GREENHOUSES

Advanced Technology for Protected Horticulture

GREENHOUSES

Advanced Technology for Protected Horticulture

Joe J. Hanan
Professor Emeritus
Colorado State University
Fort Collins, Colorado

CRC

CRC Press
Boca Raton Boston London New York Washington, D.C.

Library of Congress Cataloging-in-Publication Data

Hanan, Joe J., 1931–
Greenhouses : advanced technology for protected horticulture / Joe J. Hanan.
p. cm.
Includes bibliographical references and index.
ISBN 0-8493-1698-7 (alk. paper)
1. Greenhouse management. I. Title.
SB415.H346 1997
635.9′.823—dc21 97-31768
CIP

International Standard Book Number 0-8493-1698-7
Library of Congress Card Number 97-31768
Printed in the United States of America 1 2 3 4 5 6 7 8 9 0
Printed on acid-free paper

PREFACE

In 1989, when I retired from Colorado State University after 26 years, it was my goal to revise the 1978 *Greenhouse Management* textbook by Hanan, Holley and Goldsberry. That text, despite having read every word in it twice with my wife, was a humbling experience. Mistakes continued to appear. The revision would also be something to keep me occupied in retirement. It would be, hopefully, a significant contribution to the field in which I had spent the major portion of my life. Little did I know what that would entail! Until 1991, when writing began, I had been unable to keep up with the literature in the greenhouse field –given the University's bureaucratic requirements. As one may surmise from the publications cited here, the greenhouse literature had burgeoned. The focus and productivity had shifted to, with some exceptions, countries such as the U.K., The Netherlands, Israel and other European nations. Fortunately, the relevant publications were mostly in English. The long gestation also led to problems with the data included in the early chapters. Despite this lengthy process, with some update, most of the information in the tables is still useful for comparisons. In these six years of labor, we have seen the breakup of the Soviet Union, the continuing development of China, and the Yugoslavian chaos. So, I have left most of the tables intact, with appropriate comments in the text. One will also find that I have frequently utilized literature considered old if not decrepit (i.e., 1921). In the first place, figures so chosen illustrate a basic principle that one should understand to appreciate why things happen to a crop. Such principles do not change despite attempts to reinvent the wheel, and much basic phenomena have never been published in a book expressly tailored to greenhouse production. The wide range of conditions under which greenhouse production occurs further extended the topics discussed in the text. It is sometimes hard for an up-to-date hydroponics specialist to appreciate that most greenhouse production worldwide is in the ground.

Another reason for the book was my dissatisfaction with existing greenhouse management courses –my own included. It is my feeling that instruction depended too much on the feedback that inevitably comes from willingness to continue with existing conditions, rather than introducing new technology and advanced practices which are too frequently resisted. This desire to upgrade, and bring into the course the latest ideas, led me far afield, into areas where I had, as a horticulturist, little training. This was one of the reasons for a six-year effort in getting the first draft of some 800 pages down on paper. It required a course in soil chemistry and a dip into calculus. It also brought home the realization that training at the university level in this field was deficient. There was a proclivity to "dumb down" course requirements in the interests of retaining student numbers and promoting "diversity." Introduction of advanced techniques could not be easily made. Many students did not have the necessary background in mathematics and chemistry. It seems to me that our higher institutions were, and are, turning out too many poorly trained individuals whose contribution is limited through no fault of their own. It reflects on the greenhouse industry. Thus, one finds here an attempt to push the limits to the best of the author's abilities. It has resulted in a book which will not be easy to read. Comments to me by reviewers suggest that few will require this tome as a course textbook. I feel that R.J. Sternberg's comment (*Successful Intelligence*, 1996), that he "never was under pressure to raise the level of what he wrote, but always to lower it", is applicable to many horticultural textbooks.

A part of it may also be the fact that I interject my own comments to which some will take exception. However, I feel, on the basis of some 50 years of experience and training in the field, that I have points to make that are germane to the operation of greenhouses. To write a text as though one gives only the "facts and nothing else" is ridiculous. Even in the selection of those facts to present to a reader, the author unavoidably colors choice on the basis of his history.

The 1978 textbook contained chapters on personnel management, growth regulators, pollution and marketing. I thought I could do so here. As the chapters unfolded, however, it was soon obvious that I would be writing until the cows came home. The book would be several volumes. In the interest of keeping things to a "reasonable" length, I quit at eight chapters. One will also note that English units of measurement are not used. The text uses metric units. This was a great simplification since it reduced the number of names, made everything uniform, and avoided cumbersome conversions. By the time of retirement, I had become very tired of writing two papers of the same thing —one using English units, the other metric, with two sets of figures and tables. My own opinion is that retention of English units has done a disservice to the industry in a period of rapid change and technological advance.

As a result, I have attempted to cover the main divisions in greenhouse production: structure, radiation, temperature, water, nutrition, CO_2, and climate control. My main objective in all these chapters was to provide a viewpoint in those areas that are often publicized too much, such as soilless culture or hydroponics, and to enlarge those areas that I feel are still relatively archaic, such as nutrition. The chapter on climate control brought me into areas that, by training, I was incompetent to deal with as a mathematician. As with other chapters, detail in presentation was decided upon by my interest and competence and the need to keep things somewhat in bounds.

The last reason for undertaking this opus can best be described by quoting from Richard Mitchell's *Less Than Words Can Say*:

> "If we want to pursue extended logical thought . . . we need a discipline imposed from outside of the mind itself. Writing is that discipline. . . we have to suspect that coherent, continuous thought is impossible for those who cannot construct coherent, continuous prose.
>
> "Writing is an audacious and insolent act. When we write, we call the other members of our tribe to order.
>
> ". . . there is no other way to judge the work of a mind except through its words . . ."

I think this emphasizes the point that the act of putting all these words on paper has been a learning process by which I could clarify concepts and, more important, outline deficiencies.

Readers may decide that Hanan's mind is not something desirable, and he may be a male, chauvinist pig. That's fine. I have no objection to the use of "his" or "her," but **not** "his/her," "hiser/herer," or whatever other abomination the reader is inclined to desire in the interest of degrading language. In my creeping senility, I have my little predilections to which I hold come hell or high water. There is no need for me to be politically correct.

It was not until the end that I ran into the more unpleasant aspects of publishing in scientific fields. Readers will note extended figure legends, and although most publishers gave permission to use figures freely, they often bound them with numerous restrictions. A very few required payment for the privilege of using data that was probably obtained with taxpayers' money in the first place. For one who charged nothing, and wished to use published scientific information in the interests of a vocation, this practice required by present copyright laws does not yield the best impression, contributing nothing to writing a "scholarly" opus. It is one of the poorer aspects of our present "technological" society.

Lastly, I acknowledge the assistance I received on this voyage. First, my wife, Julia, who allowed me to turn a former bedroom into an office and purchase all the goodies one generally requires nowadays. Second, Ken Brink allowed use of departmental facilities, which was also continued under his successor Steve Wallner. At least they never said anything! Thirdly, I managed to stay on the good side of the secretaries, Judy Croissant, Gretchen Deweese, Bonnie Schilling, and technician, Ann McSay. As anyone knows, good secretaries and technicians are to be esteemed. There were a number of other people who suffered through my prose by reviewing it. These included Royal Heins, Willard Lindsay, Byron Winn, Gene Giocomelli, and Christos Olympios. The staff at Wageningen and Naaldwijk also gave of their time and publications at my request, including Hugo Challa, A.J. Udink ten Cate, Teo Gieling, and G.P.A. Bot, among others. Lastly, there were a number of people in the greenhouse industry and research institutions who provided information and data. These included George Dean, Bill Pixley, Cornelius van Bavel, A.S. Economou, M. Grafiadellis, Howard Hughes, Al Schmidt and Bernie Busch. My appreciation also extends to CRC Press for their willingness to publish a technical book with so many illustrations. One never undertakes something of this magnitude without the assistance, large or small, of many others. Undoubtedly I have left some out, but they have my gratitude.

This has been a rather long journey for something that is out-of-date before it gets published. I never expected it to go on so long. I am responsible for any mistakes or conclusions to which a reader may take exception.

BIOGRAPHICAL SKETCH

Joe J. Hanan
Professor Emeritus

For more than 50 years, Joe J. Hanan has been associated with some phase of greenhouse production, ranging from laborer to teacher, investigator and editor of technical and scientific publications in the field. He has traveled extensively and consulted with numerous organizations and commercial operators.

Receiving his B.S. Degree in Horticulture from the University of Missouri, Columbia, in 1951, he went to work as a commercial grower for $42.50 per week in southwest Missouri. However, shortly after major surgery, he enlisted for four years in the U.S. Marine Corps where he served as rifle coach, overseas, and electronics technician in the Fleet Marine Air Wing. Hanan returned to his schooling after release from active duty, matriculating at Colorado State University, receiving his Master of Science degree in 1957 in Horticulture, and his Ph.D. from Cornell University in 1963 in Floriculture. He returned as Assistant Professor to Colorado State University where he remained during the rest of his career, rising to full professor, leader of investigations and bulletin editor. He retired in 1989, and soon began work culminating in the present textbook.

Over these years, Hanan was fortunate to be associated with, and trained by, some outstanding individuals: i.e. W.D. Holley, F.B. Salisbury, A.W. Dimock, D.G. Clark, F.C. Stewart, R.W. Langhans, to mention a few. He was intimately connected with the Colorado greenhouse industry, including such individuals as Gordon Koon, Harold and Ray Crowley, C.T. Haley, Bernard Busch, the Kitiyama brothers, Homer Hill, Wayne Pearson, the Tagawa brothers, Roger Weakland, etc. His national contacts included Francis Aebi, Sr., David Thompson, Howard Hughes, L.S. Busch, D.K. Dillon, C.R. Wright, and many others too numerous to mention, including several foreign contacts and friends in industry and educational and research institutions. Many of his former students are now active in industry and agricultural institutions.

Hanan was privileged to work extensively in the Mediterranean region in Sardinia, Cyprus and Israel as well as Costa Rica. He taught short courses at the Mediterranean Agronomic Institute of Crete. These activities were in addition to travel and consultation in Europe, the U.K., Colombia, and South Africa.

During the latter 30 years, his more than 200 scientific and technical publications and two textbooks earned him four awards for outstanding research from the American Society for Horticultural Science (two Alex Laurie and two Kenneth Post), Best Friend of the Industry Award (Colorado), and the Gamma Sigma Delta Meritorious Award. He was elected Fellow of the American Society for Horticultural Science in 1988, and the Colorado Greenhouse Growers' Association elected him life member of the Association and honorary member of the Colorado Floriculture Foundation in 1989.

Hanan is presently a member of the American Society for Horticultural Science, The International Society for Horticultural Science and the honorary societies Pi Alpha Xi, Sigma Xi, Gamma Sigma Delta, and Pi Alpha Phi.

In addition to the activities outlined above, he remains active in civic affairs, being a life member of Lions Club International. Throughout his career, he took an active part in University operation and in his local church.

TABLE OF CONTENTS

LIST OF TABLES

CHAPTER 1

CHAPTER 2

CHAPTER 3

CHAPTER 4

CHAPTER 5

CHAPTER 6

CHAPTER 1

INTRODUCTION

I. DEFINITION

In this text, "greenhouse" means a structure covering ground for growing a crop that will return a profit to the owner risking time and capital. However, a "greenhouse" can mean other things to other people. A greenhouse may be a hobby, a structure for research, a profitable business enterprise or an attempt to prove a point. Because humans deal with greenhouses, greenhouses will be diverse. In a textbook of this type, it is difficult to decide when to stop and when to go into detail. My objective, as in 30 years of teaching, is to provide what I think modern students should be exposed to as graduates of greenhouse programs in institutions of higher learning. The student may not like it. Overall, I speak of greenhouses for a livelihood for an owner, which means that the structure must be **economically** practical for the particular environment (country, climate, social order, etc.) in which it stands. In this speeding, technological age, failure to communicate and to use the best available technology leads to failure.

Fig. 1-1. Low level technologies are unlikely to support sophisticated greenhouses or practices. Machines will be replaced by labor, and structures will be simple.

Greenhouses are a means of overcoming climatic adversity, using a free energy source, the sun. But, the structure also depends on climate. That is, in certain parts of the world, heating may not be required. Practices and economics, therefore, are different than in a climate where both heating and cooling are necessary. In some places, one may need supplemental irradiation since solar energy can be deficient during periods of the year. To emphasize, the practices carried out in the structures, and the structures themselves, will depend on the economic, political, and social strictures of the location. The lack of a supporting, developed technology (i.e., Fig. 1-1) may mean maximum use of cheap labor with a simple and cheap structure. On the other hand, the presence of a suitable, supporting technology, readily available capital, a good economic climate and political stability, with an educated and trained workforce, will usually mean advanced, sophisticated technology as suggested by the modern, high-rise metropolis in Fig. 1-2.

In this text, I will discuss the general relationships between practice, structure, climate and the social milieu in which the structure is placed. Unless these relationships are appreciated, the capital invested, either by an individual or group, will be wasted. This also means that specialized greenhouse structures, which often receive more than their share of publicity, will not be given a great deal of attention.

II. SOME GENERAL CHARACTERIZATIONS

Since the publication of *Greenhouse Management* by Hanan, Holley and Goldsberry in 1978, the practice of greenhouse culture has changed drastically. Shortly after publication of that textbook, a well-known individual said that all the major problems had been solved. All there was to do was to "fine-tune" the procedures. After some 40 years, one finds that such statements can subject the speaker to sudden decapitation. Emphasis on this subject is best given by some quotations from famous individuals as provided in an advertisement placed by the Electronic Data Systems Corporation in the March 8, 1996, *Wall Street Journal*:

Fig. 1-2. Advanced technologies as typified by this metropolis are likely to have a highly developed greenhouse industry. This figure and the preceding one serve to illustrate the fact that greenhouse production cannot be considered separately from the surrounding environment which includes political situation, educational attainments and standard of living as well as available technology.

"Radio has no future. Heavier-than-air flying machines are impossible. X-rays will prove to be a hoax." [William Thomson, Lord Kelvin, 1824-1907].
"Rail travel at high speeds is not possible because passengers, unable to breathe, would die of asphyxia." [Dionysius Lardner, 1793-1859].
"While theoretically and technically television may be feasible, commercially and financially I consider it an impossibility . . . " [Lee DeForest, 1873-1961].

Probably, among the biggest changes to have occurred in the last two decades following 1978 are the development of computers, the movement to large-scale production in suitable climates, combined with fast transportation, specialized structures fitted to the particular climate and mechanization; and new advances in irrigation, fertilization, and integrated pest management. All this has been aided by the development of the largest mass market in the world in the United States (Table 1-1).

Despite the diversity, we can make some general characterizations about greenhouses:

Table 1-1. Per capita consumption of floral products in selected countries in 1980 and estimated per capita consumption in the U.S.A. [Anon., 1989]. Consumption values marked with an asterisk (*) are from Nannetti, 1982.

Per capita consumption of floral products in 1980 or 1989

	$
Germany	37.72
Sweden	29.94
Switzerland	40.72
England	13.77
Holland	44.91
Japan	47.90
Italy	40.72
Belgium	31.73
France	23.95
Spain	7.19
Norway	7.19
*U.S.A.	24.77

*Target potential for per capita consumption $55.00

*Potential increase for U.S.A. $30.23

*Total potential market for products in U.S.A.: $\$6.7 \times 10^9$

A. RELATIONSHIP TO THE PARTICULAR COUNTRY

Although, historically, greenhouses were developed in regions where transportation to distant markets was prohibitive, there were restrictions to flow across political boundaries, and there was a need for protected cultivation during winters; we find that climate can substitute for technology. Greenhouses will be seen, often, in so-called "developing" nations. In fact, greenhouse production can be a significant export important to the particular country's balance-of-payments, as well as a means to provide employment for people. For example (Table 1-2), the relatively large area in greenhouse structures in small countries such as Holland or Israel suggests a greater relative importance to the economy than in the U.S. where the area under protection in ornamentals is comparable. The land area and population in the U.S. are much larger in comparison to many countries by several hundred fold (Table 1-2). The contribution to the Gross National Product by the U.S. greenhouse industry is, therefore, small. Horticulture under glass in The Netherlands provides 63% of the net contribution in the balance of payments in agriculture of that country [Meijaard, 1988], whereas the contribution by the U.S. industry is less than 1%. Meijaard [1995], however, pointed out that the share of the agricultural sector in The Netherlands' national income is about 4%.

In still other countries such as Colombia, Kenya, Mexico or Central Americas, the low GNP combined with the need to provide income for a burgeoning population makes greenhouse production for export especially attractive [e.g., Kellen, 1983; Shypula, 1981; Rochin and Nuckton, 1980; Accati, 1978]. Countries such as

Table 1-2. Comparisons of total land area, population, gross national product (GNP), GNP per capita and land area in protected ornamental production for several countries [Anon., 1982].

Country	Total area (sq.km.) (10^3)	Population (10^6)	GNP (10^9) ($)	GNP/capita ($)	Ornamental area (ha)
Australia	7770.0	14.9	161.0	10805.4	1110.5
Brazil	8806.0	123.0	286.0	2325.2	999.6
Central America	271.4	13.6	0.03	2.0	1699.7
Colombia	1138.0	28.8	36.0	1250.0	1214.1
Denmark	43.0	5.1	57.6	11294.1	385.3
France	551.7	54.0	568.0	10518.5	2023.4
Israel	20.7	3.9	16.0	4102.6	1295.0
Italy	301.2	57.2	345.0	6031.5	3429.3
Japan	372.4	117.6	1.1×10^{12}	9353.7	11877.6
Kenya	583.0	17.2	6.8	395.3	404.7
Netherlands	40.9	14.2	138.0	9718.3	4734.8
Spain	504.8	37.6	212.0	5638.3	1499.4
U.K.	244.0	55.7	525.0	9425.5	520.4
U.S.A.	35223.8	229.8	2.9×10^{12}	12619.7	5997.5
West Germany	248.6	61.7	658.0	10664.5	2598.5

Germany, France, Australia, etc. have substantial areas in greenhouses (Tables 1-2 and 1-7), but these areas are insufficient to meet the consumptive needs in those developed nations. The classic example is the Dutch thrust to advance consumption in the U.S., which has a population of 230 million and the highest standard of living. The Dutch had 63% share of the cut flower and a 51% share of the pot plants in the international export trade [Krause, 1988]. The export from The Netherlands to the U.S., and other countries, is remarkable. In 1973, per capita consumption of floral products in the U.S. was only $9.73 [Fossum, 1973]. In 1980, that had increased to $24.77 (Table 1-1), with a predicted potential of 6.7 billion dollars ($55.00 per person)[1]. Colombian imports to the U.S. now make up more than 76% of all carnations sold in the U.S., and more than 45% of all cut roses [Stewart, 1989]. Despite the negative attempts in the U.S. to restrict imports [Sullivan, 1976], development of mass marketing of floral products in the U.S. would have never reached the point it has without the constructive, farseeing competitiveness of outsiders. One must, after all, remember that ornamentals compete with other products such as candy, alcohol and tobacco. However, floral products have none of the negative attributes those products provide. The American industry has never mustered the forces to use this potential. The result has been marked changes in the U.S. industry's structure [Harris, 1982; Galante, 1986].

The classic effect of imports on an industry that was unable to make suitable adjustments in an open market is noted by Colorado's example. Beginning with the surge of floral imports in the early 1970s, the carnation production area in Colorado fell from a high of 659 thousand sq.m. in 1974 to a low of 268 thousand in 1986 as growers changed to other crops or closed their doors [Kingman, 1986]. This was at a time when the technology to produce more than 500 standard carnation cut flowers per sq.m. was available, but average production was 300 per sq.m.-yr. A few growers have remained in business with carnations because of their ability to produce maximum yields with high quality –and to market them successfully. At the same time, Colorado growers removed from their flowers the only specific product identification ever to be used for flowers in the U.S. Thus, they were unable to identify their product in the marketing system. As the growers perceived no immediate help

---

1 Currency conversions to U.S. dollars used in this text as of October, 1990 are: Dutch guilder = $0.5988, British pound = $1.996, Greek drachma = $0.006935, Spanish peseta = $0.0108.

from their research program, funds for ongoing research and development were reduced –in a period of general public fund

Table 1-3. Comparisons of major trade regions [House, 1989]. Note this information prepared just prior to breakup of the Soviet Union.

	United States	Soviet Union	Japan	European Community	China
Population (10^6)	244	284	122	324	1074
Gross Natl. Prod. (GNP) (10^9 U.S. $)	4436	2375	1607	3782	294
Per capita GNP (1987 U.S. $)	182000	8360	13180	11690	270
GNP growth rate (1987)	2.9%	0.5%	4.2%	2.9%	9.4%
Inflation (change in consumer prices)	3.7%	-0.9%	0.1%	3.1%	9.2%
Labor force (10^6)	122	155	60	143	513
Agriculture	3	34	5	12	313
Non agriculture	118	121	56	131	200
Foreign trade (10^6)					
Exports (U.S. $)	250	108	231	954*	45
Imports (U.S. $)	424	96	151	955*	40
Consumption (bbl oil equiv. per capita)	56	37	23	24	5
Oil reserves (10^9 bbl)	33	59	0.1	8	18
Oil production (10^6 bbl per day)	10	13	Negligible	3	3
Natural gas reserves (10^{12} cu.m.)	5	41	0.04	3	0.9
Coal reserves (10^9 metric tons)	264	245	1	91	170
Grain production (kg per capita)	1150	740	130	480	402
Meat production (kg per capita)	109	65	31	82	18
Life expectancy (yrs)	75	69	78	76	68
Cars per 1000	570	42	235	347	Negligible

* Data include trade between EEC members.

retrenchment in agriculture– and significant monies put into the process for restricting imports. Rather than perceiving a huge market which they could share through a positive and coordinated approach to surviving in a competitive market, the cut flower industry left it to others. Although Colorado's climate means that both heating and cooling are required, it is an ideal high solar radiation regime that allows precision environmental control for maximum productivity commensurate with high quality. The forefront of industry development, in the 1980s, shifted from the U.S. to export committed nations such as The Netherlands, Israel, etc.

Concerning the above, it is useful to compare the U.S. with other large countries of immense potential, particularly the European EEC union (Table 1-3). The EEC represents a serious challenge to the U.S. Certainly the U.S. greenhouse industry has not even begun to consider possible implications in terms of competitive ability. Shortly after the publication of Table 1-3, the situation in Russia changed completely. One deals now with a multiplicity of smaller countries whose effects on the international scene are not likely to be significant for several years given existing political problems. Also, in the last few years, another formidable trading block has been formed between Canada, Mexico and the U.S. The effect on greenhouse industry relationships and product flow are likely to be interesting.

The case of vegetable production under cover is different compared to ornamentals. Historically, protected vegetable production developed in those regions where winter conditions made such production a necessity

Table 1-4. Energy produced per hectare-year and the ratio between energy input versus that produced for various types of agricultural production [Enoch, 1978a].

Type of agriculture	GigaJoules per hectare-year output	ER (energy ratio, output/input)
Shifting cultivators, slash/burn, Congo	15.7	65.0
Nomad, Bedouin, Sinai	0.2	50.0
Subsistence farming, India	10.4	14.8
Rice, Philippines	22.9	5.5
Cereal, U.K., average farm 1970	48.6	1.9
Intensive rice, U.S.A.	84.1	1.3
Cattle and sheep, U.K., average farm, 1970/71	9.3	0.6
Dairy, U.K., average farm, 1970/71	17.3	0.4
Brussels sprouts	9.1	0.2
Heated greenhouse, rose crop, Israel	356.0	0.03
Heated greenhouse, winter lettuce (2 months), U.K.	63.6	0.002

–northern Europe, Russia, etc. – at a time when boundaries and lack of rapid transportation prevented free flow of commerce. For these reasons, protected vegetable growing can be quite significant in some of these countries. In the U.S., on the other hand, rapid, inexpensive transport of vegetables from those regions where they can be produced outdoors the entire year (Florida, California, Texas), meant that it was difficult to achieve any significant protected cultivation in vegetables. This is particularly so when one considers the price differential necessary to cover the higher production costs in greenhouses. Dalrymple [1973] stated it was cheaper to ship tomatoes from Mexico to Canada than to produce them in greenhouses in that country. Profitable vegetable producers remaining in business deal largely with a restricted market with high-income consumers, or, there is a marketing system in place that provides vegetable growers with a reasonable return. In Colorado and elsewhere, there has been a remarkable increase in vegetable production in the last 10 years. A part of this may be governmental subsidies for cogeneration and the use of Dutch technology. Once one has satisfied the basic needs of the consumer such as for food, increasing consumption is much more difficult. In recent years, the Mediterranean has seen rapid development of greenhouse vegetable production (Spain, Greece, Israel, etc.) by using cheap structures and low-cost labor in a suitable climatic region [Mavrogianopoulos, 1989; Montero et al., 1985; 1989]. This has put increasing pressure on North European producers, and the northern area in protected vegetables has tended to decline.

The greenhouse industry can be characterized as international [Staby and Robertson, 1982]. Unless countries become insular, as the U.S. was after World War I, product mobility will continue to increase. It has led to some problems such as disease and insect spread to such an extent that different localities have all the diseases that are present elsewhere. Survival of a greenhouse industry in highly advanced nations requires coordinated effort to provide a long-term research base for continued technological advance, and a regularized marketing system that includes product standards, product identification and adequate promotion. Examples of such marketing systems include the Dutch and Israelis. In Holland, production costs have decreased 2 to 3% annually over the past 10 years (corrected for inflation) with an annual increase in production of 6% although total greenhouse area decreased by 1% [Meijaard, 1995].

Table 1-5. Operational costs per sq.m.-year for various greenhouse operations. **Labor** includes payroll taxes, unemployment compensation, social security, pension, etc.. **Utilities (Util)** include fuel, power and telephone. **Supplies (Sup)** include chemicals, fertilizers, containers, plants, CO_2, etc. **Administration (Admn)** includes accounting, legal, computer, etc.. **Insurance (Ins)** includes health and catastrophic, etc.. **Marketing (Mark)** includes promotion, research, license fees, vehicles, packaging, commission, etc.. **Taxes** include income and property. Management and profit are not included. Data obtained in 1990.

Type	Labor	Util.	Sup.	Admn.	Ins.	Depr.	Mark.	Taxes	Total
Propagator, 22000 m^2, U.S.A.	28.83	9.02	21.94	1.10	2.83	11.21	5.63	----[a]	80.56
Roses, 4259 m^2, U.S.A.	30.70	8.68	4.53	2.21	1.83	4.36	20.90	2.52	75.73
Tomatoes, 1 ha, Cyprus[b]	2.21	----	2.04	----	----	0.71	----	0.56	5.52
Cucumbers, 1 ha, Cyprus[b]	2.19	----	2.43	----	----	0.73	----	0.59	5.94
Roses, 1000 m^2, Israel[c]	9.80	3.50	3.22	1.03	----	7.50	----	----	25.05
Roses, 5190 m^2, Greece[d]	12.31	7.58	2.21	0.63	0.31	2.84	4.73	0.95	31.56
Dutch operation, 8500 m^2,[e]	7.79	6.59	4.50	----	----	----	4.50	1.80	38.03

[a] Taxes not separated in information provided.
[b] Figures for total interest not taxes, from Papachristodoulou et al. [1987].
[c] Interest 11.9%.
[d] Personal communication, May, 1991, A.S. Economou.
[e] Average of data ranges as given by Meijaard, 1995. Total costs are total direct costs.

B. ENERGY, CAPITAL, AND LABOR REQUIREMENTS

Regardless of location, greenhouse production remains the most intensive agricultural process known. It is intensive in terms of labor. It is intensive in terms of capital since the erection of a covering over ground is a large investment whatever location and requires money to purchase and operate. In temperate climates, greenhouses, by the very nature of their structure, are large energy consumers. If all inputs are converted to energy terms (Table 1-4), we see that, for energy efficiency, primitive slash and burn agriculture has the highest energy return (ER=65). Most modern agriculture pursuits require more energy input than is gained. Of course, the assumption in Table 1-4 is that an energy unit from human labor is equal in value to an energy unit from fossil fuel on a physical basis. Whether it is so from a monetary standpoint is an argument I will not address. Greenhouses are the worst from an environmentalist's viewpoint, especially if supplemental irradiation is employed. Since electrical generation is about 40% efficient, and usable irradiation from the most efficient, high-intensity lamps is about 40%, then conversion of a fossil fuel to energy in a form useable by a plant is less than 20% efficient. Given also the fact that conversion of radiant energy by photosynthesis to a saleable economic unit is less than 10%, then the total process has less than a 2% efficiency! Environmentally speaking, it would be better to place all greenhouse production in regions having many clear days (e.g., arid or semiarid climates). Note in Table 1-4, that there is an improvement in efficiency at lower altitudes (Israel at 32° versus England at 55° N) [Enoch, 1978b]. Nevertheless, economics dictate in our existing context, that such agricultural pursuits are profitable, and will, therefore, be used. Nor, do I think we wish to return to a nomadic, hunter culture. The only good thing about the "good" old days is that they are gone.

Table 1-6. U.S.D.A. Floriculture Crops Report for 1986. Gross wholesale value per square meter for selected crops [Hamrick, 1987].

Crop	$ per sq.m.
Spray chrysanthemums	11.77
Standard chrysanthemums	23.33
Standard carnations	25.92
Poinsettias	26.35
Mini carnations	27.00
Cutting geraniums (pot)	30.98
Seed geraniums (pot)	32.94
Flats vegetable, bedding	34.78
Florist azalea	41.80
Potted chrysanthemums	46.01
Hybrid tea roses	49.57
Sweetheart roses	52.06
Easter lilies	52.49

A modern greenhouse in temperate climates represents a capital investment of more than $1 million per hectare exclusive of land acquisition. Replacement costs for a Dutch operation are given as about $90.00 per sq.m. Operational costs are proportionally higher (Table 1-5). For example, chemicals and fertilizers for 1 hectare in the U.S. can be as high as $13,000 per year. A general farm operation, on the other hand, would consider a fourth of that amount as excessive. Labor can exceed $300,000 per hectare. As can be noted, other expenses to operate a hectare of protected horticulture are also high in comparison to conventional farming such as wheat, maize or similar crops. A good example of labor cost significance was provided by Lukaszewska and Jablonska in 1991. They compared the number of flowers a grower had to sell to pay for one hour of work in Poland versus The Netherlands. In 1989, Polish growers had to sell 6.4 carnation flowers and 2.4 cut roses; whereas, in The Netherlands, one had to sell 88 and 66 flowers, respectively, to pay for an hour's worth of labor. One should compare the total costs per sq.m. for operations in the Mediterranean climate of Cyprus, versus those for Israel and Greece, and as contrasted to costs in a temperate, continental climate of the U.S.A. (Table 1-5). Thus, we see that the solution for growers in high-cost, developed regions is often the selection of crops with which low-cost regions cannot compete (import restrictions, transportation costs, etc.); or, the willingness to make best use of new technology to reduce growing costs and improve productivity. The differences are startling. The fact that The Netherlands remains a dominant force in the market for greenhouse products says much to their ability. With a centralized marketing system, growers in The Netherlands do not have to compete with each other, and the communications and technology transfers between growers are facilitated. In climatic regions where heating or cooling is not required, the cost of greenhouse structures can be considerably reduced. If labor is cheap, as suggested above, operational costs can be markedly lowered by labor substitution (Table 1-5). By far, the greater part of greenhouse production is in the ground, and this is cheaper compared to raised benches. The latter may be a necessity with pot plant production, or where disease must be eradicated. Hydroponics and soilless culture have continued to receive publicity since Gericke first coined the term in the 1930s. Soilless culture has received renewed interest by means of recirculating nutrient solutions or production in rockwool and similar inert materials. Unfortunately, hydroponic installations can double the initial capital investment, and require a very high level of technical competence –something that not all operators possess.

C. RETURN TO THE GROWER

If greenhouses are intensive, using large amounts of energy, they can also offer high return. Again, as with practice and structure, this return will depend upon political, economic and social considerations. For example, ornamental production usually provides the highest net return (Table 1-6). However, this means that the consumer must have discretionary income to purchase ornamentals or high-priced vegetables, and thus the major markets are restricted to developed regions with a high living standard. Vegetable production generally returns less, but profitability will vary drastically with standard of living, social practices and geographical boundaries. As with any general statement, there are exceptions to the rule. Greenhouse production of vegetable transplants to field producers is an example of a profitable service. But, compared to the total industry, transplant production is small and specialized. We see that the thrust of production is toward those countries with a suitable climate where a product can be produced cheaply, with high-speed transportation to advanced regions with high technology and high income.

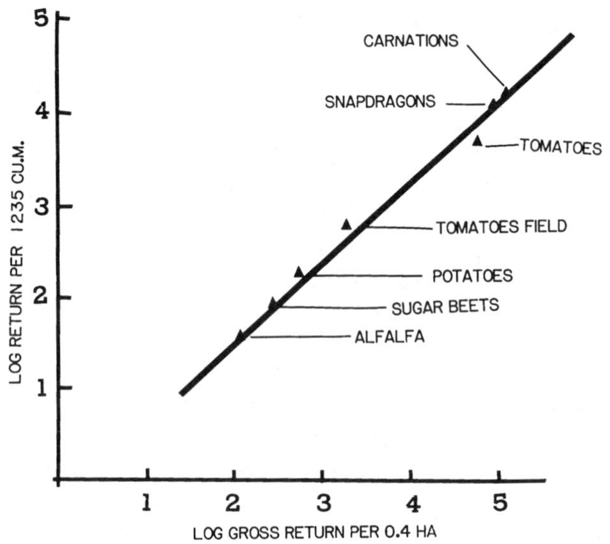

Fig. 1-3. Relationship between gross return per 0.4 hectare and return per 1235 cu.m. water. Consumptive water use for all field crops taken as 100 cm and for all greenhouse crops as 200 cm. "Consumptive" use is irrigation. Gross return in U.S. dollars [Hanan, 1967].

In some respects, greenhouse production makes maximum use of raw materials compared with conventional agriculture, and it provides opportunities to reduce pollution such as contamination of groundwater supplies from fertilizers and other chemicals. Hanan [1967] showed that the return from water use in carnation or snapdragon production was more than 10-fold that returned per unit in outdoor potato, beet or tomato cultivation (Fig. 1-3). Similar comparisons for other raw inputs to the greenhouse process have not been made. If means can be found to provide structures with the insulating qualities of normal residences, while still using solar energy, then one can conclude that greenhouse production is a means to maximize utilization of scarce materials. The cost of water in producing a carnation cut flower, however, is less than $0.01. As long as the water quality is good, it represents a minuscule part of total operational cost.

D. SIZE

Although the situation has changed somewhat in the last 20 years, one finds that most of greenhouse operations are still what we, in the U.S., call "small businesses." In fact, in some parts of the world, an area of less than 1000 m^2 represents a viable economic unit. In the U.S., a covered area of 7000 to 8000 m^2 is often considered suitable for a single-family operation. Meijaard [1995] gives an average area of 8500 sq.m. for establishments in The Netherlands. Size, however, can often be found at more than 15 to 20 hectares in some parts of the world. A small business is not necessarily bad, but such operations are commonly subject to many difficulties such as poor management, lack of technical skills, insufficient capital and the necessity to sell in a highly competitive marketing system. Here again, the relative importance of the industry to the political entity in which the industry functions (see Table 1-2 as to relative areas) can mean generous state subsidy and technical assistance through a state-supported extension service. The differences that can be found in various parts of the world are enormous. Many politicians find it expedient to subsidize agriculture, and this desire to protect their agriculture is an offshoot of previous practice when farmers were the population majority. The controversy in the GATT conferences where several countries refuse to reduce agricultural subsidies in trade agreements is a case in point [Knight et al., 1992]. In 1993, we saw problems between the U.S. and Japan, especially in Japanese farmers' attempts to restrict rice imports. This political power of agriculture is less important in developed countries where the urban population possesses the voting power and agriculturists can represent less than 5% of the total population. Unfortunately, most urban people are unorganized in terms of political input unless they belong to a particular group such as unions, elderly, etc. Agricultural interests represent an organized group, wielding considerable influence through tradition and conservatism plus entrenched contacts and practices. The humorous, cynical and sarcastic comments by O'Rourke [1991] on U.S. agricultural policy should be read to appreciate the relationships between agriculture, politics and business operations.

Table 1-7. Estimated area of glass and plastic structures in the world [Germing, 1986].

Country	Hectares
Algeria	100
Australia	1200
Austria	385
Belgium	1900
Bulgaria	800
Canada	400
China	13130
Cyprus	300
Denmark	400
Finland	350
Germany	4770
Greece	3100
Hungary	2850
Iceland	15
Ireland	200
Israel	400
Italy	17000
Japan	27300
Korea (south)	1300
Morocco	660
Netherlands	9000
New Zealand	200
Norway	225
Poland	2600
Portugal	4120
Romania	6300
Saudi Arabia	80
Spain	12150
Sweden	340
Switzerland	180
Tunisia	1050
Turkey	9000
United Kingdom	2660
U.S.A.	4000
U.S.S.R	6900
Yugoslavia	560

E. EXTENT

The industry, as a whole, represents a minuscule part of total worldwide agricultural production, as can be noted in Table 1-2 by the disparity between area in greenhouses versus total area of any country. The gross return to the local economy, however, can be significant, as discussed previously. Russia may be a special case because the total arable area under the present regime is about 13% of the total land area. Russia is 1.8 times the size of the U.S. Given Russia's climate, greenhouse production could be highly important.

Although The Netherlands were generally given first place in the list of total area in protected cultivation [i.e., Sheard, 1979], Table 1-7 shows that Japan, including soft plastics, takes first place. According to Takakura [1988], the area in protected cultivation in Japan is more than 42000 hectares with about 7500 ha in fruit culture –grapes, oranges and pears. Various authorities give different figures. The most recent available for The Netherlands [Meijaard, 1995] gives the area in vegetables as 4700 ha, with foliage plants at 1100 ha and cut flowers equal to 4200 ha. Japan has the highest per-capita consumption. An article in *The Grower* magazine [Anon., 1981] gave the total area for the U.S.S.R. as about 12000 ha as compared to 6900 ha recorded by Germing [1986]. Mavrogianopoulos [1989], for Greece, provided a value of 3700 ha, an increase of 600 ha over Germing's 1986 figure. Montero et al. [1987] gave figures of 21680, 20000 and 1200 ha for Spain, Italy, and Portugal, respectively. The total for data in Table 1-7 is 135900 ha, as compared to 180500 ha, of which 45500 ha were greenhouses, the rest tunnels, given by Meijaard [1995] for 1989. In large nations such as the U.S., China and Russia, the total land area in protected cultivation is not remarkable (Tables 1-2, 1-3).

F. SPECIAL PECULIARITY OF GREENHOUSES

When one covers a crop with a structure, very significant changes are made to the internal environment as is appreciated by anyone who has ever been inside a greenhouse. The single greatest effect, however, is reduction of wind velocity in comparison to that normally found in the field [Businger, 1966]. This is the true "greenhouse effect" as far as we are concerned in this book. Although the cover will also markedly influence energy exchange, especially outgoing radiation, this is, in respect to the influence of wind movement, minor –although the popular use of the term "greenhouse effect" refers to energy transfer through the earth's atmosphere, the atmosphere, in this case, being a transparent cover. The effect of a cover, because of reduced convective energy transfer by wind, is a marked increase in internal temperatures under clear skies, as well as changes in a number of other factors that will be discussed later.

G. QUESTIONS TO ANSWER IN STARTING A GREENHOUSE BUSINESS

One of the problems to be faced by individuals wishing to begin their own business, especially involving a greenhouse, is what questions must be answered in order to assure success. As with most small businesses, the owner frequently fails to have the necessary expertise to make technical systems work, even if most of the questions have been answered properly. In today's greenhouses, the need for knowledge and skills is much greater than ever before –and it is not always limited to business or production knowledge.

What are some questions that any entrepreneur should attempt to answer before actually committing himself to risk? The information-gathering process can be laborious. Sometimes a question cannot be answered, if at all, with reasonable degree of accuracy. In other cases, the prospective owner may have to compromise between a solution to one question and a poor solution to another question.

1. What is the market size in relationship to the crops one proposes to grow? To whom is the product to be sold and what is the general standard of living? What, if any, marketing system is available? What will be the transportation cost? What are the possible costs for grading, packaging and promotion? How is the market likely to fluctuate with changing conditions? Who will be the competitors? And, how good are those competitors?

   Over the years, all of us in the business have seen people who wish to grow plants in greenhouses, who erect the structure, borrow money to operate until the first crop comes in, and <u>then</u> worry about where and how they are going to sell it. That is getting the cart before the horse. Research on marketing is always money and time well worth spending, for one might find that a greenhouse business in the particular location would not be viable under any circumstance.

2. What are the climatic conditions of the location chosen (temperatures, available solar radiation, rainfall, snowfall, hail, wind and wind directions)? These external factors will directly influence structural type, covering, heating and cooling systems, humidity control and many other factors. Are there problems from local meteorological effects? From pollution? Chapter 2 addresses a few of these problems. They directly influence the initial capital outlay and future operating costs.

   The lack of solar radiation at certain critical periods may limit yields or require supplemental irradiation. High temperatures may mean one can dispense with heating systems, but excessive temperatures can be a problem that require different greenhouse structures. Hail, high winds and possible snow loads may require a stronger structure, investment in windbreaks and insurance against damage. Locating in areas that are not well drained in terms of air movement may increase fuel costs and reduce the frost-free season. These are but a few of the answers that can be obtained from a thorough research of climate and weather, generally available from governmental agencies and in talking with people who have lived all their lives in the same place.

3. What is the fuel and power availability? Is natural gas available? Will the grower have to use oil? If coal is available, what restrictions are there on its use? What pollution controls are required? How far away is the natural gas connection? The main electrical power supply? Chapter 4 may provide some answers. Putting a gas line underground is an expensive proposition. In extremely cold weather, the supplying gas company may shut the greenhouse off. So, is there an alternative fuel for use in an emergency? Is electrical power available

three phase or single phase? 220 or 440 V? How reliable is the electrical supply? An emergency generator may be necessary. In fact, most greenhouses in temperate climates will install a standby generator to operate water pumps, boilers and minimal air-conditioning in the event of power failure.

4. What is the water availability and what is it's quality? Does that quality vary with the year? What restrictions are there on well drilling, or on the size of the tap if the operator wants to use domestic sources?
   The failure to adequately outline and appraise water supplies, particularly in the Western U.S., is a common fault. The reason is that good water is cheap. However, the entrepreneur does not realize that a poor water is horribly expensive because it seriously limits productivity, causes problems with fertilization, and locks the grower into a substandard operation unless he can change his supply. Salt water intrusion in local wells is a serious problem in many Mediterranean countries, particularly Spain [Palmer, 1987]. Chapter 5 deals with the limits placed on water supply and what can be done. Money spent on this aspect is money well spent. The prospective grower should be aware of local water law and the likelihood of water pollution with respect to organic chemicals –largely herbicides.

5. What is the soil like in the location chosen? Is it well drained? Or, is there a high water table? Is it subject to flooding? What is the topography? Steep sites may require extensive leveling or one is faced with increased costs in product transportation around the site. Locations with high water tables may mean excess water from irrigation systems cannot be removed without expensive alterations. Can the crop be grown directly in the ground? If so, what is the fertility level and what measures will be needed to control salinity? Can the local soil be used with minimal alteration as part of a potting mixture?

6. How much land is needed now and for future expansion? How is its value expected to fluctuate? Is it in an area likely to urbanize, or to be surrounded by polluting industry? What are the zoning restrictions? Most government entities, at least in the U.S., are likely to be unfamiliar with greenhouse structures and they may or may not consider them as agriculture. Officials may attempt to enforce unrealistic requirements on structural strength, or, at some time in the future, change the zoning regulations so the establishment becomes a non-conforming use, eliminating any possible expansion. Local home owners may complain about pollution and dirt. Thus, the owner must have some knowledge of local ordinances. If the greenhouse is in the country, the grower may have to be careful of pollution and insects from local farming procedures –especially if the farmer decides to spray his crop with an herbicide.

7. Where is capital to be obtained? Unless one has a rich uncle, the individual must have money to buy, transport and erect the necessary structures; then he must have money on which to operate and pay the labor and utility bills until the first crop is sold. This is where friendliness with a local banker pays dividends, and, even better, to know what governmental programs may be available that may provide low interest loans, cash subsidies or business assistance. In some countries, an exporting company is often given special dispensation such as importing special equipment without paying prohibitive duties. In other countries, a highly developed extension service may be available that tells the operator what to do. In still others, special tax incentives are frequently offered. Unfortunately, one must then deal with a bureaucracy, and the paperwork and the need to deal with a bureaucratic mind can be almost insurmountable.

8. Quite often, a proposed operation will attempt to locate in a region where there is supposedly a ready supply of cheap labor. One can do this in some developing countries, but there may be other expenses. That is, the operator may be required, once an individual is hired, to maintain them in sickness as well as health, and firing them may be next to impossible. In still other countries, there simply may not be a source of low-priced labor. Or, entrepeneurs must compete with other employers for the available personnel. Governmental regulation may require many additional benefits, and hiring of illegal persons is often a great temptation to the businessman –on which, of course, the entrepreneur fails to pay withholding tax, retirement benefits, unemployment compensation, etc. Thus, the labor pool that will be available to the greenhouse operator must be investigated.

III. ABOUT THE FUTURE

In reviewing Section 1.2 of the book published in 1978 on greenhouse management, I find the times are just as challenging, and the needs for new thinking, new techniques and new ways of looking at our environment are just as needful. We can say, in this decade, that international horticultural trade is here to stay. No matter where products are produced, each will have their particular challenges, but insularity is not one of them. In some localities, greenhouse production is an industry, not an agricultural pursuit. It is an industry that requires all the skills and acumen of successful businessmen without that conservatism usually attributed to them, and with the ability to think in the long-term possibilities.

Although the problems of pollution and chemical dangers may not seem so important to those in developing nations, we can expect the environmental movement to continue as it has in America. Growers will be ever more subject to restrictions on use of dangerous chemicals so necessary in existing greenhouse practice. Continued development of integrated pest management can be expected, which may have unforeseen consequences on product quality to the extent that some areas probably cannot continue operation. Of course, environmental protection means increased costs since individuals can no longer utilize, or dump their refuse, in the "commons." Few realize that of the total cost of building new power plants in some countries, pollution control expenses can exceed 25% of the total. The consumer pays for this, reducing discretionary income that could buy greenhouse products.

In Chapter 2 I will briefly discuss the possibility that some day we will see greenhouses constructed with very high insulation properties. This, of course, will increase capital investment, shifting a greater advantage to milder climates. Yet, as the Dutch have shown, it is possible to survive well in a high-cost environment through the achievement of effective and uniform procedures for all entrepreneurs.

One thing we know will be with us always is taxes and governmental regulation. To say currently that we are free to run our business in the manner to which we are accustomed, come hell or high water, is to be exceedingly myopic. Particularly in developed nations, the industry is increasingly surrounded with governmental regulations and requirements. The ability to carry on a business by hanging receipts on a nail is a practical impossibility. One does not operate nowadays without taking into account the various restrictions and requirements of governmental agencies. Really, we have no one to blame but ourselves since we were the ones who initially provided dangerous places to work, abused the human condition, and polluted the environment in pursuit of profit as the sole objective, or hired illegal immigrants.

There are other objectives besides profit –which one must have if he is to stay in business. One can desire to have the best-run greenhouse, an asset to the community, a place to provide work for people, a wish to provide a product to others that gives sustenance in some fashion, or a place that contributes to the welfare of the community. Another objective might be that the grower just likes to grow plants. Why not do something one likes to do that allows a livelihood? Why not, through the medium of your business, make a contribution to society by training others and providing an example to those around one?

IV. REFERENCES

Accati, E. 1978. Ideal climate, no labour costs, make Kenyen flowers bloom. *The Grower*. 89(14), Apr. 13, 1978.

Anon. 1982. Flowers Unlimited. Coop. Verenging Verenidge Blumenveilingen Aalsmeer, The Netherlands.

Anon. 1981. Russian salad. International Round-up. *The Grower*. 96(24). Dec. 10, 1981.

Businger, J.A. 1966. The glasshouse (greenhouse) climate. *In* Physics of Plant Environment, ed. W.R. van Wijk. North-Holland Publ. Co., Amsterdam. 382 pp.

Castilla, N. et al. 1989. Alternative greenhouses for mild winter climate areas of Spain. Preliminary report. *Acta Hort.* 245:63-70.

Dalrymple, D.G. 1973. A global review of greenhouse food production. USDA, ERS Report 89, 150 pp.

Palmer, C. 1987. It's not all going Spain's way. *The Grower*. 107(23). June 4, 1987.

Enoch, H.Z. 1978a. A theory for optimilization of primary production in protected cultivation. I. Influence of aerial environment upon primary plant production. *Acta. Hort.* 76:31-43.

Enoch, H.Z. 1978b. A theory for optimilization of primary production in protected cultivation: II. Primary plant production under different outdoor light regimes. *Acta. Hort.* 76:45-57.
Fossom, T. 1973. Trends in commercial floriculture. Crop production and distribution. A statistical compendium for the United States 1945-1970. The Soc. Amer. Florists Endowment, Washington, D.C.
Galante, S.P. 1986. Smaller flower retailers wilt as competition grows intense. *The Wall Str. J.* Aug. 4, 1986.
Gericke, W.F. 1940. The Complete Guide to Soilless Gardening. Prentice Hall, New York.
Germing, G.H. 1986. The world acreage under greenhouses. *Chron. Hort.* 26:11.
Gillette, R. 1982. The Netherlands and the future of US floriculture. *Florists' Rev.* 170(4404):18-23,158-160.
Hamrick, D. 1987. The USDA floriculture crops report. *Grower Talks*. 51(3):96-100. July, 1987.
Hanan, J.J. 1967. Water utilization in greenhouses: Alternatives for agriculture in arid regions. *Proc. 3rd Ann. Amer. Water Res. Conf. San Francisco*, Nov., 1967, 160-169.
Hanan, J.J., W.D. Holley and K.L. Goldsberry. 1978. Greenhouse Management. Springer-Verlag, Heidelberg. 530 pp.
Harris, W. 1982. Flower power. *Forbes Magazine*. Oct.25, 1982.
House, K.E. 1989. For all its difficulties, U.S. stands to retain its global leadership. *The Wall Str. J.* Jan. 23, 1989.
Kellen, V. 1983. Forced growth. *Florists' Rev.* Oct. 27, 1983. 17-21.
Kingman, R. 1984. 1984 Colorado greenhouse survey. CO Greenhouse Growers Assoc., Denver, CO.
Knight, R. et al. 1992. Breaking away. *U.S. News & World Rpt.* 112(13):36-40.
Krause, W. 1988. Leading the world in exports. *The Grower.* 109(4).
Lukaszewska, A.J. and L. Jablonska. 1991. Ornamental plant production in Poland. *Chron. Hort.* 31:35-37.
Mavrogianopoulos, G. 1989. Characteristics of greenhouse construction and greenhouse production in Greece. Unpubl. MS, Agric. Univ. of Athens, Greece.
Meijarrd, D. 1988. Characteristics of export-oriented horticulture and their developments. *Acta Hort.* 223:28-34.
Meijaard, D. 1995. The greenhouse industry in The Netherlands. *In* Greenhouse Climate Control. J.C. Bakker et al eds. Wageningen Pers.
Montero, J.I., P.F. Martínez and N. Castilla. 1987. Posibilidades de mejora climatica de los invernaderos de España. *Reunion Agrimed de Cultivos Protegidos*, Barcelona, Nov., 1987.
Nannetti, M. 1982. Does potential worldwide production exceed potential worldwide demand? *Florists' Review.* 171(4438):41-44.
O'Rourke, P.J. 1991. Parliament of Whores. The Atlantic Monthly Press, N.Y. 233 pp.
Papachristodoulou, S., C. Papayiannis and G.S. Panayiotou. 1987. Norm input-output data for the main crop and livestock enterprises of Cyprus. *Agric. Econ. Rpt. 16*. Cyprus Agric. Res. Inst. Nicosia.
Rochin, R.I. and C.F. Nuckton. 1980. The Mexicans are coming. How come? *Amer. Veget. Grower*. 28(12). Dec., 1980.
Sheard, G.F. 1979. Intensive greenhouse horticulture in Western Europe overview – Industry background and situation. *Roses, Inc. Bul.* June, 1979.
Shypula, D. 1981. The flower industry in Colombia –where did it come from and where is it going? *Florists' Rev.* 167(4339):10-12, 58-59.
Staby, G.L. and J.L. Robertson. 1982. The international movement of cut flowers. *Florists' Rev.* 170(4403):10-14.
Stewart. E.L. 1989. Report of Eugene L. Stewart, esq. to the 1989 annual meeting of Roses, Inc. *Roses, Inc. Bul.* Nov., 1989.
Sullivan, G.H. 1976. Floral trade policy: Procrastination, protection, public perspective. *Florists' Rev.* 160(4148):30-32.
Takakura, T. 1988. Protected cultivation in Japan. *Acta Hort.* 230:29-37.

CHAPTER 2

STRUCTURES: LOCATIONS, STYLES, AND COVERS

I. INTRODUCTION

Greenhouses have been built on the tops of buildings (Fig. 2-1), on the sides of hills (Fig. 2-2) and buried in the ground with only the roof showing. They have been built as towers with plants moved vertically, constructed in the shape of geodesic domes (Fig. 2-3) or in other special shapes in an attempt to take advantage of solar radiation (Fig. 2-4). Greenhouses have been erected without any internal support (Fig. 2-5). However, as related by Vuoto in 1975, power failures and ice storms can lead to disaster when the roof has no supporting superstructure. One might consider low tunnels (Fig. 2-6) as a form of greenhouse. Regardless of the variations that may be found, we are interested in few styles. In addition to radiant energy transfer from the sun, one must deal with the practicality of growing plants for profit. It is the necessity to cover ground as cheaply as possible with an efficient structure, easily managed, that determines style for a particular climatic region. Although Sims [1976] stated that, theoretically, hemispherical shapes would be the most efficient structure for radiant energy transfer, there are other factors which must be considered for profitable return.

Fig. 2-1. Greenhouse for specialized use on top of an office building. Examples of such specialized structures are common.

Fig. 2-2. Commercial uneven-span greenhouses located on the side of a hill in southern France.

Of the various designs depicted in Fig. 2-7, ridge-and-furrow or gutter-connected (multispan) structures have proven most common for commercial use since they are constructed with minimum expense (Fig. 2-8). The Mansard type, originally proposed for the reason that the roof design allows the sun to penetrate at right angles for many months of the winter under British conditions [Smith, 1967], was later withdrawn from consideration since the proposed advantage was not sufficient to outweigh the increased cost of a complicated structure [Sims, 1976]. The sawtooth design is particularly applicable in warm climates, and there are many variations of this type (Fig. 2-9). If the vertical roof portions are faced south, Bailey and Critten [1985] proposed slight modifications which computer simulations showed to transmit 6% more energy as compared to conventional roofs (Fig. 2-10). The arch, quonset, or barrel designs have proven to be especially economical where the roof covering is flexible, and the reduced superstructure increases the radiation transmitted [Aldrich, 1988].

Fig. 2-3. Geodesic hobby greenhouse. Very large designs may be found in a number of botanical gardens.

It is obvious that greenhouses come in all sizes and shapes. If there is one design that could be considered the standard, it is the Dutch Venlo (Fig. 2-11). It is, probably, the world's most widely used and copied greenhouse design [Anon., 1989]. Unfortunately, some were built in warm regions with inadequate modification for the differences in climate. It took a period of years for engineers to appreciate the difficulties in exporting such structures to regions with hot climates. Furthermore, the Venlo design is not cheap and perhaps represents overbuilding for still-developing production areas where local materials are satisfactory.

Fig. 2-4. Uneven span greenhouse, facing south (right side of picture) in order to take advantage of winter solar radiation.

Fig. 2.5. Air-inflated plastic greenhouse with no supporting internal structure. House requires continuous operation of an air blower to pressurize and inflate the plastic. Air locks are necessary for entry.

The opinion has been expressed [Kruyk, 1982] that the traditional "glasshouse" industry will disappear within 30 years. So far, such rapid changes as Kruyk proposed have not appeared, especially since glass for greenhouse roofs in northern Europe remains the covering of choice in the 14 years after Kruyk's publication [Waaijenberg, 1995]. It remains to be seen if his other possibilities will, in fact, be realized.

With this brief introduction to greenhouse structures, this chapter will develop the importance of location and climate, supporting superstructures, coverings, internal arrangements, perils to the structure and, lastly, general costs involved. It will be necessary to refer to radiation transfer, ventilation, and temperature control, aspects of which will be discussed in greater detail in following sections of the text.

II. CLIMATIC CONSIDERATIONS AND LOCATION

Fig. 2-6. Low plastic tunnels for early vegetable production in mild climates. Such enclosures are common in maritime, warm regions. The main purpose is to hasten growth for early markets.

As with other agricultural operations, greenhouse production is strongly dependent upon climate and weather. In other words, if water and nutrition are readily available, arid or semi-arid regions with their large number of clear days are most productive [Hanan, 1968; Enoch, 1978]. It is the possibility of excessive temperatures in dry areas that can seriously limit growth. The impor-

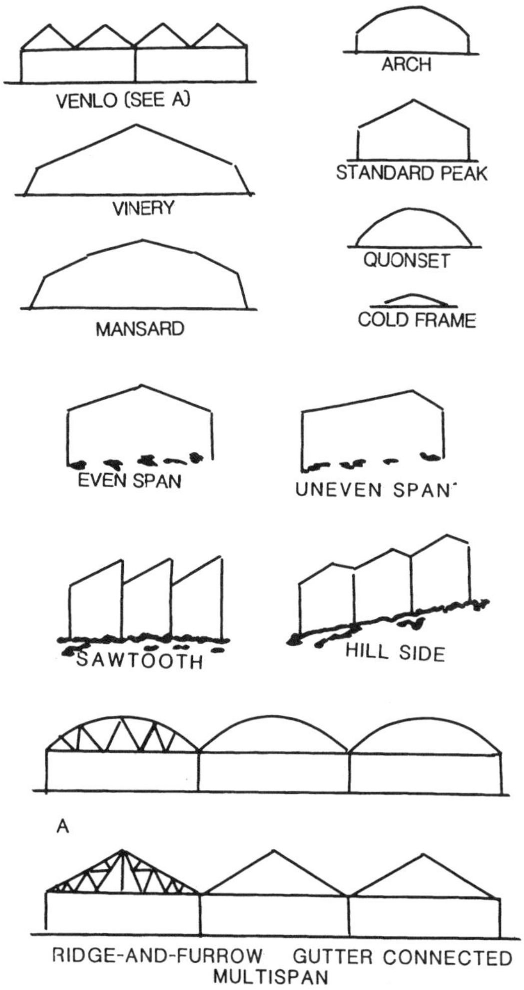

Fig. 2-7. Styles of commercial greenhouse design. The Mansard and uneven span types are not common in large part due to cost. Sawtooth types are common in mild climates. The Venlo type is considered as probably the "standard" design at least in European countries. The very wide spans such as the vinery and mansard are less common.

tance of solar radiation is readily apparent from Fig. 2-12, which shows Besemer's calculations [1966] on carnation cut-flower production as a percentage of possible sunshine. This type of relationship has been derived for other crops, and indicates that locating a greenhouse for maximum efficient return is not always a simple process. There are several regions, such as the U.S. northeastern and northwestern regions, where percentage sunshine commonly falls below 30% of that possible in midwinter. Winters in north Europe are particularly prone to low radiation, as well as regions in Central and South America. Unless supplemental irradiation is employed, temperatures must be lowered to prevent poor flower quality or failure of the crop to develop properly [Canham et al., 1969].

One should exercise care to distinguish between "climate" and "weather," the latter being the day-to-day variations as influenced by the earth's circulatory system, and which play a considerable role in operation of any business [Maunder, 1970]. Climate, however, can be considered to be determined by four factors: 1) latitude, 2) altitude, 3) whether maritime or continental, and 4) the local topography.

Fig. 2-8. Common gutter-connected (ridge-and-furrow) greenhouses found in temperate, semi-arid climates. Both standard peak and arch types present, covered with flexible, corrugated fiber-reinforced plastic (FRP).

The effect of latitude is illustrated in Fig. 2-13, showing the average solar radiation on a horizontal surface for different latitudes, north of the equator. Of course, the day length varies accordingly, so that regions greater than 40° N or S latitude will have significant variations in total solar energy regardless of local conditions such as clouds, obstructions or pollution. Similarly, the average mean temperature will change with season (Fig. 2-14), so that one can appreciate the differences in costs entailed for heating, cooling and supplemental irradiation.

Through location at high altitudes, such as the Bógota, Colombia, region, Kenya, Central America, etc., areas close to the equator can be found which provide average temperatures suitable for a particular crop with minimum structures largely for protection from rain (Fig. 2-15). By locating at sufficiently high altitudes, compensation can also be made for low sunshine. Generally, for every 30 m increase in elevation, flowering of unprotected crops is delayed one day. While total yield may be significantly reduced under such conditions, the lower costs in operation, labor and construction more than make up the difference resulting from low yields.

Latitude and altitude can be markedly modified by the presence of large bodies of water that have an ameliorating effect on climatic extremes by extending the growing season and reducing maximum and minimum temperatures. The classical examples of maritime climates include the Mediterranean basin, the California coast and San Joaquin and Sacramento Valleys, and on the leeward shores of the Great Lakes. These, and other, areas are the most intensive regions for horticultural production to be found in the world, whether protected or outdoors. Climates of England, The Netherlands, Denmark and other north European countries are also modified by the maritime influence unless, as sometimes happens, wind directions change, bringing in severe continental weather. One of the obvious differences between maritime and continental climates is illustrated by Fig. 2-16. Continental climates are characterized by extreme temperature variations over the year and diurnally. In the high plains, along the eastern slope of the Rocky Mountains, as well as in other locations, one may find diurnal temperature extremes in excess of 20 C. These changes obviously have an effect upon vegetation and greenhouse practice in terms of heating and cooling costs, protection from weather damage and requirements for advanced control systems.

Another factor to be considered in choosing a location is local topography. The effect of mountain ranges on precipitation is well known. Thus, for example, the western slopes of the Sierra Nevadas and Rocky Mountains in the U.S. are generally considered the "upslopes" with regard to the prevailing wind direction, and are the

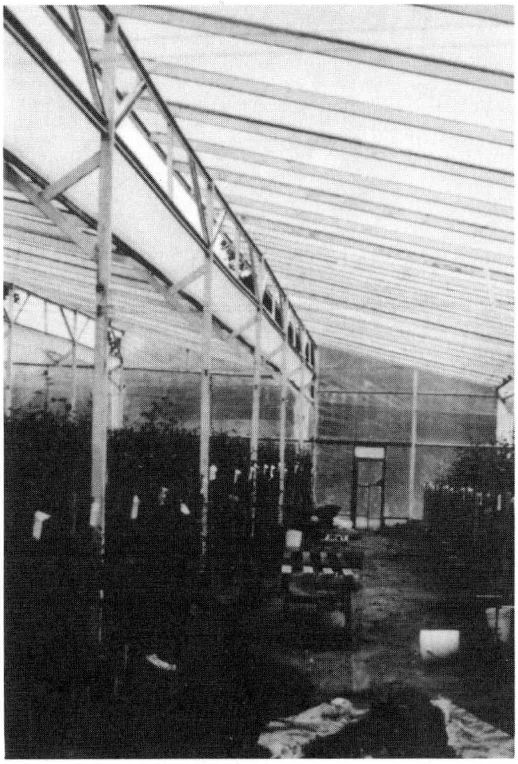

Fig. 2-9. Two types of sawtooth or lean-to structures. The upper picture, cheaply built with a wood superstructure and covered with single polyethylene layer was common to Southern California, a maritime climate. The bottom structure could be found in Central and South American countries.

wet regions where greater cloud cover can prevail. On the other hand, the leeward, or downslopes on the eastern sides are usually in a "rain shadow" so that precipitation is greatly reduced. These latter are the clear sky regions that greatly increase potentials for crop production. The same situation can be found in the Mediterranean regions (Spain, Italy, etc.), the Himalayas and South America.

Another type of topographical influence is air drainage in valleys which reduce climatic extremes. An example for southern California is given in Fig. 2-17, where air drainage at night can raise minimum temperatures by more than 2 C. This behavior of air movement accounts for the extensive orchard production in the high mountain valleys of western Colorado. Even small variations in local elevation, under appropriate weather conditions, can mean a difference of several degrees in minimum temperatures (Fig. 2-18). Railway embankments, dense tree belts, buildings, etc. can act as dams to the movement of air. The greenhouse operator can find himself ensconced in a cold air lake as suggested in Fig. 2-18.

One should be aware of the effect of urbanization on local climatic conditions. Most large cities drastically affect temperature, wind movement and precipitation. Cities are heat islands that can raise the average temperature, decrease or change wind velocity and direction, and increase precipitation. Probably more important for the greenhouse grower is the pattern of air pollution and location of polluting industries. Local meteorological records are very important to the grower since these indicate

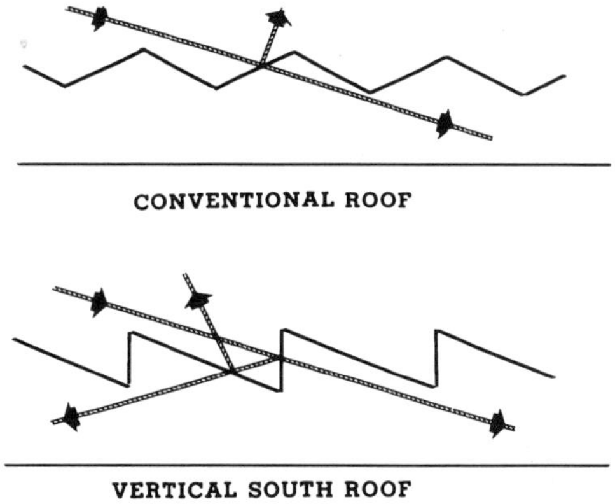

Fig. 2-10. Suggested improvement in radiation transfer by a south facing, vertical roof (sawtooth) design proposed by Bailey and Critten [1985](©1985, Int. Soc. Hort. Sci., *Acta Hort.* 170:193-199).

Fig. 2-11. Examples of the Dutch Venlo design. Note in the top picture the wide and sparse placement of roof ventilators in older styles. While adequate in northern Europe, milder climates require extensive modification, such as in the bottom picture where the house is being fitted for fan-and-evaporative pad cooling. The middle picture shows one type of overhead ventilator arrangement.

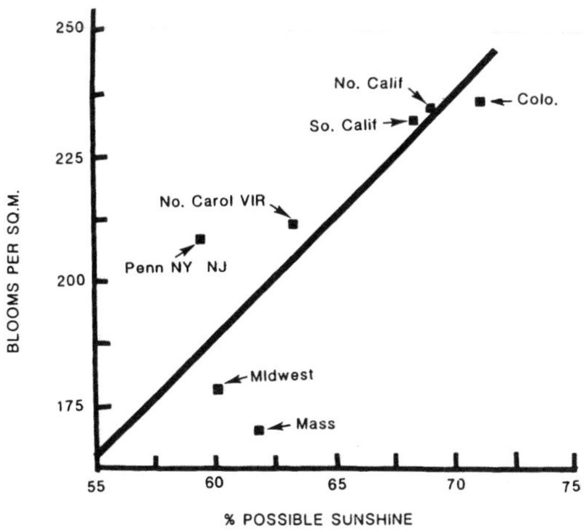

Fig. 2-12. Effect of percent of possible sunshine on carnation cut-flower production [Besemer, 1966].

costs, insurance needs, and help plan efficient operations. The text on the *Value of the Weather*, by W.J. Maunder in 1970, is especially illuminating. One of the original and most outstanding expositions of local climate is the text by Geiger [1965] on the climate near the ground. Individuals able to choose their location for greenhouses would do well to wade through this book for ideas on what to look for. For the United States, the texts by Conway and Liston [1974] and Visher [1954] provide a wealth of information in diagrammatic form on climates. Such information can probably be found for most parts of the world. Therefore, before one even thinks of siting a greenhouse, large savings in future operating costs, and higher production, can be achieved by first considering climate through perusal of meteorological records and such other local information as may be available. Of course, site selection in regard to climate may be one of those compromises one must make in order to achieve other, more important goals: for example, market, water supply, labor, etc., or the fact that the particular piece of property is available. The effects of climate on greenhouse design are pictorially summarized in Fig. 2-19. Plant production can be outdoors or in very minimal structures.

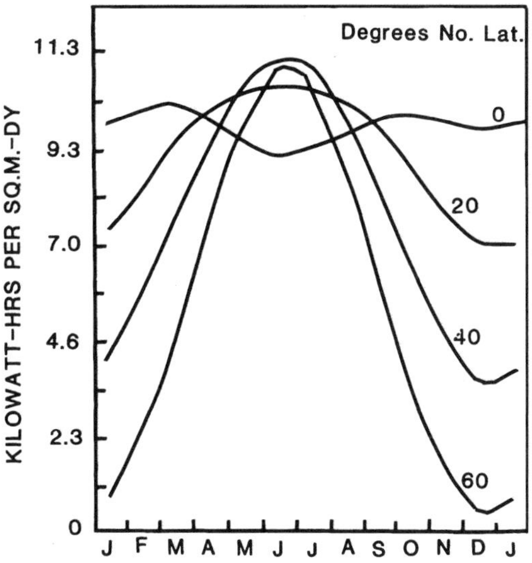

Fig. 2-13. Average solar radiation on a horizontal surface at sea level for different latitudes and seasons [Albrecht, 1951].

III. SITE SELECTION

Site selection is a matter of good thinking. Walker and Duncan [1974a], Aldrich and Bartok [1985], Sheldrake and Sayles [1974], Courter [1965] and Smith [1988], among others, have provided details. If the operation is to be retail, an improved road should be present, together with the necessary utilities such as water, fuel, telephone and electrical power. Many urban regions require that off-street parking be provided for employees. Certainly, if the greenhouse engages in retail trade, additional, paved parking will be required, depending upon the number of customers to be accommodated at any one time. One should also determine labor availability since the operation may have to compete with other industries for a limited labor supply. Several studies have been conducted on labor required. Short [1982] reported six to eight employees per hectare for Dutch greenhouses. Meijaard [1995] gives a figure of one person per 2000 m^2 or five per hectare. The number will vary with the crop grown. In 1976, Krause stated that five men per hectare were required for rose production, 6.7 for carnations and four for year-round chrysanthemums. For vegetables, Dalrymple [1973] estimated 3.5 to 9.6 man-years per hectare depending upon

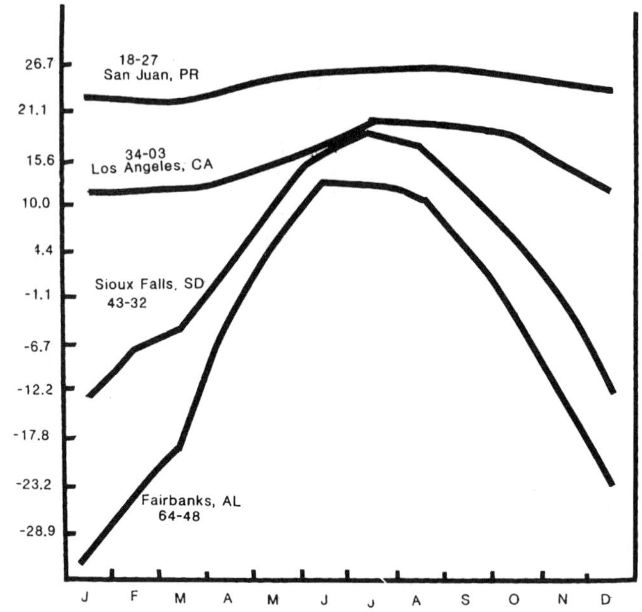

Fig. 2-14. Mean monthly temperatures for various locations. Seasonal temperature variation increases the higher the latitude [From *Climate and Agriculture*, F.W. Went, ©1957 by *Scientific American, Inc.* All rights reserved].

whether double cropping was practiced and the crop produced. William Swanekamp's [1994] discussion of the 5.6 ha Kube-Pak operation indicated 40 full-time employees, swelling to 120 during the main season –or 7 to 21 employees ha^{-1}. As mechanization continues to develop, the labor requirement will decline. However, in some countries with cheap labor, labor is a substitute for operations usually mechanized elsewhere.

The water requirement will vary with crop, structure and local climate. One will find

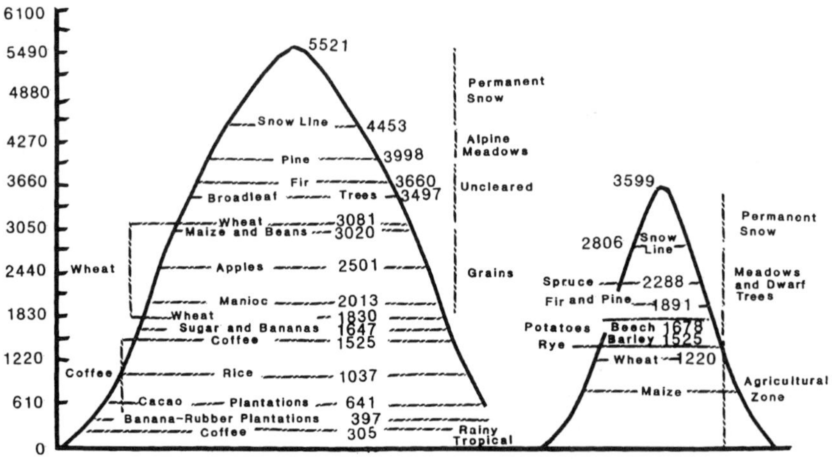

Fig. 2-15. Illustration of the effect of altitude on unprotected crop production at two different latitudes. The peak on the left is at the equator, the one on the right at about 40° N [From: *Crop Adaptation and Distribution* by C.P. Wilsie © 1962 by W.H. Freeman & Co. Used with permission].

Fig. 2-16. Mean daily temperatures in maritime (San Francisco) and continental (St. Louis) climates [From: *Crop Adaptation and Distribution* by C.P. Wilsie © 1962 by W.H. Freeman & Co. Used with permission].

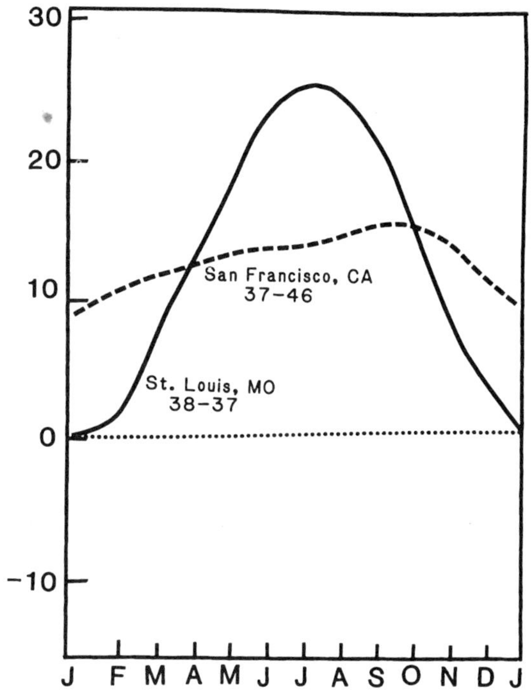

considerable variation in amounts given in the literature. It is not unusual to find recommendations of 140 m^3 ha^{-1} dy^{-1} (1 to 2 mm dy^{-1}) [Walker and Duncan, 1974a], which would suggest a maximum water supply of over 51000 m^3 ha^{-1} yr^{-1}. Note that 1 to 2 mm dy^{-1}, a 100% difference, is not unusual. A single crop of tomatoes may require 3135 m^3 ha^{-1} [Courter, 1965], and Hanan [1967] used a figure of 20000 m^3 ha^{-1} yr^{-1} (200 cm). Jasper and Hanan [1967] found that standard carnation flowers required 3.4 liters per flower. This amount did not vary regardless of irrigation regime and number of flowers produced. That is, the higher the water supply to the root system of the carnation (assuming the substrate will allow frequent irrigation), the more flowers produced. Thus, if the grower managed to grow only 300 cut flowers per m^2, the water actually used by the crop would be over 11000 m^3 ha^{-1} yr^{-1}, whereas yields of 500 per m^2 would require over 18000 m^3 ha^{-1} yr^{-1}. This situation of high water availability, in combination with soilless culture was outlined by van Nierop in 1982. He stated that the more water you allow through the root system, the better production will be. This recommendation should be taken with a grain of salt. I would, as a general rule, suggest a grower plan for a minimum water supply of at least 20000 m^3 ha^{-1} yr^{-1} up to a maximum of 50000 m^3. A grower should allow additional water supplies, especially if evaporative pads are in use. This water supply should have a maximum salinity of 700 dS m^{-1} with no debris or sand [Hanan, 1982]. Further information on water is provided in Chapters 5 and 6.

Electrical service will vary with crop and climate. If no photoperiodic or supplemental irradiation is employed, a 1000 m^2 range will require about 4 to 6 KW [Walker and Duncan, 1974a] for forced ventilation. Some engineers may consider this excessive. The electrical service entrance size for anything 1000 m^2, or smaller, recommended by Aldrich and Bartok [1985], is 60 amperes at 240 volts AC (U.S.A. 60 Hz) (14.4 KW). Over 1000 m^2, the service size should be increased to 100 amperes (three phase), and to 600 amperes with greenhouse areas above 10000 m^2. With areas of 15000 m^2, or greater, the voltage entrance should be 440 VAC, with amperage increasing to 800 A for anything approaching 30000 m^2. Where the grower is using electricity at night for photoperiod control, the same service as employed for exhaust fans (fan-and-pad evaporative cooling) can be used since neither is in use simultaneously. But, supplemental irradiation, which requires high intensity lamps, necessitates much larger service entrances. For example, if one assumes an irradiation level of 30 watts per sq.m., and a radiation source that is 40% efficient, then 75 KW may be required, which would mean a service entrance of over 300 A at 240 VAC for 1000 m^2. What the grower should do is to specify all electrical equipment that will be utilized *before* construction, and to base service and power requirements on that estimate. One should also keep in mind that supplying companies often impose a "demand" charge. Sudden demand for power will increase the electrical bill. Operators should attempt to "smooth" electrical demand. Large buildings and industrial concerns will often include computer programs especially designed to avoid, if possible, excessive high demand. In developing countries, power supplies can be limited and unreliable. A grower may combine

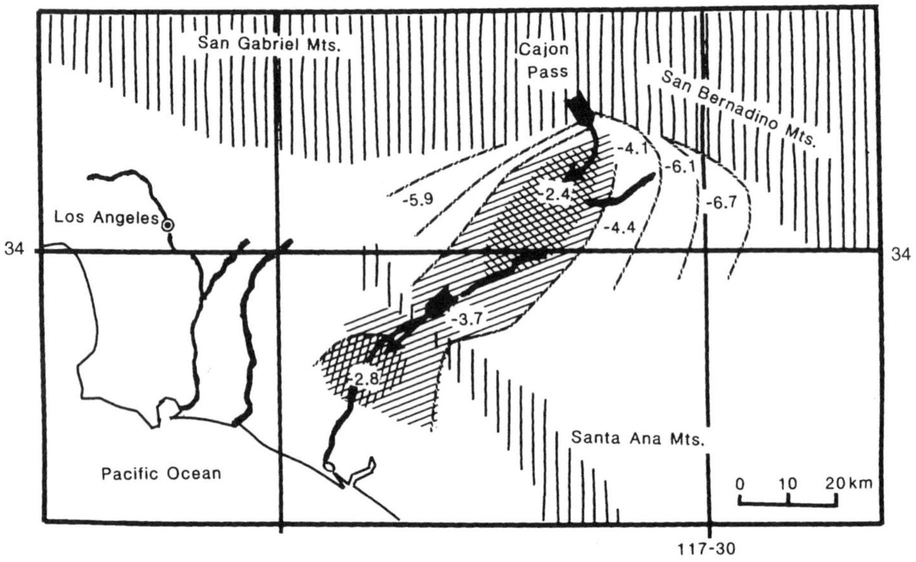

Fig. 2-17. Effect of local topography on minimum temperatures at night. Minimum temperatures are higher in the path of the main air movement down from Cajon Pass [Young, 1921]. This serves to illustrate the effects of downvalley night winds on ameliorating temperatures. Colorado experience shows that a difference of a few meters can result in severe frost damage to unprotected crops.

Fig. 2-18. Diagrammatic sketch of temperatures in low areas on still nights. Location of a structure should be on the slope within the thermal belt. The vertical scale is greatly exaggerated.

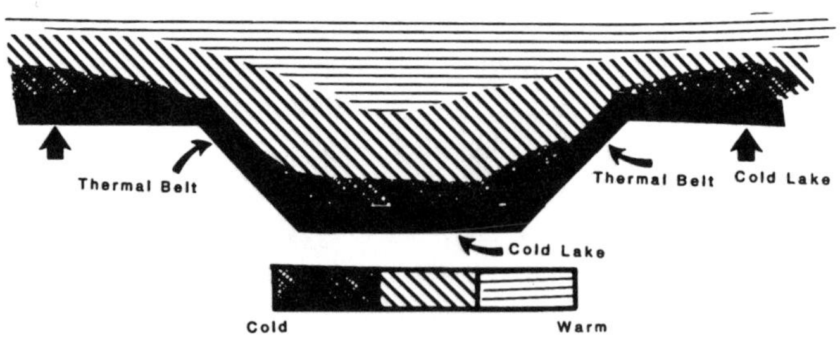

Fig. 2-19. Representation of how climate affects ornamental production. The top picture shows outdoor carnation production in Kenya in the early 1970s. The second picture is chrysanthemum cutting production in cloth houses in Florida. The third picture reveals polyethylene structures in Colombia at about 3000 m elevation. The last picture shows fully enclosed, heated and cooled greenhouses in Colorado at about 1600 m elevation, 39° N latitude.

Fig. 2-20. Problems arising from locating a greenhouse in an area with poor drainage and a high water table. Employees must wear boots and work in a disagreeable environment. Splashing will spread disease in spite of raised benches.

cogeneration, which means he provides his own generation with the waste heat used to help heat the greenhouse. Further details are given in Chapter 3. Information on fuel requirements is provided in Chapter 4.

Some grading and leveling will be required on almost any site. A perfectly level site is not desirable. Smith [1988] stated that most modern greenhouses are built with a slope of 1 in 200 to facilitate drainage. This also helps ground drainage. Beyond a 1% slope, costs of leveling will increase markedly, and any fill will require suitable compacting for the greenhouse foundations. Greenhouses built on steep slopes will delay product movement, decrease worker efficiency, and cause problems with uniform water distribution from most irrigation systems. At least considerable effort will be required in design.

Since greenhouses are labor intensive, the owner should always pay particular attention to facilitate worker movement and ease necessary tasks [Hendrix, 1975]. Obviously, if the crop is to be grown in the ground, the grower should assess depth, drainage and fertility of the native soil. In some parts of the world, operators have gone to great effort to import suitable substrates, such as sand. This, of course, increases costs significantly. With pot plant production, or the implementation of hydroponic or soilless culture, problems with the local soil can be eliminated except for the need to provide drainage from the structure. Location in areas with high water tables can lead to serious difficulty. Even if the crop is raised above ground level, in benches, surplus water can lead to considerable difficulty (Fig. 2-20). A rainfall of 100 mm means that 1000 m^3 water must be disposed of from 1 hectare.

There are several miscellaneous factors that need to be considered, such as instances of locating very expensive growing structures on land formerly occupied by petroleum industry with possible serious repercussions. Waste disposal must now be considered. California growers have had to install waste disposal and water recycling systems to avoid downstream pollution. Severe restrictions have been imposed in The Netherlands. Herbicide drainage from greenhouse operations has also led to legal suits. The grower cannot afford to neglect the quality of his waste water, nor where it goes. Some cities may have strict requirements about what can be disposed of in the local sewer system. The converse of whom you pollute is what others may do to you. There have been instances of inadvertent chlorine releases from nearby industry that have severely damaged crops, contamination of shallow wells from petroleum refineries and storage facilities, and accumulation of dust on greenhouse roofs from nearby, heavily traveled graveled roads. Urban pollution may travel up and down river valleys and pollutants may be very high along highways, especially in the early morning and late evening hours. Excessive run-off and possible flooding should always be considered in low areas. The owner cannot restrict himself to his property, but he must think about the impact his operation may have on the surrounding territory. He should be a good neighbor as well as assess dangers to his operation. As pointed out in the previous section, locating on flat land may mean you are locating in an area where there will be a "cold air lake," which will increase fuel costs and increase problems with environmental control. Locating on a slope, if there are no "dams" to impede usual air slope drainage, will help greenhouse climate control.

IV. ARRANGEMENT

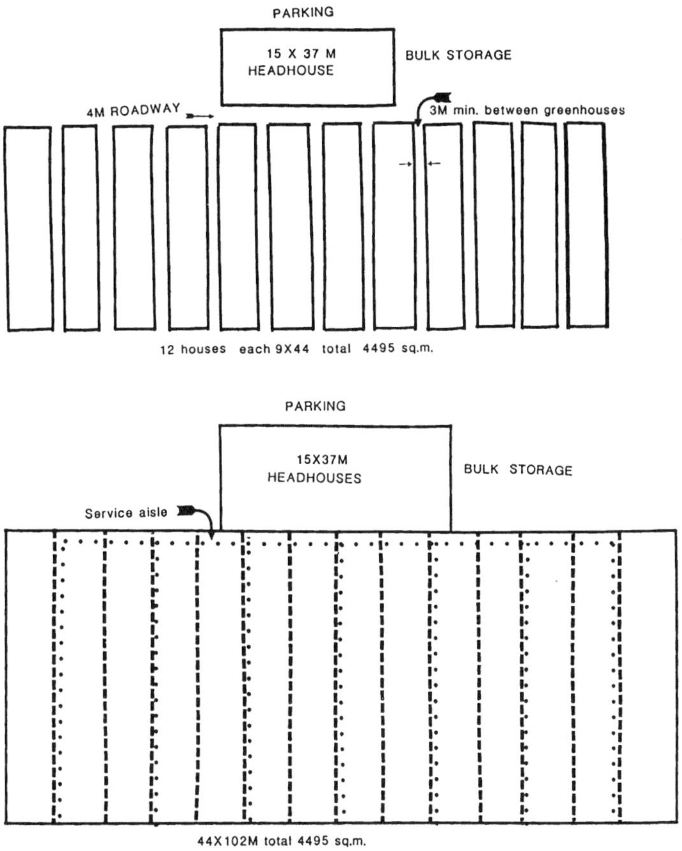

Fig. 2-21. Two systems for arranging greenhouses: separate, individual (**upper**) and gutter-connected, block (**lower**) [Aldrich and Bartok, 1985].

The arrangement of structures for retail business has been discussed by Aldrich and Bartok [1985]: 1) Locate service structures to the north of the greenhouses if possible. 2) Separate customer traffic from supplier traffic. 3) Arrange the retail area to keep customers from the production area. 4) Provide convenient access and parking to the sales area. 5) Arrange the layout so that traffic moves from clean areas to dirty ones to reduce possible disease and insect problems. 6) Locate any residence to ensure privacy. 7) Locate any windbreaks upwind, at least 30 meters from any structure. And 8), arrange the sales area so all customers must pass the cash register.

Regardless of arrangement, particular attention should be given to facilitating cleanliness. Sanitation is the first line of defense in disease and pest control. It also makes a point with employees in setting up good habits. Several of the above suggestions can be applied to wholesale operations.

There are essentially two basic methods for greenhouse arrangement, separate structures or gutter-connected (Fig. 2-21), depending upon the crop grown. The former is easily expanded or contracted by moving the houses into or out of production. This might be the best for bedding plant operations where not all houses are needed for many months of the year. Movable greenhouses have been built, some glass covered, others with light-weight plastic [Harrison and Rogers, 1982]. Moving the greenhouse allows a crop to be grown outside while the heated structure is utilized to start another crop or different species. Some bedding plant growers, by placing their plants on movable benches, move the plants outdoors in the spring by opening the ends of the greenhouses. This allows the plants to be hardened off while others are being started inside. Separate houses also allow species with unique requirements to be separated and grown without interference, and isolation of new, imported plants, is easily arranged until shown uncontaminated. If spaced far enough apart, shading from one house to the other can be avoided. One disadvantage is that surface area of the greenhouses to ground covered will be larger with greater energy requirements in the winter [Bartok, 1987]. Unless connected by covered service alleys, product and employees must move outdoors in inclement weather. Costs of materials and construction will be higher. Another possibility is increased difficulty in handling employees.

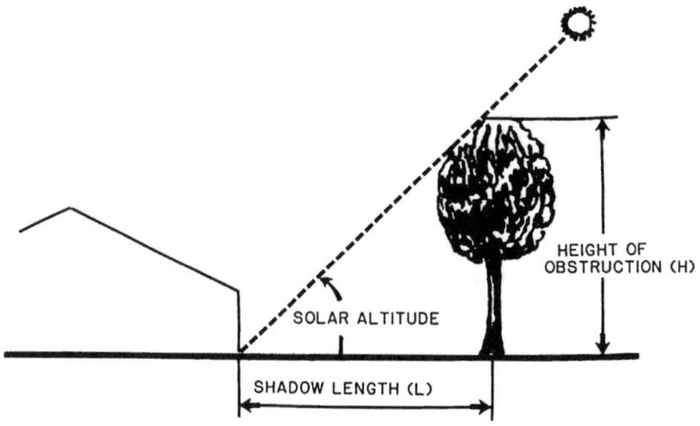

Fig. 2-22. Right triangle created by an obstruction and solar altitude. The latter can be obtained from Table 2-1. The shadow length (L) can be calculated, or the distance required between obstruction and greenhouse [Walker and Duncan, 1974a].

The gutter-connected arrangement is generally most common and cost effective. All activities are inside one building and a central heating system can easily serve most areas. The service building can be attached to the greenhouses for rapid movement. While these structures can be zoned for heating, high outside winds will cause high infiltration. Heat will tend to migrate downwind in a gutter-connected structure; and if fan-and-evaporative pad cooling is employed, the entire area covered must be cooled simultaneously. The grower will have to arrange varieties in accordance with warm and cold areas. Such houses are also prone to snow accumulation in the gutters. Where there are heavy snow loads, growers run heating lines under the gutters. With individual houses, particularly the standard peak (Fig. 2-7), snow will usually slide off. Examples of interior designs are given in Figures 2-79 through 2-84.

Particularly on southern exposures in the northern hemisphere, greenhouses should be placed far enough away to avoid shading by surrounding obstacles during the day (Fig. 2-22). As a rule the greenhouse should be located away from the obstruction 2.5 times the obstruction's height. Calculation of the necessary distance (L) can be done by multiplying the height of the obstruction by the cotangent of the sun's altitude ($L = H(\cot A)$). The solar altitude (A) for the shortest day of the year, for various latitudes, can be read from Table 2-1.

This discussion indicates that the possible arrangements are practically infinite. The operator should carefully plan his site, outlining the major traffic flows of machines, personnel and materials. Positioning of outdoor storage and service buildings in accordance with easy access and movement of product is particularly important. Far too often, the grower, perhaps limited by available land, money and site characteristics, finds himself locked into an inefficient operation if he has not spent some time to arrange structures properly. Explanation of suitable layouts is difficult to find in the horticultural literature. It would be desirable to hire a suitably experienced engineer during planning stages.

Table 2-1. Solar position (altitude) for various latitudes and solar times.

Date	Latitude	Time	Altitude
Dec. 21	24°N	0800 & 1600	15
		1200	43
	32°N	0800 & 1600	10
		1200	35
	40°N	0800 & 1600	6
		1200	27
	48°N	0900 & 1500	8
		1200	19
	56°N	0900 & 1500	2
		1200	11
	64°N	1100 & 1300	2
		1200	3
June 21	24°N	0800 & 1600	36
		1200	89
	32°N	0800 & 1600	37
		1200	82
	40°N	0800 & 1600	37
		1200	74
	48°N	0800 & 1600	37
		1200	66
	56°N	0800 & 1600	36
		1200	58
	64°N	0800 & 1600	34
		1200	50

A. SERVICE FACILITIES

While a number of activities can be carried out inside a greenhouse, even at the expense of some crop area, almost any greenhouse requires a supporting facility, the arrangement of which will depend upon crops grown, climate, heating system and whether the operation is wholesale or retail. Fig. 2-21 shows a headhouse for the operation, but it is not always necessary to place the service building to one side of the main range. For convenience, service can be situated between, or, in the center of several ranges of greenhouses –at the possible expense of some shading to the north. This may be the best compromise. Or, with a mind to future expansion, the first group of houses, with the support facility, can be placed to allow additional

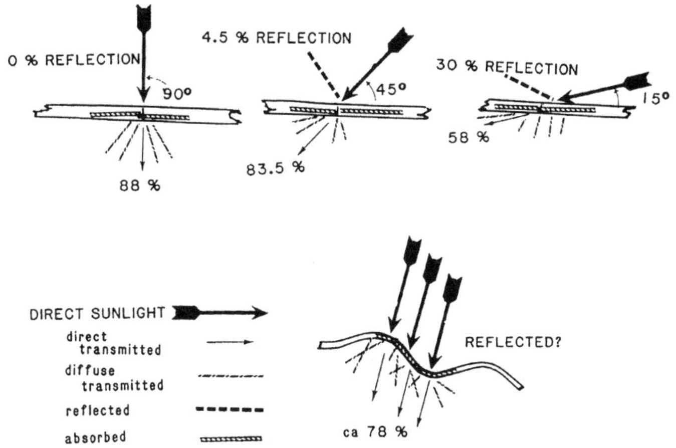

Fig. 2-23. Effect of angle of direct solar incidence on transmission of radiation for glass and FRP.

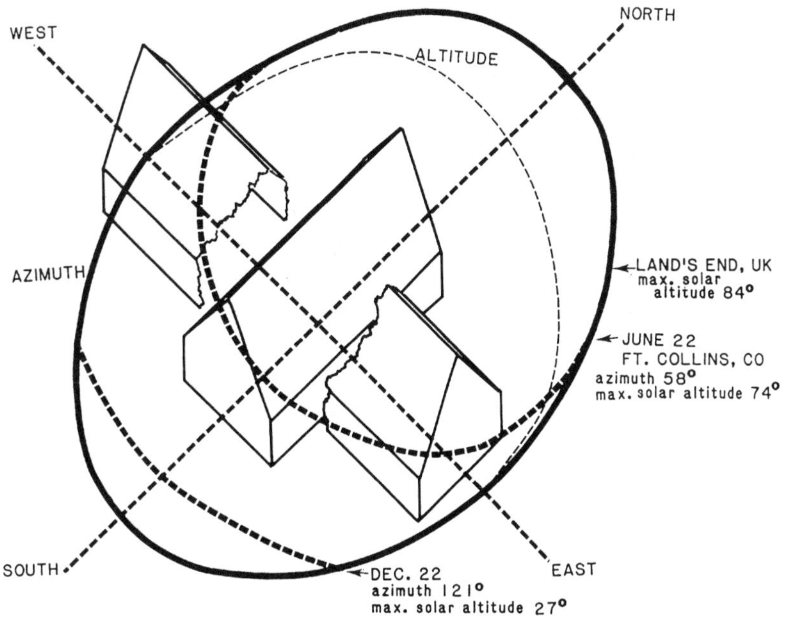

Fig. 2-24. Path of the sun on the longest and shortest days at Ft. Collins, CO (39° N) with respect to an E-W and a N-S oriented greenhouse [Hanan, 1970].

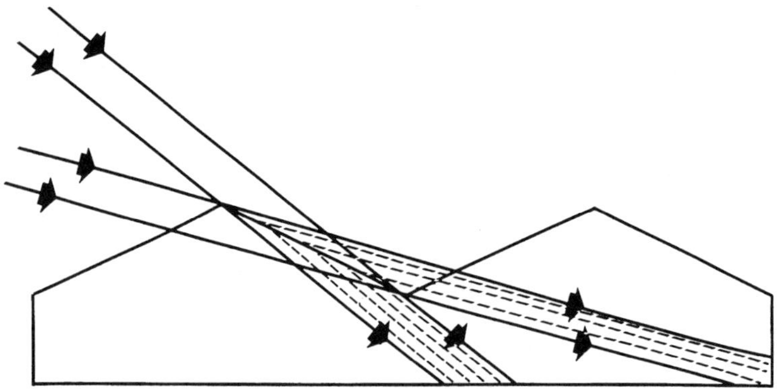

Fig. 2-25. Winter and equinoctial shading in a gutter-connected, east-west greenhouse at noon on a clear day in the U.K. [Edwards, 1965].

greenhouses on other sides. In that situation, interior planning for product flow, customer service, etc. should consider the possibility of changed flow patterns in the future. Additional expense may be contracted initially for the necessity of making the first step efficient, or making future alterations easily.

As a rule, for greenhouses covering an area up to 4000 m^2, about 15 m^2 per 100 m^2 greenhouse floor area will be needed to adequately service the operation. Between 4000 to 8000 m^2, the area per 100 m^2 can drop to 10 m^2. Over 8000, one may use the figure of 8 m^2 for every 100 m^2 greenhouse area. Of course, this will vary with several factors: e.g., if a centrally located heating plant is used instead of heaters inside the structure. If the operation forces bulbs, or species such as azaleas, it will require specialized, temperature controlled storage. An operation producing bedding plant seedlings will use specialized germination chambers. Is an area required for retail selling or buyers coming to a wholesale establishment? An indoor loading area is required for tender foliage plants in inclement weather. These, and many other variations preclude any general floor plans. However, one may outline some of the facilities that are necessary in almost any greenhouse support building in temperate climates:

1. Separate storage rooms for pesticides, fertilizers, and herbicides are required. These three general groups should never be stored together. The problems of cross-contamination, misuse and self-pollution are all too common. Pesticide and herbicide storage is locked. Fertilizers such as NH_4NO_3 are explosive. Acids such as H_2SO_4 and HNO_3 are highly corrosive. In combination with other chemicals, very unstable, explosive situations may occur. Fire and health authorities in many developed countries are very sensitive to the existence of such materials.
2. While bulk materials such as soil amendments, composts, etc. can be stored outside, space for immediate use inside a protected building might be necessary.
3. Some type of refrigerated storage is almost always required, capable of being cooled to 0 C for many cut-flower species. At such low temperatures, specialized cooling systems are required. Ordinary, walk-in coolers, capable of maintaining 5 to 8 C will not usually be adequate except for vegetables, potted plants and transplants. However, some vegetables require special equipment, such as vacuum containers for rapid lettuce cooling, cold-water sprays for other types, or special gas chambers for ripening fruit such as tomatoes.
4. A reasonably well-equipped space for machine repair and fabrication is usually necessary. Or, it can be multipurpose for potting, transplanting, etc. If a cut-flower operation, space may be needed for grading and packaging. A pot plant range will need an area for packaging and for loading, preferably inside. Vegetables will require cleaning, grading and packing equipment.
5. Showers and restrooms are required by local code and, in the U.S. the EPA mandates such areas with appropriate washing and changing areas. The grower might want to include an employee rest area or lunchroom. Clean and neat employee facilities go far to improve employee performance.
6. Spaces for heating plant (boilers, etc.), fertilizer injectors, irrigation control, temperature control (computer?), pesticide pumps and emergency generators are necessary. Electronic equipment should

not be located where possible fumes from chemicals or operations will be prevalent. Greenhouse atmospheres do not go well with unprotected electronic systems.

7. Office space is usually required for accounting, records, order taking, supervisors, etc.
8. Storage space will be required for the materials needed for cultural handling of the crop. These may be support wire for a crop, pots, packaging materials, labels, trays, *ad infinitum*.
9. In some specialized producers, the benches may actually be brought into the headhouse where they can be planted, or the product removed for sale, and then returned to the greenhouse with new plants. These may require space and equipment not commonly found.

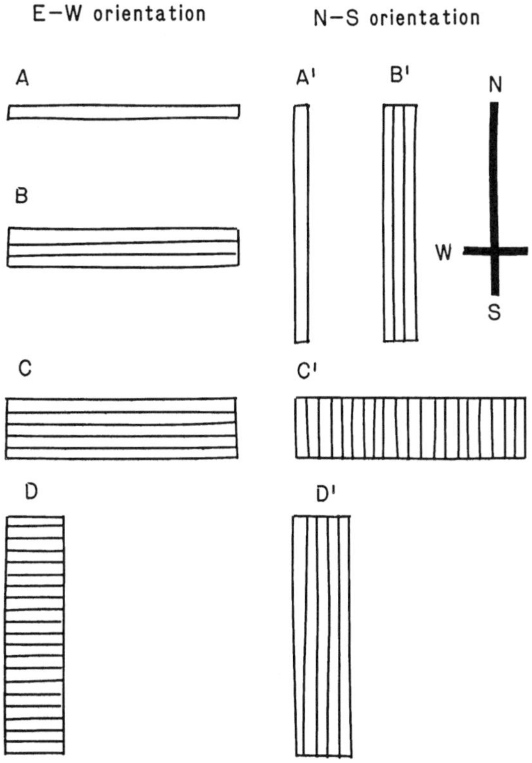

Fig. 2-26. Greenhouse arrangements examined by Kozai et al. [1978]. All N-S orientations had good radiation distribution, but low transmissivity with the exception of **C'**, which had medium transmittance to direct solar radiation. Uniformity was good in **A** but bad or worse in all other E-W configurations. Transmissivity, however, was high in **C** and **D** and very high in **A** and **B**.

Bartok [1987] suggests that steel buildings offer easy construction, low maintenance and good resale value. Wood buildings usually have lower cost. For ease in maneuvering tractors, trucks and materials, a minimum width of 7.5 m should be used. In large operations, building widths of 18 to 24 m are common.

It can be seen that greenhouses and their handling are simple neither in their construction nor in their operation. Careful thought is necessary to their successful operation, not the least of which are location, siting and arrangement. Information on design and layouts, as indicated earlier, is not commonly found in readily available literature. Contact with extension engineers, however, would be a good place to start.

IV. ORIENTATION

Fig. 2-23 shows the obvious fact that for maximum transmission of direct solar radiation into a greenhouse, the transparent covering should be at right angles to the sun's rays. This has led to considerable work by English and Dutch workers on the best orientation of common structures for maximum efficiency. Unfortunately, the sun is not fixed in position, and the relationship between direct solar radiation and the structure varies throughout the year (Fig. 2-24). At latitudes greater than 40°, the low altitudes of the sun (angular distance of the sun above the horizon) in the winter calls for some consideration. The sun, in the winter, also rises much further south. The azimuth in the morning, as measured in degrees clockwise from north, increases. The work in Europe, which is, for the most part north of 50°, indicates that east-west ridge-line orientations are best for winter conditions [Sims, 1976; Whittle and Lawrence, 1959]. Sims [1976] states that N-S orientation can have a 16% loss on bright days. Studies by Mattson and Maxwell [1971] showed that east-west greenhouses cast fewer shadows than those oriented north-south. There are differences between large, clear-span houses and smaller ones arranged as

Fig. 2-27. Note the difference in orientation of the greenhouses in the foreground versus those in the background. In this instance, in mountainous countryside, structures have had to conform with the steep topography.

multispan or gutter-connected. Small spans are likely to cast large shadows on bright days from the gutters in east-west greenhouses (Fig. 2-25). Kozai et al [1978] showed that the daily average of direct solar radiation transmissivity was higher in east-west houses when the ratio of the side wall height to the width of the span was greater than 0.5. A 10 m wide house with 4 m eaves would have a ratio of 0.4. This appears excessively high, but accounts for the trend toward higher houses. Venlo spans, however, which are usually 3 to 4 m wide, easily meet the above ratio. Transmissivity decreased as the number of spans increased in east-west, gutter-connected houses, whereas transmissivity in a north-south arrangement was independent of the number of spans. Transmissivity in a N-S house decreased when the ratio of length to width was greater than 5. E-W houses were less affected by this ratio. However, the distribution of radiation in an E-W house was less uniform than in N-S structures. The transmissivity of a single-span house was more sensitive to orientation than that of a multispan range (Fig. 2-26). Although, large, clear-span greenhouses show advantages in terms of light transmission, costs of construction are higher than multispan ranges [Sims, 1976]. For multispans, the advantages are cheaper, good land utilization, flexible floor area, good summer light, good summer crop response, and less heat utilization. Clear, or wide-span houses offer flexible use of floor area, good winter radiation, good winter crop response, and easy installation of thermal screens.

Fig. 2-28. An early style wood frame greenhouse in South America. Even though the posts are treated, they will only last four to six years.

Some authors [e.g., Courter, 1965; Sheldrake and Sayles, 1974] either dismiss the importance of orientation or neglect to mention the differences. According to Walker and Duncan [1974] and Manbeck and Aldrich [1967],

Fig. 2-29. Two relatively modern, mild climate greenhouses in South America. More durable, light-weight steel superstructures improve radiation transfer and distribution. Note the height of the gutters, a trend of all modern greenhouses. The climates in which these structures are located have high rainfall, so the roofs have steeper slopes to remove water. Ventilation is provided through the sides, ends and peaks of the roofs.

Fig. 2-30. Two early styles found in the Mediterranean. Note the shallow roof pitches, indicating low rainfall. No ventilation is through the roofs, making these rather hot houses in the summer. However, the plastic may be allowed to deteriorate, resulting in a completely open structure in the summer.

the E-W arrangement is preferable in the winter above 40 to 45° latitude. At other times of the year, and closer to the equator, N-S is better. Hanan [1970] suggested that any advantage of E-W versus N-S would disappear as one approached the equator.

Another factor to consider is that most investigations deal with direct solar radiation, available only on clear days. Smith [1988] pointed out that 70% of winter radiation in northern Europe is diffuse (less than 30% possible sunshine). The practical result is that conventional roofs represent the best compromise between good energy transmission and capital costs. In fact, an overcast sky is often considered a uniformly distributed radiation source, and it makes no difference as to orientation of the structure. The north wall and roof become just as important as the south sides in terms of available energy. Thus, opaque insulation of the north wall for purposes of energy conservation is not the best solution. The advantages of clear sky regions for maximum production, as pointed out by Hanan [1968], become more obvious. Other problems, such as the necessity to conform with the particular site, may result in the decision to disregard orientation regardless of any conceivable advantage (Fig. 2-27). This is one of the numerous compromises a grower will have to make in planning his structures.

V. SUPERSTRUCTURES

While this section deals largely with standards and designs of structures in developed countries, the article by Monteiro et al. [1989] points out differences between those regions in the process of developing an industry, and those possessing an industry for several years with a well-developed infrastructure. For example, Fig. 2-28 compares an older structure, erected with local materials versus more modern houses in Fig. 2-29 for the same general locality. Fig. 2-30 compares two styles utilized in the Mediterranean, while Fig. 2-31 shows a wooden structure erected in Central America. Note that all of these examples are unheated, covered with single-layer polyethylene. Where there is little precipitation, roofs may be nearly flat. Standards imposed by regulatory agencies may be minimal. Monteiro et al. make the point that the cost of more advanced structures does not return

Fig. 2-31. Home-grown rose greenhouse in Central America. The curved trusses are an attempt to maintain tightness on the single-layer polyethylene roof. Ventilation is through the roof and gutters only.

profitably. There is usually a need to provide more ventilation for unheated houses. Provision for ventilation has been addressed by several authors in the last decade, e.g., Dayan et al. [1986a, 1986b], Brun and Lagier [1985] and Castilla et al. [1990]. Both Monteiro et al. [1988] and Montero et al. [1985] point out rapid progress in new designs while maintaining expense as low as possible. The basic limitations of these structures for mild climates are the low temperatures and high humidities during the night. Some of these problems will be addressed in later chapters.

Modern greenhouses have progressed considerably beyond the type depicted in Fig. 2-32. Standards have been developed in several countries [Richardson, 1987; Spek, 1985; Waaijenberg, 1995; Spelman, 1988; NGMA Standards, 1985]. A general European standard is in the process of being formulated. The result has been to improve radiation distribution inside structures through new coverings, reduced superstructures, wider panes of glass, and greater heights between ground and eave (Fig. 2-33). The uniformity of the Venlo design has made it possible to mass-produce greenhouse

Fig. 2-32. A greenhouse probably built in the 1930s. Steel with wooden sash bars, narrow glass and interior supporting members, reducing interior versatility, with manual ventilation. This style is now uneconomical and completely outdated. The picture serves to emphasize the improvements made in recent years.

Fig. 2-33. Two examples of modern greenhouses in temperate climates. **Upper**: Commonly found in the Western U.S., arch type with FRP covering. **Lower**: A Danish wide span with glass covering.

components, simplify construction techniques, expedite erection and standardize heating systems, irrigatior installations and environmental control equipment (Fig. 2-11) [Anon., 1989].

A. DESIGNS AND STANDARDS

Greenhouses are lightweight structures, seldom exceeding 9 m in height. The problem is more often to holc them down than to keep them up. Windload in parts of the world (e.g., Colorado) may be more important thar snow loads. Standards for construction vary with country, although the engineering of such buildings remains much the same. Failure to maintain some standards can lead to severe damage and financial loss [Spelman, 1988].It was Spelman's conclusion that pressure by the industry on manufacturers of greenhouses tended to reduce

Table 2-2. Comparisons of design wind pressures obtained from calculation with various codes in Europe, U.K., and the U.S. [Adapted from Aldrich and Wells, 1979].

Source and method	Basic wind speed (m/s)	Design wind speed (m/s)	Design wind pressure (N m^{-2})
CP3. 3 s gust. 50 yr recurrence, isopleth map, 10 m	45	33.5	690
STL 106, maximum 3 s gust, 50 yr recurrence, 10 m	45	34	710
Gradient wind at 275 m, Scruton and Newberry	33	41	1030
Surface wind at 10 m, mean hourly speed, 50 yr recurrence, Scruton and Newberry	23	35	750
Hoxey and Wells with CP3	45	37	840
Holland, surface wind at 10 m	45	33	510
NGMA, fastest speed at 9.2 m, 50 yr recurrence	45	39	1210
ASAE, fastest speed at 9.2 m, 25 yr recurrence	49	39	935
France, basic pressure at 10 m	45	45	770

Design wind speed and pressure calculated for a 3 m height.

Fig. 2-34. Note the curvature of the bottom cord of the truss. The grower has attached cable to this cord as a means to support end posts on his rose benches, a purpose never intended for the structure.

structural strength in the interests of costs and to reduce shading. Wind speeds of 46 m s^{-1} resulted in severe and widespread damage in the U.K. Lack of attention to safety led to severe damage in The Netherlands in 1972 and 1973 [Waaijenberg, 1995].

In addition to bearing the weight of the structure, greenhouses are subject to a number of other stresses that must be taken into account, as pointed out above. In addition to the **dead load**, the designer must take into consideration **live loads, wind loads**, and **snow loads**. The requirements will vary with location, topography and local code authorities. For the U.S., Reilly's [1988] survey indicated three major model building codes. The Uniform Building code covers the area west of the Rocky Mountains; the Southern Building Code, the southern U.S., and the Building Officials Code Association, the rest of the country. Local building officials in the U.S. are the final factor in application of any specific code to the locality [Bartok and Aldrich, 1989]. In the smaller countries of Europe, a central code body provides a single authority. The British Standard is BS5502, introduced in 1981, which requires compliance with the British Standard Code of Practice, CP3 [Richardson, 1987]. The Dutch Standard is NEN3859, sometimes regarded as the "Venlo" standard [Spelman, 1988; Spek, 1985]. Unfortunately, competitive pressures and differences among code authorities can result in widely differing requirements. For example, some municipalities in the U.S. may require the structure to carry live loads exceeding 146.9 kg m^{-2}, whereas a short distance into the country, the county officials may require 73.6 kg m^{-2}. Greenhouse manufacturers have attempted to alleviate some of the more unrealistic requirements through the proposal by the National Greenhouse Manufacturers Association on greenhouse standards [NGMA, 1985]. Some of the variation in wind loads between codes and proposals can be noted in Table 2-2, as the result of a survey by Aldrich and Wells [1979]. Their calculations showed the 1972 NGMA standards to be the most conservative of those examined. This might be due to the general litigiousness of Americans, and the desire of engineers to avoid legal suits. However, it also points out that there is considerable disagreement among engineers as well as lack.

Table 2-3. Weights of common structural components used in greenhouse construction [Hanan et al., 1978].

Cross section	Material	Kg/meter
L	Angle 6X6X0.5 cm	
	Aluminum	35
	Steel	59
C	Channel 6X10X0.4 cm	
	Aluminum	196
	Steel	295
I	I-Beam (L & B) 5X8X0.4 cm	
	Aluminum	90
	Steel	171
O	Pipe 3 cm O.D.	
	STD Steel	37
	High tensile	89
□	Tubular steel 5 cm square	
	Aluminum	37
	Steel	111
⊓	Formed steel	44
▮	Wood 5X10 cm No. 1 fir	37
▮	Wood 5X15 cm No. 1 fir	77

Table 2-4. Site specific versus zoned ground snow loads (kg m^{-2}), illustrating the variation in snow loads that can exist over relatively short distances. Zoned values considered as large regions of several km^2 as compared to small areas of less than 1 to 2 km^2 [NMGA Standard, 1985].

State	Location	Zoned value	Site-specific value
California	Mt. Hamilton	24.5	220.9
Arizona	Chirachahua National Monument	49.1	147.3
Arizona	Palisade R.S.	24.5	981.8
Tennessee	Monteagle	73.6	98.2
West Virginia	Fairmont	196.4	270.0
Maryland	Edgemont	245.5	219.1
Pennsylvania	Blairsville	245.5	294.5
Vermont	Vernon	294.5	368.2

of sufficient information. Most of the data for design loads, particularly wind, comes from model studies in wind tunnels. It appears that the British were most forward in examining full-scale greenhouses for wind loads [e.g., Hoxey, 1990; Wells and Hoxey, 1980]. In any event, attempting to design structures for all possible events leads to an expensive and inefficient building. The essence of good design is to build for the most probable conditions that will be encountered, providing an efficient greenhouse at minimal cost. In such cases, some risk will always be present and cannot be avoided. The owner, in cooperation with the manufacturer, must assess those risks realistically. That is, a probability can be assigned to severe weather. On this basis, the grower can determine his risk for a particular structure. Some engineers may take exception to this viewpoint since the structure may represent only 10% of the total investment. The point is the requirement for a well-designed house with no false

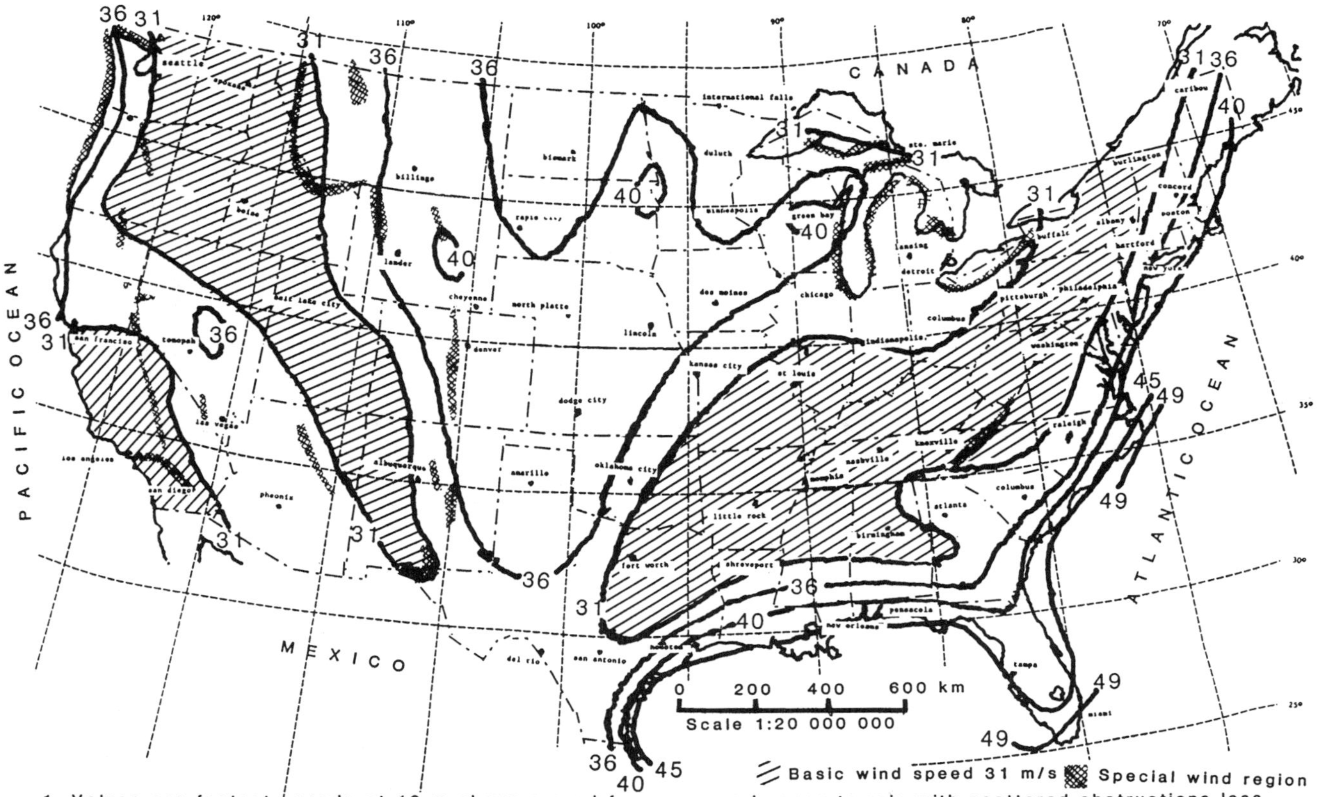

1. Values are fastest speeds at 10 m above ground for exposure in open terrain with scattered obstructions less than 10 m, flat, open country with a probability of 0.02.
2. Linear interpolation between countours acceptable.

Fig. 2-35. Basic wind speed in m/s for the conterminous U.S., adapted from ASCE Minimum design loads for buildings and other structures 7-88, 1988 (By permission of ASCE).

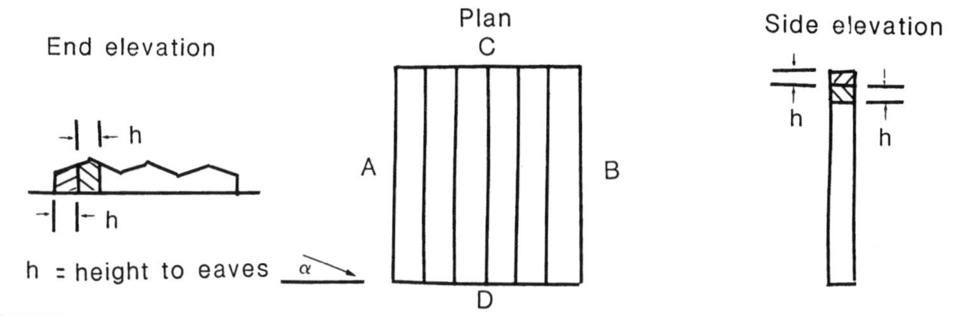

Wind angle α	Cp_e for surface A	B	C	D	Local Cp_e ▨	▧
0°	0.6	-0.4	-0.3	-0.3	-1.2	-0.6

ON GABLE ENDS C, D

SINGLE SPAN

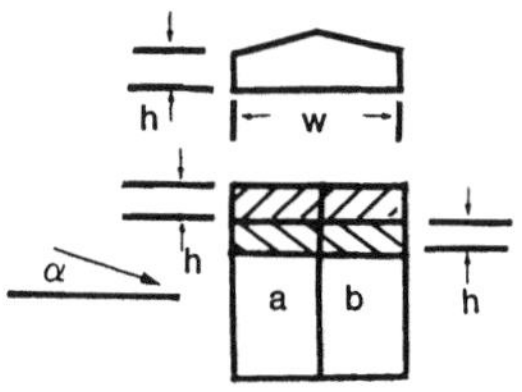

Wind angle α	Roof angle	Cp_e a	b	Local Cp_e ▨	▧
0°	26°	see "end" above	-0.6	-1.0	-0.8
				ON ROOF SLOPE b ONLY	
	20°	see "end" above	-0.6	-1.0	-0.8
				ON ROOF SLOPE b ONLY	

MULTISPAN

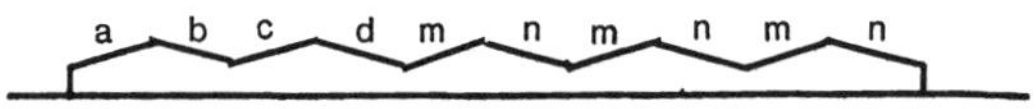

Wind angle α	Roof angle	h/w	Cp_e First spans a×	b	Second span c	d	Other spans m	n	Local Cp_e+ ▨	▧
0°	26°	0.26	0.3	-0.5	-0.5					
		0.36	-0.1	-1.2	-1.1	-0.6	-0.4	-0.5	-	-
		0.73	-0.8	-1.0	-0.7					
0°	20°	0.44	-0.85	-1.0	-0.8					

× As for single span,i.e. pressure coefficients from top figure
+ As designated for single span and to be applied to all spans

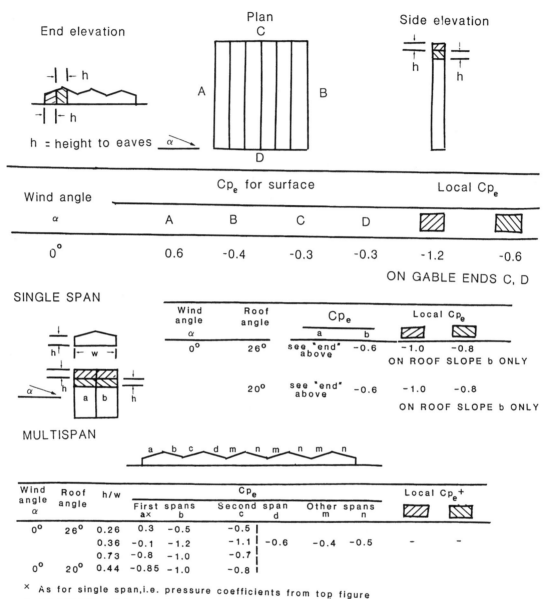

Fig. 2-36. Example of wind coefficients for walls (top) single span and multi-span roofs (bottom) adapted from Hoxey [1990]. Each coefficient is multiplied by the design wind speed and suitable constant (e.g., 0.613 in SI units, N/m^2, m/s) to arrive at the force acting upon the wall or roof. h = height of the structure. Note that coefficients for a multispan house vary with the ratio of height to width. With the exception of the first roof upwind, all others undergo a suction force. See also Fig. 4-82 for additional information on wind coefficients.

economies such as leaving out sufficient fasteners on fiberglass roofs. This was a problem in the 1970s in Colorado.

1. Design Loads

From the NGMA Standards [1985], the dead loads to be designed for are the weights of all construction material, including the covering, and all fixed equipment such as heaters, cooling equipment, electrical and lighting systems, as well as other systems when supported by the structural members. During the first energy crisis in the 1970s, growers often hung plant materials on the structure in the interests of increasing usable space. Even before this period, many cut-flower producers used the structure as a part of the bench support. These loads are often not taken into account, and can lead to considerable problems as noted in Fig. 2-34. NGMA requires such loads to be included in the dead load calculations if in place for more than 30 days. An example of weights of various structural materials used in greenhouse construction is provided in Table 2-3. The weight of film covers such as polyethylene is considered negligible (<0.5 kg m^{-2}). However, 3 mm thick glass will weigh about 8.8 kg m^{-2} and double glass over 17 kg. Weights of structured plastics will vary between 2 and 5 kg m^{-2} depending upon thickness.

Live loads are temporary loads caused by the use and occupancy of the greenhouse. Such temporary loads may be imposed during construction, repair or the weight of workmen and temporary scaffolds. Hanging objects within the house are also live loads if not remaining in place more than 30 days. The maximum live load in NGMA [1985] standards is limited to 73.6 kg m^{-2} or 722 N m^{-2}. The Dutch standard for crop loads is 150 N m^{-2}. However, all roof members such as purlins, rafters, truss members, etc. shall be capable of supporting a minimum concentrated live load of 45.5 kg applied downward and perpendicular (normal) to the roof surface at their mid-span.

Initial wind speeds for design are taken from isopleths as presented in Fig. 2-35. These are wind speeds at a 10 m height and one can interpolate between lines. Note that in some areas, special precautions must be taken as wind speeds may by too variable when presented on a map of this size. The British code also uses a map to show wind speeds for various locations. According to Spelman [1988], a wind speed of 42 m s^{-1} selected for southeast England would translate to a dynamic pressure of 500 N m^{-2}. The Dutch standard specifies a single loading figure regardless of location. For a Venlo house with 3 m eaves, the loading is also 500 N m^{-2}. A choice of 46 m s^{-1} in England, however, would increase the wind load factor to about 600 N m^{-2}. Spelman concluded that houses designed to 46 m s^{-1} would be stronger than the Venlo. More recently, Waaijenberg [1995] gave a wind and snow loading of 250 N m^{-2}, with the proviso that wind and snow loads do not act together. These points should be kept in mind when considering greenhouse selection, inasmuch as the structure designed to meet standards elsewhere might not be sufficiently strong for your location.

Once the basic wind speed has been determined, calculation of pressures on various structures can still vary considerably from one authority to another. Aldrich and Wells [1979] recommended that factors relating to ground cover, exposure or similar influence should not be used. However, Hoxey [1990] indicated that three factors are applied to the basic wind speed in the U.K.: 1) A topographical factor which takes into account local features with a value of 1.0 for level terrain. 2) A ground roughness, building size and height factor, with a typical value of 0.7 for a greenhouse 4 m high. And 3), a factor which takes account of security and time exposed. In the U.K., houses covered with soft plastic are given a value of 0.88 (10 years), and glass covered structures are assigned 0.93 (20 years). In the calculation of velocity pressures, NGMA [1985] assigns factors such as "importance," "exposure," and "gust response" that increase or reduce the final value. For example, in a large city with tall buildings, the gust response value will be greater than 2, but the exposure factor will be markedly reduced. Exposures of 4 m high greenhouses in flat, unobstructed coastal areas, will result in a velocity value of 1.20, a gust response factor of 1.15, and an "importance" coefficient of 1.05. The design wind pressures are calculated from a series of coefficients that take into account positive windward pressures and negative leeward pressures which vary with the type of structure under consideration. An example from Hoxey [1990] is provided in Fig. 2-36 for rectangular, glass-covered houses (see also Fig. 4-82). NGMA [1985] Standards use different combinations to determine the necessary coefficients. Arch roofs and polyethylene, air-inflated roofs will have different coefficients –especially as the plastic layers are likely to be fastened only at the edges. The final result is the force that the structure must withstand, and the designer plans the structure accordingly.

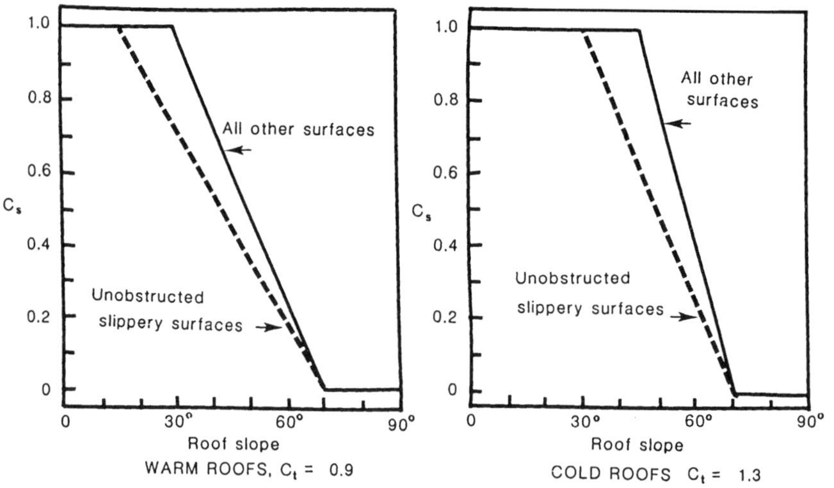

Fig. 2-37. Coefficients (C_s) for roof slopes in determining snow loads [ASCE Minimum design loads for buildings and other structures, *Amer. Soc. Civil Engr., 7-93, 1988.* By permission of *ASCE*].

Snow loads may act in conjunction with wind and dead loads, although the Dutch state differently. Live loads are not included with possible snow accumulations but can act in conjunction with wind [NGMA, 1985]. As with wind loads, calculation begins with the "ground snow load" for the particular location. Maps, similar to that shown for winds (Fig. 2-35), may be found in such sources as ANSI A58.1 [1982] or the ASCE American Standards 7-88. Unfortunately, the mountainous regions of the Western U.S., Appalachians and the northern New England regions are so variable that local meteorological data are required. Even where a zoned, ground snow load is provided, there may be site specific values that differ remarkably from the zoned value (Table 2-4). Greenhouses are remarkable due to the fact that resistance to heat transfer through roofs (R value = $m^2\,°C\,W^{-1}$) is commonly less than 1, and always less than 2 –as compared to a value of 16 for rockwool insulation in the ceiling of a home. The latter indicates a heat transfer coefficient (h) much less than 1 as compared to greenhouse roofs where h may range from 2 to nearly 10 W m^{-2}. To qualify for the reduced structure the common R values permit, the greenhouse must be continuously heated to an internal temperature of 10 C, 1 meter above floor level. Snow on double, air-inflated roofs causes the plastic layers to become one so that it behaves as a single layer in terms of the R value. Double-structured plastic roofs usually have R values less than 2. Thermal screens are not considered, and growers must beware of the possible effects by opening thermal curtains to melt snow. The code also requires a suitable alarm installation, or an attendant present, if the grower is to take advantage of the reduced R.

Fig. 2-38. Collapse of a gutter-connected, FRP covered rose greenhouse from excessive snow accumulation in the gutters. Other houses that suffered damage from the particular blizzard did not always fall immediately, but began progressive collapse within one to three days after the main snow fall. Physical snow removal from such structures is not technically feasible.

To determine the adjusted snow load from the ground snow load, the structure is considered as a flat roof with adjustment for roof exposure, thermal condition and occupancy and function of the house. The designer then considers the roof slope, unbalanced loads, snow drifts, sliding snow and unloaded portions. If the greenhouse is intermittently heated or unheated, then the flat roof design load should not be less than the ground snow load. Exposure factors in the NGMA standard vary from 0.8 for windy areas with the roof exposed on all sides to 1.2 for densely forested areas that experience little wind. The risk factor is 1.0 for houses open to the general public, and 0.8 for all others. The thermal factor for unheated houses is 1.3, 1.1 for a continuously heated double glazed structure and 0.9 for a heated, single glazed house.

From the flat roof snow load, the sloped roof, design, snow load is obtained by multiplying by a roof slope factor C (Fig. 2-37). The value will be different for arch and gutter-connected roofs. Usually, the gutter should be uninsulated with provision for heat below the gutter.

In all these calculations of loads, the idea is to establish a value for structural strength that reduces the possibility of failure to an acceptably low level. The NGMA code states that the few snow-induced failures that have occurred were when the structures were out of service. That is not completely true, and there have been occasions when snow has accumulated so rapidly that load to the point of structural collapse has happened (Fig. 2-38). O'Rourke's discussion (1997) on loads due to snow drifting suggests that gutter-connected greenhouses are prone to excessive loads between houses, especially if winds are at right angles to the longitudinal axis of the houses.

In general, the NGMA Standards [1985] are taken with slight modification from the ANSI A58.1 [1982] minimum design loads for buildings and other structures. I am not aware of full-scale tests in the U.S. similar to those performed by the British engineers over the past two decades. As noted, Dutch standards do not deal with such a wide variety of conditions common to the U.S. Similarly, French, German, etc. standards are different from those discussed here. In other parts of the world, standards, if available, are likely to be unenforced. The EEC is likely to publish uniform standards in the near future, and companies exporting structures to the EEC and surrounding countries will probably have to conform with those standards.

2. Structures, Materials and Methods

Old style greenhouses, as depicted in Fig. 2-32, are no longer being built. The trend has been to gradually higher structures, beginning with 1.8 m to the gutters in 1952 to modern-day buildings as high as 4 m to the gutters. The advantages are better radiation distribution and greater versatility in the use of the space. The Venlo design (Fig. 2-11), using a steel superstructure, was constructed in basic modules of 3.2 x 3.2 m until the early 1980s [Anon., 1989]. Currently, the most modern design is 4 x 6.4 m which permits l m wide glass, 4 mm thick. Glass 3 mm thick with glazing bars at 600 mm centers transmits about the same as 4 mm glass at 800 mm. An increased width to 1000 mm affords more radiation [Smith, 1988], built on the structure shown in Fig. 2-39. New introductions of wider glass have met resistance from builders since such large panes are difficult to handle. Another innovation has been hardened glass for ventilators, eliminating the need for a metal frame except at the hinge [Anon, 1989; Lawson, 1988]. Some manufacturers will now supply structural parts painted white over the galvanizing. Moreover, the importance of improved radiant energy supply has been emphasized by such authors as Spelman [1987], who stated the general rule that 1% more energy would provide 1% higher yield. This point is discussed further in Chapter 8, and caution should be used in applying this rule indiscriminately. It is probably correct under European winter conditions, where any irradiance improvement can be readily noted under their conditions.

Almost all greenhouses are now built as "clear" spans; that is, no supporting posts between the side columns (Fig. 2-33, 40). An anonymous author in 1989 made the comment that such clear spans of 20 m or more are fewer in number. U.S. practice is to build gutter connected structures with clear spans varying between 10 and 15 m (Figures 2-8, 33). However, almost any type, including Venlos, may be found in the U.S. The Israelis have developed a metal structure suited for their climate, described by Dayan et al. [1986], characterized by height up to 4 m or more and limited to maximum dimensions of about 30 m. This permits good side ventilation since roof ventilators on polyethylene-covered houses are seldom found in Israel.

The thrust of the Mediterranean regions, as well as other mild climate areas has been toward structures that reduce high temperatures through improved ventilation [Brun and Lagier, 1985; Castilla et al., 1990; Zabeltitz, 1985]. Even in these regions, however, the trend is toward higher buildings (Figures 2-9, 29).

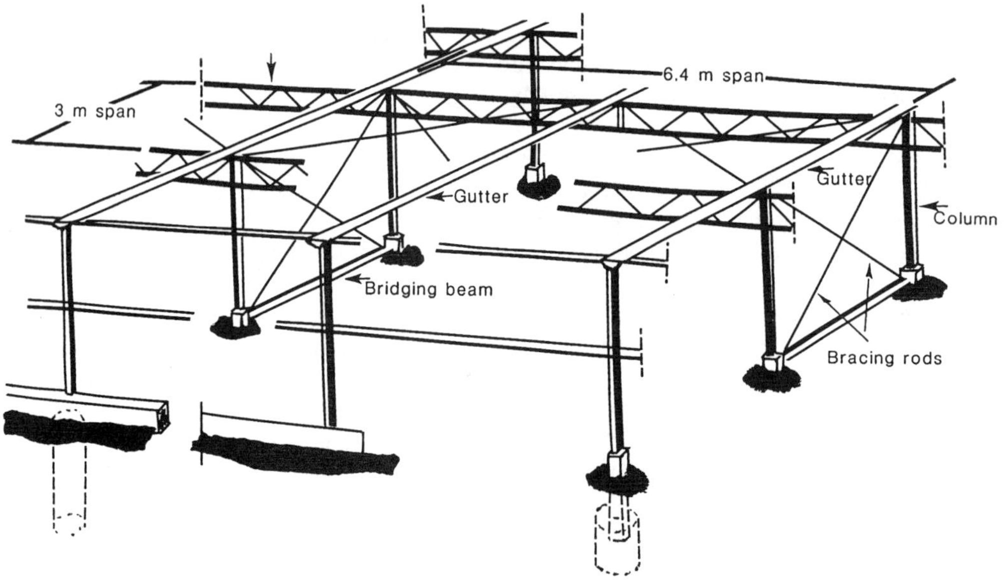

Fig. 2-39. Standard Venlo superstructure [Spek, 1985] (©1985. Int. Soc. Hort. Sci. *Acta Hort.* 170:11-23).

Fig. 2-40. Major parts of a typical, clear-span structure. Roof bars will not be used if house is covered with plastic. Arch or barrel roofs will not have a ridge in the sense of a standard peak house. Anti-sway braces are not included in this figure.

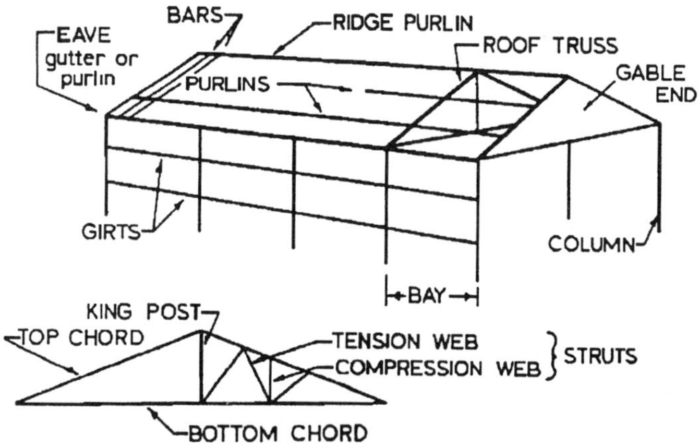

a. Materials

1. Wood

Greenhouses have been built out of most things imaginable: plastics, wood, steel, aluminum and concrete. In some instances, volatiles from plastics have been shown to be phytotoxic [Day, 1984; Scott and Wills, 1972]. In cold re gions the grower must be aware of toxic properties of not only plastics, but paints, preservatives, fumigants and other chemicals likely to be employed. Seeley [1980] discussed the problems with paints in particular. Wood, which is probably most often used in many parts of the world, is nevertheless unavailable in some locations. Concrete structures have been employed (Fig. 2-41). During the early years of this century, rot-resistant wood species such as cypress, as well as

Fig. 2-41. Greenhouse with a concrete superstructure, covered with single-layer polyethylene sandwiched between two layers of chicken wire. Note the shallow roof pitch.

redwood, were readily available in the U.S., and most components of the structure were formed from wood, particularly in direct support of the glazing (Fig. 2-42, 45). However, wood requires considerable maintenance under greenhouse conditions. In general, the ability of steel to carry greater loads with smaller cross-sections means greater structural transmissivity

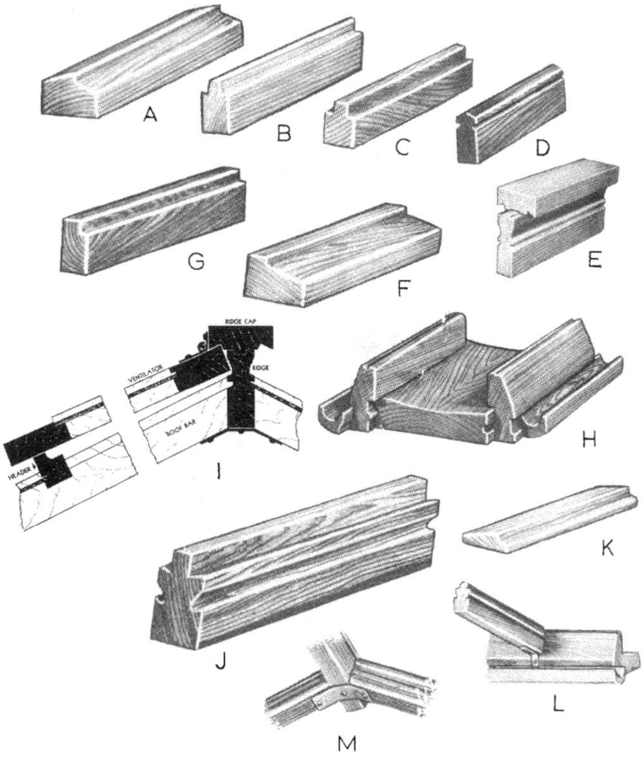

Fig. 2-42. Old style woodwork for greenhouses. **A** glazing sill; **B** gable bar; **C** side wall bar; **D** ridge; **E** ridge and cap; **F** sash sill; **D** gable bar; **H** gutter; **I** wood vent sash for roof; **J** roof bar; **K** glazing sash; **L** eave bar tie; **M** ridge bar tie [Hanan et al., 1978].

and reduced preventive care. Materials such as wood or concrete, because of their greater cross-sectional area, will cast more shade inside the house. Most modern structures use galvanized steel or aluminum.

Despite the move to metal structures, wood in the U.S. and most developing areas remains the cheapest, most readily available material for greenhouse support (Figures 2-9, 28, 30, 31, 43). There is also a large body of practical experience in most localities, so that design and construction usually proceed without need of advanced engineering. Wood also has a number of advantages in addition to ease of manipulation; it has a low thermal conductivity compared to metal, a high resistance to electric current and a low coefficient of expansion. Early, long aluminum structures sometimes gave problems with glass slippage as the house expanded and contracted with temperature changes. The problems with wood in greenhouse construction arise from the high humidities usually encountered that cause undesirable swelling with the possibility of rotting organisms attacking it. Almost always, there will be some contact with soil,

Fig. 2-43. Two types of prefabricated wood trusses for greenhouses. The lower picture shows a lean-to or sawtooth design for mild climates.

especially if the supporting posts are sunk in the ground. The choice of preservatives to resist rot is small. Creosotes and pentachlorophenol are not to be used in unventilated greenhouses. One that has been shown to be relatively harmless is copper naphthenate. For maximum protection, pressure processes are recommended since they permit greatest penetration of the preservative. Tanalith (Wolman salts) is another material for pressure

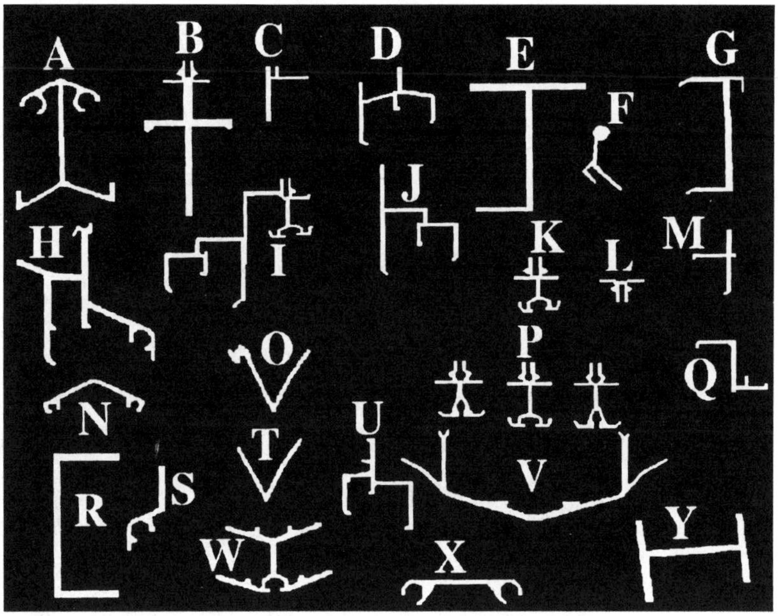

Fig. 2-44. Aluminum extrusions for greenhouses. For example: **A** ridge bar with provision for hinged ventilators either side; **I** gable end bar; **P** various sash bars for glass; **V** gutter.

or surface application that has been found safe for greenhouse use [Beese, 1978].

Engineering of wood structures has changed markedly in the past decade. Recourse to simple nomographs for designating column and beam size is not usually employed. With increased use of computers, the necessary requirements are simply entered into appropriate equations. The engineer can make several trials to arrive at the best solution for the conditions.

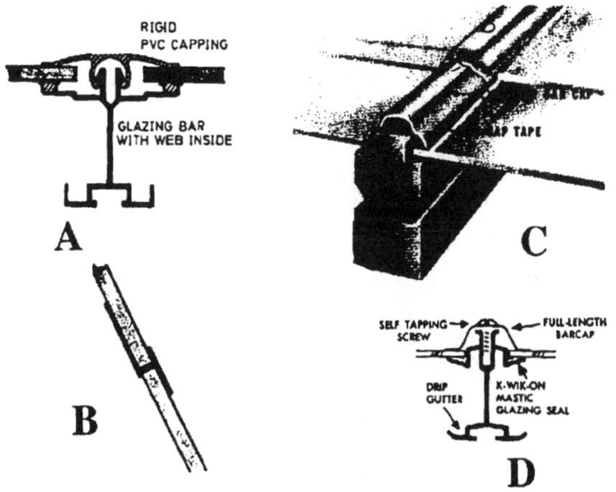

Fig. 2-45. Typical glazing bars for glass covered structures. **A** glazing bar with web inside; **B** plastic "butting strip" for roof glass; **C** a wood sash bar with bar cap; and **D** aluminum sash bar and cap [Hanan et al., 1978].

In the U.S., at least, numerous companies manufacture roof trusses to order (Fig. 2-43). The company engineers will design a truss to meet the owner's specifications and design loads [Faherty and Williamson, 1989]. For example, designating the size of column necessary to support the walls and roof depends upon whether one uses simple solid columns –which is the usual case– spaced columns or built-up columns. The relationship between length and cross-sectional area is determined by the slenderness ratio, and is further modified by whether the columns are fixed against rotation at both ends. The result is to obtain the effective length of the column. Depending upon the slenderness ratio, columns are further divided into short, intermediate or long columns, determining their behavior under load. From a knowledge of probable loads, it becomes possible to make appropriate selections for the proper size column. Calculations for beams or trusses are more complicated, and should not be indiscriminately employed by the neophyte. Fortunately, with the accumulated experience to be found in most regions, wood greenhouses are usually constructed by "rules of thumb." and, quite often, are over-designed.

Fig. 2-46. Two examples of steel pipe frame greenhouses, both covered by double polyethylene, air-inflated. The upper example used fabrication on site, originally covered with single polyethylene, sandwiched between two layers of chicken wire. The bottom was a prefabricated structure.

Since the advent of plastic covers, numerous wood designs have been proposed: for example, those by Courter [1963], Sheldrake and Sayles [1974], Hartz et al., [1981], and Walker and Duncan [1973]. A recent bibliography of plant growth facilities by Ross [1988] lists over two dozen types, ranging from hobby houses to solar-heated structures. Some of these use laminated and glued wood members. The publications from the University of Connecticut, by Bartok [1986], lists nine structures for which plans and directions may be obtained.

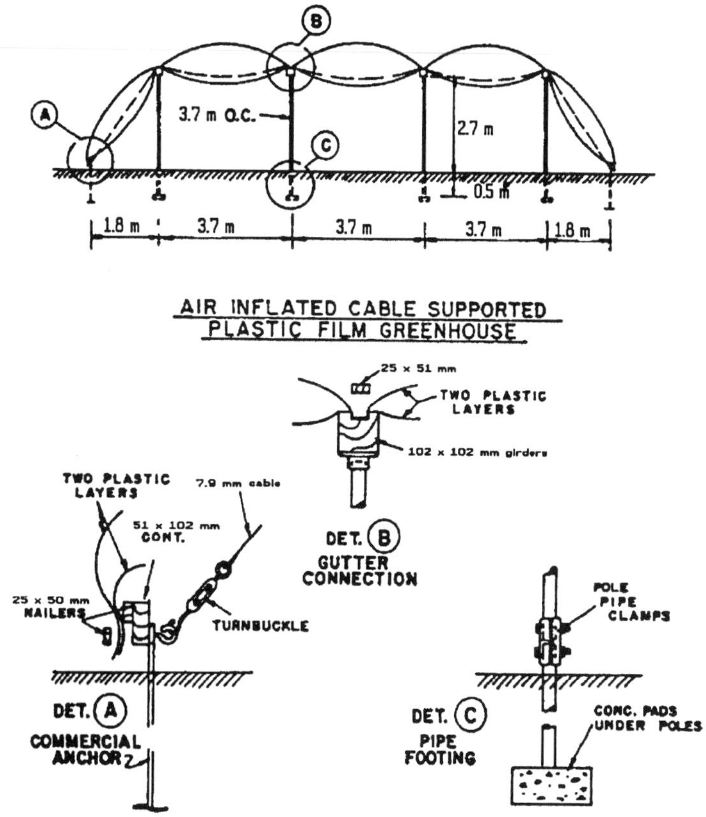

Fig. 2-47. Minimal, air-inflated, double polyethylene, cable supported greenhouse proposed by Roberts [1971].

2. Metal

Most modern, commercial greenhouses today are constructed with high tensile, galvanized steel, formed or tubular configurations, or pipe. Aluminum, because of its light weight, resistance to corrosion, and the fact that specialized units may be extruded (Fig. 2-44), is commonly found. Some care is required where joining of steel with aluminum may result in corrosion from galvanic effects. Insulation of joints may be required. While galvanized steel is very resistant to corrosion, black pipe or wrought iron can be employed if suitably painted and maintained. Galvanized pipe is preferable. Care should be used to reduce contact with fertilizer salts wherever possible. Although the use of metal sash bars for glass increases heat transfer compared to wood, they are common (Fig. 2-45).

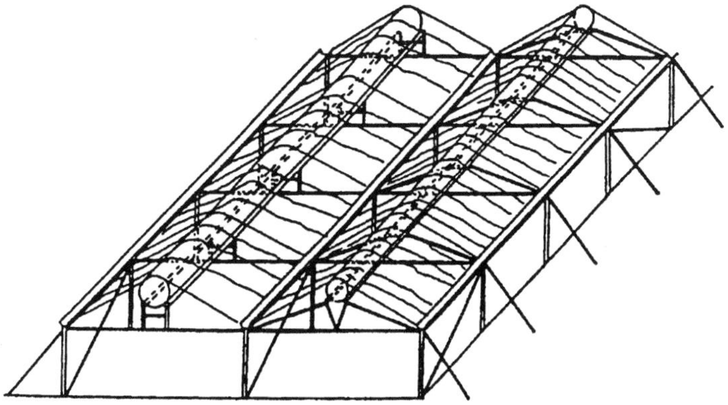

Fig. 2-48. Inexpensive, light-weight, single-layer polyethylene structure proposed by Zabeltitz [1985] (© 1985. Int. Soc. Hort. Sci. *Acta Hort.* 170:25-28). The tube at the ridge of each house, when suitably inflated, would keep the plastic tight, reducing wind damage. See also the design by Olympios [1982].

The forms that may be found for metal superstructures are almost as common as those for wood. Previous Figures 2-4, 9, 11, 19, 29 and 33 provide a representative sample. In some instances,

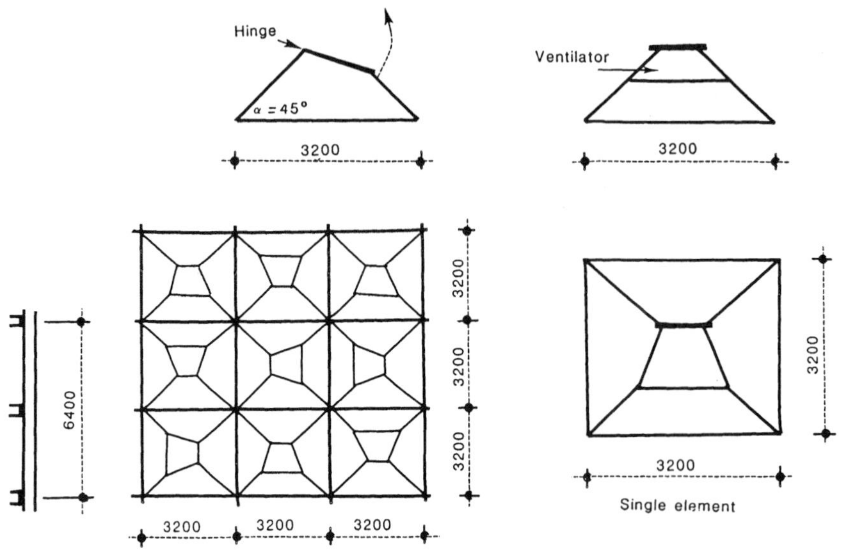

Fig. 2-49. Revised Venlo roof proposed by Spek [1985] to fit on standard superstructure (Fig. 2-39). Ventilators can be adjusted regardless of wind direction (© 1985. Int. Soc. Hort. Sci., *Acta Hort.* 170:11-23).

growers have provided their own structure by bending steel pipe to the appropriate curvature, and manufacturers commonly construct houses from pipe frames (Fig. 2-46). In many respects, metal is more versatile, with buildings ranging from minimal, taking advantage of plastic coverings (Fig. 2-47), to simple structures for mild climates (Fig. 2-48). Spek [1985] proposed a newer version of the Venlo (Fig. 2-49), building upon the supporting structure depicted in Fig. 2-39. It has been my observation over a period of 30 years that, as industry develops in a particular location, there is a gradual movement to metal structures with special provisions for the particular climate.

3. Foundations

All vertical and horizontal forces on the greenhouse structure must be transferred to the ground via the foundations. There are several configurations, ranging from simply driving the column into the soil to quite elaborate footings (Fig. 2-50). However, simple posts have limited resistance to withdrawal, and high winds have seen such structures pulled completely from the ground. Conversely, there have been examples where extremely heavy snows have forced supporting columns further into the ground without collapsing the building. The size and type of footing will depend upon the load-bearing capacity of the soil (Table 2-5), and the design of the structure (Table 2-6), as well as the actual weight that will be present. Due to problems with sulfate attack on concrete, Dutch engineers found it necessary to consider foundation replacement for up to 4 million m^2 of greenhouses. A proposed replacement, using a heavy steel plate, was made by engineers (Fig. 2-51).

VI. COVERINGS

Coverings are considered separately from superstructures since almost any covering may be placed on almost any structure. The possible exception is the Venlo design, using glass. The most common materials are listed in Tables 2-7, 8 and 9. Table 2-10 lists the properties of these coverings. There are numerous others such as polypropylene, EVA, etc. that have been tested, including colored and luminescent. The problem is that there are no rigorous legal standards to which one may refer for decision although such standards have been proposed [Hamrick, 1988; Waaijenberg, 1985; Nisen, 1976].

According to Hamrick, for example, NGMA has proposed ASTM E-903 (American Standard Testing Method) for determining PAR transmission under laboratory conditions. Thus, 3 mm thick, low iron, tempered and rolled glass transmits 92% of the outside radiation, whereas float glass transmits 89%. Data from these tests

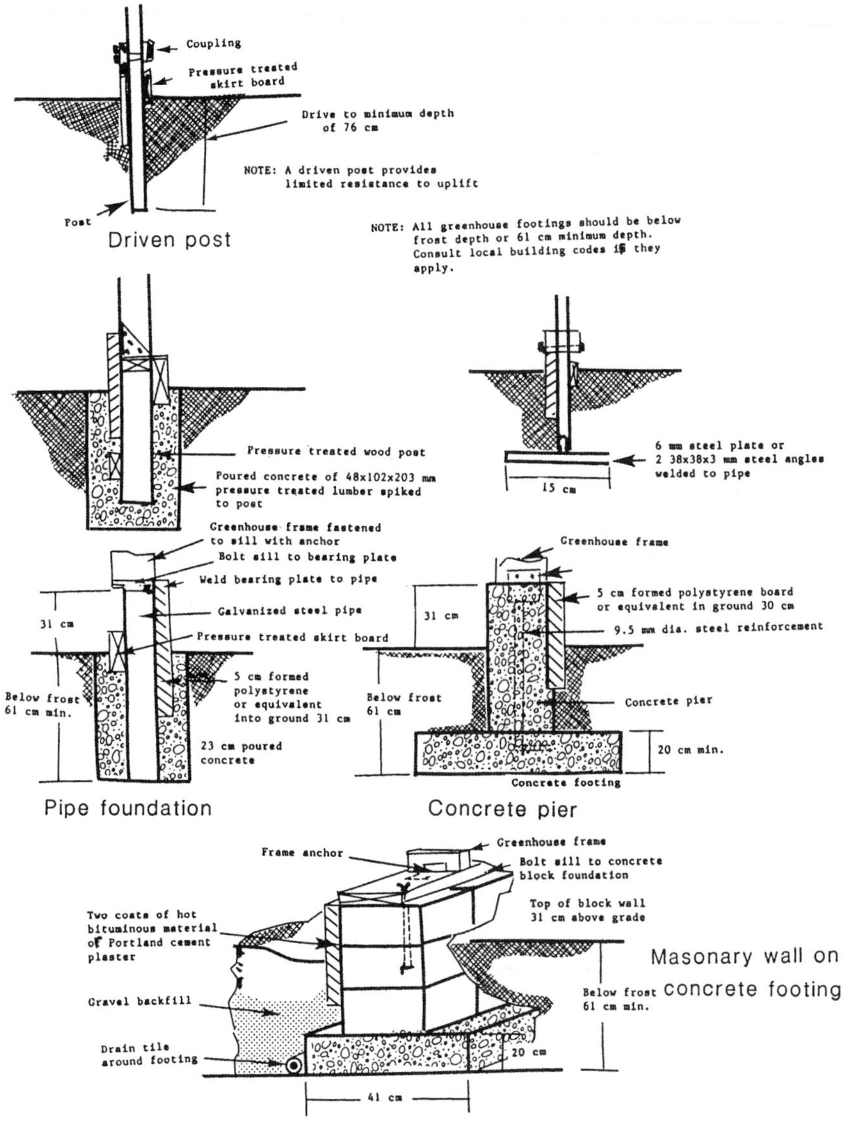

Fig. 2-50. Examples of foundations for greenhouses suggested by Aldrich and Bartok [1985]. In some areas of the U.S., galvanized columns are set in concrete with FRP placed about 15 cm into the ground on the outside.

are not likely to agree with figures presented in Table 2-8. Unfortunately, none of the several tests are legally required. It is up to the buyer to request such tests when making an approach to the manufacturer for possible purchase. Another point to be made is that these tests are carried out under laboratory conditions and may not be reliable indications of how the material will perform in real life. Using PAR transmission, the percentage resulting is of direct radiation perpendicular to the specimen surface –something that seldom occurs. The ability of a cover to transmit diffuse radiation is very important. Even where radiation measurements are carried out in the greenhouse, the investigator, in order to reduce data variability, will often place the common "point" sensor close to the covering in order to avoid any shading effects of the superstructure. When placed in the crop position, the

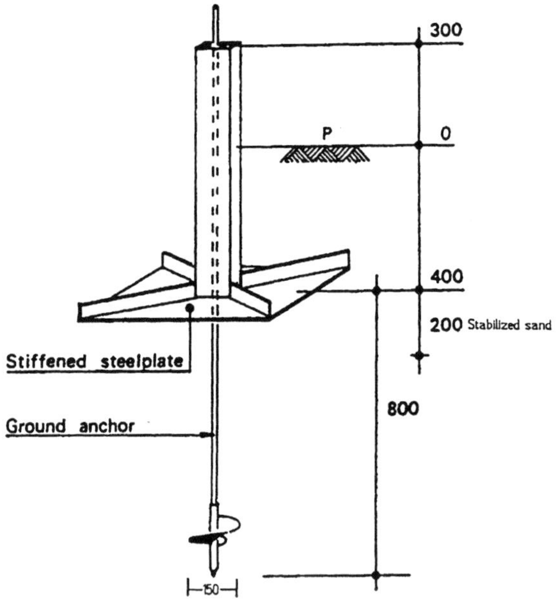

Fig. 2-51. Steel footing with ground anchor devised by Dutch engineers [Spek, 1985] (© 1985. Int. Soc. Hort. Sci., *Acta Hort.* 170:11-23).

solar variation may be more than ten times that obtained with the transducer directly under the covering. In fact, due to position and superstructure, the total radiation transmitted by all types of structures and covers will be about 60% of that outside [Hanan, 1989; Dayan et al., 1986a]. Although 60% is an average, and there are materials that perform well above that average, it is discouraging to realize that the average greenhouse is not very efficient in making use of sunshine.

A. PLASTICS

The advent of plastics (Fig. 2-52), over the past 30 years, has completely revolutionized greenhouse construction, and helped make possible the construction of cheap structures in developing countries. The article by Mark [1984] is a good exposition of their evolution and basic chemistry. The most commonly employed material in mild climates is so-called "soft" plastic or polyethylene film (PE). It is the cheapest material available and comes in wide sheets capable of covering large structures. Unfortunately, even with UV inhibitors, it seldom lasts longer than two years (Fig. 2-53), although there are claims for as much as five years. Even in temperate climates, polyethylene is common where it can be employed as a double, air-inflated covering, reducing energy loss compared to single layers of either glass or plastic. Pressurizing the plastic allows fasteners to be applied along the edges of the film and prevents flapping in the wind –the most common cause of early failure (Fig. 2-54). First use of film plastics usually involved nailing the plastic to the structure with covering strips of wood (Fig. 2-55). Degradation, due to heating on clear days, often caused initial loss at the attachment, especially as the wood strips often loosened, placing major stress at the nails. The general methods of attaching film plastics at the edges are depicted in Fig. 2-56. Other methods have been considered for holding films taut on the structure (Fig. 2-48) [Zabeltitz, 1985; Lawson, 1989; Franklin, 1983; Aldrich and Bartok, 1987]. Initially, single layers were often sandwiched between wide mesh wire to prevent flapping, but newer systems are much more efficient. Polyethylene also makes a very good emergency cover when roof damage occurs from hail or wind [Goldsberry, 1973].

Table 2-5. Load carrying capacity of some soils [Dalzell, 1972].

Soil type	Capacity (metric tons m^{-2})
Soft clay	9.8
Wet sand or firm clay	19.6
Fine, dry sand	29.4
Hard dry clay or coarse sand	39.2
Gravel	58.8
Hardpan or shale	98.0
Solid rock	No limit

Numerous studies have been carried out with different colors of transparent covers, films especially, such as films blocking infrared (IR) transmission and films capable of fluorescing at different wavelengths [Grafiadellis, 1985; Simpkins et al., 1984; Novoplansky et al., 1990; Goldsberry, 1971; Blom and Ingratta, 1985, Bowman, 1962]. More recently, Daponte [1994] discussed the use of multilayer films in order to achieve the required specifications of longevity, antifog, antidust, inhibition of certain disease development, anticreeping, etc. Deliberately tinted coverings reduce total transmission to the detriment of yield. Reasons and references for this

Table 2-6. Pier footing diameters for average soil [Aldrich and Bartok, 1985]

Greenhouse span (m)	Pier spacing (m)					
	1.2	1.8	2.4	3.1	3.7	4.9
	Pier diameter (cm)					
6.1	15	23	31	31	31	38
7.3	23	23	31	31	38	38
8.5	23	31	31	38	38	46
9.8	23	31	31	38	38	46
11.0	23	31	38	38	46	**
12.2	31	31	38	38	46	**
14.0	31	38	38	46	46	**
18.3	31	46	46	46	**	**

** Requires special design

attribute will be covered further in Chapter 3. Although coloring will usually reduce condensation, particularly on thin films, due to higher temperatures from energy absorption, the tradeoff is not worth the general yield reduction. Clear films or coverings have usually been the most advantageous. Polyethylene is transparent to IR radiation, which results in higher energy consumption and greater condensation on the inside, particularly on clear, cold nights. Blocking IR by incorporating IR absorbers in the film reduces energy loss –at least the interior temperature does not have to be as high. Such films have shown improved growth [Simpkins et al., 1984; Goldsberry et al., 1985). Grafiadellis [1985] showed a nonsignificant higher production under EVA and IR polyethylene covers. Tinted films with photoselective orange and violet colors markedly reduced yield of vegetables. In a highly teleological exposition, Novoplansky et al. [1990], exploiting the response of plants to the ratio of red/far red radiation, incorporated fluorescent dyes into plastic films. This converted the green spectrum of radiation into red wavelengths, resulting in an increased yield for gypsophilia. Whether this will prove practical remains to be seen. In any event, the testing of greenhouses and their coverings on yield is fraught with difficulty, with results significantly influenced by objectives, cultural management as well as the species being grown for testing [Dayan et al. 1986b; Ferare and Goldsberry, 1984; Elenes-Fonseca and Hanan, 1987; Bowman, 1962; Goldsberry et al., 1986]. Furthermore, the reduced infiltration of plastic-covered greenhouses relative to glass in cold climates significantly changes the interior climate [Metcalf, 1983], sometimes providing advantages in CO_2 fertilization and energy conservation, but often leading to further problems in regulating growth and preventing disease. Some of these will be discussed in following chapters.

Of the films employed in greenhouse coverings, polyvinyl fluoride (PVF) has the reputation for durability and high transmission (Tables 2-7, 8, 9). Even though available only in narrow widths, PVF has been employed in conjunction with EVA films attached to rigid frames. As in the case of double PE covers, there is a significant reduction in heat loss, but transmission in the short wavelengths is higher than double PE films. According to some authors [Franklin, 1983; Sherry and White, 1984], rose production in such double structures, using PVF as the outer covering, was significantly higher than other covers tested (Table 2-11). Other factors besides transmission and durability must be considered in such double covers. For example, using a film highly permeable to water vapor on the inside (e.g., PVF 25 μm thick) with a film highly resistant to vapor transmission (e.g., PE, UV stabilized, 180 μm thick) on the outside will lead to a highly undesirable situation of rapid condensation between the two layers [Waaijenberg, 1985]. The vapor barrier should be inside the house.

The use of double covers has been advanced by the fabrication of "structured" plastics, some semiflexible, but mostly rigid, of polycarbonate or acrylic (Fig. 2-52). Polycarbonates come in a number of configurations, with two or three layers. Acrylic has the highest transmission and durability. Acrylic laminated to polyester significantly improves weatherability. These structured covers also have the advantage, compared to glass, of requiring minimal structure for support. However, their high coefficients of thermal expansion and contraction mean

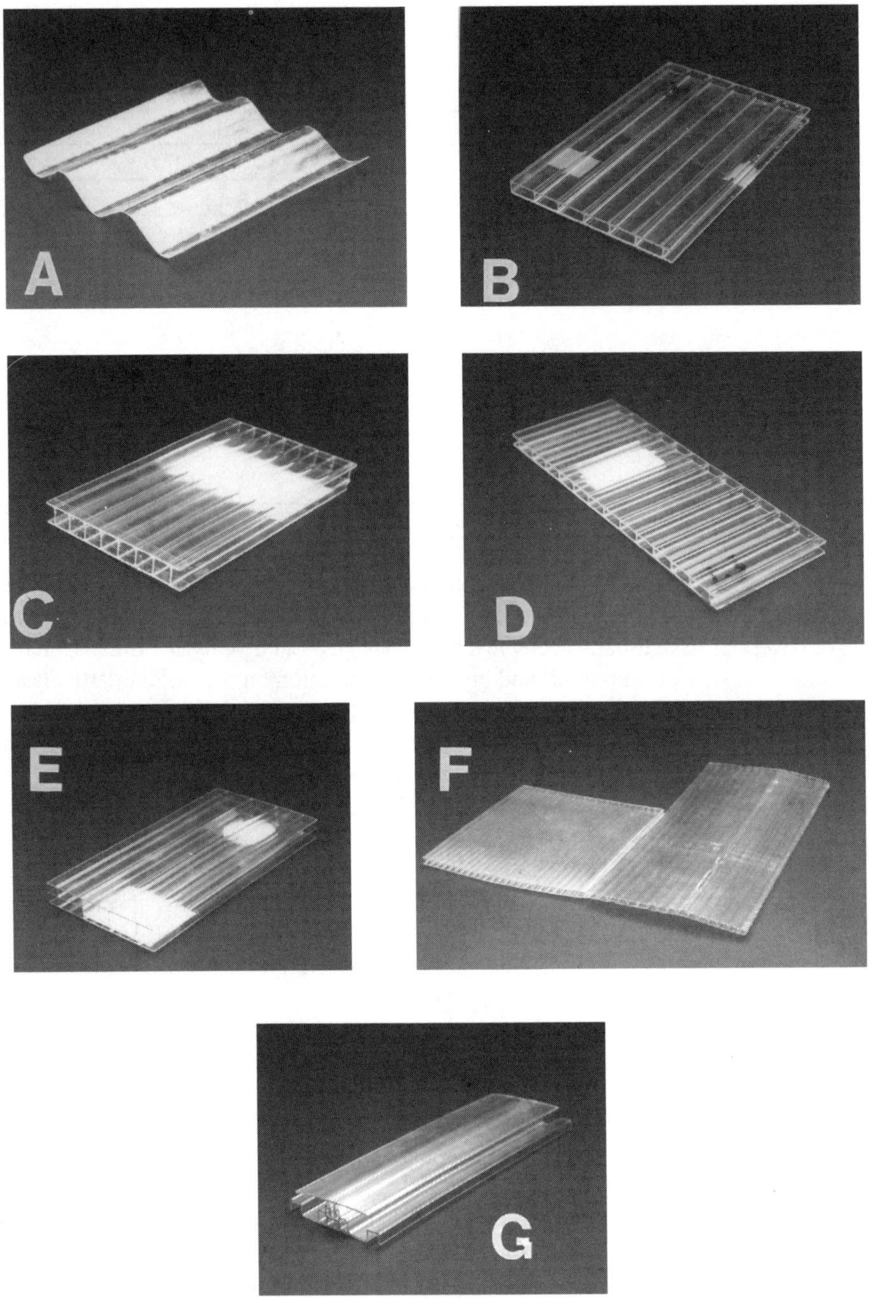

Fig. 2-52. Types of plastics for covering greenhouses. **A** clear, corrugated FRP; **B** structured acrylic, 8 mm thick; **C** structured polycarbonate, 16 mm thick; **D** structured acrylic, 8 mm thick; **E** structured, 3 layered polycarbonate; **F** 4 mm thick polycarbonate, semiflexible. Note the right piece in use for a year, visibly weathered with a crack in the top layer and the shaded area caused by the hold-down strap; **G** plastic connector for joining two structured pieces.

specializedattachments (Fig. 2-52G). Polycarbonates also have the reputation of lower radiation transmittance, and some formulations have shown rapid darkening in less than a year (Fig. 2-52F).

Table 2-7. Advantages and disadvantages of greenhouse coverings [Compiled from Aldrich, 1985; Ball, 1986; Jewett, 1985; Anon., 1981a; White, 1987].

Covering	Type	Advantage	Disadvantage
A. Glass	Soda lime	High transmittance, high weatherability, low thermal expansion, resistant to heat, UV and abrasion	Low impact resistance, high cost, heavy, reqires sash bars for installation
	Tempered low iron	Resists hail damage, larger panes possible	
	Patterned	Greater diffuse radiation	
	Double	30 to 40% reduced energy transfer	Very high cost
B. Acrylic (PMMA)	Rigid, structured, 2 layer	High transmittance, superior UV resistance and weatherability, no yellowing, light weight, easy fabrication	Easily scratched, high thermal expansion, high cost, flammable, slight embrittlement with age
C. Polycarbonate (PC)	Rigid or semiflexible, 2 layered, structured	High impact resistance, wide range of service temperatures	Poor weatherability and UV resistance, high expansion and easily scratched
D. Polyvinyl fluoride (PVF)	Film	High transmittance, resistant to UV, high impact resistance, heat shrinkable	High cost, tears easily if punctured, not available in wide widths
E. Polyvinyl chloride (PVC)	Film, rigid, corrugated or structured	High transmittance initially, available in many forms	Darkens quickly, turns black over structural members, embrittlement, low impact resistance
F. Fiber reinforced plastic (FRP)	Semiflexible, flat or corrugated	Low cost, strong, easily fabricated and installed, high impact resistance, diffuses radiation	Susceptible to UV degradation, requires PVF lamination, turns yellow, flammable, medium life
G. Polyethylene (PE)	Film with or without IR blockers and UV resistance	Lowest cost, easy to install, large sheets, high impact resistance	Short life, low heat transfer resistance, low service temperature requirements
H. Polyester	Film, laminated PMMA	High transmittance, high service temperatures, high weatherability, UV resistant	Low impact resistance, narrow sheets, UV degradable

Table 2-8. Characteristics of greenhouse coverings [Compiled from Jewett, 1985; White, 1987; Franklin,1983; Duncan and Walker, 1975; Aldrich and Bartok, 1985].

Covering	Type	Thickness (mm)	PAR transmissivity (%)[1]	Thermal transmissivity (%)[2]	Heat loss (W $m^{-2}C^{-1}$)	Lifetime (years)	Cost ($ m^{-2})
A. Glass	Soda lime	3	88	3	6.3	25+	11-22
	Tempered	3, 4	90-92	<3	6.3	25+	≤39
	Double	25	71	<3	3.0	25+	38-76
B. Acrylic (PMMA)	Structured	8, 16, 32	83	<3	3.5	20+	27-43
	Single layer	---	93	<5	---	20+	22-32
C. Polycarbonate (PC)	Structured	4, 6, 8, 16	79	2-3	3.5	5-7	19-32
	Single layer	---	87	<3	---	7-10	32-43
D. Polyvinyl fluoride (PVF)	Film	90-120 μm	92	21	5.7	>10	4-5
E. Polyvinyl chloride (PVC)	Numerous configurations		No longer recommended [Duncan and Walker, 1975]				
F. Fiber reinforced plastic (FRP)	Corrugated, PVF coated	1.2-1.8 kg m^{-2}	88	<3	5.7	10-15	9-19
G. Polyethylene (PE)	UV resistant, ±IR	100-150 μm	87	50	6.3	2-3	0.65
H. Polyester	Single layer	51-150 μm	85-88	30	---	7-10	5-11
	Laminated to PMMA	---	87	9.5		10+	5-12

[1]---PAR transmittance, photosynthetically active radiation, 400-700 nm, ratio transmitted to radiation incident on surface.

[2]---Thermal transmittance, heat transfer at wavelengths >3000 nm.

Table 2-9. Trade names and general comments on greenhouse coverings [Compiled from Anon, 1984; Jewett, 1985; Anon., 1982; O'Flaherty, 1985; Franklin, 1983; Ball, 1986; White, 1987].

Covering	Trade names	Comments
A. Glass	Sedo	Double, sealed panes, inert gas within
	Hortisave 34, Hortipane, SunMate	Double panes
	Solakleer, Solatex	Tempered panes up to 1150 mm wide by 2 m long, conventional panes 1.65 m long, 800 mm wide
B. Acrylic (PMMA)	Plexiglass, Lucite, Acrylite SDP, Rosalite,	Single layers, sheets up to 1.2 x 4.9 m
	Exolite, Degussa, Rohm	Structured, 2 layer units, 8 to 32 mm thick
C. Polycarbonate (PC)	Qualex, Exolite PC, Polygal, Lexan, Lexan PS, Polyglaz, Makrolon, Cryoflex, Dynablass, Tuffak K, Twinwall	Structured 2 or 3 layers, 4 to 16 mm thick, sheets up to 1.2 x 4.9 m, as with PMMA requires specialized fittings for expansion
D. Polyvinyl fluoride (PVF)	Tedlar	Limited to narrow widths, laminated to FRP on one side, heat shrinkable
E. Polyvinyl chloride (PVC)	Windocel, Maxolux, Rodeca, Ondex, UltraBrit, Paltought 2002, Polygal SD, Lexan Thermoclear	Variety of configurations in thin films, corrugated or structured with tongue-and-groove 40 mm thick. Limited to narrow widths
F. Fiber reinforced plastic (FRP)	Filon, Lascolite, Denverlight, Crystal Coat, Structoglas, Barclite, Glasteel	23% glass fibers in an acrylic and polyester matrix, corrugated, designated by weight per unit area, PVF coating on one side, sheets up to 1.2 x 4.9 m
G. Polyethylene (PE)	Monsanto 602, 703, Cloud Nine, Duratherm, Sunsaver, Tufflite III, Visqueen 1504, Loretex III, Fogbloc	Cloud Nine, Duratherm and Sunsaver are IR blocked. Most contain UV inhibitors. Sheets up to 12 x 67 m, 150 µm thickness better, may contain ethylvinyl acetate (EVA)
H. Polyester	Flexigard, Escolite, Hostaflow ET, Mylar, Melinex 071 and OW, Llumar	Flexigard laminated with PMMA, available only in narrow films

Table 2-10. Physical properties of some greenhouse covering materials. Abridged from Walker and Slack [1970]. Data converted to SI units.

Covering	2 C	19 C	36 C
Impact strength (load in grams required for 50% failure) ASTM Test B 1709-62T			
Polyethylene, 150 μm	358	327	327
Polyethylene, 150 μm, UV inhibitor	331	263	236
Polyvinyl chloride, 200 μm	No failure at 399 g		
Polyester, 130 μm	No failure at 399 g		
Corrugated FRP, 1.4 kg m^{-2}	No failure at 399 g		
Corrugated PVC	395	395	No failure
Glass, 3 mm	95	181	195
Load per 1 cm deflection of transversely loaded materials (kg) ASTM Test D 1502-60			
Polyethylene, 150 μm	11.3	8.0	6.8
Polyethylene, 150 μm, UV inhibitor	10.2	7.8	6.2
Polyvinyl chloride, 200 μm	8.2	5.9	3.9
Polyester, 130 μm	42.6	28.8	39.2
Corrugated FRP, 1.4 kg m^{-2}	13.2	14.2	14.5
Corrugated PVC	9.2	9.1	9.4
Glass, 3 mm	14.5	14.6	14.6
Maximum load at 15 cm deflection or to failure (kg) ASTM Test D 1502-60 continued			
Polyethylene, 150 μm	142.9	103.4	83.0
Polyethylene, 150 μm, UV inhibitor	119.7	101.6	78.5
Polyvinyl chloride, 200 μm	119.3	73.5	57.6
Polyester, 130 μm	Maximum load of 245.5 kg applied, deflection limit		
Corrugated FRP, 1.4 kg m^{-2}	53.1**	49.9**	46.9**
Corrugated PVC	48.5**	38.6**	40.8**
Glass, 3 mm	27.2**	29.5**	17.4**

** Material failed, test terminated.

Slight scratches on structured materials may lead to cracking in a relatively short time. Plastics of any kind should not be exposed to outdoor weather prior to installation. This is a factor for the grower to keep in mind when scheduling delivery of roofing materials to the construction site. Water and high sunlight on plastics leads to rapid degradation when placed in stacks outdoors, not to mention the possibility of high winds dispersing roof panels through your neighbors' windows.

Polyvinyl chloride (PVC) can also be obtained in structured forms, or in corrugated sheets or film. Early tests of PVC showed rapid embrittlement and degradation (Fig. 2-57). However, studies by Tesi and Tesi [1985] and Tesi et al. [1986] suggest that some of these difficulties have been corrected by new formulations and manufacturing methods –despite the general recommendation against PVC use by Duncan and Walker [1975] ten years earlier.

The difficulty with double-layer plastics is the real reduction in radiation transmission compared to glass, which becomes very important under limiting irradiance conditions of northern Europe and the northern U.S. In some instances, growers have installed supplemental irradiation. The relationship between yield and energy savings of two covers can be economically justified under appropriate conditions of low insolation and low outside temperatures. However, observation of European practices indicates major use of single, high transmissivity glass covers. Waaijenberg [1995] stated that larger glass panes, combined with improved structure has increased the total transmissivity of diffuse radiation from 65 to 72%. Of course, if one is growing a species that does better under low irradiance and warm temperatures, justification for a double layer is more easily obtained.

Fig. 2-53. Degradation of 150 μm thick vinyl film after 20 months of service. This first occurs over superstructure from heat buildup. Polyethylene, without UV inhibitors, usually degrades and tears in less than one year.

One of the most common covers, at least in the high radiation regions of the southwestern U.S. is corrugated, fiber-reinforced plastic (FRP). Various styles have been available (Fig. 2-52, 58), but 1.8 kg m^{-2} weight, corrugated panels, laminated on the outside with a PVF coating, have proven most satisfactory. Although some authors have presented data showing unsatisfactory results with FRP [e.g., Hasskelus and Beck, 1963], Holley and his students [Holley and Bennett, 1962; Holley, 1964; Briggs and

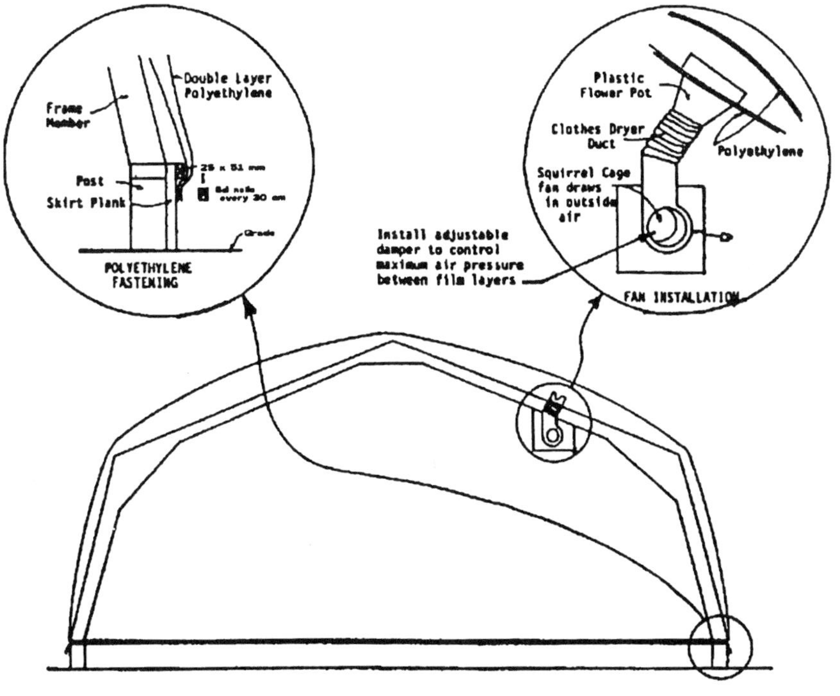

Fig. 2-54. Example of a double, air-inflated polyethylene greenhouse. The plastic is fastened only at the edges [Aldrich and Bartok, 1985]. Other methods for fastening film plastics are shown in Fig. 2-56.

Fig. 2-55. Early method of fastening soft plastics to greenhouse structure with nailed, wood strips. Subject to damage from high winds due to flapping and heat buildup under wood strips.

Holley, 1961; Spomer and Holley, 1963] showed a distinct advantage for carnation production under Colorado conditions. FRP was distinctly cheaper than glass, easier to install, and permitted less superstructure. The diffusing capabilities of FRP compared to glass, and its different behavior, resulted in improved growth of roses and carnations without the necessity of applying shading compounds to the roof as was the case with glass during the summer periods. Although impact resistant to hail (Fig. 2-59), FRP did prove to be highly flammable under appropriate conditions (Fig. 2-60). Long sheets of FRP can be applied to curved structures, and can be fixed to metal or wood superstructures, using the appropriate fasteners (Fig. 2-61). In high wind zones, damage has resulted largely from failure to adequately fasten the material to the structure. FRP must be fastened to purlins every other ridge. During the first year of installation, condensation on clear panels is high, and panels should never be fastened to lie directly on purlins since this will result in a drip line on the plants below as condensate runs toward the gutters. Application of hydrophilic coatings to the interiors of plastic covers [Delano and Raseman, 1972] aids considerably in handling condensation inside greenhouses.

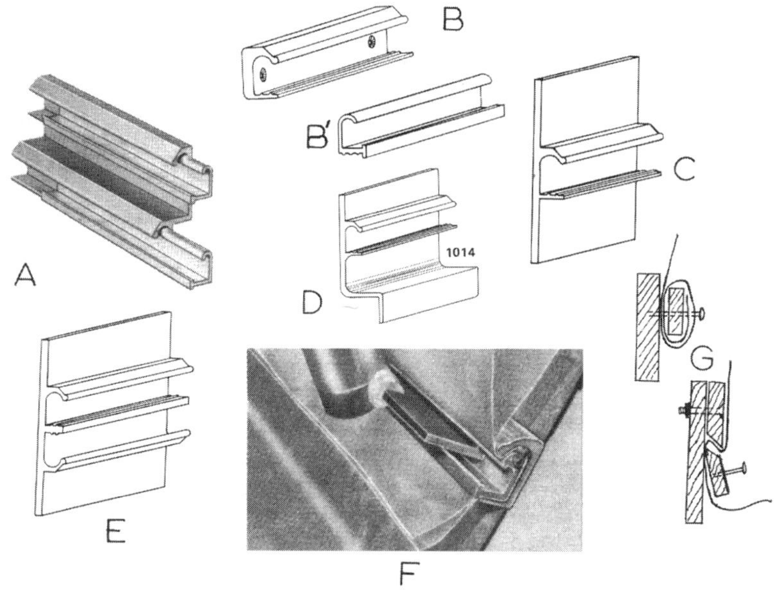

Fig. 2-56. Accessories for attaching plastic film covers to superstructures. **A** aluminum extrusion for a double layer; **B** simple extrusion with its insert **B'**; **C** extrusion for fastening at gable ends; **D** baseboard extrusion; **E** eave (roof-sidewall); **F** tool for installing inserts; **G** wooden systems.

The problem with FRP is gradual browning and weathering of the panels in relatively short periods (6 to 8 years) (Fig. 2-62), and the fact that panel surfaces can be attacked by common fungi [Durrell and Goldsberry, 1970]. Failure of the PVF lamination, or cracks from hail damage can accelerate weathering. A considerable problem is reliability in the manufacturing process. The result is that FRP must usually be replaced within seven to eight years. Refinishing has not proven satisfactory [Goldsberry and Homan, 1975; Goldsberry, 1971]. Despite the seeming advantages of FRP, at least in high radiation regions, application is not common outside of such regions in the U.S. A part of the problem may be psychological. Humans appreciate the warmth of the sun coming through glass on a cold, clear day. FRP, at high radiant intensities, definitely transmits less energy than glass, and its diffusing capabilities result in less direct radiation.

The evolution of plastics has come far since their first application in the 1950s. At that time, one was never sure the covering would be in place the next morning. On very cold mornings, these early plastic films would often shatter explosively when touched. One often had to wear a raincoat inside the structure as slight movements in the film would result in severe rainstorms. Recent improvements suggest that proper formulation may directly aid in disease control [Daponte, 1994].

Table 2-11. Effect of different roof covers on rose production. Table by Franklin [1983]. All comparisons made with glass as 100% (By permission of the British *Grower* magazine).

Comparison	Single glass	Double air-inflated PE	Twin-walled acrylic	Double PVF, air inflated
Yield	100	88	109	110
Stem weight	100	85	70	118
Stem length	100	96	91	109
Quality index	100	78	92	108

Fig. 2-57. Blackening of corrugated, rigid PVC over structural members of the greenhouse after two years in place. Picture taken in the 1960s. The entire covering has darkened, but particularly where heat has built up. Material becomes extremely brittle.

B. GLASS

Glass covers have always been, and will probably remain, an important factor in greenhouse design. With the advent of tempered glass, structural members required for support (Figs. 2-44, 45) can be reduced and spaced further apart, thus increasing usable energy inside the greenhouse. The increased impact resistance permits panes as wide as 1150 mm (Table 2-9), and newer construction makes use of ventilators without framing materials except at the hinge. There are possibilities of still stronger glass to come in the future. For example, Olcott [1963] discussed chemical hardening which permits glass to be bent sharply. Unfortunately, once strengthened, such glass can no longer be altered in any of the traditional finishing processes such as flameworking or cutting. A point is eventually reached where the weight becomes too much to be handled by workers, and a large pane would be prone to sagging under load.

The chief advantage of glass, of course, is its durability. With care and suitable maintenance, glass and metal structures will last generations. It can be cleaned and restored to original clarity with the use of hydrofluoric acid. The latter is the only chemical that will remove oily, sooty films. Acid products of combustion can react with glass to form durably bonded deposits. Other acids are not only dangerous but can corrode metal parts rapidly [Wells, 1969a; 1969b].

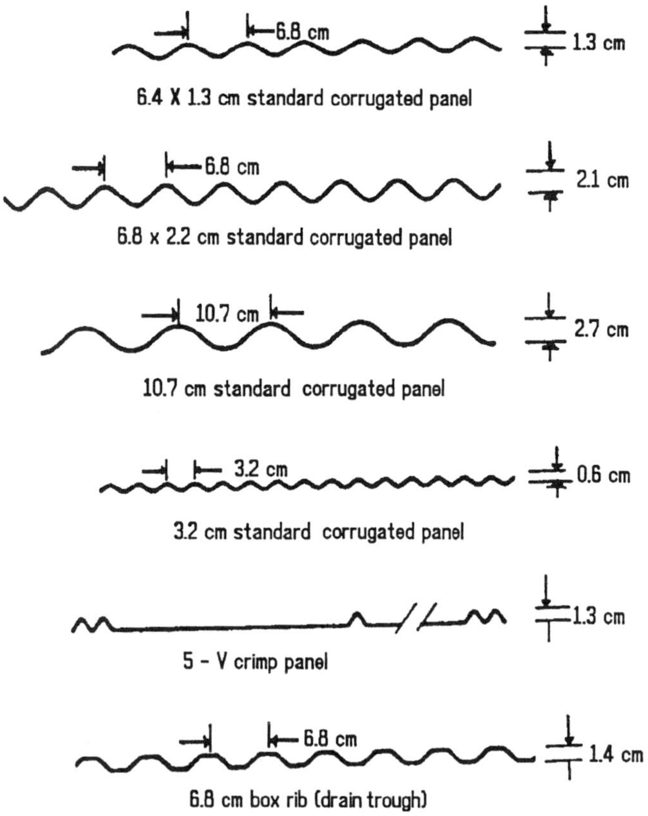

Fig. 2-58. Common configurations of fiber-reinforced plastic panels [National Bureau of Standards, PS 53-72].

Typical soda-lime, or float glass, contains about 0.1% iron, with reduced transmission compared to tempered, low iron (0.04%) glass. Special surface treatments can increase transmittance, and White [1987] predicted availability of glass at 96% transmission in a few years. White also indicated that 800 to 900 mm wide, diffusing glass can increase potential radiation at plant level to about 80%. Metallic oxide coatings have been applied to glass, which reflects long wavelength radiation. This, in combination with double layered glass, can effectively reduce heat loss by 70 to 80% compared to a single layer. Eyeglasses that darken with high light intensity are fairly common. This suggests the possibility of glass covers that automatically shade. Double layer glass, sealed at the edges, and the space filled with argon or CO_2, is quite strong but one of the most expensive greenhouse coverings. Some growers have retrofitted their structures with another glass layer (Fig. 2-63) [Anon., 1984; Gilette, 1982]. There is always, however, the problem of dirt accumulation between two layers of any material where the layers are not adequately sealed. There have been attempts to apply polyethylene either above glass, or below the sash bars. These are not satisfactory due to dirt and algae on the underneath plastic, or leakage of air through the overlaps on glass panes with an inflated top cover. As with double layered plastics, radiant transmission will be less. The tradeoff in energy saving may not outweigh the possible loss in yield.

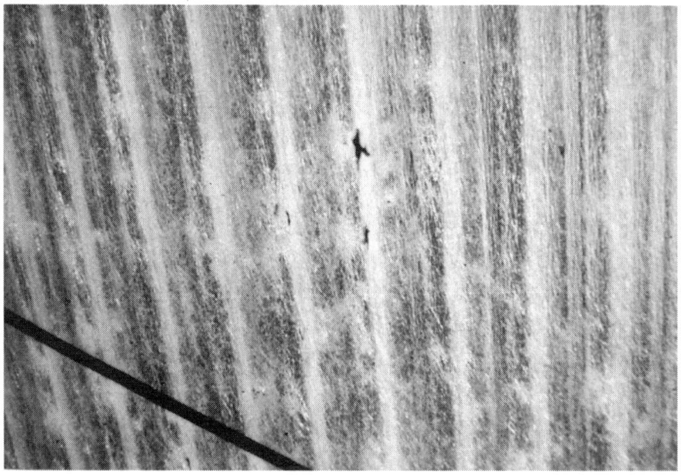

Fig. 2-59. Impact damage on FRP. Hailstones were sufficiently large to crack the PVF layer. Rapid weathering will occur. A glass-covered structure would have lost most of the panes. The dark, irregular spots show dirt accumulation.

C. SPECIAL VARIATIONS

The 1970s fuel problems led to considerable exploration of ways to decrease energy utilization. Among the more interesting was flowing $CuSO_4$ solutions between double layers [Nilsen et al., 1983; Mortensen, 1987; Tross et al., 1984]. The idea was to permit photosynthetically active radiation to reach the plants, and absorb other wavelengths. Such possibilities, including insulation between layers and its removal during the day, will be discussed further in Chapter 4. Growers have attempted opaque insulation of north walls. Spelman [1987] discussed the idea of placing double covers on the north roof, suggesting a significant improvement in an operation's gross margin. These procedures, despite some publicity, have not achieved widespread use –largely due to the increased initial cost and greater operating expense. The fact remains that, up to now, any roof transparent to solar radiation will possess a high heat conductivity (low R value) compared to a common insulated house roof. There is the possibility that, in the future, the use of unusual solids, called aerogels, might be possible [Fricke, 1988] (Fig. 2-64). These materials have been employed to reduce back radiation in solar collectors, and possess insulation qualities approaching common materials, with aerogel tiles having a conductivity 100 times smaller than fully dense silica glass. Should such materials prove economically feasible, greenhouses could be built with the insulation qualities now available to common buildings.

Fig. 2-60. Experimental test of FRP flammability. Under suitable conditions with exhaust fans on, loss of covering is rapid, usually without damage to superstructure.

VII. PERILS TO THE STRUCTURE

A. WIND

Several examples of perils to greenhouse structures have been presented already –some due to misapplication on the part of owners themselves. Damage from wind (Fig. 2-65) is probably the most common as the result of structural lightness, and the fact that greenhouse operators will sometimes try to cut corners. In regions where local topography and location can result in hurricane force winds, failure to use first-class fittings and fasteners will usually result in roof loss, not to mention the entire structure. If this occurs where temperatures can drop below freezing, the grower is likely to lose his entire crop, if the wind does not destroy it first. As discussed in sections 2.II and 2.III, recourse to climatic data and site selection is one of the major methods for eliminating future problems.

B. HAIL

There are well-known hail belts in certain regions of the world, of which eastern Colorado is one. This could be considered a climatic disadvantage, especially before the days of tempered glass. On the other hand, there are regions suitable for ornamental production where hail has never fallen in recorded history. Prior to the introduction of FRP, hail insurance or special hail covers were often required to avoid the result shown in Fig. 2-66. However, when hail can reach the size depicted in Fig. 2-67, even the most resistant covers will be pierced or severely damaged.

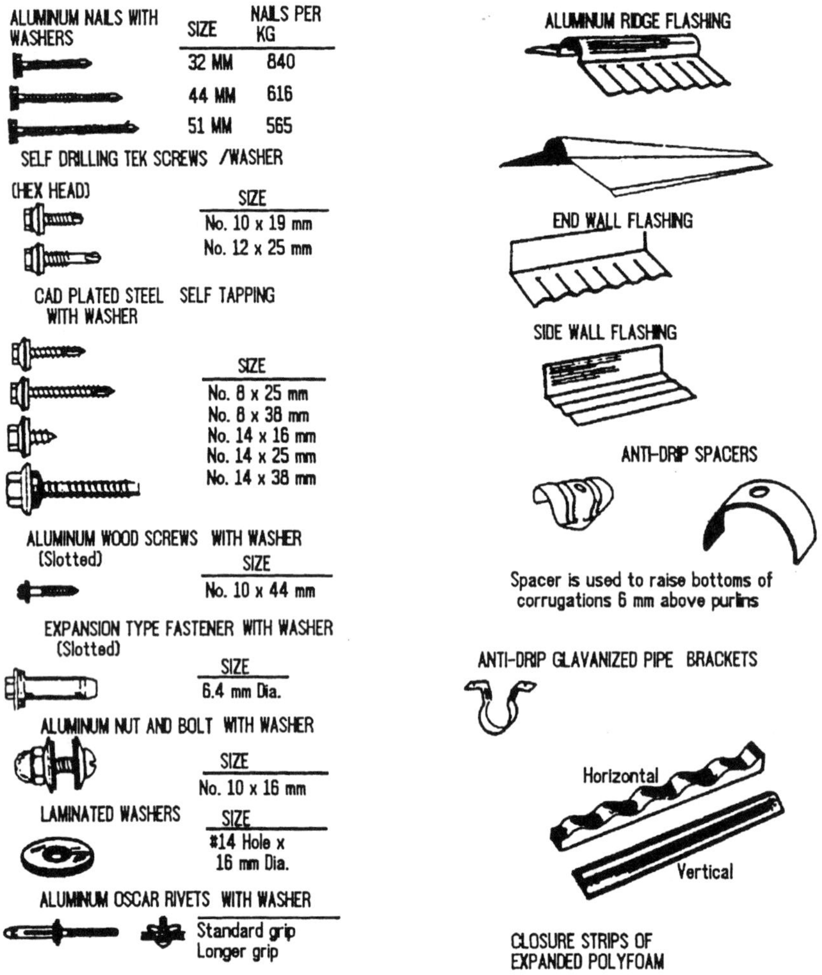

Fig. 2-61. Accessories used to attach FRP panels to superstructure [A.H. Hummert Seed Company].

C. FIRE

With the arrival of FRP, some growers thought they could dispense with hail insurance and hail screens. As indicated above, this has not always proved true; and, lurking in the future was the very serious problem of fire (Fig. 2-60, 68). Probably the first major fire occurred in Colorado when nearly 14000 m^2 of FRP covered structures burned in less than 10 minutes. In most instances, fires have resulted from carelessness, especially around dry aspen evaporative pads. When such pads are ignited, and exhaust fans are operating, the fire is carried rapidly through the house with very little damage to the superstructure. Crop loss is usually total, and the cleanup afterwards rather difficult, due to remaining glass fibers (Fig. 2-68). Acrylic covers are also highly flammable.

In an exhaustive study in the 1970s, Goldsberry [1970] [Hanan et al., 1978] reported that PVC or flame retardant panels installed as barriers were of no value, nor were sprinklers with fine droplets. Once heat has built up, decomposition of the plastic occurs rapidly and ignition is nearly spontaneous. Goldsberry summarized his investigations, placing the blame squarely on the operator as the primary cause. Among the factors were improper

Fig. 2-62. An example of severe weathering on FRP. The dark panel has never been replaced due to its position above the heater. Note the difference between the clearer panels –right versus left of center.

storage and failure to clean up combustible materials, improper electrical wiring, failure to inspect and maintain heating systems, carelessness on the part of employees, owners and children, and improper gas piping. Among his recommendations were clean facilities, proper construction and design, proper electrical systems for high moisture areas, establishment of employee education and fire drills, and coordination with the local fire department. Greenhouses often have hazardous materials of which firefighters should be aware.

D. SNOW

Problems of snow loads in the design of greenhouses have already been examined in Section 2.VI.A.1, especially the extreme variations that can occur within relatively short distances (Table 2-4). There have been

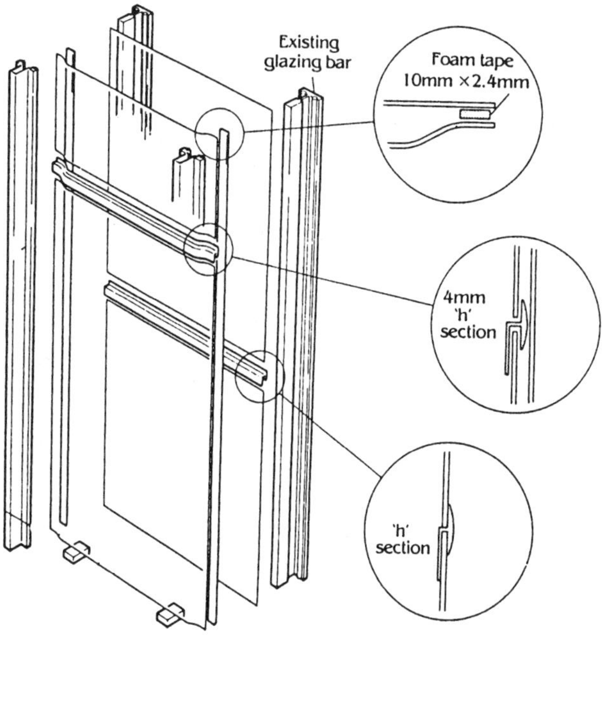

Fig. 2-63. Example of retrofitting an existing glass cover with a second layer of glass [Anon., 1984] (By permission of the British *Grower*).

examples of snow accumulating so fast that, despite the low thermal resistance of a greenhouse roof and the placement of heating pipes directly below the gutter, accumulation caused eventual collapse (Fig. 2-38). A grower may fail to get heat installed in new construction before the first snows of the winter season (Fig. 2-69). Coupled with minimal superstructure, an accumulation of less than 70 mm is sufficient to damage a house. Single, free standing houses have a distinct advantage inasmuch as snow will usually slide off the building. Most cases of snow damage have occurred with gutter-connected ranges. Snow can also be combined with high winds, leading to drifting in unexpected locations, causing severe structural loss. Although it is not economical to build for extreme conditions that might occur, suitable engineering can go far to reduce the probability of catastrophe. Some engineers feel that the total loss possible in such cases far exceeds the incremental cost of an adequate structure. The grower should thoroughly evaluate probability factors of catastrophic weather occurrences, and avoid accepting the cheapest structure. One should

Fig. 2-64. Example of an aerogel in granular form. This sample is made from sulfur dioxide and would be unsuitable due to its solubility in water. Other materials have been utilized. These granules are more than 90% air and translucent. Sandwiching such materials between two covers would more than quadruple the insulative quality of present covers.

Fig. 2-65. Two examples of severe damage from wind. In both cases, the wooden structure with columns merely buried in the ground was subject to overturning, as well as rupture of the film plastic.

Fig. 2-66. Example of hail damage to a glass-covered greenhouse. During the 1940s and 1950s, growers often employed a small mesh-screen above the roof as hail protection.

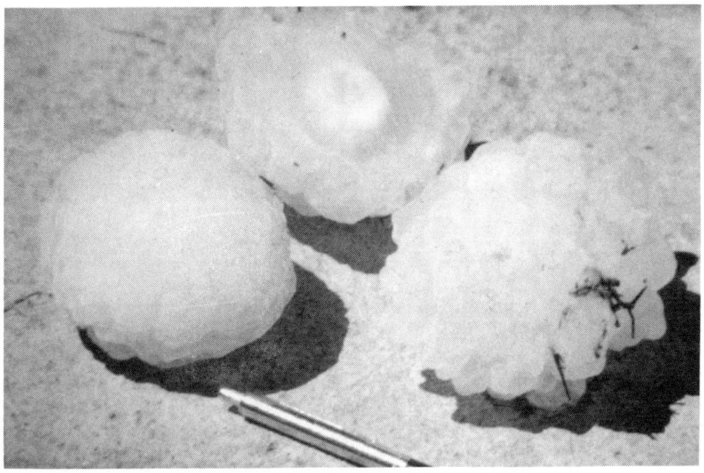

Fig. 2-67. Hailstones of this size can easily penetrate FRP as well as composition roofs underlaid by 16 mm thick plywood.

Fig. 2-68. Aftermath of a greenhouse fire. Note the remaining glass fibers which are difficult to handle. The superstructure is essentially undamaged although the crop loss appears total. This particular fire started from carelessness with a welding machine close to a dry aspen evaporative pad.

Fig. 2-69. Damage to new construction from an unexpected snow storm at a time when the heat system had not been installed. This greenhouse had a fairly substantial superstructure.

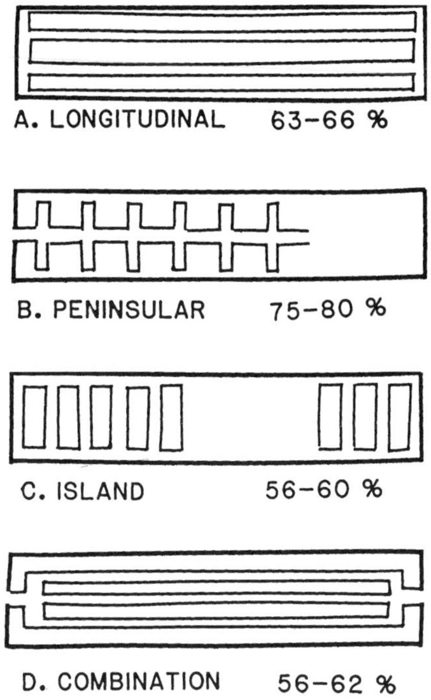

Fig. 2-70. Basic floor plans for fixed greenhouse benches [Hanan et al., 1978].

make sure that there is heat in the structure before any possibility of snow, and that suitable alarm systems are installed and operable in case of heating failure.

E. MISCELLANEOUS PERILS

As with any building, equipment or machine, there should be a regular procedure for maintenance: not only corrective with an easily perceived problem, but also preventive. Airplanes are subjected to a regular series of inspections with standard replacement of major components at stated intervals. The passenger is duly thankful when levitating at the pilot's command. There is no reason why such inspection and maintenance procedures should not be carried out with the more mundane and less perilous greenhouse. It emphatically saves money and removes stress. Even simple greenhouses in mild climates need inspection. In this age of hazardous materials handling, packing machines, etc., these operations need attention that can be traced with suitable records. Even high temperatures in plastic structures can lead to rapid degradation of the roof. Thus, it may not be the simplest procedure with empty houses to turn off ventilation and close the building. A hot greenhouse is always an empty greenhouse.

Fig. 2-71. Typical ground production of roses. This system is universal for roses and almost as common for all other cut flowers, especially in major production areas and mild climatic regions.

VIII. INTERIOR ARRANGEMENTS

Having discussed the structure and covering to some extent, one comes to how the inside of the greenhouse may be arranged for maximum efficient utilization. This planning phase should have been completed long before the structure is erected. One could calculate from the data in Table 1-5, that, to date, the percentage of total operational costs attributed to labor between operations in the U.S. and other locations is not all that different. This is a serious problem in developed countries where high labor costs exacerbate the industry's competitive position. The publication by Walker and Duncan [1974b] summarizes some common bench construction. Beginning in the 1980s, several publications [e.g., Lawson, 1989b; Ball, 1988;

Fig. 2-72. Raised longitudinal bench construction for cut flowers. These benches are constructed from redwood. Commonly, the undersupports are placed directly on concrete walkways, with the bottom boards placed lengthwise, with a minimum 6-7 mm spacing between.

Reilly, 1981a; 1981b; Hamrick, 1988] have examined and discussed new systems of mechanization, leading to more efficient product flow and concomitant reduction of labor costs.

A. LONGITUDINAL ARRANGEMENTS

The common, "old-style" arrangements of production area inside a greenhouse are presented in Fig. 2-70. Longitudinal bench arrangement, particularly for cut-flower production, is still the most common style found in both advanced and developing countries. Bench widths vary from 80 to 122 cm with aisles ranging from 46 cm for most cut-flower crops to 60 cm. In some locations, a bench may consist of a single row of plants with nearly 1 meter between each row. A bench width of 122 cm is too wide for a cut-flower crop, although suitable if pot plants are produced. Pot plant benches can be wider as long as workers can reach the center comfortably. Of course, main service aisles must be large enough to accommodate motorized traffic (1.8 to 2.5 m). cut-flower production in major production areas of the world is in the ground (Fig. 2-71), largely because of simplicity and lower expense. It is the cheapest. Raised benches are less common (Fig. 2-72) and are not often found except where they are the only way to exclude disease. Almost any material may be employed in bench construction, ranging from wood to concrete. For cut flowers, benches will be placed as close to the ground as possible in order to allow for a tall crop. For pot production, benches will be at a height for ease in access without excessive bending on the part of the laborer. Among the items that determine length of benches, are the maximum allowable temperature rise in structures using fan-and-evaporative pad cooling, and the avoidance of "backtracking" on the part of flower cutters. Cutting and picking is mostly manual. Depending upon the conveyance utilized by the worker, the ideal situation is for the cutter, or picker, to walk completely around each bench, removing everything ready for sale, without returning to the central aisle several times. The amount of time spent by workers traveling to and from their place of work can be astonishing. In some cut-flower operations, where the entire crop is removed in a short time, conveyor belts are installed above the crop so workers do not have to carry the product to the main service aisle. Special carts can be pushed or pulled down aisles so the worker does not have to backtrack. Hendrix [1975] spent considerable time discussing conveyor systems. Among those items that determine conveyor use is savings in time through quicker transport, less tiring to the worker and less damage to goods transported. Capital cost can be too high to justify equipping the whole of a production area. Also, moving a conveyor may not pay for the time spent. Hendrix mentioned their use in lettuce, pot and cut chrysanthemums, endive, celery and other potted plants that are cleared in a lump. Rail systems, using the heating pipe, are common on tomatoes and cucumbers. The disadvantages of transport systems include capital investment, many operators for correct distribution, and

Fig. 2-73. Two examples of increasing productive area for foliage plant production. Care should be used in overhead trays as noted here where the channel prevents drip on plants below.

monotony. There are, however, numerous advantages not only in labor savings. Mechanization may require a considerable increase in labor to make use of expensive machinery [Swanekamp, 1994].

Fig. 2-74. Utilization of aisle space through hanging baskets. This practice is not particularly desirable. Some shading of the crop below will occur and workers can be subjected to drip of hazardous materials employed for pest control. Dribble tube irrigation is desirable since hand watering is usually irregular and messy. Unless planned in the house design, such arrangements could put an unacceptable load on the structure.

B. PENINSULAR AND OTHER TYPES

The problem with longitudinal bench arrangement is the low area (60-65%) in actual production. The service access to the crop must be heated and maintained as any other portion of the structure, and returns little to the enterprise. Peninsular arrangement has been found particularly useful for pot plant production since the short aisles all end on a central aisle down which motorized carriers can proceed. Productive areas can be increased to 75 to 80% of the total inside area. It is possible, with fixed bench arrangement, to increase the productive area beyond 90 to 100% with certain crops such as foliage. These usually grow well with reduced irradiant intensity, and they can be located in tiered benches. Examples are shown in Fig. 2-73. Production of miscellaneous crops such as hanging baskets, while making use of normally unproductive space (Fig. 2-74), is not a viable option where the main operation is cut flowers. One or both crops usually get neglected. Hanging such weights on a structure built for cut-flower growing could result in structural damage. The other two types of fixed bench arrangement noted in Fig. 2-70 are seldom found, largely because of reduced production area. They are more likely to be utilized in botanical gardens.

Production area can be 90 to 100% of the total space with some cut-flower species, pot plants and specialized operations. Lettuce is a common one inasmuch as few cultural procedures are required between planting and harvest. Another example, with chrysanthemums (Fig. 2-75), can be achieved by transporting workers on gantries above the crop. Fortunately, *Dendranthema* can be grown very uniformly and harvested over a short period. This allows a single supporting tier, raised as the crop grows, and all operations up to harvesting are conducted from above.

C. MOVABLE ARRANGEMENTS

To go beyond this point with interior space, one must install movable benches (Fig. 2-76, 77), discussed by Reilly [1981a; 1981b] and others [Ball, 1981; Gillette, 1983; Short, 1982]. Benches are usually 1.5 to 1.8 m wide, capable of moving from side-to-side 46 cm. The space is filled with benches with room in the house for one access aisle. By moving the benches, any bench may be accessed from any side (Fig. 2-76). An example of the percentage increase in production area is given in Table 2-12. Of course, such benches are not cheap, with Reilly [1981a] suggesting a cost of about \$23.00 m^{-2}. This seems low, particularly for some more modern and elaborate systems which can also include the irrigation system. However, Ball, in 1981, reported a cost of \$30.00 m^{-2} for a bench mechanization system. Benches of this type can be upwards to 30 m in length. They have not been successful with common cut-flower production because of weight and the necessary supports which have to be quite strong to adequately maintain a heavy crop upright. Other examples of movable systems are provided in Fig. 2-78. Of course, when one retrofits an existing structure to increase usable space, he should have a profitable market for

Fig. 2-75. Example of total area production for chrysanthemum cut flowers. Note the heating pipes on the crop support, providing heat next to developing flower buds. All cultural operations are from the overhead gantry.

the increased production. Some growers have been known to forget this common sense realization that better space utilization is of no value if the new product cannot be sold.

Once one begins to think in terms of movable systems, it is a short step to systems of completely movable benches that can be rotated into and out of the growing structure. There are two

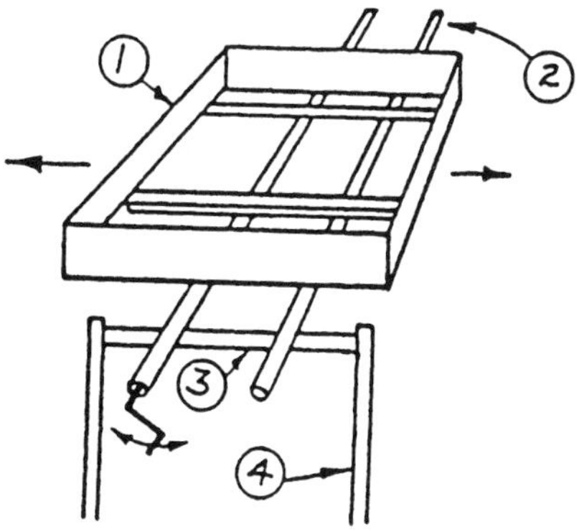

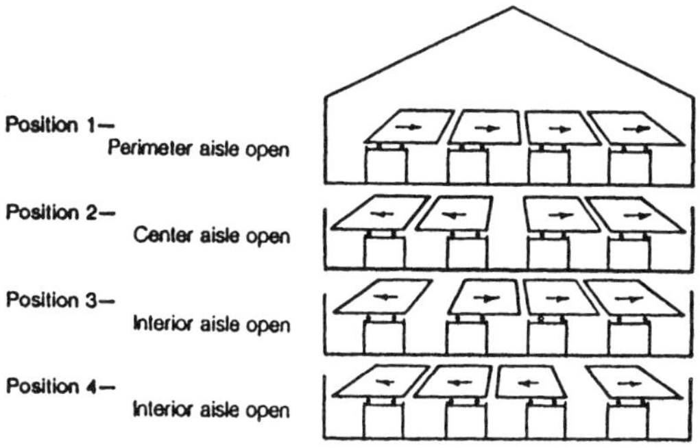

Fig. 2-76. Movable benches. **Upper**: **1**, Bench top is mounted on two continuous rollers, **2**, running the length of the bench. Rollers move laterally on the crosspiece **3**, each crosspiece supported by two uprights, **4** set in concrete. **Lower**: How one floating aisle is provided for access to all benches [Reilly, 1981a] (By permission of *Florists' Rev.*).

main schools of thought [Krause, 1981]; i.e., bringing the crop to the worker which requires crop containers capable of being moved regardless of arrangement or location, or bringing the workers to the crop through the use of gantry systems supported from the structure. The British have been most active regarding the latter, and the Dutch with the former. According to Krause [1981], the British gantry system has wider and more varied application, with the grower making his own system from standard parts. On the other hand, the Dutch system is highly mechanized for pot plant production. An example of a highly sophisticated Danish gantry system combined with movable trays is shown in Fig. 2-79. With this device, a tray may be moved from any location and carried to the packing and preparation area. Another basic arrangement, and less expensive, diagrammed by Reilly [1981b], is depicted in Fig. 2-80.

The idea behind these systems, of course, is better utilization of greenhouse space by reducing or eliminating aisles, labor savings, better attention to the crop, and putting the labor force into a better work environment.

Fig. 2-77. Two examples of movable benches. The upper photograph shows very wide platforms of expanded metal mesh; the lower are modern, plastic trays suitable for ebb-and-flow irrigation or capillary mats.

Hamrick [1988] reviewed the pros and cons of these systems, pointing out that trays must be cleaned, stacked and stored, requiring space to hold the trays. Stationary rails and tracks take up space, and overhead gantries reduce available solar energy and can interfere with thermal screens and supplemental irradiation installations. Another way around problems is to place containers directly on the floor, eliminating trays. The Dutch have tested this application (Fig. 2-81), but it can be appreciated that this is expensive. Nevertheless, growing on the floor allows maximum space versatility, and it can create a more uniform plant environment.

Whether trays, floor, gantry or combination, thorough planning is required to ensure that expensive systems operate properly. Material and product flows should be carefully planned (Fig. 2-82). Among the items one should consider, whatever the system or location [Hamrick, 1988], are:

Fig. 2-78. Upper: Simple, home-built, laterally movable platforms up to 30 m in length. This particular range was formerly a cut-flower operation. **Lower:** This pot plant grower has supported trays from the structure which are moved vertically when necessary. When lowered, they fill in the aisles between the fixed benches.

1. If one is growing in freestanding houses, make all houses the same size and length, and move production in one direction. Sometimes, by laying out a string while walking a crop through the range will help to eliminate backtracking and crossing. Note the discussion later in this section.
2. Make all doors of equal height and eliminate wherever possible.
3. Never handle single units.
4. Create buffers between every major work action in the headhouse. If actions are directly connected, the work line will always be as slow as the slowest activity. If buffered, then each function can operate independently.

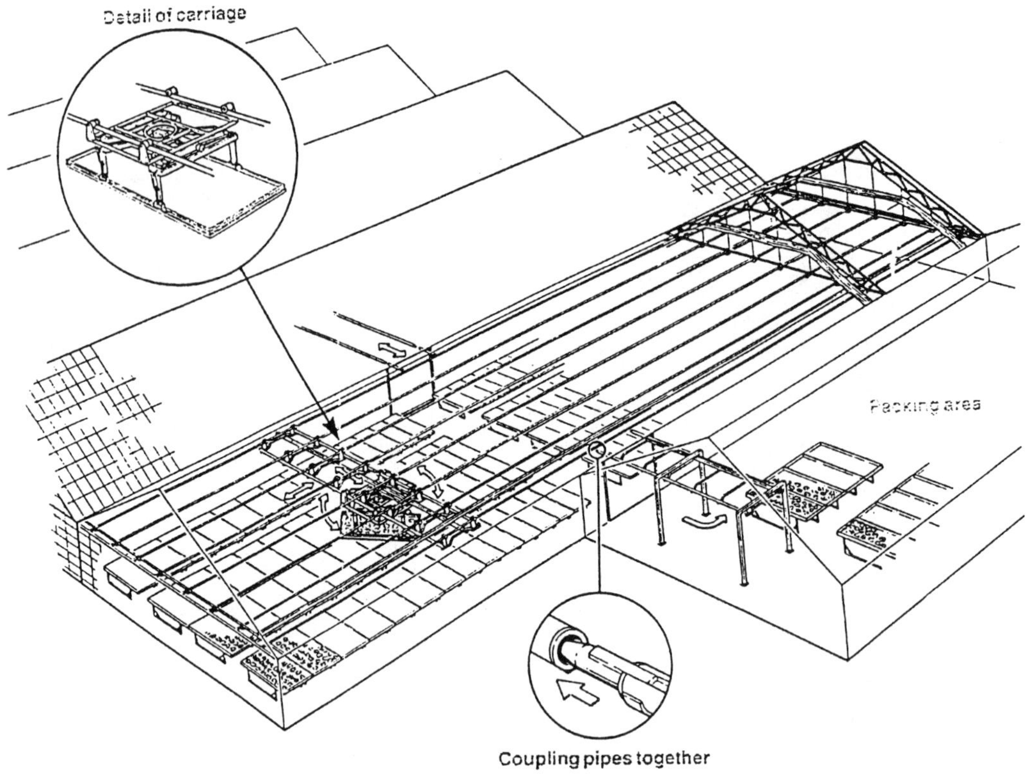

Fig. 2-79. Danish gantry system for moving trays from greenhouse to packing shed and back [Lawson, 1988b] (By permission of the British *Grower* magazine).

Fig. 2-80. Basic system for moving tray benches described by Reilly [1981b] (By permission of the *Florists' Rev.*).

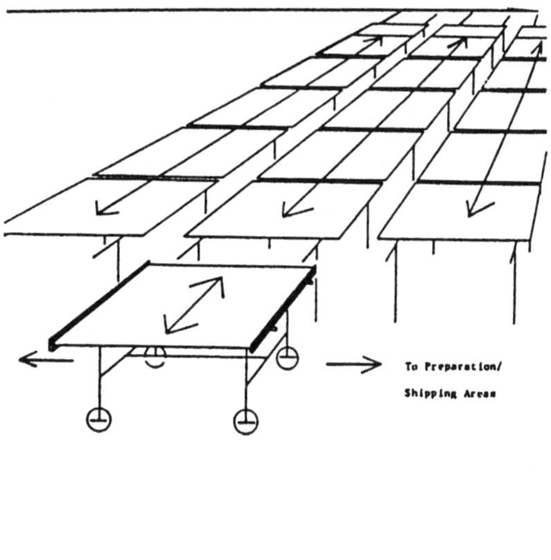

Table 2-12. Increases in production area for different width houses, comparing fixed benches with floating aisle benches [Reilly, 1981a]. The number and size of benches are given with total meters Equal signs indicate total width of preceding benches. (By permission of the *Florists' Rev.*).

Greenhouse width(m)	Fixed	Floating aisle	% increase
6	2, 1.5 m = 4.2 m, 1, 1.2 m	3, 1.8 m = 5.4 m	29
7.8	3, 1.8 m = 5.4 m	4, 1.8 m = 7.2 m	33
9.6	4, 1.8 m = 7.2 m	5, 1.8 m = 9 m	25
10.8	3, 1.5 m = 8.1 m 2, 1.8 m	2, 1.5 m = 10.2 m 4, 1.8 m	26
12.6	5, 1.5 m = 9.3 m 1, 1.8 m	2, 1.5 m = 12 m 5, 1.8 m	27

Fig. 2-81. Dutch growing floor for pot plant production. Hot water pipes are in the concrete base to control temperature at the base of the plants. Irrigation is by flooding the floor (large scale ebb-and-flow), and CO_2 injection and warm air can be introduced through the floor to move upward through the plants.

5. If many people are doing the same thing, then set up parallel work lines. Each worker can function independently, and the pace of the production line is not limited to the slowest worker.
6. The transportation system must be able to move the product faster than the activity itself.
7. Put items requiring the most activity closest to the activity station.
8. Make allowances for downtime with a backup system standing by. Plan a maintenance schedule, and set up training sessions for staff. The greenhouse organization system will have to be changed. Grafting pieces onto existing operations is unsatisfactory.

For example, Fig. 2-82A shows a propagation operation with the plants stuck and rooted in week one. The plants then proceed through the house to week eight, when they are removed and packed. In Fig. 2-82B, robotic movers, capable of moving several trays at a time, move through the house, one mover for each side. On the other hand, 2-82C shows a single direction of movement for cutting production. In both of the latter two systems, robotic movers are programmed from computers which provide staff with information on production, shipping and sales, so that the company knows exactly how many trays of what variety are available and where they are located. Likewise (Fig. 2-82D), product flow for separate houses is in one direction. In all of these, planting and packaging can be carried out in a more comfortable environment for the workers. With some crops, the greenhouse becomes a "holding" area and little or no work is carried out within the structure (Fig. 2-83).

Other systems have been devised (Fig. 2-84). The main stumbling blocks remain at the beginning and ending. Here also, machines are being manufactured to automatically stick cuttings, plant seed trays and transplant the

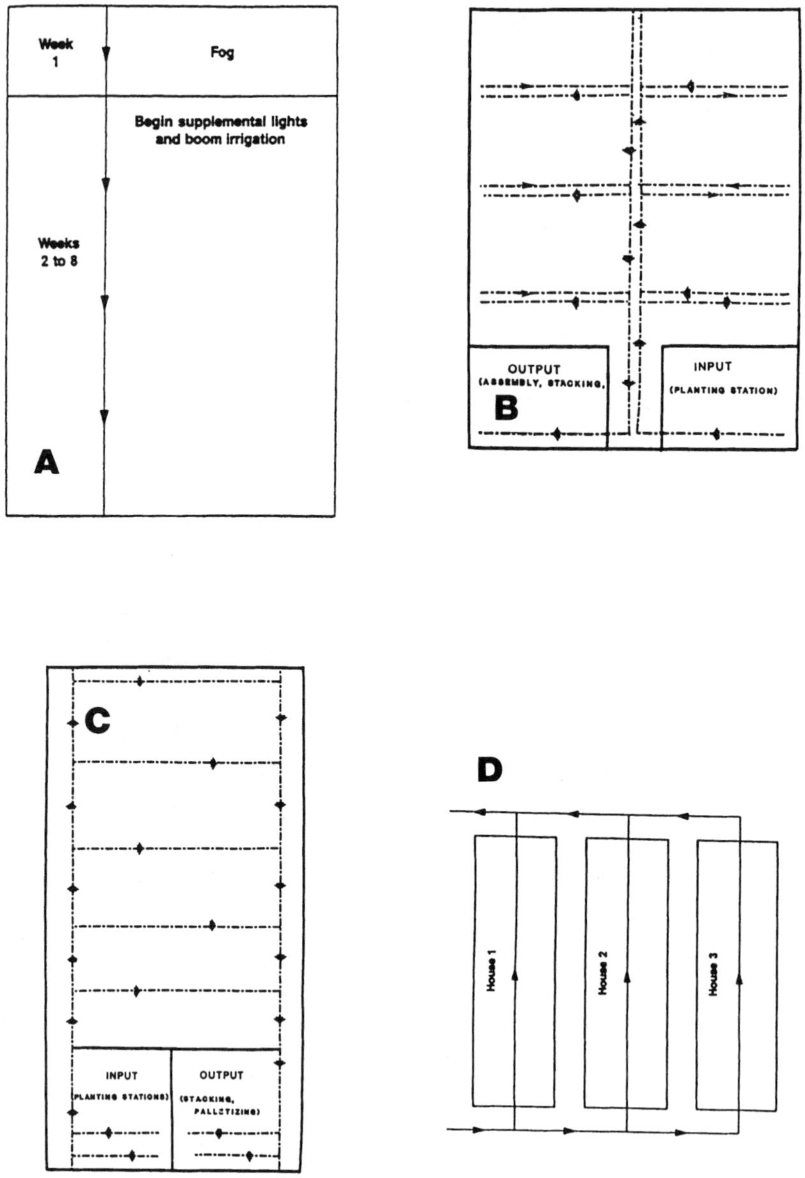

Fig. 2-82. Various types of flow patterns for product movement in greenhouses [Hamrick, 1988]. See text for more explanation (By permission of *Grower Talks*).

seedlings, and remove, grade and package the final product [e.g., Kutz et al., 1987; Anon., 1977; Royle, 1980; Young and Murphy, 1982]. Ting et al. [1991a,b] and Yang et al. [1991] have spent considerable effort on robotic transplanting and materials flow of seedlings, particularly for plug production. An example of an actual operation is shown in Fig. 2-85. The problem in the U.S. has been the diffuse nature of the greenhouse industry in a large nation. There has not been any real standardization so that advances are made by individuals, each to his own pattern. This is contrasted with the situation in several European countries where central agencies coordinate

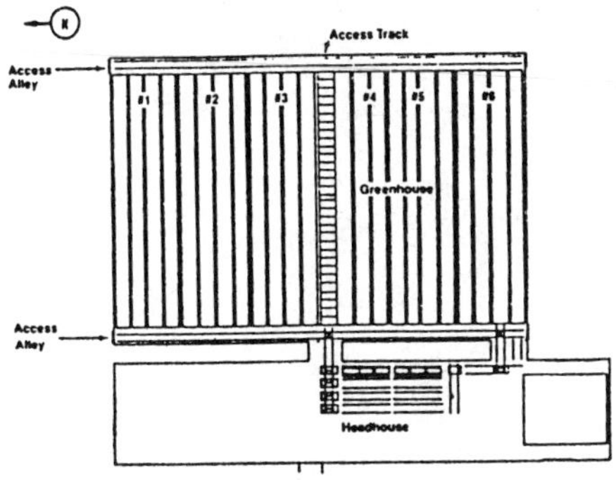

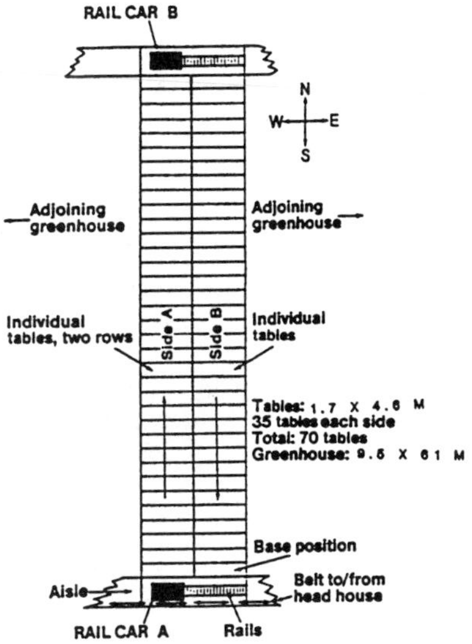

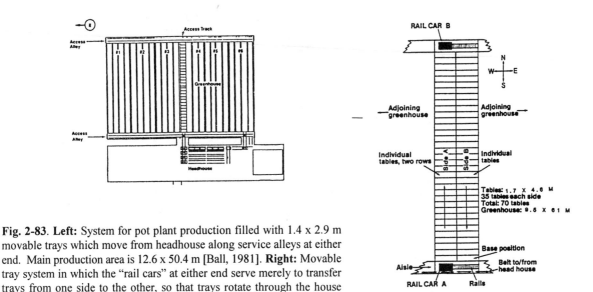

Fig. 2-83. **Left:** System for pot plant production filled with 1.4 x 2.9 m movable trays which move from headhouse along service alleys at either end. Main production area is 12.6 x 50.4 m [Ball, 1981]. **Right:** Movable tray system in which the "rail cars" at either end serve merely to transfer trays from one side to the other, so that trays rotate through the house [Ball, 1988] (By permission of *Grower Talks*).

action and provide a modicum of standardization. Several years ago, I toured a civil engineer through several greenhouses. His comment was that not one of the more than 12 operations visited were the same. An entrepreneur would, and does, have difficulty in retrieving design and setup costs to manufacture many of the foreign origin machines now found in the U.S.

While great advances have been made in pot plant and bedding plant mechanization, difficulties remain with cut-flower production. Here, the problem can be divided into three parts: 1) The fact that plant material is usually in place for a long time, and it can be tall and heavy. Thus, the support, and any moving system, must be robust if the entire planting is to be moved. Attempts to do this with tomato production have been made [Lawson, 1989; Giacomelli, 1987]. 2) The root substrate is commonly heavy, although movement toward hydroponic systems has greatly reduced this limitation. Unfortunately, the systems that have been devised, such as the nutrient film technique, do not lend themselves to transport of the type designed for small containers. And 3), as with the chrysanthemum, which has gradually evolved through re-selection and breeding programs to the type that can be single-cropped, many of our other cut-flower crops need to be bred for mechanization. This is what occurred with field tomatoes. For example, the present carnation cultivars are unsuitable for fast cropping –although single cropping has been tested several times. The species must be developed in tandem with the mechanization scheme. The problem No. 3 has never received adequate attention in the U.S., the growers often endorsing esoteric, biotechnical systems that appear as cheap substitutes –especially if the presentation is good. Breeding in the U.S. has been heavily slanted toward new varieties and disease resistance. Overhead conveyors of various types that move equipment and materials between headhouse and greenhouses are almost too numerous to mention.

The flow of material in the production process has been addressed by several authors [e.g., van Weel, 1991; Young and Murphy, 1982; Fang et al., 1990]. Young and Murphy set up a design to accommodate grading, staging, order picking and packaging of foliage plants. There were three problems: 1) Individual customer orders were generally small quantities of several species, sizes, etc. 2) A high incidence of small quantity "add-on" orders. And 3), the plants had to be graded for quality. The authors emphasized that a training phase was necessary to stop old habits. Fang et al. utilized a simulation program to examine internal transport systems, which allowed one to find potential bottlenecks in the system.

Fig. 2-84. Carnation propagator using movable trays. **Upper**: Rooted cuttings come from the greenhouses to the left and are pulled for packaging. Trays are renewed and pasteurized at the far end and return overhead for new cuttings. **Lower**: A tray of new cuttings is watered as it passes on the way into the greenhouse. No cultural operations are performed in the house.

Space allocation in pot plant ranges, growing several species, types, cultivars, etc. is a serious problem. Fortunately, however, there are means to address space allocation for the purpose of maximizing return through the use of linear programming. Heinemann [1994] discussed this procedure on an elementary level, and Fang et al. [1990] developed a program specifically for greenhouses. Basham and Hanan [1983] discussed the use of linear programming, but failed to develop it into a user-friendly system for greenhouse application. With these types of programs where space is a constraint, the mathematical procedure allows one to incorporate costs, variable labor, time requirements for sale of a particular item, prices, etc. in order to allocate space, depending upon stage of

Table 2-13. Construction costs (U.S. $) for three different sized greenhouses, gutter-connected, in North Carolina. [Abridged from Brumfield et al., 1981]. Annual interest rate 13%.

Size (m^2)[a]	**1800**	**9000**	**36000**
Land required (ha)	2	4	20
Area graded (m^2)[b]	4233	16000	52694
Earth moved (m^3)	1266	4786	15763
Land value ($/ha)	7000	3750	2500
Grading cost ($/$m^2$)	0.76	0.76	0.65
Total price land ($)	14000	15000	50000
Grading costs ($)	3175	12330	35680
Cost of land use/yr ($) (13% int.)	2233	3553	11138
Cost, Dbl, PE, air-inflated (Total)	62820	285500	1093600
Cost, glass structure (Total)	107800	488000	1884000
Cost, basic for PE structure (m^{-2})	14.58	13.50	10.74
Cost, heating (m^{-2})[c]	3.67	3.13	3.13
Cost, cooling (m^{-2})[d]	7.13	6.91	6.80
Cost, labor (m^{-2})	5.94	4.86	4.43
Cost, freight (m^{-2})	0.97	0.83	0.81
Cost PE cover (m^{-2})	1.63	1.57	1.56
Total cost PE (m^{-2})	33.91	30.89	29.48
Cost, glass structure (m^{-2})	29.92	26.78	25.81
Cost, heating (m^{-2})[c]	8.32	7.56	7.45
Cost, cooling (m^{-2})[d]	6.05	5.94	5.94
Cost, labor (m^{-2})	11.88	10.48	9.72
Cost freight (m^{-2})	2.05	1.94	1.94
Total cost, glass (m^{-2})	58.21	52.70	50.87
Service bldg., size (m^2)[e]	185	907	3889
Economy model ($)	27000	133700	503700
Premium model ($)	44000	191100	714000
Ave. cost m^{-2}	192.24	178.96	156.60
Coolers ($)	10447	21369	45207
Employees (No.)	6	22	87
Driveway and parking ($)	(gravel) 2862	(paved) 28592	(paved) 66468
Parking size (m)	19.5 x 21.6 (16 cars)	19.5 x 40.5 (30 cars)	19.5 x 40.5 (30 cars)
Concrete aisles ($)	9082	42989	179939
Gravel under beds ($)	4978	21903	81065
Benches, fixed, home built ($)	19500	97500	390000
Benches, prefabricated ($)	35880	179400	717600
Benches, movable ($)	34164	170820	683280

[a]Operations located in country, wholesale only. [b]Land nearly flat for minimal grading.
[c]Heating, using gas-fired unit heaters. [d]Cooling with evaporative pad and fan, cellulose pads.
[e]Steel, insulated building.

Table 2-14. Greenhouse construction costs for various styles [Aldrich and Bartok, 1985].

Type	Materials ($ m^{-2})	Erection labor cost ($ m^{-2})	Total cost ($ m^{-2})
Conventional glass house, concrete foundation, galvanized frame, truss roof	64.80-86.40	27.00-32.40	91.80-118.80
Gutter connected, plastic, concrete piers, galvanized frame	27.00-43.20	16.20-21.60	43.20-64.80
FRP covered, pipe arch, pipe foundation, 32 mm galvanized pipe	18.90-27.00	2.70-5.40	21.60-32.40
Steel pipe arch with PE cover, pipe foundation, 32 mm galvanized pipe	12.96-18.36	2.16-3.24	15.12-21.60
Rigid frame, wood, PE cover, wood post foundation, clear span	8.64-10.80	3.78-6.48	12.42-17.28

Note: Site preparation $2.70 to $3.78 m^{-2}, 76 mm concrete floor $7.56 to $8.64 m^{-2}, benches $16.20 to $48.60 m^{-2}.

Fig. 2-85. The effect of ground covered on relative cost per unit value of a greenhouse [Smith, 1988] (By permission of the British *Grower* magazine).

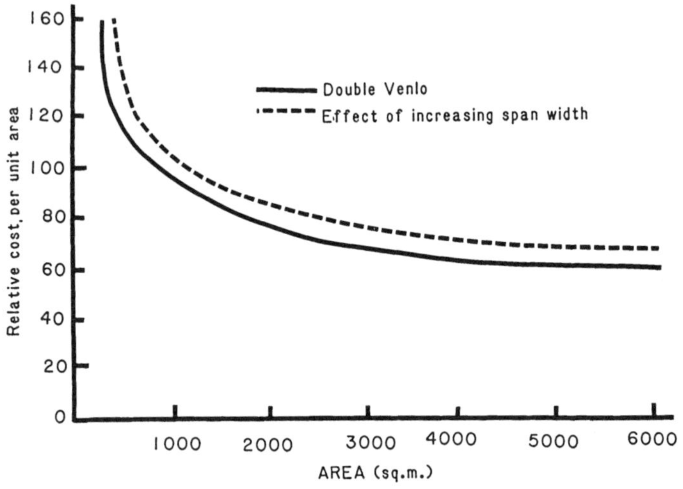

growth and time of year. The method determines the best use of space in accordance with maximization of the "objective function" or profit. From it one may obtain a sensitivity analysis; that is, if the price is increased or decreased, what is the effect on profit? Or, if one forces additional or fewer units, what is the likely outcome? Linear programs, and related procedures, have opened a new era in greenhouse management. These have also been applied, as discussed in Chapter 8, where climate is optimized. Powerful systems are available which will require changes, new thinking and greater technical competence.

IX. COSTS

There are a few caveats to consider when looking at costs of greenhouses: 1) Due to inflation and rapidly changing conditions, any figures are out-of-date before they can be published. 2) Data for industry in developed regions will have no relation to costs in less developed countries, especially when one includes the possible governmental policies and subsidies that may be available. Table 1-5 for operational costs in different countries

Table 2-15. Greenhouse construction options employed by Halsey [1975] in calculating present values by means of discounting over a 30 year period for a 4000 m^2 operation (By permission of the *Florists' Rev.*)

Option I	Venlo glass house, initial cost \$23.00 m^{-2}, annual maintenance costs = \$100, re-seal glass every 10 years at \$0.86 m^{-2}.
Option II	Steel post, extruded aluminum at \$10.69 m^{-2}, PVF laminated FRP replaced every 10 years at \$4.53 m^{-2}, removal cost at \$0.22 m^{-2}, annual upkeep at \$300.00.
Option III	Same structure as Option II, FRP without PVF lamination at \$3.13 m^{-2}, removal cost at \$0.22 m^{-2}, recover every 5 years, annual upkeep = \$100.00.
Option IV	Same structure as Option II, covering double layer, air-inflated PE at \$1.19 m^{-2}, replace every 2 years at \$0.22 m^{-2} removal costs, annual upkeep = \$100.00.
Option V	Treated pine-cypress structure with PVF laminated FRP, cover replaced every 10 years, \$7.34 m^{-2} construction, \$4.12 m^{-2} for covering, \$3.24 m^{-2} to remove old covering, annual upkeep = \$300.00.

points up the problem. And 3), information of this type is more likely to be published in the native language and not widely disseminated outside of the particular region. Nevertheless, good information, even if outdated, can be utilized as a basis of comparison to alert one to those factors that may be most important. For example, Table 2-15 compares 1975 costs for several types of construction and the means of arriving at present values. The relative differences remain valid today even if more than 20 years old. Obviously, the data must be updated. The data in Table 2-14 is 12 years old. Again, the relative differences between different construction types are meaningful.

Some figures have been given in previous pages, but probably one of the most comprehensive and detailed expositions on costs was the publication by Brumfield et al. in 1981 for North Carolina operations (Table 2-13). Aldrich and Bartok [1985] also provided a simplified schedule for materials and erection of several greenhouse types (Table 2-14). If one compares the differences in prices derived by Brumfield et al. in 1981 versus those given by Aldrich and Bartok in 1985, most of the change can be attributed to inflation. In general, glass-covered structures are likely to be most expensive, with wood superstructures and polyethylene covers the cheapest. Between, there is sufficient variation to please anyone. The data by Brumfield et al. is particularly interesting in its detail by pointing out numerous areas of expense that may not be adequately considered by the neophyte –such as preparing the land, freight costs, the service building, parking and benches. There are obvious economies of scale. That is, the larger the operation, the smaller the per unit cost as exemplified in Fig. 2-86. The high costs of small structures (i.e., hobby greenhouses) can be appreciated, and visual observation of Smith's curves suggests any new commercial structure covering less than 1000 m^2 will be prohibitively expensive, whereas beyond 4000 m^2, increased area will provide only small benefits in terms of reducing unit area expense.

Are there means by which one can make intelligent decisions as to type of structure other than the state of one's exchequer and friendliness with the local banker? One method was examined by Halsey in 1975. This is the discounted cash flow analysis that serves to yield a "present value" of a structure that may endure for upwards to 30 years. Halsey set up a series of options (Table 2-15), each option to last 30 years, including the costs expected for each year of the option's usable life. It is doubtful that any greenhouse will be recognizable after 10 years in this day and age. However, Halsey's work serves as an illustration. The cost for each year is multiplied by a "discount coefficient," which is, in effect, the percentage return on investment expected by the operator. The formula for calculating the coefficient (d_c) is:

$$d_c = \frac{1}{(1 + i)n} \qquad (1)$$

where i = *the time preference (discount) rate as a fraction (10% = 0.10), or the rate of return on investment*

n = *the year*

Table 2-16 shows the method for Option I given by Halsey in Table 2-15. If all the costs are added for the 30 years, one arrives at the "actual" value, or cash expended at the end of 30 years. The discounted cash flow analysis brings back future values to the present, working in reverse of compounding. If this procedure is carried out for each of the options in Table 2-15, for different rates of return, Table 2-17 is the result. The results show, at a 10% rate, that Option IV (double PE, air inflated) is the least cost alternative in present real values (1975). However, there is not much difference between IV and Option II, aluminum, PVF-FRP. When the timing of costs is introduced, and cash outlay discounted at 10%, Option I is the most expensive. If the owner is willing to accept 6% return on investment, Option II would be the least cost alternative (Table 2-17). According to Halsey, this illustrates that when as well as how much will affect investment decisions. With existing desk computers, the procedure for calculating "present value" shown in Table 2-16 can be carried out automatically and in a very short time, thus giving the grower the opportunity to try several scenarios.

The analysis presented here breaks down with high inflation rates. That is, greenhouses that require frequent covering replacement and more labor will be hit hardest by inflation. Assuming that the price of FRP will rise 1% per year, PVF will increase 2%, polyethylene 5%, lumber 5%, and wages 5% per year, recalculation of the present value for the options in Table 2-15 will show Option II to be the least cost alternative, followed by III, V, IV, and I last with a present value of $95586.00. Another possibility is that lower interest rates are likely to be charged for long-term structures such as Option I. The return on investment is likely to be different. That is, 12% for I, 10% for II and III, 6% for Option IV and 8% for V.

Table 2-16. Expected costs (U.S. $) and present values for Option I, Venlo style glass greenhouse, 4000 m^2, discounted at 10% [Halsey, 1975] (By permission of the *Florists' Rev.).*

Year	Discount	Actual cost	Present
0	1.0000	92783	92783
1	0.9091	100	91
2	0.8264	100	83
3	0.7513	100	75
4	0.6830	100	68
5	0.6209	100	62
6	0.5645	100	56
7	0.5132	100	51
8	0.4665	100	47
9	0.4241	100	42
10	0.3855	3585	1382
11	0.3505	100	35
12	0.3186	100	32
13	0.2897	100	29
14	0.2633	100	26
15	0.2394	100	24
16	0.2176	100	22
17	0.1978	100	20
18	0.1799	100	18
19	0.1635	100	16
20	0.1486	3585	533
21	0.1351	100	14
22	0.1228	100	12
23	0.1117	100	11
24	0.1015	100	10
25	0.0923	100	9
26	0.0839	100	8
27	0.0763	100	8
28	0.0693	100	7
29	0.0630	100	6
30	0.0573	100	6
Total		102753	95586

There are other methods that can be used in the same manner for computing present values of a structure. Assuming that the grower has sufficient information, and is deciding to expand, he can compute his "internal rate of return" which is a measure of the economic soundness and feasibility of an investment [Kirschling and Jensen, 1974; 1976]. The IRR is a measure of the potential return on capital investment in a project, based on the time flow of money into and out of the project. As with present values, it is the annual compound discount rate, which makes the present value of scheduled investments equal to the present value of expected dollar returns. The IRR is the anticipated rate of return internally produced by the project. It can be compared directly with the rate of return external to the project, or the opportunity or value cost of capital. If the IRR is greater than or equal to the opportunity cost of capital, or external rate of return, then the proposed project is economically profitable. The method, for example, can be utilized to determine the effects of

Table 2-17. Comparison of future value (U.S. $) of five greenhouse options (0% return on investment) compared with present values when discounted over a 30 year period.

Percent return on investment (Discount rate)	I (See Table 2-15)[a]	II	III	IV	V
(0%)=future value	$102753	$108752	$125777	$129478	$154055
6%	97192	82226	87530	85427	93570
8%	96271	77787	81272	78409	83345
10%	95586	74485	76627	73233	75727
12%	95072	71997	73111	69330	69981
15%	94513	69297	69263	65061	63701

[a] Greenhouse area in each option is 4000 m^2. The percentage rates are calculated by means of an appropriate formula, or can be obtained from tables in various textbooks.

wage rates, land costs, inability to sell 100% of the product, price of the product, etc. It requires, however, good record-keeping. Other systems have been discussed [e.g., Spelman, 1987].

Sometimes, the decision to invest depends on more than costs of materials and labor. It includes not only the cost of interest on capital borrowed, the period of the loan, but also the personality and experience of the banker. In a *Grower* article in 1983, the anonymous author pointed out that when a Dutch grower goes to his local bank for a loan, the banker not only knows about greenhouse growing and its prospects, but he has up-to-date cost and performance data on file, as well as auction returns to corroborate cash-flow predictions. The banker is conditioned to say "yes". In 1983, construction costs per hectare in Holland varied from $460000 to $520000, boiler installations $119760 to $131736 per ha, and land costs at $89820 per ha. The total cost was not all that much different from the U.K. The biggest advantage is the ability to obtain long-term loans (15-20 yrs.) at 7.5 to 8.5% as contrasted with the British situation of 7 year loans at 16% annual interest. The terms of the loan are likely to be the most important in making decisions to build a greenhouse.

X. REFERENCES

Albrecht, F. 1951. Intensitat und Spektralverteilung der Globalstrahlung bei klarem Himmel. *Arch. Met. Geophys. Bioklim.* 3:220-243.

Aldrich, R.A. 1988. Evaluate structure and glazing before building. *Greenhouse Grower*. 6(6):88-92. June, 1988.

Aldrich, R.A. and J.W. Bartok, Jr. 1985. Greenhouse Engineering. Agric. Eng. Dept., Univ. of CT, Storrs, CT.

Aldrich, R.A. and J.W. Bartok, Jr. 1987. Alternate methods of installing double skin glazing. *Roses, Inc. Bul.* June, 1987.

Aldrich, R.A. and D.A. Wells. 1979. Estimating wind loads on glasshouses. *Trans. ASAE*. 22:1122-1128.

Anon. 1977. The real block revolution is yet to come. *The Grower*. 88(8):329-333. Aug. 25, 1977.

Anon. 1981. Roof covers - An overview. *Grower Talks*. Sept., 1981.

Anon. 1982. Solakleer glass improves light passage. *Florists' Rev*. 171(4432):36-37. Nov. 11, 1982.

Anon. 1983. What makes the Dutch successful? *The Grower*. 105(17). Apr. 24, 1983.

Anon. 1984. Double glazing comes of age. *The Grower*. 102(21):32-33. Nov. 22, 1984.

Anon. 1989. Value of Venlo. *The Grower*. 111(4). Jan. 26, 1989.

ASCE. 1993. Minimum design loads for buildings and other structures. *Amer. Soc. Civil Engr.*, 7-93, New York.

ANSI A58.1. 1982. American National Standard minimum design loads for buildings and other structures. Amer. Nat. Standards Inst., New York. 100 pp.

Bailey, B.J. and D.L. Critten. 1985. The vertical south roof multispan greenhouse. *Acta Hort.* 170:193-199.

Ball, V. 1981. The Pan-American Plant Parrish System. *Grower Talks*. 45(2):9-17, June, 1981.

Ball, V. 1986. A fresh look at greenhouse coverings. *Grower Talks*. June, 1986.

Ball, V. 1988. Go around! A new labor-saving concept. *Grower Talks*. 52(8):14-18, 23-24. Sept., 1988.

Bartok, J.W., Jr. 1986. Information and plans for greenhouse/nursery operations. *Nat. Res. Mgmt. and Eng.*, SEG-29 Revised. Univ. of CT., Storrs.

Bartok, J.W., Jr. 1987. Plan before you build. *Greenhouse Grower*. 5(5):25-26. May, 1987.

Bartok, J.W., Jr. and R.A. Aldrich. 1989. Greenhouses and local zoning ordinances. *Proc. ASAE*, Paper No. 89-4031. Quebec, 1989.

Basham, C.W. and J.J. Hanan. 1983. Space optimization in greenhouses with linear programming. *Acta Hort.* 147:45-56.

Beese, E.J. 1978. Wood preservatives and treated lumber for use in landscape construction. *Roses, Inc. Bull.*, June, 1978.

Besemer, S.T. 1966. An economic analysis of the carnation industry in the United States. M.S. Thesis, CO State Univ., Ft. Collins. 124 pp.

Blom, T.J. and F.J. Ingratta. 1985. The use of polyethylene as greenhouse glazing in North America. *Acta Hort.* 170:69-80.

Bowman, G.E. 1962. A comparison of greenhouses covered with plastic film and with glass. *Proc. XVIth Int. Hort. Cong.*, Brussels. 1962.

Briggs, R.A. and W.D. Holley. 1961. Effects of glass and fiber-glass on carnation growth. *CO Flower Growers Assoc. Res. Bul.* 135:1-3.

Brooks, C.E.P. 1949. Climate Through the Ages. Second revised edition. Dover Publ., New York. 395 pp.

Brumfield, R.G. et al. 1981. Overhead costs of greenhouse firms differentiated by size of firm and market channel. *NC Agric. Res. Service, Tech. Bull.* 269. 89 pp.

Brun, R. and J. Lagier. 1985. A new greenhouse structure adapted to Mediterranean growing conditions. *Acta Hort.*170:37-46.

Canham, A.E., K.E. Cockshull and A.P. Hughes. 1969. Supplementary illumination of Chrysanthemums. Manuscript, Electrical Res. Sta., Univ. of Reading, Shinfield. 2pp.

Castilla, N. et al. 1989. Alternative greenhouses for mild winter climate areas of Spain. Preliminary report. *Acta Hort.* 245:63-70.

Conway, H. McKinley, Jr. and L.L. Liston. 1974. The Weather Handbook. Conway Res. Inc., Atlanta, GA. 255 pp.

Courter, J.W. 1965. Plastic greenhouses. Univ. of IL, *Coop. Ext. Serv. Circular* 905. 55 pp.

Dalrymple, D.G. 1973. A global review of greenhouse food production. Econ. Res. Serv., USDA, *Foreign Agric. Econ.* Rpt. No. 89. 150 pp.

Day, D. 1984. Plasticizer fears lead to British Standards change. *The Grower*. 102(21):3. Nov. 22, 1984.

Dalzell, J.R. 1955. Simplified Masonry Planning and Building. McGraw-Hill, NY. 312 pp.

Daponte, T. 1994. Multilayer greenhouse films. *In* Greenhouse Systems. Automation, Culture and Environment. NRAES-72. Rutgers Univ., New Brunswick, NJ.

Dayan, E. et al. 1986a. Suitability of greenhouse building types and roof cover materials for growth of export tomatoes in the Besor region of Israel. I. Effect on climatic conditions. *Biotronics*. 15:61-70.

Dayan, E. et al. 1986b. Suitability of greenhouse building types and roof cover materials for growth of export tomatoes in the Besor region of Israel. II. Effect on fresh and dry matter production. *Biotronics*. 15:71-79.

Delano, R. and C.J. Raseman. 1972. Control of condensate and light in greenhouses and solar stills. *Plasticulture*. 14:1-8. June, 1972.

Duncan, G.A. and J.N. Walker. 1975. Greenhouse coverings. Dept. Agric. Eng., Univ. KY, AEN-10, 10 pp.

Durrell, L.W. and K.L. Goldsberry. 1970. Growth of *Aureobasidium pullulans* on plastic greenhouse roofing. *Mycopath. Mycol. Appl.* 42:193-196.

Edwards, R.I. 1965. The transmission of solar radiation in vinery, propagation and clear span glasshouses. Note No. 11, NIAE, Silsoe, U.K..

Elenes-Fonseca, C. and J.J. Hanan. 1987. The effect of greenhouse cover and shading on 'Royalty' rose yield. *CO Greenhouse Growers' Assoc. Res. Bul.* 449:1-6.

Enoch, H.Z. 1978. A theory for optimilization of primary production in protected cultivation. II. Primary plant production under different outdoor light regimes. *Acta Hort.* 76:45-57.

Faherty, K.F. and T.G. Williamson. 1989. Wood Engineering and Construction Handbook. McGraw-Hill Publ. Co., New York.

Fang, W., K.C. Ting and G.A. Giacomelli. 1990a. Animated simulation of greenhouse internal transport using SIMAN/CINEMA. *Trans. ASAE.* 33:336-340.

Fang, W., K.C. Ting and G.A. Giacomelli. 1990b. Optimizing resource allocation for greenhouse potted plant production. *Trans. ASAE.* 33:1377-1382.

Ferare, J. and K.L. Goldsberry. 1984. Response of roses and pot chrysanthemums to different plastic film greenhouse covers. *CO Greenhouse Growers' Assoc. Res. Bull.* 403:1-4.

Franklin, M. 1983. Challengers to single glass. *The Grower.* 108(3). July 16, 1983.

Fricke, J. 1988. Aerogels. *Sci. Amer.* 258(5):92-97.

Geiger, R. 1965. The Climate Near the Ground. Trans. from German. Harvard Univ. Press, Cambridge, MA. 611 pp.

Giacomelli, G.A. 1987. Movable row tomato production system for the greenhouse. *Appl. Engin. in Agric.* 3:228-232.

Gillette, R. 1982. Twice the glass, half the energy and beautiful, beautiful light. *Florists' Rev.* 170(4404):40-41.

Gillette, R. 1983. Push benches where? *Grower Talks.* Apr., 1983.

Goldsberry, K.L. 1970. Flammability control of FRP greenhouse coverings. Progress Report I. *CO Flower Growers Assoc. Res. Bul.* 242.

Goldsberry, K.L. 1971. Plant response to light quality created by tinted FRP panels. *Proc. 10th Nat. Agric. Plastics Conf.*, Univ. of IL.. pp 118-123. Nov., 1971.

Goldsberry, K.L. 1973. Air-inflated system for an emergency roof. *CO Flower Growers Assoc. Res. Bull.*. 276:3-4.

Goldsberry, K.L., J.J. Hanan and C. Elenes-Fonseca. 1986. The effects of plastic glazings on rose growth in computer controlled greenhouses. *CO Greenhouse Growers' Assoc. Res.* Bull. 426:1-3.

Goldsberry, K.L. and T. Homan. 1975. Refinishing FRP greenhouse panels. *CO Flower Growers Assoc. Res.* Bull. 297:1-3.

Goldsberry, K.L. and T. van der Salm. 1985. Plant responses to plastic greenhouses. *CO Greenhouse Growers' Assoc. Res.* Bull. 419:1-3.

Grafiadellis, M. 1985. A study of greenhouse covering plastic sheets. *Acta Hort.* 170:133-142.

Halsey, L.A. 1975. Time is money for greenhouse operators. *Florists' Rev.* 156(4040):69-71, 133-134. May 8, 1975.

Hamrick, D. 1988a. The covering choice. *Grower Talks.* 51(12):64-74. April, 1988.

Hamrick, D. 1988b. The logistics of efficiency: Benches out floors in? *Grower Talks.* August, 1988.

Hanan, J.J. 1968. Advantages of clear-day climatic regions for greenhouse production. *CO Flower Growers Res.* Bull. 217:1-3.

Hanan, J.J. 1970. Observations on radiation in greenhouses. *CO Flower Growers Assoc. Res.* Bull. 239:1-4.

Hanan, J.J. 1982. Salinity III. Handling water supplies to minimize salinity problems. *CO Greenhouse Growers Assoc. Res.Bull.* 384:1-4.

Hanan, J.J. 1989. Sunlight in Colorado greenhouses. *CO Greenhouse Growers Assoc. Res.* Bull. 466:1-5.

Hanan, J.J. and F.D. Jasper. 1969. Consumptive water use in response of carnations to three irrigation regimes. *J. Amer. Soc. Hort. Sci.* 94:70-73.

Hanan, J.J., W.D. Holley and K.L. Goldsberry. 1978. Greenhouse Management. Springer-Verlag, Heildelberg. 530 pp.

Harrison, D.J. and L. Rogers. 1982. New glass house solves problems. *The Grower.* 98(6):36. Aug. 12, 1982.

Hartz, T.K., A.J. Lewis and H.A. Hughes. 1981. Performance of a modified Brace Institute greenhouse in Virginia. *HortScience.* 16:748-752.

Hasskelus, E.R. and G.E. Beck. 1963. Plant responses to light transmitted into a fiberglass reinforced plastic greenhouse. *Proc. Amer. Soc. Hort. Sci.* 82:637-644.

Heineman, P. 1994. Preliminary methods of integrating components and processes into a greenhouse system. *In* Greenhouse Systems. Automation, Culture and Environment. NRAES-72. Rutgers Univ., New Brunswick, NJ.

Hendrix, A.T.M. 1975. Transport systems in glasshouse horticulture. *Neth. J. Agric. Sci.* 23:231-237.

Holley, W.D. 1964. Type of greenhouse covering may affect CO_2 utilization by carnations. *CO Flower Growers Assoc. Res.* Bull. 172:1-3.

Holley, W.D. and B.B. Bennett. 1962. Effects of glass and fiberglass coverings on carnation growth. Second report. *CO Flower Growers Assoc. Res.* Bull. 148:1-4.

Hoxey, R.P. 1990. Greenhouse constructional design. Proc. Int. Seminar and British-Israel Workshop on Greenhouse Technology. Mar.-Apr., 1990. Bet Dagan.

Jewett, J.T. 1985. An update on greenhouse covering materials. *Roses, Inc. Bull.* Mar., 1985.

Kirschling, P.J. and F.E. Jensen. 1974. Costs and returns for bedding plants produced in three alternative greenhouse types. *NJ Agric. Expt. Sta., Dept. Agric. Econ. & Marketing.* 31 pp.

Kirschling, P.J. and F.E. Jensen. 1976. How land investment should affect greenhouse feasibility decisions. *Florists' Rev.* 159(4122):43-44, 85-87. Dec. 2, 1976.

Kozai, T., J. Goudriaan and M. Kimura. 1978. Light transmission and photosynthesis in greenhouses. Simulation Monograph, Wageningen.

Krause, W. 1976. Mainstreet glasshouseville. *The Grower.* 85(4):11

Krause, W. 1981. End of the road. *The Grower.* Oct. 1, 1981.

Kruyk, P. 1982. In 30 years most glasshouses will go. *The Grower.* July 29, 1982. Supplement. pp 15.

Kutz, L.J. et al. 1987. Robotic transplanting of bedding plants. *Trans. ASAE.* 30:587-590.

Lawson, G. 1988a. Light, height and white. *The Grower.* 109(4). Jan. 28, 1988.

Lawson, G. 1988b. A crane in the house. *The Grower.* 110(24):17-19. Dec. 15, 1988.

Lawson, G. 1989a. Systems for plastic. *The Grower.* 111(7). Feb. 16, 1989.

Lawson, G. 1989b. Tomatoes on the move. *The Grower.* 111(2). Jan. 12, 1989.

Manbeck, H.B. and R.A. Aldrich. 1967. Analytical determination of direct visible solar energy transmitted by rigid plastic greenhouses. *Trans. ASAE.* 10:564-567.

Mark, H.F. 1984. The development of plastics. *Amer. Sci.* 72:156-162.

Mattson, R.H. and T.J. Maxwell. 1971. Studies of greenhouse orientation in a heliodon. *HortScience.* 6:209-210.

Maunder, W.J. 1970. The Value of Weather. Methuen & Co., London. 388 pp.

Meijaard, D. 1995. The greenhouse industry in The Netherlands. *In* Greenhouse Climate Control. J.C. Bakker et al. Eds. Wageningen Pers.

Metcalf, E.D. 1983. Polyethylene works out best at Lee Valley. *The Grower.* Mar. 10, 1983.

Monteiro, A. et al. 1988. Evolution of and issues associated with economically useful plastic greenhouses for production in the Portuguese mild winter. *HortScience.* 23:665-668.

Montero, J.I. et al. 1985. Climate under plastic in the Almeria area. *Acta Hort.* 170:227-234.

Mortensen, L.M. 1987. Optimizing the environment. *Greenhouse Grower.* 5(5):28-32. May, 1987.

NGMA. 1985. Standards. Design loads in greenhouse structures. Ventilating and cooling greenhouses. Greenhouse heat loss. Nat. Greenhouse Manuf. Assoc. Pana, IL.

Nilsen, S., C. Dons and J. Hovland. 1983. Effect of reducing the infrared radiation of sunlight on greenhouse temperature, leaf temperature, growth and yield. *Sci. Hort.* 20:15-22.

Nisen, A. 1976. Functional photometric properties of the covering materials for ground or greenhouses. *Plasticulture.* 32:16-22.

Novoplansky, A. et al. 1990. Greenhouse cover for morphogenetic signaling. *Proc. Int. Seminar and British-Israel Workshop*, Bet-Dagan, Aug. 1990.

O'Flaherty, T. 1985. A window on plastic cladding. *The Grower.* 104(3):3-7. Jul. 18, 1985.

O'Rourke, M.J. 1997. Snow load on buildings. *Amer. Sci.* 85:64-70.

Olcott, J.S. 1963. Chemical strengthening of glass. *Science.* 140:1189-1193.

Olympios, C.M. 1982. A new multispan plastic covered greenhouse. Misc. Rpt. 5, Agric. Res. Inst., Nicosia, Cyprus. 16 pp, (Greek).

Papachristodoulou, S., C. Papayiannis and G.S. Panayiotou. 1987. Norm input-output data for the main crop and livestock enterprises of Cyprus. Agric. Econ. Rpt. 16, Agric. Res. Inst., Nicosia, Cyprus.

Reilly, A. 1981a. Floating aisles boost production. *Florists' Rev.* 168(4353):26-27, 135-136. May 7, 1981.

Reilly, A. 1981b. Using greenhouse space efficiently with rolling pallet system. *Florists' Rev.* 168(4357):32-33. Jun. 4, 1981.

Reilly, A. 1988. Understanding design loads. *Grower Talks*. 51(12):82-87. April, 1988.
Richardson, G. 1987. Computers on construction. *The Grower*. 108(3). July 16, 1987.
Roberts, W.J. 1971. Air-inflated and air-supported greenhouse structures. *Proc. 10th Natl. Agric. Plastics Conf.*, Univ. of IL, Nov., 1971. pp 103-109.
Ross, D.S. 1988. Bibliography of greenhouse and plant growth facilities. Northeast Reg. Agric. Eng. Service, Cornell Univ., Ithaca, NY.
Royle, D. 1980. Old ideas mix with new technology. *The Grower*. 93(13):25-27. Sept. 25, 1980.
Scott, K.J. and R.B.H. Wills. 1972. Ethylene produced by plastics in sunlight. *HortScience*. 7:177.
Seeley. J.G. 1980. Some precautions during greenhouse rehabilitation. *NY State Flower Industries Bull.* 119:1-2.
Sheldrake, R., Jr. and R.M. Sayles. 1974. Plastic greenhouse manual. Dept. Vegetable Crops, Cornell Univ., Ithaca, NY. 21 pp.
Short, T.H. 1982. Holland as seen by an agricultural engineer. *Florists' Rev*. 170(4409):16-19. Jun. 3, 1982.
Simpkins, J.C. et al. 1984. Testing of infrared radiation barrier greenhouse films. Departments of Biological and Agric. Eng. and Hort. and Forestry. Rutgers Univ., New Brunswick. 9 pp.
Sims, T. 1976. Seven years of research on house type and orientation. *The Grower*. 86(20):1027-1033. Nov. 11, 1976.
Sherry, W.J. and J.W. White. 1984. A comparison of greenhouse coverings. *Florists' Rev*. May 3, 1984.
Smith, B. 1988. The how, when and why of building a new glasshouse. *The Grower*. 109(19). May 12, 1988.
Smith, C.V. 1971. A contribution to glasshouse design. *Agric. Meteor*. 8:447-468.
Smith, F.G. 1967. We need to adopt new thinking on light transmission. *The Grower*. Nov. 4, 1967.
Spek, J.C. 1985. A study of greenhouse foundations - new systems. *Acta Hort*. 170:11-23.
Spelman, N. 1987. The return to investing in light. *The Grower*. 108(3). July 16, 1987.
Spelman, N. 1988. October's lesson: invest in a house built to standard. *The Grower*. 109(9). Mar. 3, 1988.
Spomer, S.A. and W.D. Holley. 1963. Carnation growth under glass, mylar and fiberglass panels. Report III. *CO Flower Growers Assoc.Res. Bull.*. 160:1-3.
Swanekamp, W. 1994. Commercial integrated production of bedding plants and potted plants. *In* Greenhouse Systems. Automation, Culture and Environment. NRAES-72. Rutgers Univ., New Brunswick, NJ.
Tesi, D. and R. Tesi. 1985. The growing of tomatoes in Italy using tunnel greenhouses clad with rigid plastics. *Plasticulture*. 68:29-38.
Tesi, D., C. Dehennau and D. Malarme. 1986. Properties of new PVC rigid sheets for greenhouses. *Plasticulture*. 71:37-46.
Ting, K.C., G.A. Giacomelli and S.J. Shen. 1990a. Robot workcell for transplanting of seedlings. Part I. Layout and materials flow. *Trans. ASAE*. 33:1005-1010.
Ting, K.C. et al. 1990b. Robot workcell for transplanting of seedlings. Part II. End-effector development. *Trans. ASAE*. 33:1013-1017.
Tross, M.J. et al. 1984. An optical liquid filter greenhouses: Numerical solution and verification of a thermodynamic model. *Acta Hort*. 148:401-409.
van Nierop, J. 1982. International advisor warns: Substrate growing is the answer to energy saving problems. *The Grower*. (Horticulture Now supplement). Jan. 21, 1982.
van Weel, P.A. 1991. Integrated crop production systems. Transport and materials handling in the greenhouse. *ASAE Symp. Automated Agriculture for the 21st century*. Chicago. 458-467.
Visher, S.S. 1954. Climatic Atlas of the United States. Harvard Univ. Press, Cambridge, MA. 403 pp.
Vuoto, V. 1975. Wright's replaces "the bubble that burst." *Florists' Rev*. May 8, 1975.
Waaijenburg, D. 1995. Construction. *In* Greenhouse Climate Control. J.C. Bakker et al., Eds. Wageningen Pers.
Waaijenburg, D. 1985. Research on greenhouse cladding materials. *Acta Hort*. 170:103-109.
Walker, J.N. and G.A. Duncan. 1973. Rigid-frame greenhouse construction. *Dept. of Agric. Eng., Univ. of KY. Bull.* AEN-15. 7 pp.
Walker, J.N. and G.A. Duncan. 1974a. Greenhouse location and orientation. *Dept. of Agric. Eng., Univ. of KY. Bull.* AEN-32. 4 pp.
Walker, J.N. and G.A. Duncan. 1974b. Greenhouse benches. *Dept. of Agric. Eng., Univ. of KY Bull.*, AEN-13. 9 pp.
Walker, J.N. and D.C. Slack. 1970. Properties of greenhouse materials. *Trans. ASAE*. 13:682-684.

Wells, D.A. 1969a. Interim report on cleaning of horticultural glass. Note E.C.4., Nat. Inst. Agric. Eng., Silsoe.

Wells, D.A. 1969b. Cleaning of horticultural glass. DN/G/2/69. NIAE, Silsoe, U.K..

Wells, D.A. and R.F. Hoxey. 1980. Measurement of wind loads on a full-scale glasshouse. *J. Wind Eng. and Ind. Aerod.* 6:139-167.

Went, F.W. 1957. Climate and agriculture. *Sci. Amer..* 196:82-94.

White, J.W. 1987. Greenhouse coverings for geraniums. *PA Flower Grower Bull.* 380:3-6.

Whittle, R.M. and W.J.C. Lawrence. 1959. The climatology of glasshouses. I. Natural illumination. *J. Agric. Eng. Res.* 4:326-340.

Wilsie, C.P. 1962. Crop Adaptation and Distribution. W.H. Freeman, San Francisco. 184 pp.

Yang, Y., K.C. Ting and G.A. Giacomelli. 1991. Factors affecting performance of sliding-needles gripper during robotic transplanting of seedlings. *Appl. Engin. in Agric.* 7:493-498.

Young, F.D. 1921. Nocturnal temperature inversion in Oregon and California. *Mo. Weather Rev.* 49:138-148.

Young, R.E. and D.D. Murphy. 1982. Centralized shipping for an ornamental greenhouse. *Trans. ASAE.* 25:872-875.

Zabeltitz, C. von. 1985. New construction of a plastic-film-greenhouse. *Acta Hort.* 170:25-28.

CHAPTER 3

RADIATION

I. INTRODUCTION

Sunlight is the determinant in greenhouse production. There is, however, more confusion about behavior and terminology dealing with radiation than almost any other factor. At the practical level, knowledge of, and application of that knowledge in the industry, is slight. It is the purpose of this chapter to define and explain radiation, and how it is used in greenhouses. In the control of plant growth, or photomorphogenesis, there are at least five types of information that may be derived from the radiation environment: 1) radiation quantity, 2) radiation quality, 3) direction of radiation, 4) duration of radiation (timing of light-dark transitions), and 5) polarization (different photoreceptor arrangement) [Smith, 1986]. Of these five groups, industry has utilized only 1) and 4) to any significant extent in design and management decisions.

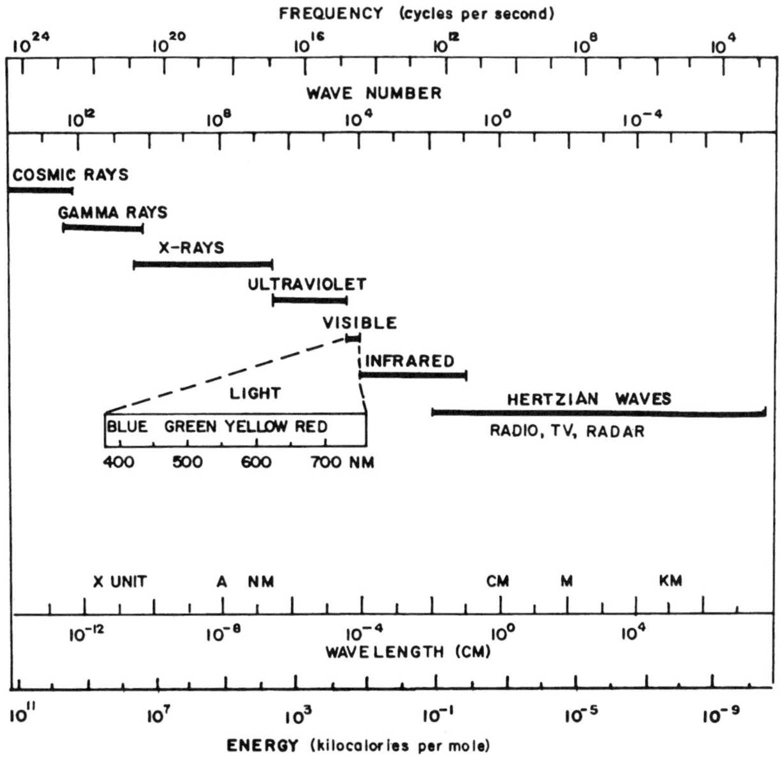

Fig. 3-1. The electromagnetic spectrum showing position of visible radiation with respect to other parts of the spectrum [Adapted from Withrow and Withrow, 1956].

Apart from the above five factors, an additional complication is that the energy regime of a greenhouse, and plants within it, should be considered simultaneously with temperature, water, CO_2 and nutrition – subjects that are discussed in the following chapters. Physiologists, and many horticulturists, often deal solely with quantum relationships to the exclusion of such factors as latent energy transfer (conversion of liquid water to a gas). It is a fact that, for well-watered plants, which is the common situation in greenhouses, more than 70% of incoming radiation is used to evaporate water (see Chapter 5). To neglect this factor leads to significant misunderstanding of the greenhouse climate and plant response. When considering photosynthesis and photomorphogenesis, one must deal with quantum interaction with receptive pigment systems in plants. The energy balance of the total operation, however, includes all radiant energy between wavelengths

300 and about 50000 nm.

Greenhouses become particularly fascinating when comparing their productivity with crops grown outdoors. For example, tomatoes grown over a 10-month period in greenhouses can produce 47 kg m^{-2} commercially, or about 4.3 kg m^{-2} dry matter, as compared with wheat which produces 2.1 kg m^{-2} dry matter of which half of the dry matter will be in the grain [Aikman, 1989].The energy content of tomato plant dry matter is 16 MJ kg^{-1} total, or the crop is trapping 67 MJ m^{-2} into photosynthate. The tomato plant converts about 7.6% of the intercepted PAR (Photosynthetically Active Radiation) energy into dry matter energy. Such high efficiencies contrast markedly with the overall photosynthetic efficiency of the biosphere. According to Gates [1980], the average gross productivity of the earth's biosphere is 12.8 MJ m^{-2} yr^{-1}, for an overall efficiency of 0.207%. Net productivity (usable gain) is only 0.086%. Crops are generally designed for high productivity compared with the natural ecosystem, and Gates estimates the average annual crop efficiency for the U.S. at 1.7%, with 4 month averages for the summer in the Central U.S. at 2.4%.

Table 3-1. Wavelength classification [Cathey and Campbell, 1980].

Classification	Wavelength
Ultraviolet (UV)	100-380 nm
UV-C	100-280 nm
UV-B	280-320 nm
UB-A	320-380 nm
Visible	380-780 nm
Infrared	780-2500 nm
Thermal	2500 + nm

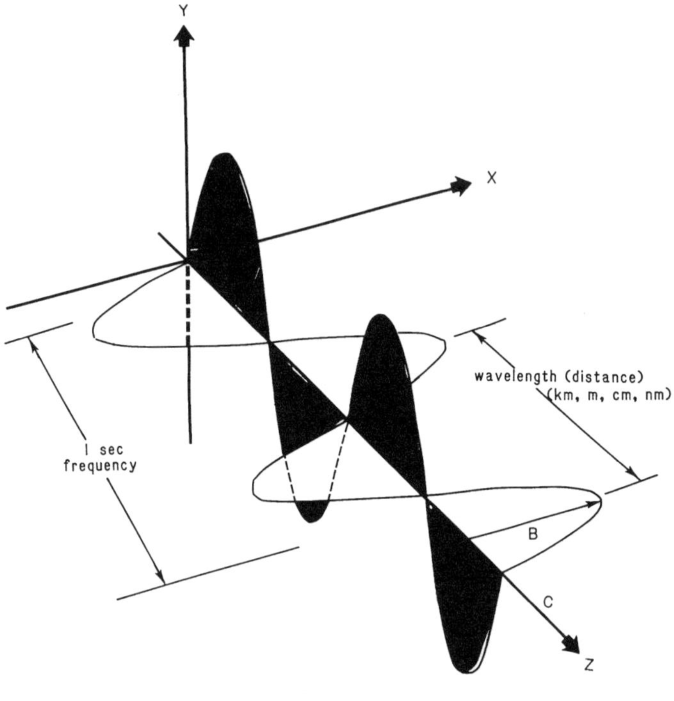

Fig. 3-2. Schematic representation of electric fields (shaded part) and magnetic fields (B) in an electromagnetic wave moving from the intersection of **y** and **x** toward **z** [From Kraml, 1986] (With kind permission from Kluwer Academic Publ.).

In the last 15 years, there has been rapid development of mathematical procedures for radiant energy transfer by many individuals and groups, especially those in the U.K., Holland and Japan. This information is coupled with further elucidation of the role of radiant energy in the photomorphogenesis of plants, which may lead to new practices, and increasing productivity. As pointed out in Chapter 1, although structures for growing plants are energy intensive, they make maximum use of raw materials with the potential for minimum pollution of the biosphere.

II. RADIATION BASICS

Summarizing, there are three major sets of measurements used to describe radiation in the greenhouse: 1) Those terms dealing with total radiation (300 to 50000 nm), with units in Joules or Watts. 2), Those terms reported in number of quanta (moles) for specified wavelengths such as PAR. And 3), illumination (photometric) terminology which is commonly used (e.g., lux, lumen). One has no choice but to employ all three as necessary, particularly as most supplemental sources (Tables 3-3, 3-4) are specified in photometric terms. Further comments on the relationships of these sets are made below.

A. THE ELECTROMAGNETIC SPECTRUM

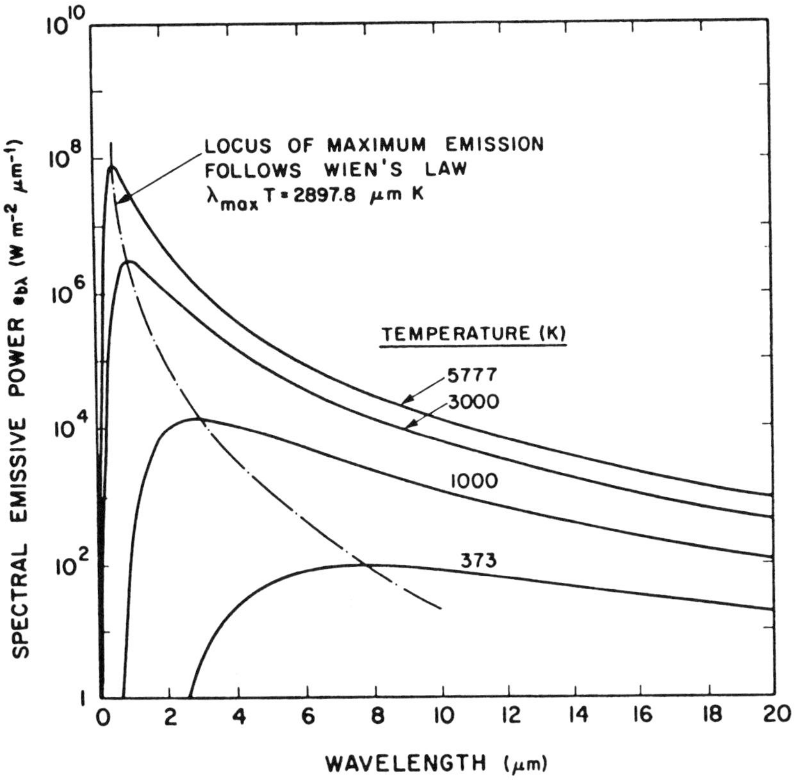

Fig. 3-3. Spectral emissive power of a perfect blackbody, $\epsilon = 1.0$, at different temperatures. The dotted line plots the points of maximum emission of the blackbody as a function of temperature [Iqbal, 1983] (By permission of Academic Press).

The radiant energy used by plants comprises a small part of the total electromagnetic spectrum shown in Fig. 3-1. In the wave theory of radiation, we can imagine propagation of energy through space as though someone threw a stone into a pond, and we are viewing the ripples (Fig. 3-2). The "quality" of this radiation, whose speed in a vacuum is a constant, can be designated by frequency (cycles per second or Hertz), wavelength (distance between identical points on the wave, (λ), or wave number, the number of waves in one centimeter (v'). These are readily converted from one to another:

$$c = \nu\lambda \tag{3.1}$$

$$\lambda = \frac{c}{\nu}, \text{ and} \tag{3.2}$$

$$\nu' = \frac{1}{\lambda} = \frac{\nu}{c} \tag{3.3}$$

where: $c = 2.998 \times 10^{10}$ *cm sec*$^{-1}$
ν = *waves per second (Hz)*
ν' = *wave number or waves per cm*
λ = *wavelength in cm* = 10^7 *nm*

Most commonly, we use wavelength to designate the "quality" with which we are dealing, usually in nanometers (nm). Thus, radiation having a visual attribute of "blue" will have a wavelength (λ) of 400 nm, whereas wavelengths of 500 nm will look green. Humans do not visually perceive radiation of shorter wavelengths than about 380 nm or longer than about 780 nm.

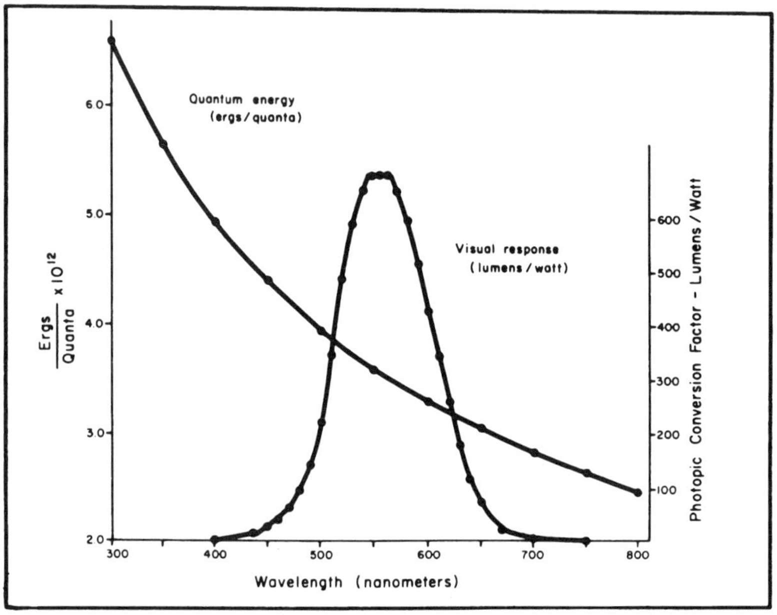

Fig. 3-4. Quantum energy as a function of wavelength and the human visual response [McFarlane, 1978] (By permission of Cornell Univ. Press).

Although the spectrum depicted in Fig. 3-1 is quite extensive, for photosynthesis and photomorphogenesis, we are interested in a very restricted portion that is arbitrarily divided into regions as noted in Table 3-1. Other authors may use different divisions depending upon their objectives. Thus, Tuller and Peterson [1988] divided their intervals into two UV bands (300-315 and 315 to 400 nm), blue-green photosynthesis (400-510 nm), green-orange (510-610 nm), red photosynthesis (610-700 nm) and IR from 700 to 1100 nm. Nijskens et al. [1985] combined the photosynthetically active regions into one band (400-700 nm), a solar UV (300-380 nm), solar visible (380-760 nm), short or solar IR (760-2500 nm) and total solar from 300 to 2500 nm. Usually, however, the photosynthetically active waveband (PAR) is taken to be 400 to 700 nm. For the action spectra for photoperiodism, the "near IR" region is 655 to 665 nm, and the "far IR" is 725 to 735 nm. Note, however, the phytochrome pigment can absorb radiation above and below the values given (Fig 3-67). In more recent times,

	Light source[a]													
	Incandescent		Fluorescent					HID discharge					Sun[b]	
Parameter	60	100	FCW	FCW	FWW	PGA	PGB	HG	HG/ DX	MH	HPS	LPS	DL	SKY
Heat losses (W per lamp)														
Conduction-convection	6.5	9.7	16	85	16	18	17	74	77	67	60	51	---	---
Ballast	0	0	6	30	6	6	6	40	40	60	70	50	---	---
Distribution of output as a fraction of electrical input power (mW/W)														
400-700 nm	57	69	204	188	199	127	146	124	131	265	261	276	429	567
580-700 nm	37	44	71	65	87	68	72	6	36	102	168	271	160	142
700-850 nm	72	81	4	4	3	2	15	8	9	32	99	37	152	96
800-850 nm	26	29	0.6	0.6	0.4	0.1	1	3	3	22	77	36	46	26
250-400 nm	1.0	1.4	9	8	6	5	10	52	21	29	2.8	0.6	64	216
400-850 nm	129	150	208	192	202	129	161	132	140	297	260	313	581	663
850-2700 nm	678	668	4	8	4	4	4	134	134	59	89	17	342	119
Thermal	83	84	305	322	209	342	324	423	439	339	271	230	13	1
Conduction-convection	109	97	344	347	349	390	370	168	175	146	128	222	0	0
Ballast	0	0	130	122	130	130	130	91	91	130	149	217	0	0

[a] **Fluorescent light sources**: **FCW** = cool white (2 sizes with similar spectral distribution); **FWW** = warm white; **PGA** = plant growth A (e.g. Gro-Lux, Plant Light, Plant Gro); **PGB** = plant growth B (e.g. Gro-Lux Wide Spectrum). **HID discharge**: **HG** = clear mercury; **HG/DX** = mercury deluxe; **MH** = metal halide (sodium, scandium, thorium, mercury, lithium iodides); **HPS** = high-pressure sodium; **LPS** = low-pressure sodium.

[b] Clear-sky summer sunlight: **DL** = daylight (includes both direct and skylight); **SKY** = skylight (does not include emission from the atmosphere). Do not add DL and SKY together.

[c] One lumen m^{-2} = one lux; mW lm^{-1} = W $m^{-2}(klx)^{-1}$.

NOTE: Unless irradiance in W m^{-2} is specified as PAR, one must assume the entire shortwave band from 300 to 3000 nm. To convert from W m^{-2} total short wave solar energy, sum mW/lm from 250 nm through thermal and divide by the mW/lm in the 400 to 700 nm band to obtain percentage of the total spectrum due to PAR. This can be multiplied by the conversion value in Table 3-4 to convert to μmol $m^{-2}s^{-1}$ PAR.

B. QUANTA

Radiation can be represented as shown in Fig. 3-2 in the "wave theory," or as bundles or packets of energy known as quanta in the "particle theory." For biological phenomena, such as human vision, photosynthesis, photoperiodism, phototropism, etc., the number of quanta of a given frequency (wavelength) associated with a particular irradiance is what interacts with the photoreceptor system to transfer energy from radiation to the system. A single quantum of radiation contains energy, **E**, proportional to the frequency of the radiation:

$$E = h\nu, \text{ or} \tag{3.5}$$

$$E = \frac{hc}{\lambda} \tag{3.6}$$

where: E = *energy in Joules (J)*
h = *Plank's constant* 6.62×10^{-34} *J s*
ν = *frequency*
c = *speed of light* = 3×10^{8} *m* s^{-1}
λ = *wavelength (m)*

The product of h x v = 19.86, so for one quantum per second, the equivalent energy rate, or power, in watts is:

$$19.86 \times 10^{-26}\, W\, m\, \lambda^{-1} \tag{3.7}$$

when wavelength is given in meters.

The result, in terms of energy per quantum is shown in Fig. 3-4. As wavelength increases, energy per quantum decreases.

C. TERMINOLOGY (Table 3-2)

1. Radiation and Illuminosity

Radiant energy (**E**) is measured in joules (**J**), whereas the passage of energy, or flux, is measured in **J** s^{-1} or Watts (**W**). It is necessary to separate terminology when speaking of "light," or illumination, as contrasted with radiant energy (W), which is applicable to any portion of the electromagnetic spectrum, including visible. The sensitivity of the human eye to radiation is given in lumens (**lm**) and varies with wavelength (Fig. 3-4). Radiant flux of 1 Watt at 555 nm wavelength corresponds to a luminous flux of 680 lm. At 600 nm, however, 1 Watt generates a luminous flux of 422 lm, whereas at 400 nm, the illumination is practically zero. The lumen does not correspond to a definite number of Watts except at a specific wavelength. Although photometric terminology has no relation to what goes on in greenhouses, the terms –such as light, illumination, lumens, lux,

Table 3-4. Conversion factors for radiometric to quanta or lux to quanta for various radiation sources. Multiply W m^{-2} by constant or divide lux by constant [Thimijan and Heins, 1983].

Light source	µmol $s^{-1}m^{-2}$ $(W\ m^{-2})^{-1}$ 400-700 nm	lx $(\mu mol\ s^{-1}m^{-2})^{-1}$ 400-700 nm	400-850 nm
Sun and sky, daylight	4.57	54	36
Blue sky only	4.24	52	41
High-pressure sodium	4.98	82	54
Metal halide	4.59	71	61
Warm-white fluorescent	4.67	76	74
Cool-white fluorescent	4.59	74	72
Plant growth A[a]	4.80	33	31
Plant growth B[a]	4.69	54	47
Incandescent	5.00	50	20
Low-pressure sodium	4.92	106	89

[a] See Table 3-3 for description

etc.– are very common. Even among photobiologists, "light" is used instead of radiation, although the individual is really talking about the number of quanta in a specific waveband.

When energy is emitted from a point source (i.e., an incandescent bulb), it spreads out as it passes through space. The radiant intensity at any point in space is designated by quanta per square meter, Watts per steradian (**W sr^{-1}**) or lumens per steradian (**lm sr^{-1}**), where the area of a steradian is determined by the distance from the source or, if we assume the source is in the center, of a sphere. Any sphere surrounding a point source contains 4 x 3.14 steradians, and since the sphere's surface area equals 12.57 x r^2, the energy per steradian will be spread over more area the further it is from the source at the center of the sphere. That is, for a uniform 100 W source, at 1 cm distance (r), the radiant intensity will be 100 W/4πsr = 7.96 W sr^{-1} or 7.96 W cm^{-2}. At the surface of a sphere 10 cm in diameter, the intensity will still be 7.96 W sr^{-1}, but the irradiance per cm^2 will be 0.0796 W since there are now 100 cm^2 in each steradian. If one imagines a set of lines that connect the boundaries of the steradian on the sphere's surface to the source (a point) at the center of the sphere, the angle between them is commonly called the "solid" angle. The radiant intensity from a point source (I) will be: $I = dF/d\tilde{\omega}$ where F = radiant flux and $\tilde{\omega}$ = the solid angle.

Radiant emittance from a source (**R**) is in Watts per square meter (**W m^{-2}**). Similarly, the density or irradiance at a receiving surface is given in **W m^{-2}**, and is designated by **Q**. For luminous emittance or illuminance, the units are lumens per square meter (**lm m^{-2}**), or commonly lux (**lx**). In dealing with supplementary irradiation, most of the artificial sources are rated by their luminous efficacy in lumens per Watt of electrical input to the radiant source (**lm W^{-1}**) (see Table 3-9). The luminous efficiency is a decimal fraction, and, at wavelengths less than 380 nm or greater than 780 nm, it is zero since we cannot see above or below those wavelengths.

2. Quantum Energy

Part of the information required to understand plant response is quantum counting. That is, the number of quanta of a given wavelength associated with a particular radiant flux. Individual quanta invoke specific photochemical responses in living systems. In Section 3.II.B, the relationship between wavelength and energy was derived and shown in Fig. 3-4 over the wavelengths which interest us the most. Especially in the PAR range of 400 to 700 nm, the number of quanta is the deciding criterion, and this is designated by the SI unit of concentration the **mole**. One mole (mol.) means the number of quanta equal to Avogadro's number (6.023 x 10^{23}). The mole is a large value, and in photomorphogenesis and photosynthesis, the common range is in micromoles (μmol.), with the rate being given in micromoles per sq. meter-second (**μmol. m^{-2}s^{-1}**). Earlier literature uses the term "microeinstein" for the same thing.

D. CONVERSIONS

The technological and scientific literature is replete with photometric terminology. Most specifications dealing with irradiation sources use photometric terms. To gain an understanding for greenhouse purposes, one is forced to convert, for example, from illumination units to quanta, lumens to Watts, etc. The 1983 publication by Thimijan and Heins was particularly detailed in providing many of these conversions. Tables 3-3 and 3-4 present this information with further discussion on radiation sources later in this chapter. Even in the scientific literature where radiometric terms are correctly employed, the units may be in **ergs**, **Watts**, **joules**, **langleys**, **calories**, and in English **BTU**s (British Thermal Units). Fortunately, in the SI system based upon 10s, the decimal points can be easily moved. For example, a milliWatt per sq.cm. is 10^{-3} x 10^4 Watts per sq.m. or 10 W m^{-2}. An erg is 10^{-7} joule, so that an erg per sec.-sq.cm. is 10^{-7} x 10^4 joules per sec.-sq.m., which is equal to a milliWatt per sq.m. (mW m^{-2}).

Conversion of lux to W m^{-2} depends upon the radiation source and the waveband of interest. From Table 3-3, the radiation per unit of luminous flux for high-pressure sodium lamps in the 400-700 nm range is 2.45 mW lm^{-1}. If the lamp provides 30 kilolux (klx) for 12 hrs, then:

$$\frac{(30000)(2.45)(12)}{(1000)(24)} = 36.8 \; W m^{-2} \; PAR \qquad (3.8)$$

The value 36.8 is an average for a 24-hour period. Otherwise, the result would be twice that given. For conversion of quantum flux to radiometric, Table 3-4 shows a conversion factor of 4.98 for the same lamp, so that 1000 μmol. s^{-1} m^{-2} would be:

$$\frac{1000\ \mu mol\, s^{-1}\, m^{-2}}{4.98\ \mu mol\, s^{-1}\, m^{-2}\, (W\, m^{-2})^{-1}} = 201\, W\, m^{-2}\, PAR \tag{3.9}$$

There may be need to convert photometric to quantum:

$$\frac{30000\ lx}{82\ lx\, (\mu mol\, s^{-1}\, m^{-2})^{-1}} = 366 \mu mol\, s^{-1}\, m^{-2} \tag{3.10}$$

Keep in mind that the above conversions are restricted to the 400-700 nm waveband. Table 3-4 also provides conversion factors from quantum to photometric units for two wavebands.

Now, much of the scientific literature dealing with photobiology (photosynthesis, photoperiodism, phototropism, etc.) reports irradiance as quantum flux (μmoles $s^{-1}m^{-2}$). Many European publications often report PAR in terms of flux density or W m^{-2} (PAR). In other words, energy specifically in the 400 to 700 nm waveband. Many publications, when referring to irradiance over an hour, day, or longer period, report energy in Joules, e.g., MJ $m^{-2}dy^{-1}$, etc. This avoids the necessity to use the term Watt-hour (1 W-hr = 3600 J) since otherwise a Watt by itself is one joule per second. Where PAR is not specified, one must assume total shortwave, or global, radiation from 300 to 3000 nm. The later is the most common meteorological designation with instruments readily available for use [WMO, 1965; Gates, 1980; Hanan, 1984]. I will show in later chapters that: 1) conversion of radiant energy in the greenhouse is largely concerned with changing water from a liquid to a vapor. 2) The proportion of radiant energy actually involved in converting CO_2 and water to carbohydrates, or changing the status of a pigment receptor in the plant, is sufficiently small as to be ignored in studies of energy balances (see Chapter 8). 3) The general instrumental error is large enough to disguise the energy used in photochemical systems. And 4), the total shortwave and thermal radiation exchange without and within a greenhouse must be considered. A report of quantum flux effectively excludes consideration of water relationships in plants. Chapter 5 will show that plant water relationships can affect plant growth. One can only conclude that report of total radiant energy remains the most viable procedure –distasteful as that may be to photobiologists.

E. MEASURING RADIATION

Instruments for measuring radiation can be separated into three classes: radiometers, photometers, and quantum meters. A few examples are provided in Fig. 3-5. Glass-covered radiometers, such as the Eppley or Kipp-Zonen, convert radiant energy to heat. The rise in temperature is measured and calibrated to provide an output in suitable units such as W m^{-2}. With glass domes, the spectrum measured is 300 to about 3000 nm with a flat response over the bandwidth. In other words, sensors converting radiant energy to heat are not considered sensitive to wavelength. The Eppley is the standard weather instrument in the U.S., while the Kipp is most common in Europe. Other transducers, such as silicon or selenium cells or photodiodes can be calibrated to measure solar radiation, but their response is nonlinear. They show considerable deviation in sensitivity as wavelength varies, and they should not be used with radiant sources different from those for which they have been calibrated. Photometers, although easy to use and cheap, are made to provide a response similar to the human eye (Fig. 3-4). They do not detect radiant energy above or below the visible wavelengths. Quantum meters for measurement of PAR are constructed to measure the radiation equally over the band 400 to 700 nm (Fig. 3-6). The difference in sensitivity over the 400 to 700 range shown in Fig. 3-6 is to compensate for lower quantum energy as wavelength increases. These are called PPF sensors for "photosynthetic photon flux" (photons referring to visible quanta). Another sensor type is called YPF or "yield photon flux," which weights the quanta in the range from 360 to 760 nm according to plant photosynthetic response. YPF sensors, according to Barnes et al. [1993], were less accurate than PPF devices under metal halide, high-pressure sodium, and low-pressure sodium

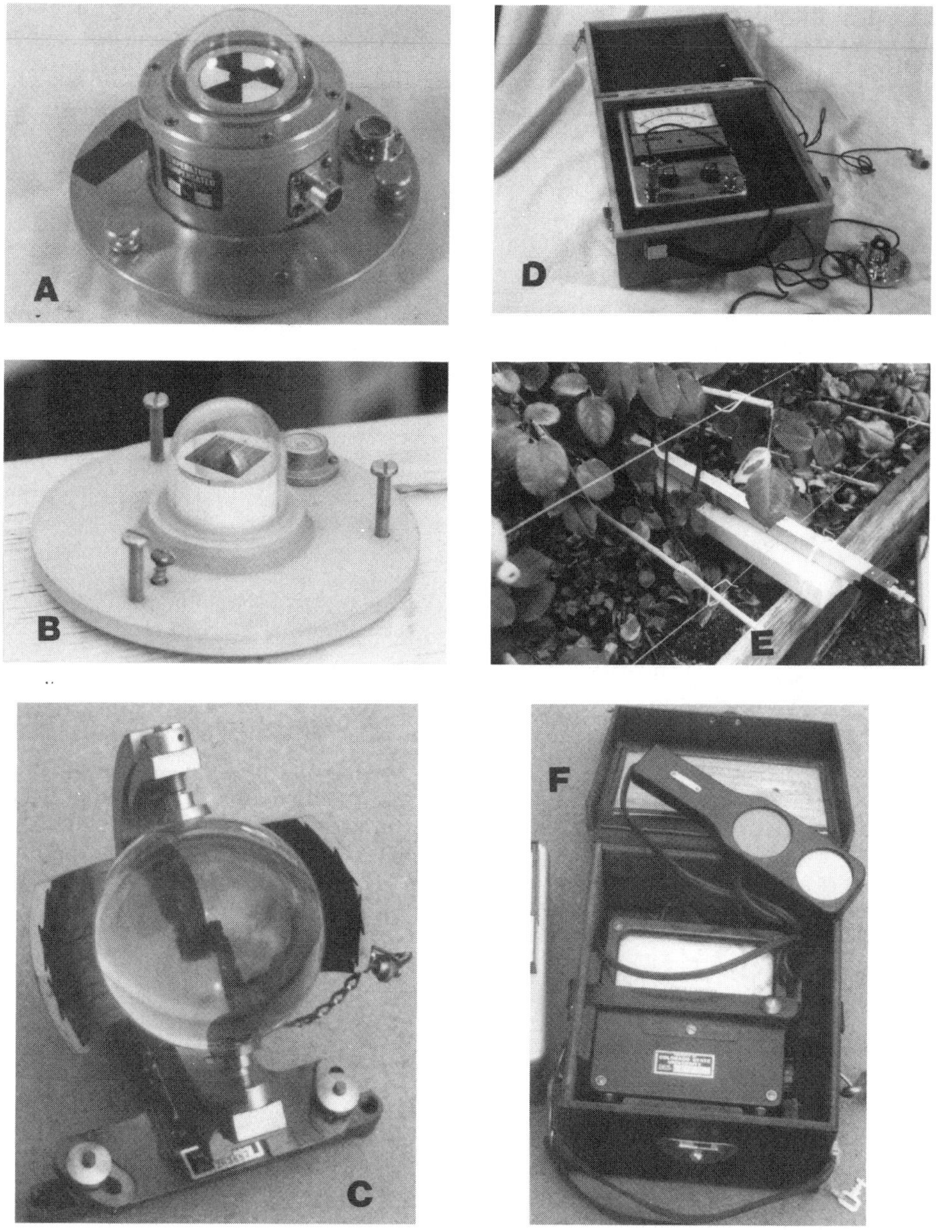

Fig. 3-5. Examples of radiation measuring instruments: **A** Eppley pyranometer for solar radiation, even response throughout range of 300 to 3000 nm, very expensive ($1000.00); **B** Silicon cell pyranometer for measuring solar radiation (ca $300.00); **C** Campbell-Stokes sunshine recorder; **D** Meter for measuring radiation, light or quanta, depending upon the probe used (probes about $200.00); **E** Linear quantum meter. Linear net radiometers can also be purchased, but are very expensive; **F** Old-style photometer calibrated in foot-candles. The reflective sensing surfaces are undesirable for anything other than normal illuminance.

lamps. The measurement accuracy for the most common quantum instruments appears to be within ±5%.

Instruments made to measure radiation are manufactured to follow the cosine law as closely as possible. Essentially, as the angle of incident radiant energy increases (closer to the horizon), the same radiant flux is spread over a larger area compared with that perpendicularly incident on the surface. The change in intensity will

Fig. 3-6. Spectral response of a quantum sensor compared with the ideal response over the PAR range of 400 to 700 nm.

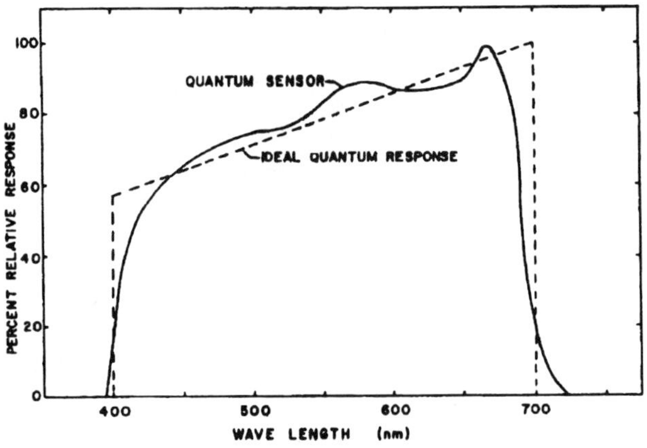

depend on the trigonometric cosine of the angle of incidence. In many instruments, a translucent lens is placed over the sensor to achieve a good cosine response. Instruments having specular, reflecting surfaces, such as the photometer in Fig. 3-5F, will not provide valid data unless the incident energy is normal (perpendicular) to the receiving surface.

Devices such as the Eppley or Kipp-Zonen pyranometers for measuring shortwave solar radiation are expensive. Silicon cell pyranometers, photometers, or quantum devices are much less costly. Cheaper devices than these are available, but are unlikely to be reliable or accurate. Common measurement of solar radiation in the field usually has errors of ±10%, using the standard Eppley depicted in Fig. 3-5A. Even first-class instruments, with a calibration directly traceable to national standards laboratories, can have errors of ±3%. A problem with most sensors is their small response surfaces that make them very susceptible to shadows from the greenhouse superstructure. Linear devices that provide a sensitivity over some distance should permit such effects to be averaged, i.e., Fig. 3-5E, or a linear net radiometer.

Fig. 3-7. Spectral distribution of direct solar radiation at sea level on a horizontal surface (air mass = 1). The solar spectrum is compared with the luminosity curve [adopted from Gates, 1965] (By permission of the Amer. Meteor. Soc.).

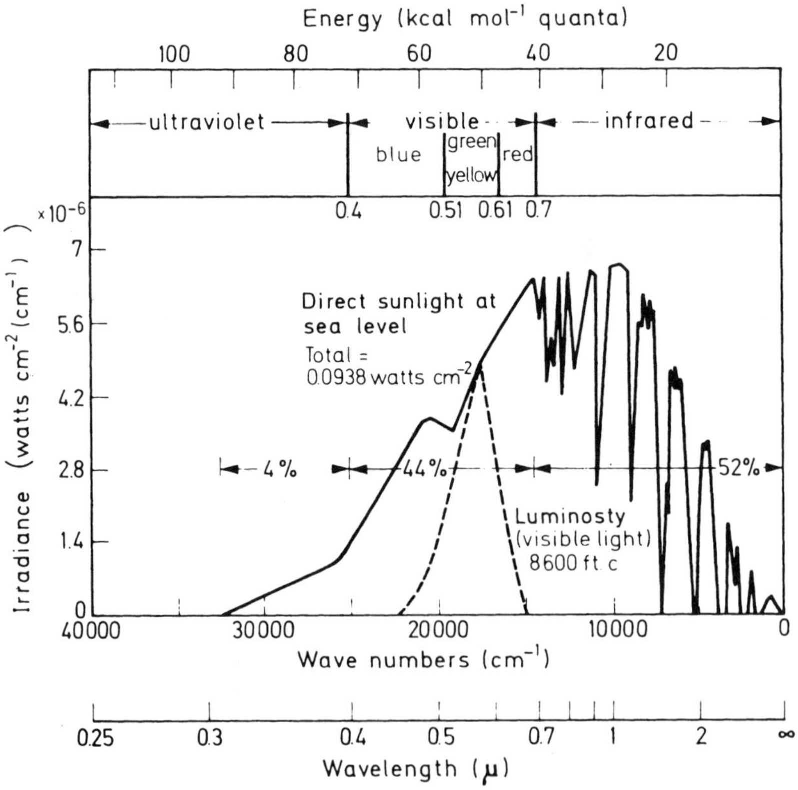

III. SOLAR RADIATION

The spectral distribution of sunlight at sea level, on a horizontal surface, is commonly plotted as a function of wavelength. This method shows the maximum monochromatic intensity to appear at 470 nm. According to Gates [1962], this is an illusion because of the method of plotting. It is correct to plot the energy as a function of frequency in which case the area under any section of the curve is directly proportional to the energy. Although I have chosen to use Gates' 1965 figure, where the curve varies with wave number (Fig. 3-7), both types of plots show that maximum solar intensity occurs just beyond 1 μm or 1000 nm. The median value occurs at 710 nm, at the upper edge of visible radiation. Fig. 3-7 shows that 52% of the solar energy reaches the earth at wavelengths beyond 700 nm, with 48% between 300 and 700 nm. Obviously, this curve is ideal, varying markedly, depending upon the depth of the atmosphere the rays must penetrate, elevation, cloudiness, etc. A scatter plot, comparing a cloudy location with a clear region, is provided in Fig. 3-8. It shows the variability that can be expected in most locations, and some of the difficulty in dealing with solar radiation. Note that, in a clear-day region, variability of direct solar radiation is increased with the greatest percent of possible global radiation exceeding 60% as contrasted to only 45 to 50% for cloudy places. The former results from fewer clouds and the higher atmospheric transparency common to arid climates. Fig. 3-9 compares the solar radiation curve outside the earth's atmosphere with the curve at sea level, cloud light, skylight, and radiation transmitted through vegetation. The latter will be referred to again in discussing the significance of the predominant far red found below vegetative canopies. The considerable variability in the curve beyond 1000 nm results from absorption of radiant energy by water vapor and CO_2 in the air.

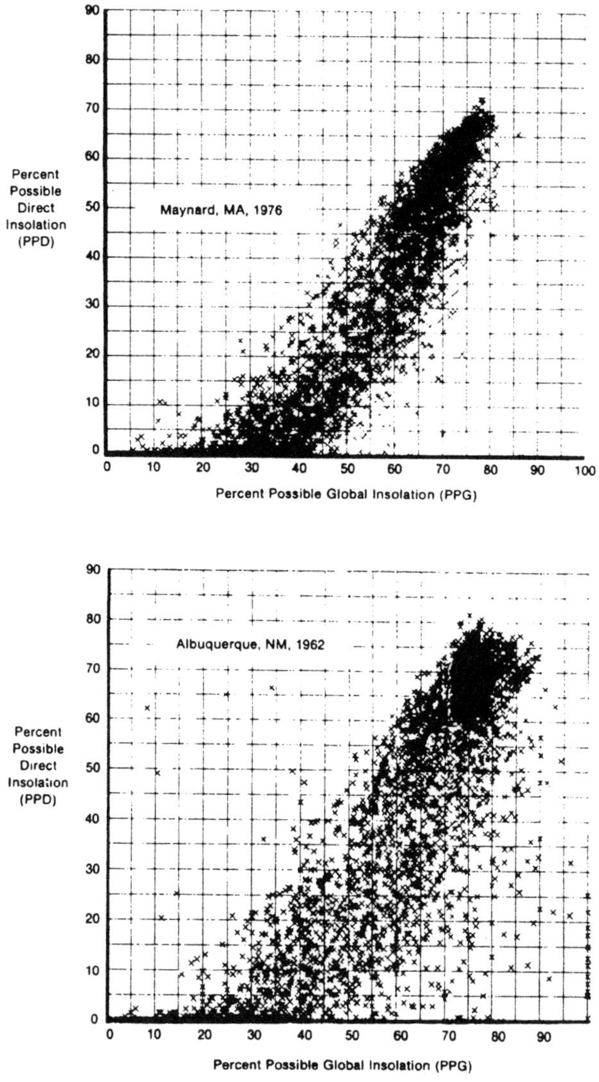

Fig. 3-8. Comparison of global and direct insolation at clear (**lower**) and cloudy (**upper**) locations. Each symbol represents an hourly average value [Knapp and Stoffel, 1982].

Instantaneous values of global radiation on a horizontal surface during the day may vary from less than 50 W m^{-2} to more than 1250 W m^{-2} (98 to 2400 μmol. $s^{-1}m^{-2}$ PAR). Typical values for different latitudes and times of the year were presented earlier in Fig. 2-13. Hourly integrals may range from 150 kJ $m^{-2}hr^{-1}$ to 3.6 MJ $m^{-2}hr^{-1}$. Total daily ranges are from 4 to 30 MJ $m^{-2}day^{-1}$ [Gates, 1980]. The solar constant (radiant intensity outside the atmosphere) is usually taken to be 1395 W m^{-2}. Under partly cloudy conditions and high elevations, intensity at the surface can often approach, and even exceed, the solar constant due to reflection from clouds. The sun emits a much broader band of electromagnetic radiation than is depicted in Figs. 3-7 and 3-9. It includes not only frequencies characterized as ultraviolet through infrared, but also microwaves and radio waves.

As for present greenhouse practice, the three most important factors concerning solar radiation are: 1) the total energy that may be available, 2) the length of the light and dark hours, and 3), the quanta in the various wavebands, influencing photosynthesis and photomorphogenesis. Evaluation of data, stating average annual solar radiation over the earth's surface show: 1) Northern Europe, including the U.K. and Russia, are at a severe climatic disadvantage as to total solar energy in comparison to such regions as the Mediterranean. 2) The Mediterranean regions, California, and the Chilean and South African western coasts have more than double the European energy, combined with a moderate maritime climate. 3) Central Australia, the Southwestern U.S., Northern Chile, the southwestern part of Africa, and Kenya north into Egypt have nearly three times the average annual energy that is available in Europe. Unfortunately, some of these latter climates have excessively high temperatures during portions of the year, or may be continental in regard to climatic variations.

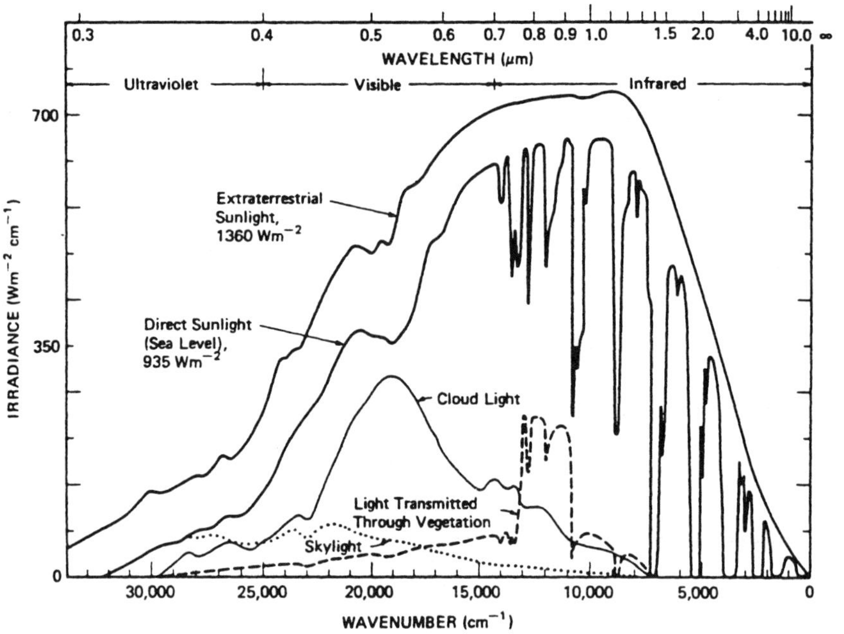

Fig. 3-9. Spectral distribution of direct sunlight, cloud light, skylight and energy transmitted through a plant canopy received on a horizontal surface [Gates, 1980] (By permission of Springer-Verlag).

It is usual for daily global radiation in December to be less than 2 MJ m^{-2} for locations such as England, Holland, and Seattle, WA, as compared to more than 10 MJ $m^{-2}day^{-1}$ for places like Phoenix, AZ, San Diego, CA, or Orlando, FL. [Anon., 1984]. Fig. 3-10 pictorially shows the differences to be expected between summer and winter in Scandinavian greenhouses.

The second factor to be considered for flowering control in greenhouses is the seasonal change in daylength, depending upon latitude of the location. Fig. 3-11 includes the critical daylength for flowering of some important ornamentals. It does not show the considerable variation that can occur among the various cultivars of a particular species. Photoperiodic response is one of the most important practical applications of radiant energy control in present industry, and greater detail will be given later in this chapter. It is sufficient to point out here that flowering is not the only important factor that may be controlled by light-dark transitions and duration.

IV. RADIATION AND THE GREENHOUSE

In Chapter 2, coverings and structures were discussed partially. In this section, radiation transfer and distribution inside greenhouses will be examined more closely. The last decade has seen considerable effort with many publications by the English, Dutch, and Japanese, with mathematical development of programs to assess radiation transmission. I am not competent to judge the particular mathematics that may have been employed. However, most of the English literature will be cited as necessary for most of it is seldom, if ever, discussed in industry trade journals. Students have virtually no contact with this information, but an appreciation of what has been done, and the sources of that information, can be vital in understanding plant behavior in high cost buildings.

A. COVERINGS AND ENERGY TRANSFER

Ideally, one needs 100% transmission of all solar wavelengths (300 to 3000 nm) at angles of incidence from zero to 90°, and zero transmittance for all thermal wavelengths beyond 3000 nm [Robbins and Spillman, 1980]. The latter qualification would greatly reduce energy loss. However, it is unlikely that such a material will ever be found. The transmittance of a greenhouse cover varies with the wavelength and incident angle of the radiation reaching it, of the refractive index of the material, and its extinction coefficient. Incidence angle is the angle (radians or degrees) between a line normal (perpendicular) to the surface and the ray (Fig. 3-12). Light rays are bent when passing from one material to another (refraction), so that when a light ray passes from air into, for example, glass, the ray is bent, and again when it passes out of the glass into the air below the pane (Fig. 3-12). The refractive index of a material is the ratio of the speed of light in a vacuum to its speed in the material, and is always greater than 1. The index for glass is 1.526, with most other greenhouse coverings varying from 1.460 for PVF to 1.520 for FRP and 1.515 for PE. Air is generally very close to 1. When passing from a medium of high refractive index to one with a low index, the incident angle can become so large that no radiation is refracted and total reflection occurs. The extinction coefficient accounts for absorption of radiation in the material with glass at 9.37 x 10^{-3} mm^{-1} as compared with 7.52 x 10^{-2} mm^{-1} for PE. Note that PE absorbs ten times as much as clear glass. When used as a double cover, the absorption can be unacceptable for maximum growth in low irradiance climates. Inclusion of IR absorbers or UV stabilizers in thin films can change properties of the material. The first stabilizers were based on nickel chelates that imparted a greenish cast to the film. These were followed by transparent "HALS" stabilizers. These were more prone to deactivation from products containing sulfur that may be used in the greenhouse. Newer products have greatly improved the ability of thin films to resist aging from UV.

Fig. 3-10. Upper: Interior of a greenhouse during December in Copenhagen, Denmark. **Lower**: Inside a greenhouse during June, Malmoe, Sweden. These illustrate the great differences in insolation during the year in European countries.

There are reports [Daponte, 1994] showing the use of plastic materials to control interference. Interference results when two rays from the same source, but traveling by different paths, "interfere" with each other when recombined. Depending upon the combination, the energy in each ray can be enhanced, or the rays may cancel each other –essentially preventing light passage. By varying thickness and refractive indices in thin films, certain wavelengths, such as green (500 nm), can be removed, leaving a red color. By selecting materials of differing refraction and thickness, one can select which wavelengths in the total spectrum to reflect. With multilayer extrusions, and the inclusion of one or more pigment types, radiation quality can be controlled. Right now, most

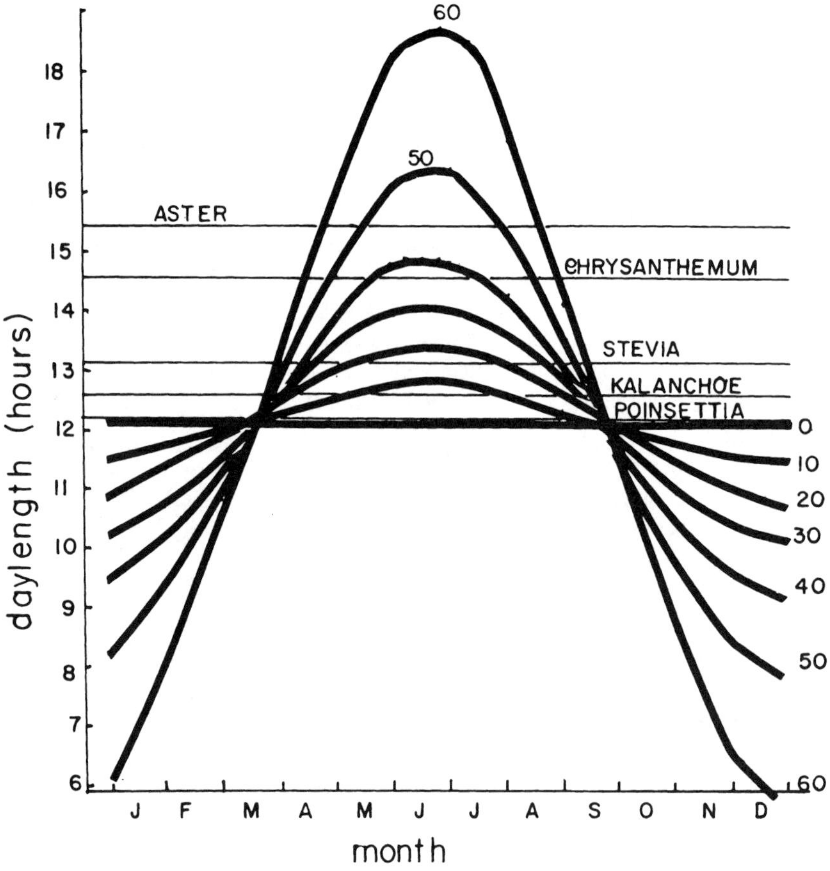

Fig. 3-11. Variation in daylength for different latitudes in the Northern Hemisphere. The horizontal lines for species, where they intersect with the curves, show the critical daylength above, or below, which the species will initiate flowers. For some species, the critical daylength is cultivar dependent and will vary from that shown at the right.

of these films discussed by Daponte are expensive.

The angle of incidence is always taken into account (Fig. 3-13). When the incidence angle changes from 0 to 90° in 10° increments, the respective transmissivity of 3 mm glass is 0.86, 0.86, 0.86, 0.85, 0.85, 0.82, 0.77, 0.65, 0.40 and 0.00 [Kozai and Kimura, 1977], where transmittance is the ratio of the energy passing through the cover to that incident upon it. Reflection of energy by the cover may decrease or increase energy into the greenhouse, depending upon incidence angle and the cover. Most materials follow responses similar to those shown in Fig. 3-13.

Fig. 3-12. Radiation transmission through one cover. The angle θ_1 is the angle of incidence and θ_2 the angle of refraction. Also shown are the possibilities for reflection [Robbins and Spillman, 1980] (By permission of the Amer. Soc. Agric. Engr.).

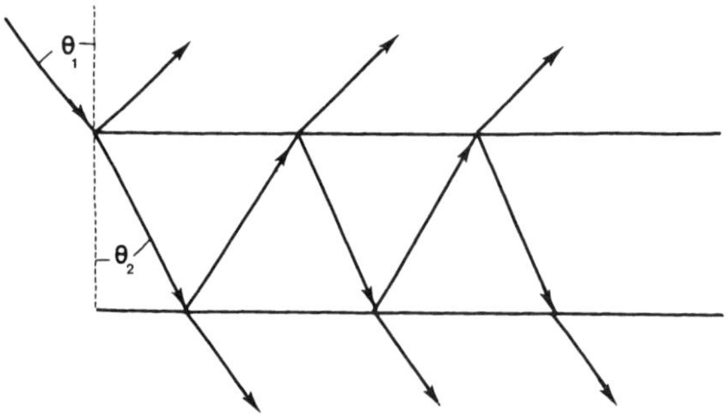

Early work by Baker and Aldrich [1964], however, showed that corrugated FRP did not follow the regular pattern suggested by Fig. 3-13. In fact, Hanan's measurements [1970] of radiation in an east-west glasshouse versus a north-south FRP cover showed widely varying slopes for a plot of outside radiation versus inside radiation (Fig. 3-14). FRP had a lower slope than glass, with the curves intersecting at R_os (outside radiation) of about 170 W m^{-2}. When glass and FRP are compared directly, FRP showed greater transmissivity than glass at R_os below 200 W m^{-2}, and lower than glass above 200. At high outside intensity, FRP has lower transmittance. Goldsberry's data [1969] (Fig. 3-15) shows distinct differences between glass and FRP on clear days. On overcast days, the differences between glass and FRP are nearly indistinguishable. Practical experience and work by Holley and his students, cited in Chapter 2, showed that under Colorado climatic conditions, FRP provides an advantage for high energy requiring crops. Under cloudy regimes at higher latitudes, the advantage of FRP *vis-a-vis* glass disappears. One can suggest that, the

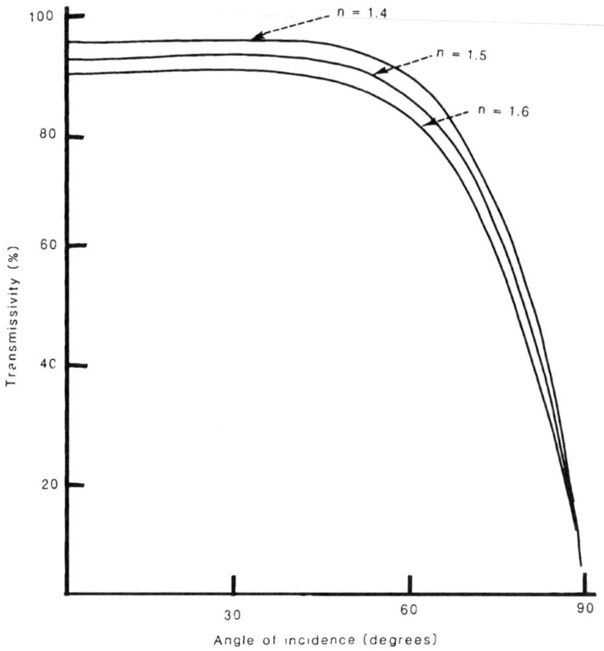

Fig. 3-13. The relationship between radiation transmission and angle of incidence for materials of different refractive indices [Bowman, 1970].

lower the available energy in a given location, the more inefficient the structure becomes in transmitting energy.

When double covers are employed, such as the structured plastics (Fig. 3-16) or double-layer PE, radiation transfer is increasingly complicated, as shown by Russell [1985]. Some of the possible reflections that have to be accounted for are depicted in Fig. 3-17. For structured materials, according to Russell, the overall transmittance of any twin-walled cover is much less than one plain sheet of the same material (8 to 20% less). The absorption coefficient in this case plays a dominant role. The smaller the number of ridges per unit wall distance, the higher the transmittance. Dependency of radiant transfer on the number of cross walls decreases as the extinction coefficient decreases. Therefore, for a virtually non-absorbing material such as PMMA, the overall transmittance is virtually independent of the number of ridges per unit length.

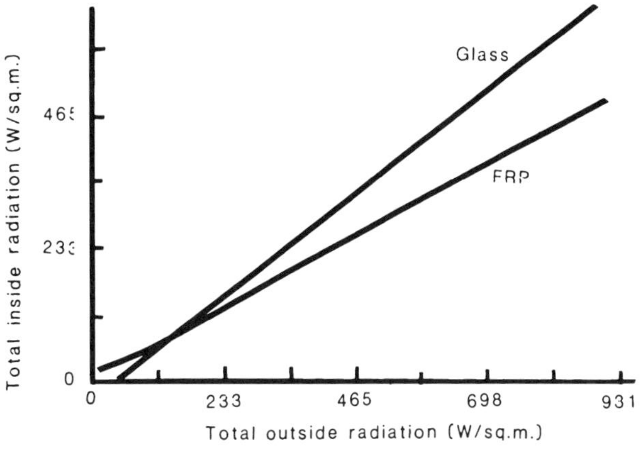

Fig. 3-14. Comparison of inside radiation as a function of outside insolation for an east-west glass house and a north-south, corrugated FRP-covered greenhouse [Hanan, 1970].

Robins and Spillman [1980] worked out formulae for double covers of varying materials, finding that FRP diffused about 21% of the transmitted radiation, so that rays passing through a sheet of flat FRP had different directional characteristics after passage. When flat FRP was employed in combination with other covers such as PVF, glass, or PE, transmittance was higher at 15° than at 0° angle of incidence. There were significant differences in transmission between clear and cloudy days. The dependence of transmittance on incidence angle was not as closely correlated for corrugated FRP as for flat FRP.

Diffuse radiation has been mentioned several times, and particularly the effect diffuse energy may have on greenhouse production. In fact, personal observation in the late 1950s in Colorado between glass and FRP covers showed striking differences in carnations when grown more than 1 year (Fig. 3-18). This was commonly attributed to better radiation distribution in the crop canopy. Although clear glass has a very low extinction

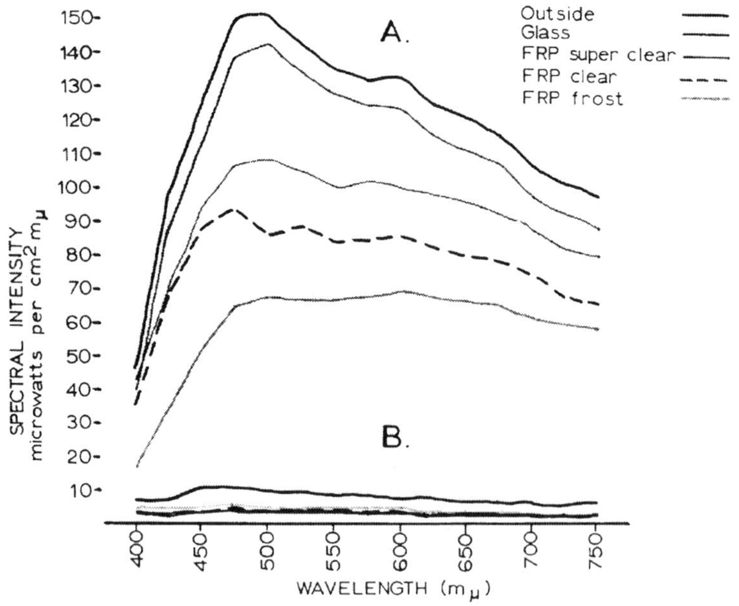

Fig. 3-15. Spectral transmission of various FRP covers versus glass compared with outside irradiance on clear days and uniformly overcast days [Goldsberry, 1969].

coefficient so that direct sunlight penetrates with little change, there remains a considerable part of the available energy, even on clear days, that is diffuse. The work by Basiaux et al. [1973] is one publication dealing with diffusion properties of different covers. It is difficult to trace accurately the path of rays emerging from materials that tend to diffuse. A material that diffuses radiation is normally called "translucent," as contrasted to "transparent" materials. A translucent body changes the direction of the incident beam, but there is no change in monochromatic radiation frequency. Basiaux et al. [1973] worked with striated glass as a diffusion cover, finding that more energy was transmitted on clear days at low solar elevations up to 30°. There may be a "light trap" effect of striated glass. The diffusion of glass is most often due to irregularities on the surface, whereas in plastic materials, diffusion is also due to the heterogeneity within the material. It is a common observation that shadows tend to disappear in a greenhouse with FRP or PE covers as contrasted to glass. Using a computerized program, Basiaux et al. found that at solar elevations above 35°, a diffusive cover caused a shading effect in strong insolation.

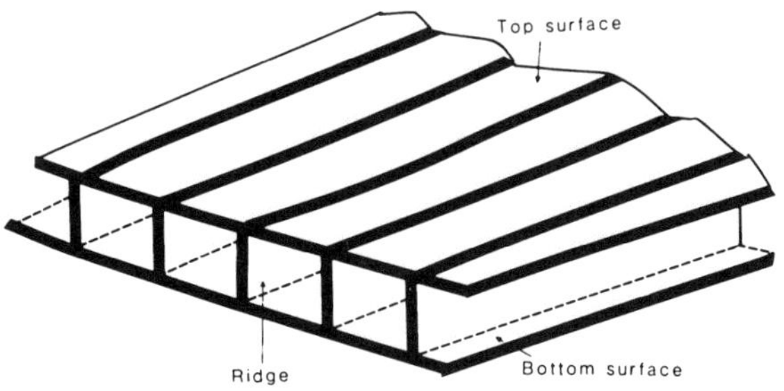

Fig. 3-16. Typical structured plastic material (twin-walled) [From Russell, 1985].

Hanan [1989], in a study of 13 different covers with measurements of global, shortwave radiation at plant level, found that diffuse and direct radiation in FRP-covered houses could vary considerably depending upon house orientation, age of the roof, and time of year. When all measurements were plotted as a function of total inside radiation (Fig. 3-19), the diffusive component leveled off above 400 W m^{-2}, although direct radiation continued to increase. It may be concluded that up to levels of 200 W m^{-2}, nearly all energy in a greenhouse is diffusive. Diffuse radiation external to the greenhouse is also higher in proportion to direct radiation. Depending upon the transmissivity of the structure, outside radiation levels above 300 W m^{-2} will be required before direct radiation begins to increase significantly. In other words, solar energy levels below 300 W m^{-2} through the midday are likely to be on days that are overcast.

The Belgians [Nijskens et al., 1985] have gone to considerable effort to categorize covering materials. Under careful laboratory procedures, they have cataloged most of the common coverings as to transmission in the solar

UV (300-380 nm), solar visible (380-760 nm), short or solar IR (760-2500 nm), total solar (300-2500 nm), and PAR (400-700 nm). They have provided values for transmittance and reflectance at various angles of incidence for these bands under clear and overcast days. Fig. 3-20 provides graphically some values they obtained for many materials. Nijskens et al. state that 85 to 90% of the incident energy is estimated to penetrate uncoated single panes. Nevertheless, coated glasses and all double covers reduced radiant transfer to 65%. For IR transmittance, PE varied from 20 to 60%, depending upon the additives incorporated. The materials differed greatly as a clear sky became cloudy. Under overcast skies, the reduced transmittance varied from 17 to 22% compared to a clear sky. None of the films tested differed greatly from that corresponding to total solar radiation in the PAR band. Figs. 3-15 and 3-21 show the transmittances for common materials over the 400 to 700 nm band.

B. STRUCTURE AND TOTAL TRANSMISSIVITY

The study of greenhouse transmissivity is complex. It is one thing to consider radiant energy transfer of the cover singly and quite another to account for efficiency of the entire structure. Figs. 3-23 through 3-33 show some results that have been obtained through actual measurement and the use of mathematical models. Transmissivity varies with: 1) shape and height of the structure, 2) design of superstructure, 3) slope of the roof, 4) orientation (i.e., N-S to E-W), 5) time of year and day, 6) proportion of diffuse and direct radiation, 7) latitude of the structure, and 8), general climatic conditions. Some of these

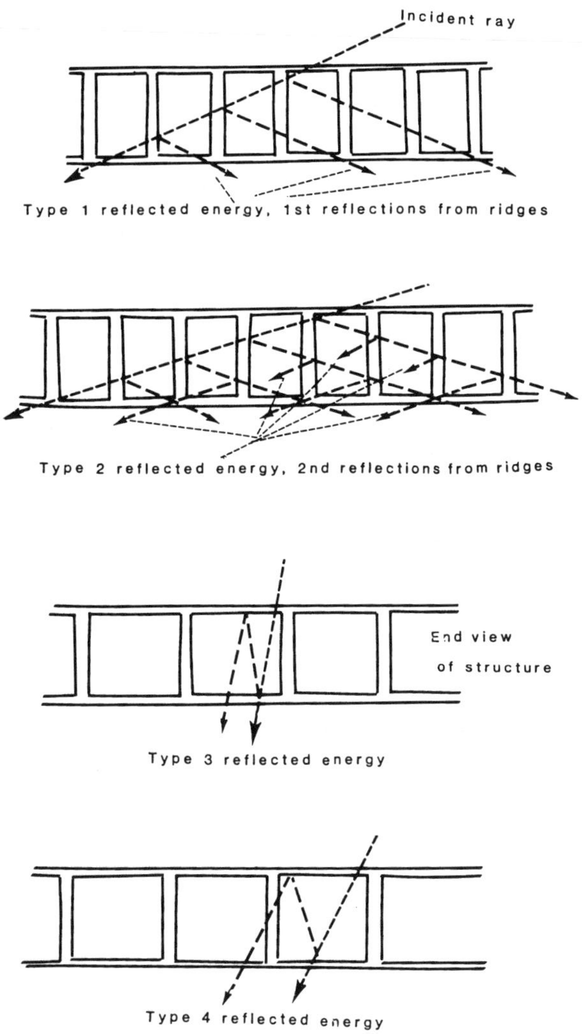

Fig. 3-17. Types of reflected radiation considered by Russell [1985] in assessing transmissivity of structured plastics.

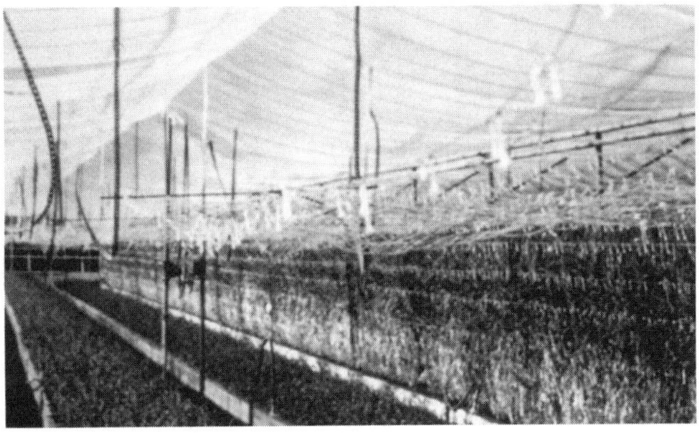

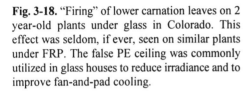

Fig. 3-18. "Firing" of lower carnation leaves on 2 year-old plants under glass in Colorado. This effect was seldom, if ever, seen on similar plants under FRP. The false PE ceiling was commonly utilized in glass houses to reduce irradiance and to improve fan-and-pad cooling.

items have been mentioned. Explaining all possibilities is not feasible.

Beginning with the early work by the British [e.g., Whittle and Lawrence, 1959; Edwards and Lake, 1965a; 1965b; Edwards and Moulsley, 1958], the pace of investigation on greenhouse transmissivity accelerated in the 1970s and 1980s with engineering studies by the Dutch and Japanese [e.g., Bot, 1983, Kozai, 1977; Kozai et al.., 1978, Kozai and Kimura, 1977; Kurata, 1983, 1990], and particularly D.L. Critten who published more than 13 papers on the subject in the 1980s [e.g., Critten, 1983 through 1987e].

Usually, to learn the overall transmission of a structure, the irradiance of the hemisphere above the greenhouse must be evaluated (Fig. 2-22). For direct sunlight, the elevation and azimuth of the sun must be determined. Even overcast sky is not a uniformly distributed source as sometimes suggested. For a given section of the sky, the elevation of that section (sin a) is taken into account [e.g., Littlefair, 1981; Critten, 1983a, 1986; Bot, 1983; Kozai, 1977]. An overcast sky has a mathematical formula:

$$L_e = \frac{L_z(1+2\ \sin\ a)}{3} \tag{3.11}$$

where L_e *= illuminance of sky at angle a above the horizon*
L_z *= illuminance of sky at the zenith*

Fig. 3-19. Total diffuse (**upper**) and total direct (**lower**) radiation as a function of total internal greenhouse global radiation at plant level in 13 greenhouses of varying covers at different times of the year. Each point represents average irradiance over 1 hour, 6 readings per hour [Hanan, 1989]. Correlation coefficient upper = 0.83, lower = 0.92. Twelve of the 13 structures were commercial, north-south oriented houses and 1 east-west.

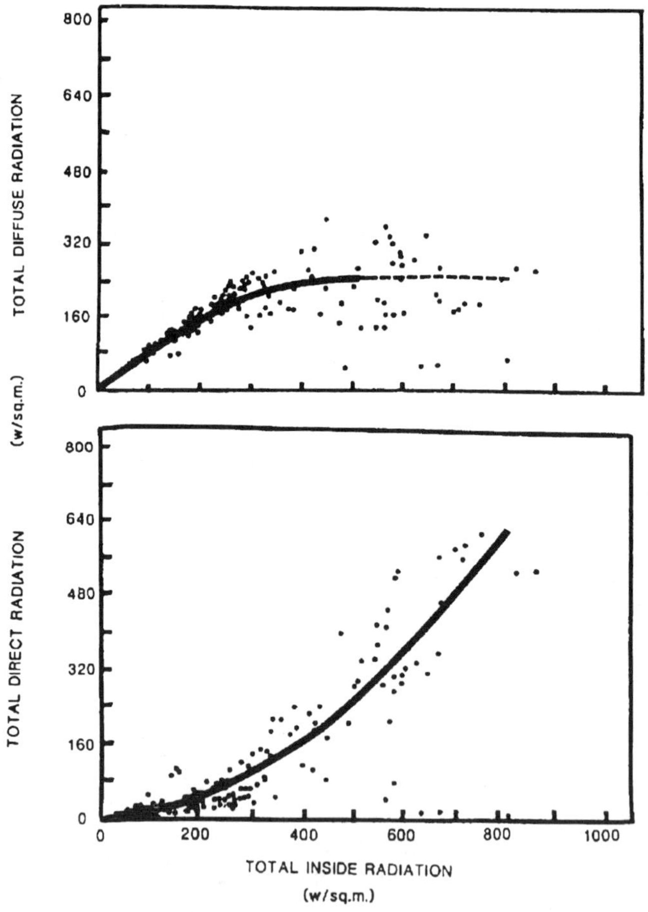

A similar analysis using odd Legendre polynomials for angle **a** and Fourier analysis for azimuth has been used by Critten [1986]. The planes of the greenhouse must be taken into account and the receiving surface inside the greenhouse considered. The three categories, sky, greenhouse, and floor (plants), are usually divided into a series of segments and each evaluated and then summed or integrated to obtain total sky irradiance (energy from all portions of the hemisphere above the greenhouse), transfer through the planes of the greenhouse

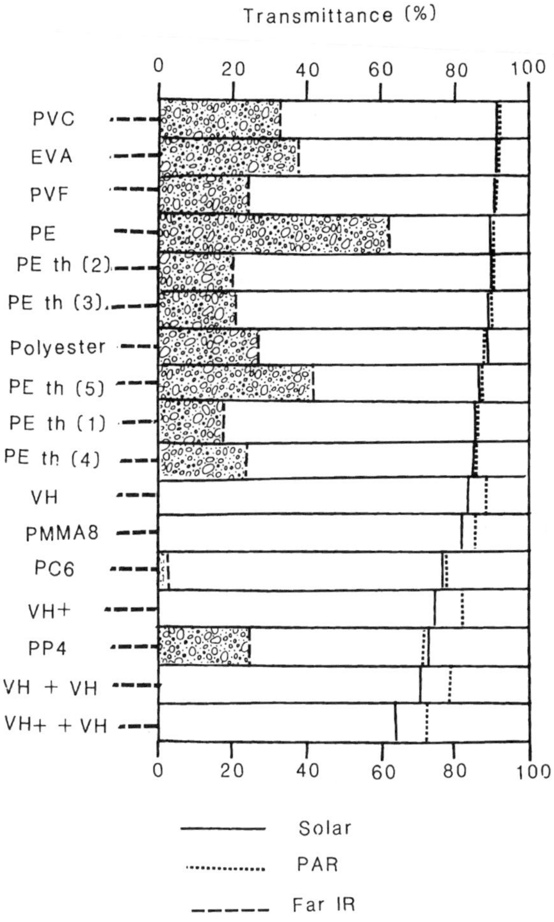

Fig. 3-20. Transmittance at 0° incidence for total solar radiation, PAR, and thermal IR for some widely used coverings. **PVC** = polyvinyl chloride, **EVA** = ethyl vinylacetate, **PVF** = polyvinyl fluoride, **PE** = polyethylene, **PE TH** = PE with IR block, **VH** = glass, **PMMA8** = poly methylmethacrylate double wall 8 mm thick, **PC6** = polycarbonate double wall 6 mm thick, **VH+** = low emissivity glass, **PP4** = polypropylene double wall 4 mm thick, **VH+VH** = double glass, **VH+ + VH** = double glass with one low emissivity. Shaded areas represent far infrared transmittance. The solid vertical lines indicate total solar transmittance, and dotted vertical lines PAR [Reprinted from *Agric. and Forest Meteor.*, Nijskens, J. et al., 1985. 35:229-242. With kind permission of Elsevier Sci.-NL, Sara Burgerhartstraat 25, 1055 KV Amsterdam, The Netherlands].

surface are and then received on the floor. Results are expressed as transmittance (T or τ), normalized on a scale of 0 to 1. Of course, clear and overcast skies represent two extremes. One is faced with picking the most common condition for the locality. One scheme, used by Bot [1983], employs a scale of 0 to 8. During the winter in Germany, heavy cloudiness on a scale of 7/8 to 8/8 is dominant. These two classes account for 80% of the time in December, January, and February.

In evaluating the energy transfer of greenhouse surfaces, one must take into account whether the house in question is single span or multispan. In the latter case, the role of reflection and shading by the superstructure becomes important, with some differences of opinion [e.g., Kurata, 1990; Critten, 1983a; 1983b; 1987a; 1987c; Bot, 1983] as to how the procedure for accounting for these factors is carried out. Occasionally, the results have shown that there is a loss in transmissivity at the edges of a house (Figs. 3-23, 3-24) which does not accord with plant response in high-sided greenhouses. Reflection has been examined by Critten [1984; 1985c; 1987b] in the improved transmissivity of vertical, south-roofed greenhouses (Fig. 2-10), generally showing a slight advantage of 8 to 10% under low-angle, direct-beam sunlight, but a 6% loss under diffuse sky conditions. Caution might be necessary in some of this European work conducted in glass-covered structures with clear spans.

As with reflections from the cover, energy loss due to structural members varies with whether the skylight conditions are diffuse or clear sky [Critten, 1987c; 1987d]. Examples of losses under diffuse radiation are presented in Table 3-5, which suggests that losses due to structure can exceed 20%, and with the cover, losses exceed 30%, depending upon the type of structure. One might conclude that fully inflatable houses with no superstructure would be an approach to the ideal. However, this is highly dangerous in severe climates with good chances of catastrophic loss. Also, most of the cover is not likely to be at the most advantageous angles for radiant penetration. If, however, the superstructure can be reduced by incorporating a part of the support in the cover, less energy loss due to shading by the supporting structure will occur –such as the use of corrugated FRP, structured plastics, or double, air-inflated PE.

Fig. 3-23 shows total transmissivity across a four-span Venlo house as compared with Critten's measurements (Fig. 3-24 bottom) for the same structural type. In each example, radiation dropped off precipitously at the edges. Note the gutter effects in a multiple span structure. With many of these studies, the author assumes a house of infinite length and an infinite number of gutter-connected houses in order to omit the edge effects. According to Critten [1985b], transmission near the gable ends of both single and multispan houses can be 3% higher than that further into the greenhouse. In small houses, Israeli work has shown enhanced productivity at the sides of the structure (Figs. 4-39, 4-40) as contrasted with the possible effects suggested by Figs. 3-23 and 3-24. The effect of N-S orientation on radiation at two times of the year is shown by Fig. 3-25 as compared to E-W spans on the shortest day of the year (Fig. 3-26). Going from the gable end, along the length of the house, radiation at certain times of the year can vary remarkably. Variation with distance from the gable will also be found on overcast days.

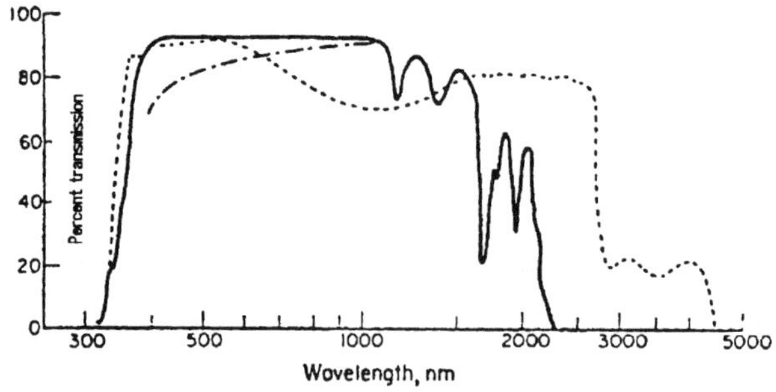

Fig. 3-21. Spectral transmission of three common cover materials [Bowman, 1970] With permission of Academic Press Ltd., London.

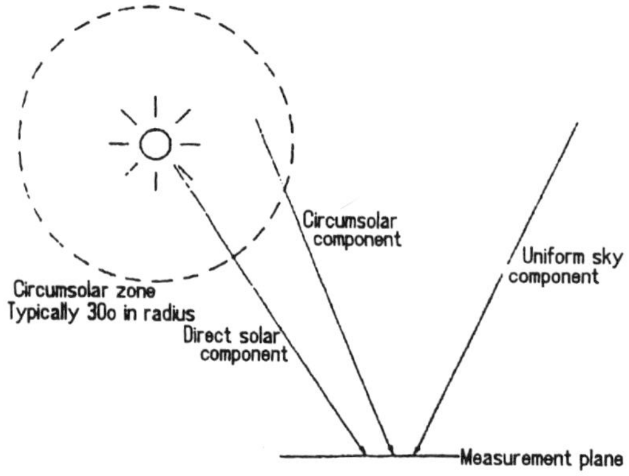

Fig. 3-22. Sky model employed by Littlefair [1981] in evaluating illuminance of the sky. Utilized by Critten [1985c] to assess sky irradiance With permission of Academic Press Ltd., London.

Examples of transmissivity as a function of roof angle, latitude and number of spans are given in Figs. 3-27 through 3-31. In the case of Fig. 3-27, transmissivity is expressed as a <u>loss</u> that decreases as the roof angle increases much beyond about 20° on an overcast day. As latitude increases (e.g., Osaka versus Amsterdam), structure orientation becomes more critical (Fig. 3-28). A height/span ratio of 0.6 could be applicable to a house 2 m wide with 1.2 m side walls, or 4 m wide with 2.4 m side walls. On the other hand, a ratio of 0.2 could be a 10 m span with side walls 2 m high. Fig. 3-29 would seem contradictory. There was not much difference in transmissivity between latitudes until the structure orientation was about 60° from E-W. Houses in Fig. 3-28, however, have differing types (wall/span), whereas those in Fig. 3-29 examined roof angles, ranging from 16 to 45° at three latitudes for a single structureless span. Note Fig. 3-30 shows that as the number of spans increases, average daily total transmissivity decreases. The effect of orientation remains pronounced. As suggested in Fig. 3-25, Fig. 3-31 shows that transmissivity with distance from the gable end can vary radically, especially as latitude increases. Many results from models assume a house without a superstructure. Fig. 3-31 shows remarkable effects of roof angle, especially at high latitudes. These latter diagrams deal with clear sky conditions, as contrasted to Fig. 3-27 under an overcast sky. One concludes that the prevailing climatic conditions are highly important in determining the most efficient greenhouse design.

In Fig. 3-32, the total transmissivity of a Venlo multispan is provided by Bot under diffuse sky conditions as functions of azimuth and altitude of the designated part of the sky. This serves to show that

Table 3-5. Calculated and measured percentage global energy losses due to greenhouse structural members at various stages of erection [From Critten, 1987c]. Conditions under diffusive radiation.

	Energy loss					
	Simpson twinspan		**Venlo**		**Robinson multispan**	
Stage	**Calculated**	**Measured**	**Calculated**	**Measured**	**Calculated**	**Measured**
Stage 1	9.1	10.0	12.8	12.3	16.8	15.2
Stage 2	18.2	20.0	22.6	---	19.3	---
Complete house	32.2	31.0	35.8	34.0	32.9	---

Note: Stage 1 = erection of all weight-bearing elements.
Stage 2 = insertion of glazing bars.
Complete house = glass covering.

Fig. 3-23. Calculated transmissivity of a 4 span Venlo. **E**, **R**, and **C** indicate house edge, ridge and center respectively. [Reprinted from *Agric. and Forest Meteor.*, Kurata, L. 1990. 52:319-331. With kind permission of Elsevier Sci.-NL, Sara Burgerhartstraat 25, 1055 KV Amsterdam, The Netherlands].

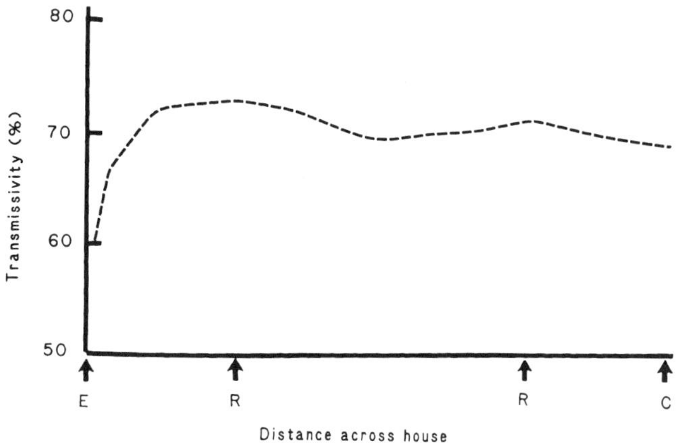

an overcast sky cannot be considered as a uniform radiant source. Since diffusive radiation represents the prevailing north European conditions, particularly in the winter, structures should be designed to permit maximum transmission under those conditions. Similar climatic areas can be found in the U.S.

Contrasting to these conditions are the data provided by Hanan for greenhouses in an arid, clear sky region (Figs. 3-33 through 3-35). All these measurements were obtained by point sensors at crop level. Part of the variability could have resulted from shading and reflection onto the sensors. On the average, about 60% of the total outside radiation was transmitted to the top of the plant canopy. In high latitudes, transmissivity can be lower than 50%. The old practice of applying shade to glass in the summer showed a very significant reduction in transmittance (Fig. 3-34). This would suggest very low energy for growth, necessary for temperature reduction before introduction of evaporative cooling and FRP covers. Structured acrylic had the highest proportion of direct radiation in the summer. The difference between acrylic and clear glass, as for diffuse and direct radiation, could be attributed to the differing superstructures. The glass structure had narrow panes supported on wood sash bars. Their shadows would affect data from point sensors. Except for an E-W, corrugated FRP cover, all other FRP-covered greenhouses were not significantly different from each other in terms of transmissivity.

C. MODIFICATIONS TO INCREASE OR MODIFY IRRADIANCE

Over the years many attempts have been made to either increase total irradiance inside greenhouses, or to modify the radiation that reaches the plant material [e.g., Aikman, 1989; Kurata, 1983; Critten, 1985a; Elwell and

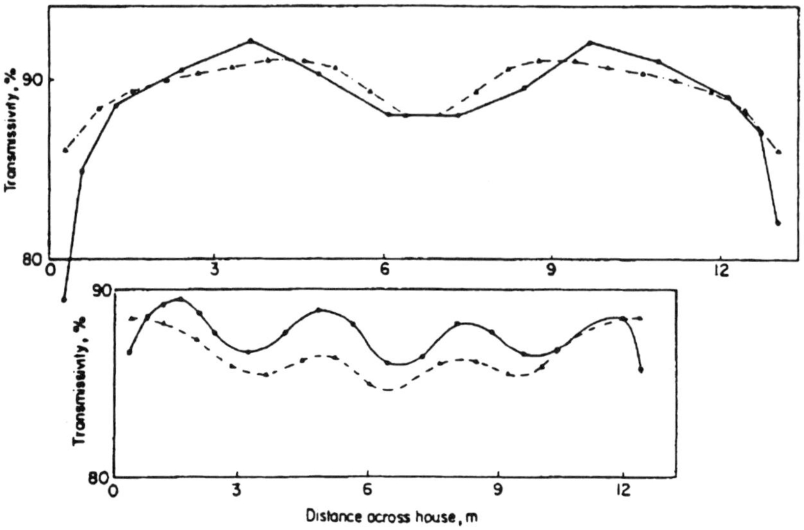

Fig. 3-24. Transmissivity of a twin-span (**top**) and a 4-span Venlo (**bottom**). Solid lines are experimentally measured, dotted lines are calculated from a model [Critten, 1983b]. Note the decrease at the sides, which was unexpected and does not accord with practical experience. With permission of Academic Press Ltd., London.

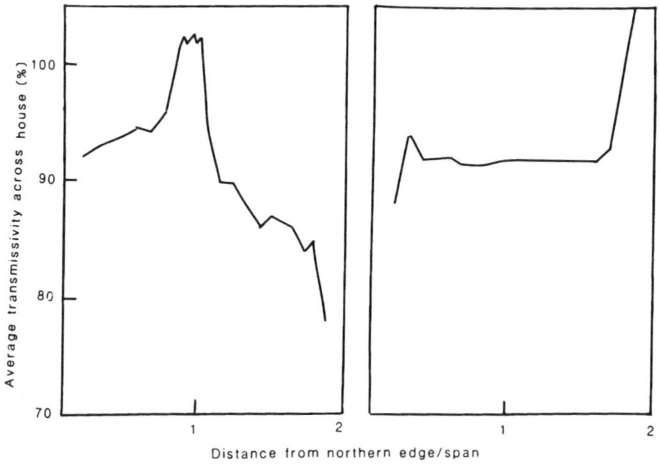

Fig. 3-25. Transmissivity along the length of a N-S, 30° roof greenhouse at midwinter (**left**) and midsummer (**right**). Length/span ratio = 2.0, wall height/span ratio = 0.2 [Critten, 1985b]. With permission of Academic Press Ltd., London.

Short, 1989; Novoplansky et al., 1990; etc.]. For example, Thomas [1978] investigated the possibility of specularly-reflecting north, back walls of E-W oriented greenhouses. This offered a significant increase in irradiance on clear days with the possibility of thoroughly insulating that part of the house. Critten [1985a] analyzed forms of high-gain, nonfocussed sunlight concentrators suspended from the ridge, finding a 20% gain over the whole house width with specialized louvers under sunny conditions. Kurata [1983] tried the use of Fresnell prisms that concentrated radiant flux. Such prisms are commonly used in signal and search lights. Aikman, in an interesting paper [1989], approached the problem as one of redistribution of the available energy; that is, to reflect energy back to that part of the canopy that may be shaded. Deriving a mathematical model,

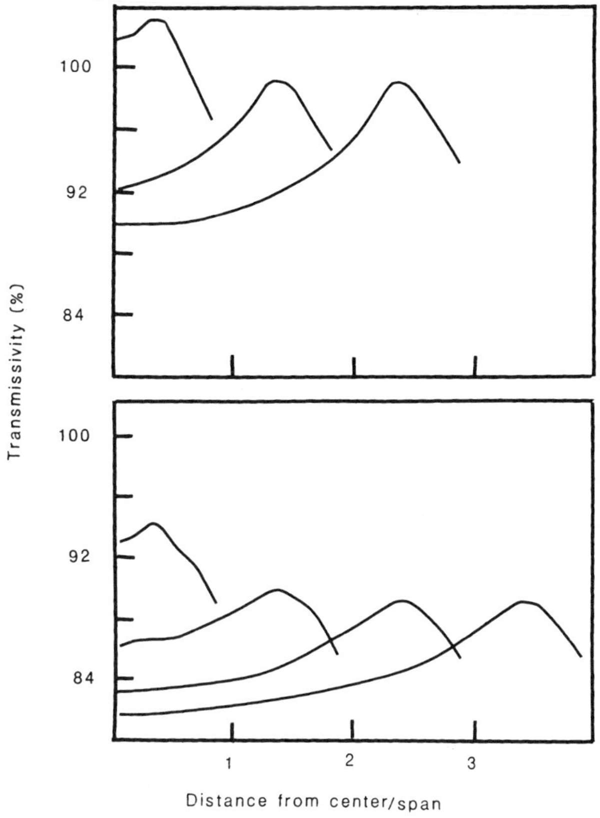

Fig.3-26. Effect of length of greenhouse on the transmissivity of an E-W multispan under clear sky on December 21. **Upper**: greenhouse with a 26° roof. **Lower**: house with a 45° roof. Wall height/span ratio = 0.4. Each curve represents different spans, south to north. No structural members in this simulation [Critten, 1985b]. With permission of Academic Press Ltd., London.

Aikman estimated that increases of 33 to 38% may be obtained by appropriately distributed irradiance. He suggested the possibility of partially reflective screens to distribute direct solar radiation more uniformly over the foliage.

Other authors, such as Elwell and Short [1989] and He et al. [1990], have examined the use of polystyrene pellets in double-glazed houses to control total radiation and also to serve as insulation at night. Over several years, the possibility of water flowing over the greenhouse roof as a means to cool and heat has been tried; with copper sulfate solutions to act as solar collectors, while allowing PAR to be transmitted [e.g., Mortensen, 1987; Tross et al., 1984; van Bavel, 1978; Nielsen et al., 1983]. Many of these will be discussed further in Chapter 4. Novoplansky et al. [1990] hit upon the novel

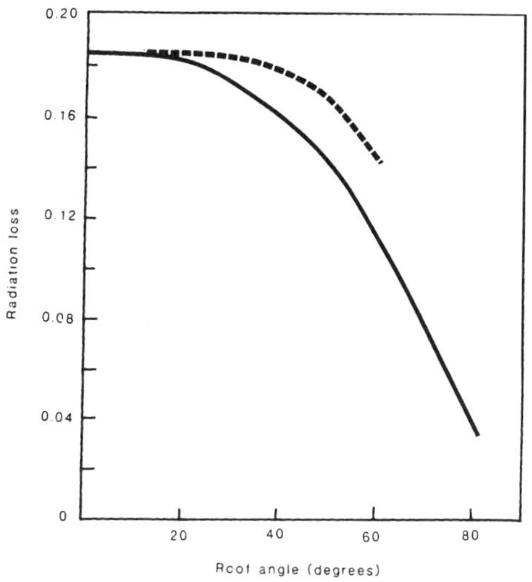

Fig. 3-27. Transmissivity **loss** as a function of roof angle for a standard overcast sky, double glazed, multispan greenhouse [Critten, 1987a]. Dashed line is for a single span. Note in this graph that loss **decreases** as roof angle **increases**. With permission of Academic Press Ltd., London.

idea of incorporating fluorescent pigments in plastic films that have the capability of absorbing green radiation and fluorescing in the red. This can have significant effects on plant morphology. Even more interesting is the understanding of blue receptors in plants and the fact that botrytis sporulation is modified by UV. Thus, Honda et al.. [1977] could show a reduction in sporulation on cucumbers and tomatoes by using a vinyl cover that limited UV to wavelengths longer than 390 nm as contrasted to a lower limit of 300 nm. Other important fungi in greenhouses are influenced by –as noted in Section 3.I– quantum ratios and counts. Use of covers to control disease and morphogenesis may be a significant advance in greenhouse practice.

Fig. 3-28. Effect of house orientation on average transmissivity at 34°39'N and 52°20'N for three wall height/span ratios on December 22 for a four span house. [Reprinted from *Agric. and Forest Meteor.*, Kozai, T. and M. Kimura, 1977. 18:339-349. With kind permission of Elsevier Sci.-NL, Sara Burgerhartstraat 25, 1055 KV Amsterdam, The Netherlands].

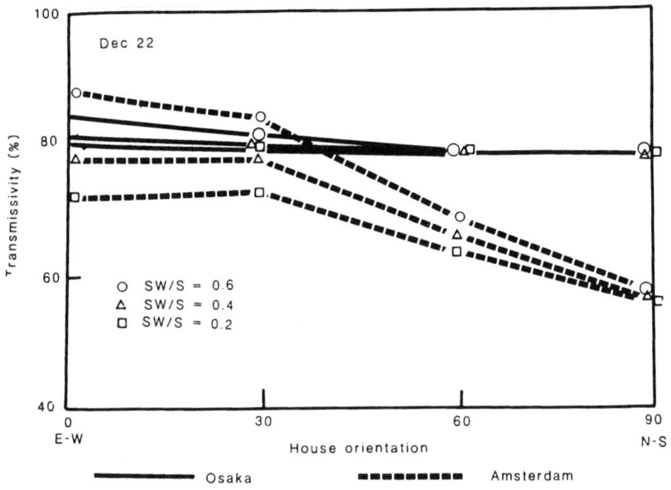

Unfortunately, many procedures enumerated above, while perhaps technically feasible are uneconomic. Reflectors have been used several times, but their maintenance is troublesome. Presence of reflectors overhead will reduce diffuse radiation, and Bailey [1984] pointed out that sunlight reflected from north walls did not make up for the loss of energy from the northern sky under cloudy conditions. The importance of energy through the north walls was shown in Chapter 2. One has to keep in mind that reflective systems work with direct radiation. Since the main objective is to increase energy under diffusive conditions, these novel ideas are less useful. Even the single, wide-span greenhouse, which was shown to have the most uniform and highest transmissivity, succumbed to the fact that multispans were more economical to build and operate. Reflection systems increase capital investment and operational costs, particularly where the possibilities of flowing water systems have been attempted through structured plastics. The latter are very expensive. Critten, in a summary of the research in this area [1993], pointed out that the designer must reduce radiation losses to the minimum by decreasing the structural content of the house, or redirecting reflected radiation so that it is not wasted.

V. THE PLANT AND RADIATION

Providing a complete text on plant physiology is not possible, and we are reduced to dealing with those aspects of radiation and growth which, based on 30 years experience, I think are relevant and interesting. In this section, we will discuss, first, energy interception by the crop. Unfortunately, radiation interception, penetration, and distribution have been sadly neglected at least in the U.S. A complete analysis would require serious attention to the agronomic literature, and a review of quantum transport in the canopy such as published by Myneni et al. [1989]. The second aspect is discussion of photosynthesis and the methods presently employed to supplement solar radiation under winter conditions. The third is photomorphogenesis where low quantum counts (low irradiance) in certain wavelengths (quantum ratios) play a dominant role in flowering as well as other aspects of greenhouse practice.

A. RADIATION INTERCEPTION BY THE CANOPY

Energy falling on a leaf may be absorbed by the pigment system such as chlorophyll or phytochrome, transmitted through the leaf, reflected, or absorbed and utilized to raise plant temperature or evaporate water. The latter are the most likely to occur with well-watered plants, and, in fact, the solar radiation curve can be closely

Fig. 3-29. Effect of house orientation at three locations on transmissivity of single span, structureless glass house with roof pitches of 16° (**top**), 32° and 45° (**bottom**), each 20 m long, clear sky. [Reprinted from *Agric. and Forest Meteor.*, Kozai, T., 1977. 18:327-338. With kind permission of Elsevier Sci.-NL, Sara Burgerhartstraat 25, 1055 KV Amsterdam, The Netherlands].

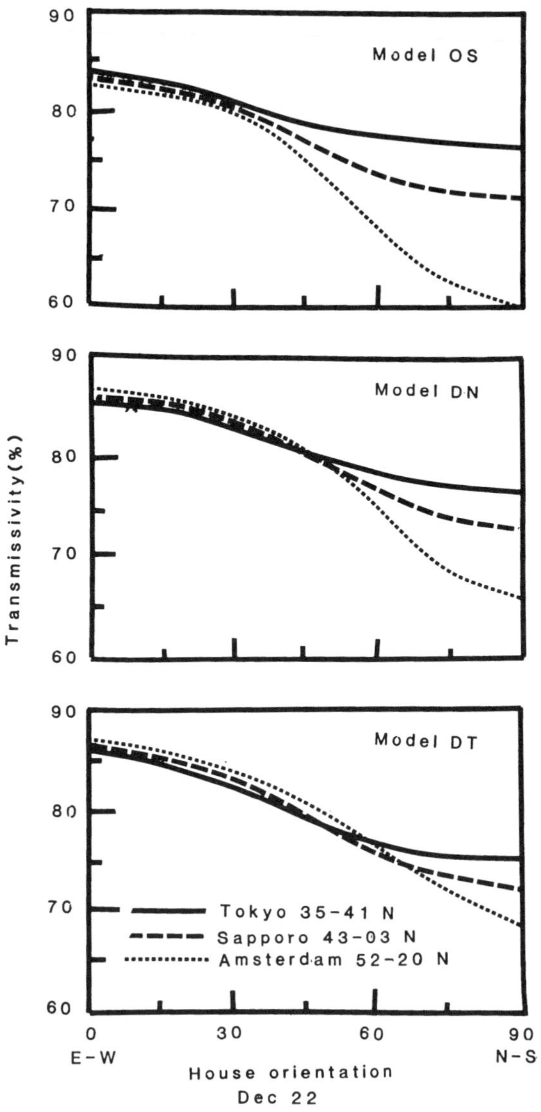

approximated by converting water loss of well-watered plants on an equivalent basis (Fig. 3-36). A significant portion of the incident energy is reflected by the vegetation –as much as 30%. However, there is a linear relationship between incident radiation and reflected (Fig. 3-37). The proportion of the total reflected from a carnation crop, expressed as a percentage of the total, appears to increase rapidly when irradiance drops much below 140 W m^{-2} [Hanan, 1970]. The process of increasing reflectivity at irradiances less than 140 may be worse under glass compared to FRP. These types of relationships between quantity and partitioning of energy under covers have not been adequately appraised. For simplicity, nearly all authors use a single value for reflectivity whatever the conditions.

One investigation of energy interception by an ornamental greenhouse crop was published by Stanhill et al. in 1973. Their results showed that 20% of the global solar radiation reaching the rose canopy was reflected, as contrasted to 28% for individual leaves. Reflection in the near-IR waveband was twice as great for the canopy with 30% of the global radiation transmitted to the floor of the house below the canopy, largely via the aisles between benches. By means of photography, they plotted the probability of a rose canopy intercepting radiation. This showed a maximum mean probability of 80% that direct solar radiation at a given angle of solar elevation would be intercepted by a rose crop in N-S oriented benches in a N-S greenhouse with diffusing glass. According to Stanhill et al., the net terrestrial radiation (thermal radiant exchange) within a glasshouse is very small, as originally shown by Businger [1966]. The net radiation balance was essentially equal to the net shortwave flux (R_{in}-R_{out}, 300-3000 nm). The photosynthetic efficiency was 1.8% when expressed on the basis of incident global energy and 3.9% when expressed by absorbed energy (net radiation). According to Hanan [1970], the ratio between net radiation and incoming, total shortwave flux within the structure could vary considerably as the result of external conditions. For example, Fig. 3-38 shows that net radiation received by a carnation crop seldom equals the inside global radiation unless the outside temperature is equal to, or exceeds, the inside temperature (April 20). The other three days, whether clear or overcast, had outside temperatures below the inside house temperature (>5 C May 20 to more than 20° below inside temperature on Dec. 12 and 17). Hanan [1970] concluded that as total outside radiation decreases in the winter months, an increasing proportion of the energy within the structure is

Fig. 3-30. Average daily total transmissivity of a standard greenhouse at 35°N on Dec. 21 with a roof angle of 30°, for single span, three spans and an infinite number of spans. [Reprinted from *Agric. and Forest Meteor.*, Kurata, K., 1990. 52:319-331. With kind permission of Elsevier Sci.-NL, Sara Burgerhartstraat 25, 1055 KV Amsterdam, The Netherlands].

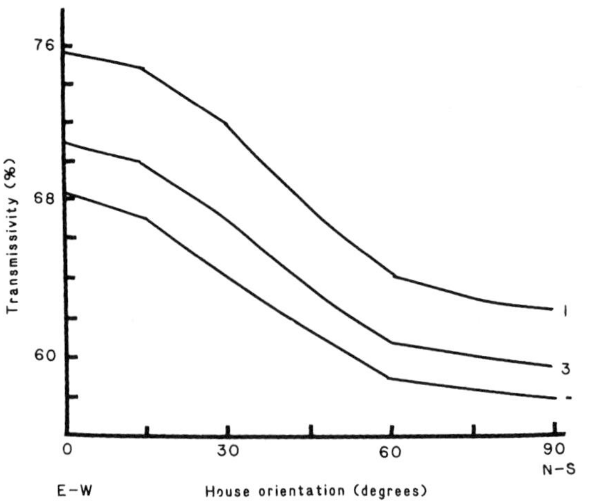

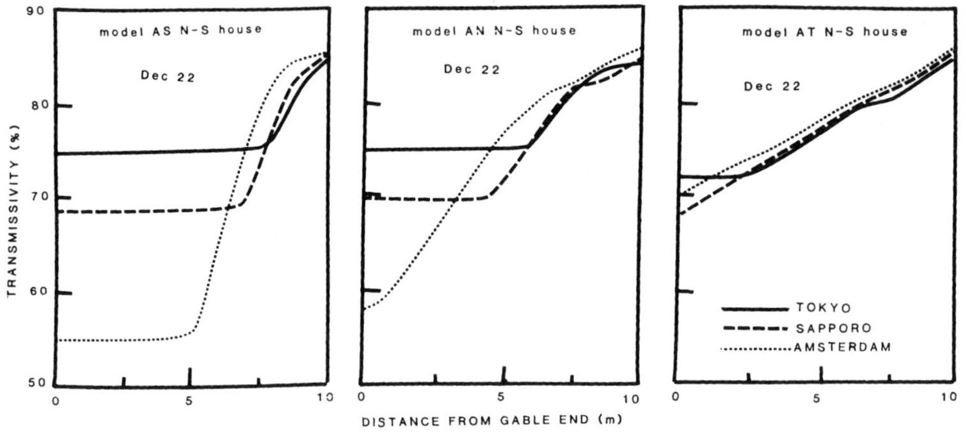

Fig. 3-31. Transmissivity gradients in N-S houses at different latitudes on Dec. 21 for structureless glass houses at roof pitches of 16° (left), 32° and 52° (right), each 10 m long. Clear sky conditions. [Reprinted from *Agric. and Forest Meteor.*, Kozai, T., 1977. 18:327-338. With kind permission of Elsevier Sci.-NL, Sara Burgerhartstraat 25, 1055 KV Amsterdam, The Netherlands].

unavailable for photosynthesis or evaporation with a larger part required to maintain plant temperature. These types of relationships have not been adequately examined.

Radiant interception by a canopy will obviously vary with internal crop arrangement (i.e., E-W versus N-S benches), planting arrangement, and stage of growth [e.g., Nederhoff, 1984; Papadopoulos and Ormrod, 1988a, b; Stoffers, 1975; Yang et al., 1990; Kasrawi, 1989]. At sometime or another during the season, the crop will not have covered the space in which planted, and the LAI (leaf area index = ratio of leaf area to ground covered) will

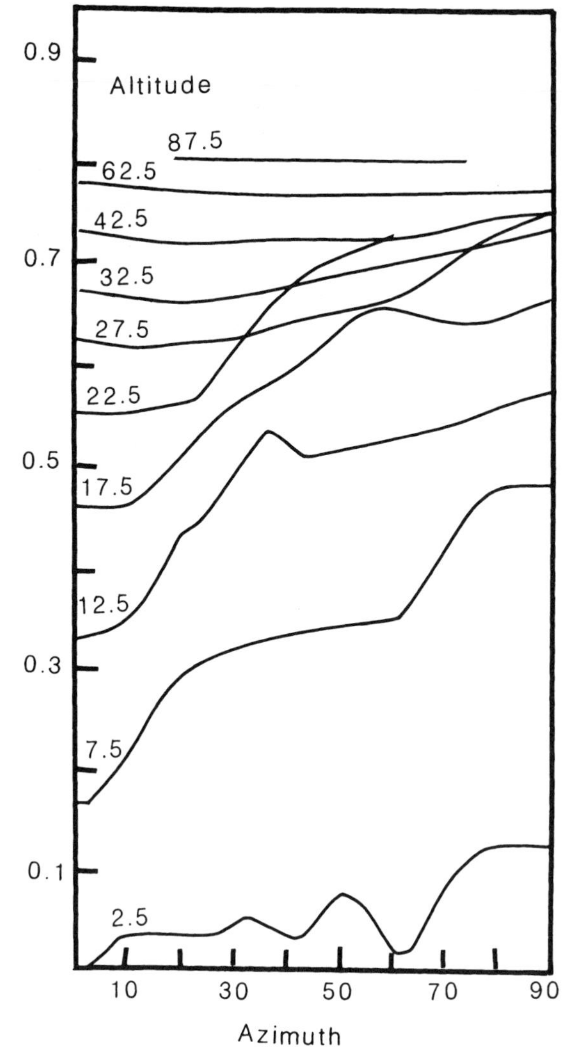

3-32. Total transmissivity of a standard Venlo multispan with roofslope of 22° at 52°N latitude under diffuse sky conditions where α = azimuth and ν = altitude of a designated portion of the diffusive sky [Bot, 1983] (By permission of the author).

be quite low. The greenhouse efficiency as for energy conversion will be correspondingly low. During initial stages of growth, or in the first year of production, with carnations, higher planting densities are likely to provide an advantage in higher yields per unit area since the production area will be fully covered with vegetation in a shorter time. On the other hand, once the crop fully occupies the space to which it is restricted, plant density is unimportant [Hanan and Heins, 1975]. In a fully dense, 1 year-old carnation crop, where height of the crop is 120 cm, shortwave radiation gradually decreases until, 12 cm in from the outside edge and 50 cm above the soil level, energy flux is practically zero [Hall and Hanan, 1976]. Production by plants in inside rows of a bench will be less than the outside rows. With some species, the grower cannot afford to

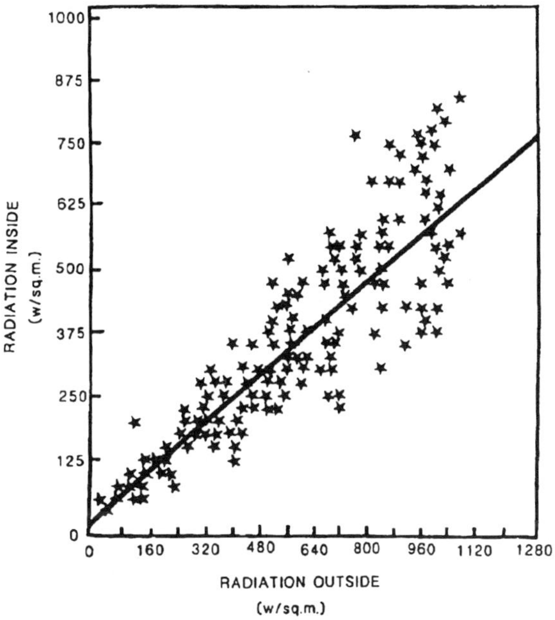

Fig. 3-33. Absolute global radiation inside 12 N-S and 1 E-W commercial greenhouses at plant level in Colorado as a function of outside radiation. Each point represents average irradiance measured every ten minutes from 1000 through 1600 hours. See Fig. 3-34 for covering types [Hanan, 1989]. Note the variability of points at high outside irradiance, resulting from shading and reflection on point sensors at crop height.

maximize LAI since too dense spacing may have undesirable effects on quality (e.g., potted chrysanthemums, poinsettias, etc.).

All greenhouse growers with any experience are familiar with the variation in growth that can occur inside structures. In a good exposition of the possibilities, Moreshet et al. [1976] examined spatial variation of roses in N-S, glass covered greenhouses, finding a significant difference along the E-W axis with a minimum at the center of the house (Figs. 3-39, 3-40). Similar variations may

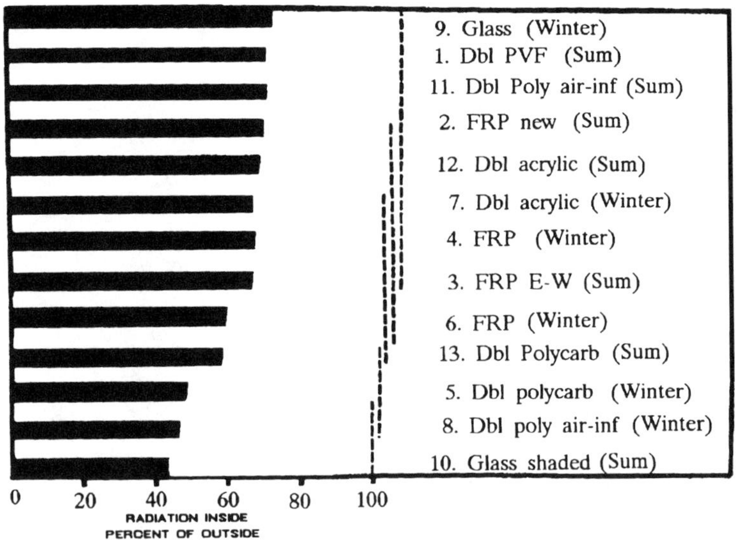

Fig. 3-34. Transmissivity of 12 N-S and 1 E-W commercial greenhouses at plant level in Colorado. Numbers 1, 4 through 8 under winter conditions; 2, 3, and, 9 through 13 in the summer. Vertical lines indicate significance at 0.05 [Hanan, 1989].

be expected with any crop grown in single spans of limited length (15 x 12.5 m in this case). As noted in Section 3.IV.B, for gutter-connected, multispan houses, the length must exceed eight span widths, with the number of spans more than three, before uniform irradiance conditions can be expected over a significant area at the center of the structure. Growth will be greater at the edges of the greenhouse unless some type of insulation is installed on the side walls (i.e., double cover).

All three factors of reflectivity, absorptance, and transmissivity of green foliage must be considered when dealing with radiant energy exchanges, especially as all three are extremely wavelength sensitive. Note, in Fig. 3-41, that greatest absorption of radiation occurs in the visible region from 300 to 700 nm. Reflectivity and transmission are greatest between 700 and 1300 nm, the region of highest solar energy intensity. Although leaves appear to us to be green, if we could "see" beyond 700 nm, the color below and within the canopy would appear dark and "red." Most higher plants exhibit this spectral behavior [Smith, 1986; Gates et al.., 1965]. Vegetation absorbs red but transmits far-red, which affects the R/FR ratio within vegetative canopies, and, consequently, the behavior of plants. An example of spectral distribution below plant canopies was given in Fig. 3-9. Another example for greenhouse conditions is provided in Fig. 3-42. Tuller and Peterson [1988] discuss the significance of their results regarding *Botrytis cinerea* infection on conifer seedlings.

The penetration of energy into the crop canopy varies with the total leaf area, arrangement and type of leaves, height of the canopy and its arrangement, and whether the radiation is diffuse or direct. There are many models to be found in the literature [e.g., Spitters, 1986; Stoffers, 1975; de Wit, 1965; Acock et al., 1978, etc.]. Assimilation (photosynthesis) is often based on calculating the assimilation of individual leaves, or clusters, or at specific levels in the canopy, which implies that energy distribution within the canopy is calculated. Since photosynthesis responds nonlinearly to radiation, the average value of direct radiation within the canopy is not suitable for calculation [Nederhoff, 1984]. A single beam of sunlight can greatly affect radiation level, depending upon its direction, leaf arrangement, and effects of interleaf shading. Under diffuse radiation, when energy is coming from all directions, an average value is acceptable. As mentioned earlier for carnations, energy gradients will be present, vertically and horizontally. In his summation of energy penetration of canopies, Critten [1993] said that all models calculate the average energy for diffuse radiation, with the direct radiation added as appropriate. He cites Nederhoff [1984] as showing that 80% of the leaves in a cucumber crop were in shade at only 25 cm below the top of the canopy. Results from Nederhoff and Papadopoulos and Ormrod [1988a; b]

Fig. 3-35. Proportion of direct (**upper**) and diffuse (**lower**) global radiation as a percentage of total inside radiation in 12 N-S and 1 E-W commercial greenhouses in Colorado. FRP in all cases was corrugated. See Fig. 3-34 for details. Values encompassed by vertical lines are not significantly different from each other [Hanan, 1989].

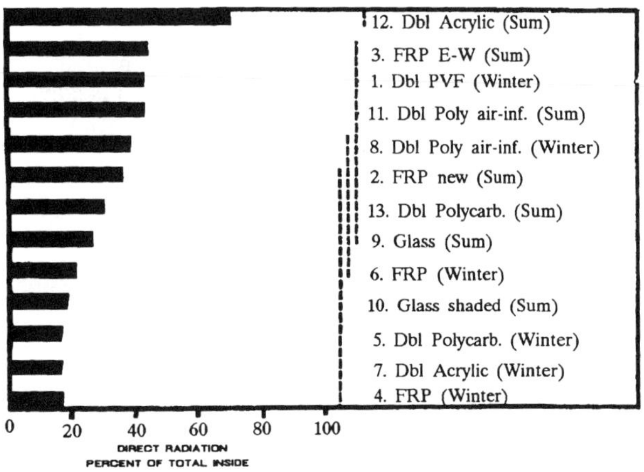

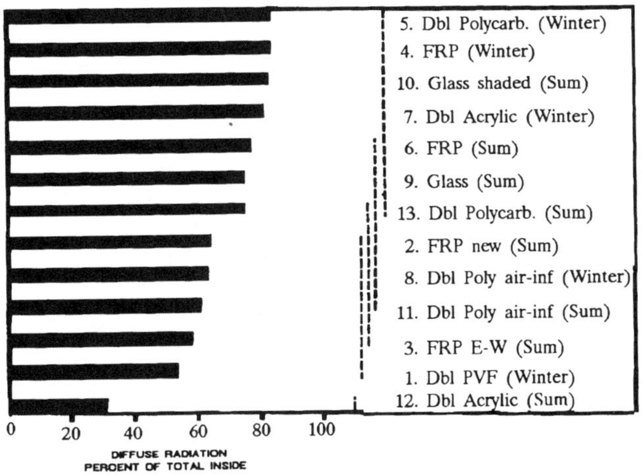

Fig. 3-36. Curve of solar radiation for 1 year in Colorado, calculated in terms of daily water loss for carnations under glass [Hanan, 1969]. Latent heat of vaporization = 2.47×10^6 J kg^{-1}.

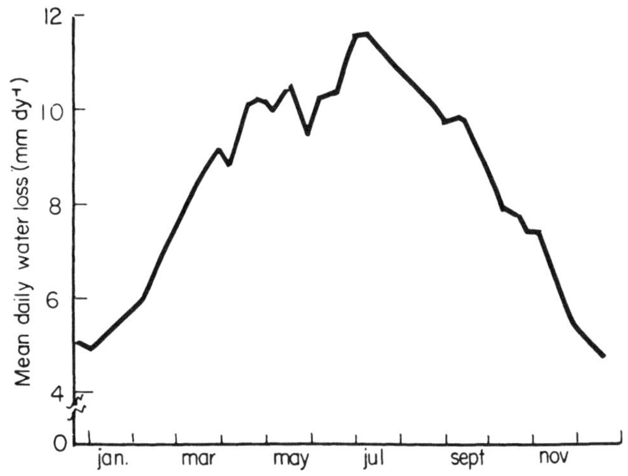

(tomatoes) showed that depth within the crop had the greatest influence on energy penetration. Whatever the initial row spacing, leaves will spread to intercept all available irradiance –given enough time. Probably the most common formula for determining energy penetration is some type of extinction coefficient or modification of Beer's law:

$$I = \frac{I_o K e^{-KL}}{1 - m} \tag{3.12}$$

where: I = *irradiance on the leaf surface (W m*$^{-2}$*)*
I_o = *irradiance at the top of the canopy (W m*$^{-2}$*)*
K = *canopy extinction coefficient*
L = *Canopy leaf area index*
e = *natural logarithm*
m = *leaf transmission coefficient*

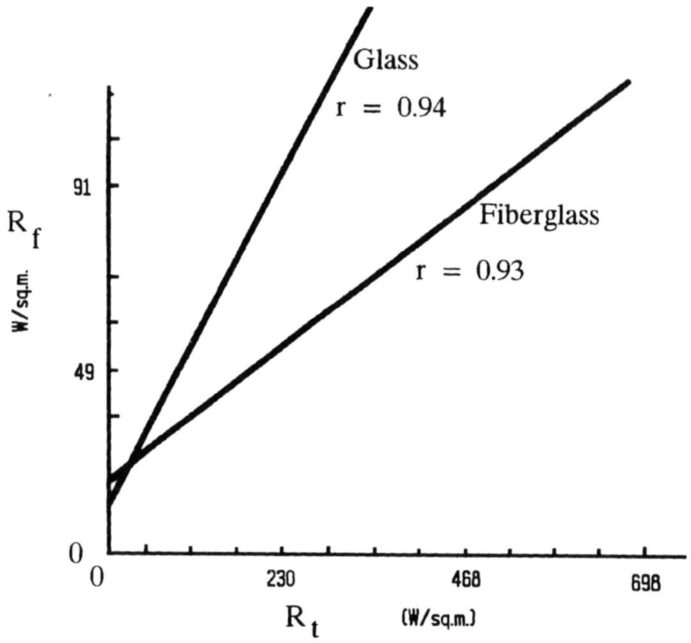

Fig. 3-37. Relationship between reflected radiation (R_f) from carnations and total short wave radiation (R_t) in glass and FRP houses [Hanan, 1970]. Data obtained with Eppley pyranometers, one upside-down above crop. Note that the lines intersect the y-axis where R_t = 0, perhaps indicative of outgoing thermal radiation from plants and heating system.

B. PHOTOSYNTHESIS AND RADIATION

1. General Relationships

The photosynthetic system in plants represents an energy gathering process for the purposes of converting CO_2 and water into a useful plant product. The photochemical system uses quantum energy on the order of 159 KJ to accomplish an ordered chemical reaction at room temperature with a high degree of efficiency. The thermal reactions in photosynthesis rarely involve energy changes higher than 42 to 63 KJ. To manipulate a package of energy two or three times this size without damage to the apparatus in a highly directed and specific way is a formidable accomplishment [Calvin and Androes, 1962].

Naturally, we would like to maximize photosynthesis by growing plants properly, such as supplementing deficient solar radiation and injecting CO_2 when possible. One should have, first, an appreciation of basic

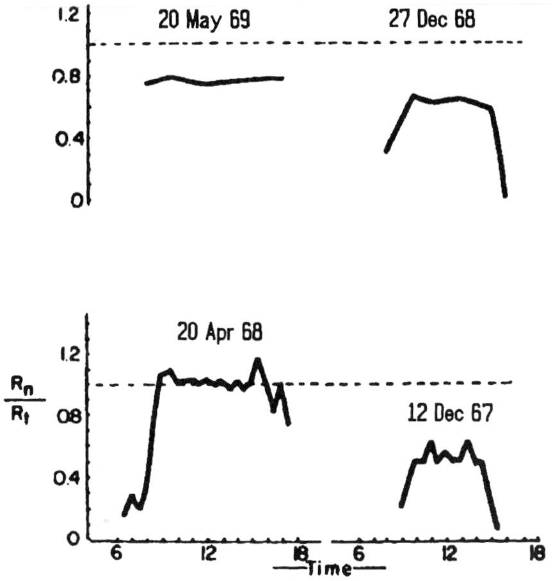

Fig. 3-38. Ratio of net (R_n) and total short-wave global radiation (R_t) above carnations in glass (**lower**) and FRP (**upper**) greenhouses for selected days [Hanan, 1970]. $\mathbf{R_n} > \mathbf{R_t}$ on April 20, suggesting, perhaps, additional thermal radiation unmeasured by the Eppley pyranometer (300-3000 nm) as contrasted to the linear, polyethylene-covered net radiometer (300-50000 nm). In general, however, total global radiation is much greater than net radiation, particularly during the cold season of December.

capabilities. Nearly all species grown commercially in greenhouses are known as C_3 plants with a few having crassulacean acid metabolism (CAM). C_3 and CAM plants are characterized, as noted in Table 3-6, by relatively low rates of CO_2 uptake at saturating irradiances, as contrasted to C_4 species that may have photosynthetic rates double that of C_3s or CAMs. Furthermore, C_3s and CAMs have high CO_2 compensation points. That is, the external CO_2 concentration that exactly equals the internal CO_2 level provided by the respiratory process. The high compensation point results from the fact that such plants use a different respiratory process in light as contrasted to the basic method that proceeds regardless of light or dark. Photorespiration is accelerated by normal oxygen concentrations (21%), low CO_2 levels, and high temperatures. In C_4 species, on the other hand, there is little loss of CO_2 due to photorespiration. It is the principal reason C_4s show higher photosynthetic rates (CO_2 uptake) at maximum irradiances. C_4 species also have a different anatomy, using different chemical pathways.

CAM species, frequently called succulents, usually open their stomata and fix CO_2 at night. Stomata may be closed during the day. CAM is favored by desert climatic conditions of hot days, high irradiance, cool nights, and dry soils. CAM species may be found in the families *Cactaceae, Orchidaceae, Bromeliaceae, Liliaceae, Euphorbiaceae,* and *Crassulaceae* [Kluge and Ting, 1978; Szarek and Ting, 1977], which include many commercial greenhouse species.

C_4 species are most often found among the grasses and sedges, including millet, sugar cane, sorghum, and maize [Downtown, 1975]. Krenzer et al. [1975] provided a listing of C_4 and C_3 species. Of the 285000 species of flowering plants, the presence of the C_4 pathway is found in roughly 0.4% of those investigated, but includes several important economic species as mentioned [Salisbury and Ross, 1985].

The photoreceptive pigments in plants can be thought of as machinery to harvest quanta. The number of quanta necessary, to fix 1 molecule of CO_2, use two molecules of water, and release 1 oxygen molecule (quantum yield), was determined by McCree [1971] over the PAR wavelength band. Fig. 3-43 illustrates that the entire 400 to 700 nm band contributes significantly with a maximal quantum yield in the red. The fact that the major pigments in the system (chlorophylls a and b) transmit in the green (Fig. 3-44), thereby making leaves appear green, disguises the contributions of other pigment molecules in transferring energy. The action spectrum in Fig. 3-44 may be compared with the quantum distribution of solar radiation on a clear day (Fig. 3-45). Under maximum efficiency, where any molecule can absorb only 1 quantum at a time, and this quantum causes excitation of only 1 electron; a minimum eight quanta are required to oxidize two water molecules and release four electrons for further work. Since there are two photosystems that work synergistically in higher plants, two quanta

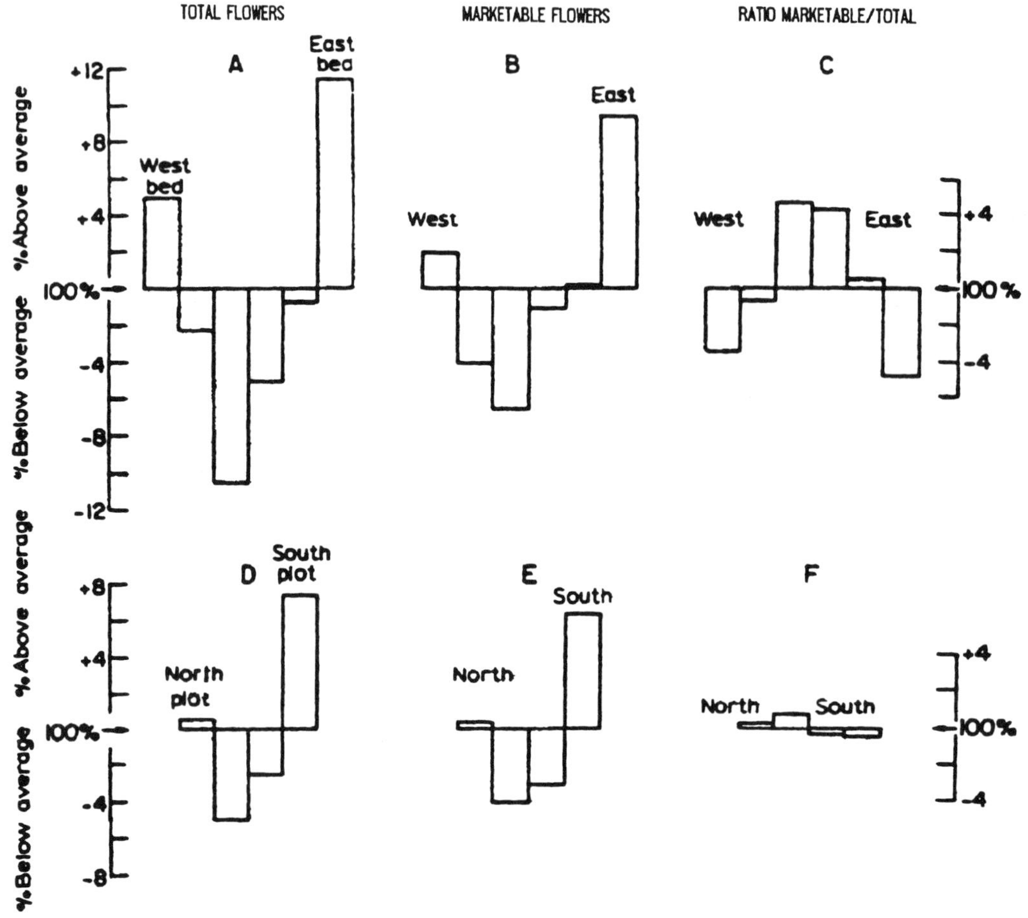

Fig. 3-39. Distribution of rose flowers along the E-W and N-S axes of 15 x 12.5 m glasshouses, averaged over the winter season. Data expressed as percent above or below the mean number of flowers over the year. [Reprinted from *Scientia Horticulturae*, Moreshet, S. et al., 1976. 5:269-276. With kind permission of Elsevier Sci.-NL, Sara Burgerhartstraat 25, 1055 KV Amsterdam, The Netherlands].

are needed initially. For leaves of many plants, 15 to 20 quanta are required, although, under ideal conditions, only 12 may be necessary [Ehleringer and Björkman, 1977]. Although the quanta in the blue wavelengths contain more energy, much of this degrades to heat before contributing to photosynthesis. Radiation within the leaf is also reflected and scattered so that a quantum may contribute energy throughout the PAR band. In fact, under appropriate conditions, energy can be several times as concentrated within a leaf as compared to that incident on the external surface. The leaf is a highly efficient quantum trap [Volgemann, 1986].

The question arises, how many quanta are available in the photosynthetically active range? For Leicester, U.K., an example is given in Fig. 3-45. The quantum irradiance and relative proportions have been put in a table (Table 3-7) by Frankland [1986]. At 52°N, according to Frankland, this corresponds to an irradiance of 200 W m^{-2} in the winter to 800 W m^{-2} in the summer. These irradiances represent total intensity between 300 and 3000 nm, so the proportion in PAR (400-700 nm) must be determined by multiplying by 42.9 % (proportion of total in PAR) and then by 4.57 (Table 3-3), or 1568 $\mu mol\ m^{-2}s^{-1}$ maximum. Frankland gives a range of 450 to 1800 $\mu mol\ m^{-2}s^{-1}$. Thimijan and Heins [1983] suggest a maximum 24-hr average for clear sky to be 552 W m^{-2} or 1082

Fig. 3-40. Seasonal iso-yield lines of total number of rose flowers per m^2 for four glasshouses each 15 x 12.5 m. Dashed lines are interpolation of yield in the aisles. [Reprinted from *Scientia Horticulturae,* Moreshet, S. et al., 1976. 5:269-276. With kind permission of Elsevier Sci.-NL, Sara Burgerhartstraat 25, 1055 KV Amsterdam, The Netherlands].

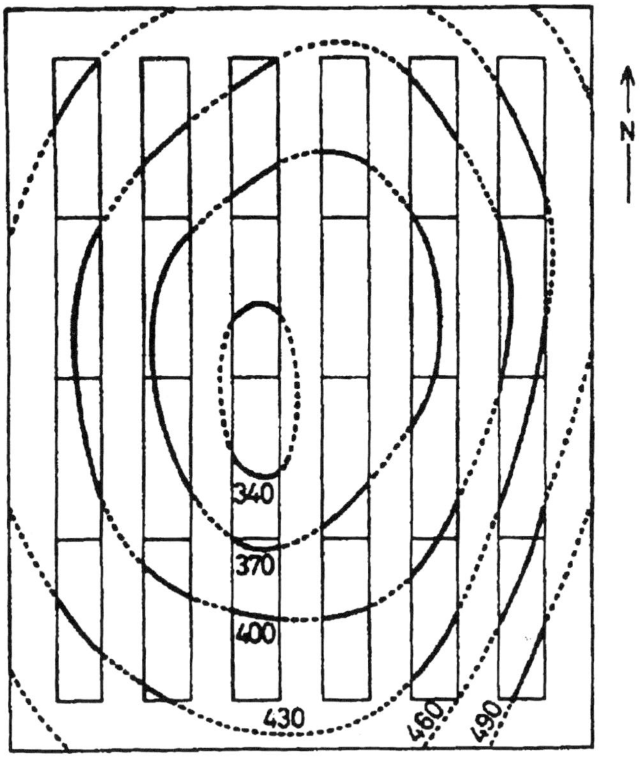

μmol $m^{-2}s^{-1}$. Salisbury and Ross [1985] suggest that, at sea level, a maximum of 900 W m^{-2} reach plants outside, with about half of this in the infrared and 5% in the UV, leaving about 400 W m^{-2} available in the 400 to 700 nm range. Under conditions of high altitude, semiarid conditions, and partially cloudy conditions (i.e., Colorado), total shortwave irradiance (300-3000 nm) can exceed 1000 W m^{-2}, or approximately 1960 μmol $m^{-2}s^{-1}$.

Unfortunately, greenhouses, on the average, transmit only 60% of the figures mentioned above (Fig. 3-33); so that, under ideal conditions of clear skies in England, the maximum quantum irradiance is

Fig. 3-41. Spectral reflectivity and transmissivity of rose leaves. Dashed line = a young red leaf; Dotted line = young green leaf; solid line = mature, dark green leaf. The difference between reflection and transmission is absorptance. Area under the curves is proportional to the flux density. This pattern is typical of most healthy green leaves of higher plants [See Gates.1980]. [Reprinted from *Agric. and Forest Meteor.*, Stanhill, G. et al.., 1973. 11:385-404. With kind permission of Elsevier Sci.-NL, Sara Burgerhartstraat 25, 1055 KV Amsterdam, The Netherlands].

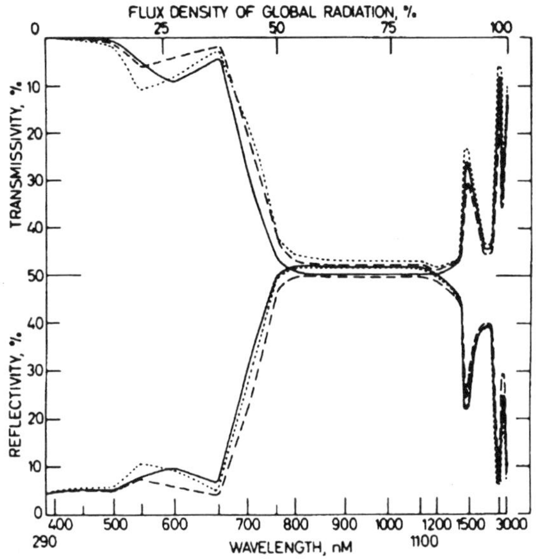

likely to be 1000 to 1100 μmol $m^{-2}s^{-1}$ –usually much less due to shading to control temperature. Earlier discussion in this chapter (Figs. 3-14 and 3-37) suggested that transmission efficiency declines more rapidly at outside global radiation levels below 200 W m^{-2}.

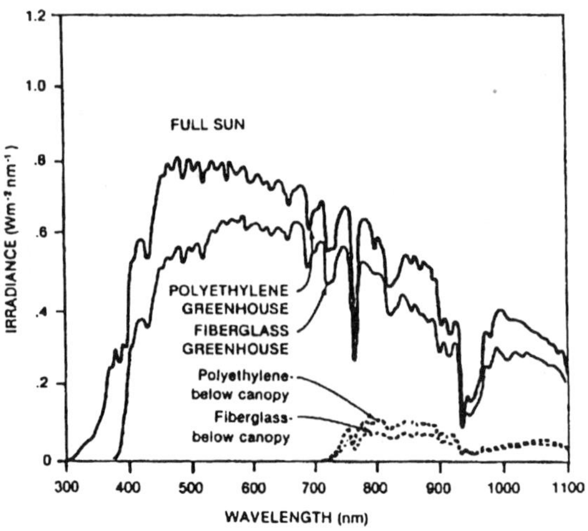

Fig. 3-42. Spectral distribution of solar irradiance above and below a conifer seedling canopy inside PE and FRP covered greenhouses on a clear day. Note the far IR below the canopy. To humans, it would appear as though there was no light. [Reprinted from *Agric. and Forest Meteor.*, Tuller, S.E. and M.J. Peterson. 1988. 44:49-65. With kind permission of Elsevier Sci.-NL, Sara Burgerhartstraat 25, 1055 KV Amsterdam, The Netherlands].

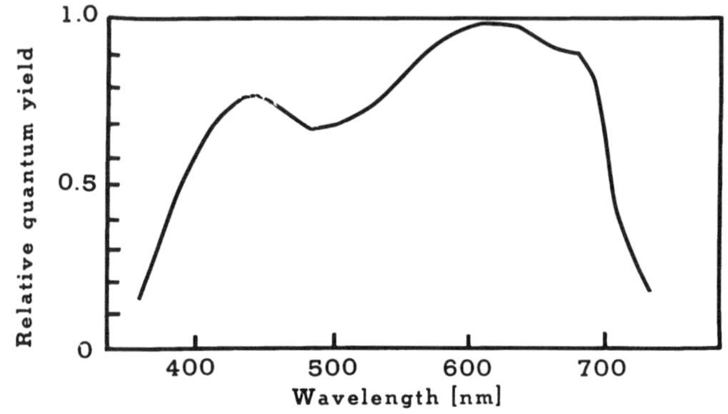

Fig. 3-43. Average quantum yield for 22 species of crop plants over the wavelength band 350 to 750 nm as determined by McCree. [Reprinted from *Agric. and Forest Meteor.*, McCree, K.J. 1972. 9:191-216. With kind permission of Elsevier Sci.-NL, Sara Burgerhartstraat 25, 1055 KV Amsterdam, The Netherlands]

Table 3-6. Some general characteristics of C_3 and C_4 plants [Goudriaan and Atjay, 1979].

	C_3	C_4
CO_2 assimilation rate in high irradiance	0.013-0.025 mmol $m^{-2}s^{-1}$	0.025-0.044 mmol $m^{-2}s^{-1}$
Optimum temperatures	20-25 C	30-35 C
CO_2 compensation point in high irradiance	50 $\mu l\ \ell^{-1}$	10 $\mu l\ \ell^{-1}$
Photorespiration	Present	Not present

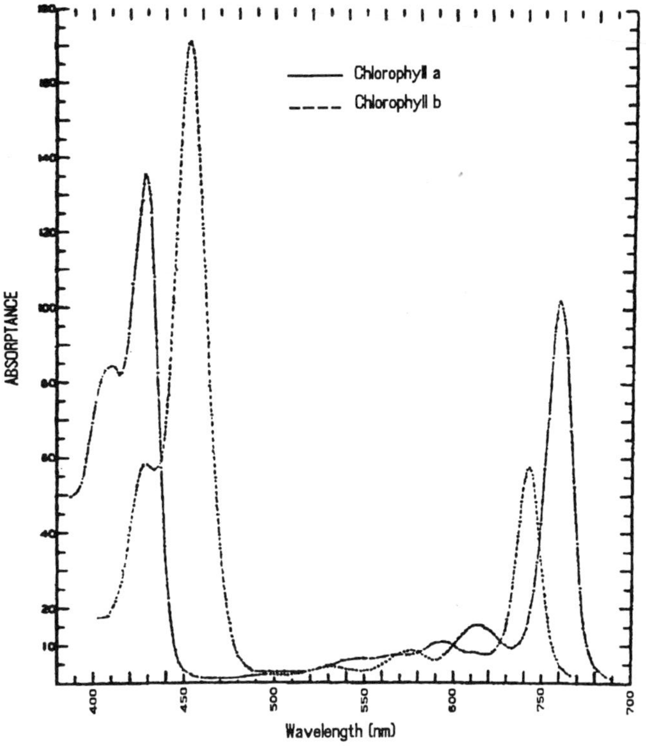

Fig. 3-44. Absorption spectra of chlorophylls a and b in ether [Comar and Zscheile, 1942] (By permission of the Amer. Soc. for Plant Physiol.).

That is, transmissivity varies with climatic conditions; 200 corresponds to 235 μmol $m^{-2}s^{-1}$ PAR inside the greenhouse. Heavily overcast days in Colorado often have R_os less than 60 W m^{-2} outside, or about 70 μmol $m^{-2}s^{-1}$ PAR. Some 30 years of observation suggest that, no matter the location, radiation will always be deficient in greenhouses during January and February in the Northern hemisphere for species such as roses, carnations, cucumbers, etc. Of course, low energy requiring plants such as African violets and other indoor decorative plants, are not influenced as strongly by low irradiances in the winter. Growers encountering much cloudy weather find that supplemental irradiation can provide a significant improvement in growth, especially in meeting early market requirements.

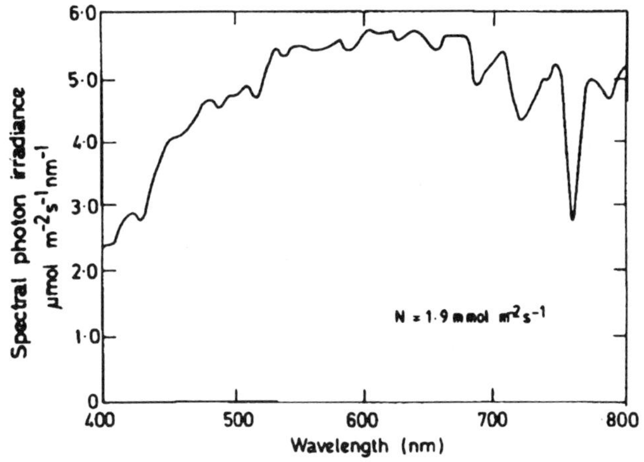

Fig. 3-45. Spectral quantum distribution of daylight on a clear day in Leicester, U.K. Plot remains qualitatively similar on cloudy days when total irradiance (N) may be reduced 10 times. [Reprinted from *Photomorphogenesis in Plants*, Smith, H. 1986. With kind permission of Kluwer Academic Publ.].

Another problem in dealing with photosynthesis is that the process is most efficient at minimum irradiance. As intensity increases, photosynthetic efficiency declines rapidly (Fig. 3-46) for a single leaf. While the curves in Fig. 3-46 deal with sugar beets, the same general relationship holds for most crop plants, as indicated in Figs. 3-47 through 3-49, which show photosynthetic response curves for cucumber, tomato, and rose respectively. One will note that authors employ a variety of units for CO_2 uptake, which I have converted to moles, assuming standard temperatures and pressures. Cucumber responds with a maximum rate of about 0.030 mmol $m^{-2}s^{-1}$

Fig. 3-46. Photosynthetic rate of a sugar beet leaf under a mercury lamp and efficiency of solar energy conversion at 30 Pa (300 μl ℓ⁻¹) CO_2 [Gaastra, 1958].

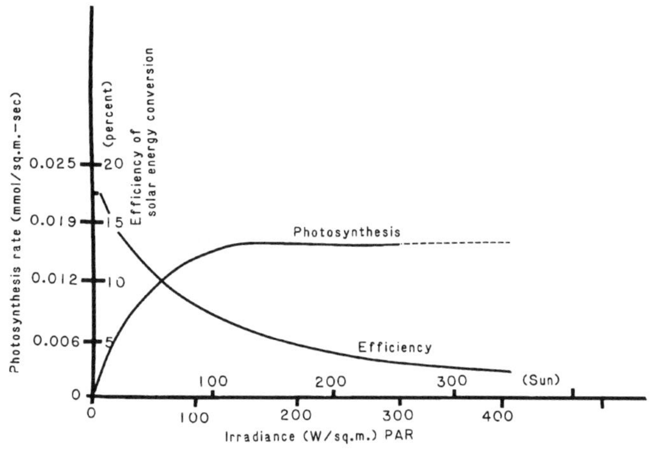

at high CO_2 concentrations (Fig. 3-47), but CO_2 uptake falls precipitately at irradiances much below 100 W m^{-2} PAR, which corresponds, roughly, to 397 μmol m^{-2}s^{-1}. Fig. 3-48 shows the common effect of previous history on potential photosynthetic rate for tomato. That is, plants acclimatized to high irradiances will do better when subjected to high intensity than those acclimatized to low intensities. Depending upon the species,

Fig. 3-47. Photosynthesis of a cucumber leaf in relation to irradiance and temperature at two CO_2 levels, under incandescent lights [Gaastra, 1962].

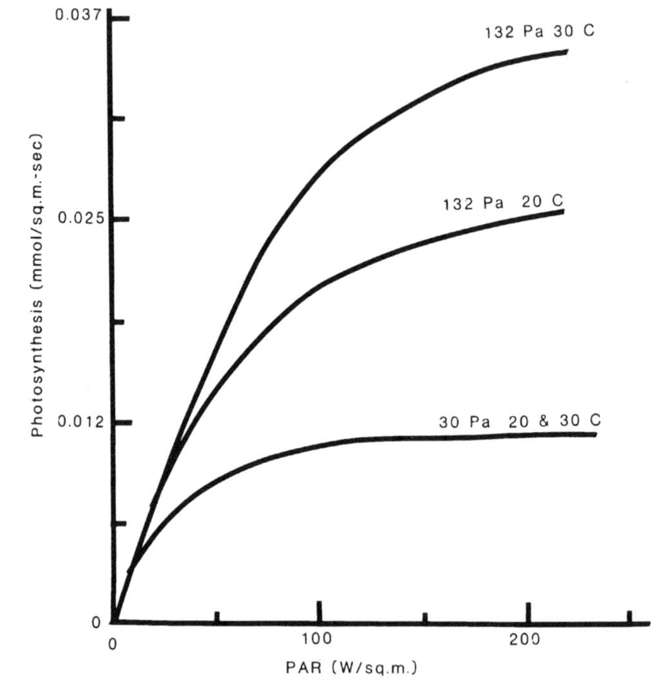

photosynthetic saturation for single leaves will occur in the region of 300 to 500 μmol m^{-2}s^{-1} at ordinary CO_2 concentrations. Bozarth et al.'s data in Fig. 3-49, however, showed saturation being approached at quantum fluxes around 500 μmol m^{-2}s^{-1} at normal CO_2 levels. CO_2 uptake appeared to be maximum at about 0.010 mmol m^{-2}s^{-1} as contrasted to a maximum of about 0.0045 found by Aiken and Hanan [1975]. The usual situations under greenhouse conditions are more likely, as shown in Fig. 3-50. Other factors that influence the rate include disease, nutrition, water supply, leaf age, etc. For the commercially important rose, Fig. 3-51 suggests the variations that can occur with age of the leaf.

One can summarize this information by pointing out that maximum photosynthetic rates will occur with higher irradiances and higher CO_2 levels, maximal water supply, optimal nutrition, and high temperatures –within reason. The main point is, however, the benefit of such factors as CO_2 fertilization will be determined primarily by available radiant energy. Furthermore, raising temperatures will frequently achieve adverse results even at high CO_2 levels if radiant energy is lacking.

Irradiances for maximum photosynthesis will be higher for a vegetative canopy compared to a single leaf. Although the upper leaves may be saturated, attenuation due to shading will mean that lower foliage will not

Table 3-7: Irradiance quantity and quality for some representative situations in the natural environment for a clear day. Leaf shade values for an LAI of 4. [Adopted from Frankland, 1986].

	Quantum irradiance µmol $m^{-2}s^{-1}$ 400-800 nm	Proportion of irradiance in the wavebands				
		Blue (400-500 nm)	Green (500-600 nm)	Red (600-700 nm)	Far-red (700-800 nm)	R/FR ratio
Direct sunlight	1700	0.23	0.26	0.26	0.25	1.04
Below leaf shade	60	0.04	0.15	0.11	0.70	0.16
Below 5 mm soil	0.002	0.01	0.05	0.17	0.76	0.22
Below 1 mm clear water	700	0.30	0.39	0.36	0.05	7.2

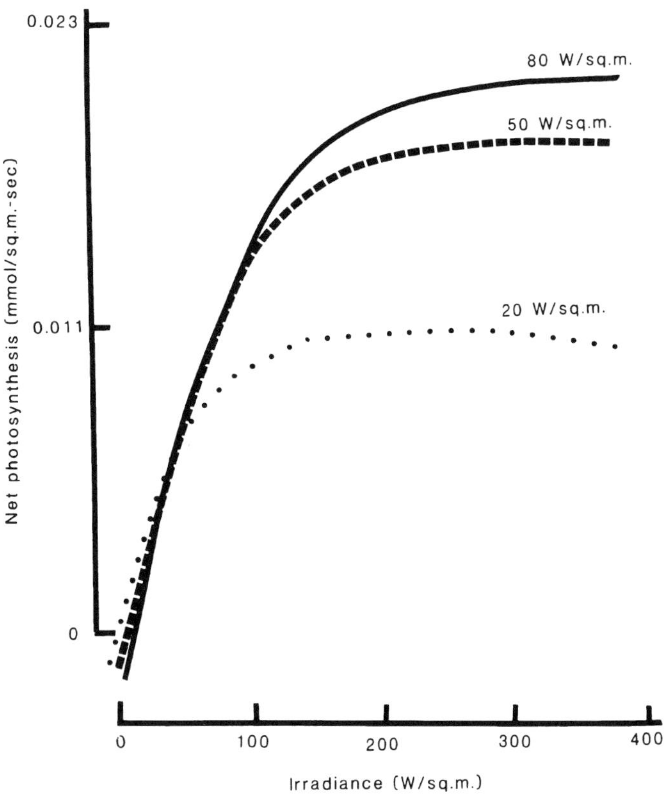

Fig. 3-48. Photosynthetic response curves for tomato leaves adapted to 3 intensity regimes of 20, 50, and 80 W m^{-2} for 16 hr at 20 C [Aikman, 1989].

reach maximum photosynthesis except at higher irradiances. Foliage area of a crop is generally shown by the leaf area index (LAI), or the ratio of total leaf area to the ground area covered. As Fig. 3-52 shows, increasing foliage area as the crop grows will result in higher total photosynthetic rates up to an LAI of about 4. LAIs as high as 8 are common, but one can usually expect the crop to fill the available ground area completely at an LAI of 4. The redistribution of energy proposed by Aikman [1989], by means of reflectors, is one possibility to improve energy distribution in a crop canopy. Attempts have been made to alter leaf arrangement of a canopy to improve energy distribution [e.g., Shannon et al., 1986], but this type of genetic alteration has not been found remarkable. Another attempt to get around the problem typified by Fig. 3-47 is to use pulsed irradiance [Kleuter et al., 1980]. By pulsing the radiation, time is provided for the chemical reactions in photosynthesis to reduce concentrations because of the quantum reactions. With a treatment of 10 Hz and 25% duty cycle, Kleuter et al. found higher apparent photosynthesis for cucumbers compared to continuous irradiation. The idea is to avoid saturation of the photochemical system. The problem in pulsing fluorescent lamps is to achieve a true square sine wave of radiation –i.e., the irradiance was not the same throughout each pulse.

Fig. 3-49. Effect of irradiance on photosynthesis in a mature rose leaf at 31 Pa (312 μl ℓ$^{-1}$) CO_2 [Bozarth et al., 1982] (By permission of the Amer. Soc. Hort. Sci.].

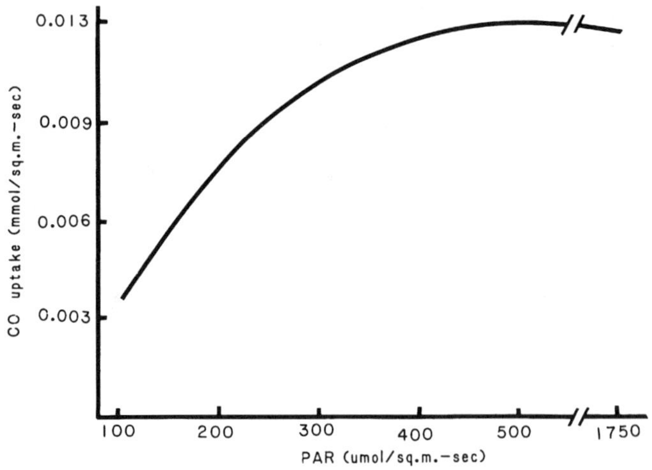

Fig. 3-50. Effect of PAR at four temperatures on predicted net photosynthesis of roses at 86 Pa (1013 μl ℓ$^{-1}$) CO_2. $R^2 = 0.88$ [Coker and Hanan, 1988].

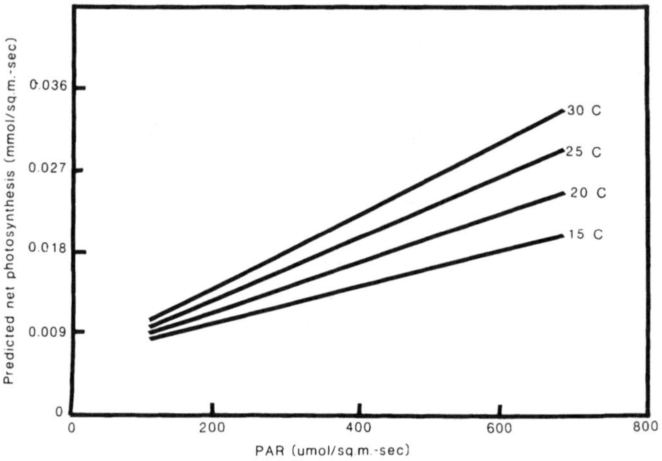

Fig. 3-51. $^{14}CO_2$ uptake of rose leaves as a function of leaf age of the first 5-leaflet leaf from the base of the flowering stem [Aikin and Hanan, 1975]. Age based upon days from loss of red color and upon days from cut of previous flower.

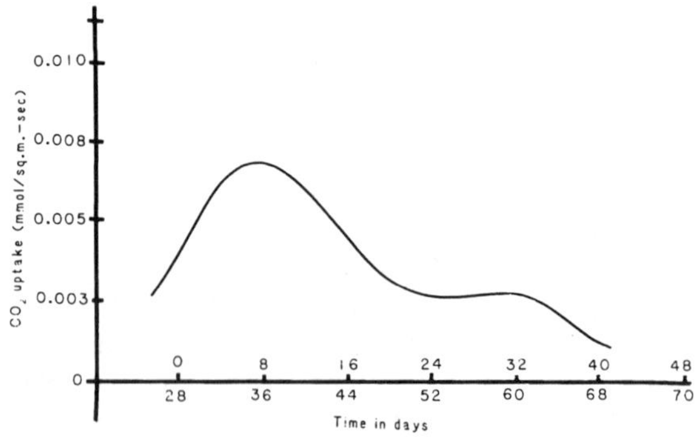

A second factor to consider is that total photosynthesis depends on intensity and time. The longer the interval, regardless of intensity, the more CO_2 reduced and the more carbohydrates manufactured. Thus, many authors [e.g., Hughes and Cockshull, 1971; Bruggink and Heuvelink, 1987; Charles-Edwards, 1978; Bierhuizen et al., 1984; Nilwik, 1980; 1981a; 1981b; Challa and Schapendonk, 1984] show total irradiance, in terms of megaJoules (MJ) per unit area for the day (MJ m^{-2} dy^{-1}), and the total photosynthesis as net assimilation rate (NAR), or the net rate of dry matter accumulation per unit leaf area for the plant or crop (g $m^{-2}dy^{-1}$). Many studies are carried out in growth chambers. The authors may explicitly state the irradiance as PAR. Even so, this results in confusion since the same units are also employed over the entire electromagnetic spectrum (300-50000 nm). It is quite easy for the reader to miss the distinction between radiation in the PAR waveband and irradiance over the entire global shortwave waveband –particularly as

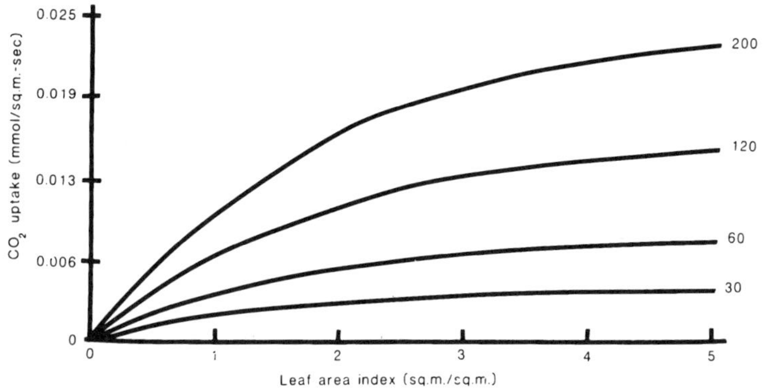

Fig.3-52: Relationship between gross photosynthesis (P) and leaf area index (LAI) at different irradiances (W m^{-2} PAR) [Challa and Schapendonk, 1984] (© 1984, Int. Soc. Hort. Sci., *Acta Hort.* 148:501-509).

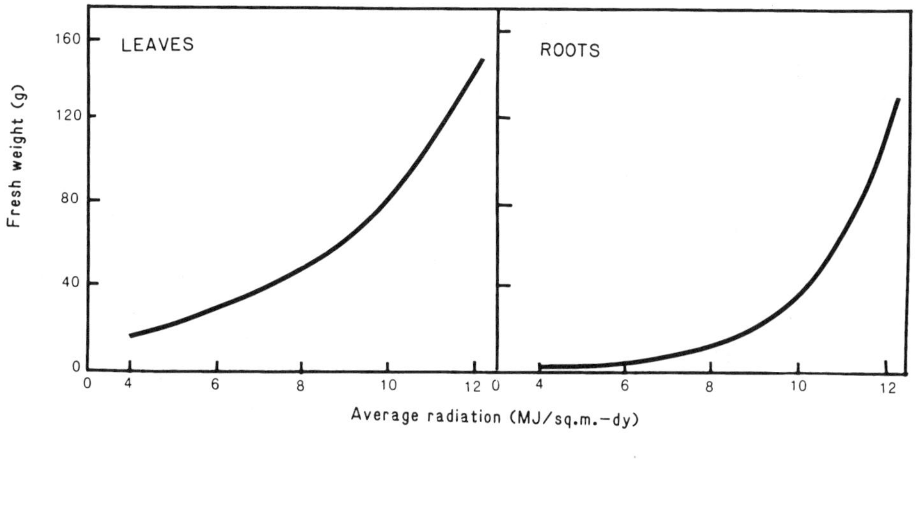

Fig. 3-53. Turnip leaf fresh weight (**left**) and root weight (**right**) after 30 days at 22/18 C day/night temperatures as a function of daily irradiance [Downs, 1985] (By permission of the Amer. Soc. Hort. Sci.).

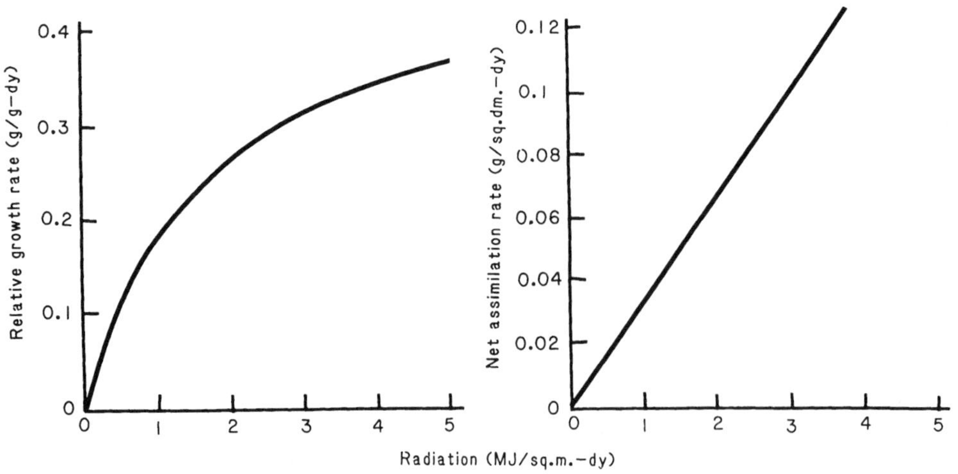

Fig. 3-54. Relative growth rate (RGR) and net assimilation rate (NAR) of cucumber plants as a function of daily PAR irradiance [Challa and Schapendonk, 1984] (© 1984, Int. Soc. Hort. Sci., *Acta Hort.* 148:501-509).

engineers and meteorologists use instruments that commonly measure irradiance between 300 and 3000 nm. Furthermore, if conversion to quantum flux is needed, the author must have clearly stated the irradiating source. For comparison purposes, the reader must use Tables 3-3 and 3-4 to do his own conversion.

Table 3-8: Abbreviations for lamps used for supplemental irradiation in greenhouses.

Lamp type	Abbreviation	Wattages
Incandescent, tungsten filament	GLS, INC	100-1000
Fluorescent	MCF, MCFR	40-200
Warm white	FWW	
Cool white	FCW	
High-intensity discharge (HID)		
Clear mercury, mercury deluxe	HG, MB/U, MBFR/U	400
Metal halide	MH, MBI	400-1000
High-pressure sodium	HPS, SON/T	400-1000
Low-pressure sodium	LPS, SOX	400-1000

For whole plants, one can find maximum photosynthetic rates varying from as little as 0.004 mmoles $m^{-2}s^{-1}$ for a pot plant specie such as *Vriesea splendens* [Bierhuizen et al., 1984] to 0.014 mmoles m^{-2} s^{-1} for chrysanthemum at 10 C and 5.4 MJ m^{-2} daily radiation integral [Acock et al., 1979]. Maximum NARs of 12.8 g $m^{-2}dy^{-1}$ were obtained by Bruggink and Heuvelink [1987] for sweet pepper, 9.8 for cucumbers, and 9.6 g $m^{-2}dy^{-1}$ for tomato. Challa [1976], for cucumbers, obtained a maximum photosynthesis rate of 0.006 mmol $m^{-2}s^{-1}$. Quite extensive investigations have been carried out by Nilwik [1980; 1981a; 1981b] for sweet peppers and on the chrysanthemum by the British researchers such as Acock et al. [1977; 1979] and Hughes and Cockshull [1971; 1972].

While there are significant genotypic differences in maximum photosynthetic rates (P_{max}), the major source of variation is attributable to differences in leaf thickness [Charles-Edwards, 1978]. Within C_3 species, according to Charles-Edwards, there are no significant differences in the ratio of photorespiration to gross photosynthesis, nor in the energy use of the leaf. While the photorespiratory loss is negligible in C_4 species, this is achieved at reduced energy utilization efficiency when compared to C_3 plants. Charles-Edwards states that the total dry matter

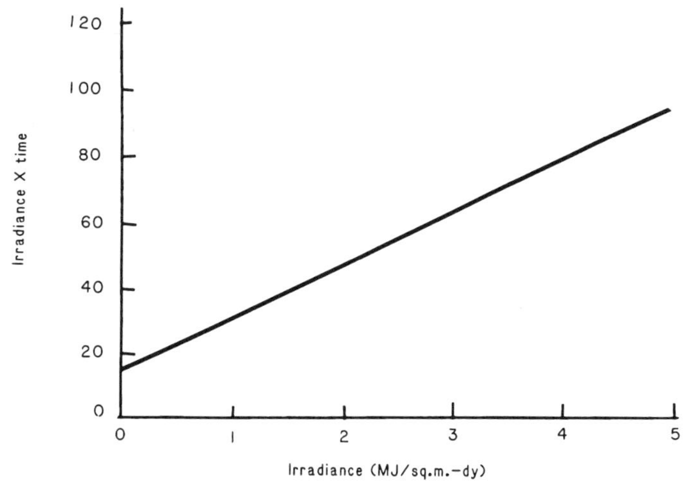

Fig. 3-55. The product of growing time (t) required to attain a reference weight and daily PAR radiation (I*t) as a function of daily irradiance (I) [Challa and Schapendonk, 1984] (© 1984, Int. Soc. Hort. Sci., *Acta Hort.* 148:501-509).

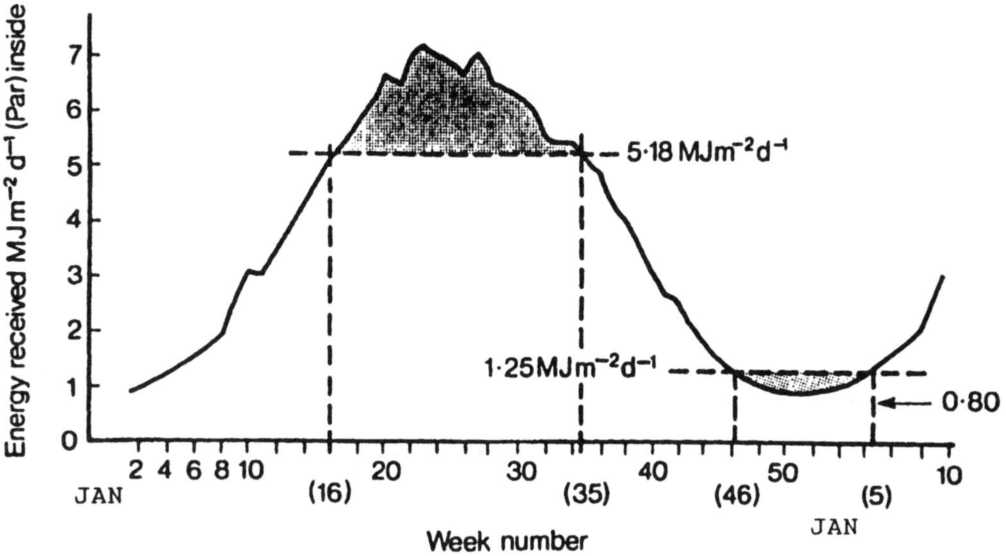

Fig. 3-56. Average daily solar radiation (PAR) inside greenhouses at Efford, U.K., 1961-81. The value 0.80 is the average irradiance (MJ $m^{-2}dy^{-1}$) at the darkest time of the year [Potter, 1987] (By permission of the British *Grower* magazine).

Table 3-9. Brief comparison of artificial irradiance sources [Bartok, 1988] (By permission of the *Greenhouse Manager*).

Irradiance source	Wattage	Ballast (Watts)	Total Watts	Average life (hr)	Initial lumens	Lumens per Watt
Incandescent	40	---	---	750	46	12
	100	---	---	to	1740	17
	200	---	---	1000	3940	20
Fluorescent CW	40	8	48	20000	3150	66
CW (cool white)	75	16	91	12000	6300	69
CW-HO (high output)	110	16	126	12000	9000	74
Metal halide	400	25	425	15000	40000	94
	1000	60	1060	10000	125000	118
High-pressure sodium	400	25	425	24000	50000	117
	1000	60	1060	24000	140000	132

production rates of C_4 plants in the U.K. are similar to those of C_3s. He concluded that there was no strong evidence that selection of plants with higher rates of P_{max}, or C_4 characteristics would lead to significant improvement in total dry matter production in the U.K. We are left with the problem of available solar energy and how we may supplement it.

2. Supplementing Solar Energy

The point has been made that yield will vary directly with available radiant energy. A typical example was provided by Downs [1985] who grew seedling turnips throughout the year under glass, finding that root growth was reduced 50% when the average irradiance per day decreased to 10.75 MJ m^{-2} (PAR) as compared to the maximum shown in Fig. 3-53, and leaf growth was reduced by half when the irradiance fell to 9.50 MJ $m^{-2}dy^{-1}$ (PAR). Challa and Schapendonk [1984] examined the effect of energy reduction on growth, distinguishing between widely spaced seedling cucumber growth and the production phase when crop photosynthesis is independent of the LAI. Obviously, the first phase of growth will markedly influence the second phase. The ratio between leaf area and plant weight influenced the saturation type curve for relative growth rate (RGR), whereas the net assimilation rate (NAR) was linearly related to daily irradiance (Fig. 3-54). The effect of radiation on growth strongly depends on the radiation intensity and is most pronounced at low irradiance levels. By multiplying the growing time required to obtain a given reference weight by the daily irradiation, Challa and Schapendonk obtained a straight line when plotted as a function of daily radiation (Fig. 3-55). According to Bruggink and Heuvelink [1987], NAR for young tomatoes, cucumber, and sweet peppers was at a maximum level when the daily light integral was about 4 MJ m^{-2}, and RGR was at about 3 MJ $m^{-2}dy^{-1}$. Other studies on sweet peppers by Nilwik [1981a; b] found highest leaf area and total dry weight at 1.67 MJ m^{-2} in a 16-hr irradiance period. Continuous radiation (24 hr) caused leaf drop. For chrysanthemums, studies have shown increasing plant weights up to daily energy levels of 3.75 MJ m^{-2} [Hughes and Cockshull, 1972]. Apparently, there is considerable variation on irradiance requirements among the various authorities, not to mention the difficulty in comparing units. I would also suspect that the results may be influenced by water relationships during the studies, which are seldom given much attention in investigations of this type.

To meet market demands, one may build more greenhouses –which merely increases the problem of surpluses during the high irradiance season, and it may not solve the problem of growing high energy-requiring crops in the winter– or, one can supplement solar energy with artificial sources. The problem faced by growers in dark climates is illustrated in Fig. 3-56. It is not simply installing lamps to meet desired irradiances since, in the summer season, supplemental irradiation is of no use. The situation is complicated by the variety of sources available (Tables 3-3, 3-8, 3-9), differences in spectral distribution, efficiency of these sources, and requirements of the species and grower objectives.

Cathey and Campbell reviewed supplemental systems in 1980. Table 3-8 illustrates the major types and their abbreviations. Tables 3-3 and 3-9 provide the characteristics of these lamps. One may note differences in the

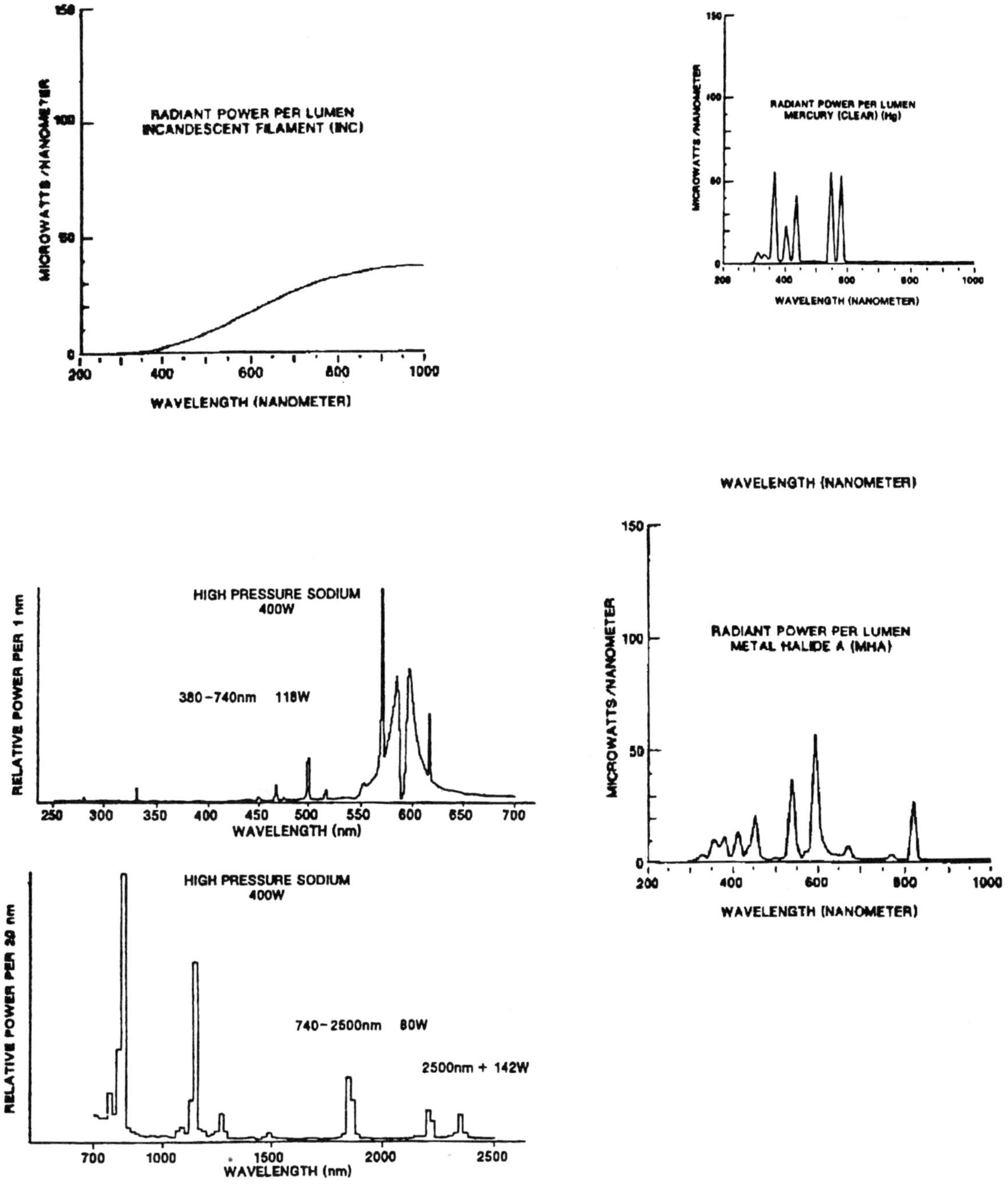

Fig. 3-57. Spectral radiant power for some horticultural lamps as a function of wavelength [Cathey and Campbell, 1980].

Table 3-10. Mounting heights (m) versus irradiance for HPS 400 W and LPS 180 W lamps [Cathey et al., 1981].

	Watts per square meter (400-850 nm)			
Lamp & wattage	6	12	24	48
HPS (400 Watt)	3	2.3	1.6	1.0
LPS (180 Watt)	2.4	1.7	1.2	0.8

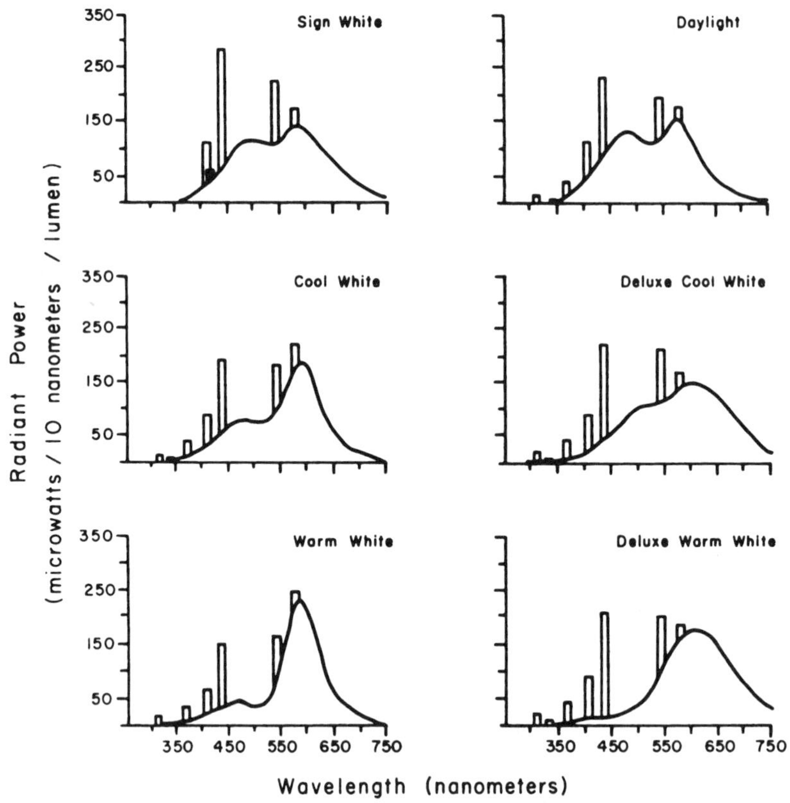

Fig. 3-58. Spectral energy distribution of some typical white fluorescent lamps [General Electric, 1970].

simplified Table 3-9 compared to 3-3, resulting from varied information sources. Manufacturers and other information for the trade have been provided by such authors as Harless [1988] and Falk [1985]. Despite the spectral variations shown in Figs. 3-57 and 3-58, any of the sources listed can be used to grow plants or supplement solar radiation. Sometimes they are used in combination, particularly incandescent with fluorescent in growth chambers. However, incandescents are not used to supplement sunlight since their efficiency is so low (0.24 to 0.40 μmol s^{-1} [electrical Watt]$^{-1}$), and most of the energy output occurs in the invisible infrared (Fig. 3-57). Fluorescents are much more efficient compared to incandescent as to visible radiation per Watt input. Similarly, the quantum flux for cool white would be 0.89 to 1.00 μmol $s^{-1}W^{-1}$. By changing the phosphors in the

Fig. 3-59. Four examples of supplementary irradiation installations for greenhouses. Note that high-intensity lamps are carried on a gantry in the lower right picture, which moves up and down the greenhouse.

tubes, various spectral distributions may be obtained (Fig. 3-60), but these have not been significantly better than ordinary cool or warm white [Biran and Kofranek, 1976]. An HPS lamp with an output of 132 lm W^{-1} would provide 1.61 μmol $s^{-1}W^{-1}$ (Table 3-9). The efficiency of lamps to convert electrical power to visible radiation is not a reliable criterion for lamp selection for plant growing [Andersen, 1986].

Overall, fluorescent lighting has been found most useful for seedling germination and initial growth or where low energy-requiring plants are grown in double tiers. Cathey and Campbell [1980] and Cathey and Thimijan [1986] have separated various lamps based on growth stage, interior survival, maintenance, propagation, etc. However, fluorescents are seldom used in greenhouses since the units and their reflectors severely reduce solar radiation by casting considerable shade.

Mercury and metal halide lamps are among those first used for high-intensity supplemental irradiation, being sufficiently efficient to be cost effective. More recently, high-pressure sodium lamps have come into increasing

Table 3-11. Irradiances (kilolux) at different horizontal and vertical distances from a single HPS 400 Watt fixture [Holcomb and Heins, 1988] (By permission of *Grower Talks*).

Horizontal location	Vertical distance from lamp (cm)				
	30	40	50	60	70
1	54.3	31.6	22.7	15.4	12.2
2	51.8	32.4	23.5	16.6	13.0
3	51.8	31.6	21.5	15.8	12.2
4	48.6	29.2	21.1	15.0	12.2
5	50.2	31.6	22.7	17.4	15.4
6	43.1	28.0	22.3	16.2	12.5
7	38.9	26.7	21.5	15.8	12.2
8	25.9	21.9	17.8	13.4	11.3
9	44.6	34.8	22.3	15.4	13.8

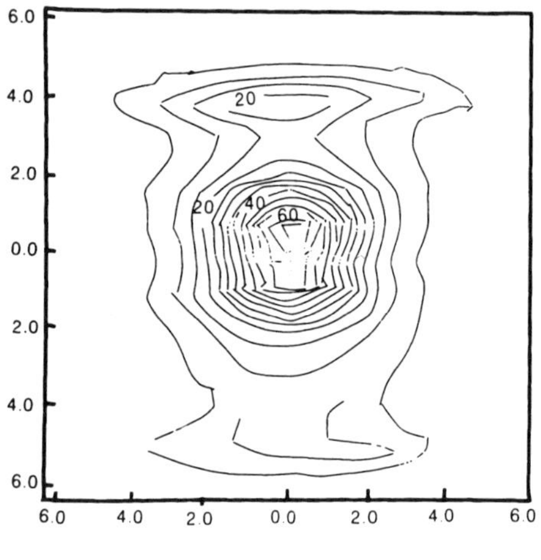

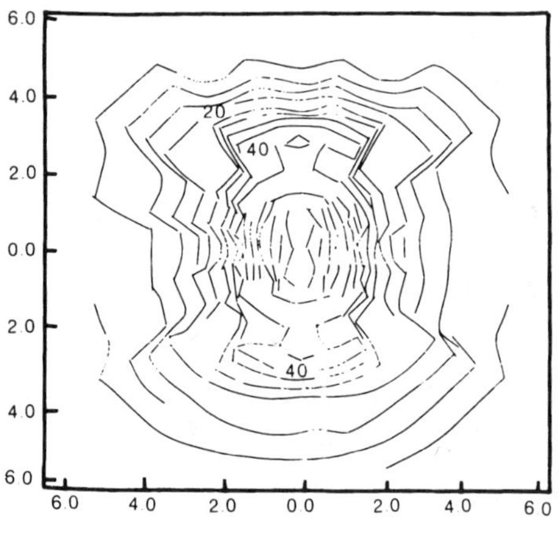

Fig. 3-60. Quantum flux (μmol $m^{-2}s^{-1}$) distribution as measured 2 m below an illuminaire at a laboratory (**left**) versus manufacturer's calculated data (**right**). Grid units are meters [Walker, 1987] (By permission of the *Greenhouse Grower*).

use, with outputs in the visible wavelengths exceeding 100 lumens/Watt, and intensities upward to 50000 lumens for high-pressure sodium. The yellow light from these lamps may not be desirable where appearance is important, but their ratings are such that size can be reduced significantly to reduce interfering with natural daylight. Examples of commercial installations with high-intensity lamps are provided in Fig. 3-58.

Given a particular lamp, the area it will irradiate will depend upon the square of the distance of the lamp from the crop –assuming it is a point source. Possible mounting heights for various wattages are provided in Table

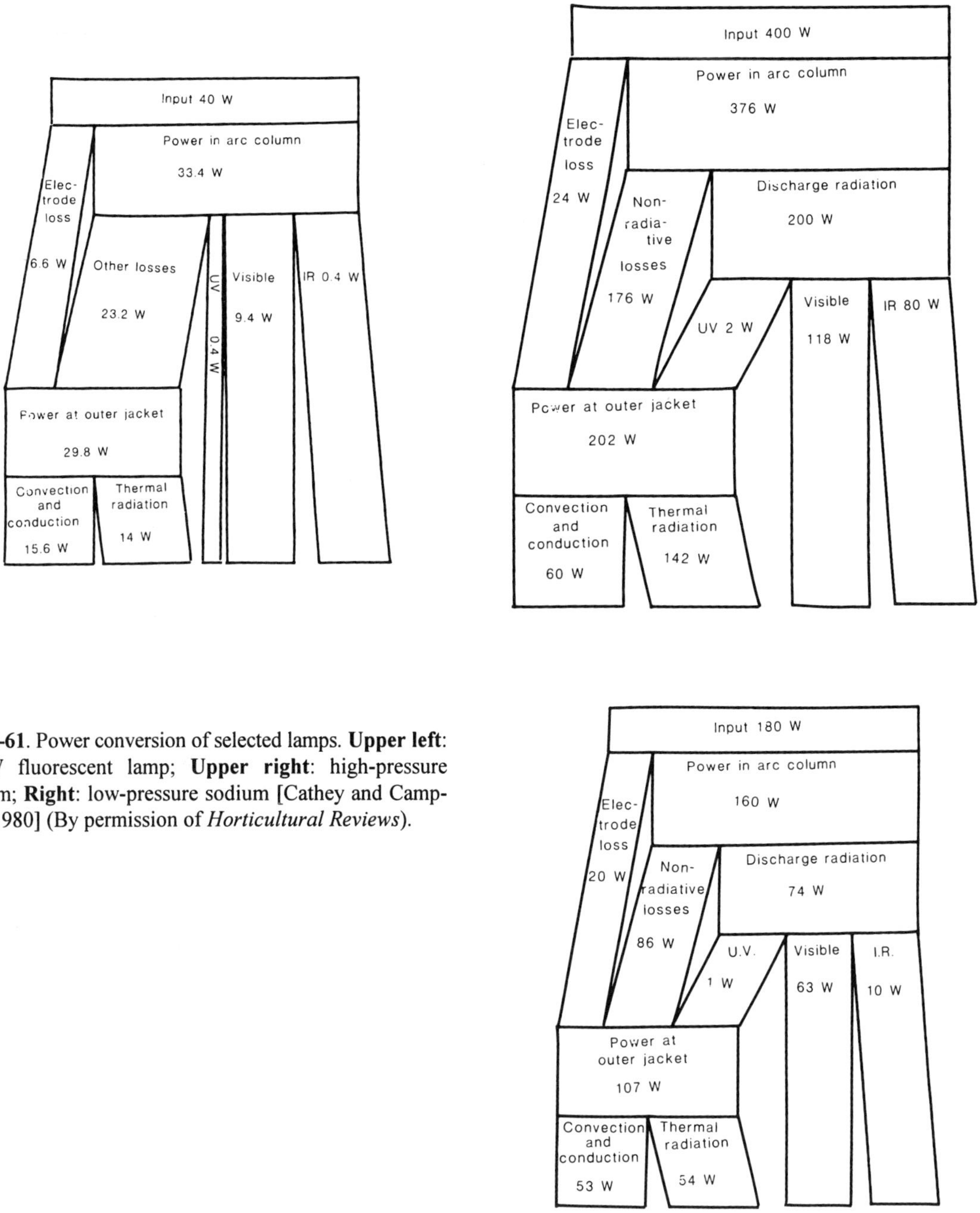

Fig. 3-61. Power conversion of selected lamps. **Upper left**: 40 W fluorescent lamp; **Upper right**: high-pressure sodium; **Right**: low-pressure sodium [Cathey and Campbell, 1980] (By permission of *Horticultural Reviews*).

Table 3-12. Suggested irradiance (W m^{-2}, 400-850 nm) and daily cycle for supplemental irradiation in greenhouses [Adopted from Cathey et al., 1981].

Plant and growth stage	Irradiance	μmol $s^{-1}m^{-2}$ Quantum PAR[a]	Duration	Time	mol $dy^{-1}m^{-2}$ Total quanta[b]
African violet (early flowering)	12-24	43-86	12-16	0600-1800 0600-2200	1.9-5.0
Ageratum (early flowering)	12-48	43-172	24	0000-2400	3.7-14.9
Begonia, fiberous rooted (early flowering)	12-24	43-86	24	0000-2400	3.7-14.9
Begonia, tuberous rooted (branching & early f-lowering)	12-24	43-86	24	0000-2400	3.7-14.9
Carnation (branching & early flowering)	12-24	43-86	16	0800-2400	3.7-14.9
Chrysanthemum (vegetative growth, branching, multi-flowering)	12-24	43-86	8	0800-1600	1.2-2.5
Cineraria (seedling, 4 wks)	6-12	22-43	24	0000-2400	1.9-3.7
Cucumber (early flowering)	12-24	43-86	24	0000-2400	3.7-7.4
Eggplant (early flowering)	12-48	43-172	24	0000-2400	3.7-14.9
Foliage (philodendron, schefflera) rapid growth	6-12 12-48	22-43	24 24	0000-2400 0000-2400	1.9-3.7
Geranium (branching, early flowering)	12-48	43-172	16	0800-2400	3.7-14.9
Gloxinia (early flowering)	6-12	43-172	24	0000-2400	2.5-9.9
Lettuce (rapid growth)	12-48	22-43	24	0000-2400	1.9-3.7
Marigold (early flowering)	12-48	43-172	24	0000-2400	3.7-14.9
Impatiens - New Guinea	12	43-172	16	0800-2400	3.7-14.9
Impatiens - Sultana (early)	12-24	43	24	0000-2400	2.5
Juniper (vegetative growth)	12-48	43-86	24	0000-2400	3.7-7.4
Pepper (early fruiting, compact)	12-24	43-172	24	0000-2400	3.7-14.9
Petunia (early and branching)	12-48	43-86	24	0000-2400	3.7-14.9
Poinsettia (vegetative) (branching & multi-flowering)	12 12-24	43 43-86	24 8	0000-2400 0800-1600	3.7 1.2-2.5
Rhododendron (vegetative)	12	43	16	0800-2400	2.5
Roses (early and rapid)	12-48	43-172	24	0000-2400	3.7-14.9
Salvia (early flowering)	12-48	43-172	24	0000-2400	3.7-14.9
Snapdragon (early flowering)	12-48	43-172	24	0000-2400	3.7-14.9
Streptocarpus (early)	12	43	16	0800-2400	2.5
Tomato (early and rapid)	12-24	43-86	16	0800-2400	2.5-5.0
Trees (deciduous, vegetative)	6	22	16	1600-0800	1.3
Zinnia (early flowering)	12-48	43-172	24	0000-2400	3.7-14.9

[a] Quantum flux calculated for an HPS lamp. Ratio mW/lm 400-700 nm to 400-850 = 0.72 (Table 3-3), conversion to μmol from Table 3-4.

[b] Total quantum flux per day, calculated as for previous footnote.

Table 3-13. Effects of supplemental irradiation on yield and quality of 'Forever Yours' roses [Tsujita, 1980].

Harvest period	Irradiance (klx)	Yield (Flowers m^{-2})	Stem length (cm)	Stem diameter (mm)	Fresh weight (g)	Maturity (days)
Jan-Mar	None	30.0	52.2	4.9	20.8	54.8
	6.8/18 hr	36.0	53.1	5.3	22.3	53.5
	10.8/18 hr	55.4	49.7	5.3	25.2	49.8
Oct-Jan	None	51.1	50.5	5.3	21.0	61.2
	6.8/18 hr	81.2	48.7	5.2	22.4	56.9
	10.8/18 hr	134.2	45.0	5.0	20.5	51.0

3-10. The distribution of intensity can be quite variable, depending upon bulb, bulb orientation and reflector type. Table 3-11 provides an example of variation between location that can occur. Fig. 3-60 is a more detailed comparison between two measurements of the same lamp type. This variability, of course, can cause serious differences in growth when provided over a long time. The grower should require energy distribution patterns produced by the individual fixture. In general, rectangular reflectors have been recommended [Falk, 1985; Harlass, 1988; Bartok, 1988], since these provide a rectangular irradiance distribution versus a circular pattern. Mist protectors can be purchased to protect the lamp from water, but these can markedly reduce energy reaching the plants. Fluorescents, mercury, metal halides, and sodium lamps require ballasts that can be attached to the lamp or placed remotely. These fixtures, and the lamps, are heavy, and the grower should investigate the ability of the structure to support the system. The wattage needed at plant level, with a distribution pattern, will determine the number of fixtures, which can cost up to $200 each. One common suggestion [Tsujita, 1981] for HPS installations has been 5 to 10 m^2 per fixture on 2.2 x 2.2 to 3.2 x 3.2 m spacings, 1.5 to 2 m above plant level (Table 3-10).

Table 3-14. The effect of a constant daily PAR irradiance given over different daily durations on growth of selected potted ornamentals. HPS supplementing natural irradiance 6.6 to 10.2 mole $m^{-2}dy^{-1}$ [Reprinted from *Scientia Horticulturae,* Gislerød et al., 1989, *The interaction of daily lighting period and light intensity on growth of some greenhouse plants.* 38:295-304. With kind permission of Elsevier Sci.-NL, Sara Burgerhartstraat 25, 1055 KV Amsterdam, The Netherlands].

Irradiance period (hr)	Dry weight (g $plant^{-1}$)	Height (cm)	Days to first flower	No. of flowers & buds
		Begonia 'Schwabenland'		
16	7.9	21.1	54.4	10.3
24	7.9	23.6	45.7	11.2
		18.7	49.3	10.9
		Kalanchoe 'Pollux'		
16	6.6	9.9	57.3	22.5
20	8.9	11.2	54.0	30.0
24	8.1	10.4	57.2	25.5
		Pelargonium 'Alex'		
16	8.0	6.4	--	2.0
20	11.4	7.5	--	4.0
24	9.5	7.7	--	5.0
		Hedera 'Svendborg' and 'Gloire de Marengo'		
16	5.9	20.0		
20	8.4	25.2		
24	7.9	24.5		

One can note in Table 3-3 that a considerable portion of electrical energy is dissipated by the ballasts as heat, plus heat from the bulbs. A detailed breakdown of energy distribution for three common lamp types was provided by Cathey and Campbell [1980] (Fig. 3-61). In a 40-W fluorescent, 23.5% of the input is distributed as visible radiation, 29.5% for a 400 W HPS lamp and 46.1 % for a 180-W LPS lamp. Thus, 29.8, 202, and 107 W, respectively will be dissipated by convection, conduction, and thermal radiation. This energy is available for heating the structure so not all is lost. This "benefit" should be figured into the grower's calculations for heating, especially as it will usually be at night.

Measuring illuminance with an inexpensive photometer is easy. However, the conversion factors will vary with the source (Tables 3-3, 3-4). For example, if one wants 6 W m^{-2} PAR at plant height, using HPS lamps, then from Table 3-3, 1 400 W HPS lamp produces 2.45 x 10^{-3} W lm^{-1} in the 400 to 700 nm PAR band. A light meter would have to read:

$$\frac{6}{2.45x10^{-3}} = 2449 lm\ m^{-2} \tag{3.13}$$

or 2.5 klx approximately, or 3.1 klx for an LPS lamp (1.92 mW/lm). These figures correspond to about 30 µmol. $m^{-2}s^{-1}$ quantum flux for both lamp types. Mounting heights for standard HPS and LPS lamps to provide the desired PAR intensity in Watts are given in Table 3-10.

Table 3-15. Typical supplemental irradiation applications in Ontario commercial greenhouses [Adopted from Tsujita, 1980].

	Irradiance (klx)	PAR (µmol $s^{-1}m^{-2}$)[a]	Duration (hrs)	Total PAR (mol $dy^{-1}m^{-2}$)[a]	Application period
Roses	2.2-5.4	27-66	12-24	1.2-5.7	Sept.-Apr.
Pot mums	2.7-5.4	33-66	12-24	1.2-5.7	lst 2 wks
Mum stock plants	2.2-3.8	27-46	12-20	1.2-3.3	Sept.-Mar.
Reiger begonia	2.2-3.2	27-39	12-24	1.2-3.4	Propagation or early stage
African violets	2.7-5.4	33-66	12	1.2	Propagation or preflowering
Bedding plants	2.7-5.4	33-66	12-24	1.2-5.7	1st 2-4 wks

[a] PAR for 400-700 nm band for an HPS lamp (Tables 3-3, 3-4), 2.45 W m^{-2} $(klx)^{-1}$.

Suggested irradiances in the 400-850 nm band for various species, with durations, are provided in Table 3-12. Given the required irradiances, the duration can be highly important. For example, with roses, intensities above 30 µmol $m^{-2}s^{-1}$ have been highly beneficial when given for 18 to 24 hr daily. A typical result is shown in Table 3-13 where 10.8 klx corresponds to about 130 µmoles $m^{-2}s^{-1}$. Roses can be irradiated continually at higher intensities with higher yields [Zieslin and Mor, 1990]. An added benefit, according to Tsujita [1980], is that supplementary irradiated roses under Canadian conditions can be grown at wider spacings, resulting in considerable savings in plant costs. Roses, however, show no photoperiodism which is the case with chrysanthemums, poinsettias, etc. Supplemental irradiation must be restricted to either early phases of vegetation growth or limited to the hours necessary to provide the required photoperiodic response. On the other hand, 20 hrs of supplemental irradiation were found by Gislerød et al. [1989] to provide the best results for many pot plant species (Table 3-14). A significant improvement in strawberry yield was also found by Ceulemans et al. [1986] with supplemental lighting. For lettuce, extending the daily radiation interval from 16 to 24 hrs resulted in doubling the weight of all loose leaf cultivars examined [Koontz and Prince, 1986]. Continuous irradiation may cause leaf drop with some species, and adverse response on tomatoes. Typical irradiation of ornamental crops under Canadian conditions is provided in Table 3-15. The accumulative energy (MJ $m^{-2}dy^{-1}$) that will be provided at different intensities and durations is given in Table 3-16.

The cost of supplemental irradiation not only depends upon initial outlays, but also the cost of electricity. Some power companies in the U.S. provide a special, interruptible rate [Walker, 1985; Brumfield and Ford, 1987]. For example, an off-peak rate in Pennsylvania was 4.13¢ for the first 750 kW-hr and 3.0627¢ for all additional kW-hr. The grower must be willing to permit interruption on 1 hour's notice. The average greenhouse operator pays about 6¢ per kW-hr. According to Walker [1985], an analysis showed that if energy costs were 40% of operating costs, and if supplemental irradiation provided half the heat, then it would be cost effective if the process increased productivity by 17%. A comparison of electrical and fuel oil heating costs is shown in Table 3-17. Still another example for Ontario, Canada, is provided in Table 3-18. Lawson [1987] stated that the annual cost of supplementary irradiation in Holland was $2.99 m^{-2}, based on a lamp life of 4 years and 8 years for the reflector, lamp holder, interest, and maintenance costs. This amounted to a cost of $49.16 for a lamp irradiating 16 m^2. Electricity costs varied from 6¢ to 12¢ per kW-hr, depending on whether the power station supplying the area was fired by gas, coal, or fuel oil. Potter [1988] gave a lamp cost of $240.00, including wiring and fitting, amortized over 8 years at 12%. For each potted chrysanthemum, Potter calculated a total running cost of about 4¢ and a capital cost of about 4¢, each lamp irradiating an area of 7.6 m^2.

Table 3-16. Irradiance per day (MJ m^{-2}dy^{-1}) (**upper**), quantum flux per day (moles m^{-2}dy^{-1}) (**lower**) for an HPS lamp at different intensities (W m^{-2}) and durations (hr).

Illuminance (klx)[a]	**Irradiance (W m^{-2})**			
	6	**12**	**24**	**48**
	2.5	**4.9**	**9.8**	**19.6**
Duration (hrs)				
8	0.17[b] 0.86[c]	0.35 1.72	0.69 3.44	1.38 6.88
12	0.26 1.29	0.52 2.58	1.04 5.16	2.07 10.32
16	0.35 1.72	0.69 3.44	1.38 6.88	2.76 13.77
20	0.43 2.15	0.86 4.31	1.73 8.60	3.46 17.21
24	0.52 2.58	1.04 5.17	2.07 10.32	4.15 20.65

[a] Conversion from Table 3-3, 2.45 mW lm^{-1}.
[b] Irradiance in PAR (400-700 nm).
[c] Conversion from Table 3-4, 4.98 μmol m^{-2}s^{-1} per W m^{-2}.

Table 3-17. Comparison of electric and heating oil costs in U.S. cents [Walker, 1985] (By permission of the *Greenhouse Grower).*

Electric cost (per kW-hr)	2.9	4.5	5.0	6.0	8.0	10.0	12.0
Heating oil (per liter)	21.0	33.0	36.0	44.0	58.0	59.0	84.0

Assumed oil heat content of 35.7 MJ ℓ$^{-1}$, heating plant efficiency of 70%

Table 3-18. Estimated yearly cost per square meter to own and operate a 400 Watt HPS unit, based on a rate of $0.03 per kilowatt-hour [Tsujita, 1980].

Illumination	**Area/lamp**	**Lighting regime**	
(klx)	**(m^2)**	**12 hr/dy**	**24 hrs/dy**
2.7	9.0	$9.36	$11.84
3.8	6.7	$12.37	$15.82
5.4	4.5	$18.61	$23.78

One recent approach to determining electrical costs has been the proposal by Heuvelink and Challa [1989]. They proposed dynamic optimization of on-off control of supplemental irradiation by means of a computer program based upon the return achieved by each gram dry weight increase in gross CO_2 assimilation (Fig. 3-62) versus the marginal costs of supplemental irradiation. The break-even point is the natural irradiance in the greenhouse at which the costs of supplemental irradiation equal the marginal production. An example of the break-even point is provided in Fig. 3-63. Rather than using time clocks to switch lights on or off at some arbitrary interval, or perhaps a simple program, which provides a delay –protecting against rapid cycling– and an arbitrary solar intensity; the program optimally adjusts on-off with the projected returns and costs. The natural energy at which both possibilities (on or off) are equal is the break-even point. Below this point, the lamps should be on. The authors examined the influence of CO_2 levels, product price, LAI, conversion efficiency and electrical cost on the break-even point. Examples are provided in Fig. 3-64 for supplemental irradiation at 7.36 W m^{-2}.

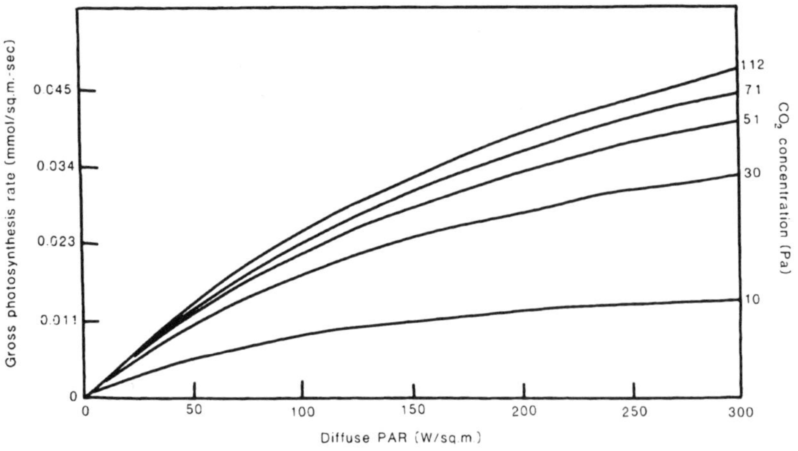

Fig. 3-62. Gross photosynthesis of a whole vegetable crop with a leaf area index of 3 as a function of diffuse PAR at different CO_2 concentrations. The range of CO_2 levels are from 100 to 1100 μl/ℓ [Heuvelink and Challa, 1989] (© 1989, Int. Soc. Hort. Sci., *Acta Hort.*, 260:401-412).

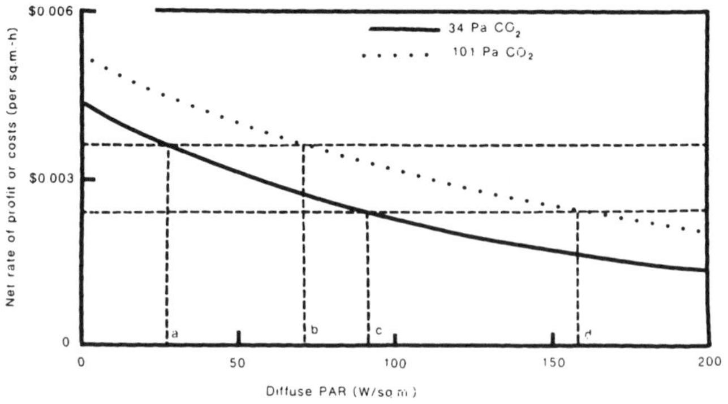

Fig. 3-63. Break-even points (**a**, **b**, **c**, and **d**) for supplemental irradiation as related to natural irradiance inside a greenhouse. Economic yield of supplemental irradiation (7.36 W m^{-2}) at two CO_2 concentrations. Electrical costs of 6 and 9¢ (kW-hr)$^{-1}$, LAI = 3, leaf temperature 20 C, crop conversion efficiency = 10 g fresh weight per gram CO_2 and product price of 0.15¢ per gram fresh weight [Heuvelink and Challa, 1989] (© 1989, Int. Soc. Hort. Sci., *Acta Hort.*, 260:401-412).

The economic value of 1 gram gross CO_2 assimilation and price of electricity are the two major parameters. Chinese cabbage was unprofitable, using this program. Cucumber, lettuce, and tomato were unprofitable for most of the year, and radish, eggplant, and sweet pepper had high potential. This program would have to be modified where the yield is fixed (i.e., pot chrysanthemums or poinsettias). The grower must obviously be able to assess future returns that can be input to the program. Other authors have published on programs for supplemental irradiation [i.e., Carrier et al., 1994; Meyer, 1989]. Carrier et al.'s program has not yet been set up on a large scale since the model requires improvement. In Meyer's development, published in 1989, heat requirements of the greenhouse caused lights to be switched on in one strategy tested.

It can be appreciated that the profitability of supplemental irradiation will depend greatly on the ability of the grower to optimize all other cultural conditions. Requirements have been stated by such authors as Shaddick [1987], Bartok [1988], and Potter [1988]. CO_2 levels should be raised, the area irradiated should be kept full of

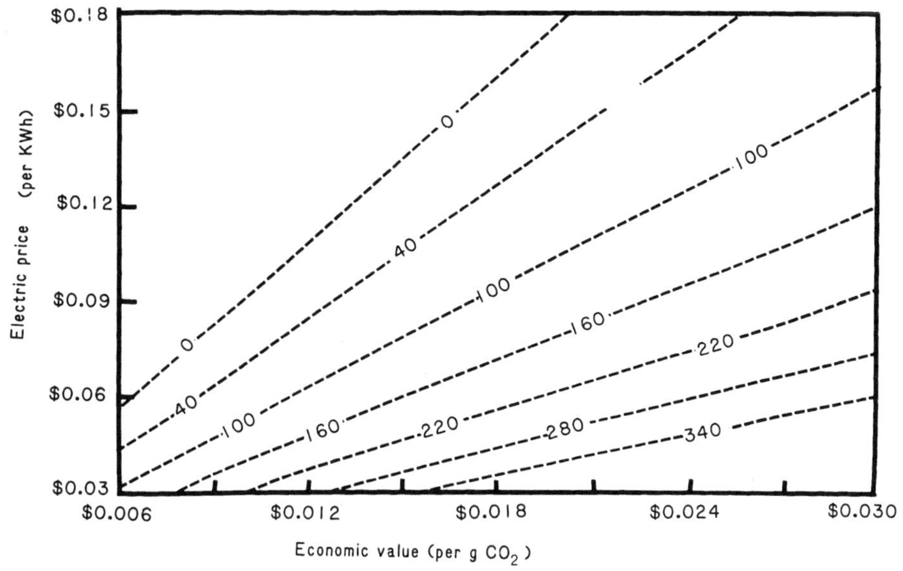

Fig. 3-64. Break-even points (values in graph) for supplemental irradiation (intensity = 7.36 W m^{-2}) as a function of economic value of gross CO_2 assimilation and electrical price. LAI = 3, leaf temperature = 20 C, and CO_2 concentration = 101 Pa (1000 μl ℓ^{-1}). 1 ct = 0.01 Dfl and 1 Dfl = $0.5988 U.S. [Heuvelink and Challa, 1989] (© 1989, Int. Soc. Hort. Sci., *Acta Hort.*, 260:401-412).

Table 3-19. Average percent transmission (dawn-to-dusk) of shade cloth fabrics located inside different plastic covered greenhouses [Goldsberry and van der Salm, 1985].

	Manufacturer's designation				
	30%	**40%**	**50%**	**60%**	**70%**
House cover	Dbl PVF	Dbl PE	Dbl PE	FRP	PVF panels
Condition	**Percent transmitted**				
Through greenhouse cover only (% of outside radiation)	73	63	65	26	68
Through shade (% of inside radiation above shade)	72	42	53	30	32
Greenhouse cover and shade (% of outside radiation)	53	27	35	27	22

plants, commensurate with plant quality, nutrition may have to be adjusted along with irrigation, and disease must be adequately controlled. In particular, the grower may find the cost of providing adequate electrical service to be expensive. For example, assume that one wishes to irradiate 1000 m^2 with 400 W HPS lamps, each covering 10 m^2. One will have 100 lamps requiring electrical service capable of supplying nearly 50 kW. At a voltage of 220, the service must supply more than 200 amperes. This is heavy service, and though the rate charge per kW-hr may be cheap (1.5 to 2.7¢ per kW), the demand charge (cost of the installation necessary to meet the requirements) may be considerable. Using Hanan's 1976 figures, a 40-kW demand would cost $127.80 per month. If the lamps were on 20 hr per day for 30 days, the total power consumption would be 24000 kW-hr. This would be a very high cost if supplemental irradiation were the only benefit.

3. Reducing Solar Energy

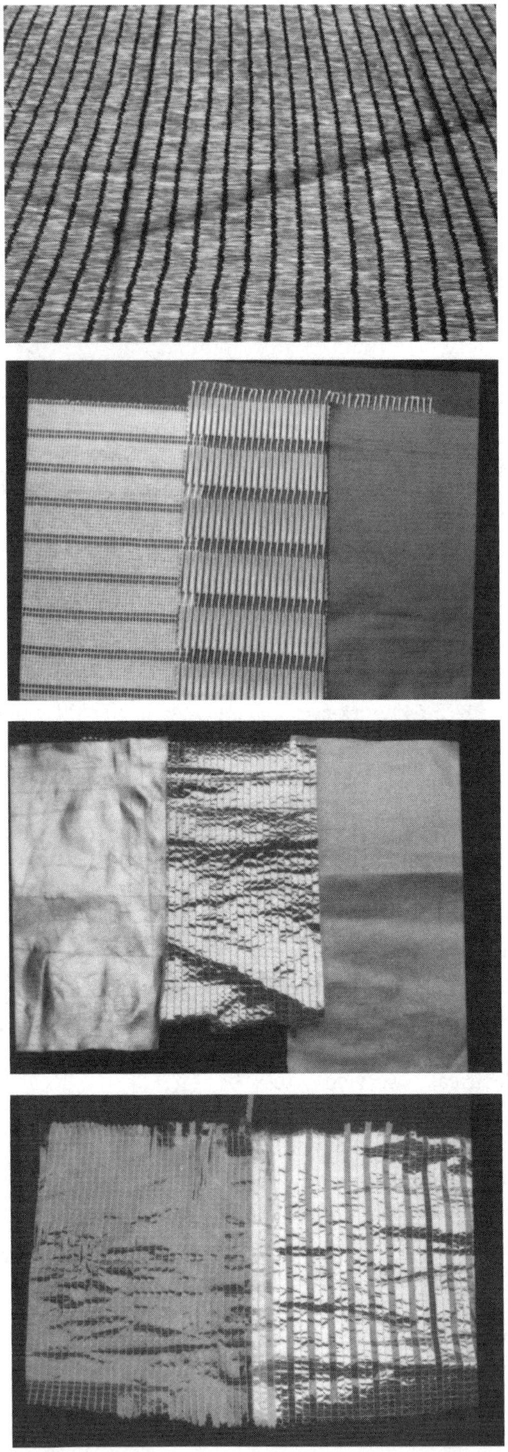

Fig. 3-65. Some examples of shading materials used with mechanical systems for opening and closing inside greenhouses. Other materials are available. Most of the materials in the bottom two pictures are employed as thermal screens, interlaid with plastic film or a foil-backed film. The top picture shows a woven plastic screen. These come in many sizes and densities. The second picture shows a metallic, woven cloth, which may be obtained in numerous densities.

There are occasions when it is necessary to reduce irradiance, especially in the summer, and with low energy-requiring plants such as many indoor decorative species. There are also times, either due to the greenhouse cover or because of high temperatures, the grower wishes to reduce intensity deliberately. There are many methods. The old way of applying whitewash on glass is cheap but difficult to control. It was noted in Fig. 3-34 and 3-35 that applying whitewash to glass roofs is likely to severely reduce intensity. Yet it was a necessity in Colorado before fan-and-evaporative pad cooling and FRP covers. Even with the newer, internal automatic shades, the grower must be careful not to significantly reduce intensity too much on high energy-requiring plants. He will also reduce yields if high temperatures are not a problem [Elenes-Fonseca and Hanan, 1987; Coker and Hanan, 1988]. Some examples of shading materials for internal covers are shown in Fig. 3-65. Some of these are also employed for energy conservation. Where shading is necessary in the summer, growers may install two separate moveable systems –one for shading, the other for energy conservation. More examples of such movable systems will be exhibited in Chapter 4.

Almost any kind of material may be used for shading, ranging from cheesecloth to those in Fig. 3-65, and intended for automatic systems. Attempts have been made to use flowing water over the greenhouse roof, or colored solutions such as copper sulfate. These can significantly alter spectral transmission as shown in Table 3-7. Further discussions of such methods are also deferred to Chapter 4. The manufacturer usually designates the percent reduction in intensity, but this will vary in actual use, especially if the shade is employed outside or inside a greenhouse (Table 3-19). According to Nijskens et al. [1985], the primary action of solar screens is not the creation of shade. The first aim is to limit an unacceptable increase in temperature resulting from high solar irradiance. The ideal solar screen should reject the sun's infrared rather than the visible or PAR. From Figs. 3-7 and 3-9, it was noted that the greatest portion of solar energy is in the near-IR. Absorption of IR by the screen will simply postpone overheating since the screen will emit heat through thermal radiation. In an examination of several materials, Nijskens et al. found that all commercial materials reduce transmittance in the visible rather than in the short (near) IR. Using their criteria, two polyester materials could be suitable, but any colored acrylic, polyethylene, or woven screens would be rejected. Considerable care is required not to overdo shading, since from Table 3-19, the results may be different from expected. As long as temperatures can be adequately controlled, and the species are dense canopies with high energy requirements, full sunlight is usually preferred.

C. PHOTOMORPHOGENESIS

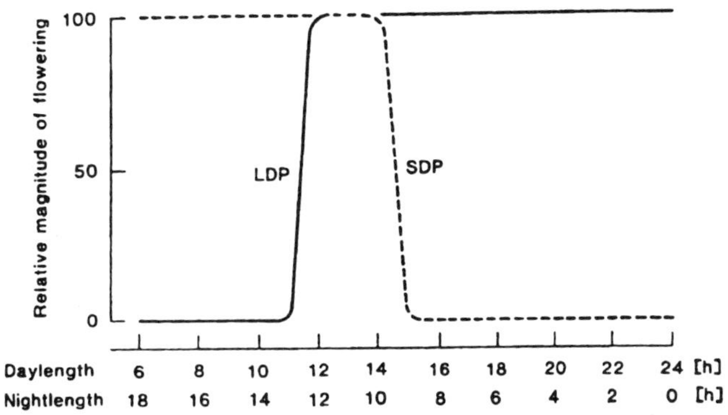

Fig. 3-66. Long-day plants (LDP) flower only when the daylength exceeds a critical duration, and short-day plants (SDP) flower only when the daylength is less than a critical duration. In this example from Vince-Prue [1986], both the LDP and SDP would flower in photoperiods between 12 and 14 hr duration (Reprinted from *Photomorphogenesis in Plants*, 1986. With kind permission of Kluwer Academic Publ.).

The photocontrol of growth occurs via photomorphogenetic pigments of which phytochrome is the best known in plants, with still unidentified blue light and UV absorbing receptors [Cosgrove, 1986]. These pigments do not provide energy or carbon for plant growth, but detect one or more conditions in the environment such as quality, quantity, direction, and duration of irradiance (Section 3.I). The photomorphogenetic pigments regulate the rate and directionality of growth of different plant organs. In the last decade, the literature shows a proliferation of photomorphogenetic responses in biological organisms, and the impression is one of considerable confusion since there is a wide response range in the plant world alone. The best exposition of photomorphogenesis is the volume edited by Kendrick and Kronenberg, *Photomorphogenesis in Plants*, published in 1986. That text is cited frequently in the following section.

1. Photocontrol

The most common use of the phytochrome system in greenhouse practice is the control of flowering through manipulation of daylength. Unfortunately, as Smith stated in 1986, we do not have a satisfactory hypothesis for the mechanism of phytochrome action in green plants. The fact remains, however, that

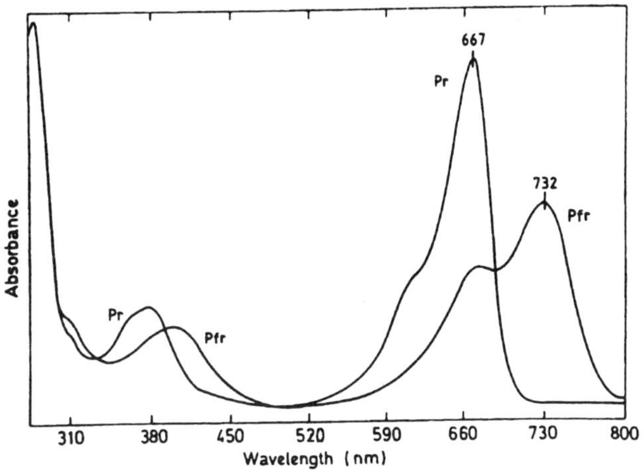

Fig. 3-67. Absorption spectra of purified phytochrome from etiolated oat seedlings. The spectrum for Pfr is an equilibrium mixture of Pr and Pfr after saturating with red light [Reprinted from *Photomorphogenesis in Plants*, Smith, H., 1986. With kind permission of Kluwer Academic Publ.].

exposure to alternating periods of light and dark is a feature of the environment, and it is the action of phytochrome in timing of this feature that we use in forcing plants in greenhouses. Vince-Prue [1986] clearly outlined the problem in that time-keeping events of the kind necessary to induce flowering –or any other process for plant survival– requires the ability to measure duration of day or night. There must be a clock to measure passage of time and a photoreceptor (phytochrome) to distinguish between light and dark.

Based on the work by Beltsville investigators, Hendricks, Borthwick, and others, two major categories of responses to alternating light and dark were found: short-day plants (SDP) when flowering occurs or is accelerated in short days, and long-day plants (LDP) where flowering occurs or is accelerated by long days. Day-neutral plants were those where daylength does not play a regulatory role in flowering –but, other photomorphogenic reactions may be important to the grower, depending upon his objectives. The characteristics of these photoperiodic responses are depicted diagrammatically in Fig. 3-66. Mechanisms are such that the plant can identify the period when reproduction can be carried out for specie survival. For example, chrysanthemum requires a shorter day for flower development than for flower initiation, thus ensuring that normal flowers occur only in the autumn. In actuality, it is the length of the dark period measured, since, for a range of species, response is independent of the duration of the preceding irradiance. The duration of the dark period is precisely timed and the trigger for the time measurement is the transfer from light to darkness.

According to Vince-Prue, most of the experimental evidence supports an internal clock with a 24-hr period. The clock can be reset by exposure to radiation, the precise response depending upon the phase of the rhythm when irradiance is presented. This is analogous to the problems humans have when faced with jet-lag. The individual can reset his clock to the local time quicker by exposure to high-inensityillumination. The rhythms found in plants are called circadian cycles (about 1 day). The rhythm drifts since a period is not exactly 24 hr, but can be entrained to a 24-hr cycle by a periodic signal, usually light. So with plants, the ability to respond to a dark period can be modified, or even eliminated, by the accompanying photoperiod. Some energy is necessary before a subsequent dark period is effective. Although some photosynthesis must obviously occur if the plant is to survive, the conclusion is that the light requirement in photoperiodic induction in SDPs is not photosynthetic. While some SDPs will respond to a single cycle of photoperiod-darkness, the majority require a succession of sufficiently long dark periods, and must be preceded by a photoperiod whose duration is not critical. The phase of the circadian timer, at least in SDPs, is set at dusk by the photoperiod. During darkness, the timer establishes phases of sensitivity to light, and the photoperiodic response of the plant depends on whether the inducible phase of the rhythm has been reached before exposure to light occurs (dawn).

The situation with LDPs is different. LDPs flower only when the daylength exceeds a certain critical value in a 24-hr cycle (Fig. 3-66). LDPs show a negligible response to a brief night-break and a strong flowering response to long daily exposure to light [Vince-Prue, 1986]. Sometimes, a few species appear to respond as though controlled by dark processes as in SDPs (e.g., Fuschia), or a dark dominant response. This contrasts with a light-dominant response where the flowering depends on the accumulated irradiance (light integral), regardless of whether it is given continuously or intermittently (i.e., carnation). This is not the only variation that can be encountered in LDPs. Light dominant plants appear to have an action spectrum with maximum effect around 710 to 720 nm, exhibit increased flowering when FR is added to a day extension with R, and a change in responsiveness to R and FR during each daily cycle. It is not, however, possible to be certain that phytochrome is the photoreceptor although effectiveness of mixtures of R and FR support the idea. In SDPs, flowering is inhibited by any mixture of R+FR, but in LDPs, the maximum flowering response is at intermediate R:FR ratios and decreases markedly as the R content of the radiation is increased.

There are other responses such as bulbing, elongation, abscission, tuberization, dormancy, sex expression, stomatal regulation, chloroplast orientation, lateral branching, and so forth. In fact, although Salisbury [1982] used the term "qualitative" to denote species with an absolute requirement as contrasted to "quantitative" species, Smith [1986] stated that nearly all phytochrome effects are quantitative in nature. Phytochrome cannot be considered solely as a "switching" agent. A problem encountered in the study of the phytochrome system is the fact that concentration of the pigment is very low in green plants. The presence of chlorophyll interferes with spectral analysis. The result is that much of our knowledge of phytochrome has been obtained from etiolated plants or plants subjected to abnormal conditions that do not always tell us how to use the system practically. The absorption spectrum shown in Fig. 3-67, for example, was obtained from etiolated oat seedlings. Note that the pigment also absorbs in the blue and UV wavelengths.

From the simple system of switching between the red (Pr) and far-red (Pfr) forms of phytochrome in vogue 30 years ago, the possibilities of what is involved have gradually become more complicated. One of the more recent schemes was proposed by Vierstra and Quail [1986], suggesting not only a direct connection with the phytochrome gene, but a feedback mechanism (Fig. 3-68). Not only can the different forms be obtained by suitable irradiance with appropriate wavelengths, but both, including phytochrome mRNA, are subject to degradation. Each process, gene → mRNA → Pr and mRNA → mRNA′, Pr → Pr′, Pfr → Pfr′, has its own rate constant, so that reactions do not proceed at the same velocity in all cases. Similarly, there are intermediate compounds in the change from Pr to Pfr, and they are not the same in the path Pr → Pfr as in Pfr → Pr. All these reactions not only continue at different rates, but are also influenced by concentration and temperature. It is simply unrealistic to assume that one merely has to irradiate a green plant with Pr (600 to 700 nm) or with Pfr (710 to 780 nm) to achieve the desired objective. It has now been shown that there can be times during the photoperiodic cycle when Pfr can be either inhibitory or promotive in the flowering of SDPs and LDPs, depending upon the species. In response to radiant intensity alone, investigators have distinguished between high irradiance responses (HIRs), such as hypocotyl lengthening in etiolated seedlings, low fluence responses (1 to 10^3 µmol $m^{-2}s^{-1}$)(LFRs), with which greenhouse operators are most familiar, and very low fluence responses (VLFRs, 10^{-4} to 9 x 10^{-1} µmol m^{-2} s^{-1}).

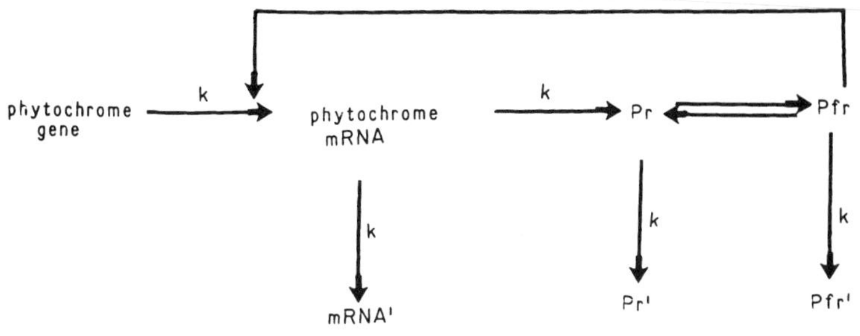

Fig. 3-68. Phytochrome system suggested by Vierstra and Quail [1986]. mRNA′, Pr′, and Pfr′ are degradation products. The symbol "k" refers to rate constants (Reprinted from *Photomorphogenesis in Plants*, Vierstra, R.D. and P.H. Quail, 1986. With kind permission of Kluwer Academic Publ.).

Although it is often assumed that Pfr is the active agent in the process, some authorities have proposed that it is the ratio Pfr/Pr or Pfr/P (total phytochrome) that determines response and measures the ratio of R/FR radiation. Many radiation sources, including the sun, contain both red (R) and far-red (FR) quanta. It would be logical to assume that the R/FR ratio in the radiance from the source (designated by the Greek letter zeta, ζ) would influence the ratio Pfr/P (designated as Ø). This latter ratio, under conditions of steady irradiance, would be constant.

Investigators such as Smith and Holmes [1977] and Holmes and Smith [1977] have shown Pfr/P to be the possible trigger for elongation of certain shade-avoiding plants. This ratio was 0.54 above a canopy and 0.04 below the canopy (see Figs. 3-9 and 3-42). The possibilities of using this ratio in greenhouses have been examined by Novoplansky et al. [1990] and cited earlier in this chapter. The possibilities of the R/FR ratio have also been examined over the day to determine if there are changes in the ratio that could signal the end of the day [Hughes et al., 1984]. An example of this spectral distribution is provided in Fig. 3-69. Hughes et al. concluded that although substantial changes in spectral composition occurred during twilight, they were less capable of providing reliable and accurate time signals than the absolute irradiance. One can note the high value in the blue at the beginning and end of the day, and there is a small, measurable drop in R/FR. Smith [1986] suggested the possibility of a B/UV photoreceptor to detect end-of-day signals.

There is now a consensus that there are blue (B) and UV-B photoreceptors in many organisms. Senger and Schmidt [1986] listed 28 responses, particularly in the fungi, as well as 10 photoreversible responses in the R/FR given by Kronenberg and Kendrick [1986]. Some of these possibilities, from the standpoint of disease control, have already been mentioned.

The variations in photoperiodic responses that can exist are almost as many as the number of species. An

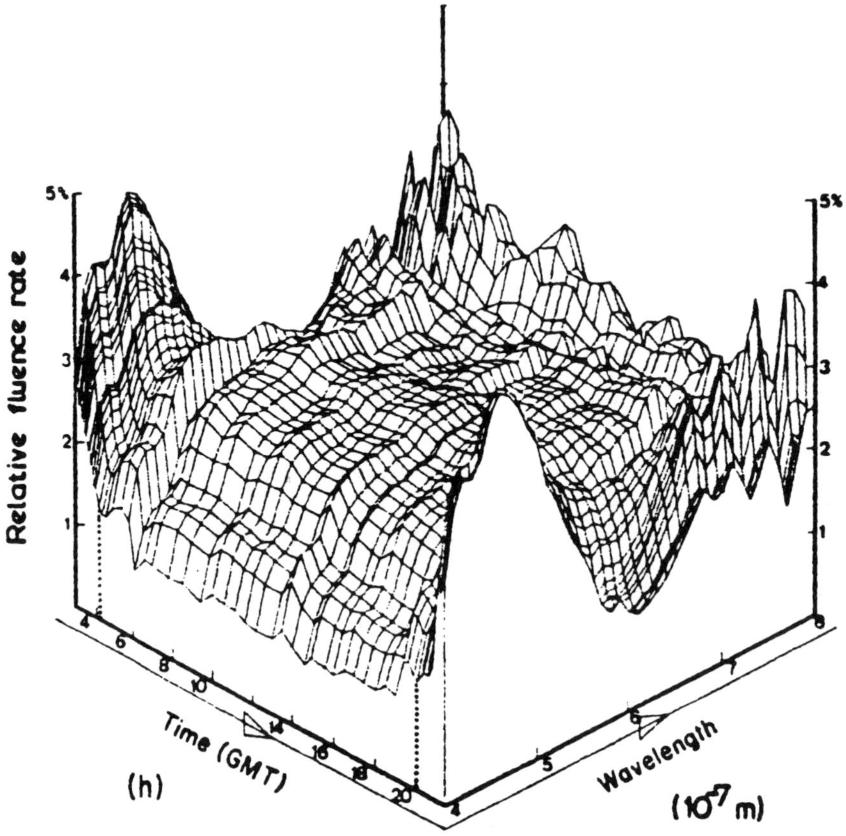

Fig. 3-69. Spectral quality surface for daylight. Spectroradiometric scans of the 400 to 800 nm spectral band from twilight to dusk at 10 nm intervals for July 7, 1981 plotted 3-dimensionally as a function of relative quantum irradiance versus time versus wavelength. Minor fluctuations are cloud cover variations, but note the marked increase in blue at dawn and at dusk [Hughes et al., 1984] (With permission of Blackwell Science, Ltd., *Plant, Cell and Environment*).

abbreviated Table 3-20 from Salisbury [1982] lists some of those of interest to greenhouse growers. Table 3-20 deals with the flowering response, generally the most common purpose of photoperiodic manipulation in greenhouses. Other responses are of interest and require knowledge of the individual specie. Only one or two examples are provided here in the interest of allowable space. For example, Soffe et al. [1977] examined effects of photoperiod on some vegetables. The extension of the photoperiod with a FR-rich source (incandescent at 3 W m^{-2}) resulted in distinct photomorphogenic changes with significant increases in leaf area and dry weight of lettuce, celery, beetroot, and spinach. All plants received 12 hr of irradiance at 115 W m-2. 3 W would be equivalent to about 15 µmol $m^{-2}s^{-1}$ in the 400 to 700 nm band (Table 3-4), suggesting photocontrol, but certainly not similar to the usual critical night-break. In celery, the problem is to prevent bolting. Roelofse and Hand [1989] found that flower initiation and bolting eventually occurred regardless of daylength regime, but short-day regimes induced a more rosette-like habit, shorter leaves and more side shoots. Chrysanthemums, carnations, snapdragons, and many other species also behave similarly in that flowering will eventually occur no matter the photoperiod, as contrasted to other species where a single light-dark cycle is sufficient to induce or prohibit a response (Table 3-20). Chrysanthemum is one of the more important species for which photoperiodic control is practiced on a large scale. The species is not very sensitive when a night-break is given to prevent flowering. But, Horridge and Cockshull [1989] showed that a 4 hr night-break (0.84 W m^{-2} PAR) could be used to take advantage of off-peak electrical rates, as long as the uninterrupted dark period before or after night-break was no longer than 8 to 9 hr in length. The greatest number of leaves was formed, with maximal delay of flower initiation, when the night-break was preceded by 4 to 8 hr of darkness. This contrasts with the carnation, which requires continuous dusk-to-dawn irradiation for maximum acceleration of

Table 3-20. Twenty-five photoperiodic response types with selected representative species. Plants marked with an asterisk appear in more than one category, illustrating variability with a given specie. This list has some notable omissions of greenhouse crops [Abridged from Salisbury, 1982] (With permission of *Horticultural Reviews)*.

SHORT-DAY PLANTS

1. Absolute short-day plants

Amaranthus caudatus L., love-lies-bleeding

Bryophyllum pinnatum (Lam) Pers., bryophyllum

Cattleya trianae Linden & Rehb. f., cattleya orchid

*Dendranthema grandiflora** Tzvelev., chrysanthemum

*Cosmos sulphureus** Cav., yellow cosmos, 'Klondike'

Glycine max L., soybean

*Impatiens balsamina** K, balsam impatiens

*Ipomoea batatus** (L.) Lam., sweet potato

Kalanchoe blossfeldiana Poellniz, kalanchoe

*Oryza sativa** L., rice

*Zea mays** L., maize or corn

2. SDP at high temperature, quantitative at low temperatures

Fragaria X ananassa Duch., strawberry

3. SDP at high temperature, day-neutral at low temperature

*Nicotiana tabacum** L., tobacco

*Salvia splendens** F. Sellow ex Roem. & Schult, scarlet sage

4. SDP at low temperature; day-neutral at high temperatures

*Cosmos sulphureus** Cav., yellow cosmos

5. SDP at high temperature; LDP at low temperatures

Euphorbia pulcherrima L., poinsettia (For greenhouses, considered only as SDP at all temperatures: Hanan)

Ipomoea purpurea (L.) Lam., morning glory

6. Quantitative short-day plant

Cannabis sativa L., hemp or marijuana

Capsicum frutescens L., bush redpepper

*Dendranthema grandiflora** Tzvelev., chrysanthemum

Cosmos bipinnatus Cav., common cosmos

*Helianthus annus** L., sunflower

*Helianthus tuberosus** L., Jerusalem artichoke

*Ipomoea batatus** (L.) Lam., sweet potato

*Oryza sativa** L., rice

Solanum tuberosum L., potato

Zinnia X hybrida, zinnia

7. Quantitative SDP, require or are accelerated by low-temperature vernalization

*Allium cepa** L., onion

*Dendranthema grandiflora** Tzvelev., chrysanthemum

8. Quantitative SDP at high temperature; day-neutral at low temperature

*Salvia splendens** F. Sellow ex Roem. & Schult., scarlet sage

Zygocactus truncatus (Haw.) K. Schum., Christmas cactus

LONG-DAY PLANTS

9. Absolute long-day plants

Agrostis palustris Huds., bentgrass

Anagallis arvensis L., scarlet pimpernel

Anethum graveolens L., dill

Chrysanthemum maximum Ramond, chrysanthemum

Dianthus caryophyllus L., carnation (Only wild-type, LD accelerates commercial varieties: Hanan)

Fuschsia X hybrida Hort. ex Vilm., fuchsia
Hibiscus syriacus L., hibiscus
Mentha piperita L., peppermint
Nicotiana sylvestris Speg. & Comes., tobacco
Phlox paniculata Lyon ex Pursh., summer phlox
Raphanus sativus L., radish
Rudbeckia hirta L., black-eyed susan
Sedum spectabile Boreau. and *S. telephium* L., stonecrops
*Spinacia oleracea** L., spinach

10. LDP, require or are accelerated by low-temperature vernalization
Anagallis tenella, pimpernel
Beta vulgaris Moq., common beet, sugarbeet
Dianthus spp., pinks
Oenothera spp., evening primroses
Saxifraga hypnoides L., moss saxifrage
*Spinacia oleracea** L., spinach
*Triticum aestivum** L., winter wheat

11. LDP at low temperature; quantitative LDP at high temperature
Brassica pekinensis Lour., Chinese cabbage

12. LDP at high temperature, day-neutral at low temperatures
Cichorium intybus L., chicory
Tropaeolum majus L., common nasturtium

13. LDP at low temperature, day-neutral at high temperatures
Delphiniuum X cultorum Voss, larkspur
Rudbeckia bicolor Butt., coneflower

14. LDP, vernalization will substitute (at least partly) for long-days
Silene armeria L., sweetwilliam silene

15. Quantitative long-day plants
Brassica campestris L., bird rape
Camellia japonica L., common camellia
*Dianthus barbatus** L., sweetwilliam
*Nicotiana tabacum** L., tobacco
*Scrophularia arguta**, figwort
*Triticum aestivum** L., spring strain, wheat

16. Quantitative LDP, require or are accelerated by low-temperature vernalization
Campanula spp., bellflowers
*Dianthus barbatus** L., sweetwilliam
Digitalis purpurea L., foxglove
*Iberis intermedia**, candytuft 'Durandii'
Lychnis coronaria (L.) Desr., rose campion

17. Quantitative LDP at high temperature; day-neutral at low temperature
Antirrhinum majus L., common snapdragon
Begonia semperflorens Link & Otto, begonia
Centaurea cyanus L., cornflower
Petunia X hybrida Hort. Vilm.-Andr., petunia
*Poa pratensis** L., bluegrass

DUAL DAYLENGTH PLANTS

18. Long-short-day plants
Aloe albiflora Guillaum, aloe
Bryophyllum crenatum (medic.) O. Kuntze, bryophyllum

Cestrum nocturnum L. (at 23 C, day-neutral at 24 C), night-blooming jasmine

19. Short-long-day plants

Escheveria harmsii Macbr., echeveria

20. Short-long-day plants, require or are accelerated by low-temperature vernalization

*Poa pratensis** L., bluegrass variety

21. Short-long-day plants, low-temperature substitutes for the SD effect and, after low temperature, plants respond as LDP

Campanula medium L., Canterbury bells

*Iberis intermedia**, candytuft 'Durandii'

INTERMEDIATE-DAY PLANTS

22. Plants flower when days are neither too short nor too long

Coleus X hybridus Voss, coleus 'Autumn'

Cyperus rotundus L., purple nutsedge

Saccharum spontaneum L., sugarcane

AMBIOPHOTOPERIODIC PLANTS

23. Plants quantitatively inhibited by intermediate daylengths

Madia eleans D. Don ex Lindl., tarweed

DAY-NEUTRAL PLANTS

24. These plants have the least flowering response to daylength. They flower about the same time under all daylengths, but some may be promoted by high or low temperature or by temperature alternation.

Calendula officinalis L., pot marigold calendula

Fagopyrum spp., buckwheats

Fragaria vesca L., alpine strawberry

Gardenia jasminoides, capejasmine

Gomphrina globosa L., globe amaranth

*Helianthus annuus** L., sunflower

Helianthus tuberosus L., Jerusalem artichoke

Ilex aquifolium L., holly

*Impatiens balsamina** L., impatiens

*Lunaria annua** L., dollarplant

Lycopersicon esculentum Mill., tomato

*Nicotiana tabacum** L., tobacco

*Oryza sativa** L., rice

Poa annua L., annual bluegrass

Rhododendron obtusum (Lindl.) Planch., Hiryu azalea

*Scrophularia arguta** and *S. peregrina*, figworts

*Solanum tuberosum** L., potato

*Zea mays** L., maize or corn

25. Day-neutral plants require or are accelerated by low-temperature vernalization

*Allium cepa** L., onion

Daucus carota L., carrot

Euphorbia lathyris L., caper euphorbia

*Lunaria annua** L., biennial strain, dollarplant

Pyrethrum cinerariifolium Trevir., dalmation pyrethrum

Saxifraga rotundifolia L., broadleaf saxifrage

*Scrophularia alata** and *S. vernalis*, figworts

Vicia faba L., broadbean

Fig. 3-70. Examples of photoperiodic effects on plants. A Aster (left) short days, (right) elongation and eventual flowering under long days. **B** Petunia (left) long days, (right) compact growth under short days. **C** Snapdragon (left) short days, (right) long days. **D** Chrysanthemum (left) short days, (right) long days. **E** Kalanchoe (left) long days, (right) short days. **F** Ageratum (left) long days, (right) short days. **G** Poinsettia (left) short days, (right) long days. **H** Park oats (left) long days, (right) short days.

Table 3-21. Some common lighting practices for responsive greenhouse crops, generally based on locations above 40° N latitude. Fluorescent lamps may also be employed. Actual practice may vary considerably with cultivar and objectives of grower [From Hanan et al., 1978]. Note: See also Larson [1980].

Species	Long day/ Short day	Hr light (dark) required	Intensity		Remarks
			Lux	W m^{-2}	
Aster *Callistephus chinensis*	LD	16	50-100	0.45-0.9	LD required for elongation, preceding flowering, can use 4 hr night-break. Minimum temperature 10 C.
Azalea *Rhododendron* spp.	LD	16	270-430	2.4-3.9	LD promotes stem length and increases leaf size, supplemental irradiation.
Carnation *Dianthus caryophyllus*	LD	24	50-100	0.45-0.9	Dusk-to-dawn at 6 to 7 expanded leaf pairs, for 2 to 4 weeks. Promotes elongation.
Chrysanthemum *Dendranthema grandiflora*	SD	(11)	50-200	0.45-1.8	SD required for flowering at 16 C. Note comments in text on chrysanthemum. Cultivars are divided into response groups; see Kofranek [1980], Crater [1980], and Gloeckner [1989].
Gloxinia *Sinningia speciosa*	LD	16	100-200	0.9-1.8	Hastens flowering, daylength extension.
Kalanchoe *Kalanchoe blossfeldiana*	SD	(11.5)	50-100	0.45-0.9	SD required for flowering, 4 hr night-break to prevent.
Easter lily *Lilium longiflorum*	LD	16	160	1.4	Hastens flowering, at emergence, later elongation, 2 to 4 hr night-break. May substitute for vernalization . See Wilkins [1980].
Poinsettia *Euphorbia pulcherrima*	SD	(12)	10-50	0.1-0.45	SD to flower, LD plant at temperatures below 10 C.
Fuchsia *Fuchsia hybrida*	LD	16	160	1.4	Daylight extension to hasten flowering.
Snapdragon *Antirrhinum majus*	LD	16	50-100	0.45-0.9	Hastens flowering, night-break 4 hr at 5 to 10 leaf pairs.
Stephanotis *Sephanotis floribunda*	LD	14	50-100	0.45-0.9	LD to flower, 4 hr night-break, night temperatures of 18 C.
Strawberry *Fragaria* spp.	LD	16	50-160	0.45-1.4	LD to promote runner formation and flowering, prevent dormancy at night temperatures of 6 C or lower.
Calceolaria *Calceolaria herbeohybrida*	LD	16	50-100	0.45-0.9	Earlier flowering, 8 hr daylength extension or 5 hr night-break, temperature below 10 C to initiate flowers.
Cineraria *Sinningia cruentus*	LD	16	50-100	0.45-0.9	Responds similarly to calceolaria.
Centaurea *Centaurea* spp.	LD	16	50-100	0.45-0.9	Requires LD to flower, day-neutral at low temperatures.
Camellia *Camellia japonica*	LD	16	100-200	0.9-1.8	LD promotes flowering at high temperatures, minimum night temperature of 16 C for flower initiation.
Begonia *Begonia* spp.	SD	(12)	50-160	0.45-1.4	Cv 'Lorraine' flowers on SD, 'Elatior' goes dormant, tuberous rooted produce tubers on SD.
Christmas cactus *Zygocactus truncatus*	SD	(15)	100	0.9	Flowers on SD, 4 wk period.

Species	Long day/ Short day	Hr light (dark) required	Intensity		Remarks
			Lux	W m^{-2}	
Tall marigold *Tagetes erecta*	SD	(15)	100	0.9	Hastens flowering on SD.

flowering. On the other hand, Kalanchoe, which is an SDP species, exhibits a rhythm in its flowering response with a 24-hr periodicity. All too few economically important greenhouse crops have been subjected to basic studies on photoperiodism. Much of our knowledge comes from such species as cocklebur (*Xanthium*) or goosefoot (*Chenopodium*). Undoubtedly, cultural procedures could be improved with additional investigation based on good science. Other examples of photoperiodic response for a few species are provided in Fig. 3-70.

2. Daylength Control

Table 3-21 provides additional information of practical use in controlling the flowering of many important greenhouse species. Table 3-22 provides actual hours and minutes of daylight throughout the year for various latitudes in the Northern Hemisphere. To convert to the Southern Hemisphere, enter a date about 6 months earlier or later than the one needed. Reference may also be made to Fig. 3-11, which graphically shows the daylength variation. When, and how much, days must obviously be shortened or lengthened, varies with season and latitude. Thus, in the summer, there is no need to lengthen the day on chrysanthemums to prevent flowering at latitudes much above 30° N as the days will be sufficiently long (nights are short). On the other hand, the grower must artificially shorten the daylength if flower development is to occur. Fig. 3-11 shows the limits for chrysanthemums and also for other species. The figure also shows that at latitudes less than 30°, chrysanthemum plants will flower at anytime of the year, and it is only necessary to lengthen the day to maintain vegetative growth.

Incandescent light is commonly used to achieve long days with the irradiances shown in Table 3-21. In general, a long night is broken into shorter periods by turning on lights during the middle of the night (i.e., 2200 to 0200), depending upon the species. Incandescents are cheap and easily utilized, and the irradiance required is low (LFR), in the range of less than 1 to 3 W m^{-2}. One 60-W bulb, with a reflector, every 1.2 to 1.5 m above a standard bench is sufficient. For two beds, a single row of 100 W bulbs can be placed every 1.9 m. For a 6 m wide house, a single row of 300 W reflector bulbs every 3 m will be sufficient. Examples of arrangements and bulbs are shown in Fig. 3-71. With some plants, the cost can be reduced by breaking the dark period into shorter cycles (cyclic lighting), and arranging the control so that a fifth of the total area is illuminated at any one time. As worked out by Cathey and Borthwick [1961], each of the five sections receives 6 minutes of light every 30 min (20% of the time). It may be necessary, depending upon cultivar with chrysanthemums, to increase intensity. The authors suggested 110 lx with the above schedule. If the frequency is increased to 3 sec per minute for 4 hr, the intensity should be increased to 220 lx. This type of night-break has not been used to a great extent, most growers preferring to use a simple system to provide continuous illumination during the night-break. Langhans [1980] provided a simple table, giving spacing and wattage (Table 3-23). However, where the entire house is in the same crop, the entire area may be illuminated from a single row of lights in the center, or one can use a high-inensity lamp such as a xenon lamp to provide the required intensity.

Shortening the daylength is more complicated since curtains are necessary to reduce illumination below 5 to 10 lux. There are some photoperiodic responses to even lower illuminance, but this seems the maximum allowed to obtain the desired darkness with most commercial species. Obviously, the material must be opaque, and the methods are numerous, depending upon area to be covered, house, and bench arrangement. Labor can be reduced by using mechanical systems (Fig. 3-72) to open or close the cloth. Under high solar intensity, summer conditions, an excessive heat buildup under the cover can seriously interfere with the desired objective. Using fans to pull air through under the curtain to remove heat may be necessary, or one is forced to close the curtains later in the day and open them later the following morning. The general practice is to close at 1700 hrs and open at 0800; but with automatic systems, the time can be varied as necessary. Other covering materials that are black on the inside and reflective on the outside can help reduce temperature, and can be used as thermal screening. On a small scale, growers often employ black plastic as a cover since it is much cheaper than the usual

Fig. 3-71. Various examples of incandescent and fluorescent lighting for photoperiodic response. **Upper**: Light to lengthen days in Florida to prevent initiation. **Middle**: Internal reflectarized bulb with aluminized, opaque cloth. **Lower left**: 200 W bulb with internal reflector. **Lower right**: Traveling fluorescent tubes. This is a type of cyclic lighting.

Table 3-22. Duration of daylight for different dates and latitudes, Northern Hemisphere. 0.5 to 1 hr must be added to arrive at photoperiodically active light. [Abridged from List, 1966].

Day of month	Hours-minutes											
	Jan	Feb	Mar	Apr	May	Jun	Jul	Aug	Sept	Oct	Nov	Dec
Latitude 10° N												
1	11-33	11-42	11-56	12-14	12-29	12-40	12-42	12-33	12-18	12-02	11-47	11-36
17	11-37	11-50	12-05	12-22	12-35	12-42	12-38	12-26	12-09	11-53	11-40	11-32
Latitude 20° N												
1	10-57	11-16	11-45	12-20	12-52	13-16	13-19	13-02	12-32	11-57	11-25	11-00
17	11-05	11-32	12-03	12-38	13-07	13-20	13-12	12-48	12-13	11-40	11-10	10-56
Latitude 30° N												
1	10-15	10-46	11-33	12-29	13-20	13-57	14-03	13-34	12-46	11-53	10-59	10-22
17	10-27	11-12	12-02	12-57	13-42	14-04	13-52	13-11	12-18	11-25	10-36	10-14
Latitude 40° N												
1	09-23	10-10	11-18	12-39	13-54	14-49	14-58	14-16	13-05	11-47	10-29	09-33
17	09-42	10-47	12-00	13-20	14-27	15-00	14-42	13-41	12-24	11-06	09-55	09-20
Latitude 50° N												
1	08-10	09-20	10-58	12-55	14-41	16-04	16-18	15-14	13-31	11-39	09-48	08-24
17	08-38	10-15	11-58	13-53	15-30	16-22	15-53	14-23	12-32	10-40	08-58	08-06
Latitude 60° N												
1	06-03	08-00	10-28	13-17	15-58	18-17	18-43	16-51	14-10	11-28	08-43	06-28
17	06-51	09-23	11-55	14-45	17-18	18-50	17-57	15-30	12-44	10-02	07-24	05-54

Fig. 3-72. Various types of black cloth systems for photoperiodic control. **Upper**: Note exhaust fans for cooling under the cloth. **Middle**: Black cloth in a parked position. There are systems that do a better job of parking to reduce overhead obstruction. **Lower**: System partially closed. Such systems may also be used for energy conservation, serving a dual purpose. See Chapter 4.

Table 3-23. Lamp height and wattage for 100 lux [Adapted from Langhans, 1980].

Bench width (m)	Height from bench to bottom of reflector (m)	Spacing (m)	Lamp wattage (W)
0.9	0.6	0.9	25
1.2	0.8	1.2	40
1.5	1.0	1.5	50
1.8	1.2	1.8	75

opaque materials. This is a dangerous practice since plastic is impermeable to water vapor exchange compared with cloth. Excessively high humidities are likely to occur, and, coupled with high temperatures, can lead to severe damage from foliar diseases. Light leaks through tears and failure to close around the edges can cause problems if not delay in the timing schedule. During the early years of photoperiodic control, turning on lights to check temperatures, ventilators, etc., or outside street lights (e.g., sodium vapor) were known to delay crops, particularly poinsettias. Furthermore, shade curtains, and also photoperiodic or high-inensity lights are more objects to be placed above the crop, thus reducing natural irradiance. Choice, especially of shade curtains, thermal screens, etc., should be done so that there is minimum obstruction when they are opened or "parked." Failure to maintain the greenhouse adequately can lead to leakage and additional condensation on the curtains below the roof, which can cause sagging and damage to the system. While automatic systems are great labor savers, and they can be controlled with present computers, they require close attention to detail. Backyard mechanics may be fine with Model A cars, but greenhouses are rapidly growing more complicated. They require knowledgeable and trained technicians in order to ensure reliable and efficient operation.

VI. REFERENCES

Acock, D., D.A. Charles-Edwards and A.R. Hearn. 1977. Growth response of a *Chrysanthemum* crop to environment. I. Experimental techniques. *Ann. Bot.* 41:41-48.

Acock, B., D.A. Charles-Edwards and S.Sawyer. 1979. Growth response of a Chrysanthemum crop to the environment. III. Effects of radiation and temperature on dry matter partitioning and photosynthesis. *Ann. Bot.* 44:289-300.

Aiken, W.J. and J.J. Hanan. 1975. Photosynthesis in the rose: Effect of light intensity, water potential and leaf age. *J. Amer. Soc. Hort. Sci.* 100:551-553.

Aikman, D.P. 1989. Potential increase in photosynthetic efficiency from the redistribution of solar radiation in a crop. *J. Expt. Bot.* 40:855-864.

Andersen, A. 1986. Comparison of fluorescent lamps as an energy source for production of tomato plants in a controlled environment. *Sci. Hort.* 28:11-18.

Anon. 1984. Solar radiation in Europe. *CO Greenhouse Growers Assoc. Res. Bull.* 406:4.

Baker, J.H. and R.A. Aldrich. 1964. Light transmission of rigid plastics. *Paper No. 64-432.* Ann. Mtg. ASAE, Ft. Collins, CO

Barnes, C. et al.. 1993. Accuracy of quantum sensors measuring yield photon flux and photosynthetic photon flux. *HortScience.* 28:1197-1200.

Bartok, J.W. Jr. 1988. High intensity discharge lighting. *Greenhouse Manager.* 6(9):171-177. Jan., 1988.

Basiaux, P., J. Deltour and A. Nisen. 1973. Effect of diffusion properties of greenhouse covers on light balance in the shelters. *Agric. Meteor.* 11:357-372.

Bierhuizen, J.F., J.M. Bierhuizen and G.F.P. Martakis. 1984. The effect of light and CO_2 on photosynthesis of various pot plants. *Gartenbauwissenschaft.* 49:251-257.

Biran, I. and A.M. Kofranek. 1976. Evaluation of fluorescent lamps as an energy source for plant growth. *J. Amer. Soc. Hort. Sci.* 101:625-628.

Björn, L.O. 1986. Introduction. *In* Photomorphogenesis of Plants. R.E.Kendrick and G.H.M. Kronenberg eds. Martinus Nijhoff, Boston. pp 3-14.

Bot, G.P.A. 1983. Greenhouse climate: From physical processes to a dynamic model. Ph.D. Dissertation, Wageningen. 240 pp.

Bowman, G.E. 1970. The transmission of diffuse light by a sloping roof. *J. Agric. Eng. Res.* 15:100-105.

Bozarth, C.S., R.A. Kennedy and K.A. Schekel. 1982. The effects of leaf age on photosynthesis in rose. *J. Amer. Soc. Hort. Sci.* 107:707-715.

Bruggink, G.T. and E. Heuvelink. 1987. Influence of light on the growth of young tomato, cucumber and sweet pepper plants in the greenhouse: Effects on relative growth rate, net assimilation rate and leaf area ratio. *Sci. Hort.* 31:161-174.

Brumfield, R. and P. Ford. 1987. Lights at the right price. *Greenhouse Grower.* 5(11):75-76. Nov., 1987.

Businger, J.A. 1966. The glasshouse (greenhouse) climate. *In* Physics of Plant Environment. W.R. van Wijk ed. North Holland Publ. Co., Amsterdam. pp 277-318.

Calvin, M. and G.M. Androes. 1962. Primary quantum conversion in photosynthesis. *Science.* 138:867-873.

Carrier, M., A. Gosselin and L. Gauthier. 1944. Description of a crop growth model for the management of supplemental lighting in greenhouses. *HortTechnology.* 4:383-389.

Cathey, H.M. and H.W. Borthwick. 1961. Cyclic lighting for controlling flowering of Chrysanthemums. *Proc. Amer. Hort. Sci.* 78:542-552.

Cathey, H.M. and L.E. Campbell. 1980. Light and lighting systems for horticultural plants. *Hort. Rev.* 2:491-537.

Cathey, H.M. and R.W. Thimijan. 1986. USDA's universal system for lamps and lighting. *Ohio Florists' Assoc. Bul.* 685:4-8.

Cathey, H.M., L.E. Campbell and R.W. Thimijan. 1981. Strategies for supplemental lighting in greenhouses: How to boost the growth of plants. *Ohio Florists' Bull.* 621:8-9.

Ceulemans, R. et al.. 1986. Effects of supplemental irradiation with HID lamps, and NFT gutter size on gas exchange, plant morphology and yield of strawberry plants. *Sci. Hort.* 28:71-83.

Challa, H. and A.H.C.M. Schapendonk. 1984. Quantification of effects of light reduction in greenhouses on yield. *Acta Hort.* 148:501-509.

Charles-Edwards, D.A. 1978. An analysis of the photosynthesis and productivity of vegetative crops in the United Kingdom. *Ann. Bot.* 42:717-731.

Coker, F.A. and J.J. Hanan. 1988a. The effect of shading on 'Samantha' roses. *CO Greenhouse Growers Assoc. Res. Bull.* 455:1-5.

Coker, F.A. and J.J. Hanan. 1988b. CO_2 uptake by 'Samantha' roses. *CO Greenhouse Growers Res. Bull.* 456:1-3.

Comar, C.L. and F.P. Zscheile. 1942. Analysis of plant extracts for chlorophylls a and b by a photoelectric spectrophotometric method. *Plant Physiol.* 17:198-209.

Cosgrove, D.J. 1986. Photomodulation of growth. *In* Photomorphogenesis in Plants. R.E. Kendrick and G.H.M. Kronenberg ed. Martinus Nijhoff, Boston. pp 341-366.

Crater, G.D. 1980, Pot mums. *In* Introduction to Floriculture. R.A. Larson ed. Academic Press, NY. pp. 261-285.

Critten, D.L. 1983a. A computer model to calculate the daily light integral and transmissivity of a greenhouse. *J. Agric. Eng. Res.* 28:61-76.

Critten, D.L. 1983b. The evaluation of a computer model to calculate the daily light integral and transmissivity of a greenhouse. *J. Agric. Eng. Res.* 28:545-563.

Critten, D.L. 1984. The effect of geometric configuration on the light transmission of greenhouses. *J. Agric. Eng. Res.* 29:199-206.

Critten, D.L. 1985a. The use of reflectors in venetian blinds to enhance irradiance in greenhouses. *Solar Energy.* 34:83-92.

Critten, D.L. 1985b. The effect of house length on the light transmissivity of single and multispan greenhouses. *J. Agric. Eng. Res.* 32:163-172.

Critten, D.L. 1985c. A theoretical assessment of the transmissivity of conventional symmetric roofed multispan E-W greenhouses compared with vertical south roofed greenhouses under natural irradiance conditions. *J. Agric. Eng. Res.* 32:173-183.

Critten, D.L. 1986. A general analysis of light transmission in greenhouses. *J. Agric. Eng. Res.* 33:289-302.

Critten, D.L. 1987a. An approximate theory for reflective losses from infinitely long greenhouses and plastic tunnels under diffuse light. *J. Agric. Eng. Res*. 38:47-56.
Critten, D.L. 1987b. The transmissivity of E-W aligned models of a conventional and clerestory greenhouse under natural winter irradiance. *J. Agric. Eng. Res*. 37:129-139.
Critten, D.L. 1987c. Light transmission losses due to structural members in multispan greenhouses under diffuse skylight conditions. *J. Agric. Eng. Res*. 38:193-207.
Critten, D.L. 1987d. Light transmission losses due to structural members in multispans under direct light conditions. *J. Agric. Eng. Res*. 38:209-215.
Critten, D.L. 1987e. An improved theory for reflective losses from infinitely long greenhouses. *J. Agric. Eng. Res*. 38:301-311.
Critten, D.L. 1993. A review of light transmission into greenhouse crops. *Acta Hort*. 328:9-31.
Downs, R.J. 1985. Irradiance and plant growth in greenhouses during winter. *HortScience*. 20:1125-1127.
Downtown, W.J.S. 1975. The occurrence of C_4 photosynthesis among plants. *Photosynthetica*. 9:96-105.
Edwards, R.I. and J.V. Lake. 1965a. Transmission of solar radiation in a large-span east-west glasshouse. II. Distinction between the direct and diffuse components of the incident radiation. *J. Agric. Eng. Res*. 10:125-131.
Edwards, R.I. and J.V. Lake. 1965b. The transmission of solar radiation in a small east-west glasshouse glazed with diffusing glass. *J. Agric. Eng. Res*. 10:197-201.
Edwards, R.I. and L.J. Moulsley. 1958. Preliminary measurements of the distribution of light in a glasshouse. *J. Agric. Eng. Res*. 3:69-75.
Elenes-Fonseca, C. and J.J. Hanan. 1987. The effect of greenhouse cover and shading on 'Royalty' rose yield. *CO Greenhouse Growers Assoc. Res.* Bull. 449:1-6.
Elwell, D.L. and T.H. Short. 1989. Control of electrostatic effects in a polystyrene pellet variable shading greenhouse. *Trans. ASAE*. 32:2117-2122.
Ehleringer, J. and O. Björkman. 1977. Quantum yields for CO_2 uptake in C_3 and C_4 plants. *Plant Physiol*. 59:86-90.
Falk, N.K. 1985. Lighten up. *Greenhouse Manager*. Oct., 1985.
Frankland, B. 1986. Perception of light quality. *In* Photomorphogenesis in Plants. R.E. Kendrick and G.H.M. Kronenberg eds. Martinus Nijhoff, Boston. pp 219-235.
Gaastra, P. 1958. Light energy conversion in field crops in comparison with the photosynthetic efficiency under laboratory conditions. *Med. Landbouw. Wageningen*. 58:1-12.
Gaastra, P. 1962. Photosynthesis of leaves and field crops. *Netherlands J. Agric. Sci*. 10:311-323.
Gates, D.M. 1965. Heat, radiant and sensible. Chapter 1. Radiant energy, its receipt and disposal. *Meteor. Monog*. 6:1-25.
Gates, D.M. 1980. Biophysical Ecology. Springer-Verlag, New York. 611 pp.
Gates, D.M. et al.. 1965. Spectral properties of plants. *Appl. Optics*. 4:11-20.
General Electric. 1970. Fluorescent lamps. TP-111.
Gislerød, H.R., I.M. Eidsten and L.M. Mortensen. 1989. The interaction of daily lighting period and light intensity on growth of some greenhouse plants. *Sci. Hort*. 38:295-304.
Gloeckner, Inc. 1989. Chrysanthemum manual. Gloeckner, Inc., NY. 118 pp.
Goldsberry, K.L. 1969. Spectral transmission of greenhouse glass. *CO Flower Growers Assoc. Res. Bull*. 208:3-4.
Goldsberry, K.L. and T. van der Salm. 1985. Shading fabrics: Transmission characteristics under plastic covers. *CO Greenhouse Growers Assoc. Res.* Bull. 416:1-4.
Goudriaan, J. and G.L. Atjay. 1979. The possible effects of increased CO_2 on photosynthesis. *In* The Global Carbon Cycle, B. Bolin et al. eds, Wiley, New York.
Hall, A. and J.J. Hanan. 1976. Measurement of total light energy in a carnation bench. *CO Flower Growers Assoc. Res. Bull.* 308:2.
Hanan, J.J. 1969. Water loss and stress in carnations grown under glass. *CO Flower Growers Assoc. Res. Bull.* 233:1-3.
Hanan, J.J. 1970. Some observations on radiation in greenhouses. *CO Flower Growers Assoc. Res.* Bull. 239:1-4.
Hanan, J.J. 1976. The cost of electricity. *CO Flower Growers Assoc. Res. Bull*. 311:1-3.

Hanan, J.J. 1989. Sunlight in Colorado greenhouses. *CO Greenhouse Growers Assoc. Res. Bull.* 466:1-5.
Hanan, J.J. and R. Heins. 1975. Effect of plant density on two years of carnation production. *CO Flower Growers Assoc. Res.* Bull. 302:1-3.
Hanan, J.J., W.D. Holley and K.L. Goldsberry. 1978. Greenhouse Management. Springer-Verlag, Heildelberg. 530 pp.
Harless, S. 1988. See the light. *Greenhouse Manager*. 7(6). Oct., 1988.
He, L., T.H. Short and X. Yang. 1990. Theoretical analysis of solar radiation transmission through a double-walled, acrylic, pellet-insulated, greenhouse glazing. *Trans. ASAE*. 33:657-664.
Heuvelink, E. and H. Challa. 1989. Dynamic optimization of artificial lighting in greenhouses. *Acta Hort.* 260:401-412.
Holcomb, E.J. and R.D. Heins. 1988. Growing green. *Growers' Talks*. Oct., 1988.
Holmes, M.G. and H. Smith. 1977. The function of phytochrome in the natural environment. IV. Light quality and plant development. *Photochem. and Photobiol.* 25:551-557.
Honda, Y., T. Toki and T. Yunoki. 1977. Control of gray mold of greenhouse cucumber and tomato by inhibiting sporulation. *Plant Disease Rptr*. 61:1041-1044.
Horridge, J.S. and K.E. Cockshull. 1989. The effect of the timing of a night-break on flower initiation in *Chrysanthemum morifolium* Ramat. *J. Hort. Sci.* 64:183-188.
Hughes, A.P. and K.E. Cockshull. 1971. The effect of light intensity and carbon dioxide concentration of the growth of *Chrysanthemum morifolium* cv Bright Golden Anne. *Ann. Bot.* 35:899-914.
Hughes, A.P. and K.E. Cockshull. 1972. Further effects of light intensity, carbon dioxide concentration, and day temperature on the growth of *Chrysanthemum morifolium* cv Bright Golden Anne in controlled environments. *Ann Bot.* 36:533-550.
Hughes, J.E. et al.. 1984. Photoperiodic time signals during twilight. *Plant, Cell and Environ.* 7:269-277.
Iqbal, M. 1983. An Introduction to Solar Radiation. Academic Press, NY. 390 pp.
Klueter, H.H. et al.. 1980. Photosynthesis in cucumbers with pulsed or continuous light. *Trans. ASAE.* 23:437-442.
Kluge, M. and I.P. Ting. 1978. Crassulacean Acid Metabolism: Analysis of an Ecological Adaptation. Springer-Verlag, NY.
Knapp, C.L. and T.L. Stoffel. 1982. Direct normal solar radiation data manual. Solar Energy Res. Inst., Golden, CO. 69 pp.
Kofranek, A.M. 1980. Cut chrysanthemums. *In* Introduction to Floriculture. R.A. Larson ed. Academic Press, NY, pp. 3-45.
Koontz, H.V. and R.P. Prince. 1986. Effect of 16 and 24 hours daily radiation (light) on lettuce growth. *HortScience*. 21:123-124.
Kozai, T. 1977. Direct solar light transmission into single-span greenhouses. *Agric. Meteor.* 18:327-338.
Kozai, T. and M. Kimura. 1977. Direct solar light transmission into multispan greenhouses. *Agric. Meteor.* 18:339-349.
Kozai, T., J. Goudriaan and M. Kimura. 1978. Light transmission and photosynthesis in greenhouses. Simulation Monograph, Wageningen. 99 pp.
Kraml, M. 1986. Light direction and polarization. *In* Photomorphogenesis in Plants. R.E. Kendrick and G.H.M. Kronenberg ed. Martinus Nijhoff, Boston. pp. 237-267.
Krenzer, E.G., D.N. Moss and R.K. Crookston. 1975. Carbon dioxide compensation points of flowering plants. *Plant Physiol.* 56:194-206.
Kronenberg, G.H.M. and R.E. Kendrick. 1986. The physiology of action. *In* Photomorphogenesis in Plants. R.E. Kendrick and G.H.M. Kronenberg eds. Martinus Nijhoff, Boston. pp 99-114.
Kurata, K. 1983. Studies on improvement of the light environment in greenhouses. (1) Application of a Fresnel prism to greenhouse covering. J. Agric. Meteor. 39:103-106 (Japanese).
Kurata, K. 1990. Role of reflection in light transmissivity of greenhouses. *Agric. and Forest Meteor.* 52:319-331.
Langhans, R.W. 1980. Greenhouse Management. Halcyon Press, Ithaca, NY. 239 pp.
Landsberg, H.E. 1961. Solar radiation at the earth's surface. *Solar Energy*. 5:95-98.
Larson, R.A. ed. 1980. Introduction to Floriculture. Academic Press, NY. 607 pp.
Lawson, G. 1987. Light up for stronger stems and better colour. *The Grower.* 108(12):15-21. Sept. 17, 1987.

List, R.J. 1966. Smithsonian Meteorological Tables. *Smith. Misc. Collection* Vol. 114, Publ. 4014. Washington, D.C. 527 pp.

Littlefair, P.J. 1981. The luminance distribution of an average sky. *Lighting Research and Technology*. 13:192-198.

McCree, K.J. 1972. The action spectrum, absorptance and quantum yield of photosynthesis in crop plants. *Agric. Meteor.* 9:191-216.

McFarlane, J.C. 1978. Light. *In* A Growth Chamber Manual. R.W. Langhans ed. Cornell Univ. Press, Ithaca, NY. 222 pp.

Meyer, J. 1989. Evaluation of artificial light systems. *Acta Hort.* 245:370-376.

Moreshet, S., Z. Plaut and N. Zieslin. 1976. Spatial variation in glasshouse rose flower production in relation to solar radiation. *Sci. Hort.* 5:269-276.

Mortensen, L.M. 1987. Optimizing the environment. *Greenhouse Grower*. 5(5):28-32. May, 1987.

Myneni, R.B., J. Ross and G. Asrar. 1989. A review on the theory of photon transport in leaf canopies. *Agric. and Forest Meteor.* 45:1-153.

Nederhoff, E.M. 1984. Light interception of a cucumber crop at different stages of growth. *Acta Hort.* 148:525-534.

Nijskens, J. et al.. 1985. Radiation transfer through covering materials, solar and thermal screens of greenhouses. *Agric. and Forest Meteor*. 35:229-242.

Nilsen, S., C. Dons and K. Hovland. 1983. Effect of reducing the infrared radiation of sunlight on greenhouse temperature, leaf temperature, growth and yield. *Sci. Hort.* 20:15-22.

Nilwik, H.J.M. 1980a. Photosynthesis of whole sweet pepper plants. 1. Response to irradiance and temperature as influenced by cultivation conditions. *Photosynthetica.* 14:373-381.

Nilwik, H.J.M. 1980b. Photosynthesis of whole sweet pepper plants. 2. Response to CO_2 concentration, irradiance and temperature as influenced by cultivation conditions. *Photosynthetica.* 14:382-391.

Nilwik, H.J.M. 1981a. Growth analysis of sweet pepper (*Capsicum annuum* L.) 1. The influence of irradiance and temperature under glasshouse conditions in the winter. *Ann. Bot.* 48:129-136.

Nilwik, H.J.M. 1981b. Growth analysis of sweet pepper (*Capsicum annuum* L.) 2. Interacting effects of irradiance, temperature and plant age in controlled conditions. *Ann. Bot.* 48:137-145.

Novoplansky, A. et la. 1990. Greenhouse cover for morphogenetic signaling. *Proc. Int. Seminar and British-Israel Workshop*. Bet-Dagan. Aug., 1990.

Papadopoulos, A. and D.P. Ormrod. 1988a. Plant spacing effects on light interception by greenhouse tomatoes. *Can. J. Plant Sci.* 68:1197-1208.

Papadopoulos, A. and D.P. Ormrod. 1988b. Plant effects on photosynthesis and transpiration of the greenhouse tomato. *Can. J. Plant Sci.* 68:1209-1218.

Potter, R. 1987. Brightening up bleak mid-winter. *The Grower*. 108(13). Sept. 24, 1987.

Robbins, F.V. and C.K. Spillman. 1980. Solar energy transmission through two transparent covers. *Trans. ASAE.* 23:1224-1231.

Roelofse, E.W. and D.W. Hand. 1981. The effect of daylength on the development of glasshouse celery. *J. Hort. Sci.* 64:283-292.

Russell, R.W.J. 1985. An analysis of the light transmittance of twin-walled materials. *J. Agric. Eng. Res.* 31:31-53.

Salisbury, F.B. 1982. Photoperiodism. *Hort. Rev.* 4:66-105.

Salisbury, F.B. and C.W. Ross. 1985. Plant Physiology. Wadsworth Publ. Co., Belmont, CA. 540 pp.

Senger, H. and W. Schmidt. 1986. Diversity of photoreceptors. *In* Photomorphogenesis in Plants. R.E. Kendrick and G.H.M. Kronenberg eds. Matinus Nijhoff, Boston. pp 137-183.

Shaddick, C. 1987. Moerman's expand lighting to 2½ acres of glass. *The Grower*. 108(9), Nov. 5, 1987.

Shannon, W.M., W.C. Bausch and C.M.U. Neale. 1986. Spectral response of corn under varying canopy geometry. *ASAE Paper No. 86-3516*. St. Joseph, MI.

Smith, H. 1986. The light environment. *In* Photomorphogenesis in Plants. R.E. Kendrick and G.H.M. Kronenberg ed. Martinus Nijhoff, Boston. pp 187-217.

Smith, H. and M.G. Holmes. 1977. The function of phytochrome in the natural environment. III. Measurement and calculation of phytochrome photoequilibria. *Photochem. and Photobiol.* 25:547-550.

Soffe, R.W., J.R. Lenton and G.F.J. Milford. 1977. Effects of photoperiod on some vegetable species. *Ann. Appl. Biol.* 85:411-415.

Spitters, C.J.T. 1986. Separating the diffuse and direct component of global radiation and its implications for modeling canopy photosynthesis. Part II. Calculation of canopy photosynthesis. *Agric. and Forest Meteor.* 38:231-242.

Stanhill, G. et al.. 1973. The radiation balance of a glasshouse rose crop. *Agric. Meteor.* 11:385-404.

Szarek, S.R. and I.P. Ting. 1977. The occurrence of crassulacean acid metabolism among plants. *Photosynthetica.* 11:330-342.

Thimijan, R.W. and R.D. Heins. 1983. Photometric, radiometric, and quantum light units of measure: A review of procedures for interconversion. *HortScience.* 18:818-822.

Thomas, R.B. 1978. The use of specularly-reflecting back walls in greenhouses. *J. Agric. Eng. Res.* 23:85-97.

Tross, M.J. et al.. 1984. An optical liquid filter greenhouse: Numerical solution and verification of a thermodynamic model. *Acta Hort.* 148:401-409.

Tsujita, M.J. 1980. High pressure sodium lamps help Canadian growers. *Roses, Inc. Bull.*, June, 1980.

Tsujita, M.J. 1981. How we do it in Canada –Crops and lamps. *Ohio Florists' Assoc.* Bull. 621:6-7.

Tuller, S.E. and M.J. Peterson. 1988. The solar radiation environment of greenhouse-grown Douglas-fir seedlings. *Agric. and For. Meteor.* 44:49-65.

Van Bavel, C.H.M. 1978. Projecting crop growth in a fluid-roof solar greenhouse. *Acta Hort.* 87:301-310.

Vierstra, R.D. and P.H. Quail. 1986. The protein. *In* Photomorphogenesis in Plants. R.E. Kendrick and G.H.M. Kronenberg ed. Martinus Nijhoff, Boston. pp 35-60.

Vince-Prue, D. 1986. The duration of light and photoperiodic responses. *In* Photomorphogenesis in Plants. R.E. Kendrick and G.H.M. Kronenberg eds. Martinus Nijhoff, Boston. pp. 269-305.

Volgelmann, T.C. 1986. Light within the plant. *In* Photomorphogenesis in Plants. R.E. Kendrick and G.H.M. Kronenberg eds. Martinus Nijhoff, Boston. pp. 307-337.

Walker, P.N. 1985. Save with electric lights. *Greenhouse Grower.* Dec., 1985. pp 18-20.

Walker, P.N. 1987. Select the right lights for your greenhouse. *Greenhouse Grower.* 5(5):90-94. May, 1987.

Wang, J.W. and C.M.M. Felton. 1983. Instruments for Physical Environmental Measurements. Vol. I. Kendall/Hunt Publ. Co., Dubuque, IA. 378 pp.

Withrow, R.B. and A.P. Withrow. 1956. General control and measurement of visible and near-visible radiant energy. *In* Radiation Biology, Vol. III, A. Holleander, ed. McGraw-Hill, New York.

Whittle, R.M. and W.J.C. Lawrence. 1959. The climatology of glasshouses. I. Natural illumination. *J. Agric. Eng. Res.* 4:326-340.

Wilkins, H.F. 1980. Easter lilies. *In* Introduction to Floriculture. R.A. Larson ed. Academic Press, NY. pp. 327-352.

World Meteorological Organization. 1965. Guide to meteorological instruments and observing practices. WMO No. 8. TP3, Supplement 5.

Zieslin, N. and Y. Mor. 1990. Light on roses. A review. *Sci. Hort.* 43:1-14.

CHAPTER 4

TEMPERATURE

I. INTRODUCTION

In this chapter, the objectives are to review the growth and the systems that implement temperature manipulation. Temperature control effects of temperature on will be left to Chapter 8. There are a number of reasons for this approach. Prior to the 1960s, temperature was the parameter best capable of being managed to some degree by the grower. Textbooks dealing with the subject, even in the 1970s, tended to treat each major climate parameter as a separate entity. Those of us fortunate to be in this field over the past 40 years have seen what amounts to a revolution in instrumentation and control, with a greatly enhanced appreciation of the relationships between surrounding technology and corresponding advances in greenhouse culture. For example, Korns and Holley [1962] attempted automated temperature control with solar irradiance, to be followed later by Holley and Juengling [1963] with CO_2 as an additional parameter. Unfortunately, the equipment available was bulky, slow, inaccurate, and subject to breakdown. The potential for this type of climate management was never introduced into commercial practice until the advent in the late 1970s of desktop computers with sufficient memory. Similarly, Hanan [Unpublished, 1965] constructed an automatic irrigator based upon solar energy accumulation. Again, equipment available was inadequate. In fact, during the work leading to Hanan and Jasper's 1967 publication, measuring and recording equipment was all electromechanical and subject to severe temperature effects on accuracy. By the time this equipment was installed, the same apparatus was all solid-state –in less than 2 years. Thus, the thought may be present but limited by supporting technology.

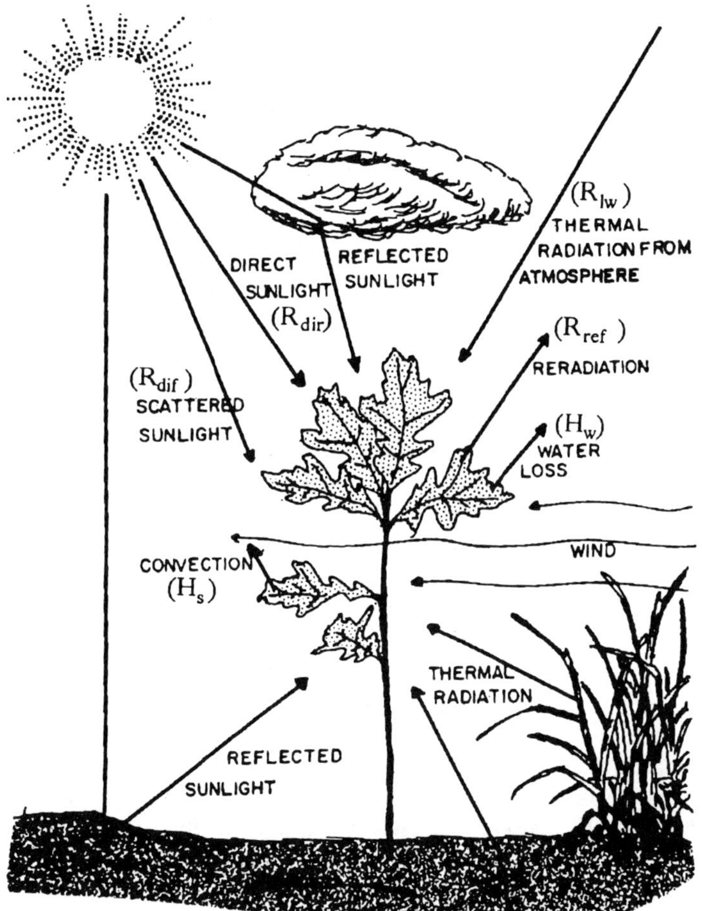

Fig. 4-1. Energy fluxes for a single plant in the open.

The present situation is shown by the numerous symposia on climate control, energy utilization, and modeling that have been published since the 1970s (e.g., *Acta Horticulturae* Volumes 106, 147, 162, 174, 229, etc.). In particular, the European workers have been in the forefront of such subjects as plant and greenhouse modeling, with integration into the economic decisions that a greenhouse manager is called upon to make. It would be nonsense to consider temperature control as a separate issue. Climate control in greenhouses has now reached a level of complexity and sophistication that requires separate attention –apart from how the output may be implemented in the greenhouse.

II. THE PLANT AND THERMOMORPHOGENESIS

A. PLANT TEMPERATURE

The temperature a plant actually achieves is a function of several factors –aside from the temperature of the air surrounding the plant. To take the simplest case (Fig. 4-1), plant temperature (T_p) will be a function of solar radiation, thermal radiation exchange, sensible heat convection, and conversion of water from a liquid to a vapor or transpiration. At equilibrium –which seldom occurs– the sum of these factors can be formalized:

$$R_{dir} + R_{dif} - R_{ref} \pm R_{lw} \pm H_s - H_w - P + M \pm S = 0 \quad (W\,m^{-2}) \tag{4.1}$$

where: R_{dir} = *direct solar radiation, including that reflected from clouds and surroundings*
R_{dif} = *diffuse solar radiation*
R_{lw} = *longwave thermal radiation exchange between sky and plant, and from surroundings*
R_{ref} = *reflected shortwave radiation*
H_s = *sensible heat exchange between plant and air*
H_w = *latent heat exchange due to transpiration*
P = *energy used in photosynthesis*
M = *energy released from respiration*
S = *energy stored in the plant*

R_{dir} and R_{dif} include radiation that may be reflected from the clouds, ground or nearby objects. The radiation reflected by plants is often considered constant (20-30%) [Penman et al.,1967], although data from Hanan [1970a] (Fig. 3-37) would suggest that plant reflectivity can vary, depending upon cover and outside irradiance (R_{tot}). When R_{ref} is subtracted, and R_{lw} accounted for, the net radiation (R_n) is what is actually available to plants. Inasmuch as only 4 to 5% of R_n is typically utilized in photosynthesis, the energy involved, together with that liberated in respiration is generally ignored. Even where efficiency may approach 8% [Aikman, 1989], the usual error expected in field radiation measurements (±10%) [Hanan, 1984; Gates, 1980] more than disguises contributions resulting from P. Stanghellini [1983d] suggested a ±5% error in radiation measurements made on a tomato crop. Gieling and Schurer [1995] state that uncertainty can be kept within ±5% if coverings are cleaned monthly and the instruments calibrated every two years. The storage term (S) is often neglected in long term measurements (>24 hr), and few [e.g., Stanghellini, 1983b; 1983d; 1987; Albright et al., 1985] have considered the storage term of vegetative canopy and ground in many of the formulations of the energy budget. In short term analyses for purposes of climate control, the effect of storage must be included.

The term H_s can be written [Campbell, 1977]:

$$H_s = \rho C_p \frac{(T_p - T_a)}{r_H} \quad (W\,m^{-2}) \tag{4.2}$$

where: ρ = *air density (1.29 kg m^{-3}) at STP*

C_p = specific heat of air ($1.01x10^3$ J $kg^{-1}K^{-1}$). The product of ρC_p = the volumetric heat capacity of air (1200 J $m^{-3}K^{-1}$)
T_p = plant temperature (K)
T_a = air temperature (K)
r_H = resistance to heat transfer (s m^{-1})

H_w follows Stanghellini [1987]:

$$H_w = \frac{\rho C_p}{\gamma r_v}(e_o - e_a) \qquad (W m^{-2}) \tag{4.3}$$

where: γ = psychrometric constant (0.66 Pa K^{-1})
e_o and e_a = vapor pressure (Pa) at the evaporating surface and air respectively

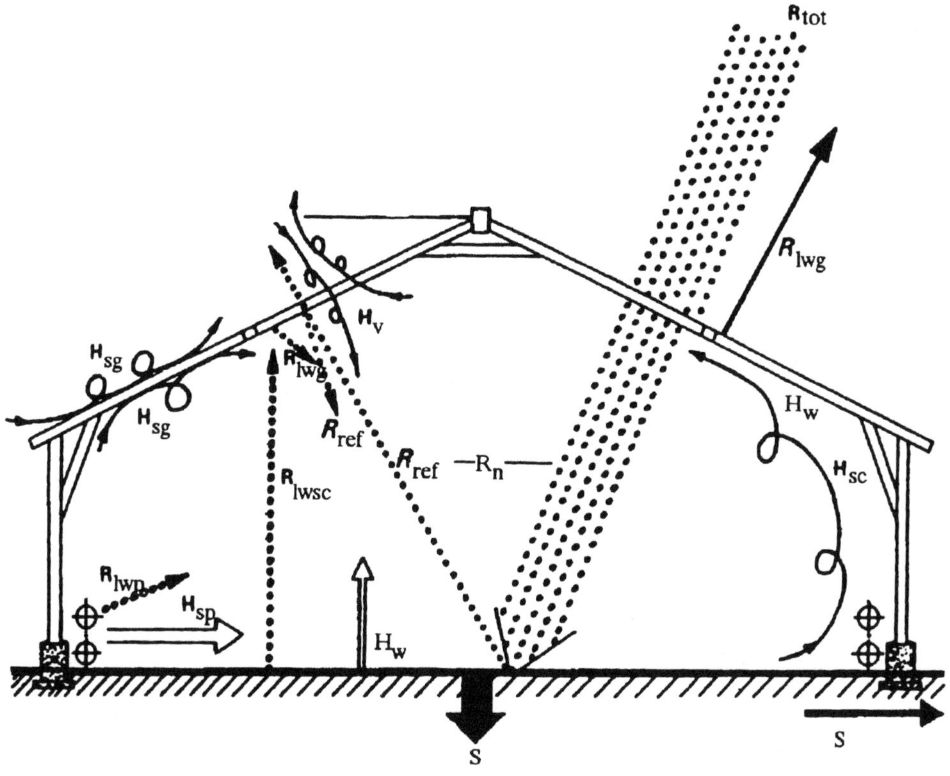

Fig. 4-2. Energy fluxes in a naturally ventilated, hot water heated greenhouse with a fully mature crop. The heat storage term includes the crop mass [Modified from Stanghellini, 1983c] (With permission of the author).

Further discussion of vapor flux from the plant will be left to Chapter 5. But, it is sufficient to emphasize again that, for well-watered plants, an average 70% of net radiation (R_n) is utilized to evaporate water, whereas the respective contributions of photosynthesis and respiration in the energy balance Eq. 4.1 can be safely ignored. In climate control of the greenhouse, this major form of energy partitioning as latent heat cannot be dismissed. There is a serious dichotomy in the literature when those interested in photosynthesis, growth, and development utilize PAR terms (μmol $m^{-2}s^{-1}$ or W m^{-2} PAR) versus those interested in total energy distribution, which must take into account not only the 400 to 700 nm PAR region, but also those wavelengths beyond this restricted region. When the radiation integral is so important to growth (Chapter 3), unfortunate confusion can arise when that integral is provided in µmoles versus Watts. If the latter, one is never sure, unless clearly stated, that W m^{-2} is meant to be PAR or total spectrum. When dealing with an energy balance, the units must include the full spectrum .

The remaining major term in Eq. 4.1 is that due to longwave, thermal radiation exchange between the plant and sky and surrounding objects (e.g., heating pipes, a cold greenhouse cover, etc.). Some individuals, such as Stanghellini [1987], include incoming thermal radiation in R_n over the range 300 to 40000 nm. However, since most radiation instrumentation restricts irradiance measurement to 300 to 3000 nm, I choose to consider R_{lw} separately:

$$R_{lw} = \epsilon\sigma(T_p^4 - T_s^4) \qquad (W\,m^{-2}) \tag{4.4}$$

where: ϵ = *emissivity, generally taken to be 0.95*
σ = *Stefan-Boltzmann constant (5.67x10^{-6} W m^{-2}K^{-4})*
T_p = *absolute radiating surface temperature (K)*
T_a = *absolute temperature of absorbing surface (K)*

This is a simplified equation since the determination of surface areas involved (view factors) may be difficult. For a plant outdoors, the hemisphere of the sky above it is relatively simple, contrasted to an enclosure.

When the plant is enclosed in a structure, the situation becomes much more complicated since now there is the interaction of energy fluxes between a structure and vegetative canopy, and between the greenhouse and external climate. The situation is shown in Fig. 4-2, where Stanghellini [1983c] has depicted a naturally ventilated greenhouse heated with hot water pipes. I have modified the diagram to fit the units used here. Obviously, energy fluxes will vary depending upon whether the greenhouse is heated with hot air, warm water, infrared, etc., or employs some type of forced ventilation. The major point to emphasize is that climate in a greenhouse should never be considered without a vegetative canopy. While there has been work of value dealing with empty structures [e.g., Albright et al., 1985; Silva and Rosa, 1987], it is not the normal state of a commercial range. A greenhouse and its crop should be treated as an entity:

$$R_n + R_{lwp} + H_{sp} = \pm H_{sg} \pm R_{lwg} \pm H_v \pm S \qquad (W\,m^{-2}) \tag{4.5}$$

where: R_{lwp} = *longwave radiation from heating system*
H_{sp} = *sensible heat exchange from heating system*
H_{sg} = *sensible heat exchange by convection and conduction through walls and covers*
R_{lwg} = *longwave radiation from cover to sky,* H_v = *sensible and latent heat exchange through ventilators*
S = *energy flux to and from ground and canopy*

There are several simplifying assumptions in this formula, compared to many to be found in the literature [e.g., Bot, 1983; Stanghellini, 1987; ASHRAE, 1978; Walker, 1965; Garzoli and Blackwell, 1973]. For example, R_{lwsc}, H_{sc}, and H_w are neglected. Longwave radiation (R_{lwsg}) from the soil and crop will be trapped within the greenhouse, especially if the cover is glass (see Section 4.IV.B.4 and Chapter 3). Exchanges involving sensible heat from the crop (H_{sc}) and latent heat exchange from the crop or on the glass (H_w, condensation) are both involved in energy exchange by ventilation or infiltration (H_v). Although total radiation (R_{tot}) is easier to measure outdoors, a significant proportion will be reflected, and a smaller amount will be absorbed in the cover. A portion of that reflected from the crop (R_{ref}) may pass through the cover and be lost. However, this could be compensated for by a measurement of net radiation (R_n). The latter is possible with existing instruments, and the variability in such internal measurements due to the greenhouse structure can be reduced by using a linear net radiometer to remove the effect of shading by the superstructure. Although there are few investigators reported using this method, it should be encouraged. This leaves the major heat fluxes of a greenhouse to H_{sg}, H_v and R_{lwg}. In all of these, temperature differences are a major driving force, that must be understood for efficient climate control. However, Chapter **5** will show the close interaction between an interior climate and the crop. Simplification of the whole process may be possible.

B. THE EFFECTS OF TEMPERATURE

One should remember that temperature manipulation in greenhouses is a function of: 1) the external climate and weather, 2) the systems and fuel available, and 3) the objectives of the grower. Thus, in some locations where heating is minimal or nonexistent, the grower may have to compromise with objectives in the sense that yield is delayed and quality less than desired. The economics of energy costs for heating may be prohibitive. On the other hand, the grower may be unable to maintain optimal temperatures because of low sunlight or the structure has insufficient heating or cooling capacity. Other objectives of the grower include height manipulation, timing for a good market, vegetative growth rather than reproduction, germination, rooting of vegetative cuttings or floral induction for future forcing. Obviously, the so-called "optimum" temperature will vary with these objectives and the plant species.

1. General Considerations

The temperature range for greenhouse plant production is relatively narrow, i.e., from about 10 to 30 C for almost all species with some exceptions in the case of specialized application. Table 4-2 provides some optimal temperatures as a starting point for discussion. These recommendations have, in large part, been determined empirically and are usually suitable where the old-style thermostatic control is employed. We assume the temperatures, whether day (DT) or night (NT), are constant. At present, many of the recommendations have been replaced by sophisticated temperature manipulations, some of which are discussed later. Many species of tropical origin will perform adequately when DT = NT, but many others (e.g., tomatoes, cucumbers, carnations, chrysanthemums, etc.) perform better with alternating environment where NT < DT. There are at least two good reasons why NTs should be lower than DTs: 1) As nearly 75% of the energy loss by a greenhouse occurs at night, lowering temperatures at night represents an energy conservation measure. And 2), a lower night temperature reduces respiratory loss of food reserves –slower development occurs. Although Went [1957], summarizing the work at the Earheart Laboratory in California, suggested night temperatures to be most important in growth of most temperate origin plants, this has been found to be dependent upon a number of other parameters [Erwin and Heins, 1995].

Having given a set of temperatures for a given species (Table 4-1), it is necessary to qualify the recommendations:

1. Optimum temperatures will generally decrease with age, with germinating seeds and young seedlings having the highest ranges [Went, 1957; Slack and Hand, 1983; Hanan, 1959; etc.]. In many cases, researchers divide growth into a number of phases in order to simplify. For example, Slack and Hand [1983] divided cucumber growth into four stages –germination, preplanting, early post planting, and late post planting. White and Warrington [1988] divided geranium growth into seedling emergence, appearance of a visible bud after 7th leaf, and from bud to anthesis, while Karlsson et al. [1989b] utilized four stages in chrysanthemums –start of short days to a visible bud, a visible bud to disbud, disbud to color, and color to full flowering. In many respects, such staging simplifies the approach to examining thermomorphogenic effects since the objectives in each stage may be different and one stage may affect the succeeding stage [Karlsson et al., 1989b; c]. The actual average temperature change as the crop grows through the various stages toward harvest will obviously vary with species and with the particular cultivar. For this reason alone, one can appreciate the necessity for a uniform vegetative canopy. In other

Table 4-1. Suggested temperatures (C) for various greenhouse crops Adapted from Hanan et al., [1978], Ball [1985], ASHRAE [1989], and Wittwer and Honma [1979]. Note the exceptions discussed in the text. Many of these recommendations may be out of date in light of recent developments.

Plant	Temperature: Day	Temperature: Special	Temperature: Night	Remarks
Alstroemeria			18	Until floral induction
		5		2-4 weeks for floral induction
		16		Maximum soil temperature, flowering will cease above this
		13		Soil temperature for continuous flowering
Alyssum		21		Germination
	3-5+		4-10	Day temperatures above night
Amaryllis			16-21	Higher temperature for forcing, lower when in flower
Anemone	16-17	15	6-8	Maximum temperature for germination. Grow as cool as possible
Aquilegia		21-29		Germination
Asparagus		21-29		Germination
Aster	3-5+		10-12	Day temperature 3-5 C higher
Azalea		4-10	16-18	4-10 preripening temperature under lights. Day temperatures for forcing 5-8 higher
Bedding plants		21		Germination for ageratum, alyssum, aster, begonia, calendula, celosia, cosmos, dahlia, gaillardia, impatiens, lobelia, marigold, portulaca, salvia, zinnia
		18		Germination for centaurea, coleus, pansy, phlox, verbena
		16		Germination for hollyhock
		13		Gemination for sweet pea
	3-5+		4-10	3-5 higher day temperatures for calendula, larkspur, lobelia, snapdragon, stock, pansy.
	3-5+		10-16	Ageratum, aster, begonia, centaurea, coleus, dianthus, geranium, nierembergia, petunia
	3-5+		13-18	Browallia, celosia, dahlia, impatiens, marigold, salvia, verbena, zinnia
	3-5+		16-18	Kochia
			10-13	Temperature until transplanted for alyssum, calendula
			16	Temperature until transplanted for aster, balsam, browallia, coleus, geranium, impatiens, marigold, petunia, portulaca, salvia, verbena, zinnia
			10	Temperature until transplanted for dianthus, larkspur, lobelia, pansy, snapdragon, stock
			13	Temperature until transplanted for centaurea, dahlia, nierembergia, phlox
Begonia		18, 21	13-16	Germination temperatures for tuberous (18) and fibrous (21), 16 to first flowers then 13 for tuberous
			17-21	Non-stop fibrous begonias
			21-22 18	Reiger begonias, 18 for better color and flower initiation
Bulbs		5-9	16	9 for precooling, 5 until forcing, 16 for forcing, varying with required marketing period. Tulips, daffodils, hyacinths, crocus
Caladium		21		Storage in dry air prior to planting
		27-29	21-24	27-29 = soil temperature for forcing, minimum 18
Calceolaria	3-5+	18-21	9-10	18-21 for germination, 3-5 C higher day temperature

Plant	Temperature			Remarks
	Day	Special	Night	
Carnation	16-22		10-13 13-16	Temperature varies with age, irradiance, whether in soil or inert media and if CO_2 is employed. Ventilation usually at 22 C maximum, 13-16 for summer
Chrysanthemum	17-21		16-17	Minimum night temperature for flower initiation. Can be grown at lower temperatures otherwise. Will vary with irradiance, height control, substrate, and CO_2 use
Cineraria		21	9-10	Will not initiate flowers above 16 C
Cucumber	26-29	28	21	28 = germination temperature
Cyclamen		16-18	17-20 10-11	Higher temperature in first 5-6 wks, then lower temperature until flowering, 16-18 C germination
Easter lilies		4-7		Precooling, varies with variety
		2-17		Rooting temperatures, depending upon date of Easter and pretreatment
	18-21		16	Optimum temperatures vary with year and cultural technique. Leaf counting technique important
		0.11+ 0.09ADT		Formula for controlling leaf unfolding rate, ADT = average daily temperature, Karlsson et al. [1988]
Exacum	24-26		16-18	
Foliage plants	35	21-24	18	Maximum day and minimum night temperatures for stock plants. 21-24 for propagation
	24		18	Best temperature for wide number of species
Freesia		31, 16		31 C for ripening, 16 and below for initiation, 13 recommended for later process
		16-18		Seed germination
			13-17	Beginning at 17 C in summer and fall, dropping to 13 following winter and spring
Fuchsia		18-24	13	Seed germination at 18-24 C
Geranium	18, 20-24		13-17	Lower temperatures for cloudy days, higher on sunny days or night following clear days
Gerbera	21-27		9-16	Varies with climate and cultural conditions
Gloxinia	24	18-21	18	18-21 for seeded germination, water temperature above 10 C
Hibiscus	22-23	22-24	20-21	22-24 for rooting, temperatures for Mar through Sept., dropping 1-2 C Oct through Feb
Hydrangea		2,11	13-16	2 = long-term storage, 11 C for releasing dormancy
Iris		32,10	16	32 for 10 days to initiation and 10 to hasten for forcing at 16
Kalanchoe			16-21	
Lettuce	17-24		14,10-12	14 for Bibb lettuce, Grand rapids 10-12. Lower day temperatures on cloudy days higher on clear days
Poinsettia	27-32		18	For stock plants
	27		18,16	Initiation and development, 27 not to be exceeded, 16 = final temperature for near Christmas
Rose	21-29		16-17	Varies with higher temperatures under clear days, reduce to 24 C if possible when no CO_2 injection
Saintpaulia	21-23		20-21	
Snapdragon		18-21	16-10	Germination, start at 16 first 10 days and reduce gradually to 10 at flowering

Plant	Temperature			Remarks
	Day	Special	Night	
Sweet pepper	7-9+	21-21.5		21-21.5 average daily temperature for yield (ADT), 7-9 day - night temperature difference (DIF), Bakker and van Uffelen [1988]
Tomato	21-26	29	14-17	Temperatures vary according to sunlight, 29 = germination
			13-18	For furnishing transplants

words, each plant should be as close as possible to its neighbors in terms of age, morphology and treatment. Otherwise, it becomes quite impractical to attempt precision climate control. In commercial ranges, consideration of temperature in respect to stage may be impossible if several plants are in various stages such as may occur with a continuous chrysanthemum program.

2. The optimum temperature will vary with available solar radiation. In Figs. 3-52 and 3-53, higher irradiances result in greater photosynthetic activity and, according to Holder and Hand [1988], each additional 100 MJ m^{-2} radiation from the start of picking is equivalent to a yield of 1.98 kg m^{-2} tomato fruit under U.K. conditions. On chrysanthemum, Hughes and Cockshull [1972] stated that with increasing irradiance, temperatures could rise to 24 C. For the same species, Karlsson et al. [1989c] found that high DT or NT combined with low irradiance prevented flower initiation under short-day conditions. There were no flower buds present after 100 short days when the photosynthetic irradiance integral was 1.8 mols $m^{-2}dy^{-1}$. From Table 3-4, this is equivalent to 4.6 W m^{-2} clear sky radiation (400-700 nm) over a 24 hr period or 0.4 MJ $m^{-2}dy^{-1}$. Flowering occurred under all DT-NT combinations when plants received 21.6 mols $m^{-2}dy^{-1}$ (52.7 W m^{-2} or 4.6 MJ $m^{-2}dy^{-1}$). The combined effect of radiation and age of individual rose leaves on photosynthesis may be noted in Fig. 4-3. Unfortunately, maximum photosynthetic rates are usually found at relatively higher temperatures (e.g., 24 to 29 C for sweet pepper [Nilwik, 1980], and these temperatures may not be best for maximum yield and quality [Bakker and van Uffelen, 1988].

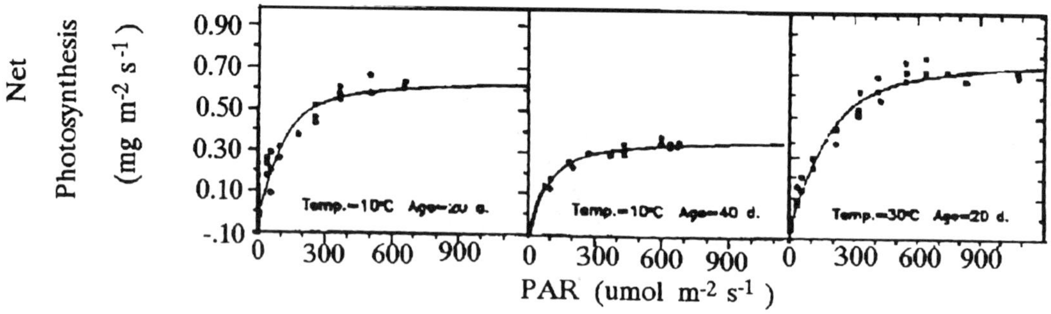

Fig. 4-3. Effect of irradiance at two temperatures and two leaf ages on net photosynthesis of individual rose leaves [Pasian and Lieth, 1989]. Note: These types of curves for other temperatures and leaf ages were not as distinctly different (With permission of the Amer. Soc. Hort. Sci.).

3. If the CO_2 concentration can be increased, then the optimum temperature is also increased. In fact, low CO_2 levels in closed greenhouses can be a serious growth limiting factor. The combination of high CO_2 and higher temperatures go hand-in-hand (Figs. 3-48, 3-49, and 4-4). This relationship is particularly apt to clear-day regimes in the winter where the sun can be allowed to heat the greenhouse up to 30 C as long as CO_2 is increased correspondingly [Hanan, 1973b]. In Colorado, a greenhouse may be closed permanently from November through March as long as CO_2 is maintained above 100 Pa (1000 ppm)

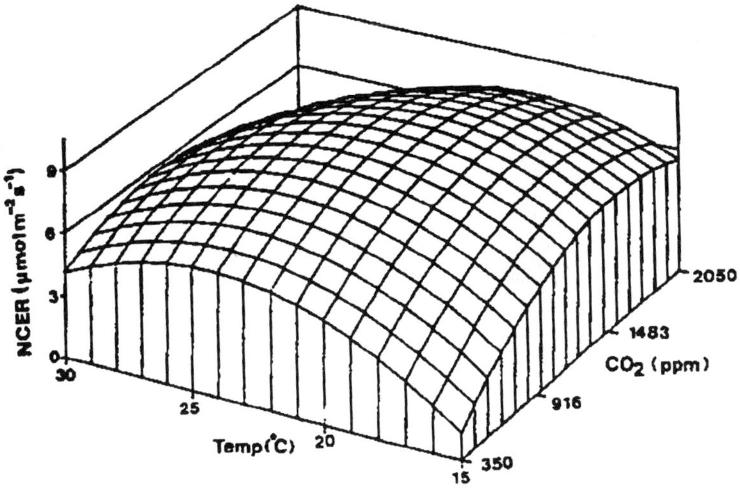

Fig. 4-4. Effect of temperature and CO_2 concentration on net carbon exchange rate (NCER) of 'Samantha' roses. Irradiance at 450 µmol $m^{-2}s^{-1}$ [Jiao et al., 1988].

for roses, which is possible if the outside temperature is at least 16 C below the inside. This practice is not appropriate for all species, nor for human comfort. Under low irradiance, the benefits of high CO_2 are reduced, and the optimum temperature is correspondingly lowered. The possibility for CO_2 enrichment of the kind employed in cold, arid climates has been discussed by Cockshull [1985] and will be considered later (Chapter 7).

4. One must carefully specify the objective for an optimum temperature, which varies depending upon the particular process (Fig. 4-5). A temperature for maximum sugar translocation may be completely different from the best temperature for root growth. The fact that maximum photosynthesis may occur at temperatures higher than that suitable for maximum yields has already been mentioned. The literature is replete with a jargon such as NAR (net assimilation rate), NCER (net carbon exchange ratio), LAR (leaf area ratio), and SLW (specific leaf weight); whereas, other authors refer to leaf unfolding rates, stem elongation rates, visible bud index, production of lateral branches, number of flowers, and days to harvest, among others. Some of these are defined in Appendix A. The point to be made is that each of these processes may have explicit optimal temperatures that are influenced by preceding treatments. In almost all of these investigations, a basic assumption is that water is freely available to the plant and little regard is given to internal plant water potential. The latter can strongly

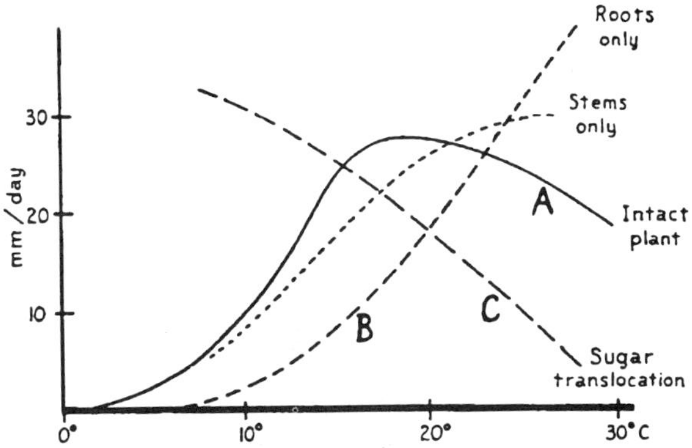

Fig. 4-5. Rate of growth of different plant parts as a function of the night temperature for tomato [Went, 1956] (With permission of the *American Scientist*).

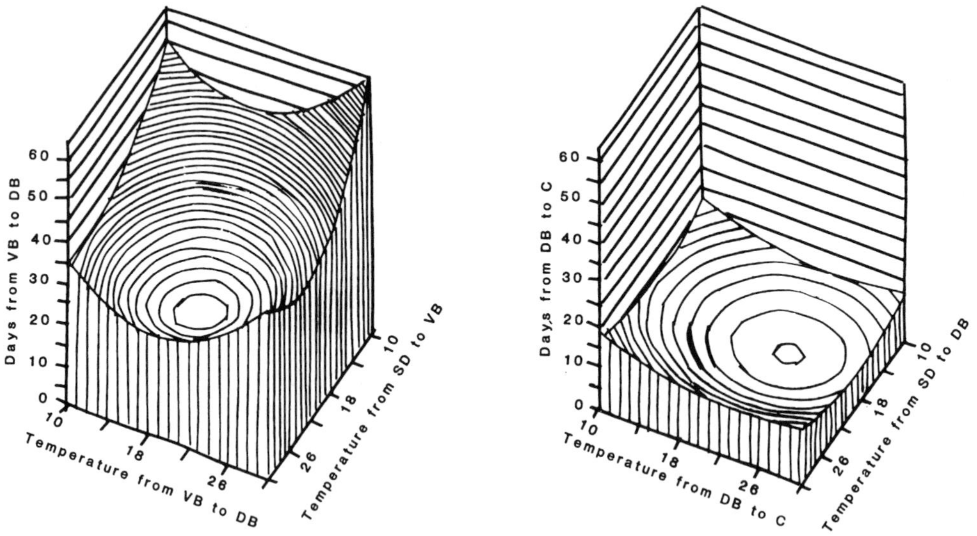

Fig. 4-6. Effect of temperature and time of exposure on development rate in chrysanthemum. **Left:** The effect during start of short days to visible bud (SD to VB) and from visible bud to disbud (VB to DB) on development in the phase VB to DB. **Right:** The effect during SD to DB (2 succeding phases) and disbud to color (DB to C) on number of days required for DB to C [Karlsson et al., 1989b] (With permission of the Amer. Soc. Hort. Sci.).

influence experimental results. Nevertheless, available publications are seriously deficient in this regard.

5. Response to a particular temperature is often predicated on what happened previously. The effect of previous history has been reinforced in recent years by Karlsson et al.'s [1989b] work on chrysanthemums (Fig. 4-6). Delay during visible bud to disbud was greater at 20 C if in the previous stage, start of short days to visible bud, plants were subjected to 10 C as contrasted with those plants subjected to 30 C. On the other hand, Hurd and Enoch [1976] found that mean temperature over the previous 24 hr had virtually no effect on current photosynthesis of spray carnations at 20 C. With sweet pepper [Bakker, 1989], unfavorable temperature treatments can result in an unfavorable leaf area/fruit ratio, which will significantly affect mean fruit weight at harvest. Not only will temperature usually affect following growth and response to temperature, but other factors such as nutrition, CO_2 levels, and water stress can modify morphology in such a fashion that response to a given treatment in following growth phases will be different. In any case, one must be careful to separate photosynthesis and dry weight gain *per se* from such processes as "growth" or "development" –the former such as elongation as contrasted to floral initiation and differentiation for the latter. Other examples can be given.

These five statements, when taken together, appear to imply that a correct target temperature for the grower's objectives is not possible. Table 4-1 appears misleading in most respects. That is not the case, as will be noted in this and later chapters.

2. Temperature Extremes

The effect of temperature can be both obvious and very subtle. Some of the more obvious effects are noted in Figs. 4-7 and 4-8. In most instances, these are examples of improper heating systems or failure to follow through with appropriate procedures. Several problems result from low temperatures commonly encountered in mild climates in unheated greenhouses. In particular, vegetable crops subjected to lower than optimal temperatures often have an increase in malformed fruit. On carnations, lack of control combined with low temperatures increases calyx splitting to the point that bands must be used to prevent flower shattering. Even though appropriate temperature control can improve quality significantly under most situations, the economics

Fig. 4-7. Examples of improper temperature control. **Upper left**: bullheads in roses caused by excessive and rapid temperature variation between heating and cooling. **Upper right**: Bullheads in carnations caused by grower failing to raise the day temperature during a long period of dark, cold weather. Intended as an energy conservation measure. **Lower left**: A poinsettia crop in which can be barely discerned in this picture, the plants in the foreground are fully in flower, whereas those in the background are still green. An example of poor heat distribution. **Lower right**: Effect of improper heat distribution on stock. Note plants to the right in full flower with those in the foreground actually damaged from hot air (latter courtesy of Jay S. Koths).

Fig. 4-8. Two obvious examples of improper temperature. **Left:** Chrysanthemums subjected to low temperatures at time of flower initiation, resulting in by-passed flowers and crown buds. Flowers should be on the same level. **Right:** Snapdragon subjected to high temperatures on one side of the spike, resulting in an abnormal and low quality flower (latter courtesy of Jay S. Koths).

of the locality may preclude the additional capital and operating expense. Examples to be found in mild climates include the fact that at temperatures below 11 C, hibiscus will develop chilling injury and will not survive below 5 C [Karlsson et al., 1989]. Cucumbers grown at temperatures of 15 DT and 10 NT will reduce the number of high-quality fruits. Depending upon the market and energy costs, the best financial result could be obtained from their treatment with the highest heating costs [Slack and Hand, 1983; Challa, 1978a]. In chrysanthemums, there is an optimal temperature for flower development of 18 C. Increasing the temperature above 24 C will reduce flower size [Erwin and Heins, 1988]. Temperatures below 10 C or above 30 C will markedly delay flowering in chrysanthemum [Karlsson et al., 1989b]. The problem of low temperature effects on time to yield and product quality is one of the most serious problems faced in unheated greenhouses. Solutions to improve the situation commonly involve methods to add heat as cheaply as possible.

3. The Temperature Integral

Went [1957] summarized the general effects of temperature by suggesting that such plants as the tomato were highly thermoperiodic, with the greatest influence on growth occurring at night. In the past decade, experimental evidence now shows that the majority of greenhouse species are capable of integrating temperature over a 24 hr period or longer. In other words, changes in temperatures during a 24-hr cycle may cause the same developmental rate so long as the 24-hr mean (ADT) is numerically the same within a restricted range of about 10 to 25 C. There are numerous examples, one of the first being Slack and Hand's [1983] results with cucumber (Fig. 4-9). The results are not always straight lines (Figs. 4-10 through 4-13), depending on where one is starting from. Above the base temperature and below the optimum temperature, straight lines are most common. These types of relationships have been found for cucumber, Easter lilies, poinsettias, chrysanthemums, hibiscus, sweet pepper, tomatoes, geraniums, as well as others. The application may be made to stem growth, leaf area development, days from planting to first harvest, fruit yield, final yields, lateral branch number, etc. The

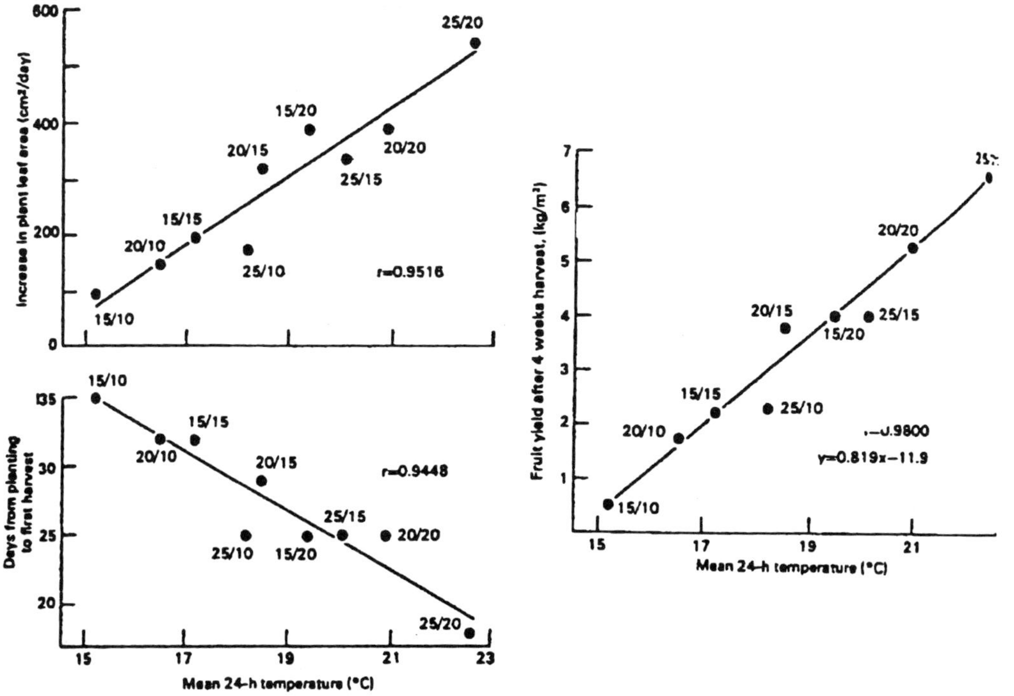

Fig. 4-9. Effect of ADT in the early post-planting period of cucumbers on rate of increase in leaf area (**upper left**), time from planting to first harvest (**lower left**), and yield after 4 weeks harvesting (**right**). Plants sown in January. Numbers on each point are the nominal DT/NT treatments [Adapted from Slack and Hand, 1983] (Reproduced by permission of the *Journal of Horticultural Science,* 58:567-572).

possibilities in the case of temperature control are numerous, and, as discussed later, capable of being modeled through statistical analyses. There are some complications. Challa and Brouwer [1985] found that after a bright day, temperature had a much stronger effect on leaf area ratio (LAR) than after a dull day. Challa and Brouwer [1985] concluded that plants do not respond in a simple fashion to integrated temperature, and indicated that a detailed knowledge of the diurnal course of growth is required. Krug and Liebig [1980] pointed out that there was a clear relationship between irradiance and the 24-hr ADT. Acock et al. [1979] stated that, in the case of dry matter gain by chrysanthemums, chrysanthemum growth was known to be unresponsive to temperature. A more recent study by Karlsson and Heins [1992] shows that dry matter accumulation varies with the particular part of the plant (roots, stems, leaves, flowers) and growth stage.

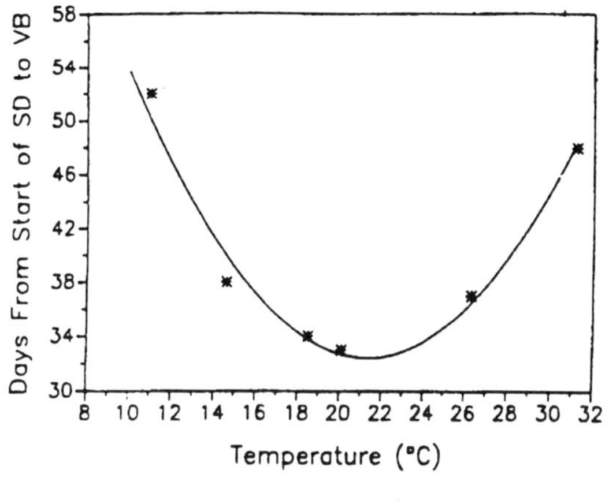

Fig. 4-10. Effect of ADT on development in phase 1 (start of short days to visible bud) in chrysanthemum [Karlsson et al., 1989b].

Fig. 4-11. Leaf unfolding rate of hibiscus in response to ADT [Karlsson, Heins, and Hackman, 1989].

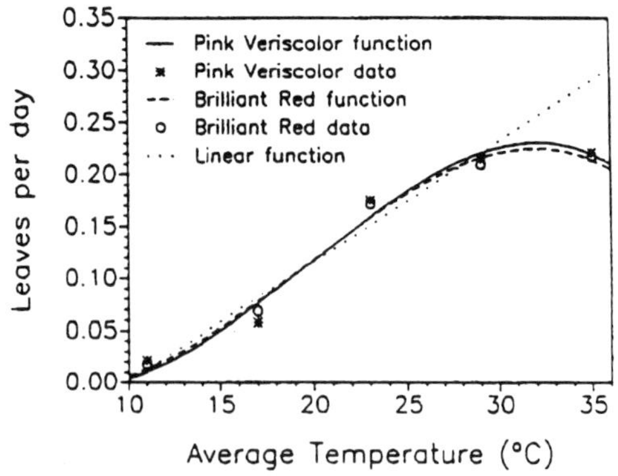

Fig. 4-12. Effect of ADT and photoperiod on number of lateral branches of *Fuschia x hybrida* after 78 days grown under long (LD) and short (SD) days. [Erwin et al., 1991] (With permission of the Amer. Soc. Hort. Sci.).

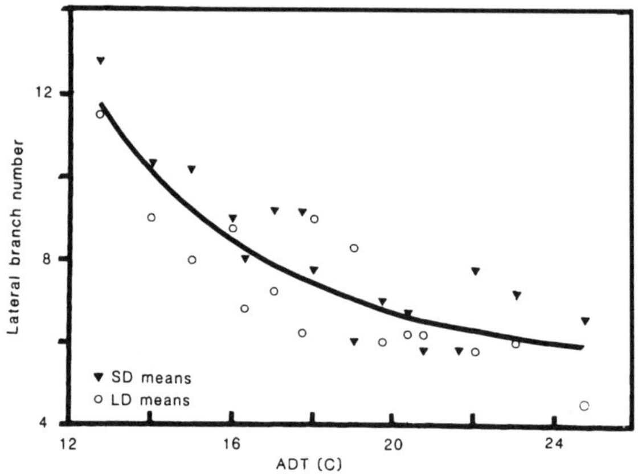

The possibilities of ADT in greenhouse control open a number of new approaches to precision climate control. This was shown by Langhans et al. [1981] where, with chrysanthemums and lettuce, the night temperature was varied in a number of ways compared to a constant NT. There were no observable differences in size, flower quality, or in time to flower with chrysanthemum as long as the average temperature was the same –within a limited temperature range. In another study by Hurd and Graves [1984] on tomatoes, 24 hr regimes were generated algebraically and compared to plants grown in constant temperatures with adjustments to maintain the same ADT in all treatments. Plants in the so-called square-wave treatment were similar to those in the "fluctuating" treatment. Stem lengths were 7% shorter in the "wind" treatment, which adjusted temperature in accordance with external wind velocity. However, the overall impression was the treatments similarity and the "wind" treatment resulted in 14% less accumulative fuel consumption. The ability of the tomato, and others, to integrate temperatures increases flexibility in modifying the diurnal temperature pattern for fuel saving, increased yield, faster growth, and reduction in the number of records required to be maintained. The effect of variations in temperature during various day or night periods often depends upon when a particular regime is instituted. Erwin et al. [1989a] stated that stem elongation in poinsettia can be reduced markedly by cooling only during the first 2 hours of the day. Erwin and Heins' [1995] summary indicated that there are more sensitive periods during the day and night. The rate of stem elongation is greatest at the end of the dark period and the beginning of the day. In an analysis of starch content in cucumber under north European winter conditions, Challa [1978a] showed that starch content dropped to starvation levels after 12 hr in the dark (Fig. 4-14). This caused a marked reduction in respiratory rate. By decreasing the temperature to 12 C in the starvation phase, growth was increased compared to standard conditions of 25 C DT/NT with 8 hr of irradiance. When the same temperature was given for the whole night (22 C), growth rate was not affected but development was reduced. At 10 C, plants showed chilling

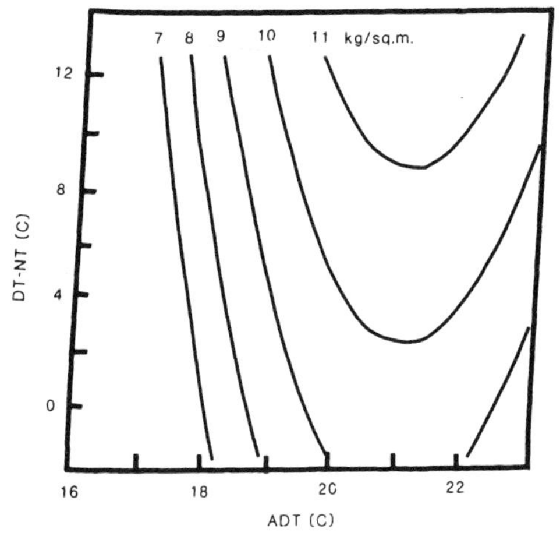

Fig. 4-13. Final yield (kg m^{-2}) of class 1 fruit of sweet pepper as a function of ADT and DT/NT amplitude or DIF [Bakker and van Uffelen, 1988).

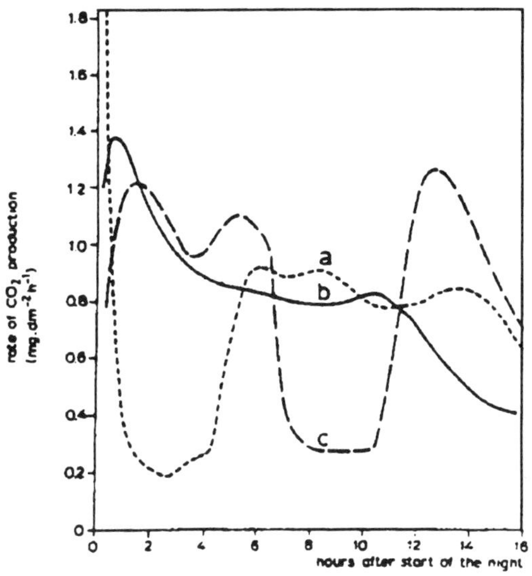

Fig. 4-14. Cucumber respiration in the dark when cold treatment given in the first 4 hr (**a**), in the middle of the night (**b**), compared to a constant 25 C (**c**) [Challa, 1978] (© 1978, Int. Soc. Hort. Sci., *Acta Hort.*, 76:147-150).

damage; 4 hr of 12 C at onset of dark caused a corresponding shift of the starvation phase. A low temperature in the middle of the night was slightly less successful in shifting the starvation phase (Fig. 4-14). Generally, the application of ADT and DIF under European conditions has not been as successful as in the U.S. [Erwin and Heins, 1995]. This difference has been attributed to the climatic differences between the two regions.

4. Day and Night Temperatures

The various effects of night temperatures versus day temperatures on growth and development of greenhouse plants have been a subject of considerable importance for many years. In general, it was not until research techniques and facilities became available that one could look at DT/NT effects separately and in combination with ADT. What one finds is that the difference between DT and NT (DIF) determines such factors as stem elongation, internode elongation, and leaf orientation [e.g., Erwin et al., 1989a; b; 1991; Karlsson et al., 1989a; Erwin and Heins, 1988; Blacquiere, 1991; Berghage and Heins, 1991, etc.]. On the other hand, DIF may be relatively minor and not related to such things as individual fruit growth in sweet pepper, or the total number of fruits per plant [Bakker, 1989]. DIF (DT-NT) has been proposed as a means of controlling stem and internode elongation in Easter lily [Erwin et al., 1989] (Fig. 4-15), poinsettia [Berghage and Heins, 1991] (Fig. 4-16), and fuschia [Erwin et al., 1991] (Fig. 4-17). There is an interaction between DIF and ADT illustrated in Fig. 4-18. Fig. 4-19 shows that response in regard to flower area will vary with the irradiance integral. As noted in Fig. 4-17, there can be a distinct photoperiodic reaction. According to Erwin et al. [1991], the effects of DIF has been noted on *Streptocarpus nobilis*, *Xanthiuum strimium*, *Lycopersicum esculentum*, *Zea mays*, *Salvia splendens*, *Impatiens walleriana*, *Nephrolepis exaltata*, as well as *Lilium, Euphorbia, Dendranthema* and *Campanula.* Plants grown under positive DIF (DT>NT) appear morphologically similar to those

Fig. 4-15. Effect of DIF on plant height of Easter Lily [Erwin et al., 1989] (With permission of the *Amer. J. Bot.*)

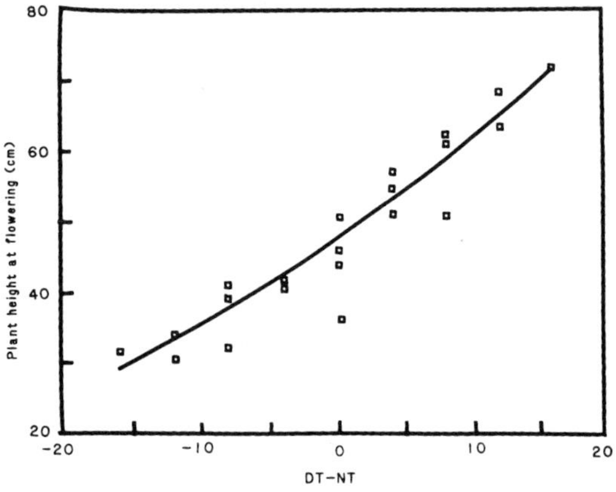

Fig. 4-16. Effect of DIF on internode length of second lateral shoot of poinsettia [Berghage and Heins, 1991] (With permission of the Amer. Soc. Hort. Sci.).

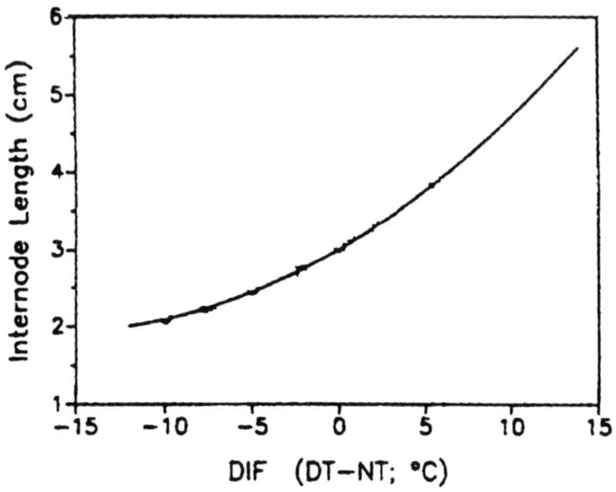

Fig. 4-17. Effect of DIF on internode length of *Fuschia x hybrida* grown under long days and short days [Erwin et al., 1991] (With permission of the Amer. Soc. Hort. Sci.).

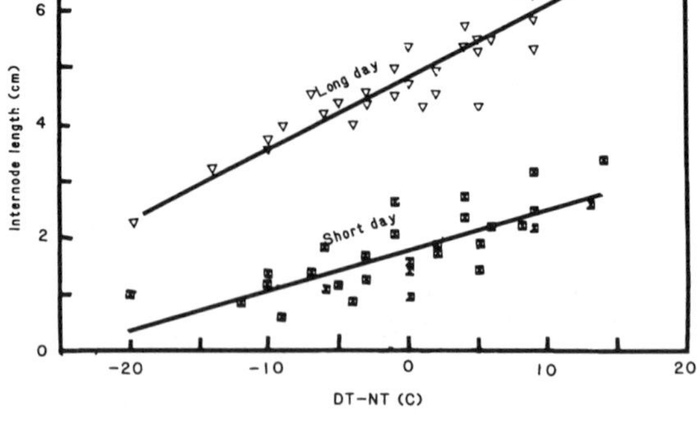

Fig. 4-18. Effect of DIF and ADT on internode length of chrysanthemum [Karlsson et al., 1989a] (With permission of the Amer. Soc. Hort. Sci.).

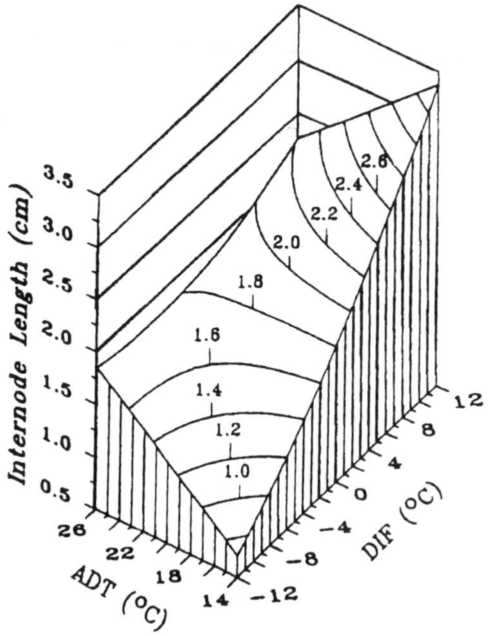

irradiated with a low R/FR ratio, whereas those grown under a negative DIF (DT<NT) are similar to those irradiated with a high R/FR ratio. Plants subjected to negative DIF often show chlorosis, and such situations may not be necessary in height control inasmuch as a change from + to 0 has a more marked effect than a change from 0 to –. A growth retardant can be included, although the data suggest that DIF application could reduce or eliminate chemical usage. Obviously, if DT is decreased without changing NT, the ADT will be less, and there can be a delay unless NT is

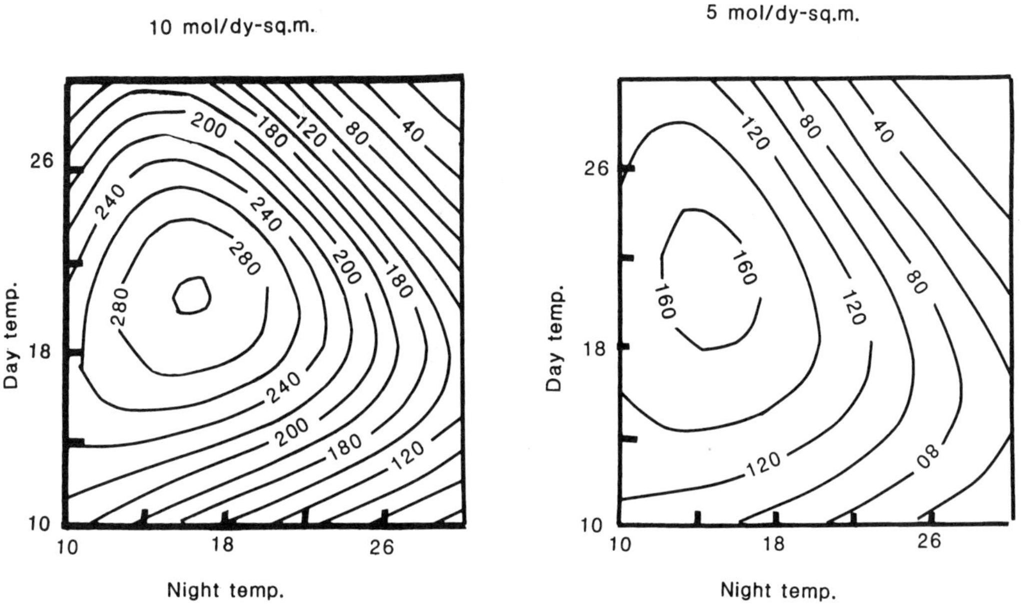

Fig. 4-19. Effect of DT and NT on total flower area (cm^2) for chrysanthemum grown under two daily irradiance integrals (10 and 5) [Reprinted from *Scientia Horticulturae*, 39:257-267, Karlsson, M.G. et al., ©1989. With kind permission of Elsevier Sci.-NL, Sara Burgerhartstraat 25, 1055 KV Amsterdam, The Netherlands].

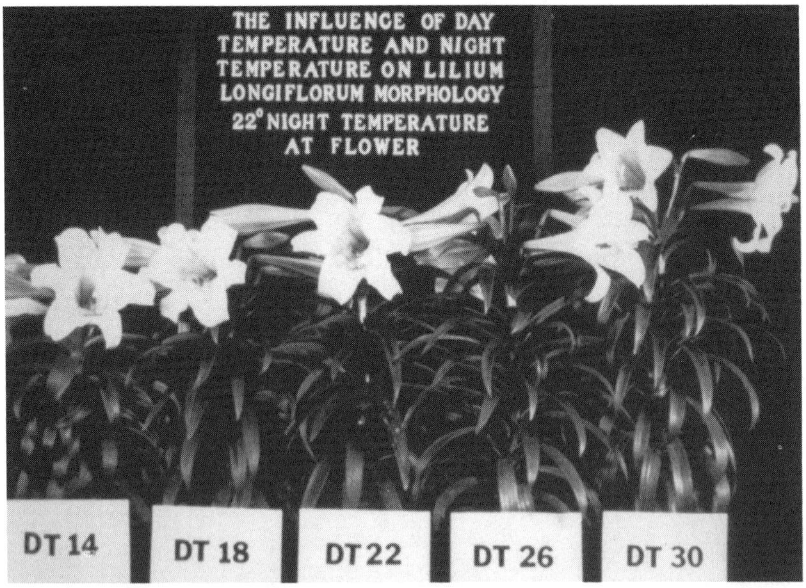

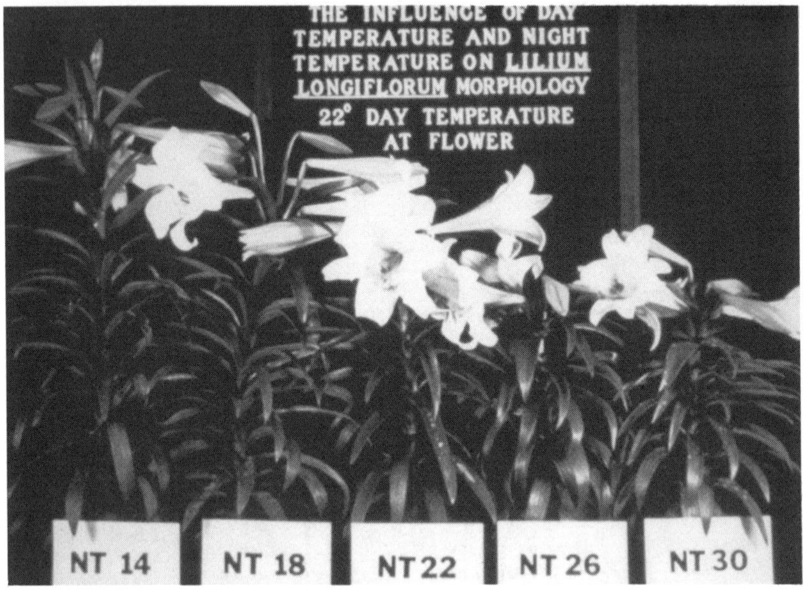

Fig. 4-20. Upper: Effect of increasing DT at constant NT as compared to increasing NT at a constant DT (**Lower**) on Easter lily [Courtesy of R.D. Heins, 1991].

correspondingly adjusted to maintain the same ADT. There are distinct effects of DT versus NT as noted in Fig. 4-20, which shows that increasing temperatures in the day increase elongation, but it is reversed at night if temperature is increased with a constant day temperature. Similar effects may be noted in a number of species, but the net result of both day and night temperature is related to DIF.

5. A General Summary

Certainly, the use of such temperature regimes as one can envision is hardly applicable using old-style analog instrumentation. Nor can they be applied where temperature control cannot be maintained accurately and precisely. However, the potential exists for digital computer applications that are only now under investigation.

In most instances, the effects of ADT and DIF –within a limited range (10 to 26 C)– can be subjected to statistical analysis with the generation of mathematical formulae that can be used in appropriate computer programs to manipulate temperature. There are several examples such as prediction of days to flower from start of short days for chrysanthemums by Karlsson et al. [1989c]:

$$Days\ to\ flower = 243.05 \exp(a1) PPF^{-0.17215} \tag{4.6}$$

$$\text{where:} \quad a1 = (-1.1588 \times 10^{-2} DT \cdot NT) + (2.1748 \times 10^{-4} DT \cdot NT^2) + (3.2368 \times 10^{-4} DT^2 \cdot NT) + (-8.5909 \times 10^{-8} DT^2 \cdot NT^2) \tag{4.7}$$

While these equations look rather complex, their solution is rapid and easily carried out on present-day computers, and the appropriate steps taken to incorporate them into ongoing programs. Another example, and more sophisticated, is the use of Richard's functions to predict internode length in poinsettias [Berghage and Heins, 1991]:

$$IL_t = P_1 (1 + P_2 e^{P_2 - P_{3t}})^{-\frac{1}{P_4}} \tag{4.8}$$

where: IL = internode length, time (t) after pinching
P_1 through P_4 = estimated parameters

The parameters can be estimated from suitable statistical programs, such as the one below for parameter P_1 for internode 1:

$$P_{1(1)} = 4.647 + 0.111(DIF) + 0.00348(DIF)^2 \tag{4.9}$$

The equation forms a static model that can be used to predict internode length with a given DT and NT. P_1 = final internode length asymptote, P_2 = when the function begins to increase on a time axis, P_3 = rate of increase, and P_4 = the inflection point. In this case, DT and NT must be held constant. This represents, to my knowledge, the first publication of the use of a Richard's function for a greenhouse crop in American literature. However, there are a number of such functions, derived by several authors, purporting to predict growth curves (see Chapter **8**). Through suitable integration, a dynamic model was constructed by Berghage and Heins:

$$L_{(1+DT)} = L_t + \int_t^{t+DT} (R(x))\,dx \tag{4.10}$$

where: L = internode length
DT = time step used in model (1 dy)
R = Richard's function

Not all formulations must be this complicated. Karlsson et al. [1988] quantified leaf unfolding rates on Easter lily:

$$leaves\ per\ day = -0.1052 + 0.0940(ADT) \tag{4.11}$$

This equation necessarily assumes a straight-line response to average daily temperature (ADT), and is valid only over a narrow temperature range. A slightly more complex equation, statistically derived by Bakker [1989] for sweet pepper is:

$$Y = a(ADT) + b(ADT)^2 + c(DIF) + d \tag{4.12}$$

where: Y = variable such as number of flowers, number of fruits, etc.
a, b, and c = constants generated through the appropriate statistical program

As descriptions of a process, the above formulae are what one calls a "black box" [Udink ten Cate, 1980; Bot, 1989, Challa, 1985]. They merely describe statistically what happens with a given input over a limited range. Since they do not describe the basic process (mechanistic approach), extrapolation to conditions outside of the range covering the original data is dangerous. For unheated greenhouses where the temperature commonly drops below 10 C, and rises above 26 C, models similar to those cited here cannot be utilized. Practical application of DIF is covered in a series of compiled reprints from the *Greenhouse Grower* magazine [Lieberth, 1990]. Unfortunately, the following Chapter 5 will show that water relationships are highly influential in plant response. Given the fact that most stem elongation occurs at night –due to lower water stress– it may be appreciated that the results of existing research on temperature effects may well be correlated with internal plant water potential. This interaction has yet to receive sufficient attention.

The possibilities for eventual automatic control of the greenhouse, based upon economic decisions of the grower, should be a reality in the next few years. Even in regions where temperature may be uncontrolled, and at the whim of weather, an appropriate model should permit the grower to predict crop outcome.

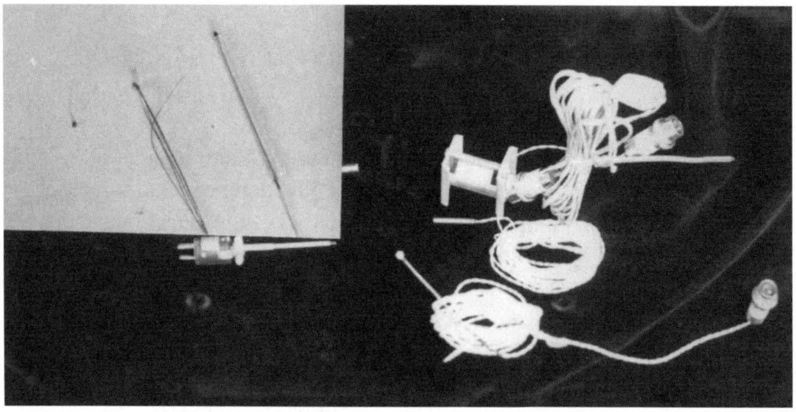

Fig. 4-21. Examples of electrical resistance sensors for temperature measurement. All of these shown here are thermistors with prices ranging from a few cents to nearly $100. Temperature range is limited with lower accuracy and stability compared to RTDs, but sensitivity can be very high. That is, they can detect very minute changes [Hanan, 1984].

III. TEMPERATURE AND ITS MEASUREMENT

Temperature is the degree of sensible heat, or intensity, whereas heat is a form of energy [Hanan, 1984]. Temperature can be thought of as a driving force for energy transfer, and it is, probably, the most common measurement made in greenhouses. Temperature, aside from its most obvious effect on plant growth and development, can be the basis for measurement of numerous other climatic values required for proper greenhouse operation, such as humidity, radiation, wind velocity, or CO_2 concentration. That is, the measurement of humidity often involves a determination of temperature in some fashion to arrive at a value for the amount of water in air (see Chapter 5). Thermal pyranometers essentially measure temperature differences. Low wind velocities are sometimes measured in terms of temperature differences. The property of temperature can be inferred from a number of thermally induced physical attributes such as changes in pressure,

Table 4-2. Thermal response times of various parts of a typical greenhouse according to Bot [1989]. The time response is the number of seconds required to reach 36.8% of the final equilibrium value.

Part of greenhouse	Thermal capacity ($J\ m^{-2}K^{-1}$)	Thermal resistance ($K\ W^{-1}m^{-2}$)	Specific time response (s)
Greenhouse air	3000	0.1	300
Cover (4 mm glass)	10000	0.04	400
Crop canopy	3000	0.05	150
Soil (1 cm thick)	25000	0.01	250
Heating pipes (55 mm diameter per meter, water)	8500	1.25	10000

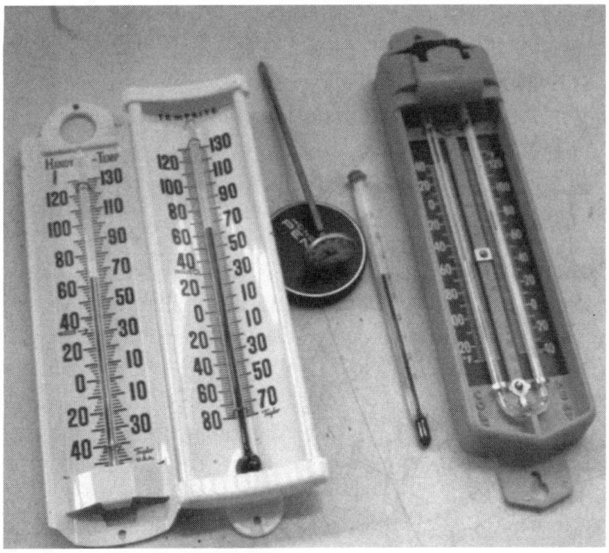

Fig. 4-22. Examples of liquid-in-glass and mechanical thermometers. The two left instruments contain alcohol and probably are the cheapest devices available. The instrument on the right is a maximum-minimum mercury thermometer. All of these are initially accurate to ± one scale division, but inaccuracy may exceed ±2.0 C after some years of use without calibration (Hanan, 1975). This is particularly true where the scale is separate from the liquid column.

volume, electrical resistance, etc. Unfortunately, a temperature indicator shows only the temperature of the sensor. It is for the user to decide how this relates to the temperature he wishes to know. In greenhouses, one not only must pay attention to accuracy and precision, but also to the manner in which the measurement is made. Unless care is taken, the results are likely to be garbage.

All temperature measurements made in greenhouses today, especially if computers are the basis for control, use some type of electrical property such as resistance or the voltage output from the junction of dissimilar metals (thermocouples). Examples of electrical sensors that may be found are provided in Fig. 4-21. Cheaper instruments, such as liquid-in-glass thermometers (Fig. 4-22), remain common. Devices, such as thermometers are only guaranteed accurate to ± 1 scale division. Thus, if one has an alcohol thermometer marked in divisions of 2 degrees, which reads 10 C, then the actual temperature could be 8 or 12 C. Basic electrical devices used for calibrating other instruments are generally accurate to within ±0.10 C. That is, ±0.10 C in reference to the "true" value –whatever that "true" value may be. Generally, the true value is determined, for example, by the National Bureau of Standards in the U.S. "Precision," on the other hand, is how close a series of measurements made by the same instrument are to each other. High precision is certainly not indicative of "accuracy." A liquid-in-glass instrument (Fig. 4-22) can be bought for less than $20.00, but electrical sensors, with accuracy to within ±0.1 C are likely to cost much more, exclusive of the metering equipment, especially if such sensors are RTDs (resistance temperature detectors). The latter have been recommended by Gieling and Schurer [1995], stating that uncertainty should not be more than ±0.3 C. Solidstate resistance devices such as thermistors, while extremely sensitive and precise (>±0.001 C), and often cheaper, will usually have an accuracy around ±1.0 C. Their calibration curve is curvilinear as contrasted with a linear RTD calibration. By the time one includes the errors found in amplification

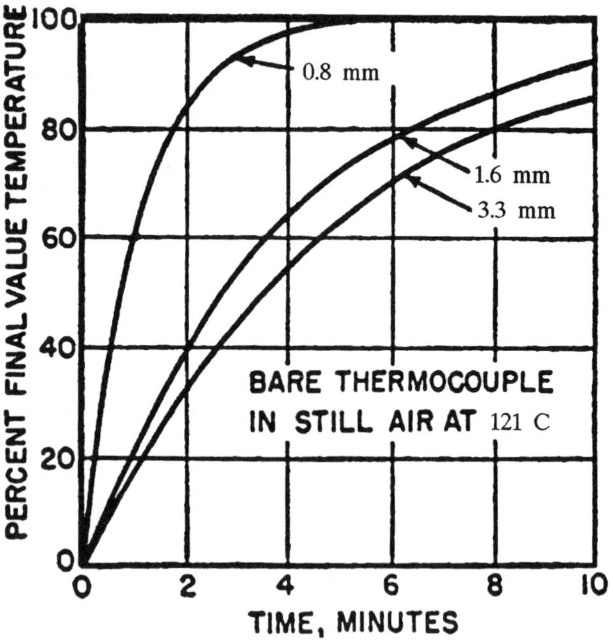

Fig. 4-23. Response of different size, bare thermocouples in still air as the result of a change in ambient temperature [Wilson and Evans, 1957].

and recording equipment, an accuracy of ±0.3 C, when using thermocouple systems, can be expensive [Hewlett-Packard, 1983].

It is very easy for present-day digital computer systems to provide an average value from several hundred measurements, obtained in less than a second, with a large number of digits to the right of the decimal point. Such information, in the case of temperature, even to two decimal places (0.00), is meaningless. Although, one will find in the literature data giving temperatures to the nearest tenth, with considerable emphasis on half-degree C increments as highly significant; about the best the grower can expect in terms of absolute accuracy of a temperature reading is ±0.5 C. The digital readout of most instrumentation will provide values to ± one digit (e.g., ±0.1 C), but, unless a lot of money is spent, anything better than ±0.5 is not realistic for common greenhouse use. Secondly, computers can overload one with data. As a general rule, the thermal mass of a greenhouse is such that climatic changes are relatively slow [Bot, 1980; 1989]. Frequency of the variation is likely to be less than 0.01 Hz or cycles per second (>100 seconds per cycle). In other words, significant changes in temperature are commonly on the order of several minutes. Measurements made too frequently are statistically unreliable since each datum depends on the previous measurement. The values are not random nor independent of each other. They do not follow the theoretical Poisson distribution curve for randomly sampled data so that statistical analyses are not valid. A temperature measurement every minute, and averaged over a 10 to 15 minute period, can be adequate to determine temperatures in greenhouses [Hanan et al., 1987]. Other parameters may have to be determined at different frequencies. For example, Bot [1980] suggested a radiation measurement each second. In 1989, Bot indicated methods for determining response times of a greenhouse, calculating values from the thermal capacity and thermal resistance of several parts of the usual greenhouse (Table 4-2). The response time (1/e) of a canopy may be 150 s to more than several hours for a soil 15 cm deep. For a first-order reaction (i.e., equation for a straight line), with a simple step response, the result is remarkably similar to what happens with individual sensors as discussed below.

There is another important factor in arriving at an indication of true temperature. That is the response time of the sensor. Due to inertia, any device placed in a new environment requires time to reach the new equilibrium. The rate of response to a change depends upon the physical properties of the sensor and its environment, as well as relevant dynamic properties (e.g., shape, size, etc.). The time it takes for a sensor to achieve a new equilibrium is referred to as the "time constant," "response time," or "lag," the coefficient (τ) having dimensions in seconds. τ specifies the time remaining for completion of an adjustment to a change such as temperature as 1/e or 0.368 (36.8%). The adjustment already completed is 1-(1/e) or 0.632 (63.2%). The symbol "e" is the natural logarithm. The time constant, thus, does not say how long it takes for the sensor to reach the actual new temperature, but how long to complete 63.2% of the change. An example of response of thermocouples to a step change in temperature is provided in Fig. 4-23. The smallest diameter wire of 0.8 mm has a response of about 1 min in still air, whereas a larger couple with 3.3 mm diameter wire (quite large) requires more than 4 min to reach 63%

Fig. 4-24. Two examples of poor controller mounting in greenhouses. In still air, response time will be very slow and sensors will be influenced by radiation and, on the right, heat conduction through the mounting will occur.

Fig. 4-25. Examples of shielded and aspirated sensor enclosures. The center example is a section of PVC irrigation pipe with an air velocity, from bottom to top, between 6 to 7 m s^{-1}. Note that all of these can be adjusted in height.

of the final value. If a computer acquisition system were to take a reading every minute, a 0.8 mm thermocouple would be closer to the true value than the 3.3 mm couple. Especially if temperature was varying rapidly, none of the sensors in this example would likely read a true value in still air. Probably the slowest response times one may expect are with alcohol thermometers in still air –more than 30 min. In water, mercury-filled thermometers will have a response time between 50 and 60 sec. Instruments that provide rapid readings are quite likely to be less accurate than a common thermometer read every 10 min –as long as the temperature

Fig. 4-26. Example of a remote sensing, infrared thermometer that is focused on the greenhouse vegetative canopy to the right of the picture. The field-of-view of such an instrument should be >30°. This device is mechanically operated, but newer devices are much smaller and solid-state.

is not changing rapidly. The response time of any instrument will, of course, depend upon the medium in which it is immersed. A rapid flow of air over the sensor will decrease its response time as heat will be convected more rapidly to and from the sensor. Or, if placed in water, which has a much higher heat capacity than air, response time will be further reduced.

The final aspect to consider in measuring temperatures in greenhouses is the location and manner in which the sensor is protected from undesirable influences such as radiation or heat exchange due to conduction in the sensor's mounting. After all, we wish to know the true air temperature, which is a matter of convective energy exchange between sensor and air. Energy exchange due to radiation or conduction would increase error. Two examples of undesirable situations are indicated in Fig. 4-24. Even though shaded, the left sensor is placed on bare ground where the radiation interception is greatest (a flat surface), resulting in excessively high temperatures on clear days. In the other case, solar radiation on the back of the mounting will cause heat to be conducted to the thermostats. In both cases, the sensors are in still air. Response time will be extremely slow. Temperature sensors must be completely shielded from radiation and aspirated by sucking air over the sensor at velocities of at least 0.5 m s^{-1}. If humidity is being measured with a wet bulb in the same aspirated shelter, the air velocity should be on the order of 3 to 4 m s^{-1} (Fig. 4-25). Good sensor mounting is of no use, however, if the instrument is located in the wrong place. Greenhouses have large volumes, and temperature distribution is likely to be uneven, especially if the heating system is hot air. Therefore, the sensor must be located where the action is and in the most representative region. As a general rule, sensors should be located at flower height on ornamentals, or where the fruit is being produced on vegetables. A better idea is to measure several places in the house or zone to provide a reasonable average, depending on the total area. The common factor of a single sensor per management zone is discussed further in Chapter 8. It shall be seen in Chapter 5 that analyses of transpiration rates require good spatial determination of several climate parameters, especially if the vegetative canopy is not continuous and uniform. Further discussion on temperature variations to be expected is provided in Section 4.IV.

To summarize: 1) Any temperature transducer must be shielded and aspirated. 2) The sensor must be located properly in respect to the crop. 3) Data must be accumulated with due regard to the change (time constant). And 4), accuracy of the sensor and its attendant equipment should be known –it usually is not.

Of course, one should not assume that the plant temperature will be the same as the air temperature. This is seldom the case, and plant temperature can often vary significantly from air temperature, especially under high solar radiation and low air movement. Ideally, we need an instrument that gives a temperature of the average plant canopy. Such instruments are available (Fig. 4-26). Unfortunately, they are expensive, not likely to be as accurate as commonly available transducers, and are not, up to now, sufficiently reliable. Calibration requires special equipment, and accuracy can be influenced by foliage emissivity and reflection in the greenhouse. Nevertheless, this would be a good solution to greenhouse climate control in terms of Hashimoto's approach to a "speaking" plant [Hashimoto et al., 1985].

IV. TEMPERATURE MANIPULATION IN THE GREENHOUSE

In this section, I shall discuss the equipment and methods employed to control temperature. First, under consideration is heating to maintain a given minimum temperature, the second, the means to limit excessive temperatures.

A. HEAT LOSSES AND REQUIREMENTS OF THE STRUCTURE

In Eq.4.5, Section IV.B.1, an energy balance statement of a greenhouse was provided. For adequate heating, one must estimate energy loss resulting from H_{sg} (convection and conduction), H_w (condensation), H_v (infiltration and ventilation), and R_{lwg} (thermal radiation loss); and balance this with H_{sp}, the sensible and radiant heat exchange from the heating system. There are a number of formulations for calculating H_{sg} among which the simplest is standard for U.S. conditions from ASHRAE [1987]:

$$H_{sg} = hA(T_i - T_o) + 0.5VN(T_i - T_o), \quad (W) \qquad (4.13)$$

where h = *heat exchange coefficient* ($W\ m^{-2}K^{-1}$)
A = *exposed surface area of the greenhouse* (m^2)
T_i = *inside air temperature* (K)
T_o = *outside air temperature* (K)
V = *internal house volume* (m^3)

Table 4-3. Natural air exchanges per hour for greenhouses [ASAE, 1988].

Construction system	Exchanges
New construction, glass or fiberglass	0.75-1.5
New construction, double layer plastic film	0.5-1.0
Old construction, glass, good maintenance	1-2
Old construction, glass, poor condition	2-4

Table 4-4. Approximate heat transfer coefficients for greenhouse covers [ASAE, 1988].

Covering	h value $W\ m^{-2}K^{-1}$
Single glass (sealed)	6.3
Single plastic	6.8
Single fiberglass	6.8
Dbl plastic, polyethylene	4.0
Rigid dbl-wall PMMA (acrylic)	3.0
Dbl glass (sealed)	3.0
Single glass & thermal screen	3.0
Dbl plastic & thermal screen	2.5

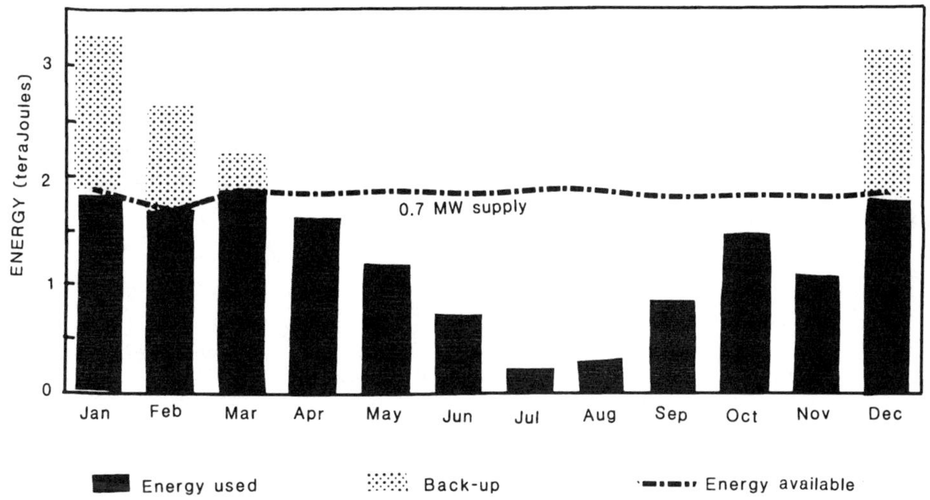

Fig. 4-27. Variation in heating requirements per month for a 1 ha, glass-covered greenhouse in the U.K. The energy supply is sized for 25% of maximum demand with top up from a standby boiler [Ellis, 1990].

N = number of air exchanges per hour

Simpler relationships are provided by Eq. 8.13 and 8.14 (Chapter 8).

The first term on the right (Eq. 4.13) represents energy loss due to radiation, conduction, and convection through the roof and walls of the greenhouse, whereas the last term on the right is energy loss due to exchange of air through cracks in the structure, or infiltration (Table 4-3). Numerous variations of Eq. 4.13 may be found in the literature. In some, a correction is made for solar heating [e.g., Bailey, 1990; Schockert and Zabeltitz, 1980]. In others, infiltration is ignored or combined with other factors [e.g., Seginer and Kants, 1986; Albright et al., 1985]. The values for infiltration in Table 4-3 are necessarily approximate. Hanan [1974] found rates for commercial ranges that varied from 0.34 air exchanges hr^{-1} for a steam-heated, plastic film cladded house to 0.96 hr^{-1} for an old, steam-heated, glass-covered range with windspeeds below 4.5 m s^{-1}, and outside temperatures below freezing. On the other hand, at wind speeds between 13 and 22 m s^{-1}, exchange rates in old houses exceeded 7 hr^{-1}. It is easier to heat an old, glass greenhouse at outside temperatures approaching -20 C, than at warmer conditions, inasmuch as all cracks are frozen shut. Infiltration is reduced to very low levels. Such situations in cold climates are extremely dangerous where combustion occurs inside the structure (natural gas unit heaters) and no provision is made for outside air to reach the combustion chambers. With modern structures, especially those covered with plastic, I would suggest air exchanges rates as low as 0.5 hr^{-1} under most conditions. Hanan [1974] also found that the infiltration rate increased with a greater temperature differential and if the heating system used forced hot air.

Table 4-5. Calculated heat transfer coefficients (h) for a variety of covers under clear and overcast skies, wind speed = 4.0 m s^{-1}, ΔT = 30 C [Adapted from Nijskens et al, 1984].

	h (W $m^{-2}K^{-1}$)	
Materials	**Clear sky**	**Over-cast sky**
Polyethylene single (PE)	9.0	7.2
Ethylvinylacetate single (EVA)	7.8	6.6
Polyvinyl chloride single (PVC)	7.6	6.4
Reinforced PVC single	6.6	5.8
Standard glass, 4 mm single	6.1	5.5
H^+ glass single layer outside[1]	5.4	5.2
H^+ glass single layer inside	4.4	3.9
PE + PE double	6.4	4.8
Polypropylene 4 mm double	5.9	4.8
Standard glass + PE double	4.3	3.9
Polycarbonate 6 mm double (PC)	3.5	3.2
Acrylic (PMMA) 8 mm double	3.4	3.0
Standard glass double	3.1	2.8
H^+ glass outside + glass[1] (double)	2.8	2.7
Glass + H^+ glass inside[1] (double)	2.5	2.2
Glass + H^+ glass between[1] (double)	2.3	2.1

[1] H^+ refers to glass with a low emissivity layer with the layer on the <u>outside</u> surface, on the <u>inside</u> surface or <u>between</u> the double panes on the outside of the inner glass.

One of the most detailed investigations on infiltration was Okada and Takakura's development [1973]. According to these authors, "ventilation rate" has some ambiguity and it would be better for comparisons between structures to base infiltration on volume change per unit cover area (roof and walls). They presented a formula based upon wind velocity and temperature difference inside to out:

$$V_g = av + b\sqrt{T_i - T_o} \tag{4.14}$$

where: V_g = *volume of air change rate per unit cover area* ($m^3m^{-2}hr^{-1}$)
a and b = empirical constants
v = *wind speed* (m s^{-1})
$T_i - T_o$ = *temperature difference inside to outside (C)*

Okada and Takakura's evaluation resulted in the formula $V_g = 0.44v + 0.14\sqrt{\Delta T}$. The calculated values were "reasonably" close to the observed volume changes per hour (N) until external wind speeds exceeded 4.7 m s^{-1}. Of particular interest was their discussion of the need to measure air velocity outside of the boundary conditions of the structure (see Section 5.IV.B.3). The authors proceeded to calculate the heat transfer coefficient as the result of infiltration, including that due to latent heat as well as sensible heat:

$$h_{inf}^{*} = \left[Cp + \frac{L\Delta q}{\Delta T}\right]\rho V_g \tag{4.15}$$

where: h_{inf}^{*} = *heat transfer coefficient due to infiltration (kJ m*$^{-2}$*hr*$^{-1}$*K*$^{-1}$*)*
Cp = *specific heat of dry air (kJ kg*$^{-1}$*K*$^{-1}$*)*
L = *heat of vaporization (kJ kg*$^{-1}$*)*
Δq = *specific humidity difference, inside to outside (kg kg*$^{-1}$*)*
ΔT = *temperature difference, inside to outside (K)*
ρ = *air density (kg m*$^{-3}$*)*
V_g = *volume air exchange per unit cover area (m*^{3}*m*$^{-2}$*hr*$^{-1}$*)*

It was Okada and Takakura's conclusion that moisture transfer (latent heat) shared the heating loss to more than 50% of the total heating load due to infiltration. Observed infiltration rates for a glass house ranged from 0.21 hr^{-1} at v = 0.2 m s^{-1} to 1.38 at v = 4.8. The transfer coefficient varied with windspeed, ranging from less than 0.6 W m^{-2}K^{-1} to more than 2.2 W at 6 m s^{-1} windspeed. The values increased with increasing volumetric heat capacity of the air (more vapor) and increasing temperature difference.

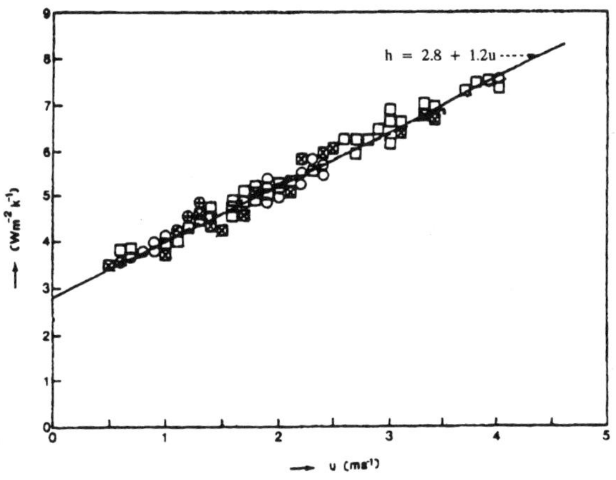

Fig. 4-28. Heat transfer coefficient between cover and outside air of a Venlo glass house as a function of windspeed [Bot, 1983] (With permission of the author).

Energy lost at night, when $T_o < T_i$, is directly proportional to the total surface area of the structure and the temperature difference (Eq. 4.13). This type of relationship may be found in numerous publications. However, it represents peak power requirements necessary to maintain a given internal temperature. It does not make most efficient use of heating equipment since the maximum requirement may be necessary only a few times in the year. The problem was discussed by Ellis [1990], and his data (Fig. 4-27) shows that the energy requirement may vary from less than 0.5 x 10^{12} J in July to more than 3.0 x 10^{12} J per month in January and February in the U.K. In the continental regions of the U.S., the difference between summer and winter is much greater. With a -8 C outside temperature, and a 20 C inside requirement, van de Braak [1995] pointed out that a standard greenhouse in the Netherlands will require 246 W m^{-2} heating capacity. This is with a single glass cover. A constant supply, sized for maximum demand is poorly utilized. Equipment is commonly designed to operate

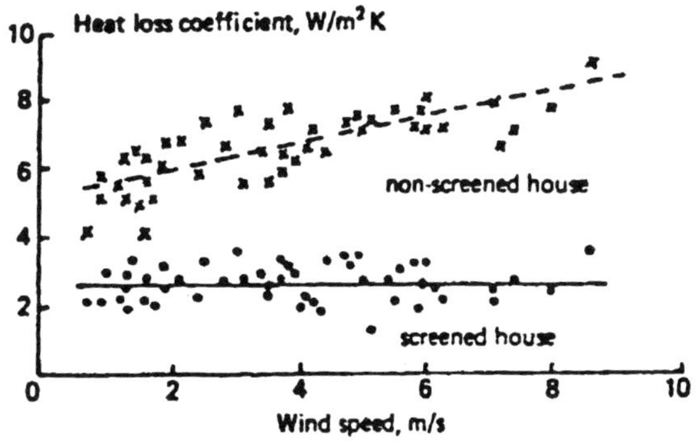

Fig. 4-29. Heat loss coefficients as a function of windspeed, comparing a glass house with one fitted with a thermal screen [Bailey, 1981] (©1981, Int. Soc. Hort. Sci., *Acta Hort.*, 115:663-670).

most efficiently at maximum ratings. Experience in Colorado has indicated that a grower may reduce the requirements imposed by Eq. 4.13 by nearly a third. This will sometimes result in subnormal temperatures in the winter. But, one, or even three nights of low temperatures does not appear to cause marked delay in carnation growth. Ellis suggested that better utilization can be had by an energy supply sized for 25% of the maximum load. It can still provide 65% of the energy required (Fig 4-27). Backup equipment sized to provide 75% of the maximum will provide additionally for very cold periods and serve in an emergency such as failure of the main system.

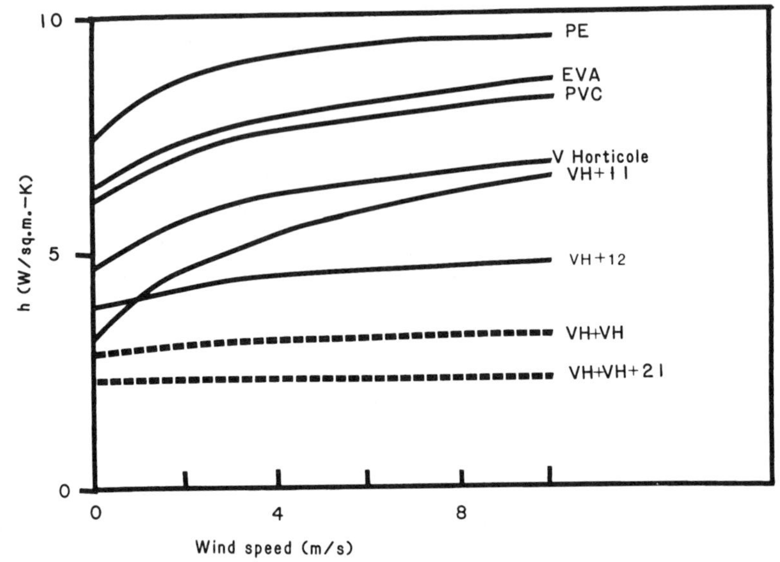

Fig. 4-30. Calculated heat transfer coefficients as a function of wind speed for covers. See Table 4-5 for code designations [Reprinted from *Agric. and Forest Meteor.*, Nijksen et al., ©1984, *Heat transfer through covering materials of greenhouses*, 33:193-214. With kind permission of Elsevier Sci.-NL, Sara Burgerhartstraat 25. 1055 KV Amsterdam, The Netherlands].

In the American literature, standard works provide heat transfer coefficients for the most common claddings (e.g., Table 4-4), leaving the impression that such values are constants. Aside from the thermal conductivity of the material itself, the heat transfer coefficient (h) of any material is not constant and varies markedly with conditions inside and outside the structure. For example, the detailed work of Nijskens et al. [1984] clearly showed the large differences between the extremes of overcast versus clear skies –especially coverings that transmit wavelengths beyond 3000 nm (Table 4-5). A clear sky is much colder than a cloudy sky, hence thermal radiation will be greater even if the covering does not transmit thermal radiation. The coefficients in Table 4-5 show a 20% difference between clear and overcast skies for single layer PE –a material common to mild climates– versus 25% for double PE, assuming no condensation. Bot and van de Braak [1995] mentioned the difficulty of determining the radiative exchange between the cover and the hemisphere above the greenhouse. To overcome this, a sky temperature (T_{sky}) is defined as the temperature of a black hemisphere according to the Stefan-Boltzmann law (Eq. 4.4). This can be measured directly, using a pyrgeometer. I have not seen this in use in any commercial application in the U.S. Several have examined thermal radiation [e.g., Garzoli and Blackwell, 1987; Silva and Rosa, 1987; Maher and O'Flaherty, 1973; Bot, 1983; Walker and Walton, 1971; etc.]. It is the Japanese who have contributed most to theoretical development of energy transfer in greenhouses [e.g., Takakura, 1967, 1968; Okada, 1980]. The publications by Takakura, especially, represent one of the most detailed, mathematical expositions I have seen for greenhouses. Takakura's model for unheated structures showed that the phase lag of the inside air temperature change behind the outside temperature change increased with the increase of the ratio of floor area to total glass surface area (B) and with the decrease in the ratio of air volume to floor area (μ). The factor of surface-to-ground area was also discussed by Tantau [1980], except he reversed the ratio. Large multiple-span houses have higher ratios (B) (less surface area versus ground covered) than single-span houses. This combined with smaller μ (greater volume versus ground covered) will slow the temperature change inside as the outside temperature varies; so that under appropriate conditions, the interior of an unheated, glass-covered greenhouse may be as much as 2 hr behind the temperature change outside. Such properties of the greenhouse will have a bearing on the types of heating and cooling systems and their control.

A second factor changing the transfer coefficient (h) are the conditions within and without the greenhouses in regard to convective transfer. This has been examined by numerous investigators [e.g., Nijskens et al., 1984; Bot, 1983; Bailey, 1990; Seginer and Krantz, 1986; Whittle and Lawrence, 1960a; c; Albright et al., 1985; etc.]. For outside the structure, heat loss is commonly assumed to increase linearly with increasing wind speed (Fig. 4-28, 4-29). However, this variation may not occur if the structure is fitted with a thermal screen. Others have provided data to show nonlinearity with considerable variation among cladding materials (Fig. 4-30) as a function of wind speed. Most authors assume a linear relationship [e.g., Albright et al., 1985]:

$$h = 5.0 + 0.5V \tag{4.16}$$

where: $5.0 = h$ *at 0 velocity*
V *= wind velocity* ($m\ s^{-1}$)

Bot [1983] found that wind direction made little difference on the regression model as noted by the close cluster of the various points in Fig. 4-28. The Dutch have a considerable advantage since their research is conducted in structures that are most common in their industry. In actuality, the transfer coefficient can only be considered as linear at night with constant temperatures. Okada's investigation [1980] showed the heat transfer coefficient to vary with the temperature difference and placement of the heating system (Fig. 4-31). During the day, solar radiation definitely results in a curvilinear relationship where total heat consumption followed the relationship [Gonzales and Hanan, 1988]:

$$\ln(Gas) = a(T_i - T_o) - bR_{in} + cV + d \tag{4.17}$$

where: $ln(Gas)$ *= natural logarithm of total natural gas consumption* ($m^3 hr^{-1}$)
T_o *and* T_i *= mean outside and inside temperature (C)*
R_{in} *= inside short wave radiation (300-3000 nm)* ($kJ\ m^{-2} hr^{-1}$)
V *= mean wind velocity* ($m\ s^{-1}$)
a, b, c, and d = statistically determined constants which vary with the greenhouse cover

The differences between gas consumption for 89 m^2 greenhouses, with and without thermal screens, with two claddings, are depicted in Fig. 4-32. Note the differences in the ordinate scales on the night-versus-day gas consumption. Unfortunately, Gonzales and Hanan did not carry these results to a logical conclusion in regard to the heat transfer coefficient. Fig. 4-32 does show that there is a significant effect of external wind velocity on energy consumption, even with thermal screens and a double covering (PVF SHD). It is doubtful that the curves in Fig. 4-32 (day) for screened and unscreened houses are significantly different from each other as the screens were opened during the day.

An approach to *in situ* determination of coefficients has been made by Albright et al. [1985] and Seginar and Kantz [1986]. This process can be accomplished although, unfortunately, the latter authors used an empty structure. The mathematics and instrumentation are such that it may not become a common procedure for several years.

On the inside of the cover, the same problems with regard to wind speed will occur. In almost all publications, the prevailing condition is to assume convective transfer with air velocities below 0.2 $m\ s^{-1}$ [Stanghellini, 1983a]. This is certainly not the case with horizontal air flow [Koths, 1974], or where hot air heating with forced circulation is employed. I have not seen any data dealing with heat transfer under these conditions. However, the inside condition is commonly complicated by condensation on the cladding. Whenever the inside cover surface temperature is below the dewpoint of the inside air, water will condense. The formation of a water layer has startling effects on energy transfer. As a thin film of water is opaque to thermal radiation, infrared transmission by a polyethylene cover, for example, drops. The situation is complicated if the cover has

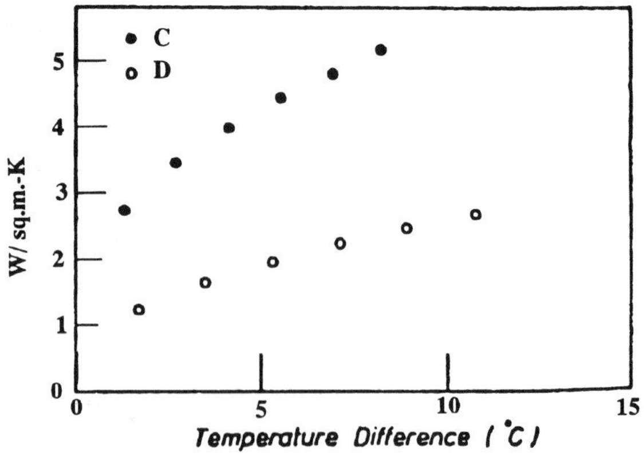

Fig. 4-31. Heat transfer coefficients as calculated by Okada [1980] as a function of the temperature differences between inside and outside the greenhouse. **C** are values for heating pipe stacked vertically along the outside wall, and **D** for heating pipe arranged horizontally across the width of the house close to the ground.

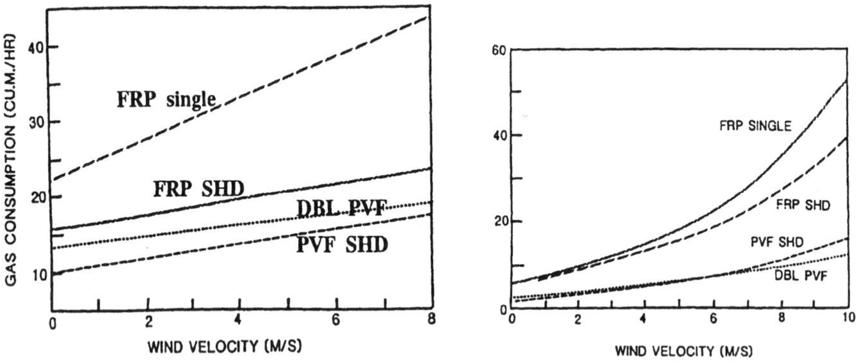

Fig. 4-32. Effect of outside wind velocity on total, hourly natural gas consumption of four, 98 m^2, quonset-shaped houses, two covered with FRP, two with double-layer PVF, during the night (**left**) and during the day (**right**) [Gonzales and Hanan, 1988]. One house of each cover fitted with an automatic thermal screen (SHD). Screens were open during the day, closing and opening at 50 W m^{-2}, ±30 W m^{-2} toggle, inside irradiance.

a low emissivity layer on the inner side. Condensation usually increases due to the lower temperature of the cover and emissivity of water is commonly close to 1.0 [Nijskens et al., 1984]. If a low emissivity layer is outside, relative convective loss will be increased due to a higher surface temperature, and in rain, the same situation occurs as inside. Also, as 1 kg water requires 2.47 x 10^6 J to convert liquid to a gas (latent heat of vaporization), this same amount will be released upon condensation at the inner roof surface. Eq. 8.19 and 8.20 show the relationships with regard to condensation.

The situation of condensation was approached by Bakker [1986] with a proposed vapor balance model. The flux of water vapor to a cover was assumed to be mainly free convection and characterized by a mass transfer coefficient, a_m:

$$M_c = \frac{a_m(q_i - q_g)}{\rho} \tag{4.18}$$

where: M_c = *mass transfer due to condensation (g* $m^{-2}s^{-1}$*),* ρ *= air density (kg* m^{-3}*)*
q_i *and* q_g *= specific humidity inside and inside of the glass cover (g* kg^{-1}*)*

The coefficient, a_m, was estimated from the heat transfer coefficient, h, through the use of dimensionless numbers:

$$a_m = \left[\frac{h}{\rho Cp}\right]\left[\frac{Sc}{Pr}\right]^{(1-b)}, \tag{4.19}$$

$$Nu = 0.12 Gr^{0.33} Pr^{0.33} \quad and, \tag{4.20}$$

$$h = \frac{0.12\lambda}{\ell(g\beta(T_g - T_i)(\frac{\ell^3}{v^2})^{0.33}0.71^{0.33}} \tag{4.21}$$

where: a_m = $m\ s^{-1}$
h = $W\ m^{-2}K^{-1}$
Sc, Pr, Nu, and Gr = Schmidt, Prandtl, Nusselt and Grashoff numbers respectively
λ = *thermal conductivity (*$2.2 x 10^{-2}\ W\ m^{-1}K^{-1}$*)*
ℓ = *characteristic length (1.7 m, distance ridge-to-gutter, Venlo)*
g = *acceleration of gravity (*$0.980\ m\ s^{-2}$*)*
β = *thermal expansion coefficient (*$3.6 x 10^{-3}\ K^{-1}$*)*
v = *kinematic viscosity (*$1.4 x 10^{-5}\ m^2 s^{-1}$*)*

When these equations are solved, the mass transfer coefficient becomes a function of the difference between the greenhouse air temperature and the glass temperature:

$$a_m = \frac{1.33}{(\rho Cp)(T_i - T_g)^{0.33}} \tag{4.22}$$

Using this model, Bakker [1986] found a maximum vapor loss by condensation of about 0.75 kg $m^{-2}dy^{-1}$, which was a slight underestimation of the measured value. At low transpiration rates, the errors could be as much as 30%. The estimated maximum amount of condensation on a glass roof was about 25 g m^{-2}. Further discussion on this subject is provided by Stanghellini [1995], cited in Chapter 8.

In an earlier publication, de Graff [1981] found a linear relationship between nighttime transpiration and amount of heating energy expressed in degree-hours per night. For 300 degree-hours, the equivalent amount of latent heat was 3.4 MJ or 1.4 mm water. De Graff stated that about 20% of the heating energy was utilized for evaporation (see Section 5.IV). The evaluation of the heating transfer coefficient, therefore, cannot be made on the basis of sensible heat transfer alone. The energy requirement of a structure must include latent heat transfers due to infiltration and condensation, as well as wind speeds, the actual temperature differential, and radiation.

This emphasizes that calculation of heat loss of a greenhouse is not as simple as standard engineering texts seem to imply. ASAE [1988] states that the recommended h value (Eq. 4.13) is nearly a constant, whether covered with a single or double layer. From the previous discussion, this is not at all correct. Secondly, ASAE

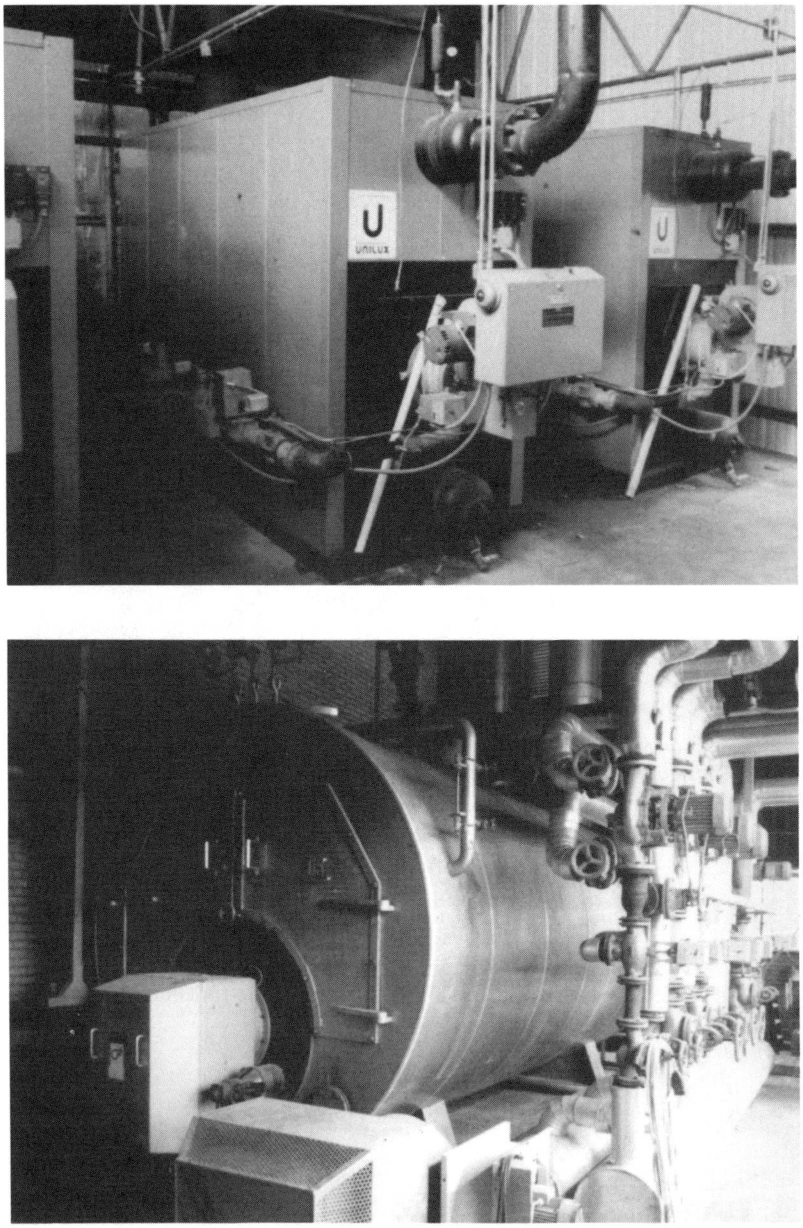

Fig. 4-33. Two examples of a central energy supply, both hot water. In the upper picture, several small units allow a measure of flexibility according to load requirements. In the bottom, the hot water circulating pumps are on the right. Similar steam installations of varying sizes may be found with fuels ranging from coal to wood.

[1988] recommends that the outside design temperature be the lowest temperature for 99% of the time. In Colorado, a -20 C outside temperature is not unusual, which for roses would result in a temperature differential of 37 C. Design of a heating system for this extreme case underlines the need for careful consideration of heating requirements for a particular locality. Furthermore, with suitable models, it becomes possible to make maximum use of computers to calculate ongoing energy requirements for maximum efficient utilization of energy and maximum production. Attempts have already been made by incorporating outside wind velocities to reduce energy requirements without markedly altering plant growth [Bailey, 1990; Hurd and Graves, 1984]. Some results have indicated a 10% increase in net income on lettuce, and a 14% savings in fuel costs on tomatoes. According to Butters [1977], a gradual change in accordance with growth stage in chrysanthemum could reduce energy consumption by 10%. Other trials included reducing temperatures part of the night or during one night of the week. These manipulations cannot be easily accomplished with standard thermostats or analog controls and require digital machines capable of the necessary calculations and decisions within a very short time (Chapter 8).

B. HEATING SYSTEMS AND ENERGY SOURCES

1. Heating Systems for Greenhouses

There are four primary systems for maintaining temperatures in greenhouses: 1) steam, 2) hot water, 3) hot air, or 4) infrared. The first three may be used in combination (e.g., steam conversion to hot water, steam or hot water conversion to hot air, etc.). Infrared heating is not utilized on a wide basis due to some other problems. In the U.S., steam has been employed historically although hot water is being increasingly applied, particularly for ground heating. Hot air units are quite common in the U.S. since the method is often the cheapest, particularly where natural gas is available for combustion within the structure. Hot water is the most common in Europe. Steam and hot water are often supplied to the structure from a central plant even though this is the most expensive initial cost installation (Fig. 4-32).

There are a number of advantages to a central main energy supply [ASHRAE, 1984]: 1) A central plant offers greater flexibility to use alternative energy sources. 2) The method uses less space in the buildings served. 3) They usually require less total installed capacity than dispersed units. 4) Partial load performance might be much more efficient than many isolated small systems. 5) Instrumentation and control are generally of better quality and more comprehensive, improving efficiency and reliability. And 6), energy conservation apparatus usually has a longer service life and reduced cost. The steam or hot water is then conveyed to the various greenhouses where final dispersal takes place. Central units as in Fig. 4-33 usually come prepackaged with controls and burners installed by the factory. The burners are often capable of handling either oil or natural gas, with modulating controls so combustion can vary with demand. Main supply to the greenhouses is an expensive heating process and most are elevated in the U.S. However, the Dutch are moving most of their main supply lines underground despite the increased cost [Anon, 1989a]. Central boilers are about 10% more efficient than unit heaters and are often quickly convertible from oil to gas [Anon, 1983]. Operation of a central heat supply requires trained personnel, or quick availability of such people. In the U.S., this may be difficult although there are various manuals that can be used for training [i.e., Dyer and Maples, 1979; Payne, 1985].

Table 4-6. Some typical efficiencies of various steam boiler types [White, 1982].

Boiler type	CO_2 (%)	Exhaust gas temperature (C)	Combustion efficiency (%)
Economic packaged (3 pass)	13	232	85
3 pass economic	11.5	260	82
2 pass Scotch marine	11	288	79
Cast iron sectional	10	316	77
Vertical, firetube type	10	371	75
Lancashire (and similar)	9	343	74
Vertical, cross tube type	8	538	58

04-34

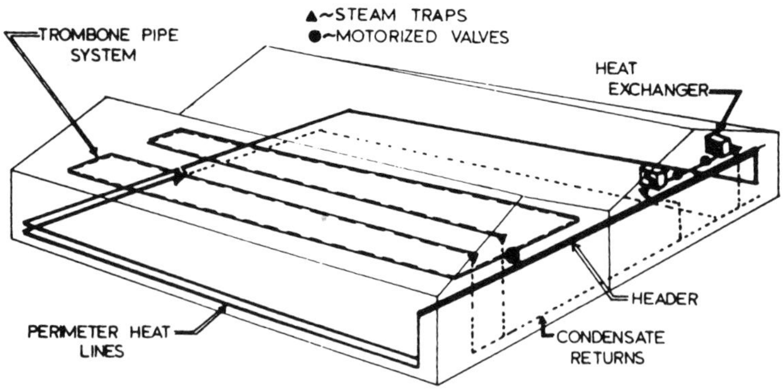

Fig. 4-34. Arrangement of heating surfaces in a greenhouse, suitable for steam or hot water. Perimeter heat is nearly always required; overhead, above crop, becoming less common. Note hot air converters to right, common in the U.S.

a. Steam

Steam heating systems offer a number of advantages and some disadvantages [ASHRAE, 1984]: 1) Steam flows through the system without the need for external sources such as pumps. 2) System components can be repaired or replaced without need to drain and refill the system. 3) System temperature can be controlled by varying the pressure. 4) Steam is distributed with little variation in temperature. 5) It is a constant-temperature heat transfer medium. It requires considerable energy to convert a kilogram of water to vapor, and this heat (ca. 2.26 MJ kg^{-1} at 100 C, 0 gauge pressure) is released when condensed. Steam is of particular value where the medium must travel long distances and where intermittent changes in a heat load are required. For greenhouses, steam availability is of particular usefulness for pasteurization of containers and root substrates. In climates such as Colorado, another advantage considered by some is the rapidity with which a load can be handled. In cold weather and partially cloudy skies, steam availability has the potential of quickly compensating for sudden changes in irradiance. However, the high temperature of radiating surfaces generally preclude placing pipes low to the ground where it could come into contact with the crop and workers. Typical types of steam boilers and their efficiencies have been discussed by White (Table 4-6).

Table 4-7. Heat loss from bare steel horizontal pipe, air temperature at 21 C, Watts per linear meter [Stamper and Koral, 1979].

Size	38 C	66 C	99 C	116 C
25 mm	18	59	118	154
32 mm	22	72	146	190
38 mm	26	82	166	215
51 mm	32	100	204	264

Table 4-8. Heat emission of horizontal coils on a wall, steam at 7 kPa, 102 C, air at 21 C, Watts per linear meter of coil [ASHRAE, 1988].

	25 mm	32 mm	38 mm
Single row	127	156	178
2 rows	242	300	335
4 rows	423	524	592

In general, radiation surfaces are arranged linearly around the walls (perimeter heating), and overhead or close to the ground across the house (Fig. 4-34). Note also, steam may be converted to hot air for distribution. But, some perimeter heat should always be employed, and in snowy climates, a heating pipe is often included under the gutter to melt snow. The arrangement of pipes above the crop was common prior to 1970 since this got an immovable system out of the way for cultural operations. Although Whittle and Lawrence [1960c] stated there was no significant difference between high- and low-level pipes, it is generally conceded that high-level radiating

Table 4-9. Steam and hot water ratings for bare steel and copper-aluminum, finned heating elements. Minimum water flow 0.9 m s^{-1}. Watts per linear meter [Ted Reed Thermal, Inc., 1985].

Description	Rows of pipe	Mounting height[a]	Steam[b]	Hot water[c]			
				115 C	104 C	93 C	82 C
Steel pipe, 32 mm 112 fins/m	1	25 cm	1153	1442	1210	990	798
	2	41 cm	2026	2594	2181	1788	1432
Steel pipe, 32 mm, 138 fins/m	1	25 cm	1326	1662	1393	1144	913
	2	41 cm	2383	2989	2508	2057	1643
Steel pipe, 51 mm, 112 fins/m	1	25 cm	1259	1576	1317	1076	865
	2	41 cm	2153	2700	2268	1855	1490
Steel pipe, 51 mm, 138 fins/m	1	25 cm	1336	1672	1403	1153	923
	2	41 cm	2249	2816	2364	1932	1557
Copper tube, 32 mm, 112 fins-/m	1	25 cm	1393	1739	1461	1201	961
	2	41 cm	2441	3046	2566	2095	1682
Copper tube, 32 mm, 138 fins-/m	1	25 cm	1566	1960	1643	1345	1076
	2	41 cm	2672	3344	2806	2297	1845

[a] Distance from floor to top of element.
[b] Steam at 7 kPa, 102 C, and air temperature at 18 C, Watts per linear meter.
[c] Minimum water flow 0.9 m s^{-1}, air temperature at 18 C, Watts per linear meter.

surfaces will increase the temperature difference across the roof, leading to increased heat loss [Bond, 1982; Okada, 1980]. High-level pipes will certainly interfere with thermal and shade screens, and this was a problem in retrofitting existing structures for energy conservation. High-level piping also reduces irradiance. The grower should always strive to keep the area above the crop as uncluttered as possible. Where the grower uses hot air convectors, perimeter heating should also be installed. There have been numerous instances of crop damage in severe weather where all heat was introduced above the vegetation. Even though a thermostat was placed at 1.5 m above ground level, cold air flowing from the walls froze carnations planted in the ground. This is particularly apt to occur with unit heaters and polytube distribution.

Different sized pipes for linear heating are generally rated in Watts per linear meter. Tables 4-7 through 4-11 provide this information for bare steel pipe and finned steel or copper-aluminum pipe under a variety of conditions. Steam ratings are all for steam at 7 kPa gauge pressure, temperature of 102 C to air at 21 C. Obviously, steam temperature will be higher at higher pressures, but most steam heating systems seldom, if ever, exceed 100 kPa except at the boiler and main distribution system if distances to the heated area are great; 51 mm diameter pipe is common. Note also the difference in thermal exchange between bare pipe and finned pipe. Despite the higher cost, the greater exchange capacity of finned piping greatly reduces the amount required. The greater heat transfer, however, tends to introduce "hot spots" with less uniformity in temperature distribution.

With steam, as with hot water, heating lines, and controls should be arranged so that a line is hot the entire length of the greenhouse. Hot lines only in part of the house can result in large temperature differentials. According to ASAE [1988], houses 9 m or less in width need only perimeter lines, whereas in wider houses, sufficient perimeter lines are installed to meet sidewall losses and the remaining lines spaced across the house, 3 m above the crop, or one line under each raised bench, sufficient to meet roof losses. If hot air convectors are employed, then sufficient units with appropriate capacity are installed above the crop. For Dutch conditions, van de Braak [1995] states the most common arrangement for vegetables are four 51 mm pipes per 3.2 m bay, 5 to 10 cm above the soil surface. They are combined in two pairs, 40 cm apart at a mutual distance of 1.6 m. The pipe can be used for a rail system for internal transport. In other arrangements, the pipes are attached to benches, adjustable with crop height (Fig. 2-75), overhead, and around the sides.

Steam installations require devices to prevent steam from entering the condensate return lines, which usually drain by gravity to a central sump for eventual return to the boiler. The devices on the end of each heat line or exchanger are called steam traps, which permit condensate to pass but not steam. Considerable heat will remain in the condensate. Return lines should be insulated so the condensate will be hot as possible when returned to the boiler. Provision is also required for expansion of a long steam line to prevent buckling of either the line or structure.

b. Hot Water

This system has been the choice in Europe for a number of years, and is becoming increasingly important in the U.S. Such systems are extremely versatile, generally operating at pressures below 1102 kPa and temperatures below 121 C. Water is continuously circulated (minimum 0.9 m s^{-1}) in heating lines under benches, along the outer walls, or close to the ground, as noted in Fig. 4-35. By means of 3-way or 4-way valves (Fig. 4-36), the temperature in the terminal elements may be maintained at a level sufficient to meet the system requirements. And, as hot water generally circulates in loops, heat is uniformly distributed the length of the house. The design temperature of supply and return water is commonly 90 and 70 C, respectively [van de Braak, 1995]. On-off systems for water circulation in the U.S. are common and less expensive than modulated control.

As temperatures are seldom high enough to cause injury, heating elements may be distributed through the crop (Fig. 2-76, 4-35), or in or on the ground (Figs. 2-82, 4-37, 4-38). ASHRAE [1987] states that under-floor heating can provide 25 to 30% of peak heating requirements in northern houses or about 50 to 65 W m^{-2} for a bare floor. This is reduced by about 25% when potted plants or seedling trays are set directly on the floor. Monk [1987] suggested up to 40% of the annual heating load may be supplied by root zone heating under Ohio conditions. Floor heating installations range from placement of heating loops permanently in the floor of the structure to the use of 6 mm EPDM, synthetic rubber tubes directly under the plants (e.g., Fig. 4-38). For low-temperature floor requirements (<40 C), polyethylene, polybutylene, or polyvinyl chloride piping can be utilized.

Fig. 4-35. Three examples of hot water piping arrangement, underneath benches (top), perimeter heating (middle), and just above the ground (bottom). Pipes in the bottom picture are also employed for movable picking carts in tomato or cucumber production. Pipes may be painted as long as a latex-based paint is used. Painting pipes aluminum will reduce heat output up to nearly 30%. Note also the flexible connections in the background in the bottom picture. These pipes may be moved out to the overhead for planting operations. Typical Dutch Venlo heating arrangement

Commonly, 13 or 19 mm diameter, PE pipes are buried in porous concrete or gravel, on 30 to 40 cm centers; 6 mm EPDM tubes may be purchased in a twinwall configuration, creating an even temperature throughout the entire loop (Fig. 4-36). Maximum length for 13 mm loops is about 60 m, for 19 mm about 120 m [Aldrich and Bartok, 1985].

New types of heating elements have been developed to replace steel pipe (Fig. 4-39). According to Krause [1982], each length of the aluminum pipe replaces two lengths of standard steel heating pipe, with a reduction in weight to less than 10% of the original. No welding or regular heavy pipe fittings are required. The special design of this piping is that the surface area is the same as 51 mm diameter steel pipe, but water volume is much less so that response to the control system is far quicker [Anon., 1989a]. With these systems, the heating elements may be easily positioned out of the way of cultural operations when necessary.

Table 4-10. Steam and hot water ratings for copper-aluminum, 11 x 9 cm finned tube, 131 fins/m, 27 cm height, Watts per linear meter [Sterling Radiator, 1984].

Tube size (mm)	Steam (7 kPa, 102 C to 21 C)	Hot water (flow = 0.9 m s^{-1}) 93 C	82 C	71 C
19	1140	980	788	605
25	1182	1019	817	625
38	1163	999	798	615

It is obvious that the control system where floor heating is involved is more complex since such a system must be separated from the main terminal elements where temperatures can be much higher. Even ground-level systems may be split into several circuits, each with its own control, separate from perimeter heat, etc., which ensures uniform and level temperatures through the crop area. Hot water systems react slower than steam or hot air, and the time response, especially in the case of soil heating, has to be taken into account (Table 4-2).

Table 4-11. Heat transfer (watts per linear meter) for various temperature differences between pipe and greenhouse air [van de Braak, 1995].

Temperature difference (C)	Steel pipe diameter (mm) 51	33.2	26.4	Plastic tube diameter (mm) 25
20	34	23	18	14
30	55	38	31	24
10	15	10	8	6
40	77	53	44	35
50	101	71	58	46
60	128	90	73	--
70	156	108	90	--
80	185	129	107	--

The thermal capacities of linear heating elements given in Tables 4-7 through 4-11 do not include any separation between heat loss through sensible heat versus radiation. The value for radiation is not considered significant in American literature. Dutch experience, however, indicates an important effect of thermal radiation from the heating pipes on transpiration. This is especially significant under conditions of low solar irradiance (refer to Figs. 4-40 and 5-93). Stanghellini [1983b; c] concluded that there appeared to be a strict coupling between heating regime and transpiration. Data from de Graff [1980], Okada and Takakura [1973], and Bakker [1986] were cited in the previous section. Businger [1966] showed the relationship between pipe temperature and coefficients of heat transfer (Fig. 4-39). However, we are most interested in pipe temperatures less than 100 C in the case of hot water systems. For convective transfer, Bot's [1983] data (Fig. 4-41) is the only information I am aware of that is directly applicable to the greenhouse condition. Table 4-11 provides additional information on heat transfer at different temperature differentials for steel pipe and plastic. Under winter conditions in northern climates, it appears safe to assume that a significant portion of energy input for maintaining appropriate temperatures goes to latent heat –i.e., evapotranspiration. Whereas, in the case of steam and hot water systems, the energy source is partly radiative; in hot air systems, which have no radiative component, transpiration may be a partial function of advective energy exchange such as occurs in the field with dry winds.

c. Hot Air

This method, whether using steam or hot water as the primary source or combustion inside the house, has one fundamental advantage over steam or hot water. It is cheap. Arrangement and styles are innumerable, with the least expensive usually being natural gas convectors where that fuel is available. Figs. 4-42 and 4-43 provide examples employed in both temperate and mild climates. In any case, combustion within the greenhouse is dangerous, especially if units depend upon house air for combustion. Problems with air pollution from burners in plastic houses have been noted before 1963 [Moore and Jones, 1963], especially if the heaters depend upon

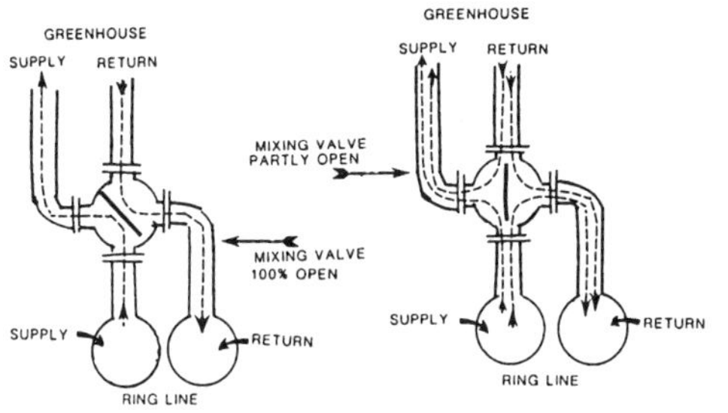

Fig. 4-36. Examples of 3- and 4-way valves for hot water heating systems. **Right:** A 3-way valve with attendant circulating pump. As the valve opens, more water circulates through the boiler for additional heat. **Left:** A 4-way valve in two positions. Water continually circulates in the main ring line, which connects to the boiler and, in the heating loops, in the greenhouse. Depending upon valve position, additional hot water is bled into the heating line for more heat with cold water returning to the ring line.

convective air exchange and are equipped with continuously burning pilot lights. Hanan [1973a] showed that ethylene will be found as a by-product of combustion regardless of precautions to ensure complete burning. Forced air heaters, which do not make use of some type of distribution system (e.g., burners for CO_2 supply), will show a gradual increase in ethylene concentration in the house. Hanan recommended at least 1 cm^2 free air opening for each 400 kJ hr^{-1} output from a burner which is double the National Fire Protection Code requirements in the U.S. Unfortunately, forced air heaters, which do not use a distribution system, usually set up independent circulation patterns for each heater so that the heater should have its own air supply piped directly to it from outside. There have been numerous instances in Colorado of pollution damage in severe weather where the burner depends upon infiltration and interior air supply for combustion. Burners are available where the combustion chamber is totally enclosed with separate air supply from the outside with electronic ignition. Although more expensive, these types offer safety even in climates where external temperatures seldom go below 0 C. Convective, hot air heaters have been shown to provide extremely nonuniform heat distribution in structures of any size [Meneses and Monteiro, 1990b].

Fig. 4-37. EPDM, 6 mm diameter twin-wall tubing laid directly on the ground for heating.

Unit heaters with internal combustion are about 10% less efficient than centrally located boilers; however, they take less capital cost to buy and install [Anon., 1983]. Any horizontally discharging heater located above the crop (Figs. 4-42, 4-43), will have the coldest region directly under the heater, and if all the units are in one location, this will ensure a significant temperature differential along the house length –even if the heaters are fitted with distribution tubes. Heaters should be installed at both ends of a house, even though this increases cost. Downward discharge heaters (Fig. 4-43, C, and D) will also have marked temperature differentials, especially if the house is fitted with perimeter heat. The discharge pattern will interfere with normal convective patterns [ASAE, 1988].

The best installations from the standpoint of temperature uniformity use polyethylene tubes attached to a heater, or in some other fashion (Fig. 4-42A, C, and D), to distribute heat uniformly the length of a house. Information from such individuals as Parsons [1975] gives workable solutions for sizing tubing to distribute heat and for cooling purposes. Unfortunately, information of the type provided by Bailey [1974] on perforated ducts is not readily available in the technical literature. Static pressure, temperature, and air discharge along uniformly perforated polytubes are neither uniform nor linear. Pressure could be maintained uniform by reducing tube size with distance from the heater. Practically, such tubes are difficult to manufacture, and the principal idea is uniform heat output in a system where temperature drops with distance. It is necessary to increase the amount of air discharged progressively along the tube and space the discharge holes to produce the required distribution of air. For example, Fig. 4-44 shows the variation in air discharge along a uniformly perforated duct from a 31 cm layflat tube with 2.5 cm holes. An important ratio is the total area of discharge holes to cross-sectional area of the tube or aperture ratio (Fig. 4-45). Bailey concluded that the minimum static pressure does not always occur at the duct entrance, and a compromise must be made between an aperture ratio to give a large air flow at low pressure and too low a static pressure for good inflation of the tube. Bailey suggested an aperture ratio of two as an upper limit. Unfortunately, the requirement for heating is a uniform heat output, which means the air discharge rate must increase with distance from the heater, whereas, for ventilation, the discharge should be uniform. In many practical situations in the U.S., the grower experiments with hole spacing in his particular situation to obtain satisfactory conditions, depending upon whether the tube is for heating or ventilation. Experience in Colorado indicates that hole spacing should be increased toward the distal end in order to avoid large temperature differentials the length of the greenhouse.

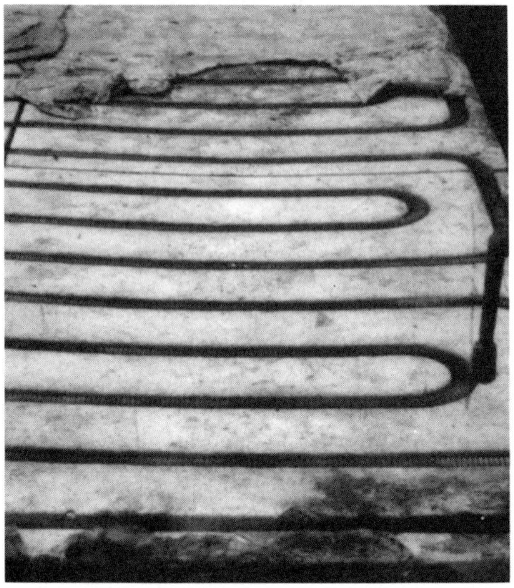

Fig. 4-38. Two examples of small hot water tubing laid directly under the crop on raised benches.

Fig. 4-39. Pear-shaped aluminum, hot-water heating pipe with the same thermal capacity as 51 mm diameter steel pipe.

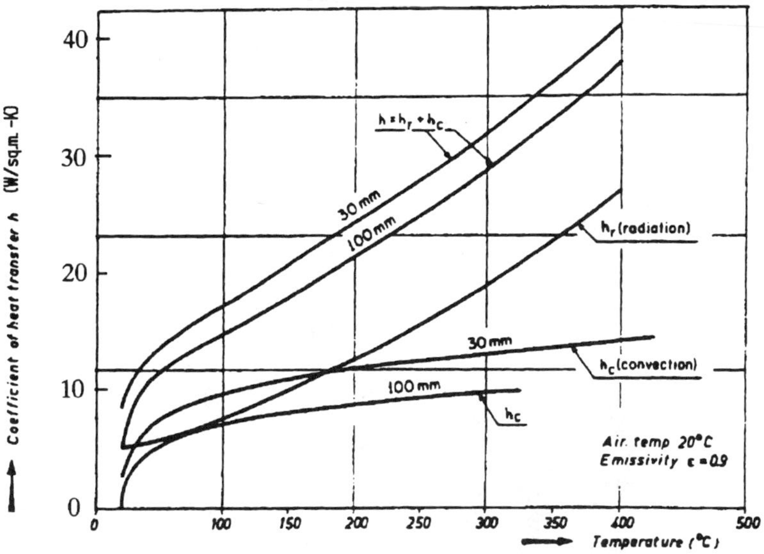

Fig. 4-40. Effect of pipe temperature on convective (h_c) and radiative (h_r) heat transfer coefficients of steel pipe (30 and 100 mm diameter) [Businger, 1966].

I have not seen information published pertaining to this requirement. It is, usually, much simpler to buy layflat tubing already perforated. In any event, lengths should not exceed 30 m, and the heater fan capacity should be sized to handle a perforated duct.

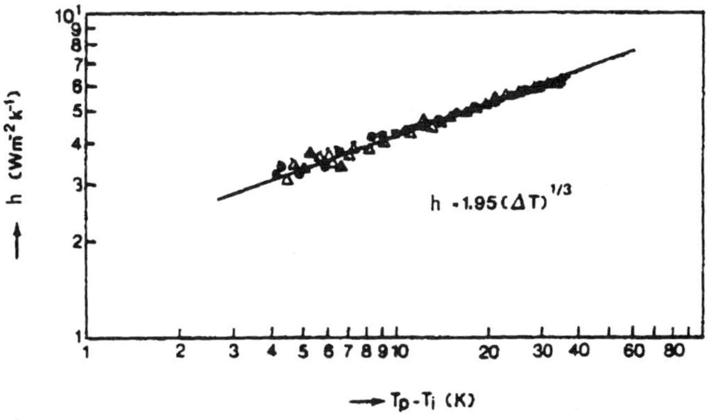

Fig. 4-41. Convective heat transfer coefficient between heating pipes and air as a function of the temperature difference between pipes and air [Bot, 1983] (With permission of the author).

d. Infrared

Such systems which depend upon high temperatures of a surface to radiate thermally to colder objects enjoyed considerable popularity in the 1970s (Fig. 4-46). For commercial ranges, the fuel is propane, more commonly natural gas. The system is arranged overhead to provide an even distribution of radiation to the vegetative canopy below. For houses 9 m or more wide, at least two lines of radiating pipe should be installed; 10cm diameter black pipe is the standard radiating device, with a 40 cm wide reflector shield above the pipe. Burners will generally be spaced about 6 m apart, and temperatures of the pipe may vary from above 400 to 200 C just before each successive burner, to less than 150 C at the exhaust vacuum blower. The claims in terms of fuel saving have ranged from 30% [Blom and Ingratta, 1981; Rotz and Heins, 1982] to more than 60% [Youngsman, 1978] in comparison to conventionally heated structures. The system seems to work best on "two-dimensional" crops (e.g., potted chrysanthemums, poinsettias, etc.) versus "three-dimensional" canopies such

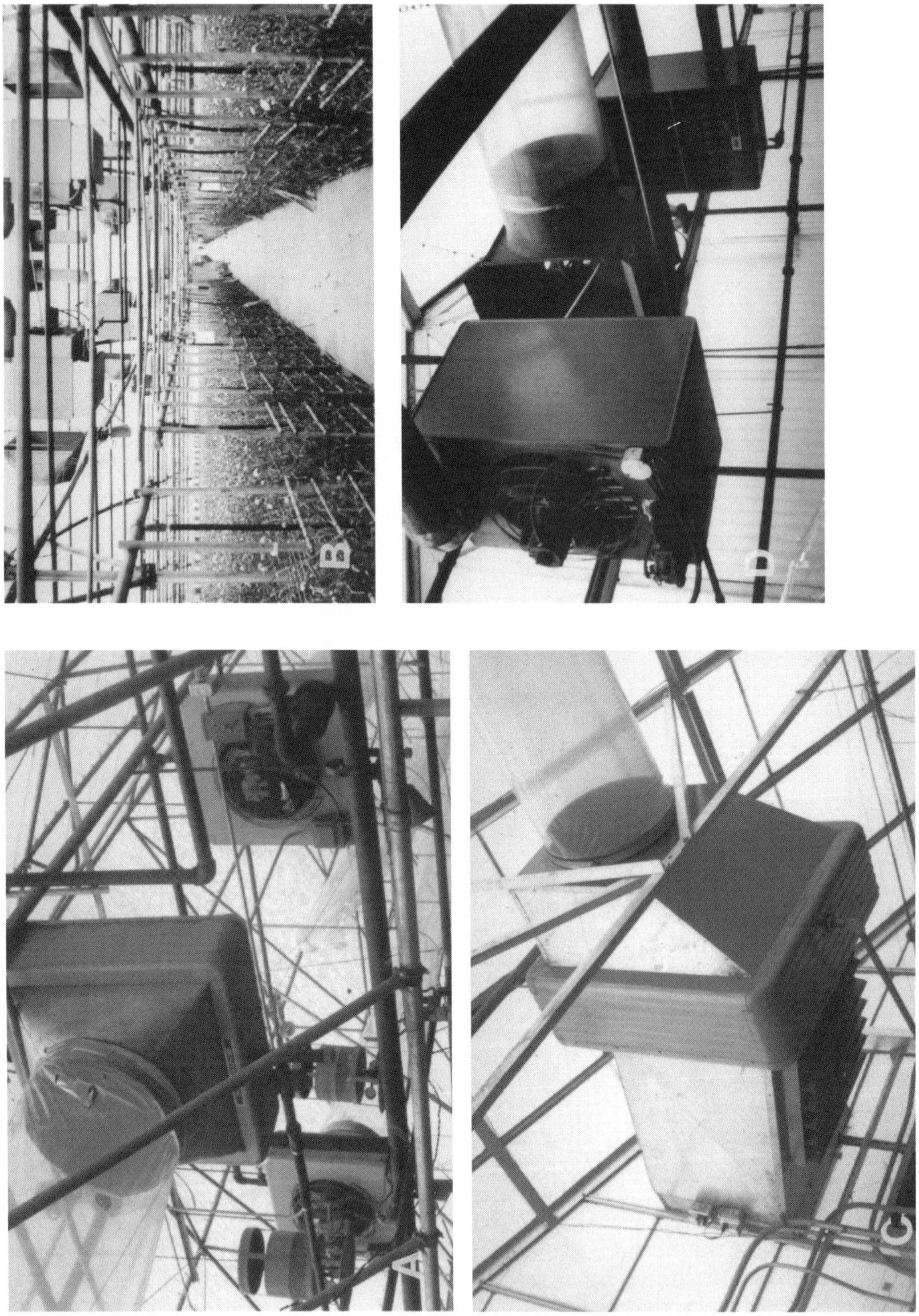

Fig. 4-42. Examples of hot air convectors, all using polyethylene tubing for heat distribution. **A** and **C** are steam convectors; **B** a **D** use natural gas.

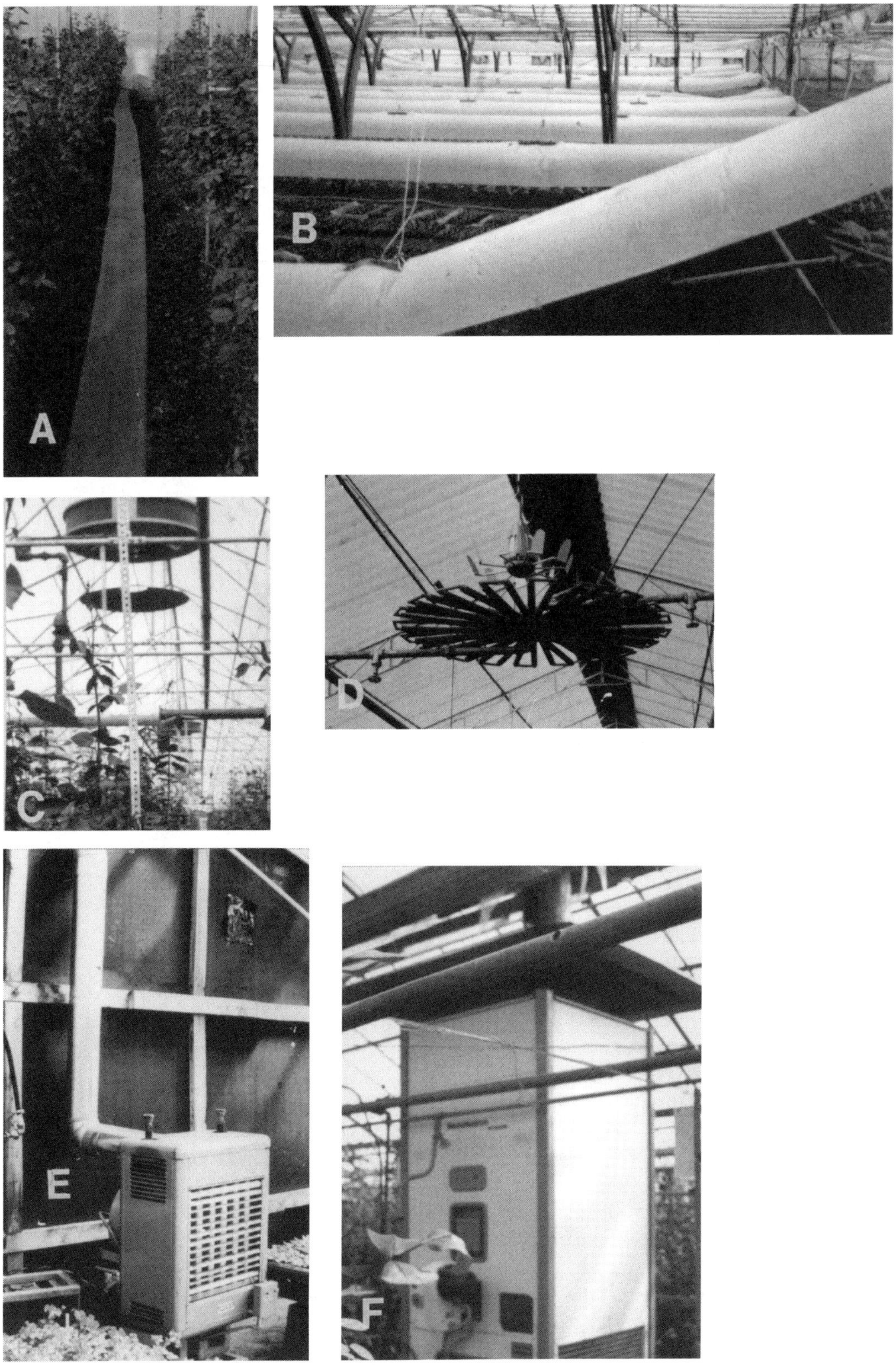

Fig. 4-43. Additional hot air systems. **A** and **B** are examples of polytube distribution; **C** and **D** are not recommended; **E** and **F** will be found in mild climates.

roses, cucumbers, tomatoes, etc. The system is on-off so that temperatures can change rapidly. Blom and Ingratta [1981] found significant variation in vegetative temperatures depending upon location. Problems with distribution have also been emphasized by Koths [1985]. According to Smith [1987], there does not appear to be any great advantage over conventional piped systems for vegetable production in the U.K. and Europe. According to Challa [1980], in Holland, the system leads to considerable problems due to wide temperature variations between leaves directly in the path of radiation versus those shaded, with condensation sometimes on those shaded. The thermal radiation also increases transpiration, which can result in higher stress. There have also been reports of excessive elongation, and, of course, thermal screens may be impractical with the heating system above the crop (at least 1.5 to 3 m). While IR systems apparently work better in mild climates –there are a number installed in California– no one has actively investigated the type of temperature control that should be employed. At the present time, these systems appear to have lost favor, at least in the European literature.

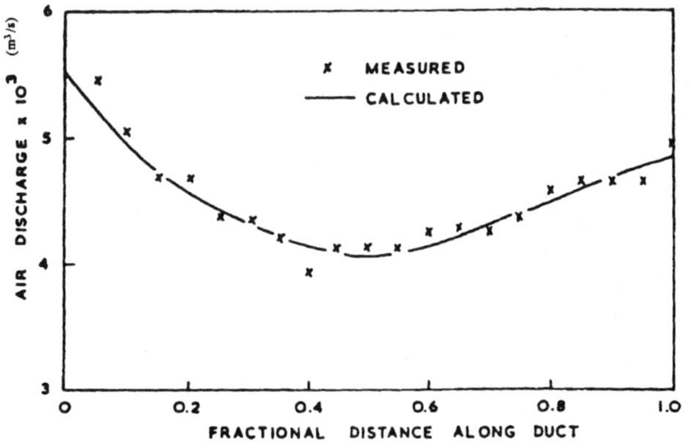

Fig. 4-44. Variation of air discharge rate along a 31 cm layflat tube, uniformly perforated with 2.5 cm holes [Bailey, 1974].

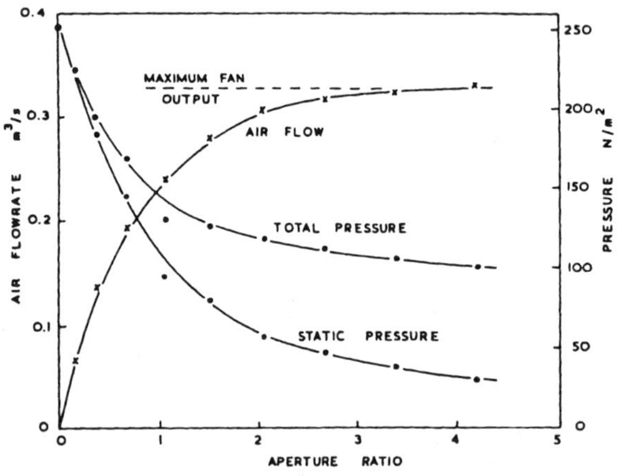

Fig. 4-45. Variation of inlet air flow conditions as a function of the aperture ratio in a uniformly perforated tube [Bailey, 1974].

2. Low-Temperature Systems

The previous section dealt with heating systems conveying energy at high temperatures (>80 C), suitable for temperate climates where external temperatures are commonly below 0 C. Some of these may be employed in mild climates such as California or the Mediterranean (Fig. 4-43, A, E, and F). But, even in mild climates, combustion inside a greenhouse, for reasons given above, is not desirable. Experience over several years in the Mediterranean region has shown a gradual progression to combustion outside the structure, and in most instances, these have been hot air methods using inflatable tubes for distribution inside the house (Fig. 4-43A, 4-47). A particular problem in exporting, mild climate regions is the close relationship between market price and heating costs where high energy fossil fuels are employed. For example, the study by Vakis and Photiades [1990] showed that conventional oil heating systems markedly improved yield and quality, with earlier harvests, for Cyprus. In cold seasons, the extra cost of heating was more than made up by the extra income. On the other hand, in a mild winter season, the extra cost was not covered due to lower product prices. Fossil fuel heating was definitely uneconomical during the 1970s fuel crisis. That has changed, and for export, heating is required in Cyprus from December to April, sufficient to maintain 15 C for high-temperature crops such as cucumbers, tomatoes, etc.

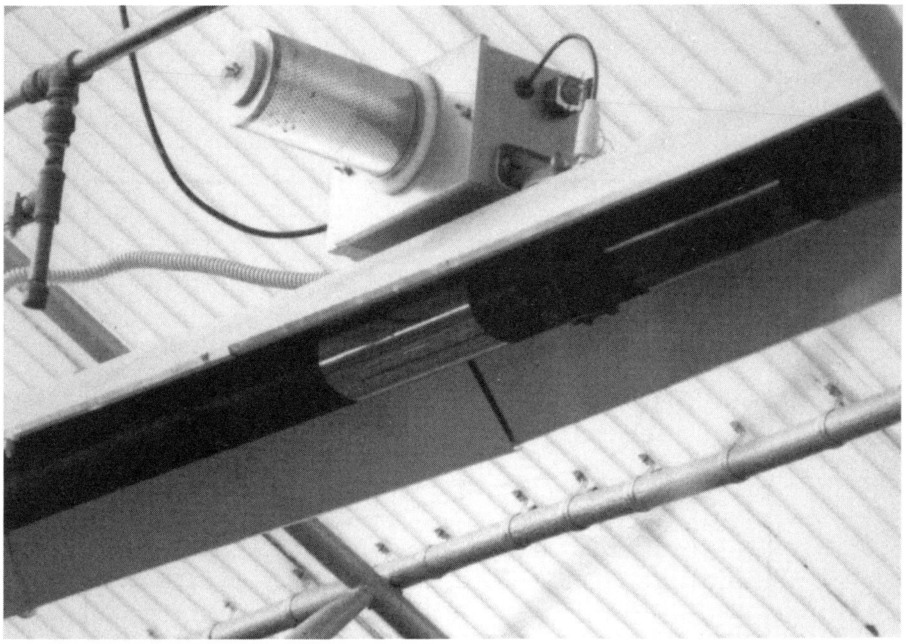

Fig. 4-46. An infrared heating system with reflector, showing one of several, in-line gas burners.

Fig. 4-47. Hot air system used in a mild climate. Combustion outside the greenhouse with distribution through PE tubing.

In mild climates, it is possible to employ low temperature energy sources such as geothermal, solar energy or waste heat that would be unusable in colder regions. These are supplies at temperatures below 70 C. At 40 C, almost four times as much pipe is required to give the same amount of heat as at 80 C and an air temperature of 20 C [Ellis, 1990]. Obviously, conventional exchange methods would be uneconomical, and the success of a particular system is very closely related to cost. Also, the situation may be more of an avoidance of damaging conditions than to maintain what would be considered an optimum temperature. The typical situation that often occurs is depicted in Fig. 4-48 for a major frost event. A passive system such as solar tubes (Fig. 4-49) provides a gain of several degrees over a house without heating, and in a high light region, could be sufficient to produce a high quality product. Passive solar heating, due to its low cost, has found utilization in Israel and Greece [Grafiadellis et al., 1990; Grafiadellis and Kyritsis, 1981; Grafiadellis, 1984; Mattas et al., 1990; Segal et al., 1990b]. Usually, 25 cm PE tube, with 0.2 mm wall thickness, and lengths up to 30 m, are laid between rows and filled with water (Fig. 4-49). In some instances, a black foil is laid under the tube, or the tubes are placed with a space between them to form a container for the plant row. The advantages and cost of this system in comparison to others can be

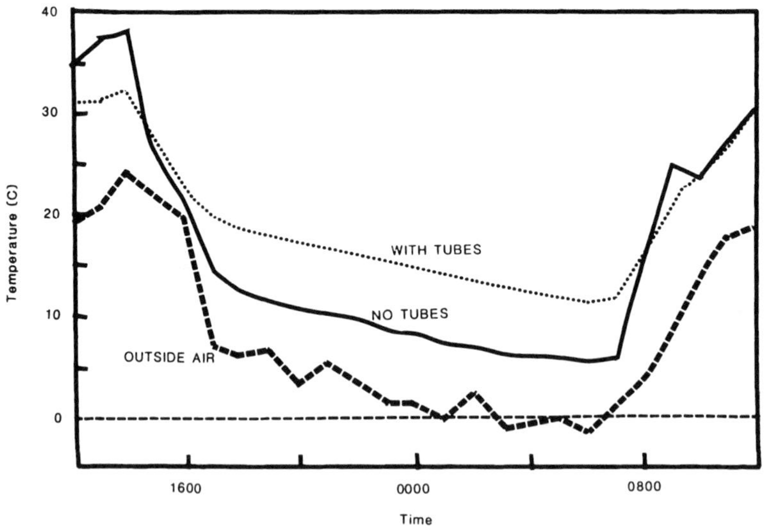

Fig. 4-48. Ambient temperatures compared with those in an unheated greenhouse and one containing passive solar tubes on a night of frost in Israel [Segal et al., 1990b].

noted in Table 4-12. Other systems have been devised to allow for adequate heat exchange (Fig. 4-50). These are more expensive than passive solar tubes, and, as with tubes, may take up valuable floor space, reduce structural transmissivity, or shade the plants. The popularity of low-temperature systems in mild climates can be appreciated by the fact that the cheapest hot air system in the U.S. is about $7.50 per sq.m., nearly $20.00 for a hot water system, and more than $21.00 per sq.m. for a steam system.

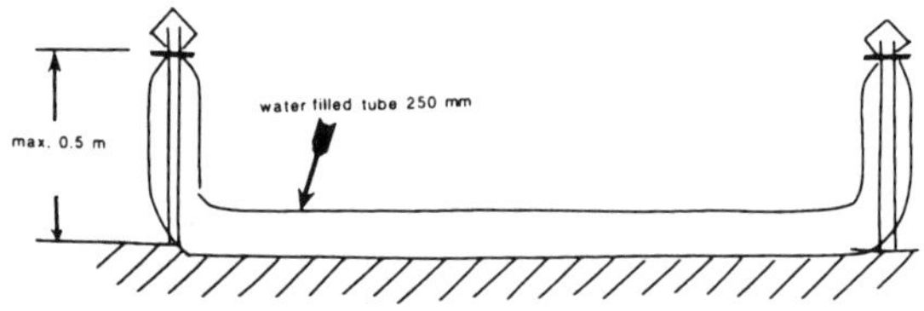

Fig. 4-49. Typical solar tube, containing water, for storing solar energy and raising internal temperatures [Segal et al., 1990a].

3. Energy Sources for Greenhouses

The possible energy sources for greenhouses are numerous and, sometimes, novel. Many have been investigated, particularly since the energy crisis in the early 1970s. They may be listed:

1. Fossil fuels which include: natural gas, propane, butane, coal, oil, and wood,
2. Solar energy
3. Geothermal
4. Wind
5. Cogeneration
6. Electricity
7. Heat pumps
8. Heat from decomposition (biothermal)
9. Waste heat from power plants, industrial processes such as whiskey distilleries, poultry operations, and dairies

In fact, there was once a newspaper report in Oregon where the grower used rabbits in great numbers to provide most of the heat for the greenhouse.

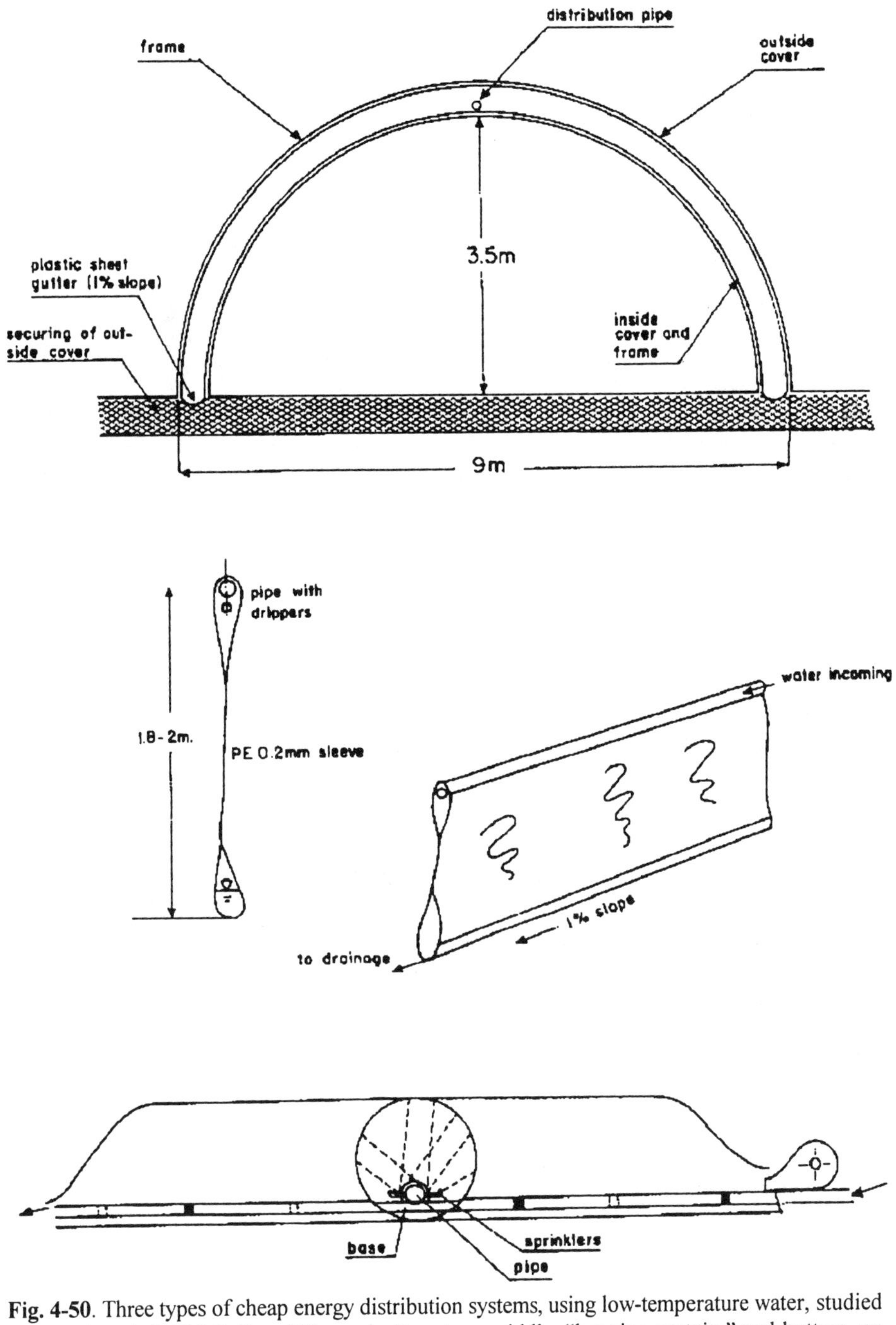

Fig. 4-50. Three types of cheap energy distribution systems, using low-temperature water, studied by Segal et al. [1990a]. Top, "Hungarian" system; middle, "hanging curtain;" and bottom, an inflated floor tube.

Fig. 4-51. Two examples of hot water solar heat collectors intended as demonstration for solar heating. **Top:** Two of these units for a 600 m^2 greenhouse. **Bottom:** A collector to heat 100 m^2.

Table 4-12. Comparison of heat exchangers for greenhouses, applicable to mild climates [Segal et al., 1990a].

	Hot air	**Hot water**	**Hungarian**	**Hanging water**	**Inflated tubes**	**Passive solar**
Heating power	Sufficient	Sufficient	Medium	Sufficient	Sufficient	Small
Obstruction to solar	None	None	High	Small	None	None
Need for electricity	High	Small	Small	Small	Small	None
Maximum fluid temperature	90 C +	90 C	50 C	50 C	50 C	<40 C
Use of corrosive and sedimentive fluids	Impossible	Light	Minimum	Possible	Possible	Irrelevant
Average transfer coefficient (W/m^2- C)	20 +	10	2.3	4.3	3.2	ca 10
Maximum exchange area (m^2/m^2 floor). Allow space for crop & work	Unlimited	0.5-0.7	ca 1.4	2.5	1.25	0.4
System cost ($\$/m^2$ exchange area)	Expensive	ca 7	1.2	1.5	2.2	0.25
Cost of energy transfer ($/W- C)	1 +	0.7	0.4	0.35	0.6	Very small
Convenience for workers	Very	Convenient	Convenient	Not convenient	Not convenient	Not convenient

a.. Fossil Fuels

The characteristics of the most common fuels are listed in Table 4-13. Of these, natural gas is the cleanest and most convenient when available. The actual composition varies with the source. The Dutch have now shifted to calorific values as the basis for pricing the commodity as compared to the common practice of price per unit volume. Where natural gas is unavailable, there are several grades of fuel oil as noted in Table 4-13. Even when natural gas is available, fuel oil is commonly used as a standby in an emergency. Burners are generally provided that allow rapid switching between the two fuels. In general, purchase, installation, and operating expenses of gas fired unit heaters are low [Welles, 1975] –about two-thirds that of steam or hot water. Natural gas is delivered as it is consumed, whereas oil and coal require appropriate storage facilities. With some oils such as No. 6

Table 4-13. Typical heating values of various fuels [ASHRAE, 1981] and wood [Corder, 1973].

Fuel type	Remarks	Specific gravity or mass	Heating value (MJ)
Natural gas, type II, high methane	Composition depends on source, odorless and colorless	0.590-0.614 g g^{-1}	37.6-39.9 MJ m^{-3}
Propane	Boiling point -40 C, storage in liquid state	1.52 g g^{-1}	50.2 MJ kg^{-1} 93.2 MJ m^{-3}
Butane	Boiling point 0 C, storage in liquid state	2.01 g g^{-1}	49.3 MJ kg^{-1} 119.2 MJ m^{-3}
Fuel oil, No. 1	Light distillate, high volatility	0.883-0.800 kg ℓ^{-1}	38.2-37.0 MJ ℓ^{-1}
No. 2	For use with pressure-atomizing gun burners, most in domestic and medium capacity burners	0.874-0.834 kg ℓ^{-1}	39.5-38.2 MJ ℓ^{-1}
No. 5	Preheating may be required	0.951-0.921 kg ℓ^{-1}	41.8-40.9 MJ ℓ^{-1}
No. 6, Bunker C	High viscosity, preheating required for pumping and atomizing	1.012-0.965 kg ℓ^{-1}	43.5-42.2 MJ ℓ^{-1}
Coal, anthracite	Clean, dense, hard, non-caking, burns smokelessly and uniformly		29.5 MJ kg^{-1}
bituminous, medium volatile	Large range in grade and properties, may be friable and dusty, may cake, subject to improper firing		32.6 MJ kg^{-1}
subituminous B	High in moisture when mined, likely to ignite spontaneously when stored, non-caking, ignites quickly		20.9 MJ kg^{-1}
lignite	Woody structure, high moisture, low heating value, liable to spontaneous combustion, non-caking, little smoke or soot		16.0 MJ kg^{-1}
Wood, 20 % moisture content, dry	Heating value reduced with increasing moisture. Density varies 304 to 593 kg/m^3	Douglas fir Bark Ponderosa pine Lodgepole pine Bark Red Alder	20.8 MJ kg^{-1} 23.0 MJ kg^{-1} 21.2 MJ kg^{-1} 19.5 MJ kg^{-1} 25.1 MJ kg^{-1} 18.6 MJ kg^{-1}

(Bunker C), installations require heating in order to flow and for proper atomization at the burner. Coal and wood may also require storage to prevent freezing. For the latter, increased moisture content will reduce heating potential. In some areas, coal is the lowest cost fuel source. Although this may seem significant, it requires more labor to burn coal, produces large quantities of ash, and may require expensive pollution control equipment. Operation of solid fuel systems is much more difficult, inasmuch as combustion cannot be turned on or off as with natural gas or oil [Roberts et al., 1985].

Table 4-14. Fuel costs for comparative net heat production from different fuels [Adapted from White and Aldrich, 1980].

Net heat production per dollar (MJ)	Fuel oil[a] cost/liter	Electric[b] cost/kW-hr	Natural gas[c] cost/m^3	Propane[d] cost/liter	Bituminous coal[e] cost/ton
340.8	0.076	0.010	0.076	0.052	40.18
253.2	0.103	0.014	0.103	0.070	54.04
246.9	0.105	0.014	0.105	0.072	55.43
240.5	0.108	0.015	0.106	0.074	56.81
235.3	0.111	0.015	0.116	0.076	58.19
219.4	0.128	0.016	0.119	0.081	62.36
197.3	0.132	0.018	0.132	0.090	69.28
178.4	0.145	0.020	0.145	0.099	76.21
164.6	0.158	0.021	0.158	0.108	83.14
149.8	0.174	0.024	0.174	0.119	91.45
106.6	0.245	0.033	0.246	0.168	128.87
80.2	0.326	0.045	0.328	0.224	171.82
64.4	0.403	0.055	0.404	0.276	212.01

[a] Fuel oil, liter 39.9 MJ ℓ^{-1} at 65% efficiency.
[b] Electric, kWhr 36.0 MJ kW-hr^{-1} at 100% efficiency.
[c] Natural gas, m^3 37.3 MJ m^{-3} at 70% efficiency.
[d] Propane, liter 25.5 MJ ℓ^{-1} at 70% efficiency.
[e] Coal, metric ton 248.8 MJ t^{-1} 55% efficiency.

Wood may include sawdust, wood chips, so-called "hogged" fuel from lumbering operations, and sometimes pellets or wafers. According to Corder [1973], the largest single use of wood in the world is for fuel, especially in undevel oped countries. In the U.S.; fuelwood con-sumption reached a peak in 1875 and then declined to the point where wood accounts for about 1% of the national use for energy (1970). Grantham and Ellis [1974] also discussed the potential of wood for energy. They pointed out that per capita consumption of fossil fuels in the U.S. is about six times that of all forest products. The nation's entire annual timber harvest would contribute relatively little to total energy requirements. Even straw has been employed where it can be obtained cheaply [Anon, 1982b]. Wood burning, as with coal, requires special units for maximum efficient combustion, which will increase the initial cost as well as expenses for installation and operation. Under special conditions of a cheap and readily available supply, materials such as olive husks [Meneses and Monteiro, 1990b] have been utilized. One could also consider pecan hulls, rice hulls, seaweed, etc.

Fuel cost is the second largest variable expense in temperate climates. The ability to compare costs with alternative energy sources is a valuable adjunct to greenhouse operation. Unfortunately, oil prices are based upon cost per unit volume of fluid, while coal may be priced per ton. In each case, the potential energy is different, and conversion efficiency to usable energy in the greenhouse will vary. Tables 4-14 and 4-15 provide various comparisons of fuel cost versus energy obtained per dollar and approximate costs of fuel per hectare greenhouse area in the northeastern U.S. One may also calculate the cost per unit energy for comparison:

$$\text{cost for X units of energy (J)} = \frac{(\text{No. of units } (\ell,\ \text{ton},\ m^3))(\text{price per unit})}{(\text{energy content per unit})(\text{efficiency})} \tag{4.23}$$

For example, a ton of coal costing \$50.00 per metric ton, containing 29.5 x 10^9 joules per ton will have an efficiency of 70%. The cost per gigaJoule (10^9) will be \$2.38. Similarly, No. 2 oil at \$0.26 per liter, containing 39.5 MJ per liter, at 80% efficiency, will cost \$8.22 per GJ. Natural gas at \$0.13 per cubic meter and 37.6 MJ/m^3, 85% efficiency, will cost \$4.06 per GJ. Efficiencies of energy supplies will be found in the literature to vary from 55% for coal to 100% for electricity. Even electricity, however, if losses in transmission and generation are taken into account, will be considerably less than 1.00.

Table 4-15. Estimated heating costs per hectare for some ornamental crops, northeast U.S. [Adapted from White and Aldrich, 1980].

	Night temperature (C)	Fuel cost per ha-yr at $ 0.21 ℓ^{-1}	Liters No. 2 oil $m^{-2}yr^{-1}$	Cost ($) per m^2 area - year			
				0.18 ℓ^{-1}	0.21 ℓ^{-1}	0.24 ℓ^{-1}	0.26 ℓ^{-1}
Carnations and Snapdragons	10-12	130680	61.3	11.41	12.91	14.20	16.14
Chrysanthemums and Roses	17	172240	81.8	15.06	17.22	19.37	21.52
African violets and Foliage	20 to 21	226510	106.3	19.58	22.38	25.18	27.98
Azalea, Begonia, Cyclamen, Geranium, Gloxinia, and Poinsettia	18	200875	94.0	17.43	19.80	23.38	24.75
Hydrangea, Kalanchoe, and Lily	16	165530	77.7	14.42	16.36	18.51	20.44

b. Solar Energy

As with most of the energy sources other than fossil fuels, solar energy was extensively studied during the decade or so after the energy crisis in 1974. At first glance, one might believe a greenhouse to be a "natural" structure for solar heating. It is not so for a number of reasons: 1) The structure loses energy rapidly so that requirements are far higher compared to conventional, insulated buildings. 2) Most of the net radiation is utilized for evaporation. 3) Plant requirements preclude high temperatures, which would be more efficient in energy trapping and transfer. And 4), as noted in Chapter 2, a greenhouse is not particularly efficient as to transmissivity. With the exception of passive solar tubes in mild climates discussed above, solar energy is not now employed to any great extent in greenhouse heating. Examples of so-called pilot tests are provided in Fig. 4-51.

Numerous textbooks have been written on solar heating. The literature is extensive [e.g., Hoecker, 1979; Rotz et al., 1982; Kozai et al., 1986; Willits et al., 1985; Zabeltitz, 1986; Hunn, 1979; Mears et al., 1978; etc.]. Solar energy has been popularized and quite often the reports are controversial [Bond, 1983; Hood, 1975; Tweedell, 1980; etc.]. Some publications have tended to use the truth with penurious frugality. When reading the published articles, one should keep in mind the climatic area for which the results are intended. For example, the information already cited from Grafiadellis and co-workers, Baille [1990], Seginer and Albright [1981], etc. deal with systems most likely to be applied in mild climates. As discussed under low-temperature heating systems, the objective may be more to avoid extremes than to maintain temperatures at optimum. As pointed out by Hoecker [1979], effective solar heating was 40 times greater at Miami, FL, compared to Pittsburgh or International Falls, MN. But, this might not be relevant to heating applications due to the much warmer temperatures in Miami. Zabeltitz [1986] summarized by stating that only 15 to 25% of the annual heat requirement in Germany could be saved by solar energy. About 65 to 75 liters of oil are needed to heat a greenhouse in Hannover. On the other hand, in southern regions (Mediterranean), 60 to 100% of the annual heat requirement can be covered by solar energy, but the total annual heat consumption is only 5 to 10 liters oil per m^2. In a consideration of solar availability for space heating in coldest weather, Asbury et al. [1979] showed that all but one of the eight sites studied required a peak-day backup energy requirement in excess of 85% of the peak-day energy requirement of a conventional (nonsolar) heating system. The exception was Albuquerque, NM. These results, in effect, state that solar energy systems in most of the U.S., especially for greenhouses, require a second system for backup –essentially doubling the cost of the heating system. As with other alternative energy sources, the collapse of oil prices has reduced the intensive study of solar energy for greenhouses.

Another variation in passive solar heating is the use of the ground as heat storage for the greenhouse by circulating greenhouse air through pipes buried in the soil [Mavroyanopoulos and Kyritsis, 1986; Boulard et al., 1989]. In a 150 m^2 greenhouse, Mavroyanopoulos and Kyritsis buried 20, 20 cm diameter aluminum tubes at a depth of 2 m. Greenhouse air was blown through these pipes whenever the interior air temperature dropped below 12 C or exceeded 28 C. The results were that an earth-air heat exchanger could be used at the latitude of Athens, Greece, from November to March with no ventilation, with the fan requiring only 20% of the total energy supplied to the structure by the ground. In a more detailed study in southern France, Boulard et al. buried 12.5 cm diameter, perforated pipe at two levels of 80 and 50 cm. This system could maintain a 7 to 9 C temperature differential, inside-outside, in March and April. Auxiliary heating was only 20% of the whole heating season requirement for lettuce followed by tomatoes. A particular problem, however, was the fact that such a storage system dehumidified the greenhouse air in the daytime and vice-versa at night. The nighttime humidification was the main disadvantage of the system. The average sensible plus latent heat was 95 W m^{-2} for storage and 48 W m^{-2} for night extraction. Latent heat exchange comprised about 30% of the total heat extraction.

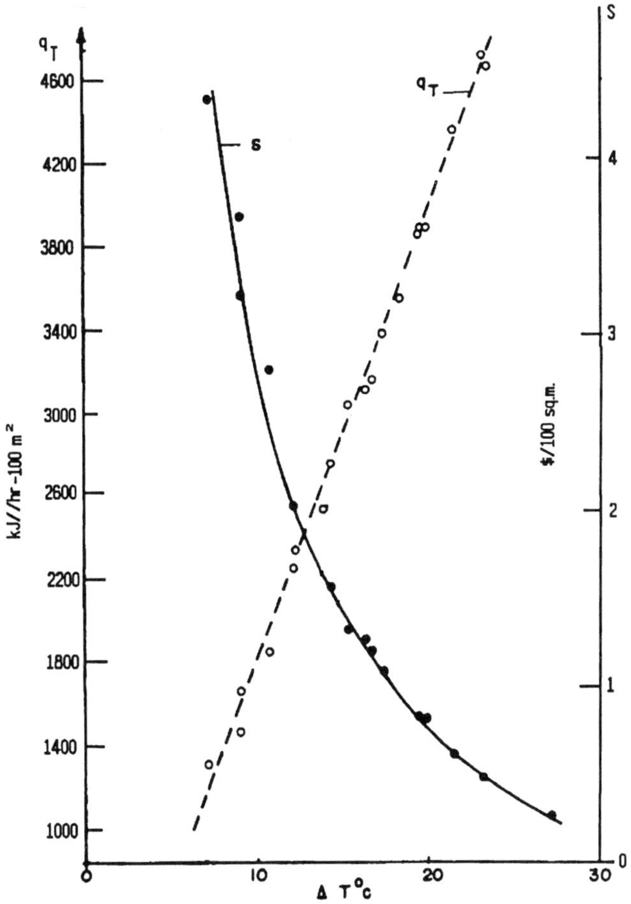

Fig. 4-52. Heat flux (q_T) per 100 m of 16 mm PE pipe and cost per 100 m^2 of greenhouse as a function of the temperature difference [Reprinted from *Energy in Agriculture*, 6:27-34, Tal et al, ©1987. With kind permission from Elsevier Sci. Ltd., The Boulevard, Langford Lane, Kidlington 0X5 1GB, U.K.].

Another interesting process for heating is the use of phase-change materials to store heat. Such chemicals as Glauber salt ($Na_2SO_4.10H_2O$), polyethylene glycol, and calcium chloride hexahydrate liquify above certain temperatures and recrystallize when their temperature is lowered. In changing to a liquid, the material absorbs heat and releases it when freezing, just as water will give up heat when it freezes. The PCM, such as the patented compound Serrolithe, is packaged in plastic containers over which the air from the greenhouse is blown during the day when temperatures are above 20 to 25 C, and again at night when the temperature drops below 15 to 20 C. The enthalpy of fusion (freezing) is 150 kJ kg^{-1} [Boulard et al., 1990; Tesi and Malarme, 1987; Ginzburg et al., 1989]. According to Boulard et al., in southern France, temperature lift in a 176 m^2 house covered with a structured PC before mid-February was insufficient to melt the PCM, and auxiliary heating was required. On the other hand, after late April, ventilation was required, which prevented additional stored heat. During March and April, however, storage could exceed capacity of heating due to solar gain on the storage facility itself. During the same period, the PCM system could maintain the air temperature 10 C higher than ambient. In Israel, Ginzburg et al. were able to maintain interior minimum temperature 16 C above the outside minimum temperature from November to April. PCMs have also been used for humidity control (Section 5.IV.B.2)

Rotz et al. [1982] concluded those attempts to use the greenhouse as a means to collect excess solar heat had the least potential. In effect, in temperate climates, passive solar heat collection is nonsense. Whereas, if the system is cheap enough, it can be applicable in warm regions (e.g., solar tubes).

c. Geothermal

These systems have received increasing attention despite the drop in fossil fuel cost. Many countries have hot water sources that may be close to the earth's surface or can be obtained by drilling. In many cases, the water may be low temperature (<40 C), which is particularly useful in mild climates [e.g., Behnke and Lambert, 1982; Tal et al., 1987; Kacem, 1989; Nikita-Martzopoulou, 1990]. Political units in many areas are likely to subsidize research and practical tests –as with solar energy. The Idaho National Engineering Laboratory is one example of a state program for geothermal development [Kunze, 1976].

One particular success story in the U.S. involved an operation receiving a subsidy for the first well (1524 m deep), which produced water at 65 C and nearly 1 ℓ s^{-1} flow. Unfortunately, the Environmental Protection Agency required the water to be re-injected, and the company had to convince the EPA that the water was clean enough to discharge into a local river. A second well, in another location, provided an artesian source, flowing more than 6 ℓ s^{-1} at 92 C, and could be pumped up to 47 ℓ s^{-1} without affecting the source. A second well re-injected the water [Wright, 1981]. The results reported by Wright point to a number of problems with geothermal sources. The supply may be highly corrosive, and governmental agencies may require suitable disposal of the effluent. The supply is seldom directly usable for irrigation. Secondly, for deep sources, there is the cost of drilling a well and attendant piping, along with probable corrosion problems. For both shallow and deep geothermal sources, pumping may be required, which increases cost. Thirdly, the location may be remote, and the operator is faced with problems of utilities and transportation. According to Howland [1983], problems of finding geothermal water in close juxtaposition with good water for irrigation are serious. And, the geothermal sources usually occur in isolated areas. Howland [1986] also pointed out that cutting fuel costs 50 to 75% did not provide for a profitable operation if labor and transportation ate up the difference.

Despite the caveats, geothermal energy has received increasing attention in Europe and the Mediterranean. According to Popovski [1988], Hungary has more than 130 ha in geothermally heated houses, followed by Yugoslavia with 60.7 ha. A number of other countries also have significant areas in geothermal heating. Examples of the "Hungarian" system (Fig. 4-49) have been described by Dixon [1985] for using low-temperature water. Heat flux and piping requirements for 58 C water and ambient temperature of 5 C have been given by Tal et al. [1987] (Fig. 4-52, Table 4-16), using PE piping laid on the ground surface. Such applications appear to be successful, especially in those regions with little or no fossil fuel resources.

d. Cogeneration

This is the simultaneous generation of heat and electricity. Due to changes in the public utilities in the U.S., power companies must offer to purchase electricity from qualified co-generators under the Regulatory and Policy Act of 1978. For almost every utility, the marginal energy source for power generation is the "avoided" cost based on the cost of natural gas [Jenkins, 1985]. To receive subsidies for development of cogeneration, the waste heat must be beneficially utilized. The constraint is [Jenkins. 1985]:

$$\frac{(E_p + \frac{1}{2}E_t)}{E_f} > 0.425 \tag{4.24}$$

where: E_p = *sum of total annual electrical energy produced*
E_t = *thermal energy used in the greenhouse*
E_f = *total annual fuel energy consumed by the engine*

Table 4-16. Black PE piping for geothermal water at 58 C, ambient temperature at 5 C, minimum inside 12 C, calculated by Tal et al. [1987] for 1000 m^2 of greenhouse.

Pipe dimensions		Number of pipes	Supplied heat (GJ/hr)	Pressure drop (kPa)	Flow rate (m^3/hr)	Outlet temperature (°C)
Diameter (mm)	Length (m)					
16	150	40	416.2	12.2	5.6	38.2
		44	414.5	7.0	4.6	33.8
		48	412.8	4.6	4.0	30.3
16	300	20	415.5	86.3	5.5	38.0
		23	413.8	40.4	4.2	31.8
		25	411.8	27.9	3.7	28.6
16	600	12	412.8	238.7	3.9	30.5
		14	408.8	130.5	3.3	24.8
		16	404.8	84.3	3.0	21.2
12	150	56	416.6	33.1	4.5	33.6
		64	418.6	20.3	3.7	28.4
		64	418.6	20.3	3.7	28.4
12	300	30	415.3	191.5	4.0	30.7
		33	413.2	128.8	3.6	27.2
		35	410.9	93.4	3.3	24.5

This constraint limits the size of the engine that can be used and requires careful matching of the thermal demand in the greenhouse and the electrical production of the generator. Unfortunately, the peak electrical demand may occur in the summer when heating is not required. The ratio will be less than 0.425 and the cogenerator would not qualify for avoided cost rates. Cogeneration, according to Jenkins, can result in substantial savings. 105.5 MJ of natural gas used in a cogeneration plant will produce 8.8 kW-hr (31.7 MJ) of electricity and 36.9 MJ of useful heat for the greenhouse, assuming a thermal efficiency for the engine of 30% and half of the engine waste heat is recovered. To produce the same amount of energy separately would require 158.3 MJ or 50% more gas. Jenkins [1985] concluded that cogeneration appeared to offer substantial potential for reducing heating costs, whereas heat pumps were marginally attractive where electrical rates are low and annual heat demand high. Lawson [1984] has described a Dutch "total energy" system wherein the waste heat from electrical generators is utilized. To avoid loss of heat when the generators must be run in the summer for electricity, the grower has installed a wind turbine that has saved about 20% in energy compared with the previous year. That is, the wind turbine allows the generators to be shut down when there is no need for heat. Van de Braak [1995] pointed out that the overall efficiency of cogeneration can be 85 to 90%. A cogenerator produces heat and electricity at a ratio of two to one. About 500 Dutch growers are using heat or electricity of a cogenerator.

Fig. 4-53. Ground-level heating system in southern France, using low-temperature water from a heat pump.

e. Heat Pumps

These have been thoroughly examined by a number of authors [Jenkins, 1985; Kozai, 1986; Smith, 1982; Kyritsis et al., 1988; White, 1982; Rotz et al., 1981]. Essentially, a heat pump operates similarly to a refrigeration system or an air conditioner. For heating purposes, the system extracts heat from the air outside, ground, or water well, and pumps it into the house. The compressor with its refrigerant, reverses the usual situation of heat flow to a colder region. The ability of a heat pump to extract energy is determined by its coefficient of performance (COP), which is the ratio of energy delivered to the electrical energy used by the system. With source temperatures in the range of 16 to 20 C, COP is typically 3.5 to 4. When the temperature falls to -12 to -18 C, the ratio drops to about 1, and the heat pump has no advantage over electrical resistance heating [Jenkins, 1985]. If COP is 3, and the electrical utility is generating and transmitting energy at 25% efficiency, then the heat pump is equivalent to a natural gas boiler operating at 75% thermal efficiency. White [1982] states that heat pumps can offer an average of 20% savings over conventional cooling systems and central electric resistance heating. For maximum efficiency, a heat pump should be used year-around, cooling in the summer as well as heating in the winter. An example of a low-temperature heating system in France, using a heat pump, is provided in Fig. 4-53. However, the initial capital costs, and costs of operation, have led a number of individuals to conclude that heat pumps for greenhouses are uneconomical at the present time [Jenkins, 1985; Smith, 1982; Kyritsis et al., 1988]. The latter conclusion is somewhat contrary to results published by Rotz et al. in 1981. Under their projected price scenarios, a hybrid system with gas or electrical heaters as backup could be operated at lower cost. But the most critical element in the economic analysis was the life of the heat pump equipment. Reducing life from 20 to 10 years increased costs markedly. The system was no longer competitive with natural gas.

f. Biothermal

Biothermal systems depend upon decomposition of vegetable products, manure, or garbage to produce methane that can be employed to heat the greenhouse conventionally. Heat from manure decomposition was a common practice in cold frames 50 years ago. The use of the term biomass is the more-or-less conventional combustion of vegetable products such as wood, crop residues, fruit pits, nut shells, etc. [Jenkins, 1985]. These methods are less convenient than natural gas, and they may lead to problems from emissions and wastes due to high ash contents of some fuels. Heating values may be highly variable with large amounts of dirt. Investigations have been made into the use of methane from waste disposal facilities. However, few of these systems have achieved widespread utilization. Paybacks calculated by Jenkins indicated 2 to 4 years for biomass systems, 4 to 6 years for cogeneration and from 6 to more than 10 years for heat pumps. Anaerobic decomposition in garbage dumps, which produces methane (biogas), has been looked into. None of these have achieved widespread use.

g. Wind

Energy from wind is another frequently investigated possibility. However, the variability prevents large-scale applications to greenhouses –although one has been mentioned as a supplement to existing systems [Lawson, 1984]. Apparently, the U.S. is the leader in electrical generation from wind with about 1400 MW being produced in California. Wind energy is also making a significant contribution in Denmark and investigations are being carried on in the U.K. and the Netherlands. As with solar energy, the particular site must have sufficient wind speed, and a backup in case of calm conditions. As with other sources, the continuing low cost of fossil fuels has greatly reduced the incentives for continued interest.

h. Waste Heat

The large waste heat from industrial processes, particularly electrical generating plants has been given much attention. In fact, according to Winspear [1974], one large base load generating station has enough reject heat to fill the needs of the entire greenhouse industry in the U.K. The prime problem with this, as with other alternative energy sources, is the generally low temperatures obtained if the heat is taken from the final cooling loop at the generating plant; which varies, depending upon the plant load; and the fact that the electrical plant may not be located in a desirable spot from the standpoint of greenhouse operation. Nevertheless, there have been a number of projects examined in recent years. One of the best known to American industry is the installation in Minnesota described by Ashley [1979]. A 2123 m^2, air-inflated double PE structure is heated with condenser cooling water with a 29 C temperature at full electrical load. Design criteria for the greenhouse heating system were based on a -34 C ambient temperature, 10 C inside, with a calculated heat loss of 2.1 GJ hr^{-1}. Some problems occurred with fouling of the heat exchangers. Because of the relatively low temperature differences, significant electrical energy is required for pumping. In another 4000 m^2 greenhouse, the cost to operate and deliver warm water was about $15000 per year. The economic feasibility is most dependent upon: 1) distance between the source and greenhouse, 2) climatic conditions, 3) electric rates, 4) land cost, and 5), distance to market. According to Widmer [1979], the power company must guarantee a relatively steady water flow for 10 years or longer. The electrical plant must be a base load, not a peaking producer. Results have shown that a 24 C minimum water is adequate for short periods when ambient temperatures are above 3 C.

A larger installation is in Yorkshire, U.K., consisting of 12 ha of tomato producing glass houses [Anon, 1981; Long, 1988]. Cooling water from the electrical plant varies between 27 and 35 C, with the enterprise the result of a joint venture between a state monopoly and a large private concern. This was the first venture by the latter into greenhouse production. The heat costs, according to Long [1988], are about 25% of current commercial fuel costs. However, the added overhead charges for pumping large water volumes raises cost to nearer 50% of conventional methods. Only 3 to 4 degrees C heat is extracted, so the water must be cooled further for power station use. A steam generation unit is available for backup in case of interruptions to the water supply, which is via 900 and 1200 cm diameter mains. The French also started a waste heat system at St. Laurent Des Eaux [Anon, 1976a]. This 3000 m^2, plastic structure uses 56 $\ell\ s^{-1}$ water at 20 C. The temperature is raised to 30 C and circulated through layflat PE tubes laid on the ground and sloped to flow by gravity.

A number of other investigators have examined the feasibility of waste heat from power plants [e.g., Rotz and Aldrich, 1979; Walker et al., 1987; Walker, 1978; Heard et al., 1989]. In a computer examination of five water heating systems, using reject heat, Heard et al. found none of them profitable for the conditions of the study. The costs of pipelines alone exceeded the energy cost saving for a location near Pittsburgh. In a literature review for the Four Corners Regional Commission, Colorado, Goldsberry [1975] suggested that if energy equal to 35% of the heat input to the generator was extracted for heating at 121 C, only 30% of the heat input would be wasted to the condenser versus 60% –assuming a 40% efficiency for electrical generation. Electrical production might be reduced 10%, but overall efficiency of both electrical generation and greenhouse operation would be raised significantly. So far, this possibility has not received much attention from the power industry.

i. Summary

By and large, most of the alternative energy sources discussed here have not proven economically feasible given present fossil fuel prices and conditions. The exceptions to this are cheap systems intended for mild climates such as the Mediterranean, Florida, California, etc. One will find in the technical and popular literature many articles that purport solution of environmental and energy concerns. All these energy sources are technically feasible, but usually fail to account for the practical economic restraints. Particularly, in the case of greenhouses, one must always consider exterior climatic conditions and the technical surround of the local community.

4. Energy Conservation

With the fuel crisis in the 1970s, industry and political entities were rudely shocked. In a few short years, a mass of publications resulted that dealt with energy reduction and alternative energy sources. It was instructive to observe the approach made to the problem by two different, but highly developed technologies. Krause [1982] pointed out that while many British growers were talking of reducing growing temperatures and altering cropping schedules to directly influence fuel costs, the Dutch were considering increasing capital investments on a massive scale in order to reduce the cost per unit produced without substantially changing their cropping programs. Thus, the Dutch looked at long-term capital investments versus the British short-term solution. The difference between Dutch and English operations was emphasized by Kooistra in 1986. Between 1975 and 1985, there was a 4% increase in yield per unit area per year in the Netherlands. The energy consumption per unit of product was reduced by 50%. Combined with the introduction of soilless culture or artificial substrates, this was the first introduction of a computer in 1975. In 1985, more than 4000 computer systems were in operation. Use of soilless cultures such as NFT and rockwool, combined with appropriate nutrition, provided production improvement varying between 10 and 15%. Thus, the Dutch were able to compete profitably with imports, which contrasted sharply with the situation in the U.S. Although the agricultural research system in the U.S. far outweighs the Dutch in terms of facilities and personnel, the system is diffuse and individualistic, each entity suffering from budget restraints and competition. This is complicated by a disorganized industry seeking only short-term solutions for immediate profitability.

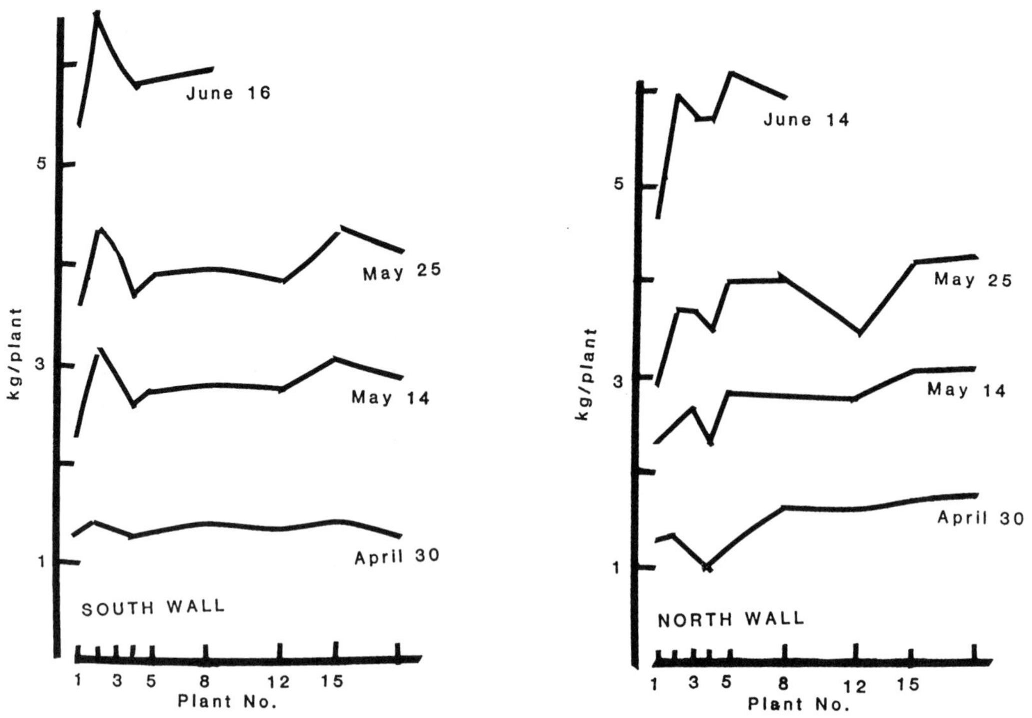

Fig. 4-54. Production of plant 1 through 20 behind the south wall (**left**) and north wall (**right**) of a greenhouse at different times of the year. Single glass plus plastic film [Buitelaar et al., 1984] (©1984, Int. Soc. Hort. Sci., *Acta Hort.*, 148:511-517).

The literature is replete with the savings that could be had by lowering temperatures [e.g., Bond, 1981; Ross et al., 1978; Anon, 1974]. However, the ability to save 10% by lowering the temperature 1 C from 18 C is a statement of fact and not a recommendation (Winspear, 1974]. In fact, these types of measures do more damage than good since they are based upon insufficient knowledge of plant physiology discussed in Section 4.II of this chapter.

a. Double Cladding

One approach to energy saving involved double covers, which have already been discussed several places in this book (Tables 2-7 through 2-10, Fig. 2-53, 2-55, 2-64). While the savings are significant (Tables 4-4, 4-5, Fig. 4-29 through 4-32), it is instructional to observe the European scene where single covers are retained with efforts to maintain, and even improve, irradiance with a distinct shift to thermal screens. Structured materials are often installed on the side and end walls. Use of insulation on side walls may cause a significant irradiance reduction that affects yield but not flowering rate [Buitelaar et al., 1984]. Buitelaar et al. examined the effects of different insulation materials such as double glass versus glass plus polystyrene (opaque) on north and south walls (Fig. 4-54). They found less of an effect of insulative materials on the north wall versus the south wall, with the greatest differences in the summer in terms of yield reduction. Insulative materials, even when transparent, did have a marked effect on yield of the first plants inside the wall. The common practice of fully insulating the north wall with opaque materials will definitely reduce growth in temperate climates.

Considerable effort in the U.S. and U.K. has been extended on fuel saving features of double covers, with much discussion of "R" values of materials, or the resistance to heat transfer (m^2 C W^{-1}). R is the reciprocal of

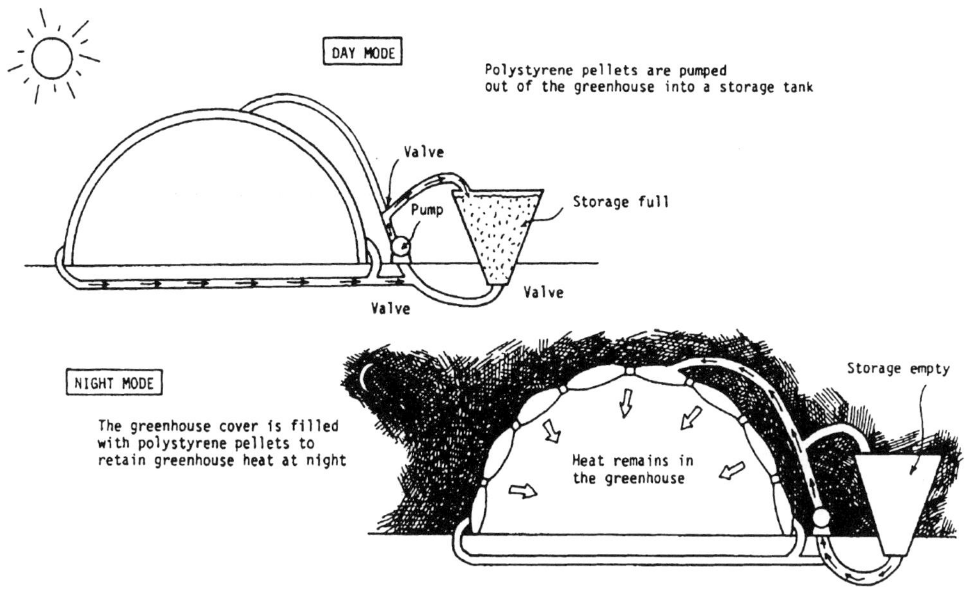

Fig. 4-55. Sketch of a polystyrene pellet insulated, double film-covered greenhouse [Roberts et al., 1985].

conductivity (h). Waaijenburg [1984a; b], in reporting investigations on cladding materials, has suggested the use of the ratio of h value of the test material to the h value of single glass as the insulative index (I_d). Thus, if $I_d = 0.6$, then the heat consumption of the greenhouse with the covering being tested will be 60% of that of a comparable single glazing. Or, there will be a heat savings of 40%.

Other attempts to reduce energy loss through the greenhouse roof have employed polystyrene pellets that are blown between double covers at night and removed during the day [He et al., 1990; Short and Shah, 1981; Elwell and Short, 1989, Roberts et al., 1985] (Fig. 4-55). The potential heat savings, according to Roberts et al., are 60 to 90% annually. Obviously (Fig. 4-55), such a system requires considerable ancillary equipment with large storage for the pellets –assuming one has solved the problems of uniformly distributing the pellets between the double cover, overcoming static electricity and effects of any moisture in the pellet medium. Other attempts have included blowing fire foam between greenhouse layers [Cunningham, 1983; Roberts et al., 1985]. A liquid foaming agent is foamed under pressure and blown between the plastic film layers. The foam is continuously generated at night and the liquid is drained by gravity. A problem is that freezing temperatures break down most foams. Another idea has been to actually enclose the structure with movable, insulated panels (another building). Glass may also be purchased with low emissivity films. Halleux et al. [1985] found that radiative heat loss was reduced to 40% of its common value, and the radiation from the crop to cover by 10%. For cloudy sky conditions, the advantages of low emissivity glass are markedly reduced. Nijskens et al. [1984] showed that the presence of water (condensation) on the film will drastically change emissivity, and condensation can be worsened. The advantages become illusory with the change in emissivity. Low emissivity on the outer side of the inner layer of a double glass provided the best results.

In summary, reports from the U.K. [Norman, 1983; Mitchell, 1982] indicate problems with double covers that could be due to reasons other than reduced radiation (e.g., reduced CO_2 concentration, reduced nutrient uptake, internal air pollution, etc.). Van Winden et al. [1984], in six experiments with cucumber, found a yield reduction of 4 to 13% compared to single glass. The results varied with season, with tomatoes showing a yield under double glass 10 to 15, and 4 to 13% less in comparison with yield under single glass in the spring and autumn. Steinbuch and van de Vooren [1984] showed that species such as begonia and African violet do well under double covers, but chrysanthemums required a longer period to flower. Quality was definitely reduced on roses. Bulbous crops such as iris and tulip were not easy to grow in insulated houses. Nor was it possible to achieve the same average temperature under the different covers. Mitchell makes the point that there is no virtue in energy saving *per se*. Saving fuel or labor at the expense of yield or quality has many times proven to be counterproductive.

b. Thermal Screens

The problems of double covers, and high heat losses through PE claddings, led to the development of thermal screens that totally enclose the crop and separate it from the cold exterior cover (Figs. 4-56, 4-57, 3-65, 3-72). A number of years were required to achieve suitable systems and appropriate screening materials (Table 4-17). Materials for rolling screens should be nonshrinkable, or not more than 2% [Bakker and van Holsteijn, 1995]. Bakker and Holsteijn report that screening materials are made from three raw materials –polythene, polyester, or acrylic. Polypropylene and polyamide cellulose have little durability. Reports have indicated a saving up to 60% compared to conventional, unscreened houses [Bailey, 1981] at night and a seasonal reduction of 25 to 30%. The effect of various screening materials may be noted in Table 4-17 and Fig. 4-58. According to Krause [1982], the Dutch are now employing two screens and sometimes three, one or more of which may be utilized for shading. This practice may be more beneficial than the real energy savings.

Table 4-17. Thermal performance (W $m^{-2}K^{-1}$) of selected insulating curtain materials in a commercial greenhouse [Roberts et al., 1981].

Material	Transfer coefficient (h) based on roof area
Reemay Sunbonded polyester	3.59
Double knit cloth	3.53
Black PE, drilled	2.70
Reinforced PE	2.59
Black PE over Reemay	2.49
Prefabricated aluminized vinyl	2.38
Aluminized vinyl (worn)	2.18
Polypropylene 97% shade	2.16
Experimental black PE, drilled	2.10
Foylon	1.93

The articles, especially in the British *Grower*, made interesting reading during the late 1970s and early 1980s. Two reports in 1978 [Anon, 1978a; b] gave full recommendation for thermal screens with considerable information on retrofitting existing structures. But, following articles [e.g., Krause, 1980; Butters, 1980; Royle, 1980a; b] raised serious questions about the yield loss that generally occurred with screens. In particular, Royle reported that yield was reduced more under an impermeable screen. Attention was being given to higher humidities occurring under screens. Krause stated that experiments with present growing systems showed that yield losses quite outweighed the benefit of fuel saving. Monetary losses of up to 12% were reported on tomatoes or $16557 per hectare in gross return for a savings of $12475 per hectare in fuel costs. At the end of 5 weeks of picking cucumbers, the loss reported was 15%. At one nursery, the loss in irradiance with parked screens was 4%, but the crop loss was closer to 9%. Problems could also be arising from lower temperatures under thermal screens. By 1989, reports [Anon, 1989b] indicated that nearly all Dutch growers had moved to thermal screens, and as the result of severe fires, several Dutch nurseries were removing structured acrylic covers and replacing with glass. One important conclusion was that thermal screens altered the interior climate to such an extent that everyone had to start over. Temperature patterns are changed completely with development of vertical gradients [Butters, 1980] about 30 cm above the floor, with changes in irrigation frequency and higher humidities. Butters reported higher yields after 4 weeks of picking in the new screened houses. Thus, the installation of thermal screens required completely different environmental practices –as well as the research necessary to provide efficient screening materials and equipment [Fenton, 1983].

Similar studies have been carried out in other locations such as Denmark [Amsen, 1986], Canada [Staley et al., 1986; Arinze et al., 1986], and the U.S. [Aldrich, 1980; Rebuck et al., 1977; Seginer and Albright, 1980]. Research has been active in France and the Mediterranean as well. In Canada and the U.S., considerable attention has been given to problems of snow accumulation on cold roofs and the temperature drop when screens are opened in the morning, causing an increased heat load. According to Staley et al. [1986], under British Columbia conditions, thermal screens can be a very profitable investment. For commercial-sized structures (4560 m^2), with No. 2 fuel oil at $0.28 ℓ^{-1}, any curtain system costing $40.00 m^{-2} or less would be profitable with savings of 43%. For a colder, continental climate such as Ontario, the net present value (NPV) (future cash flows of costs and benefits during life of system discounted to determine present values) ranged from $11.00 to $46.00 m^{-2}, depending upon installation costs. If the fuel savings were increased to 50% from 43%, the NPVs would be

Fig. 4-56. Various examples of thermal screens retrofitted in existing greenhouses. Note that **B** uses an infrared heating system. **C** has two curtains, the lower a shade screen as is **D**.

\$27.00 to \$74.00 m^{-2} with maintenance costs much less than \$4.50 m^{-2}. Rebuck et al., in 1977, quoted several sources as indicating a payback ranging from 8 years to less than 2 years.

There are a number of details to be kept in mind [Roberts et al., 1981; Ross et al., 1978; Bailey, 1981]: 1) Variations in performance caused by changes in weather and mechanical sealing of the edges dominate differences due to the curtain material. 2) The possibility of condensation collecting on the material should be avoided by using porous material or drilling the material with 2 mm diameter holes on 15 or 30 cm spacings. This latter practice may now be obsolete. 3) Performance of thermal screens is improved by making them nonporous with an aluminized layer on the upper surface. 4) The edges of closure must be well sealed to prevent air leakage. 5) The shade should be "parked" during the day in as small a space as possible to avoid loss of solar energy.

Fig. 4-57. Further examples of horizontal thermal screens, some also being used for shading. **Left**: Note two screens at right angles to each other. **Right**: An opaque screen in the parked position.

Aluminized vinyl and experimental PE were more bulky than Foylon or double-knit cloth (Table 4-17), with the 97% shade material most bulky. 6) There is a strong correlation between a material's ability to drain condensate and durability. And 7), the curtain should be drawn horizontally with the curtain area minimized and all heating equipment within the curtain enclosure.

According to Bailey, there was a reduction in tomato yield with screens during the first year of experimentation. When the side screens were removed during the day in the second year, crop performance in

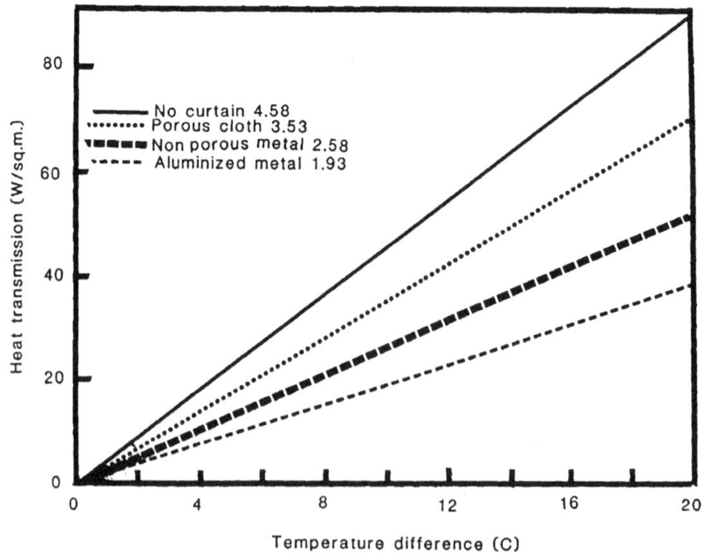

Fig. 4-58. The effects of various curtain types on the heat transmission as a function of temperature difference inside to out. The transfer coefficient (h) (W m^{-2} K^{-1}) is given for each material [Roberts et al, 1981].

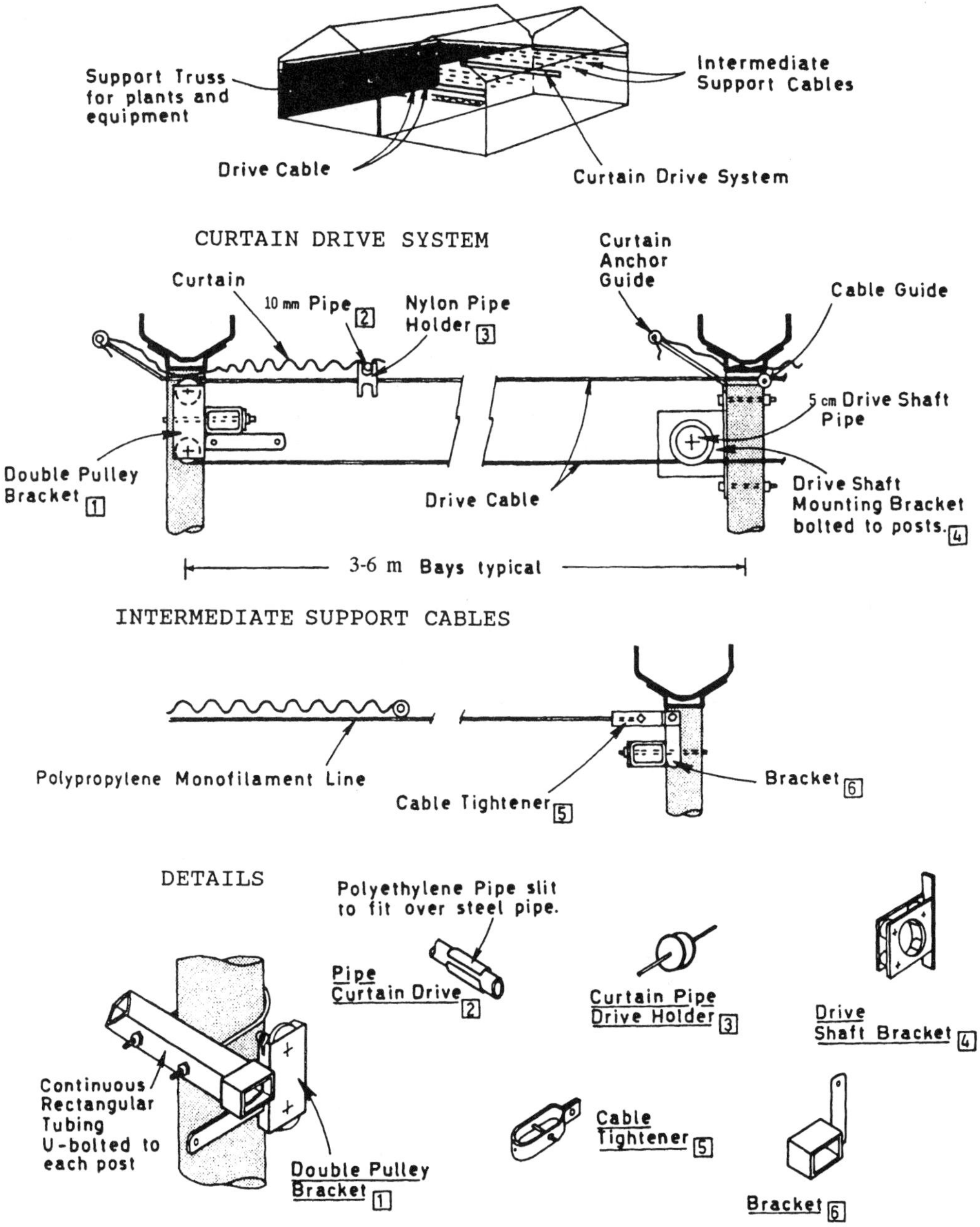

Fig. 4-59. Details for horizontal thermal screen in clear span, gutter-connected greenhouses [Ross et al., 1978].

both screened and unscreened houses were comparable. An example of a thermal screen setup is provided in Fig. 4-59. Numerous other types are available. The screen moves from gutter-to-gutter above the crop supports. In many earlier houses, the trusses were utilized as crop supports and obviously required the screens to move between the trusses as typified by Fig. 4-56A, B. In tight systems, temperatures above the screen are likely to be quite low. Thus, the opening device should have some means of "cracking" the openings in the morning until

Table 4-18. Predicted cost of greenhouse heating with various curtain control methods [Seginer and Albright, 1980] (With permission of the ASAE).

Control method	Heating cost $/ha-yr	Loss from optimum[a] $/ha-yr	Loss from optimum[a] %
Without curtain	100800	44000	100.0
With curtain			
1) Optimal (computer model)	56800	0	0.0
2) Monthly best outside radiation (R_o)[b]	56900	100	0.2
3) $R_o = 33$ W/m^2	57400	600	1.1
4) Best day length	60800	4000	7.0
5) $R_o = 0$	62400	5600	9.9
6) One hr off best day length	68100	11300	19.9

[a] Set points for opening and closing determined by proposed model and used for basis of comparison for other methods.
[b] Calculated "best" outside radiation level for curtain opening and closing, which varied with time-of-year. R_o for 3) and 5) fixed at levels given.

temperatures above and below the screen have opportunity to equalize without damage to the crop. However, this is easily solved.

In general, curtain control is based upon timed opening and closing. Seginer and Albright [1980] were among the first investigators to attempt computer control, basing opening and closing upon the return for fuel savings versus costs of delaying crop growth, with the idea that the best time to open or close a thermal curtain was when the potential gain from photosynthesis when opened equaled the potential gain from heat conservation when closed. Their predicted costs, comparing timed, radiation, and "optimal" are given in Table 4-18. The optimal control considered the tradeoff between photosynthesis, respiration, and fuel requirements with and without the thermal screen. Their conclusions were that it was less expensive to base opening and closing in New York state on the radiation level inside the greenhouse with the sensor located above the screen. For outside radiation the best single preset radiation level was 33 ± 10 W m^{-2}. The harvest of a 4 month chrysanthemum crop was delayed 8 hours which would not be noted

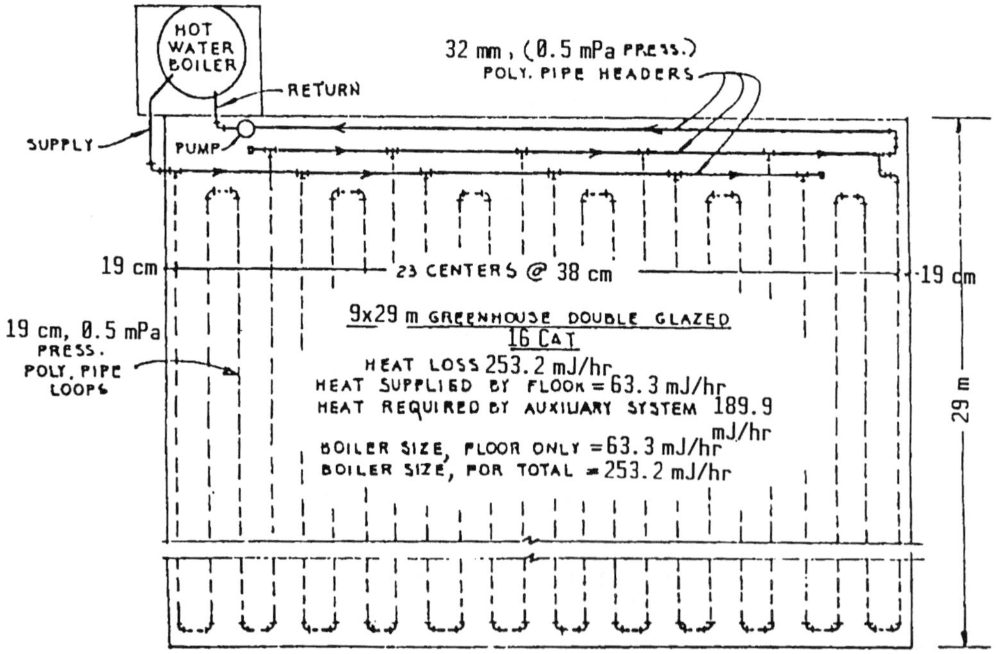

Fig. 4-60. Floor plan of a greenhouse using hot water heat in polyethylene pipe [Adapted from Roberts, 1983].

Fig. 4-61. Nomographs calculated by Puri for designing ground heating systems using hot water. All piping 25 mm diameter. The top graph is for water 5 C above floor temperature (ΔT_f) and air temperature 10 C below floor temperature (ΔT_g). The remaining two are conditions for air temperature 5 and 2.5 C below floor temperature. A = pipe spacing (cm), B = floor depth (cm), D_d = pipe diameter [Reprinted from *Solar Energy*, 28:469-481, Puri, V.M., ©1982. With kind permission of Elsevier Sci. Ltd., The Boulevard, Langford Lane, Kidlington 0X5 1GB, U.K.].

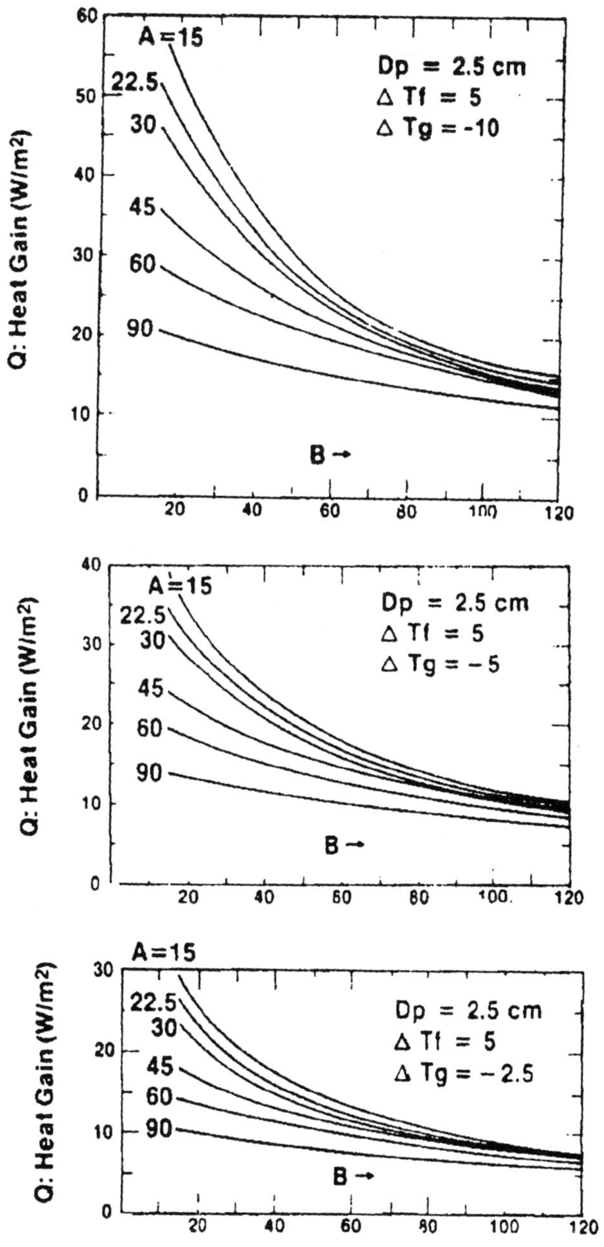

under practical conditions. The radiation method had the advantage of saving \$3500 $ha^{-1}yr^{-1}$ over a time-clock operated system. Seginer and Albright noted that best solar radiation levels are lower for the winter months, with error in setting being more important in the winter. Of course, variability of solar radiation requires that either an average radiation set point be employed, or, more practically, a toggle method is used that prevents constant screen movement until the irradiance has increased or decreased significantly from the initial value. Such a system was employed by Hanan et al. [1987] to switch day-night regimes and prevent excessive operation of the curtain. Delaying curtain opening has the advantage of reducing initial cooling. Computer control can also incorporate staged opening to prevent a sudden drop of cold air on the crop below. Or, one might consider utilization of DIF temperatures to control height. Certainly, a variable setting in accordance with season is more reliable than a time clock that may not be re-set when necessary. Many of the newer systems are not described in the scientific literature. There has been a marked improvement in control systems and materials.

c. Surface Heating

Flowing water over the surface of a greenhouse as a means to either cool or heat the interior has been examined for a long period. Harold Gray, at Cornell University, in the 1940s examined flowing water possibilities. It was employed by the Phytotron in California [Went, 1957] to help cool the inside. More recently, Morris et al. [1958] examined the limitation of temperature by use of a water film. According to them, it was not practical to maintain on the glass a water film thick enough to absorb a significant amount of solar radiation,

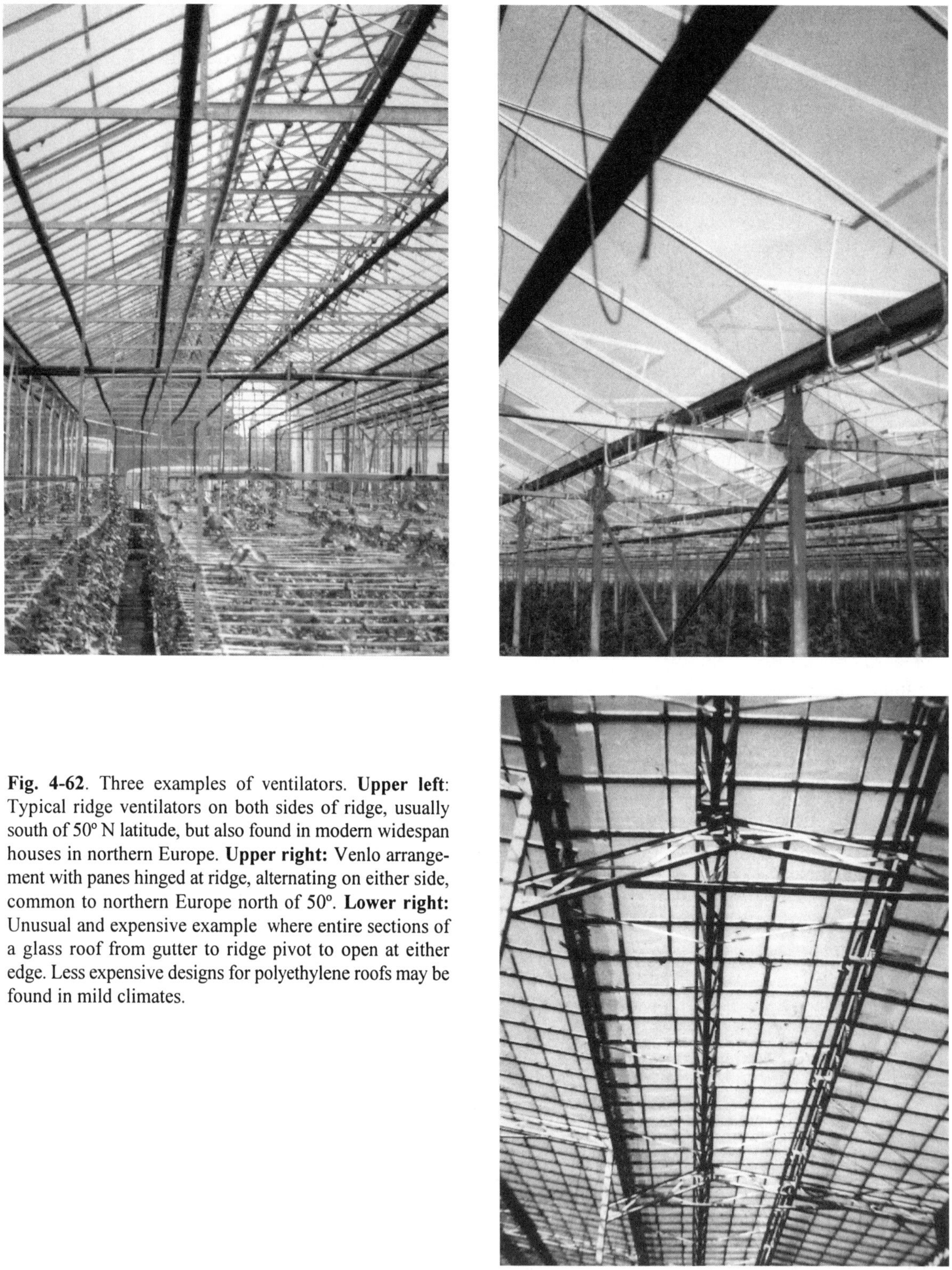

Fig. 4-62. Three examples of ventilators. **Upper left**: Typical ridge ventilators on both sides of ridge, usually south of 50° N latitude, but also found in modern widespan houses in northern Europe. **Upper right:** Venlo arrangement with panes hinged at ridge, alternating on either side, common to northern Europe north of 50°. **Lower right:** Unusual and expensive example where entire sections of a glass roof from gutter to ridge pivot to open at either edge. Less expensive designs for polyethylene roofs may be found in mild climates.

although a reduction of the inside temperature could be obtained. More to the issue, possibilities were enhanced during the fuel crisis to use waste heat from water to heat the greenhouse. Studies were carried out in the late 1970s and early 1980s by Walker and co-workers [Heinemann and Walker, 1986; Walker et al., 1982; 1983; Walker, 1982; Lazarus et al., 1981]. Starting with a small experimental greenhouse in 1977, Walker [1978] found that for an outside temperature of -6 C and a water temperature of 30 C, the inside climate could be maintained at 15 C with no increase in humidity. This amounted to a waste heat input of 5.9 kW m^{-2} versus a conventional system input of 0.31 kW m^{-2}. Such a system is extremely inefficient, but since there are few commercially attractive alternative uses for this waste heat from large generating plants, the costs for greenhouse heating could be reasonable. In larger houses, at a rate of 0.4 ℓ $s^{-1}m^{-2}$, 81% of the surface area was covered, and this heating system provided a 26% energy savings over electrical resistance heating methods. In general, a glass house was better than FRP cladding in transferring heat.

An alternative to surface heating of the house by flowing water is the development of the so-called "fluid roof" by van Bavel and co-workers [van Bavel and Damagnez, 1978; van Bavel et al., 1980; 1981; van Bavel and Sadler, 1979]. The idea is to circulate water containing 3% copper chloride between the layers of a double roof. Transmission is greatest between about 350 and 700 nm for thin or less concentrated solutions. Therefore, energy required for photosynthesis will be available while the remainder is absorbed and can be used to heat the greenhouse at night. The approach by the investigators has been to simulate the effects, using a suitable simulation language. To the best of my knowledge, the authors have not actually tested plant response under suitable conditions, even though the system has also been proposed as a means of reducing excessive temperatures in mild climates.

Such a system has been tested by Nilsen et al. [1983] under Scandinavian conditions. The idea here was to reduce the infrared radiative portion of solar radiation. Unfortunately, northern European weather conditions are completely different from more southerly latitudes. Although, according to Nilsen et al., the method produced an almost optimal room temperature, and reduced the leaf temperature, the overall reduced irradiance did not outweigh the leaf temperature reduction in terms of growth. The authors also pointed out that it was not possible to produce the same flow rates in all channels of the structured cover. Weather during most of the experiment was rainy and cloudy and the yield of tomatoes was not significantly increased by the fluid roof. A model of an "optical liquid filter" greenhouse was examined by Tross et al. [1984] in Israel. This was a system very similar to that proposed by van Bavel. According to van Bavel et al. [1980], the fluid roof system offered considerable promise under Texas conditions, and they dismissed evaporative cooling as no solution [van Bavel et al., 1980; 1981]. The latter conclusion has not been substantiated in respect to practical experience with fan-and-pad cooling systems in the arid regions. Furthermore, the effects on spectral quality would undoubtedly have an effect on photomorphogenesis.

d. Ground Heating

This was discussed in Section 4.IV.B.2, particularly as it is applicable to mild climates. As noted, suitable ground heating can supply up to 25 to 30% of the total heat load in temperate climates. The main limitation, of course, is the maximum temperature allowable –especially when plants are set directly on the ground or grown in the soil. Some species will benefit from warm soil, others may not, and still other species may require lower temperatures during various growth stages (i.e., alstroemeria). However, the ability to maintain a reasonable ground temperature is certainly a good energy conservation method as it reduces the temperature differential in the upper greenhouse air. A number of individuals have examined ground heating [e.g., Shen and Mears, 1977; Roberts, 1983; Puri, 1982; Boulard et al., 1989; Otten et al., 1989; Kruyk, 1976; Mavroyanopoulos and Kyritsis, 1986]. An example of an installation is provided in Fig. 4-60. According to Roberts [1983], the system can be "dry," that is, the piping may be embedded in porous concrete and any water drains away; or, it can be "wet" where the floor is lined with a biocide-treated vinyl liner, containing 23 to 30 cm of stone and a porous concrete cap. The dry floor, although less expensive to install, delivers only 50% of the heat that the wet floor system will deliver. In most cases, especially where the water table is high, an insulation layer must be under the floor. Flooded floors have received attention in the Netherlands [Lawson, 1985]. In some instances, heat pipe is glued directly under flooded movable benches. Or, the plants are placed directly on a concrete floor, which can be flooded for irrigation (Fig. 2-82). Lawson makes the point that all possible surfaces should be dry as heat can be used for evaporation, which is a waste.

A heated floor is particularly useful when waste heat is available in the range of 32 to 38 C. Standard 80 psi (551 kPa) PE pipe made from a virgin resin with Nylon fittings is employed. Double clamps are recommended. Kruyk [1976] states that Dutch growers employ 15 mm diameter polypropylene piping, buried at 40 cm. Loops are limited to about 80 m length. The deep depth allows usual cultivation. Some growers in the U.S. utilize polybutylene, which has a higher temperature rating, and temperatures up to 50 C may be reached without damage. However, assuming that for most purposes, the maximum temperature of a floor is about 20 C for most plant species, water at lower temperatures can be employed. Under-floor heating provides a much larger area of heat input [Anon, 1982a], with the result that there are no horizontal temperature gradients. A certain portion of the heated water can be circulated several days before the crop is planted to bring the mass up to desired temperature. Thereafter, the system is a constant heat source of about 63 W m^{-2} [Giacomelli, Personal Communication 1992]. Puri [1982] has developed a series of nomographs for designing floor heating systems. A selected few of the more likely nomographs are in Fig. 4-61. Note that ΔT_f and ΔT_g are temperature differentials. $\Delta T_f = T_f - T$, where T_f is the fluid temperature in the piping and T = floor temperature. $\Delta T_g = T_g - T$, where T_g is the air temperature. If one assumes a floor temperature of 20 C, then the heating fluid temperature is 25 C and the air temperature is 10 C (Fig. 4-60, top).

Otten et al. [1989] examined ground heating affects on tomato production grown in NFT, rockwool, and soil. Apparently, the NFT troughs acted as solar collectors, which reduced yield due to high water temperature. Otten et al. stated that plants grown in good soil at cool night temperatures, and fertilized through a drip irrigation system, could produce yields similar to those obtained with NFT and rockwool systems. But, the percentage of large fruits was less from plants grown in soil. Rootzone heating did not improve performance of tomato plants in NFT, but was beneficial for plants in rockwool grown at cool night temperatures in the spring. Energy savings of at least 30% were obtained by reducing the night air temperature to about 11 C with rootzone heating on tomatoes without reducing yield. For plants in NFT, no rootzone heating was required and yields and quality were improved at 11 C air temperature.

The 1995 text by the editors Bakker et al., does not give floor heating a great deal of attention. Van de Braak [1995] considered the slow response of a floor heating system to be a significant disadvantage, and van Meurs [1995] states that the efficiencies of heated floors are very low. Because of large thermal mass, slow response time may mean that ventilation during the day will be required since the floor cools down slowly, even if maintained at the night setting. I suspect that the success of heated floors is closely related to climatic conditions, particularly in regard to radiation. Clear-day regimes would increase difficulties in control due to relatively rapid changes in the radiant heat load.

e. Miscellaneous Energy Conservation Methods

The energy crisis marvelously concentrated growers' attention on common procedures to ensure efficient energy use. It was not unusual prior to the 1970s for growers to leave out closure strips on FRP covers in order to reduce condensation common to such claddings in the first 2 or 3 years. That abruptly changed. Winspear [1974] also pointed out several procedures to follow: more frequent boiler inspections, fixing faulty steam traps capable of wasting 16 kg steam per hour, one pair of uninsulated flanges on a 10 cm diameter steam line will consume nearly 400 liters of oil yearly, and 30 cm of uninsulated steam line will consume 203 kg oil per year. Methods were undertaken to wring as much energy as possible from the combustion process [e.g., Anon, 1977], such as recovering heat from the exhaust gases, prewarming the combustion air, etc. Studies were carried out on means to more efficiently distribute air within the greenhouse [Tesi and Tosi, 1985]. Adequate attention was finally given to accuracy of temperature measurement since an error of 0.5 C over a sufficient period could seriously affect growth and increase fuel consumption needlessly [e.g., Anon, 1976b]. Windbreaks also received attention [Freeman, 1976; Botacchi, 1977; Nelson, 1979]. Even air from abandoned mines was investigated as a means to heat and cool greenhouses [Ewen et al., 1980].

All these procedures to reduce energy loss and improve efficiency should be standard and common. The grower should maintain a logbook to make sure that timely maintenance is accomplished. However, most of

Fig. 4-63. **Upper:** Two examples of side ventilation in more developed areas of the Mediterranean. Double PE with no ridge vents, restriction to 30 m maximum between sides. **Lower left:** Common arrangement for southern California. **Lower right:** Probably the least expensive ventilating system in the Mediterranean.

those methods for conserving energy and making use of solar radiation are very climate sensitive. They are also subject to various problems, as listed in Section 4.IV.B.2. So far, none of the novel methods to utilize solar radiation are commonly accepted practice. They also have the disadvantage that one cannot predetermine temperature control in a manner to ensure maximal yield and quality.

f. Summary

Methods to tighten the structure to reduce infiltration and heat loss, installation of thermal screens, and use of double covers have a remarkable effect on an interior climate. Humidity is likely to be higher, CO_2 concentration lower during the day, and the interior is less influenced by exterior conditions. Failure to ensure maximum structural transmissivity is likely to lead to an unacceptable yield loss and crop delay. Diseases are likely to increase (i.e., botrytis, mildew, rust, etc.), and physiological diseases such as blossom end rot on tomatoes can be more frequent [Watkinson, 1975]. Kooistra [1986] stated that the application of thermal screens caused greater vertical and horizontal temperature differences. Adaptations to the heating systems were necessary, with increased use of CO_2 injection. Thus, these procedures must be implemented slowly with all the various ramifications carefully worked out before success can be assured.

As a final point with regard to all these alternative energy methods and ways to conserve energy, it might be well to mention a remark made to me several years ago by a very intelligent grower who kept meticulous books. It was the fact that, although energy prices had increased more than 400% in the past decade, his costs as a percentage of the total cost of growing had only increased 4%. This relationship looms large when considering capital investment in novel techniques.

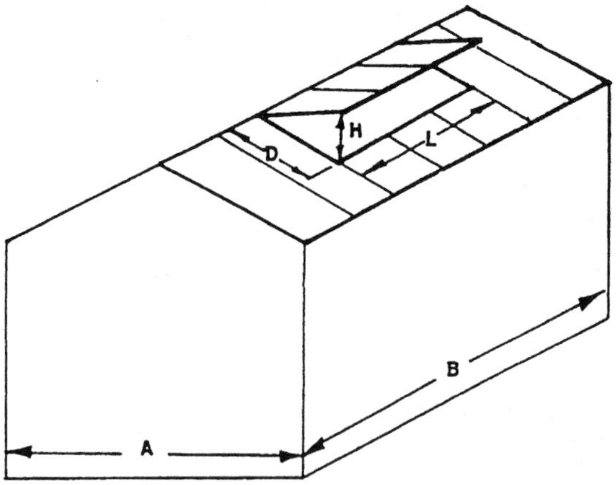

Fig. 4-64. Calculating percentage ventilation. The 'mouth' area is (H x D) + (L x H) and 'throat' area is (D x L). Take the smaller times number of vents and divide by (A x B) [Smith, 1988] (With permission of the British *Grower* magazine).

C. VENTILATION AND COOLING

1. Natural Ventilation

The principal, and least expensive method used in greenhouses to prevent excessive temperatures is natural ventilation, caused by pressure differences or natural buoyancy forces through ventilators arranged on the top of the structure, on the sides, or both. Examples of various arrangements may be noted in Figs. 2-9, 11, 19, 27, 29-32, 44, and 63. Still, other examples are presented in Figs. 4-62 and 4-63. The important item to note from these examples are the variations, which depend upon climate and local technology.

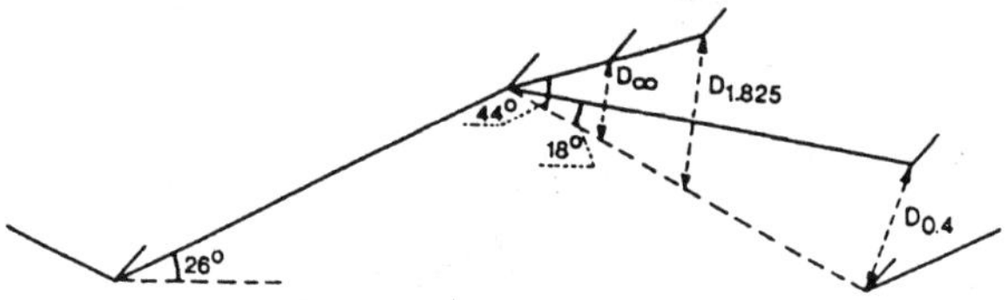

Fig. 4-65. Typical Dutch Venlo ventilator arrangement where the vent is half or the full distance between ridge and gutter. The aspect ratio (length/height) of 0.4 means that ventilation will increase faster compared to the ventilator with an aspect ratio of 1.825. For ventilator angles of 18 and 44°, ventilation between the two is comparable [Bot, 1983] (With permission of the author).

a. Northern Methods

For example, the common design for northern Europe (Netherlands, Germany, Belgium, etc.) and the U.K. is Figs. 2-11 and 4-62. The ventilators, especially on the roof are arranged individually, alternating with each other on opposite sides of the ridge (Fig. 4-62). In calculation of air movement through such ventilators,

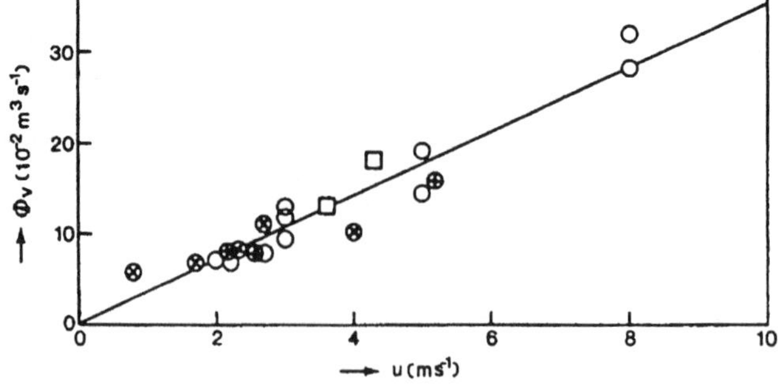

Fig. 4-66. Flux of air (Φ_v) as a function of wind speed (u) through a Venlo ventilator when open 10% of full or 4.4° [Bot, 1983] (With permission of the author).

the effect of the triangular ends is considered [Bot, 1983]. According to Smith [1988], ventilator area (see Fig. 4-64 for calculating area as percentage of total floor area) should be about 16%. Almost all Dutch houses are built as large multispan, gutter-connected structures that preclude the use of side ventilation, especially where the sides are permanently sealed by double walls. Under normal conditions, only the leeward vents are opened [Bot, 1983] (Fig. 4-65). This helps to avoid sudden drafts on the crop within, particularly in cold weather. For this type of opening, for any given opening angle, the air flow is linearly proportional to the wind speed (e.g., Fig. 4-66). However, typical behavior of ventilators hinged on one side is indicated in Fig. 4-67, showing that initial opening is more critical than larger variations at wider opening angles. Under Dutch conditions, buoyancy effects due to temperature differences are relatively minor. Ventilation effects due to wind will be dominant if $3u > (\Delta T)^{1/2}$, where u = wind speed (m s^{-1}), and ΔT = temperature difference between inside and outside. Temperature effect increases with the square root of ΔT [Bot, 1983]. The side areas are dominant for air flux at opening angles less than 10°. Cold air will generally flow in on the bottom side of an opening, and warm air will flow out the top half of an opening. An extreme situation is where $\Delta T > 25$ C, in which case the outside wind speed must be more than 1.7 m s^{-1} in order to be dominant. Under leeward ventilation there appears to be no large effect of wind direction with this vent arrangement.

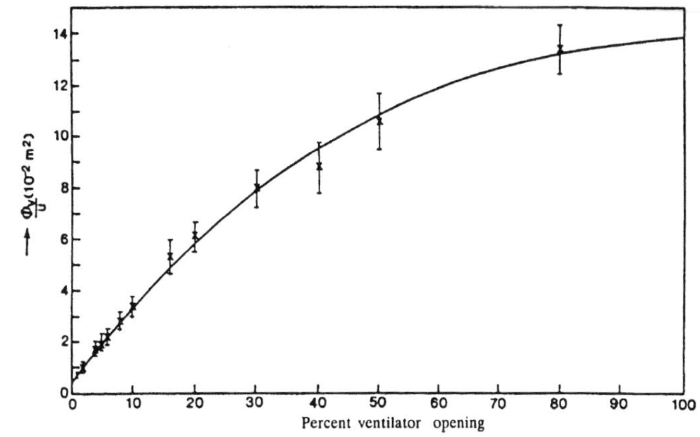

Fig. 4-67. Ventilation flux divided by wind speed (Φ_v/u) as a function of ventilator opening (r) for a Venlo arrangement [Bot, 1983] (With permission of the author).

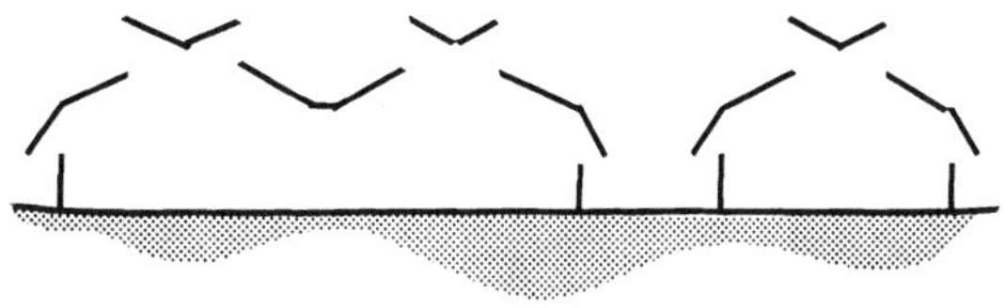

Fig. 4-68. Ventilator openings for double- and single-span greenhouses, running the length of the structure.

b. Middle Latitudes

As one proceeds further south, ventilator design frequently changes to continuous arrangements the full length of the structure (Figs. 2-67, 4-62, 4-68). In fact, although highly expensive, the entire roof may be arranged in sections for ventilation (Fig. 4-62). Continuous ventilators are common in southern Europe, the U.S., and Japan. Wide-span houses in Northern Europe will also be found with continuous vents. American recommendations suggest 15 to 25% ventilation area [ASAE, 1988]. Assuming a single-span structure 10 x 30 m with a vent width of 80 cm, double vents on the ridge and sides, the "throat" area would be 96 m^2 or 32%. The situation indicated in Fig. 4-63, for a Maritime climate, a 30 x 30 m multispan, 3 m to the gutters, with all four sides capable of being opened to a maximum 2 m, would be equivalent to 27%. Side ventilation, in the case of single- or double-span structures, has the capability of quadrupling ventilation rates when combined with top ventilators [Bot, 1983]. The effects of total ventilator opening as a percent of floor area for roof vents only and both side and roof vents may be noted in Fig. 4-69. White [1975] did not find a mathematical model that could be used to predict a temperature rise under New Zealand conditions. He suggested that only small benefits could be expected from fitting greenhouses with openings much larger than 30% of the floor area.

The Japanese, particularly, have been instrumental in investigating natural ventilation in wind tunnels and in the field [Sase et al., 1980; Kozai et al., 1980; Kozai and Sase, 1978; Sase, 1989; Sase et al., 1984]. A simulation model for glass houses up to four spans was tried by Kozai and Sase [1978]. Their results showed that for outside wind conditions less than 2 m s^{-1}, the number of air changes was mainly dependent upon the inside-to-outside temperature difference. Above 2 m s^{-1}, the number of air changes (N) was approximately proportional to wind speed and the number of spans (Fig. 4-70). The method used by Kozai and Sase [1978] could not be applied to varying wind speed and natural convection conditions.

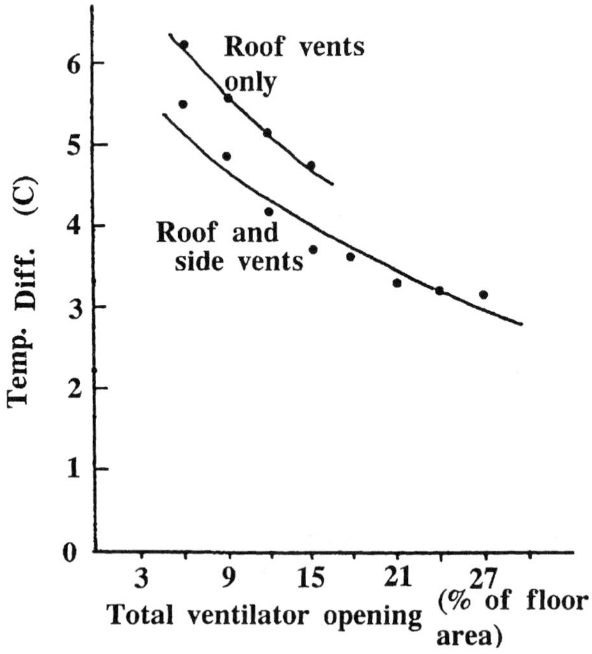

Fig. 4-69. Observed inside-to-outside temperature difference for New Zealand greenhouses as a function of % ventilator opening [White, 1975] (©1975, Int. Soc. Hort. Sci., *Acta Hort.*, 46:63-70).

For single- and double-span models in a wind tunnel (Fig. 4-68), Sase et al. [1980] computed pressure coefficients for various ventilators as functions of ventilator angle and wind direction. These coefficients are a measure of resistance of the opening to air flow, and are used to compute the number of air changes per hour, which is the common method of expressing a ventilation rate. Pressure coefficients of side ventilators were dependent only on wind direction and not opening angle. Coefficients also changed with manipulation of the various ventilators. Up to windspeeds of approximately 1 to 2 m s^{-1}, air exchange in single-span houses did not appear to vary remarkably (Fig. 4-71), but with a continued wind velocity increase, air changes increased linearly except where the wind angle was parallel to the greenhouse length, or 90° [Kozai et al., 1980]. There are no openings in the gable ends. Although 60 air changes per hour [Walker, 1965] are generally considered necessary to avoid heating above the outside air temperature, this number will not be reached until external wind speeds at 0° (normal to ridge) exceed 1 m s^{-1} in a house with a windward ridge vent opened to 40° and both side vents are opened (Fig. 4-71, case B). Information of this type has been utilized by a number of authors to program ventilator control [e.g., Sase et al., 1984; 1985; Kozai et al., 1980]. To achieve exchange rates of 60 hr^{-1}, side ventilation is necessary, or fan ventilation. The side ventilator area must be equal to the ridge ventilator area for maximum ventilation [Whittle and Lawrence, 1960a]. According to Fuchs [1990], if 28 C inside is considered the critical, maximum temperature allowable, the exchange rate must be equal to 30 hr^{-1}. The variable nature of winds greatly complicates ventilator control. Unless provision is made to obtain reliable measurements, as well as incorporating suitable delays in ventilator movement [e.g., van de Vooren and Strijbosch, 1980], ventilators will be constantly moving –much to the detriment of machinery and performance of control algorithms. Vent control systems should include overrides to adjust a ventilator angle in rain and high winds (see Chapter 8). These latter adjustments prevent rain from entering the house and avoid the possibility of structural damage.

Almost all calculations on ventilation employ volumetric units such as total exchanges per hour or cubic meters per second. While an exchange rate greater than 30 hr^{-1} would imply a significant air velocity within the structure, it is not directly related to velocity over the plant canopy. The latter are especially important under high irradiances. Hanan [1970b] showed for carnation flowers (whose petals have no stomata), that the rise in petal

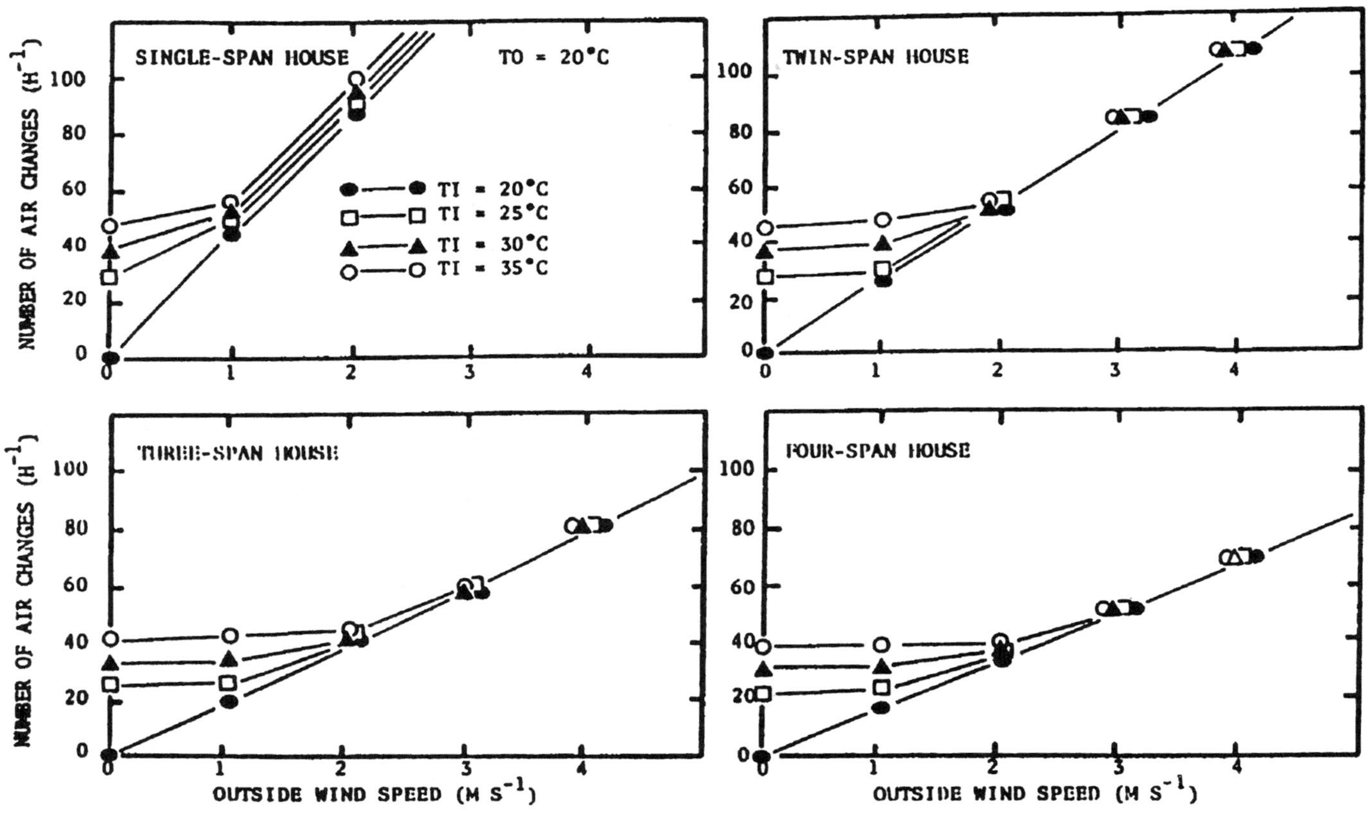

Fig. 4-70. The effect of outside wind speed and temperature differential ($T_I - T_o$) on hourly air exchanges for single and multispan greenhouses as calculated by Kozai and Sase [1978] (©1978, Int. Soc. Hort. Sci., *Acta Hort.*, 87:39-49).

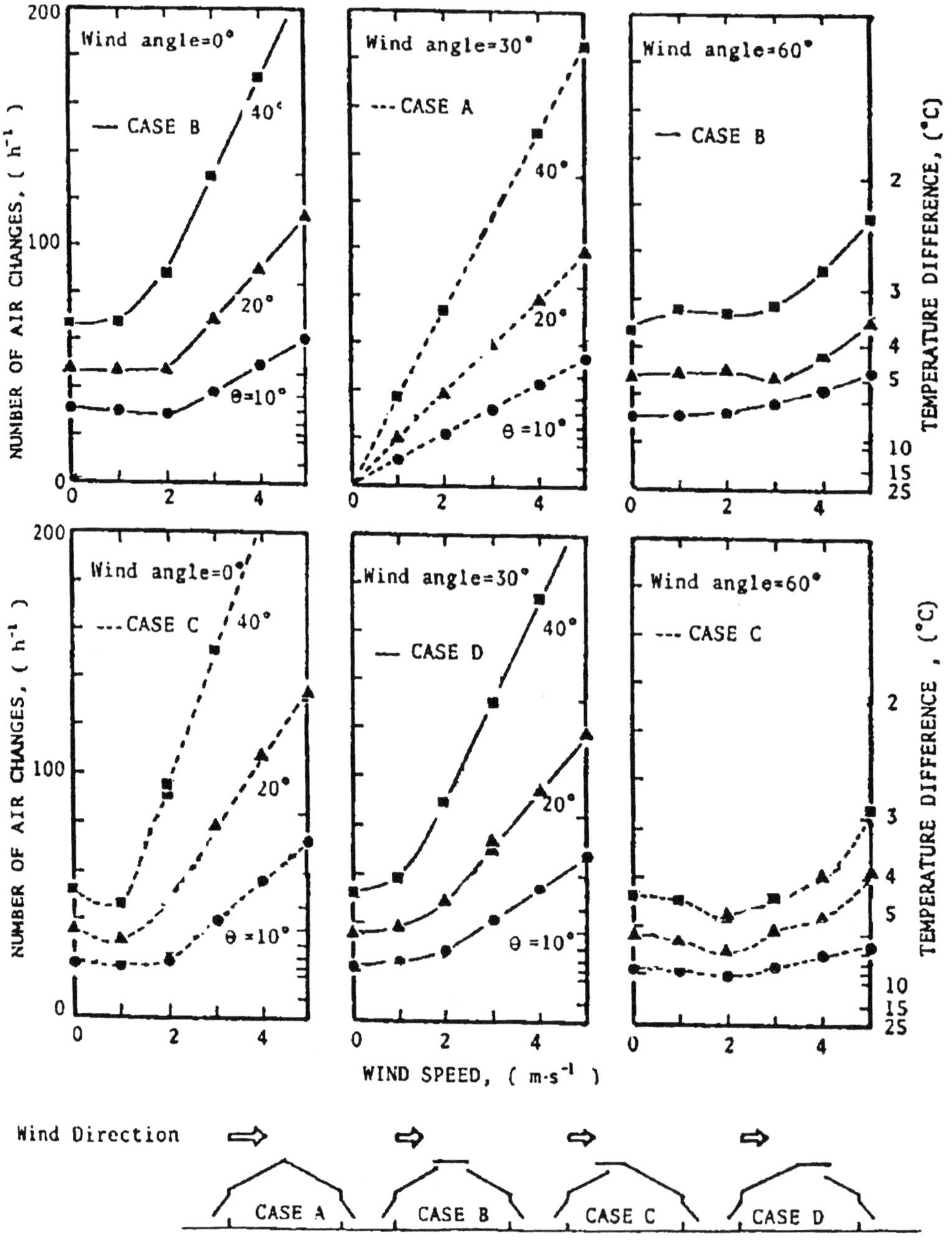

Fig. 4-71. Effect of wind speed, wind angle, and vent opening angle on air exchanges and ΔT for a 7.2 x 27 m, single-span greenhouse. Vent positions for cases shown below [Adapted from Kozai et al., 1980] (©1980, Int. Soc. Hort. Sci., *Acta Hort.,* 106:125-136).

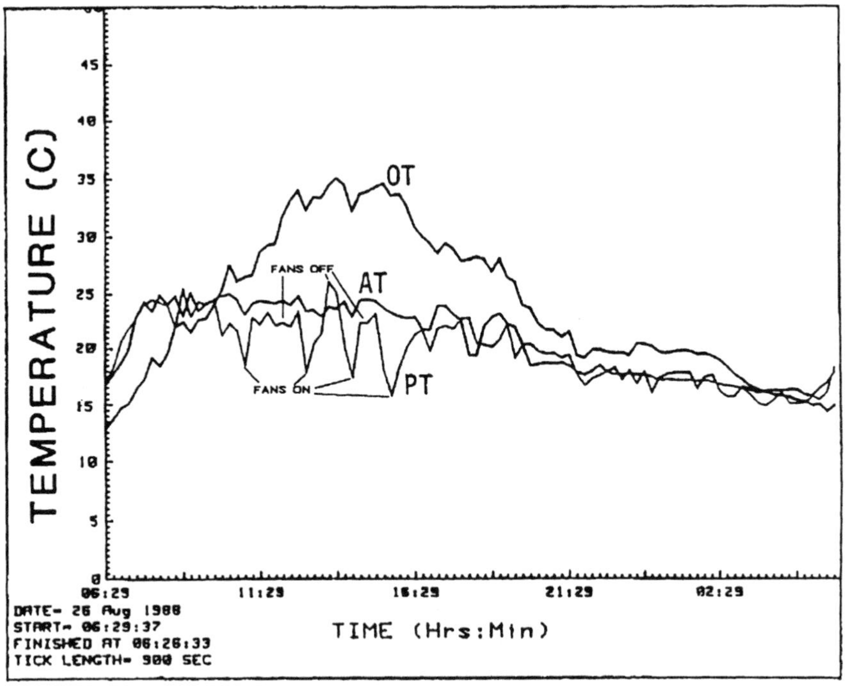

Fig. 4-72. Inside rose canopy (**PT**) and air temperature (**AT**) in an FRP-covered greenhouse compared with outside air temperature (**OT**) on a clear summer day with fan-and-evaporative pad cooling [Hanan, 1988].

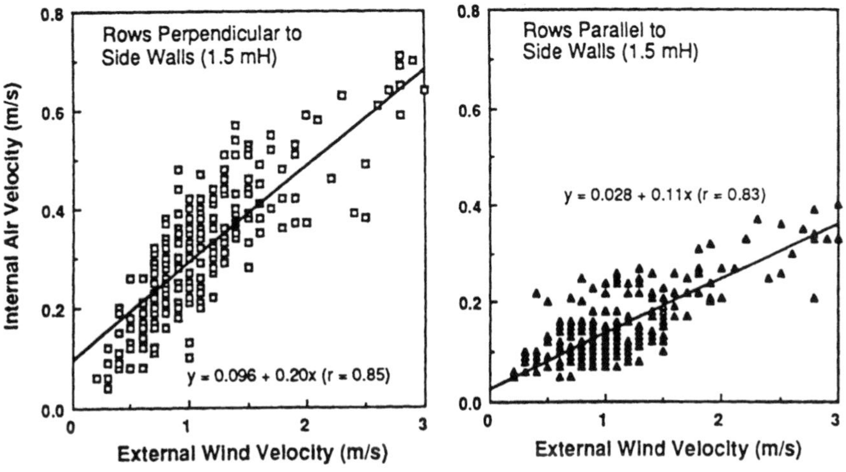

Fig. 4-73. Effect of external wind velocity on inside velocity, depending upon plant arrangement, wind angle at 0°, side and ridge ventilators of a twin-span house opened to 50° [Sase, 1980].

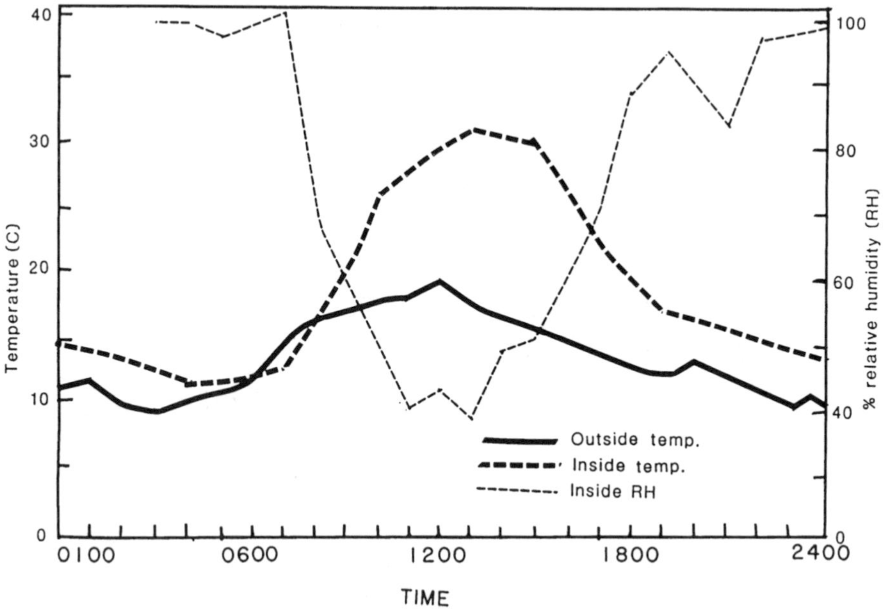

Fig. 4-74. Typical relative humidity and temperature variation over a 24 hr period in an unheated, PE-covered greenhouse in the Spanish Almeria region on a spring day [Montero et al., 1985] (©1985, Int. Soc. Hort. Sci., *Acta Hort.*, 170:227-234).

temperature above air temperature was abrupt at wind velocities below 0.2 m s^{-1}. High irradiances of 1000 W m^{-2} can mean red carnation flower temperature exceeding air temperature by 7 C at wind speeds of 1.5 m s^{-1}. Hanan recommended that, under high solar irradiance, wind speeds over the flower buds should not be less than 1.0 m s^{-1}. Even for an entire transpiring canopy, changes in air velocity between 0 and 0.5 m s^{-1} can mean rapid fluctuations in plant temperature exceeding 5 C (Fig. 4-72). In the only study I am aware of in regard to effects of plant arrangement on internal wind velocity, Sase [1989] showed that, for plant rows parallel to the side walls in a twin-span house, external wind velocity had to exceed 1 m s^{-1} before interior velocity approached 0.2. And, 0.5 m s^{-1} was never reached, even at exterior wind velocities of 3 m s^{-1} –with the greenhouse ventilating at maximum (Fig. 4-73). If plant rows were perpendicular to the side walls, inside velocity usually exceeded 0.2 m s^{-1} when outside air velocity reached 1.0 m s^{-1},

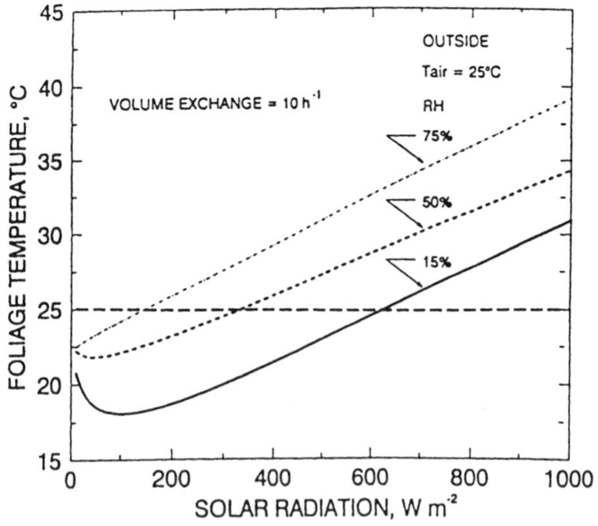

Fig. 4-75. Foliage temperature in an unheated, 1000 m^2 greenhouse a function of solar irradiance for different relative humidities at a low exchange rate of 10 hr^{-1} as calculated by Fuchs (1990).

Fig. 4-76. Effect of air exchange rate on air and foliage temperature in an unheated, 1000 m^2 greenhouse at two outside air temperatures and two relative humidities (15% = arid, 75% = humid, tropical) as calculated by Fuchs [1990].

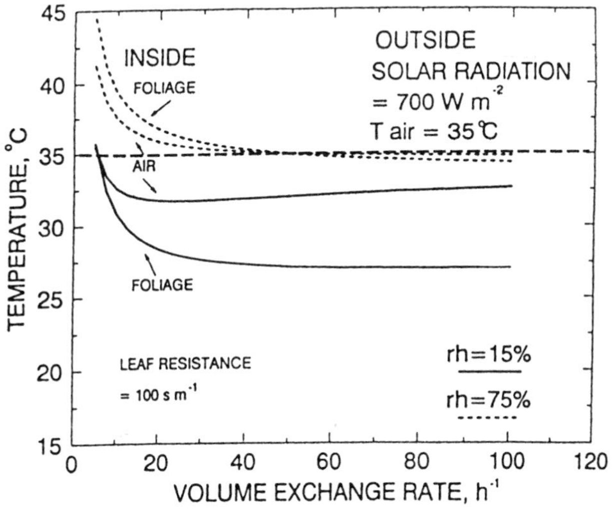

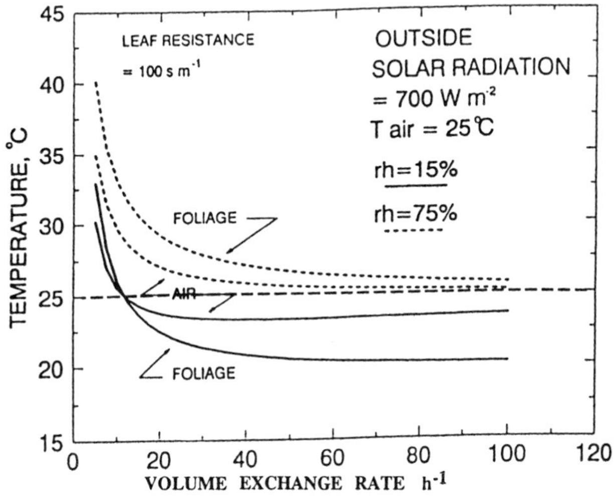

with all ventilators opened to 50°. Air movement and mixing within a greenhouse has a direct influence on energy exchange of the vegetation. This aspect of wind velocity over the plant canopy versus ventilation exchange rate for the structure is not adequately addressed by most of the energy balances described in the literature. Ventilator operation is particularly critical for species such as roses where slight drafts through improper vent operation (especially the side vents) can result in powdery mildew epidemics. Air movement and slight temperature drops under critical conditions can provide conditions for rapid germination of such propagules as powdery mildew conidia (see Section 5.IV).

c. Mild Climates

The ultimate structure for warm climates has been depicted in Figs. 2-19 (3rd from top) and 2-29. These are high structures with ventilation at the ridge and gutters as well as open sides. The high roof pitch is indicative of heavy rainfall. The objectives, of course, are to limit day temperatures and protect the crop from rain. Other examples of less elaborate systems are provided in Figs. 2-9, 28, 30, 31, 44, and 4-63. Structures should be a minimum of 2 m at the eaves or gutters in order to allow maximum side ventilation. Newer structures can be 4.5 m to the gutter. Low structures (i.e., 2 m to the ridge), with limited ventilation will be subject to excessive temperatures unless the roofs are removed during the summer. This is a common procedure in the Mediterranean where crops for export are grown through the winter. Roofs may be very shallow, without ventilators (e.g., Figs. 2-30, 2-42). The difficulty with these low houses –which may be the effect of cost and available local materials– are excessive temperatures during the late spring, summer, and early fall, with high humidities at night. An example of conditions is provided in Fig. 4-74. The variation in relative humidity is due to temperature change since absolute humidity seldom varies much over a 24 hr period. The attempt to retain heat in the common PE cladded houses under such conditions is a mistake, and can lead to severe disease problems. Houses should be allowed to ventilate regardless of outside temperatures. It is usual to find internal temperatures lower than outside on clear nights [Castilla et al., 1988; 1989; Montero et al., 1985; Baille, 1988; Hanan et al., 1979; Meneses and Monteiro, 1990a; b]. Total ventilation area should, probably, never be less than 30%.

2. Evaporative Cooling Methods

I have stated several times that a significant portion of radiant energy is utilized to convert water to a vapor. In Section 4.IV.B, it was pointed out that the energy from the heating system is utilized in this fashion to varying extents. The situation is even more marked in the summer when cooling is required and solar radiation is greatest. Examination of energy balances of empty greenhouses [e.g., Silva and Rosa, 1987; Jimenez and Casas-Vazquez, 1978; Montero et al., 1990a] can be considered as an abnormal situation. The only approach to an empty house is when the area is being prepared for planting or the small plants do not cover a significant portion of the ground area. Even here, syringing, or some type of mist system may be utilized to raise humidity and disperse heat by converting more to latent heat. The effect of a plant canopy on lowering air temperature in greenhouses is well understood by growers. The difference can amount to more than 30 C between a dry house and one containing an actively growing crop. It also emphasizes the problems of using a greenhouse as a passive solar heater.

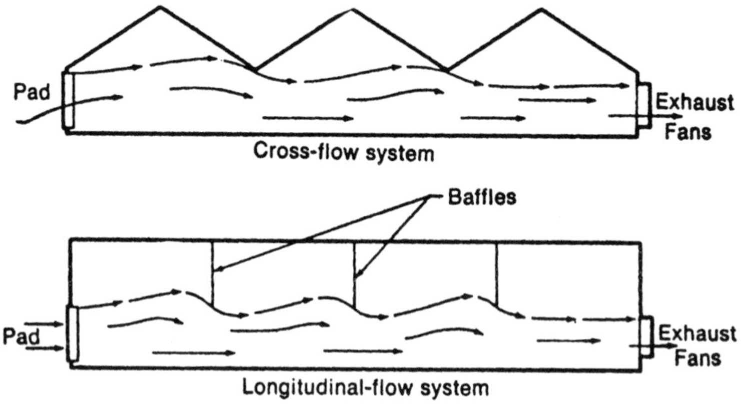

Fig. 4-77. Common cross or longitudinal fan-and-evaporative pad arrangements for cooling greenhouses. Vertical baffles are not always recommended.

One of the best and most recent articles on effects of transpiration on cooling has been by Fuchs [1990], utilizing a simplified approach to calculating sensible and latent heat exchange for an unheated greenhouse covering 1000 m^2. As with most such formulae, several assumptions are made to reduce complexity. These are discussed in Chapter 5. However, Fuchs' results are particularly interesting. As humidity is reduced (arid regions), foliage temperature is also reduced because of increased transpiration (Figs. 4-75, 4-76), and at air exchange rates less than 10 hr^{-1}, the rapid increase in foliage temperature conforms with results obtained by Hanan [1970b] for wind velocities less than 0.2 m s^{-1} (Fig. 4-76). The indications at high humidities would be typical of warm, humid climates where transpiration would be reduced and energy loss is mostly through sensible heat exchange. Crop cooling would not be sufficient to create a climate suitable for tomatoes, even at high air exchange rates. The situation for maximum cooling under warm, arid climates requires freely available water. Restriction to water supply is probably unavoidable inasmuch as most greenhouse production, especially cut-flower and vegetable, is produced in native soils. Water restriction is often required to prevent deficient aeration in the substrate.

a.. Fan-and-Evaporative Pad Cooling

Since a transpiring crop is usually insufficient to prevent excessive temperatures during the summer, other methods can be employed to increase conversion to latent heat. Probably the most common is misting the crop, especially cuttings, as devised by Langhans in the mid-1950s. The other major advance was the introduction of evaporative cooling by DeWerth in 1954. In the same year, one large commercial range in the Denver region converted to the system. In the following year, more than 75% of the commercial ranges in Denver had converted to fan-and-pad cooling. Exhaust fans arranged at one end of the structure, pull air through a wet pad at the opposite end (Fig. 4-77). The pad is essentially a large psychrometer, with maximum cooling dependent upon the wet bulb temperature (T_{wb}). This makes the system very good for high radiation, dry climates. Examples of commercial fan-and-pad arrangements are shown in Figs. 4-78 and 4-79. The general recommendations [ASAE, 1988] state a basic requirement of 0.04 $m^3\ s^{-1}m^{-2}$ air movement or about one complete air exchange per minute, which allows for a 4.5 to 5.5 C temperature rise from pads to fan over a distance of 30 m. For a standard 9 x 30 m greenhouse with a volume of 1080 m^3, this would amount to a total 10.8 $m^3\ s^{-1}$ air volume, or a complete air exchange of 0.6 min^{-1}. The velocity would be 0.3 m s^{-1}. The exhaust fans should be sized to deliver 10.8 $m^3\ s^{-1}$ at 0.03 kPa with guards and louvers in place. Distances between pads should never exceed 50 m since the temperature rise can easily exceed 10 C (Fig. 4-93). At distances less than 30 m, volume must be increased by:

$$F = \frac{5.5}{\sqrt{D}} \qquad (4.25)$$

where: F = multiplication factor
D = distance between pads and fans (m)

For example, in the hypothetical greenhouse above, if the fan arrangement is to pull across the house (9 m) versus the length (30 m), the air velocity is reduced to 0.1 m s^{-1}. Eq. 4.25 requires 1.8 x 10.8 = 19.8 $m^3\ s^{-1}$. The air exchange would be 1.1 min^{-1} for a velocity of 0.2 m s^{-1}. Short distances between pads and exhaust fans are much more difficult to cool, and require a disproportionate increase in electrical consumption. Electricity must be cheap and in reliable supply. Jamieson, in 1971, made the comment

Fig. 4-78. Two examples of exhaust fans on commercial ranges. Bottom picture shows a ventilator arrangement with evaporative pad inside.

Fig. 4-79. Examples of evaporative pads, the left picture most commonly found with shredded aspen material hung vertically. The upper right is a rigid cellulose material. The bottom right is V-shaped to increase evaporative area. Pads may also be horizontally arranged, necessary where vertical area is insufficient to meet requirements.

that the cost for fan-and-pad cooling was \$4400 to \$5900 per hectare, with electrical costs of \$592 to \$986 per hectare for the U.K. [Morris, 1971]. Other corrections for altitude and solar intensity will be required since fan efficiency will be significantly reduced above 600 m and light levels above 54 klx will increase the heat load. Under Colorado conditions, a volume requirement of 0.06 to 0.07 $m^3\ s^{-1}m^{-2}$ floor area is usual.

Cost comparisons for modified evaporative cooling and refrigeration were explored by Wolfe [1970] for U.K. conditions. Assuming that a relative humidity of 90% was acceptable, the capacity of a refrigeration plant for 80% of 630 W m^{-2} solar radiation would be about 60 tons (205 kW). Wolfe's calculations resulted in a total cost of \$27944 per year –far in excess of the estimated financial return. For evaporative cooling, using an indirect, recirculating system with one or two stages, Wolfe estimated \$15968 yr^{-1} for a single stage and \$31936 for the two stages. Both were in excess of financial return despite improvements from better temperature control and continuous CO_2 enrichment. Experience has shown that pad-to-fan evaporative cooling is successful given the appropriate climatic conditions. In the U.K., however, direct evaporative cooling has been tried and largely abandoned [Wolfe, 1970].

Other methods for using evaporative pad cooling have been proposed [Zieslin and Gazit, 1978]. That is, the location of the fans just inside of the pads and blowing the air into a structure opened at the sides. According to the authors, a positive pressure creates a more uniform climate compared to the conventional system. This benefit is attributed to the greater mixing as contrasted with the laminar air flow and gradual temperature increase from pads to fans with a suction system. Experience with ducted positive pressure systems in the U.S. generally leads to marked temperature differentials in the house, and Zieslin and Gazit's proposal does not seem to have received wide utilization. A rather novel adaptation of evaporative cooling [Cunningham and Thompson, 1988] has been a thermally driven system for structures in arid regions. Air flow depends upon buoyancy forces through appropriate towers at either end of a greenhouse. The intake chimney is 7.5 m high, and entering air is cooled by a wet pad, with the heavier, cool air falling to the floor and flowing into the house and out the opposite end tower. Neither has this system been adapted to any great extent in the industry.

Table 4-19. Recommended air velocity through pad materials [ASAE, 1988].

Type	Air velocity through pad ($m\ s^{-1}$)
Aspen fiber, mounted vertically, 50-100 mm thick	0.75
Aspen fiber mounted horizontally, 50-100 mm thick	1.0
Corrugated cellulose, 100 mm thick	1.25
Corrugated cellulose, 150 mm thick	1.75

Table 4-20. Water flow and sump capacities for vertically mounted cooling pad materials [ASAE, 1988].

Pad type & thickness	Flow rate per lineal length ($\ell\ min^{-1}\ m^{-1}$)	Sump capacity per unit pad area ($\ell\ m^{-2}$)
Aspen fiber, 50-100 mm	4	20
Aspen fiber, desert conditions	5	20
Corrugated cellulose, 100 mm	6	30
Corrugated cellulose, 150 mm	10	40

The evaporative pads are sized to permit adequate air movement (Table 4-19), with the cheapest material being shredded aspen fiber. The standard will handle between 0.7 and 0.8 m^3 air $s^{-1}m^{-2}$ of pad area. Aspen pads must be replaced yearly, are a fire hazard when dry, and quite often sag in the holder mounts, with large holes often appearing after some use (Fig. 4-80). Another problem is a water supply with high salinity, leading to a blockage. A bleed-off rate should be established of about 0.2 $\ell\ min^{-1}m^{-2}$ pad area to avoid salt buildup. Bleed-off as a function of salt content in the water has not been found in the literature. Cellulose fibers do not sag and can be used for several years. However, they have a high initial investment. Vertical water flow rates and sump capacities are listed in Table 4-20 for vertically mounted pads. The water is distributed at the top of the pad with holes bored in the top of the pipe, which reduces plugging compared to holes in the bottom, with clean-out plugs at the ends of the distribution pipe. The return water must be filtered, with ASAE [1988] suggesting a 2 mm square mesh screen. Some growers install a drain pipe in the recovery trough and fill the trough with fine gravel, coupled with filters such as discarded panty hose.

Heat and mass transfer coefficients for horizontal and vertically mounted aspen pads were evaluated by Kimball et al. in 1977. There was no significant effect of water temperature over a range of 12 C, but the thinner the pad the more efficient; that is, the closer the actual air temperature to the wet bulb temperature after passing through the pad. The heat transfer coefficient for 25 mm thick pads was about five times larger than for 127 mm thick pads. There is a practical problem with 25 mm thick pads. Such thin materials would be difficult to maintain under commercial conditions even if they were available for purchase. The statistically derived formula for liquid film transfer was suitable for either horizontal or vertical pads, based on an optimum air flow of 1.07 kg $s^{-1}m^{-2}$ (ca. 0.89 $m^3\ s^{-1}\ m^{-2}$).

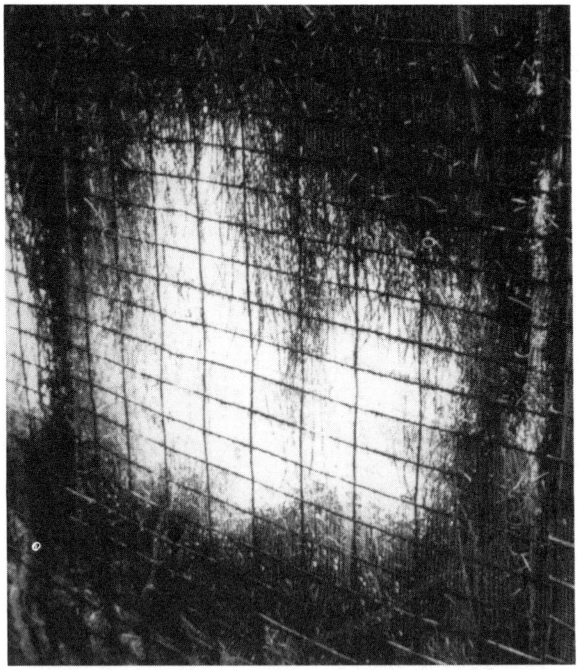

Fig. 4-80. Upper: Common problems with vertically mounted aspen fiber, evaporative pads. To avoid this, growers will often tightly compress the pad between wire mesh (Fig. 4-79), which increases wind resistance. **Lower:** Salt and dirt build-up on aspen fiber pads. This greatly reduces pad efficiency, and can be prevented by a suitable bleed-off of water from the system. Attempts to use such pads 2 years in succession is a poor practice.

In most cases, single-hinged ventilators are installed external to the pads with a maximum design velocity of 1.8 m s^{-1}. External ventilator mounting allows water to remain in the pad system for operation in the spring and fall. However, initial opening of these ventilators is highly critical, with nearly 100% of total capacity being achieved well before the vent is opened 50% of its maximum distance. Small changes in opening are required to avoid large air fluctuations with fan operation, and to allow the system to "catch up" with the temperature change. Water is continuously circulated over the pads, and can be simply controlled with a photoelectric cell and outside thermostat. The setting in arid climates is generally around 13 C. This avoids the possibility of a wet bulb temperature below freezing. For crops susceptible to disease at high humidities, water in the pads should be turned off some time before cooling fans shut off so the greenhouse interior can be dried before sunset. Or, the fans should be forced on for a suitable drying cycle.

The pad-and-fan system efficiency is usually given to be about 85%, and under low humidity conditions (<20% relative humidity), it is capable of cooling air more than 10 C below ambient temperature conditions. ASAE [1988] recommends a maximum distance of 7.5 m between exhaust fans, but this is commonly much closer (Fig. 4-78). Types of pressure actuated openings on fans other than those shown in Fig. 4-78 can be installed. Low-pressure, high-volume propeller fans are generally selected. An efficient greenhouse exhaust fan for cooling should deliver 8.5 to 9.5 m^3 s^{-1} against a 0.03 kPa static pressure (1.2 m diameter) [Roberts, 1988]. An example of a fan characteristic curve is provided in Fig. 4-81. In any event, these should be covered in the winter in northern, continental climates in order to avoid excessive energy loss and possible freeze damage to plants close to the fans. Exhaust fans are usually staged, with one or two being used for cold weather ventilation through perforated PE tubes. The induced suction automatically inflates the tubes through louvers in the gable end walls (Fig. 4-78). As inside temperature rises, more fans come on, and the ventilators may be employed to prevent rapid temperature drop that occurs when water begins to flow in the evaporative pads.

Albright [1995] proposed a ventilator inlet control to create a negative pressure that limits wind effects. This involves the determination of positive and negative pressure coefficients around structures, similar to those illustrated in Fig. 2-37 and in Fig. 4-82. As an example, at a wind speed of 6.6 m s^{-1}, the downwind suction pressure on the house where the coefficient is -0.6 would be -16 Pa (0.16 cm water column). Unless exhaust fans can create a greater suction than -16 Pa, inlets located on the downwind sides act as outlets. Fan cutoff is often at a static pressure difference of nearly 100 Pa, which could cause structural problems and difficulty in opening and closing doors. Due to the usual construction, it is unlikely that static pressures above 25 Pa (0.25 cm) can be produced [Albright, 1995]. Ventilation fans can be expected to operate near the free-air intercept (Fig. 4-81) so that air flow rate is relatively insensitive to pressure difference. The usual rule is that fans alone control ventilation rate and inlets alone control the air's entering speed, distribution, and mixing; 25 Pa represents a fairly high pressure. The most likely range is less than 7 Pa (8 mm water column). Albright suggested that control of greenhouse inlets be separated from the temperature control of fans by measuring the pressure differential between a reference point outside and that inside, and using this difference to control air entry within a prescribed range, preventing reverse air flow through the inlets. This would also serve to reduce problems mentioned above with rapid ventilator movement when first opened. The system buffers the negative effects of wind, but the greenhouse must be very tight since any crack becomes a possible inlet. Leaks subvert any attempts to control ventilation through modulating the inlets.

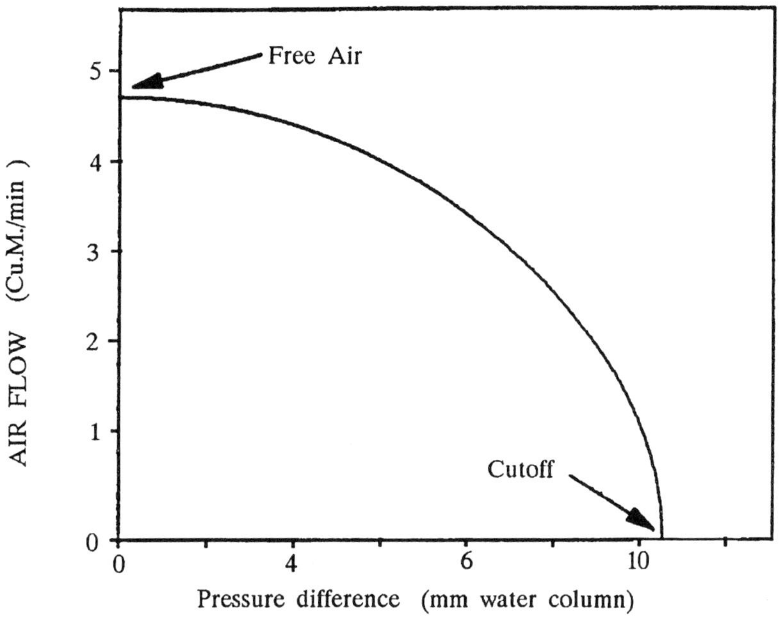

Fig. 4-81. Typical shape of a fan-characteristic curve, showing free air and cut-off points. This graph is the reverse of the usual form [Albright, 1995] (With permission of the Amer. Soc. Hort. Sci.).

The mathematical relationships between evaporative cooling and the greenhouse energy budget have been examined by a number of authors [i.e., Walker and Cotter, 1968; Morris, 1971; Seginer and Livne, 1978; Landsberg et al., 1979; etc.]. Both Walker and Cotter and Morris showed that fan-and-pad cooling is not likely to be efficient in humid climates where the relative humidity is generally above 60% –as is the usual case in northern Europe and the U.K. (see Wolfe's comments above). Morris [1971] discussed the problems of air dividing into two air streams, one passing through the foliage, the other bypassing the canopy. Personal experience has shown this bypassing to be a common situation with dense crops arranged in benches. A 1 year-old carnation crop is as effective as a polyethylene sheet in diverting air flow. Morris also considered the possibility of recirculation to reduce temperature gradients, and to avoid cold drafts near the inlet. This has not been employed to any great extent.

One of the more interesting mathematical analyses of fan-and-pad cooling was carried out by Landsberg et al. in 1979. The assumptions common to these types of analyses will be discussed to greater extent in Chapter 5. The final equation allowed prediction of average inside air temperature for a 6 x 18 m, glass-covered house with 108 m^2 floor area, and a volume of 300 m^3:

$$T_h = \frac{\left[\dfrac{A\beta R_n}{1+\beta} + \left(hA_c + \dfrac{2F\rho c_p\gamma}{\gamma+\Delta}\right)T_a + \dfrac{2F\rho c_p \Delta T_d}{\gamma+\Delta}\right]}{(hA_c + 2F\rho c_p)} \tag{4.26}$$

where: T_h = *average inside air temperature (C)*
A = *floor area (108 m^2)*
β = *Bowen ratio* = H_Q/LE *(see Section 5.IV.B.4 for the definition of the Bowen ratio)*
R_n = *net radiation on greenhouse floor (W m^{-2})*
h = *transfer coefficient for sensible heat transfer of house (W $m^{-2}C^{-1}$)*
A_c = *surface area of greenhouse (215 m^2)*
F = *air flow rate through house (m^3 s^{-1})*
ρ = *air density (1.2 kg m^{-3})*
c_p = *specific heat of air (1.01x10^3 J $kg^{-1}C^{-1}$)*
ΔT_d = *change in dewpoint temperature (C)*
Δ = *change in saturated vapor pressure per °C*
γ = *psychrometric constant (0.66 mb C^{-1})*

Landsberg et al. calculated the results for ventilation rates of 0.0077, 0.0157, 0.023, and 0.046 m^3 $s^{-1}m^{-2}$ floor area. Assuming the system was arranged the long distance (18 m), these values would correspond to windspeeds of 0.05, 0.1, 0.15, and 0.3 m s^{-1}. Only the latter could be considered comparable to commercial situations. The assumption is made that the pad reduces air temperature to T_{wb}, or 100% efficient, and that a freely transpiring crop covers the floor area with the Bowen ratio = γ/Δ or 0.33. Changing the Bowen ratio markedly effects the temperature rise in the house (e.g., 0.33 to 1.0 = 100% increase in temperature rise relative to 0.33). Note also in Section 5.IV.B.4, the questions that can be raised about the use of a ratio originally defined for use over water surfaces and later applied to outdoor vegetation.

A different approach to the problem of assessing fan-and-pad cooling was made by Seginer and Livne [1978], comparing dry canopies with wet canopies and the introduction of an evaporative pad. In this case, the ventilation rate and power required to maintain a certain canopy temperature was significantly smaller if the ceiling height was reduced, preferably to the top of the hypothetical rose canopy. Obviously, which the authors acknowledge, this is not feasible. Personal experience has also shown that tender shoots directly under a ceiling are likely to be cooked under conditions of low wind movement and clear skies. The authors state that it is impractical to utilize exhaust fans with a net power requirement above about 100 W m^{-1}. The units are not defined explicitly, and I assume this means Watts per linear meter of house length.

b. Other Evaporative Methods

Several methods have already been mentioned in this section –such as mist propagation (Fig. 4-84). The canopy is actually wetted, which effectively maintains a free water surface and actual evapotranspiration equals the potential evapotranspiration. The resultant high humidity also tends to reduce transpiration from the wetted leaves. Canopy wetting has been examined by Seginer and Livne [1978] and Cohen et al. [1983]. Whether in naturally ventilated or forced ventilated greenhouses, canopy wetting greatly reduces plant temperature. Cohen et al. also looked at cooling provided by water on the glass roof. This procedure was investigated by Harold Gray at Cornell University in the 1940s, and utilized by Went at the Pasadena phytotron. In my estimation, the practical possibilities of this practice, which are still being examined, have not been proven. The ability of wetting foliage for cooling purposes depends upon the species and the temperature to be maintained within the house. Thus, misting is not often encountered with cool crops (see Section 5.IV.B.2, Figs. 5-80 and 5-81), especially as, for example, air is saturated at 1.23 kPa at 10 C versus 2.34 kPa at 20 C. Therefore, control of relative humidity is much more critical at lower temperatures compared to warmer conditions. One must also consider the possibility of disease where the foliage may be constantly wet.

The other possibility is to employ high-pressure systems (5-6 MPa) to provide an extremely fine mist (i.e., 10 μ diameter), allowing essentially a fog that tends to remain in the air. Evaporative cooling occurs above the crop with minimal wetting to the foliage (Fig. 4-84). A heavy fog also reduces solar intensity. Such a system is expensive, requiring heavy pumps, piping, and special nozzles (Fig. 4-85), very clean water, and it has a high electrical consumption. The usual water supplies commonly contain sufficient salts to seriously plug equipment and damage foliage. Rain water is an acceptable source with at least two 5 μm filters at the pump and each nozzle fitted with a 40 μm filter. One of the few studies on the more modern systems has been conducted by Montero et al. [1991] under Spanish conditions. The system was able to maintain temperatures very near ambient even under humid conditions. The authors concluded that it is important to provide a capacity to maintain high humidity in well ventilated houses (>30 exchanges hr^{-1}). Nozzle density in this study was 1 per 10 m^2. The ability of a high-pressure fog system to reduce internal temperature below ambient depends upon outside humidity as well as ventilation rate. The advantages of this system were greater uniformity compared to pad-and-fan, no requirement for forced ventilation or the demand for a "tight" structure. These systems, as others, can be misused –e.g., rust on roses, etc. I have seen no information on control devices for these systems, although one can use either humidistats or thermostats, or connect them into more sophisticated apparatus (Chapter 8).

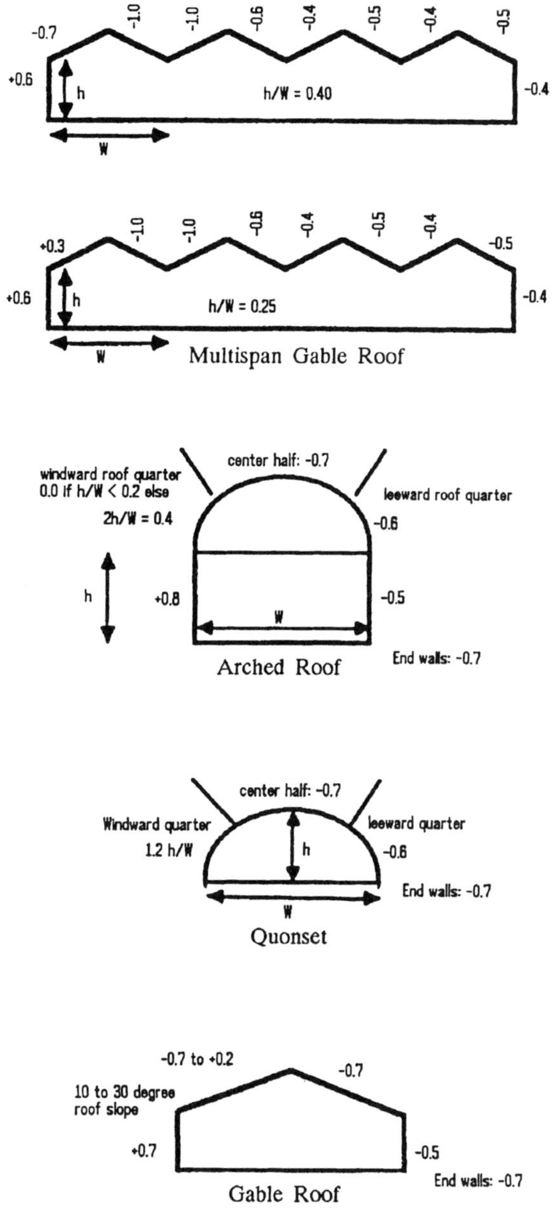

Fig. 4-82. Greenhouse shapes and wind pressure coefficients used by Albright [1990] in air flow control (With permission of the Amer. Soc. Hort. Sci.).

3. Temperature Variations Inside Greenhouses

In Chapter 2 (Section II.E) and Chapter 3 (Sections III.D.2 and III.C.1), the variations of radiant energy inside a structure and its interception by a vegetative canopy were discussed. As a general rule, highest temperature will be found where irradiance interception is greatest –this may be the ground, some distance within the upper regions of a dense canopy, or portions of the structure. One can hardly expect temperatures throughout a commercial range to be the same, given the many arrangements and crop species that may be encountered, and the continual change in solar intensity outside that occurs over a year. Roses, for example, may

Fig. 4-83. Top: Low pressure misting system for vegetative propagation, often controlled by a timing device. **Bottom:** High-pressure fog system in operation. The latter can provide a fog so thick that workers get lost.

transmit a significant part of radiant energy to the ground surface below the canopy [Stanhill et al., 1973], whereas a 2-year-old carnation crop is practically impenetrable in the center of a bench [Hall and Hanan, 1976]. The presence of aisles means that at least some energy will reach the ground in the walk area, and that the sides of the crop in benches will intercept significant energy. Despite this, a common basic assumption, explicitly or by default, is often made that the interior climate is homogeneous [e.g., Bakker, 1986; Garzoli and Blackwell, 1973; Duncan et al., 1981; etc.]. The canopy may be considered as a single large leaf [e.g., Fuchs, 1990] or a flat plane [e.g., Maher and O'Flaherty, 1973]. The investigator may divide the house into ground, crop, air above the crop, and the structure as separate compartments for purposes of evaluating energy flow, each compartment being homogenous [e.g., Sadler and van Bavel, 1984]. The variations that may be found in the literature for purposes of simplifying mathematical approaches to energy flows are many. It is a fact that the greenhouse environment is extremely heterogeneous or nonuniform. On this basis, the common practice of a single radiation or temperature measurement on which a sophisticated and expensive control is based (i.e., computers) makes little sense –even though acceptable crops

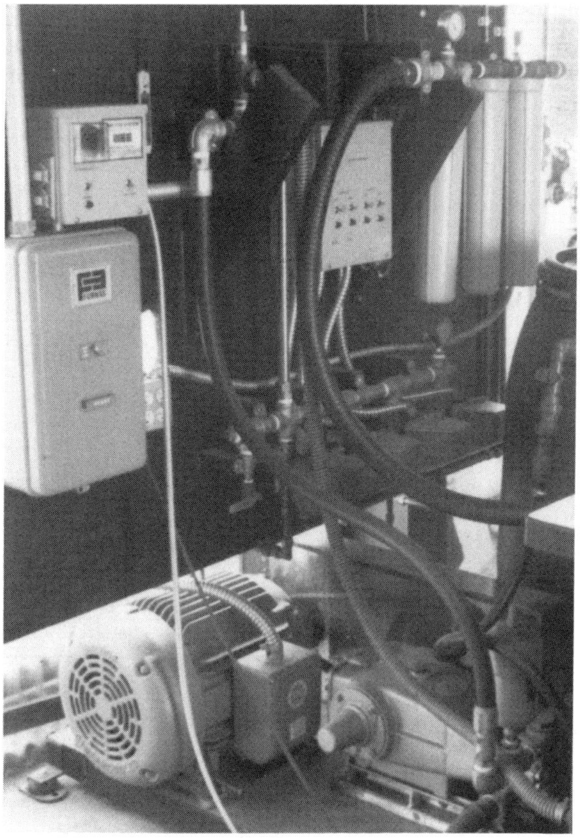

Fig. 4-84. Pump and controls for a high-pressure fog system, operating at 6 to 7 MPa pressure.

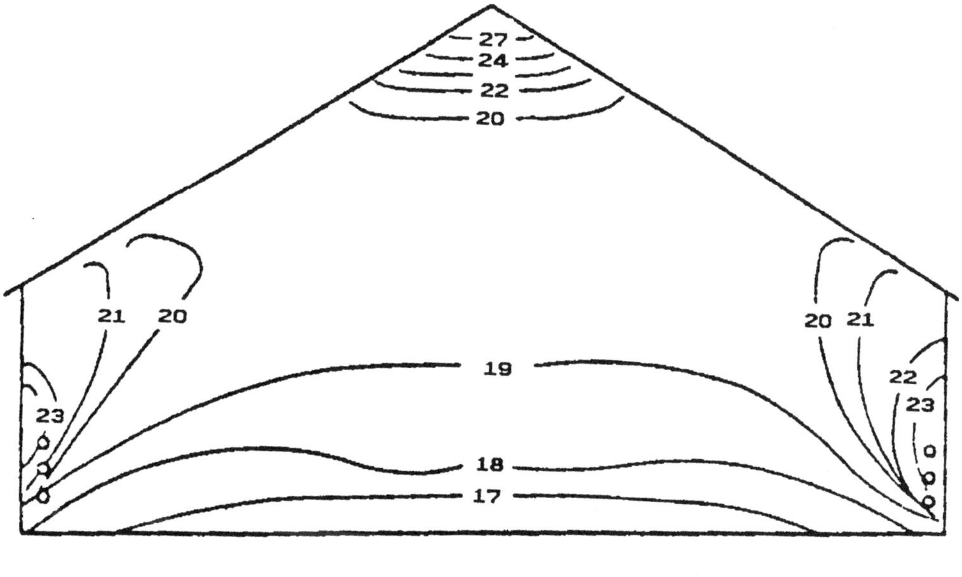

Fig. 4-85. Cross-section of a greenhouse with perimeter pipe heating showing the temperature profile that may be expected [ASHRAE, 1978].

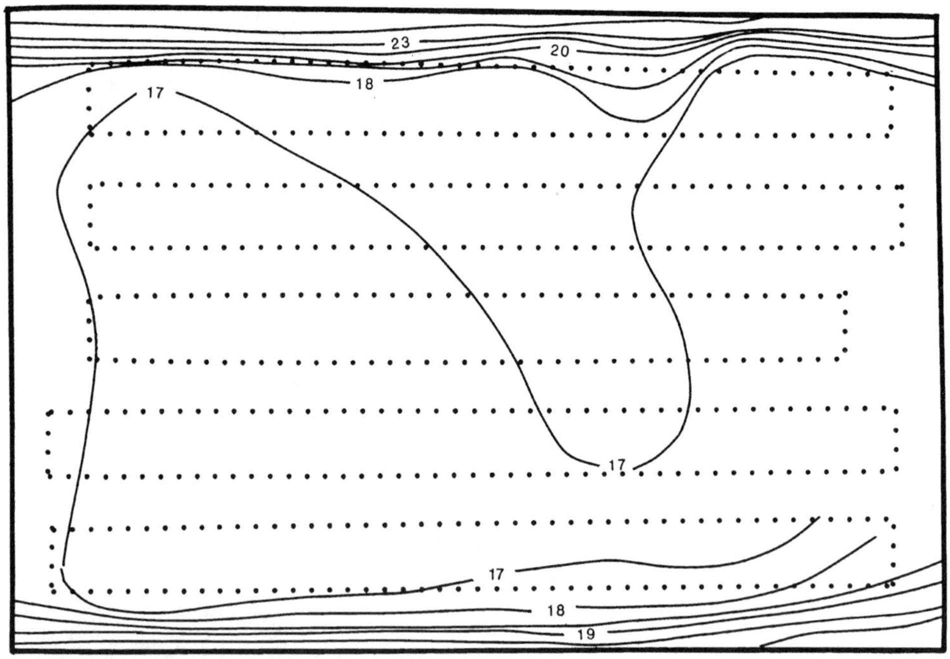

Fig. 4-86. Horizontal temperature pattern produced by steam coils on the sidewalls without forced air circulation [Carpenter and Bark, 1967] (With permission of the *Florists' Rev.*).

can be produced with single sensors per zone. For that fact, "acceptable" crops have always been produced since there is a greenhouse industry. This does not mean that practices today will be useful tomorrow for "acceptable" greenhouse production. The types of equipment and methods to control environment can make nonuniformity so great that plant response becomes obvious. Net return to the enterprise can be reduced accordingly.

Probably the closest to homogeneity one can achieve is at night or in overcast conditions with mild heating loads –common to northern Europe and the U.K.– where the heating system is hot water or steam uniformly distributed across the house near the floor, or the floor itself supplies the major portion of energy. Such uniformity was found by Whittle and Lawrence [1960b] in the U.K. In empty, unheated greenhouses, no

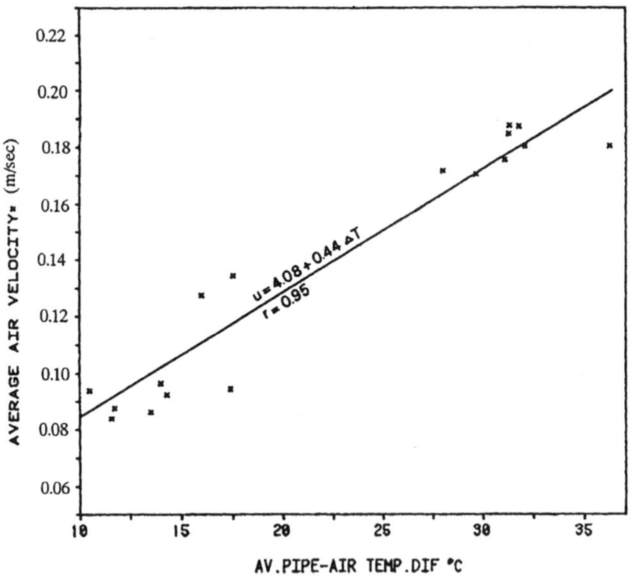

Fig. 4-87. Air velocity as a function of the difference in temperature between pipe surface and the air measured 0.8 m above the ground [Stanghellini, 1983a] (With the permission of the author).

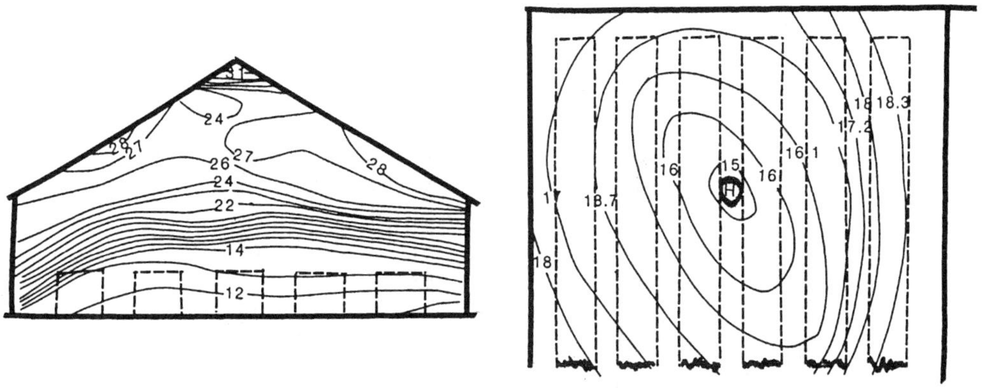

Fig. 4-88. Vertical (**left**) and horizontal (**right**) temperature patterns produced by a vertically mounted hot air heater at the location noted as "H" on the left [Carpenter and Bark, 1967] (With permission of the *Florists' Rev.*).

vertical gradients could be established, with the difference between ends (ca. 15 m) less than 1.1 C. In heated houses (outside temperatures 0 to 7 C), temperatures inside were within 0.8 C of that desired 95% of the time. On one clear night with an outside minimum of -7 C, the house having ground-level steam piping had differences of 0.1 and -0.8 C between 0.3 and 3.3 m vertically. On the other hand, in houses with pipes at 2.25 m above the ground, the maximum vertical gradient was 3 C. The data, according to Whittle and Lawrence, suggested that perimeter heating plays an important part in good temperature distribution, with pipes at the ground level resulting in negligible temperature gradients. Results by Okada [1980] have already been cited; to the effect that low, horizontally arranged pipes will provide the most uniform temperature profile within a structure. Uniformly heated floors are good in heat distribution until the heat load begins to require additional energy above the floor. The situation can be improved by using only as much radiative piping as necessary to meet requirements. The latter situation of meeting minimum requirements is best achieved by using hot water, inasmuch as the water temperature can be varied at will without having to turn off entire piping loops.

Examples of temperature gradients that can be encountered with perimeter steam heat are provided in Figs. 4-85 and 4-86. For such systems, heat transfer, in addition to the radiative component, is usually free convection, laminar flow, with low, dimensionless Reynold's numbers and relatively high Grashof values ($Re \approx 350$, $Gr \approx 10^5$-10^6), at air velocities around 0.1 m s^{-1}. A number of authors have discussed dimensionless numbers [e.g., Gates, 1980; Campbell, 1971; Bot and van de Braak, 1995]. These numbers, of which there are several in addition to the two mentioned, serve to relate the numerous factors involved in energy and mass transfer.

Stanghellini [1983a] stated that the Grashof number is a measure of the importance of buoyancy due to the difference in temperature between the pipe surface and the air around it, while the Reynolds number describes the effect of the pipe surface itself on turbulence in the air. Gates [1980] stated, as a general rule, that air flows of 0.1 m s^{-1} or greater result in forced convection, and less than 0.1, result in free convection. In either case, turbulent flow does not occur. Stanghellini [1983a] found that only under extremes of temperature differences exceeding 30 C between pipe and air are wind speeds likely to approach 0.2 m s^{-1}. At low temperature differences, air velocity may be as low as 0.08 m s^{-1} (Fig. 4-87). Thus, pipe systems, if suitably designed and operated, will provide highly uniform conditions where air movement will be forced convection at very low wind speeds. However, under extreme heat loads ($T_o < -10$ C), vertical temperature gradients of more than 10 C have been observed between the ground and 1.5 m above. Furthermore, due to heat migration in a large multispan range, resulting from outside wind, and the variation in heating resulting from the solar passage on a clear day,

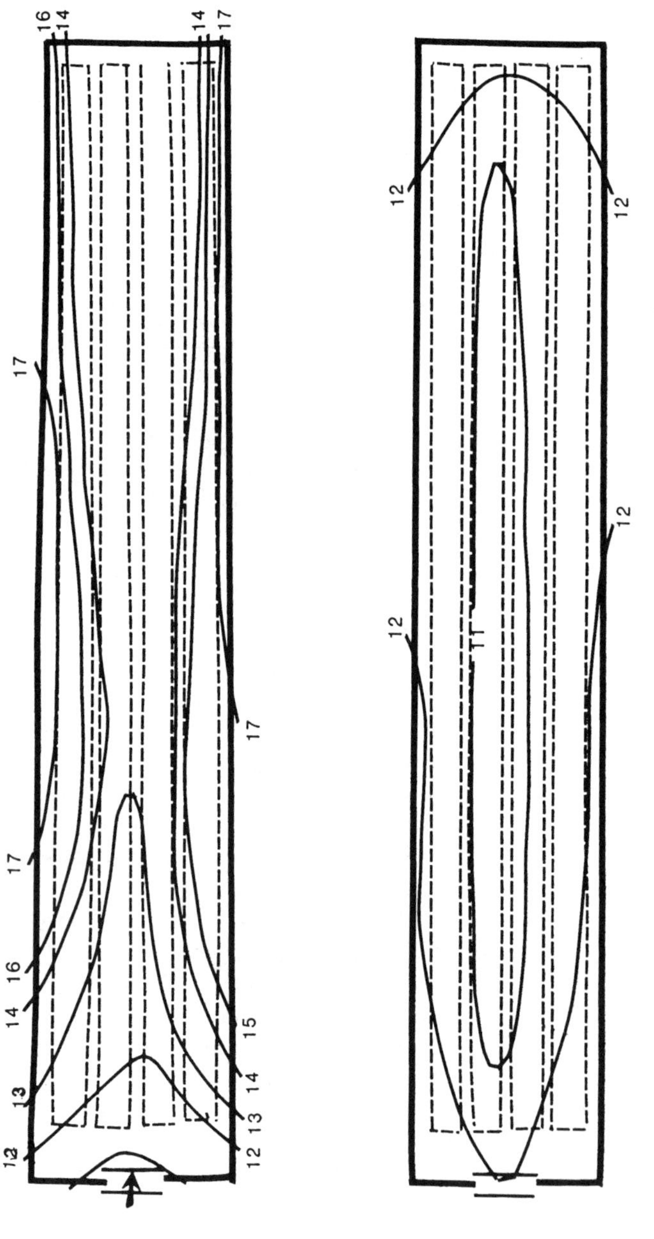

Fig. 4-89. Horizontal temperature patterns at bud height (+1 m) when a horizontally mounted, PE ducted heater **(top)** is on and off **(bottom)** [Bark and Carpenter, 1967; see also Bark and Carpenter, 1969] (With permission of the *Florists' Rev.*).

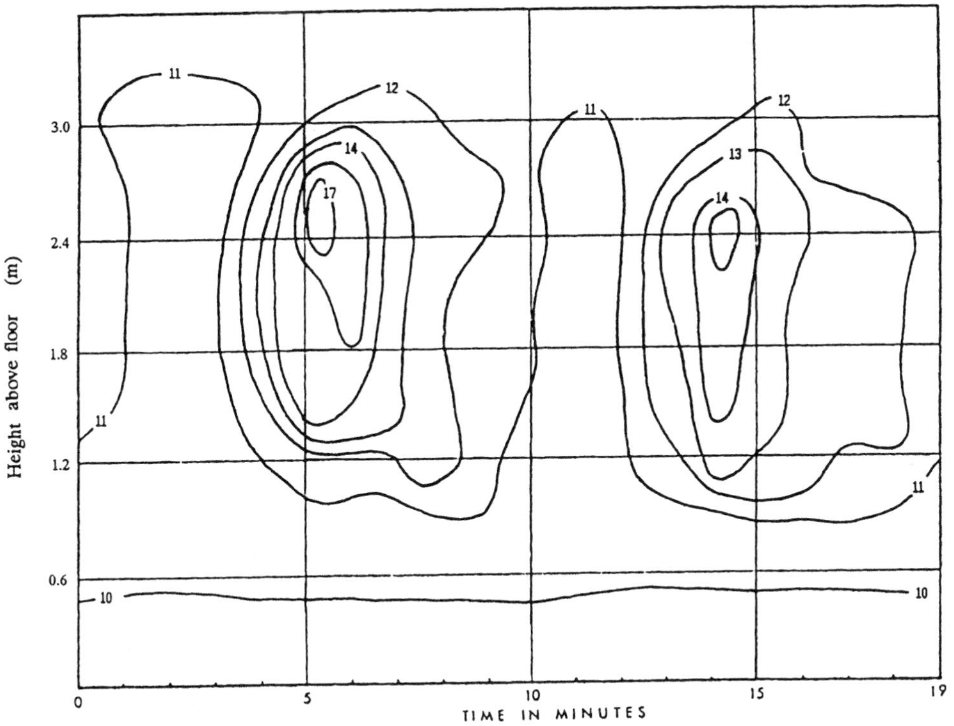

Fig. 4-90. Variation of vertical temperature with time from the floor to 3.6 m above in the center of a house using a horizontal heater with perforated PE tubing attached [Bark and Carpenter, 1969] (With permission of the *Florsts' Rev.*).

large blocks must be zoned, each zone with its own climate control system (single sensor). I have seen the east side of a large north-south oriented range cooling at midmorning while the west side was still heating –and vice-versa in the afternoon.

Temperature distribution of bench-top hot water systems was examined by Jenkins et al. [1988]. With perimeter systems, strong convection cells were formed along the walls, rising toward the ceiling before mixing in the canopy. These convection cells were not formed with a bench-top system, but high-temperature regions were directly over the benches, with the temperature distribution being non-uniform. Local convection, according to the authors, may have been disrupted by the benches.

During the 1960s and 1970s, considerable attention was given to forced circulation of the air within greenhouses, especially with cheaper hot air systems and PE tube ventilation. Unfortunately, most of these air circulation systems, especially overhead, vertically mounted heaters (Fig. 4-43C, D) are likely to increase temperature heterogeneity –particularly if combined with perimeter piping. In any event, non-ducted unit heaters should never be employed in commercial ranges [Meneses and Monteiro, 1990]. Even if fitted with blowers or exhaust fans, large temperature variations can be expected [Bailey, 1974]. For horizontally mounted, non-ducted heaters above the crop, a cool region will be found directly under and behind the heater, with high temperatures in the main discharge. Such heaters tend to set up individual air circulation patterns. Examples of problems with vertically discharging, hot air heaters are provided in Fig. 4-88, and for horizontally mounted, PE ducted heaters in Figs. 4-89 and 4-90. In all of these examples, the heaters and PE tubes are mounted above the crop. Fig. 4-90 shows the high temperatures that can occur some distance above ground level. Under severe winter conditions, in high houses, this can lead to marked, vertical temperature gradients. Invariably, the region directly under the heater will be cooler. If all heaters in wide span, or multispan ranges are located in the central aisle and blowing toward the ends –as is often the case– that area will be cold compared to the far end of the structure (Fig. 4-91). Attempts to reduce this variation with additional, high velocity fans are likely to be unsuccessful inasmuch as the velocity necessary for turbulent mixing is not usually reached. An example of wind speeds likely to be

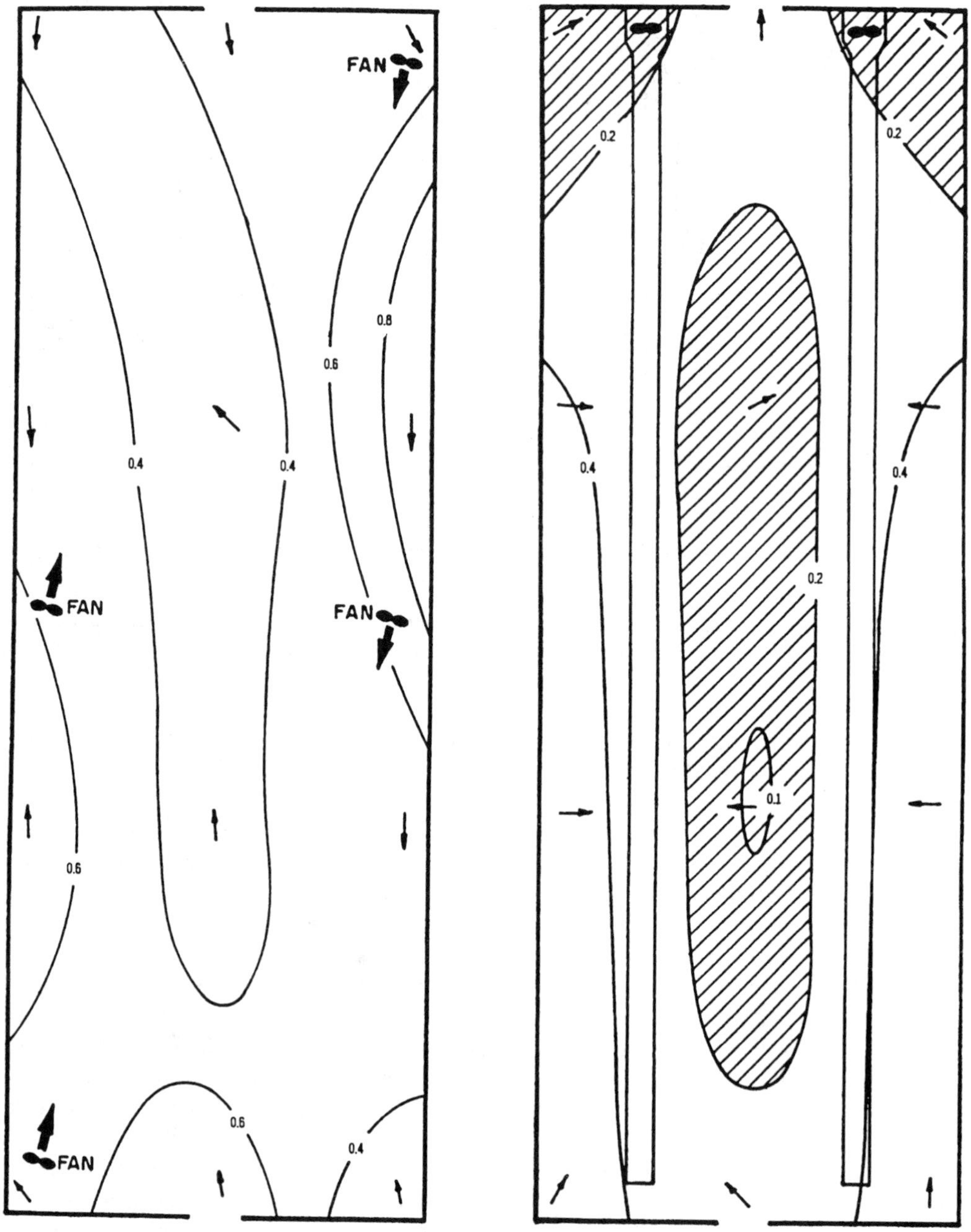

Fig. 4-91. Air circulation 30 cm above ground level for a horizontal convection circulation system using individual fans pointed toward the center of the house (**left**), and two overhead perforated PE tubes (**right**) [Walker and Duncan, 1973]. The values on the isolines are windspeeds (m/s). The small arrows indicate wind direction, while the hatched areas enclose regions of equal velocities.

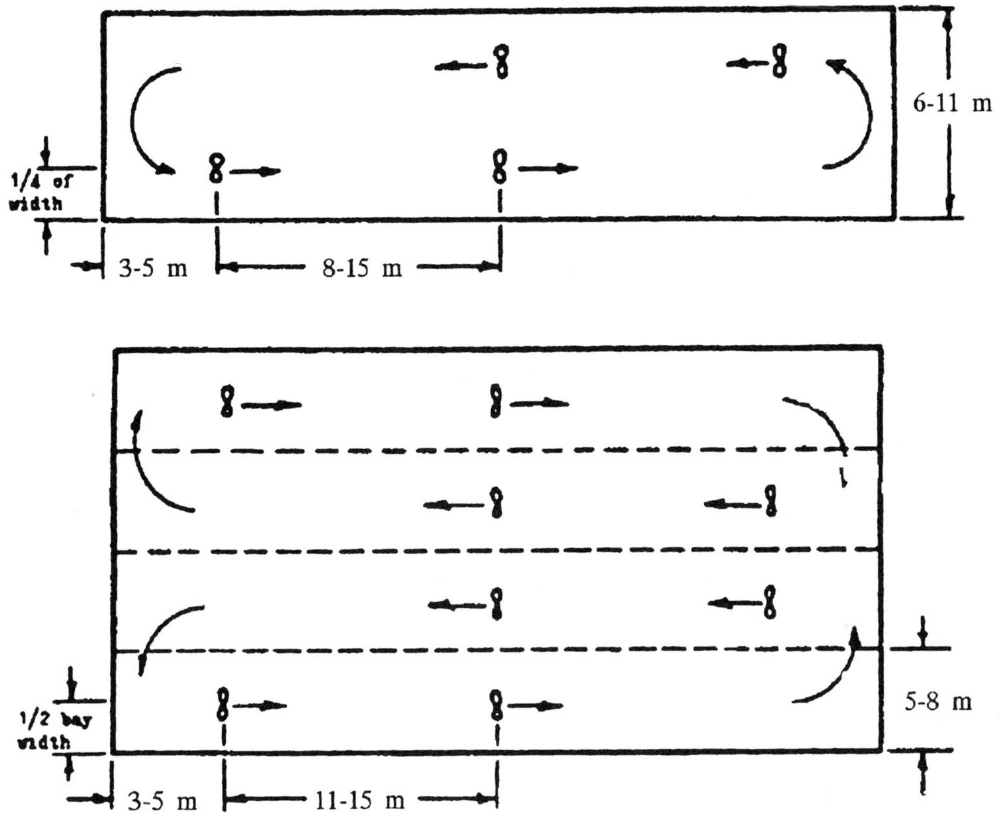

Fig. 4-92. Illustration of fan placement for horizontal air flow in a single span (**upper**) and gutter-connected houses (**lower**), using 40 cm diameter fans [Bartok, 1988] (With permission of the *Greenhouse Manager*).

encountered with non-ducted and PE ducted fans is provided in Fig. 4-91. One will note that air velocities can vary dramatically. Temperatures may vary in a similar fashion.

When air moves at slow speeds, it tends to continue in the same path upon which initially directed, with remarkably little mixing with the surrounding air into which the stream is injected. The flow is laminar and heat transmission may be pure, free convection in the vertical direction (from hot pipes) or forced convection (horizontally or vertically). At certain critical velocities as exemplified by the Re and Gr numbers, the flow is turbulent, mixes with the surrounding air rapidly, and energy transmission will be much larger and quicker. This is the objective of small holes in PE ducting. Mixing occurs within a few diameters of the holes and hot –or cold– air will not directly impinge on the crop. Another example of practical use was the Earheart Plant Research Laboratory where air was introduced through 9 mm wide slots in the floor at 1.6 m s^{-1}. Turbulence and mixing occurred almost immediately and rose evenly through the whole greenhouse without channeling [Went, 1957]. Fig. 4-91 shows that the perforated tube distribution results in a more even air velocity throughout a house compared with individual fans.

Fig. 4-87 shows that air velocities in greenhouses can easily be less than 0.1 m s^{-1}. The literature suggests [e.g., Coker and Hanan, 1988] that aerodynamic resistance to water loss and CO_2 uptake can far outweigh leaf resistance of well-watered plants, leading to serious difficulty with CO_2 supply within the vegetative canopy. The ASHRAE Fundamentals Handbook [1989] states that air velocities in the range of 0.5 to 0.7 m s^{-1} are optimum for plant growth under controlled conditions, that air speeds across the leaf of 0.03 to 0.1 m s^{-1} increase CO_2 uptake, and that velocities above 1.0 m s^{-1} inhibit growth. A more detailed discussion on wind velocities is provided in Chapter 5. With the possible exception of fan-and-pad cooling, or high outside wind velocities with natural ventilation, air velocities in the bulk greenhouse air will seldom approach 0.5 m s^{-1}, unless forced air circulation is employed –leading to wide temperature variations inside.

As a means to get around the problems discussed above, Koths [1969; 1974] proposed the idea of uniformly moving all the air within the greenhouse in a well established circulatory pattern above the plant canopy at speeds in the range of 0.2 m s^{-1}. As the flow would be laminar, without turbulence, and since the entire air mass would be in motion; the result would be a uniform temperature throughout the inside with improved CO_2 supply to the canopy when heating is employed. The method for single and multiple span ranges is shown in Fig. 4-92. Koths applied the acronym "HAF" to this method, or horizontal air flow. Continuously operating, split capacitor fans of 100 to 400 W, with high-pitch blades are mounted above the crop, at least 30 cm below the roof and 2 to 3 m above the floor [Bartok, 1988]. When located more than 2.1 m above the floor, blade guards are not required. Fan diameter may range from 46 cm up to 80 cm, with the blade matched to the motor. In an evaluation of HAF in several commercial ranges, Brugger et al. [1987] found an airflow between 0.13 to 0.73 m^3 $min^{-1}m^{-2}$ floor area, which was less than the recommended rate of 0.9 to 1.1 m^3 $min^{-1}m^{-2}$. Velocities within the plant canopy ranged from 0 to 0.5 m s^{-1}. Obviously, air movement of this type will increase heat transfer from hot piping, and through the roof and sides of the house. This system has been popular in the U.S..

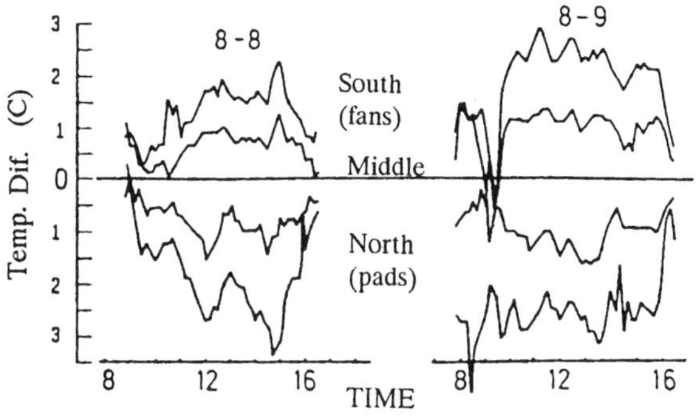

Fig. 4-93. Differentials between temperatures at the center of a 30 m long house and four locations at different distances from the evaporative pad located at the north end, under summer conditions in Colorado with carnations in their second year [Hanan, 1973].

Other conditions that can result in highly significant temperature differentials in a greenhouse are the temperature rises that can occur in fan-and-pad cooling (Fig. 4-93). Interestingly, the biggest temperature rise in this situation commonly occurs in the first one-third to one-half of the distance from the evaporative pads. Wolfe and Cotton [1975] also examined airflow and temperature distribution where exhaust fans were used. Their results confirmed the recommendations of the ASAE [1988]. Ventilation rates in excess of 0.04 m^3 m^{-2} s^{-1} did not produce further substantial reduction of the temperature rise in the greenhouse. Their observations showed a deep air stream moving directly from inlet to fans, the pattern being determined by the configuration of the air inlet and that of the roof.

Since temperature gradients of this type can markedly affect growth, the grower often compensates by locating cultivars tolerating warmer temperatures near the exhaust fans, and those doing better at cooler temperatures close to the pad. Similar disposition of species and cultivars may be made if the observed temperature differences are stable. The variations that may be found in the U.S. were noted by Carpenter et al. [1976]. These contrast remarkably with the situation in Europe of uniform construction, equipment, and practices.

V. REFERENCES

Acock, B., D.A. Charles-Edwards and S. Sawyer. 1979. Growth response of a chrysanthemum crop to the environment. III. Effects of radiation and temperature on dry matter partitioning and photosynthesis. *Ann. Bot.* 44:289-300.

Albright, L.D. et al.. 1985. *In situ* thermal calibration of unventilated greenhouses. *J. Agric. Eng. Res.* 31:265-281.

Albright, L.D. 1990. Environment control for animals and plants.. *In* Agric., Food and Biol. Syst., ASAE, Joseph, MI.

Albright, L.D. 1995. Controlling greenhouse ventilation inlets by pressure difference. *HortTechnology*. 5:260-264.

Aikman, D.P. 1989. Potential increase in photosynthetic efficiency from the redistribution of solar radiation in a crop. *J. Expt. Bot.* 40:855-864.

Aldrich, R.A. 1980. Evaluating thermal blanket materials for greenhouses. *CT Greenhouse Newsletter*. 97:13-16.

Aldrich, R.A. and J.W. Bartok, Jr. 1958. Greenhouse Engineering. *Dept. of Agric. Eng.*, Univ. of CT, Storrs.

Amsen, M.G. 1986. Thermal screens in greenhouses: Diurnal variations in heat consumption. *J. Agric. Eng. Res.* 33:79-82.

Anon. 1974. No loss of yield with carnation regime run five degrees lower. *The Grower*. 82(25):1156.

Anon. 1976a. Industrial waste heat used by French project. *The Grower*. 85(2):79-81.

Anon. 1976b. Gloucester glasshouse growers are not getting temperatures right. *The Grower*. 85(14):724.

Anon. 1977. Heat rescued from combustion gases. *The Grower*. 87(6):305.

Anon. 1978a. No doubt about value of thermal screens. *The Grower*. 89(16):974-975. Apr. 27, 1978.

Anon. 1978b. Screens and speed save fuel. *The Grower*. 89(18):1012. May 4, 1978.

Anon. 1981. Reject heat glass will save 1m gal of oil a year. *The Grower*. 95(26):36-37.

Anon. 1982a. Inertia problems in underfloor heating. *The Grower*. 96(20):31-33.

Anon. 1982b. Tomato grower slashes heating fuel bill. *The Grower*. 98(29):27-29. July 29, 1982.

Anon. 1983. Unit heaters or central boilers. *Grower Talks*. April, 1983.

Anon. 1989a. Innovations in heating. *The Grower*. 111(4).

Anon. 1989b. Screen and clad. *The Grower*. 111(4). Jan. 26, 1989.

Arinze, E.A., G.J. Schoenau and R.W. Besant. 1986. Experimental and computer performance evaluation of a movable thermal insulation for energy conservation in greenhouses. *J. Agric. Eng. Res.* 34:97-113.

Ashley, G.C. 1979. Heating greenhouses in northern climates with power plant reject heat. Symp. Proc. *HortScience*. 14:155-160.

ASAE 1988. ASAE Engineering Practice. Heating, ventilating and cooling greenhouses. Hahn, R.H. and E.E. Rosentreter, eds. 35th edition. ASAE, St. Joseph, MI.

Asbury, J.G., C. Maslowski and R.O. Mueller. 1979. Solar availability for winter space heating: An analysis of SOLMET data, 1953 to 1975. *Science*. 206:679-681.

ASHRAE. 1978. Applications Handbook. Chapter 22. Environmental control for animals and plants. ASHRAE, Atlanta, GA.

ASHRAE. 1981. Handbook of Fundamentals. Chapter 2, Heat transfer. Chapter 5, Psychrometrics. Chapter 15, Combustion and fuels. ASHRAE, Atlanta, GA.

ASHRAE. 1984. Systems Handbook. ASHRAE, Atlanta, GA.

ASHRAE. 1987. HVAC Handbook. Chapter 37, Environment for plants and animals. Plants: Greenhouses and other facilities. ASHRAE, Atlanta, GA.

ASHRAE. 1989. Fundamentals Handbook. Chap. 9, Environmental control for plants and animals. ASHRAE, Atlanta, GA.

ASCE. 1988. Standard minimum design loads for buildings and other structures 7-88. Amer. Soc. Civil Engineers, Washington, DC.

Bailey, B.J. 1974. Glasshouse air heating: Pressure changes and flow distribution along perforated air-distribution ducts. *NIAE Departmental Note* DN/G/391/2101, Silsoe.

Bailey, B.J. 1974. Temperature distribution in a multi-span glasshouse with free-discharge air heaters. *NIAE Dept. Note* DN/G/410/2101, Silsoe. 11 pp.

Bailey, B.J. 1981. The evaluation of thermal screens in glasshouses on commercial nurseries. *Acta Hort.* 115:663-670.

Bailey, B.J. 1990. Greenhouse models and climate control. *Proc. Int. Sem. and British-Israel Workshop on Greenhouse Technology*. Bet-Dagan., Aug., 1990.

Baille, A. 1988. Greenhouse microclimate and its management in mild winter climates. *ISHS Symp. on Protected Cultivation of Ornamentals in Mild Winter Climates*. 18-21 Oct. 1988. 16 pp.

Baille, A. 1990. Renewable energy sources and heat storage systems in greenhouses. *EEC Report on Energy Savings in Greenhouses*.

Bakker, J.C. 1986. Measurement of canopy transpiration or evapotranspiration in greenhouses by means of a simple vapour balance model. *Agric. and Forest Meteor.* 37:133-141.

Bakker, J.C. 1989. The effects of temperature on flowering, fruit set and fruit development of glasshouse sweet pepper (*Capsicum annuum* L.). *J. Hort. Sci.* 64:313-320.

Bakker, J.C. and G.P.A. van Holsteijn. 1995. Screens. *In* Greenhouse Climate Control. J.C. Bakker et al. eds. Wageningen Pers.

Bakker, J.C. and J.A.M. van Uffelen. 1988. The effects of diurnal temperature regimes on growth and yield of glasshouse sweet pepper. *Neth. J. Agric. Sci.* 36:201-208.

Ball, V. ed. 1985. Ball Red Book. Reston Publ. Co., Reston, VA., 14th edition.

Bark, L.D. and W.J. Carpenter. 1967. Temperature patterns in greenhouse heating. *Florists' Review.* 139(3609):17-19.

Bark, L.D. and W.J. Carpenter. 1969. Greenhouse heating with horizontal unit heaters and convection tubing. *Florists' Review.* 143(3711):44-45, 98-99, 112-114. 143(3713):18-19, 60-61.

Bartok, J.W. Jr. 1988. Horizontal air flow. *Greenhouse Manager.* 6(10):107-112.

Behnke, M.A. and B. Lambert. 1982. Sunshine foliage world burns no fuel for heat. *Florists' Rev.* 168(4353):144-146.

Blacquiere, T. 1991. Thermo- and photomorphogenesis for ornamentals. *Chronica Hort.* 31:3-4.

Blom, J. and F.J. Ingratta. 1981. The use of low intensity infrared for greenhouse heating in southern Ontario. *Acta Hort.* 115:205-216.

Boulard, T., E. Razafinjohany and A. Baille. 1989. Heat and water vapour transfer in a greenhouse with an underground heat storage system. Part I. Experimental results. *Agric. and Forest Meteor.* 45:175-184.

Boulard, T. et al.. 1990 Performance of a greenhouse heating system with a phase change material. *Agric. and Forest Meteor.* 52:303-318.

Bond, T.E. 1981. A multitude of energy-saving tips for cost conscious greenhouse operators. *Florists' Rev.* 169(4386):29-32, 104-107.

Bond, T.E. 1982. Greenhouse energy notes - 5: Energy conservation ideas. *Flower & Nursery Rpt.* Spring, 1982.

Bond, T.E. 1983. Time to cast light on solar research. *Southern Florists and Nurseryman.* 96(10):25-27.

Bot, G.P.A. 1980. Validation of a dynamical model of greenhouse climate. *Acta Hort.* 106:149-158.

Bot, G.P.A. 1983. Greenhouse climate: From physical processes to a dynamic model. Ph.D. Dissertation, Wageningen. 240 pp.

Bot, G.P.A. 1989. Greenhouse simulation models. *Acta Hort.* 245:315-325.

Bot, G.P.A. and N.J. van de Braak. 1995. Transport phenomena. *In* Greenhouse Climate Control. J.C. Bakker et al. eds. Wageningen Pers.

Botacchi, A.C. 1977. Living windbreaks. *Pennsylvania Flower Growers Bull.* 296:1, 7-8.

Brugger, M.F., T.H. Short and W.L. Bauerle. 1987. An evaluation of horizontal air flow in six commercial greenhouses. Paper No. 87-4020. *ASAE, 1987 Summer Meeting*, Baltimore. 9 pp.

Buitelaar, K. et al.. 1984. Effects of different insulation materials in glasshouse walls on growth and production of tomatoes. *Acta Hort.* 148:511-517.

Businger, J.A. 1966. The glasshouse (greenhouse) climate. *In* Physics of Plant Environment, W.R. van Wijk ed. North-Holland Publ. Co., Amsterdam. 382 pp.

Butters, R. 1977. Fuel saving night regime. *The Grower.* 88(9):386-387.

Butters, R. 1980. Venlo trials for the north. *The Grower.* 3(25):14-21. Jun. 19, 1980.

Campbell, G.S. 1977. An Introduction to Environmental Biophysics. Springer-Verlag, New York. 159 pp.

Carpenter, W.J. and L.D. Bark. 1967. Temperature patterns in greenhouse heating. *Florists' Review.* 139(3611):21-22; 139(3612):28, 92-93; 139(3613):98-101.

Carpenter, W.J., R.A. Mecklenburg and W.H. Carlson. 1976. Greenhouse heating efficiency. *Florists' Review.* 157(4075):31-32, 70-73.

Castilla, N., F. Elías and E. Fereres. 1988. Caracterizacion decondiciones climaticas y de relaciones suelo-agua-raiz en el cultivo enarenado del tomate en invernadero en Almeria. *In* Invest. Agric. Prod. Prot. Veg., Instituto Nacional de Investigaciones Agrariea, Spain. In Print.

Castilla, N. et al.. 1989. Alternative greenhouses for mild wind climate areas of Spain. Preliminary report. *Acta Hort.* 245:63-70.

Challa, H. 1978. Programming of night temperature in relation to the diurnal pattern of the physiological status of the plant. *Acta Hort.* 76:147-150.

Challa, H. 1980. Stralingsverwarming en plantengroei. *Tuinderij.* 19:30-31.

Challa, H. 1985. Report of the working party "Crop growth models". *Acta Hort.* 174:169-175.

Challla, H. and P. Brouwer. 1985. Growth of young cucumber plants under different diurnal temperature patterns. *Acta Hort.* 174:211-217.

Cockshull, K.E. 1985. Greenhouse climate and crop response. *Chronica Hort.* 25:49-51.

Cohen, Y., G. Stanhill and M. Fuchs. 1983. An experimental comparison of evaporative cooling in a naturally ventilated glasshouse due to wetting the outer roof and inner crop surfaces. *Agric. Meteor.* 28:239-251.

Coker, F.A. and J.J. Hanan. 1988. CO_2 uptake by 'Samantha' roses. *CO Greenhouse Growers' Assoc. Res.* Bull. 456:1-3.

Corder, S.E. 1973. Wood and bark as fuel. *Res. Bull. 14.* Forest Res. Lab., School of Forestry, Oregon State Univ., Corvallis. 28 pp.

Cunningham, W. 1983. Liquid foam costs fuel costs. *Greenhouse Grower.* 1(9):31-32. Sept., 1983.

Cunningham, W.. and T.L. Thompson. 1988. Passive greenhouse cooling. *Greenhouse Grower.* 6(4):18-20. Apr., 1988.

de Graff, R. and J. van den Ende. 1981. Transpiration and evapotranspiration of the glasshouse crops. *Acta Hort.* 119:147-158.

DeWerth, A.F. 1954. A practical method of cooling greenhouses. *Texas Agric. Expt. Sta. Prog. Rpt.* 1729.

Dixon, G. 1985. Heating with hot water under the skin. The Grower. June 13, 1985.

Duncan, G.A., O.J. Loewer, Jr. and D.G. Colliver. 1981. Simulation of energy flows in a greenhouse: Magnitudes and conservation potential. *Trans. ASAE.* 24:1014-1021.

Dyer, D.F. and G. Maples. 1979. Measuring and improving the efficiency of boilers. Boiler Efficiency Inst., Auburn, AL.

Elwell, D.L. and T.H. Short. 1989. Control of electrostatic effects in a polystyrene pellet variable shading greenhouse glazing. *Trans. ASAE.* 32:2117-2122.

Ellis, R.G. 1990. Low temperature heating of greenhouses. *Proc. Int. Sem. and British-Israel Workshop on Greenhouse Technology.* Bet-Dagan. Aug., 1990.

Erwin, J.E. and R.D. Heins. 1988. New concepts on the effects of day & night temperature on plant growth. *Canadian Florist.* 83(1):42-54.

Erwin, J.E. and R.D. Heins. 1995. Thermomorphogenic responses in stem and leaf development. *HortScience.* 30:940-949.

Erwin, J.E., R.D. Heins and M.G. Karlsson. 1989. Thermomorphogenesis in *Lilium longiflorum*. *Amer. J. Bot.* 76:47-52.

Erwin, J.E., R.D. Heins and R. Moe. 1991. Temperature and photoperiod effects on *Fuchsia x hybrida* morphology. *J. Amer. Soc. Hort Sci.* 116:955-960.

Erwin, J. et al.. 1989a. How can temperatures be used to control plant stem elongation. *MN State Florists Bull.* 38:1-14.

Erwin, J. et al.. 1989b. Do cool days/warm nights work with plugs? You bet! Ball *Grower Talks.* 52(11):46-52.

Ewen, L.A., J.N. Walker and J.W. Buxton. 1980. Environment in a greenhouse thermally buffered with ground-conditioned air. *Trans. ASAE.* 23:965-970.

Fenton, S.P. 1983. Developing effective screening techniques for the future. *The Grower.* Jan. 6, 1983. pp 19-21.

Freeman, B. 1976. Artificial windbreaks in horticulture. *Plasticulture.* 32:45-53.

Fuchs, M. 1990. Effect of transpiration on greenhouse cooling. *Proc. Int. Sem. and British-Israel Workshop on Greenhouse Technology.* Bet-Dagan. Aug. 1990.

Garzoli, K and J. Blackwell. 1973. The response of a glasshouse to high solar radiation and ambient temperature. *J. Agric. Eng. Res.* 18:205-216.

Gates, D.M. 1962. Energy Exchange in the Biosphere. Harper and Row, New York. 151 pp.

Gates, D.M. 1980. Biophysical Ecology. Springer-Verlag, New York. 611 pp.

Ginsburg, C. et al.. 1989. The growth of roses in a greenhouse air-conditioned by phase-change material. *Hassadeh.* 66:742-747. (Translation R. Shillo in Roses, Inc. Bull. Oct. 1989).

Goldsberry, K.L. 1975. Literature review on the utilization of thermal discharge water from electric generating plants. *Progress Rpt. I. Four Corners Regional Commission, Farmington, NM. FCRC No.* 252-336-083. 39 pp.

Gonzales, D. and J.J. Hanan. 1988. Effect of radiation, wind velocity and temperature differential on natural gas consumption of greenhouses. Preliminary report. *CO Greenhouse Growers Assoc. Res.* Bull. 454:1-4.

Grafiadellis, M. 1984. Development of solar systems for heating greenhouses. *Acta Hort.* 154:223-231.

Grafiadellis, M. and S. Kyritsis. 1981. Heating greenhouses with solar energy. *Acta Hort.* 115:553-563.

Grafiadellis, M., G. Spanomitsios and K. Mattas. 1990. Recent developments introduced in the passive solar system for heating greenhouses. *Acta Hort.* 263:111-117.

Grantham, J.B. and T.H. Ellis. 1974. Potentials of wood for producing energy. *J. of Forestry.* Sept., 1974. 552-556.

Hall, A. and J.J. Hanan. 1976. Measurement of total light energy in a carnation bench. *CO Flower Growers' Assoc. Res.* Bull. 308:2.

Halleux, D. et al.. 1985. Dynamic simulation of heat fluxes and temperatures in horticultural and low emissivity glass-covered greenhouses. *Acta Hort.* 170:91-96.

Hanan, J.J. 1959. Influence of day temperatures on growth and flowering of carnations. *Proc. Amer. Soc. Hort. Sci.* 74:692-703.

Hanan, J.J. 1970a. Some observations on radiation in greenhouses. *CO Flower Growers' Assoc. Res.* Bull. 239:1-4.

Hanan, J.J. 1970b. Statistical analysis of flower temperatures in the carnation. *J. Amer. Soc. Hort. Sci.* 95:68-73.

Hanan, J.J. 1973a. Ethylene pollution from combustion in greenhouses. *HortScience.* 8:23-24.

Hanan, J.J. 1973b. Ventilation temperatures, CO_2 levels and rose production. *CO Flower Growers' Assoc. Res. Bull.* 279:1-5.

Hanan, J.J. 1974. Air infiltration in greenhouses. *CO Flower Growers' Assoc. Res. Bull.* 286:1-4.

Hanan, J.J. 1975. Accuracy of common thermometers. *CO Flower Growers' Assoc. Res. Bull.* 298:1-3.

Hanan, J.J. 1984. Plant Environmental Measurement. Bookmakers Guild, Longmont, CO.

Hanan, J.J. 1988. Summer greenhouse climate in Colorado. *CO Greenhouse Growers' Assoc. Res. Bull.* 461:1-6.

Hanan, J.J. and F.D. Jasper. 1967. Water utilization by carnations. *CO Flower Growers' Assoc. Res. Bull.* 204:1-4.

Hanan, J.J., F.A. Coker and K.L. Goldsberry. 1987. A climate control system for greenhouse research. *HortScience.* 22:704-708.

Hanan, J.J., W.D. Holley and K.L. Goldsberry. 1978. Greenhouse Management. Springer-Verlag, Heildelberg.

Hanan, J.J., D. Moon and W.D. Holley. 1973. Summer quality in carnation benches. *CO Flower Growers' Assoc. Res. Bull.* 276:1-2.

Hanan, J.J., C. Olympios and C. Pittas. 1979. Observations on temperatures, humidity and light in polyethylene-covered greenhouses. *CO Flower Growers' Assoc. Res. Bull.* 351:1-3.

Hashimoto, Y, T. Morimoto and T. Fukuyama. 1985. Some speaking plant approach to the synthesis of control system in the greenhouse. *Acta Hort.* 174:219-226.

He, L., T.H. Short and X. Yang. 1990. Theoretical analysis of solar radiation transmission through a double-walled, acrylic, pellet-insulated, greenhouse glazing. *Trans. ASAE.* 33:657-664.

Heard, L.R. et al.. 1989. Comparison of five reject water heating systems for a proposed greenhouse at the Bruce Mansfield power plant: A case study. *Appl. Eng. in Agric.* 5:102-108.

Heinemann, P.H. and P.N. Walker. 1986. Modeling heat loss from surface heated greenhouse water. *Trans. ASAE.* 29:1379-1384.

Hewlett-Packard. 1983. 3497A data acquisition/control unit. Hewlett-Packard Instruments, Palo Alto, CA. 25 pp.

Hoecker, W.H. 1979. Relative effective solar space heating over the United States obtained from southward-tilted solar collectors. *Roses, Inc. Bull.* Feb., 1979.

Holder, R. and D. Hand. 1988. Biological aspects of energy saving in protected cultivation. *Chronica Hort.* 28:25.

Holley, W.D. and C. Juengling. 1963. Effects of additional CO_2 and automated day temperatures on carnations. *CO Flower Growers' Assoc. Res. Bull.* 162:1-4.

Hood, P.C. 1975. Clouds of doubt hang over a solar furnace. *The Nat. Observer*. Oct. 25, 1975.

Howland, J.E. 1983. A sobering update on geothermal greenhouses. *Southern Florist & Nurseryman*. 96(23):8.

Howland, J.E. 1986. Geothermal: A strong business comes first. *Greenhouse Manager*. Oct., 1986. pp 108-109.

Hughes, A.P. and K.E. Cockshull. 1972. Further effects of light intensity, carbon dioxide concentration, and day temperature on the growth of *Chrysanthemum morifolium* cv. Bright Golden Anne in controlled environments. *Ann. Bot.* 36:533-550.

Hunn, B.D. 1985. Performance and cost of a hybrid passive/active solar house. *Trans. ASAE*. 85:445-457.

Hurd, R.G. and C.J. Graves. 1984. The influence of different temperature patterns having the same integral on the earliness and yield of tomatoes. *Acta Hort*. 148:547-554.

Hurd, R.G. and H.Z. Enoch. 1976. Effect of night temperature on photosynthesis, transpiration and growth of spray carnations. *J. Expt. Bot.* 27:695-703.

Jenkins, B.M. 1985. Alternative greenhouse heating systems. *Roses, Inc. Bull.* June, 1985.

Jenkins, B.M., R.M. Sachs and G.W. Forister. 1988. A comparison of bench-top and perimeter heating of greenhouses. *Calif. Agric*. 42(1):13-15.

Jiao, J., M.J. Tsujita and B. Grodzinski. 1988. Predicting growth of Samantha roses at different light, temperature and CO_2 levels based on net carbon exchange. *Acta Hort*. 230:435-442.

Jimenez, J.I. and J. Casas-Vazquez. 1978. An experimental study of micrometeorological modifications in a glasshouse during summer. *Agric. Meteor*. 19:337-348.

Karlsson, M.G. and R.D. Heins. 1992. Chrysanthemum dry matter partitioning patterns along irradiance and temperature gradients. *Can. J. Plant Sci.* 72:307-316.

Karlsson, M.G., R.D. Heins and M.E. Hackman. 1989. Temperature controlled leaf unfolding rate in Hibiscus. Unpubl. MS., Mich. State Univ., East Lansing. 5 pp.

Karlsson, M.G., R.D. Heins and J.F. Irwin. 1988. Quantifying temperature-controlled leaf unfolding rates in 'Nellie White' Easter Lily. *J. Amer. Soc. Hort. Sci.* 113:70-74.

Karlsson, M.G. et al.. 1989a. Temperature and photosynthetic photon flux influence chrysanthemum shoot development and flower initiation under short-day conditions. *J. Amer. Soc. Hort. Sci.* 114:158-163.

Karlsson, M.G. et al.. 1989b. Development rate during four phases of Chrysanthemum growth as determined by preceding and prevailing temperatures. *J. Amer. Soc. Hort. Sci.* 114:234-240.

Karlsson, M.G. et al.. 1989c. Irradiance and temperature effects on time of development and flower size in chrysanthemum. *Sci. Hort.* 39:257-267.

Kacem, C. 1989. Heating greenhouses with geothermal energy: A study of three geothermal water distribution systems. MS. Thesis. Med. Agron. Inst., Chania, Greece.

Kimball, B.A., D.S. Benham and F. Wiersma. 1977. Heat and mass transfer coefficients for water and air in aspen excelsior pads. *Trans. ASAE*. 20:509-514.

Kooistra, E. 1984. Energy saving temperature regimes for vegetable growing. *Acta Hort*. 148:561-566.

Kooistra, E. 1986. Developments in the control of growing conditions in Dutch glasshouse horticulture. *Netherlands J. Agric. Sci.* 3:381-385.

Korns, C.H. and W.D. Holley. 1962. Effects of temperature automation on carnations. *CO Flower Growers' Assoc. Res. Bull.* 150:1-6.

Koths, J.S. 1969. Air movement within greenhouses. *CT Greenhouse Newsletter*. 26:7-8.

Koths, J.S. 1974. Effectiveness of horizontal air flow (HAF) in greenhouses. *CT Greenhouse Newsletter*. 56:17-21.

Koths, J.S. 1985. Radiant energy heating warms greenhouses if evenly distributed. *Roses, Inc. Bull.*, Oct., 1985.

Kozai, T. 1985. Ideas of greenhouse climate control based on knowledge engineering techniques. *Acta Hort*. 174:365-373.

Kozai, T. 1986. Thermal performance of an oil engine driven heat pump for greenhouse heating. *J. Agric. Eng. Res*. 35:25-37.

Kozai, T. and S. Sase. 1978. A simulation of natural ventilation for a multi-span greenhouse. *Acta Hort*. 87:39-49.

Kozai, T., S. Sase and M. Nara. 1980. A modelling approach to greenhouse ventilation control. *Acta Hort*. 106:125-136.

Kozai, T., K. Shida and I. Watanabe. 1986. Thermal performance of a solar greenhouse with watertanks for heat storage and heat exchange. *J. Agric. Eng. Res.* 33:141-153.

Krause, W. 1980. Rising fuel costs may outweigh crop loss. *The Grower.* 93(3):19. Jan. 17, 1980.

Krause, W. 1982. All eyes on energy saving. *The Grower.* 97:26-36. 11 Mar. 1982.

Krug, H. and H.P. Liebig. 1980. Diurnal thermoperiodism of the cucumber. *Acta Hort.* 118:83-89.

Kruyk, P.A. 1976. Soil heating makes a comeback. *The Grower* (Supplement). 85(4):14-15.

Kunze, J.F. 1976. Idaho geothermal development projects. *Ann. Rpt. for 1976.* INEL, Energy Res. & Develop. Adm. Boise, ID.

Kyritsis, S. et al.. 1988. A solar assisted heat pump for greenhouse heating in a typical Mediterranean climate. *Rpt. of the French-Israeli Symp. on Greenhouse Technology.* Bet Dagan, Israel. 1987.

Landsberg, J.J., B. White and M.R. Thorpe. 1979. Computer analysis of the efficacy of evaporative cooling for glasshouses in high energy environments. *J. Agric. Eng. Res.* 24:29-39.

Langhans, R.W. 1954. Mist propagation and growing. *NY State Flower Growers' Bull.* 103:1-3.

Langhans, R.W., M. Wolfe and L.D. Albright. 1981. Use of average night temperatures for plant growth for potential energy savings. *Acta Hort.* 115:31-37.

Lawson, G. 1982. For energy ideas, viva la France! *Greenhouse Manager.* 1(4):36-37.

Lawson, G. 1984. Dutch total energy system takes off. *The Grower.* Feb. 16, 1984.

Lawson, G. 1985. Flooded floors show promise. *The Grower.* 103(5):41-43.

Lazarus, S.S., J.B. Braden and P.N. Walker. 1981. Water-blanket greenhouses: The economics of using waste heat. *IL Res.* 23:8-9.

Lieberth, J.A. ed. 1990. Control plant growth with temperature. Reprints from the *Greenhouse Grower*, Meister Publishing, Willoughby, OH. 34 pp.

Long, E. 1988. £2.5m expansion at Europe's largest tomato growing complex. *The Grower.* 109(25):25-28.

Maher, M.J. and T. O'Flaherty. 1973. An analysis of greenhouse climate. *J. Agric. Eng. Res.* 18:197-203.

Mattas, K. et al.. 1990. Evaluating the effectiveness of the passive solar system for heating greenhouses. *Acta Hort.* 263:97-101.

Mavroyanopoulos, G.N. and S. Kyritsis. 1986. The performance of a greenhouse heated by an earth-air heat exchanger. *Agric. and Forest Meteor.* 36:263-268.

Mears, D.R. et al.. 1978. Rutgers' solar project explores greenhouse heating alternatives. *Florists' Review.* 162(4204):22-24, 62-65.

Meneses, J.F. and A. Monteiro. 1990a. Permanent ventilation in non-heated greenhouses to reduce *Botrytis* on tomatoes. *Proc. Int. Sem. and British-Israel Workshop on Greenhouse Technology.* Bet-Dagan. Aug., 1990.

Meneses, J.F. and A.A. Monteiro. 1990b. Ducted-air heating systems in greenhouses: Experimental results. *Acta Hort.* 263:285-292.

Mitchell, T. 1982. No virtue in energy saving alone: *The Grower* (Supplement). Jan. 21, 1982.

Monk, G.J. and J.M. Molnar. 1987. Energy-efficient greenhouses. *Horticultural Rev.* 9:1-52.

Monteiro, A.A. 1990. Greenhouses for mild-winter climates: Goals and restraints. *Acta Hort.* 263:21-32.

Montero, J.I. et al.. 1990. Natural ventilation in polyethylene greenhouses with and without shading screens. *Proc. Int. Sem. and British-Israel Workshop on Greenhouse Technology.* Bet-Dagan. Aug., 1990.

Montero, J.I., A. Anton and I. Segal. 1991. Cooling of greenhouses in Mediterranean climate: Shading, natural ventilation and fogging. Unpubl. MS, IRTA, Cabrila Center, Barcelona. 20 pp.

Montero, J.I. et al.. 1985. Climate under plastic in the Almeria area. *Acta Hort.* 170:227-234.

Moore, E.L. and T.N. Jones. 1963. Heating plastic greenhouses. *Mississippi Agric. Expt. Sta.* Bull. 666, State College, MI. 11 pp.

Morris, L.G. 1971. The theory of fan ventilation. *Acta Hort.* 22:74-85.

Morris, L.G. et al.. 1958. The limitation of maximum temperature in a glasshouse by use of a water film on the roof. *J. Agric. Eng. Res.* 3:121-130.

Nelson, P.V. 1979. Artificial windbreak material can protect crops, structures. *NC Flower Growers'* Bull. Oct-Dec, 1979.

Nikita-Martzopoulou, C. 1990. Greenhouse heating systems with geothermal energy of low enthalpy in Greece. *Acta Hort.* 263:183-189.

Nilsen, S., D. Christian and K. Hovland. 1983. Effect of reducing the infrared radiation of sunlight on greenhouse temperature, leaf temperature, growth and yield. *Sci. Hort.* 20:15-22.

Nilwik, H.J.M. 1980. Photosynthesis of whole sweet pepper plants. 2. Response to CO_2 concentration, irradiance and temperature as influenced by cultivation conditions. *Photosynthetica.* 14:382-391.

Nijskens, J. et al.. 1984. Heat transfer through covering materials of greenhouses. *Agric. and Forest Meteor.* 33:193-214.

Norman, B. 1983. Twin skin saver disappoints at Efford EHS. *The Grower.* 100(15). Oct. 13, 1983.

Okada, M. 1980. The heating load of greenhouses. *J. Agric. Meteor.* (Japanese). 35:235-242.

Okada, M. and T. Takakura. 1973. Guide and data for greenhouse air conditioning. 3. Heat loss due to air infiltration of heated greenhouse. *J. Agric. Meteor.* (Japanese). 28:11-18.

Otten, L. et al.. 1989. Energy saving hydroponic greenhouse pilot project: Energy aspects. *Canadian Agric. Eng.* 31:147-152.

Parsons, R.A. 1975. Sizing perforated polyethylene tubes for greenhouse ventilation and heating. *Canadian Florist.* 70(4):29-31.

Pasian, C.C. and J.H. Lieth. 1989. Analysis of the response of net photosynthesis of rose leaves of varying ages to photosynthetically active radiation and temperature. *J. Amer. Soc. Hort. Sci.* 114:581-586.

Payne, F.W. 1985. Efficient Boiler Operations Sourcebook. The Fairmont Press, Inc., Atlanta, GA. 221 pp.

Penman, H.L., D.E. Angus and C.H.M. van Bavel. 1967. Microclimatic factors affecting evaporation and transpiration. *In* Irrigation of Agricultural Lands. R.M. Hagan et al., Eds. No. 11. Amer. Soc. Agron., Madison, WI. 1180 pp.

Popovski, K. 1988. Greenhouse heating with geothermal energy. *Chronica Hort.* 28:19-20.

Puri, V.M. 1982. Greenhouse floor heating system optimization using long-term thermal performance design curves. *Solar Energy.* 28:469-481.

Rebuck, S.M., R.A. Aldrich and J.W. White. 1977. Internal curtains for energy conservation in greenhouses. *Trans. ASAE.* 20:732-734.

Roberts, W.J. 1983. Bury that heat! *Greenhouse Grower.* Oct., 1983. pp 12-14.

Roberts, W.J. 1988. Selecting the right fan. *Greenhouse Grower.* 6(4):16-17. Apr., 1988.

Roberts, W.J. et al.. 1981. Progress in movable blanket insulation systems for greenhouses. *Acta Hort.* 115:685-692.

Roberts, W.J. et al.. 1985. Energy conservation for commercial greenhouses. *NRAES*-3. Univ. of CT., Storrs, CT. 40 pp.

Ross, D.S. et al.. 1978. Energy conservation and solar heating for greenhouses. *NRAES*-3, Univ. of CT, Storrs, CT. 48 pp.

Rotz, C.A. and R.A. Aldrich. 1979. Feasibility of greenhouse heating in Pennsylvania with power plant waste heat. *Trans. ASAE.* 22:1375-1381.

Rotz, C.A. and R.D. Heins. 1982. Evaluation of infrared heating in a Michigan greenhouse. *Trans. ASAE.* 25:402-407.

Rotz, C.A., R.A. Aldrich and J.W. White. 1982. The use of thermal insulation and solar heat in the greenhouse. *Florida Foliage.* Feb., 1982. pp 59-67.

Rotz, C.A., R.D. Heins and D.E. Cochell. 1981. Heat pumps for temperature control in Michigan greenhouses. *ASAE Paper No. 81-3007.* Orlando, FL. 18 pp.

Royle, D. 1980a. Wye choice based on humidity. *The Grower.* 93(22)

Royle, D. 1980b. Screens still raising doubts. *The Grower.* Mar. 27, 1980. p 4.

Sadler, F.J. and C.H.M. van Bavel. 1984. Simulation and measurement of energy partition in a fluid-roof greenhouse. *Agric. and Forest Meteor.* 33:1-13.

Sase, S. 1989. The effects of plant arrangement on air flow characteristics in a naturally ventilated glasshouse. *Acta Hort.* 245:429-435.

Sase, S., T. Takakura and M. Nara. 1984. Wind tunnel testing on airflow and temperature distribution of a naturally ventilated greenhouse. *Acta Hort.* 148:329-336,

Sase, S. et al.. 1980. Ventilation of greenhouses. I. Wind tunnel measurements of pressure and discharge coefficients for a single span greenhouse. *J. Agric. Meteor.* (Japanese). 36:3-12. (English summary).

Schockert, K. and C. von Zabeltitz. 1980. Energy consumption of greenhouses. *Acta Hort.* 106:21-26.

Segal, I. et al.. 1990a. Developing and testing of heat exchangers for greenhouses utilizing very low temperature geothermal water. *Proc. Int. Sem. and British-Israel Workshop on Greenhouse Technology*. Bet-Dagan, Aug., 1990.

Segal, I. et. al. 1990b. Passive solar heating of greenhouses utilizing water tubes. *Proc. Int. Sem. and British-Israel Workshop on Greenhouse Technology*. Bet-Dagan, Aug., 1990.

Seginer, I. and L.D. Albright. 1980. Rational operation of greenhouse thermal-curtains. *Trans ASAE*. 23:1240-1245.

Seginer, I. and D. Kantz. 1986. *In situ* determination of transfer coefficients for heat and water vapour in a small greenhouse. *J. Agric. Eng. Res*. 35:39-54.

Seginer, I. and A. Livne. 1978. Effect of ceiling height on the power requirement of forced ventilation in greenhouses. *Acta Hort*. 87:51-67.

Shen, H. and D.R. Mears. 1977. Computer simulation of warm floor greenhouse heating. *ASAE Paper No. 77-4532*. St. Joseph, MI. 9 pp.

Short, T.H. and S.A. Shah. 1981. A portable polystyrene-pellet insulation system for greenhouses. *Trans. ASAE*. 24:1291-1295.

Silva, A.M. and R. Rosa. 1987. Radiative heat loss inside a greenhouse. *J. Agric. Eng. Res*. 37:155-162.

Slack, G. and D.W. Hand. 1983. The effect of day and night temperatures on growth, development and yield of glasshouse cucumbers. *J. Hort. Sci*. 58:567-572.

Smith, B. 1982. Heat pumps and geothermal energy may be too expensive. *The Grower* (Supplement). Jan. 21, 1982.

Smith, B. 1987. Heating choice changes to gas, water or air. *The Grower*. 108(3):13-19: July 16, 1987, supplement.

Smith, B. 1988. The how, when and why of building a new glasshouse. *The Grower*. 109(19):27-33. May 12, 1988.

Staley, L.M., G.J. Monk and J.M. Molnar. 1986. The influence of thermal curtains on energy utilization in glass greenhouses. *J. Agric. Eng. Res*. 33:127-139.

Stamper, E. and R.L. Koral eds. 1979. Handbook of Air Conditioning, Heating and Ventilating. 3rd edition. Industrial Press, NY.

Stanghellini, C. 1983a. Calculation of the amount of energy released by heating pipes in a greenhouse and its allocation between convection and radiation. *Res. Rpt. 83-3*. IMAG, Wageningen. 20 pp.

Stanghellini, C.1983b. Evaporation of a greenhouse crop and its relationship to the supply of heat. *Res. Rpt. 83-6*. IMAG, Wageningen. 31 pp.

Stanghellini, C. 1983c. Forcing functions in greenhouse climate and their effect on transpiration of crops. *Res. Rpt. 83-4*. IMAG, Wageningen. 55 pp.

Stanghellini, C. 1983d. Radiation absorbed by a tomato crop in a greenhouse. *Res. Rpt. 83-5*. IMAG, Wageningen. 23 pp.

Stanghellini, C. 1987. Transpiration of greenhouse crops. Ph.D. Dissertation, IMAG, Wageningen. 150 pp.

Stanhill, G. et al.. 1973. The radiation balance of a glasshouse rose crop. *Agric. Meteor*. 11:385-404.

Steinbuch, F. and J. van de Vooren. 1984. Production and quality of cutflowers and potplants grown in greenhouses covered with energy saving double layer materials. *Acta Hort*. 148:555-560.

Sterling Radiator. 1984. "Classic" Elements. Catalog AR-5. Sterling Radiator, Westfield, MA.

Takakura, T. 1967. Predicting air temperatures in the glasshouse (I). *J. Meteor. Soc. Japan*. 45:40-52.

Takakura, T. 1968. Predicting air temperatures in the glasshouse (II). *J. Meteor. Soc. Japan*. 46:36-44.

Tal, A. et al.. 1987. Design procedure for a greenhouse space heating system utilizing geothermal warm water. *Energy in Agric*. 6:27-34.

Tantau, H.J. 1980. Inventory of more significant parameters which can be useful to characterize energy requirements for heating. *Acta Hort*. 107:99-102.

Ted Reed Thermal, Inc. 1985. Commercial fin tube radiation. Catalog TFF-85. Ted Reed Thermal, Inc., West Kingston, RI.

Tesi, R. and D. Tosi. 1985. Risparmo energetico in serra con impiego di schermature e destratificatori. *Colture Protette*. 14:57-62. (English summary).

Tesi, R. and D. Malarme. 1987. Risparmio energetico in serra con l'impiego di "Serrolithe". *Colture Protette.* 16:57-61. (English summary).

Tweedell, R. 1980. Solar dreamers: Heat 'quiet truths'. *The Denver Post.* May 4, 1980.

Udink ten Cate, A.J. 1980. Remarks on greenhouse climate control models. *Acta Hort.* 106:43-47.

Vakis, N.J. and I. Photiades. 1990. Effects of heating methods and temperature regimes on the production of greenhouse crops. Agric. Res. Inst., Cyprus. Unpublished MS.

van Bavel, C.H.M. and J. Damagnez. 1978. A simulation model for energy storage and savings of a fluid-roof solar greenhouse. *Acta Hort.* 76:229-236.

van Bavel, C.H.M. and E.J. Sadler. 1979. A computer simulation program for analyzing energy transformations in a solar greenhouse. *Dept. of Soil and Crop Sci.*, Texas A&M Univ., College Station. 75 pp.

van Bavel, C.H.M., E.J. Sadler and J. Damagnez. 1980. Analysis of heat and water stress of plants in greenhouses. *Acta Hort.* 107:71-78.

van Bavel, C.H.M., E.J. Sadler and J. Damagnez. 1981. Cooling greenhouse crops in a Mediterranean climate. *Acta Hort.* 115:527-536.

van de Braak, N.J. 1995. Heating equipment. *In* Greenhouse Climate Control. J.C. Bakker et al. eds. Wageningen Pers.

van de Vooren, J. and T. Strijbosch. 1980. Glasshouse ventilation control. *Acta Hort.* 106:117-123.

van Winden, C.M.M., J.A.M. van Uffelen and G.W.H. Welles. 1984. Comparison of the effect of single and double glass greenhouses on environmental factors and production of vegetables. *Acta Hort.* 148:567-573.

Waaijenberg, D. 1984a. Strength and durability of greenhouse cladding materials. *Acta Hort.* 148:657-662.

Waaijenberg, D. 1984b. Research on insulating greenhouse cladding materials. *Acta Hort.* 184:651-656.

Walker, J.N. 1965. Predicting temperatures in ventilated greenhouses. *Trans. ASAE.* 8:445-448.

Walker, J.N. and D.J. Cotter. 1968. Cooling of greenhouses with various water evaporation systems. *Trans. ASAE.* 11:116-119.

Walker, J.N. and L.R. Walton. 1971. Effect of condensation on greenhouse heat requirement. *Trans. ASAE.* 14:282-284.

Walker, J.N., and G.A. Duncan. 1973. Air circulation in greenhouses. *Dept. of Agric. Eng.*, Univ. of KY, AEN-18, 9 pp.

Walker, P.N. 1978. Surface heating greenhouses with power plant cooling water. *Trans. ASAE.* 21:322-324.

Walker, P.N. 1982. An experimental surface-heated greenhouse. *Trans. ASAE.* 25:1022-1025.

Walker, P.N., S.S. Lazarus and J.B. Braden. 1982. Surface-heating greenhouse: Microeconomics. *Trans ASAE.* 25:408-412.

Walker, P.N., S.S. Lazarus and J.B. Braden. 1983. Surface heated greenhouses: Algorithms for modulating flow. *Trans. ASAE.* 26:170-174.

Walker, P.N. et al.. 1987. Unit heaters for utilizing reject warm water in greenhouses: Control algorithms. *Trans. ASAE.* 30:1116-1118.

Watkinson, A. 1975. Insulating fully creates other problems. *The Grower.* 84(4):611-613.

Welles, D.G. 1975. What to look for when choosing a system for heating the greenhouse. *Florists' Review.* 156(4040):67-68, 131.

Went, F.W. 1956. The role of environment in plant growth. *Amer. Sci.* 44:378-398.

Went, F.W. 1957. The Experimental Control of Plant Growth. Chronica Botanica, Waltham, MA. 343 pp.

White, J.W. 1982. Energy conservation and alternative energy sources for commercial greenhouses. Part IV. *Roses, Inc. Bull.* Jan. and Feb., 1982.

White, J.W. and R.A. Aldrich. 1980. Greenhouse energy conservation. PA State Univ., University Park, PA. 27 pp.

White, J.W. and I.J. Warrington. 1988. Temperature and light integral effects on growth and flowering of hybrid geraniums. *J. Amer. Soc. Hort. Sci.* 113:354-359.

White, R.A.J. 1975. Effect of ventilation on maximum air temperatures in twelve identical glasshouses. *Acta Hort.* 46:63-70.

Willits, D.H., P. Chandra and M.M. Peet. 1985. Modelling solar energy storage systems for greenhouses. *J. Agric. Eng. Res.* 32:73-93.

Wilson, R.E. and J.P. Evans. 1957. Thermocouples. *In* Process Instruments and Control Handbook. D.M. Considine, ed. McGraw-Hill, New York

Whittle, R.M. and W.J.C. Lawrence. 1960a. The climatology of glasshouses. II. Ventilation. *J. Agric. Eng. Res.* 5:36-41.

Whittle, R.M. and W.J.C. Lawrence. 1960b. The climatology of glasshouses. III. Air temperature. *J. Agric. Eng. Res.* 5:165-178.

Whittle, R.M. and W.J.C. Lawrence. 1960c. The climatology of glasshouses. V. The heat consumption of glasshouses. *J. Agric. Eng. Res.* 5:399-405.

Widmer, R.E. 1979. Commercial greenhouse heating with reject heat from electric generating plants. *MN State Florists* Bull. 1(6):1-4.

Winspear, K.W. 1974. Glasshouse engineering research and the energy crisis. *Nat. Glasshouse Energy Conf.*, Littlehampton. Oct. 1974. 12 pp.

Wittwer, S.H. and S. Honma. 1979. Greenhouse Tomatoes, Lettuce and Cucumbers. Mich. State Univ. Press, East Lansing. 225 pp.

Wolfe, J.S. 1970. Feasibility and economics of conditioning recirculated greenhouse air by means of evaporative cooling. *Note No. 46/2104.* NIAE, Silsoe, U.K. 18 pp.

Wolfe, J.S. and R.F. Cotton. 1975. Airflow and temperature distribution in greenhouses with fan ventilation. *Acta Hort.* 46:71-89.

Wright, C.R. 1981. Utah Roses, Inc. geothermal heat project. *Roses, Inc. Bull*, Sept., 1981. pp 55-57.

Youngsman, J. 1978. Infrared heating for greenhouses. *Ohio State Florists' Assoc. Bull.* 587:1-4.

Zabeltitz, C. von. 1986. Greenhouse heating with solar energy. *Energy in Agric.* 5:111-120.

Zieslin, N. and M. Gazit. 1978. Greenhouse cooling with a positive pressure pad-and-fan system. *Gartenbauwissenschaft.* 43:185-187.

CHAPTER 5

WATER

I. INTRODUCTION

Most of the solar radiation received by a well-watered crop evaporates water. The proportion can be as high as 87% of the global radiation on a rose canopy under arid conditions [Stanhill and Albers, 1974], and as low as 50% under northern European conditions [van der Post et al, 1974]. The energy used to transpire water from outdoor crops can exceed the incident net radiation if energy is brought into the canopy from outside the field (dry surroundings), i.e., "advective" energy [Gates, 1980; Woodward and Sheehy, 1983; etc.]. Under dry conditions, plants may transpire more than 99% of all water absorbed from the soil [Plaut and Moreshet, 1973]. While I have given the general figure of 70% of total shortwave radiation, Chang [1968] suggested 80 to 90% of net radiation because some advected energy also exists in a humid climate. The result is that very high correlations can be obtained between net radiation and evapotranspiration [e.g., Hanan, 1970b], and such relationships can be employed for scheduling irrigations (see Fig. 3-38). An example of transpiration versus net radiation for a greenhouse crop is provided in Fig. 5-1.

Fig. 5-1. An example of the effect of total solar radiation inside the greenhouse (R'_C) on transpiration of a tomato crop. Similar relationships can be found for almost any well watered crop. In this case radiation has been expressed in equivalent terms of water depth [Reprinted from *Agri. Meteor.*, 3:187-196, Lake, J.V.., ©1966. With kind permission of Elsevier Sci.-NL, Sara Burgerhartstraat 25, 1055 KV Amsterdam, The Netherlands].

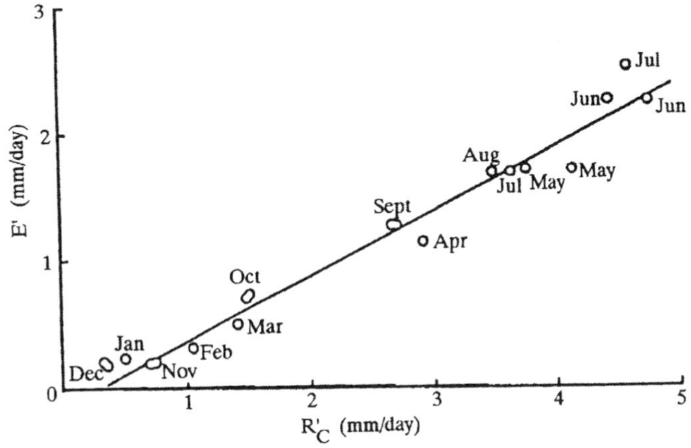

In effect, a vegetative canopy is a complex, flexible piping system, transferring water from the root environment to the air above it. The quantities of water are large –for example, to produce a 36 g standard carnation flower requires nearly 4 kg water during growth to harvest [Hanan and Jasper, 1969]. A single stem *Antirrhinum majus* (Snapdragon) may require 5 kg of water under New York State conditions [Hanan and Langhans, 1964b]. Other authors have shown as much as 600 kg water are required to produce 1 dry kg of maize seed. Boyer [1985] stated that a rapidly transpiring sunflower leaf will turn over its water content in 20 minutes. In each case, the change in state of a kilogram from liquid to vapor represents the disposal of about 2.47 MJ. The literature shows that deliberate restriction of water to a plant will generally reduce growth, and the higher the transpiration rate in well-watered crops, the greater the yield (Fig. 5-2). The maximum efficient use of water by economic plants is when water is freely available, and transpiration can continue at the maximum potential rate as determined by the energy supply [see Lemon, 1965, for a discussion on water use efficiency]. Under these conditions of maximally available water, the amount transpired by a vegetative canopy that completely covers the land surface (LAI ≈ 4) will equal the evaporation rate from a wet surface. According to Slatyer [1967a], crop yield is maximal only when water is freely supplied throughout the life of the crop. There will be little difference

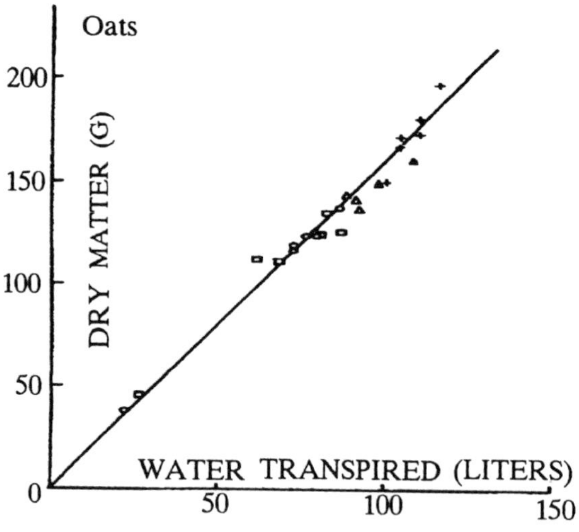

Fig. 5-2. Typical example of the relationship between transpiration and yield of several oat varieties [Arkley, 1963]. Although this example deals with a field crop in regard to dry matter production, the same general relationship holds for most economic crops whether inside a greenhouse or outside (With permission of *Hilgardia)*.

in transpiration rate between herbaceous species commonly grown in greenhouses (e.g., lettuce, tomato, cucumber, carnation, rose, *ad infinitum*) provided they are not severely stressed and completely cover the ground [de Graff and van den Ende, 1981].

In Chapters 1 and 2, the influence of radiation and temperature as features of the local climate was emphasized. Climatic regions with many clear days and mild temperatures allow maximum efficient use of greenhouses for crop production. Unfortunately, such regions also guarantee high potential evapotranspiration. These same arid regions experience moisture deficits with annual precipitation usually less than 500 to 700 mm; also subject to such severe variations that an average precipitation figure has little meaning. The relationship between aridity and man was well discussed in 1965 by the American Association for Advancement of Science Publication No. 74. In that review, Duisberg [1965] quotes Koenig that "the use of irrigation in the arid lands of the U.S. is not an appropriate use of that valuable resource, water . . . " In fact, the appropriation of water by those able to pay for it (industry and domestic) usually leaves agriculture on the short end of the stick. However, I have already shown (Fig. 1-3) that the return to a greenhouse operation for each water unit consumed places greenhouse production in the enviable position as a strong competitor for available water resources. Hanan [1967] suggested that greenhouse production in arid regions represents the most efficient use of a scarce resource –if that resource is of suitable quality. However, one thing to keep in mind is that a greenhouse itself represents a desert [Enoch, 1986]. Serious problems can arise when land is covered so that natural precipitation can no longer be employed to control salinity –apart from the fact that the grower must supply water in some fashion.

Fig. 5-3. Example of guttation from a young rose plant growing in rockwool. This is indicative of a positive water potential. The phenomenon usually occurs in the early morning hours or under very low irradiances.

The removal of water from a soil matrix, or from a solution containing solutes, requires a force greater than the force tending to keep the water in place. A still greater force is necessary to move that water to evaporation sites in leaves and stems, with the greatest difference in pressure arising from the significant differences in water vapor concentrations between the leaf and surrounding air. The grower is faced with problems of water supply

Fig. 5-4. Statistical relationship between evapotranspiration and internal water potential in carnations in a greenhouse [Hanan, 1970a]. Actively transpiring and photosynthesizing plants will invariably experience some degree of negative water potential.

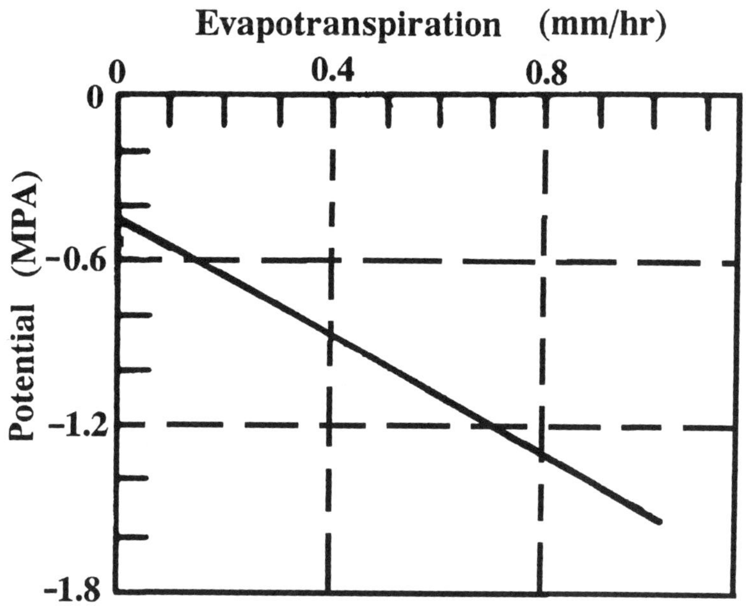

and water demand with the pipe (plant) between. If the suction becomes great enough, the pipe will collapse (plant wilts). Between that disaster and maximum turgor, the status of water inside the plants making up the vegetative canopy, in respect to pressure and flow rate, will markedly influence growth. This chapter, therefore, is divided into three principal sections: 1) the effect of water stress on growth, 2) the supply of water to the root system, and 3) factors influencing the demand (transpiration). Problems of water quality and salinity will be discussed in Chapter 6 along with nutrition. I make the assumption, until Chapter 6, that osmotic forces caused by solutes in the water supply can be neglected –largely to simplify and reduce the length of this chapter.

II. WATER AND PLANT RESPONSE

One can look upon the movement of water within a vegetative canopy as the response to a high pressure in the soil, pumping the water into the atmosphere –similar to pumping water to the top of a tall building. In the plant system, however, the pressures and chemical activities of water in this system will be less than zero. In other words, flow is in response to greater negative pressures common to the atmosphere surrounding the crop. The highest pressures, although still negative, will be in the root substrate. Turner and Begg [1981] delineate the problem clearly in stating that water moves only from sites of high potential to those of low potential (the atmosphere), and to extract water from the substrate against gravity and resistance to liquid flow in the vascular system, the potential energy of water in the plant must be less than that of the water in the substrate. Thus, water deficits occur in the tissues of all transpiring plants as an inevitable consequence of the water flow. The lower water potential of the atmosphere provides the driving force for movement out of the tissues, through the stomata, and into the air. There are occasions when pressures and activities can be higher than zero (Fig. 5-3), leading to the common observation of guttation by plants usually during early morning periods and when the water supply is freely available. This is not common when plants are actively growing and photosynthesizing. Invariably, there will be stress, or negative pressures, in green plants under the conditions provided in greenhouses. In fact, over a considerable range of transpiration and internal water potentials, the relationship can be closely approximated by a straight line (Fig. 5-4) under greenhouse conditions [Hanan, 1970a].

A. BASICS AND UNITS OF WATER POTENTIAL

The status of water in the soil-plant-air system is based upon thermodynamic relationships and the ideal gas law that allows one to compare the energy difference between a suitable reference and actuality. The last 30 years have seen a revolution in terminology and definitions that allows an individual to consider the entire system using uniform units. A simplified derivation developed here is based upon Slatyer's discussion [1967a, b] and others [Boyer, 1969; Campbell, 1977; Gates, 1980; etc.]. The procedure makes the assumption that behavior of water, water vapor, and the solutes that may be present is ideal. For most purposes, deviations from ideal behavior do not cause remarkable error.

It is not possible to know the absolute energy of a system such as water in plant or soil, but it is acceptable to determine the difference between what is actually the situation and some reference point, such as, in this case, pure, free water at standard temperature and pressure:

$$\Delta\mu = \mu_w - \mu_w^o \tag{5.1}$$

where: $\Delta\mu$ = *the energy difference (erg mol^{-1})*
μ_w = *the chemical potential of the water in the system*
μ_w^o = *the chemical potential of pure, free water at the same temperature*

$\Delta\mu$ represents the work involved in moving 1 mole of water from a pool of pure water to some arbitrary point in the system at constant temperature and pressure. A difference in the value between two locations shows that water is not in equilibrium, so there will be a tendency for water to flow toward the region where $\Delta\mu$ is lower. Rather than using this cumbersome quality to denote the activity of water, one defines:

$$\psi = \frac{(\mu_w - \mu_w^o)}{\overline{V}_w} \tag{5.2}$$

where: ψ = *chemical potential (erg cm^{-3})*
$\overline{V}_w$ = *molal volume of water (cm^3 mol^{-1}) (ca 18.0 cm^{3}mol^{-1})*

Dimensionally, the Greek symbol, ψ, is equivalent to pressure units since 1 erg cm^{-3} = 1 dyne cm^{-2}. Therefore, 0.1 megapascal (MPa) = 1 bar = 10^6 dyne cm^{-2} = 0.987 atmospheres (atm). The megapascal is coming into increasing use, and atmospheres will be found in the physiological literature before the 1950s. Water potential may also be found expressed in Joules per kilogram (J kg^{-1} = 0.01 bar) or per unit volume of water (J m^{-3}).

If the term $\mu_w^o = 0$, $\Delta\mu$ is thermodynamically equivalent to:

$$\Delta\mu = RT \ln \frac{e}{e^o} = RT \ln a_w = RT \ln \phi_w N_w \tag{5.3}$$

where: R = *perfect gas constant (8.314 x 10^7 erg mol^{-1}K^{-1})*
T = *temperature (K)*
e = *vapor pressure of water in air (Pa)*
e^o = *vapor pressure of pure free water at the same temperature and pressure, or saturation (Pa)*
ln = *the natural logarithm*
a_w = *chemical activity of water*
ϕ_w = *chemical activity coefficient*
N_w = *mole fraction of water*

The water potential of a cell, tissue, plant or soil can be related to the relative vapor pressure (e/e^o):

$$\psi = \frac{\Delta\mu}{\overline{V}_w} = \frac{RT}{\overline{V}_w} \ln \frac{e}{e^o} \tag{5.4}$$

The quantity $RT/\bar{V}_w$ is 135.0 MPa at 20 C. If the relative humidity of the air is 1.0, then ln 1.0 = zero and the water potential of saturated air is zero. At RH = 0.99, ψ = -1.36 MPa, and at 0.50 RH –which is not at all uncommon for the bulk air– ψ = -93.6 MPa (see Table 5-1). Some typical values of water potential were provided by Nobel [1991] (Table 5-2). In each case from soil to air, the water potential (ψ) becomes less. The water movement in one direction is inevitable.

Table 5-1. Relationship between relative humidity and water potential.

Relative humidity (%)	Water potential (MPa)
100.0	0.00
99.6	-0.54
99.0	-1.36
96.0	-5.51
90.0	-14.2
50.0	-93.6
0.0	$-\infty$

The chemical potential of water in plants is affected not only by the hydrostatic pressure (tension), but by the colligative effects of solutes and the interaction with matrices of solids (cell walls) and macromolecules [Hsiao, 1973]. Colligative effects include freezing point depression, boiling point elevation, lowering of the vapor pressure of the solvent, and osmotic pressure. The effects of gravity for most cases in greenhouse practice are small enough to ignore. One can write:

$$\psi_t = \psi_p + \psi_o + \psi_m \tag{5.5}$$

where: ψ_t = *the total water potential (MPa)*
ψ_p = *potential due to pressure*
ψ_o = *potential due to osmotic pressure*
ψ_m = *potential due to matric pressure*

Because of fairly rigid walls, large hydrostatic pressures can exist in plants. Thus, ψ_p is important in support of the plant and in the movement of water and solutes in the xylem and phloem. One of the most important colligative properties is the behavior of water surrounding a cell containing solutes separated by a semipermeable membrane. Since the presence of solutes in the cell lowers potential, water will tend to move from its higher concentration outside into the cell, increasing the cell's turgor pressure. This is one of the major factors influencing cell enlargement and the ultimate length of flower stems and fruit size in greenhouse crops. The greater the concentration of solutes, the more negative is ln a_w, or the activity of water, and the larger the turgor pressure caused by water movement into the cell. ψ_o is always negative. The term ψ_m, or matric potential, represents interactions of water with interfaces such as the retention of water by colloidal particles, membranes, etc. This reduces the thermodynamic activity of water. According to Boyer [1967], in the species he studied, matric potentials were correlated with the amount of cell wall present and the volume of water outside the leaf protoplasts. While significant, ψ_m is often ignored in view of the importance of ψ_p and ψ_o in determining ψ_t and ultimate growth response. This allows one to state the two extreme conditions found in plants. As $\psi_t = \psi_p + \psi_o$, then when the plant is fully turgid, $\psi_p = \psi_o$ and

Table 5-2. Some representative values of water potentials in the soil-plant-air system [Adapted from Nobel, 1991].

Location	ψ_t (MPa)	ψ_p (MPa)	ψ_o (MPa)
Soil, 0.1 m below ground and 10 mm from root	-0.3	-0.2	-0.1
Soil adjacent to root	-0.5	-0.4	-0.1
Xylem of root near ground surface	-0.6	-0.5	-0.1
Xylem in leaf at 10 m above ground	-0.8	-0.8	-0.1
Vacuole of mesophyll cell in leaf at 10 m	-0.8	-0.2	-1.1
Cell wall of mesophyll cell at 10 m	-0.8	-0.4	-0.5
Air in cell wall pores	-0.8		
Air just inside stomata at 95% RH	-6.9		
Air just outside stomata at 60% RH	-70.0		
Air outside boundary layer at 50% RH	-93.6		

$\psi_t = 0$. On the other hand, when completely wilted, $\psi_t = \psi_o$ and $\psi_p = 0$.

One may also write an equation similar to Eq. 5.5 for the soil system:

$$\psi_t = \psi_m + \psi_o \qquad (5.6)$$

The gravitational and electrical potentials are ignored. Contrasting with the plant system, matric potential in soils is the most important potential one deals with unless saline conditions are encountered. The fact that the matric potential is 0 in flowing hydroponic systems is one of their principal advantages.

B. EFFECTS OF STRESS ON GROWTH

Water movement is governed by two factors: the driving force and the conductance of the flow path [Boyer, 1985]. There are two types of water flow one must deal with: 1) mass flow of water for transpiration, largely through the xylem, and 2) flow through the tissue from the xylem to individual cells. Water enters the root in response to a lower potential, and moves across the endodermis, which can form one of the major resistances to flow. Water movement is largely mass flow, whereas solute uptake (essential ions, etc.) is largely an active process. The conductivity of the plant as a system will vary with the flow rate. Since ions can be accumulated against a concentration gradient by expenditure of metabolic energy, under low transpiration rates, water can still move into the roots and cells in response to osmotic relationships. Thus, at low flow rates, the endodermis is not a rate-limiting barrier. Resistance to water flow is also high in the leaf mesophyll. However, under high transpiration rates, conductance appears to decrease in several species, and this results from cavitation and formation of embolisms that effectively block the xylem tissue [Tyree and Sperry, 1989]. Little research on economic greenhouse plants regarding cavitation has been carried out, although Hanan [1987] reported on preliminary work with roses. The breakage (cavitation) appears to begin at water potentials around -0.8 to -1.0 MPa, increasing in rate to about -2.0 MPa, and then decreasing as most of the xylem elements have formed embolisms. According to Tyree and Sperry [1989], one important role of stomatal regulation is to prevent catastrophic xylem embolism while pressing water conduction through stems to the maximum limit.

A general relationship that relates water flow to potential gradients and resistance to flow was outlined by Stanhill and Vaadia [1967], based upon concepts originally proposed by van den Honert in 1948:

$$Q_{s,r} = -\frac{\psi_r - \psi_s}{r_s + r_r} \qquad (5.7)$$

$$Q_{r,l} = -\frac{\psi_l - \psi_r}{r_r + r_l} \qquad (5.8)$$

$$Q_{l,a} = \frac{\psi_a - \psi_l}{r_l + r_a} \qquad (5.9)$$

where: Q = water flux
ψ = water potential
r = resistance and, the subscripts s, r, l, and a refer to the soil, root, leaf, and air respectively

These equations have been discussed by Slatyer [1967a] and Plaut and Moreshet [1973]. They show that the flow of water in the plant depends on a potential gradient and resistance to that flow. Since xylem resistance is almost negligible compared to the soil-root and leaf-air resistance, it is often ignored. If the transpiration rate is constant, then according to Newman [1974]:

$$Q = \frac{\psi_r - \psi_s}{r_s + r_r} = \frac{\psi_l - \psi_r}{r_r + r_l} = \frac{\psi_a - \psi_l}{r_l + r_a} \qquad (5.10)$$

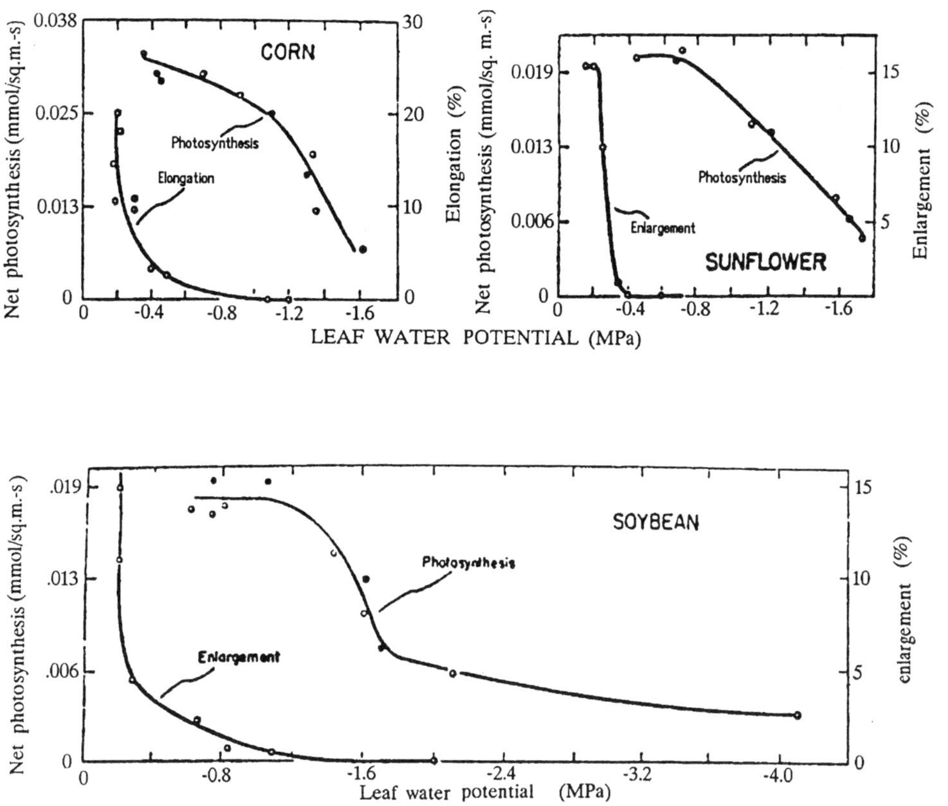

Fig. 5-5. Effect of leaf water potential on leaf enlargement and photosynthesis in corn, soybean, and sunflower [adapted from Boyer, 1970a] (With permission of the *J. of Plant Physiology*).

There is no indication of the water that is directly used in cell uptake for enlargement [See Fiscus et al., 1983]. It was Honert's purpose to show that r_a is always much smaller than r_s and r_p. Although Newman states the latter are generally accepted, it will be shown that there are exceptions. There are other criticisms to Eq.s 5.7 through 5.10 that are not considered here. A more recent discussion by Fiscus et al. [1983] deals with the relationships between the water taken up by individual cells versus that evaporated to the air. Except over rigorously defined ranges, the relationships are seldom linear.

A basic factor in plant growth is cellular enlargement, which requires a high enough turgor to extend the cell walls, and a low enough water potential to provide water for the enlargement process [Boyer, 1985]. The osmotic potential must be sufficiently low to allow both these requirements to be met simultaneously. Enlarging cells require a water potential of -0.1 to -0.15 MPa below the water potential of the external solution in order to supply water. Solute uptake must be continuous if dilution of cell solution is to be prevented as cell enlargement occurs and more water enters the cell. The importance of enlargement can be learned by the fact that in such species as carnation, all cells in a stem are formed at six to seven visible leaf pairs when the branch is about 10 to 15 cm long. Thereafter, stem length strictly depends on cell enlargement. This situation is applicable to nearly all economic crops grown in greenhouses, whether vegetables or ornamentals. Boyer's work, carried out in the early 1970s [1970a; b; 1971a; b], provides several examples (Fig. 5-5), which include the effect of water potential on photosynthesis. The results show that elongation is extremely sensitive to water stress, with significant reductions at leaf water potentials around -0.3 to -0.4 MPa. Reduced photosynthesis was not usually noted until

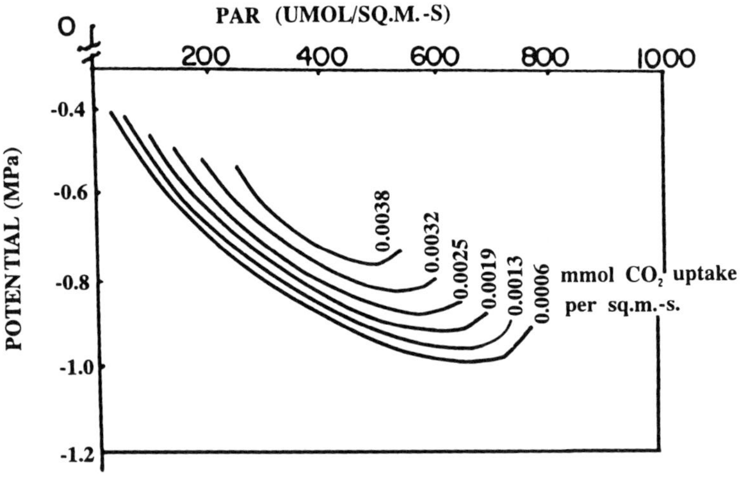

Fig. 5-6. Response contour plot of $^{14}CO_2$ uptake of mature 'Forever Yours' rose leaves at 54 Pa CO_2 concentration as a function of ψ_t and PAR. Curves connect points to equal $^{14}CO_2$ uptake. Statistical correlation = 0.94 [Aikin and Hanan, 1975].

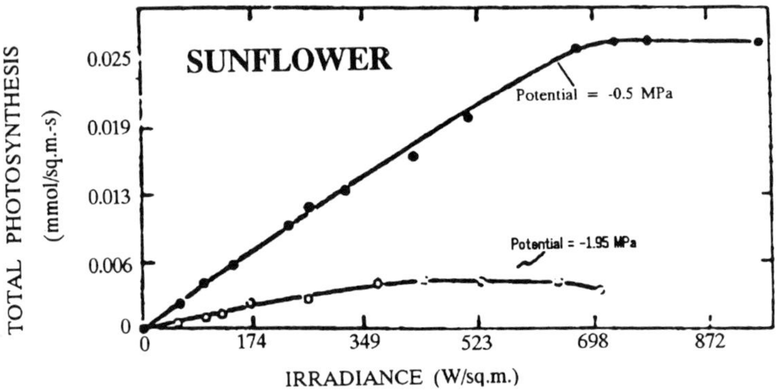

Fig. 5-7. Photosynthetic rate at different irradiances in an intact sunflower leaf at different potentials [adapted from Boyer, 1971b] (With permission of the *J. of Plant Physiology*).

potentials less than -0.6 to -1.2 MPa were obtained. Photosynthesis reduction may be due to stomatal closure, and one may surmise that most elongation and enlargement of plants occurs at night since, during the day, water potentials are low enough to prevent rapid cell enlargement –if any (see Fig. 5-4).

A basic study on one greenhouse crop was carried out by Aikin and Hanan in 1975. The results (Fig. 5-6) showed close correlation between PAR and water potential and CO_2 uptake, with potentials of -0.4 MPa and lower having significant effects on photosynthesis. Increasing quantum flux tended to compensate for lower water potentials, although, obviously, there was a close interaction between irradiance and resultant ψ_t. Boyer's work [1971a; b] on sunflower showed a remarkable effect of lower leaf water potential (ψ_l) on photosynthesis (Fig. 5-7). According to Boyer, there was no photosynthetic inhibition resulting from stomatal closure in sunflower. On the other hand, some individuals say that photosynthetic inhibition at low ψ_ts is the result of stomatal closure, preventing CO_2 uptake [i.e., Behboudian and van Holsteijn, 1977; van Holsteijn et al., 1977].

There are many other physiological processes in plants that can be affected by water stress. These have been reviewed by many authors [e.g., Boyer, 1985; Turner and Begg, 1981; Hsiao, 1973; etc.]. However, we are concerned with the economic return in greenhouse production. In practical terms, the effect of internal plant water potentials may be obviously severe (e.g., Fig. 5-8), where, through error, the grower failed to keep track of what

Fig. 5-8. Two examples of severe water stress encountered over short time periods. The roses on the left were growing in an inert medium and the irrigation system inadvertently shut off for 1 day. On the right, the grower failed to irrigate snapdragons in soil on a bright day, resulting in floret abortion on the spike.

Fig. 5-9. Carnations on the left subjected to greater stress as the result of lower irrigation frequency in an inert medium compared to the right. Responses to photoperiod, CO_2 fertilization, or higher temperature by the plants on the right will be greater than by those on the left.

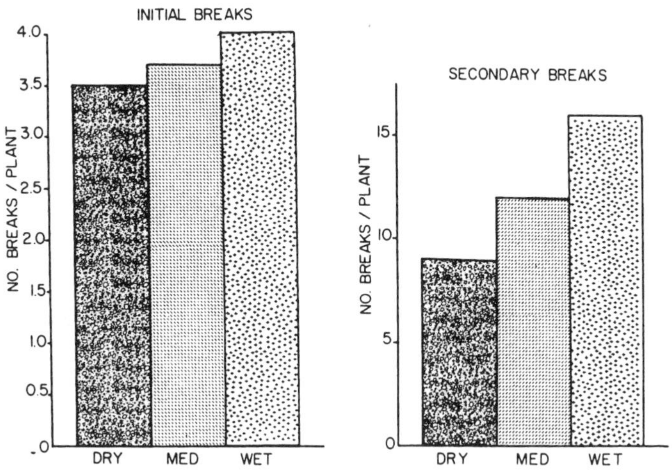

Fig. 5-10. Effect of water stress on side breaks produced by carnations for initial breaks from a single pinch (left) and subsequent breaks on the resulting stems for the second crop (right) [Hanan and Jasper, 1969].

Fig. 5-11. Effect of higher water stress (lower potential) on African violet and Boston fern (left) compared to the two plants on the right. Stress varied by changing relative humidity. Plants grown in an environmental chamber. Note that both treatments are saleable. The differences are not great.

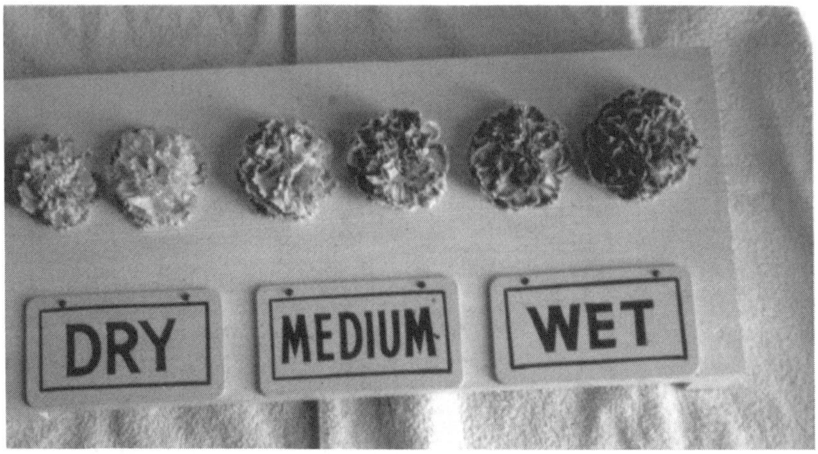

Fig. 5-12. Unusual effect of water stress on dye uptake in carnations. Cut flowers subjected to lower potentials probably have smaller diameter xylem vessels, which would restrict dye movement to a greater degree. Flowers on the left typical of "dry" growers.

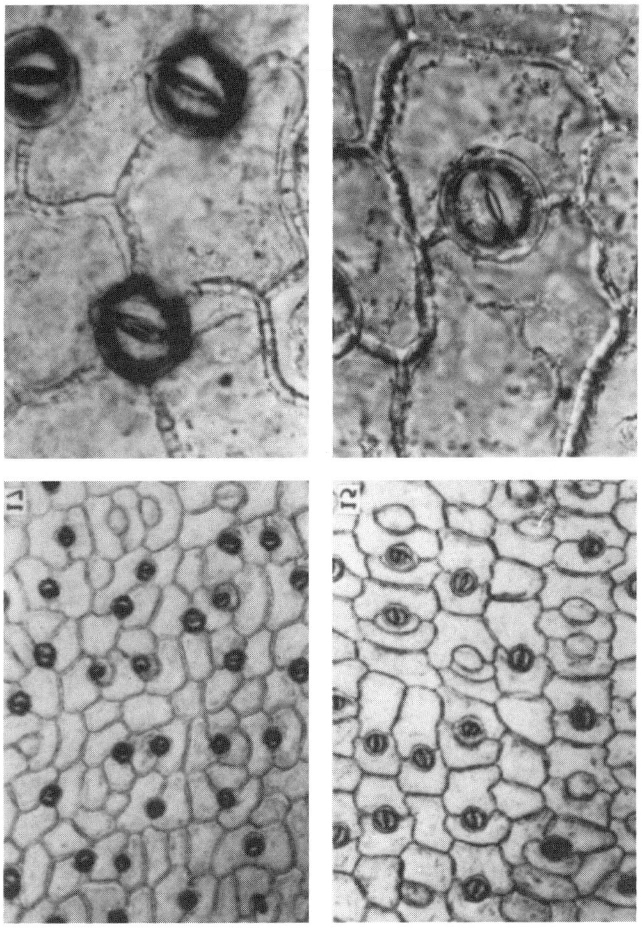

Fig. 5-13. Effect of lower internal water potential (left) on carnation stomata and stomatal density. Examples on left show smaller epidermal cells, with stomata set deeper within the tissue compared to those on the right, which were grown in an inert medium. This serves to illustrate that these plants subjected to an identical environment will not respond similarly.

was going on. Or, the results can be more subtle but just as significant as to economic return. For example, short stem lengths, due to low water supply (Fig. 5-9), will reduce the number of side breaks (i.e., roses, carnations, etc.) (Fig. 5-10), and result in smaller and fewer cucumbers, tomatoes, etc. Excessive demand for water in a dry environment will cause smaller plant sizes (Fig. 5-11). These are common results in commercial production. The effects can be highly subtle (Fig. 5-12). Subjecting plants to continued high stress results in morphological changes that will influence future responses to the environment (Fig. 5-13). With indoor decorative plants, acclimatization is highly important if the consumer's purchase is to provide satisfaction in usual dark and dry homes under winter conditions. As noted in Fig. 5-13, plants possess the ability to accommodate to water potentials imposed over extended periods, particularly in regard to drought resistance and response to saline conditions. Compare the carnations in Fig. 5-14 with those in Fig. 5-9. One might state with truth that the carnation is both drought resistant and tolerant to saline conditions –because the plant survives. However, the examples shown in Fig. 5-14 are utterly uneconomic in Colorado or California. Although specific effects of radiation and temperature on plant response have been discussed in Chapters 3 and 4, it is obvious that the practices to achieve the desired results will not be possible if the plants are similar to those typified in Fig. 5-14.

Nearly all the examples shown above deal with a single species. Nevertheless, the same general responses may be found for most important greenhouse crops [e.g., Friis-Nielsen, 1969; Phene et al., 1992; Plaut et al., 1973; Plaut and Zieslin, 1974; Letey and Blank, 1961; Hanan, 1969a; b; Brouwer, 1963; Peters and Runkles, 1967; etc.]. Judging from what has just been reviewed, one might conclude that water stress should be minimized wherever possible. This is not always desirable. Growing plants under conditions to prevent stress may result in abnormal growth [Heydecker et al., 1970], increased brittleness in some cultivars of species such as carnations [Hanan, 1972], fruit cracking [Gerhardt and Smith, 1945], or problems of calcium nutrition in tomatoes, leading to blossom-end rot [Bakker, 1990a; Tibbitts, 1979]. In fact, stress may be desirable with some crops (for example, Fig. 5-15), to tailor the crop to market requirements or to ensure the crop withstands rough handling. According to Boyer [1969], water potentials between -0.1 and -0.4 MPa are characteristic of the growth process. The tissue does not reach zero as long as growth can occur.

C. MEASUREMENT OF WATER POTENTIAL

The ability to measure internal plant water potential on a continuous, real-time basis would be an unparalleled advance in greenhouse climate control. ψ_t represents the integration of water supply and water

Fig. 5-14. Two examples of carnations adapted to saline conditions for survival. Note narrow leaves, short stems, and small flowers in the bench on the left. The left-hand plants were benched June and flowered for Christmas. With low salinity, such plants would flower in September and return crop in December in an arid region. The single plant on the right is nearly 1 year old. When subjected to concentrated solutes in the external substrate, many species can accumulate solutes against a concentration gradient that lowers the cellular water potential below that of the solution outside [Slatyer, 1961; Russell and Shorrocks, 1959]. Water will then flow into the cell in response. Obviously, however, there will be a lasting effect on growth and economic return. Note the salt crystals on some of the lower leaves, indicative of the high salts in the irrigation water.

Fig. 5-15. Effect of decreasing water stress (left to right) on spike formation and size in single-stem snapdragons. Under very low stress, floret spacing may be too great.

demand. Plant response to ψ_t is just as remarkable as is response to radiation, temperature, and nutrition. Adequate measurement of the internal water status represents the most serious challenge to environmental control.

Of the several laboratory procedures employed today, the psychrometric and pressure chamber methods are most common (Fig. 5-16, 5-17). The psychrometric system was devised originally by Spanner [1951] and Richards and Ogata [1958], and is based upon the relationship presented in Eq. 5.4. In the Spanner system, leaf tissue samples are placed in the chamber (Fig. 5-16, left) and the thermocouple cooled until water condenses on the junction. This is then allowed to evaporate as a miniature wet-bulb psychrometer, the amount of the temperature depression providing the ratio e/e^o, where e^o is

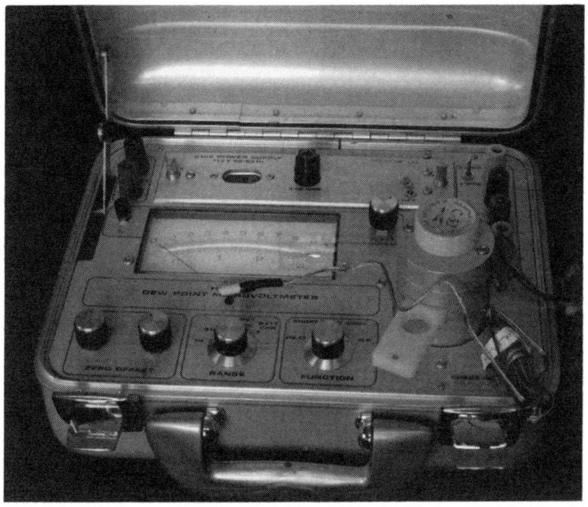

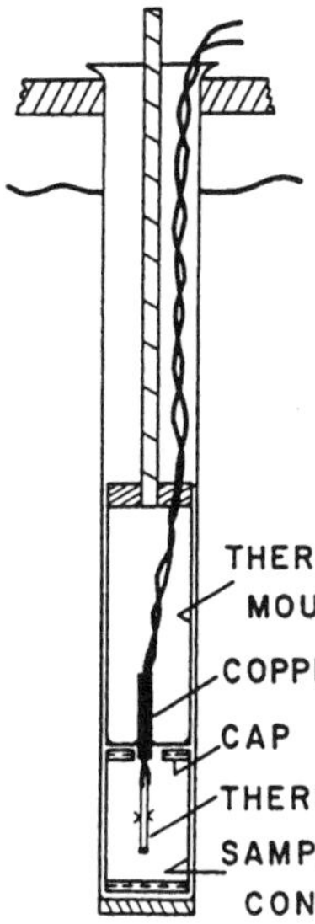

Fig. 5-16. **(Left)**. Older model of a thermocouple psychrometer with the sample chamber sitting on top to the right. This device uses Peltier cooling or dewpoint mode. The **right** diagram shows a system in which a drop of water is placed on the thermocouple and the chamber submerged in a water bath for temperature control.

Fig. 5-17. Pressure bomb with plant sample enclosed in a plastic bag to prevent drying. Pressure is gradually increased until water can be seen in the xylem of the cut end. Problems include compression of a stem with soft tissues, and the stem cannot be re-cut.

the saturation vapor pressure at the same temperature (see Table 5-1). The method is elegant as for its thermodynamic relationships, but extremely difficult in its requirements for appropriate temperature control and protocol [Boyer, 1969; Turner, 1981; Hanan, 1984]. Devices have been manufactured for determining ψ_t in the field without destructive sampling. However, in 1984, Shackel found that exposure of the psychrometer and sampled leaf to moderate irradiances resulted in substantial errors, combined with the difficulty in achieving isothermal conditions in the instrumentation. The range of e/e^o varies between 0.988 and 1.0 for most soil and tissue systems. For a change in a potential of 0.1 MPa, the amplification and recording system must be capable of showing differences less than ±1.0 μV, or about 5 to 6 μV per MPa. Newer instruments have been greatly improved, and for sampling under greenhouse conditions, the commonly high potentials allow rapid equilibration for readings.

The second common method, the pressure chamber revised by Scholander et al. [1965] (Fig. 5-17), allows an average potential to be obtained on an entire shoot by applying a counter pressure to the sample sufficient to force water back through the xylem to the exposed, cut stem projecting out of the chamber. Although there are difficulties with this method [e.g., Barrs et al., 1970; Ike et al.,

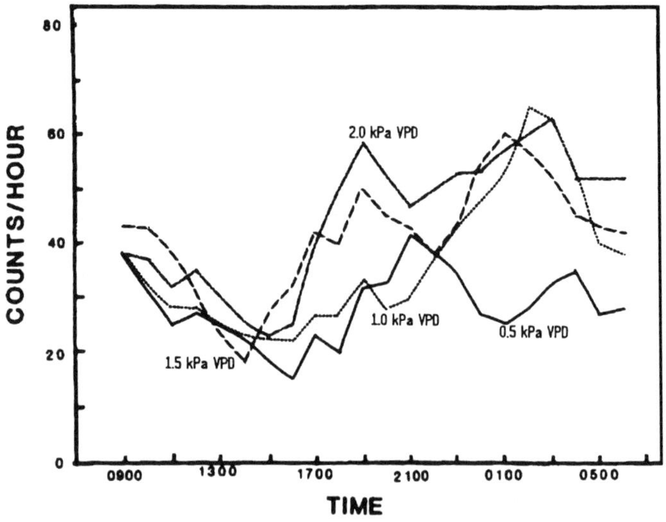

Fig. 5-18. Hourly cavitation rates in single stems of 'Forever Yours' roses grown in rockwool and controlled over successive 24 hr periods at different vapor pressure deficits (VPD). Curves subjected to moving means smoothing process with resultant loss of first and last data points [Hanan, 1987].

Table 5-3. Approximate classification of methods employed in greenhouses for water and nutrient supply of economic crops with some general characteristics of each.

Type	Remarks
Field soils	Mineral, specific gravity ≈ 2.65, bulk density (BD) >0.7 g cm^{-3}, infiltration < 30 cm hr^{-1}, porosity <60%, moisture holding capacity <30%, fertility usually low, variable cation exchange capacity (CEC), unsaturated hydraulic conductivity relatively high compared to all others.
Modified soils	Mixtures with field soils, BD 0.3 to 0.6 g cm^{-3}, infiltration 100 to 700 cm hr^{-1}, porosity 60 to 80%, MHC >30 %, CEC variable, fertility variable, hydraulic conductivity practically zero at potentials <-0.03 MPa.
Soilless media	BD <0.1 to 0.3 g cm^{-3}, infiltration >700 cm hr^{-1}, porosity 80 to 95%, MHC >50%, hydraulic conductivity zero at potentials <-0.001 MPa, CEC dependent upon medium, fertility from zero to low.
Hydroponics (recirculating systems, NFT, rockwool, sand, gravel, etc.)	May not have any substrate, CEC usually zero, infiltration, BD and MHC not applicable. Hydraulic conductivity not applicable.
Aeroponics	Roots enclosed and sprayed with water and nutrient mixture.

1978], the method is much simpler than the thermocouple psychrometer and reasonably accurate with the proper procedure.

Simpler systems include the relative turgidity method where tissue samples are brought to equilibrium in water. The difference between fresh and dry weight is divided by the difference between turgid and dry weight, giving the relative water content. Still other systems examined include beta-ray gauging [Challa, 1975], changes in stem diameter with linear variable transformers [Namken et al., 1969], determining sap flow with heat pulses [Marshall, 1958; Valancogne and Nasr, 1989], correlations with stomatal conductances, nuclear magnetic resonance (NMR) [van As et al., 1985], pressure probes capable of showing potentials within single cells, or the recording of cavitations in plant stems (Fig. 5-18). All these methods have severe problems from the standpoint of a practical grower. The most common is destructive sampling. Other systems require expensive equipment, and all require close attention to appropriate sampling of representative plants. The environmental diversity of

Fig. 5-19. Schematic of a root cross-section in a substrate, suggesting the arrangement of soil particles and resulting voids filled with air or water.

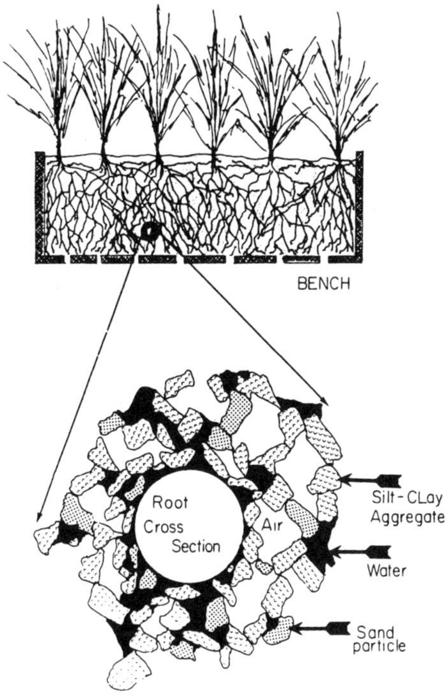

a greenhouse means that selecting a typical specimen that would give the grower assurance that results are typical of the vegetative canopy will be difficult.

A number of approaches have been made to indirect methods such as determining the vegetative canopy temperature with infrared thermometry (Fig. 4-6). The idea is that as stress increases, transpiration will be reduced, and the canopy temperature will increase. With a suitable field-of-view, most of the canopy can be sensed by the instrument, providing an integrated mean temperature of the vegetation. Irrigation scheduling with IR thermometry in the field has been intensively investigated by many individuals [e.g., Walker and Hatfield, 1983; Aston and van Bavel, 1972; Glenn et al., 1989; Weigand and Namken, 1966; Jackson et al., 1977; Clawson and Blad, 1982; etc.]. A number of ideas have been introduced such as stress degree days (SDD), canopy temperature variability (CTV), and temperature stress day (TSD). Usually, canopy temperature measurements are made at fixed times during the day and the difference between leaf and air temperature correlated with stress or comparisons made with a well-watered crop as a control. In the latter case, Clawson and Blad [1982] found the onset of stress when the range of six measurements in a plot exceeded the well-watered control by 0.7 C. Jackson et al. [1977] concluded that if SDD became positive, stress was occurring and the field should be irrigated. In other words, the decision to irrigate was whether SDD was positive or negative.

Fig. 5-20. General relationship between bulk density (g cm^{-3}) and porosity (air and water filled voids) for various substrates, taken from Hanan et al. [1981], White [1974], and Johnson [1968].

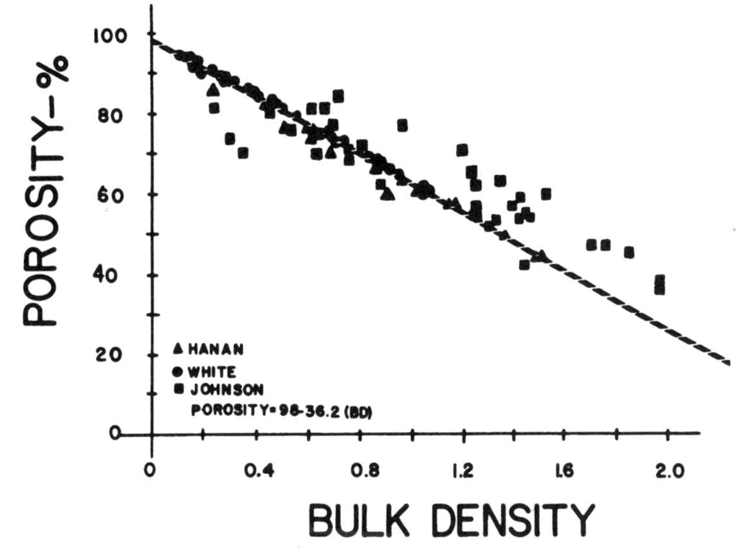

The range of temperatures and factors influencing canopy temperature has been found quite variable. In at least one instance, the author concluded that IR thermometry could not be employed for forest trees without knowledge of the specific physiological behavior of the species [Körner, 1985]. Where the temperature differences are small, measurement accuracy of the instrumentation can be a chief difficulty, requiring complicated calibration procedures [Fuchs and Tanner, 1966]. In their discussion, Walker and Hatfield [1983] stated that stress was more easily detected at high vapor pressure deficits. If VPD was less than 840 Pa, then the corresponding change in the leaf-to-air temperature difference was only 1.0 C, which approached the limit of measurement accuracy. These findings suggested that IR methods were not suitable for crop stress determinations in cool or humid areas. In his discussion of leaf temperature and transpiration, Gates [1964] concluded that a large change in transpiration must take place to produce a significant change in energy balance and in the resulting leaf temperature. In Chapter 4, Eq. 4.1 showed that many factors can determine the energy balance of a plant. In a definitive study of greenhouse transpiration, Stanghellini [1987] stated that transpiration rate and temperature excess are explicit functions of five variables: net radiation and heat storage, air temperature, saturation deficit, and external and internal resistances to vapor flow. In view of the comments by Gates, Stanghellini, and Walker and Hatfield, it is doubtful that canopy temperature measurements are of any use in assessing plant water potential in greenhouses.

This section ends on a negative note in the sense that all known methods for determining plant water potential are not of immediate use to greenhouse growers –although its measurement would advance efficient production to a greater extent than any other climate control parameter.

III. WATER SUPPLY

A particular problem in greenhouse production is the number of variations available to the grower in providing water and nutrition to an economic crop. As means to outline the major methods, I have arranged them into categories with some brief remarks (Table 5-3). The progression is from the simplest and cheapest method of production to the most expensive with the highest technical requirements. Most greenhouse cut-flower and vegetable production is in the ground (field soils), especially in developing countries. Bedding and pot plant production requires soil modification of some type to be successful. In the Netherlands, van Os [1982a] predicted a total area of about 500 ha in soilless culture for 1982, which was more than double the amount in 1981, or roughly 5% of the total greenhouse area. According to Roe [1986], hydroponics (NFT) accounted for some 34 ha in the U.K. in 1985 while production in rockwool accounted for about 85 ha. The Dutch have mandated that by 2000, all cultivation must take place in closed circuits. By 1994, 30% of all vegetable production must be with a closed recirculation system [Meijaard, 1995]. Iwata [1992] gave a figure of about 373 ha of vegetable and ornamental production in some type of soilless culture, or about 1% of the total Japanese growing area. Capital costs of rockwool and NFT were given by Roe [1986] as more than \$37000 ha^{-1} for rockwool and more than \$62000 ha^{-1} for NFT. Operating costs were about \$23000 ha^{-1} for rockwool renewed yearly and \$19000 ha^{-1} for NFT. The difference in capital requirements between ground production in field soils and hydroponic systems is highly significant.

Although there appears a move toward soilless recirculation systems as exemplified by The Netherlands, few growers have the technical competence required for advanced hydroponic systems in spite of numerous "how-to" publications. Despite all the publicity various advanced systems have received in recent years, the main thrust of greenhouse production remains in the ground or in modified soils.

Another aspect of root substrates –or the lack of them– is that plants can be grown in any mixture or in any situation that permits root growth and an adequate supply of water, oxygen, and nutrients. The decision is economically based, biased toward the least expensive and least technically demanding if resultant yields permit suitable profits. There are situations where more expensive systems are required, either because suitable soil is unavailable or the competitive situation requires the additional production usually achievable with most hydroponic methods.

A. THE ROOT SYSTEM

Maximum water uptake by plant roots generally occurs within the first 1 to 4 cm of the elongating root tip [Boyer, 1985; Slavík, 1965]. In soils, the absorbing surface may be increased by root hair production, but root hairs are seldom seen in flowing hydroponic culture. It may be, despite the progressive suberization of the root epidermis behind the main absorption zone, that considerable water uptake can occur throughout the root system in hydroponics. Water moves from the epidermis through the loose cells of the cortex, through the single layer of cells in the cylinder called the endodermis, to the central vascular cylinder in which lies the xylem. Some

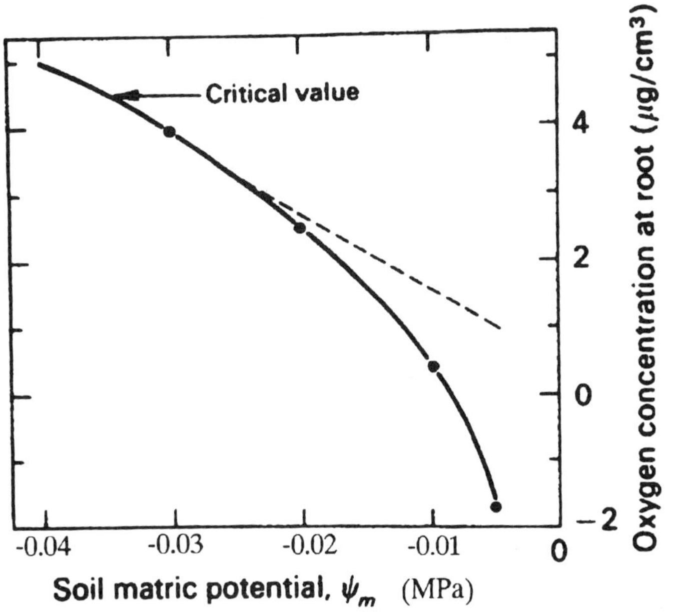

Fig. 5-21. Oxygen concentration at the root surface as influenced by matric potential, assuming that all roots consume oxygen at the same rate regardless of aeration status. If the uptake rate does decrease, the dotted line would be more likely. [Adapted by Taylor and Ashcroft, 1972, from Weigand and Lemon's data, 1958]. Note the curve was derived for field soils. The "critical value" is the minimum O_2 level for root metabolism (From *Physical Edaphology: The Physics of Irrigated and Nonirrigated Soils* by Taylor and Ashcroft, ©1972 by W.H. Freeman and Co. Used with permission).

authors [e.g., Boyer, 1985] consider the endodermis a major restriction to water uptake. However, severe restriction can also occur in the soil matrix [Gardner, 1960], and also the transfer from the cell matrix in the leaf to the outside air. Transpiration is driven by water evaporating from the microfibril pores in the cell wall matrix of the leaf [Nobel, 1991]. As indicated (Eq. 5.7-5.9), water moves according to a potential gradient, and in soils, the situation of the root with respect to soil particles and water is suggested in Fig. 5-19. One factor determining water availability in soils is continuous root growth to new water sources since it is doubtful, even in field soils with high rates of hydraulic conductivity, that the root's sphere of influence extends much beyond 2 to 3 cm from the root axis [Gardner, 1960]. In porous mixtures, the distance is probably less than a few millimeters. Root extension can vary between 2 and 5 cm per day [Nakayama and van Bavel, 1963], with depths to 10 m for alfalfa or 3 m for sugar beets [Slavlk, 1965]. Some plants have been found to possess more than 600 km of roots. Thus, under usual conditions, the plant will produce sufficient root surfaces to meet transpiration requirements. Under drought conditions, the root-top ratio will increase (greater root production). It is a common observation in greenhouses that plants often wilt on bright days following several dark days in which the root-shoot ratio has decreased under low water demand, to the point that demand cannot be met on the bright day until the system has had time to readjust. Obviously, a living, growing, non-diseased root system, with sufficient absorbing surface area, is required for adequate water supply. The statement, attributed several years ago to some well-known individuals, that one does not sell the root system, unfairly de-emphasizes the role played by a healthy root system in greenhouse production.

B. FACTORS LIMITING ROOT ACTIVITY

One of the more easily measured soil parameters is bulk density (BD) (mass of soil plus voids per unit volume, g cm^{-3}). One can find a distinct relationship between BD and total porosity of substrates as indicated in Fig. 5-20, prepared from many sources, showing the ranges of BD and porosity (total volume of water and air) one may find with most greenhouse media. Table 5-3 provides the usual values one may encounter for the different major classifications. From Fig. 5-19, one can also discern that water must be in contact with the root if it is to be absorbed in meaningful quantities. Vapor transfer in the soil matrix is, for greenhouse culture, insignificant. The fact that oxygen diffusion in water is about 10^4 times slower than in air says that the thickness of the water film is more important than an indication of porosity as shown in Fig. 5-20, or the amount of air-

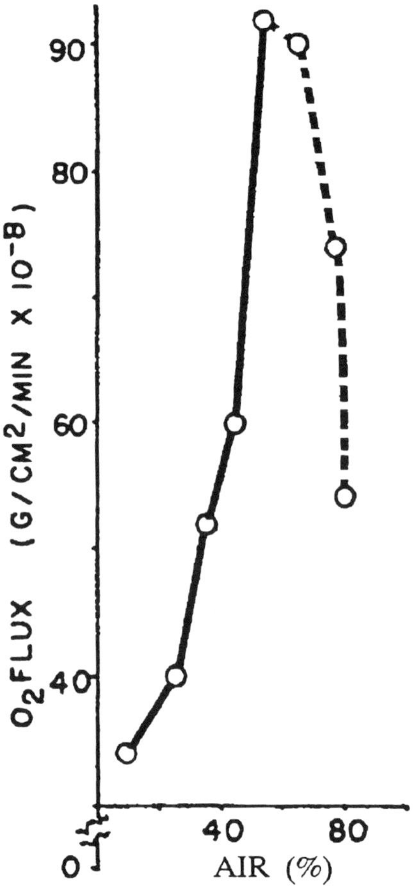

Fig. 5-22. Oxygen flux rates to a platinum microelectrode simulating the plant root in a peat moss-perlite medium as influenced by air porosity. Snapdragons were the test crop [Hanan and Langhans, 1964b]. The decrease in flux rate beyond about 55% was likely due to rupture of water films about the electrode, preventing electron flow in the measurement.

filled voids at maximum water content. Soil with no air-filled voids is a poor candidate for most economic greenhouse plants. As an example, Bugbee and Frink [1986] examined air porosity over a range from 0.01 to 0.336 cm^3 cm^{-3} for a peat-vermiculite medium, showing that air porosities less than 0.113 cm^3 cm^{-3} (11.3%) limited growth of many greenhouse pot plants. At 0.336 cm^3 cm^{-3}, growth was also reduced. I suggest the latter might have been inadequate water supply. However, Hanan [1964a] showed that the air-filled voids in shallow (<30 cm) greenhouse media are usually close to atmospheric oxygen concentrations, and his conclusions were to the effect that the thickness of the water film was more important than air porosity. Generally, a root substrate should have a minimum 0.10 to 0.15 cm^3 cm^{-3} air porosity (10 to 15%) at maximum water content [Hanan, 1964a; Hanan and Langhans, 1963].

The influence of water films around the root on oxygen supply has been extensively examined by California workers [e.g., Letey et al., 1964; 1962; 1961; Stolzy et al., 1961; Rickman et al., 1965], using the platinum microelectrode for simulating a plant root as devised by Lemon and Erickson [1955]. The critical oxygen concentration as a function of soil matric potential is plotted in Fig. 5-21, showing that, at soil water potentials most commonly approaching the situation in greenhouses, oxygen concentration at the root can be close to zero. A typical result with a greenhouse medium such as peat moss-perlite may be noted in Fig. 5-22. Although the California investigators settled on a minimum oxygen flux rate of 20 x 10^{-8} g oxygen per cm^2 min^{-1} as necessary for reasonable plant growth in the field, Hanan and Langhans [1963] felt that a value of 30 x 10^{-8} g cm^{-2} min^{-1} would be safer. Oxygen supply can be just as critical in some hydroponic systems such as the nutrient film technique (NFT) or aeroponics. Cooper [1979] states that the flowing nutrient solution should be only a few millimeters deep so that most of the root system is above the solution. According to Soffer and Burger [1988], oxygen dissolved in the water is essential to root formation and growth. Whether grown in thin films or bulk solution cultures, some attention should be given to adequately aerating the root system, either by bubbling air through the solution or vigorously stirring it.

Low oxygen supply is particularly critical when root systems are subjected to limited root growth by compaction (Fig. 5-23). As bulk density increases (Fig. 5-24), root growth is correspondingly slowed. There can be a synergistic effect, especially in field soils, between oxygen supply and high soil strength. Many soils shrink and swell with changes in water content. Shrinkage as the soil dries will restrict root growth (Fig. 5-25). Other systems for determining soil strength and root restriction have been examined [e.g., Barley et al., 1965; Taylor and Gardner, 1963; Phillips and Kirkham, 1962]. The data show, however, that as a reasonable estimate, bulk densities above 0.9 g cm^{-3} are very likely to have deleterious effects on root penetration. Field soil used for production is particularly susceptible to compaction as equipment is run over it. Hardpans are likely to form with time, restricting water movement and root penetration although the upper 30 cm are carefully cultivated.

Several other factors can limit root activity, such as inadequate water and nutrition supply and disease. These will be discussed in the next sections.

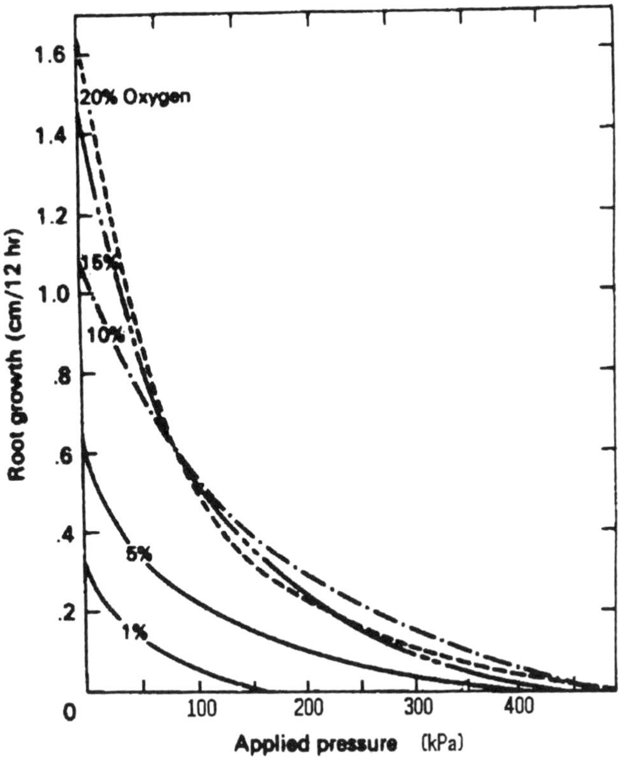

Fig. 5-23. Effect of applied pressure on root growth at different oxygen concentrations. The pressure would be indicative of restriction to growth [Gill and Miller, 1957]. Roots grown in an ideal mixture to which the indicated pressures were applied (With permission of the *Amer. Soc. of Agronomy*).

C. FIELD SOILS

Apart from the physical structure of a soil, three factors determine water availability in any medium: 1) total volume, 2) moisture content, and 3) unsaturated hydraulic conductivity.

1. Water Content and Water Potential

Fig. 5-19 shows pore spaces in most media to be highly irregular, and it is the size of these pores, and their connections, that determine the ability of substrates to retain water against gravitational forces. Water will also be absorbed on the surfaces of the particles so that surface area will also be an important factor. For

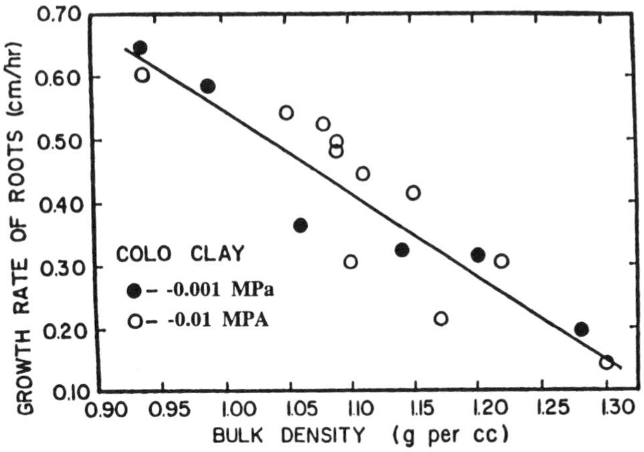

Fig. 5-24. Relationship between bulk density and root growth rate at 2 soil water potentials in a Colo clay [Phillips and Kirkham, 1962] (With permission of the *Amer. Soc. of Agronomy*).

example, if $1 cm^{-3}$ ($6\ cm^2$ area) is subdivided sufficiently (each particle 1 μm on a side), the total surface area of the resultant 10^{12} particles will be $6\ m^2$. Depending upon the type of clay in a soil, particle sizes will vary from 0.01 to 5.0 μm [Brady, 1974]. $1\ cm^3$ of 0.01 μm clay colloids will have an external surface area equivalent to 6 ha. The external surface area of 1 g of clay is 1000 times that of 1 g of course sand. To remove water from these surfaces, and from the soil pores, requires work, expressed as matric water potential (Eq. 5.6), which must be higher than ψ_t in the plant if water is to flow into the plant system. The relationship between soil water potential and water content can be plotted as a soil water characteristic curve (Fig. 5-26). Unfortunately, such curves are merely approximate, depending upon whether the curve was plotted on a wetting cycle or a drying cycle. Because of pore size variation, emptying a pore can require a higher suction force (lower potential) than filling the same pore –the variation called hysteresis. A clay soil may contain more water than a sandy soil, but most of it may be unavailable due to the low potentials required to remove it from the soil matrix. The curves in Fig. 5-26 are characteristic of many field soils, the shapes at high potentials largely due to pore characteristics. Similar curves may be obtained for any porous substrate. There is no general relationship expressing the water content of all soils to water potential, nor is there a unique relation for any given soil [Gardner, 1965b].

Fig. 5-25. Effect of bulk density and water potential on root penetration in an Amarillo fine sandy loam soil. The differences in BD between this figure and Fig. 5-24 indicate that soil type has a considerable influence on soil strength. Bulk density in this respect is not a good indicator of compaction [Williams and Wilkins 1963, Taylor, S.A. and H.R. Gardner. *Soil Science,* 96:153-156].

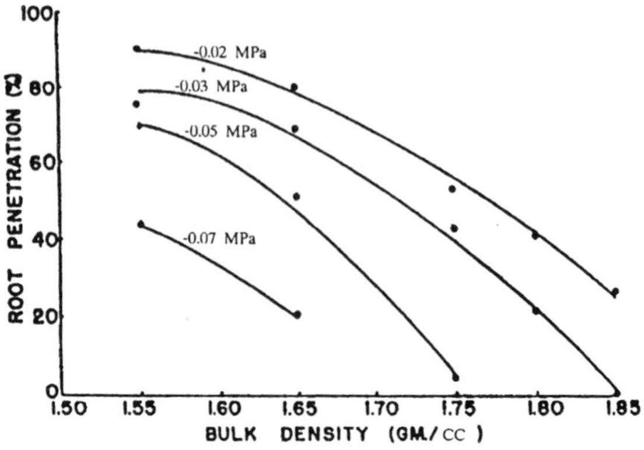

Fig. 5-26. Water characteristic curves for several typical field soils. Note moisture content based upon mass of dry soil. These curves are very general for broad soil classifications [From: *Physical Edaphology: The Physics of Irrigated and Nonirrigated Soils*, by Taylor and Ashcroft, ©1972, by W.H. Freeman and Co. Used with permission].

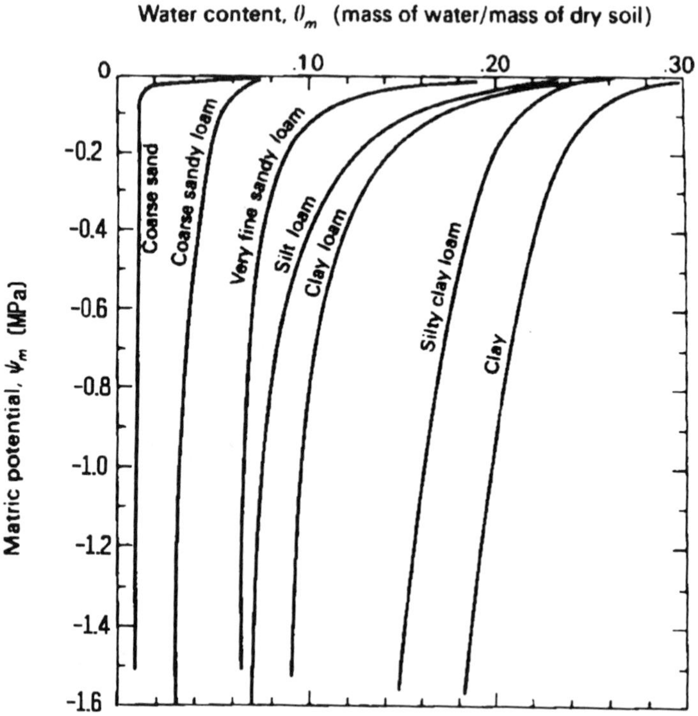

Note, in Fig. 5-26, that moisture content is expressed as grams per gram dry soil. With the high specific gravity of most mineral soils, drying a soil sample at 105 C, and determining the weight loss, is a common and simple gravimetric procedure which accords closely to $cm^3\ cm^{-3}$ on a volume basis. However, this can lead to serious problems with high organic soils and most modified or soilless culture substrates. Moisture content of such soils, expressed on the basis of dry weight, can result in moisture contents as high as 3000% [Boelter and Blake, 1964]. Moisture content should be expressed on a volume basis that requires undisturbed soil cores of known volume. When done so, Fig. 5-27 compares a typical mineral field soil with some highly organic peats.

Many soils in the field are generally considered at maximum moisture holding capacity at matric suctions around -0.02 to -0.03 MPa. The older literature calls this field capacity, with the lower limit of water availability at -1.5 MPa (permanent wilting point). As pointed out by Slatyer [1967a] and others [Turner and Begg, 1981; Ritchie, 1981a; b; Slavík, 1965; Hadas, 1973; etc.], these limits of so-called freely available water are difficult

Fig. 5-27. Water characteristic curves of three peats compared with a mineral soil with moisture contents expressed as volume water per unit bulk volume soil [Boelter and Blake, 1964]. If the values for peat were given in mass per unit mass, the wet portions of the curves would be far in excess of 1.0 (>100 %) (With permission of the *Amer. Soc. of Agronomy).*

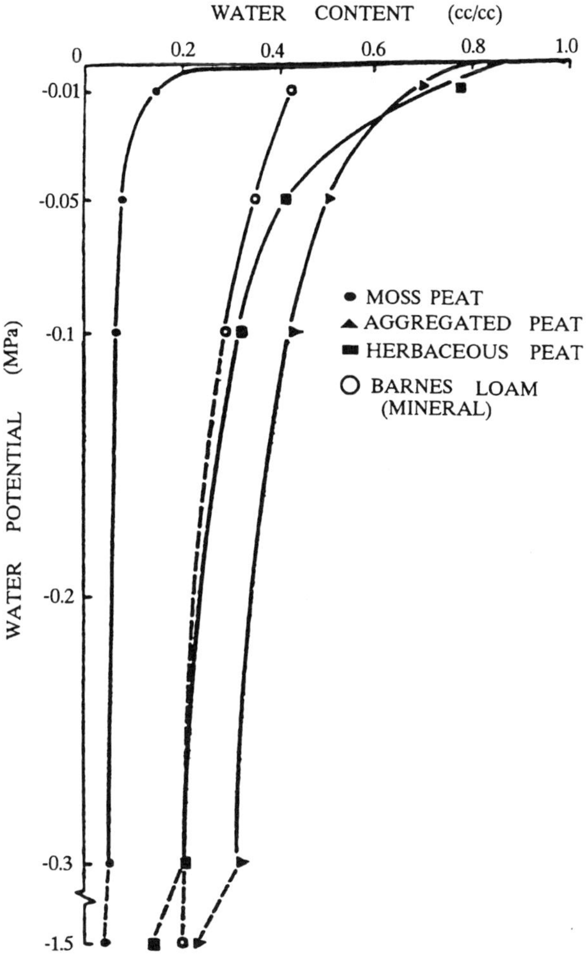

to establish. While useful in the field, there are other factors to consider such as the transpiration rate, soil water conductivity, and extension of the root system. This is especially so in the greenhouse with modified soils or soilless culture where the minimum medium water potential is seldom less than -0.03 to -0.05 MPa. The highest potential at maximum moisture content of greenhouse soils can be more than -0.001 MPa. One may note in Figs. 5-26 and 5-27 that, for several substrates such as sand and peats, a large part of the water is removed at -0.03 MPa; whereas, for silt and clay soils, sufficient water may be retained at -0.03 MPa to reduce aeration. These curves also emphasize the criticality of irrigation timing of common mixtures employed in greenhouse production, and why growers are generally more confident with field soils –because failure to supply water until Monday, rather than irrigating Saturday or Sunday, is less likely to affect crop growth.

2. Hydraulic Conductivity

For several years, following Viehmeyer and Hendrickson's work in California, water held in the soil was considered as freely available to the plant between field capacity and PWP. However, Gardner's work in the 1960s showed that problems of **rate** of water flow to the root system with the transpiration rate could result in water restriction even at maximum moisture holding capacity [Gardner, 1960; Gardner and Nieman, 1964; Gardner and Ehlig, 1962a, b]. Gardner developed formulae based upon the root as a linear sink, showing that when the soil suction was above -0.06 MPa, the impedance to water flow was largely in the plant, while if suction was less than -0.1 to -0.2 MPa, the soil became the limiting factor. In more recent publications [e.g., Stanhill and Vaadia, 1967; Hadas, 1973], the authors use the terms resistance or conductivity. In any case, typical curves relating soil water conductivity with soil water potential are shown in Fig. 5-28. Unsaturated hydraulic conductivity in most field soils declines rapidly at potentials less than -0.01 MPa as the water is removed from the pores and water flow must be more tortuous through a greater distance. In highly porous media, such as peat moss, conductivity can be nearly zero at soil water potentials as high as -0.001 MPa. In fact, starting small seedlings is often easier (restricted root system) in a heavy soil compared with a deep, porous medium. As maximum root growth has not been established by the seedling, severe stress can occur in porous peat media although water can be squeezed from a sample by hand.

The relationship between suction at the root surface and distance from the root is depicted in Fig. 5-29 as calculated by Gardner[1963] at a constant water uptake of 0.1 cm^3 dy^{-1}. The effect of water uptake rates on suction at the plant root is also shown for 2 soils in Fig. 5-30 –indicating that soil type plays a marked role.

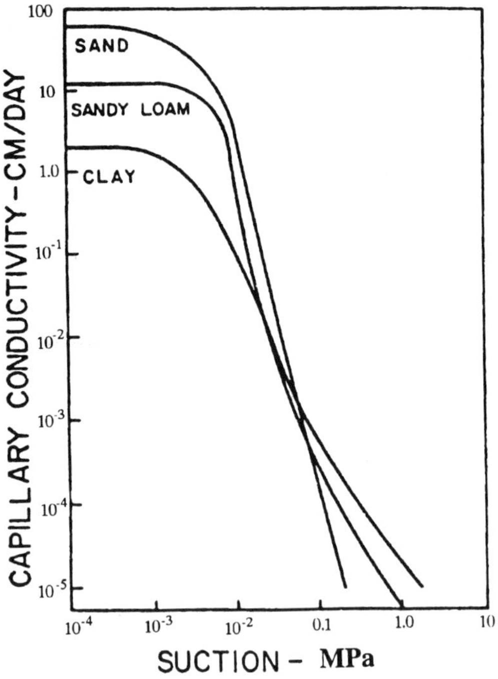

Fig. 5-28. Capillary conductivity in three field soils as a function of matric water potential. Note the logarithmic scale on the left axis [Gardner, 1965]. Between -0.001 and -0.01 MPa, conductivity rate decreases dramatically (With permission of the *Amer. Soc. of Agronomy*).

The practical implication of the relationship between moisture content and conductivity, and their effects on transpiration, was shown by Denmead and Shaw [1962] (Fig. 5-31). As water demand increases, soil water content must be higher if there is to be no restriction to water demand, and, consequently, increased plant stress. A grower in field soils must constantly balance transpiration rates with soil water content without causing oxygen deficiencies. Since this is difficult at best, crops in field soils will usually be subjected to higher water stress with concomitant reduced growth compared to modified soils or hydroponic systems. The relationship between soil water potential, plant water potential, and calculated water potential at the plant root is suggested in Fig. 5-32 as the soil dries out over a period of several days. Previous discussion (Section 5.II.B) showed that plant water potentials of -0.1 MPa are capable of significantly affecting elongation or cell enlargement. Potentials much higher than those at which wilting occurs will reduce growth.

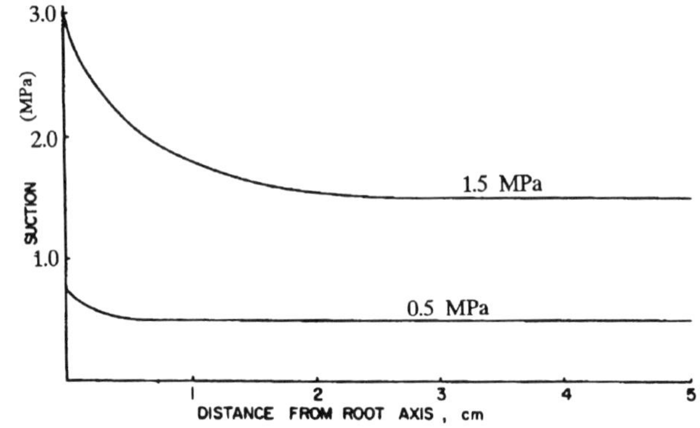

Fig. 5-29. Calculated soil water suction as a function of distance from the root, using capillary conductivity values to provide a constant water uptake of 0.1 cm^3 dy^{-1} cm^{-1} root length when initial soil water potential was -0.5 or -1.5 MPa [Williams and Wilkins, ©1960, Gardner, W.R. *Soil Science*, 89:63-73].

3. Infiltration

Infiltration refers to the penetration of water into a soil from rainfall or irrigation. In general, when first wetted, water will flow into the matrix at a high rate. Afterwards, the rate becomes less, particularly if the soil particles swell upon wetting. For outside conditions, an infiltration rate is important since this will determine the runoff and the wetting depth for a given amount of water on the surface of the soil. The infiltration rate is a major distinguishing characteristic between field soils and modified soil or soilless mixtures. The rate seldom much exceeds 30 cm water depth per hour for field soils (Fig. 5-33), whereas some soilless mixtures such as peat-moss and perlite can have infiltration rates greater than 1000 cm hr^{-1}.

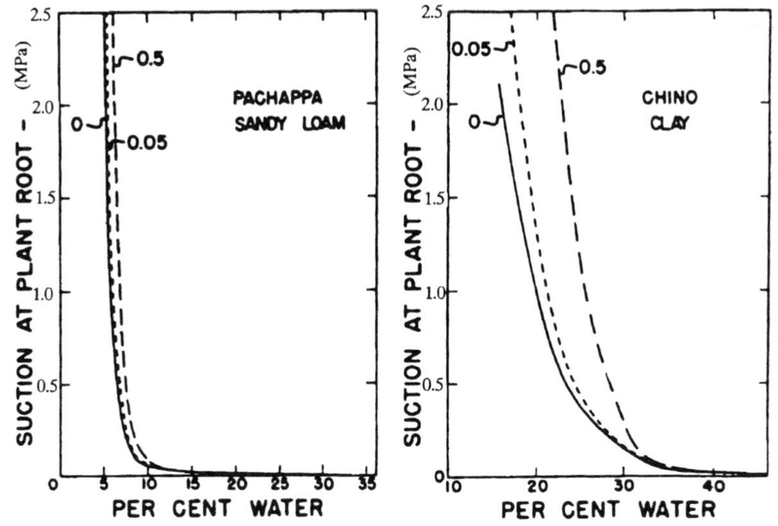

Fig. 5-30. Water potential at the plant root as a function of soil water content at various rates of water uptake ($cm^3dy^{-1}cm^{-1}$ root length) for 2 different field soils. Note the percent water scales for the 2 soils are different [Williams and Wilkins, ©1960, Gardner, W.R., *Soil Science*, 89:63-70].

Flood irrigation is seldom practiced in greenhouse cultivation –even if one assumes that one can flood a highly porous, freely draining substrate. With trickle irrigation systems, the low infiltration rates of field soils are not often a significant problem.

Fig. 5-31. Actual transpiration rate of a corn crop as influenced by soil moisture content [Denmead and Shaw, 1962]. Note that a low water demand allows the soil to become relatively dry before affecting the transpiration rate through possible stomatal closure. This figure shows clearly the relationship between soil water supply and the water demand as they affect the rate of water loss. The transpiration reduction indicates water stress in the plant (With permission of the *Amer. Soc. of Agronomy*).

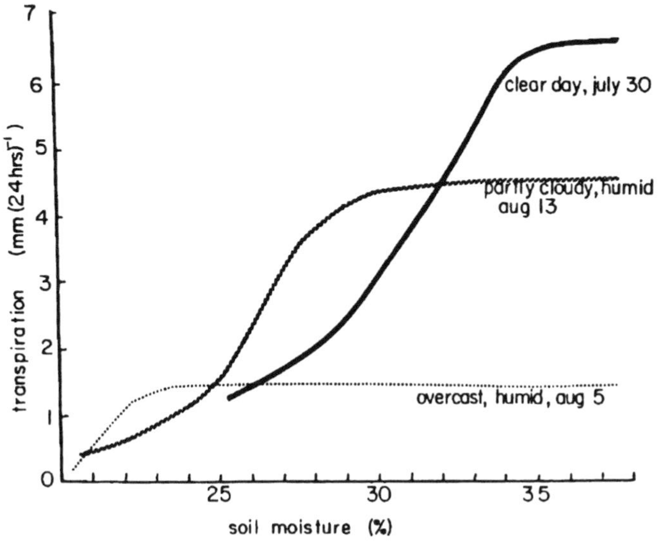

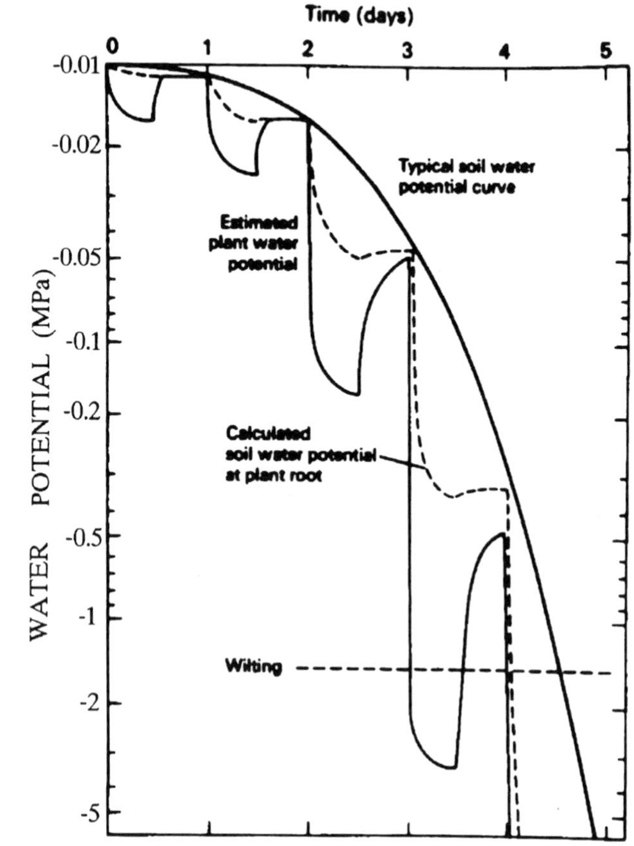

Fig. 5-32. Daily changes in water potentials of plant and soil as the soil dries. The curve for plant ψ_t is based on the assumption that transpiration begins at sunrise and continues at an average rate for 12 hr, then stops, based on data from Taylor, 1964 [From: *Physical Edaphology: The Physics of Irrigated and Nonirrigated Soils* by Taylor and Ashcroft ©1972 by W.H. Freeman and Co. Used with permission]

4. Advantages and Disadvantages of Field Soils

Growing in a field soil has three advantages compared with more esoteric procedures: 1) it is there to be used, 2) assuming no high-density regions, the volume is such as to allow unrestricted root growth, and 3) it is least expensive to grow in. As an example, assume a modified soil mixture in a pot containing 500 cm^3 with an 80% porosity, a depth of 13 cm, so that at container capacity, the moisture holding capacity is 50% or 200 cm^3 of water. Of this water, 100 cm^3 can be removed before plant water potential declines to unacceptable levels. On the other hand, assume rose plants are in the ground at a density of 30 m^{-2}. The soil porosity is 40%, maximum moisture holding capacity is 30% and the soil depth is 30 cm. Each plant's root system will have access to about 10000 cm^{-3}, containing 3000 cm^3 water. Even if only one-third of the total water is readily available for each rose root system, 1000 cm^3 water are there to be extracted, especially with a rapidly extending root system that is essentially unconfined. The same situation is also applicable to nutrition. The point is that growing in a "large" volume as represented by field soils requires remarkably different cultural practices such as reduced irrigation frequency and lower fertilizer applications.

A particular problem is the wide diversity of soil types that may be found, ranging from the typical red, alkaline clay soils of arid regions to black, acidic, and often highly organic, soils of humid climatic areas. There is no standardization. Sometimes, where the native soil is totally unacceptable, soils or sands may be imported. Since a hectare of soil, 30 cm deep, with a bulk density of 0.7 g cm^{-3} will weigh 2100 metric tons, the cost is usually prohibitive. This large volume also leads to other difficulties in any attempt to modify the soil. As a rule-of-thumb, significant physical changes in a soil are made when at least one-third to one-half of the final volume consists of whatever amendment used. Attempts to change drainage and aeration with a few truckloads will seldom provide benefits. In fact, adding sand to an adobe clay soil is likely to turn it into cement. A grower, therefore, using a field soil, while less likely to make a catastrophic mistake, is inherently limited to lower yields compared to the potential of modified substrates or hydroponics. Significant changes in physical and chemical properties of field soils are much more difficult to achieve compared with limited volume container production.

5. Disease Problems in Field Soils

A frequently overriding disadvantage to growing in the ground is the fact that the soil is unconfined. It is

Fig. 5-33. Infiltration curves of several soils: **A**. Honeoye gravely silt loam; **B**. Aiken clay loam; **C**. Crown sandy loam; **D**. Palouse silt loam; **E**. Cecil clay loam , using Free et al's data [1940]. Infiltration is in cm of water depth, showing that all of these soils have infiltration rates of about 30 cm hr^{-1} (Honeoye) or less [From: *Physical Edaphology: The Physics of Irrigated and Nonirrigated Soils* by Taylor and Ashcroft, ©1972 by W.H. Freeman and Co. Used with permission].

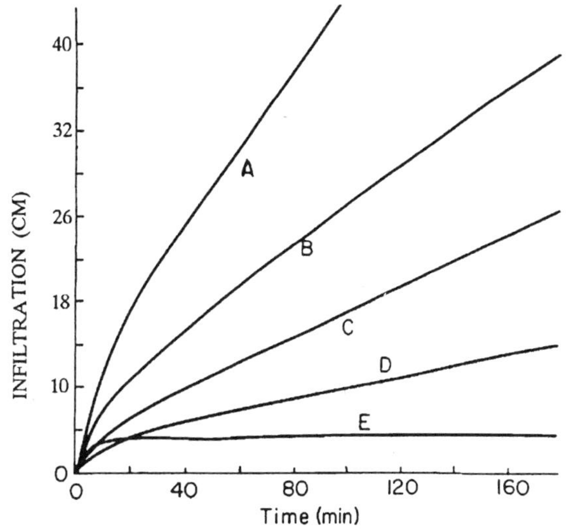

extremely difficult to control many soil-borne, pathogenic organisms since the substrate volume, for all practical purposes, is infinite. Standard procedures for chemical sterilization or steam pasteurization are suitable for a short term only. With usual systems of injecting steam or chemicals into the upper surface, it is doubtful that eradication goes deeper than 15 cm inasmuch as there is no way to displace the air ahead of the steam front. I have observed tomato production in ground soils where nematodes were a serious problem. Despite all efforts of the grower to deep steam his soil, the crop always showed damage after a few months. Similar problems occur with such chemicals as methyl bromide. In some areas, growers will bury pipes 40 to 45 cm deep in the soil, and by using vacuum pumps, they will suck steam downward through the soil. Or, they may inject steam through the buried pipes. Such large volumes are expensive to pasteurize, and, even then, only adequate for one season's growing period. Experience in Colorado has shown that, with a clean stock program, carnations can be grown in virgin soils up to 6 years. Afterwards, the grower must raise culture above the ground level in benches to control disease in the common 2- and 3-year cultural program. In most regions, carnations may be grown less than a year (winter), so that a partial sterilization treatment can be successful. The soil may be treated yearly. Interestingly enough, such difficulties as prevalent with carnations, chrysanthemums, tomatoes, etc., do not appear to apply to rose production in the ground where the same plants may be in place several years. Disease control problems in the ground are one reason for the intensive research and breeding programs in Europe on disease resistance. It is not my purpose here to give an extensive discussion on disease control. However, one can refer to the excellent two-volume text on diseases of floral crops, edited by Strider [1985].

D. WATER MOVEMENT INTO SOILS AND IRRIGATION FREQUENCY

Water, when applied to the soil surface, does not flow instantaneously to all parts of the soil mass. The situation, especially where the surface is flooded from a point source (trickle system), is depicted in Figs. 5-34 and 5-35. Under the influence of gravitation, hydraulic pressure, and the attraction of surfaces and surface tensions in small pores, water will flow through a uniform medium in a well-defined wetting front. The rapidity will depend upon the physical characteristics of the soil and water flow rate from the applicator. In highly porous media, the wetting front may move mostly vertically and little horizontally. Numerous examples of methods to calculate wetting areas under point sources may be found in the agronomic literature [e.g., Risse and Chesness, 1989; Levin et al., 1979; Clothier et al., 1985; Raats, 1971; etc.]. With trickle irrigation systems, emitter spacing is critical. The situations as shown in Figs. 5-34, 5-35 are ideal, but they serve to show that it takes time to wet a soil mass. When, say, the upper 30 cm are wetted, and water application ceases, water may continue to move downward in a field soil and be lost from the main area of root activity. A danger, of course, is insufficient water application to wet the entire soil mass, so parts of the root system may not receive sufficient water.

Fig. 5-34 shows a situation where a more porous medium underlies the soil. This is particularly important with greenhouse cultivation in shallow, freely draining layers. As noted in the second picture of Fig. 5-35, the wetting front does not move into the coarser layer immediately. It will not do so until the water potential at the interface is raised sufficiently. Once the criterion is met, water may then move very rapidly as exemplified in the bottom picture. In Fig. 5-35, the opposite situation of a clay lens is inserted into the soil layer. Here, water will move readily into the clay lens, but as conductivity is much lower, the rate can be quite slow, as noted in the

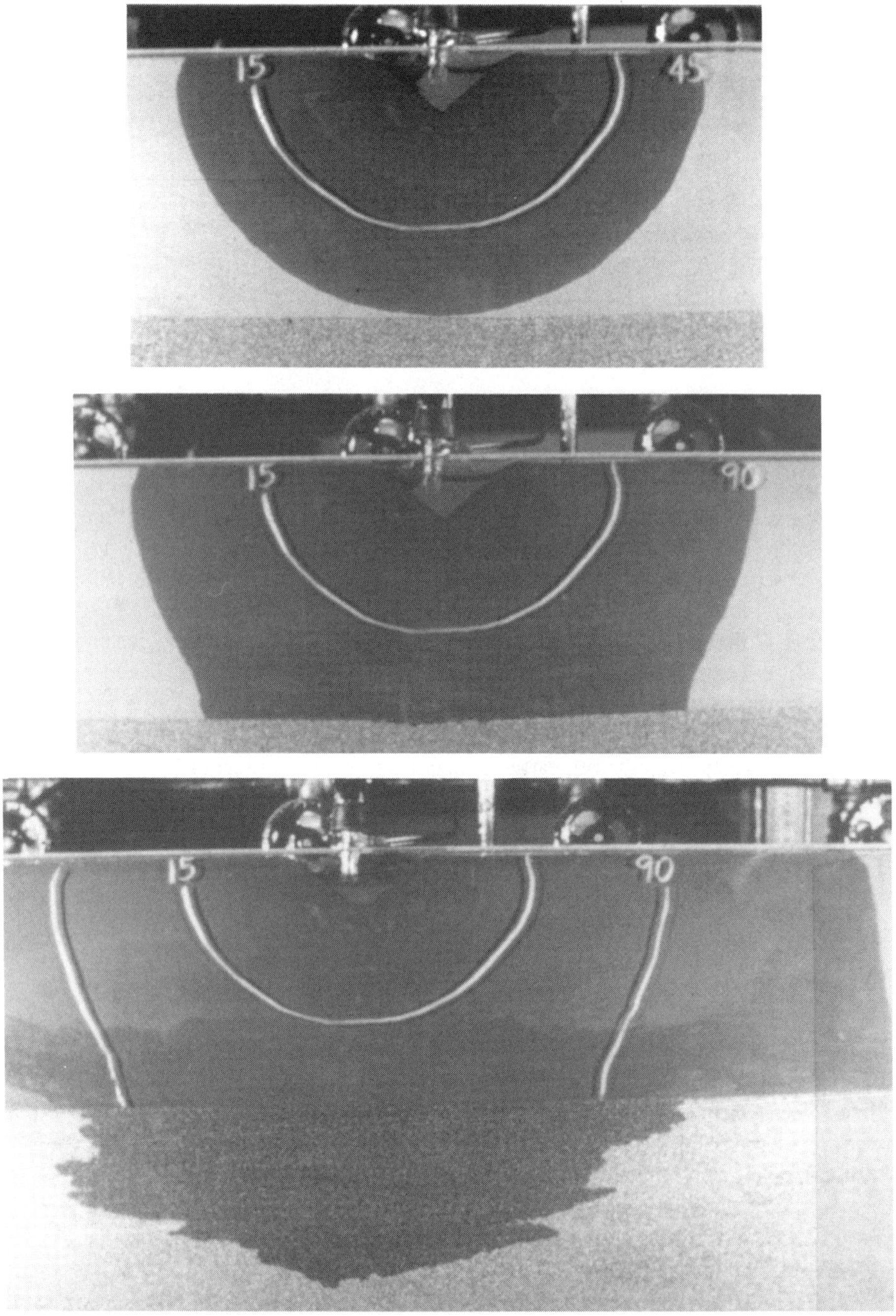

Fig. 5-34. Penetration of water into a uniform soil from a point source, underlain by a more porous, sand mixture. Note in the middle picture, water does not move immediately into the sand, but then moves rapidly when suitable conditions are reached (i.e., pressure at interface zero or greater) [Courtesy of W.H. Gardner].

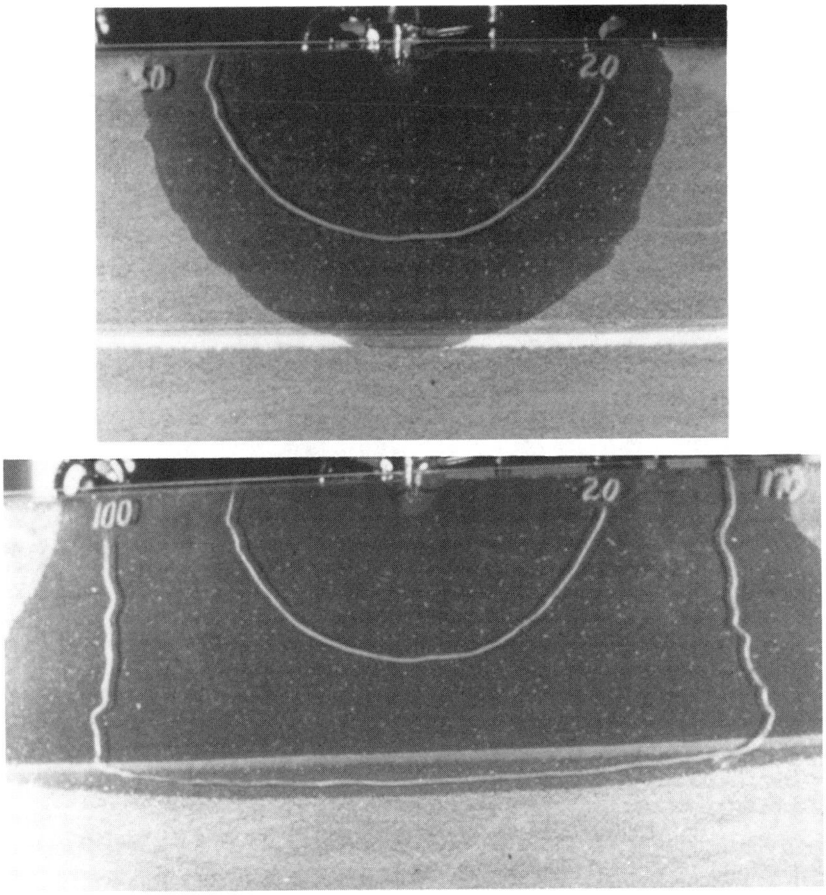

Fig. 5-35. Water penetration into a uniform soil from a point source. A clay lens is inserted. Water penetrates into the clay easily, but moves slower (conductivity less) [Courtesy of W.H. Gardner].

bottom picture. The point to be made is that a substrate, in which plants are established, cannot be maintained at a constant moisture level for any period [see Brady, 1974; Gardner, 1965b; etc.]. One is reduced to growing a plant in a constantly changing situation with respect to water and nutrient availability. The internal plant water potential will change accordingly. One can only wet a soil mass to its maximum moisture-holding capacity and allow it to dry to some predetermined water content as measured with a suitable device. Each irrigation interval will tend to follow the pattern outlined in Fig. 5-32. Experiments that attempt to examine plant response as the result of irrigation treatment cannot be duplicated because the water demand may change from one trial to another as the result of weather and climatic changes. The grower can only control water supply, once the plant is established in the soil –or any medium– by how often he waters the crop or, at what minimum water content or soil moisture tension he has established. The assumption is made that the entire soil mass is brought to maximum water content at each irrigation. Attempting to reduce water by restricting application sufficient to wet the entire root mass is dangerous for the reasons outlined above and, in addition, salinity problems are much more likely to occur.

Growing "wetter" is irrigating more often so the substrate does not dry out as much (Fig. 5-36). Under winter conditions, allowing the substrate to dry to relatively low soil matric potentials may mean that the crop receives no water or nutrients for long periods. On the other hand, under high irradiance conditions of the summer, it may be practically impossible to maintain maximum moisture content in a limited volume soil mass. An actual situation of irrigation frequencies, under commercial conditions for three growers, was examined by Hanan [1981]. By using devices to record the matric suction, the potentials were measured weekly throughout the winter period for first-year carnations (Fig. 5-37). There were marked differences in yield between the growers with No.1 (top) having the lowest yields, No. 2 (middle) was noted in the industry for very high yields, and No. 3

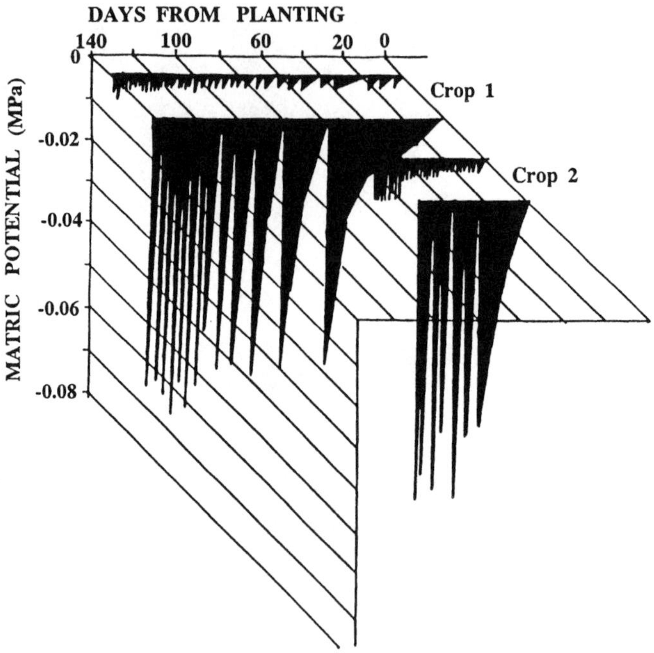

Fig. 5-36. Crop 1 of single-stem snapdragons planted in November and grown to flowering, while crop 2 was planted in June. Both crops watered when matric suction, read once daily, was -0.003 MPa or -0.06 MPa [Hanan, 1963, unpublished data]. Note the plants in Crop 1, watered at -0.06 MPa, went for nearly 60 days before the first irrigation under New York State conditions. On the other hand, during the last days of crop 2, matric suction was less than -0.003 MPa even though watered daily. These curves serve to emphasize the varying conditions to which plants are subjected regardless of how the grower applies water in surface irrigation.

(bottom) known as a "dry" grower with lower yields on the average but not so low as No. 1. The soils in each greenhouse were different. No. 2 used tensiometers to keep track of his soil moisture content and obviously varied his watering practice according to his estimate of weather conditions. In addition, No. 2's soil drained rapidly, as was noted by the fact that the matric suction rapidly increased. All growers applied water until drainage occurred.

The situations described above are applicable to any medium, even hydroponics, where a substrate is employed. The differences are a matter of degree which can be highly critical under certain conditions.

E. MEASURING SOIL MOISTURE CONTENT

In field soils, and often in modified soils, knowledge of water content is a useful adjunct to efficient irrigation of greenhouse plants –although it is surprising how few growers carry out the process on a regular basis. The gravimetric method, using various sampling tools as shown in Fig. 5-38, has already been described. Water content (θ_m) is expressed as mass of water per mass of dry soil. As pointed out, with highly organic soils, the practice can result in ridiculous values of moisture content. An undisturbed core sample should be taken of known volume (Fig. 5-38) and water volume expressed as the water ratio (θ_v) or volume of water per unit volume. If the bulk density is known, or assumed, the volume basis may be obtained from mass basis figures by use of the formula:

$$\theta_{vb} = \left(\frac{BD}{D_w}\right)\theta_{dw} \tag{5.11}$$

where: DB = bulk density of the soil (g cm^{-3})
D_w = density of water, usually taken as 1
θ_{dw} = water content on a mass basis

Mineral soils are usually dried at 105 C; however, this may lead to loss of constituents in highly organic

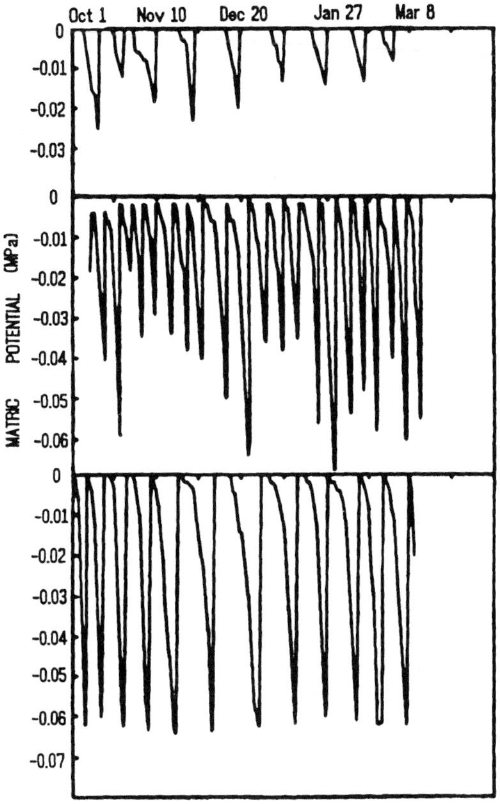

Fig. 5-37. Weekly matric suctions of soils in freely draining, shallow layers in which carnations were planted in June and irrigated by a peripheral "Gates" spray system. Three different commercial ranges in Colorado, each grower following his own standard procedure [Hanan, 1981]. At each watering, sufficient water applied to leach with automatic fertilizer injection. Tensiometers were read once weekly.

soils. Gardner [1965a] suggested 50 C as more appropriate in drying such soils. Gardner also stated that water content of stony or gravelly soils, both on a mass and volume basis, may be misleading. Large rocks occupy appreciable volume and contribute markedly to mass without a commensurate contribution to porosity or water capacity of the soil.

Another method for direct determination of water volume is the neutron thermalization method (Fig. 5-39). This has been the apparatus of choice for field research, but it is expensive. It has never been employed in commercial greenhouse practice to my knowledge. In recent years, "time-domain reflectometry" has received considerable attention [e.g., Topp et al., 1980; Topp et al., 1984; Topp and Davis, 1985; Brisco et al., 1992; Nadler et al., 1991; Hilhorst et al., 1992; etc.]. In effect, one measures the dielectric constant of the soil between 2 parallel probes (waveguides) inserted into the medium. One may also consider the probes to be two sides of a capacitor, with the soil forming the insulation between them. Since water content is the largest determinant of the dielectric constant, the latter can be related to moisture content when suitable probes and microwave frequencies are employed. Topp et al. [1980] showed that calibration for different soils was unnecessary and that soil texture, density, salt content, and temperature did not affect the TDR measurement of liquid water. Portable, hand-held probes have been designed [Topp et al., 1984; Briscoe et al., 1992; Hilhorst et al., 1992]. Pelletier and Tan [1993] have reported on the use of TDR to measure wetting patterns in peach orchards. Hilhorst et al. [1992] have examined TDR and capacitance methods for soils and rockwool in greenhouses. Their results appeared encouraging.

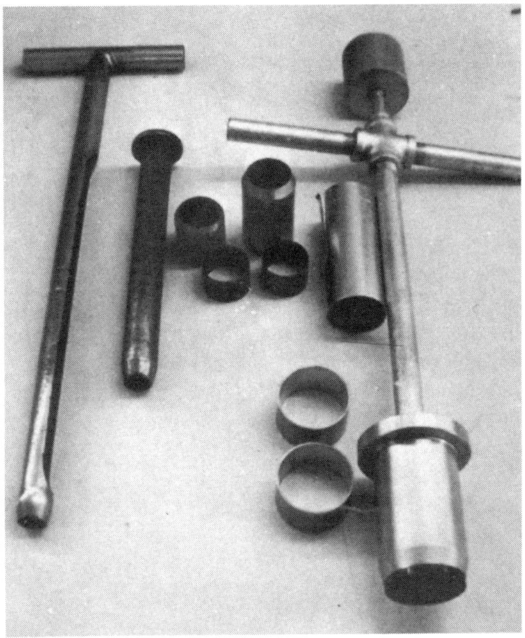

Fig. 5-38. Several examples of soil sampling devices. The small tubes in the center with rings are for extracting undisturbed samples, the rings holding the sample and removed from the outer covering. The two on the left are commonly empioyed for nutrient analysis. The one on the right is employed for field soils for undisturbed samples of known volume.

Fig. 5-39. Neutron thermalization probe. System depends upon slowing of neutrons emitted by a radioactive source by water molecules. With this instrument, the probe is lowered into an access tube in the ground for the measurement.

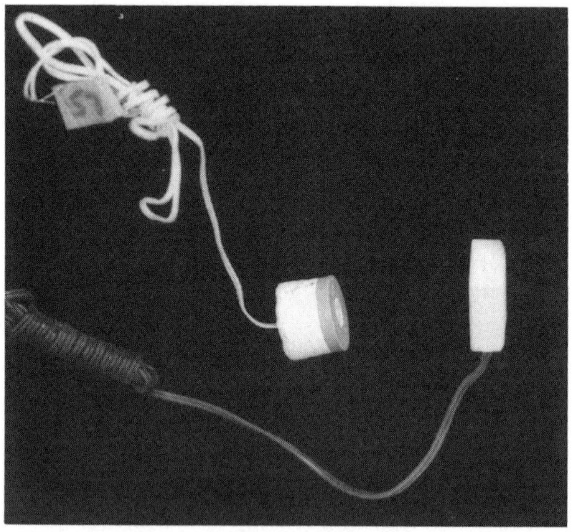

Fig. 5-40. Gypsum (Bouyoucos) blocks for electrical resistance measurement of soil water. Simple and easy to use but difficult to interpret, operating outside the ranges usually encountered in greenhouses.

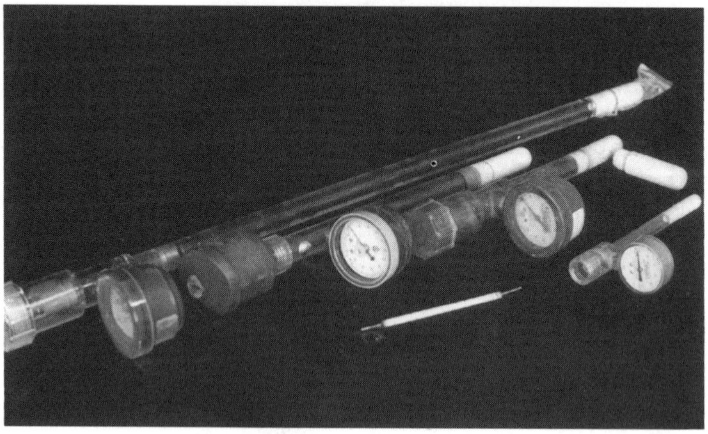

Fig. 5-41. Common styles of tensiometers to be found in greenhouse practice. As the medium dries out, water is extracted from the water-filled instrument through the ceramic cup, resulting in a corresponding reading on the vacuum gauge. Other styles may use mercury manometers.

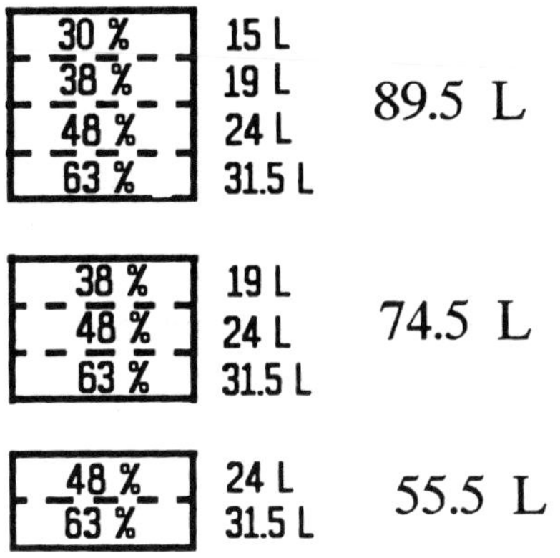

Fig. 5-42. Moisture content in each 5 cm layer of three soils, top 20 cm deep, middle 15 cm deep, and bottom 10 cm deep. The total water content for each layer is calculated for a square meter of soil [Adapted from Hanan, 1964, unpublished data]. Note, as depth is reduced, one is growing in wetter soil. However, with less volume, the substrate may dry out faster in the summer.

Other methods have been frequently discussed by many authors [e.g., Gardner, 1965a; Holmes et al., 1967; Taylor and Ashcroft, 1972; Slatyer, 1967a; etc.]. The thermocouple psychrometer can be employed to detect total soil water potential, using small diameter, porous ceramic tubes with the thermocouple inside. This allows the atmosphere inside the tube to come into equilibrium with the soil environment.

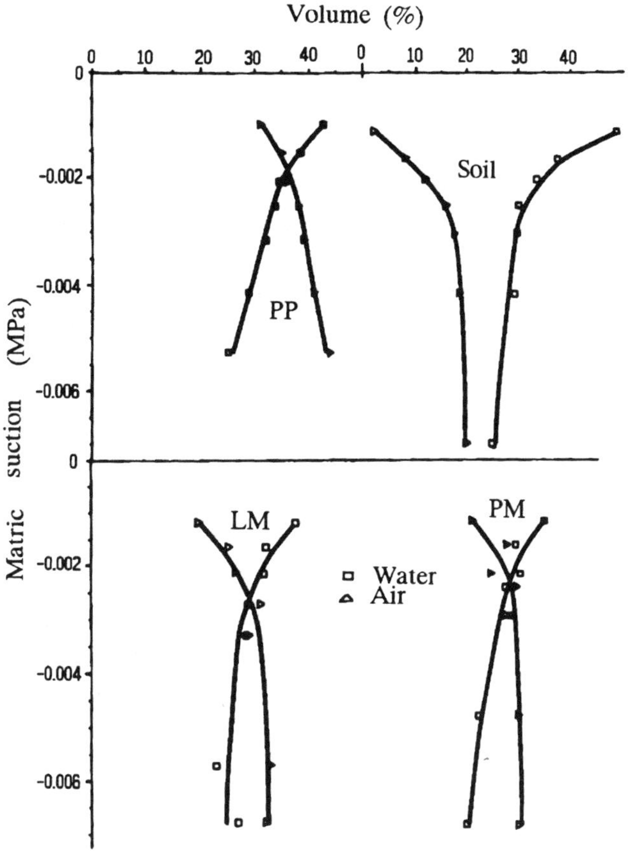

Fig. 5-43. Air and water content of various root media at different soil suctions. These are results from actual samples taken from the surface to 4 cm depth in soil layers up to 70 cm deep so that the highest potential represents a soil 10 cm deep, -0.002 MPa = 20 cm, etc. **Soil** = silt loam field soil, **PM** = 1-1-1 mixture by volume of soil, sand, and peat moss, **PP** = 1-1 mixture of peat moss and perlite, and **LM** = 1-1-1 mixture of soil, sand, and leaf mold [Hanan, 1963, unpublished data]. -0.001 MPa is approximately equal to a 10 cm water column.

More common for practical purposes is electrical resistance (Fig. 5-40) or capacitance systems and tensiometric devices (Fig. 5-41). The latter permit a direct measurement of matric potential, but they are not useful at potentials lower than -0.07 MPa. Since this is the maximum value for most greenhouse conditions, tensiometers are the most often used in greenhouse practice. Simple electrical resistance systems are not useful for greenhouse application since they operate best below -0.1 MPa. Inexpensive devices that one may buy in local stores have no application as they are usually inaccurate and markedly influenced by the electrical conductivity of the soil solution. This is not so with TDR systems. Tensiometers may be equipped with electrical contacts to initiate irrigation at specific matric water potentials. Their principal problem in highly porous media is inadequate contact between the medium and ceramic cup of the tensiometer. They are of no use in gravel media. A particular problem in greenhouse cut-flower production is losing the instrument in the crop, plus failing to service instruments at regular intervals. If the devices are to be used for showing water content, they must be calibrated for the particular substrate. Another

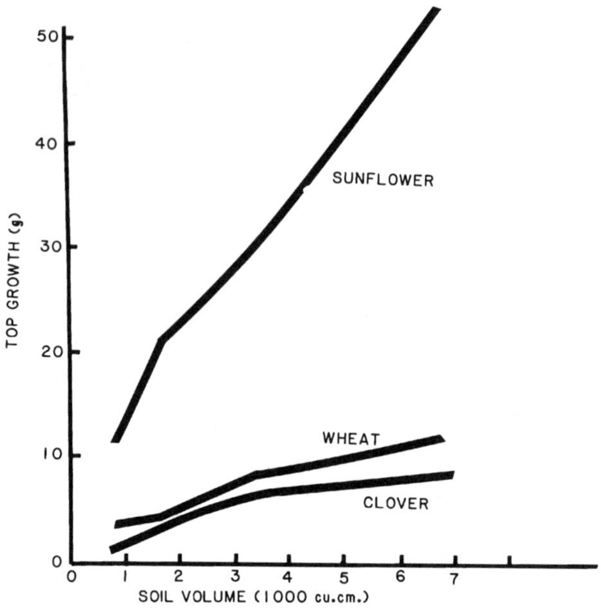

Fig. 5-44. Top growth of plants in various soil volumes [Adapted from Stevenson, 1967] (With permission of the *Canadian J. Soil Sci.*).

difficulty is the variation to be found in most field soils, and the fact that while there may be an accurate indication of water suction in the bulk soil, this value may not be true for actual suction at the root-soil interface (Fig. 5-29). In limited-volume containers, roots quickly surround the cup so that there should be close correspondence between suction of the bulk soil and suction at the root surface.

In hydroponic systems, of course, none of the above methods –except for thermocouple psychrometers– are of any value. Flowing solutions eliminate matric potential, and irrigation of soilless media is usually frequent enough to preclude any instrument indications. Much of the literature uses the term "well-watered" in the sense that there was, in the experiment, no restriction to water uptake by a root system. Typically, this may be a way to get around the difficulty of specifying the root environment. The author did not really know. I think much of the scientific research on greenhouse production is seriously deficient in this respect –to which the author has made his own illimitable contributions.

F. MODIFIED SOILS AND SOILLESS MEDIA

Before the 1950s, container production in greenhouses could be characterized by continuous repotting (shifting) of the crop to larger-sized containers as the plants grew. Plants were started in very small clay pots, and shifting the crop was a major process in greenhouse culture. In this manner, problems of aeration and drainage with the existing potting soils were avoided as the large plant removed water rapidly. "Pot-binding," overgrowth of the root system, was a major problem. Several publications dealt with the subject of pot-binding before the 1950s [i.e., Knight, 1944]. Each year, carnation soils were removed from the bench, the benches repaired, and new field soil brought into the greenhouse for the next crop. Manure was a standard additive when used, although studies on the use of other organic additives were carried out in the 1930s [i.e., Sprague and Marrero, [1931]. At a time when hourly wages were less than $1.00, many of these practices appeared cost effective. However, as salaries increased with inflation, combined with movement toward single potting of major crops, development of new irrigation systems, steam pasteurization, and new practices in fertilization, extensive investigations were carried out on soil modification and the use of various additives to achieve adequate plant growth in minimum time. These studies gathered momentum in the 1960s through the 1980s.

1. Water Relationships

When using modified soils or soilless substrates, we assume a freely draining, shallow medium of limited volume. That is, there is no restriction to water flow from the underside of the medium into the air. This is analogous to the situation shown in Fig. 5-34 where the water was moving into a more porous soil. Under these conditions, water will not drain from the medium until the potential is equal to, or exceeds zero. Richards [1950] called this situation one of the laws of soil moisture, and it has come to be called "Richards' outflow law." In 1964, Hanan and Langhans and White published on the behavior of soil mixtures in shallow layers. If the entire mass (pot, flat or bench) has been brought to its maximum moisture-holding capacity, then, at equilibrium, after drainage has ceased, the suction at the bottom soil interface will be zero. At the upper surface, suction can be expressed in terms of the soil depth and the water matric potential will be equal to the soil depth in centimeters

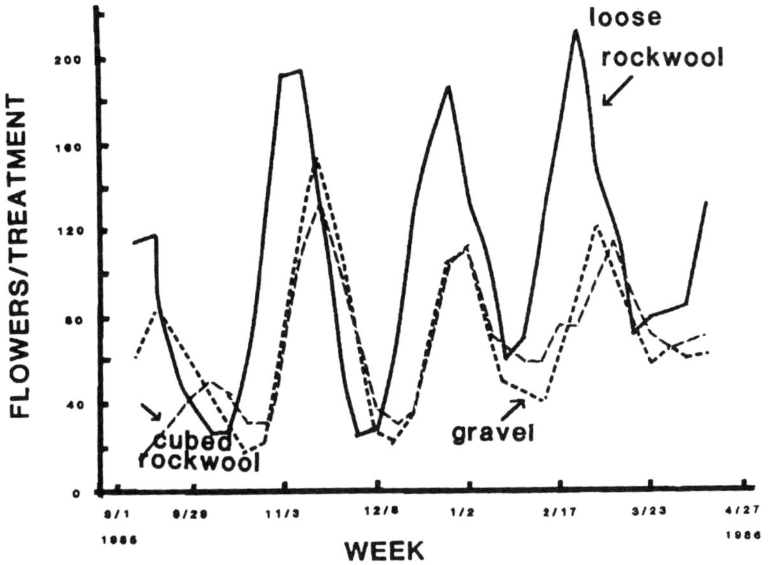

Fig. 5-45. Yield of cut roses grown in 20 cm deep media of loose rockwool or gravel, compared with 8 cm deep rockwool cubes, all watered by a trickle system with identical fertilization automatically injected [Hanan, 1986].

of water column. If matric suction measurements are made at any point above the bottom of the soil, the result will be equal to the distance above the bottom the measurement was made. This basic principle has remarkable implications for crop culture in greenhouses. Assuming a soil depth of 20 cm, at maximum moisture holding capacity, the suction at the upper surface will be 20 cm water or -0.002 MPa matric potential. If 10 cm deep, the matric potential will be -0.001 MPa. The values are ten times higher than normally prevalent under field conditions (-0.01 to -0.03 MPa). In this respect, **any** field soil **must** be modified if it is to be utilized for container production at depths less than 30 cm. The situation as depth decreased was discussed by Hanan [1981], showing that, although total water volume will often decrease as soil volume decreases, there will be an increase in water content in $cm^3 cm^{-3}$, or percentage. The operator is growing in wetter soil (Fig. 5-42). Choice and handling of substrates for shallow containers, especially plug seedlings, become highly critical to avoid deficient aeration under winter conditions

Table 5-4. John Innes compost. Loam selected should have enough clay to be slightly greasy when smeared without stickiness, from a good pasture, 2-7 % humus. This to be composted in alternating layers of 12 cm loam and 5 cm strawy manure, allowing 6 month's time before using [Adapted from Lawrence and Newell, 1952].

	Ingredients, parts by volume	**Fertilizer ingredients, per m^3**
Seed compost	2 loam	1.2 kg 20% superphosphate
	1 peat moss[a]	0.6 kg calcium carbonate[c]
	1 sand[b]	
Potting compost	7 loam	1.2 kg 20% superphosphate
	3 peat moss	0.6 kg potassium sulfate
	2 sand	0.6 kg calcium carbonate

[a] Canadian or German sphagnum or equivalent.
[b] Coarse sand.
[c] Lawrence and Newell refer to this as "fine chalk."

of low water demand and small plants. Hanan and Langhans [1964] used the term "bench capacity" to denote the maximum moisture holding abilities of shallow, freely draining media. However, "container capacity," coined by White [1964], has gained wide acceptance.

The effect of depth and matric suction on air porosity and moisture content was examined by Hanan in the early 1960s. Fig. 5-43 compares relationships between a field soil and some standard amendments and mixtures. In most field soils placed in shallow containers, air porosity can approach zero in the depths at which they are normally used. It can also be appreciated that amendments to soils should be capable of providing sufficiently large pores to permit drainage at reduced suctions. Uniform mixtures of small particle sizes are to be avoided since there may be insufficient suction to empty the pores and the entire mass remains saturated after irrigation [Spomer, 1976]. However, Steiner [1968] approved using uniformly sized media as a means to avoid deficient aeration. Some soilless materials are particularly dangerous in this respect when used in shallow containers (e.g., fine sand). That is, if the pore spaces are small enough, a shallow layer may remain saturated.

2. Effects of Volume

As the size of the container decreases, adjustments must be made to other cultural procedures if adverse effects on plant growth are to be avoided. This problem is well known by agronomists conducting research on field crops in small containers [Baker and Woodruff, 1963; Cook and Millar, 1946]. Maize grown in small containers may require up to 20 times the amount of phosphorous usually applied under field conditions. Similar requirements have been noted for nitrogen and potassium. A study by Stevenson [1967] showed root and top growth to vary consistently and quantitatively with changing soil volume (Fig. 5-44). Stevenson postulated that above a certain root density, individual roots could interfere with each other through competition. Many studies have been carried out by horticulturists [e.g., Ruff et al., 1987; Tilt et al., 1987; Latimer, 1991; Wells and Postlethwaite, 1969]. Every time, growth was related to container size, with reduced growth the smaller the container. However, one must be careful in comparing soil volumes since depth will influence available water as outlined above, and provision of adequate nutrition on a comparative basis is extremely difficult. Hanan [1986] compared rose production in loose rockwool and gravel in benches at 20 cm depth with rockwool in cubes each 15 x 100 cm, 8 cm deep. These treatments were all watered and fertilized at the same rates. The results in Fig. 5-45 show that watering practice and fertilizer must be adjusted since other experiments have shown much better performance than those shown in the figure for blocked rockwool and gravel. Generally, the smaller the substrate volume, the higher must be the fertilization rate plus irrigation frequency if growth is not to be checked. Kemble et al.'s [1994] studies on a compact growth tomato in cell volumes of 3.3, 27, 37.1 and 80 cm^3 showed a general reduction in time from sowing to anthesis as container volume increased. Flowering was delayed more than 2 weeks when grown in 3.3 cm^3 cells. Delay was 5 to 7 days in the intermediate sizes compared to 80 cm^3. To date, most recommendations to be found in the literature are empirically determined for the particular container sizes, substrate, and local climatic conditions.

3. Amendments for Soil Mixes

a. Standard Mixes

Through the 1950s and early 1970s, considerable interest was generated by the John Innes composts (Table 5-4), U.C. Mixes (Table 5-5), and the Cornell peat-lite mixes (Table 5-6). These were attempts to provide standard, reproducible modified soil and soilless media that were readily available. Bunt [1988] has given three definitions of "composts": 1) plant material that has undergone decomposition, 2) animal manure and straw used as a basis for mushroom production, and 3) mixtures of organic and mineral materials for container production. Usually, the word "compost" has been reserved for mixtures based upon mineral soils. The difficulty with composts is that not everyone has available the required soils, and, as noted previously, field soils are exceedingly diverse. Furthermore, composting is difficult in arid regions. Since their development, almost every type of amendment has been investigated and found "suitable" for container production provided the appropriate precautions are taken in preparation and fertilization. Even ground-up automobile tires have been examined [Bowman et al., 1994]. Zinc toxicity was a major problem with the latter. Many examples of investigations on substrates may be found in the *Acta Horticulturae* Volumes, 99 and 133.

Table 5-5. Basic U.C. mixes, using fine sand and peat moss. Sand should consist of particles between 0.5-0.05 mm, with silt plus clay not exceeding 15%, and coarse sand not exceeding 12-15%. Peat moss finely ground and pathogen-free, Canadian, German sphagnum or equivalent [Adapted from Baker, 1957].

	Ingredients, (% by volume)		Bulk density ($g\ cm^{-3}$)		Maximum water content (% by volume)	Fertilizers for indefinite storage (per m^3)
Mix	Sand	Peat moss	At max[a]	Oven dry		
A	100	0	1.87	1.42	43	0.2 kg potassium nitrate 0.1 kg potassium sulfate 1.1 kg 20% superphosphate 0.7 kg dolomitic lime 1.1 kg calcium sulfate (gypsum)
B	75	25	1.68	1.22	46	0.2 kg potassium nitrate 0.1 kg potassium sulfate 1.1 kg 20% superphosphate 2.0 kg dolomitic lime 0.6 kg calcium carbonate 0.6 kg calcium sulfate (gypsum)
C	50	50	1.50	1.01	48	0.1 kg potassium nitrate 0.1 kg potassium sulfate 1.1 kg 20% superphosphate 3.4 kg dolomitic lime 1.1 kg calcium carbonate
D	25	75	1.06	0.54	51	0.1 kg potassium nitrate 0.1 kg potassium sulfate 0.9 kg 20% superphosphate 2.3 kg dolomitic lime 1.8 kg calcium carbonate
E	0	100	0.69	0.11	59	0.2 kg potassium nitrate 0.5 kg 20 % superphosphate 1.1 kg dolomitic lime 2.3 kg calcium carbonate

[a] Weight at maximum water content for a 15 cm column of a fine sand of the Oakley series and Canadian peat moss. Also applicable to maximum water content.

b. Amendments

A general listing of several common amendments are given in Table 5-7, with general comments. Often, an amendment is material that is readily available. In the Southeastern U.S., this may be pine bark. In Spain, it could be olive husks and in Cyprus, one might use pecan hulls. In other cases, the amendment is the substrate and the medium is soilless culture. If inert, with no organic material, the dividing line between traditional culture and hydroponics is indistinguishable. A more detailed outline of amendments and mixtures is given in Table 5-8, with bulk densities ranging from 0.10 to 1.68 $g\ cm^{-3}$, air porosities from 1.8 to 54.7%, and container capacities (18

Table 5-6. Peat-lite Cornell mixes [Adapted from Cornell Recommendations, 1974].

Mixture	Ingredients	Fertilization per m^3						Wetting agent[b]
		Lime-stone	20% superphos-phate	Calcium nitrate	Trace[a]	Iron sulfate	10-10-10	
A	50% sphagnum peat moss[c] 50% vermiculite	3 kg[d]	1.2 kg	0.6 kg	43 g			64 g
B	50% sphagnum peat moss 50% perlite	3 kg					4.8 kg	64 g
Foliage plants	50% sphagnum peat moss							
	25% vermiculite 25% perlite	4.8 kg	1.2 kg	0.6 kg KNO_3	43 g	16 g	1.5 kg	64 g
Epiphytic mix	1/3rd sphagnum peat moss							
	1/3rd perlite 1/3rd douglas fir bark, fine ground, 3-13 mm	4.2 kg	2.4 kg	0.6 kg KNO_3	43 g	11 g	1.5 kg	64 g

[a] Fritted trace element mix [b] Wetting agents such as Aqua-Gro, Ethomid or Triton B-1956, etc.
[c] Canadian, German or equivalent. Note: Allow for 20% shrinkage [d] Only "ground" limestone specified

cm deep) 15 to 53%. Section 5.III.B recommended air porosities of 10 to 15% minimum, although higher values (20%) have been suggested for species such as azaleas, ferns, and orchids [Faber and Hoitink, 1983]. Air porosities less than 5% have definitely shown damage to foliage plants [Conover and Poole, 1977]. The decision on what material to be used is most often based on cost, although that criterion is not always the best. The variation at different ratios of a given series of components was examined by White [1974], and these are presented in Table 5-9. Although total porosity gradually increases as amendments are added to a basic soil, marked, significant changes in porosities, water content, and infiltration rates do not appear until the amendment is nearly 50% or more of the final volume (Table 5-10).

c. Handling Mixtures

There are many things to consider with modifying soils and using soilless media. Adding small particles (i.e., fine sand) to a course medium such as peat moss will reduce total volume of the final material. That is, 2 m^3 plus 2 m^3 do not add to 4 m^3. Small amounts of fine material will fill large pores, reducing air porosity. This is probably the reason for Steiner's [1968] comments on variable-sized amendments. For an increase in porosity, the fine amendment may have to exceed more than 50% of the final volume –called the "threshold proportion" by Spomer [1975]. Note the air porosities in Table 5-8 where fine sand, sandy loam, or clay loam are used with various amendments. A second problem with some media such as peat moss is shrinkage and wettability. Up to a 30% volume loss in peat moss mixtures may be expected over time. In a Monte Carlo simulation procedure, Burés et al. [1993] found maximum shrinkage when the proportion of coarse pine bark particles and sand mixtures ranged from 50 to 70% of the volume. When dry, peat moss can be exceedingly difficult to wet. This is the reason for the wetting agents employed in the peat-lite mixes (Table 5-7). On the other hand, with such materials as rockwool in shallow layers, a non-wetting agent may be required so that surface tension in the pores is reduced and the material drains adequately. Lack of drainage with very uniform materials in shallow layers (i.e., sand)

Table 5-7. Listing of amendments used for soil modification and soilless culture [Derived from a variety of sources: Bunt, 1976; 1988; Nelson, 1972; Johnson, 1968; White, 1974; etc.].

Amendment	Remarks
Coal ash	Heavy, can contain a number of heavy metals (B, Ba, Co, Cr, Ni, Pb, Si, etc.), pH 4.5 to 12.0, variable in particle size and mineral content.
Corfuna	Proprietary, ground oil cake and seaweed, fermented and inoculated with beneficial bacteria, an animal manure substitute. Made in France.
Expanded or calcined aggregates	Extruded clay heated to 1000 to 1100 C, 0.3-0.7 g cm^{-3}, internal porosity 40-50%, no CEC, washing necessary before use, number of proprietary names: Waylite, Idealite, Freelite, Haydite, Turface, etc.
Expanded polystyrene flakes	4-12 mm particle size, chemically neutral, no decomposition, 0.02 g cm^{-3}, water not absorbed in flake, proprietary name "Styromull."
Gravel	Washed with no carbonates, particles <0.5 mm less than 5% of total, 0.5 to 5 mm diameter, can grow in 20 mm diameter gravel with special precautions, heavy.
Hardwood bark	Several species with 40% cellulose, rapid decomposition, requires composting to eliminate phytotoxicity, 0.5 to 12.8 mm particle size, often suppressive to root rot organisms.
Leaf mold	Rapid decomposition, can be contaminated by herbicides and NaCl as well as oil residues when picked up by street-cleaning operations.
Limestone gravel	See Schwarz and Vaadia [1969]. Can be used as substrate with special handling. Otherwise, calcium source (Chap. 6).
Animal manure	Highly variable, low nutrient containing, liable to contain weed seeds and subject to high ammonia concentrations.
Pecan hulls	Example of locally available material, upwards to 20 mm size.
Peat moss, sphagnum	Several species, loose and bulky, CEC 110-130 meq/100 g, pH 3.5-4.0, 0.06-1.0 g cm^{-3}.
Peat moss, sedge	*Carex* spp., some Phragmites, higher nutrients than sphagnum, darker, more decomposed, higher CEC than sphagnum.
Perlite	Expanded aluminosilicate, volcanic origin, high porosity, cellular structure but enclosed and unavailable for water storage, 0.1 g cm^{-3}, no CEC, aluminum toxicity below pH 5.0, some fluorides.
Phenolic resin foam	Open structure, pH about 7.0, no decomposition, no CEC, 0.03 g cm^{-3}, water holding capacity 70%.
Polyurethane foam	Open structure, pH about 7.0, 0.01 g cm^{-3}, no decomposition, no CEC.
Pumice (Scoria)	Volcanic alumino-silicate, some K^+ and Na^+, highly porous, 0.48 g cm^{-3}, breaks easily on handling.
Rice hulls	0.1 to 0.2 g cm^{-3}, 80% porosity.
Rockwool	Aluminosilicate, heated to 1500 C and spun to fibers, 97% porosity, no CEC, 0.65 g cm^{-3}, will remain saturated at <5 cm depth unless treated.
Sand	Fine sand 0.05 to 0.5 mm diameter, no carbonates, 1.6 g cm^{-3}. Do not use "soft" sand (eroded, rounded grains). Use "sharp" sand.
Sawdust	Must be composted, rapid decomposition, 0.19 g cm^{-3}, 78% porosity.
Sewage sludge and municipal wastes	Use not more than 30% of final volume. Liable to contain heavy metals (Cr, Zn, Ni, Cd, etc.). Processed sewage, distinct from "nightsoil."
Softwood bark	*Pinus* spp. mostly, 0.25-0.45 g cm^{-3}, internal porosity 43%, 2.4 mm or smaller diameter, lower cellulose and slower decomposition than hardwoods.
Straw	Should be composted, rapid decomposition. Used in bales as cucumber substrate in northern Europe.
Urea formaldehyde foam resin	Absorbs water, 0.1 to 0.3 g cm^{-3}, degrades and releases N slowly, pH 3.0, shredded to 12-13 mm particles.
Vermiculite	Al-Fe-Mg silicate, heated to 1000 C for 1 min to expand laminated plating, 80% porosity, 0.08 g cm^{-3}, particle size up to 6 mm, CEC 100-150 meq/100 g, 5-8 % K^+, 9-12% Mg^{+2}, pH 6.0-6.8, also alkaline types, pH 9.3-9.7. Will lose structure gradually.

Table 5-8. Physical properties of amendments, soils, and mixtures [Condensed from Johnson, 1968]. Values determined for substrate depths of 18 cm.

Material	Bulk density		Moisture capacity (% volume)	Total porosity (%)	Air porosity (%)
	Dry ($g\ cm^{-3}$)	Wet ($g\ cm^{-3}$)			
Bark, fir (0-3 mm)	0.23	0.62	38	70	32
Bark, fir (3-16 mm)	0.19	0.34	15	70	55
Bark, redwood (10 mm)	0.13	0.44	31	80	50
Loam, clay	0.95	1.51	55	60	5
Loam, sandy	1.58	1.95	36	38	2
Peat sedge, AP4	0.21	0.74	52	69	17
Peat, sedge, BD	0.26	0.95	69	77	8
Peat moss, hypnum	0.19	0.79	59	72	12
Peat moss, sphagnum	0.11	0.70	59	84	25
Perlite (1.0-2.0 mm)	0.09	0.52	43	76	33
Perlite (6-8 mm)	0.10	0.29	20	74	54
Pumice (2-9 mm)	0.46	0.87	41	62	21
Pumice (8-16 mm)	0.48	0.74	26	71	45
Rice hulls	0.10	0.23	12	82	69
Sand, builders	1.68	1.95	27	36	9
Sand, fine B	1.44	1.83	39	45	6
Sawdust, cedar	0.21	0.60	38	81	43
Sawdust, redwood	0.18	0.68	49	77	28
Vermiculite (0-5 mm)	0.11	0.65	53	81	28
50-50 clay loam with					
Peat sedge	0.66	1.23	57	62	5
Peat moss, hypnum	0.63	1.23	60	66	6
Peat moss, sphagnum	0.55	1.18	61	71	10
Sand, builders	1.28	1.69	41	47	6
Sand, fine	1.32	1.74	42	47	7
50-50 mixture, sandy loam with					
Peat sedge	0.90	1.41	50	54	4
Peat moss, hypnum	0.94	1.45	50	54	4
Peat moss, sphagnum	0.87	1.41	53	59	6
Sawdust, redwood	0.80	1.33	53	63	10

Material	Bulk density Dry (g cm^{-3})	Bulk density Wet (g cm^{-3})	Moisture capacity (% volume)	Total porosity (%)	Air porosity (%)
50-50 mixture, fine sand with					
Bark, fir (0-3 mm)	0.86	1.23	37	55	15
Bark, redwood	0.94	1.38	44	57	13
Peat, sedge	0.93	1.43	49	54	5
Peat moss, sphagnum	0.75	1.23	47	57	9
Perlite (2-5 mm)	0.86	1.29	43	52	8
Pumice (2-3 mm)	1.05	1.43	38	42	5
Sawdust, redwood	0.93	1.31	41	53	12
50-50 mixture, peat moss with					
Perlite (5-6 mm)	0.11	0.63	51	75	24

can be a danger. Under summer conditions of high irradiance, this is not usually a problem due to rapid plant water uptake. However, in winter –e.g., production of bedding plant seedlings in January– saturated substrates are likely to give difficulty. Maximum infiltration rates for modified soils and soilless culture can easily exceed 900 cm hr^{-1} [Hanan et al., 1981]. Under such conditions, water from an irrigation system such as a trickle, or drip, will not travel far horizontally (Figs. 5-34, 5-35), nor will unsaturated hydraulic conductivity be significant. Problems will arise with practices to leach such media for salinity control, and portions of the root system may not be sufficiently watered.

Despite the usual increase in water content of modified substrates in shallow containers, water availability remains a problem that has been approached by incorporating water absorbing polymers (hydrogels) in the mixture. Hydrogels can absorb 20 to 1000 times their own weight in water, and are categorized into three main families [James and Richards, 1986]: starch copolymers, polyacrylamides, and polyvinylalcohols. These have been promoted as amendments to increase the water available to plants and as a water conservation method [e.g., Jensen, 1971; Wang, 1987]. Unfortunately, hydrogel performance can be influenced by many factors, such as the iron source employed, medium, plant species, etc. [James and Richards, 1986]. As ion concentration in the water increases, amount of water absorption decreases [Orzolek, 1993]. They have also been considered as soil conditioners to improve structure [Wallace et al., 1986]. Letey et al. [1992] conclude that water-absorbing polymers do not conserve water. The quantity of extra water held in the container depends on the nature of the medium and size of the container, and tests of polymers did not change transpiration rates –which were the same in all treatments. Time between irrigation was extended, but only about 1 day out of 7. The only factor altered is storage capacity, with greater retention in coarse-textured soils. Polymers do not alter water demand. Orzolek [1993], however, suggested increasing use of these materials in plant production, painting an optimistic outlook.

Most producers in the U.S. mix their own substrates, using machines such as used cement mixers. These can be equipped with steam attachments for pasteurization. The variations one may find are extensive. There have been incidents where, for example, the mixing operator ran his tractor over a scoria pile, rapidly reducing the material to fines that resulted in a poor mixture. In another case, the grower chose a cheaper material which, in preparation, had a large proportion of small particles. This also led to difficulties in crop culture. There is an obvious moral to this story. Problems of product quality in the U.S. are difficult to assess. Harless [1984] listed 54 different potting soils sold in the U.S., not to mention suppliers of peat moss, perlite, vermiculite, etc. In the Netherlands, making composts is considered a highly technical operation, and the major research stations work directly with the operators to ensure quality and uniformity. To be listed by the research station, the mixture must meet the approval of a compost specialist, for which the manufacturer pays a levy on a voluntary basis [Boertje and Bik, 1975]. This type of cooperation is unavailable in the U.S.

Table 5-9. Effect of mixture proportions on some physical characteristics of a modified soil [Adapted from White, 1974]. Values for a 17 cm deep column.

Mixture (soil-perlite-peat moss)	Bulk density ($g\ cm^{-3}$)	Total porosity (%)	Maximum water capacity (%)	Air porosity (%)	Percolation rate ($cm\ hr^{-1}$)
10-0-0	1.15	57	44	13	4
9-1-0	1.15	57	42	15	5
9-0-1	1.05	61	44	17	5
8-1-1	1.03	61	46	15	7
7-3-0	1.03	62	42	20	51
7-0-3	0.93	65	41	24	39
7-1-2	0.85	68	46	22	36
7-2-1	0.90	66	45	22	49
6-1-3	0.72	73	44	28	30
6-2-2	0.82	69	41	28	31
6-3-1	0.86	68	44	24	35
5-5-0	0.82	69	42	27	20
5-0-5	0.69	73	48	26	100
3-7-0	0.68	74	40	34	133
3-0-7	0.48	81	57	24	148
3-6-1	0.54	79	40	39	108
3-1-6	0.45	83	53	27	123
2-7-1	0.46	82	39	43	>152
2-1-7	0.38	85	64	21	>152
2-6-2	0.40	84	42	42	>152
2-2-6	0.36	86	54	32	>152
1-9-0	0.40	84	40	44	>152
1-8-1	0.31	88	38	50	>152
1-7-2	0.30	88	46	42	>152
1-6-3	0.29	88	43	45	>152
1-3-6	0.26	89	56	33	>152
1-2-7	0.27	89	64	25	>152
1-1-8	0.27	89	65	24	>152
1-0-9	0.22	91	69	23	>152
0-10-0	0.18	92	37	56	>152
0-9-1	0.17	93	39	54	>152
0-7-3	0.14	94	44	50	>152
0-5-5	0.14	93	52	42	>152
0-3-7	0.12	94	53	41	>152
0-1-9	0.18	90	65	25	>152
0-0-10	0.10	94	64	31	>152

Table 5-10. Effect of adding peat moss to a silt loam on pore space and infiltration [From White, 1964].

Silt loam-peat moss proportions	Total porosity (%)	Water content for 17 cm column (%)	Air porosity (%)	Infiltration rate (cm hr^{-1})
10-0	57	44	13	4
9-1	61	44	17	5
7-3	65	41	24	39
5-5	73	48	26	100
3-7	81	57	24	148
1-9	91	69	23	>152
0-10	94	64	31	>152

Fig. 5-46. Bag culture in a greenhouse. Common materials are peat moss, perlite, modified soils, etc. The bags are easily handled and are sterile when purchased.

The literature is replete with investigations on the use of, and plant productivity in, various mixtures, using different handling procedures. Examples include several from Volume 99 of *Acta Horticulturae* [e.g., Strojny, 1980; Wilson, 1980; Feigin et al., 1980; Verdonck et al., 1980; etc.]. Other examples include studies by Bearce [1986], Greenwood et al. [1978], Poole and Waters [1972], and the articles on sewage sludge by Hornick et al. [1984] and Gouin [1993]. As with examination of plant response to irrigation frequencies or methods, most of the results are limited to the climatic region in which the investigations were carried out. They might not be capable of duplication in successive years as the result of different weather conditions. Interestingly enough, Simitchiev et al. [1983] showed quality of tomato fruits grown on soil or rockwool did not differ significantly. These were tested on many animals and possible aberrations in function of bodily organs were not evident. Thus, the "quality," or nutritive value, of products from plants grown in soil, soilless media, or hydroponics are not different if they are grown properly. The substantive conclusion from all these studies is that economic plants can be grown in about anything. The criterion is economic practicality in the particular location.

One advantage most of the newer practices have is that the medium is usually sterile when received, or can be easily pasteurized or renewed. Examples are shown in Figs. 5-46 and 5-47. Bagged culture is common in Europe, and rockwool (Fig. 5-47) is a major medium in the Netherlands. Diagrammatic examples of rockwool block arrangements are provided in Fig. 5-48.

d. Phytotoxicities and Contaminants

Material selected for an amendment, or to grow in as a soilless medium, should not be toxic or be contaminated with diseases, pests, weeds, or unwanted chemicals. This is not usually a problem with materials heated in manufacture (e.g., expanded aggregates, rockwool, perlite, vermiculite, etc.). However, sometimes peat moss, especially manure, sludge, etc., can contain substances that result in disaster for the grower. A particular problem with hardwood bark is phytotoxicity when used fresh. Hardwood must be composted for at least 6 months to avoid toxicity [Faber and Hoitink, 1983], although Mazuri et al. [1975] suggested a 13-month

Fig. 5-47. Growing tomatoes on rockwool cubes. Plants are started in the cube to the left-center of the picture and then set on the larger cube, which is covered by the white plastic. Note the irrigation emitter in the cube.

incubation period. Hardwood bark also possesses the ability to suppress some pathogenic organisms. Lawson and Horst [1984] stated that losses caused by *Phytophthora* spp. were lower in tree bark media than in peat-based substrates. Substitution of all or a major portion of peat with composted hardwood bark eliminated rhododendron root rot as a problem in many nurseries. Many other factors also influence the ability of media to suppress certain diseases such as *Phytophthora*, *Pythium*, *Rhizoctonia*, *Thielaviopsis*, *Verticillium,* and *Fusarium*. Nematodes may also be suppressed [Malek and Gartner, 1975]. According to Faber and Hoitink [1983], soil fungicides do not need to be used if the container medium contains at least 50% composted hardwood bark. Unfortunately, bark continues to decompose, and some barks, particularly if fine sawdust, rapidly decompose, tying up nitrogen as the result of high microbial demand. Higher levels of nitrogen fertilization are required. Pine bark is quite common in the southern U.S., and many individuals [e.g., Nelson, 1972; Love and Nelson, 1972; Love, 1978; Tilt et al., 1987] have investigated its use in various combinations with other materials. Love [1978] suggested particle sizes from about 0.6 to 9.5 mm as suitable. Particle sizes from >6.4 to <0.106 mm were investigated by Tilt et al. [1987], with their conclusion that the key to top growth response may not lie with size of container or type of media, but in the matching of the medium and container geometry with plant growth habit. Liptay and Edwards [1994] are some of the few who have examined container geometry. With tomatoes, 2 to 4 cm high cell widths were decreased incrementally from 1.36 x 1.36 cm to an elongated rectangle of 1.74 x 1.06 cm. More narrow cells caused seedling height to decrease, with shortest plants in cells 0.3 x 5.14 x 3 cm deep. Liptay and Edward's data show that cell depths of 3 to 4 cm resulted in greatest height, fresh and dry weights. Smoothness of the inner cell surface affected root growth, but did not affect shoot growth. The authors state that a rough surface texture reduced root growth, but their data does not always confirm this on an individual basis. Roots normally proliferate at the interface between the soil mass and container which I attribute to better oxygenation at that interface. Other phytotoxicity problems have been noted with mixes containing polyurethane foam [Wees and Donnelly, 1992]. Washing with water before use has been recommended. With foliage plants, fluoride levels are a problem [Peterson, 1975]. Fluoride can accumulate in plant tissues, and sources include the water supply, superphosphate fertilizers, and some soil amendments, including bark, peat moss, perlite, and others.

There is a continuing problem with herbicide contamination, the sources of which are often novel. Field soils in particular may be contaminated from continuous herbicide use. There have been problems from fertilizer contamination due to the mixing equipment or improper manufacturing procedures. A herbicide may be stored or transported and the floors and walls of the container are sometimes contaminated, resulting in contamination

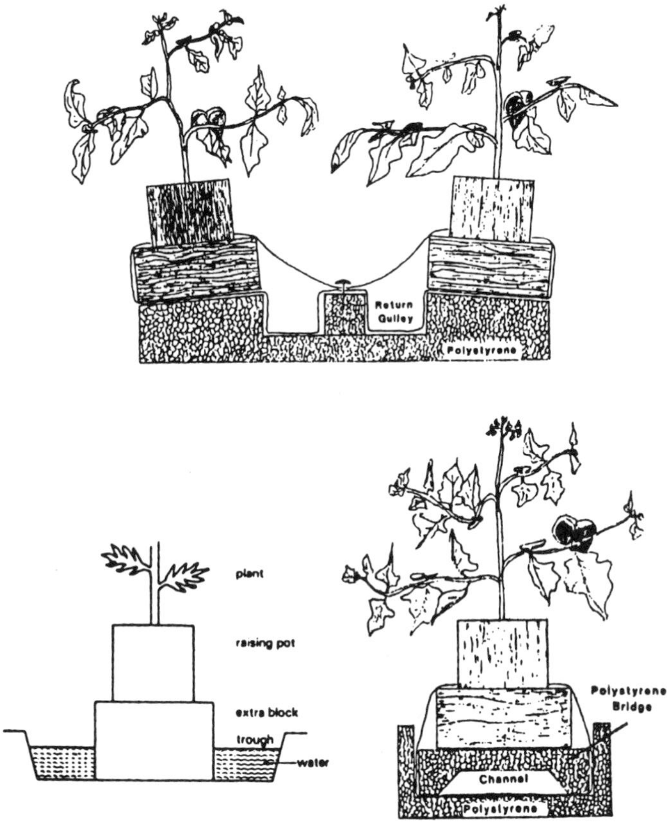

Fig. 5-48. Three examples of rockwool block arrangement. Those with polystyrene bridges from Leatherland [1987]; the stationary water system from van Os [1982]. See comments under hydroponics.

of future fertilizer or amendments stored in the same location. The many examples emphasize that herbicides should always be stored separately. The grower must always be aware of the history of his substrates. Quite often, the grower himself is his own worst enemy by using unmarked containers, possessing chemicals he does not require, failing to store herbicides separately, using the same equipment for application of pesticides and herbicides, and many other sins and omissions. Attempting to identify a herbicide before use is infeasible. Neither are there good means to eliminate a herbicide once it becomes apparent [Peterson, 1979]. Attempts have been made with massive additions to the substrate of activated charcoal or sugar; but these are of little value except as placebos.

e. Predicting Physical Properties

One of the greater difficulties in handling modified soils and artificial substrates is predicting the outcome. In fact, according to Johnson [1980], there are nearly as many methods for assessing composts as there are

Table 5-11. Optimum ranges for container mixture properties as given by Jenkins and Jarrell [1989], taken from the literature. See text for reservations of the author on these values.

Property	Units	Optimum range	Factors affecting optimum range in addition to plant type
Bulk density	g cm^{-3}	0.15-1.3	Texture, compaction, structure
Porosity	% volume	60-75	Depth, texture, structure
Air porosity	% volume	10-20	Depth, texture, structure
Container capacity	% volume	50-65	Depth, texture, structure
Available water	% volume	>30	Depth, texture, structure
Saturated hydraulic conductivity	cm hr^{-1}	>5	Texture, compaction, structure
pH	pH	5.0-6.0	Organic matter content
Cation exchange capacity	mol (+)/m^3 x 10	>10	

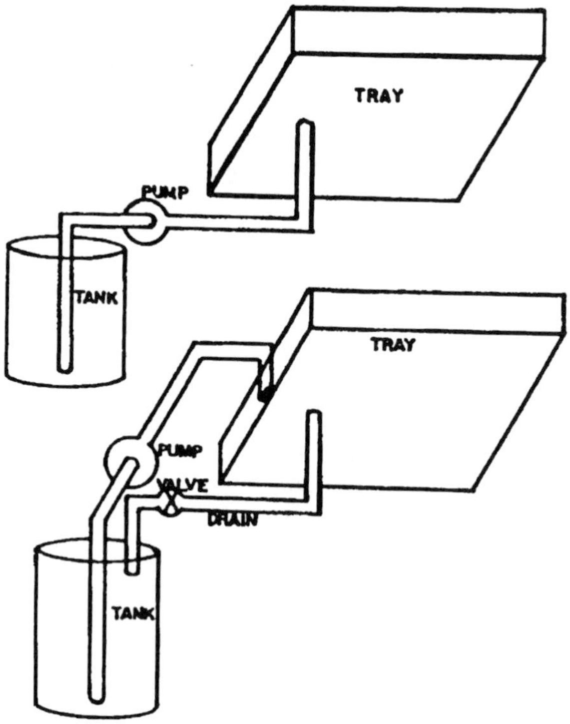

Fig. 5-49. Two diagrammatic examples of "classical" hydroponic systems, using a gravel substrate in water-tight benches with periodic flooding and draining [Hanan and Holley, 1974].

researchers. In 1978, Verdonck et al. compared results from seven laboratories to find that the figures were tremendously different for the same substrates. No correlation could be calculated between the different methods. It was Verdonck et al's conclusion that one could not provide a single method for determination of physical properties that satisfied all demands. More recently, a series of papers by Jenkins and Jarrell [1989] and Milks et al. [1989a, b, c] have attempted to solve problems of predicting physical properties of container mixtures.

Jenkins and Jarrell [1989] employed an additive model that assumed that a mixture property was the weighted sum of the properties contributed by the individual components, testing 24 combinations of sandy loam soil, sand, bark, and perlite. Their results showed considerable variation in predictions of bulk density, total and air porosity, container capacity, available water, saturated hydraulic conductivity, pH, and CEC. Their results contrast with the simulations of Burés et al. [1993], which suggest that additive models may not be appropriate. Of course, in my opinion, the use of the term "available" water is not valid in consideration of the water relationships of plants for the reasons discussed in Section 5.II.B and subsequent sections. I do not agree entirely with Jenkins and Jarrell's table of optimum ranges for container mixture properties (Table 5-11). For example, I believe the upper range of bulk density is too high, and any value less than 30 cm hr^{-1} for saturated hydraulic conductivity is too low. Lastly, a weighted sum does not account for the fact that mixing finer particles with larger particles results in a smaller volume.

Fig. 5-50. Example of a commercial hydroponic installation growing chrysanthemums in steel tanks of vermiculite rotated through the range. This operation received considerable publicity in the 1940s and 1950s. The picture was taken in the late 1960s. Interestingly enough, there was a commercial establishment within 1 km, growing the same genus in the ground which looked better.

The approach of Milks et al. [1989a, b, c] was to evaluate various statistical models for their ability to describe the soil water content versus the water potential relationship. A nonlinear moisture retention function was combined with container geometry in an "equilibrium capacity variable" (ECV) that provided predictions of porosity, air porosity, container capacity, available and unavailable water, and a solid fraction. In view of the wide variations of mixtures in which plants can be economically grown, one might question the usefulness of these types of models. As pointed out, moisture retention curves of soils are subject to hysteresis and can be considered as unique to the particular medium [Gardner, 1965b]. Some means of predicting physical properties, especially for plug culture (depths <5 cm), is devoutly wished. Milks et al. state that using their model permits any medium to be evaluated for performance in any container. Their methods, however, would require the services of a reputable laboratory. They are not, right now, to be undertaken by the average greenhouse operator.

In summary, the state-of-the-art in handling restricted volume, shallow, freely draining substrates can be related to some general rules that are usually necessary if acceptable plant growth is to be obtained. Some of these discussed in the text are repeated here:

1. Any field soil must be modified in some fashion.
2. Any amendment should comprise one-third to one-half of the final volume.
3. Materials to be used in a container mixture should be sterile, or capable of being pasteurized, free from contaminants, weed seed, or disease.
4. Vermiculite, when used singly as a medium, is likely to loose its structure over time, and it is not considered a suitable material for plug culture.
5. The use of fine particles such as sand (<0.1 mm) will fill larger pores in course media such as peat moss, resulting in loss of pore space and increased moisture retention.
6. Perlite is usually too light and requires addition of heavier materials for container stability. Peat moss as a part of a mixture requires upward pH adjustment to avoid aluminum toxicity, resulting from its release from perlite at low pH.
7. Rockwool (loose) should not be used at depths less than 5 cm, and behaves like sand when employed as an additive.
8. Do not compact mixtures more than necessary to stabilize the plant.
9. Store media properly, protected against contaminants, and do not add highly soluble nutrients before sterilization, or if the medium is to be stored for long periods.
10. Use of gravel (3-5 mm) to amend substrates is unlikely to be satisfactory.
11. Do not use "soft" (rounded) sand, or any sand containing quantities of calcium and magnesium carbonate.
12. Specify German, Canadian or Finish peats with some vegetation recognizable in the bale. Local peat mosses are likely to be very fine and mostly decomposed.
13. A grower requires that any material purchased for use as a substrate have the same properties from one lot to another, or from year to year.

G. HYDROPONICS

Hydroponics as defined here is the culture of a crop without any soil. If substrates are employed, they are usually inert with very little, if any, nutritive or cation exchange capacity. The lack of buffering capacity is a major distinction of hydroponics. Therefore, any substrate employed follows the same behavior as outlined previously for soils, modified soils, and soilless mixtures. Only in solution culture, or flowing solutions, is matric potential, hydraulic conductivity, air porosity, and moisture capacity irrelevant. Many individuals think of culture in peat moss or vermiculite as a form of hydroponics since an appropriate, complete nutrient application is required. However, both materials have considerable CEC, which means that they are "buffered" against changes in pH and nutritive content. Steiner [1968] did not include peat moss under the heading of soilless culture –contrary to Penningsfeld [1978].

1. Systems

Historically, water culture has been undertaken in research since the 17th century, with considerable publicity engendered by Gericke in the 1930s. Only in the last 2 decades have commercial producers begun

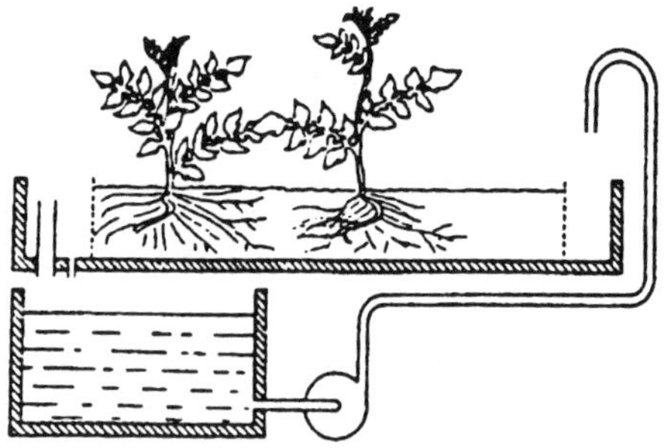

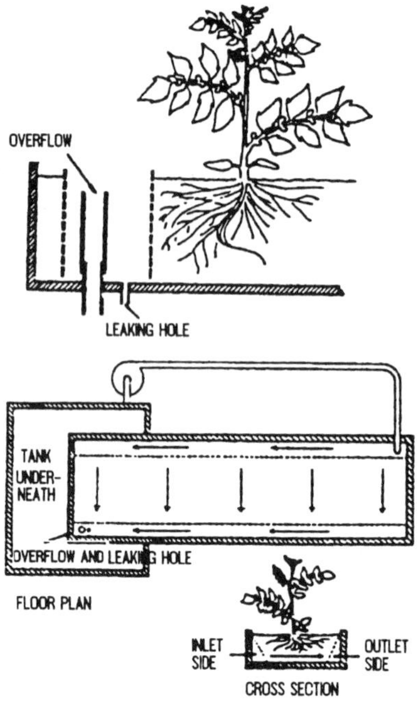

Fig. 5-51. Filippo's [1947] system for adequately flushing large-scale hydroponic systems, as presented by Steiner [1968]. Certainly an expensive system, seldom implemented on a large commercial scale.

to set up true hydroponics commercially [Steiner, 1968; Bunt, 1982; Bragg and Chambers, 1988; Sonneveld, 1980; etc.]. According to van Winden [1988], 90% or more of the Dutch production of tomato, cucumber, sweet pepper, and eggplant was in rockwool. There has been a regular progression of techniques, procedures, and methods that can fairly astound one in their diversity and the claims found in innumerable articles on the subject. Amusingly, one well-publicized effort resulted in political repercussions and the resignation of the premier in Newfoundland, Canada [Lamphier, 1989].

a. "Classical" Hydroponics

"Classical" hydroponics requires the use of watertight containers, storage tanks, pumps, and a substrate, as illustrated in Fig. 5-49 and 5-50. Although one may cycle the system several times daily (flood and drain), the substrate, usually gravel, may have extremely inflected water characteristic curves (see Figs. 5-26 and 5-27). Very severe stress can occur if the flooding frequency is not adequately regulated (Fig. 5-8) according to water demand. One is also faced with the problem of oxygenation of the root system as discussed by Steiner [1968] in some detail. Water saturated with oxygen contains about 10 mg ℓ^{-1} O_2 at 15 C and 9 mg ℓ^{-1} at 20 to 21 C. A tomato plant in 5 liters of water at 20 C has only about 45 mg of O_2 on which to draw, sufficient for about 2 hr on a sunny day. At 18 C, the diffusion rate of O_2 in water is 1.9 x 10^{-7} mg $s^{-1}cm^{-2}cm^{-1}$ distance for a difference in

Fig. 5-52. Example of an inert, gravel medium for carnations, using a drip irrigation system. Spray systems were found to physically damage the plants. Fertilized at each watering with surplus going to waste.

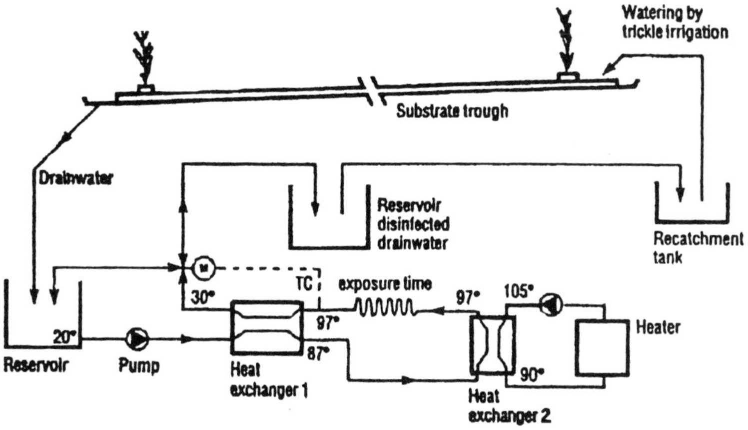

Fig. 5-53. Diagram of a recirculating, sterilization system for rockwool culture, as studied in The Netherlands [Lawson, 1989].

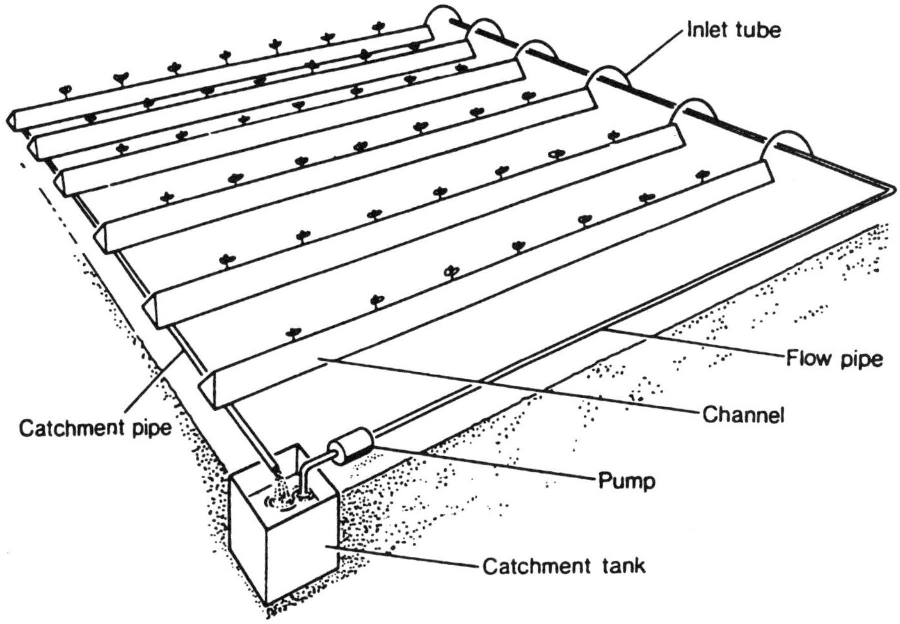

Fig. 5-54. Basic layout of a nutrient film, hydroponic, continuously circulating growing system [Cooper, 1979] (With permission of the British *Grower*).

partial pressure of 21.3 kPa (21% O_2) to zero at a hypothetical rootlet. Steiner calculated that, in a gravel substrate with an air porosity of 30%, the O_2 diffusion rate to the roots would be 0.42 mg $s^{-1}cm^{-2}cm^{-1}$. Since considerable water would be left behind after each watering, the oxygen supply could be severely limited. It was for this reason that Steiner recommended uniform-sized gravel for the best oxygen supply since mixed sizes increase the water film thickness retained about the root system. However, if the pore spaces of the gravel are too small for the depth, the substrate is likely to remain saturated. Once drained, very little water may remain for the plants to use. For oxygenation, Steiner suggested flooding every 75 to 150 minutes for carnation and tomato, respectively (particle sizes 4 to 12 mm diameter). To ensure complete replenishment of the substrate solution, Filippo's [1947] idea of moving he incoming flow uniiformly across the narrow width of the bench (Fig. 5-51) was strongly advocated. Solutions should be circulated and aerated vigorously.

The problem with these systems, other than the physical operation of any medium discussed above, is their cost and the fact that installation locks the operation into a single system that requires further high capital outlay for change. This was particularly true in the early years of this century when most installations were concrete (Fig. 5-50). Even plastic did not reduce installation costs sufficiently. A variation on this method that enjoyed brief popularity in Colorado was merely replacing the soil with gravel in existing raised benches (Fig. 5-52), and allowing the fertilizer solution to run to waste [Holley and Parker, 1968]. Here again, several problems cropped up apart from the growers inexperience with the necessity to provide an adequate nutrition program. The high irrigation frequency with existing spray systems (peripheral Gates) showed actual physical damage on the plants at existing water pressures [Hanan and Duke, 1970], not to mention problems of high humidity and increased foliar disease on the interior vegetative canopy. A final problem is the fact that each crop leaves behind part of its root system when the bench is pulled and replanted. Thus, after a few years, the grower winds up with a gravelly soil with the requirement to adjust irrigation frequency accordingly. Replacing the substrate in a 20 cm deep, long bench is a laborious and expensive proposition. This system, however, did provide earlier production of carnation cut-flower s with a distinct improvement in flower quality under Colorado conditions.

b. Bag and Block Culture

With the development of inexpensive, trickle irrigation systems in the 1960s, there was a considerable movement to growing cucumbers in straw bales. Plants were started in peat blocks and placed on top of the bales, each with its own trickle applicator. Straw bales were cheap, the floor needed no particular leveling, and automatic fertilizer injectors were readily available. This was followed by bags filled with peat moss, perlite, and finally rockwool blocks (Figs. 5-46 and 5-48). In all these, the irrigation water was allowed to run to waste. This waste of surplus water is a simplification of the nutrient program since the nutrients applied through the irrigation system are always the same. Detailed chemical analysis is not usually required on a regular basis. The problems of adequately monitoring a recirculated solution are bypassed. The material is sterile when received and can be reused by steaming at a fraction of the cost of steaming soil. Van Os [1982a] stated that soil pasteurization cost about $1.38 per cubic meter soil, and bags or rockwool could be resteamed for a fraction of that price. Verwer [1978] gave a figure of $898 per 1000 m^2 ground area, while an article in the *Grower* [Anon, 1981] gave a figure of $5433 per hectare for soil steaming. Clayton [1981] stated a value more than $15000 per hectare for sheet steaming of year-around chrysanthemum culture in the ground. Other materials have been used in bag culture such as polyurethane and polyphenol foams, flocks, mats, barks, vermiculite, and polymers. The latter material has particular problems, but initial results appeared encouraging [Anon, 1989]. The greenhouse floor is often covered with an insulating polystyrene sheet with another, white polyethylene sheet, over the entire floor. The polyethylene bags filled with the material are slit for drainage. With perlite, the procedure is different [Hall, 1986]. The bags are wrapped with an outer sheet. A rigid, polystyrene dam is placed under the outer sheet, between each bag. This forms a 6 cm deep puddle for each bag and the solution flows over the last dam in the row to waste. Woven bags for better aeration have been shown to provide an improved yield [Stirling, 1981]. One problem with peat moss is its yearly availability and variation in quality that sometimes occurs [Reilly, 1980].

Although rockwool can be used in the loose form –note that rockwool for insulation may have been treated with oil to reduce dust– [Lee, 1985], its most popular use is in blocks treated with a phenolic resin to bind the fibers and with a wetting agent. Blocks come in various dimensions, depending upon the crop, usually 7.5 to 8 cm deep, sleeved or unsleeved, and have become particularly popular in the Netherlands (Fig. 5-47) [Blaabjerg, 1983; van Os, 1982a, b; Bragg and Chambers, 1988; Verwer, 1978; Kitchener, 1978; Holcomb, 1982; Sonneveld, 1980; etc.]. Rockwool can be reused for several years, with some growers placing new blocks over the used when the old become thin enough [Drakes, 1982]. I believe the rockwool system has been the first use of a hydroponic system on a large commercial scale with innumerable variations (Fig. 5-48).

All these methods, including recirculating systems, are dependant upon high water quality [Kitchener, 1978; Krause, 1983; Sonneveld and Voogt, 1975; Verwer, 1978; Anon., 1978b; etc.]. As will be discussed in Chapter 6, this means a basic water supply with the least salt content possible. Hanan [1982] suggested a maximum electrical conductivity of 700 $\mu S\ cm^{-1}$ in the supply for minimum problems in handling salinity and nutrition

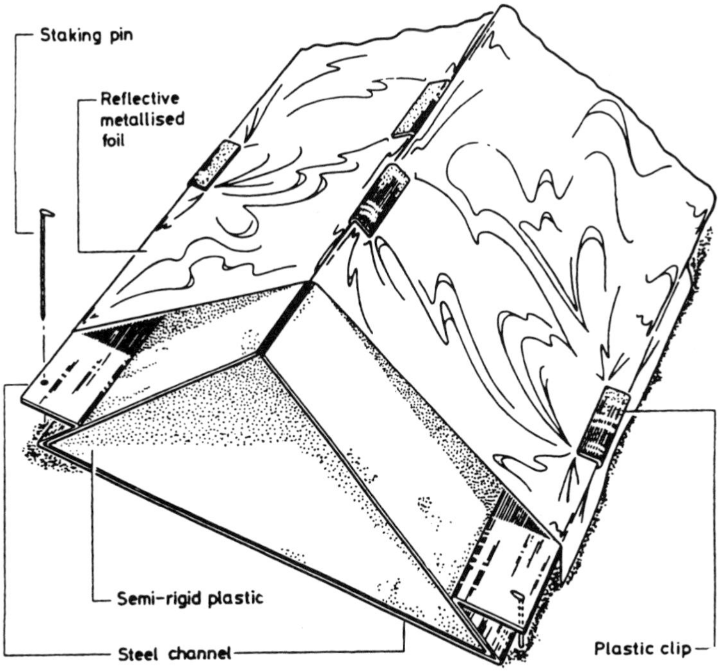

Fig. 5-55. Detailed section of an NFT channel according to Cooper [1979]. The reflective cover is often necessary to prevent excessive temperatures under high irradiance conditions (With permission of the British *Grower*).

programs. Levels below this value are to be sought even if paying more for water is necessary. Bag culture systems and rockwool block cultures are difficult to leach, and saline water supplies require more water to be used if maximum yields are to be obtained. Currently, concerns about the return of these waste waters to the ground water have aroused considerable emphasis on recirculation of bag cultures and rockwool as suggested in Fig. 5-48, with extensive research being carried on in the Netherlands to continuously sterilize the recirculated solution from rockwool culture (Fig. 5-53) [Lawson, 1989]. The Dutch are also heavily investing in mechanical operations, especially lettuce harvesting, where troughs, covers, planting, and eventual harvesting are carried out by machines [Lawson, 1984].

c. Recirculating

There are two approaches to continuous, recirculating systems: 1) use large water volumes (up to 50 mm depth [Kitchener, 1978]), with vigorous aeration for purposes of helping to buffer the solution (4 ℓ m^{-2}), and provide a safety factor in water supply; or 2), employ thin water films continuously flowing over some of the root system, and which may or may not have a large-volume bulk solution. In fact, most of the solution may be flowing through the channels –about 65 m^3 per hectare versus 2 m^3 for storage [Lauder, 1977]. The latter require particular care for emergencies that are likely to interrupt water supply. The best, well-known system is the British Nutrient Film Technique (NFT) attributed to Cooper [1979] as the principal originator. A typical layout for NFT is shown in Fig. 5-54. Cooper spends considerable effort in outlining the requirements of an NFT channel (Figs. 5-55, 5-56, 5-57). Slopes, ranging from 1:200 at the minimum to 1:30, have been examined [Anon., 1978a; Wilson, 1978; Cooper, 1974; Verwer, 1978; etc.], there being no limit to steepness according to Cooper [1979]. A 1:100 slope probably represents a valid minimum for practical purposes. With a slope of 1:50, or steeper, Hurd [1978] did not observe low oxygen levels or ethylene buildup in the root mat. A steady, uniform

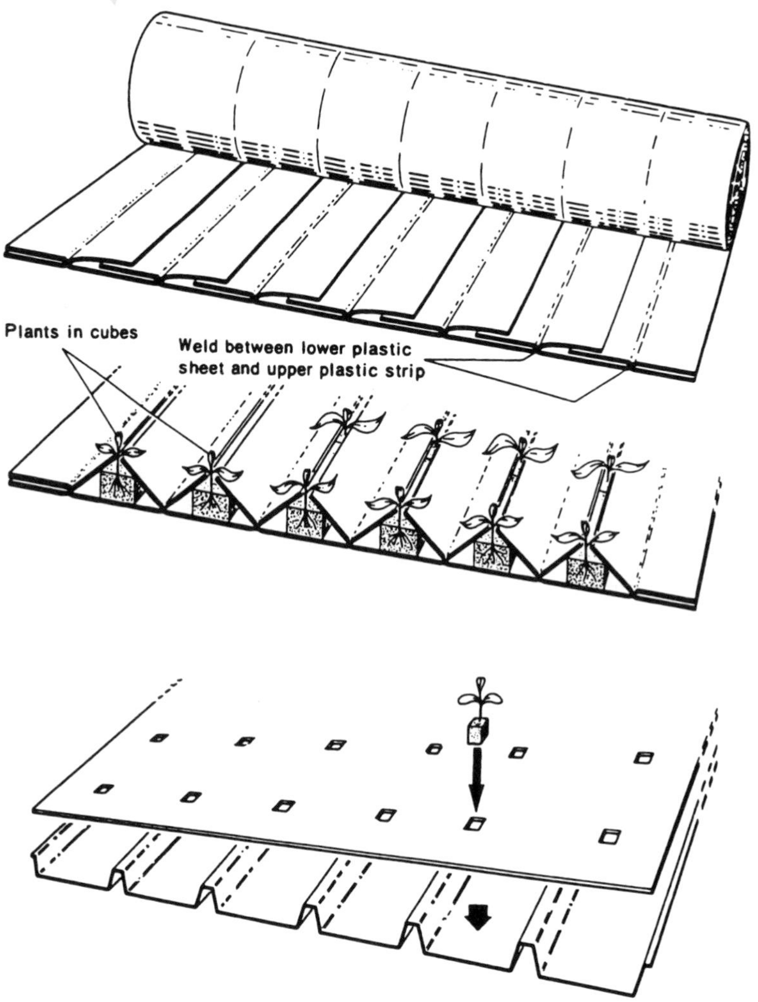

Fig. 5-56. Example of a multiple channel NFT system for crops such as lettuce, capable of being mechanized [Cooper, 1979] (With permission of the British *Grower*).

slope, with a rigidity sufficient to bridge slight depressions is required. The troughs must not change shape with time –a problem with some plastics. A flat channel, with a very slight concavity of the center is needed to ensure that the solution flows down the center of the channel, eliminating the need for a capillary mat. However, if the solution meanders across the channel, a thin, nondegrading capillary mat can be used to ensure that young plants receive water. A base width of 23 cm is suggested. This will provide sufficient room for the root mat without undue resistance to water flow and consequent damming of the water. Channels can be elevated to avoid stoop labor, or placed on movable tables. Sometimes, the entire floor has been concreted and epoxied, forming the necessary channels for, say, a lettuce production facility [Lauder, 1977a; c].

The critical factor is the water depth as it flows down the channel. Cooper [1979] stated that depth should not be more than a few millimeters. On the other hand, Verwer [1978] limited film depth to no more than 2 mm as contrasted to 10 mm attributed to Hurd [Anon, 1978a]. The root mat may be 2 to 3 cm deep on a fully grown crop [Hurd, 1978], so that a large portion of the root system is exposed to the air. Published water flow rates

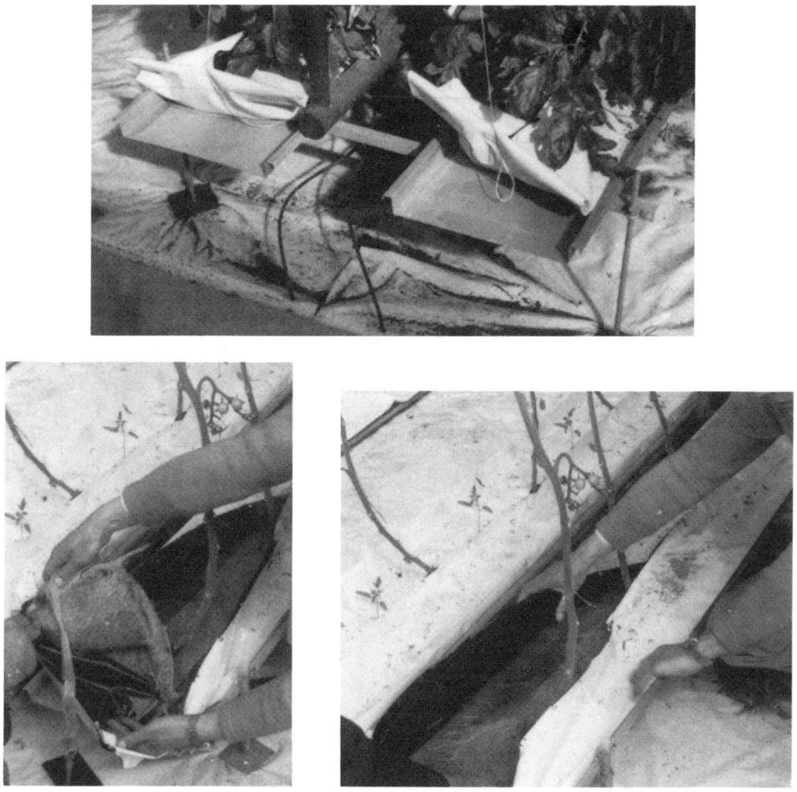

Fig. 5-57. Three examples of NFT, the bottom two showing the root mat usually formed on a crop such as tomatoes.

have been 1ℓ m^{-1} per channel [Wilson, 1978], or 3 ℓ $m^{-2}hr^{-1}$ [van Os, 1982]. According to Hurd [1978], maximum transpiration demand is about 0.8 ℓ per plant per day. For a typical 0.4 ha installation and 12000 plants, this rate will result in one complete change of the nutrient solution per day (total volume in circulation about 9 m^3). Maximum channel lengths of up to 40 m have been used. Intermittent circulation of the solution has been examined [Graves and Hurd, 1983] as a means to control growth. This may be highly drastic and negates the whole purpose of a freely available water supply. I should emphasize that most of the investigative work on NFT has been under northern European climatic conditions. Such practices as stopping the solution periodically might be more practical under the low winter irradiance conditions found in Europe. Investigations of the type cited here have not been published for high solar irradiances found in dry climates. It is my belief, for reasons already given, that clear-day regions would provide maximum return for high-cost, highly technical installations of the types described here. In the original NFT system, plants were started in peat blocks and later planted to the channels. The problem with peat was the necessity to provide a filter in the system to remove peat particles washed from the blocks. A filtering system should probably be installed in any case. Later, starting cubes included plastic and rockwool. An interesting sidelight on some more subtle technical difficulties to be found in these systems is the

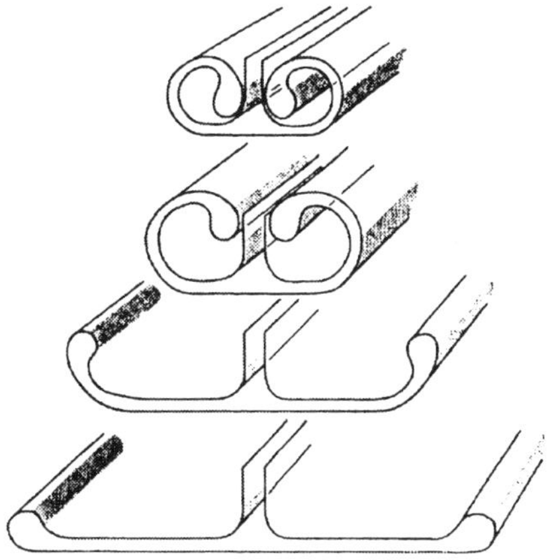

Fig. 5-58. Rolled, plastic NFT troughs that adjust with root growth [Mossman, 1980] (With permission of the British *Grower*).

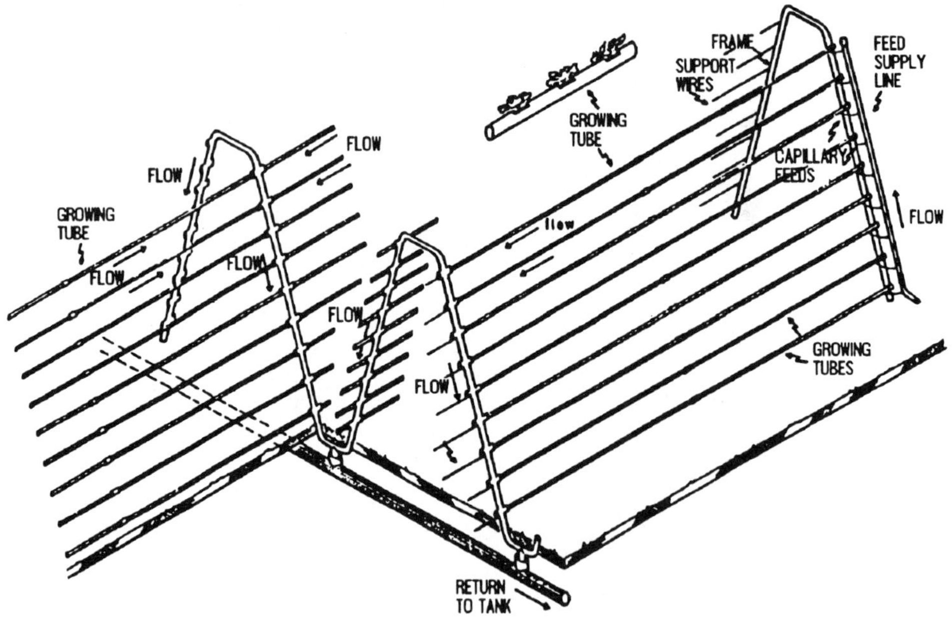

Fig. 5-59. Archway NFT system proposed by Morgan and Tan [1983] for high-density lettuce production (©1983, Intern. Soc. Hort. Sci., *Acta Hort.*, 133:39-46).

Fig. 5-60. NFT system proposed by Giacomelli et al. [1983] for year-around tomato cropping. The channel is supported by an overhead cable with solution returning to the catchment tank in the center between the supporting posts. The latter are movable from side-to-side (©1983, Int. Soc. Hort. Sci., *Acta Hort.*, 133:89-102).

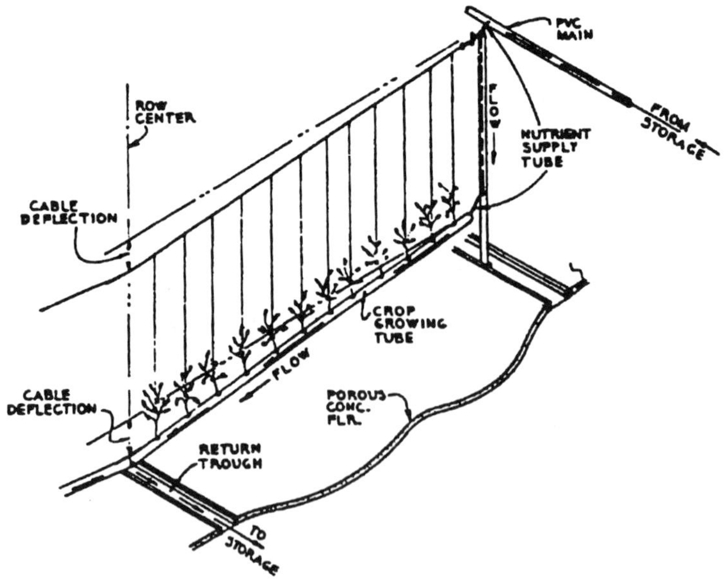

fact that rockwool blocks can be vertically or horizontally grained (fibers running at different orientations). Clayton [1981] stated that separating horizontally grained rockwool starting blocks severely damaged the root systems because the roots grew out the side of the blocks as contrasted to the vertically grained cubes.

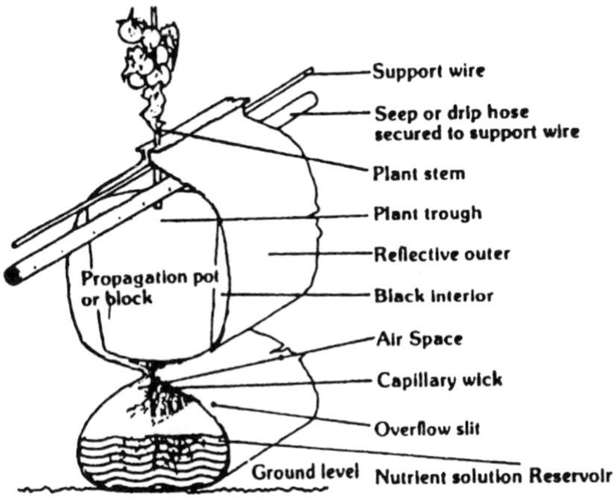

Fig 5-61. One of several variations of NFT proposed in the last decade [Anon., 1982a]. See Fig. 5-48 for other examples of combinations of substrate and recirculated water (With permission of the *Amer. Vegetable Grower*).

Nelson and Fonteno [1991] have also examined rockwool, showing that the various slab types investigated had released most water at -0.004 MPa. There were considerable differences in physical properties of various proprietary slabs. Direct rooting of chrysanthemums was not successful and the effect of horizontal or vertical graining was not remarkable. A system has been developed to divide the root system [Anon., 1985]. At planting time, the new plant is placed upon an upside-down, V-shaped channel located in the NFT trough, covered with a capillary mat. The solution can be run down one side for a period, then switched to the other side of the V. There is no need of a starting cube that may put trash into the system. Still another modification developed by Cooper [Mossman, 1980] has been a rolled channel that adapts to the size of the root mat (Fig. 5-58).

Since the introduction of NFT, many modifications of the original system have been devised. For example, Fig. 5-59 shows a proposal by Morgan and Tan [1983] for high-density lettuce production, whereas Fig. 5-60 is a cable supported system developed by Giacomelli et al. [1983] for continuous cropping. Other combinations and permutations may be suggested by Fig. 5-48. Fig. 5-61 is still another method, seeking to simplify problems of slope, film depth control, etc.

Of course, water temperature in recirculating systems should be controlled. Recommendations have ranged from 20 to 22 C for such crops as tomato and cucumber [Clayton, 1981; Anon., 1978a; van Nierop, 1982] –although one may find in the trade literature statements to the effect that "heating the solution to encourage extra growth shows no benefits as for extra yield" [Anon., 1979b]. It is equally valid, however, that failure to heat can result in loss. Production of high temperature-requiring crops such as tomatoes and cucumbers often demand heating of the substrate in northern climates. It is customary in bag and rockwool systems to run 1 or 2 hot water pipes under each row [Verwer, 1978]. In Chapter 4, this point was discussed from the standpoint of energy savings by putting heat in low and perhaps even reducing the air temperature. According to some authors [i.e., Mossman, 1980; van Nierop, 1982: Verwer, 1978], the energy savings can be substantial without greatly affecting yields of most crops, especially if a thermal screen is employed. Results reported have ranged from 24 to 50% lower energy consumption compared to running the air temperature at recommended values [Anon., 1978a; van Nierop, 1982]. Proper control of substrate and air temperature intimates a suitable, sophisticated control system (Chapter 8).

2. Phytotoxicities, Contaminants and Disease

The lack of buffering in most of the newer methods of greenhouse culture means that contamination can be quick and disastrous. Some types of plastics can release toxic materials [Cooper, 1979] that are rapidly circulated throughout the system. The small water volumes, however, suggest that if the problem is diagnosed soon enough, the system can be drained, flushed, and refilled –although this is wasteful. This is not true with infectious disease organisms where a single infected plant may release causative organisms to the rest of the culture. This was a dominant point in the early years of NFT [e.g., Davies, 1981; Anon., 1979c; Evans, 1977; Price, 1980; Lawson and Dienelt, 1988; etc.]. A system running water to waste at least avoids recirculating a contaminant.

a. Disease Problems

Disease problems were particularly emphasized with the appearance of root death on tomatoes at the stage of first fruit production. Hurd [1978], however, pointed out that this also is at the time when leaves are removed, not to mention the fact that fruit formation acts as a highly preferential sink for photosynthates in the plant so that roots may not receive sufficient respiratory substrates. Root death apparently represents a normal procedure with tomato growth as presently practiced in most greenhouses. Hurd [1978] makes the point that the shoot-root ratio in the tomato is a constant, and root death is apparently a response to this physiological requirement. Given the heterogeneity of root systems, death and new generation is to be expected, as noted in Section 5.III.A Unfortunately, the presence of dead and decaying tissues is likely to have repercussions as for other organisms present in the culture –i.e., *Pythium* spp. Another, physiological disease common to NFT cultures is "collar rot" of tomatoes. Wilson [1978] pointed out that this resulted from salt accumulation on the necks of the plants, resulting in stem damage. Phosphate concentrations in solution should be limited to 0.5 meq/ℓ, pH rise should be limited and controlled, and the salts washed off with clean water.

Besides *Pythium*, many other organisms are important such as *Phytopthora*, *Fusarium*, *Colletotrichum*, *Pyrenochaeta*, *Didymella*, and the bacterial organisms *Erwinia* and *Corynebacterium*. The spread of Big Vein virus in lettuce has been attributed to the fungus *Olpidium* [Price, 1980], and is a particular problem in NFT [Birch, 1980]. In fact, more than 35 microorganism species have been isolated from recirculating hydroponics, not to mention algae. The presence, multiplication and spread of fungal flora in NFT systems is common and normal [Vanachter et al., 1983]. *Pythium* spp., as the result of their ubiquitousness, can be a continuing difficulty, especially with young seedlings. It is not exceptional to isolate *Pythium* from ordinary water supplies. Even if the installation is sterilized with steam or urea formaldehyde, it may be reinfected. Work by Hockenhull and Funck-Jensen [1983] showed that severity can be related to the presence of suitable substrates upon which the normally saprophytic fungus can feed. NFT systems do not usually provide much of this substrate although addition of sucrose could enhance pathogenicity of the organism. It is a well-known fact that roots exude many proteinaceous compounds that can provide a substrate for fungal growth. Certainly, a dead and decaying root system would also provide a suitable substrate. Thus, *Pythium*-related damage is often worse in soils or highly moist substrates such as peat moss [Funck-Jensen and Hockenhull, 1983]. The idea is to prevent *Pythium* from building up to a sufficiently high inoculum potential to successfully infect the host by means of competition from other organisms, keeping the solution clean or using some system of continuous sterilization of the solution.

The reports of Vanachter et al. [1983] and Staunton and Cormican [1978] have shown that many of the above organisms are not any more dangerous in NFT than in any other substrate or method. *Fusarium oxysporum f. dianthii* is just as bad in NFT, peat moss, rockwool, or soil.

b. Genetic Resistance

Although there have been numerous calls for resistant carnation varieties by the trade, the requirements to incorporate resistant genes in the highly heterozygous carnation have been intractable to the present, as contrasted to some vegetable species. It has been my observation over 30 years, that carnations can be grown disease-free continuously (up to 5 yr) if the propagative material is disease-free, the substrate is suitably isolated, pasteurized, and precautions are taken to prevent inoculation. The problem in most commercial establishments is that the surroundings have already been infected with many pathogens. It is easy to reinfect isolated cultures through carelessness and the failure to install the culture with appropriate precautions. Hanan et al. [1963] showed it possible to eliminate *Pythium* from soil cultures of the highly susceptible *Antirrhinum* (snapdragon) through appropriate sanitation procedures. It may be that the new gene manipulation techniques will be successful in providing resistance in ornamentals.

c. Chemical Control

The possibilities of fungicidal applications in recirculating solutions have been investigated [Anon., 1979a; Vanachter et al., 1983; Evans, 1977; etc.]. With food crops, this can be a dangerous practice and likely to be severely regulated by governmental agencies. The use of chlorine and copper as disinfectants has been suggested, but this is inadvisable as effective concentrations would also be phytotoxic to many economic species –not to mention interfering with nutrition control. Exposure of the solution to ultraviolet light has been suggested [Evans, 1977; Lawson and Dienelt, 1988]. Although Lawson and Dienelt state this has been successful in controlling *Pythium* root rot of spinach, iron must be added to the solution after each treatment. It is also expensive, and the system proposed in Fig. 5-53 may be the best from the standpoint of temperature control, cleaning solutions, and disease control. The situation emphasizes the role of primary sanitation and clean stock in disease control regardless of cultural method. Good sanitation is generally the cheapest way to go.

Table 5-12. Comparative costs of soil and rockwool production for cucumbers and tomatoes. Verwer [1978] give figures based upon 1000 m^2, which I have converted to $U.S. per m^2, using the monetary conversion given in footnote 1, Chapter 1 (©1978, Int. Soc. Hort. Sci., *Acta Hort.*, 82:141-147).

	Cucumbers		Tomatoes	
	Soil	**Rockwool**	**Soil**	**Rockwool**
Steam pasteurization	0.90	--	0.90	--
Rockwool medium	--	1.35	--	2.19
Cow manure	0.11	--	0.07	--
Plants	0.69	0.78	0.96	1.10
Substrate heating	0.66	0.30	--	--
Nutrition	0.06	0.45	0.10	0.45
Watering system	0.06	0.14	0.10	0.22
Fertilizer analysis	0.04	0.24	0.04	0.24
Removing old crop	0.05	0.11	0.05	0.16
Extra labor, 7.3 hr	--	0.07	--	0.21
Total costs m^{-2} ($ U.S.)	2.57	3.43	2.22	5.07

3. Costs

Some comparisons between soil culture and hydroponics have been given in the introduction of this section, along with some general costs of sheet steaming ground soils (Section 5.III.G.1). Verwer 's 1978 figures are more thorough in comparing soil versus rockwool blocks (Table 5-12), with more recent data provided by van Os (Table 5-13). Rockwool culture in the Netherlands has been tremendously popular with the introduction of NFT resisted. Roe's figures [1986] are fairly detailed in comparisons between rockwool and NFT. However, Table 5-14 compares capital investment for peat bags, rockwool, and NFT, and continuous flow rockwool systems (Fig. 5-48).

Table 5-13. Investment, annual cost, and total extra costs for several culture systems [van Os, 1982b], based upon a 10000 m^2 nursery, converted here to $U.S. per m^2 (With permission of the British *Grower*).

Substrate	Investment	Annual cost	Extra costs compared to soil
Rockwool	6.53	1.90	1.77
Stationary water	7.04	2.22	1.46
Rockwool in circulating water	8.11	2.80	2.39
Peat bags	4.58	1.71	0.78
NFT	9.76	2.69	2.58

Numerous figures may be found elsewhere in the literature. Sieur and Brungardt [1977] quote a number of authorities that a 400 m^2, hydroponic greenhouse can be raised for a capital investment between $30 and $38 m^{-2}, although other figures go up to $86 m^{-2}. These data are at variance with those given in Tables 2-13 and 2-14 where, depending upon the materials and type of construction, greenhouse costs may vary from $12.42 m^{-2} to more than $190 m^{-2} for houses complete with raised benches and soil culture. Such figures as given by Sieur and Brungardt can be misleading. Describing a 1 ha NFT installation in the U.K., an anonymous author [Anon., 1977] states that adaptation of NFT cost $8.87 m^{-2} with the inclusion of a $16000 laboratory for nutritional control. Cooper [1986], however, touting the use of NFT, provided a mean combined labor and fuel cost of $11 m^{-2} for NFT versus $13 m^{-2} for rockwool. Cooper [1977] also stated fertilizer costs for NFT as $0.54 m^{-2} versus $0.37 m^{-2} for vegetables grown in soil. In fact, fertilizer and water costs in greenhouse production are generally less than 10% of the total operational cost. Although water cost can exceed $1.00 m^{-3}, or $0.50 m^{-2} [Krause, 1983], looking to save expenses in this area can be a mistake. According to a 1982 article in the *American Vegetable Grower*, consumables for hydroponic systems ranged from $1.02 to $1.45 m^{-2}, and durables (over a 7 yr period) would range from $2.40 to $6.30 m^{-2}. Krause [1983] stated that although previous costs were calculated at $0.50 to $1.00 m^{-2} more than growing in soil, the actual costs in 1983 were closer to $0.10 m^{-2}. Building a rockwool system into a newly erected glass house in Holland actually produced a $0.50 m^{-2} saving. This was because of subsidies of 10% provided by the government for installation of rockwool on the grounds that the system saves energy.

The variable reports cited can leave one thoroughly confused. There is nothing to say about which of the publications may be most reliable since the articles may lack sufficient detail concerning methodology and data collection. Suffice it to conclude that hydroponic systems will be more expensive than growing in the ground. The grower must assess capital and operating costs versus the supposed improvement in yield and quality of the product and the proposed net return as a consequence. The general impression I received from many scientific and popular publications leads to a cynical outlook that much of the thrust for hydroponics has been over publicized.

Table 5-14. Capital invested, net annual capital costs, and net extra costs for four different hydroponic cultures, in \$U.S. m^{-2} [Anon., 1981]. For the U.K. (With permission of the British *Grower*).

	Peat bags	**Rockwool**	**NFT**	**Rockwool continuous flow**
Leveling	0.22	0.22	0.22	0.22
Pumps, meters	0.89	0.89	0.89	0.89
PVC pipes, filters	0.44	0.44	0.44	0.44
Water heating	--	--	0.08	0.08
Drip irrigation	0.67	0.67	--	0.67
Substrate warming	--	0.89	--	0.89
Peat bags	1.18	--	--	--
Rockwool	--	1.33	--	1.33
Gullies (channels)	--	--	4.44	1.11
Total \$ m^{-2}	3.39	4.43	6.17	4.72
Net annual capital costs	1.15	1.24	1.21	1.34
Net extra costs per year	0.99	1.73	2.21	2.00

4. Yields

Quite often, the high yields attributable to hydroponics are heralded to the point of being misleading. The tremendous increase in yields as the result of esoteric practices results from the wrong comparison. A number of authors [e.g., Albery et al., 1985; Sieur and Brungardt, 1977] make the comparison between hydroponics and field culture. Sufficient information has been given in preceding chapters to emphasize that greenhouse culture **is intensive**. The fact that one may produce year-round in a greenhouse versus 3 or 4 months in the field in temperate climates is bound to result in huge yield differences in favor of greenhouses despite the cultural procedure. Whereas, some more sophisticated hydroponic systems have often been disappointing and questionable for ornamentals and vegetable production [e.g., Anon., 1978b; van der Hoeven, 1980; Waterfield, 1981; Crane, 1981; Dungey, 1983; Anon., 1982c; d; etc.].

Many things can influence yields in any substrate, ranging from the cropping method to how much of the greenhouse is in production –i.e., the extent of the vegetative canopy. Few if any authors state whether yields per unit area include the whole greenhouse (corner-to-corner) or just the portion actually planted to crop. As an example, Wilcox [1980] emphasized the importance of available solar radiation in accordance with production scheduling (Table 5-15), showing that, in the Midwest and northern U.S., profitability of an operation could be drastically modified by planting time and production period. Schippers [1982] discussed the possibilities of inter-cropping tomatoes (Table 5-16) and methods of rearranging the crop as it grows to increase productivity. Some better comparisons are Cooper's [1986] use of the U.K. Grower Tomato Performance Recording Scheme (Table 5-17) to compare rockwool and NFT. It can be noted that improvement of yield in NFT varied widely, depending upon the length of the harvest period with an overall advantage to NFT of 3% higher yield per unit area after 6 months cropping. Figures specifying yields as high as 44 kg m^{-2} tomatoes (+200 tons/acre) usually fail to specify time in production, actual area, plant density, and location. A still better comparison is van Os' data presented in Table 5-18. Typically, NFT was not particularly interesting –in a region where technical skill is high and readily available. One of the best discussions was Mary Peet's presentation in 1985 (Fig. 5-63). Depending upon whether yields are compared on a per-plant basis, per unit area, or per unit area per week, one can get widely varying results. The latter comparison, on a weekly basis, shows that growers in soil compare quite favorably with worldwide production averages in the various cultural systems for tomatoes. Peet [1985] states that when different systems are compared in a single study so that weeks of cropping and plant densities are the same, yields are also similar. Further, she states that the surest way to increase yield per plant is to grow longer. The

Table 5-15. Hydroponic tomato yields and production costs per m^2 of midwestern U.S. growers for a non-optimum schedule versus an "optimum" scheduling in the northern U.S. [Adapted from Wilcox, 1980] (With permission from the *Amer. Vegetable Grower*).

Month	Accumulated expense ($/m²)	Fruit yields (kg/m²)	Accumulated value ($/m²)	Profit or loss ($/m²)
Non-optimum, winter production schedule of six midwestern growers				
August	3.69			
September	5.54			
October	7.96			
November	10.76	4.3		-4.46
December	13.88	2.0	9.20	-4.68
January	17.05	0.9	10.58	-6.56
February	20.31	0.7	11.55	-8.76
March	23.29	0.5	12.24	-11.06
April	25.80	1.6	14.63	-11.19
May	27.85	1.5	16.77	-11.09
Optimum scheduling in northern U.S. for tomatoes				
February	1.67			
March	4.72			
April	7.30			
May	9.42	0.8		
June	11.38	5.7	9.56	-1.82
July	13.33	5.7	17.91	4.58
August	15.28	6.2	26.96	11.67
September	17.24	4.9	34.15	16.91
October	19.63	3.1	38.77	19.11
November	22.54	0.6	39.61	17.07
December	25.80	0.5	40.36	14.56

surest way to increase yield per unit area is to optimize plant densities.

The problems of greenhouse, hydroponic vegetable production in the U.S. were pointed out in Chapter 1. They have also been discussed by Gerber [1984] and Killenberg [1986]. The publicity has attracted many individuals who, besides lacking the necessary technical skills, may also lack the managerial talents necessary in any business. Gerber emphasized that a small operation (1000 plants) does not provide incentive for automation, although such an operation can survive by supplying local and specialized markets with a high-quality product. Gerber stated that a single greenhouse can keep one individual employed at a minimum wage with perhaps a profit of $1000 to $2000 per year in Illinois. Killenberg, for the northeast U.S. and lettuce operations, suggested that the average operation covered less than 600 m^2 and required an investment of $28000 to build. The sales were mostly to high-income earners (>$25000 per year). It is my observation that there is a serious lack of interaction between growers and extension-research institutions. There have been a number of failures. Schippers [1986] took exception to some of Killenberg's remarks. The main reason, according to Schippers, to grow winter lettuce is to protect a market, with the higher production in spring and early summer for profitability. It was his experience that about 1.5 million heads of lettuce $ha^{-1}yr^{-1}$ was a respectable goal with lettuce in hydroponics and, with a variable spacing method, 1.9 million heads $ha^{-1}yr^{-1}$ was possible. Schippers stated that hydroponics as a technique "offered no technical problems at all" and did not need much research. In another article, Schippers [1981] described a lettuce-producing facility expected to achieve a weekly summer yield of 6000 heads per hectare in hydroponics in Massachusetts.

The purpose of all this discussion is to illustrate representative yields and costs for hydroponic systems for greenhouse culture. That is, the systems used for greenhouse culture should be compared within the context of greenhouse operations –although Phene et al. [1992] have proposed that 200 tons per hectare of processing tomatoes may be achievable commercially in the field without the benefits of hydroponic systems. In any event, improvements are directly related to the competence of the grower and his ability to obtain and operate the required equipment. Improvements are incremental, generally within the range of 10 to 15% or less. My point in spending this time and effort on cultural procedures for supplying water is that choice and success is largely dependent upon many factors (climate, technology, economics, political). The fact that a particular system is touted as the solution to all problems is misleading. As I will point out in Chapter 6, there is a long way to go, for, in my estimate, the existing systems for nutrient control in many greenhouses are archaic.

Table 5-16. Kg per plant tomato yield under various cropping patterns of 32 wks for continuous, 20 wks total for two crops, and 28 wks for intercropped trial [Schippers, 1982] (With permission of the *Amer. Vegetable Grower*).

Variety		Total
Tropic	Continuous crop	9.3
	Spring plus fall crops	8.6
	Spring with interplanted fall crop	8.6
Sobeto		
	Continuous crop	10.1
	Spring plus fall crops	11.2
	Spring with interplanted fall crop	11.3

Table 5-17. Cumulative tomato yields over 6 months for plants in rockwool and NFT [(Cooper, 1986]. (metric tons per hectare). Ten growers (With permission of the British *Grower*).

Month	Rockwool mean	NFT mean
Feb	0.85	1.59
Mar	14.66	15.31
Apr	45.19	50.18
May	98.06	104.98
June	146.30	157.76
July	215.56	222.88

Note: to convert to kg m^{-2} move decimal 1 place to the left.

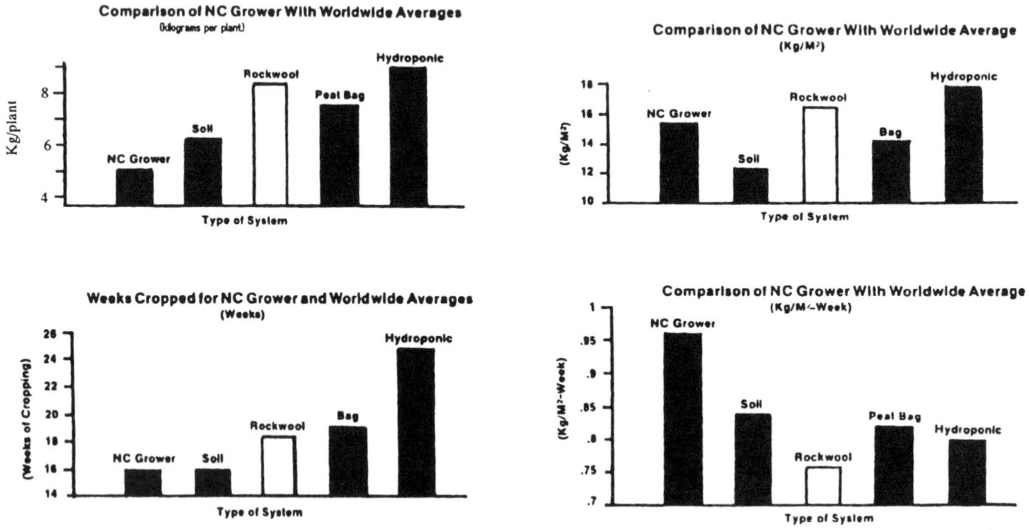

Fig. 5-62. Comparisons of North Carolina (U.S.) growers in soil with worldwide averages: **Upper left,** kg per plant; **Upper right**, kg m^{-2}; **Lower left**, number of weeks cropped; **Lower right**, kg m^{-2}wk^{-1} [Peet, 1985] (With permission of the *Amer. Vegetable Grower*).

Table 5-18. Experiments in the Netherlands comparing growing systems for a number of vegetable crops [Adapted from van Os, 1982a].

Location	Substrate	Variety	Sowing date	Last harvest date	Yield (kg m^{-2})
		Tomatoes			
Vleuten	Rockwool	Sonatine	1 Nov	28 July	15.4
	Rockwool in stationary water				15.4
	Foambags				14.8
	Peat bags				15.4
Vleuten	Rockwool	Dombito	5 Nov	7 July	13.7
	Rockwool in stationary water				12.8
	Soil				10.2
Bleiswijk	Rockwool in circulating water	Sonatine	26 Oct	15 July	18.8
	NFT				19.5
	Soil				16.0
	Rockwool		29 Oct	14 July	18.1
	Rockwool in stationary water				16.8
	Peat bags				16.3
	NFT				17.7
Wageningen	Rockwool		8 Dec	17 Sept	19.8
	Rockwool in circulating water				18.4
Westland-Z	Rockwool		4 Dec	17 July	20.7
	NFT				19.2
	Soil				19.5
		Peppers			
Naaldwijk	Rockwool	Bruinsama	?	4 Sept	12.7
	NFT				11.5
Wateringen	NFT		7 Oct	9 Sept	12.2
	Soil				10.5
		Eggplants			
?	Rockwool	?	14 Feb	15 Sept	16.5
		Cucumbers			
Sappemeer	Rockwool	Corona	15 Dec	2 June	18.8
	Foambags				17.6
Pijnacker	Rockwool		1 Dec	30 July	45.1
	Rockwool in stationary water				41.9
Vleuten	Rockwool		5 Dec	27 June	30.5
	Foambags in stationary water				28.1
	Foambags				29.5
Utrecht	Rockwool	Primio	14 Dec	15 July	35.0
	Rockwool in stationary water				34.0

H. IRRIGATION SYSTEMS

There are several ways to apply water to substrates. The selections depend upon technology, labor and material costs, and the cultural procedures used in the greenhouses. These methods are listed.

1. **Flood** (furrow): Application of free water to medium's surface and allowed to flow over the surface until reaching the furthest distance from the source.
2. **Hand watering**: (Fig. 5-63D). Generally the cheapest method in terms of equipment, but labor costly. Application of high volumes in a short period. Still utilized under special circumstances.
3. **Sprinkler**: (Figs. 5-63A and E, 5-64). Lower rates of application, increased capital investment. With possible exception of Fig. 5-63A, seldom used for cut-flower production. Most common use for small units on a dense spacing (Fig. 5-64 right).
4. **Capillary** (sub-irrigation): (Figs. 5-65 and 5-66). May also be employed with **ebb-and-flow** systems (Fig. 5-67), most commonly for pot plant production. Capital intensive. Can be considered in conjunction with spaghetti systems.
5. **Trickle** (**Drip**): (Figs. 5-63B, C, and F, 5-68, 5-60). Most expensive as to equipment but widely used for cut-flower and vegetable production in greenhouses and in the field. Low water application rates, infiltration not usually a problem. Spaghetti systems may be considered a variation of trickle systems.

Fig. 5-63. **A**: Gates peripheral spray system with nozzles spraying toward bench center, spaced on 45 to 60 cm distances, pressures limited to 35 kPa at the nozzle. **B**: Porous wall, drip tubing, low pressures, low application rates. **C**: Sown plastic tubes with capillary supply (Drip). **D**: Hand watering with a hose breaker to reduce velocity and force. **E**: A lateral spray system that moves up and down the walks between benches. Gradually going out of use. **F**: Twin-wall drip tubing on a gravel substrate.

1. Primitive Systems

Flooding in greenhouses is usually where cultural conditions are primitive, structures are unheated, and labor is cheap. It is never used for maximal production as the industry develops. Hand watering, which was most common up through the early 1950s, rapidly went out of vogue with the increase in labor costs and development of plastic pipe and fittings. Fries [1979] calculated the cost. If 12 ℓ m^{-2} water is applied at a rate of 19 ℓ min^{-1} from the hose, and the labor spends 80% of the time in actual application, then for 7300 m^2 in a 1 ha house and a labor cost of \$3.00 hr^{-1}, the total labor cost for 1 watering will be \$277.00. Obviously, one man cannot water 1 ha without taking several days. In fact, if the crop must be watered every 2 days, then six waterers will be required for a cost of \$50552 per year. There will be times when the watering cycle can be more than a week, and other times when the crop must receive water every day. Fries compared this with system costs for a number of advanced installations (Table 5-19). For a 2 to 5 year depreciation of the systems, figured at 10% interest on the outstanding balance, the yearly cost is much less then using hose watering with \$3.00 hr^{-1} labor.

Hand watering is still employed for emergencies, particularly with the planting of a new crop, or where it is necessary to establish capillarity in mat watering systems. However, hose watering as a standard procedure is seldom satisfactory since its success depends upon the skill and care of the laborer. Hose ends are also very good means to inoculate sterile substrates with pathogens –especially as most growers leave hoses lying on the ground. The method is the application of a large volume of water in a short time –which may, or may not, adequately soak the substrate, depending upon infiltration rates. It will disturb soils, reducing water and air permeability at the soil surface. With automated systems, one can afford to pay \$10.00 hr^{-1} or more to oversee system operation with skilled labor.

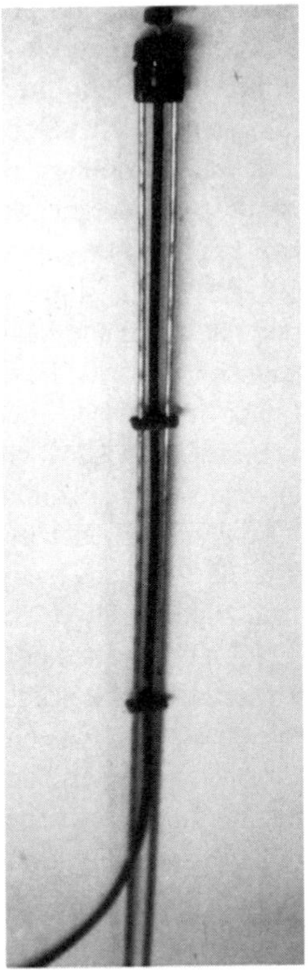

Fig. 5-64. (**Left**) An early (1960s) setup of rotating sprinklers for carnations. Not a good system and no longer in use in Israel. (**Right**) A "spray stake" with capillary tubing for controlling water flow. Sprays in a 360° pattern, for closely spaced small units.

2. Sprinkler Methods

Sprinkler systems, especially the Gates peripheral system (Fig. 5-63A) and variations of that system, were popular in the 1960s. They were the first real application of new materials in the greenhouse to reduce labor costs and improve growth. Overhead sprinkler systems as typified by Fig. 5-64 (left) are not recommended for ornamentals –especially if the water is salty. If water quality is good, some operations still employ overhead sprinkling. It is instructive, however, that over a decade, overhead sprinkling disappeared in Israel and elsewhere. Spray stakes, on the other hand, are still common where the production units are small and at close spacing (<15 x 15 cm). These systems should not be confused with fog or mist systems used for humidity control or propagation (to be covered in the next section). It was found, in hydroponic substrates particularly, that these systems could cause physical damage to herbaceous plant foliage, plus spreading foliar diseases. This was especially true if the grower watered late in the day during the winter so that water remained on the foliage into the night hours. Comparisons were made with other methods by van den Ende and de Graaf [1974]. Sprinkler systems not only wetted the bench surfaces on vegetable crops, but also the walkways, distributing water often into areas where there were no roots. Thus, according to these authors, sprinkler systems were less satisfactory than trickle or drip methods. Sprinklers that were located close to the ground, resulted in lower yields on tomatoes over 2 years of trials. Peripheral sprinkler systems may still be commonly found on crops that possess few leaves on the lower stems and are woody –such as roses. Even here, however, crown gall can be spread by splashing water.

3. Sub-irrigation

Capillary mat systems (Figures 5-65, 5-66), along with ebb-and-flow methods (Fig. 5-67), have become increasingly popular for potted plant production on a large scale. Early work on constant water levels was carried out at Cornell University in the 1940s by Post and others. This was followed by the capillary sand bench developed at NIAE in the U.K. [Anon., 1962; Wells and Soffe, 1962]. The latter were expensive and difficult to build. With the development of suitable mats that could be easily laid over the bench, without the need for water tightness, capillary mats became more popular. According to Larson and Hilliard [1977], algal formation can result in unpleasant odors, especially if the mat dries out. Fungus gnats can also be a problem. Bjerre [1983], however, showed that a pierced, black polyethylene sheet laid over the mat would prevent algal growth. This also helps to prevent root growth in the mat, with evaporation three to five times less than an uncovered mat. Pots must be flat based and heavy enough to establish contact with the mat. Hose watering will initially be required to establish capillarity. Harbaugh and Stanley [1985] do not recommend fertilization with the water on capillary mats since this tends to encourage root growth into the mat. Fertilizer injection is liable to lead to a salt buildup on the upper surface to a greater extent compared with other systems. Slow-release fertilizers, incorporated into the substrate before planting, appear suitable. The mat should be kept moist always, requiring three to six applications daily. The bench must be without depressions or high spots. A crowned bench, or one sloping 2 to 5 cm across the bench, may be used to eliminate wet or dry spots. Giacomelli [personal communication, 1992] stated that root zone heating is not to be used.

Ebb-and-flow methods (Fig. 5-67) (ebb-and-flood according to Fynn, 1994) are another means of applying water from below the substrate. Fertilized solutions may be employed with a considerable reduction in water loss due to evaporation [Holcomb et al., 1992]. Water is pumped into the bench tray from a storage tank and allowed to drain back by gravity. Trays are flooded to a depth of 15 to 18 mm and remain flooded until upper soil surfaces in containers are moist. The geometry of the benches should be to ensure even wetting of all pots at the same time and for the same duration [Fynn, 1994]. The water should rise to pot level without splashing and without encountering other pots. Similarly, the water should drain without passing other pots. Holcomb et al. [1992] did not find a significant difference in water and fertilizer use between ebb-and-flow systems and drip methods. Dole et al. [1994] found that ebb-and-flow used the least amount of water, with the least run off, as contrasted to capillary mats which used the most water with the greatest run off. Poinsettia production was most satisfactory when grown with ebb-and-flow.

4. Trickle or Drip Systems

Trickle or drip systems, which are the most expensive, were originally developed in Israel where cultivation occurred in heavy soils, water had to be conserved, and the salinity was due mostly to high NaCl levels [Blass,

1969] (Fig. 5-68). Examples of various trickle systems are provided in Figs. 5-63B, C, and F, 5-69, and 5-70. Generally, I distinguish between trickle and drip methods in that water supply of the former systems are usually

Fig. 5-65. Two examples of a capillary mat for pot culture. Spaghetti tubing is utilized in the upper picture to wet the mat uniformly. In the bottom, a black plastic sheet, perforated to allow contact with the mat and pots eliminates algae growth. One can see designs on the plastic for uniform spacing of the crop.

at a higher rate than drip methods. However, many authors use the two terms interchangeably.

The ease of forming plastic shapes led to a proliferation of emitter types and methods. Moser [1979] listed more than 20 types on the market, dividing them into long path emitters, orifice emitters, combination long path and orifice emitters, and one porous wall type. Moser provided a detailed description of each kind. Sanders [1992], discussing tape systems, divided them into laminar and turbulent flow types (Fig. 5-69). Turbulent rate of 2 to 4 ℓ hr^{-1} at 100 kPa pressure devices are a more recent development from in line emitters (Fig. 5-68). These use the friction from turning a corner to reduce pressure, with turbulence helping to clean emitter paths. Lengths for tape-type laterals in the field range from 100 to 130 m, with outlet spacings ranging from 4 to 120 cm,

Fig. 5-66. Example of pot culture on a capillary mat covered with black plastic. This system appears to be most favorable for small units in large numbers, especially on plastic, movable benches.

depending upon conditions employed. Some systems are pressure compensated so they can be used on undulating ground, with lengths of run up to 180 m, and an emitter application rate of 2 to 4 liters per hour at 100 kPa pressure. Emitters, often purchased individually, can be self-cleaning types that flush themselves at low pressure at the beginning and end of the irrigation cycle. Others can be adjusted in the field to different flow rates. Tapes have become popular in the U.S. for cut-flower and vegetable production, whereas individual emitters appear more common in such regions as the Mediterranean (Fig. 5-70). Chapin [1971] listed eight basic systems: flow resistance tube, spiral resistance dripper, porous wall hose, small orifice dripper, adjustable flow dripper, self-cleaning orifice dripper, multiple outlet loop system, and a two-pressure hose (tape). The flow resistance tubes are capillary tubes common in spaghetti systems and one of the earliest forms of drip irrigation in the U.S. A number of proprietary systems were available in Europe in the late 1960s.

Fig. 5-67. Small scale ebb-and-flow benches for potted crops. The tanks at the ends store the solution, which is pumped at need into the trays and allowed to drain back.

Table 5-19. System cost and cost per year for three different greenhouse irrigation systems. Adapted from Fries [1979] with figures converted to a per-hectare basis, 73% area in production (With permission from the *Amer. Vegetable Grower*).

	Total system cost			
System	**Installation labor cost**[a]	**Material cost**	**Controller cost**	**Total cost**
Spaghetti, 106 zone	$8949	$8949	$5030	$22930
Capillary, 106 zone, mat	$9109	$9109	$5031	$23249
Spray stake, 106 zone	$8765	$8765	$5031	$22553
	Cost per year (10% interest)			
	Total initial investment	**Average useful life**		**Amortized cost per year**
Spaghetti, 106 zone	$22930	5 years		$5732
Capillary mat, 106 zone	$23249	2 years mat, 5 years equipment		$10108
Spray stake, 106 zone	$22553	5 years		$5640

[a] Labor cost estimated at 2 times material cost (controller).

Fig. 5-68. One of the first true drip emitters developed by Blass [1969] in Israel. Water pressure and flow control is by an internal spiral of predetermined length with the water outlet at one end. This can be inserted into the feeder line as necessary.

The principle of drip irrigation is to supply water at very low rates, in the region of maximum root activity. Both these types and sub-irrigation systems do not wet the foliage, thereby significantly reducing disease problems, and leading to a considerable savings in water consumption [Harbaugh and Stanley, 1985; Harbaugh et al., 1986; Chapin, 1971; Hall, 1980; etc.]. Usually, yields and irrigation efficiencies have been increased with water saving greater than 30%. This also applies to fertilizer consumption where it is injected into the irrigation water. Application rates for dripper systems are usually 1.2 to 4.5 ℓ hr^{-1} per emitter, whereas spray stake systems are typically 6.8 ℓ hr^{-1} [Bebb, 1980]. Bebb distinguishes a trickle tube (capillary, spaghetti tubes, microtubes) as having average rates of 1.34 ℓ hr^{-1}. At these low rates, larger areas can be irrigated simultaneously. Usually, application rates must exceed 6 to 8 ℓ hr^{-1} before low infiltration rates in field soils become a problem [Keller and Karmeli, 1974]. As infiltration rates increase, with corresponding decreases in hydraulic conductivity, emitter density must be increased to ensure adequate substrate wetting. Thus, for example, in gravel hydroponics, a tape system with emitters at 4 cm intervals, 2 laterals between each row, is required as compared to 1 lateral with emitters at 60 cm in a heavy soil (Table 5-20). With proper system design, there should not be any differences in plant response as the result of culture or cultivar [Harbaugh et al., 1986]. Seasonal water use will vary with differences in cropping time, which is influenced by the production method and cultivar choice.

Technical aspects and design of drip systems have been examined by many individuals [e.g., Moser, 1979; Keller and Karmeli, 1974; Wu and Gitlin, 1983; Sanders, 1992; etc.]. Wu and Gitlin speak of "deficit" irrigation, capable of increasing application efficiency to 100%. This does not appear applicable to the greenhouse

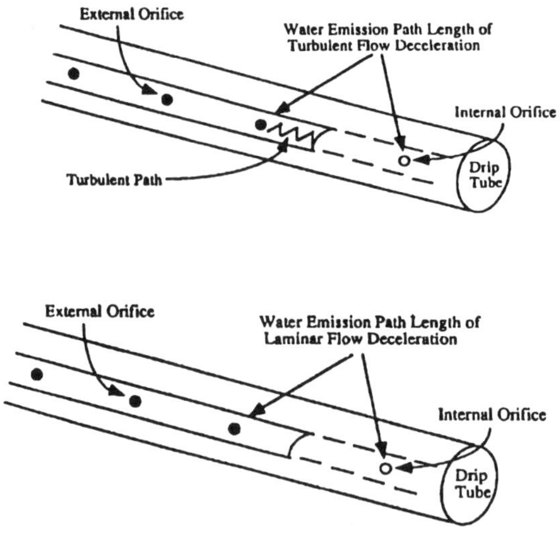

Fig. 5-69. Cross-sections of laminar flow (lower) and turbulent flow (upper) drip tapes as given by Sanders [1992]. Internal orifice diameter is less than external orifices to reduce clogging (With permission of the *Amer. Soc. Hort. Sci.*).

situation where it is the limitation of the substrate in terms of aeration that may be the limiting factor for soils. A number of formulae have been determined for emitter performance, which is not only dependant upon pressure, but also upon temperature. Emitters can be characterized by:

$$q = K_d + H^x \tag{5.12}$$

where: q = *emitter discharge* ($\ell\ hr^{-1}$)
K_d = *a constant, characterizing the emitter*
H = *pressure head at which emitter operates (m)*
x = *an exponent characterized by the flow regime*

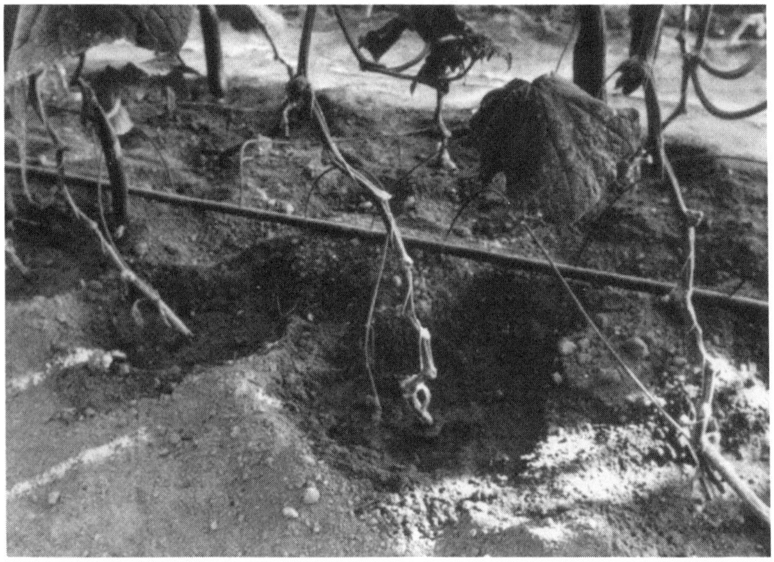

Fig. 5-70. Use of capillary tubes for drip irrigation in greenhouse vegetable production. In-line emitters may also be employed, installed according to plant spacing.

For fully turbulent flows, x = 0.5, and for laminar flows, x = 1.0. The exponent will also vary with the amount of flow regulation provided by the emitter. Friction will cause decreasing pressures that will decrease discharge along the lateral, as will changes in head due to slope variations. However, a general rule-of-thumb is to limit flow differential so the minimum emitter discharge is at least 90% of the average discharge. This may be well within practicality [Keller and Karmeli, 1974]. Not only is emitter discharge very sensitive to pressure fluctuations, but discharge will vary with aging, plugging and slow clogging, and deposits in addition to temperature. Pressure regulation is generally achieved at the head of a manifold (supply to the laterals), an example of which for a cut-flower bench is provided in Fig. 5-71. The most economic division of the allowable head loss is about 55% in the lateral and 45% in the manifold. It is obvious from this short exposition that system design requires the services of a competent engineer. Bartok [1986] has suggested some maximum daily water requirements, which are presented in Table 5-21. Other values have been presented in Chapters 1 and 2 as well as this one, keeping in mind that actual water use may vary drastically from these values as a function of plant size, irradiance, and greenhouse climatic conditions. Prevatt et al [1992] compared costs for outdoor vegetable

Table 5-20. Percentage of soil wetted by various discharges and spacings for emission points in a straight line, applying 40 mm of water per cycle over the wetted area [Keller and Karmeli, 1974] (With permission of the *Amer. Soc. Agric. Engr.*).

Effective spacing between laterals (m)	Effective emission point discharge rate[a]														
	<1.5 ℓ hr^{-1}			2 ℓ hr^{-1}			4 ℓ hr^{-1}			8 ℓ hr^{-1}			>12 ℓ hr^{-1}		
	Soil texture and recommended emission point spacing on lateral (m)[b]														
	C 0.2	M 0.5	F 0.9	C 0.3	M 0.7	F 1.0	C 0.6	M 1.0	F 1.3	C 1.0	M 1.3	F 1.7	C 1.3	M 1.6	F 2.0
	Percentage of soil wetted[c]														
0.8	38	88	100	50	100	100	100	100	100	100	100	100	100	100	100
1.0	33	70	100	40	80	100	80	100	100	100	100	100	100	100	100
1.2	25	58	92	33	67	100	67	100	100	100	100	100	100	100	100
1.5	20	47	73	26	53	80	53	80	100	80	100	100	100	100	100
2.0	15	35	55	20	40	60	40	60	80	60	80	100	80	100	100
2.5	12	28	44	16	32	48	32	48	64	48	64	80	64	80	100
3.0	10	23	37	13	26	40	26	40	53	40	53	67	53	67	80
3.5	9	20	31	11	23	34	23	34	46	34	46	57	46	57	68
4.0	8	18	28	10	20	30	20	30	40	30	40	50	40	50	60
4.5	7	16	24	9	18	26	18	26	36	26	36	44	36	44	53
5.0	6	14	22	8	16	24	16	24	32	24	32	40	32	40	48
5.5	5	12	18	7	14	20	14	20	27	20	27	34	27	34	40

[a] Where relatively short pulses of irrigation area applied, effective emission point discharge rate should be reduced to about half of instantaneous rate.

[b] Texture of soil designated by **C**, coarse; **M**, medium; and **F**, fine. Emission point spacing is equal to about 80% of the largest diameter of the wetted area of the soil underlying the point.

[c] The percentage of soil wetted is based on the area of the horizontal section about 0.3 m beneath the soil surface.

Fig. 5-71. Example of a fixed pressure regulator at the head of a manifold, regulating pressure to a series of laterals in a greenhouse bench. Variations are numerous, depending upon requirements.

production in sandy soils for 2 types of sub-irrigation and a drip system. Table 5-22 presents Prevatt et al's figures for the drip system on a per-hectare basis. The authors did not recommend a surface drip system for field vegetable production. This would not be applicable to greenhouse conditions.

The principal problem with drip systems is emitter clogging. Good filtration is essential [Hall, 1980; Bebb, 1980; Long, 1987; Royle, 1980; Bucks et al., 1979; 1981]. The items that can lead to clogging are outlined in Table 5-23, while Table 5-24 is a system devised by Bucks et al. to classify irrigation waters for drip methods. Particle sizes that are likely to be important are listed in Table 5-25, along with equivalent screen mesh numbers. There is no proven, practical method for determining whether a user will encounter clogging problems. Commonly, domestic water supplies that have been filtered and chlorinated are least troublesome, but treatment of agricultural water to this quality can be impractical. This is not the usual case in greenhouse production. Also, municipal supplies can contain suspended materials that not only clog drip systems (emitter orifice diameter can be as small as 0.3 mm), but can damage equipment such as mist pumps and injector systems when utilized over a long period (Fig. 5-72). If the sum of the three factors in Table 5-24 is less than 10, little problem can be anticipated; 10 to 20 suggests some problem, and 20 to 30 is a severe problem. There can be large fluctuations in water quality as will be discussed in Chapter 6.

Table 5-21. Estimated maximum daily water requirements (ℓ m^{-2}) for a variety of greenhouse conditions [Adapted from Bartok, 1986] (With permission of the *Greenhouse Manager*).

Crop	Liters water m^{-2}	Remarks
Bench crops	16.2	Based on twice daily watering
Bedding plants	20.5	Based on twice daily watering
Pot plants	20.5	
Mums, hydrangeas	61.5	Based on 3 times daily watering
Roses	29.1	
Tomatoes	10.8	Per m^2 of bed, watered every other day

Filters for suspended solids include screen, centrifugal, media (sand or gravel packs), or gravity (settling basins). Usually, growers may use 2 filters in succession such as a sand filter followed by a screen or centrifugal filter. Each type has its own advantages and disadvantages, and one should follow the manufacturer's recommendation for the particular system. A homemade sand filter has been described by Roberts and O'Hern [1993], using No. 20 crushed silica sand and yielding an effective 200 mesh filter bed. If this is not available, filtration at a size of one-tenth the diameter of the emitter's smallest opening is a guideline [Bucks et al., 1979; 1981]. Filter capacity should be large enough to permit the rated flow without frequent cleaning. A pressure loss of about 69 kPa is allowable before cleaning is required. Filtration units should be designed with at least 20 to 30% extra capacity.

Sometimes even worse than solids clogging a drip system is precipitation. At least in the southwest U.S., most shallow wells contain calcium and magnesium carbonates with high pHs. Above pH 7.5, precipitation will occur in the filter, tubing, or emitter (Fig. 5-73). Acidification is generally recommended to reduce pH to about 6.5 [Royle, 1980; Bucks et al., 1979]. Below 6.0, corrosion will occur on metal surfaces. Acids such as HNO_3 and H_3PO_4 are most commonly used. In both cases, they also supply an essential element. Precipitation can occur with phosphoric acid if high enough in concentration in waters containing calcium. Also, phosphate is not required in very high concentrations for most crops (ca. 0.5 to 1.0 meq/ℓ). As most hard waters contain sufficient sulfate for crop requirements, sulfuric acid is not recommended for acidification although it is usually cheaper. While pH is lowered, salinity is unnecessarily increased. Gaseous chlorine will also acidify water. Chlorination for bacterial control, however, is not suggested when water has 0.4 mg or more dissolved iron as the chemical reaction will result in iron oxide that can precipitate, clogging emitters. Some alternative chemicals to control bacteria and algae are xylene permanganate, ozone, quaternary ammonium salts, copper salts, acrolein, hydrogen peroxide, bromine and iodine. Some of these may be too costly or phytotoxic. Ozone is a good bactericide but has no residual effect. Furuta et al. [1976] suggested 1 mg ℓ^{-1} copper to keep algae from growing, and this rate will not injure most crops. Hydrogen peroxide was recommended by Royle [1980] to get rid of algae and bacteria, as well as other organic matter. Unfortunately, concentrated hydrogen peroxide attacks brass, butyl, neoprene, steel, and fertilizers. Quaternary ammonium salts and bromine have been most commonly employed.

Table 5-22. Annual fixed and variable costs for a vegetable drip system for use on sandy soils in Florida [Adapted from Prevatt et al., 1992] (With permission of the *Amer. Soc. Hort. Sci.*).

Item	Cost
Annual fixed costs ($U.S. ha^{-1})	
Depreciation	139
Interest	125
Repairs	83
Taxes	33
Insurance	13
Total annual fixed costs	393
Total annual fixed costs per hectare	640
Total annual fixed cost per 100 m	12
Variable costs ($U.S. ha^{-1})	
Chemical treatment	36
Drip tube	259
Drip tube installation	18
Irrigation maintenance labor	28
Irrigation manager	12
Pumping (diesel fuel)	55
Operating interest	24
Total variable costs/season	497
Total variable costs/season-hectare	810
Total variable costs/100 linear m	15

Note: Costs originally based upon a 40.5 ha field, which the author has reduced to 1 ha, 61 cm emitter spacing, 1 lateral in each plant bed.

Most drip and trickle systems require periodic flushing. This can be accomplished automatically at the start and end of irrigation cycles, or manually, using some of the above materials. Various methods of flushing have been described [Royle, 1980; Anon., 1988; Bucks et al., 1979]. Very elaborate schemes have been outlined [Long, 1987], in which the water supply is vigorously oxygenated, filtered, and with automatic pH control. In field production, a drip system is often buried. It was this type recommended by Prevatt et al. [1992] and Phene et al. [1992] for maximizing field production. Such practices have not been observed in greenhouse production. It has been my experience in small substrate volumes that emitters can be blocked by root growth into the orifices.

5. Aeroponics

The procedure of enclosing a root system, and supplying water and nutrients with a mist system, has been examined for several years. Aeroponics have not gained widespread use, however, since technical requirements are advanced. Water quality requirements are very high, and the system must be very reliable. The method, of course, is not applicable to products sold that require a substrate. For cut-flower s or vegetables, the supporting system must be rearranged. As a rule, the method has been used mostly for research. While I have seen commercial use in many locations, aeroponics do not appear to have remained in place for extended periods or expanded greatly. Apparently the improvement in net return makes it difficult to warrant the initial investment and operating costs.

Table 5-23. Principal physical, chemical, and biological contributors to clogging of drip irrigation systems [Bucks et al., 1979] (With permission of the *Amer. Vegetable Grower*).

A. Physical: suspended solids	B. Chemical: precipitation	C. Biological: bacteria and algae
1. Organic	1. Calcium or magnesium carbonate	1. Filaments
a. Aquatic plants (phytoplankton/algae	2. Calcium sulfate	2. Slimes
b. Aquatic animals (zooplankton)	3. Heavy metal hydroxides, oxides, carbonates, silicates, sulfides	3. Microbial deposits
c. Bacteria	4. Fertilizers	a. Iron
2. Inorganic	a. Phosphate	b. Sulfur
a. Sand	b. Aqueous ammonia	c. Manganese
b. Silt	c. Iron, zinc, copper, manganese	
c. Clay		

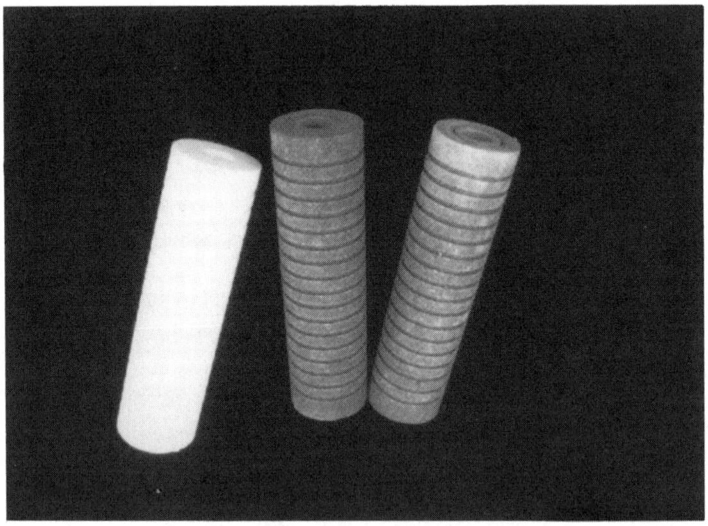

Fig. 5-72. Example of suspended solids in a domestic water supply as noted in the 2 right filters after 1 month in service compared to a new filter on the left. The solids removed were very abrasive and could not be seen in a glass of water. This caused damage to mechanical equipment and blockage of drip tapes.

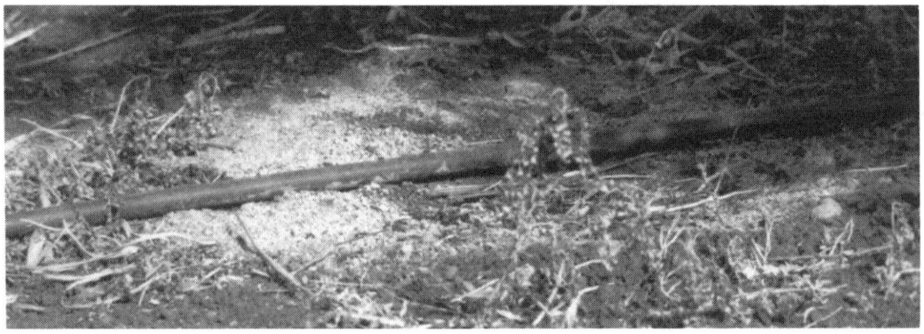

Fig. 5-73. Precipitation of calcium and magnesium carbonates by a drip emitter under arid conditions in the Mediterranean.

I. AUTOMATIC CONTROL OF IRRIGATION

Over several years, investigators have attempted design of irrigation systems to automatically initiate watering, based upon some parameter such as soil water potential [Wells, 1966; Wells and Soffe, 1961] or weight of the container [Dwyer et al., 1987]. Models have been set up to simulate the water balance of the substrate [Michels and Feyen, 1984], using tensiometers. Others have employed tensiometers with Penman's evapotransiration formula [Norrie et al., 1994a, b]. Tensiometers were studied by Post and others in the 1940s, but never became widely utilized. Most scheduling is based upon grower experience and observation. The literature dealing with outdoor irrigation scheduling is enormous, some of which will be cited in the following section. Even with NFT, Hurd and Graves [1981] have attempted to control plant growth through intermittent flow of the nutrient water film. This is an on-off method which could be dangerous in high solar regions and dry climates. On-off methods, as with temperature control, may not provide sufficient precision.

Table 5-24. System for classifying irrigation waters used in drip systems [Bucks et al., 1979].

	Physical	Chemical[a] (max. mg/ℓ)		Biological
Rating	Suspended solids (max mg/ℓ)	Dissolved[b] solids	Iron and/or manganese	Bacteria[c] populations (max. No./ml)
0	<10	<100	<0.1	<100
1	20	200	0.2	1000
2	30	300	0.3	2000
3	40	400	0.4	3000
4	50	500	0.5	4000
5	60	600	0.6	5000
6	80	800	0.7	10000
7	100	1000	0.8	20000
8	120	1200	0.9	30000
9	140	1400	1.0	40000
10	>160	>1600	>1.1	>50000

[a] Tentative chemical classification based on highest rating for either dissolved solids, soluble iron or manganese.
[b] If water pH 7.5 or greater, rating increased by 2.
[c] If water contains an abundant snail population, rating increased by 4. Bacteria populations do not reflect increased algae and microbial nutrients.

Table 5-25. Soil particle size in microns and screen mesh numbers [Bucks et al., 1979].

Soil particle size	Microns[a]	Screen mesh No.
Very coarse sand	1000-2000	18-10
Coarse sand	500-1000	35-18
Medum sand	250-500	60-35
Fine sand	100-250	160-60
Very fine sand	50-100	270-160
Silt	2-50	
Clay	<2	

[a] To find equivalent mm, move a decimal point to the left three places.

It is my guess that attempts to control substrate moisture content will not be completely successful for greenhouse culture. The whole thrust of research and practical application cited in this chapter has been to make water maximally available to a healthy root system. Recirculating systems, such as NFT, seem to be an approach to the ideal where nutrients can eventually be subject to precision control, and utilized as a means to modify growth according to market demand and the particular climatic conditions. If water is freely available, then it becomes a matter of controlling the water demand, considerably simplifying plant growth control. This may seem at variance with the increasing requirements for pollution control, especially in such regions as California and The Netherlands [Hasek et al., 1986; Biernbaum, 1992; Whitesides, 1989]. Such problems will be discussed in Chapter 6 where NO_3^- runoff from the operation is a particularly serious difficulty. The whole thrust of greenhouse operation up to the 1980s was overuse of water supplies in controlling salinity and maintaining maximal water supply in many modified soils and inert mixtures. In developed countries, greenhouses can be shut down if compliance with local pollution control regulations is not carried out by the operation successfully.

IV. WATER DEMAND

Only in recent years have concerted efforts been made to place problems peculiar to greenhouses on a firm basis concerning water loss. Very little of this has made its way into the practical literature. Instead, especially in the American literature, a simplistic approach dealing almost entirely with humidity relations as they affect growth is common. The vital relationship between water supply and demand has not been adequately studied for greenhouse practice. We have already shown in this chapter that water uptake, movement through the plant, its loss from leaves and consequent water potentials imposed in the plant can have serious repercussions in terms of growth, quality, and timing of economic greenhouse crops. The control of plant water potential (ψ_t) cannot be accomplished through water supply alone except to restrict supply –either deliberately (bedding plants) or inadvertently. Resultant plant stress, where water supply is reduced, will be determined by the rate at which water is lost through transpiration. As noted in previous chapters, especially Chapter 3, solar energy variation is enough to seriously complicate the process of obtaining meaningful results. Add to this the climatic variations of wind, humidity and temperature; empirical data obviously derived from local research has limited application. Principal investigators in field research have frequently commented on the limits of empirically obtained data [e.g., Tanner, 1966; Monteith, 1981], using statistically derived relationships that have little relationship to physical principles. These comments are also applicable to greenhouses. Furthermore, water loss complicates humidity control and disease relationships.

Table 5-26. Conversion factors for some common units used in expressing transpiration. To convert, multiply the quantity in the left column by the value in the desired right column. Adapted from Nobel [1991] and modified by the author.

Transpiration units	mmol $m^{-2}s^{-1}$	mg $m^{-2}s^{-1}$	MW m^{-2a}
μmol $cm^{-2}s^{-1}$	10	180.2	441.5
mol $m^{-2}h^{-1}$	0.278	5.01	12.3
μg $cm^{-2}s^{-1}$	0.555	10	24.5
μg $cm^{-2}min^{-1}$	9.25×10^{-3}	0.1667	0.4
mg $dm^{-2}min^{-1}$	9.25×10^{-2}	1.667	4.1
g $dm^{-2}h^{-1}$	1.542	27.8	68.1
kg $m^{-2}h^{-1}$	15.42	278	681.1
mm depth $m^{-2}h^{-1}$	15.42	278	681.1

[a] Latent heat of vaporization at 20 C = 2.45 MJ kg^{-1}.

Table 5-27. Summarization of van der Post et al's [1974] calculations of a yearly energy budget for a typical Dutch Venlo greenhouse (©1974, Int. Soc. Hort. Sci., *Acta Hort.*, 35:13-22).

	Energy ($J m^{-2}$)
Total outside global radiation	3.6 GJ
Total inside global radiation	2.5 GJ
Shortwave reflected radiation	349.0 MJ
Back longwave, thermal radiation	321.1 MJ
Latent heat of vaporization (466 mm)	1.1 GJ
Sensible heat	758.4 MJ
Heating energy (early tomatoes, autumn lettuce)	1.7 GJ
Loss due to air exchange and ventilation	1.1 GJ
Net longwave, thermal radiation loss	251.4 MJ
Contribution by heating to latent heat (188 mm)	460.9 MJ

To understand what is going on with the crop, one must appreciate the role of stomata, the techniques of expressing and measuring humidity, and the mathematical relationships that attempt to explain the physical basis of momentum, heat, and mass transport between a vegetative canopy and its surround. The special conditions of a greenhouse emphasized in Chapter 1 as to wind movement, places the physics of transpiration in a region where it is very difficult to make use of existing relationships derived outdoors. Assumptions for the purpose of simplification are numerous.

Table 5-28. Actual water supply and calculated water consumption in mm equivalent depth as determined by van der Post et al. [1974] for Dutch conditions. The author has converted these to equivalent energy of MJ m^{-2} in parenthesis (©1974, Int. Soc. Hort. Sci., *Acta Hort.*, 35:13-22) .

Period	Actual water supply				Evapotranspiration of tomatoes according to:	
	Tomato	Cucumber	Rose	Carnation	Energy balance	1966-68 measurements
May-Sept., 1968	380 (931)	450 (1103)	490 (1200)	470 (1151)	380 (931)	390 (956)
Feb.-Sept., 1969	620 (1519)	750 (1838)	770 (1886)	710 (1740)	620 (1519)	610 (1495)
12 Jan.-13 Aug., 1972	600 (1470)				560 (1372)	530 (1299)
17 Jan.-6 Aug., 1972	530 (1299)				510 (1250)	520 (1274)

A. TRANSPIRATION UNITS AND QUANTITIES

In modern greenhouses, the ground is likely to be completely covered with white plastic, concreted, or even a heating system installed in the ground. Heat flux to and from storage in the ground will be significant on a short term basis [Stanghellini, 1987], whereas evaporation from that surface can be nearly zero. The term "evapotranspiration" (ET), one of the most common words found in the literature dealing with field crops, which takes into account the evaporation from wet soil and transpiration, may not be suitable for some greenhouse situations. I prefer to use "transpiration" (E), ignoring for now water loss from exposed soil or the floor. Of course, if a misting system wets the crop surface, evaporation of free water can be a highly significant influence on the internal greenhouse climate.

The transpiration units commonly employed and various conversion factors are provided in Table 5-26. However, any of these units may be converted to equivalent radiant energy terms (W m^{-2} = 1 J $s^{-1}m^{-2}$) by multiplying by the latent heat of vaporization, (L) 2.45 MJ kg^{-1} water at 20 C. This is the energy required to convert liquid water to vapor. The value is temperature dependant, varying from 2.50 at 0 C to 2.41 MJ kg^{-1} at 40 C. This range is less than 4%, so that the use of the value at 20 C under conditions found in the greenhouse will not introduce a significant error. As previous discussions in Chapters 3 and 4, dealing with energy, were in W m^{-2}, this conversion permits considerable simplification in water and energy calculations. It is also common practice to express water loss and application in terms of equivalent depth (mm). Thus, a sheet of water, 1 mm deep, over 1 m^2 is equivalent to 0.001 m^3 m^{-2} or 1 kg water m^{-2}.

We can look at some figures given previously for water use by greenhouse plants. For example, Fig. 1-3 employs an equivalent water depth of 200 cm yearly for such crops as carnations, snapdragons, and tomatoes in the greenhouse, or about 4.9 GJ $yr^{-1}m^{-2}$. In Section 2.III, several suggestions as to water requirements were cited such as 34.3 MJ $m^{-2}dy^{-1}$ (140 m^3 $ha^{-1}dy^{-1}$) for a vegetable crop, or a minimum yearly water supply of 4.9 GJ m^{-2} ranging up to the equivalent of 5.1 GJ m^{-2} (51000 m^3 ha^{-1}). If the total global radiation available at Fort Collins, CO, were expressed as equivalent energy in water converted to vapor, the rate of 29.4 MJ $m^{-2}dy^{-1}$ could be reached (12 mm dy^{-1}, Fig. 3-38) during the summer. In actuality, maximum evapotranspiration rates outdoors for most crops that completely cover the soil can range from 12.3 MJ $m^{-2}dy^{-1}$ to nearly 20 MJ $m^{-2}dy^{-1}$ (Fig. 5-31), whereas in the greenhouse, under U.K. conditions, less than 7 MJ $m^{-2}dy^{-1}$ may be found in the summer (Fig. 5-1) in regard to water use. Section 5.I showed the equivalent of 9.8 MJ to grow a carnation flower, 12.3 MJ for a snapdragon flower, and 1.5 GJ to produce 1 kg of dry maize. Based upon the recommendations given by Bartok (Table 5-21), the equivalent energy required (assuming all is converted to vapor) ranges from 26.5 to 150.7 MJ $m^{-2}dy^{-1}$. A considerable part of the water applied in irrigation is lost through drainage, or otherwise wasted.

One of the first good summations of transpiration by greenhouse crops under Dutch conditions was van der Post et al's 1974 presentation. Using Penman's mathematical development of evapotranspiration [1948], the

Table 5-29. Variables and parameters needed for an estimate of transpiration or temperature of a greenhouse plant canopy, according to the frequency with which they must be known, and the entities they determine [Adapted from Stanghellini, 1987].

	Net radiation (R_n)	**External resistance[1] (r_e)**	**Internal resistance[2] (r_i)**	**Rate of thermal storage (S)**
To be known beforehand	**Radiation extinction coefficient** for shortwave (k_s) **Reflectance** of a dense canopy (ρ_∞) **Reflectance** of the soil surface (ρ_s) **Path width** (p) **Width** of crop row (w) **Height** of crop row (h)	**Characteristic dimension** (ℓ)	**Minimum resistance** (r_{min}) Temperature that minimizes resistance (T_m)	**Density** of leaf tissue (ρ_t) **Specific heat** of leaf tissue (ρ_t) Average **leaf thickness** (d)
Once daily	**Leaf area index** (LAI)			
With the same frequency as surface temperature of leaf (T_l) and latent heat of vaporization (LE) (transpiration)	**Shortwave irradiance** (R_t) **Temperature** of pipe heating system (T_p) **Temperature** of greenhouse cover (T_c) **Temperature** of soil surface (T_s)	**Air temperature** (T_a) **Wind speed** (u)	R_t CO_2 concentration ($[CO_2]$) **Vapor pressure** of air (e_a)	

[1] Resistance to heat and vapor transport external to the canopy as contrasted to footnote 2.
[2] Resistance within the canopy, from the evaporating sites in the leaves.

authors derived probably one of the first energy balances of typical Dutch Venlo structures for an entire year (Table 5-27). Their studies of actual transpiration for a number of greenhouse crops are presented in Table 5-28, with the equivalent energy required. Although Table 5-28 shows differences between crops for water use based on that supplied, it is the consensus that once a well-watered crop covers the soil area completely –certainly by the time LAI = 4.0, there will be little difference between species as to transpiration rate [de Graaf, 1981; Gates and Hanks, 1967, Thornthwaite and Hare, 1965; etc.]. In fact, the development of mathematical formulae to deal with the energy balance of a crop usually makes the assumption that the vegetative canopy completely covers the ground, is uniform and homogenous, and little reference is made regarding species. For one-dimensional considerations of energy exchange, there must be no horizontal gradients [Tanner, 1966]. This is seldom the case in greenhouses, especially during initial growing phases. Such simplifications can be carried too far [Thornthwaite and Hare, 1965].

B. TRANSPIRATION

Fig. 5-1 showed that most of the water loss is closely linked to available solar radiation. The early British work [Morris et al., 1957; Neale, 1955; Lake et al., 1966; Rothwell and Jones, 1961] showed the close correlation with radiant energy. Many statistical correlations have been made between transpiration and radiation, the equations often being straight lines [e.g., Stanhill and Albers, 1974; Hanan, 1970b, Lake et al., 1966], with correlations often better than 0.90. Occasionally, the relationship may be curvilinear [e.g., de Graff, 1981].

The efficiency with which energy is dissipated as latent heat of vaporization depends not only upon radiation, but includes temperature, humidity, and transport properties along the path of vapor from the liquid water in the plant to the free atmosphere [Fuchs, 1986]. The parameters and variables required for an estimate of transpiration or plant temperature for one derivation were tabulated by Stanghellini [1987], for example, and are reproduced in Table 5-29. Stanghellini states that if one is interested in "reasonable" estimates of transpiration and plant temperature, the parameters in Table 5-29 may be reduced to six. There are many approaches in which the calculations are simplified and rapid with present-day computers. Stanghellini's method is one of many one may find in the literature. It is the most useful method when the objective is to show water demand over a short interval, to be used to help control the greenhouse climate and internal plant water stress. If one is interested only in daily, or longer periods, greater simplification can be employed. Heat storage in the crop and ground can usually be ignored. This section is broken into several parts: 1) the role of radiation. 2) the role of humidity. 3) factors in the vapor pathway from leaf to air, and 4) examples of calculating transpiration rate.

1. Radiation

Usually, authors assume constant transmissivity, reflectivity, and longwave re-radiation in arriving at a suitable radiant energy value (net radiation, R_n) per unit area (m^2). This assumes that transmissivity of a greenhouse is constant, which we know from Chapter 3 is far from the case. To summarize briefly, radiation inside will vary across the house and along its length (Figs. 3-24 to 3-26). It will vary with distance across a span and span placement in multispan ranges (Figs. 3-27, 3-28 and 3-32), from season to season (Figs. 3-26 and 3-28), with orientation of the structure (Fig. 3-30) and with latitude (Figs. 3-30, 3-31 and 3-33). Even in so-called diffuse conditions (overcast skies), radiation inside a structure will vary with altitude and azimuth of the energy stream from a particular part of the sky (Fig. 3-34). Also, crop reflectivity from the canopy may be significantly different under various covers (Fig. 3-39). Thus, in modeling and controlling greenhouse climate, one of the most serious problems is determining the total radiant energy available within the house. Although Section 3.IV.B briefly discussed the various approaches to the problem, this important aspect has not been adequately addressed. According to Kimball [1986], the variations of transmissivity due to solar altitude and azimuth are small compared to the more dramatic effects of a shade cloth. Solar angle effects were ignored in his energy balance program for greenhouses. Not many investigators follow Kimball's procedure. Examples of detailed expositions of greenhouse transmissivity (τ_g) are Bot's dissertation [1983] and the numerous papers by British and Japanese workers cited in Chapter 3.

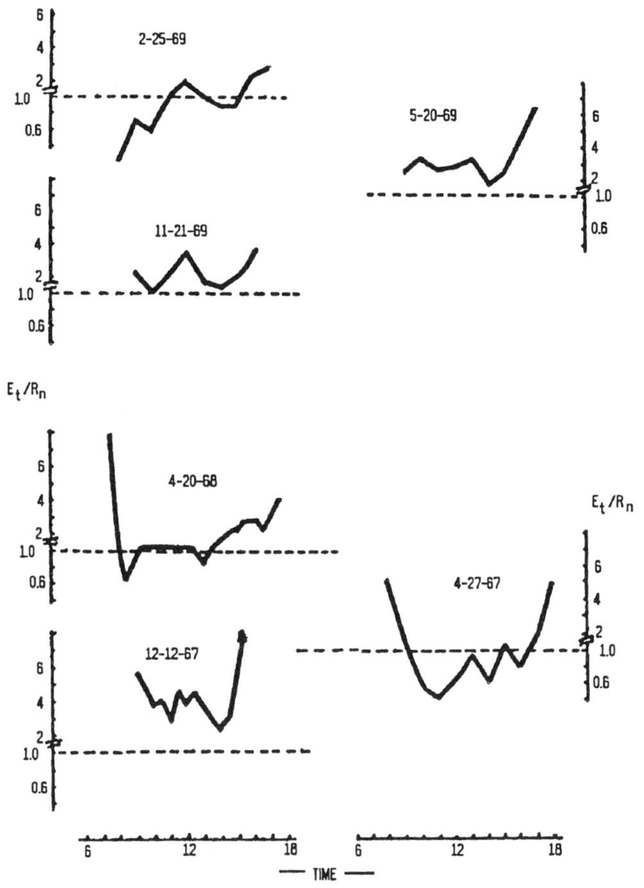

Fig. 5-74. Value of the ratio of evapotranspiration of carnations to net radiation for selected days in Colorado [Hanan, 1970b]. Note the different scale at values <1.0, and the variability of the ratio.

Having thus arrived at a transmissivity value for the structure, one is still faced with determining the net radiation absorbed by a crop canopy. For a single leaf, I have borrowed from Stanghellini [1987]:

$$R_n = (1 - \tau_s - \rho_s)R_s + R_{lw} - \sigma T_l^4 \tag{5.13}$$

where: R_n *= net radiation (W m*$^{-2}$*)*

τ_s = *shortwave transmissivity of the leaf*
ρ_s = *shortwave reflectivity of the leaf*
R_s = *shortwave radiation (W m^{-2})*
R_{lw} = *longwave radiation (W m^{-2})*
σ = *Stefan-Boltzmann constant (W m^{-2}K^{-1})*
T_l = *Leaf temperature (K)*

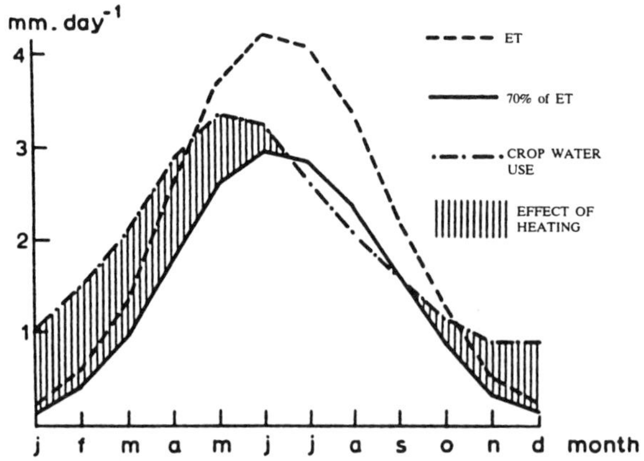

Fig. 5-75. Influence of global radiation and heating energy on water consumption of tomatoes in a Venlo house [van der Post et al., 1974]. T is the calculated evapotranspiration according to Penman's formula outside the house. 70% is the calculated amount inside the house (©1974, Int. Soc. Hort. Sci., *Acta Hort.*, 35:13-22).

Many variations of this formula may be found in the literature [e.g., Fuchs, 1990; Landsberg et al., 1979; Gates, 1980; etc.]. This formulation shows that net radiation will also depend on vegetation temperature (T_l). Accounting merely for R_s and R_{lw} is insufficient. Net radiation will vary with the temperature of the cover, which, in cold climates such as Colorado,

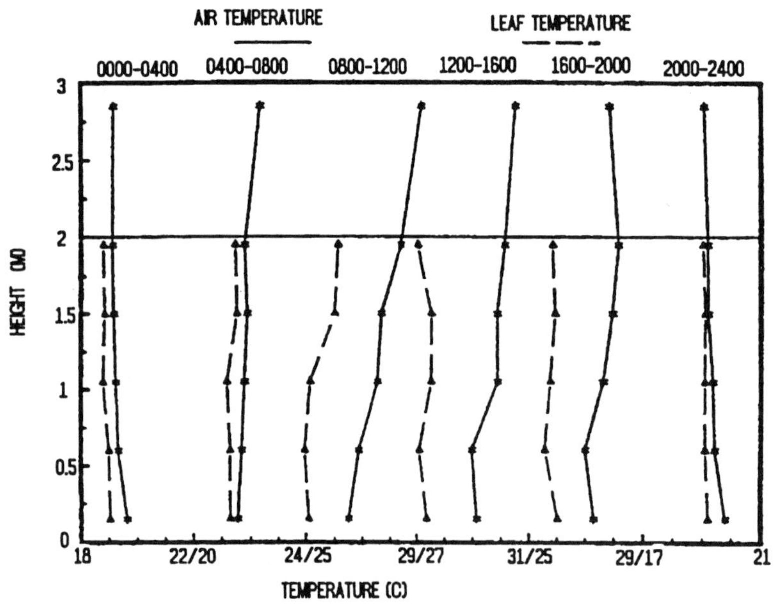

Fig. 5-76. Vertical profiles of leaf and air temperature of a greenhouse cucumber crop at various periods of the day. The horizontal line is the top of the canopy [Reprinted from *Agric. and Forest Meteor,* 1990, Yang, X. et al. 51:197-209. With kind permission of Elsevier Sci.-NL, Sarah Burgerhartstraat 25, 1055 KV Amsterdam, The Netherlands].

can have a significant influence; typified by several examples in Fig. 5-74, of the ratio between E_t and R_n (see Eqs. 3.4 and 4.5). Other factors can influence E_t/R_n. It was Hanan's conclusion [1970b] that a considerable part of the energy for water loss, where the ratio exceeded one, was from the heating system. Not until summer conditions and moderate outside temperatures was the ratio less than 1, or did net radiation exceed E_t. T_l will obviously be influenced by the heating system. The longwave radiation (R_{lw}) can also be due partly to a high temperature heating system (i.e., infrared). The contribution of the heating system was briefly discussed in Chapter 4, and has been addressed by Stanghellini [1981], van der Post et al. [1974] (Fig 5-75), and de Graaf

[1981]. De Graaf concluded that 20% of the heating energy was used in evaporation under Dutch conditions. Stanghellini [1987] stated that longwave radiation from a conventional pipe heating system did not contribute significantly to water loss.

Almost invariably, authors begin derivation of the energy balance of plants with consideration of the simplest case –a single leaf [e.g., Gates, 1980; Monteith, 1973; Woodward and Sheehy, 1983; Campbell, 1977; etc.]. To avoid difficult mathematical problems, many authors simply assume that the vegetative canopy represents a single large leaf, and mathematically average net radiation over the total leaf area index (LAI) [e.g., Fuchs, 1990; Penman, 1948; Penman and Schofield, 1951; Black et al., 1970]. Alternatively, the region of interest may be divided into 1 or more layers, each contributing to the overall energy balance and through which energy and mass transport can occur [e.g., Kimball, 1973; 1983; Sadler and van Bavel, 1984; Kindelan, 1980; Cooper and Fuller, 1983; etc.]. In other cases, the crop is included in the greenhouse and the energy and transport fluxes are developed for the total entity [e.g., Levit and Gaspar, 1988; Bot, 1980; Duncan et al., 1981; Chandra et al., 1981; etc.]. Investigators may develop detailed models of energy relationships for individual plants [e.g., Lieth and Reynolds, 1988]. Stanghellini [1987] developed special models for tomato and cucumber crops in greenhouses, based upon row and aisle width and plant height. In their development of meteorological data for measuring transpiration, Fuchs et al. [1987], for an outdoor crop of horizontally uniform foliage, computed sunlit leaf area and shade area based upon a solar zenith angle, and developed a relationship:

$$R_n = (0.55)^n R_d L^* R_{lw} \tag{5.14}$$

where: n *= an empirical factor, zero to one, accounting for radiation entrapment resulting from multiple scattering between sunlit leaves*

R_d *= direct global radiation*

L^* *= a function of LAI and mean horizontal area of shade cast by a unit leaf area*

R_{lw} *= net thermal radiation, calculated according to Campbell [1977]*

Eq. 5.14 is one of many mathematical models of varying complexity that can be found. There is little uniformity and no standard. Tanner and Fuchs' [1968] discussion of the situation is worth repeating. The sensible heat produced at the upper leaves of a canopy, under strong radiation, is transferred downward through the canopy to shaded leaves through a large temperature gradient. The resulting variations within the canopy can be remarkable, although within a greenhouse the gradients may not be as large as in the field (Fig. 5-76). Geiger's account [1965] documents well the variations not only in temperature, but also CO_2 concentration, humidity, and wind velocities that can occur within a vegetative canopy. Other descriptions include Lemon et al. [1971], Chang [1968], and Loomis and Williams [1969]. Tanner and Fuchs further state that the humidity anywhere in the canopy is uncertain except within the leaves –which varies with temperature. The strength of radiation heat sinks and sources vary spatially and temporally. Simple systems –such a flat plane, single leaf, or even resistance networks to suggest transport– are unrealistic. Canopy models that consider spatial distribution of heat sources and transfer coefficients are necessary.

More discussion of some of the foregoing will be given in Chapter 8. However, the situations of a vegetative canopy within greenhouses are obviously far more complex than can be found in a large field of uniform maize, rye, etc. Some simplification is necessary, and it may be that a single coefficient could be used to relate net radiation interception as a consequence of growth from first planting to harvest, as well as allowing for the many arrangements that can be found. There is a great deal of work to be done on this aspect alone before we arrive at relationships that are applicable to all greenhouse conditions.

2. Humidity

Atmospheric levels of water vapor significantly influence growth and development. The rate of growth, composition, and form that a plant attains, according to Tibbitts [1979], are controlled by humidity. This is an oversimplification of the factors influencing plant response. Humidity also has a direct effect on disease. A thorough understanding of how vapor behaves in air should be part of the "nuts-and-bolts" any grower has in his technical arsenal.

a. Definitions

Water vapor in air is treated as any other gas. At the concentrations, pressures and temperatures of the greenhouse, its behavior can be considered as following the perfect gas law:

$$PV = nRT \tag{5.15}$$

where: P = *pressure (kPa)*
V = *volume* (m^3)
n = *number of moles concentration*
R = *perfect gas constant* ($8.314\ J\ mol^{-1} K^{-1}$)
T = *temperature (K)*

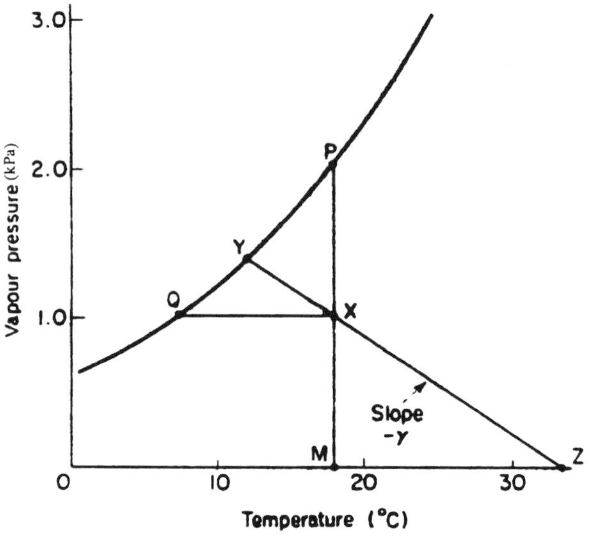

Fig. 5-77. Relationships between dry bulb temperature (T_{db}), wet bulb temperature (T_{wb}), virtual temperature (T_v), vapor pressure (e), and dew point (T_{dp}). **X** represents air at 18 C and 1.0 kPa vapor pressure. The line **YXZ**, with a slope of -γ, gives T_{wb} from **Y** (12 C), T_v from **Z** (33.3 C). The line **QX** provides T_{dp} from **Q** (7.1 C). The line **XP** gives e_s from **P** (2.1 kPa) [Adapted from Monteith, 1973].

The fact that nothing is ideal will not cause significant error. The relationship in Eq. 5.15 is applicable to any gas such as, for example, CO_2, and the following exposition will be referred to again in Chapter 7. From this equation, the methods and terminology dealing with vapor concentration can be related to each other:

1. **Vapor pressure** (e): kilopascals, millibars. The sum of all partial pressures of all components in the air will equal the total atmospheric pressure.
2. **Absolute humidity** (χ): g m^{-3}, kg m^{-3}, mass of vapor per unit volume (m/V).
3. **Specific humidity** (q): g kg^{-1} moist air, mass of vapor per unit mass of air plus vapor (m_v/m_{av}).
4. **Mixing ratio** (r): g kg^{-1} dry air, mass of vapor per unit mass of dry air (m_v/m_a).
5. **Virtual temperature** (T_v): degrees K, the temperature at which dry air would have the same density as a moist air sample at an actual temperature T.
6. **Relative humidity** (RH): percent or decimal ratio, 0 to 1; the ratio between actual vapor concentration and the vapor concentration at saturation for any particular temperature ($\chi_a/\chi_{s(T)}$, $e_a/e_{s(T)}$.
7. **Dew point** (T_{dp}): degrees C, the temperature to which unsaturated air must be cooled to produce saturation; that is, $e_a = e_{s(T)}$.

8. **Mole** (n): m or mol, a mole of any substance contains the same number of molecules (6.023×10^{23}), and at sea level (101.3 kPa) and 273 K (0 C) (STP), 1 mole of a gas will occupy 22.414 ℓ.
9. **Activity** (a): the ratio between actual concentration and concentration at some base level such as pure, free water at STP.
10. **Saturation deficit** or **vapor pressure deficit** (VPD): kPa, the difference between saturation and actual vapor pressure or absolute humidity and VPD = $e_{s(T)}(1\text{-}RH) \approx \Delta(T\text{-}T_{dp})$, where Δ = the slope of the saturation vapor pressure curve (Fig. 5-77), and is evaluated at a mean temperature of $(T + T_{dp})/2$.
11. **Wet bulb temperature** T_{wb}: degrees K, the temperature of a wet surface exposed to standard conditions and at equilibrium.
12. **Dry bulb temperature** T_{db}: degrees K, the actual air temperature.
13. **Mole fraction** N_v: dimensionless number between 0 and 1, representing the ratio of the number of moles of water to the total number of moles of all species in the system.

The relationships of some previous definitions are depicted diagrammatically in Fig. 5-77. The heavy curve, **Q-P**, shows the relationship between temperature and vapor pressure (e_s) for saturated air (100% RH). The velocity of random movement of the molecules will be a direct function of temperature. Water vapor occupies space independent of other gases that may be present. At any given temperature, the same amount of vapor can be present in any empty volume of space as in an equal volume of air. The saturation vapor pressure on **QYP** (Δ = de/dT) (Pa K^{-1}) can be calculated (Table 5-30) [Woodward and Sheehy, 1983]:

$$\ln e_s = A - \frac{B}{T} - C \ln T \tag{5.16}$$

where: A = 57.96 Pa
B = 6731.0 Pa K
C = 4.796 Pa

Note the marked increase in water vapor as temperature increases. At a temperature of 18 C and a vapor pressure (e_a) of 1 kPa (point **X**), a wet bulb thermometer will be cooled, under standard conditions, to point **Y** (12 C). In practical terms, this is the lowest air temperature that can be obtained with a fan-and-pad cooling system operating at 100% efficiency. **P** represents the vapor pressure if that same air parcel were saturated at the same temperature (2.06 kPa). **Q** is the temperature to which the sample would have to be cooled to be saturated (T_{dp}), and **Z** represents the virtual temperature (T_v) of the sample, or the temperature to which dry air would have to be raised in order to have the same density as the moist air at 18 C (33.3 C). Any relationship for dry air into which T_v is substituted is a relationship applicable to moist air.

The absolute humidity (χ) is related to vapor pressure by Eq. 5.17, and in a similar fashion, Eq. 5.18 provides the solution for specific humidity (q). The solution for mixing ratio (**r**) is given in Eq. 5.19. Alternatively, another solution for χ is Eq. 5.20. Mixing ratios and specific humidities are not common in plant studies, their principal differences being that **q** is based upon unit mass of <u>moist</u> air as contrasted with **r** which is based upon a unit mass of <u>dry</u> air. There is little difference in values between **q** and **r** (Table 5-31).

$$\chi = \frac{e_a M_v}{RT} = m/V = \left[\frac{e_a}{461.7T}\right] \tag{5.17}$$

where: χ *= kg m^-3, absolute humidity*
e_a *= N m^-2 = Pa, vapor pressure*
M_v *= molecular weight of water, 0.018 kg mol^-1*
m = mass, kg
V = volume, m^-3
R = perfect gas constant, 8.314 J mol^-1 K^-1
T = temperature (K)

$$q = \frac{\frac{M_v e_a}{M_a}}{(P - e_a) + \left(\frac{M_v}{M_a}\right) e_a} = \frac{0.622 e_a}{(P - e_a) + 0.622 e_a} = \frac{\chi}{\chi + \rho_a} \tag{5.18}$$

where: q = *specific humidity, kg kg*$^{-1}$ *moist air*
M_a = *molecular weight of air (the ratio,* $M_v/M_a = \epsilon = 0.622$*), 0.029 kg mol*$^{-1}$
P = *atmospheric pressure, 101.3 kPa*
ρ_a = *density of dry air, 1.292 kg m*$^{-3}$

$$r = \frac{\frac{M_v e_a}{RT}}{(P - e_a)\frac{M_a}{RT}} = \frac{0.622 e_a}{P - e_a} = \frac{\chi}{\rho_a} \tag{5.19}$$

where r = *the ratio of mass of vapor to mass of dry air, kg kg*$^{-1}$

$$\chi = \frac{\rho_a \epsilon e_a}{P - e_a} \tag{5.20}$$

The mean molecular weight of dry air, as noted, is 0.029 kg mol^{-1}, with a mean density of 1.292 kg m^{-3}. However, the average water vapor density is about 0.005 to 0.01 kg m^{-3}. Although highly important, water vapor is a small part of the atmosphere. Table 5-31 compares the various definitions for a 50% relative humidity. The virtual temperature may be calculated from Eq. 5.21:

$$T_v = \frac{T}{\left[1 - \frac{e_a}{P}(1 - \epsilon)\right]} \tag{5.21}$$

where: T_v = *virtual temperature, °C*
ϵ = *ratio of molecular weights of water and air, 0.622*

If e_s is taken as the reference for pure free water, e_0, then it will be seen that the ratio e_a/e_s is uniquely related to the chemical activity of water (a_w) and mole fraction (N_w) as given previously by Eqs. 5.3 and 5.4 in Section 5.II.A Some examples of water potential were provided in Table 5-1. Activity and mole fraction are seldom employed in formulae dealing with E_t.

Relative humidity (RH), which is the most common unit in practice is a “non-conservative” value, dependent upon two factors –water vapor present and the temperature of the sample (Eq. 5.22):

$$RH = 100\left[\frac{e_a}{e_s}\right]_{(T)} \tag{5.22}$$

One may calculate RH from χ as well as **e**, or **q**, or **r**, without great differences in the result. However, List [1966] uses the mixing ratio (r). One of the most common devices to equate RH with T_{db}, T_{wb}, e_s, and r is the psychrometric chart (Fig. 5-78). This is a nomograph frequently employed by refrigeration and air-conditioning

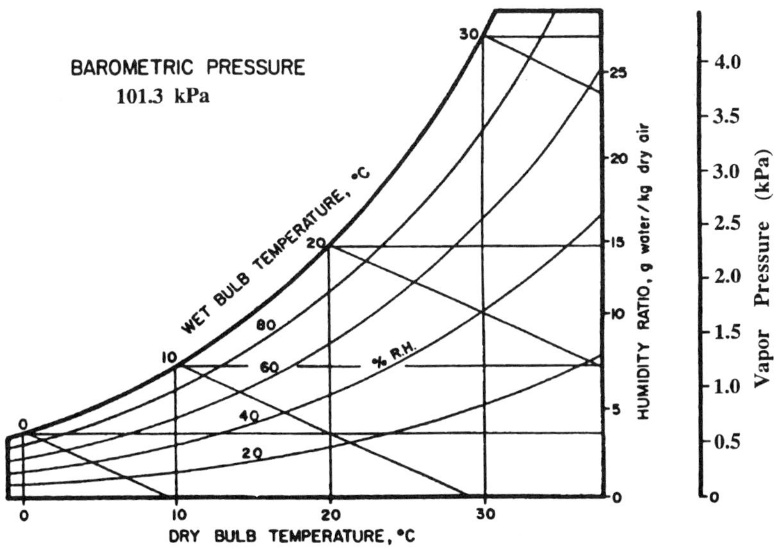

Fig. 5-78. Psychrometric chart often utilized to determine relative humidity (RH), vapor pressure (e) and mixing ratio (r) as functions of dry bulb and wet bulb temperatures [Gaffney, 1977] (With permission of *Amer. Soc. Hort. Sci.*).

engineers in installation design. However, computers can do this equally well to greater precision. Ways to show relationships between RH, T_{db}, and VPD are presented in Figs. 5-79 and 5-80. These curves serve to illustrate that common use of relative humidity for control is subject to serious errors. For one thing, a leaf can easily lose water to air at 100% RH if the leaf temperature is higher than the air temperature (greater e_s) and we assume the internal spaces of the leaf are saturated. Secondly, since air holds less vapor at lower temperatures, very slight changes in e_a or T_{db} can result in large changes in RH (Fig. 5-80). Thus, misting devices are seldom employed at temperatures around 10 C since wild fluctuations in RH can occur. At temperatures above 20 C (common for such species as roses), control based on RH is more readily accomplished. Because control of water loss involves the concentration gradient between leaf and air, we are more interested in the vapor pressure deficit (VPD), and Fig. 5-80 shows that much below T_{db} = 12 C, VPDs of 1.0 kPa or greater are impossible –as contrasted with the situation at dry bulb temperatures above 15 C.

b. Humidity Measurement

Fig. 5-81 shows 2 instruments frequently used to measure humidity by determining the dry bulb and wet bulb temperatures of the air, or psychrometry. A standard instrument, known as the Assman psychrometer is employed in research. Although Fig. 5-81 shows mercury thermometers as the sensors, other devices such as thermocouples or resistance devices can be employed as long as a rigorous protocol is followed. This is the use of clean wicks, distilled water, sufficient time for equilibrium, adequate shielding from radiation, matched ther- mometers and, above all, adequate aspiration at an air velocity not less than 3 m s^{-1}. The latter requirement means that these instruments cannot be used where the environment –such as a canopy– should not be disturbed by extraneous and rapid air movement. If the conditions are fulfilled, then the actual humidity can be found from T_{db} and T_{wb} measurements, using an ordinary pocket calculator:

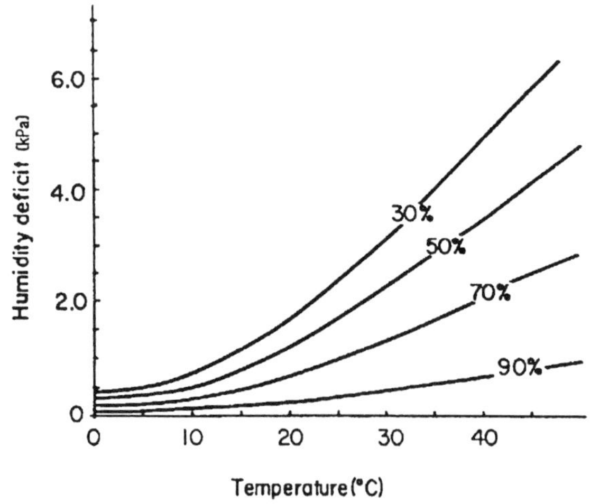

Fig. 5-79. Relationship between air temperature (T_{db}) and the "humidity deficit" (vapor pressure deficit VPD) at various relative humidities (RH). In this case, deficit is the difference between actual (e_a) amount of vapor present and what would be present at saturation (e_s) [Hanan, 1985].

Table 5-30. Absolute temperature, saturation vapor pressure (e_s) and slope of saturation vapor pressure curve (Δ) in kPa [Adapted from Monteith, 1973].

Temperature		e_s	Δ
C	K	(kPa)	(Pa K^{-1})
8	281.2	1.072	73
10	283.2	1.227	83
12	285.2	1.402	93
14	287.2	1.598	104
16	289.2	1.817	117
18	291.2	2.063	130
20	293.2	2.337	145
22	295.2	2.643	162
24	297.2	2.983	179
26	299.2	3.361	199
28	301.2	3.780	221
30	303.2	4.243	244
32	305.2	4.755	269
34	307.2	5.320	297
36	309.2	5.942	327
38	311.2	6.626	357
40	313.2	7.378	394

Table 5-31. Amount of water in air using various definitions of humidity for a relative humidity of 50% at 20 C [Woodward and Sheehy, 1983].

Density of water vapor (χ) (kg m^{-3})	Vapor pressure deficit (kPa)	Specific humidity (kg kg^{-1})	Mixing ratio (kg kg^{-1})	Dewpoint temperature (C)
0.00865	1.171	0.00713	0.00718	8.7

Fig. 5-80. Variation of relative humidity (RH) at fixed vapor pressure deficits (VPD) as a function of air temperature (T_{db}). VPDs in excess of 1.0 kPa cannot be obtained at temperatures much below 12 C [Hanan, 1991].

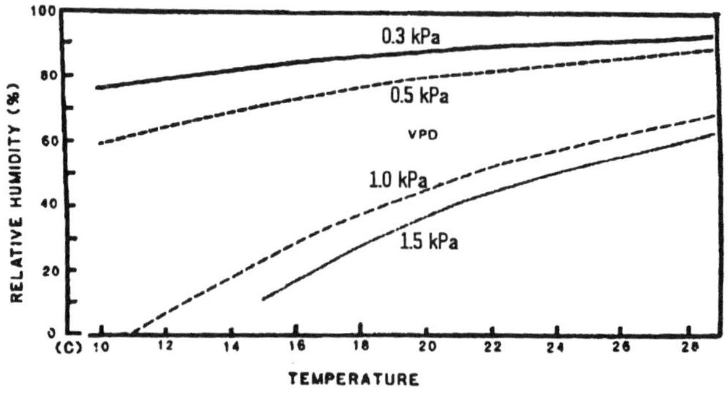

$$e_a = e_s - \gamma P(T_{db} - T_{wb}) \tag{5.23}$$

where: e_a = actual vapor pressure, kPa
e_s = vapor pressure at saturation at T_{wb}, which may be calculated from Eq. 5.16
γ = psychrometric constant, 65.8 Pa C^{-1} at 20 C
P = atmospheric pressure, 101.3 kPa at sea level

A form utilized by the National Weather Service is:

$$e_a = e_s - 0.00066(1 + 0.00115T_{wb})(T_{db} - T_{wb})P \tag{5.24}$$

where e_s = the saturation vapor pressure at the wet bulb temperature (T_{wb})

The psychrometric constant, γ, is:

$$\gamma = \frac{c_p \rho}{L \epsilon} \tag{5.25}$$

where: c_p = specific heat of air at constant pressure, $1.01x10^3$ J $kg^{-1}K^{-1}$
ρ = density of air including vapor, 1.292 kg m^{-3}
L = latent heat of vaporization, 2.45 MJ kg^{-1}
ϵ = M_v/M_a = 0.622

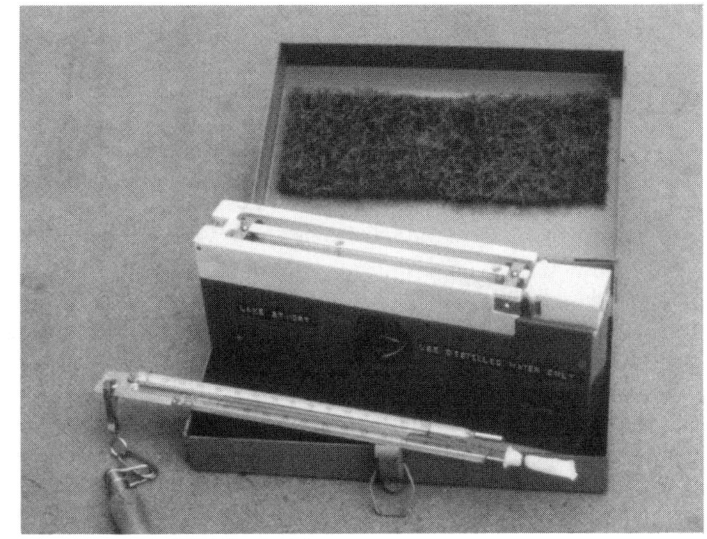

Fig. 5-81. Two types of common psychrometers. Note the wick on the bottom, sling instrument that is whirled to provide adequate ventilation. The enclosed device has an internal fan.

According to Monteith [1973], the slope of the saturated vapor pressure curve (Δ) (Fig. 5-77) and the psychrometric constant (γ) are regarded as weighting factors that determine the partitioning of radiant energy between evaporation (latent heat) and convection (sensible heat). Monteith stated that partitioning depends upon the ratio of Δ/γ as determined by the physical properties of water as a liquid and as a vapor. Both the sum of Δ and γ, and the ratio of the two, figure prominently in equations modeling vapor and heat transfer from plants. Table 5-32 compares the numerical value of Δ/γ with temperature.

There are many other devices for measuring humidity. One, very common, is the hair element (Fig. 5-82), which changes length with a change in humidity. Cheap versions of these devices may be bought in local stores, but are seldom satisfactory. Even under the best of conditions, error may be more than ±10% RH and seldom less than ±20%. They cannot be relied upon at RHs much below 30% or above 80%. At extremes, calibration is likely to shift permanently. Specially treated paper may also be employed, particularly for humidistats. Other systems include electrical methods that measure resistance of the element with changes in humidity (LiCl Dunmore cells, sulfonated polysterene, etc.), infrared gas analyzers, coulometric hygrometry, absorption of water in a desiccant, and so forth. One of the more interesting has been the development of capacitance probes (Fig. 5-83) in the last decade. These have been shown to have very high linearity over the entire range from 0 to 100% RH, and considerable precision with fast response. The device shown in Fig. 5-83 has been used inside greenhouses and outside under Colorado conditions. Inside probes require monthly calibration since the high humidities cause calibration drift. However, outside probes have functioned satisfactorily for as much as a year without significant change in calibration. Reference may be made to Hanan [1984], Rosenberg et al. [1983], and Wang and Felton [1983] for further details on hygrometry as applied to plant-atmosphere systems.

Table 5-32. Comparison of the psychrometer constant (γ) and the dimensionless ratio of the slope of the saturation (e_s) curve and γ with temperature [Stanghellini, 1987; Monteith, 1973].

Air temperature (C)	Δ/γ	Psychrometer constant, γ (Pa C^{-1})
0	.67	64.6
5	0.92	64.9
10	1.23	65.2
15	1.64	65.5
20	2.14	65.8
25	2.78	66.2
30	3.57	66.5
35	4.53	66.8
40	5.70	67.1

c. Humidity Variation

The general behavior of humidity in a greenhouse over time is a direct function of whether a heating system is present; the type of heating and ventilation system;and the outside climatic conditions of humidity, temperature, and wind. The outside conditions will often determine the methods and ability to control inside humidity. For example,

Fig. 5-82. Example of a mechanical thermohygrograph for recording both air temperature and relative humidity. A spare hair element is in the box at the bottom of the picture. Such devices are not suitable for greenhouse environmental measurements without special precautions.

Fig. 5-84 shows the general progression of absolute humidity with season in an arid, continental climate as denoted by the dewpoint temperature (T_{dp}). Obviously, maintaining low humidity for disease control during the winter is less of a problem, especially with hot air heating systems, than in a humid maritime climate where seasonal changes in absolute humidity are likely to be unremarkable.

On the other hand, absolute vapor concentration seldom varies significantly over a 24 hr period (Fig. 5-85). While the outside absolute humidity (χ) is below the inside, humidity reduction inside can be accomplished by ventilation. The use of relative humidity as an indicator, however, can show extreme variations (Fig. 5-85). In humid climates and unheated structures, it is usual for RH to reach 100% inside as temperature drops in the evening (Fig. 4-72). If outside χ is as high, or higher than χ inside, little can be done for humidity reduction without heating simultaneously with ventilation. In temperate and arid regions, disease problems are likely to be more severe in the spring and fall [Cobb et al., 1978]. Even in dry climates, the decrease in temperature where night setback is practiced is likely to result in high RHs, especially if evaporative pads are not shut off early so the greenhouse can dry out. Both Figs. 5-85 and 5-86 consider conditions at different times of the year in Colorado. Note in Fig. 5-85 the variation in outside RH as contrasted to the constancy of

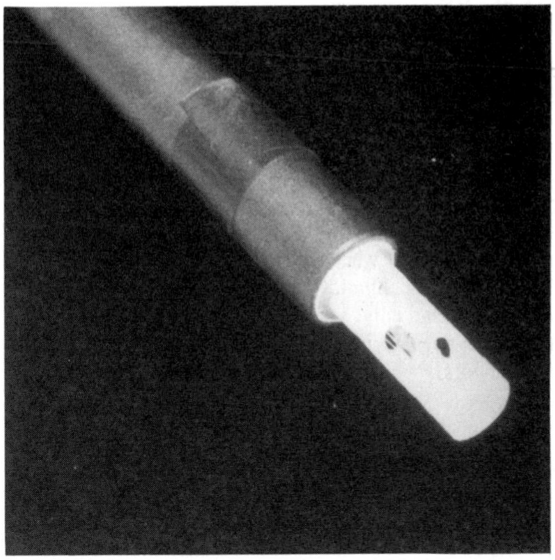

Fig. 5-83. Example of the business end of a capacitance humidity probe with the protecting filter removed. This particular device has been found to work well with digital acquisition systems for climate control. They do require a separate power supply, periodic calibration, and they are expensive.

outside vapor pressure. The variation in outside temperature, inside temperature, and plant temperature with fan-and-pad cooling may be noted in Fig. 4-72. This is sufficient to show how RH may vary with temperature when there is almost no change in absolute humidity. The VPD record presented in Fig. 5-86 was obtained during a period when outside temperatures were below -24 C, and outside e_a less than 0.1 kPa. Under these conditions, with no humidification, RHs inside can drop below 40% even during the night, especially in hot-air heated houses. If the houses had contained carnations, the VPD of 1.5 kPa would have been impossible due to the lower air temperature necessary for a cool crop.

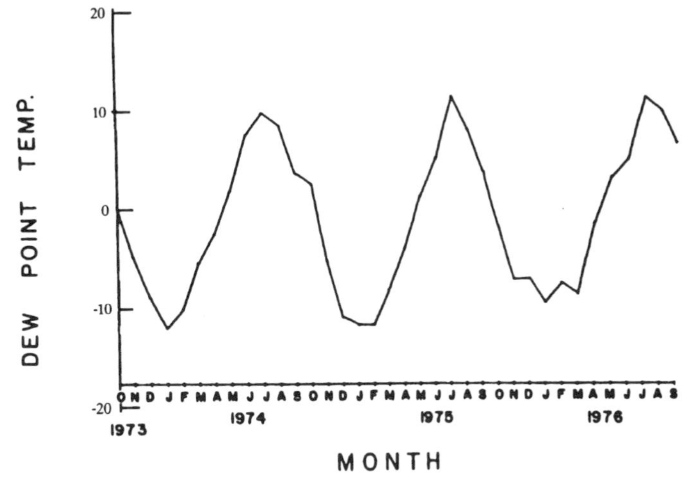

Fig. 5-84. Mean monthly dewpoint temperatures (T_{dp}) from Oct. 1973, through Sept. 1976, Ft. Collins, CO [Cobb et al., 1978].

In a more humid region, such as Kentucky, humidity outside may be between 90 and 100% RH [Walker and Duncan, 1973]. However, even for Kentucky, which is among the regions with the highest RH, values above 90% are only an average of 1.6 hr per day. Even here, positive ventilation of a house will often limit the duration of high humidity. In most climates in the winter, dehumidification may also occur due to condensation on the cover. Whenever it is sufficiently cold enough to reduce the cover temperature below T_{dp}, condensation will occur. As noted in Chapter 4, condensation can result in a significant decrease in humidity and release of energy. The possible benefits of condensation for humidity reduction are limited when energy-conservation methods such as thermal blankets and double covers are employed, resulting in a lower infiltration rate and warmer inside cover temperatures. Conservation measures have often led to considerable difficulty with increased disease until growers learned to manipulate the environment appropriately [Anon, 1986; Van Meurs and Gieling, 1980; de Graaf, 1985; etc.]. Plants can still evaporate water at night although stomata may be closed, besides vapor contribution from soil surfaces, etc.

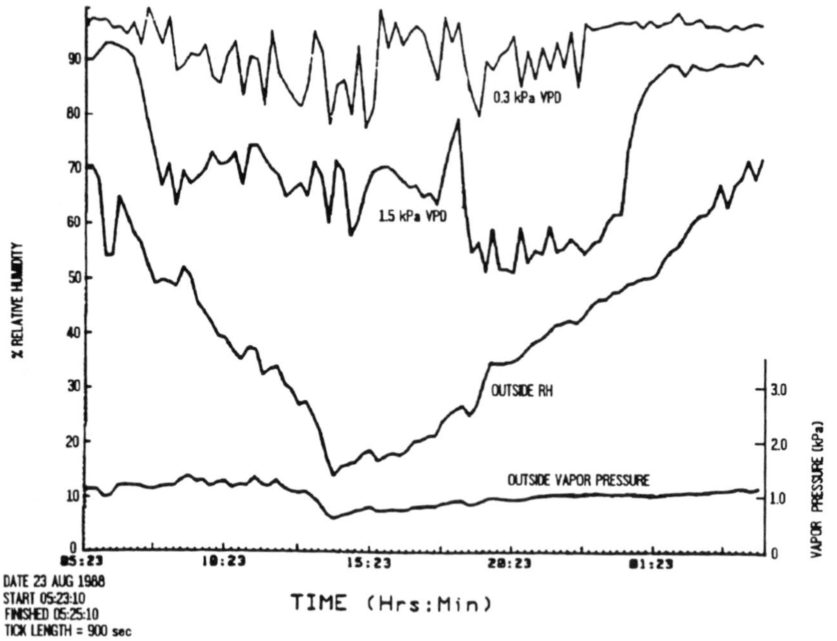

Fig. 5-85. Relative humidity inside and outside a greenhouse, and outside vapor pressure in Colorado during the early fall [Hanan, 1988]. Roses grown at 16 C night and 21 C day minimum. Fan-and-pad cooling. Over a 24-hr period, outside absolute humidity varies little, whereas outside relative humidity decreases markedly as temperature rises during the daylight hours.

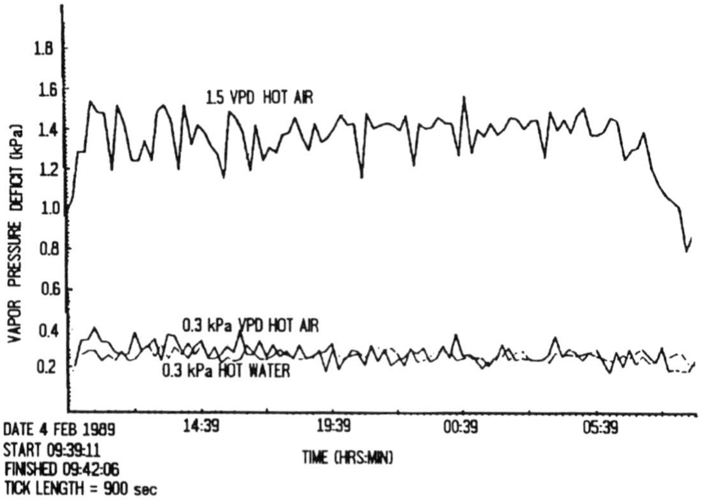

Fig. 5-86. Vapor pressure deficits inside greenhouses controlled at 0.3 and 1.5 kPa VPD during the winter in Colorado [Hanan, 1989]. Roses grown at 16/21 C night/day, respectively.

Problems of high humidity are particularly troublesome within a dense canopy. Air movement is usually very low (<0.01 m s^{-1}) unless some type of air circulation is employed such as HAF, or the benches are constructed to permit natural circulation upward through a canopy. For these reasons, watering with spray devices should not occur in late winter afternoons since the inside foliage is unlikely to dry out sufficiently. Evaporative pads should be turned off 1 to 2 hr before dark so that exhaust fans can dry pads and foliage. Closing plastic houses in Mediterranean climates, under the misleading idea of conserving energy, is a poor practice. The house should remain open.

A problem under clear-night conditions, especially where heat is not used in mild climates, is the radiation loss from the canopy to a cold cover and sky. This, quite frequently, can reduce leaf temperature below the dewpoint, resulting in condensation on the leaf (Fig. 5-87) and increased susceptibility to foliar diseases.

d. Humidity Control

A number of suggestions have been made in the previous section regarding manipulation of humidity. Almost any of the devices discussed in the section on measurement may be employed to attempt humidity regulation. Electrical systems are best used for computer input. Most of these methods can be employed for

Fig. 5-87. Condensation of water vapor noted in the early morning on cucumber. This usually occurs during clear sky conditions, when provision is not made to dry the greenhouse the previous evening or during the night. This condition is particularly conducive to infection and damage from pathogens. This situation particularly prevalent in polyethylene-covered houses in dry climates.

direct settings of relative humidity. Nevertheless, as I have suggested, and discuss later, RH manipulation can be difficult due to the relationships of temperature and actual vapor concentration. Furthermore, the effect of humidity is best given in terms of the vapor pressure deficit (VPD) between the leaf and bulk air, as noted previously in Eq. 4-3. The assumption is made, under greenhouse conditions, that the internal leaf spaces are saturated. If leaf temperature is known, or can be estimated, or is assumed equal to the air temperature, then e_s can be calculated and e_a of the surrounding air subtracted to provide VPD in kPa. As noted later, and in Chapter 8, the necessary calculations are easily and rapidly carried out by present-day tabletop computers.

Any greenhouse that encloses a full vegetative canopy is likely to have problems with excessive humidity conditions or VPDs close to zero. There are essentially three methods to control this situation: 1) ventilation, 2) cover condensation, and 3) heating, or combination of heating and ventilation. A fourth, presented by Assaf [1990a], Assaf et al. [1990b], and Seginer and Kantz [1989], involves the use of a concentrated brine solution with appropriate circulation systems. Vapor is absorbed from the air circulated over the brine, and the latent heat released is transferred to the air. To recycle, when the brine solution is sufficiently diluted, the solution is heated and the fresh water released. According to Flaherty [1988], 2000 hr or more of continuous operation generally results in a cheaper system as contrasted to fan ventilation for humidity control.

According to Walker and Cotter [1968], warm-weather humidity can be defined as three different problems. 1) High humidity at night when no solar heating is available, and when the external temperature (T_o) is near or above the desired inside temperature (T_i). 2) The humidity condition when solar heating is available, but inadequate to cause ventilation. And 3) low humidity when solar heating is adequate to result in ventilation. In the first condition, ventilation must be maintained whenever T_o is above T_i. In the second case, the problem can be described:

$$E = \frac{F}{\nu}(r_i - r_o) \tag{5.26}$$

where: E = *moisture added by transpiration*
F = *air exchange rate,* $m^3\ s^{-1}$
ν = *humid volume in house,* $m^3\ kg^{-1}$ *dry air*
r_i = *mixing ratio inside,* $kg\ kg^{-1}$
r_o = *mixing ratio outside the house*

The combination of solar-induced temperature rise and increased E tend to compensate for each other. Only slight changes in RH usually occur in humid climates. However, in conditions of high solar radiation, and full ventilation, undesirably low humidities may be the result –especially during initial crop establishment when the canopy is not fully developed with an LAI much less than 1. This can be a particular problem in arid climates.

In those cases where both heating and ventilation are necessary, Augsburger and Powell [1986] calculated requirements for two conditions of humidity and temperature, and related the results to actual costs. In both examples, the house covered 937 m^2 with a volume of 3387 m^3. In the first case, the house maintained at 24 C,

Fig. 5-88. Effect of vapor pressure deficit (VPD) on leaf area of 'Royalty' plants (per plant) and 'Red Success' roses [Harkess and Hanan, 1988].

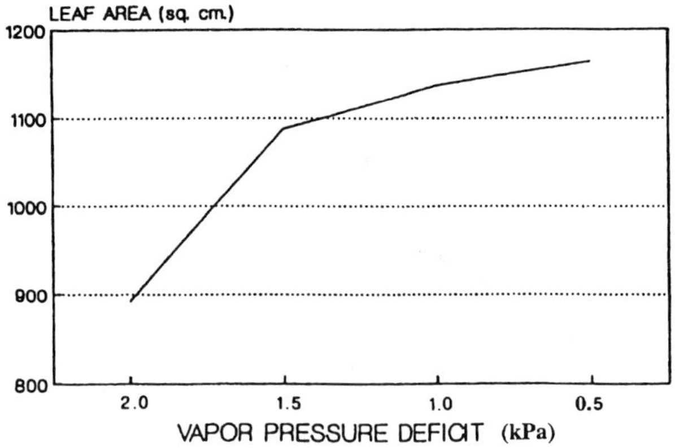

required reduction from 95% RH to 70%, using outside air at 10 C and 100% RH (raining). At 95% RH, the inside air contained 17.7 g water per kg air. Greenhouse air at 70% would contain 13 g vapor kg^{-1}. The outside air held 7.6 g kg^{-1} even though saturated. Here, 2397 kg of air would have to be exchanged, which with a 1.2 m diameter exhaust fan, would require 4 min of operation for about 0.5¢ at \$0.08 $(kW\text{-}hr)^{-1}$; 60% of the total greenhouse air would be exchanged and 34.4 MJ energy required to heat the new air. At a gas cost of 17.8¢ m^{-3}, the heating cost would be about 20¢, for a total cost of 20.5¢ per cycle. Assuming a drying cycle every 30 min for half the day, the cost per day would be about \$5.00.

In their second example, Augsburger and Powell assumed a house temperature of 20 C, RH = 85%, and it was desired to reduce humidity to 70%, using outside air at 10 C and 80% RH. This required an exchange of 1392 m^{-3} to remove 8.6 kg vapor. The same size fan as in the first example would require 2.5 min, and 16.8 MJ of energy would be necessary to heat the 10 C air, for a total cost of 10.3¢. Assuming three cycles per hour over 12 hours for 150 days, the total yearly cost would be about \$5.91 m^{-2}. These calculations make no allowance for the fact that the crop would be transpiring (either night or day), so there would be continual addition to the vapor in the house air, as indicated in Eq. 5.26. Ventilation and heating would be required over a greater period than given in these two examples. Nevertheless, the method shows a practical application of humidity control in temperate climates. The costs do not seem excessive when given the fact that disease problems certainly reduce profitability. The most difficult and expensive situation is when the outside temperatures and humidities closely correspond to inside. Without significantly raising the inside temperature, ventilation is not sufficient. Then, a brine dehumidifier as described by Assaf [1990a] might be useful.

Walker and Cotter [1968] and Walker and Duncan [1973] suggested that small amounts of heat addition are possible to reduce relative humidity –such as the use of the heater on a cyclic timer, heating intermittently for 3 or 4 minutes. This was also suggested by Ross [1974], even though the air temperature will be raised above the desired setpoint. The ventilation setpoint might have to be adjusted to avoid undesirable interaction between heating and cooling. With computers, this type of dehumidification can be easily carried out. Hanan et al. [1987], in their description of a computer system, controlled minimum VPD at 0.5 kPa for roses. Note: this method completely disregards RH and operates without incorporating temperature. Under Colorado conditions, the requirement simultaneously to heat and ventilate was limited to the spring and fall, seldom occurring beyond the early morning hours. At 20 C, a VPD of 0.5 kPa would result in 78% RH. At 10 C, this same VPD would result in a 58% RH. A VPD minimum control point of 0.2 kPa would be more suitable (84% RH). If it is desired to maintain humidity based on RH, then the VPD should be adjusted according to temperature. Other investigators [e.g., Matthews and Saffell, 1986] have studied computer control of humidity, using psychrometers for determining T_{db} and T_{wb}. Further discussion of this facet of climate control will be in Chapter 8.

As a rule, relative humidities greater than 90% should be prevented. There are special instances for mist propagation, seedling establishment, and new plantings where it is necessary to reduce transpiration and plant temperature by deliberately wetting the foliage or employing a high-pressure fog system, as outlined in Chapter 4. These occasions, however, are special and not often employed when the crop has developed a full canopy and is approaching harvest. An exception to not using fog or mist on fully developed crops would be roses and some decorative foliage plants.

Addition of water vapor to the greenhouse is much simpler. In arid regions, using evaporative pads for cooling, relative humidities below 60% are seldom encountered. Where, due to the necessity as outlined above, additional water has to be added, various types of misting systems are available, ranging from low pressure methods (e.g., 500 kPa) up to fogging systems operating above 5 MPa (Fig. 4-84). The higher the pressure, the more expensive the system, with increased requirements for clean water supplies. The urgent criterion, however, is the use of low salinity water –preferably below 0.5 mS cm^{-1} electrical conductivity.

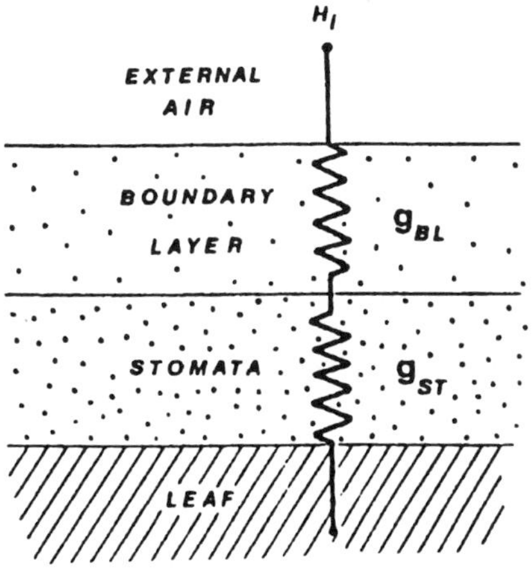

Fig. 5-89. A simple circuit from Aubinet et al. [1989], showing the major conductances from within the leaf to the external or bulk air. This diagram does not take into account pathways for vapor through the cuticle, from both sides of the leaf, or the fact that the boundary layer is usually much more complex. (Reprinted from *Agric. and Forest Meteor.*, Aubinet, M. et al., 1989. 48:21-44. With kind permission of Elsevier Sci.-NL, Sara Burgerhartstraat 25, 1055 KV Amsterdam, The Netherlands).

e. Effect of Humidity on Growth

Humidity effects are difficult to examine, largely due to the interactions between vapor concentration and temperature, and the greater difficulty of accurate measurement. As pointed out, basing results on relative humidity (RH) fails to take into account the basic physical principles of water loss from plants. This is not to denigrate some work by Norwegian and Dutch workers in particular [Gislerød and Nelson, 1989; Mortensen and Gislerød, 1990; Gislerød and Mortensen, 1990, 1991; van de Sanden, 1985; Bakker, 1989; 1990a; b; c]. These workers are usually careful to specify the VPD for the experimental conditions and Bakker dispensed with RH completely.

The review by Grange and Hand in 1987 concluded that humidities between 1.0 and 0.2 kPa VPD do not affect the physiology and development of horticultural crops. At VPDs greater than 1.0 kPa, dry weight can be reduced in many species such as *Begonia*, *Saintpaulia*, *Dendranthema*, *Euphorbia*, etc., whereas others such as *Rosa*, cucumber, and lettuce were not affected [Gislerød and Mortensen, 1991]. Low VPDs (0.11 kPa) and high CO_2 levels increased relative growth rate (RGR) and leaf area [Gislerød and Nelson, 1989]. The latter authors showed that stomatal opening in chrysanthemum (*Dendranthema*) was decreased slightly at high CO_2 levels, but was increased by high humidity. Leaf area may be the dimension most greatly affected by high humidities. As noted by Harkess and Hanan (Fig. 5-88), at VPDs of 1.5 kPa, leaf area in roses was markedly reduced. Significant effects over a range of 0.2 to 1.0 kPa VPD have been noted on transpiration rates of most plants, i.e., a decrease at high humidities as expected. Over the mean range of 0.30 to 0.78 kPa VPD, Bakker [1989] did not find any detrimental effects on growth of sweet pepper, nor on eggplant (24 hr mean, 0.34 to 0.99 kPa VPD) [Bakker, 1990b]. In tomato, however, Bakker [1990a] found that calcium deficiency and leaf area reduction occurred at continuously high humidity (0.2 kPa VPD as contrasted to 0.8 kPa). Bakker et al. [1987] found vegetative growth in cucumbers to be enhanced by high day or night humidity, but there was no effect on an early yield.

The problems encountered at high humidities have been examined by Gislerød and Mortensen [1991; 1990] and Bakker [1990a]. The situation was summarized by Grange and Hand [1987], in which VPDs less than 0.3 to 0.2 kPa can result in calcium deficiencies in tomatoes with increased blossom end rot of both tomatoes and peppers. Similar disorders are found in "tip-burn" of strawberries, chicory, and lettuce leaves, "tip-scorch" of chrysanthemums, and "blackheart" of cauliflower, celery, and Brussels sprouts. Lowered uptake of such macronutrients as K, N, Ca, P, and Mg has been noted in a number of species at high humidity [Aikman and Houter, 1990]. Increasing solution concentrations will tend to counteract these problems resulting from high humidity. "Glassiness" in lettuce and "edema" in geraniums has often been observed as the result of high humidities under winter radiation conditions. It should be noted that most of the recent work on humidity has been carried out in north European countries (Oslo, Norway, at 60° north) where winter conditions are characterized by extremely low radiation values (see Chapter 3), and energy conservation measures tend to exacerbate high humidity regimes.

Grange and Hand state the optimum humidity levels for controlling plant diseases are poorly defined. In some cases, it is known that pathogens such as *Puccinia* require free water to germinate, whereas the oidia of the powdery mildew organism, *Sphaerotheca*, requires RHs above 95% but below 100% –because free water may cause death of the spores. Thus, disease control and suitable accomplishment of biological control require close attendance to greenhouse humidity level.

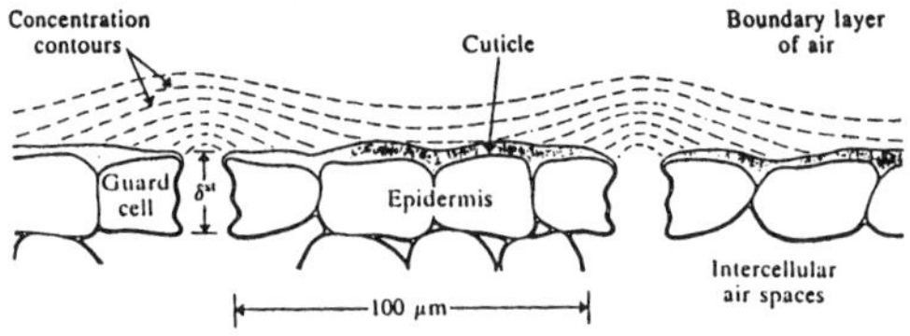

Fig. 5-90. Enlarged cross-section of a leaf epidermal layer, showing stomata and approximate dimensions between stomata on an average leaf.

Fig. 5-91. The effect of stomatal width on transpiration in *Zebrina* in wind (upper curve) and still air (lower curve) [Bange, 1953]. The lower curve probably more closely approaches greenhouse conditions. Bange stated the wind was estimated at several meters per second for the upper curve (With permission of *Acta Bot. Neerl.*).

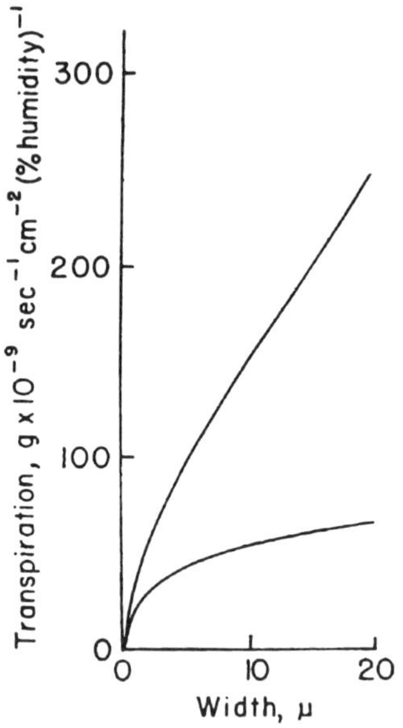

3. Factors of the Vapor Pathway

To this point, two parameters of water demand have been briefly examined: 1) the energy supply (radiation), and 2) the concentration gradient between leaf and air (vapor pressure deficit). The third major factor is the resistance to molecular movement from the evaporating sites in the leaf to the bulk air. The situation for a very simple case is outlined in Fig. 5-89. There are many variations of increasing complexity to be found [e.g., Waggoner and Reifsnyder, 1968; Moreshet et al., 1968; Lynn and Carlson, 1990; etc.]. As can be appreciated, there are two sides to a leaf, and there may be stomata on both sides that may not behave similarly [Kanemasu and Tanner, 1969a, b]. Water loss through the cuticle occurs in parallel with the stomata. So if one uses electrical resistance as an analogy to examine vapor flow, there will be resistances in parallel and in series. This practice has been employed since the 1900s in units of seconds per meter (s m^{-1}, s cm^{-1}, or m^2s mol^{-1}). In recent years, there is increasing emphasis on units of conductivity –the reciprocal of resistance (m s^{-1}, mol $m^{-2}s^{-1}$). Fig. 5-89 uses conductance. Fiscus et al., in 1983, suggested that the Ohm's law analogy be abandoned, if for no other reason that one expects linearity with resistance measurements. Linearity is probably the exception with biological systems. Nobel [1991] states that, in electrical terms, resistivity is a fundamental property of a material with units of ohms m^{-1}. Resistance (R), on the other hand, is expressed in ohms. Resistance may vary from some lower limit to infinity, and the material transported in unit time (flux density) is inversely proportional to resistance, but directly proportional to conductance. Furthermore, conductance varies between zero and some maximum value. Nobel concludes that conductance terminology is often more convenient for relating plant response to environmental factors. As so much of the literature in this section uses both resistance and conductance terms, we will use both where convenient.

There are a number of factors to be considered in this section. First, there is effect of resistance on vapor flux from the mesophyll through the cuticle and stomata of the canopy. Second, there are the circumstances of the boundary layer that are peculiar to greenhouses as to wind movement, turbulent air flow, and the interaction between the greenhouse environment and plant canopy. Lastly, there is a need to put together these major factors of energy, gradient and resistance (conductance) in some fashion that allows one to predict: 1) the effect on plant water potential (stress), 2) the greenhouse environment, and 3) to control greenhouse environment to the betterment of profitability.

Table 5-33. Some representative values of conductance and resistance of leaves and stomata [Adapted from Nobel, 1991; Seginer et al., 1990; Seginer, 1984].

	Conductance ($mm\ s^{-1}$)	Resistance ($s\ m^{-1}$)
Stomata		
Open	4 to 20	160 to 800
Closed	0	∞
Cuticle		
General	0.1 to 0.4	2500 to 10000
Cucumber	0.7 to 3.2	310 to 1350
Tomato	0.8	1220
Pepper	0.8	1220
Eggplant	0.5	1920
Dieffenbachia	0.5	2220
Dracena	0.1	19920
F. benjamina	0.2	4300
Dendranthema	0.2	4200
Rose	0.4	2400

a. Physiological Regulation of Vapor Flow

Fig. 5-13 showed an example of the pores that will be found in green plants in leaf surfaces. These openings are necessary for CO_2 uptake, so they are open during the day and usually closed at night –except for plants with a crassulacean metabolism. With opening, there will be, necessarily, an outward flow of water molecules if the potential of the water in air is lower than the potential inside the leaves. Fig. 5-90 is an enlarged diagram of a small leaf section with typical stomata. The morphological variations that can be found in the plant world are enormous. The behavior of stomata has been reviewed and examined by many individuals [e.g., Rashcke, 1975; Zelitch, 1967; 1969; Ofir et al., 1968; Lee and Gates, 1964; Shaer and van Bavel, 1987; etc.]. Stomata open in response to low CO_2 concentrations within the leaf at low radiant intensities as photosynthesis begins in the mornings. The opening width of stomata is seldom constant, with diurnal rhythms often superimposed. The important factor to consider is that, from the standpoint of maximum productivity, stomata should be opened fully when there is sufficient energy for photosynthesis. The effect of stomatal aperture on transpiration has been examined by several authors. One example is presented in Fig. 5-91. The lower curve in still air is particularly important since it closely approximates greenhouse conditions. The lower curve shows that the effect of stomatal closure at zero, or low wind speeds, will not be pronounced until the stomata are nearly closed.

Nobel [1991] summarized general values of resistances and conductances found in the literature. These may be noted in Table 5-33. Raschke [1975] showed that resistance of stomatal pores can be as low as 100 $s\ m^{-1}$, although Table 5-33 gives a value of 160 $s\ m^{-1}$. Slatyer [1966] stated that typical open stomata values can range from less than 10 $s\ m^{-1}$ to more than 100 $s\ m^{-1}$. An assumption that the interior of the leaf is always saturated at the particular leaf temperature is based on the fact that (Eq. 5.4), at 99% RH inside the leaf, the water potential will be -1.4 MPa. At 95%, the potential would be -6.9 MPa. Studies on resistance within the leaf [e.g., Slatyer, 1966] suggest that drainage of the interfibrillar spaces as the result of low water supply would require -1.5 MPa for spaces roughly 100 mμ, and -15 MPa for spaces of 10 mμ diameter. Jarvis and Slatyer [1970] presented evidence that the mesophyll cell walls of cotton could influence transpiration rates. The authors state that the relative water vapor pressure at the mesophyll surfaces equal to e_s at T_l cannot be supported under many conditions. However, in plant communities where the total leaf area greatly exceeds ground area, the relative significance of internal factors is much less than is the case with a single leaf. These values show that humidities low enough to significantly affect the evaporative process within the leaf will not be found in unwilted, well-watered, herbaceous crops. There was considerable doubt about the possibility of non-stomatal regulation of vapor flow within the leaf. However, the appreciation of xylem cavitation in the 1970s suggests that embolisms within the xylem can be a significant factor in water supply to the leaves since the xylem volume for water movement is reduced [Tyree and Sperry, 1989]. Fiscus et al. [1983] point out that the water system can be depicted as a hollow tube topped by a fibrous matrix as the evaporating surface. Evaporation from this matrix will result in increased curvature of the liquid surface, leading to lower potential in the liquid phase. For well-watered greenhouse plants, the assumption of saturation at any given leaf temperature appears valid, and one need deal only with stomatal and cuticular resistance.

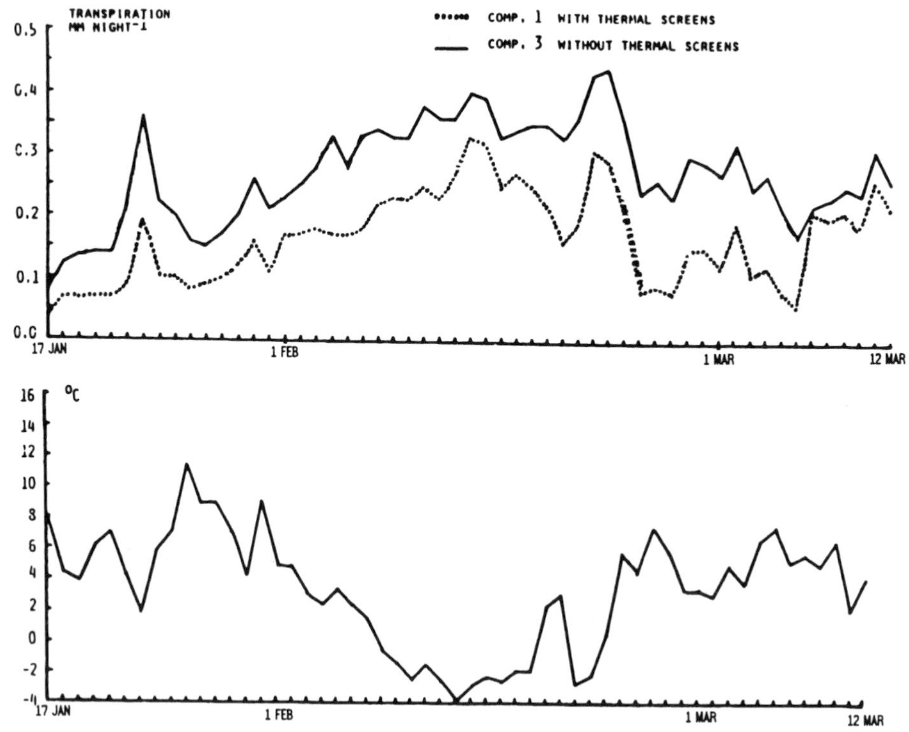

Fig. 5-92. Nighttime transpiration of tomatoes in a Dutch greenhouse. Comparison is made between treatments with and without thermal screens and with outside temperature [de Graaf, 1985] (©1985, Int. Soc. Hort. Sci., *Acta Hort.*, 174:57-66).

Cuticle resistance can be 10 to 100 times higher than fully opened stomata. Assaf [1989] cited Zhao et al. as showing that resistance at night for chrysanthemums was 4300 s m^{-1} under IR heating but 1700 s m^{-1} for plants grown in houses with convective heating systems. On the other hand, Panter and Hanan [1987] found no significant differences in daytime stomatal resistances between *Calceolaria* plants grown in IR-heated structures versus convective regimes. Table 5-33 shows that cuticular resistance to water vapor flux can be quite variable, ranging from as low as 310 s m^{-1} for cucumber to more than 19000 for *Dracena*. An example of actual water loss at night under European conditions is given in Fig. 5-92, showing a close correspondence of transpiration with outside temperature. This figure emphasizes the climatic situation north of 50° latitude where the length of the dark period in winter, combined with high heating requirements, results in more water lost at night than during the day though the stomata are closed [Stanghellini, 1981a; de Graff, 1981; 1985]. The idea that transpiration ceases at night, an assumption used in some models found in the literature, is nonsensical.

A factor causing stomata to close is low plant water potential (ψ_t). Low potentials may be caused by failure in water supply or excessive transpiration loss. Certainly, when the plant wilts, stomata will close. From the greenhouse operator's viewpoint, the ideal situation is for stomata to be fully open through the day. Fortunately, there is a "critical" water potential (ψ_t) above which the stomata will remain open and transpiration will be the consequence of energy input, vapor gradient, and canopy resistance. Leaf resistance is assumed to be a constant, total resistance changing only when stomatal closure occurs during the dark hours. This, in fact, is an assumption made by Lynn and Carlson [1990], citing several authors to support their contention. For many reasons discussed by Aubinet et al. [1989], the impact of stomatal conductance on transpiration rate of greenhouse crops is markedly reduced during the day. Development of models, according to Aubinet et al., that incorporate stomatal conductance is a superfluous sophistication.

Examples of the relationships between internal water potential and leaf resistance are given in Figs. 5-93 and 5-94, which accord with Raschke's conclusion [1975] that the critical threshold varies between -0.7 and -1.8

Fig. 5-93. The relationship between stomatal resistance and leaf water potential for upper and lower leaf surfaces of a bean plant (*Phaseolus vulgaris*) [Kanemasu and Tanner, 1969a]. The critical potential for the lower surface is about -1.1 MPa as contrasted to the upper surface (adaxial) with rapidly increasing resistance below -0.8 MPa (With permission of the *J. Plant Physiology*).

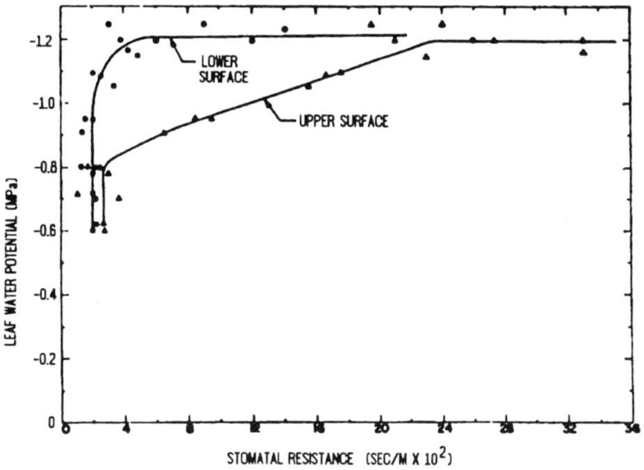

MPa. Numerous field studies [i.e., Cox and Boersma, 1967; Moreshet and Stanhill, 1965; van Bavel, 1967; Ofir et al., 1968; etc.] have shown that transpiration proceeds at the maximum determined by available energy if water is freely available. Some data on the relationship between availability and transpiration rate were presented in Fig. 5-31. The discussion in Section 5.II.B showed that many plant functions such as elongation, photosynthesis, etc. can be significantly reduced at potentials closer to zero than these "critical" potentials for stomatal closure. Bedding plant growers often withhold water for height control. Still, I wonder if other procedures could be used to control height (e.g., suitable timing of germination, etc.), rather than possibly increasing the crop period in an expensive greenhouse as the result of slower growth.

From this brief review, one may conclude that the range of water potentials for most greenhouse plants should not be lower than -1.0 MPa to ensure full stomatal opening for maximum CO_2 uptake. The whole thrust of crop culture should be to control climate to maximize water loss by ensuring open stomata. This serves to emphasize the importance of some day measuring leaf water potential of the crop on a real-time basis.

b. External Resistance

Because of frictional drag exerted by an object immersed in a moving fluid –such as air– wind speed will increase with increasing distance from the object as the result of the fluid's viscosity. At the surface, such as a leaf, molecules will stick to the surface, and their velocity will be zero. The transition zone from zero to full velocity is called the boundary layer (Fig. 5-89). For a single leaf, the situation can be approximated by a flat plate, so that, under equilibrium conditions, a logarithmic profile of increasing air velocity may be achieved with distance from the surface (Fig. 5-96). If heat and vapor are being transferred to the surrounding fluid, there will be temperature and vapor gradient profiles. These profiles from zero to full air velocity can be roughly less than a millimeter to as much as 1 centimeter in depth. The exact situation will vary with numerous parameters, such as dimensions of the leaf, distances from the leading edge, thickness, temperature, surface roughness, etc., plus velocity of the bulk air. For a canopy, the boundary layer resistance is sometimes called aerodynamic resistance. Some authors may also employ the term "external" resistance as contrasted to "stomatal," "leaf," "epidermal," or "internal" resistance.

Usually, the fluid flow immediately next to the surface will be laminar. That is, there is a streamline flow in which 1 particle follows another in an orderly fashion. With temperature or vapor gradients, heat or vapor will be transferred by molecular conduction. Depending upon conditions, laminar flow will cease at some point above the surface, resulting in turbulent flow in which the smooth character of the flow breaks down. There are many eddies and irregularities generated as the air passes over the object. The distinction between laminar and turbulent flow may be easily seen by the smoke rising from a cigarette, a campfire, etc. under calm conditions. The effect of turbulent versus laminar flow is to increase the rates of heat and vapor conduction from a surface such as a plant leaf.

Transfer across the boundary layer can be treated as resistance or conductance, as noted previously for stomata. Eqs. 4.2 and 4.3 used r_H and r_V in that brief discussion of sensible and latent heat transfer. Boundary layers and resistance to turbulent transfer will obviously be entirely different for a full canopy versus

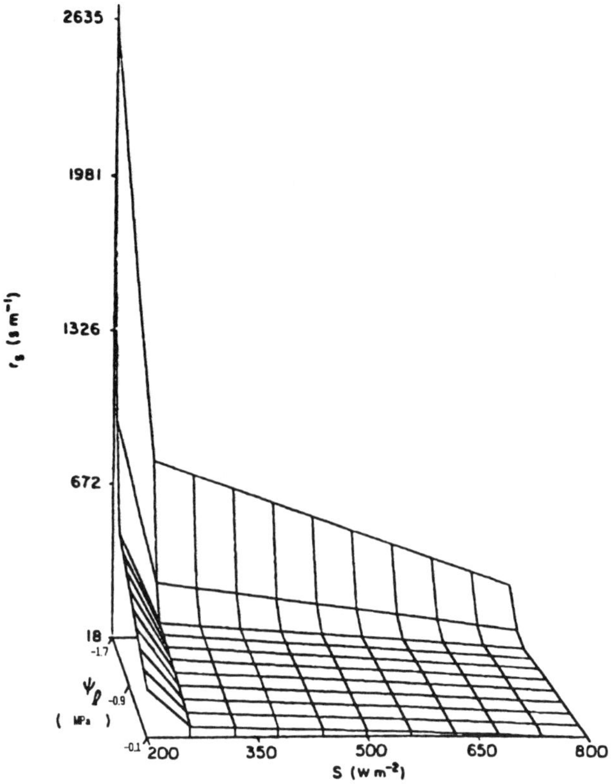

Fig. 5-94. Three-dimensional graph showing the effect of solar flux and leaf water potential on stomatal resistance. Calculated from a mathematical model proposed by Lynn and Carlson. [Reprinted from *Agric. and Forest Meteor.*, 1990, Lynn, B.H. and T.N. Carlson. 52:5-43. With kind permission of Elsevier Sci.-NL, Sara Burgerhartstraat 25, 1055 KV Amsterdam, The Netherlands].

a single leaf. Aubinet et al. [1989] and Jarvis and McNaughton [1986] concluded that disagreement among authors about the role of stomata versus aerodynamic resistance is a question of scale. The importance of stomatal regulation of water loss progressively decreases when one scales upward from a single leaf, to a plant, and to the entire canopy. Some difficulties of the greenhouse climate were admirably set out by Stanghellini [1987]. The resistance to vertical heat exchange for a field crop is usually estimated by the assumption that the canopy is immersed in a boundary layer, characterized by a well-defined, logarithmic profile of wind speed. Such an assumption requires that the upwind distance (fetch) be at least 50 to 100 times the instrument height above the crop surface [Tanner, 1960; Kanemasu et al., 1979]. If one can place an anemometer 1 m above the crop in the greenhouse, this would mean the instruments for measuring windspeed, temperature and humidity must be 100 m downwind –a rather impractical situation, assuming that there is a regular wind movement in one direction!

c. Wind Movement

Wind velocity influences several features of the greenhouse climate besides canopy resistance to vapor transfer. Evaporative-pad cooling, with exhaust fans, represents one situation in which a rapid wind movement in one direction can prevail for a significant period. The maximum wind speeds for a standard cooling system pulling air 30 m is about 0.3 m s^{-1}. Invariably, wind speed will be lower than 0.1 m s^{-1} when the structure is tightly closed, steam or hot water heating employed, and outside temperature only slightly below the inside air temperature. Air movement was discussed in Sections 4.IV.C.2 and 4.IV.D, with the maximum wind speed as the result of pipe temperature differentials at 0.2 m s^{-1} (Fig. 4-88). Wind speeds of 0.5 m s^{-1} can be represented as maximum for nearly all cases and will result in leaf fluttering, markedly altering the boundary layer. Other examples of wind speeds determined experimentally were presented in Figs. 4-69 and 4-71. Wind velocities within most greenhouses with pipe heating will be less than 0.1 m s^{-1}. Hanan et al. [1978] suggested an average of 0.04 m s^{-1} for interiors, whereas Stanghellini [1981a] employed a value of 0.12 m s^{-1}. Wind speeds in the field are usually about 1 to 5 m s^{-1}. Unfortunately, air velocities found in greenhouses are difficult to measure [Hanan, 1984]. Special instruments are required which lack robustness required for ordinary greenhouse use. One item, simplifying consideration of energy transfer between canopy and atmosphere in a structure, is the fact that one can neglect "momentum" energy transfer. That is, there is insufficient wind velocity to cause significant frictional energy transfer to the canopy.

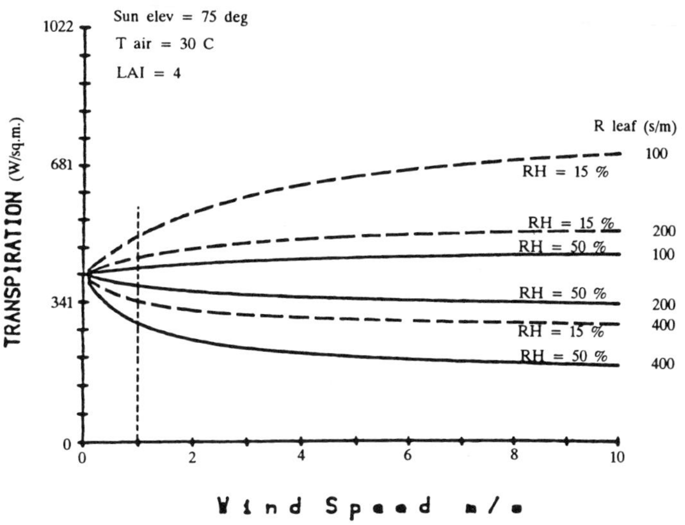

Fig. 5-95. The effect of windspeed on transpiration of a cotton crop [Fuchs, 1986]. Transpiration, given by Fuchs in mm hr^{-1}, has been converted to latent heat. The vertical bar at 1 m s^{-1} serves to emphasize the air velocities common to greenhouses are below 1 m s^{-1} –more commonly <0.5.

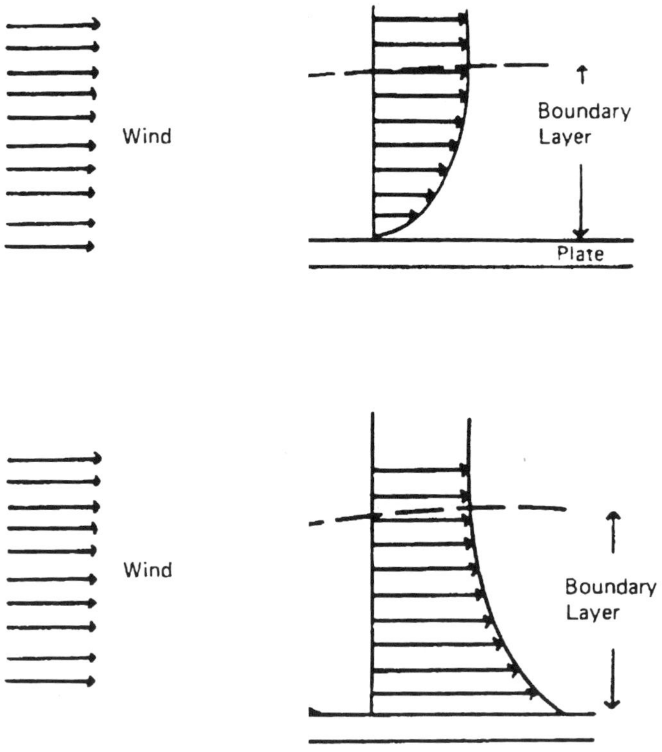

Fig. 5-96. Wind speeds above a smooth plate some distance downwind from the leading edge. The lower diagram shows the temperature profile with height when the plate is warmer than the air. This illustrates the logarithmic wind profile commonly found outdoors when calculating energy fluxes of a plant canopy.

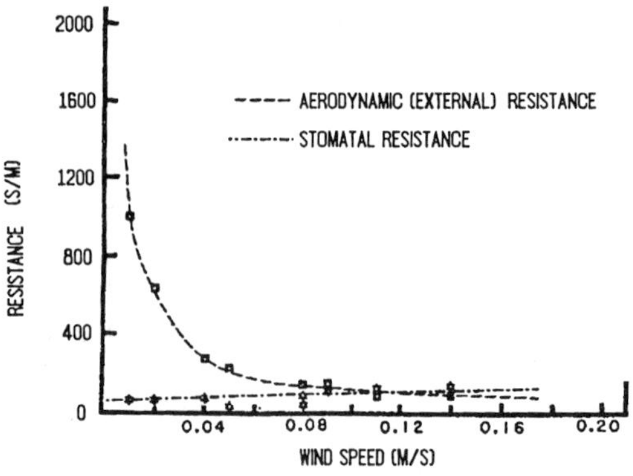

Fig. 5-97. Comparison of calculated aerodynamic resistance with epidermal resistance of rose leaves. Formula for aerodynamic resistance from Grace [1981]. Data for roses from Coker and Hanan [1988].

The effects of wind velocity on transpiration of greenhouse crops are suggested in Fig. 5-95, using a mathematical model defined by Fuchs et al. [1987]. The wind speeds common to greenhouses are in the region where the slopes of the curves change most rapidly. It will be noted that the effect of wind speed on plant water loss is not straightforward. At low humidity and low resistance (<200 s m^{-1}, Fig. 5-95), transpiration will increase with higher wind speed. However, with moderate resistance, transpiration is almost independent of air velocity, and at high resistance ($\geq$400 s m^{-1}), water loss decreases with wind speed. The result at low resistance and humidity is because leaf temperature is below the air temperature, and at high humidity and resistance, solar radiation will heat the leaf above air temperature. As wind speed increases, more heat is carried away from the leaf, reducing the energy available for evaporation, thereby reducing transpiration (reduced VPD). In reality, the convergence of the lines to a single point where water loss is dependent only on air temperature and radiation is not as precise as indicated. At zero wind speed and positive R_n, the leaf temperature is above the air temperature which causes free convection and buoyancy forces.

The main effect of these low wind speeds is that heat and mass transport cannot be assigned solely as due to free or forced convection. Quite often, buoyancy effects as discussed by Assaf [1989] may be predominant. Many authors have estimated external or aerodynamic resistance of leaves and vegetative canopies. Using an empirically derived equation, Stanghellini [1987], –for leaves with a characteristic dimension of 20 cm (cucumber) and leaf temperature equal to air temperature– calculated an external resistance of about 450 s m^{-1} at 0.01 m s^{-1} air velocity, decreasing to about 200 s m^{-1} at 0.05 m s^{-1}. Monteith [1981] used a boundary layer resistance of 40 s m^{-1} at wind speeds between 1 and 3 m s^{-1} for plants in the field. In an investigation of relative roles of stomatal versus external resistance, Shaer and van Bavel [1987] calculated external resistance of a tomato canopy in small greenhouses as approximately 200 s m^{-1} (conductance = 5 mm s^{-1}). They concluded that external resistance was the dominating variable in water loss, being 2 to three times larger than epidermal resistance (mesophyll, stomatal, and cuticular) even at a CO_2 level of 109 Pa. My calculations of air velocity in Shaer and van Bavel's miniature greenhouses resulted in a value of 0.04 m s^{-1}. For rose plants at night, Seginer [1984] calculated external resistances of 190 to 400 s m^{-1}. In still another study, Seginer et al. [1990] estimated external resistances for various species, obtaining values ranging from 80 to 300 s m^{-1}. Part of this work by Israeli investigators was conducted in a very small greenhouse with convective heating.

The reason for some problems in determining relative importance of stomatal and external resistances in mass transport of a canopy may be noted in Fig. 5-97. Using a formula given by Grace [1981] for estimating aerodynamic resistance, Fig. 5-97 shows that below air velocities of about 0.08 m s^{-1}, external resistance of the canopy rapidly increases, far exceeding the stomatal resistance of roses –the latter remarkably constant regardless of air velocity. Experimental verification of this aerodynamic resistance curve for greenhouse conditions has not been carried out. But, the result appears logical for conditions commonly encountered in greenhouses, and is consistent with values found in the literature. For greenhouse conditions, small variations of wind velocity in and about the canopy would be critical concerning water loss, humidity, and temperature variations.

d. Coupling of Atmosphere and Transpiration

The argument for stomatal control was examined by Jarvis [1985] and McNaughton and Jarvis [1983]. Jarvis states that the large drop of potential between evaporating sites in the leaf and the atmosphere –commonly on the order of 10 MPa (50% RH outdoors)– cannot be sustained across mesophyll walls. The controlling resistance lies in the gas phase between the water-air interface within the leaf and some reference location in the bulk ambient air. Stomatal control may be important for single leaves, or even single plants. Still, these conclusions can be wholly misleading when applied to an entire vegetative canopy. Jarvis [1985] cites several publications to show that feedback between plants and their immediate aerial environment can diminish or eliminate dependence of transpiration on stomatal conductance. Penman and Schofield [1951] concluded that under still or nearly still conditions (closed greenhouses), stomatal control is very slight. The stomata must be nearly closed before rates of water loss and CO_2 assimilation are appreciably reduced (Fig. 5-91).

Leaf boundary layer resistances are 10 to 100 times greater than stomatal resistances during the day, depending upon leaf size and ventilation rate. The extent to which water loss from a crop is controlled by stomata depends upon the degree of coupling between the leaves and the atmosphere. Jarvis quotes Monteith as stating that two systems are said to be coupled when they can exchange force, momentum, energy, or mass. The first two, for greenhouse conditions, can be ignored. How much coupling between crop and atmosphere was designated the "omega" factor (Ω) by Jarvis, with a range of values between 0 and 1. When Ω is small (close to 0), the ambient VPD is imposed at the leaf surface (very thin boundary layers). The total conductance from leaves in a canopy to some reference point above the canopy is smaller than the conductances across boundary layers of individual leaves. Short, aerodynamically smooth canopies, according to Jarvis, are largely decoupled from the atmosphere above them (large Ω). In this situation, the tendency for transpiration to depart from the equilibrium rate as determined by radiation interception is opposed by a change in VPD in the opposite direction. The effect of low wind speeds discussed previously is to increase Ω, so that greenhouse crops are decoupled from the atmosphere. Jarvis concluded that the boundary layer conductances of crops in greenhouses may be at least 10 times smaller than the outdoors –or resistances are correspondingly greater.

Jarvis' development of the coupling factor was a significant contribution to understanding greenhouse crop behavior. Further development of coupling and greenhouse climate by Aubinet et al. [1989] is equally important. Most mathematical models of transpiration, especially those outdoors, require some measurement of wind speed, temperature, and humidity above the plant canopy at some height "z," representing atmospheric conditions above the canopy boundary layer. This is not true since, according to Aubinet et al., the greenhouse climate is both influenced by the canopy, that in turn, influences water loss from the canopy. There is no height "z" in the greenhouse that can be considered as uninfluenced by the conditions of the canopy. The authors call this aspect a "hydrological" relationship with a negative feedback. The true limit of the boundary layer in greenhouse crops is the outer skin of the structure. Three factors reduce the influence of stomatal conductance on water loss: 1) the "damping" effect of the air layer next to the leaves, as outlined in the previous section; i.e., high resistance to sensible and latent heat transfer compared with the usual leaf resistance; 2) the "thermal negative feedback" resulting from the fact that changes in transpiration will modify the energy balance of the crop and by that change its temperature; e.g., an increase in transpiration will result in cooling, reducing water loss; and 3) the "hydrological" feedback, meaning that a variation in transpiration directly modifies the water balance of the greenhouse air.

Both 2) and 3) operate in the sense of a negative feedback system, analogous to feedback in electrical control systems, so that any perturbation of the flow rate will be insignificant if the resultant gain in the feedback loop is negative and very large. That is, if transpiration suddenly changes, factors 2) and 3) will tend to reduce that change, depending upon the magnification, or gain, of the feedback circuit. Aubinet et al. define a value, C, which is the global environmental gain of the system, taking into account the three effects listed. All three effects lessen the impact of stomatal resistance on transpiration, with C always positive. Zero is reached when there is no stomatal control of transpiration. If the thermal and hydrological feedbacks are negligible, and the boundary layer of individual leaves small, than C approaches 1. The stomata control water loss. A larger LAI increases the importance of the feedback systems, and these will be more important when the ventilation rate is small, disappearing if the house is well ventilated. Jarvis' omega factor did not take into account the hydrological feedback discussed by Aubinet et al. since that factor is unimportant outside a greenhouse.

There are two conditions one needs to consider concerning decoupling of greenhouse canopies: 1) the photoperiod when stomata are fully open, the canopy is decoupled, and transpiration will be a direct function of net radiation –unless the house is being ventilated; and 2) during the night when stomata are closed and transpiration is largely cuticular. Coupling of the canopy with the interior climate will increase (Ω large, C small), since the gain of the thermal and hydrological feedback systems will be reduced as the result of the large increase in leaf resistance with closed stomata.

4. Calculation of Transpiration Rate

Methods for determining transpiration rates in greenhouses have received marked attention the last 2 decades. The methods employed for outdoor crops as listed by Kanemasu et al. [1979] include eddy-correlation, aerodynamic techniques, Bowen ratio approaches, and combination or energy balance methods. Fuchs [1973] separated methods as to energy balance, mass and heat transport, momentum transport and turbulent mixing, aerodynamic and the Bowen ratio method. All these methods are essentially one-dimensional without horizontal divergence (advection) of sensible and latent heat through the canopy. Horizontal movement of air through a canopy will obviously occur, especially with convective heating and forced ventilation cooling systems, depending upon the crop status (LAI, arrangement, species, etc.). Tanner [1960] discussed problems resulting from inhomogeneity of the surface, and effects of widely spaced row crops, which can induce large differences in horizontal variations. This drastically increases requirements for spatial sampling. Kaneumasu cites Thom as suggesting that horizontal heat fluxes as large as 100 W m^{-2} for horizontal temperature gradients of 1 C can be present. Advective effects on transpiration have been examined by many authors, among them Abdel-Aziz et al. [1964], who found that low estimates of water loss using the Penman formula could only be attributed to advective energy carried by wind, especially at night. Three-dimensional approaches to the problem as discussed by Kanemasu et al. and Tanner are probably better applied to the greenhouse situation as compared to field conditions. However, for simplicity, only one-dimensional, vertical heat and mass transport will be considered here.

All of the methods mentioned above require some measurement of various properties of the air above the canopy. Even with Bowen ratio, Penman and Penman-Montieth combinations, such measurement remains a major requirement. As noted previously, most of the listed methods require the development of an adequate profile above the canopy, which means an adequate fetch –or placement of the measuring instruments some distance downwind from the leading edge of the crop. Kanemasu et al. [1979] state that although the Bowen ratio can be determined without knowledge of surface properties, wind speeds, and height above the surface, the use of the method requires an assumption of a horizontally uniform surface for a distance 100 to 1000 times the greatest height of the measurement. We shall see that the Bowen ratio has been commonly used in greenhouse investigations of transpiration even though this requirement for fetch cannot be achieved.

Of the methods listed above, the Penman-Monteith combination method and Bowen ratio have been the most commonly applied to greenhouse conditions. Monteith [1981] stated that Penman's contribution [Penman, 1948, 1952; Penman and Schofield, 1951] to the study of natural evaporation has been immense. Many variations of the basic Penman equation have been published in the literature [e.g., Bakker, 1986; Stanghellini, 1987; Penman et al., 1967; Black et al., 1970; Lemon et al., 1971; Fuchs et al., 1987; Tanner and Fuchs, 1968; etc.].

a. Combination Formulae (Penman-Monteith Energy Balance)

Eq. 4.1 presented the factors in the energy balance of a plant at equilibrium. The radiation factors can be combined to give net radiation (R_n), which can be equated to radiation received, less that radiation reflected, plus or minus the longwave exchange (Eqs. 5.12 and 5.13). By neglecting **P** and **M** in Eq. 4.1, the plant's energy balance can be equated to R_n [Tanner and Fuchs, 1968]:

$$R_n = H + E + S \tag{5.27}$$

where: $R_n = W\ m^{-2}$

H = *sensible heat flux density,* $W\ m^{-2}$

E = *latent heat flux density,* $W\ m^{-2}$

S = *soil and crop heat flux density,* $W\ m^{-2}$

Both R_n and S can be measured by net radiometers for the former (Section 3.II.F), and appropriate heat flux discs for the ground, and estimated for the vegetation as discussed by Stanghellini [1987] for the latter. **H** and **E** can be equated as follows:

$$H = \rho C_p h(T_o - T_z) \tag{5.28}$$

$$E = \left(\frac{L\rho\epsilon}{P}\right) h(e_o - e_z) \tag{5.29}$$

where: ρ *= air density, kg* m^{-3}
C_p *= specific heat of air, J* $kg^{-1}K^{-1}$
h *= transfer coefficient for heat and water vapor from surface to height z, m* s^{-1}
T_o *= surface temperature, K*
T_z *= air temperature at height z, K*
L *= latent heat of vaporization, J* kg^{-1}
ϵ *= ratio of mole weights of water and air, 0.622*
P *= barometric pressure, kPa*
e_o *= water vapor pressure at the surface, kPa*
e_z *= water vapor pressure at height z, kPa*

From Eqs. 5.27 through 5.29, there can be derived an equation for **E**, assuming that the surface is saturated at T_o:

$$E = \left[\frac{\Delta}{(\Delta + \gamma)}\right]\left[(R_n - S) + \left(\frac{\rho C_p}{\Delta}\right) h(e_z^* - e_z)\right] \tag{5.30}$$

where: Δ *= slope of the saturation vapor pressure curve versus temperature = de/d**T**, Pa* K^{-1} $\boldsymbol{T} = (T_o+T_z)/2$
γ *= psychrometric constant, Pa* K^{-1}
e_z^* *= saturated vapor pressure at* T_z
e_z *= vapor pressure at height z*

The transfer coefficient, h, can be found for a one-dimensional surface with an adequate fetch and a suitable wind profile [Tanner and Fuchs, 1968]. None of these are likely to be found in a greenhouse. This formula also presumes that water loss proceeds at its maximum potential rate.

The combination approach can also incorporate resistance (or conductance) in cases where **h** is unknown. If one assumes that the internal or stomatal resistance of a well-watered crop is constant, then one deals only with external or aerodynamic resistance (r_a) of the crop between the evaporating surfaces and height **z** (see Eqs. 4.2 and 4.3). The resistance r_a = 1/h and is the same for both heat and vapor. From Stanghellini [1987]:

$$E = \frac{\frac{\Delta}{\gamma}(R_n - S) + \frac{2LAI\rho C_p}{\gamma r_a}(e_a^* - e_a)}{1 + \frac{\Delta}{\gamma} + \frac{1}{r_a}} \tag{5.31}$$

where: e_a^* *and* e_a *= saturated and actual vapor pressures of the air*
2LAI = twice the leaf area index of the crop

The use of 2(LAI) is equivalent to stating that all the canopy leaves are wired in parallel, and behave as a leaf of unit area. If, as stated by Fuchs et al. [1987], the conductance of leaves is proportional to the photosynthetic quantum flux density, then an internal resistance should be included in Eq. 5.31.

b. Bowen Ratio

If one ignores energy storage in the ground and crop (**S**), then R_n = E+H. Furthermore, if the transport constants for heat and vapor are equal, then from Penman et al. [1967]:

$$\frac{H}{E} = \left(\frac{C_p P}{\epsilon L}\right)\left(\frac{T_1 - T_2}{e_1 - e_2}\right) = \gamma \frac{T_1 - T_2}{e_1 - e_2} = \beta \tag{5.32}$$

*where the subscripts **1** and **2** refer to temperature and vapor pressures measured at heights z_1 and z_2, and β = the Bowen ratio.*

If the formula R_n = E+H is combined with Eq. 5.32, a deceptively simple formula is obtained:

$$E = \frac{R_n}{(1 + \beta)} \tag{5.33}$$

The larger the ratio, the more sensible heat lost per unit of latent heat. Eq. 5.33 requires no measurement of wind speed or surface properties, and involves only the ratio of the temperature gradient to the vapor pressure gradient between any two levels of measurement. Over homogenous, irrigated areas, β will range from 0.2 to 0.5. If there is strong advection, β will be negative, and at -1, Eq. 5.33 is meaningless [Penman et al., 1967]. Kanemasu et al's [1979] statement on an adequate fetch has been given. Furthermore, a very basic assumption required for this relationship is that the diffusivities of sensible and latent heat are equal. Nevertheless, the Bowen ratio has been applied in greenhouses [e.g., Cohen et al., 1983]. With no crops and dry surfaces, ratios of 3.5 and 5.5, during the day were calculated –value of γ ignored. If the soil surface was saturated, β = 0.5, dropping to less than -1 during the period mist sprayers were in operation. In houses with growing plants, β was usually low, suggesting that most of the energy was partitioned to latent heat versus sensible heat. A variation of the previous formulae was given in Chapter 4, Eq. 4.19, in which the Bowen ratio was employed to evaluate fan-and-pad cooling [Landsberg et al., 1979]. Seginer et al. [1990] and Seginer and Kantz [1989] defined a "ventilation" Bowen ratio where the temperature and humidity differences were measured from inside to outside the greenhouse. Their calculation resulted in a ratio limited to 1.1 to 1.5 at night. Based upon my reading of the field literature, these uses of the Bowen ratio, as originally defined, would be wrong. Calling the ratio by another name to avoid confusion would be better.

c. Water Relations at Night

Section 5.IV.B.3 discussed resistance factors of the vapor pathway with Table 5-33 giving some data on cuticular resistance of various species. Although stomates may be closed at night, transpiration is not zero unless the leaf temperature is below the dewpoint. De Graff [1985] showed that water loss during winter nights could exceed daytime totals under Dutch conditions (Fig. 5-92). Evaporation rates could approach 0.03 mm hr^{-1} at night, which is equivalent to 20 W m^{-2}. Under Israeli conditions, Seginer [1984] found rates of about 0.015 mm hr^{-1} (10 W m^{-2}) for roses. The fact that such water loss at night, especially in mild climates, can be highly critical from the standpoint of disease control has been discussed, with the suggestion that mathematical formulae for water loss at night are separated from the equations given above for daytime periods. At such low transpiration rates, existing formulae can be imprecise, resulting in errors as large as ±50%.

The situation at night has been extensively investigated by Seginer [1980, 1984], Seginer et al. [1990], Seginer and Kantz [1989], and Assaf [1989]. Although Seginer [1984] states that nighttime transpiration is at least 10 times smaller than that in the daytime, his conclusion that its contribution to total water loss is of no significance cannot be accepted under all climatic conditions (Section 5.IV.B.3). In a situation where a dehumidifier can be employed, Seginer and Kantz [1989] developed a model based on six balance equations for sensible heat, latent heat, the ceiling, the roof outside, plant canopy, and the dehumidifier. From these equations, the energy gain and loss of the greenhouse is:

$$G = H_h + H_g = H_h + b\mu LE_d = H_o + R_{nrs} + H_v + LE_v \tag{5.34}$$

where: G = *heat loss from greenhouse,* $W\ m^{-2}$
H_h = *sensible heat flux from heater,* $W\ m^{-2}$
H_g = *sensible heat flux due to regeneration of dehumidifier brine,* $W\ m^{-2}$
b = *factor pertaining to dehumidifier = 1.5*
μ = *regeneration fraction for dehumidifier*
L = *heat of vaporization,* $J\ kg^{-1}$
E_d = *vapor flux from dehumidifier,* $kg\ m^{-2}s^{-1}$
H_o = *sensible heat flux of outside air,* $W\ m^{-2}$
R_{nrs} = *net radiation from roof to sky,* $W\ m^{-2}$
H_v = *sensible heat flux due to ventilation,* $W\ m^{-2}$
E_v = *vapor flux due to ventilation,* $W\ m^{-2}$

A further series of equations dealt with evaporation and conduction fluxes, convective fluxes, and net radiation fluxes. Of interest was the use of an equation to set the upper relative humidity value:

$$RH_p = \frac{q_i}{q_p^*} \tag{5.35}$$

where: RH_p = *the relative humidity above which dew will form on the canopy*
q_i = *specific humidity of the air,* $kg\ kg^{-1}$
q_p^* = *specific humidity at saturation at the canopy temperature,* $kg\ kg^{-1}$

Seginer and Kantz used a value of $RH_p = 0.90$, the maximum permissible RH equal to, or below which, no dew could form on the canopy. The conclusions of these authors were: 1) Condensation on the ceiling as a means to reduce humidity was preferable to other methods. 2) Up to a critical value of canopy conductance (**g***), energy consumption was hardly affected by canopy conductance (**g**). When, however, conductance exceeded the critical value, the energy loss through ventilation was proportional to the difference g - g*, and the ventilation Bowen ratio. 3) The critical conductance value depended on external weather conditions –g* high for cold weather and low for mild weather. 4) A greater fraction of total energy can be saved in mild climates by using a dehumidifier. Below g*, condensation will occur and the dehumidifier saving will be zero. 5), a dehumidifier is more likely to succeed with crops that have high conductances (low cuticular resistances), and will be limited to the beginning and ending of the heating season in cold climates.

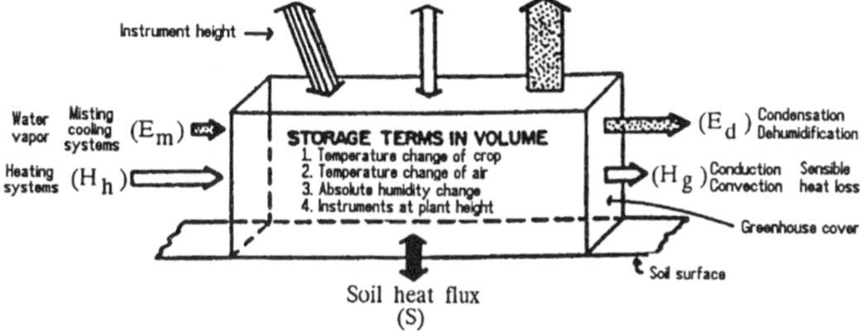

Fig. 5-98. Modification of Tanner's [1960] energy balance of a field crop volume. The volume is the greenhouse in this case, with the covering analogous to the epidermis of a leaf. The boundary layer is around the house outside. The author has not determined if this is a reasonable approach to solving the energy budget of a crop-filled structure (With permission of the *Amer. Soc. Agron.*).

Assaf [1991], who introduced the regenerative brine dehumidifier, has taken exception to some of Seginer's and Stanghellini's methods, concluding that his work [1991] in commercial houses supports the assumption that nighttime transpiration is independent of VPD. Nighttime transpiration is induced largely by convective buoyancy fluxes.

5. Summary

Having very briefly summarized information on water demand, the question a grower may very well raise is: "So what?" Testing of the various mathematical models given in the previous section generally requires specialized equipment, such as weighable lysimeters, that permit actual water loss to be measured under conditions closely approximating real life [e.g., Hanan and Huffsmith, 1967; Stanhill and Albers, 1974; Yang et al., 1990; de Graff, 1981; Meijer et al., 1985; etc.]. The requirements of such equipment make it doubtful that such devices will be common to commercial ranges. Van Meurs and Stanghellini, however, described an off-the-shelf electronic balance for measuring transpiration in greenhouses [1992]. More important, what can be done with the resulting information?

A start was made by Stanghellini [1987] and van Meurs and Stanghellini [1989] in predicting transpiration, and setting up a "transpiration setpoint" that could be used to modify existing temperature and humidity setpoints to optimize "targets" for economical greenhouse operation. This procedure of controlling water demand is only halfway home. One should use an estimate of transpiration to control plant water potential. This has been discussed by such authors as Hopmans [1981], Marcelis [1989] and Bruggink et al. [1988]. Models have been devised to predict water potential based on transpiration rates, with some very good discussion by Fiscus et al. [1983]. Although Jolliet and Bailey [1992] tested five models of tomato transpiration, finding that simplified Penman, Stanghellini, and Jolliet models gave reasonable accuracy, there are a number of problems that will be deferred to Chapter 8 for more detailed discussion.

Considering the greenhouse *in toto* might be better, without trying to separate the crop from the inside space, in a manner similar to that outlined by Tanner [1960] in his consideration of the complete energy balance of a field crop (Fig. 5-98). This is less complex than Fig. 4-8. Approaches to this are suggested in Aubinet et al's [1989] and Seginer et al's [1990] developments. Profile theory can be applied outside the greenhouse with vapor and sensible heat transport the internal conductance. Transport through the boundary layer outside of the structure becomes the external conductance or resistance. The greenhouse cover is the epidermis of the leaf with the ventilators assuming the role of stomata. To the best of my knowledge, the possibility that this idea is a mathematically and physically better undertaking has not been investigated.

V. REFERENCES

Abdel-Aziz, M.H., S.A. Taylor and G.L. Ashcroft. 1964. Influence of advective energy on transpiration. *Agron. J.* 56:139-142.

Aikin, W.J. and J.J. Hanan. 1975. Photosynthesis in the rose; effect of light intensity, water potential and leaf age. *J. Amer. Soc. Hort. Sci.* 100:551-553.

Aikman, D.P. and G. Houter. 1990. Influence of radiation and humidity on transpiration: Implications for calcium levels in tomato leaves. *J. Hort. Sci.* 65:245-253.

Alberry, W.J., B.G.D. Haggett and L.R. Svanberg. 1985. The development of sensors for hydroponics. *Biosensors*. 1:369-397.

Anon. 1962. Automatic watering of pot plants with the N.I.A.E. capillary bench. Ministry of Agric. Fisheries and Food. U.K.. STL/16. 6 pp.

Anon. 1977. Research is lagging behind NFT's development –Cooper. *The Grower*. 87(13):735-736.

Anon. 1978a. Water losses important in NFT systems symposium told. *The Grower*. 88(17):836-837.

Anon. 1978b. Water quality vital for NFT success. *The Grower*. 89(15):880.

Anon. 1979a. Identifying NFT root death causes. *The Grower*. 92(17):28.

Anon. 1979b. Yield and quality not in the bag. *The Grower*. 92(18):20-22

Anon. 1979c. Root death a matter of culture, not disease. *The Grower*. 92(18):22.

Anon. 1979d. Artificial media: Cost of comparison of preparing your own soil mixes. *Southern Florist & Nurseryman*. May 25, 1979.

Anon. 1981. Alternative media costings for glasshouse vegetables. *The Grower*. 95(9):15.
Anon. 1982a. Hydroponics take root in Europe. *Amer. Vegetable Grower*. 30(11):94.
Anon. 1982c. Carnations on rockwool. *The Grower*. 97(21):20.
Anon. 1982d. A rockwool success story. *The Grower*. 98(6):27-29.
Anon. 1985. Simplifying NFT for the rest of the world. *The Grower*. 103(20). May 16, 1985.
Anon. 1986. Humidity is complicating the fuel saving equation. *The Grower*. 105(18). May 1, 1986.
Anon. 1988. Susceptible to blockage. *The Grower*. 110(17). Oct. 27, 1988.
Anon. 1989. Tomatoes in bags of jelly. *The Grower*. 111(18):10-12.
Arkley, R.J. 1963. Relationships between plant growth and transpiration. *Hilgardia*. 34:559-584.
Assaf, G. 1991. Night time transpiration in commercial greenhouses. *Acta Hort*. 287:495-504.
Assaf, G. et al. 1990. Night transpiration in rose greenhouses and its energy balance implications. Unpublished MS. Ormat Turbines, Ltd., Yavne, Israel. 51 pp.
Aston, A.R. and C.H.M. van Bavel. 1972. Soil surface water depletion and leaf temperature. *Agron. J.* 64:368-373.
Aubinet, M., J. Deltour and D. de Halleux. 1989. Stomatal regulation in greenhouse crops: Analysis and simulation. *Agric. and Forest Meteor*. 48:21-44.
Augsburger, N.D. and C.C. Powell. 1986. Greenhouse ventilation: Some basics on controlling humidity. *Grower Talks*. Feb., 1986. 70-71.
Baker, K.F. ed. 1957. The U.C. system for producing healthy container-grown plants. Manual 23. CA Agric. Expt. Sta. 332 pp.
Baker, D.E. and C.M. Woodruff. 1962. Influence of volume of soil per plant upon growth and uptake of phosphorous by corn from soil treated with different amounts of phosphorous. *Soil Sci*. 94:409-412.
Bakker, J.C. 1986. Measurement of canopy transpiration or evapotranspiration in greenhouses by means of a simple vapour balance model. *Agric. and Forest Meteor*. 37:133-141.
Bakker, J.C. 1989. The effects of air humidity on growth and fruit production of sweet pepper (*Capsicum annuum* L.). *J. Hort. Sci*. 64:41-46.
Bakker, J.C. 1990a. Effects of day and night humidity on yield and fruit quality of glasshouse tomatoes (*Lycopersicon esculentum* Mill.). *J. Hort. Sci*. 65:323-331.
Bakker, J.C. 1990b. Effect of humidity on growth and quality of eggplant (*Solanum melongena* L.). *J. Hort. Sci.* 65:747-753.
Bakker, J.C., G.W.H. Welles and J.A.M. van Uffelen. 1987. The effects of day and night humidities on yield and quality of glasshouse cucumbers. *J. Hort. Sci*. 62:363-370.
Bange, G.G.J. 1953. On the quantitative explanation of stomatal transpiration. *Acta Bot. Neerl*. 2:225-297.
Barley, K.P., D.A. Farrell and E.L. Greagen. 1965. The influence of soil strength on the penetration of a loam by plant roots. *Austr. J. Soil Res*. 3:69-79.
Barrs, H.D. et al. 1970. Comparisons of leaf water potential and xylem water potential in tomato plants. *Austr. J. Biol. Sci*. 23:485-487.
Bartok, J.W., Jr. 1986. Where there's water, there's a way. *Greenhouse Manager*. June, 1986. 96-105.
Bearce, B.C. 1986. Looking for alternative growing media? *Greenhouse Grower*. Oct., 1986. 44-48.
Bebb, D. 1980. Trickle or drip? *The Grower* (Supplement). 94(19):22
Behboudian, M.H. and H.M.C. van Holsteijn. 1977. Water relations of lettuce. I. Internal physical aspects of two cultivars. *Sci. Hort*. 7:9-17.
Biernbaum, J.A. 1992. Root-zone management of greenhouse container-grown crops to control water and fertilizer use. *HortTechnology*. 2:127-132.
Birch, P. 1980. Big vein beaten but wilt persists. *The Grower*. 93(6):18-22. Feb. 7, 1980.
Bjerre, H. 1982. Pot plant growing on a capillary mat covered with a perforated polyethylene foil. *Acta Hort.* 133:161-164.
Blaabjerg, J. 1983. Physical and chemical composition of the rockwool medium Grodan@ and its fields of application and extension. *Acta Hort*. 133:53-57.
Black, T.A., C.B. Tanner and W.R. Gardner. 1970. Evapotranspiration from a snap bean crop. *Agron. J.* 62:66-69.
Blass, S. 1969. Drip irrigation. Water Works, Tel-Aviv. 19 pp. (Translated from Hebrew)

Boelter, D.H. and G.R. Blake. 1964. Importance of volumetric expression of water contents of organic soils. *Proc. Soil Sci. Soc. Amer.* 28:176-178.

Boertje, G.A. and R.A. Bik. 1975. Potting substrates in the Netherlands. *Acta Hort.* 50:135-142.

Bot, G.P.A. 1980. Validation of a dynamical model of greenhouse climate. *Acta Hort.* 106:149-158.

Bot, G.P.A. 1983. Greenhouse climate: from physical processes to a dynamic model. Ph.D. Diss. Wageningen. 240 pp.

Boyer, J.S. 1967. Matric potential of leaves. *Plant Physiol.* 42:213-217.

Bowman, D.C., R.Y. Evans and L.L. Dodge. 1994. Growth of chrysanthemum with ground automobile tires used as a container soil amendment. *HortScience.* 29:774-776.

Boyer, J.S. 1969. Measurement of the water status of plants. *Ann. Rev. Plant Physiol.* 20:351-364.

Boyer, J.S. 1970a. Leaf enlargement and metabolic rates in corn, soybean and sunflower at various leaf water potentials. *Plant Physiol.* 46:233-235.

Boyer, J.S. 1970b. Differing sensitivity of photosynthesis to low leaf water potentials in corn and soybean. *Plant Physiol.* 46:236-239.

Boyer, J.S. 1971a. Recovery of photosynthesis in sunflower after a period of low leaf water potential. *Plant Physiol.* 47:816-820.

Boyer, J.S. 1971b. Nonstomatal inhibition of photosynthesis in sunflower at low leaf water potentials and high light intensities. *Plant Physiol.* 48:532-536.

Boyer, J.S. 1985. Water transport. *Ann. Rev. Plant Physiol.* 36:473-516.

Brady, N.C. 1974. The Nature and Properties of Soils. 8th ed. Macmillan Publ. Co., New York. 639 pp.

Bragg, N. and B. Chambers. 1988. Soilless media from peat to hydroponics. *The Grower* (Supplement). 106(20).

Brisco, D. et al. 1992. Soil moisture measurement using portable dielectric probes and time domain reflectrome try. *Water Resour. Res.* 28:1339-1346.

Brouwer, R. 1963. The influence of the suction tension of the nutrient solutions on growth, transpiration and diffusion pressure deficit of bean leaves (*Phaseolus vulgaris*). *Acta Bot. Neerl.* 12:248-261.

Bruggink, G.T., H.E. Schouwink and T.H. Gieling. 1988. Modelling of water potential and water uptake rate of tomato plants in the greenhouse: Preliminary results. *Acta Hort.* 229:177-185.

Bugbee, G.J. and C.R. Frink. 1986. Aeration of potting media and plant growth. *Soil Sci.* 141:438-441.

Bucks, D.A., F.S. Nakayama and R.G. Gilbert. 1979. Trickle irrigation water quality and preventive maintenance. *Agric. Water Mgmt.* 2:149-162.

Bucks, D.A., F.S. Nakayama and R.G. Gilbert. 1981. Is your trickle fickle? *Amer. Veg. Grower.* 29(4):8-10, 68.

Bunt, A.C. 1961. Some physical properties of pot-plant composts and their effect on plant growth. I. Bulky physical conditioners. *Plant and Soil.* 14:322-332.

Bunt, A.C. 1982. Pot waxing loam waning. *The Grower.* 97(6):24-29.

Bunt, A.C. 1988. Media and Mixes for Container-Grown Plants. Unwin Hyman, London. 309 pp.

Burés, S. et al. 1993. Computer simulation of volume shrinkage after mixing container media components. *J. Amer. Soc. Hort. Sci.* 118:757-761.

Campbell, G.S. 1977. An Introduction to Environmental Physics. Springer-Verlag, Heidelberg. 159 pp.

Challa, H. 1975. A critical study of β-gauging for water content variation measurement in plant leaves. *Newsletter on the Application of Nuclear Methods in Biology and Agriculture.* 4:9-12.

Chandra, P., L.D. Albright and N.R. Scott. 1981. A time dependent analysis of greenhouse thermal environment. *Trans. ASAE.* 24:442-449.

Chang, Jen-Hu. 1968. Climate and Agriculture. Aldine Publ. Co., Chicago. 304 pp.

Chapin, R.D. 1971. Drip irrigation in the United States. *Int. Experts Panel on Irrigation.* Herzilya-on-Sea, Israel, 1971. 8 pp.

Clawson, K.L. and B.L. Blad. 1982. Infrared thermometery for scheduling irrigation of corn. *Agron. J.* 74:311-316.

Clayton, A. 1981. Steaming out for NFT chrysanth production. *The Grower.* 96(8):23-28.

Clothier, B., D. Scotter and E. Harper. 1985. Three-dimensional infiltration and trickle irrigation. *Trans. ASAE.* 28:497-501.

Cobb, G.S., J.J. Hanan and R. Baker. 1978. Environmental factors affecting rose powdery mildew in greenhouses. *HortScience.* 13:464-466.

Cohen, Y., G. Stanhill and M. Fuchs. 1983. An experimental comparison of evaporative cooling in a naturally ventilated glasshouse due to wetting the outer roof and inner crop soil surfaces. *Agric. Meteor*. 28:239-251.
Coker, F.A. and J.J. Hanan. 1988. CO_2 uptake by 'Samantha' roses. *CO Greenhouse Growers' Assoc. Res. Bul.* 456:1-3.
Conover, C.A. and R.T. Poole. 1977. Characteristics of selected peats. *FL Foliage Grower*. 14:1-6.
Cook, R.L. and C.E. Millar. 1946. Some techniques which help to make greenhouse investigations comparable with field plot experiments. *Soil. Sci. Soc. Amer. Proc*. 11:298-304.
Cooper, A. 1974. Soil? Who needs it? *Amer. Veg. Grower*. 22:18-20. August, 1974.
Cooper, A. 1977. Research is lagging behind NFT's development. *The Grower*. 87(13):735-736.
Cooper, A. 1979. The ABC of NFT. Grower Books, London. 181 pp.
Cooper, A. 1986. Which is the winning side in the media war? *The Grower*. 101(15). Sept. 25, 1986.
Cooper, P.I. and R.J. Fuller. 1983. A transient model of the interaction between crop, environment and greenhouse structure for predicting crop yield and energy consumption. *J. Agric. Eng. Res*. 28:401-417.
Cornell Recommendations. 1974. NY State College Agric. and Life Sci. Part I. Ithaca, NY. 60 pp.
Cox, L.M. and L. Boersma. 1967. Transpiration as a function of soil temperature and soil water stress. *Plant Physiol*. 42:550-556.
Crane, R. 1981. New NFT not cheap and simple. *The Grower*. 95(14):11.
Davies, J.M.L. 1981. Alternative wilt controls. *The Grower*. 95(3):18-19.
Denmead, O.T. and R.H. Shaw. 1962. Availability of soil water to plants as affected by soil moisture content and meteorological conditions. *Agron. J*. 54:385-390.
Dole, J.M., J.C. Cole and S.L. von Broembsen. 1994. Growth of poinsettias, nutrient leaching, and water-use efficiency respond to irrigation methods. *HortScience*. 29:858-864.
Drakes, D. 1982. Rockwool: Experiments at Stockbridge House. *The Grower* (Supplement). Jan/Feb., 1982.
Dungey, N.O. 1983. Assessing the future of NFT and rockwool. *The Grower* (Supplement). Nov/Dec., 1983.
Duisberg, P.C. 1963. Challenge of the future. *In* Aridity and Man. Hodge, C. and P.C. Duisberg eds. AAAS Publ. No. 74, Washington, D.C. 461-481.
Duncan, G.A., O.J. Loewer, Jr. and D.G. Colliver. 1981. Simulation of energy flows in a greenhouse: Magnitudes and conservation potential. *Trans. ASAE*. 24:1014-1021.
Dwyer, L.M., D.W. Stewart and D. Balchin. 1987. Accurately monitoring and maintaining soil water in greenhouse containers. *Can. Agric. Engin*. 29:89-91.
Enoch, H.Z. 1986. Climate and protected cultivation. *Acta Hort*. 176:11-19.
Evans, S.G. 1977. Disease risks affect entire crop in NFT systems. *The Grower*. 88(24):1233-1239.
Faber, W.R. and H.A.J. Hoitink. 1983. Critical properties of successful container media. *Ohio Florists' Bull*. 641:2-5.
Feigin, A., S. Dasberg and Z. Singer. 1980. Rose culture in scoria. *Acta Hort*. 99:131-138.
Filippo, H. 1947. Verschillen tussen grind-, water- en grondcultuur. *Vakblad voor Bloemisterij*. 3:8-9.
Fiscus, E.L., A. Klute and M.R. Kaufmann. 1983. An interpretation of some whole plant water transport phenomena. *Plant Physiol*. 71:810-817.
Flaherty, A. 1988. A gentle gust of warm, dry air. *Grower*. 110:(21):19-20.
Free, G.R., G.M. Browning and G.W. Musgrave. 1940. Relative infiltration and related physical characteristics of certain soils. USDA Tech. Bul. 729. U.S. Government Printing Office, Washington, D.C.
Fries, H.H. 1979. Cost savings from automated watering. *Amer. Veg. Grower*. 27(2):21, 24, 38, 76.
Friss-Nielsen, B. 1960. Vandfordampning I vÆksthuse. *Horticultura*. 14:207-221. (English summary).
Fuchs, M. 1973. The estimation of evapotranspiration. *In* Arid Zone Irrigation. Yaron, B. et al. Eds. Springer-Verlag, Heidelberg. 241-247.
Fuchs, M. 1986. Determining transpiration of crops from meteorological and agronomic data. *Proc. 4th Int. Conf. on Irrigation*. Tel-Aviv. 83-95.
Fuchs, M. 1990. Effect of transpiration on greenhouse cooling. *Proc. Int. Seminar and British-Israel Workshop on Greenhouse Technology*. Segal, I. ed Bet-Dagan.
Fuchs, M. and C.B. Tanner. 1966. Infrared thermometry of vegetation. *Agron. J*. 58:597-601.
Fuchs, M., Y. Cohen and S. Moreshet. 1987. Determining transpiration from meteorological data and crop characteristics for irrigation management. *Irrig. Sci*. 8:91-99.

Funck-Jensen, D. and J. Hockenshull. 1982. The influence of some factors on the severity of *Pythium* root rot of lettuce in soilless (hydroponic) growing systems. *Acta Hort.* 133:129-136.

Furuta, T. et al. 1976. Observations on drip irrigation for cut-flower s in greenhouses. *Florists' Rev.* 158(4093):94.

Fynn, R.P. 1994. Water and nutrient delivery - Ebb and flood. *In* Greenhouse Systems. Automation, Culture, and Environment. *NRAES-72,* Rutger Univ., New Brunswick, NJ.

Gaffney, J.J. 1977. Humidity: Basic principles and measurement techniques. *HortScience.* 13:551-555.

Gardner, W.R. 1960. Dynamic aspects of water availability to plants. *Soil Sci.* 89:63-73.

Gardner, W.R. 1965. Soil water movement and root absorption. *In* Plant Environment and Efficient Water Use. Pierre, W.H. et al. eds. Amer. Soc. Agron., Madison, WI. 127-149.

Gardner, W.R. and C.F. Ehlig. 1962a. Impedance to water movement in soil and plant. *Science.* 138:522-523.

Gardner, W.R. and C.F. Ehlig. 1962b. Some observations on the movement of water to plant roots. *Agron. J.* 54:453-456.

Gardner, W.R. and R.H. Nieman. 1964. Lower limit of water availability to plants. *Science.* 143:1460-1462.

Gates, D.M. 1964. Leaf temperature and transpiration. *Agron. J.* 56:273-277.

Gates, D.M. 1980. Biophysical Ecology. Springer-Verlag, Heidelberg. 611 pp.

Gates, D.M. and R.J. Hanks. 1967. Plant factors affecting transpiration. *In* Irrigation of Agricultural Lands. Hagan, R.M. et al. eds. *Mono. No. 11.* Amer. Soc. Agron., Madison, WI. 506-533.

Geiger, R. 1965. The Climate Near the Ground. German trans. Harvard Univ. Press, Cambridge, MA. 611 pp.

Gerber, J.M. 1984. Hydroponics, Hype and Hope. *Amer. Veg. Grower.* 32(11):62-63.

Gerhardt, F. and E. Smith. 1945. Cracking and decay of Bing cherries as related to the presence of moisture on the surface of fruit. *Proc. Amer. Soc. Hort. Sci.* 46:191-198.

Giacomelli, G.A. et al. 1982. A cable supported NFT tomato production system for the greenhouse. *Acta Hort.* 133:89-102.

Gill, W.R. and R.D. Miller. 1957. A method for study of the influence of mechanical impedance and aeration of the growth of seedling roots. *Proc. Soil Sci. Soc. Amer.* 20:154-157.

Gislerød, H.R. and P.V. Nelson. 1989. The interaction of relative air humidity and carbon dioxide enrichment in the growth of *Chrysanthemum X morifolium* Ramat. *Sci. Hort.* 38:305-313.

Gislerød, H.R. and L.M. Mortensen. 1990. Relative humidity and nutrient concentration affect nutrient uptake and growth of *Begonia x hiemalis*. *HortScience.* 25:524-526.

Gislerød, H.R. and L.M. Mortensen. 1991. Air humidity and nutrient concentration affect nutrient uptake and growth of some greenhouse plants. *Acta Hort.* 294:141-146.

Glenn, D.M. et al. 1989. Estimation of peach tree water use using infrared thermometry. *J. Amer. Soc. Hort. Sci.* 114:737-741.

Gouin, F.R. 1993. Utilization of sewage sludge compost in horticulture. *HortTechnology.* 3:161-163.

Graaf, R. de. 1981. Transpiration and evapotranspiration of glasshouse crops. *Acta Hort.* 119:147-157.

Graaf, R. de. 1985. The influence of thermal screening and moisture gap on the transpiration of glasshouse tomatoes during the night. *Acta Hort.* 174:57-66.

Graaf, R. de and J. van den Ende. 1981. Transpiration and evapotranspiration of the glasshouse crops. *Acta Hort.* 119:147-158.

Grace, J. 1981. Plants and wind. *In* Plants and Their Atmospheric Environment. Grace, J. et al. eds. Blackwell Scientific, Boston.

Grange, R.I. and D.W. Hand. 1987. A review of the effects of atmospheric humidity on the growth of horticultural crops. *J. Hort. Sci.* 62:125-133.

Graves, C.J. and R.G. Hurd. 1982. Intermittent solution circulation in the nutrient film technique. *Acta Hort.* 133:47-52.

Greenwood, G.E., G.D. Coorts and R.R. Maleike. 1978. Research conducted to determine value of various soil amendments. *Amer. Nurseryman.* 158(8):12-13, 77.

Hadas, A. 1973. Water retention and flow in soils. *In* Arid Zone Irrigation. Yaron, B. et al. eds. Springer-Verlag, Heidelberg. 89-109.

Hall, B.J. 1980. The nuts and bolts of drip. *Amer. Veg. Grower.* 28(4):8-9.

Hall, D. 1986. Perlite - a third choice in hydroponics. *The Grower* (Supplement). 106(20):9-10.

Hanan, J.J. 1964. Oxygen and carbon dioxide concentrations in greenhouse soil-air. *Proc. Amer. Soc. Hort. Sci.* 84:648-652.

Hanan, J.J. 1967. Water utilization in greenhouses: Alternatives for agriculture in arid regions. *Proc. 3rd Ann. Amer. Water Res. Conf.* San Francisco, Nov., 1967. 160-169.

Hanan, J.J. 1969a. Water loss and stress in carnations grown under glass. *CO Flower Growers' Assoc. Res. Bull.* 233:1-3.

Hanan. J.J. 1969b. Carnation stomates –Comparisons between glass and fiberglass. *CO Flower Growers' Assoc. Res. Bull.* 233:3-5.

Hanan, J.J. 1970a. Aspects of water and nutritional relationships in greenhouse production. *Proc. 18th Int. Hort. Cong.*, Tel-Aviv.

Hanan, J.J. 1970b. Water loss from carnations grown under fiberglass. *CO Flower Growers' Assoc. Res. Bul*l. 236:1-3.

Hanan, J.J. 1972. Repercussions from water stress. *HortScience*. 7:108-112.

Hanan, J.J. 1981. Irrigation study on carnations. *CO Greenhouse Growers' Assoc. Res. Bul.* 375:1-2.

Hanan, J.J. 1982. Salinity III. Handling water supplies to minimize salinity problems. *CO Greenhouse Growers' Assoc. Res. Bull.* 384:1-4.

Hanan, J.J. 1984. Plant Environmental Measurement. Bookmakers Guild, Longmont, CO. 326 pp.

Hanan, J.J. 1985. Advances in humidity measurement. *CO Greenhouse Growers' Assoc. Res. Bul*l. 420:4-5.

Hanan, J.J. 1986. Yield and quality of 'Samantha' roses in three inert media. *CO Greenhouse Growers' Assoc. Res. Bull.* 435:1-2.

Hanan, J.J. 1987. So speaks the rose. *CO Greenhouse Growers' Assoc. Res. Bull.* 442:1-3.

Hanan, J.J. 1988. Summer greenhouse climate in Colorado. *CO Greenhouse Growers' Assoc. Res. Bull.* 461:1-6.

Hanan, J.J. 1989. Winter humidity in Colorado greenhouses. *CO Greenhouse Growers' Assoc. Res. Bull.* 470:1-4.

Hanan, J.J. 1991. The influence of greenhouses on internal climate with special reference to Mediterranean regions. *Acta Hort.* 287:23-34.

Hanan, J.J. and H.R. Duke. 1970. The Gates-type greenhouse irrigation system: Design and problems. *CO Flower Growers' Assoc. Res. Bul*l. 245:1-4.

Hanan, J.J. and W.D. Holley. 1974. Hydroponics. CO State Univ. Expt. Sta. *General Series* 941. 21 pp.

Hanan, J.J. and R. Huffsmith. 1967. Weighable lysimeter and recording system used in greenhouse evapotranspiration studies. *Proc. Amer. Soc. Hort. Sci.* 91:691-698.

Hanan, J.J. and F.D. Jasper. 1969. Consumptive water use in response of carnations to three irrigation regimes. *J. Amer. Soc. Hort. Sci.* 94:70-73.

Hanan, J.J. and R. W. Langhans. 1963. Aeration adequacy in greenhouse soils. *NY State Flower Growers' Bull.* 213:1-4.

Hanan, J.J. and R.W. Langhans. 1964a. Control of moisture content in greenhouse soils. *Agron. J.* 56:191-194.

Hanan, J.J. and R.W. Langhans. 1964b. Soil water content and the growth and flowering of snapdragons. *Proc. Amer. Soc. Hort. Sci.* 84:613-623.

Hanan, J.J., R.W. Langhans and A.W. Dimock. 1963. *Pythium* and soil aeration. *Proc. Amer. Soc. Hort. Sci.* 82:574-582.

Hanan, J.J., F.A. Coker and K.L. Goldsberry. 1987. A climate control system for greenhouse research. *HortScience*. 22:704-708.

Hanan, J.J., W.D. Holley and K.L. Goldsberry. 1978. Greenhouse Management. Springer-Verlag, Heidelberg. 530 pp.

Hanan, J.J., C. Olympios and C. Pittas. 1981. Bulk density, porosity, percolation and salinity control in shallow, freely draining, potting soils. *J. Amer. Soc. Hort. Sci.* 106:742-746.

Harbaugh, B..K. and C.D. Stanley. 1985. Guidelines for the use of capillary mat, spaghetti tube and trickle irrigation systems for floricultural crops. *Foliage Digest*. 8(8):1-4.

Harbaugh, B.K., C.D. Stanley and J.F. Price. 1986. Interactive effects of trickle irrigation rates, cultivars, and culture on cut chrysanthemum. *HortScience*. 21:94-95.

Harkess, R.L. and J.J. Hanan. 1988. Effects of humidity on rose yield, average stem length and leaf area. *CO Greenhouse Growers' Assoc. Res. Bull.* 460:1-3.

Harless, S. 1984. Uncover answers to media guessing game. *Greenhouse Manager.* 3(5):102-107.

Hasek, R.F., R.H. Sciaroni and R.L. Branson. 1986. Water conservation and recyling in ornamentals production. *HortScience*. 21:35-38.

Heydecker, W., O.P. Pareek and T. Sivanavatam. 1970. Major effects of relative humidity on plant performance. *Proc. 18th Int. Hort. Congr.*, Tel-Aviv. 1:96-97.

Hilhorst, M.A., J. Groenwold and J.F. de Groot. 1992. Water content measurements in soil and rockwool substrates: dielectric sensors for automatic *in situ* measurements. *Acta Hort.* 304:209-218.

Hockenhull, J. and D. Funck-Jensen. 1982. Is damping-off, caused by *Pythium*, less of a problem in hydroponics than in traditional growing systems? *Acta Hort.* 133:137-145.

Holcomb, E.J. 1982. Hydroponics. *Florists' Rev.* 11, 60-62. Aug. 12, 1982.

Holcomb, E.J. et al. 1992. Efficiency of fertigation programs for Baltic Ivy and Asiatic Lily. *HortTechnology*. 2:43-46.

Holley, W.D. and J. Parker. 1968. Inert media compared for carnation growing. *CO Flower Growers' Assoc. Res. Bull.* 215:1-3.

Holmes, J.W., S.A. Taylor and S.J. Richards. 1967. Measurement of soil water. *In* Irrigation of Agricultural Lands. Hagan, R.M. et al. eds. Mono. No. 11. Amer. Soc. Agron., Madison, WI. 275-303.

Hopmans, P.A.M. 1981. *In situ* plant water relations monitoring for greenhouse climate control with computers. *Acta Hort.* 119:137-145.

Hornick, S.B. et al. 1984. Sewage sludge compost for soil improvement and plant growth. *Florists' Rev.* Oct. 18, 1984.

Hsiao, T.C. 1973. Plant responses to water stress. *Ann. Rev. Plant Physiol.* 24:519-570.

Hurd, R.G. 1978. The root and its environment in the nutrient film technique of water culture. *Acta Hort.* 82:87-97.

Hurd, R. and C. Graves. 1981. Controlling the water supply. *The Grower*. 96(19):17-22.

Ike, I.F., G.W. Thurtell and K.R. Stevenson. 1978. Evaluation of the pressure chamber technique for measuring of leaf water potential in cassava (*Manihot* species). *Can. J. Bot.* 56:1638-1641.

Iwata, M. 1992. Soilless culture of vegetables and flowers. *Chronica Hort.* 32:31.

Jackson, R.D., R.J. Reginato and S.B. Idso. 1977. Wheat canopy temperature: A practical tool for evaluating water requirements. *Water Resour. Res.* 13:651-656.

James, E.A. and D. Richards. 1986. The influence of iron source on the water-holding properties of potting media amended with water-absorbing polymers. *Sci. Hort.* 28:201-208.

Jarvis, P.G. 1985. Coupling of transpiration to the atmosphere in horticultural crops: The omega factor. *Acta Hort.* 171:187-205.

Jarvis, P.G. and R.O. Slatyer. 1970. The role of the mesophyll cell wall in leaf transpiration. *Planta.* 90:303-322.

Jenkins, J.R. and W.M. Jarrell. 1989. Predicting physical and chemical properties of container mixtures. *HortScience*. 24:292-295.

Jensen, M.H. 1971. A hydrophilic polymer as a soil amendment. *Proc. 10th National Agric. Plastics Conf.*, Chicago, 1971. 11 pp.

Jolliet, O. and B.J. Bailey. 1992. The effect of climate on tomato transpiration in greenhouses: measurements and models comparison. *Agric. and Forest Meteor.* 58:43-62.

Johnson, E.W. 1980. Comparison of methods of analysis for loamless composts. *Acta Hort.* 99:197-204.

Johnson, P. 1968. Horticultural and agricultural uses of sawdust and soil amendments. Tech. Bull. National City, CA. 46 pp.

Kanemasu, E.T. and C.B. Tanner. 1969a. Stomatal diffusion resistance of snap beans. I. Influence of leaf-water potential. *Plant Physiol.* 44:1547-1552.

Kanemasu, E.T. and C.B. Tanner. 1969b. Stomatal diffusion resistance of snap beans. II. Effect of light. *Plant Physiol.* 44:1542-1546.

Kanemasu, E.T. et al. 1979. Techniques for calculating energy and mass fluxes. *In* Modification of the Aerial Environment of Crops. Barfield, B.J. and J.F. Gerber eds. ASAE Mono. No. 2. Amer. Soc. Agric. Eng., St. Joseph, MI. 156-182.

Keller, J. and D. Karmeli. 1974. Trickle irrigation design parameters. *Trans. ASAE.* 17:678-684.

Kemble, J.M. et al. 1994. Root cell volume affects growth of compact-growth-habit tomato transplants. *HortScience*. 29:261-262.

Killenberg, C. 1983. Take a look at hydroponic realities. *Amer. Veg. Grower*. Nov., 1983.

Kimball, B.A. 1973. Simulation of the energy balance of a greenhouse. *Agric. Meteor*. 11:243-260.

Kimball, B.A. 1986. A modular energy balance program including subroutines for greenhouses and other latent heat devices. USDA, ARS-33. 356 pp. U.S. Govn. Printing Office, Washington, D.C.

Kindelan, M. 1980. Dynamic modeling of greenhouse environment. *Trans. ASAE*. 23:1233-1239.

Kitchener, H.M. 1978. Rockwool or NFT is basic choice in modern forms of hydroponics. *The Grower*. 89(14):761-763.

Kitchener, H.M. 1981. Rockwool promise for wilt control. *The Grower*. 95(3):17-21.

Knight, A.T. 1944. Studies of pot-binding of greenhouse plants. *MI State College. Tech. Bull*. 191. East Lansing, MI. 51 pp.

Körner, C. 1985. Humidity response in forest trees: Precautions in thermal scanning surveys. *Arch. Met. Geoph. Biocl., Ser. B*. 36:83-98.

Krause, W. 1983. Rockwool development. *The Grower*. Apr. 21, 1983.

Lake, J.V. et al. 1966. Seasonal variation in the transpiration of glasshouse plants. *Agric. Meteor*. 3:187-196.

Lamphier, G. 1989. Political careers can rise or fall in many ways; here's a new one. *Wall St. J*. Jan. 27, 1989.

Landsberg, J.J., B. White and M.R. Thorpe. 1979. Computer analysis of the efficacy of evaporative cooling for glasshouses in high energy environments. *J. Agric. Eng. Res*. 24:29-39.

Larson, R.A. and B.G. Hilliard. 1977. Mat watering: A progress report. *NC Flower Growers' Bull*. 21:1-2.

Latimer, J.G. 1991. Container size and shape influence growth and landscape performance of marigold seedlings. *HortScience*. 26:124-129.

Lauder, K. 1977a. NFT pioneer - Four years on. *The Grower*. 87(25):1310-1312.

Lauder, K. 1977b. Hydroponic pot plants now over their teething trouble. *The Grower*. 88(18):888-900.

Lauder, K. 1977c. Lettuce on concrete. *The Grower* (Supplement). Feb. 24, 1977.

Lawrence, W.J.C. and J. Newell. 1952. Seed and Potting Composts. Allen and Unwin, London. 176 pp.

Lawson, G. 1984. Dutch make lettuce harvesting easy. *The Grower*. Feb. 16, 1984, 32-33.

Lawson, G. 1989. Fine tuning. *The Grower*. 111(4). Jan. 26, 1989.

Lawson, R.H. and R.K. Horst. 1984. More search needed on bark compost usage. *Greenhouse Manager*. May, 1984. 105-112.

Lawson, R.H. and M.M. Dienelt. 1988. A hazard of hydroponics. *Greenhouse Manager*. 7(1):125-128.

Leatherland, M. 1987. Recycling on rockwool by merging with NFT. *The Grower* (Supplement). 106(20):15-17.

Lee, C.W. 1985. Hydroponic production of greenhouse crops in rockwool. Unpublished MS, CO State Univ., Ft. Collins. 9 pp.

Lee, R. and D.M. Gates. 1964. Diffusion resistance in leaves as related to their stomatal anatomy and micro-structure. *Amer. J. Bot*. 51:963-975.

Lemon, E.R. 1965. Energy conservation and water use efficiency in plants. *In* Plant Environment and Efficient Water Use. Pierre, W.H. et al. eds. Amer. Soc. Agron., Madison, WI. 28-48.

Lemon, E.R. and A.E. Erickson. 1955. Principle of the platinum micro-electrode as a method of characterizing soil aeration. *Soil Sci*. 79:383-392.

Lemon, E., D.W. Stewart and R.W. Shawcroft. 1971. The sun's work in a cornfield. *Science*. 174:371-378.

Letey, J. and G.B. Blank. 1961. Influence of environment on the vegetative growth of plants watered at various soil suctions. *Agron. J*. 53:151-153.

Letey, J., P.R. Clark and C. Armhein. 1992. Water-sorbing polymers do not conserve water. *Calif. Agric*. 46:9-10.

Letey, J. et al. 1961. Plant growth, water use and nutritional response to rhizosphere differentials of oxygen concentration. *Soil Sci. Soc. Amer. Proc*. 25:183-186.

Letey, J. et al. 1962. Influence of oxygen diffusion rate on sunflower growth at various soil and air temperatures. *Agron. J*. 54:316-319.

Letey, J. et al. 1964. Measurement of oxygen diffusion rates with the platinum microelectrode. *Hilgardia*. 35:545-576.

Levin, I., P.C. van Rooyen and F.C. van Rooyen. 1979. The effect of discharge rate and intermittent water application by point-source irrigation on the soil moisture distribution pattern. *Soil Sci. Soc. Amer. J*. 48:8-16.

Levitt, H.J. and R. Gaspar. 1988. Energy budget for greenhouses in humid-temperature climate. *Agric. and Forest Meteor*. 42:241-254.

Lieth, J.H. and J.F. Reynolds. 1988. A plant growth model for controlled-environment conditions incorporating canopy structure and development: Application to snap bean. *Photosynthetica*. 22:190-204.

Liptay, A. and D. Edwards. 1994. Tomato seedling growth in response to variation in root container shape. *HortScience*. 29:633-635.

List, R.J. 1966. Smithsonian Meteorological Tables. *Smithsonian Misc. Collections* 114, Washington, D.C. 527 pp.

Loomis, R.S. and W.A. Williams. 1969. Productivity and the morphology of crop stands: Patterns with leaves. *In* Physiological Aspects of Crop Yield. Eastin, J.D. et al. ed. Amer. Soc. Agron., Madison, WI. pp 27-47.

Long, E. 1987. Chalking up a clean water supply. *The Grower*. 108(16). Oct. 15, 1987.

Love, J.W. 1978. Basics of bark media. *Ohio Florists' Assoc. Bull*. 584:7-8.

Love, J.W. and P. Nelson. 1972. Current thoughts on pine bark. *NC Flower Growers' Bull*. 16:1-3.

Lynn, B.H. and T.N. Carlson. 1990. A stomatal resistance model illustrating plant vs. external control of transpiration. *Agric. and Forest Meteor*. 52:5-43.

Malek, R.B. and J.B. Gartner. 1975. Hardwood bark as a soil amendment for suppression of plant parasitic nematodes on container-grown plants. *HortScience*. 10:33-35.

Marcelis, L.F.M. 1989. Simulation of plant-water relations and photosynthesis of greenhouse crops. *Sci. Hort*. 41:9-18.

Marshall, D.C. 1958. Measurement of sap flow in conifers by heat transport. *Plant Physiol*. 33:385-396.

Matthews, R.B. and R.A. Saffell. 1986. Computer control of humidity in experimental greenhouses. *J. Agric. Engin. Res*. 33:213-221.

Mazuri, A.R. , T.D. Hughes and J.B. Gartner. 1975. Physical properties of hardwood bark growth media. *HortScience*. 10:30-33.

McNaughton, K.G. and P.G. Jarvis. 1983. Predicting effects of vegetation changes on transpiration and evaporation. *In* Water Deficits and Plant Growth. T.T. Kozlowski ed. Vol. VII. Academic Press, New York.

Meijaard, D. 1995. The greenhouse industry in The Netherlands. *In* Greenhouse Climate Control. J.C. Bakker et al. Eds. Wageningen Pers.

Meijer, J. et al. 1985. Development and application of a sensitive, high precision weighing lysimeter for use in greenhouses. *J. Agric. Engin. Res*. 32:321-336.

Michels, P. and J. Feyen. 1984. Automatic control of irrigation in greenhouses by simulation of the water balance in the root zone. *J. Agric. Engin. Res*. 29:223-230.

Milks, R.R., W.C. Fonteno and R.A. Larson. 1989a. Hydrology of horticultural substrates: I. Mathematical models for moisture characteristics of horticultural container media. *J. Amer. Soc. Hort. Sci.* 114:48-52.

Milks, R.R., W.C. Fonteno and R.A. Larson. 1989b. Hydrology of horticultural substrates: II. Predicting physical properties of media in containers. *J. Amer. Soc. Hort. Sci.* 114:53-56.

Milks, R.R., W.C. Fonteno and R.A. Larson. 1989c. Hydrology of horticultural substrates: III. Predicting air and water content of limited-volume plug cells. *J. Amer. Soc. Hort. Sci.* 114:57-61.

Monteith, J.L. 1973. Principles of Environmental Physics. Amer. Elsevier, New York. 241 pp.

Monteith, J.L. 1981. Evaporation and surface temperature. *Quart. J. Royal Meteor. Soc*. 107:1-27.

Moreshet, S. and G. Stanhill. 1965. The relationship between leaf resistance to mass air flow and infiltration score in the cotton crop. *Ann. Bot*., N.S. 29:625-633.

Moreshet, S., D. Koller and G. Stanhill. 1968. The partitioning of resistances to gaseous diffusion in the leaf epidermis and the boundary layer. *Ann. Bot*. 32:695-701.

Morgan, J.V. and A.L. Tan. 1982. Greenhouse lettuce production at high densities in hydroponics. *Acta Hort*. 133:39-46.

Morris, L.G. and J.V. Lake. 1962. The water loss from plants and soil under glass. *Adv. in Hort. Sci. and Its Application*. 3:323-328.

Morris, L.G., F.E. Neale and J.D. Postlethwaite. 1957. The transpiration of glasshouse crops, and its relationship to the incoming solar radiation. *J. Agric. Eng. Res*. 2:111-122.

Mortensen, L.M. and H.R. Gislerød. 1990. Effects of air humidity and supplementary lighting on foliage plants. *Sci. Hort*. 44:301-308.

Moser, E. 1979. Technically-oriented research on drip-irrigation equipment for special crops. *Acta Hort.* 89:37-45.

Mossman, P. 1980. The second generation. *The Grower.* 94(3):17-20.

Nadler, A., S. Dasberg and I. Lapid. 1991. Time domain reflectrometry of water content and electrical conductivity of layered soil columns. *Soil Sci. Soc. Amer. J.* 55:938-943.

Nakayama, F.S. and C.H.M. van Bavel. 1963. Root activity distribution patterns of sorghum and soil moisture conditions. *Agron. J.* 55:271-274.

Namken, L.N., J.F. Bartholic and J.R. Runkles. 1971. Water stress and stem radial contraction of cotton plants (*Gossypium hirsutum* L.) under field conditions. *Agron. J.* 63:623-627.

Neale, F.E. 1953. Transpiration of glasshouse tomatoes, lettuce and carnations. *Netherlands J. Agric. Sci.* 3:48-60.

Nelson, P.V. 1972. Greenhouse media. The use of Cofuna, Floramull, Pinebark and Styromull. *NC Agric. Expt. Sta, Tech. Bull.* No. 206. 17 pp.

Nelson, P.V. and W.C. Fonteno. 1991. Physical analysis of rockwool slabs and effects of fiber orientation, irrigation frequency and propagation technique on chrysanthemum production. *J. Plant Nutrition.* 14:853-866.

Newman, E.I. 1974. Root and soil water relations. *In* The Plant Root and Its Environment. E.W. Carson ed. Univ. Press of VA., Charlottesville. 363-440.

Nobel, P.S.. 1991. Physicochemical and Environmental Plant Physiology. Academic Press, New York. 635 pp.

Norrie, J., M.E.D. Graham and A. Gosselin. 1994a. Potential evapotranspiration as a means of predicting irrigation timing in greenhouse tomatoes grown in peat bags. *J. Amer. Soc. Hort. Sci.* 119:163-168.

Norrie, J. et al. 1994b. Improvements in automatic irrigation of peat-grown greenhouse tomatoes. *HortTechnology.* 4:154-159.

Ofir, M., E. Shmueli and S. Moreshet. 1968. Stomatal infiltration measurements as an indicator of the water requirement and timing of irrigation for cotton. *Exptl. Agric.* 4:325-333.

Orzolek, M.D. 1993. Use of hydrophylic polymers in horticulture. *HortTechnology.* 3:41-44.

Panter, K.K. and J.J. Hanan. 1987. Stomatal resistance and water potential of calceolaria grown under two different heating systems. *J. Amer. Soc. Hort. Sci.* 112:637-641.

Peet, M.M. 1985. Grow longer and boost density to beef up yields. *Amer. Veg. Grower.*33(4):88-89.

Pelletier, G. and C.S. Tan. 1993. Determining irrigation wetting patterns using time domain reflectometry. *HortScience.* 28:338-339.

Penman, H.L. 1948. Natural evaporation from open water, bare soil and grass. *Proc. Royal Soc., London (A).* 193:120-145.

Penman, H.L. 1952. The physical basis of irrigation control. *13th Int. Hort. Congress.* Tel-Aviv. 2:913-924.

Penman, H.L. and R.K. Schofield. 1951. Some physical aspects of assimilation and transpiration. *Symp. Soc. Expt. Biol.* 5:115-129.

Penman, H.L., D.E. Angus and C.H.M. van Bavel. 1967. Microclimatic factors affecting evaporation and transpiration. *In* Irrigation of Agricultural Lands. Hagan, R.M. et al. eds. Mono. No. 11. Amer. Soc. Agron., Madison, WI. 483-505.

Penningsfeld, F. 1978. Substrates for protected cropping. *Acta Hort.* 82:13-21.

Peterson, J.C. 1975. Toxic fluoride levels in growing media. *Commercial Flower Notes.*, Univ of GA, Athens. Oct., 1975.

Peterson, J.C. 1979. Handling soil containing residual herbicides. *Roses, Inc. Bull.* June, 1979.

Phene, C.J., R.B. Hutmacher and K.R. Davis. 1992. Two hundred tons per hectare of processing tomatoes –Can we reach it? *HortTechnology.* 2:16-22.

Phillips, R.E. and D. Kirkham. 1962. Mechanical impedance and corn seedling root growth. *Proc. Soil Sci. Soc. Amer.* 26:319-322.

Plaut, Z. and S. Moreshet. 1973. Transport of water in plant-atmosphere system. *In* Arid Zone Irrigation. Yaron, B. et al. eds. Springer-Verlag, Heidelberg. 123-141.

Plaut, Z. and N. Zieslin. 1974. Productivity of greenhouse roses following changes in soil moisture and soil air regimes. *Sci. Hort.* 2:137-143.

Plaut, Z., N. Zieslin and I. Arnon. 1973. The influence of moisture regime on greenhouse rose production in various growth media. *Sci. Hort.* 1:239-250.
Poole, R.T. and W.E. Waters. 1972. Evaluation of various potting media for growth of foliage plants. *FL State Hort. Soc.* 85:395-398.
Prevatt, J.W., G.A. Clark and C.D. Stanley. 1992. A comparative cost analysis of vegetable irrigation systems. *HortTechnology*. 2:91-94.
Price, D. 1980. Disease control: So far so good. *The Grower* (Supplement). Feb. 21, 1980. 104-107.
Raschke, K. 1979. Stomatal action. *Ann. Rev. Plant Physiol.* 26:309-340.
Raats, P.A.C. 1971. Steady infiltration from point sources, cavities and basins. *Soil Sci. Soc. Amer. Proc.* 35:689-694.
Reilly, A. 1980. From bog to bench – a tale of sphagnum peat moss. *Florists' Rev.* 167(4328):46-47, 87.
Richards, L.A. 1950 Laws of soil moisture. *Trans. Amer. Geophys. Union.* 31:750-756.
Richards, L.A. and G. Ogata. 1958. Thermocouple for vapor measurement in biological and soil systems at high humidity. *Science*. 128:1089-1090.
Risse, L.M. and J.L. Chesness. 1989. A simplified design procedure to determine the wetted radius for a trickle emitter. *Trans. ASAE.* 32:1909-1914.
Rickman, R.W., J.Letey and L.H. Stolzy. 1965. Soil compaction effects on oxygen diffusion rates and plant growth. *Calif. Agric.* 19:4-6.
Ritchie, J.T. 1981. Soil water availability. *Plant and Soil.* 58:327-338.
Ritchie, J.T. 1981. Water dynamics in the soil-plant-atmosphere systems. *Plant and Soil.* 58:81-96.
Roberts, B.W. and C.W. O'Hern. 1993. Inexpensive sand filters for drip irrigation systems. *HortTechnology*. 3:85-89.
Roe, J. 1986. Rockwool or NFT? *The Grower*. 105(8). Feb. 20, 186.
Rosenberg, N..J., B.L. Blad and S.B. Verma. 1983. Microclimate. The Biological Environment. John Wiley & Sons, New York. 495 pp.
Ross, D.S. 1974. Maintain greenhouse humidity for good plant growth. *Florists' Rev.* 166(4314):22-23.
Rothwell, J.B. and D.A.G. Jones. 1961. The water requirement of tomatoes in relation to solar radiation. *Expt. Hort.* 5:25-30.
Royle, D. 1980. Acid keeps trickle flowing. *The Grower*. Nov. 6, 1980. 25-28.
Ruff, M.S. et al. 1987. Restricted root zone volume: Influence on growth and development of tomato. *J. Amer. Soc. Hort. Sci.* 112:763-769.
Russell, R.S. and V.M. Shorrocks. 1959. The relationship between transpiration and the absorption of inorganic ions by intact plants. *J. Exptl. Bot.* 10:301-316.
Sadler, E.J. and C.H.M. van Bavel. 1984. Simulation and measurement of energy partition in a fluid-roof greenhouse. *Agric. and Forest Meteor.* 33:1-13.
Sanders, D.C. 1992. Drip-irrigation system component and design considerations for vegetable crops. *HortTechnology*. 2:25-26.
Schippers, P.A. 1981. Commercial hydroponic production. *Amer. Veg. Grower*. 29(10):9-10.
Schippers, P.A. 1982. Developments in hydroponic tomato growing. *Amer. Veg. Grower*. June, 1982. 26-28.
Schippers, P.A. 1986. Another look at hydroponic realities. *Amer. Veg. Grower*. May, 1986. 22-25.
Scholander, P.F. et al. 1965. Sap pressure in vascular plants. *Science*. 148:339-346.
Schwarz, M. and Y. Vaadia. 1969. Limestone gravel as growth medium in hydroponics. *Plant and Soil*. 31:122-128.
Seginer, I. 1980. Optimizing greenhouse operation for best aerial environment. *Acta Hort.* 106:169-178.
Seginer, I. 1984. On the night transpiration of greenhouse roses under glass or plastic cover. *Agric. Meteor.* 30:257-268.
Seginer, I. and D. Kantz. 1989. Night-time use of dehumidifiers in greenhouses: an analysis. *J. Agric. Engin. Res.* 110:1-18.
Seginer, I. et al. 1990. Night-time transpiration in greenhouses. *Sci. Hort.* 41:265-276.
Shackel, K.A. 1984. Theoretical and experimental errors for *in situ* measurements of plant water potential. *Plant Physiol.* 75:766-772.

Shaer, Y.A. and C.H.M. van Bavel. 1987. Relative role of stomatal and aerodynamic resistances in transpiration of a tomato crop in a CO_2-enriched greenhouse. *Agric. and Forest Meteor.* 41:77-85.

Sieur, H.A. and S. Brungardt. 1977. Hydroponics: Where's it at? Part I. *The American Packer* (Supplement). April, 1977. pp 10B-11B.

Simitchiev, H. et al. 1982. Biological effect of greenhouse tomatoes grown on rockwool. *Acta Hort.* 133:59-65.

Slatyer, R.O. 1961. Effects of several osmotic substrates on the water relationships of tomato. *Aust. J. Biol. Sci.* 14:519-540.

Slatyer, R.O. 1966. Some physical aspects of internal control of leaf transpiration. *Agric. Meteor.* 3:281-292.

Slatyer, R.O. 1967a. Plant-Water Relationships. Academic Press, New York. 366 pp.

Slatyer, R.O. 1967b. Terminology for cell and tissue water relations. *Z. Pflanzenphysiol.* 56:91-94.

Slavik, B. 1965. Supply of water to plants. *Meteor. Monog.* 6:149-162.

Soffer, H. and D.W. Burger. 1988. Effects of dissolved oxygen concentrations in aero-hydroponics on the formation and growth of adventitious roots. *J. Amer. Soc. Hort. Sci.* 113:218-221.

Sonneveld, C. 1980. Growing cucumbers and tomatoes in rockwool. *Proc. 5th Int. Cong. on Soilless Culture.* Wageningen. 253-262.

Sonneveld, C. and S.J. Voogt. 1975. Peat substrate as a growing medium for cucumbers. *Acta Hort.* 50:45-52.

Spanner, D.C. 1951. The Peltier effect and its use in the measurement of suction pressure. *J. Expt. Bot.* 2:145-168.

Spomer, L.A. 1975. Soil amendment (or why does 4 bushels plus 10 bushels equal only 10 bushels?). *IL State Florists' Bull.* 355:7-8.

Spomer, L.A. 1976. What is a good soil amendment? *IL State Florists' Assoc. Bull.* 367:10-11.

Sprague, H.B. and J.F. Marreno. 1931. The effect of various sources of organic matter on the properties of soils as determined by physical measurements and plant growth. *Soil Sci.* 32:35-50.

Stanghellini, C. 1981a. Estimation of energy requirement for evaporation in greenhouses. *Acta Hort.* 115:693-699.

Stanghellini, C. 1981b. Evapotranspiration and energy consumption in greenhouses. *Acta Hort.* 119:273-279.

Stanghellini, C. 1987. Transpiration of greenhouse crops. Ph.D. Diss., IMAG, Wageningen. 150 pp.

Stanhill, G. and J.S. Albers. 1974. Solar radiation and water loss from greenhouse roses. *J. Amer. Soc. Hort. Sci.* 99:107-110.

Stanhill, G. and Y. Vaadia. 1967. Factors affecting plant responses to soil water. *In* Irrigation of Agricultural Lands. Hagan, R.M. et al. ed. *Mono. No. 11.* Amer. Soc. Agron., Madison, WI. 446-457.

Staunton, W.P. and T.P. Cormican. 1978. The behavior of tomato pathogens in a hydroponic system. *Acta Hort.* 82:133-135.

Steiner, A.A. 1968. Soilless culture. *Proc. 6th Colloq. Int. Potash Inst.*, Florence. 324-341.

Stevenson, D.S. 1967. Effective soil volume and its importance to root and top growth of plants. *Can. J. Soil Sci.* 47:163-174.

Stirling, C.E. 1981. Bags of air in woven modules. *The Grower.* 95(6):17-18.

Stolzy, L.H. et al. 1961. Root growth and diffusion rates as functions of oxygen concentration. *Soil Sci. Soc. Amer. Proc.* 25:463-467.

Strider, L. ed. 1985a. Diseases of Floral Crops. Vol. I. Praeger Scientific, New York. 638 pp.

Strider, L. ed. 1985b. Diseases of Floral Crops. Vol. II. Praeger Scientific, New York. 579 pp.

Strojny, Z. 1980. Use of container method in the growing of gerberas in different substrates. *Acta Hort.* 99:243-248.

Tanner, C.B. 1960. Energy balance approach to evapotranspiration from crops. *Soil Sci. Soc. Amer. Proc.* 24:2-9.

Tanner, C.B. 1966. Comparison of energy balance and mass transport methods for measuring evaporation. *Proc. Evapotranspiration and its role in water resources management.* ASAE, St. Joseph, MI. 45-48.

Tanner, C.B. and M. Fuchs. 1968. Evaporation from unsaturated surfaces: A generalized combination method. *J. Geophys. Res.* 73:1299-1304.

Taylor, H.M. and H.R. Gardner. 1963. Penetration of cotton seedling taproots as influenced by bulk density, moisture content, and strength of the soil. *Soil Sci.* 96:153-156.

Taylor, S.A. 1964. Water condition and flow in the soil-plant-atmosphere system. *In* Brady, N.C. et al. Eds. Forage Crop Symposia. ASA Spec. Publ. 5. Amer. Soc. Agron., Madison, WI. 81-107.

Taylor, S.A. and G.L. Ashcroft. 1972. Physical Edaphology. W.H. Freeman and Co., San Francisco. 533 pp.
Thornthwaite, C.W. and F.K. Hare. 1965. The loss of water to air. *Meteor. Monog.* 6:163-180.
Tibbitts, T.W. 1979. Humidity and plants. *BioScience*. 29:358-363.
Tilt, K.M., T.E. Bilderback and W.C. Fonteno. 1987. Particle size and container size effects on growth of three ornamental species. *J. Amer. Soc. Hort. Sci.* 112:981-984.
Topp, G.C. and J.L. Davis. 1985. Measurement of soil water content using time-domain reflectrometry (TDR): A field evaluation. *Soil Sci. Soc. Amer. J.* 49:19-24.
Topp, G.C., J.L. Davis and A.P. Annan. 1980. Electromagentic determination of soil water content: Measurements in coaxial transmission lines. *Water Resour. Res.* 16:574-582.
Topp, G.C. et al. 1984. The measurement of soil water content using a portable TDR hand probe. *Can. J. Soil Sci.* 64:313-321.
Turner, N.C. 1981. Techniques and experimental approaches for the measurement of plant water status. *Plant and Soil*. 58:339-366.
Turner, N.C. and J.E. Begg. 1981. Plant-water relations and adaptation to stress. *Plant and Soil*. 58:97-131.
Tyree, M.T. and J.S. Sperry. 1989. Vulnerability of xylem to cavitation and embolism. *Ann. Rev. Plant Physiol.* 40:19-38.
Valancogne, C. and Z. Nasr. 1989. Measuring sap flow in the stem of small trees by a heat balance method. *HortScience*. 24:383-385.
Vanachter, A., E. van Wambeke and C. van Assche. 1982. Potential danger for infection and spread of root disease of tomatoes in hydroponics. *Acta Hort.* 133:119-128.
van As, H., T.J. Schaafsma and J. Blaakmeer. 1985. Application of NMR to water flow and balance in plants. *Acta Hort.* 174:491-495.
van Bavel, C.H.M. 1967. Changes in canopy resistance to water loss from alfalfa induced by soil water depletion. *Agric. Meteor.* 4:165-176.
van de Sanden, P.A.C.M. 1985. Effect of air humidity on growth and water exchange of cucumber-seedlings. *Acta Hort.* 174:259-267.
van den Ende, J. and R. de Graaf. 1974. Comparison of methods of water supply to hothouse tomatoes. *Acta Hort.* 35:61-70.
van den Honert, T.H. 1948. Water transport in plants as a catenary process. *Discuss. Faraday Soc.* 3:146-153.
van der Post, C.J., J.J. van Schie and R. de Graaf. 1974. Energy balance and water supply in glasshouses in the west-Netherlands. *Acta Hort.* 35:13-22.
van der Hoeven, A.P. 1980. Yield comparisons between chrysanthemums grown in nutrient film and in glasshouse soil. *Ann. Rpt. 1980.* Naaldwijk. The Netherlands. 78-79.
van Holsteijn, H.M.C., M.H. Behboudian and H.C.M.L. Bongers. 1977. Water relations of lettuce. II. Effects of drought on gas exchange properties of two cultivars. *Sci. Hort.* 7:19-26.
van Meurs, W.T.M. and T.H. Gieling. 1980. A research strategy to solve air humidity problems in greenhouses caused by use of energy-saving measures. *Acta Hort.* 115:1-6.
van Meurs, W.T.M. and C. Stanghellini. 1989. A transpiration-based climate control algorithm. *Acta Hort.* 245:476-481.
van Meurs, W.T.M. and C. Stanghellini. 1992. Use of an off-the-shelf electronic balance for monitoring crop transpiration in greenhouses. *Acta Hort.* 304:219-225.
van Nierop, J. 1982. International adviser warns: Substrate growing is the answer to energy saving problems. *The Grower* (Supplement). Jan. 21, 1982. 17-19.
van Os, E.A. 1982a. Dutch developments in soilless culture. *Outlook in Agric.* 11:165-171.
van Os, E.A. 1982b. Rockwool, peat or NFT - Assessing the system. *The Grower* (Suppl.) Jan. 21, 1982.
van Winden, C.M.M. 1988. Soilless culture technique and its relation to greenhouse climate. *Acta Hort.* 229:125-132.
Verdonck, O.F., I.M. Cappaert and M.F. de Boodt. 1978. Physical characterization of horticultural substrates. *Acta Hort.* 82:191-198.
Verdonck, O.F., D. de Vleeschauwer and M. de Boodt. 1980. Growing ornamental plants in inert substrates. *Acta Hort.* 99:113-118.

Verwer, F.L.J.A.W. 1978. Research and results with horticultural crops grown in rockwool and nutrient film. *Acta Hort.* 82:141-147.

Waggoner, P.E. and W.E. Reifsnyder. 1968. Simulation of temperature, humidity and evaporation profiles in a leaf canopy. *J. Appl. Meteor.* 7:400-409.

Walker, J.N. and D.J. Cotter. 1968. Control of humidity in greenhouses during warm weather. *Trans. ASAE.* 11:267-269.

Walker, J.N. and G.A. Duncan. 1973. Greenhouse humidity control. Dept. Agric. Engin. Univ. of KY. *AEN-19.* 4 pp.

Walker, G.K. and J.L. Hatfield. 1983. Stress measurement using foliage temperature. *Agron. J.* 75:623-629.

Wallace, A., G.A. Wallace and J.W. Cha. 1986. Mechanisms involved in soil conditioning by polymers. *Soil Sci.* 141:381-386.

Wang, J.Y. and C.M.M. Felton. 1983. Instruments for Physical Environmental Measurements. Vol. 1. Kendall-Hunt, Dubuque, IA. 378 pp.

Wang, Yin-Tung. 1987. Driving your soil to drink. *Greenhouse Manager.* 6(3):115-121.

Waterfield, A.E. 1981. NFT –No bonuses so far. *The Grower.* 95(8):22-23.

Wees, D. and D. Donnelly. 1992. Reduction of phytotoxicity in polyurethane foam-containing potting media. *HortScience.* 27:225-227.

Wiegand, C.L. and E.R. Lemon. 1958. A field study of some plant-soil relations in aeration. *Proc. Soil Sci. Soc. Amer.* 22:216-221.

Weigand, C.L. and L.N. Namken. 1966. Influences of plant moisture stress, solar radiation, and air temperature on cotton leaf temperature. *Agron. J.* 58:582-586.

Wells, D.A. 1966. Pressure switch control for irrigated benches and beds. NIAE Note No. 9. Silsoe, U.K. 6 pp.

Wells, D.A. and J.D. Postlethwaite. 1969. Capillary irrigation: The effect of soil volume on tomato production. *NIAE Note No. 32.* Silsoe, U.K. 9 pp.

Wells, D.A. and R. Soffe. 1961. A tensiometer for control of glasshouse irrigation. *J. Agric. Engin. Res.* 6:16-26.

Wells, D.A. and R. Soffe. 1962. A bench method for the automatic watering by capillarity of plants grown in pots. *J. Agric. Engin. Res.* 7:42-46.

White, J.W. 1964. The concept of "container capacity" and its application to soil-moisture-fertility regimes in the production of container-grown crops. Ph.D. Diss., PA State Univ. 193 pp.

White, J.W. 1974. Criteria for selection of growing media for greenhouse crops. *Florists' Rev.* 155(4009):28-39.

Whitesides, R. 1989. El Modeno Gardens: Innovative solutions to California's irrigation runoff restrictions. *Grower Talks.* 52(9):28-36.

Wilcox, G.E. 1980. High hopes for hydroponics. *Amer. Veg. Grower.* 28(11):11-13.

Wilson, G.C.S. 1978a. Collar rot in N.F.T. grown tomatoes. *Acta Hort.* 82:151-152.

Wilson, G.C.S. 1978b. A simple method of getting the desired slope in N.F.T. gulleys. *Acta Hort.* 82:149-150.

Wilson, G.C.S. 1980. Perlite system for tomato production. *Acta Hort.* 99:159-166.

Woodward, F.I. and J.E. Sheehy. 1983. Principles and Measurements in Environmental Biology. Butterworths, London. 263 pp.

Wu, I.P. and H.M. Gitlin. 1983. Drip irrigation application efficiency and schedules. *Trans. ASAE.* 26:92-99.

Yang, X. et al. 1990. Transpiration, leaf temperature and stomatal resistance of a greenhouse cucumber crop. *Agric. and Forest Meteor.* 51:197-209.

Zelitch, I. 1967. Control of leaf stomata –Their role in transpiration and photosynthesis. *Amer. Sci.* 55:472-485.

Zelitch, I. 1969. Stomatal control. *Ann. Rev. Plant Physiol.* 20:329-350.

CHAPTER 6

NUTRITION

I. INTRODUCTION

Table 6-1 lists the elements essential for normal growth of most plant species. In hydroponics, these are usually the only elements considered. The micronutrients are found in micromolar (μM) concentrations (0.000001 mole ℓ^{-1}), whereas the macronutrients are usually required in concentrations of millimoles (mM) (0.001 mole ℓ^{-1}), or 10 to 1000 times the required concentration range of micronutrients. The relative abundance of these essential elements, expressed as parts per million (ppm), may be noted as the average concentrations found in plant tissue (Table 6-1). Under the heading of forms available to plants, one notes that the essential elements occur in ionic forms with positive or negative electrical charge. In a bulk nutrient solution, electrical neutrality is required, meaning that the total positively charged cations must balance the negatively charged anions, and it is in the ionic form that most nutrients are absorbed [Hiatt and Leggett, 1974]. While there can be direct exchange of absorbed ions between soil particles and the root system [Barber, 1962], most nutrient uptake is from the soil solution [Lindsay, 1972b]. Crops require greater amounts of nutrients than the soil solution contains at any one time [Dean, 1957]. Hydroponic systems are an exception.

Table 6-1. Elements commonly considered as essential for plant growth. Ionic forms available to plants are suggested. [Adapted from Salisbury and Ross, 1985, and Brady, 1974]. Note that the common essential elements carbon, hydrogen and oxygen are not considered. Atomic weights rounded to nearest whole number[1]. Forms available to plants are usually combined with water.

Element	Chemical symbol	Forms available to plants	Atomic weight of element	Concentration in dry tissue (ppm)
		Micronutrients		
Nickel[2]	Ni	Ni^{2+}	59	0.06
Molybedenum	Mo	MoO_4^{2-}	96	0.1
Copper	Cu	Cu^+, Cu^{2+}	64	6
Zinc	Zn	Zn^{2+}	65	20
Manganese	Mn	Mn^{2+}, Mn^{3+}, Mn^{4+}	55	50
Boron	B	$H_3BO_3^o$, BO_3^{3-}	11	20
Iron	Fe	Fe^{2+}, Fe^{3+}	56	100
Chlorine	Cl	Cl^-	35	100
		Macronutrients		
Sulfur	S	SO_4^{2-}	32	1000
Phosphorous	P	$H_2PO_4^-$, HPO_4^{2-}, PO_4^{3-}	31	2000
Magnesium	Mg	Mg^{2+}	24	2000
Calcium	Ca	Ca^{2+}	40	5000
Potassium	K	K^+	39	10000
Nitrogen	N	NO_3^-, NH_4^+, NO_2^-	14	15000

[1] Many of these elements may be found in other combinations in solution, or as neutral, undissociated molecules.
[2] From Kochian [1991]; Welch et al. [1991].

As will be noted, there may be other chemical combinations with the essential elements in solution. Whether or not these also can be taken up by plants remains to be decided. For example, both Fe^{2+} and Fe^{3+} can react with water (hydrolyze) so that each forms five different compounds or oxides —the latter largely determining iron concentration in solution. Fe^{3+} can also react with other elements such as chloride, bromide, fluoride, nitrate, sulfate and phosphate to form as many as 13 different complexes, some neutral, others with varying electrical charge. These do not include the various reduction-oxidation reactions that can occur with iron [Lindsay, 1979].

Table 6-2. Other elements often found in higher plants. Plant response has been observed with some of these but essentiality not proven. Others are essential to mammals or happen to be present in the soil solution. Formulae to which a superscript **o** is attached are undissociated, neutral molecules [From Lindsay's modified data base for MINTEQA2/PRODEFA2, Allison et al., 1991][1]. "Forms in solution" are usually combined with one or more water molecules.

Element	Chemical symbol	Forms in solution available to plants	Atomic weight
Aluminum	Al	Al^{3+}	27
Antimony	Sb	$Sb(OH)_6^-$	122
Barium	Ba	Ba^{2+}	137
Bromine	Br	Br^-	80
Cadmium	Cd	Cd^{2+}	112
Chromium	Cr	Cr^{2+}, $Cr(OH)_2^+$, CrO_4^{2-}	52
Cobalt	Co	Co^{2+}, Co^{3+}	59
Fluorine	F	F^-	19
Iodine	I	I^-	127
Lead	Pb	Pb^{2+}	207
Lithium	Li	Li^+	7
Mercury	Hg	Hg_2^{2+}	201
Rubidium	Rb	Rb^+	85
Selenium	Se	HSe^-, $HSeO_3^-$, SeO_4^{2-}	79
Silicon	Si	$H_3SiO_4^-$, $H_2SiO_4^{2-}$, $HSiO_4^{3-}$, SiO_4^{4-}, $H_4SiO_4^o$	28
Silver	Ag	Ag^+	108
Sodium	Na	Na^+	23
Strontium	Sr	Sr^{2+}	88
Thallium	Tl	Tl^+	204
Uranium	U	U^{3+}, U^{4+}, UO_2^+, UO_2^{2+}	238
Vanadium	V	V^{2+}, V^{3+}, VO^{2+}, VO_2^+	51

[1] These elements may be found in other combinations in solution in ionic form or as neutral molecules.

The common practice of referring to fertilizer application as the use of "phosphorus," "potassium," or "nitrogen" is not correct, though one will find this the most common procedure in American horticultural literature. Not one of these three macronutrients exists as elemental N, P, or K in nature. Nitrogen, as it exists as N_2 in the air, is unavailable to plants. To be taken up by plants, nitrogen must be in the nitrate (NO_3^-), ammonium (NH_4^+), or nitrite (NO_2^-) forms. The application of "200 ppm N," as recommended, is incomplete and confusing. Furthermore, fertilizer analyses list the major forms (N, P, and K) as percent N, P_2O_5, and K_2O present in the mixture. The percent contents listed on a fertilizer bag do not have a direct relationship to practical plant nutrition. The statement of actual percentage N, P, and K by most authors merely confuses the issue.

A second point to be made is that the essential forms given in Table 6-1 can be found in most soils with many other combinations as presented in Table 6-2. If the element is present in solution, it will usually be found in the plant, whether or not the particular substance performs any useful function in plant metabolism [Salisbury and Ross, 1985]. Essentialities for those in Table 6-2 have not been proven, but there have been many reports on significant growth responses to silica, selenium, and sodium [e.g., Salisbury and Ross, 1985; Adams, 1987; Welch et al., 1991]. Bear classified selenium as a toxic element in 1957 –which it is often to animals feeding upon Se accumulators in semiarid regions of the U.S. On the other hand, Adriano et al. [1987] cite the literature to the effect that selenium deficiencies occur in New Zealand and Scandinavian countries. In Finland, Se is added to fertilizer formulations. Adriano et al. also mention titanium, nickel, and chromium as showing beneficial effects in plant growth. The latter may be due to interactions of these metals with essential elements. Foy [1992], reviewing the literature on factors limiting plant growth, mentioned yield responses to silicon application by sugar cane, rice, horsetail, cucumber, and tomato. Silicon can also be related in some parts of the world to a high incidence of esophageal cancer or a protective mechanism in plants against grazing. In regions of high rainfall, soil acidification leads to clay mineral decomposition and loss of soluble Si from the soil profile. Asher [1991] stated that more than half of the elements in the Periodic Table are known to occur in plant tissues. He included so-called "beneficial" elements such as sodium, aluminum, silicon, nickel, cobalt, lanthanum and cerium.

Cobalt is required by bacteria for nitrogen fixation in root nodules on legumes. Nickel is necessary in the enzyme that breaks down urea to CO_2 and NH_4^+. Sodium, nickel, vanadium, fluorine, chromium, tin, iodine, cobalt, and selenium are required by higher animals. It may be that the amounts of some forms in Table 6-2 are necessary in concentrations too low to be detected by existing analytical methods and equipment. In recent years, on the other hand, many of these elements have been considered as waste hazards in the biosphere. With few exceptions, none of these elements have been adequately studied in greenhouse production. One cannot find any mention in the horticultural literature of the use of the elements listed in Table 6-2 in hydroponic systems as a regular practice. In this chapter, the necessity to deal with chemical terminology will be required if improved techniques are to be developed along with practical application. Considerable attention will be given to terminology and units employed along with speciation (composition) of substrate solutions, particularly concerning soilless culture and hydroponics. We are rapidly approaching the day when one can predict solution composition, depending upon water quality, plus the other factors that influence solution makeup. Also note that existing recommendations are empirically determined for the conditions of the local environment –especially with respect to soil type and weathering characteristics.

It is my purpose in this chapter to present a different approach to plant nutrition from that found in the present technical literature. Empirically obtained data are useful if the appropriate caveats are kept in mind. Note the extensive citing of information provided by W.L. Lindsay and his students obtained over several years. Use of this information requires terminology and units not often found in the applied literature. Nevertheless, adequate nutritional control in greenhouse production represents one of the greatest opportunities for significant advances in the future.

II. UNITS AND TERMINOLOGY

As noted above, recommendations for fertilizer application are subject to confusion as amounts may be expressed in different chemical forms at different times. Proprietary, complete analysis fertilizers are readily available and easy to use. Recommendations may suggest so many ppm, mg per unit area, or volume, of a substrate for a fertilizer. At the risk of redundancy, the analysis required on a bag provides percentages of N, P_2O_5, and K_2O. It need not mention the form in which the elements are present in the mixture. Some authors convert the values to elemental form as N, P, and K. For example, a 20-20-20 soluble fertilizer contains 20% N, 8.8% P, and 16.6% K when converted by using molecular weight factors listed in Table 6-3. Unfortunately, none of these elemental forms are available to plants, nor is there any attention to the requirement for electrical neutrality. The possibility of conversion to other forms in the soil or hydroponic solution is not often considered. Seldom is any attention paid to the presence of solids that may dissolve or precipitate the added nutrients in a way that makes them unavailable, or completely change the chemistry of the substrate. We will see that the chemical forms (stoichiometry) of the essential nutrients in solution can markedly alter growth. Many individuals continuously refer to ammonium nitrogen (NH_4^+) and nitrate nitrogen (NO_3^-). This, I think, is an unnecessary redundancy when one could simply say ammonium or nitrate and be done with it. One may also note that few authors, if any, pay attention to the forms in which any element such as those listed in Tables 6-1 and 6-2 are present in substrates and taken up by plants. Carbon in its various combinations (CO_3^{2-}, HCO_3^-, $H_2CO_3^o$), for example, will invariably be present in soils or hydroponic solutions despite the grower's wishes in the matter. For these reasons, and others, I have chosen to review the basic chemical methods for stating concentrations, activities, acidities-alkalinities, and oxidation-reductions, and to use this terminology through the remainder of the chapter.

Table 6-3. Factors for converting chemical forms. Multiplication of the amount in the left column (ppm, kg, g, etc.) by the value in the center will provide the equivalent of the material in the right column.

Multiply this material	By this number	To obtain this equivalent amount
Ca^{2+} (calcium)	1.399	CaO (calcium oxide, lime)
Ca^{2+}	2.496	$CaCO_3$ (calcium carbonate, calcite)
CaO (lime)	0.715	Ca^{2+}
CaO	1.784	$CaCO_3$
$CaCO_3$	0.400	Ca^{2+}
K^+ (potassium)	1.204	K_2O (potash)
K_2O (potash)	0.830	K^+
KCl (muriate of potash)	0.632	K_2O
K_2O	1.583	KCl (potassium chloride, muriate of potash)
K_2SO_4 (potassium sulfate)	0.541	K_2O
K_2O	1.849	K_2SO_4
Mg^{2+} (magnesium)	1.658	MgO (magnesium oxide, periclase)
MgO	0.603	Mg^{2+}
$MgSO_4$ (magnesium sulfate)	0.335	MgO
MgO	2.981	$MgSO_4$
MgO	2.092	$MgCO_3$ (calcium carbonate, magnesite)
N (nitrogen)	4.425	NO_3^- (nitrate)
N	1.286	NH_4^+ (ammonium)
N	6.067	$NaNO_3$ (sodium nitrate)
N	7.218	KNO_3 (potassium nitrate)
N	4.717	$(NH_4)_2SO_4$ (ammonium sulfate)
NO_3^-	0.226	N
NH_4^+	0.778	N
$(NH_4)_2SO_4$	0.212	N
$NaNO_3$	0.165	N
P (phosphorus)	2.288	P_2O_5 (phosphorus pentoxide)
P_2O_5	0.437	P
$H_2PO_4^-$ (orthophosphate)	0.320	P
P	3.131	$H_2PO_4^-$
S (sulfur)	2.994	SO_4^{2-}
SO_4^{2-}	0.334	S

A. CONCENTRATION

There are a number of ways to express the amount of substance in a solvent, based upon some fundamental principles of physics and chemistry [e.g., Sienko and Plane, 1957; Petrucci, 1977; Crockford and Knight, 1964]:

1. Molarity (M)

This is the amount of solute per **unit volume** of solvent (in our case water). It is the term that will be used

Table 6-4. Ionic species often found in soil or hydroponic solutions with formula weights. Values are deliberately rounded to the nearest whole number. These represent a selection from Lindsay [1979]. The list is not intended to be inclusive since actual presence depends upon several factors. Compounds followed by a superscript zero (°) are neutral, undissociated molecules. These may be combined with one or more water molecules.

Compound	Formula weight	Compound	Formula weight	Compound	Formula weight
Aluminum Al^{3+}	27	Hydroxide OH^-	17	Nitrogen NO_3^-	62
$Al(OH)_3^o$	78	Iron Fe^{2+}, Fe^{3+}	56	NH_4^+	18
$Al(OH)_2^+$	61	$Fe(OH)_2^+$	90	NO_2^-	46
$AlOH^{2+}$	44	$Fe(OH)_3^o$	107	Phosphorus	
				PO_4^{3-}	95
$Al_2(OH)_2^{4+}$	88	$FeOH^{2+}$	73	$H_3PO_4^o$	98
AlF_3^o	84	FeF^{2+}	75	HPO_4^{2-}	96
AlF^{2+}	46	$FeCl^{2+}$	92	$H_2PO_4^-$	97
AlF_2^+	65	$FeHPO_4^{2+}$	88	$H_2P_2O_7^{2-}$	176
$AlSO_4^+$	60	$FeOH^+$	73	Potassium K^+	39
$Al(SO_4)_2^-$	221	$FeSO_4^o$	153	KSO_4^-	136
Boron		$Fe(OH)_2^o$	90	Silicon	
$H_3BO_3^o$	62			$H_4SiO_4^o$	96
BO_3^{3-}	59	$FeH_2PO_4^+$	153	Sodium Na^+	23
Calcium Ca^{2+}	40	Magnesium Mg^{2+}	24	$NaSO_4^-$	119
$CaSO_4^o$	137	$MgSO_4^o$	121	Sulfur	
				SO_4^{2-}	96
$CaHPO_4^o$	136	$MgHPO_4^o$	120	Zinc Zn^{2+}	65
$CaPO_4^-$	135	$MgCO_3^o$	84	$ZnOH^+$	82
Copper Cu^+, Cu^{2+}	64	Manganese Mn^{2+}, Mn^{3+}, Mn^{4+}	55	$Zn(OH)_2^o$	99
$CuOH^+$	81	$MnOH^+$	72	$ZnSO_4^o$	161
$Cu(OH)_2^o$	98	$Mn_2(OH)_3^+$	161		
$CuSO_4^o$	161	Mn_2OH^{3+}	127		
Carbon		$MnOH^{2+}$	72		
CO_3^{2-}	60				
$H_2CO_3^o$	62	Molybdenum			
		MoO_4^{2-}	160		
HCO_3^-	61	$HMoO_4^-$	161		
Hydrogen H^+	1	$H_2MoO_4^o$	162		

in this text. A 1 M solution contains one gram-molecular weight of solute (expressed as grams) **per liter (ℓ)**[1] of solution. Formula weights of some more common substances with which the grower deals are given in Tables 6-4 and 6-5. Table 6-4 is a selection of those compounds and ionic forms that are commonly present in the substrate solution. Methods for dealing with these will be discussed later. Table 6-5 lists those high analysis chemicals used for fertilization purposes. Other sources will be noted later (Tables 6-27, 6-28). Since concentrations of macronutrients are in the millimole (mM) range, the formula weights are taken as milligrams (mg) rather than grams (1 mg = 0.001 g). Thus, 80 mg of ammonium nitrate (NH_4NO_3) dissolved in water made up to 1 ℓ will provide a concentration of 1 mM NH_4NO_3.

[1] The SI convention gives the spelling of "liter" as "litre" and "meter" as "metre". The American spelling will be used here.

Table 6-5. Chemical fertilizers often used in greenhouse production and in hydroponics[e] for supplying macronutrients.

Compound	Formula[a]	Analysis[b]	Formula weight[c]	Equivalent weight[c]	Effect on acidity	Solubility (g 100 ml^{-1})	Other materials[d]
Aluminum sulfate	$Al_2(SO_4)_3 \cdot 18H_2O$	0-0-0	667	222	Very acid	Soluble	14% S
Ammonium chloride	NH_4Cl	25-0-0	54	54	Acid	40	---
Ammonium nitrate	NH_4NO_3	34-0-0	80	80	Acid	118	---
Ammonium polyphosphate	$(NH_4)_3H_4P_2O_7$	15-62-0	232	---	Very acid	Soluble	---
Ammonium sulfate	$(NH_4)_2SO_4$	20-0-0	132	66	Very acid	71	24% S
Basic slag	$Ca_4P_2O_9$	0-21-0	366	---	Basic	0+	44% Ca
Calcium-ammonium-nitrate	Used in place of NH_4NO_3 which can be explosive. 15.5 % N or higher, variable.						
Calcium carbonate (Limestone, Calcite)	$CaCO_3$	0-0-0	100	50	Basic	0+[e]	40% Ca
Calcium cyanamide	$CaCN_2$	20-0-0	80	---	Basic	Decomposes	---
Calcium hydroxide (hydrated lime)	$Ca(OH)_2$	0-0-0	74	37	Basic	0+	60-80% Ca
Calcium metaphosphate	$Ca(PO_3)_2$	0-64-0	198	---	Basic	0+	20% Ca
Calcium monophosphate	$Ca(H_2PO_4)_2 \cdot H_2O$	0-56-0	252	126	Basic	Decomposes	16% Ca
Calcium nitrate	$Ca(NO_3)_2 \cdot 4H_2O$	15-0-0	236	118	Basic	102	17% Ca
Calcium sulfate (gypsum)	$CaSO_4 \cdot 2H_2O$	0-0-0	172	86	Neutral	0+	23% Ca 19% S
Dicalcium phosphate dihydrate (DCPD, brushite)	$CaHPO_4 \cdot 2H_2O$	0-41-0	172	---	Basic	0+	23% Ca
Diammonium phosphate (Diammonphos, DMP)	$(NH_4)_2HPO_4$	21-53-0	132	66	Acid	43	---
Dolomite (dolomitic limestone)	$CaMg(CO_3)_2$	0-0-0	184	---	Basic	0+	22% Ca 13% Mg
Magnesium-ammonium-phosphate	$MgNH_4PO_4 \cdot H_2O$	8-46-0	155	---	Acid	0+	13% Mg
Magnesium sulfate (epsom salts)	$MgSO_4 \cdot 7H_2O$	0-0-0	247	123	Neutral	71	10% Mg 13% S
Magnesium nitrate	$Mg(NO_3)_2 \cdot 6H_2O$	11-0-0	256	128	Neutral	42	10% Mg
Monoammonium phosphate (Ammophos A, MAP)	$NH_4H_2PO_4$	11-48-0	115	115	Acid	23	1.4% Ca 2.6% S
Nitric acid	HNO_3	22-0-0	63	63	Very acid	Soluble	---
Phosphoric acid	H_3PO_4	0-52-0	98	33	Very acid	548	---
Potassium carbonate	K_2CO_3	0-0-66	138	69	Basic	112	---

Compound	Formula[a]	Analysis[b]	Formula weight[c]	Equivalent weight[c]	Effect on acidity	Solubility (g 100 ml^{-1})	Other materials[d]
Potassium chloride (muriate of potash)	KCl	0-0-62	75	75	Neutral	35	47% Cl
Potassium diphosphate	K_2HPO_4	0-41-54	174	87	Basic	167	---
Potassium metaphosphate	KPO_3	0-57-39	118	---	Acid	0+	---
Potassium monophosphate	KH_2PO_4	8-53-34	120	120	Basic	33	---
Potassium nitrate (saltpeter)	KNO_3	13-0-44	101	101	Basic	13	---
Potassium sulfate	K_2SO_4	0-0-53	174	87	Neutral	7	18% S
Pyrophosphoric acid[f]	$H_4P_2O_7$	0-80-0	178	45	Acid	709	---
Sodium nitrate	$NaNO_3$	16-0-0	85	85	Basic	73	27% Na
Superphosphate (DCP)	$CaH_4(PO_4)_2$	0-20-0	234	78	Neutral	2	18% Ca 12% S
Superphosphoric acid	H_3PO_4+$H_4P_2O_7$	0-76-0	Will solidify at low temperatures.				
Sulfur	S	0-0-0	32	---	Acid	0+	---
Treble superphosphate[g]	$CaH_4(PO_4)_2$	0-42-0	234	78	Neutral	2	12% Ca
Urea	$CO(NH_2)_2$	45-0-0	60	30	Acid	78	---

[a] Formulae may differ due to manufacturing process, addition or removal of water, etc.
[b] Percentage N, P_2O_5 , and K_2O.
[c] Formula and equivalent weights rounded to nearest whole number.
[d] These will vary with source, manufacturing process, etc.
[e] For practical use in hydroponics, solubility should exceed 10 g (100 ml)$^{-1}$. Solubility at 0 C, with one or two exceptions.
[f] Continued removal of water can increase P_2O_5 in excess of 85% and can be combined with ammonium and potassium.
[g] Contains no gypsum.

2. Molality (m)

This is the amount of solute per unit mass of solvent. A 1 m solution contains one gram-molecular weight of solute in **1 kilogram** (kg) of water. In both molarity and molality, a 1 M or 1 m solution of a substance that does not dissociate when dissolved will provide 6.02 x 10^{23} molecules or particles. Such a substance could be sugar (sucrose) which has a formula weight of 342. If 342 mg of sucrose are dissolved and made up to 1 ℓ, the concentration is 1 mM and contains 6.02 x 10^{20} particles. The same applies to a 1 mM solution in 1 kg of water.

Molality is unaffected by changes in temperature, whereas a rise in temperature will cause an increase in volume, changing molarity. Molarity (moles of solute per ℓ of solution) is a useful, very common unit but is not recommended by the international unit convention. Molality (moles of substance per kilogram of solvent) is a legitimate SI (Le Système International d'Unités) unit. The analogous SI concentration unit is mole per cubic meter (M m^{-3}). One M m^{-3} = 1 M per 1000 ℓ = 1 mM ℓ^{-1}. The multiplicity of units makes it difficult to compare results of various authorities. In this chapter, the author employs mole (M ℓ^{-1}), meaning moles per liter, commonly expressed as a logarithm to base ten.

Because the temperature range in greenhouses is narrow, and the concentrations of substances in the soil solution are invariably in the mM range or lower, there is not much to choose between molality and molarity. The number of particles in solution directly bears on the colligative properties of solution as briefly discussed in Chapter 5. Thus, as the particle number in solution increases, the freezing point will be lowered, boiling point elevated, osmotic pressure increased, and vapor pressure above the solution decreased. However, a 5 mM solution of sucrose will have the same colligative properties as a 5 mM solution of formaldehyde, fructose, glycerol, or naphthalene. They all have the same number of molecules in solution.

However, one notes that most of the chemical forms listed in Tables 6-1 through 6-4 are ionic species with varying electrical charge. For example, potassium nitrate, a very common fertilizer used in greenhouse production is highly soluble and dissociates nearly completely in solution. That is, the dissociation reaction goes to the right completely in very dilute solutions:

$$KNO_3 \rightleftharpoons K^+ + NO_3^- \tag{6.1}$$

If one wishes a common recommendation of 6 mM KNO_3 to be delivered in the irrigation supply, then (101)(6) = 606 mg KNO_3 are to be dissolved in each ℓ. Assuming complete dissociation, for each molecule there will be two ionic particles, or double the number of particles, and each species will have a concentration of 6 mM. In very dilute solutions, the effect on colligative properties will be to double, for example, the freezing point depression.

Another common chemical is calcium nitrate, which dissociates:

$$Ca(NO_3)_2{\cdot}4H_2O \rightleftharpoons Ca^{2+} + 2NO_3^- + 4H_2O \tag{6.2}$$

There are now three particles for each molecule of $Ca(NO_3)_2{\cdot}4H_2O$, one a positive calcium ion and two negative nitrate ions. The attached water molecules do not significantly change the concentration of water in 1 ℓ which is about 55.5 M. If one needs a 6 mM concentration of calcium nitrate, then (236)(6) = 1416 mg ℓ^{-1} will be required. But, there will be 6 mM of Ca^{2+} and (2)(6) = 12 mM of NO_3^- in solution, assuming complete dissociation of the 1416 mg calcium nitrate. The effects of electrical charges, concentration, and possible hydrolysis of the ions (attached water molecules), etc. will be discussed shortly.

Often, fertilizer recommendations are given in parts per million (ppm), which are equivalent to milligrams per ℓ or micrograms per gram (µg g^{-1}). The moles per ℓ of the substance can be calculated by dividing the mg dissolved in 1 ℓ by the formula weight. Unfortunately, one will find in the literature recommendations to apply 200 mg ℓ^{-1} nitrogen (N) without any specification as to a chemical source. As elemental N does not exist, one must convert the recommendation to its equivalent in a physically possible compound. This can be done, using the conversion factors in Table 6-3. This means that if 200 ppm N is from a nitrate source, then (4.425)(200) =

885 mg NO_3^-, or 885/62 = 14.3 mM NO_3^- (Table 6-3). If the grower uses ammonium nitrate, then (14.3)(80) = 1144 mg ℓ^{-1} NH_4NO_3 (Table 6-5). The grower will also add 14.3 mM NH_4^+. If he uses calcium nitrate, 7.2 mM of Ca^{2+} will be added. If the recommendation is for an ammonium source, then (200)(1.286) = 257 mg NH_4^+ or 257/18 = 14.3 mM NH_4^+ and 1144 mg of NH_4NO_3 will also provide 14.3 mM NO_3^-. These results must be adjusted downward to avoid excessive fertilization. If the grower decides to use ammonium sulfate ($(NH_4)_2SO_4$), he will need (14.3)(132/2) = 944 mg ℓ^{-1}, and 7.2 mM of SO_4^{2-} will be provided besides ammonium. Note the two ammonium ions per molecule when dissociated, requires division of the formula weight by two. Otherwise, when the salt dissolves, each molecule will provide two ammonium ions, and a 14.3 mM solution of $(NH_4)_2SO_4$ would have 28.6 mM NH_4^+.

One fact is immediately apparent. Fertilization with any element, N, K, or P, will require an appropriate compound, and a complementary ion of opposite charge will also be added (i.e., $KCl \rightarrow K^+ + Cl^-$, $NH_4H_2PO_4 \rightarrow NH_4^+ + H_2PO_4^-$, etc.). Most existing, practical publications seldom mention these physical realities in fertilizer recommendations. Even in recent scientific literature, the reader will find the use of ppm or mg ℓ^{-1} of the element (e.g., N) in solution. Hopefully, if the author shows the ionic form in which the element is supplied, the actual concentration of the ion can be calculated. For example, the concentration of nitrate to which plants may be exposed in the root substrate can be lower than 0.02 mg ℓ^{-1} nitrogen. If it is assumed that the nitrate form is the sole source, this may be converted to molarity units: (0.02 x 4.425)/62 = 0.00143 mM = 1.43 μM = 0.00000143 M = $10^{-5.84}$ M. In some publications, the requirement for the reader to convert to a common unit for comparison purposes is laborious.

B. ACTIVITY

As noted, many chemicals dissolved in a substrate solution are most often in ionic form with an electrical charge. Salts may exist in an ionized form, but behave dependently with each other. Each positive ion will be surrounded by a cluster of negative anions, and each negative ion by a cluster in which positive cations predominate. In an electrical field, the mobility of an ion will be reduced as well as reducing the magnitude of the colligative properties of the solution. Typically, single ions such as aluminum (Al^{3+}), iron (Fe^{3+}, Fe^{2+}), manganese (Mn^{2+}), etc. may be hydrolyzed as suggested in Fig. 6-1. The number of water molecules attached to the ion will vary with the hydrogen ion activity (pH). The effect is again to reduce ion mobility. The water molecules are commonly omitted in reaction formulae for greater simplicity. Thus, each ion type in solution has a total concentration based upon the amount of solute dissolved, and an "effective" concentration called the "activity" that takes into account interionic attractions. If activity is used, solution properties can usually be predicted. The value is expressed in the same terms as concentration (moles/liter), but the ratio between activity and concentration (activity coefficient, γ) is always less than 1. An activity coefficient can be calculated from a knowledge of the ionic strength (μ), valencies (number of electrical charges on the ion), and effective size of the ion. Only in very dilute, or "ideal," solutions does γ approach 1 where activity equals concentration. The activity coefficient can be below 0.002 for ions having concentrations near 0.1 M.

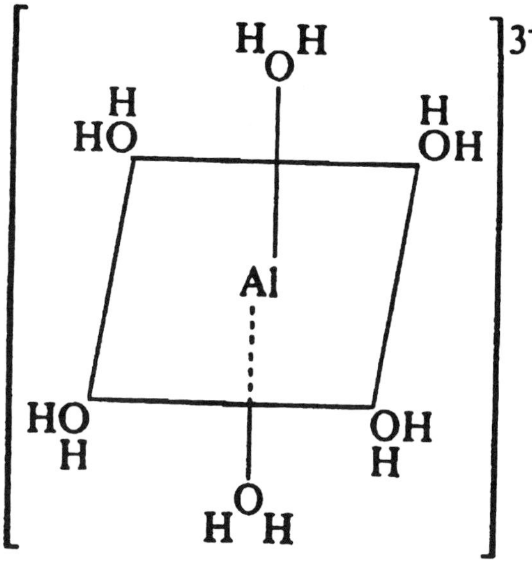

Fig. 6-1. The aluminum hexahydronium ion ($Al(H_2O)_6^{3+}$). As pH decreases, hydrogen protons are removed. Below pH 5.0, free Al^{3+} is predominant in solution [*Chemical Equilibria in Soils*, Lindsay, W.L., ©1979. Reprinted by permission from John Wiley & Sons, Inc.].

As the result of the factors outlined above, most chemical equilibria are expressed as "activity" rather than concentration. For example, the equilibrium reaction of calcite ($CaCO_3$), a common mineral in alkaline soils is:

$$CaCO_3 + 2H^+ \leftrightharpoons Ca^{2+} + CO_2(g) + H_2O \tag{6.3}$$

And the equilibrium constant, K°, of this reaction at 25 C is:

$$\frac{(Ca^{2+})(CO_2(g))(H_2O)}{(CaCO_3)(H^+)^2} = 10^{9.74} \tag{6.4}$$

The parentheses, ( ), denote "activity." As activities of $CaCO_3$ and H_2O are considered fixed, they can be set equal to 1. If the activities of the products divided by the reactants are given in logarithms to the base 10, the activity of calcium becomes:

$$\begin{aligned} -\log_{10}(Ca^{2+}) &= 9.74 - 2pH - \log_{10}(CO_2(g)) \quad or \\ pCa^{2+} &= 9.74 - 2pH - pCO_2 \end{aligned} \tag{6.5}$$

Lowercase "p" denotes that the value is a negative logarithm, and the concentration of CO_2 is given in atmospheres. The important thing to note is that the activity of calcium in solution depends on both pH and CO_2 concentration, where the dissolving mineral is calcite.

While a grower is unlikely ever to use the complex equations for calculating activities and chemical equilibria, these factors are utilized in computer programs for determining soil solution, thermodynamic equilibria based upon soil type, acidity, and reduction-oxidation potentials. In hydroponic systems, such programs open the way to control solution concentrations automatically as illustrated later in this chapter.

C. EQUIVALENCY

Equivalency in chemical terms refers to the equality of combining capacity or equal valence –the latter referring to the number of positive or negative charges on an ion. A potassium ion (K^+) has a valency of 1, whereas a magnesium ion (Mg^{2+}) has a valency of 2, and aluminum's valency (Al^{3+}) is 3. For compounds to have equal combining capacity, the formula weight is divided by the valency to obtain equivalent weight, often called the gram-equivalent weight (Table 6-5). One gram-atomic weight of hydrogen, or any other ion that will combine with or displace this amount of hydrogen, is another definition of equivalent in chemical terms. For monovalent ions such as Na^+, K^+, NH_4^+ and Cl^-, the equivalent weight and atomic weight are the same since they can react with 1 H^+ ion. Divalent cations such as Ca^{2+} or Mg^{2+} can replace two H^+ ions.

If 80 g ammonium nitrate (NH_4NO_3) are dissolved in 1 ℓ, there will be 1 equivalent per ℓ each of NH_4^+ and NO_3^- since the ions in solution each have 1 charge. On the other hand, the formula weight of calcium nitrate ($Ca(NO_3)_2{\cdot}4H_2O$) must be divided by two to give an equivalent weight of 118 g ℓ^{-1}. The Ca^{2+} ion has two charges and there are two nitrate ions, each with a valency of 1. One equivalent weight of nitric acid (HNO_3) will react with 1 equivalent weight of calcium carbonate ($CaCO_3$). As an equivalent is much too large, milliequivalents are employed as with millimoles for concentration. Six meq ℓ^{-1} KNO_3 requires (6)(101) or 606 mg in 1 ℓ, and will provide 6 meq ℓ^{-1} K^+ and 6 meq ℓ^{-1} NO_3^-. This is the same amount as required for 6 mM calculated previously. However, the equivalent weight of magnesium sulfate ($MgSO_4{\cdot}7H_2O$) is 247/2 or 123 mg (Table 6-5). A 1 mM solution would require 247 mg in 1 ℓ. A 1 mM solution of magnesium sulfate, would contain 2 meq ℓ^{-1} Mg^{2+} and 2 meq ℓ^{-1} SO_4^{2-}. Therefore, a 1 mM solution of magnesium sulfate would not be "equivalent" to a 1 mM solution of potassium nitrate (KNO_3) as to their chemical combining capacity.

The advantage to employing milliequivalents (meq) is the ability to consider the essential elements in the forms required by plants and to balance the solution electrically. Steiner, in the early 1960s, employed these units to derive physically possible solutions [Steiner, 1961; 1966; 1968]. In the same decade, students of W.D. Holley at Colorado State University used this system in detailed studies of carnation and rose nutrition [Green et al., 1971; 1973; Hartman, 1971; Hughes, 1975; Hughes and Hanan, 1977; etc.]. The results were later summarized by Hanan in a series of papers published in the 1980s [Hanan, 1982a; b; 1984; 1988]. While other authors have also employed equivalents, the system has not achieved widespread use in the industry, although manipulation of raw water supplies is considerably simplified.

D. ACIDITY

Water has the properties of both an acid and a base. Water molecules can donate protons (H^+) in a reaction with a base such as ammonia and accept protons in a reaction with acids such as hydrochloric or acetic acids. There are a few ions even in pure water, producing the equilibrium:

$$H_2O + H_2O \rightleftharpoons H_3O^+ + OH^- \tag{6.6}$$

where H_3O^+ is called the hydronium ion. For simplicity, it is usually written as H^+. The equilibrium constant of this reaction is:

$$(H^+)(OH^-) = 10^{-14} \tag{6.7}$$

The activity of water in Eq. 6.7 is 1. In pure water, the H^+ activity must equal the OH^- activity so that each must have a value of 10^{-7} moles/liter to satisfy the equilibrium constant 10^{-14}. If either H^+ or OH^- activity is known, the other can be determined using Eq. 6.7. Whether a solution is acidic or basic depends upon which ion is in greater concentration. The addition of an acidic compound will result in a greater concentration of hydrogen ions compared to hydroxide ions. Conversely, the relative proportions will favor OH^- if, for example, sodium hydroxide (NaOH) is added to the solution. The acidity or basicity of a solution is described by giving the hydrogen ion activity that may be expressed in moles/liter or pH where:

Table 6-6. Relationship between hydrogen ion concentration, pH, and pOH.

H^+ concentration (moles/liter)		**pH**	**pOH**	
1	1	0	14	Acid solutions
0.1	1×10^{-1}	1	13	
0.01	1×10^{-2}	2	12	
0.001	1×10^{-3}	3	11	
0.0001	1×10^{-4}	4	10	
0.00001	1×10^{-5}	5	9	
0.000001	1×10^{-6}	6	8	
0.0000001	1×10^{-7}	7	7	Neutral point
0.00000001	1×10^{-8}	8	6	
0.000000001	1×10^{-9}	9	5	
0.0000000001	1×10^{-10}	10	4	
0.000000000001	1×10^{-12}	12	2	
0.00000000000001	1×10^{-14}	14	0	Basic solution

$$pH = -\log(H^+) = \log\frac{1}{(H^+)} \tag{6.8}$$

In practice, the activity and concentration of the H^+ ion are not usually very different, and the pH as calculated from either will be almost identical. The relationship between concentration, pH, and pOH is presented in Table 6-6, which shows that as pH decreases by 1 unit, hydrogen ion concentration increases tenfold. At pH = 0, (H^+) = 1 mole ℓ^{-1}.

Two examples of the effect of pH may be examined for the present. In Fig. 6-2, iron equilibria calculated by Lindsay [1974] shows that the major cations in solution, Fe^{3+} and Fe^{2+}, can be reduced to levels approaching 1 ion per ℓ at pHs greater than 7 due to precipitation as iron oxide. Below activities of 10^{-6} moles (0.000001 M or 1 μM) to 10^{-8} Fe, there will be insufficient soluble iron in solution to meet plant requirements. Another effect of pH on the major orthophosphate ions in solution under ideal conditions is provided in Fig. 6-3. At pH 7.2, the ions $H_2PO_4^-$ and HPO_4^{2-} are equal in activity. As we shall see, pH has a major influence on ion availability of essential nutrients with which a grower must deal for successful plant production. A marked change in pH suggests a radical modification in soil environment. However, as stated by Moore [1974], plants will generally do well over a wide pH range (5-7). Allaway [1957] gives a pH range of 4 to 9. Note, however, that Moore published his review nearly 20 years after Allaway. The effects of pH are often indirect, affecting nutrient availability. It is this relationship between pH and nutrient availability that justifies frequent pH measurements and makes tables of pH preferences of plants often useful [Allaway, 1957].

E. OXIDATION-REDUCTION

In redox reactions, electrons (e^-) are exchanged. Oxidation of a substance occurs when electrons are lost, such as in rusting of metals, the release of electrons from the zinc anode in a flashlight battery, or the lead anodes in a car battery. Simultaneously, there must be reduction, or gain of electrons by a substance, such as occurs at the negative cathodes in batteries. Examples of redox reactions common to soils are:

$$Fe^{3+} + e^- \rightleftharpoons Fe^{2+} \quad (6.9)$$

$$Fe_3O_4(magnetite) + 8H^+ + 2e^- \rightleftharpoons 3Fe^{2+} + 4H_2O \quad (6.10)$$

$$Mn^{4+} + e^- \rightleftharpoons Mn^{3+} \quad (6.11)$$

$$MnO_4^{2-} + 8H^+ + 4e^- \rightleftharpoons Mn^{2+} + 4H_2O \quad (6.12)$$

$$Cu^{2+} + e^- \rightleftharpoons Cu^+ \quad (6.13)$$

$$NO_3^- + 2H^+ + 2e^- \rightleftharpoons NO_2^- + H_2O \quad (6.14)$$

The oxidation-reduction potential, or redox potential, is a measure of the electron availability in a chemical or biological system. Respiration by root systems and microorganisms is an oxidative process. There will be an abundance of electrons, especially close to the root surface. This is coupled with a corresponding decrease in oxygen concentration and an increase in CO_2 level as respiration goes on, and may be exacerbated by a high moisture content (see Section 5.III).

The term "potential" is commonly used since the method employed is to measure voltage differences between a platinum electrode inserted into the soil and a suitable reference electrode such as a calomel or silver-silver chloride cell. Potentials (Eh) greater than 400 mV are characteristic of well-drained substrates, whereas in flooded soils with a high carbon content, Eh can be below -180 mV –or the environment is highly reduced. The use of this measurement in explaining greenhouse conditions of the substrate has seldom been employed. There have not been any publications on the subject as applied to greenhouse culture, to my knowledge, in the last 30 years. More recently, in the agronomic ℓ ature, efforts by Lindsay, his students, and contemporaries [e.g., Lindsay, 1979; 1983; Lindsay and Schwab, 1982; Lindsay et al., 1989; Schwab and Lindsay, 1983a; b; Sajwan and Lindsay, 1986; Boyle and Lindsay, 1986; Norvell, 1991, etc.) have been made to use redox data fully. The influence of redox on trace element availability is particularly important. Rather than using Eh, a voltage, these investigators preferred using the negative logarithm of electron activity (*pe*), which allows one to use M ℓ^{-1} in calculating equilibria. Potential measurements can be converted to *pe* by means of the formula:

$$pe = \frac{Eh(mV)}{59.2} \quad (6.15)$$

To fully designate the redox status, pH and *pe* are combined to provide a value for *pe* + pH. The limit on the reduced side (flooded soils and high carbon) is determined by the equilibrium of hydrogen ions with electrons to form hydrogen gas. Where the H^+ activity is 1 M ℓ^{-1} and the partial pressure of hydrogen gas is 1 atmosphere, *pe* + pH = 0. If the pH is 6.0, then *pe* = -6. On the opposite side, in a fully oxidized condition at 1 atmosphere partial pressure for oxygen, *pe* + pH = 20.78. Again, if the pH is 6.0, *pe* = 14.78. Most soils lie in the *pe* + pH range of approximately 4 to 17.

It would be expected that most well-drained greenhouse soils and NFT systems will be highly oxidized, or at least pe + pH would exceed 10. Lindsay [1991] suggested a range for field conditions of 12 to 16 for well-aerated soils. The platinum electrode, unfortunately, is highly unstable in oxidized soils. A redox measurement of many greenhouse substrates is likely to be unreliable. As will be noted in Section 6.IV.A, however, the region close to an active root system also has the highest microbial activity compared with the bulk soil or solution. The uptake of oxygen and its consequent lower concentration as it diffuses through this highly active region, combined with increased CO_2 levels, are likely to significantly lower pe + pH. Examples of the combined effect of pH and redox on solubility of phosphorus and iron in solution are presented in Fig. 6-4. It will be noted that greatest availability of phosphorus and iron occurs under reduced conditions at low pHs. According to Lindsay [1991], if soil-Fe alone controlled iron solubility, pe + pH would have to be less than 10 at pH 7.0, or 8.0 at pH 8.0 in order for Fe^{2+} activity to exceed 10^{-8} M required by plants [Schwab and Lindsay, 1989]. As will be noted later, some plant species can reduce redox levels in solution below pe + pH 7. They are "iron-efficient" plants. Provision of iron, particularly, is a serious problem in greenhouse production where pH values are likely to range from 5.5 to 7.5 and conditions favor an oxidized environment.

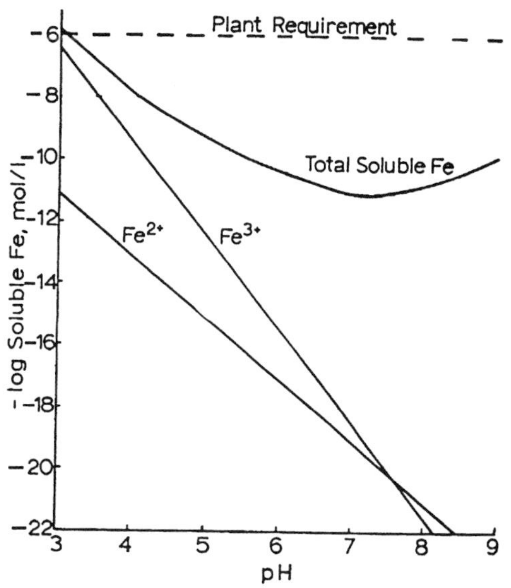

Fig. 6-2. Concentration of Fe^{3+}, Fe^{2+}, and total soluble iron in a solution in equilibrium with iron oxide and 0.2 atmospheres O_2 [Lindsay, 1974].

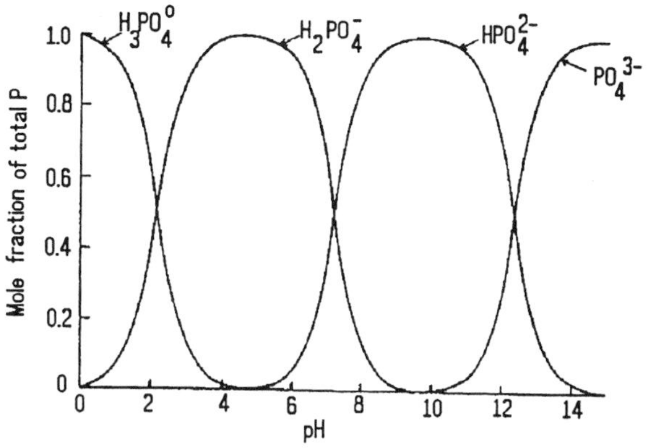

Fig. 6-3. The effect of pH on the distribution of major orthophosphate ions in solution. At pH 7.2, 50% of phosphate ions are $H_2PO_4^-$ and 50% are HPO_4^{2-} [*Chemical Equilibria in Soils*, Lindsay, ©1979. Reprinted by permission of John Wiley & Sons, Inc.].

Usually, the oxidation states of an element have an important bearing on its mobility and availability. Manganese, selenium, mercury, arsenic, chromium, copper, and iron are sensitive to redox potential in soils [Adriano et al.,1987]. As most greenhouse production is still in field soils, better distribution of agronomic information for practical application in the industry would be worthwhile. To the best of my knowledge, examination of the immediate root environment regarding redox conditions in greenhouse substrates, or hydroponic systems, has never been carried out. Despite the measurement difficulty, this is a serious lack of basic information that could prevent full automation of greenhouse production.

F. ADSORPTION-DESORPTION IN SOILS

1. Intensity versus Capacity

There are two primary factors in nutrition that determine nutrient supply: 1) intensity and 2) capacity. The first refers to the concentration of nutrients in the substrate solution. The second is the ability of the substrate to replenish a particular nutrient as it is removed from the solution. Cation exchange capacity (CEC) is one measure of the capacity factor or the ability to hold exchangeable cations. The effect of processes of weathering on

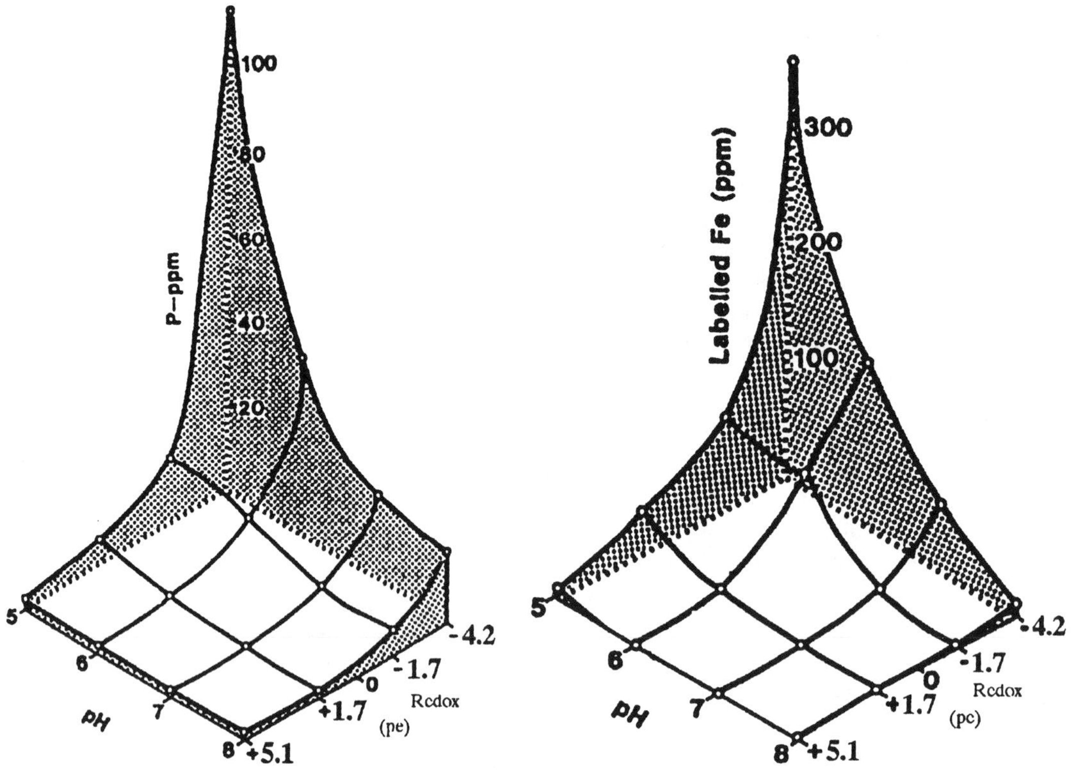

Fig. 6-4. The solubility of phosphorous (left) and iron (right) as influenced by pH and *pe*, where pH = -log(H^+) and *pe* = -log(e^-). Adapted from Reddy and Patrick [1983] (With permission of the *Amer. Soc. of Agronomy)*.

intensity and capacity factors were diagrammed by Lindsay in Fig. 6-5 for three hypothetical soil minerals, **A**, **B**, and **C**. For most potting substrates in greenhouse culture, the duration of the crop is generally short with rapid growth. The capacity of most substrates to replenish the substrate solution is similar to mineral **C**, or less. Certainly in hydroponics, the only storage factor is bulk solution. According to Moore [1974], the potential quantity of any ionic species is fixed, whereas in a soil system, additional ions can be brought into solution. Ions can be "desorbed" from the soil matrix. Fertilization causes "adsorption" of the ions back onto the soil matrix. The capacity ranges from the hydroponic extreme to production in highly organic field soils.

In Fig. 6-5, the secondary mineral **A** dissolves most rapidly and maintains solution concentration at a high level, well above the critical level required by plants growing in it. Minerals **B** and **C**, being supersaturated, will slowly precipitate until **A** disappears, and the solution concentration then drops to level **b**. **B** dissolves, maintaining the concentration at level **b**, still well above the critical plant level, and mineral **C** continues to precipitate. When, however, **B** disappears, **C** remains as the final weathering product and the soil solution drops well below the critical level. Plants growing in this "poor" soil grow either slowly or not at all. Fertilization will be required.

2. The Soil System

To appreciate the importance of the soil's capacity to store and release essential nutrients, we need some background on the colloidal soil system. The makeup of clay and humus in a soil, as influenced by primary minerals and climate, plus prior cultivation and fertilization practices, will determine the cultural procedures available to the grower and set the basic fertility level of the soil. The variation in soils from one geographical region to another is seldom, if ever, discussed in applied greenhouse practice. Warrick [1981] emphasized the problem of soil variability, with more than 10000 soil series defined in the U.S. alone. The recommendations

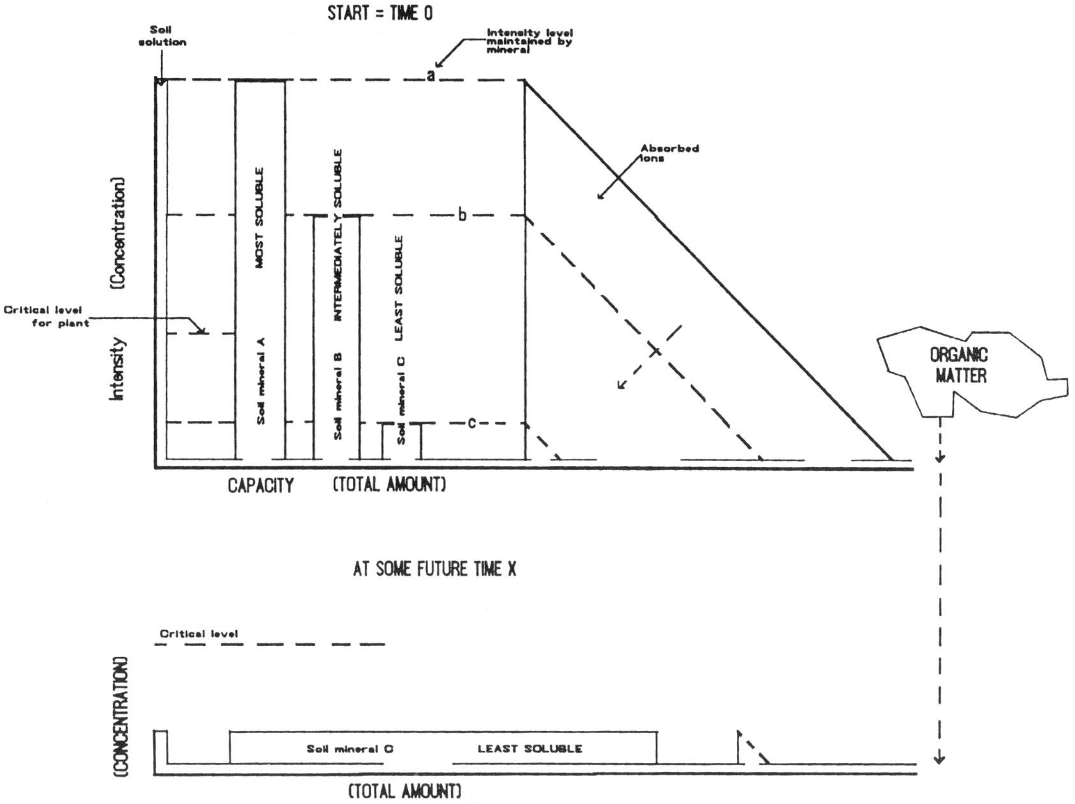

Fig. 6-5. The influence of soil minerals **A**, **B**, and **C** on solubility and availability of nutrients in a substrate. [*Chemical Equilibria in Soils*, Lindsay, W.L., ©1979. Reprinted by permission of John Wiley & Sons, Inc.].

from research stations deal largely with local conditions of soil genesis, type, and climate. Unless the grower has had direct experience in different climatic regions, he is usually unaware of what might happen if, for example, he applies a recommendation obtained from experiments in a temperate humid region to an arid or tropical soil. The results can be startling and often expensive.

The most active portions of the soil are those in colloidal state with two distinct types –clay minerals of various kinds and decomposed organic matter or humus. Clays, composed of secondary minerals and amorphous materials, differ from the components of sand and silt. The clays are products of weathering and are not found in unaltered rocks [Dean, 1957]. Two groups of clays are recognized: silicate clays characteristic of temperate regions and iron and aluminum hydrous oxide clays found in the semitropics and tropics. The hydrous oxides are mostly compounds of iron and aluminum. They play a role in phosphate fixation, influencing phosphorous availability to plants.

All clay particles, because of their small size, expose a large surface area, and in some clays, a large internal surface may be present (Fig. 6-6). Comments on this surface area were made in Chapter 5. The clay micelles usually carry negative charges. Consequently, thousands of positively charged cations are attracted to each colloid. Water molecules, carried by the adsorbed cations, are also present. For humid region clays, the ions in order of their numbers are H^+, Al^{3+}, and Ca^{2+}, followed by Mg^{2+}, and finally K^+ and Na^+. For well-drained, arid and semiarid soils, the order is usually Ca^{2+} and Mg^{2+}, Na^+ and K^+, and H^+ last. The vertical dashed line in Fig. 6-6 arbitrarily distinguishes the inner adsorbed ionic layer versus the outer ions mostly free in solution.

When drainage is reduced in arid regions, alkaline salts accumulate and adsorbed sodium ions are likely to be dominant, whereas in humid regions an aluminum-hydrogen clay results. Most silicate clays are

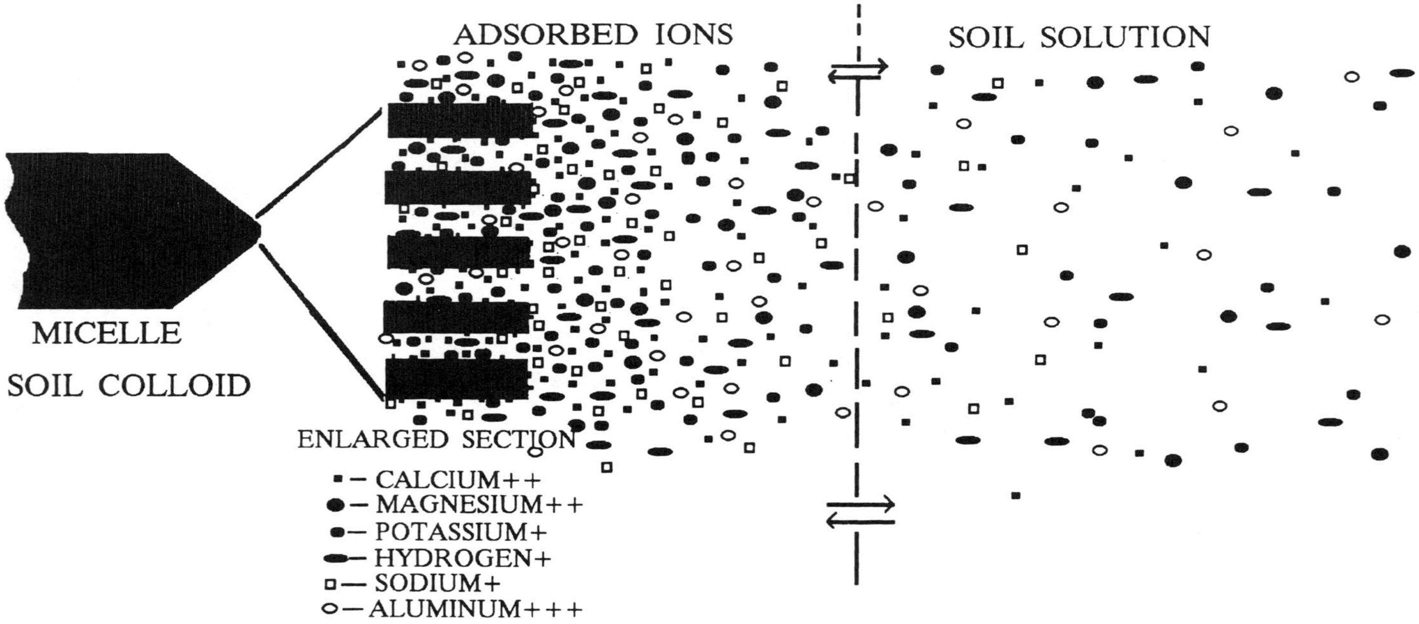

Fig. 6-6. Diagrammatic representation of the positive cation swarm about a negative clay micelle. An equilibrium between those ions adsorbed on the colloid and the ions in solution exists unless: 1) the plant removes some from solution (i.e., Ca^{2+}), and additional ions are released from the micelle to reestablish equilibrium, with perhaps H^+ ions being adsorbed in the place of the metal cation, or 2) the farmer fertilizes with a chemical such as KCl. In that case, K^+ ions can be adsorbed on the micelle.

Table 6-7. Some properties of three types of silicate clays [Brady, 1974].

Property	Clay type		
	Montmorillonite	**Illite**	**Kaolinite**
Size (μm)	0.01-1.0	0.1-2.0	0.1-5.0
Specific surface (m^2/g)	700-800	100-120	5-20
External surface	High	Medium	Low
Internal surface	Very high	Medium	None
Cohesion, plasticity	High	Medium	Low
Swelling capacity	High	Medium	Low
Cation exchange capacity (meq/100 g)	80-100	15-40	3-15

Table 6-8. Representative selection of soil minerals with approximate "type" formulae. This list is not meant to be complete [Adapted from Lindsay, 1979].

Mineral	Approx. formula	Mineral	Approx. formula
Albanite	MnS	Albite	$NaAlSi_3O_8$
Aluminum hydroxide (amorphous)	$Al(OH)_3$	Brucite	$Mg(OH)_2$
Brushite (DCPD)	$CaHPO_4 \cdot 2H_2O$	Calcite	$CaCO_3$
Chlorite	$Mg_5Al_2Si_3O_{10}(OH)_8$	Cupric ferrite	α-$CuFe_2O_4$
Cuprous ferrite	α-$Cu_2Fe_2O_4$	Dolomite	$CaMg(CO_3)_2$
Fayalite	Fe_2SiO_4	Franklinite	$ZnFe_2O_4$
Fluorapatite	$Ca_5(PO_4)_3F$	Fluorite	CaF_2
Fluorphlogopite	$KMg_3AlSi_3O_{10}F_2$	Gibbsite	$Al(OH)_3$
Goethite	α-$FeOOH$	Gypsum	$CaSO_4 \cdot 2H_2O$
Hemitite	$\frac{1}{2}\alpha$-Fe_2O_3	Hydroxyapatite(HA)	$Ca_5(PO_4)_3OH$
Illite	$K_{.6}Mg_{.3}Al_{2.3}Si_{3.5}O_{10}(OH)_2$		
Jarosite	$KFe_3(SO_4)_2(OH)_6$	Kaolinite	$Al_2Si_2O_5(OH)_4$
K-taranakite	$H_6K_3Al_5(PO_4)_8 \cdot 18H_2O$	Maghemite	$\frac{1}{2}\Delta$-Fe_2O_3
Magnesite	$MgCO_3$	Magnetite	Fe_3O_4
Malachite	$Cu_2(OH)_2CO_3$	Manganite	γ-$MnOOH$
Manganosite	MnO	Microcline	$KAlSi_3O_8$
Mg-montmorillonite	$Mg_{.2}(Si_{3.8}Al_{1.7}Fe(III)_{.2}Mg_{.3})O_{10}(OH)_2$		
Muscovite	$KAl_2(AlSi_3O_{10})(OH)_2$	Octacalcium phosphate (OCP)	$Ca_4H(PO_4)_3 \cdot 2.5H_2O$
Otavite	$CdCO_3$	Periclase	MgO
Pyrite	FeS_2	Pyrolusite	β-MnO_2
Pyrophyllite	$Al_2Si_4O_{10}(OH)_2$	Pyroxene	$CaAl_2SiO_6$
Quartz	SiO_2	Rhodochrosite	$MnCO_3$
Siderite	$FeCO_3$	Smithsonite	$ZnCO_3$
Strengite	$FePO_4 \cdot 2H_2O$	Talc	$Mg_3Si_4O_{10}(OH)_2$
Vermiculite	$(Mg_{2.7}Fe(II)Fe(III)_{.5}Ca_{.1}K_{.1})Si_{2.9}Al_{1.1}O_{10}(OH)_2$		
Vivianite	$Fe_3(PO_4)_2 \cdot 8H_2O$	Wairakite	$CaAl_2Si_4O_{12} \cdot 2H_2O$

aluminosilicates. Of the clay groups in humid regions, kaolinite, montmorillonite, and illite are the most common minerals, with some of their properties given in Table 6-7. Kaolinite represents a more advanced stage of weathering under acid conditions. Vermiculite will also be found with a cation adsorption capacity exceeding others. Other various clays can be found in diverse mixtures. Some clay minerals show positive as well as negative charges, which makes possible exchange between the colloids and anions such as phosphate, sulfate, chloride, and nitrate. Negatively charged ions, however, are often considered "mobile" in soils with the effect that anions are more easily leached. Hydrous oxide clays are often intermixed with silicate clays in temperate regions, and they are commonly dominant in tropic regions as the most advanced stages of weathering with typical red and yellow soils. In the latter, gibbsite, geothite, and limonite are prevalent. A list of some more important minerals often found in the solid state in soils is given in Table 6-8.

The previous paragraphs dealt with inorganic minerals found in soils. The diversity and variation that can be found as typified in Tables 6-7 and 6-8 merely illustrate the fact that a grower should thoroughly investigate this aspect before beginning any particular cultural procedure. Local agricultural research stations are usually the primary source of information. An example of the differences between humid and semiarid region soils is given by Brady [1974] in Table 6-9.

Table 6-9. Data for representative mineral soils, comparing humid and semiarid temperate regions, as given by Brady [1974].

Characteristics	Humid region	Semiarid region
Exchangeable calcium (meq/100 g)	6-9	13-16
Other exchangeable bases (meq/100 g)	2-3	6-8
Exchangeable hydrogen and/or aluminum (meq/100 g)	4-6	1-2
Cation exchange capacity (meq/100 g)	12-18	20-26
Base saturation (%)[a]	66.6	95 and 92
Probable pH	5.6-5.8	~7

[a] Most cations, except aluminum and hydrogen, neutralize soil acidity, and the proportion of the CEC occupied by these bases is called the "percentage base saturation."

The second major component of soils, organic matter, is considered to have a colloidal structure similar to clay. However, the principal difference between organic matter and inorganic colloids is that the complex is composed mostly of carbon, hydrogen, and oxygen as compared to aluminum, silicon, and oxygen for silicate clays. The cation exchange capacity of the product of decomposition of organic matter, humus, far exceeds any of the clay materials and is not as stable as clay. Organic matter is a potential source of nitrogen, phosphorous, and sulfur. It contains more than 95% of the total nitrogen in soils, 5 to 60% of the total phosphorous, and 10 to 80% of the total sulfur. Mineral soils vary in organic matter from trace to as much as 20 or 30%. Probably 3 to 5% is the average organic matter content of mineral soils. A high humus content for greenhouse composts is not always desirable –although organic soils are noted for intensive vegetable production. There is a wide array of organic compounds involved and a series of yellow to black substances called humic and fulvic acids [Stevenson, 1991]. These compounds form many complexes with cations, that is, a group of related units of which the degree and nature of the relationship may be imperfectly understood. Peat moss, which is the most common organic material for greenhouse production, is only partially decomposed. The physical properties of humus do not make desirable potting mixtures for shallow containers, although a decomposed black sphagnum peat is utilized in The Netherlands for bedding plants [Boertje, 1980]. Organic matter and clay minerals are the broad solid phases of a soil or compost substrate.

The colloidal system represents a storage and exchange facility between soil and substrate solution. Taken from Brady [1974], the situation as to exchange is noted:

$$\begin{matrix} Ca_{40} \\ Al_{20} \\ H_{20} \\ M_{20} \end{matrix}[MICELLE] + 5H_2CO_3 \rightleftharpoons \begin{matrix} Ca_{38} \\ Al_{20} \\ H_{25} \\ M_{19} \end{matrix}[MICELLE] + 2Ca(HCO_3)_2 + M(HCO_3) \quad (6.16)$$

For this example, the cations are monovalent in the ratio of 40, 20, 20, 20 per micelle, with "M" representing other metal cations. In humid regions where calcium may be leached, the reaction will tend to the right with hydrogen ions entering and calcium and other bases (M) being forced out of the exchange into solution where they can be taken up by root systems, or they are leached from the soil. This exchange between soil solution and the colloidal complex is chemically equivalent and the storage size of a soil is called its cation exchange capacity or CEC. The use of CEC is a simplification of the many reactions causing adsorption and desorption in soils. The subject has been discussed by Harter [1991], and consists mostly of laboratory procedures to measure cation retention. For brevity, we will deal mostly with CEC as a means to evaluate nutrient capacities.

3. Expression for CEC

CEC is expressed in equivalency or milliequivalents per 100 grams soil (meq/100 g). As noted under the section on *Equivalency*, monovalent ions such as Na^+, K^+, NH_4^+, and Cl^- are the same since they can replace or react with 1 H^+ ion. Divalent cations such as Ca^{2+} and Mg^{2+} can replace 2 H^+ ions, so their atomic weight must be divided by two to obtain equivalent weight. If a clay has a CEC of 10 meq/100 g, it can exchange 10 mg of H^+, or its equivalent, for each 100 grams of clay. For a hectare, 15 cm deep, this clay could exchange 220 kg H^+. This has practical implications since 1 meq H^+ can be replaced on the colloids by 1 meq $CaCO_3$ in limestone. As the molecular weight of $CaCO_3$ = 100, then 100/2 = 50 mg is required to replace 1 mg H^+, 1100 kg limestone per hectare would be necessary. CEC is usually determined at pH 7 since it increases with higher pH. As will be noted later, CEC can vary widely. Three examples are given in Table 6-7. For greenhouse practice, some authors [e.g., Bunt, 1988; Biernbaum and Argo, 1993, Joiner et al.,1981] have suggested that the CEC of peat mosses and related substrates should be expressed based on 100 ml rather than 100 grams (Table 6-10). Otherwise, the CEC of such materials will be much higher than they should be as the result of their low bulk density. Bunt also suggested that CEC for potting mixtures should be determined at the pH in which plants are grown –about 5.0 to 5.5 for those containing high amounts of peat moss. In effect, CEC indicates the buffering capacity of a soil as to how much fertilizer or lime is required to change the colloidal complex significantly. It sets the basic fertility reserve

Table 6-10. Cation exchange capacities of some minerals and soil mixes [Bunt, 1988; Biernbaum and Argo, 1993; Brady, 1974].

Material	Cation exchange capacity	
	meq $(100\ g)^{-1}$	meq $(100\ m\ell)^{-1}$
Humus	200	100
Vermiculite	150	4
Montmorillonite	100	100
Illite	30	
Kaolinite	10	
Fine clay	56-63	
Coarse clay	22-52	
Silt	3-7	
Sand	2-5	0
Sandy loam	20-40	--
Perlite	1.5	
Sphagnum peat	100-120	10-30
50% Sphagnum, 50% sand	8	5
75% Sphagnum, 25% sand	18	7
50% Sedge, 50% sand	21	17
75% Sedge, 25% sand	40	22
John Innes compost	9	8
25% Peat, 75% perlite	11	1
50% Vermiculite, 50% peat	141	32
66% Pine bark, 33% perlite	24	5

of the soil and is one of the easier tests to make on soil and related mixtures.

In most soils, the principal cation adsorbed on the colloidal complex will be Ca^{2+} –unless the pH is low. If one has a substrate with 6 meq exchangeable calcium, possessing a total CEC of 8 meq, there is a high probability of readily available calcium for plant growth. But, if the total CEC is 30 meq, with the same amount of exchangeable Ca^{2+}, the situation is completely reversed. With some crops, the base saturation of the soil should approach 90% Ca^{2+}. In some soils, the presence of 20% sodium on the exchange complex represents serious saline conditions and the loss of physical structure or deflocculation (puddling). From this discussion, one may conclude that the greater the CEC of a substrate, the more resistant it will be to chemical change. On the other hand, substrates consisting of plastics, rockwool, gravel, etc., or flowing hydroponic solutions such as NFT, will have no cation exchange capacity. Chemical additions to such cultures will change the root environment immediately. Failure to provide adequate nutrition on a continuous and timely basis will often show an adverse plant response in less than a week. A grower will often decide to produce in a buffered substrate since there is room, usually, for mistakes.

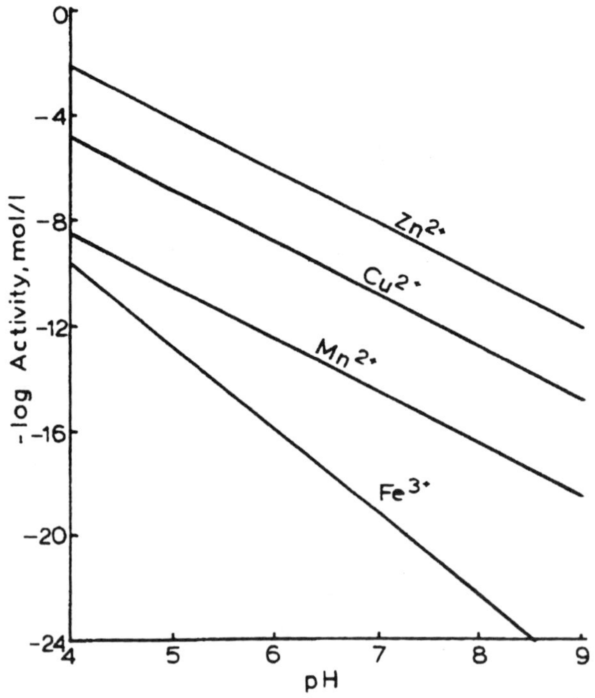

Fig. 6-7. Activity of Zn^{2+}, Cu^{2+} and Fe^{3+} as measured in soils, and Mn^{2+} calculated in equilibrium with MnO_2 [Lindsay, 1972a].

G. MASS EXCHANGE EQUILIBRIA

Another approach to determining nutrient availability of individual ions is the methodology described by Lindsay in his 1979 textbook. Under suitable experimental conditions [e.g., Norvell, 1970; Norvell and Lindsay, 1969; Lindsay, 1981; 1991; etc.], the equilibrium exchange between the soil matrix and solution was actually determined for specific ions --although the controlling solid phase was unknown. From this information, an equilibrium constant for the reaction could be calculated. Examples of such reactions are:

$$SiO_2(soil) + 2H_2O \rightleftharpoons H_4SiO_4^o \qquad (6.17)$$

$$Fe(OH)_3(soil) + 3H^+ \rightleftharpoons Fe^{3+} + 3H_2O \qquad (6.18)$$

$$Ca(soil) \rightleftharpoons Ca^{2+} \qquad (6.19)$$

$$Mg(soil) \rightleftharpoons Mg^{2+} \qquad (6.20)$$

$$Zn(soil) + 2H^+ \rightleftharpoons Zn^{2+} \qquad (6.21)$$

$$Cu(soil) + 2H^+ \rightleftharpoons Cu^{2+} \qquad (6.22)$$

Once the appropriate equilibrium constants are known, these can be combined for given pH, *pe* and CO_2 levels to calculate the theoretical level of the respective ions in solution. These data can be placed into software programs such as MINTEQA2/PRODEFA2 [Allison et al.,1991] and the speciation of a given substrate or hydroponic solution calculated in a short time. Lindsay and Ajwa [1994] have discussed the various computer programs for soil solution speciation and their use in teaching soil chemistry. However, with present computerization of greenhouses, it appears to me that these programs can be used on-line to predict need, and quantities required, in the grower's fertilization program.

As an example in a simple case, assume that Zn^{2+} in a soil solution is being controlled by the soil zinc, Eq. 6.21. The equilibrium constant of this reaction is $10^{5.8}$. Since this is equal to $(Zn^{2+})/(H^{+})^2$, then:

$$\log Zn^{2+} = 5.8 - 2pH \tag{6.23}$$

At a pH of 6.0, the concentration of Zn^{2+} in solution would be 5.8 - 12 = $10^{-6.2}$ or 0.00000063 M (0.63 μM). At pH = 8.0, $(Zn^{2+}) = 10^{-11.2}$ or 6.3 x 10^{-12} M (0.0063 μM). A 2 unit rise in pH (100-fold decrease in hydrogen ion concentration) will cause a 100-fold decrease in concentration of the zinc ion in the soil solution. Several examples, including zinc, are presented in Fig. 6-7. Note that a 10-fold decrease in hydrogen ion concentration (1 pH unit) will cause a 1000-fold decrease in Fe^{3+} activity.

Table 6-11. Conductivity units presently employed for designating salinity of a solution [Adapted from Bunt, 1988].

Units	Value	Ways of writing	Remarks
μmho cm^{-1} (micromho per cm) μSiemen (μS) cm^{-1} (microSiemen)	mho x 10^{-6} Siemen x 10^{-6}	0.000001 mho 1000 μmho = 0.001 mho 0.000001 Siemen 1000 μS = 0.001 Siemen	Units used in U.K. by ADAS and are common in U.S.literature
dS m^{-1} (deciSiemen per m)	Siemen x 10^{-3}	0.001 Siemen 1 dS m^{-1} = 1000 μS cm^{-1} = 1 mmho cm^{-1}	These units are coming into wider use in recent years
mmho cm^{-1} (millimho per cm)	mho x 10^{-3}	0.001 mho 1 mmho cm^{-1} = 1000 μmho 1 mmho cm^{-1} = 1 dS m^{-1}	Units are common in older U.S. literature Newer models read from 0.1 to 10 mho x 10^{-4}
mho cm^{-1} x 10^{-5}	mho x 10^{-5}	0.00001 mho 10 mhos x 10^{-5} = 100 μmhos = 100 μS = 0.1 mmhos = 0.1 dS	In the U.S., older models of equipment read from 10-1000 mho x 10^{-5}
SC (specific conductivity)	Siemen x 10^{-5}	0.00001 Siemen	Reported as such in Ireland
CF (conductivity factor)	mho or Siemen x 10^{-4}	0.0001 mho or Siemen CF10 = 1 mmho	Many meters in U.K. calibrated to read from 0.1 to 100 mho x 10^{-4}

H. ELECTRICAL CONDUCTIVITY (EC)

EC is the reciprocal of electrical resistance. Under standard conditions, a determination of EC is one of the simplest tests to make in solutions extracted from soils, or to make continuous measurements in hydroponic systems. The readings, in units of conductivity (Table 6-11) provide information on the general fertility level of the substrate solution and whether salinity problems are likely to exist. It is probably the most common, quick test employed. The unit of electrical resistance is the ohm, and its reciprocal, conductance, is the mho. Conductance, as measured between platinum electrodes in a suitably extracted solution, is proportional to the surface

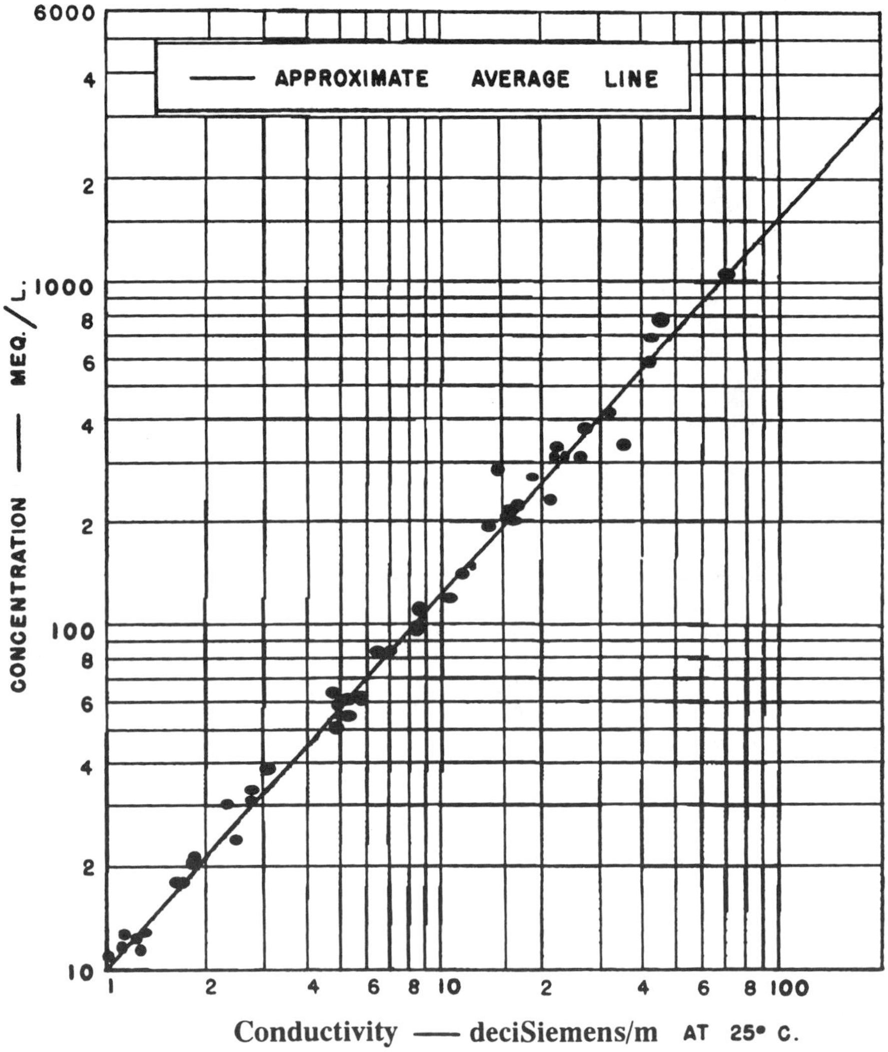

Fig. 6-8. Milliequivalents per liter concentration of several saturated paste extracts of soils as a function of electrical conductivity [Richards, 1954].

area of the electrodes and inversely proportional to the distance between them. In a given conductance cell, the measurement will depend upon ionic strength and concentration, increasing with increasing solute concentration. Temperature will also influence the reading. As noted in Table 6-11, the units can be confusing, particularly with the shift to deciSiemens per meter in recent years. Total salinity has also been expressed as ppm. The relationship between EC and solute concentration of extracts from a number of soils is given in Fig. 6-8 as an average. In actuality, EC will vary with the particular salt: e.g., $MgSO_4$, $CaSO_4$, and $NaHCO_3$ have lower conductivities than other salts at equivalent concentrations [Richards, 1954]. It will be noticed that soil extract solutions generally have an EC more than 10 times that found in irrigation water (Fig. 6-9). Soils with saturated paste extracts greater than 4 dS/m are generally considered saline. Irrigation supplies with ECs below 0.75 dS/m are usually good, whereas above 2.25 dS/m, the water supply is seldom satisfactory [Richards, 1954]. Greater detail will be given later in the section on salinity (Section 6.V).

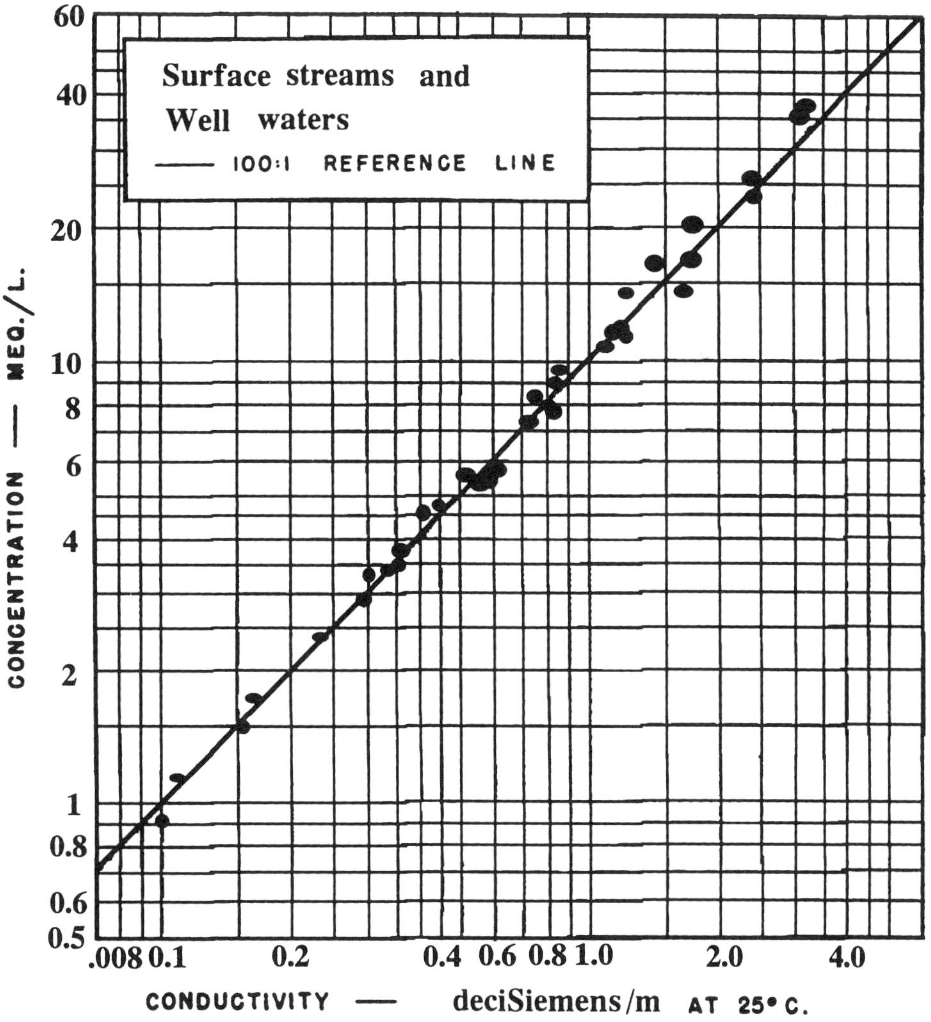

Fig. 6-9. Milliequivalents per liter concentration of several surface streams and well waters as a function of electrical conductivity [Richards, 1954].

I. COMMENTS ON LOGARITHMS

As noted in the first section, nutrient concentrations range from the micromole to millimole, where 1 µM = 0.000001 M or 10^{-6}, and 1 mM = 0.001 M or 10^{-3} M. The negative exponent can also be written as a logarithm to the base 10, or -6.0 and -3.0, respectively. Nearly always, the log of concentrations we deal with in plant production will be negative. Equilibrium constants (K°) are usually expressed as logarithms. This method has some advantages. Addition of logarithms is multiplication, subtraction is division, and raising a number to some exponential value is multiplication of the exponent [i.e., $(H^+)^2 = 2(-\log_{10}(H^+)) = 2pH$]. Addition and subtraction are much easier than multiplication and division. Present, inexpensive pocket calculators allow easy conversion of numbers to logs and vice versa. As an illustration, suppose one wishes to learn the activity of Fe^{2+} in solution where $Fe(OH)_3$(soil) is the solid furnishing ferrous ions:

$$Fe(OH_3)(soil) + 3H^+ + e^- \leftrightharpoons Fe^{2+} + 3H_2O \qquad (6.24)$$

Under standard conditions, the equilibrium constant (log K°) = 15.74 [Lindsay, 1979], and the equilibrium state can be expressed as:

$$\frac{(Fe^{2+})}{(H^+)^3(e^-)} = 10^{15.74} \tag{6.25}$$

If Eq. 6.25 is expressed in logarithms, then:

$$\log(Fe^{2+}) = 15.74 - (pe + pH) - 2pH \tag{6.26}$$

If we assume a pH of 6.5 and a *pe* of 8, then Eq.6.26 yields log(Fe^{2+}) = 15.74 - (8 + 6.5) - 2(6.5), or $10^{-11.76}$. The concentration of Fe^{2+} will be 1.74 x 10^{-12} M. Without other extenuating circumstances, an iron deficiency will result. Compare this with the previous calculations on soil zinc. In contrast to iron, *pe* does not affect Zn^{2+} concentration.

III. NUTRIENTS IN THE PLANT AND TISSUE ANALYSIS

There are three major parts to greenhouse nutrition control: 1) visual diagnosis of the crop, 2) tissue analysis, and 3) soil analysis. Visual observation is most important in assessing the general well-being of a crop in greenhouses. It is a difficult practice, requiring complete familiarity with the crop and good observational ability. This can only be achieved with practice and the opportunity to observe crop growth under a variety of conditions over a lengthy period. Classroom instruction and pictures do not provide this ability to judge acceptable growth. Expert systems for diagnostic purposes are not readily available. It is hard to do where the crop does not exhibit obvious symptoms, as shown in Figs. 6-10 and 6-11. Nevertheless, I have seen crop experts, after a thorough walk-through of a range, make general remarks as to crop status from which recommendations can be made to the grower for improvement. Growers may not always have the opportunity to judge growth in greenhouses outside their immediate region. In the example of air pollution over a wide area, growers can readily fail to note subnormal crop performance unless the damage resulting is acute.

I will not attempt to discuss visual symptomatology, nor outline the roles of the various essential elements except as necessary. The number of species produced under greenhouse conditions requires a separate textbook for such an undertaking, and familiarity with performance of the particular species under a variety of conditions. One will find in the ℓ ature [e.g., Joiner, 1983; Holcomb and White, 1982; Kofranek and Lunt, 1975; White, 1976; 1987; Nelson, 1978; Mengel and Kirkby, 1982; etc.] attention given to the role of essential elements in plant metabolism and general-to-detailed descriptions of deficiency symptoms. Micronutrient toxicities and macronutrient excesses are more likely to be important in greenhouse production as the result of high fertilization rates. Considerable attention is often given to these aspects. Several texts consist mostly of pictures of typical deficiency and excess symptoms as exemplified in Figs. 6-10 and 6-11. Wallace's 1953 text, and Smilde and van Eysinga's [1968] small book on tomatoes, are good examples of nutrient deficiencies and toxicities. Almost every technical publication and textbook on crop production will provide some examples from the authors' personal files.

Item 3), soil diagnosis, will be discussed in Section 6.IV. In this section, tissue analysis, as well as a review of the behavior of essential elements in the uptake process, will be discussed. The tissue concentration of essential nutrients in the plant, as determined under rigidly controlled conditions, is one principal means to address fertilization requirements and practices.

A. DEFINITIONS OF NUTRIENT REQUIREMENTS

Essentiality of a nutrient, for many authors, is based upon the element's requirement for the plant to survive and reproduce – often called the "critical" level, or range. For the greenhouse grower, this is not a desirable criterion for profitable production. Obvious deficiency symptoms as outlined or pictured by the previous authors, or in Figs. 6-10 and 6-11, show lost control of the fertilization program. Since fertilizer costs can be less than

Fig. 6-10. Upper picture could be excess salinity or nitrogen deficiency in carnations. The middle is typical boron deficiency, whereas the lower is zinc deficiency in roses.

2% of the total production cost, failure by a grower in this area is poor management.

Tissue analysis is a principal method for fertilizer control, particularly in hydroponic systems. Walworth and Sumner [1988], however, emphasized that successful interpretation is dependent on the accuracy of standard values (norms) used for comparison with analytical results. Furthermore, one needs an appreciation of how ionic forms are taken up and transported through the plant to interpret results properly.

According to Loneragan [1968], there is marked variation among plants as to growth related to nutrient concentrations in substrate solutions or in their tissue. Confusion occurs as the term "nutrient requirement" is often used for two purposes: 1) the least concentration required in the substrate for maximum growth, and 2) the minimal concentration within the plant required for maximum growth. From their work with agronomic crops [1969a, b], Loneragan and Snowball found that a plant with a high tissue requirement may have a low requirement in the solution. The uptake rate can be fast compared with other uptake rates. Thus, nutrient solution concentrations do not always parallel plant tissue concentration. In any case, "nutrient requirement" cannot be equated with minimum concentration for survival, nor, in our case, as Loneragan defined it [1968]: as the least concentration at maximal growth. Loneragan defined the "functional" nutrient requirement as the minimal tissue concentration that does not limit growth. Loneragan and Snowball [1969a] defined the "minimal functional requirement" for calcium as that concentration in plant tops that remained constant while yield increased substantially.

Both Bates [1971] and Jones [1985] used the term "critical" concentration. Bates considered the critical level as the tissue concentration beyond which further application of a nutrient does not return a profit. Such a definition is more applicable to the field as compared to greenhouses since fertilizer costs in field agriculture are a much larger part of total production cost. For greenhouse production, excess fertilization is common. Jones criticized both "critical" and so-called "standard" values because they are single values and do not describe the range possible for sustaining maximum growth (see Table 6-12). Expressing elemental concentration from just above the appearance of deficiency symptoms to that point beyond which toxicity occurs due to excess is the more common practice. Jones calls this the "sufficiency range." Sometimes an author may call the critical value that tissue concentration resulting in a 10% yield reduction (Fig. 6-12).

Fig. 6-11. Upper is a severe case of HCO_3^- toxicity in rose, leading to an apparent Fe deficiency. The lower (courtesy of A.C. Bunt) is a case of Fe deficiency in chrysanthemum, which was corrected.

Typical of the curves showing yield response to nutrient tissue concentration is Fig. 6-12. Differences in detail of these curves for a particular species should be kept in mind. Major differences, according to Jones

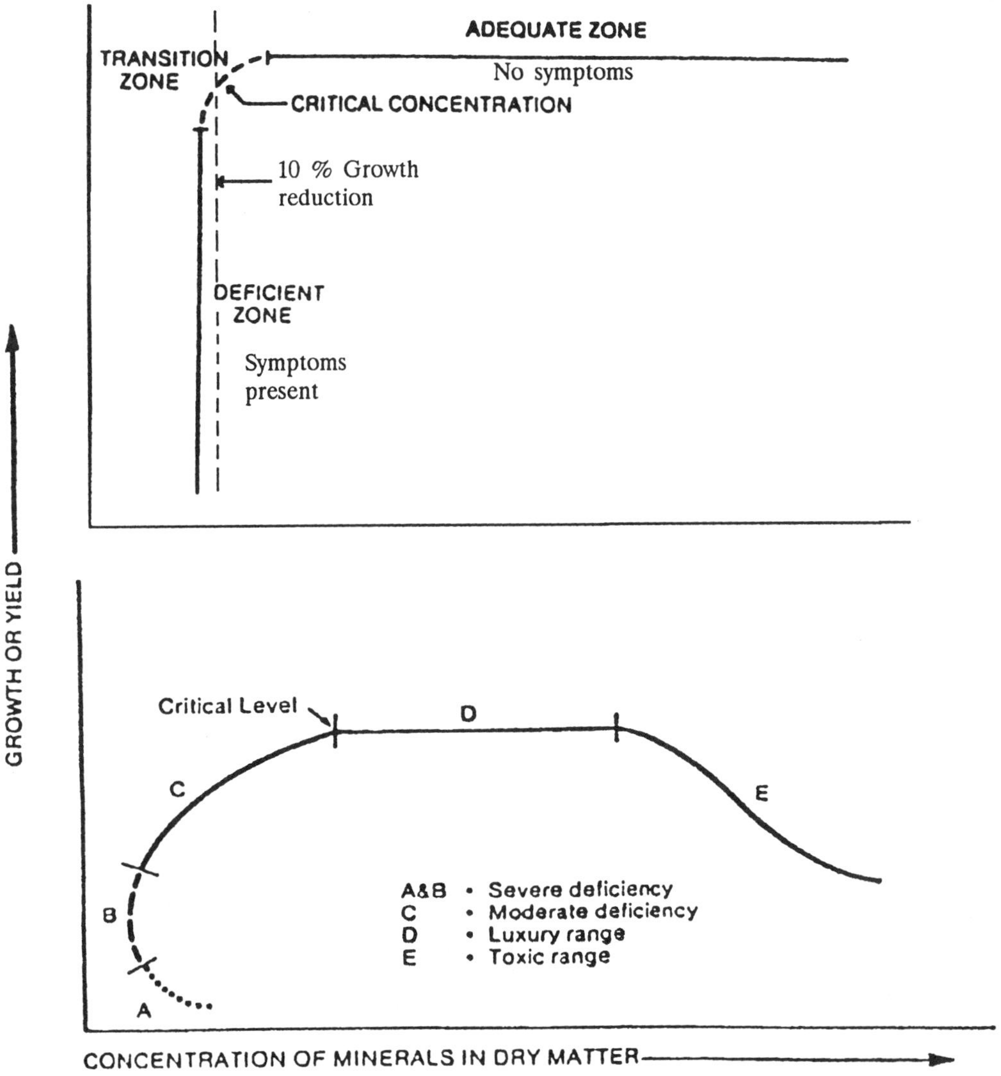

Fig. 6-12. General relationship between plant growth or yield and elemental content of the plant. [Upper graph from Ulrich, 1961 (©1961, *Amer. Inst. of Biological Sciences*), and Ulrich and Hill, 1973 (With permission of the *Amer. Soc. of Agronomy*). Lower taken from Smith, 1962 (With permission from the *Ann. Rev. of Plant Physiology*, Volume 13, ©1962, by Annual Review, Inc.)].

[1985], are in the slopes as elemental concentration increases from deficiency to sufficiency. The lower, C-shaped curve in Fig. 6-12 is called the "Steenbjerg effect." This has been shown to occur with very severe nutrient deficiencies that retard physiological development. Jones concluded that the lower figure best represented the relationship between growth and macronutrients, while the upper curve was the best for micronutrients. Bates, however, stated that plant analysis has a very limited value unless the C-shaped curve can be avoided. He believed that the Steenbjerg effect could be eliminated by careful selection of the tissue to be sampled. The upper graph does not show the possible toxic effects common to micronutrients at high concentrations. The range in sufficiency for micronutrients can be narrow. On the other hand, Walworth and Sumner [1988] bluntly state that the Steenbjerg effect is of little theoretical value. The problem, according to the latter authors, is that the determination of a "critical" nutrient concentration must be conducted under conditions where all other deficiencies are absent and plant growth conditions are optimal. Such a condition is seldom, if at all, possible in the field but may be approached in controlled greenhouse environments under suitable climatic conditions. An example of "critical" levels, below which deficiencies appear, for copper in chrysanthemums is Nelson's 1971 publication. An excellent article on micronutrient toxicity levels in geranium was published by Lee et al. in 1996. I emphasize that such critical levels may be far different from an optimal range for a maximal yield.

In their review of the literature, Walworth and Sumner find that tremendous of data are required to amounts of data establish "boundary" conditions that will ensure that yield maxima are reached at various nutrient concentrations. From a generalized model, presented in their Fig. 2, I have redrawn Sumner and Farina's model as suggesting the actual situation about nutrition and crop yield (Fig. 6-13). Using more than 6000

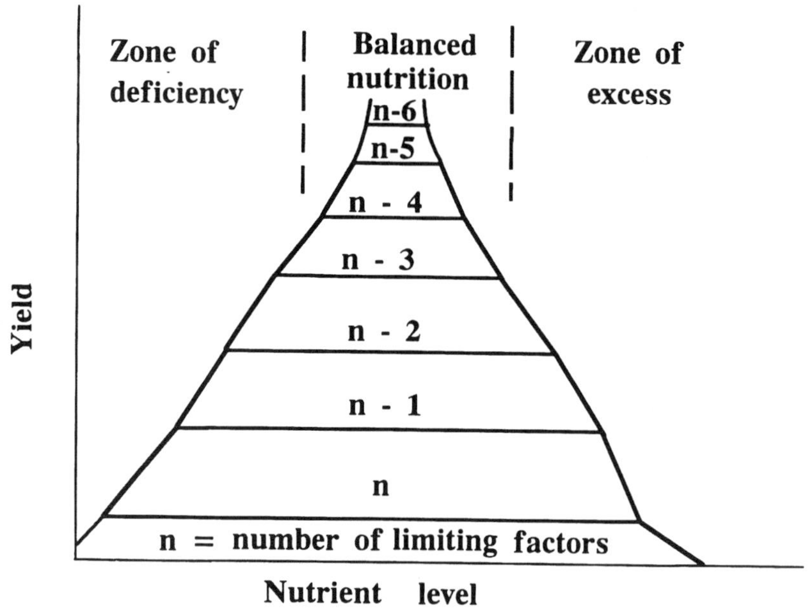

Fig. 6-13. Proposed relationship between nutrient levels in plant tissue versus yield, and as modified by other limiting factors [Sumner and Farina, 1986] (With permission of Springer-Verlag).

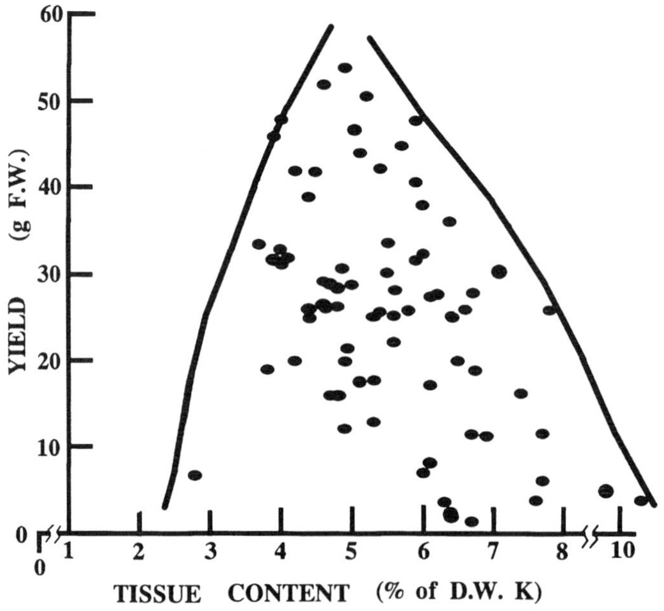

Fig. 6-14. Yield in grams fresh weight of *Salvia splendens* plants as a function of the percentage dry weight of potassium in the tissue. Data taken from the dissertation by Jeong [1990]. The heavy, sloping lines are boundaries drawn by hand. The points represent plants grown in a variety of substrates, fertilization solutions, and different times of the year.

Table 6-12. Nutrient content values published for a number of greenhouse species. Macronutrients in percent of dry weight, micronutrients in parts per million.

Plant species	Na	Ca	Mg	K	N	NO_3	P	S	Fe	Mn	B	Mo	Zn	Cu	Reference
Adiantum caudatum		0.2-0.3	0.2-0.4	2.0-3.0	1.5-2.5		0.40-0.80								Poole et al.,1976
Aechmea fasciata		0.5-1.0	0.4-0.8	1.5-2.5	1.5-2.0		0.40-0.70								Poole et al.,1976
Ageratum houstonianum[k]	0.03	1.6	0.7	4.0	4.9	1.7	0.8		160	100	39		43	18	Jeong, 1990
Ageratum mexicanum				3.12	4.75		0.75								Dight, 1977
Alyssum maritinum				4.28	4.37		0.51								Dight, 1977
Antirrhinum majus (snapdragon)		1.7	0.69	5.6	5.9		0.28-0.33								Boodley, 1962
	0.04[k]	1.0	0.7	3.6	5.0		0.6		120	70	18		26	4	Jeong, 1990
Antirrhinum nanum				3.54	4.81		0.52								Dight, 1977
Anthurium Andraeanum		1.2-1.7	0.7-1.0	1.0-2.3	1.6-2.1		0.28-0.33								Poole et al.,1976
Apium graveolens (celery)				4.0	0.5		0.2								Jones, 1985
Aphelandra squarrosa		0.2-0.4	0.5-1.0	1.0-2.0	2.0-3.0		0.2-0.4								Poole et al.,1976
Asparagus retrofractus		0.1-0.3	0.1-0.3	2.0-3.0	1.5-2.5		0.3-0.5								Poole et al.,1976
Aubrieta hybrid				4.84	4.37		0.26								Dight, 1977
Rhododendron (azalea)		0.45-1.6[a]	0.2-0.5	1.0-1.6	2.0-3.0		0.2-0.5								Bunt, 1988
		0.22	0.17	0.80	2.0		0.19				17				Twigg and Link, 1951
Begonia x semperflorens-cultorum[11]	0.14	0.9	0.7	3.4	4.0	0.08	0.6		328	120	32		36	16	Jeong, 1990
				3.00	4.17		0.21								Dight, 1977
Brassica actinophylla		1.0-1.5	0.3-0.6	2.5-3.5	2.5-3.5		0.20-0.35								Poole et al.,1976
Brassica capitata (cabbage)				2.0[b]	0.5		0.25								Jones, 1985
Calceolaria rugosa				3.92	4.58		0.44								Dight, 1977
Callistephus chinensis				4.16	4.11		0.53								Dight, 1977
Celosia cristata[k]	0.05	1.8	0.9	3.6	4.2	0.72	0.5		81	80	28		19	4	Jeong, 1990
Chamaedorea elegans		0.4-1.0	0.3-0.4	1.0-2.0	2.5-3.0		0.20-0.30								Poole et al.,1976
Chrysalidocarpus lutescens		1.0-1.5	0.3-0.6	1.0-2.0	1.5-2.5		0.10-0.20								Poole et al.,1976
Chlorophytum comosum		1.0-2.0	0.5-1.5	3.5-5.0	1.5-2.5		0.10-0.20								Poole et al.,1976
Coleus blumei[k]	0.1	1.2	0.7	4.8	4.6	1.17	1.3		142	153	33		32	10	Jeong, 1990
Coffea arabica		0.5-1.0	0.3-0.5	2.0-3.0	2.5-3.5		0.15-0.25								Poole et al.,1976
Cucumis Melo (sweet melon)					0.1										Jones, 1985
				3.0	0.5		0.15								Jones, 1985
Cucumis sativus (cucumber)					0.2[c]										Jones, 1985
		0.46-0.57[m]	0.56-0.58	3.55-4.31	6.42-6.54		0.61-0.68								Ingestad, 1973

Plant species	Na	Ca	Mg	K	N	NO_3	P	S	Fe	Mn	B	Mo	Zn	Cu	Reference
Dahlia variabilis				4.00	4.06	0.62									Dight, 1977
Dendranthema x grandiflorum (chrysanthemum)		0.46	0.06	2.15	4.50		0.17				20				Kofranek and Lunt, 1975
		0.8	0.4	1.5-5.5	4.0-5.0	0.8	0.5-6.5		50-100	50-150	20-30		25-100	5-10	Boodley, 1964
														7	Nelson, 1971
		1.0-2.0	0.35-0.65	4.5-6.5	4.0-6.5		0.2-1.0								Bunt, 1988
		0.5-4.6	0.14-1.5	3.5-10.0	4.5-6.0		0.26-1.15			195-260	25-200		7.3	10	Kofranek, 1980
Dianthus caryophyllus (carnation)		1.0-1.5	0.2-0.4	2.9-3.3	2.0-4.0	0.5-0.7	0.2-0.35	0.1+	50-100	50-150	25-100		25-100	5-10	Hanan, 1975
		1.0-2.0	0.25-0.5	2.5-6.0	3.2-5.2		0.2-0.3								Bunt, 1988
		1.0-2.0	0.24-0.50	2.5-6.0	3.2-5.2		0.2-0.3		50-150	100-300	30-100		25-75	10-30	Mastalerz, 1977
Dianthus gigantens				4.16	4.27		0.57								Dight, 1977
Dizygotheca elegantissima		0.5-1.0	0.2-0.3	1.5-2.5	2.0-2.5		0.40-0.80								Poole et al.,1976
Dracaena deremensis		1.5-2.0	0.3-0.6	3.0-4.0	2.0-3.0		0.2-0.3								Poole et al.,1976
Dracaena fragrans		1.0-2.0	0.5-1.0	1.0-2.0	2.0-3.0		0.15-0.25								Poole et al.,1976
Dracaena Sanderana		1.5-2.5	0.3-0.6	2.0-3.0	2.5-3.5		0.20-0.30								Poole et al.,1976
Dracaena surculosa		1.0-1.5	0.3-0.5	1.0-2.0	1.5-2.5		0.20-0.30								Poole et al.,1976
Euphorbia pulcherrima (poinsettia)	195 ppm								236	176	40		64	8	Boodley, 1974
		0.7-2.0	0.4-1.0	1.5-3.5	4.0-6.0		0.3-0.7		100-300	45-300	30-300		25-60	2-10	Ecke, 1971
		0.7-2.0	0.4-1.0	1.5-3.5	4.0-6.0		0.3-0.7		100-500	100-200	30-100		25-60	6-15	Mastalerz, 1977
Epipremnum aureum		1.0-1.5	0.3-0.6	3.0-4.5	2.5-3.5		0.20-0.35								Poole et al.,1976
Ficus benjamina		2.0-3.0	0.4-0.8	1.0-1.5	1.8-2.5		0.10-0.20								Poole et al.,1976
Ficus elastica		0.3-0.5	0.2-0.4	0.6-1.0	1.3-1.6		0.1-0.2								Poole et al.,1976
Foliage plants[d]		0.6-1.5	0.35-0.5	2.5-4.5	2.5-4.0		0.15-0.3		50-100	50-100			20-50	5-15	Joiner et al.,1981
General[a]		0.5-1.5	0.35-0.55	3.5-4.5	2.5-4.5		0.2-0.3		75-125	50-100	25-100		25-100	5-15	Joiner, 1983
Geranium		0.81-1.2	0.2-0.52	2.5-4.5	3.3-4.8		0.40-0.67		70-268	42-174	30-280		8-40	7-16	Holcomb and White, 1982
Impatiens holstii[k]	0.22	1.8	0.7	3.1	4.4	0.87	0.7		191	69	24		34	25	Jeong, 1990
Impatiens sultani				3.54	5.59		0.70								Dight, 1977

Plant species	Na	Ca	Mg	K	N	NO_3	P	S	Fe	Mn	B	Mo	Zn	Cu	Reference
Kalanchoe	0.02	4.43	0.25	3.25	2.90		0.32	0.59							Hansen, 1978
Latuca sativa (lettuce)				2.0	0.4		0.2								Jones, 1985
		1.4-2.0	0.5-0.7	6.0-8.0	2.5-4.0		0.4-0.6				25-45				Geraldson et al.,1973
	0.20[n]	1.62	0.54	9.64	3.87		0.47		175	53	31		58	10	Sanchez et al.,1991
		1.2-2.0[15]	0.4-0.8	4.8-8.0	4.0-5.0		0.35-0.60		80-200	25-130	20-60		30-60	5-10	Sanchez et al.,1991
Lilium longiflorum (Easter lily)		0.81-1.2	0.20-0.52	2.5-4.5	3.3-4.8		0.40-0.67		70-268	42-174	30-280		8-40	7-16	Mastalerz, 1977
Lobelia erines[k]	0.22	1.1	0.6	4.5	4.2	0.39	0.7		316	107	24		42	20	Jeong, 1990
				3.70	3.77		0.35								Dight, 1977
Lycopersicon esculentum (tomato)				1.5	0.2		0.15								Jones, 1985
		2.5-7.2	1.0-1.5	2.5-6.6		0.5-1.5	0.44-0.66		155-193[e]	27-168[e]			48-66[e]		Smilde and Eysinga, 1968
									94-819[f]	43-239[f]			201-458[f]		Smilde and Eysinga, 1968
		2.5	0.5	5.5	4.8		0.5	1.6	90	350	35	0.5	80	15	Winsor, 1973
	0.01-0.40	1.25-3.0	0.3-1.0	6.0-10.0	2.5-3.5		0.5-1.0		20-100	50-100	20-40	1-5	20-200	5-25	Wittwer and Honma, 1979
													526-1489[e]		Chapman, 1960
										1000-2000[g]					Millikan, 1951
											300-500				Brennan and Shive, 1948
													65-198		Chapman, 1960
		4.0-6.0	0.6-0.9	2.5-4.0	3.0-6.0		0.5-0.8			60-100	40-80		15-30	4-8	Geraldson et al.,1973
Maranta leuconeura Kerchoviana		0.5-1.5	0.5-1.0	3.0-4.5	2.0-3.0		0.20-0.30								Poole et al.,1976
Matthiola incana				3.54	3.64		0.39-0.48								Dight, 1977
Mesembryanthenum criniflorum				4.84	4.04		0.39								Dight, 1977
Monstera deliciosa		0.4-1.0	0.3-0.6	3.0-4.5	2.5-3.5		0.20-0.35								Poole et al.,1976
Nemesia strumosa				4.58	4.32		0.37								Dight, 1977

Plant species	Na	Ca	Mg	K	N	NO_3	P	S	Fe	Mn	B	Mo	Zn	Cu	Reference
Peperomia	2[h]	56	14	267	100[h]		17		15	1	0.2	0.1			Hansen, 1978
Pelargonium		0.77	0.14	0.62	2.4		0.28				18				Kofranek and Lunt, 1975
Petunia hybrida		0.66-1.73[i]	0.31-0.60	6.31-3.52	3.04-3.83		0.55-0.69		42-691	93-186	22-36		28-65	5-14	White, 1976
	0.39[k]	1.3	0.7	4.9	4.0	0.51	0.6		144	78	27		29	11	Jeong, 1990
Petunia multiflora				4.16	4.32		0.37								Dight, 1977
Philodendron scandans oxycardium		0.5-1.5	0.3-0.6	3.0-4.5	2.0-3.0		0.15-0.25								Poole et al.,1976
Phlox drummondii				4.24	4.23		0.60								Dight, 1977
Raphanus sativus (radish)					0.5										Jones, 1985
Rhododendron		0.14	0.39	0.30	1.0		0.11				15				Kofranek and Lunt, 1975
Rosa		1.0-1.5	0.25-0.35	1.8-3.0	3.0-5.0		0.2-0.3		50-150	30-250	30-60		15-50	5-15	Boodley and White, 1969
		1.0-1.6	0.28-0.32	2.0-2.5	3.0-3.5	0.1-0.3	0.28-0.34	0.16-0.21	80-120	70-120	40-60		20-40	7-15	Sadasivaiah and Holley, 1973
Saintpaulia (African violet)	44[h]	56	23	302	100[h]		27	12	0.9	0.2	0.1				Hansen, 1978
Salvia splendens[11]	0.27	1.5	0.7	3.6	5.1	0.65	0.8		159	155	32		35	7	Jeong, 1990
				3.24	4.48		0.72								Dight, 1977
Sansevieria trifasciata		1.0-1.6	0.3-0.6	2.0-3.0	1.7-3.0		0.15-0.3								Poole et al.,1976
Stromanthe amabilis		0.1-0.2	0.3-0.5	3.0-4.0	2.5-3.0		0.2-0.5								Poole et al.,1976
Syngonium podophyllum		0.4-1.0	0.3-0.6	3.0-4.5	2.5-3.5		0.2-0.3								Poole et al.,1976
Spinacia oleracea (spinach)					0.045-0.17										Jones, 1985
Tagetes erecta				3.84	4.23		0.60								Dight, 1977
Tagetes patula[k]	0.01	1.7	0.7	3.6	5.3	0.37	0.8		304	147	39		35	5	Jeong, 1990
				3.68	3.98		0.60								Dight, 1977
Viola tricolor				4.16	4.46		0.53								Dight, 1977
Zinnia elegans[k]	0.01	1.7	0.6	4.3	4.2	0.49	0.7		103	89	63		34	8	Jeong, 1990

Plant species	Na	Ca	Mg	K	N	NO_3	P	S	Fe	Mn	B	Mo	Zn	Cu	Reference

[a] Critical levels for deficiency symptoms.
[b] Levels below which deficiency symptoms appear.
[c] Levels for 10% growth reduction
[d] B levels not determined, S deficiency never seen, F toxicity common.
[e] Values for plants growing in marine clay soils (The Netherlands).
[f] Plants growing in river clay soils (The Netherlands).
[g] Values represent toxicity levels.
[h] Values are relative to N set at 100.
[i] Ranges for plants grown in four different root substrates.
[j] Critical levels below which growth reduction occurs.
[k] Values are for plants grown in peat-lite, and irrigated with complete fertilizer solutions with NH_4/NO_3 fixed at a ratio of 50:50, pH at 6.5, total concentration at 50 meq/l. For plants growing best in a series of treatments of differing pHs, substrates and NH_4/NO_3 ratios.
[l] Averages for all species and conditions from the ℓ ature as reviewed by Joiner (1983).
[m] Cucumber seedlings in solution, varying nutrient proportions.
[n] DRIS norms established from 3316 observations of crisphead lettuce in Florida.
[o] Sufficiency ranges for crisphead lettuce in Florida, using outermost sound leaf.

analyses for maize, Sumner [1977a] plotted the resultant data to give figures remarkably similar to Fig. 6-13. In another paper by Walworth et al.[1986], more than 8000 tissue samples for maize were used to provide boundary conditions. Use of boundary lines in analysis of biological data can be traced to Webb [1972], based upon the supposition that there are response limits to various factors influencing organisms. A boundary line occurs that provides a basis for judging population performance.

As the number of limiting factors is reduced (n - 6), the **range** of nutrient "sufficiency" in the plant decreases, until, under optimal growing conditions, the limits can be close together. The boundary conditions, as shown by the sloping vertical lines on either side in Fig. 6-13, may be skewed, resulting from the fact that the data is bounded on one side by zero and unbounded on the other side. Tissue concentrations cannot drop below zero, and there is no limit on maximum concentrations. These problems of limiting factors affecting elemental tissue levels for maximum plant response can be expected to be highly variable under field conditions. However, greenhouses should be close to optimum in situations where the internal environment can be controlled. Unfortunately, what is "high yield" with a greenhouse crop may be mostly subjective or ill-defined –especially regarding quality of fixed yield crops.

When the boundary conditions are established as shown by Walworth and Sumner [1987; 1988], Sumner [1977a], and Walworth et al.[1986] for crops such as alfalfa, maize, citrus, oats, peaches, soybeans, sugar cane, etc.; one may achieve a diagram such as in Fig. 6-13, which permits designation of nutrient "norms." To the best of my knowledge, such norms have yet to be established in greenhouse crops –with the possible exception of field-grown head lettuce [Sanchez et al.,1991]. The interesting point about these data is that the "norms" so derived are remarkably constant regardless of a locality. The result is particularly applicable to maize that used data from the U.S., Canada, Africa, South America, and France [Walworth et al.,1986]. There are exceptions. The outcome opens the possibility that such norms established for greenhouse crops would be applicable no matter whether, for example, carnations were grown in the ground in unheated greenhouses in the Mediterranean, or in environmentally controlled structures, in rockwool, in Colorado.

The published greenhouse data on tissue levels are largely what one finds in Table 6-12. These are insufficient to show clearly what Fig. 6-13 suggests. However, by using Jeong's data on bedding plants [1990], I have plotted the potassium content of *Salvia splendens* as an example of what happens when sufficient data are available (Fig. 6-14). In my estimate, the figure shows clearly the validity of Sumner and Farina's model in Fig. 6-13. The result shows that while a tissue analysis may suggest an optimum elemental level, that does not mean that yield will be high. Yield may be limited by another factor such as water availability, temperature, irradiance, etc. This emphasizes the fact that tissue analyses can only show whether nutrient levels may be limiting –not that they are the major factor causing low yield. The possibilities of this approach to greenhouse nutrition are interesting, with historical roots extending back to Blackman's classical paper on limiting factors [1905].

In summation, the terms "critical value," "sufficiency range," or "norm" require close attention to definition and application as defined by a particular author. It will be discussed later, given present methods of expressing nutrient content, that values in Table 6-12 are subject to many restrictions. A perusal of many fertilizer application articles in the horticultural literature shows little change in outmoded and incorrect approaches.

B. PUBLISHED NUTRIENT VALUES IN TISSUE (TABLE 6-12)

In view of the previous discussion, single values given in Table 6-12 require the assumption that the value represents a "critical" level below which deficiency symptoms are likely to occur, or there will be a significant yield reduction. Exceptions are noted in the footnotes of that table. Where ranges for an element are given, one may assume a "sufficiency" range for "healthy" plants –with the same exception as noted. There are at least two "general" ranges, covering most species, and for foliage plants, presented by Joiner et al.[1981] and Joiner [1983]. One may note that these "general" ranges for many plant species do not always agree with those published for a particular species within that group. As there is no way of deciding when a difference is significant, one must study the original publication to make a decision. For example, the single values cited from Jeong [1990] were taken from the highest fresh weights produced in his Experiment One for each species. When these concentrations were plotted for *Salvia splendens* in Fig. 6-14, most of them were grouped in the center of the figure and did not represent the highest yields obtained in Jeong's experimentation. The validity of plotting tissue analyses to establish "boundary" limits is proved.

One may also note that where several information sources are cited, they may differ from each other, suggesting problems in limiting factors as drawn in Fig. 6-13. For *Dendranthema*, Bunt's ranges for Ca, Mg, K, and N are higher than those given by Mastalerz. Smilde and Eysinga give two sets of micronutrient values for *Lycopersicon*, depending upon whether the plants were grown in marine clay soils or river clay soils. Whether both sets represent maximum, comparable yields is open to question. One may note the difference in range for potassium as published by Smilde and Eysinga versus Wittwer and Honma's for tomatoes. On the other hand, Hansen expressed all his values for tissue content of *Saintpaulia* as relative to nitrogen when N equaled 100. While we shall see that the ratio between ions is important, this method for nutrient content expression leaves one without means of comparison with other species. In contrast, the macronutrient values for *Dianthus caryophyllus*, as given by Bunt, Hanan, and Mastalerz, are not markedly different. The data attributed to Jones [1985] represent an onset of deficiency symptoms or 10% yield reduction. This may account for some low elemental values compared to others for the same species.

I have deliberately not cited data from field cultivation of vegetables except Sanchez et al.'s 1991 DRIS norms and sufficiency ranges for crisphead lettuce in Florida. Environmental control is nonexistent, and the range of suitable substrate nutrient concentrations will be much wider as compared with usual greenhouse practice. Plotting available data as in Fig. 6-13 might prove highly illuminating. However, Soltanpour [personal communication, 1994] stated that attempting to correct nutrition to meet the tissue norms showed by the boundary method for maximum yields is often uneconomic in field production. In many areas, other factors limit yields with consequent lower sufficiency limits. The grower does not need to correct nutrient concentrations to coincide with tissue levels for maximum yields. Certainly, the requirement to define maximum tissue nutrient levels in field crops would be less urgent compared to the greenhouse practice.

Note that sampling procedures have not been mentioned. The time, and place, of sampling on the plant are critical and subject to strict requirements. These will be discussed later.

C. FACTORS AFFECTING PLANT NUTRIENT CONCENTRATIONS

One may list the variables that can be limiting factors in both tissue nutrient levels and plant response. Some may be indirect as for nutrient availability from the root substrate:

1. Individual ion concentration in the substrate
2. Capacity of a substrate to supply nutrients
3. Soil moisture content, and, in hydroponic solutions, conditions of stirring and water flow
4. Total solution concentration or salinity
5. Aeration
6. Temperature
7. Physiological age of the plant
8. Part of plant sampled
9. Species and cultivar
10. Transpiration rate
11. Nutrient interactions
12. Nutrient source
13. Organic anion concentration

The above list is neither in order of importance nor inclusive, suggesting complexity is sufficient, and the problems one may encounter in interpreting the tissue levels presented in Table 6-12. Some of these will be discussed in greater detail in the following sections of this chapter.

D. PLANT NUTRIENT UPTAKE AND IONIC INTERACTIONS

Absorption of ions by plants is influenced by three processes: 1) diffusion of ions from the soil to the root as the result of either concentration or electrical gradients; 2) mass flow of water as the result of transpiration; this can occur in the flow from soil to root, and is the principal means of transport in the xylem; and 3) selective ionic absorption by the plant. For the present, we assume that the ions are at the root endodermis or the plasma membrane of the root cells, and we are concerned with the interactions and processes by which ions are taken into the plant. One problem that is sometimes difficult to learn in the literature, is the differentiation between uptake by individual cells of the root versus uptake that is largely transferred to the shoots via the xylem vessels.

1. Systems for Uptake by Roots

Ionic uptake is not completely understood. It does require an active metabolism by the root system with appropriate energy expenditure. According to Clarkson [1985], ion uptake represents about 20% of the maintenance respiration of plants. This does not include costs resulting from root exudation and association with mycorrhizas, or the energy required to grow a root system. Nutrients are inorganic compounds, and the incorporation of these substances is a basic process of plant metabolism [Mengel, 1974b]. Cations and anions may be absorbed independently and not in equal quantities. Monovalent ions are taken up more readily than divalent ions, and ions with small diameters will be absorbed faster than ions with large diameters. Plants are negatively charged in relation to the outer medium (e.g., a surplus of electrons), and since electroneutrality must be maintained within reasonable limits, ionic relationships are very important in nutrition [Hiatt and Leggett, 1974]. As the result of the electrical gradient, a passive force exists that acts to draw cations (NH_4^+, K^+, Ca^{2+}, Mg^{2+}, etc.) into the cell and inhibits passive entry of anions (NO_3^-, SO_4^{2-}, $H_2PO_4^-$, etc.). Kochian's calculations [1991] show that, for a monovalent ion, a 10-fold accumulation gradient inside the cell can be balanced by a membrane potential of -59 mV. For a divalent cation, the same accumulation gradient can be balanced by a potential of -29 mV. As root cells often maintain membrane potentials of -120 to -180 mV, there is a large electrical driving force for passive divalent ion uptake. Invoking active transport processes is not necessary.

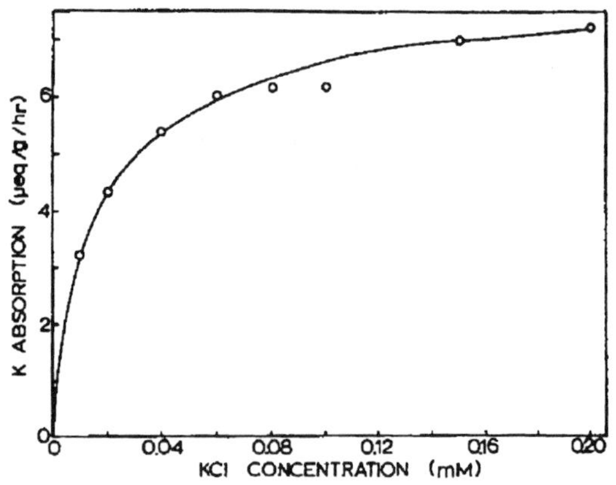

Fig. 6-15. Effect of KCl concentration on K^+ absorption in the range of system I [Hiatt and Leggett, 1974].

However, nitrate (NO_3^-) is generally the ion absorbed in largest amounts, although ammonium (NH_4^+) can also be an important nitrogen source. Since a principal characteristic of nutrient relationships is the much higher concentration of essential elements within the plant versus that in the root medium (in field soils particularly); then there must be an active, metabolic process by which anions –and some cations– are accumulated against a concentration gradient. In particular, the mechanism for K^+ uptake has been investigated extensively, showing uptake to be very efficient although Ca^{2+} concentration in the soil solution may be ten times higher. Similar mechanisms are considered to exist for NO_3^- and $H_2PO_4^-$ [Mengel, 1974b]. The process varies with the external ion concentration, and at the levels generally found in field soils (<1 mM, 10^{-3} M), active uptake is dominant with K^+ concentrations in root cells sometimes 2 x 10^4 times higher than

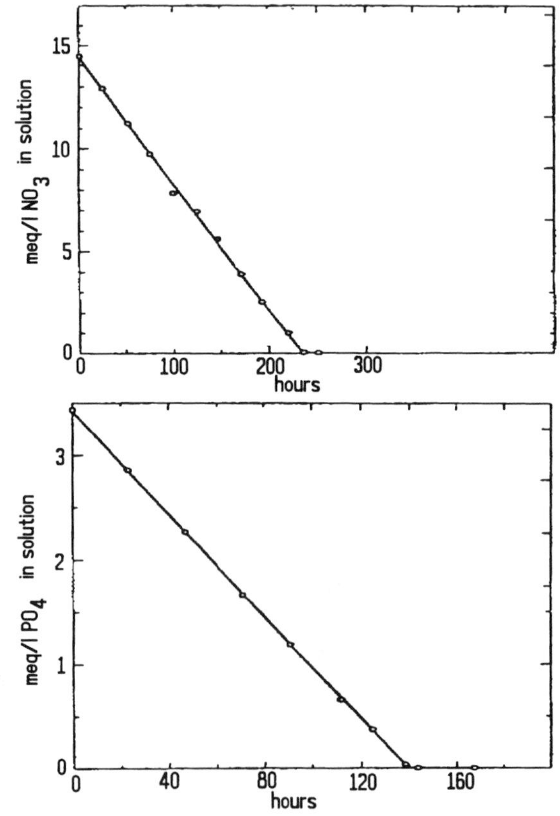

Fig. 6-16. Linear absorption of NO_3 (upper) and PO_4 (lower) by rye plants as a function of time [Olsen, 1950] (©1950, with permission of *Physiogia Plantarum*).

in the external solution [Mengel, 1974a]. Clarkson [1985] stated that natural soils contain the major plant nutrients at levels of 10^{-6} to 10^{-3} M. Within the upper range (10^{-3} M), maximum growth rate, as set by such factors as radiant intensity, temperature, etc., can be sustained. Russell and Shorrocks [1959] showed that when the nutrient status of plant and external solution was low, the concentration in the transpiration stream could exceed that of the external solution by factors exceeding 100. The transpiration rate under such conditions did not appear to affect nutrient transfer to the shoots.

The above is accumulation against an electrical gradient, sometimes called system I, with the maximum rate approaching a limiting rate in the range where the system operates. Fig. 6-15, from Hiatt and Leggett [1974], is a typical response for potassium. Again, Clarkson [1985] stated that although nutrient flux continues above the concentrations that saturate system I, it is not necessary for growth and can cause luxury consumption. This conclusion seems at odds with the general observations of fertilizer levels and plant response common to greenhouse production. Another example, where particular care was taken to ensure a minimum boundary zone at the root interface, is Fig. 6-16, resulting from a series of papers by Olsen [1950; 1953a; b; c]. Olsen concluded that rate of absorption of nitrate and phosphate was independent of solution concentration unless below 0.003 meq ℓ^{-1}. The latter is equal to $10^{-5.5}$ M NO_3^- or $H_2PO_4^-$. For calcium, Loneragan and Snowball [1969b] found that plants did not accumulate Ca^{2+} at concentrations of 0.3 μM ($10^{-6.5}$ M) in the flowing solution. Increasing the solution concentration from 0.3 to 2.5 μM ($10^{-5.6}$ M) increased yield greatly, while Ca levels in the tops remained the same. Uptake of micronutrients has not been investigated as much since the total amount is almost negligible compared to the macronutrients. However, Schwab and Lindsay [1983a] showed a distinct minimum, limiting Fe^{2+} concentration, below which Fe^{2+} was not accumulated (Fig. 6-17).

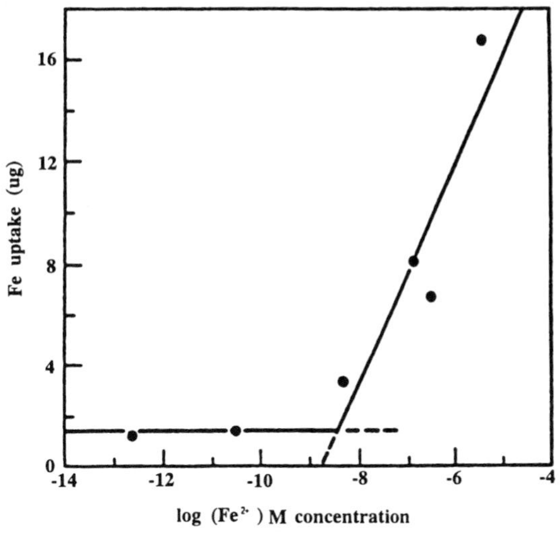

Fig. 6-17. Effect of Fe^{2+} concentration on Fe uptake by rice plants [Schwab and Lindsay, 1983a] (With permission of the *Amer. Soc. of Agronomy).*

Ingestad [1982] stated that most information about nutrient uptake relates to concentrations greater than requirements. Maintaining very low concentrations common to field soils is very difficult unless the concentrations at the plasma membrane interface are maintained, such as in flowing nutrient solutions. If care is taken in this aspect, Ingestad [1982] reported that uptake rates for nitrates have been found down to levels of 10^{-6} to $10^{-4.7}$ M (0.001 to 0.020 mmol ℓ^{-1}). With flowing solutions and constant inlet concentrations, Edwards and Asher [1974] found the uptake of nitrate sufficient to maintain maximum growth of wheat at a nitrate level of $10^{-5.3}$ M (0.005 mM/ℓ) and a flow rate of 1 ℓ min^{-1}. In essence, these results show that one cannot separate concentration from the rate at which the nutrient is supplied. When one takes into account supply rates, then concentration can be much lower than commonly encountered. This conclusion has, obviously, marked implication in hydroponic systems, especially NFT.

At solution concentrations above 1 mM, uptake may be passive, sometimes called system II (see comments above on electrical potentials), which operates with system I. Absorption by system I is considered to operate independently of the accompanying "counter" ion (e.g., K_2SO_4 where the counter ion is SO_4^{2-}). Whereas with system II, uptake is strongly influenced by the counter ion with the requirement that cations and anions be absorbed in equivalent quantities. Uptake by system II, according to Hiatt and Leggett [1974], does not result in accumulation of ions against a concentration gradient. While the range of solution concentrations in field soils (i.e., greenhouse production in the ground) may be predominantly system I, it is likely that system II operates in limited substrate volumes and hydroponic systems for greenhouses. Recommended application rates of the macronutrients for greenhouse culture exceed 1 mM, and are often 10 to 15 mM ($10^{-1.8}$ M) for NO_3^- levels. Under these high nutrient conditions, ion transfer to the shoots varies closely with the transpiration rate, and the internal concentration may be close to that of the external solution [Russell and Shorrocks, 1959]. In many species, Loneragan and Snowball [1969a] found that Ca solution concentrations from 10 to 1000 μM (10^{-3} M) markedly increased Ca levels in the plant tops but did not affect yield.

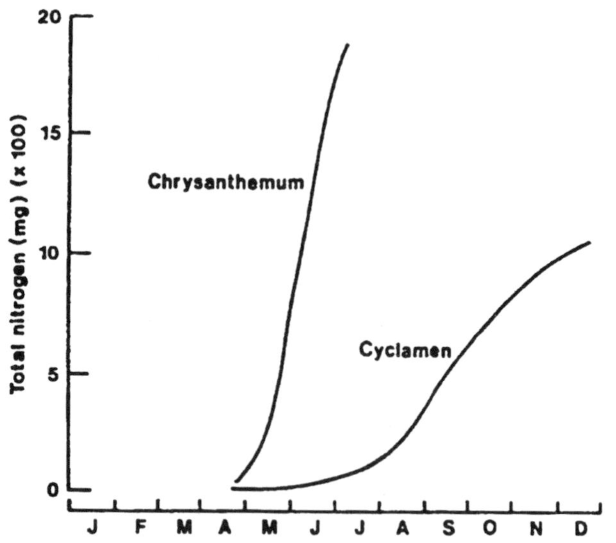

Fig. 6-18. N uptake by chrysanthemum grown 10 weeks in the summer versus a cyclamen grown over 1 year [Bunt, 1976].

2. Ion Competition and Antagonism

The presence of several ions in solution can cause competition between cations and anions. Cation-anion competition can occur after absorption. For either cations or anions, an increased concentration of one will often competitively inhibit the uptake of others, with the effect greater on ions of similar size. Potassium, rubidium, ammonium, and cesium are examples of such ions. As K^+ and NH_4^+ are normally present, competition will occur between these two. Typical of such relationships was the examination of chrysanthemum critical levels in fresh tissue by Nelson and Hsieh [1971]. They showed that below NH_4/K ratios of 0.025 to 0.026 meq, NH_4^+ injury never occurred. Above these ratios, injury to chrysanthemum was always observed. Competition between ions may be simply competition for the same uptake sites in the cell membranes. Adriano et al. [1987] listed a number of antagonistic couples such as phosphorous-zinc, cadmium-zinc, sulfur-selenium, and phosphorous-arsenic. Synergistic effects also occur, such as with calcium that will enhance absorption of small K^+ ions, and decrease absorption of ions with large hydrated diameters such as sodium. Sodium can also replace potassium to a limited extent. On the other hand, calcium can inhibit magnesium absorption in some plants such as barley and maize.

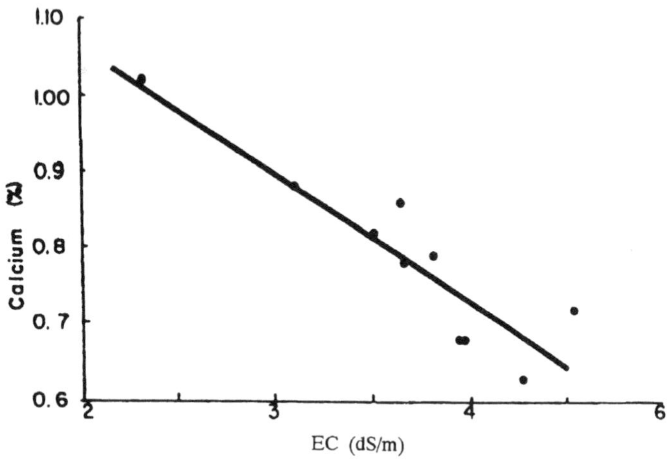

Fig. 6-19. Effect of total soluble salts on Ca levels in carnations [Hanan, 1975].

There are distinct differences in ion absorption and accumulation between species and cultivars. A good example of differences between species is Bunt's comparison between chrysanthemum and cyclamen in Fig. 6-18 concerning nitrogen accumulation. *Dendranthema* is considered a "heavy" feeder compared to cyclamen. The differences between species are so great that making general assumptions is difficult. Some plants will selectively accumulate sodium, others selenium, still others silicon. The general effect of total ionic concentration on Ca^{2+} uptake by carnations is illustrated in Fig. 6-19. Most authors stress the differences to be found between species and even cultivars within a species. In particular, much of the basic information in the agronomical ℓ ature deals with monocotyledons such as grasses, maize, etc. Many monocots are grown in greenhouse production (i.e., decorative foliage plants). However, most important species (i.e., tomato, cucumber, chrysanthemum, etc.) are dicotyledons. There can be sharp division between such groups as for nutritional requirements and behavior. In a study of the effect of micronutrients on macronutrient uptake, El Kholi [1961] concluded that there was no specific effect on macronutrient uptake common to the three plant species examined (oat, lucerne, tomato). This might have been due simply to the much lower concentrations of trace elements versus macronutrients. In oats, Yosida [1964] decided that both magnesium and potassium had a mechanism of uptake that was specific for the ion. Uptake via these mechanisms was unaffected by the concentration of other ions in solution. However, the ratios of the ions absorbed were determined by the composition of the solution and not by the solution concentration. Vlamis and Williams [1962] found that manganese toxicity in barley grown in Hoagland's solution could be eliminated by: 1) decreasing the manganese concentration, 2) adding 10 ppm silicon to the solution, or 3) increasing the concentration of Ca^{2+}, Mg^{2+}, NH_4^+, NO_3^-, or SO_4^{2-}. On the other hand, the presence of phosphate increased Mn uptake. Raising the pH of soils by adding lime will also reduce Mn uptake.

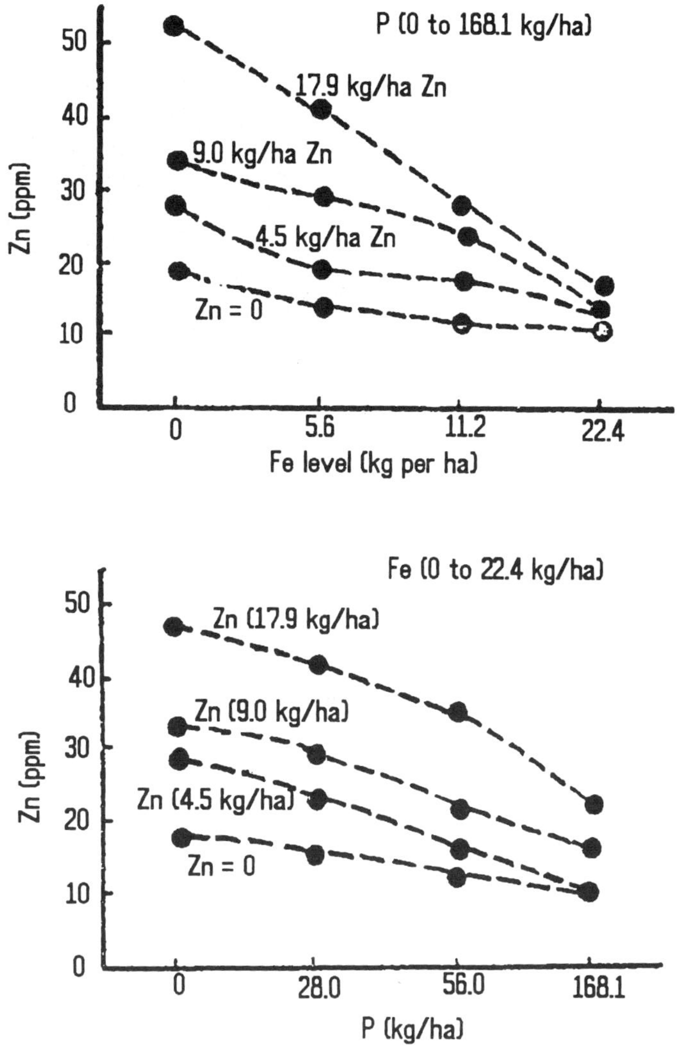

Fig. 6-20. (**Upper**) The effect of Fe levels on Zn content, and (**lower**) the effect of phosphate level on Zn uptake of maize [Lindsay et al., 1963].

The problems of iron and zinc deficiencies in Colorado soils were examined by Lindsay et al. [1963], showing interaction between iron, zinc and phosphate (Fig. 6-20). Increasing phosphate or iron levels suppressed zinc uptake in maize. There was antagonism between iron, zinc, and phosphate.

The ionic relationships to be found in the plant kingdom are numerous and complex. This complexity requires that nutrition be adjusted for the particular cultivar. For the present, Mengel [1974b] assumed that no active uptake mechanism exists for Ca^{2+} and Mg^{2+}. Active uptake systems do exist for K^+, NO_3^-, NH_4^+, $H_2PO_4^-$, Cl^- and SO_4^{2-}. Active uptake systems for other cations such as Fe^{2+}, Mn^{2+}, etc. are also likely to be present. However, the effects of micronutrients on macronutrient uptake are highly variable [El Kohli, 1961].

3. Ammonium and Nitrate Relationships

The relationships between ammonium and nitrate in nutrition are particularly important. Most plants do well solely on NO_3^-. On the other hand, Jeong [1990] found that *Ageratum* exhibited a tolerance to NH_4^+ and some toxicity to NO_3^-. As a rule, however, a 1:1 ratio of NH_4^+/NO_3^- produced the highest fresh weight in the seven bedding plant species examined. Similar results have been obtained at other ratios and with different species

Fig. 6-21. General effect of NO_3^- supply on lettuce growth. Far right = no NO_3^- applied, with increasing amounts toward the left.

[e.g., Winsor and Massey, 1978; Ingestad, 1970; 1971; 1973; etc.]. With maize, however, Blair et al. [1970] found no difference between plants grown with ammonium or nitrate supply when at low concentrations. The maximum concentration of NH_4^+ used by Blair et al. was 2 mM, while nitrate was supplied at 1 mM. Such low concentrations would result, in a commercial greenhouse situation, in unacceptable plant response with existing application systems –despite Clarkson's [1985] and Ingestad's [1982] conclusions.

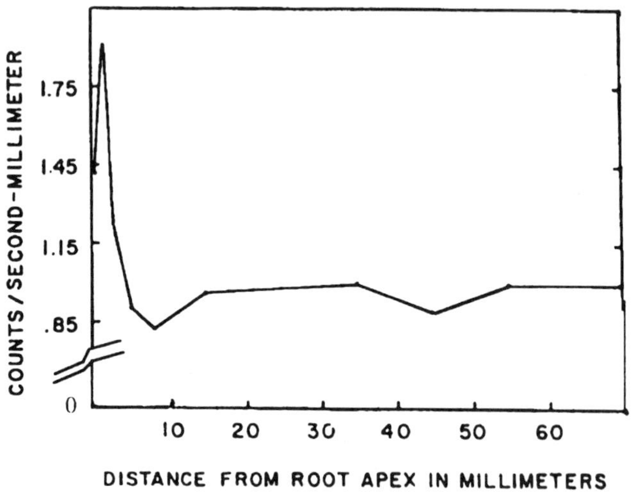

Fig. 6-22. Amount of ^{32}P accumulated in various regions of cotton roots. The curve is an average of 10 roots with an average length of 72 mm [Canning and Kramer, 1958] (With permission of the *Botanical Soc. of Amer.).*

Fertilizers using ammonium as the nitrogen supply are cheaper than nitrate fertilizers. Therefore, growers will tend to use such sources to the point that outright ammonium toxicity can occur. For reasons of phytotoxicity to plants and humans, the presence of ammonia (NH_3) in greenhouses is forbidden –although this is one of the more common nitrogen sources for field culture.

The literature was cited by Hofstra and Koch-Bosma [1970] as showing NH_4^+ to be a respiratory inhibitor. These investigators also employed low nutrient solution concentrations on tomatoes (0.5 mM NO_3^-, 0.5 mM NH_4^+ = $10^{-3.3}$ M). Their results showed, however, a distinct seasonal difference on the responses of ammonium- and nitrate-fed plants. In the summer, leaf growth of ammonium-fed plants was inferior to nitrate-fed plants. In the spring and autumn, ammonium did not have any harmful effect.

Plants have a high demand for nitrogen, largely through the uptake of NO_3^-. The "classical" effect of increasing NO_3^- concentrations on general growth of most species may be noted in Fig. 6-21. While NO_3^- toxicity has been observed (whiptail in cauliflower), excessive NO_3^- is more likely to result in "overgrowth" or excessive vegetation (Fig. 6-21), with concomitant problems of ground water contamination by nitrate (Section 6.V.C). Several types of ammonium toxicity, on the other hand, have been observed, often associated with ammonium accumulation in fumigated or steam-pasteurized soils. According to Barker and Mills [1980], plants that can convert NH_4^+ to organic nitrogen have a much greater tolerance to ammonium nutrition. If, however, ammonium reaches the shoots, biochemistry and physiology are greatly disrupted. Much of the energy production of the plant must go into carbon skeletons for ammonium incorporation and its detoxification.

While 1:1 ratios of NO_3^-:NH_4^+ are usually safe, ratios of 5:1 are usual. Greater ratios may be recommended or, levels of 10 mM NO_3^- to 2 mM NH_4^+ [e.g., Hanan and Holley, 1975; Sadasivaiah and Holley, 1973; Jones, 1983; etc.]. However, it equally common to find numerous hydroponic solutions without any ammonium [e.g., Steiner, 1961; 1968; Hoagland and Arnon, 1950; Cooper, 1979; etc.]. Hydroponic nutrition will be discussed in greater detail in Section 6.5.

4. Organic Anion Content

To establish equilibrium between anions and cations, NO_3^- must be balanced by an equivalent cation uptake or release of anions [Mengel, 1974a]. Thus, an NO_3^- supply results in release of bicarbonate (HCO_3^-) ions, increasing the external pH. On the other hand, uptake of NH_4^+ causes H^+ release, significantly reducing pH. NO_3^- uptake stimulates organic anion formation (oxalate, malate, citrate, etc.) within the plant [van Tuil, 1965]. Since the organic anions synthesized within the cell, at usual pHs, are 80 to 90% dissociated, there will be a considerable attraction for cations [Hiatt and Leggett, 1974] (see previous comments on electrical gradients). Van Tuil [1965] found that, with plants cultivated in nitrate-containing solutions, the accumulation of K^+, Na^+, Mg^{2+}, and Ca^{2+}, denoted as "**C**," minus the accumulated NO_3^-, Cl^-, $H_2PO_4^-$, and SO_4^{2-}, or "**A**," would give one the organic salt content of the plant material (C-A). If nitrates are replaced by ammonium, the C-A content is reduced, ammonium inhibiting uptake of other cations. Usually, the C-A value is constant over large changes in salt content –expressed as meq kg^{-1} dry weight. According to van Tuil, the normal C-A content for perennial ryegrass is 1000 meq kg^{-1}, whereas with sugar beets, the C-A value is 3500 meq kg^{-1}.

C-A contents have seldom been evaluated for greenhouse crops. Noteworthy was the work by Green [1967a, b] and Green et al. [1971] for carnations. The total organic anion concentration of carnation leaf tissue was highly correlated with growth. Optimal growth was invariably associated with a normal C-A content of about 1700 meq kg^{-1}. Hiatt and Leggett [1974] stated that the association of inorganic ions with the internal organic ions (i.e., malate) reduces ionic activity, so that ions may not accumulate against a concentration gradient. The interaction of organic and inorganic ions is a major determinant of ion accumulation levels and ion distribution in plants. The diagnostic flow chart devised by Green et al. [1971], wherein nutrition can be adjusted according to C-A and nutrient levels, has not been applied to any great degree in greenhouse production. This is unfortunate since evaluation of C-A content would be another parameter in tissue analysis for diagnostic purposes. In an examination of 16 species, which included lettuce and spinach, Noggle [1966] found that a lower yield was associated with plants that had higher total inorganic anion concentration. With one exception (lettuce), lower yields were associated with lower organic anion concentration. Chloride-treated plants always contained more inorganic anions than sulfate-treated plants, the former being associated with lower yields.

Nitrate, potassium, and chloride are taken up very rapidly whereas ions such as sulfate and calcium are taken up more slowly. The difference means that plants remove cations and anions in unequal amounts from solution. These differences are balanced by the accumulation or degradation of the internal organic anions such as malate [Mengel and Kirkby, 1982], or the release to the external medium of protons (H^+) or carbonate anions (HCO_3^-). For example, if K_2SO_4 is applied, the more rapid K^+ uptake causes a release of H^+, and malate is accumulated. Conversely, when $CaCl_2$ is applied, anion uptake exceeds Ca^{2+} absorption, HCO_3^- is released and malate level drops. Similar responses with NO_3^- and NH_4^+ were mentioned previously.

5. The Root and Transpiration

A very large root surface is generally exposed to the substrate environment (Chap. 5.III.A), and how much of this surface participates in ion uptake is important. Individual ions differ markedly in behavior. Mengel and Kirkby [1982] stated that movement of Ca^{2+} is restricted to young plant parts, whereas K^+ and $H_2PO_4^-$ absorption and movement to the shoots takes place readily throughout the root length. Ca^{2+} uptake is in the root tip region. A typical example, using a radioactive isotope of phosphorous with cotton as the test plant, is shown in Fig. 6-22. In the three species tested (cotton, maize and pea), heavy upward translocation from the point of supply occurred, with much more in regions 10 to 50 mm behind the root tip [Canning and Kramer, 1958]. The peak uptake in the rapidly dividing root apex may have been due more to rapidly growing cells than to export to the rest of the plant

(Fig. 6-22).

NO_3^-, $H_2PO_4^-$, and K^+ are usually transported rapidly to the upper parts of the plant, depending upon uptake and transpiration rates, while Ca^{2+}, Mg^{2+}, and SO_4^{2-} are transported at lower rates. The anions NO_3^-, $H_2PO_4^-$ and SO_4^{2-} are rapidly incorporated into the organic composition of the plant. They can be moved through the phloem and redirected in the plant, either as ions or as part of the carbohydrate system, which includes amino acids and other products. The fact that nitrogen deficiency first appears in the older parts of most species shows mobilization and removal of nitrogen to the younger, growing parts of the plant. Cl^-, Na^+, and K^+ can also be redirected as univalent ions in the phloem system.

Iron, and particularly calcium, on the other hand, are generally fixed so that deficiencies usually appear first in the actively growing regions. The lower picture in Fig. 6-11, is a classical example of iron deficiency, which, on correction, allows the plant to resume acceptable growth. Ca^{2+} is always transported to the growing meristems, and, as shown by Loneragan and Snowball [1969a, b], will not be translocated from older tissue under a deficiency situation. As these authors showed, plants transferred from 10^{-3} M calcium solutions to $10^{-6.5}$ M developed calcium deficiency though their tops had three to ten times the calcium concentration of plants grown continuously in solutions of $10^{-5.6}$ to 10^{-5} M. Loneragan and Snowball [1969a] suggested that variations in calcium supply under which deficiency develops partly account for the wide range of critical values found in the ℓ ature (Table 6-12). The average Ca^{2+} "functional requirement" found for 30 species by Loneragan and Snowball was 0.1 to 0.2%. Table 6-12 shows critical or sufficiency levels given for many greenhouse-grown species to be much higher.

Section 5.IV.B.2 elaborated on calcium deficiencies of many plants under high humidity and low solar radiation. Geraldson [1954; 1955] examined problems of blossom-end rot of tomatoes and blackheart of celery, showing that high substrate calcium levels did not always mean sufficient calcium assimilation. Appropriately timed foliar applications of $10^{-1.4}$ M $CaCl_2$ to tomatoes, or $10^{-0.6}$ M directly to the hearts of celery, were sufficient to prevent the disorders. In a rockwool slab system, Sonneveld and Voogt [1991] found that Ca deficiency symptoms could appear within 2 to 3 weeks after the start of a deficient treatment in tomatoes. They found that tomato fruits were most sensitive to Ca deficiency 7 to 10 days after flowering. Temporary disturbances of calcium supply in rockwool systems were likely to have both short-term and long-term effects. Similar calcium problems have been found in ornamentals such as the poinsettia concerning stem strength and bract necrosis [Lawton, 1989; Harbaugh and Woltz, 1989].

It is apparent that solution concentrations of the respective essential ions are not always good predictors of sufficiency in the plant. The nutrition (uptake) can be altered not only by concentration but also by the irrigation treatment. Increasing concentration will tend to offset a low irrigation frequency (Fig. 6-23, **left**). Conversely, increasing irrigation frequency, will tend to offset a low nutrient solution concentration (Fig. 6-23, **right**). Obviously, such cultural procedures will be limited by the substrate and weather conditions. In flowing nutrient solutions, where the boundary layer is as thin as possible, quite low nutrient concentrations may be satisfactory. Reducing transpiration by raising humidity has been shown to reduce uptake of K^+, Ca^{2+}, and Mg^{2+} in tomato, nitrogen in *Nephrolepis*, phosphorous in begonia, and K^+ and Ca^{2+} in chrysanthemum [Gislerød and Mortensen, 1991].

There is a distinction between ions that may be absorbed by the individual cells of the root and movement through the "free space" of the root –at least up to the endodermis. The latter refers to ion diffusion through the cell walls and the intercellular spaces. Ions can be absorbed by cells and move from cell to cell. At the root surface, the epidermis is often covered by a polysaccharide gel material (mucigel), which can influence water and ionic relationships in the rhizosphere [Kochian, 1991]. Also, within the cell walls, there are many organic groups that could bind micronutrients very tightly. Kochian states that the primary site of ion entry into roots is still not clearly outlined. There might be several ways by which ions can be transferred through the cortex, across the endodermis, and unloaded into the xylem for transfer to shoots. These considerations emphasize that flowing solutions that reduce the boundary layer at the root surface do not take into account the fact that diffusion processes still occur within the root itself.

E. NUTRIENT DISTRIBUTION AND TISSUE SAMPLING

It can be expected from the previous discussion that nutrient contents throughout a plant will not be uniform. An example of the variation to be found in tomato is exhibited in Fig. 6-24. Nutrient content will also vary with

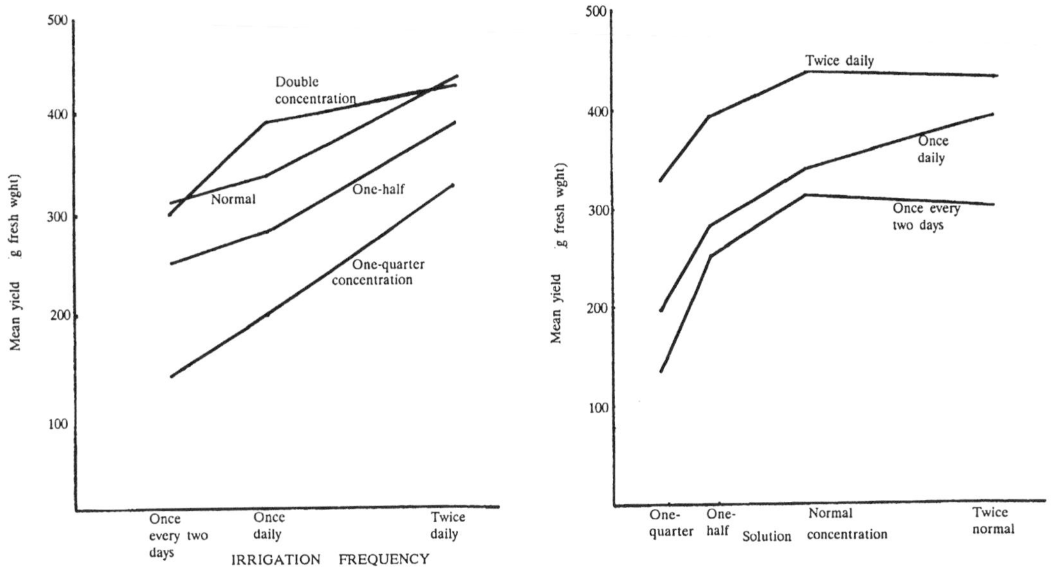

Fig. 6-23. Interaction of solution concentration and irrigation frequency on fresh weight of carnations grown in perlite with nutrients injected at each irrigation [Hartman, 1971].

age [Boodley, 1974], and as the plant ages, dry weight will increase. Since nutrient content is expressed by percent of dry weight, values at maturity will tend to be lower. According to Walworth and Sumner [1988], nutrient content should be based upon fresh weight. However, the inconvenience in storing and transporting wet samples has resulted in failure to adopt a fresh weight convention. Commonly, leaves are sampled for tissue analysis, and even here, Dole and Wilkins [1991] showed the elemental content in poinsettia leaves would vary with nodal position. One of the better expositions on plant sampling was Jones' review of soil testing and plant analysis [1985]. He pointed out that mixing of plant parts, or the taking of whole plants, is not a recommended procedure. Nutrients can accumulate in leaf margins, and Jones suggested removal of leaf margins to avoid this problem. However, this greatly increases handling, and with some species, would not be feasible.With many greenhouse crops, the short period in the greenhouse means that only one tissue analysis can be taken for future guidance since the crop will probably be marketed before any corrective action can be taken. In other words, tissue analysis becomes a postmortem. For most crops, sampling may be at, or just before, flowering. Plants damaged by insects or diseases, or mechanically damaged, should not be selected. Obviously, steps should be taken to remove any soil, dust, or chemicals that may have been applied –such as pesticides containing copper, zinc, etc. Jones [1985] recommended washing fresh tissue with a 0.1 to 0.3% detergent solution followed by a distilled water rinse. Iron and manganese concentrations may be most affected by washing.

Once obtained and treated as suggested above, tissue must be dried. Jones suggested drying at 80 C. However, Section 5.III.E cited Gardner as recommending 50 C for drying organic soils to avoid loss of organic material. I would suggest that 50 C is probably safer. One will also find investigators using 70 to 75 C for drying in more recent ℓ ature [e.g., Frick and Mitchell, 1993; Hood et al., 1993]. So, one cannot be completely sure of investigational results. Once dried, tissue should be stored in a moisture-free atmosphere until processing by the testing laboratory. If tissue can be taken to the laboratory immediately after sampling, the laboratory can provide drying and remaining sample preparation. If there is any delay, however, the sample should be dried immediately before shipping.

Sampling procedures for many greenhouse crops are given in Table 6-13. As will be noted later, problems in variance of the physiological age and sampling position on the plant can be reduced by the manner of diagnosis. Sumner [1977b] stated that if one tissue type is sampled, diagnosis by the DRIS system is likely to be independent of the leaf position in maize. Diagnosis will be discussed in the following section. Usually, tissue may be taken from 20 to 100 plants for a representative sample. This requires some judgment on the grower's

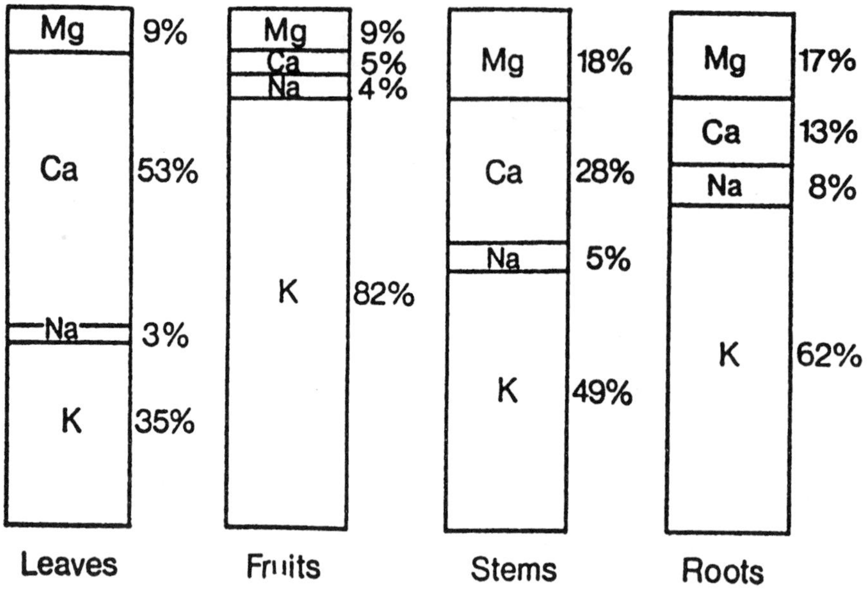

Fig. 6-24. Percentages of cations (K + Na + Ca + Mg = 100%) in different parts of tomatoes [Mengel, 1974a].

part. For example, plants of the same cultivar, if grown under uniform conditions of temperature, fertilization, substrate, etc., may only require one sample taken randomly in the crop. Single samples may not be desirable initially. Some idea of the variability to be expected between so-called "identical" treatments should be available. This may require as many as ten or more samples from identically treated plants. It allows calculation of a suitable standard error. The grower may then decide the reliability of the data obtained from what he considers a "uniform" crop. On the other hand, plants growing in the ground may require several samples to obtain a reasonable average because of soil variability common to most field soils. Grower observations of plant behavior in different parts of the greenhouse can suggest appropriate sampling patterns. About 5 g dried tissue are required. Assuming a dry weight content of 10%, then 50 g fresh tissue will be necessary –which can be more than can be comfortably held in two hands. It will be noted in Table 6-13 that some recommendations are different from each other. Those underlined are suggested by the author. Usually, a general recommendation is to sample the youngest, fully mature, and expanded leaves on a flowering stem. Unfortunately, most greenhouse operations have failed to take full advantage of foliar analysis –in part due to the higher costs compared to soil analysis. However, laboratory analyses of plant tissues are more standardized and reliable than many soil tests. That is, comparison of results between laboratories for foliar analyses are more likely to agree.

F. DIAGNOSIS FROM TISSUE SAMPLING

In previous sections we found that elemental content of plant tissue will vary with age and position on the plant and also with interactions that affect nutrient uptake and distribution. In their description of the DRIS system, Walworth and Sumner [1987] cite the literature to the effect that N, P, K, S, Zn, Mn and B concentrations decrease with advancing maturity in field crops such as alfalfa, whereas Ca and Mg increase to early flowering and then decrease. In potato, P, K, and Zn decrease as the plant ages. Even assuming that the proper procedures have been carried out on a timely basis, foliar analysis cannot provide information about fertilizer requirements or plant response to fertilizers. The fact that there is a suboptimal level of potassium does not say how much fertilizer must be applied or how the plant will respond [Walworth and Sumner, 1988]. The dynamic nature of elements within a plant means that use of values presented in Table 6-12 requires strict attention to the sampling method and time, as shown in Table 6-13. A sample must be taken at precisely the right time and from tissue that

Table 6-13. Plant tissue sampling procedures for greenhouse crops [adapted from Jones, 1985; also Witter and Honma, 1979; Winsor, 1973; Boodley, 1962; White, 1982; 1987; Nelson, 1978; Ecke, 1976].

Crop	Growth stage	Part to sample
Tomato	Immediately below last developing flower cluster	Petioles of leaves
	At fruiting	5th leaf below top
	Prior to or during fruit set 1st leaf above a fruit cluster[a]	Young plants, leaves adjacent to 2nd and 3rd clusters, or old plants, leaves from 4th to 6th clusters, or the 1st leaf above a cluster
Leaf crops, lettuce, spinach, etc.	Midgrowth	Youngest mature leaf, or wrapper leaf on head lettuce
Snapdragons	At flower	Most recent mature leaves on flowering stem
Geraniums	At flower	1st & 2nd fully mature, expanded leaves on a shoot
Melons (water, cucumber)	Early stages of growth prior to fruit set	Mature leaves near base of plant on main stem
Roses	During flower production	Upper mature leaves on flowering stem
	Flower calyx just opening	2 uppermost 5-leaflet leaves including petioles
	Flowering stem, pea size to when showing color	Entire lst 5-leaflet leaf, including petiole, counting from top of shoot
Chrysanthemum	5-6 wks after planting or after pinching	Youngest fully expanded leaves, 1/3 of distance down the stem
	Prior to or at flowering	Upper mature leaves on flowering stems
Carnations	Unpinched plants, single stem	4th or 5th leaf pairs from base of plants
	Pinched plants	5th and 6th leaf pairs from top of resulting laterals
Poinsettias	Prior to or at flowering	Most recently mature, fully expanded leaves
		Youngest mature leaves including petiole

[a] Underlined recommendations are suggested by the author.

is physiologically identical to that from which the critical value or sufficiency range was determined.

A number of approaches have been made to interpret foliar analyses. Of these, three will be discussed in detail: the Critical Value Approach (CVA), the Sufficiency Range (SR), and the Diagnosis and Recommendation Integrated System (DRIS). The latter, although developed initially in the late 1960s, has not, to my knowledge, been applied to crops in greenhouse production. Most of the available literature advances the use of DRIS, but there remain problems in practicality [Soltanpour, personal communication, 1994]. It may be a system that would advance nutritional practices in greenhouses.

1. Critical Value Approach and Sufficiency Range

Diagnosis using the CVA and SR methods requires that the composition of the tissue be compared with a value determined at the same growth stage (Tables 6-12, 6-13). In both the CVA and SR systems, problems with determining the appropriate physiological age, and the variation in nutrient content on a dry matter basis with age, is their greatest disadvantage. The SR method may decrease diagnostic precision because the limits are too wide [Walworth and Sumner, 1988].

Some of the above problems were discussed by Holland [1966]. He pointed out that a nitrogen deficiency may retard growth without affecting K uptake, which could result in an increased potassium level. There exists a dynamic balance within the plant in which single elemental analyses may not successfully suggest which

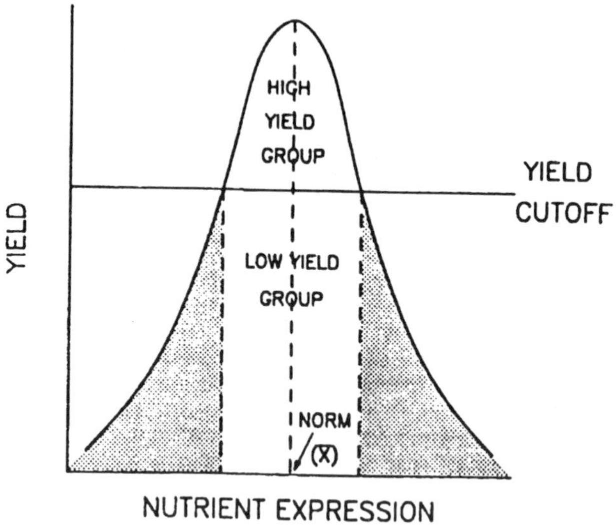

Fig 6-25. Schematic diagram of data used for DRIS evaluation. Shaded areas are values fromunhealthy tissue due to insufficiency, excess, or imbalance and not used for determining a norm value. Only those results from high yield group utilized [From Walworth and Sumner, 1988].

element is insufficient or in excess. Commonly, elemental concentration is compared with some yield index such as dry weight of seed, fresh weight of the whole plant or fruit production, or number of flowers produced. Typically, studies of this type are often called fertilizer response studies. From these, generating a mathematical relationship by statistical means is common. Such a curve may be divided into deficient, transition, and adequate zones [Ulrich and Hills, 1967] (Fig. 6-12). However, this is applicable only to a single element, whereas the relationship is actually affected by other elements as mentioned above, or the fact, for example, that the critical P level depends upon the N level. The so-called "univariate" approach (a single element), according to Holland [1966], merely produces a catalog of means, standard errors, and significance levels. It contributes nothing to interpreting the results. Holland [1966] proposed a "multivariate" approach, capable of defining restrictions without limit to the number of elements involved. Holland also developed the use of nutrient ratios as suggested by Beaufils [1971; 1973]. Terman and Nelson [1976], however, pointed out that all of the simple linear (rectilinear) or curvilinear (quadratic), multiple regression models –which are calculated by common statistical procedures– are in error if they give a positive yield intercept. (The curve crosses the yield axis at some point above zero at 0% nutrient.) Several studies have shown that the critical level of one nutrient can vary with levels of other nutrients. Terman and Nelson concluded that multiple rate, one- or two-variable fertilizer experiments, summarized by the usual quadratic regression models, are not realistically or biologically meaningful.

The problems outlined above are some principal reasons why CVA or SR methods have not achieved wide use in the greenhouse industry –not to mention higher costs. Diagnoses of adequacy or insufficiency in crops with CVA and SR methods have the advantage of simplicity. Walworth and Sumner [1988] concluded that since many CVA and SR values may have been determined under suboptimal conditions, it is impossible to infer what might have occurred had yields been higher. A CVA nutrient concentration based on yield curves from such experiments may not be applicable elsewhere. Nevertheless, the CVA and SR methods remain the most common practice in greenhouse nutritional control.

2. The DRIS System for Foliar Diagnosis

Two items in application of the DRIS system must be disposed of. The first is the development of "norms" to which foliar analyses are to be compared. This generally follows the "boundary line" definition discussed briefly in Section 6.III.A. The second is calculation of the necessary non-dimensional "indices" used to diagnose nutrient deficiency or sufficiency, and the relative order of their importance. This will be difficult and necessarily brief since there is no body of information in the ℓ ature that is applicable to greenhouse practice. Development and description of the methods for field crops may be found in articles by Beaufils [1971; 1973], Sumner [1977a; b; c; 1979], Sumner and Farina [1986], Walworth and Sumner [1987; 1988], and Walworth et al. [1986]. Many other articles may be found in the ℓ ature, but the only publication I know of in the horticultural ℓ ature that deals with what may be considered a greenhouse crop, employing DRIS, is the article by Sanchez et al. [1991], in which DRIS is used to evaluate the nutritional status of head lettuce.

a. Norm Development

The designation of boundary lines was illustrated in Fig. 6-14. Such figures are plotted for each element being considered (N, P, K, Ca, Mg, etc.), and for each ratio to be evaluated (e.g., N/P, N/K, P/K, etc.). Use of ratios eliminates the need for dry weight in the denominator:

$$\frac{N\%}{P\%} = \frac{(\frac{100N}{DM})}{(\frac{100P}{DM})} = (\frac{100N}{DM}) \; x \; (\frac{DM}{100P}) = \frac{N}{P} \tag{6.27}$$

As a result, the ratios are dimensionless. The variation in dry weight as the plant matures is eliminated. If the ratio is properly chosen, variation with time and location of the elements in tissue is also reduced. Data in a survey technique are considerable (8000+ for maize). For crisphead lettuce, 3316 observations from various field experiments were used to develop norms [Sanchez et al.,1991]. These may be skewed so that the data do not develop the "normal" or Gaussian distribution. To reduce the skewness, the data are divided into high-yield groups and low-yield groups (Fig. 6-25). The high-yield data correspond to "good" growers, with a distribution that usually allows one to calculate a mean and coefficient of variability (CV), which is used to "weight" the mean, or define the contribution of the ratio to the index to be calculated:

$$CV = \frac{100s}{\bar{x}} \quad where \quad s = \sqrt{\frac{s^2}{n}} \quad and \quad s^2 = \frac{\sum x_i^2 - \frac{(\sum x_i)^2}{n}}{n-1} \tag{6.28}$$

Here; $\bar{x}$ = the average of the data, Σx_i^2 = the sum of each datum squared, $(\Sigma x_i)^2$ = the total sum of all xs squared and, n = the number of observations (data). The symbol "s" is the standard deviation of a sample, standard error, or standard error of a mean while s^2 = the variance. This is an estimate of the variation from a mean of a population. Of course, the mean is the sum of all observations divided by the number of observations. The CV is a relative measure (percent) as contrasted with the standard deviation, which remains in the same units as the observations.

The standard deviation of the mean, $\bar{x}$, gives the reliability of the mean ($\bar{x}$), but there is no indication that the true value must be within the limits of $\bar{x} \pm s$. However, if the errors are randomly distributed, 95.4% of the errors will lie within $\bar{x} \pm 2s$, and 99.7% of the errors within $\bar{x} \pm 3s$.

A partial listing of DRIS norms for head lettuce is given in Table 6-14. Sanchez et al.expressed the nutritional data in all possible ratios and reciprocal ratios plus elemental as the ratios %/DM (Dry Matter). The elemental ratios are listed in Table 6-12, along with the sufficiency ranges. Not all of the calculated DRIS norms fall within the SA ranges. Although Sumner and others have discussed incorporation of other environmental factors in DRIS analyses, this has not been carried out. There is the problem that crop yields are often limited by factors other than nutrient concentrations and ratios. This would place sufficiency ranges lower on the diagram in Fig. 6-25. The critical value for deficiency, or minimum sufficiency, would be lower. Under such conditions, it would make little sense to attempt fertilization to bring concentrations to the optimum for maximum yield since a maximum yield would never be obtained given the particular circumstances.

b. Index Derivation

For each pair of nutrients there are three forms: e.g., N/P, P/N, and N x P. The selection may be made by choosing the ratio that has the largest variance ratio: that is s_N/s_P versus s_P/s_N. For a product (NxP), a new value of A = 1/P is defined. As N/(1/P) = N x P, then one has N/A as the product N x P. Only one expression is used to relate each nutrient pair. Learning from a single ratio such as N/P whether N or P is too high relative to the other or too low is not possible. A combined yield-quality index was used to separate the high yielding population versus number of heads alone. This requirement would be necessary for most greenhouse crops since quality would determine the number of flowers, pots, or fruits marketable. Sulfur was not included in the evaluations.

In the calculation of the indices, a simplified case of N, P, and K will be used from Sanchez et al.'s data, although, obviously, from Table 6-14, calculation for all essential elements may be used. There was insufficient data in Jeong's dissertation [1991] to determine such indices for the bedding plants Jeong used. Data for other important greenhouse crops are unavailable at this time.

Since we want determination of N, P, and K, three equations are required:

$$N\ index = \frac{[f(N/P) + f(K/N)]}{2} \tag{6.29}$$

$$P\ index = \frac{[-f(N/P) + f(100P/K)]}{2} \tag{6.30}$$

$$K\ index = \frac{[f(100P/K) - f(K/N)]}{2} \tag{6.31}$$

A value must be calculated for each of the above functions N/P, 100P/K, and K/N. This requires two equations, depending upon whether the measured nutrient ratio (i.e., N/P) is smaller or larger than the norm value determined from the boundary line survey (i.e., n/p):

if $N/P \geq n/p$ *then*

$$for\ example\ \ f(N/P) = \left(\frac{N/P}{n/p} - 1\right)\frac{1000}{CV} \tag{6.32}$$

<u>*OR*</u>

if $N/P \leq n/p$ *then*

$$for\ example\ f(N/P) = \left(1 - \frac{n/p}{N/P}\right)\frac{1000}{CV} \tag{6.33}$$

The appropriate equations (6.32, 6.33) are calculated for each nutrient ratio, which allows the indices to be determined as shown in Eqs. 6.29 to 6.31. If the actual analysis for (N/P) = (n/p) = 1, then f(N/P) = 0. N/P is optimum. However, if N/P is greater than n/p, then f(N/P) is a positive number. If less than n/p, f(N/P) will be negative. The 1000 multiplier is 100 x 10, with value 10 being included to give the resultant indices convenient numbers.

Assume that a foliar analysis of head lettuce provides us with N/P = 4.87/0.29 = 16.79, 100P/K = 29/7.43 = 3.90, and K/N = 7.43/4.87 = 1.53. Then:

$$(1)\ N/P \geq n/p:\ \frac{16.79}{9.06} - 1 = 0.85:\ CV = \frac{(100)(3.04)}{9.06} = 33.55 \tag{6.34}$$

$$f(N/P) = 0.85\left(\frac{1000}{33.55}\right) = +25$$

$$(2)\ 100P/K \leq 100p/k:\ 1 - \frac{5.56}{3.90} = -0.42:\ CV = \frac{(100)(2.59)}{5.56} = 46.58 \tag{6.35}$$

$$f(100P/K) = -0.42\left(\frac{1000}{46.58}\right) = -9$$

$$(3)\ K/N \leq k/n:\ 1 - \frac{2.62}{1.53} = -0.71:\ CV = \frac{(100)(1.19)}{2.62} = 45.42 \tag{6.36}$$

$$f(K/N) = -0.71\left(\frac{1000}{45.42}\right) = -16$$

The values +25, -9, and -16 can now be substituted into Eqs. 6.29 through 6.31:

$$N\ index = \frac{25 + (-16)}{2} = 5 \tag{6.37}$$

$$P\ index = \frac{-25 + (-9)}{2} = -17 \tag{6.38}$$

$$K\ index = \frac{-9 - (-16)}{2} = 4 \tag{6.39}$$

The above solutions show that phosphorous is deficient, whereas nitrogen and potassium are sufficient. In more complex cases, it is necessary that the computations outlined above be carried out in a suitable software program, and that a database of foliar analyses be built up for determination of norm values for greenhouse crops. Sanchez et al. [1991] carried out the above process for five macronutrients and four micronutrients, concluding that DRIS was a useful tool in diagnosing the nutritional status of lettuce. Although the norm values were obtained outdoors, there is no reason that such values would not be applicable to crops in protected cultivation.

The application of DRIS to field and orchard crops has found increasing use in the past decade. Several modifications have been proposed [e.g., Parent and Dafir, 1992]. One of the more recent summaries on plant tissue analyses has been Jones' [1991]. It is, according to Jones, doubtful that DRIS interpretation will ever be used exclusively. It is a complicated procedure for which there is insufficient data on greenhouse crops by which one can determine norm values. Jones, [1991] cited several publications suggesting that DRIS provided no better diagnoses than the usual CVA or SR methods. For Colorado field conditions, Soltanpour [personal communication, 1994] felt that DRIS had no application. There were instances where an excess of an element, such as zinc, indicated deficiencies of the others when that was not so. Thus, Soltanpour felt that DRIS represented an overly complicated method where other methods were more suitable. However, the system has been utilized by foreign governmental agencies as a means to handle chemical requirements of their cropping systems. Not having had direct experience with DRIS, offering conclusions is difficult for me. Still, the use of boundary definitions as illustrated in Figs. 6-13, 6-14 and 6-25 appears to me to be an excellent development for greenhouse use given the much higher yields usual to such production systems.

IV. NUTRIENT SUPPLY

Aung [1974] stated that the root system remains, from a physiological and practical standpoint, the primary site of ion uptake. Foliar application of nutrients may be a satisfactory method of dealing with certain problems of deficiency, but such practice is minor. In this section, we deal with the complexity of nutrient availability and the methods of soil analyses and practices that allow maximum profitability. Roots, as specialized organs, are in constant competition for available energy. Root growth is synchronized with the morphological state of the shoot. In the vegetative stage, shoot, and root growth continue linearly, but with the shoot at a faster rate. With the approach of flowering and fruiting, root growth slows or ceases abruptly as the result of energy shortage from the shoot. There is a decrease in the shoot-root ratio during reproduction. Physical and chemical factors, plus pruning, moisture, and radiant energy can modify the shoot-root ratio.

The presence of roots in the substrate –even in hydroponics– can markedly affect nutrient availability. In Section 5.III, water supply, considerable attention was given to factors limiting root activity and the influences on water uptake. In this section, we continue the discussion of root growth and environmental factors as they affect nutrient supply. We shall also consider some common practices about how nutrients are applied in greenhouse production. This is consideration of concentrations required in the substrate for maximal response according to a grower's objectives as contrasted with the previous discussion of nutrient concentrations in the foliage.

A. THE ROOT ENVIRONMENT

The root, protected by the root cap, grows through the soil, or substrate, so that it continuously moves into regions that have yet to be depleted of nutrients. Wadleigh [1957] reported on a study of a single rye plant grown in 28 ℓ s soil for 4 months. This one plant had 13.8 million roots with a length of more than 620 km, and a surface area of 237 m^2. The root hairs numbered about 14 billion with a combined length more than 10600 km and a surface area of 397 m^2. A remarkable amount of linear growth occurs with an average root system unless something restricts growth. The volume intercepted varies with species and soil conditions from 0.1 to 2% of the upper 15 cm soil layer [Barber, 1974] (see Section 5.III.B). A depletion shell is formed about the root as ionic uptake occurs. Depletion of, for example, phosphorous, will occur, beginning at the root tip, and extending back through the root hairs that begin several millimeters behind the root tip. The root apex and elongation zone will have several hours for nutrient uptake before root hairs begin to emerge. A depletion shell is already formed by the time root hairs emerge with an outer limit between 100 to 200 μm from the root surface [Clarkson, 1985]. This volume, close to the root, possessing characteristics often totally different from the bulk substrate, will

Table 6-14. Partial reproduction of Sanchez et al.'s Table 1, showing DRIS norms for crisphead lettuce with standard errors [1991].

Nutrient ratio	Mean	s
N/P	9.06	3.04
N/Mg	8.18	3.18
10N/Mg	1.09	0.64
100P/K	5.56	2.59
P/Mg	1.00	0.49
P/Na	2.86	2.36
1000P/Fe	3.42	2.01
1000P/Zn	9.24	3.79
100P/Mn	1.67	1.50
100P/B	1.39	0.83
K/N	2.62	1.19
0.1K/Mg	1.95	0.73
10K/Mn	2.96	2.22
10K/B	3.75	1.21
10Ca/N	4.41	2.09
10Ca/K	1.93	1.12
Ca/Mg	3.23	1.34
100Ca/Mn	4.62	3.98
100Ca/B	5.35	2.46
100Mg/Mn	1.43	1.15
100Mg/B	1.63	0.60
100Na/N	5.45	3.85
100Na/K	1.88	1.57
10Na/Ca	1.64	0.95
10Na/Mg	4.34	3.61
1000Na/Fe	2.21	1.56
1000Na/Zn	4.55	2.80
1000Na/Mn	4.62	3.76
1000Na/B	6.53	4.46
0.1Fe/N	4.62	2.88
0.1Fe/K	2.09	1.41
0.01Fe/Ca	1.18	0.70
0.01Fe/Mg	3.68	2.66
Fe/Zn	3.23	1.62
Fe/Mn	4.35	2.93
Fe/B	5.45	4.28
0.1Zn/N	1.56	1.38
Zn/K	6.80	4.06
0.1Zn/Ca	4.29	3.36
0.01Zn/Mg	1.24	0.85
Zn/Mn	1.61	1.33
Zn/B	1.97	1.83

be considered in the following paragraphs. The volume is commonly called the "rhizosphere," and the reactions that can occur within it often account for the various plant responses that serve to confound even knowledgeable growers. One may note that this volume about the root is much smaller than the value given for a sphere of influence for water uptake in Section 5.III.A.

In a nutrient-rich zone, growth response, even to highly mobile forms as NO_3^-, results in the formation of more lateral roots. Apparently, roots tend to elongate at rates commensurate with the rates at which most mobile ions diffuse [Uren and Reisenauer, 1988; Baldwin et al.,1972]. Root growth is a principal means of nutrient acquisition in unconfined soils as found in ground culture.

The effect of opportunistic root growth on nutrient uptake in usually "low-level" fertility field soils is highly important in greenhouse culture of cut flowers and some vegetables. Most production of such crops in the Mediterranean region, Colombia, Kenya, and the U.S. is in essentially unconfined substrates. This situation should be taken into account when considering most of the fertilizer recommendations that deal largely with limited volume, highly porous substrates (Section 5.III.F). Depending upon the volume to which the root system is restricted, depletion shells will eventually overlap so that, in container production, a root system will compete with itself for available nutrients. Even in flow-through culture systems, reducing volume available for root growth will reduce growth, assuming equal nutrient concentrations. The effects of substrate volume were discussed in Section 5.III.F.2. Personal observation over a period of 30 years suggests, for example, that roses do better in a large volume as contrasted with small rockwool blocks or peat bags. It is sometimes easier to increase volume rather than attempting to increase nutrient concentration with its consequent dangers of high salinity. Before the 1960s, a significant portion of commercial rose growing in the U.S. was in raised, wooden benches. That changed so that nearly all cut-flower rose production is in the ground now. There are situations, from an economic viewpoint as well as safety, that require production in unconfined substrates. The levels of nutrients such as phosphorous and potassium must be higher in a pot because of competition. Bray [1963] stated that pot experiments were valueless for establishing fertilizer requirements for field-grown crops. Nevertheless, a significant part of greenhouse production is in small containers. The grower must be careful to separate the different conditions in handling fertilizer adjustments.

1. Ion Movement

With ion uptake by a root, concentration of a particular element will decrease to a minimum at the root-absorbing surfaces. Concentration gradients occur, and these, combined with mass water flow, cause cation, anion, and soluble organic molecules to move toward the root. Movement of the root into spaces formerly occupied by soil particles causes an increase in bulk density so that steeper concentration gradients may occur [Barber, 1974]. The movement of ions in response to a concentration gradient has been discussed in detail by Noble [1991]. For small molecules, the time to cross 50 µm –a typical cell width– would be about 0.6 sec. For 1 meter, however, the time for diffusion would be 8 years. One has to keep in mind the scale involved when interpreting results. With roots in intimate contact with soil particles, and within the rhizosphere, movement of molecules and ions will be rapid when one considers the dimensions encountered. The diffusion rate can obviously be increased by increasing concentration in the bulk soil, or by reducing the distance the ion must move to reach the root. Boundary layers in which there is no turbulent mixing occur at the interface between any fluid and a solid. The thickness of an unstirred layer can be reduced but never eliminated. Vigorous stirring and aeration of plants in fluid hydroponic systems are means to reduce the diffusion resistance, and this is operative in such methods as the Nutrient Film Technique (NFT) [i.e., Barber, 1962; Olsen, 1953b].

Increasing the moisture content of a substrate will increase the cross-sectional area for diffusion by reducing tortuosity (Fig. 6-23). Up to a maximum, increasing bulk density will also increase diffusion rates. These phenomena probably account for the general observation by growers of enhanced plant response when clay is added to a high organic or porous substrate. Where the nutrients are supplied by mass flow, the transpiration rate and solution concentration will influence the nutrient amount supplied. Where the salt concentration is low ($<10^{-3}$ M), it is generally accepted that transpiration rates will not affect nutrient supply to the root. There are insufficient ions to be supplied and additional ions must be provided by diffusion [Barber, 1962]. At high salt levels, as generally encountered in greenhouses, mass flow can significantly affect nutrient supply. Under these conditions, the transpiration rate will influence plant nutrition. The relationship between high and low soil salt concentrations was illustrated by Barber [1962], reproduced in Table 6-15. When converted to equivalent NO_3^-,

$H_2PO_4^-$, and SO_4^{2-}, the concentrations in the low salt treatments ranged from $10^{-0.5}$ (S) to $10^{-4.1}$ M (K) as contrasted to $10^{0.3}$ (S) to $10^{-2.4}$ M (K) for the high salt levels. Plants appear able to regulate ion uptake depending upon external ion concentration. Soil analyses, therefore, are not always indicative of what is occurring in the plant. In the winter conditions of northern Europe and U.S., the low transpiration rate is, in effect, a reduction in nutrient supply as contrasted to the situation in summer conditions or arid regions such as Colorado or southern California. Sometimes, where the movement of ions to the root exceeds the uptake rate, accumulation at the root interface may occur. Examples are the accumulation of iron oxides or calcium that can precipitate as $CaSO_4$ or $CaCO_3$ [Barber, 1974].

Table 6-15. Relationship between ion concentration in the soil solution versus that within *Zea mays*, compiled by Barber [Barber, S.A. *1962. Soil Science*, 93:39-49].

Ion	Concentration (ppm)			Ratio of plant content to lowest and highest soil solution levels	
	Soil solution		*Zea mays* content		
	Low	High	Average	Low	High
Calcium	8	450	2200	275	4.9
Potassium	3	156	20000	6666	128
Magnesium	3	204	1800	600	8.8
Nitrogen	6	1700	15000	2500	8.8
Phosphorus	0.3	7.2	2000	6000	278
Sulfur	118	655	1700	155	2.6

The cation exchange capacity of the substrate, on the other hand, reduces the diffusion rate as compared to that in water. When ions diffuse, they diffuse either as a cation-anion pair or as "counter diffusion" in which an ion such as H^+ may diffuse in one direction and Ca^{2+} moves in the opposite direction. Removal of cations (e.g., K^+, Ca^{2+}, etc.) from the negatively charged clay or humus micelles requires replacement with an equivalent number of ions (Fig. 6-6). Inorganic ions released by a root are in exchange for cations and anions absorbed by the root (see also Section 4.III.D.4). Significant ionic release from roots will occur, according to Barber [1974], when cation and anion absorption are unequal. When more cations are absorbed than anions, then H^+ is released –such as happens when NH_4^+ is the predominant nitrogen supply. When more anions are absorbed, HCO_3^- ions can be released [Riley and Barber, 1969; Moore, 1974; Marschner and Römheld, 1983]. The pH near the root changes significantly. The situation varies with species and zones along the root plus the fact that more than one species may be grown together, resulting in a different pH response [Marschner and Römheld, 1983]. The general effect of ammonium and nitrate uptake on pH, which has been repeated several times in this chapter, should not be taken to be universal under all conditions. Thus, Grinsted et al. [1982] found that growing rape in a P-free solution with only NO_3^- as the nitrogen supply resulted in a pH decrease of 2.4 units, later increasing toward the end of their experiment. pH changes greater than one unit in the rhizosphere are usual.

The ionic and molecular exchange in the rhizosphere, therefore, can affect acidity so that nutrient availability can be increased or decreased (Fig. 6-7), depending upon the soil minerals available and relevant concentrations [Barber, 1974; Moore, 1974; Lindsay, 1972b; 1984]. In fact, a lower pH in the rhizosphere can go far to increase phosphorous and iron availability. Both elements are considered immobile in soils as compared to nitrate. There will also be striking effects of pH on manganese, zinc, copper, calcium, etc. availability that will be examined later. Quite often, a nutrient insufficiency may cause a corresponding plant reaction that tends to compensate for the insufficiency. Examples of such reactions have been cited by Korcak [1987]. The work by Lang et al. [1990] showed that *Ficus benjamina* could reduce iron at the cell wall that was four times greater than that of *F. marginata*. This ability was stimulated by low Fe concentrations on *F. benjamina* but not *F. marginata*.

2. Microbial Population in the Rhizosphere

Included in the rhizosphere, defined above, is a population of microorganisms that are more abundant than in the soil remote from the root. Not only may pH be different by as much as 1 to 2 units, but microscopic examination reveals that many parts of the root can be covered by bacteria 10 to 40 cells deep [Rovira and Davey, 1974]. Fungi and bacteria are often found arranged in lines along the root axis and mycorrhizal fungi may traverse all the defined zones from the outer bulk soil to several cells deep within the cortex. The change in microbial density with distance from the root of blue lupine is shown in Table 6-16. As the root grows through the soil, there is successive colonization of the root from the soil. The root tip, according to Rovira and Davey [1974], is nearly without fungi with few bacteria, and the root proximity stimulates spore germination of many nonpathogenic species. The major proliferation, as cited by Uren and Reisenauer [1988], takes place in 2 zones near the root hairs where lateral roots emerge and cells of the epidermis and cortex degenerate. Bacteria, actinomycetes, and fungi of the rhizosphere affect the host plant through their influence on nutrient availability, growth, root morphology, and nutrient uptake processes. For example, a high microbial population means that oxygen uptake for respiration will cause a lower O_2 concentration, and, conversely, the CO_2 level will be higher in the rhizosphere. These will affect redox potentials as well as nutrient solubilities in the soil solution.

Table 6-16. The rhizosphere microbial population of 18-dy-old *Lupinus angustifolius* seedlings [adapted from Rovira and Davey, 1974].

Distance from root (mm)	Microorganisms (1000s per g oven dried soil)		
	Bacteria	**Streptomyces**	**Fungi**
Root surface	159000	46700	355
0-3	49000	15500	176
3-6	38000	11400	170
9-12	37400	11800	130
15-18	34170	10100	117
80	27300	9100	91

As will be noted in the next section, part of the reason for high microbial activity close to a root is the significant root exudation of many compounds (Table 6-17). The concern with root death and disease organisms in hydroponics, discussed in Section 5.III.G.2, is due mostly to the failure to appreciate the normal microbial root population. Using the microorganisms might be better than to attempt sterile culture as suggested by the complex apparatus in Fig. 5-54.

3. Root exudation

The many microbes in the rhizosphere are partly or wholly dependent upon carbon compounds that leak or are exuded from the roots. An outline of the materials is given in Table 6-17. The compounds that come from intact root cells include, besides mucilage and other organic compounds, carbon dioxide, ethylene, nutrient ions, bicarbonate ions, protons, and electrons. These may either leak from epidermal cells (diffusates), be actively excreted (excretions), or be actively secreted (secretions) [Uren and Reisenauer, 1988]. Release of organic products is increased when high-salt roots are suddenly exposed to distilled water, acidic solutions, low calcium concentrations, iron deficiencies, or low phosphorous levels, etc. Generally, anything that increases plant growth will increase exudation.

Table 6-17. General summation of root exudates according to Uren and Reisenauer [1988].

Product	Compound
Diffusates	Sugars, organic acids, amino acids, water, inorganic ions, oxygen, riboflavin, etc.
Excretions	CO_2, HCO_3^-, protons, electrons, ethylene, etc.
Secretions	Mucilage, protons, electrons, enzymes, siderophores, allelopathic compounds, etc.
Root debris	Root cap cells, cell contents, etc.

The release of materials by a root represents a marked drain on plant resources. As noted by Barber and Martin [1976], cereals growing without microbes released material equivalent to 7 to 13% of the total dry matter production over a period of 3 weeks. In unsterilized soil, the losses increased to 18 to 25%. Barber and Martin also suggested that roots respond to dry situations by exuding additional mucilage. Clarkson [1985] concluded that the many carbon compounds lost by roots can range from 2 to 20% of the carbon fixed in the photosynthetic process. There is a functional equilibrium between the shoot and root system. A significant change in rate of shoot growth is transmitted to the roots and likewise, changes in root growth affect shoot growth. Brown and Scott [1984] conclude that cultural systems that minimize the amount of photosynthetic energy the crop must expend on the root system should lead to higher economic yields.

The layer of mucilage, starting in the root cap zone, is about 1 to 10 μm thick. The mucilage helps ensure good contact with the substrate and helps water and ionic transfer. Root cap cells can be sloughed off, yet remain viable, still producing mucilage for up to 3 weeks. Although plants adequately supplied with water and nutrients in sterile culture do not need microorganisms, few plants complete their life cycle without infestation with microorganisms. Under commercial conditions, certainly, complete sterility is highly unlikely.

The direct effects of root exudates involve increased solubility and uptake of immobile nutrients such as Fe, Mn, and P. Most iron is insoluble. For it to move to the relevant membranes, iron must be in solution. Protons and electrons, under suitable conditions, can contribute to the mobilization of iron. Most manganese in soils exists as very insoluble hydrous oxides. Again, both protons and electrons must be involved. In the case of phosphorous, the release of organic acids, enzymes such as phosphatases, and general acidification of the rhizosphere act to increase P availability. Indirect effects include the influences on the microbial population and the effects of these organisms on nutrient solubility, or the release of siderophores that chelate and mobilize Fe.

These factors of ion mobility, microbial population, and root exudation also have marked effects on soil-borne pathogens. Robson and Abbott [1989] showed that *Pythium* root rots are likely to be most severe in conditions where the host plant is under greatest stress. For *F.oxysporum* on tomatoes, plants grown in acidic conditions before inoculation are more severely affected than those grown in alkaline conditions before inoculation and then transferred to acidic conditions.

Even in hydroponic systems, exudation from roots must occur. It would appear that these exudates could be utilized as a means to improve growth if we knew more about them.

4. Mychorrizae and Symbiotic Relationships

Many fungal organisms form highly intimate and mutually beneficial relationships with plant roots –symbiosis– and are classified as mychorrizae. They are obligate parasites. VA mychorrizae have never been grown separately in culture nor have they been seen to reproduce sexually. They depend upon the host plant to give them sufficient carbon for growth. There are three general classes [Maronek et al., 1981; Gerdemann, 1974]: 1) "Ectomycorrhizae," which are most common among forest and ornamental tree species and form a thick weft or sheath of hyphal strands around the feeder roots known as the "mantle." The mantle replaces root hairs with fungal strands, greatly increasing absorptive root surfaces, and it can reach outward from the root by several meters. Hyphae also penetrate into the intercellular spaces of the root cortex, forming an interconnecting network. 2) "Endomychorrizae," the most common "vesicular-arbuscular" (VA) mychorrizae, called so as the result of the specialized vesicles and arbuscules formed within the cortical root cells. These are the most widely distributed, being found in many herbaceous shrub and tree species. There are no discernible morphological changes in the external root structure, and a loose hyphal network radiates outward several centimeters or more from the root. And 3) "Ectendomychorrizae," which exhibit characteristics of the previous two types. Koide and Schreiner [1992] cite the literature to the effect that 85 to 90% of the known seed-bearing plants (angiosperms) form this symbiosis even though there are only 120 described species of VA mychorrizae.

VA mychorrizae have been examined on Easter lily [Ames and Linderman, 1977; 1978], strawberry [Holevas, 1966], poinsettia [Barrows and Roncadori, 1977], tomatoes, cucumbers, etc. Orchids, especially, are known to require mychorrizal infection for survival, and woody species such as citrus require such infections for healthy growth. The review by Maronek et al. [1981] is particularly useful to horticulturists. Overall, the beneficial effects of symbiosis with higher plants occur in substrates with low fertility levels. Improved survival and growth of plants in landscapes or newly reclaimed land areas have been the most frequent claims [e.g., Johnson and Crews, 1979; Crews et al., 1978]. In particular, improved phosphorus nutrition has been found with the ability of VA mychorrizae to transfer phosphorus to its host. There is evidence for marked improvement in copper and zinc nutrition in some species [Clarkson, 1985]. The obvious principal benefit of mychorrizae is the expansion of hyphae far beyond the nutrient-depleted shells immediately about the roots. High fertilizer concentrations apparently reduce benefits from mychorrizae to the host plant in terms of improved nutrition, as well as reducing mychorrizal infection. Since most greenhouse container production is in pasteurized or sterilized substrates, at high nutrient concentrations, it is doubtful that deliberate mychorrizal infection as a standard practice in greenhouse production will be found cost effective. To my knowledge, direct comparisons, applicable to commercial conditions, on the benefits of deliberate mychorrizal infection have not been adequately tested.

B. BASIC FERTILITY RELATIONSHIPS

The wide range of conditions in greenhouse production makes it difficult to provide simple recommendations that are applicable in all circumstances. Crop production may be: 1) over long periods in unconfined soils, 2) in limited volumes (container production), and in substrates ranging from highly organic to essentially inert with periods ranging from a few weeks to several months (i.e., seedling transplants versus slow growing, decorative plants), and 3) in various hydroponic systems, ranging from those in which an inert substrate is periodically flushed to continuous flow systems in which the solution may be expended or reused. Any division between a true hydroponics culture and many container practices disappears.

All systems require consideration of carbonate equilibria and chelation, the latter principally with micronutrients. The standard practice of "fertigation" or injection of nutrients into the irrigation supply or nutrient solution can be applied to all systems with some variation. This latter practice, which serves as a semiautomatic fertilization procedure, simplifies cultural practice. On the other hand, liming practices are important in field soils and in container production where the substrate can be highly organic or consist partly of mineral soils.

1. Natural Fertility Levels

In an unconfined field soil, one deals with a large mass and volume. A 15 cm plow layer contains 1500 m^3 ha^{-1} and weighs more than 1000 tons, depending upon bulk density. Though a crop may be restricted to longitudinal beds, we can assume that root growth is essentially unrestricted and concentration of nutrients in the soil solution is low ($<10^{-3}$ M). Given these conditions, root extension for nutrient acquisition, at least during early crop growth stages, will be important. The volume to be dealt with means that significant physical modification can only be achieved over several years with continuous addition of amendments which may be bark, wood chips, manure, leaf mold, or such others as listed in Table 5-7 (Sections 5.III.C.4 and 5.III.F.3). Often, additions used will be those most readily available and cheap. This eliminates peat moss, vermiculite, perlite, etc. since many of these require special handling and may have to be shipped long distances. Also, as the organic content is raised, the rate at which carbon is lost also increases. The maintenance of humus at a high level is difficult and expensive. Brady [1974]

Table 6-18. Total content of various elements in soils [Abridged from Lindsay, 1979].

Element	Common range for soils (ppm)	Selected average for soils	
		ppm	Log Molar concentration at 10% moisture
Aluminum (Al)	10000-300000	71000	1.42
Arsenic (As)	1-50	5	-3.18
Boron (B)	2-100	10	-2.03
Carbon (C)		20000	1.22
Calcium (Ca)	7000-500000	13700	0.53
Chlorine (Cl)	20-900	100	-1.55
Cobalt (Co)	1-40	8	-2.87
Copper (Cu)	2-100	30	-2.33
Fluorine (F)	10-4000	200	-0.98
Iron (Fe)	7000-550000	38000	0.83
Iodine (I)	0.1-40	5	-3.40
Potassium (K)	400-30000	8300	0.33
Magnesium (Mg)	600-6000	5000	0.31
Manganese (Mn)	20-3000	600	-0.96
Molybdenum (Mo)	0.2-5	2	-3.68
Nitrogen (N)	200-4000	1400	0.00
Sodium (Na)	750-7500	6300	0.44
Nickel (Ni)	5-500	40	-2.17
Oxygen (O)		490000	2.49
Phosphorous (P)	200-5000	600	-0.71
Sulfur (S)	30-10000	700	-0.66
Selenium (Se)	0.1-2	0.3	-4.42
Silicon (Si)	230000-350000	320000	2.06
Titanium (Ti)	1000-10000	4000	-0.08
Zinc (Zn)	10-300	50	-2.12

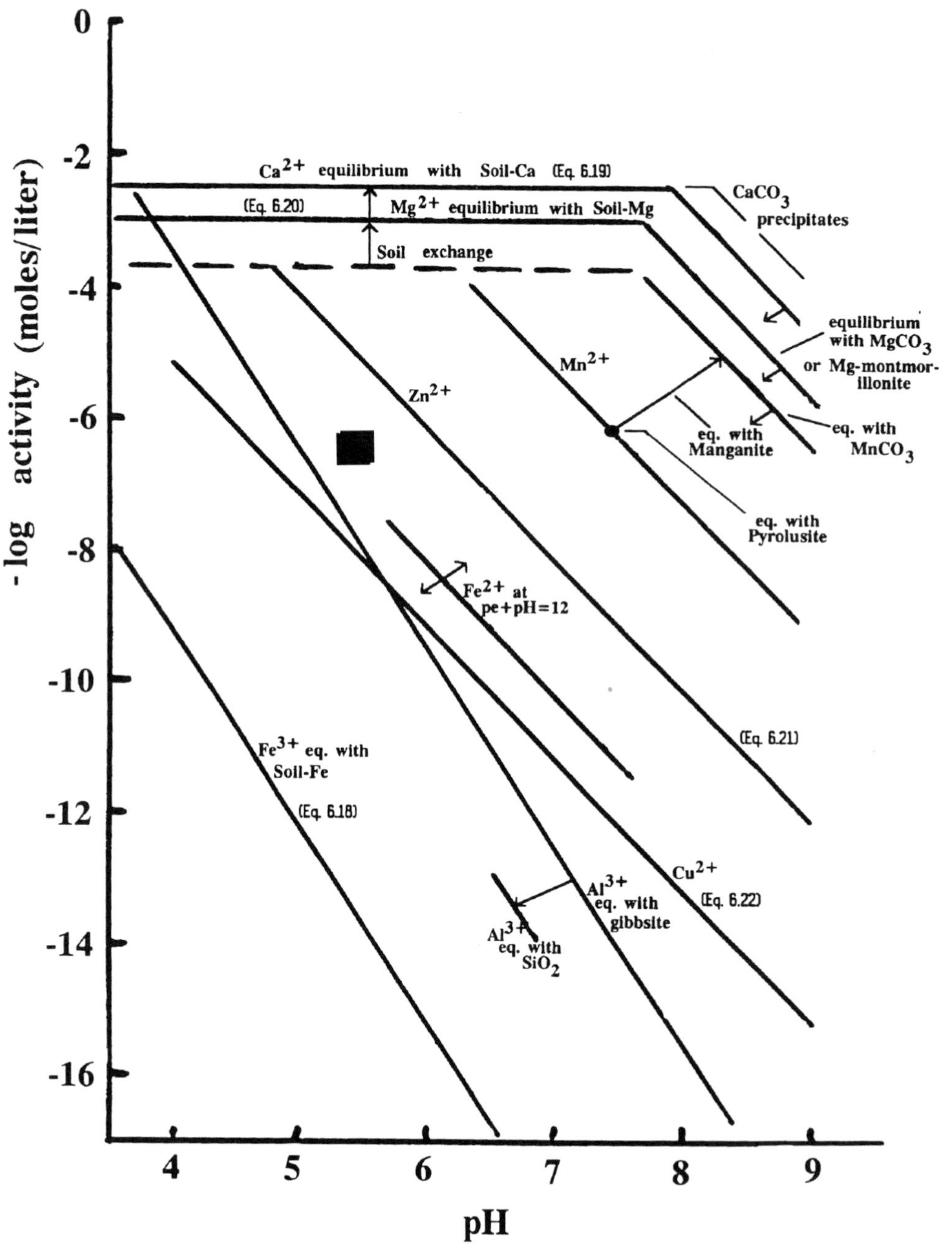

Fig. 6-26. The effect of pH on activities of several essential nutrients in soil solution. Adapted from unpublished data provided by Lindsay [1993] as a general summary of the more important ion equilibria relationships to be found in soils.

stated that it was unwise to hold organic matter above a level consistent with crop yields that pay best. Contrary to the situation of summer crop production –at least in temperate climates– continuous production, such as with roses, may occur, requiring a continuous fertilization program that varies with the season. In cut chrysanthemum culture, pasteurization and replanting of three or more crops per year will put a serious drain on soil fertility. Carnations may also be grown for more than 1 year if soil-borne pathogens are not a great problem. Numerous other cut-flower species are commonly grown in the ground and may be present for several years.

As a result, there is a restriction upon what can be done –both in terms of physical and chemical change. The grower should have some idea of basic natural fertility level of the soil, and means to improve fertility. If one looks at the total amount of the various nutrients in the soil, concentrations can be quite high. The abridged Table 6-18, from Lindsay [1979], provides an approximation of the average total elemental concentration in soils. The final column is the maximum level in the soil solution if all the elements were to dissolve in water present at 10% of the dry weight of the soil. Expressing concentration as a logarithm permits calculation of actual elemental concentration and moisture content. If a test shows a concentration of 300 ppm copper, then the ratio $(300/30) = 10^{1}$ can be added to $10^{-2.33}$ M to give $10^{-1.33}$ M for the maximum concentration of Cu at 10% moisture content. If the moisture content is 40% on a dry weight basis, then the ratio $(10/40) = 10^{-0.60}$ can be added to -1.33 to give $10^{-1.93}$ M. This is the maximum concentration of Cu in the soil containing 300 ppm Cu at 40% moisture, or 0.0117 M = 11.7 mM ℓ^{-1} . The formula can be written:

$$log[Cu^{2+}] = -2.33 + log(300/30) + log(0.1/0.4) \tag{6.40}$$

Note, in Table 6-18, the high total levels of aluminum, calcium, iron, and silicon. Although the total amounts of these and other elements are more than enough to meet plant requirements, the amounts in solution are important. Thus, one important criterion of soil tests is to determine amounts readily available to the plant. Even though Ca appears high for most soils in Table 6-18, a level of $10^{-2.5}$ M (0.003 M) is common in solution. At low pHs, Al^{3+} may be high enough to be toxic. Silicon leaches from soils and where a soil is highly weathered in humid regions, Si can be very low. Responses to silicon fertilization have been observed [Foy, 1992; Takahashi and Miyake, 1982].

a. Examples of pH Relationships

In Section 6.II, it was explained that nutrient concentration in a natural soil will depend upon the minerals present which vary with soil origin and climate (Fig. 6-5). If the controlling mineral is known, then solution concentration at equilibrium can be calculated. Examples were provided in Eqs. 6.17 through 6.22 and Fig. 6-7. Fig. 6-7 showed the activity of different micronutrients as a function of pH, whereas Fig. 6-4 showed the variation in iron and phosphorous as influenced by both pH and *pe* (redox). Fig. 6-2 compared the plant requirement for iron (10^{-6} M) with the total soluble iron to be expected in solution as influenced by pH.

The theoretical equilibrium concentrations of the ions Fe^{3+}, Fe^{2+}, Zn^{2+}, Mn^{2+}, and Cu^{2+} are repeated in Fig. 6-26. They differ from those in Figs. 6-2 and 6-7 as the result of different minerals and conditions. In practice, the concentrations as a function of pH depicted in Fig. 6-26 are not exact since equilibrium conditions are not likely. The lines do provide a starting point. Note that solution activities of Ca^{2+} and Mg^{2+} remain constant up to pH values between 7.5 and 8.0. Between pH 7.5 and 8.0, gypsum ($CaSO_4 \cdot 2H_2O$) and calcite ($CaCO_3$) can coexist. Above pH 8.0, calcite precipitates and both Ca^{2+} (Eq. 6.4) and Mg^{2+} levels drop rapidly below the usual concentrations of $10^{-2.5}$ for Ca^{2+} and 10^{-3} for Mg^{2+}. These are levels of 3 and 1 mM Ca^{2+} and Mg^{2+}, respectively. Ca^{2+} and Mg^{2+} solution levels in calcareous soils are controlled largely by the solubility of their carbonates and the CO_2 concentration [Lindsay and Moreno, 1960]. The effect of the basic mineral on ion concentrations, for Mn^{2+}, is shown by a shift upward when in equilibrium with manganite (γ-MnOOH) versus pyrolusite (β-MnO_2), and a still higher level with rhodochrosite ($MnCO_3$). In the latter case, the CO_2 level in the soil will be instrumental in controlling Mn^{2+}, whereas with manganite and pyrolusite, the redox potential will be important.

The formula for calculating the Fe^{2+} line is:

$$log\,Fe^{2+} = 15.74 - (pe + pH) - 2pH \qquad (6.41)$$

Therefore, this line will shift back and forth as determined by the redox potential *pe* + pH. If the pH is 7.0, *pe* will equal 5 (*pe* + pH = 12). The concentration of Fe^{2+} at pH 7.0 is $10^{-10.3}$ or 0.00005 μM Fe^{2+}. Iron level is insufficient (see Fig. 6-2). On the other hand, if pH drops 2 units to 5.0 (a 100-fold increase in hydrogen ion level), the Fe^{2+} concentration will be $10^{-8.26}$ M or 5 μM –a 100000-fold increase in Fe^{2+}. Several authors state that plants take up Fe^{2+} only [e.g., Barker and Mills, 1980]. One notes, however, that the Fe^{3+} concentration is nowhere high enough to be of direct importance in iron nutrition. The fact that only Fe^{2+} is taken up is largely due to the low concentration of Fe^{3+}.

Influence of a primary mineral can also be noted by the shift in Al^{3+} line, depending upon whether the ion is in equilibrium with gibbsite (γ-$Al(OH)_3$) or quartz (α-SiO_2). The effect of pH on molybdenum would be opposite to those equilibria shown in Fig. 6-26. Concentration of MoO_4^{2-} increases with increasing pH so that at pH 5, activity will be $10^{-8.4}$, versus 10^{-6} M at pH 8 (0.004 versus 1.0 μM). The relationships of boron are not well understood. Lindsay [1978; 1991] indicated activity levels for $H_3BO_3^{\circ}$ of 10^{-5} in 1978 and $10^{-5.5}$ in 1991, the latter based on newer information, regardless of the pH between 5 and 11. The ions $H_2BO_3^-$, HBO_3^{2-}, and BO_3^{3-} are present in solution at pHs above 10: otherwise the neutral molecule $H_3BO_3^{\circ}$ is dominant.

One may note that the usual range of micronutrients in soil solutions as calculated by the various authorities will vary from about 10^{-4} to less than 10^{-16} M ℓ^{-1}. Deficiencies will generally occur when concentrations drop to 10^{-6} or 10^{-8} M –a 100-fold concentration difference. Unless added, levels of Ca^{2+} and Mg^{2+} will be less than 10^{-2} M.

b. Nitrogen Relationships:

In most cropping systems, Baker and Mills [1980] stated that available nitrogen is the most limiting factor in crop growth, complicated by the fact that less than 50% of the nitrogen fertilizer applied may be used. More than 90% of the N in the surface soil layers is organically combined [Stevenson, 1991], although some authors suggest 95% or more [i.e., Legg and Meisinger, 1982]. Nitrogen is also one of the most widely distributed elements in nature, with the highest amount fixed in rocks and sediments.

A general perspective of the nitrogen cycle as modified in greenhouse production is given in Fig. 6-27. About 2 to 3% of the nitrogen is released yearly from the organic nitrogen pool by the process known as "mineralization." The product is NH_4^+, of which large amounts are used by the soil microbial population, some by the crop, some by fixation in clay minerals and organic matter, or it may be lost as NH_3 to the air. In well-aerated soils, with an abundance of basic cations, "nitrification" occurs, in which NH_4^+ is converted to NO_2^- by the bacterium *Nitrosomonas* and very shortly to NO_3^- by the bacterium *Nitrobacter*. Thus, NO_2^-, which is highly phytotoxic, does not accumulate. Other organisms can oxidize the ammonium ion. Although most plants can assimilate NH_4^+, previous discussion (Section 6.III.D.3) showed that NO_3^- is commonly the form used. Small amounts of many kinds of salts stimulate nitrification, whereas the application of large amounts of NH_4^+ fertilizers to alkaline soils have been found to depress the second step, allowing toxic accumulation of NO_2^-. Similarly, urea may cause adverse effects since it is converted quickly to ammonium ion. All these compounds can be lost to the atmosphere as N_2, N_2O, NO, and NH_3. In the case where nitrate is reduced, the process is called "denitrification." Brady [1974] states that rapid losses can occur with heavy application of urea or ammonium fertilizers, especially in alkaline soils. In greenhouse practice, particularly where soil fumigation or steam pasteurization is practiced, buildup of NH_4^+ can occur since the organisms in nitrification are killed. Heavy use of manures can be particularly dangerous in this respect.

When organic matter is added to a soil, especially materials having a high carbon/nitrogen (C/N) ratio, the microbial population, during decomposition, will claim most of the ammonium and nitrate for their own use. This is part of the "immobilization" possible in soils. Nitrogen deficiencies can readily occur until after the carbonaceous matter has been decomposed sufficiently to reduce the influence of the microorganisms. In fact, I have seen severe damage from nitrogen excess when the large microbial population dies off after decomposing a fine material such as sawdust or straw where the C/N ratio can range from 50:1 to 100:1. Well-balanced nutritional conditions are represented by a C/N ratio of about 25 [Jansson and Persson, 1981]. Decomposition

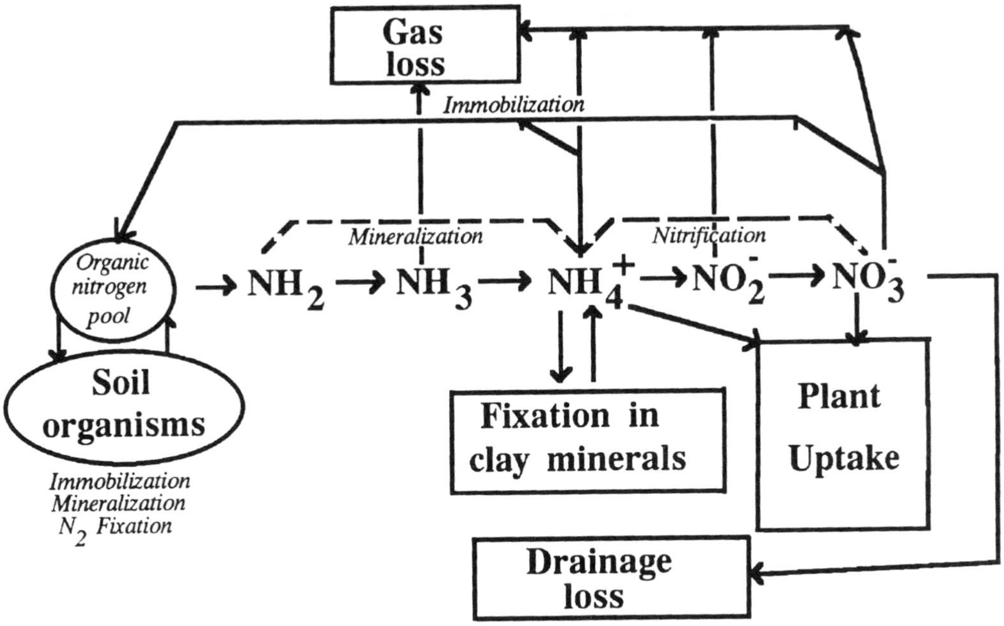

Fig. 6-27. General soil nitrogen cycle. Additions to soil nitrogen in greenhouses by rainfall are unavailable. Nitrogen fixation by bacteria unlikely to be of any importance in most greenhouse production.

Fig. 6-28. Schematic of changes that occur in pH and nitrogen when organic-type fertilizers are used in peat-sand substrates. The amount of nitrites will depend upon the rapidity of the conversion to nitrate by bacteria. Excessive ammonium combined with pasteurization can result in phytotoxicities [Adapted from Bunt, 1976].

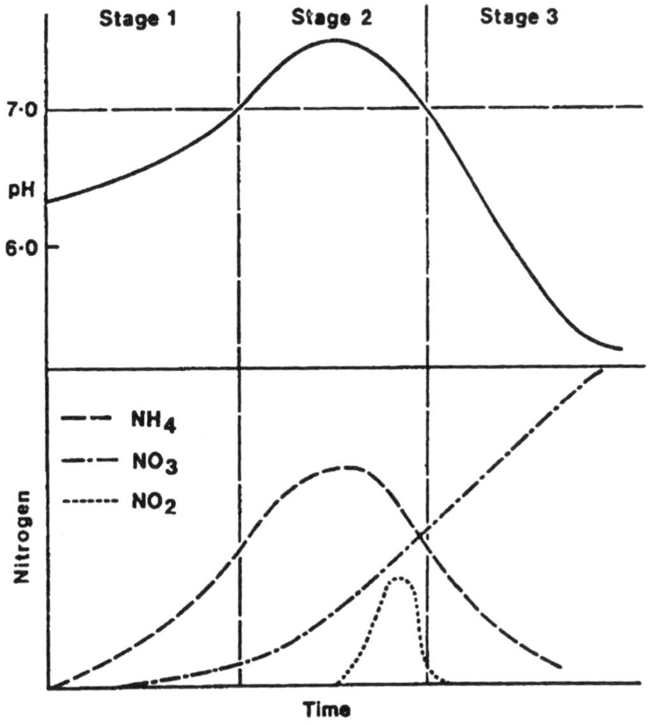

of organic materials, especially peat moss, can be very rapid under greenhouse conditions. The microbial population may vary drastically, with marked effect on nitrate and ammonium concentrations in the soil solution. A diagram showing the possible changes in composts is presented in Fig 6-28.

The mobility of NH_4^+ in soils is restricted since it will be attracted to negative colloids, and it can be fixed almost permanently on some clay minerals such as vermiculite, illite, and montmorillinites. Nitrate, on the other hand, is highly mobile and readily lost from the soil. This has led to consideration of nitrate toxicities in ground water in several parts of the world. This will be addressed later.

Although there can be no rain to cause leaching in greenhouses, excess irrigation, combined with high NO_3^- application, can drastically reduce the efficiency of NO_3^- uptake by a crop.

As discussed earlier (Sections 6.III.C.1, 6.III.D.3), plants can extract from solutions ions that are in very low concentrations. One figure given for NO_3^- was 0.003 mM or $10^{-5.5}$ M. Mengel and Kirkby [1982] summarized the situation of nitrogen in the soil solution by stating that levels of NH_4^+ and NO_3^- can change rapidly –particularly with respect to nitrate. NO_3^- can be as high as 20 to 30 mM after fertilizer application ($10^{-1.5}$ to $10^{-1.7}$ M). In fertile soils, however, the range may be 2 to 20 mM, depending upon the rate of mineralization and plant uptake. Clement et al.[1978] stated that roots of agricultural crops could be exposed to nitrogen levels ranging from 1.43 µM to 1430 mM, or $10^{-5.6}$ to $10^{0.2}$ M. It will be noted later that, in greenhouse production, nitrate and ammonium concentrations are almost always around 10^{-2} M or higher. This results, especially, from plant requirements where they are grown in small volumes or in hydroponics.

c. Phosphorus Relationships

Phosphorus is the second most important soil fertility problem in the world [Lindsay et al., 1989] today. It is only a minor component in soils with a large part in the organic fraction that is still poorly characterized. The element has unique solution properties, occurring as "orthophosphate" ions (Fig. 6-29) absorbed by plants ($H_2PO_4^-$, HPO_4^{2-}), and they are highly reactive. The result is a very low solution concentration under usual conditions, requiring rapid replenishment from the solid phases. Peaslee [1981] stated that phosphate ions may diffuse over a distance of 0.2 to 2 mm to the roots, with solution concentrations ranging from 1 to 2 µM ℓ^{-1} (10^{-5} - 10^{-6} M), depending upon the pH. In restricted volumes, the phosphorous application may have to be increased more than 20 times the usual field practice (see Section 5.III.F.2 and Tables 5-4 through 5-6). Soils exhibit sorption and desorption buffering properties by removing phosphate from solution during enrichment and releasing it during depletion. When soluble sources are applied, the phosphorous is often "fixed" or rendered unavailable to plants. As a result, the phosphorous added in fertilizers exceeds that removed by crops by more than 24% [Brady, 1974], and in some areas, additions of P may be more than triple the removal by crops. In contrast to nitrates, P is relatively immobile and seldom found in any significant concentrations in the leach water.

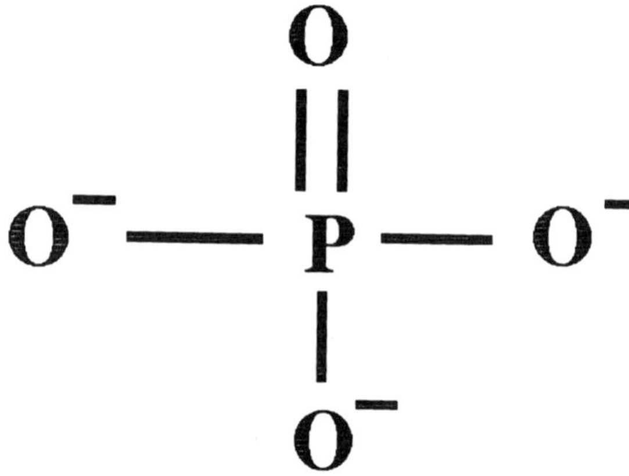

Fig. 6-29. The orthophosphate ion with P surrounded by four O in a tetrahedral arrangement.

Lindsay et al. [1989] list 65 materials formed from the reaction of phosphate fertilizers. A short list is provided in Table 6-19 as an example of the bewildering number of minerals and compounds. The application of phosphates (e.g., superphosphate, Table 6-5) results in interesting and complex chemical reactions as investigated by Lindsay and others in the 1960s and 1970s [Lindsay, 1959; Lindsay and Taylor, 1960; Lindsay and DeMent, 1961; Watanabe et al., 1965; Lindsay and Stanford, 1960; etc.]. The situation close to a phosphate fertilizer pellet is diagrammatically presented in Fig. 6-30. In the presence of a fertilizer granule, water is condensed and moves into the pellet, which dissolves the chemical. The pH can be reduced below 2.0 so that numerous aluminum, iron, calcium, ammonium, and other compounds are formed with the phosphate. Lindsay and Taylor [1960] divided these chemicals into those formed at pHs below 4.0 and those above 6.0. Products formed from polyphosphate fertilizers are a third group. As the solution, moving outward from the pellet, is neutralized, complex phosphate products are precipitated. DCPD is one of the major initial compounds formed, and where MCP (monocalcium phosphate) is applied, there is sufficient calcium to precipitate half the phosphorous applied. Above pH 6.5, DCP (dicalcium phosphate) converts to OCP. Octacalcium phosphate is unstable and gradually converts to apatite in neutral or alkaline soils, becoming less available. Lindsay and co-investigators have also documented many equilibria between phosphate ions and the numerous minerals to be found in various soils [e.g., Lindsay and Moreno, 1960; Boyle and Lindsay, 1986; Lindsay, 1979; etc.]. The reactions are strongly influenced by pH and pe, with some examples presented in Figs. 6-3 and 6-4. Separation as to general types of compounds can be based on whether the soil is acidic or alkaline and the basic reaction of the fertilizer. Lindsay and Taylor [1960] classified diammonium phosphate (DMP) and dipotassium phosphate (DKP) as in the neutral or alkaline class as compared to MCP or MAP (monoammonium phosphate). Depending upon the relative concentrations of the respective reactants, phosphorus can be rendered unavailable, or P can tie up other essential elements. Most of the reactions elucidated are extremely slow [Lindsay and Moreno, 1960], with biological reactions responsible for frequent changes in phosphate equilibria.

Table 6-19. Abridged table of compounds formed from reaction of phosphate fertilizers with soils [Adapted from Lindsay et al., 1989; Lindsay and Taylor, 1960; Lindsay, 1979].

Name	Acronym	Chemical formula
Variscite		$AlPO_4{\cdot}2H_2O$
NH_4-taranakite		$Al_5(NH_4)_3H_6(PO_4)_8{\cdot}18H_2O$
K-taranakite		$Al_5K_3H_6(PO_4)_8{\cdot}18H_2O$
Monetite (dicalcium phosphate anhydrous)	DCPA	$CaHPO_4$
Brushite (dicalcium phosphate dihydrate)	DCPD	$CaHPO_4{\cdot}2H_2O$
Octocalcium phosphate	OCP	$Ca_8H_2(PO_4)_6{\cdot}5H_2O$
Hydroxyapatite	HA	$Ca_{10}(PO_4)_6(OH)_2$
Fluorapatite	FA	$Ca_{10}(PO_4)_6F_2$
Strengite		$FePO_4{\cdot}2H_2O$
Vivianite		$Fe_3(PO_4)_2{\cdot}8H_2O$
Struvite		$MgNH_4PO_4{\cdot}6H_2O$
Reddingite		$Mn_3(PO_4)_2{\cdot}3H_2O$
β-Tricalcium phosphate	β-TCP	β-$Ca_3(PO_4)_2$
Hydrogen ammonium iron phosphate		$H_8NH_4Fe_3(PO_4)_6{\cdot}6H_2O$

In general, the materials used in loamless substrates do not provide a natural phosphorus supply, nor do they have the ability to fix or retain phosphorus to the same extent as mineral soils [Bunt, 1988]. Greater attention must be given to maintaining phosphate levels in greenhouse mixes, with care taken to ensure solubility in the $H_2PO_4^-$ and HPO_4^{2-} forms. High concentrations encountered in mixing tanks for fertigation systems can result in precipitation of calcium phosphates and others. This can also occur in alkaline irrigation supplies. The grower winds up with a sludge in the bottom of his mixing tanks, or a stopped-up trickle irrigation system.

d. Potassium Relationships

The greatest part of potassium is bound on secondary clay minerals whose particle sizes are less than 2 μm in diameter. Soils rich in clay are generally rich in potassium [Mengel and Kirkby, 1982]. Organic soils are likely to have a much lower K content. Except for soils of a sandy nature, most mineral soils are comparatively rich in K. The main source of K^+ comes from weathering of potassium-containing minerals such as vermiculite, illite, montmorillonite, etc. As with other elements, the relationships between soil minerals and plant roots can be divided into two or three major sections (Fig. 6-31): 1) K fixed as the structural element of soil minerals, 2) K^+ adsorbed in the exchangeable form on the soil colloids, and 3) K^+ present in the soil solution. The situation with respect to K parallels P and N by the fact that large portions are insoluble and relatively unavailable to plants. Lime applications are likely to increase K fixation, and K in well-limed soils is not as likely to be leached out as K in acid soils. Luxury consumption by plants can occur with K^+, so that light and frequent applications are usually superior compared to heavier and less frequent. Fertigation methods are an excellent means of potassium fertilization. In container production, K in the tissue of pot plants is likely to be of the same amount as N, and experiencing luxury consumption of K is usual.

Replacement of K by Na can occur –up to a point. The extent to which substitution can be made, however, depends on the potential for Na^+ uptake. There have been reports of the use of NaCl in solution culture, with an increase in tomato yield under U.K. winter conditions [e.g., Adams, 1987]. Similar reports were cited by Mengel and Kirkby [1982]. However, high Na^+ levels have been shown to increase tipburn on lettuce

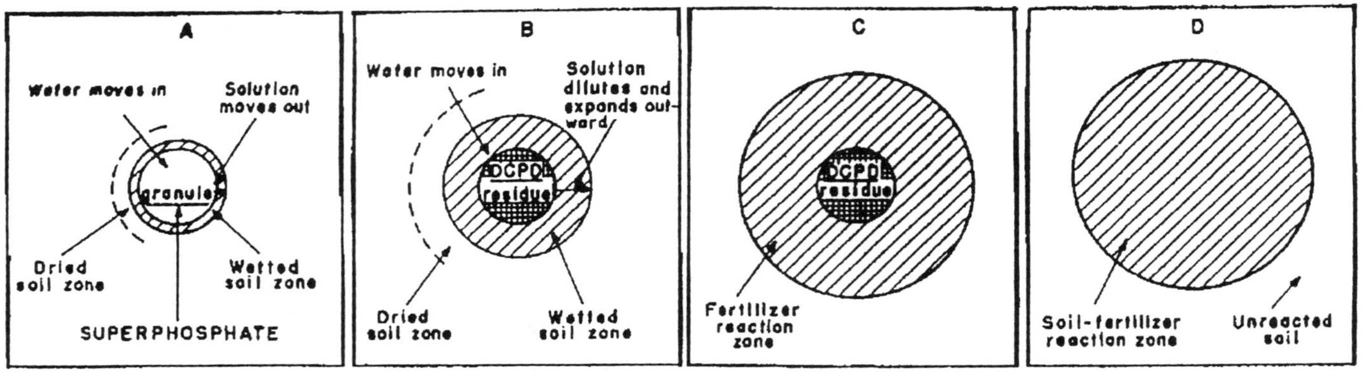

Fig. 6-30. Situation over time for a superphosphate granule in a soil. Water is condensed and moves into the granule, dissolving it within 6 to 72 hr. In acid soils the major reaction product is DCPD at pHs below 2.0. The solution dissolves large quantities of Al, Fe, Mn, Ca, K, etc. As the front moves outward, the pH gradually rises and much of the applied phosphorous is precipitated in combination with Al, Fe, and Ca. In alkaline soils, the fertilizer particle dissolves to form basic calcium phosphates [Adapted from Lindsay and Stanford, 1960].

[Sonneveld and Mook, 1983] and accumulate in roses to the point that leaf margins may burn [Hughes and Hanan, 1977]. Overall, under semiarid conditions of Colorado, minimum solution concentration is needed without sodium [Hanan, 1982b; 1984]. Schekel [1971] recommended that if the ratio of K^+ to Na^+ was 1:1, carnations would tolerate Na^+ up to maximum levels of 6 mM of either ion in solution. The problem with most studies of this type is the tendency to ignore the accompanying ion. There is considerable difference about whether Na^+ is supplied as NaCl, $NaHCO_3$, or another salt containing sodium [Sonneveld and Mook, 1983]. The behavior of the accompanying anion can be completely different in soil versus a hydroponics system. Sodium minerals are too soluble to precipitate in soils, and either accumulate as soluble salts in poorly drained soils or are leached, whereas potassium can be re-fixed on clay minerals [Lindsay, 1979]. Overall, the combinations of Na^+ and K^+ with anions (e.g., SO_4^{2-}, Cl^-, HCO_3^-, OH^- and CO_3^{2-}) in most mineral soils are not important.

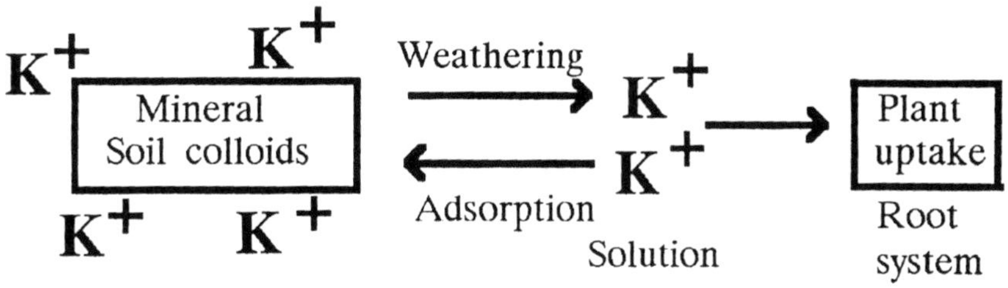

Fig. 6-31. Potassium relationships in soil.

e. Sulfur Relationships

SO_4^{2-} is the principal ionic sulfur form in solution and is not as strongly bound to soil particles as phosphate –although the S content of plants is on the same order as P (Table 6-12). Also, significant amounts of S are taken up from the air or from fertilizers containing sulfur. The atmosphere in industrialized areas is usually high in SO_2, and although sulfur dioxide is considered an air pollutant, it is also a source of sulfur when oxidized to SO_4^{2-}. Sulfur deficiencies have been noted in several areas, and responses to sulfur applications have been noted [Mengel and Kirkby, 1982]. The movement to high analysis fertilizers, air pollution control, and replacement of sulfur-containing pesticides has increased the occurrence of sulfur deficiencies. Lindsay [1979] states that the solubility of SO_4^{2-} in soils is limited by the solubility of gypsum ($CaSO_4{\cdot}2H_2O$). When in equilibrium with soil-Ca at $10^{-2.5}$ M, SO_4^{2-} is $10^{-2.1}$ M (7.9 mM). As Ca^{2+} activity is reduced by the precipitation of $CaCO_3$ at pHs approaching 8.0, SO_4^{2-} activity can increase. Fortunately, plants are more tolerant to SO_4^{2-} excess compared to chlorides. Schekel [1971] reported that carnations would produce an acceptable yield at 20 mM/ℓ SO_4^{2-} in hydroponic systems, as contrasted with Cl^- concentrations at 2 mM. Generally, in greenhouse production, sulfate fertilizers and elemental sulfur are used as means to reduce pH. This will be discussed later.

C. CARBONATES AND pH CONTROL

1. Carbonate and CO_2

Root substrates and also hydroponic systems can be considered as "open" with regard to carbonates since CO_2 can escape from solution or return to precipitate carbonate materials. CO_2 dissolves in water to form dissolved CO_2^o and undissociated carbonic acid, $H_2CO_3^o$. In solution, carbonic acid dissociates to give:

$$H_2CO_3^o \rightleftharpoons H^+ + HCO_3^- \qquad (6.42)$$

At pH 6.36, the ratio of (HCO_3^-) to ($H_2CO_3^o$) is 1. The ratio increases 10-fold for each 10-fold increase in pH (one unit), and vice versa with a decrease in pH. The bicarbonate ion also dissociates to H^+ and CO_3^{2-}. At pH 10.33, the ratio of these two is one with a change similar to that for $H_2CO_3^-$ and $H_2CO_3^o$ when pH changes. As the result of CO_2(g) levels in the substrate and pH, the concentrations of CO_3^{2-} and HCO_3^- can be calculated for a CO_2-H_2O system:

$$\log CO_3^{2-} = -18.15 + 2pH + \log CO_2(g) \quad (6.43)$$

$$\log HCO_3^- = -7.82 + pH + \log CO_2(g) \quad (6.44)$$

Table 6-20 provides examples of a carbonate in solution at different CO_2 concentrations, and Fig. 6-32 shows the effect of pH on the fraction of various carbonate ions. The table and figure serve to emphasize that carbonate in some form will always be present in solutions, whatever the system the grower uses. Due to the release of CO_2 by roots and microorganisms, CO_2 levels will always be higher in field soils compared to the atmosphere (0.030 kPa). Helyar and Porter [1989] suggest the partial pressure of CO_2 in soils to be 5 to 20 times higher than in the atmosphere. Lindsay [1979] usually increases CO_2 ten times. In shallow, porous greenhouse substrates, differences between the atmosphere and gas in the soil pores may not be detectable [Hanan, 1964]. Nevertheless, in the root rhizospheres, and microsites in the substrate, CO_2 in any system <u>must</u> be higher (see Fig. 5-22). CO_2 levels in hydroponic systems have not been thoroughly examined to my knowledge. Hurd [1978] did discuss the root environment of NFT systems, as did Steiner [1968]. Neither of these authors, however, cited any ℓ ature giving actual CO_2 levels to be expected close to the roots of crop plants in hydroponic systems.

Table 6-20. Distribution of carbonate species in solution as a function of gaseous CO_2 concentration [Adapted from Lindsay, 1979].

CO_2(g) (atm)	CO_2(g) (kPa)	log CO_2(g) (atm)	log $H_2CO_3^o$ (M)	log HCO_3^- (M)	log CO_3^{2-} (M)
0.0003	0.030	-3.52	-4.98	pH-11.34	2pH-21.67
0.003	0.303	-2.52	-3.98	pH-10.34	2pH-20.67
0.01	1.01	-2.00	-3.46	pH-9.82	2pH-20.15
0.1	10.1	-1.00	-2.46	pH-8.82	2pH-19.15
1.0	101.0	0.00	-1.46	pH-7.82	2pH-18.15

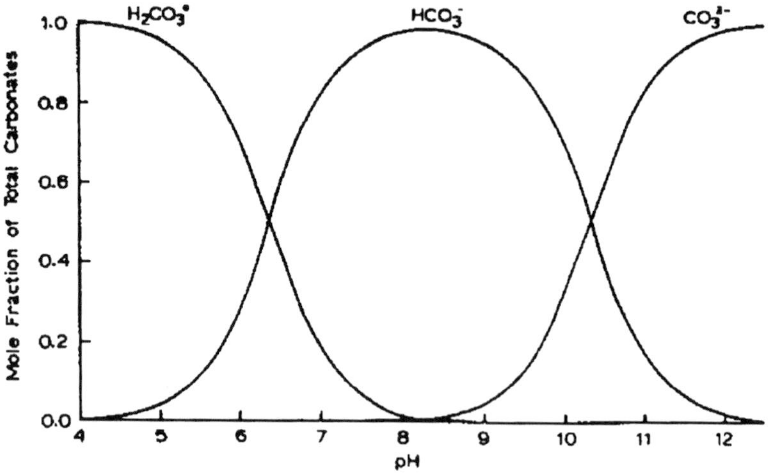

Fig. 6-32. The effect of pH on distribution of carbonate ions in solution [*Chemical Equlibria in Soils*, Lindsay, W.L., ©1979. Reprinted by permission of John Wiley & Sons, Inc.].

2. pH Control

a. Raising pH:

Most temperate soils are generally acid, and weathering will result in continuous acidification to the point that aluminum, manganese, or hydrogen toxicities may occur, plus deficiencies of phosphorous, molybdenum, and calcium [Ritchie, 1989]. Even though, in greenhouse production, the soil is not subject to rainfall, high irrigation rates, combined with some fertilizer types, can cause rapid acidification (Table 6-5).

There are exceptions in greenhouse practice such as bluing of hydrangeas where highly acidic chemicals (aluminum sulfate) are used to reduce pH deliberately. However, it is a common practice to lime soils to correct highly acid conditions. It is necessary in container production where the substrate may use media combinations incorporating peat moss. It is not the objective to apply alkaline fertilizers such as limestone to the point of neutrality (pH = 7.0) since this can depress yield. Limestone is primarily applied to correct toxicities associated with acid soils [Cregan et al., 1989]. Liming not only neutralizes soil acidity, but replenishes exchangeable Ca^{2+} [Lindsay, 1979]:

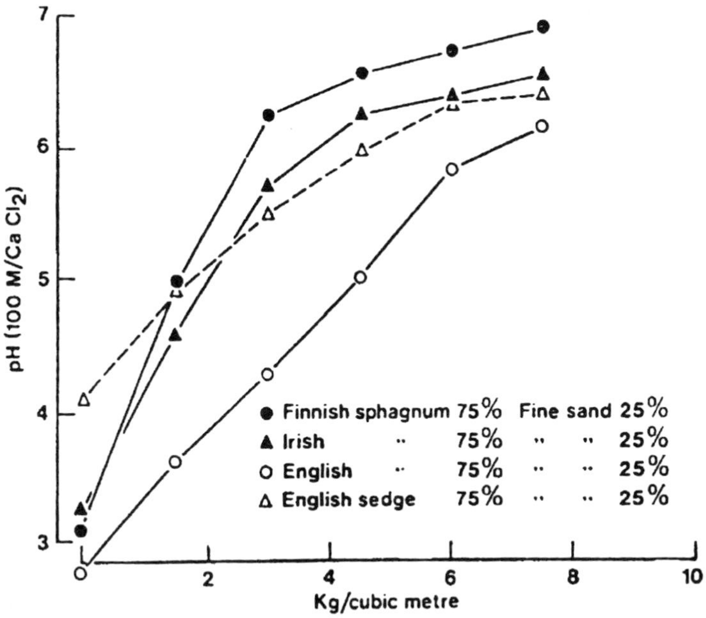

Fig. 6-33. Lime requirement of four peat-sand substrates made with different peat moss sources. Liming material was equal parts of $CaCO_3$ and $CaMg(CO_3)_2$ [Adapted from Bunt, 1988].

$$CaCO_3(c) + [2H^+\text{-}exchange] \rightarrow [Ca^{2+}\text{-}exchange] + CO_2(g) + H_2O \tag{6.45}$$

$$CaCO_3(c) + [0.66Al^{3+}\text{-}exchange] + 2H_2O \rightarrow [Ca^{2+}\text{-}exchange] + 0.66Al(OH)_3(s) + CO_2(g) + H_2O \tag{6.46}$$

The result is an equation:

$$\log Ca^{2+} + 2pH = 9.74 - \log CO_2(g) \tag{6.47}$$

which can be rearranged:

$$pH - 0.5pCa = 4.87 - 0.5\log CO_2(g) \tag{6.48}$$

Eq. 6.48 is called the lime potential relationship, where pCa = -log Ca^{2+} and CO_2 is given in atmospheres pressure. According to Mengel and Kirkby [1982], the usual application rates for mineral soils are about 4 to 6 tons per hectare over 3 to 5 years. The amounts required not only depend upon soil pH (actual acidity), but also on the H^+ adsorbed to the exchangeable system (colloids) (potential acidity). Some people call the latter the "reserve acidity." Obviously, the higher the cation exchange capacity of a soil, the more likely that an acid soil will have a high reserve acidity.

Limestone is a calcium carbonate ($CaCO_3$) or calcite. Another common liming material is dolomitic limestone ($CaMg(CO_3)_2$) (Table 6-5). Still others that are occasionally used are burned lime (quicklime, CaO) and hydroxide of lime ($Ca(OH)_2$, slaked lime). The average purity of crushed limestone is about 94% [Brady, 1974]. CaO and $Ca(OH)_2$ are powder forms and readily available. Of course, other chemicals give a basic reaction in solution, a number of them given in Table 6-5. Those with sufficient solubility to be used in fertigation methods or hydroponics can cause the solution pH to move upward gradually, as contrasted with the reaction rate in traditional liming materials as noted. The major changes in peat-based substrates can occur over a period of 2 days and be essentially complete in 14 days [Williams et al., 1988]. The medium perlite, some types of vermiculite, and many organic fertilizers (i.e., manure, wood ashes, etc.) can also contribute to liming responses of substrates into which they are incorporated. On the other hand, peat moss in perlite, used for rooting purposes, requires about 0.5 kg powdered calcium carbonate per 10 m^2. Otherwise, the rapid pH drop will cause aluminum to be released from the perlite with consequent phytotoxicity. Fig. 6-33 shows the lime requirement for different peat-sand materials. Tables 5-4 through 5-6 show that some type of liming material is invariably mixed into the substrates. Bunt [1988] presented rates of limestone application ranging from 0.9 kg m^{-3} for Finnish sphagnum to 2.0 kg m^{-3} for English sphagnum to raise pH by 1 unit. Liming reactions, however, are not easily predictable, requiring tests to decide desirable application rates.

The application to soil of two different liming materials in chemically equivalent quantities does not always mean that equivalent results will be attained. If both limestones, there may be differences in hardness and particle size. The latter are important since the finer the particle size the more rapid its reaction rate. Cregan et al. [1989] state that ground limestone reaches 100% efficiency at a fineness of about 150 μm. Experimentally, limestone that passes 60- to 100-mesh screens (250-150 μm) is the largest particle size that is 100% efficient. Brady [1974] gives the requirement that all material must pass a 10-mesh screen (1.8 mm diameter) and at least 50% of the pulverized limestone must pass a 100-mesh screen (0.15 mm diameter). Such material is called "fine" limestone. Grinding to a smaller size is not usually economical. There have been occasions when 2- to 3-cm diameter stone limestone has been used in gravel hydroponic beds. Applications must be mixed uniformly in the substrate. This is particularly important in infertile, acid soils. The target pH is 5.5 to 7.0 [Cregan et al.,1989], although Bunt [1988] recommended for organic soils a pH range of 5.0 to 5.5. Deliberate liming to neutrality (7.0) is excessive and expensive. The idea is to retard or slow acidification.

b. Lowering pH

When calcium carbonate is present in soil, Lindsay [1979] states that it has a dominating influence on many soil properties. Most calcareous (semiarid or arid) soils have a pH range of 7.3 to 8.5. The reaction showing the dissolution of $CaCO_3$ was given in Eq. 6.3 (Section 6.II.B), and the resulting formulae, expressing Ca^{2+} activity as functions of pH and CO_2 levels, appear in Eqs. 6.47 and 6.48. What can be done, practically, depends upon calcite-gypsum equilibria and CO_2 concentrations.

It is doubtful that saline-sodic or sodic soils will be encountered in greenhouse practice. The latter do not contain a great amount of neutral soluble salts, the phytotoxicity arising from sodium and hydroxyl ions, resulting in pHs above 10.0. The former usually has more than 15% of the CEC saturated with Na^+, with a saturated paste extract above 4 dS m^{-1} EC, and a pH below 8.5. These types of soils are extremely difficult to manage and are uneconomical for greenhouse production requirements. Saline soils, however, may be encountered –sometimes from the mismanagement of fertilizer programs. Brady [1974] attributes these types with high neutral soluble salts, less than 15% of their CEC is saturated with Na^+, and the electrical conductivity (saturated paste extract) is above 4 dS m^{-1}. With saline soils, the excess salts (mostly chlorides and sulfates of sodium, calcium and magnesium) can be leached with no appreciable rise in pH. Obviously, the irrigation supply must be low in sodium. High applications of gypsum are the usual means to reclaim saline or saline-sodic soils. The latter should not be leached with water low in calcium or magnesium since the pH will rise rapidly and a tight, impervious soil structure results. The neutral salt gypsum ($CaSO_4{\cdot}2H_2O$) is often recommended at several tons per hectare, mixed into the soil, to help reclaim saline or sodic soils. If calcium is required without a change in pH, calcium sulfate is recommended. Gypsum changes the caustic alkali carbonates into milder sulfates (e.g., Na_2SO_4) that can be leached.

In soils with pH above 7.0, calcite, as noted previously, precipitates. Fig. 6-26 showed the relationship with pH, and Fig. 6-34 examines the relationship in greater detail, including the effect of CO_2. As CO_2 increases, the calcite line shifts to the left. At a very high CO_2 level of 30.3 kPa, calcite could exist at pH 5.0 if (Ca^{2+}) = 1.0 M ℓ^{-1}. At the usual atmospheric level (30.3 Pa), the Ca^{2+} activity would have to be several moles. This is a situation impossible in productive soils. One also notes that at any given CO_2 concentration, the effect of raising or lowering pH moves along the calcite equilibrium line. Surprising to some, calcium chloride ($CaCl_2$) is more effective than an equal amount of hydrochloric acid (HCl) in lowering pH. $CaCl_2$ raises Ca^{2+} activity, precipitates calcite, absorbs more CO_2 from the air, forming carbonic acid that dissociates and lowers pH [Lindsay and Ajwa, 1993]. In the pH range at which gypsum and calcite can coexist (7.5-8.0), adding H_2SO_4 merely causes calcite to dissolve and gypsum to precipitate [Lindsay, 1979]:

$$H_2SO_4 + CaCO_3(calcite) + H_2O \rightleftharpoons CaSO_4{\cdot}2H_2O(gypsum) + CO_2(g) \tag{6.49}$$

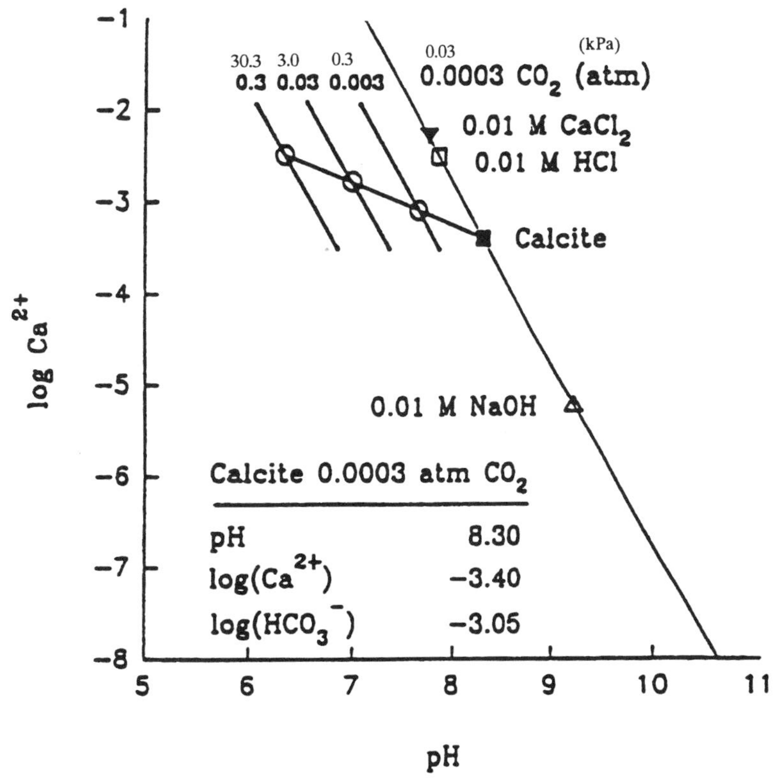

Fig. 6-34. Ca^{2+} activity versus pH for equilibrium with $CaCO_3$ at various CO_2 concentrations. At (CO_2) = 30.3 Pa, and at equilibrium (dark square), the pH = 8.30, $(Ca^{2+}) = 10^{-3.40}$ and $(HCO_3^-) = 10^{-3.05}$ M ℓ^{-1} [Lindsay and Ajwa, 1993]. Gypsum not included in this calculation.

I have seen growers attempt acidification of calcareous soils with sulfuric acid –much to the detriment of metal piping and pumps. One can calculate the equilibrium of calcite-gypsum as pH = 7.8 for 30.3 Pa CO_2. This will vary due to other components in the soil, but the addition of CaO, $CaCO_3$, $CaSO_4{\cdot}H_2O$, and H_2SO_4, to a soil containing gypsum and calcite will not affect the pH of that soil.

A list of materials used to acidify soils is given in Table 6-21. Depending upon the particular situation, some of these, as suggested above, may not be desirable. Ferrous sulfate and elemental sulfur are generally recommended. Sulfur is oxidized in the medium to H_2SO_4 by the bacterium *Thiobacillus*, hence the reaction is slow. As potting substrates may have low amounts of calcite, the pH lowering follows the curves given in Fig. 6-35 for selected substrates. Tables 5-4 through 5-6 provide suggested amounts of $CaCO_3$ for acid media. It can be appreciated that most published recommendations for container production come from experience in temperate regions with acid soils. Where the grower uses a calcareous soil as part of the substrate, the recommendations will be changed. Because conditions are so variable, the grower must experiment or rely on the local agricultural experiment station for suggestions.

For hydroponic systems where CEC is, to all purposes, nonexistent, any of the acids or soluble acid materials listed in Table 6-21 can be used to adjust pH. For upward adjustment, bases such as sodium hydroxide (NaOH), potassium hydroxide (KOH), or ammonium hydroxide (NH_4OH) can be utilized. Cooper [1979] devoted considerable space to pH adjustment in NFT solutions, suggesting for most crops that pH should be controlled between 6.0 and 6.5. In actuality, controlling pH in the range suggested is extremely difficult. The previous

Section 6.IV.C.1 emphasized that CO_2 will dissolve in water, and remarkable buffering of the solution will occur. For example, Table 6-22 is a selected list of hydroponic solutions checked for electrical neutrality. Hoagland's solutions are probably the most commonly employed in biological research at one-half the concentrations given. If the data for each solution in Table 6-22 are submitted into Allison et al.'s [1991] chemical equilibrium program MINTEQA2/-PRODEFA2, using Lindsay's modifications, the equilibrium pHs of all solutions lie between 7.0 and 7.2. These salts have been added to pure water and the equilibrium carbonic acid concentration ($H_2CO_3^o$), where the gaseous CO_2 level is set to 0.0003 atm (30.3 Pa), is $10^{-4.992}$ M (10.2 μM). The concentration where the CO_2 pressure is assumed 0.003 atm (303 Pa) is $10^{-3.992}$ (101.9 μM). These concentrations agree well with those given in Table 6-20.

The effect of adding an acid such as HCl, or a base such as NaOH, is depicted in Fig. 6-36. Not until almost 100 μM of acid has been added does pH begin to drop, decreasing more rapidly so that at 800 μM HCl, the equilibrium pH is almost 5.0. The buffering effect in raising pH with NaOH is more difficult. An addition of 800 μM base does not raise pH significantly. At the point where sufficient acid or base is added, the pH control becomes difficult. Most authors [e.g., Cooper, 1979; Jones, 1983] suggest the use of automatic injection systems for pH measurement and control. To calculate the HCl necessary, Fig. 6-36 shows that about 400 μM will be required or 0.0004 M ℓ^{-1} for a pH near 6.5. At a formula weight of 55, the amount in grams will be (55)(0.0004) = 0.022 g. The purity of the liquid acid is 90% so that an additional 10% must be added, or 0.022 + 0.0022 = 0.0242 g. The density of $HCl{\cdot}H_2O$ is 1.48. The required volume is 0.0242/1.48 = 0.016 ml ℓ^{-1}. In an NFT system, having a liquid capacity of 30 m^3, (30000)(0.016) = 491 ml HCl. Any of the acids or bases employed are highly caustic. Suitable safety procedures are required such as goggles, rubber gloves, and aprons. Be sure **never** to add water directly

Table 6-21. Inorganic and organic materials used to reduce pH.

Name	Chemical formula
Inorganic	
Aluminum sulfate	$Al_2(SO_4)_3$
Ammonium nitrate	NH_4NO_3
Ammonium polyphosphate	$(NH_4)_3H_4P_2O_7$
Ammonium sulfate	$(NH_4)_2SO_4$
Nitric acid	HNO_3
Phosphoric acid	H_3PO_4
Pyrophosphoric acid	$H_4P_2O_7$
Sulfuric acid	H_2SO_4
Elemental sulfur	S
Ferrous sulfate	$FeSO_4.H_2O$
Calcium chloride	$CaCl_2$
Organic	
Leaf mold	
Pine needles	
Tanbark	
Sawdust	
Peat moss	

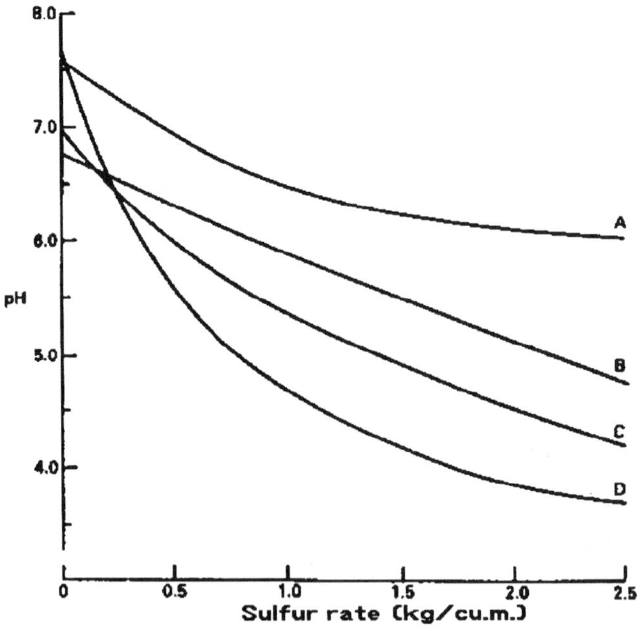

Fig. 6-35. Changes in media pH with rates of sulfur application. **A** = a spent mushroom compost, **B** = old tomato modules (sedge peat), **C** = old tomato modules (sphagnum peat, half decomposed), **D** = old tomato modules (undecomposed sphagnum peat) [Bunt, 1988].

to a concentrated acid or base. The opposite procedure is required, with suitable precautions to prevent spattering.

It should be emphasized that these comments deal with salts dissolved in pure water, and the assumed CO_2 partial pressure of 0.003 atm is correct. Undoubtedly, the equilibrium pH will change with different fertilizer salts used in making up the nutrient solution. For example, phosphorus could be supplied in phosphoric acid. Then, the formulae in Table 6-22 would have to include the H^+ ion. Furthermore, plants would be reducing concentrations of the various ions at differing rates, and they would be releasing many substances to the solution. Intermittent solution testing for pH, and hand calculation, would be laborious.

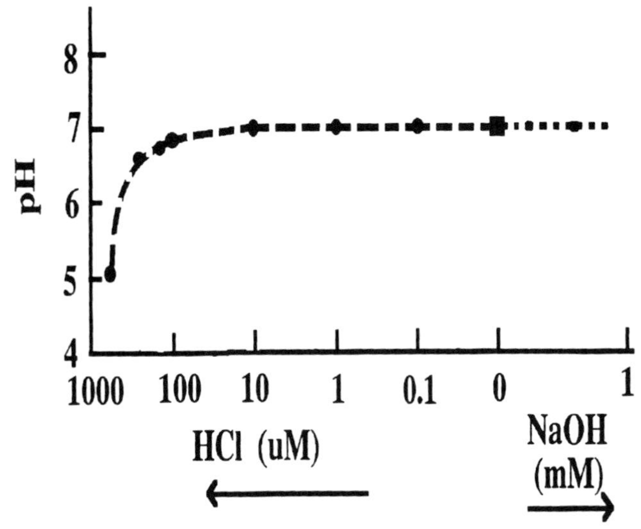

Fig. 6-36. The effect of adding HCl or NaOH to the carnation solution given in Table 6-22. The square dot is the equilibrium pH at a CO_2 pressure of 0.003 atm (303 Pa).

3. pH Measurement

While there are simple systems using dye indicators to show pH, the most common instrument capable of being used as a controller in a suitable injection system is the pH meter. Instruments with either analog or digital output, portable, and reading to three or more significant digits can be obtained. Glass electrodes are immersed in the sample, and the hydrogen ion concentration is balanced against a standard cell that behaves similarly to the standard hydrogen electrode.

However, there are difficulties. The conventional method of making a pH test on a suspension of the medium and distilled water (one part soil to 1 part water) may give pH readings that are much higher than those actually experienced by the plants growing in the medium. Bunt [1988] suggests a 1:2.5 mixture. The amount of soil is often 5 g in 5 ml water or $CaCl_2$ reagent and shaken for 10 to 30 minutes. The procedure should be the same in any series of tests. In fact, Helyar and Porter [1989] state that water is not an appropriate liquid in which to suspend soil for measuring pH. The deceptively simple process has been the center of many controversies [Ritchie, 1989]. One records a higher pH when electrodes are placed in the supernatant solution than when immersed in the sedimented soil particles. Variations in pH can be attributed to soil variability, seasonal changes, soil-liquid ratio, and the type and concentration of the ions in the extracting solution. Jones [1985] states that inaccurate meter calibration and electrode problems are common. Combination electrodes are not well suited for pH determinations. To avoid some problems with distilled water, particularly salt content variability and maintaining soil in flocculated conditions, the extracting solution should contain 0.01 M $CaCl_2$, which gives a pH similar to the saturated paste method [Bunt, 1988; Jones, 1985]. A 1 Normal (equivalent) HCl solution will also mask differences in salt concentrations. Although pH measurements can be duplicated with high precision, attempting to read an output to more than ±0.1 unit is useless. This is contrary to Jones' [1985] recommendation to read pH to the nearest tenth. In fact, in my estimation, one does very well to read or control pH to within one-half pH unit in hydroponic systems, soils, or container substrates. Despite these limitations, Brady [1974] stated that a great deal more information may be obtained from a pH determination than any other single analytical value.

D. CHELATION AND MICRONUTRIENTS

A number of micronutrient formulae are given in Table 6-22. The amounts are critical since the range between deficiency and toxicity can be narrow compared with ranges of the macronutrients. More extensive micronutrient formulations have been published [Hewitt, 1966; Jones, 1983]; but the value of these for commer-

Table 6-22. Selection of nutrient solutions used for experimental and commercial applications. Units are in micromoles per liter ($\mu M\ \ell^{-1}$). Information taken from a variety of sources as given in footnotes.

Ion	Hoagland 1[a]	Hoagland 2[a]	Modified Hoagland[b]	Johnson's 0.25 strength[c]	Steiner's formula[d]	NFT solution[e]	Carnations[f]	Roses[g]	Tomatoes[h]	Cucumber[h]	Pepper[h]
NH_4^+	---	1000	400	500	---	---	2500	1000	500	500	---
K^+	6000	6000	2400	1500	13151	7964	6000	4100	7000	5500	6000
Ca^{2+}	5000	4000	1600	1000	822	4250	1500	3000	3750	3500	3750
Mg^{2+}	2000	2000	800	250	822	2077	1000	500	1000	1000	1250
$H_2PO_4^-$	1000	1000	400	51	822	2192	1100	1000	1500	1250	1250
NO_3^-	15000	14000	5600	3500	13151	14272	10400	9000	10500	11750	12250
SO_4^{2-}	2000	2000	800	300	1233	2077	1000	500	2500	1000	1250
Fe^{3+}	15	15	100	25	7[k]	32	15	15	10	10	10
Mn^{2+}	9.1	9.1	4.56	5	9	36	9.1	9.1	10	10	10
Zn^{2+}	0.8	0.8	1.54	2	0.34	1.5	0.8	0.8	4	4	4
Cu^{2+}	0.32	0.32	0.315	0.5	0.08	1.6	0.32	0.32	0.5	0.5	0.5
BO_3^{3-}	18.2	18.2	23.1	137	18.1	11	18.2	18.2	20	20	20
MoO_4^{2-}	0.1	0.1	0.104	0.2	0.21	0.3	0.1	0.1	0.5	0.5	0.5
Cl^-		9.12									
Chelating agent	15[i]	15[i]	100[j]	25[j]	7[j]	32[j]	15[i]	15[i]	10[j]	10[j]	10[j]

[a] Hoagland and Arnon, 1950
[b] Halvorson and Lindsay, 1972. Includes trace elements
[c] Norvell, 1991. Includes trace elements
[d] Steiner, 1961. "Balanced," "Ideal" solution
[e] Cooper, 1979. Includes trace elements
[f] Hartman and Holley, 1968
[g] Sadasivaiah and Holley, 1973
[h] Welleman and Verwer, 1983
[i] FeNaDTPA, Hoagland and Arnon, 1950, with exception of DTPA in place of $FeSO_4$
[j] FeNaEDTA
[k] Trace elements from Steiner, 1966

cial use remains to be determined. Very often, a sufficient trace element can be a contaminant in the fertilizer (organic or inorganic). Zinc from galvanized pipes or copper from copper piping can dissolve sufficiently to provide the necessary Zn^{3+} or Cu^{2+}. Piping replacement with plastic has often resulted in zinc or copper deficiencies in the crop. Additions of nickel, cobalt, arsenic, barium, cadmium, etc. have yet to be shown necessary under commercial conditions. With the possible exceptions of sodium and selenium, most of the difficulty of making up solutions as given by Hewitt [1966] is not worth the effort. Proprietary mixtures of the more important chemicals can be purchased. Individual sources are noted in Table 6-23 for the common micronutrients. Of course, most of these are required in such small amounts that there can be difficulty in uniform application to soils and container substrates. Dissolving the chemicals in water and applying through the irrigation system, or suitable injection equipment, represents the easiest and safest method. Hoagland and Arnon's trace element mixture is usually made up in a concentrated solution and then an aliquot taken for final dilution. If chelates are used, the stock solution requires refrigeration. Many salts can be difficult to dissolve, boron in particular. There have been instances of a grower placing powdered borax in his concentrate tank and winding up with boron deficiency in carnations. That problem can be eliminated by making sure that each salt used is dissolved in hot water before adding it to a stock solution or into a concentrate tank.

Table 6-23. Inorganic compounds capable of being used as micronutrient sources [Adapted from Mortvedt, 1991].[a]

Micronutrient	Chemical name	Chemical formula	Solubility
Boron	Anhydrous borax	$Na_2B_4O_7$	Soluble
	Fertilizer borate	$Na_2B_4O_7{\cdot}5H_2O$	Soluble
	Borax	$Na_2B_4O_7{\cdot}10H_2O$	Soluble
	Boric acid	H_3BO_3	Soluble
	Colemanite	$Ca_2B_6O_{11}{\cdot}5H_2O$	Slightly soluble
Copper	Cupric sulfate	$CuSO_4{\cdot}H_2O$	Soluble
	Blue vitrol	$CuSO_4{\cdot}5H_2O$	Soluble
	Cupric oxide	CuO	Insoluble
Iron	Ferrous sulfate	$FeSO_4{\cdot}H_2O$	Soluble
	Ferrous sulfate	$FeSO_4{\cdot}7H_2O$	Soluble
	Ferric sulfate	$Fe_2(SO_4)_3{\cdot}9H_2O$	Soluble
	Ferrous-ammonium sulfate	$FeSO_4{\cdot}(NH_4)_2SO_4$	Soluble
Manganese	Manganous sulfate	$MnSO_4{\cdot}xH_2O$	Soluble
	Manganese dichloride	$MnCl_2{\cdot}4H_2O$	Soluble
	Manganous carbonate	$MnCO_3$	Insoluble
	Manganous oxide	MnO	Insoluble
	Manganese oxysulfate	Variable	Variable
Molybdenum	Sodium molybdate	Na_2MoO_4	Soluble
	Sodium molybdate	$Na_2MoO_4{\cdot}2H_2O$	Soluble
	Ammonium molybdate	$(NH_4)_2MoO_4$	Soluble
	Molybdic anhydride	MoO_3	Slightly soluble
	Calcium molybdate	$CaMoO_4$	Insoluble
Zinc	Zinc sulfate	$ZnSO_4{\cdot}H_2O$	Soluble
	Zinc sulfate	$ZnSO_4{\cdot}7H_2O$	Soluble
	Zinc chloride	$ZnCl_2$	Soluble
		$ZnSO_4{\cdot}4Zn(OH)_2$	Slightly soluble
	Zinc carbonate	$ZnCO_3$	Insoluble
	Zinc oxide	ZnO	Insoluble
	Zinc oxysulfate	Variable	Variable

[a] Note comments in the text on dissolving many of these compounds, especially if making stock solutions.

Previous sections have made clear that provision of the necessary trace elements in solution does not mean that they will be available to the root system. This is particularly true with respect to iron in greenhouse production. Ferrous sulfate, recommended by Hoagland and Arnon [1950] in their original publication, usually precipitates. The effect of pH on solution concentrations of Fe^{3+} and other trace elements was shown by Figs. 6-2, 6-7, 6-16, and 6-17. There are various means available to plants to overcome the low concentrations common to the pHs usually recommended in the bulk soil. For example, the pH in the rhizosphere can be lower than in

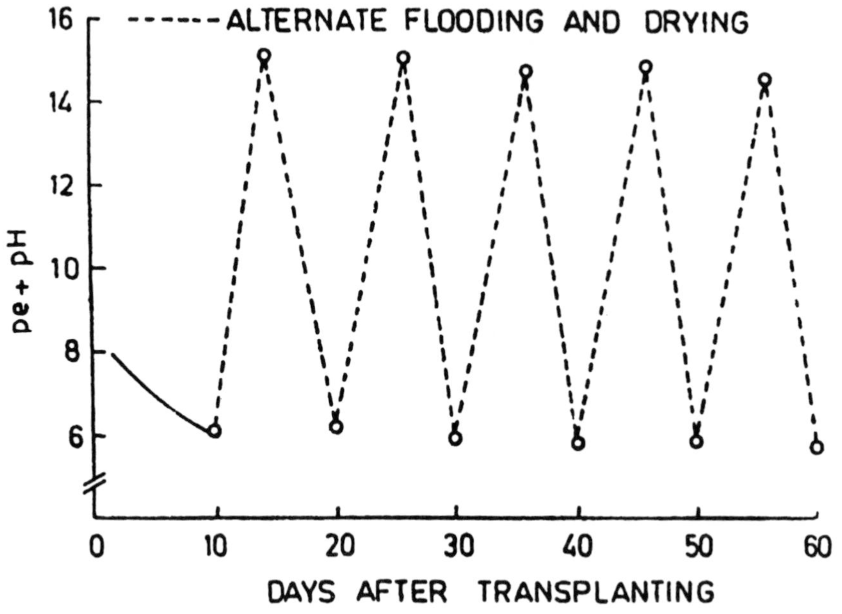

Fig. 6-37. The effect of alternate flooding and drying on soil redox of a mineral rice paddy soil 2 cm below the soil-water interface [Sajwan and Lindsay, 1986] (With permission of the *Amer. Soc. of Agron*).

Table 6-24. Common chelating agents with names and abbreviations [Adapted from Norvell, 1991; Lindsay, 1983].

Abbreviations and acronyms	Chemical name
BPDS	Bathophenanthrolinedisulfonic acid
CIT	Citric acid [(COOH)CH_2C(OH)(COOH)CH_2COOH]
CDTA (DCTA)	*trans*-1,2-Cyclohexylenetrinitrilotetraacetic acid
DTPA (Fe 330)	Diethylenetrinitrilopentaacetic acid
EDDHA (EHPG, APCA, Fe 138)	Ethylenediiminobis(2-hydroxyphenyl)acetic acid
EDMA	Ethylenediaminemonoacetic acid
EDDA	Ethylenediamine-N,N'-diacetic acid
EDTA (Sequestrene, versone)	Ethylenedinitrilotetraacetic acid
ED3A	Ethylenedinitrilotriacetic acid
EGTA	Ethylenebis(oxyethylenetrinitrilo)tetraacetic acid
HBED	N,N'-bis(2-hydroxybenzyl)ethylenedinitrilo-N,N'-diacetic acid
HEDTA (HEEDTA, Versonal, Perm. Green)	N-(2-hydroxyethyl)ethylenedinitrilotracetic acid
HIDA (HEIDA)	N-(2-hydroxyethyl)iminodiacetic acid
IDA	Iminodiacetic acid
NTA	Nitrilotriacetic acid

the bulk substrate. The effect of alternate flooding and drying on the redox potential of a mineral soil was noted by Sajwan and Lindsay [1986] (Fig. 6-37). Both lower pH and lower *pe* values will increase micronutrient activities in the substrate solution –except molybdenum. Respiratory processes by microorganisms and roots will provide conditions more suitable for iron availability. Also, plants and microorganisms can produce chelating substances that complex with the metal ion, keeping the ion in solution and making it more available. Such substances are called siderophores. The term "chelate" refers to the fact that the ion is attached to the chelate –or "ligand"– at several points. The affinity of the ligand for the ion is greater than competing anions that tend to precipitate the micronutrient. Reference has already been made to so-called iron efficient plants that can shift metabolism when iron is deficient, thereby enhancing iron supply. There are also synthetic chelates, organic molecules that have been especially beneficial in improving plant nutrition over the past 3 decades. A list of these is provided in Table 6-24. The compounds EDTA, DTPA, EDDHA, and HEDTA are the most common and readily available for commercial greenhouse use. These compounds can chelate with calcium, copper, manganese, and zinc, as well as iron. Fe330 (DTPA) is a green material most often purchased, while Fe138 (EDDHA) is dark red and recommended for alkaline conditions. Lindsay and co-workers have been particularly instrumental in explaining the complexities of chelation in soils and hydroponic solutions [e.g., Lindsay, 1984; 1991; Halvorson and Lindsay, 1972; Lindsay and Norvell, 1969; Lindsay and Schwab, 1982; Schwab and Lindsay, 1983a; b; Sadiq and Lindsay, 1989; etc.]. The effect of EDTA, for example, on the diffusion of zinc into a simulated root was examined by Elgawhary et al. [1970], showing greatly enhanced accumulation of the zinc complex at concentrations ranging from 100 to 1000 μM NaEDTA (Fig. 6-38).

The development of computer programs to calculate equilibrium levels has helped explain many results. Halvorson and Lindsay [1972] showed the ability of various chelates to complex with iron (Fig. 6-39). The high equilibrium constant for the reaction between Fe^{3+} and the ligand EDDHA means that a high pH will not result in displacement of the iron from

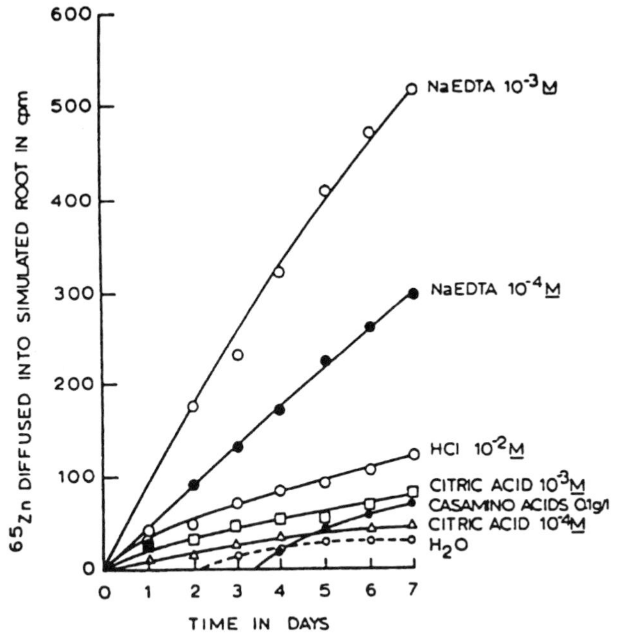

Fig. 6-38. The effect of various complexing agents and acids on diffusion of Zn into a simulated root over 7 days [Elgawhary et al., 1970] (With permission of the *Amer. Soc. of Agronomy)*.

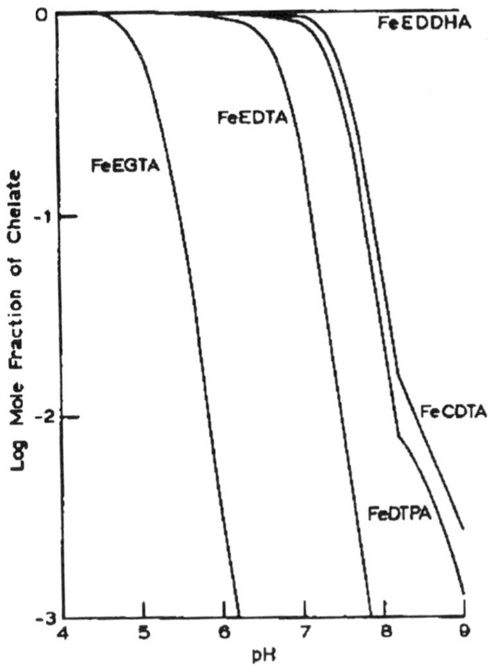

Fig. 6-39. Effect of pH on the ability of various chelates to hold iron in a modified Hoagland solution [Halvorson and Lindsay, 1972] (With permission of the *Amer. Soc. of Agronomy)*.

EDDHA. For EDTA, however, a pH above 7.0 will displace Fe^{3+}, which then precipitates as an unavailable hydrous-ferric oxide. This is the reason for the common recommendation of EDDHA in alkaline soils. Unfortunately, the situation is not that simple. Halvorson [1971] showed actual results on growing maize (Fig. 6-40). All four Fe chelates could supply Fe at pH 5.3, but at pH 7.5, only FeEDTA was effective. At 7.5, Fe was displaced to form $CaEGTA^{2-}$ and insoluble iron oxide. The poor response with FeDTPA at 7.5 resulted from zinc deficiency and not iron, since spraying the plants with Zn corrected the low maize yield. DTPA combines strongly with Zn above pH 7.5. Poor response from FeEDDHA at high pH in solution culture resulted from the presence of a free chelating agent that depressed the Fe^{3+} activity. Even when EDDHA exceeds Fe by only 0.5%, the free Fe^{3+} can decrease by some millionfold [Lindsay, 1972c]. Thus, excessive chelating agents, particularly EDDHA can cause undesirable results. Usually, ligands are not taken into the root. However, they are separated from Fe^{3+} at the root and Fe^{3+} is reduced to Fe^{2+} [Simons et al., 1962; Lindsay, 1972c]. Uptake of the ligand can occur, and in the case of EDDHA, the ability of this chelate to complex with iron can result in a severe deficiency although the plant may have more than enough iron [Simonds et al., 1962]. This strong affinity of EDDHA for iron has also been implicated in *Fusarium* suppressive soils. Scher et al. [1984] suggested that this gave a selective advantage to siderophore-producing antagonists such as *Pseudomonas putida*, resulting in Fe deficiency in *Fusarium*.

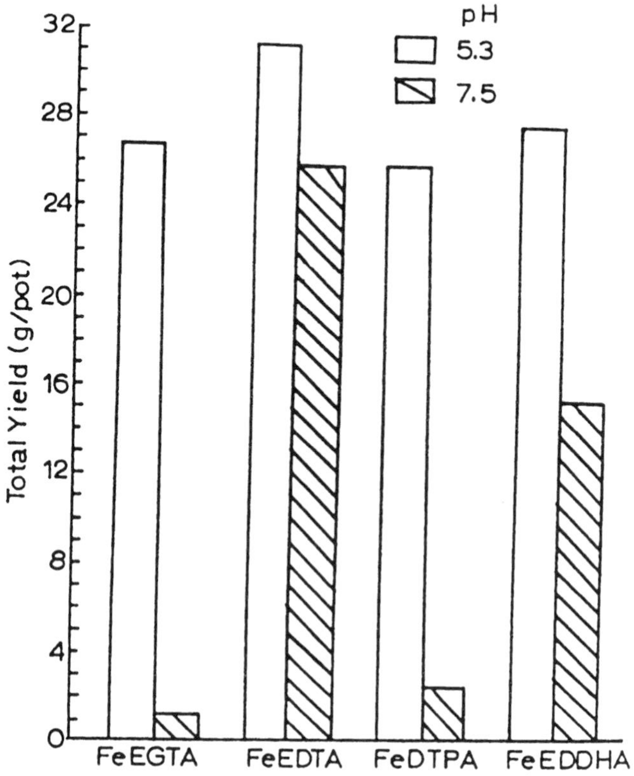

Fig. 6-40. The effect of different chelates and pH on growth of *Zea mays* [Halvorson, 1971].

Whether in soils or nutrient solutions, the chemical relationships to be found can be highly complex. Lindsay and Norvell [1969] developed "mole-fraction" diagrams for EDTA and DTPA in soils where the competing ions were Zn^{2+}, Fe^{3+}, Ca^{2+}, and H^+ at 0.003 atm CO_2 (Fig. 6-41). When Zn^{2+} is included as a competing ion, ZnL^{2-} becomes a major complex between pH 6 and 7. FeL^{2-} is dominant below pH 6 and CaL^{2-} above 7. This helps explain the results in Fig. 6-40 at high pH. For DTPA (right side), Zn_2L^- is dominant below pH 4.5, FeL^{2-} between 4.5 and 6.4, and ZnL^{3-} above 6.4.

Fig. 6-42 is similar to Fig. 6-41 except that the curves show mole fractions for a modified Hoagland's solution in equilibrium with hydrous-ferric oxide and 0.0003 atm CO_2. Depending upon pH, EDTA and DTPA also complex with manganese and copper. These two were not included in Fig. 6-41 for soils. EDTA chelates iron effectively to about pH 6.5. The zinc is nearly all chelated above 6.5, the copper above 5.5, and Mn^{2+} above 6.5. The amount of magnesium chelated is significant above pH 8.2. While not shown here, the effectiveness of FeEDTA in supplying iron to plants may decrease with increased zinc concentration. The right diagram in Fig. 6-42, shows that FeDTPA can be an effective iron supply up to pH 7.8. What is interesting about these diagrams is the number of complexes that may be present in soil solutions and in nutrient solutions. The relationships of chelated ions in solution are not simple, and the addition of a synthetic chelate sets in train a whole order of reactions that influence plant growth.

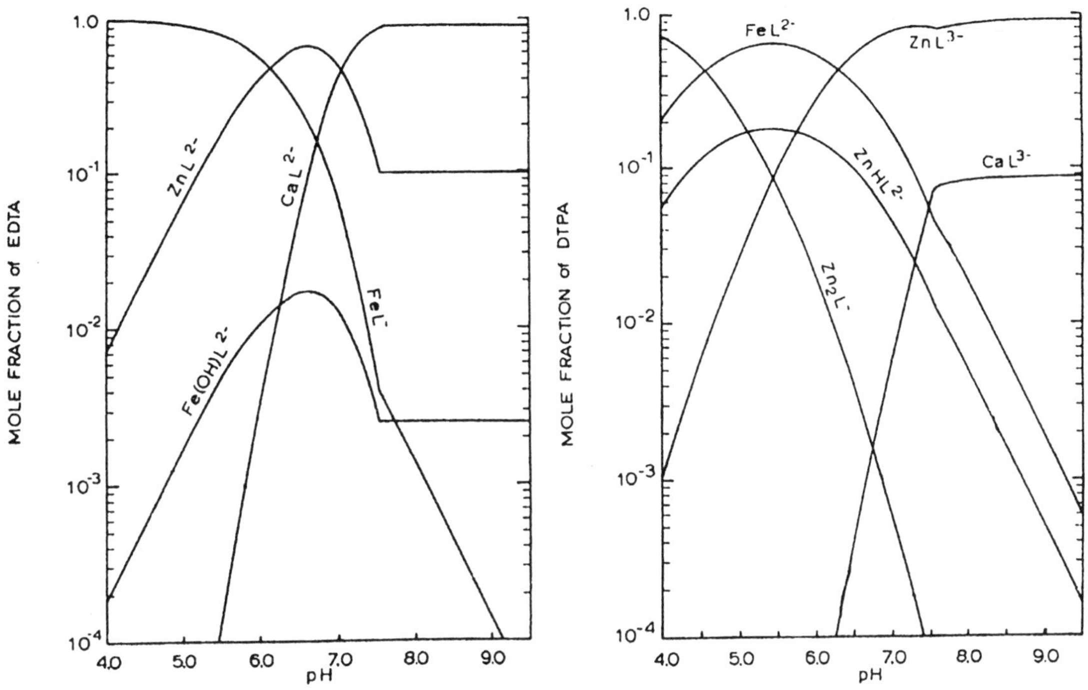

Fig. 6-41. The proportion of EDTA (left) and DTPA (right) attached to various metal ions in soils where CO_2 gaseous concentration is 0.003 atm [Modified from Lindsay and Norvell, 1969] (With permission of the *Amer. Soc. of Aronomy)*.

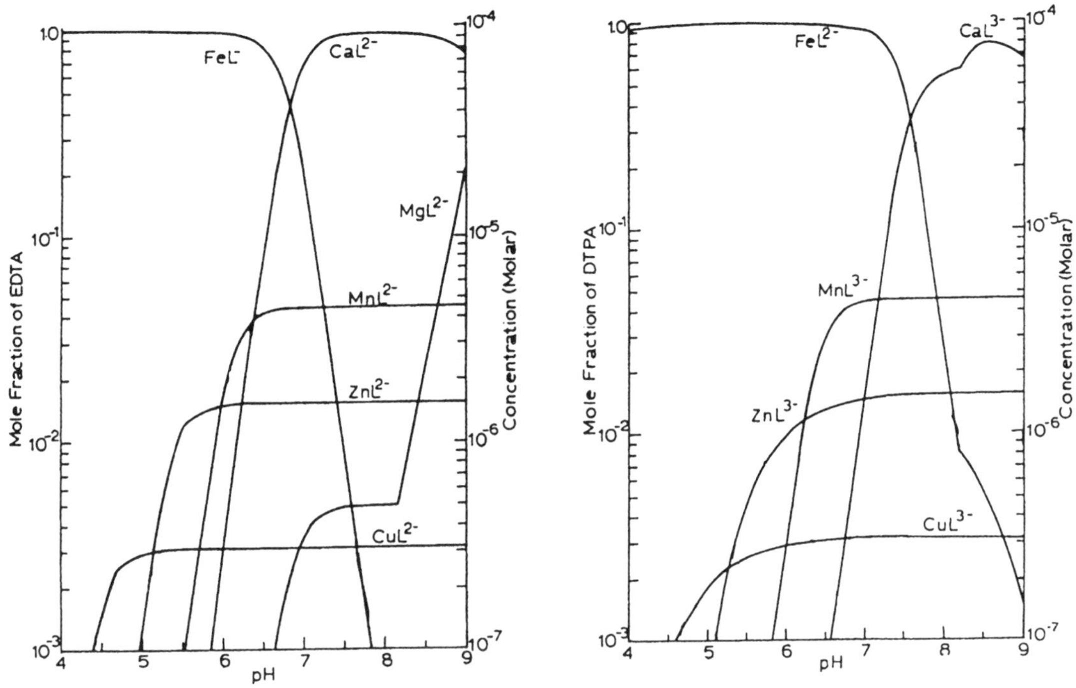

Fig. 6-42. Amount of EDTA (left) and DTPA (right) attached to various metal ions in Hoagland's modified solution in equilibrium with ferric-oxide and 0.0003 atm CO_2 [Modified from Halvorson and Lindsay, 1972] (With permission of the *Amer. Soc. of Agronomy)*.

In a more recent review, Norvell [1991] showed that many other metal ions (i.e., Al^{3+}, Co^{2+}, Cd^{2+}, Pb^{2+}, Hg^{2+}, etc.) have been found to chelate with many of the compounds listed in Table 6-24, depending upon the acidity or alkalinity of the soil. These have not been shown to have significant impact on greenhouse practice –largely since no one in the business has actively investigated the relationships. For nutrient solutions, Norvell used the GEOCHEM computer program to evaluate Johnson's 0.25-strength nutrient solution (Table 6-22), producing several diagrams similar to Figs. 6-41 and 6-42. These types of studies have not been carried forward by greenhouse researchers to the best of my knowledge. Johnson's solution is not commonly employed in greenhouse hydroponics, and some of Norvell's comments, while highly interesting, have not been examined in detail under commercial conditions.

Norvell [1991] stated that introduction of chelated Fe causes high pH sensitivity of all micronutrient ions. Concentrations of free metal ions may change 100- or 1000-fold in poorly buffered nutrient solutions. The failure of chelating agents to buffer micronutrient ions is, according to Norvell, because the chelate buffer is added in one form. The insoluble hydrous Fe(III) oxides largely control Fe^{3+} concentration whether or not a chelate is present –unless sufficient chelate is added to complex most of the Fe, Cu, Zn, etc. With a "modest" excess of EDTA, the free ligand is buffered by equilibria between Ca^{2+} and Ca-EDTA. The activities of other ions (Fe, Cu, Zn, and some Mn) are also buffered and freed from disturbance by the pH-dependent solubility of iron oxides. Inclusion of 25 µM EDTA, in addition to 25 µM ℓ^{-1} Fe-EDTA, produces nearly constant levels of metal EDTA chelates and constant activities of all metal ions from about pH 4.5 to nearly 7.5 –where iron oxides begin to precipitate despite additional EDTA. Iron phosphates may also precipitate in nutrient solutions with iron oxides. Citing the ℓ ature, Norvell [1991] rejected the use of chelating agents CIT, EGTA, HIDA, EDDHA, NTA, and HBED since they could not maintain Zn^{2+} in the desired range. In contrast, EDTA, CDTA, HEDTA, or DTPA appeared able to provide adequate buffering in a convenient range for Zn^{2+}.

To provide an example of the ions in a nutrient solution for roses (Table 6-22), MINTEQA2/PRODEFA2 has been used to calculate ion distribution for three different situations at pH 6.5 and 0.003 atm CO_2, with the results presented in Table 6-25. Table 6-26 gives the percentage distribution for the same examples. One will immediately note (Table 6-25) that none of the added nutrients remain at the initial concentrations. There is a whole host of complexes that form in solution –some 65 of those printed out by MINTEQA2/PRODEFA2 were not included in Table 6-25. Table 6-26 is more interesting. Increasing the concentration of DTPA results in chelation with calcium, increases manganese chelation from 1 to 29%, and reduces DTPA chelation with iron by some 40%. On the other hand, 99% of EDDHA chelates with iron. The partitioning of phosphate does not change with chelating agent. With additional DTPA (30 µM), 100% of the Fe^{2+} and Fe^{3+} ions will be chelated, whereas 100% of Fe^{3+} only combines with EDDHA. Better than 50% of Fe^{2+} remains as the free ion in solution.

Table 6-25. Examples of effect of chelates in a rose nutrient solution (Table 6-22) by the MINTEQA2/PRODEFA2 computer program. Ions with concentrations below 0.01 µM ℓ^{-1} ignored. Chemical formulae with "o" are neutral aqueous molecules. $Fe(OH)_3$ allowed to precipitate, in equilibrium with 0.003 atm CO_2 and pH fixed at 6.5.

Chemical formula	15 µM Fe^{3+} and 15 µM DTPA	15 µM Fe^{3+} and 30 µM DTPA	15 µM Fe^{3+} and 15 µM EDDHA
NH_4^+	879	879	879
K^+	3614	3614	3614
Ca^{2+}	1690	1690	1690
Mg^{2+}	277	277	277
NO_3^-	7943	7943	7943
SO_4^{2-}	240	240	240
Mn^{2+}	5.3	0.12	5.4
Zn^{2+}	0.03	<0.01	0.39
Cu^{2+}	<0.01	<0.01	0.05

Chemical formula	15 μM Fe^{3+} and 15 μM DTPA	15 μM Fe^{3+} and 30 μM DTPA	15 μM Fe^{3+} and 15 μM EDDHA
$H_3BO_3^o$	18.2	18.2	18.2
MoO_4^{2-}	0.06	0.06	0.06
$MnDTPA^{3-}$	0.06	2.84	---
$MnDTPA^{2-}$	<0.01	0.10	---
$H_2BO_3^-$	0.03	0.03	0.03
NH_3^o	1.58	1.58	1.58
$NH_4SO_4^-$	2.72	2.72	2.72
$MgHCO_3^+$	0.46	<0.01	0.46
$MgSO_4^o$	11.3	11.3	11.3
$MgPO_4^-$	0.17	0.17	0.17
$MgH_2PO_4^+$	25.6	25.6	25.6
$CaHCO_3^+$	2.51	2.51	2.51
$CaCO_3^o$	0.05	0.05	0.05
$CaSO_4^o$	82.8	82.8	82.8
$CaHPO_4^o$	106	106	106
$CaPO_4^-$	0.78	0.78	0.78
$CaH_2PO_4^+$	24.2	24.2	24.2
KSO_4^-	6.11	6.11	6.11
$KHPO_4^-$	0.79	0.79	0.79
$MnSO_4^o$	0.23	<0.01	0.23
$MnHCO_3^+$	0.01	<0.01	0.01
$ZnOH^+$	<0.01	<0.01	0.02
$ZnSO_4^o$	<0.01	<0.01	0.02
$ZnHPO_4^o$	<0.01	<0.01	0.09
HCO_3^-	144	144	144
$H_2CO_3^o$	102	102	102
HPO_4^{2-}	114	114	114
$H_2PO_4^-$	570	570	570
$H_3PO_4^o$	0.03	0.03	0.03
$CaDTPA^{3-}$	<0.01	0.64	---
$CaHDTPA^{2-}$	<0.01	1.14	---
Ca_2DTPA^-	<0.01	0.94	---
$MgHDTPA^{2-}$	<0.01	0.04	---
$FeDTPA^{2-}$	8.34	9.08	---
$FeEDDHA^-$			13.2
$FeHDTPA^-$	0.03	0.03	---
$ZnDTPA^{3-}$	0.18	0.20	---
$ZnHDTPA^{2-}$	0.10	0.11	---
$CuDTPA^{3-}$	0.10	0.10	---
$CuHEDDHA^-$			0.18

Table 6-26. Calculated percent distribution of various ions in a rose solution (Table 6-22) and treated as in Table 6-25 to two chelates and differing concentrations when originally mixed. Conditions the same as given in Table 6-25 legend. Data from Lindsay's modification of Allison et al.'s MINTEQA2/PRODEFA2 program [1991]. "o" = a neutral, undissociated molecule.

Major ion	15 µM Fe^{3+} & 15 µM DTPA		15 µM Fe^{3+} & 30 µM DTPA		15 µM Fe^{3+} & 15 µM EDDHA	
	Complex	Percent of major ion	Complex	Percent of major ion	Complex	Percent of major ion
$DTPA^{5-}$	$MnDTPA^{3-}$	1	$MnDTPA^{3-}$	29		
	$FeDTPA^{2-}$	92	$FeDTPA^{2-}$	50		
	$ZnDTPA^{3-}$	4	$ZnDTPA^{3-}$	2		
	$ZnHDTPA^{2-}$	1				
	$CuDTPA^{3-}$	2	$CuDTPA^{3-}$	1		
			$CaDTPA^{3-}$	7		
			$CaHDTPA^{2-}$	6		
			Ca_2DTPA^{-}	4		
$EDDHA^{4-}$					$FeEDDHA^{-}$	99
					$CuHEDDHA^{-}$	1
PO_4^{3-}	$MgHPO_4^{o}$	3	$MgHPO_4^{o}$	3	$MgHPO_4^{o}$	3
	$CaHPO_4^{o}$	11	$CaHPO_4^{o}$	11	$CaHPO_4^{o}$	11
	$CaH_2PO_4^{+}$	3	$CaH_2PO_4^{+}$	3	$CaH_2PO_4^{+}$	3
	HPO_4^{2-}	19	HPO_4^{2-}	19	HPO_4^{2-}	19
	$H_2PO_4^{-}$	64	$H_2PO_4^{-}$	64	$H_2PO_4^{-}$	64
Fe^{2+}	Fe^{2+}	48			Fe^{2+}	53
	$FeOH^{+}$	19			$FeOH^{+}$	21
	$FeSO_4^{o}$	1			$FeSO_4^{o}$	1
	$FeH_2PO_4^{+}$	10			$FeH_2PO_4^{+}$	10
	$FeHPO_4^{o}$	13			$FeHPO_4^{o}$	15
	$Fe(II)DTPA^{3-}$	7	$Fe(II)DTPA^{3-}$	87		
	$Fe(II)HDTPA^{2-}$	1	$Fe(II)DTPA^{2-}$	13		
Fe^{3+}	$FeDTPA^{2-}$	100	$FeDTPA^{2-}$	100	$FeEDDHA^{-}$	100
Mn^{2+}	Mn^{2+}	95	Mn^{2+}	2	Mn^{2+}	97
	$MnDTPA^{3-}$	2	$MnDTPA^{3-}$	96	$MnSO_4^{o}$	3
	$MnSO_4^{o}$	3	$MnHDTPA^{2-}$	2		
Zn^{2+}	Zn^{2+}	6			Zn^{2+}	80
	$ZnDTPA^{3-}$	71	$ZnDTPA^{3-}$	77	$ZnOH^{+}$	4
	$ZnHDTPA^{2-}$	21	$ZnHDTPA^{2-}$	23	$ZnSO_4^{o}$	3
					$ZnH_2PO_4^{+}$	1
					$ZnHPO_4^{o}$	11
Cu^{2+}	$CuDTPA^{3-}$	95	$CuDTPA^{3-}$	95	Cu^{2+}	23
	$CuHDTPA^{2-}$	5	$CuHDTPA^{2-}$	5	$CuCO_3^{o}$	2
					$CuHPO_4^{o}$	3
					$CuOH^{+}$	1
					$Cu(OH)_2^{o}$	2
					$CuHEDDHA^{-}$	64
					CuH_2EDDHA^{o}	3

Adding more DTPA also increases chelation of manganese, zinc, and copper. EDDHA, however, complexes about 67% of the copper and none of the manganese and zinc. There might be no buffering of the latter two in solutions with EDDHA. The remainder of the macronutrients generally exist as free ions above 90%, according to the computer program. These tables, showing possible equilibrium status of common nutrient solutions, are the first I have seen in the horticultural ℓ ature. My objective here is to illustrate the possibilities that exist in modern greenhouse nutrition, and the fact that most investigators fail to acknowledge any awareness of what goes on in their fertilization studies. Published results are, therefore, severely limited in scope and reliability.

Remember that synthetic chelates are also organic compounds. They are subject to breakdown by microorganisms. Furthermore, microorganisms require the same micronutrients as higher plants. Competition occurs, analogous to nitrogen immobilization when crop residues with wide C/N ratios are applied [Stevenson, 1991]. Sonneveld and Voogt [1980] reported that the sudden drop of manganese in recirculating hydroponics that often occurs, is due to manganese-oxidizing bacteria. The Mn content of the nutrient solution is not always a reliable measure. Clay mineral soils, in combination with organic matter, are major components involved in micronutrient retention. Organic matter contribution is highest when the predominant clay mineral is kaolinite and lowest when it is montmorillonite [Stevenson, 1991]. Mooraghan and Mascagni [1991], however, point out that when boron, for example, is released from soil minerals, it can be rapidly leached because of its nonionic nature. In contrast, increasing soil organic matter can cause copper deficiency due to complexing of copper in insoluble organic forms.

E. MACRONUTRIENT FERTILIZATION AND CONTROL

Readers will not see in this text specific fertilizer recommendations for particular species except as examples. A perusal of the most recent issues of the horticultural ℓ ature [1996] will show that most articles deal with treatment conditions as to N, P, and K. For reasons expressed at the start of this chapter, I do not agree with the existing methodology. There is little uniformity. Typical of specific fertilizer recommendations for crops is Joiner's [1983] review of ornamental fertilization in the greenhouse industry. One may also refer to publications that deal with a specific crop or several crops for greater detail. It is my opinion that fertilizer recommendations should be given in terms of millimoles per liter (mmol ℓ^{-1}) or per cubic meter (mmol m^{-3}), or millequivalents per ℓ (meq ℓ^{-1}), and the recipe should deal with those ionic forms available to plants (e.g., NO_3^-, NH_4^+, $H_2PO_4^-$, K^+, etc.).

1. Inorganic Fertilizers

Inorganic fertilizers as listed in Table 6-5 have many advantages. First, those used in agriculture are relatively pure, particularly compounds highly soluble and employed in hydroponics. Secondly, their high analysis of macronutrients means that they can be shipped long distances at low cost. Thirdly, the grower can be assured of a uniform, consistent source. Concerning the last, some care is required that handling procedures have not contaminated the product. There have been examples of bagging machinery contaminated with weed killers, resulting in crop damage. Or, placement of fertilizers on surfaces, or in containers, having had herbicides stored previously can cause significant damage. Movement of 2,4-D or Ureabor into amendments or fertilizer bags can happen. The grower finds a crop damage pattern in his greenhouses that is difficult to diagnose since there is neither record of where the particular bags were stored, nor which bags were used on the particular damaged location. Fertilizer and herbicide storages should be distinct and well separated. Another problem is the application control. Uneven application can result in uneven plant response, or through miscalculation, excess is applied, leading to salinity or specific phytotoxicities. As will be discussed further in Section 6.III.F, soil testing, an attempt to correct a perceived deficiency in one step is dangerous. The secret to a good fertilizer program is one of gradual progression based upon observation, and tissue and soil analysis. Many greenhouse cropping patterns are too short to correct nutrition –especially if readily observable– before the product is to be sold. The grower must set up a change from one crop to the next. This requires experience and good judgment since soil and tissue analyses are usually after the fact. Inorganic compounds are also rated according to their contribution to salinity and equivalent acidity or basicity. For the former, the compound is compared with $NaNO_3$ as a reference. KCl, for example, has a salt index of 114, as compared to superphosphate with an index of 10. Most compounds used in hydroponics have salt indices ranging from 8 for calcium sulfate to 104 for ammonium nitrate. Acidity is expressed as the equivalent weight of $CaCO_3$ per 45 kg of the material. In some countries, the com parison is made to CaO. Using $CaCO_3$, the neutralizing value of CaO is 178, MgO 250, and $MgCO_3$ 119.

a. Nitrogen

In field soils, nitrogen applied in fertilizers undergoes the same kinds of reactions as does nitrogen released from plant residues. Fertilizer nitrogen can be present as nitrate, ammonia, and urea [Brady, 1974]. Urea is subject to ammonification, nitrification, and utilization. Ammonium can be oxidized to nitrate, fixed by the soil colloids, or used (Fig. 6-27). Nitrate can be lost by volatilization, leaching, or use by microbes and plants. Ammonium-containing fertilizers increase acidity. In any given climatic condition, soils develop a "normal" nitrogen content, and attempts to raise this content materially higher will be accompanied by waste. Of the fertilizers given in Table 6-5, ammonium nitrate can be explosive and its direct use is prohibited in some countries unless mixed with limestone (nitrochalk). Mengel and Kirkby [1982] cite some rates of application ranging from 90 to 135 kg N ha^{-1} in the field. In container production, recommendations for fertilizer additions were given in Tables 5-4 through 5-6. Note also that some amendments contain significant nitrogen and other macronutrients (Table 5-7). There is some danger if a mixture is stored for lengthy periods. The salts may dissolve, increasing salinity over time. Composts to which fertilizer has been added should be used shortly after mixing. I have observed considerable crop damage if the grower steam-pasteurizes following the application of highly soluble, or some slow-release, fertilizers. About the only inorganic materials that can be safely steamed are limestone and 20% superphosphate. As will be noted later, macronutrients can be applied automatically with irrigation.

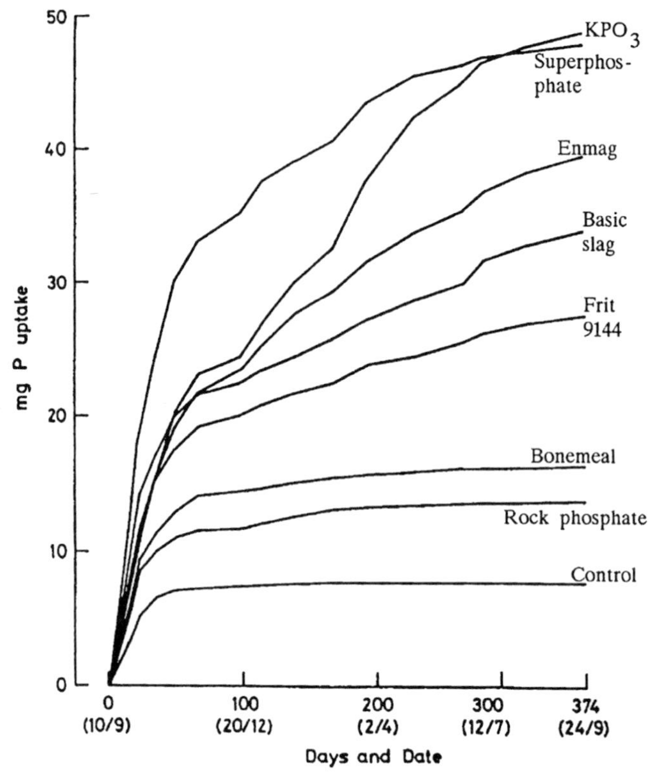

Fig. 6-43. Uptake of P by rye grass when fertilized by different P sources [Bunt, 1980] (©1980, Int. Soc. Hort. Sci., *Acta Hort.*, 99:25-32).

b. Phosphorus

The total amount of phosphorus in soils compares favorably with nitrogen, but it is much lower than potassium, calcium, or magnesium, and much of it is unavailable to plants (Table 6-18). Brady [1974] states that the tonnage of phosphorous-supplying materials exceeds all except the nitrogen carriers. Phosphorus can complex with aluminum, calcium, iron, and many other materials that can depress P availability. According to Mengel and Kirkby [1982], most P uptake by plants results from root growth to unexplored areas –which is unavailable in container production or hydroponics. Soils containing iron oxides and clay are particularly prone to strong phosphate fixation and often require very high fertilizer applications. Application rates up to 100 kg P ha^{-1} may be found in the ℓ ature. Rates of 0.5 to 1.0 kg superphosphate per 10 m^2 are usual when applied to raised soil benches. Of the phosphorus sources listed in Table 6-5, the superphosphates are most commonly employed for greenhouse production in the ground or benches as a preplant addition. Availability is determined by granule size plus chemical form. An example of phosphorus uptake by grass for several sources is given in Fig. 6-43. Both ground rock phosphate (apatite) and bonemeal (organic) are generally rejected due to their low availability

Table 6-27. Some common organic fertilizers [Hanan et al.,1978].

Material	Acidity	Analysis[1]	Remarks[2]
Activated sludge	Acid	4-6, 2-4, 0	Can contain heavy metals. Rates 2 kg per 10 m^2. Some are proprietary.
Beet sugar residue	Basic	3-4, 0, 10	Mostly calcium, micronutrient traces.
Castor pumice	Acid	5-6, 0, 0	2 kg per 10 m^2, poisonous to animals.
Cocoa shell meal	Basic	5% total	Used as conditioner in some fertilizers.
Cocoa tankage	Basic	4, 1.5, 2	May contain 20% lime.
Dried blood	Acid	9-14, 0, 0	1 kg per 10 m^2, N readily available.
Steamed bone meal	Basic	2, 25-30, 0	2 kg per 10 m^2, 20-30% calcium.
Fish scrap	Acid	9, 7, trace	1-2 kg per 10 m^2.
Garbage tankage	Basic	Variable	
Guano	Acid	12, 11, 2	Contains 8% calcium. If leached, may have 25% P_2O_5.
Hoof and horn meal	---	13, 0, 0	2 kg per 10 m^2.
Kelp (seaweed)	---	2, 1, 4-13	1-1.5 kg per 10 m^2.
Cottonseed meal	Acid	7, 2-3, 2	2 kg per 10 m^2.
Oyster shells	Basic	---	31-36% calcium, liming material.
Process tankage	Acid	4-12, 0, 0	Sometimes a conditioner in fertilizer.
Rapeseed meal	---	6, 2, 2	
Ground rice hulls	---	0.5, 0.2, 0.5	Very light and bulky.
Shrimp bran	Basic	7, 4, 0	57% calcium.
Soybean meal	---	6, 1, 2	Oil removed, mostly feed for animals.
Tobacco stems	Basic	2, 0, 6	3% Ca and 1% Cl.
Animal tankage	Basic	7, 10, 0	Principally in animal feeds.
Vermiculite	Basic	0-2, trace, 2-3	Mostly a soil conditioner.
Wood ashes	Basic	0, 0, 5	23% Ca, 2 kg per 10 m^2.

[1] Analysis for N, P_2O_5, and K_2O.
[2] Rates for a bench 15 cm deep.

[Bunt, 1988]. As with other nutrients, phosphorous can be applied in the irrigation water. Suggestions will be given later.

In general, most greenhouse species can tolerate high P levels unless the concentrations affect the availability of other nutrients such as iron. However, the use of organic-based potting substrates, coupled with some species, will cause toxicity problems. Among sensitive species mentioned by Bunt [1988] are the *Proteaceae* and some hardy flowering shrubs such as *Cytisus x praecox*. Orthophosphates at application rates above 50 mg ℓ^{-1} have been mentioned as causing damage to sensitive species. Tomato and chrysanthemum, on the other hand, are highly tolerant to excess phosphorus. P fertilizers can have sufficient fluorides present to cause phytotoxicity in many ornamental foliage plants. This can be avoided by using phosphoric acid in place of superphosphate, avoiding water with more than 0.25 ppm F, and raising the mix pH to 6.0 to 6.5.

c. Potassium

Of the inorganic fertilizers given in Table 6-5, potassium chloride (muriate of potash) is not generally recommended for greenhouse production since it contributes chloride and increases salinity compared with other compounds. It is usually much cheaper than other sources. Vermiculite is the only material in loamless mixes containing sufficient potassium. For others, K must usually be added. In field soils, the main source is usually from weathering of K-containing minerals. Mengel and Kirkby [1982] suggest an application range of 40 to 250 kg K ha^{-1} yr^{-1}. The response to K uptake depends to a considerable extent on the N nutrition level. Interactions between K and N were examined in previous sections. Deficiencies often appear where K^+ has been leached from light sandy soils or highly leached lateritic soils. Organic soils and peats are usually low in K.

2. Organic Fertilizers

The organic fertilizers listed in Table 6-27 are not widely used in modern greenhouse production –except for vermiculite and one or two others used as soil or fertilizer conditioners. They have two distinct disadvantages: 1) A low nutrient level, which means that considerable quantities may be required to supply sufficient nutrient. Shipping organic fertilizers any distance is expensive in terms of fertilizer content. 2). The analysis is likely to be variable, so the grower is not assured of a consistent nutrient supply. This is particularly true in respect to manures. Manures can cause ammonia toxicities, especially if steamed. There is the aesthetic problem with a product sold to the general consumer. Although salinity problems are not likely to arise, the high organic content of most materials can fix nutrients, resulting in deficiencies. Rapid decomposition can also tie up nitrogen. Organic substances can also be considered as slow-release fertilizers since nitrogen mineralization and release requires microbial action. The benefits of organic fertilizers are sometimes more likely to be derived from their effects on the substrate physical characteristics –such as better aeration and water relationships. No matter the organic content, nutrient availability requires the ions to be present in solution. The nutrient supply is the same as that from an inorganic source [Brady, 1974; Mengel and Kirkby, 1982].

3. Controlled or Slow-Release Fertilizers

The possibility of applying fertilizers that release nutrients gradually has intrigued growers for many years. Once-only application, either by mixing in the substrate prior to planting, or as a top dressing, would certainly reduce labor costs. Where container plants are sold, inclusion of a nutrient supply can enhance keepability and consumer satisfaction. Most organic fertilizers, mentioned in the previous section, and some inorganic compounds such as superphosphate, potassium metaphosphate, sulfur, etc. could be considered as slow release, with rapidity controlled by particle size. However, materials deliberately manufactured to provide essential elements over significant periods are given in Table 6-28.

Of the types listed in Table 6-28, magnesium-ammonium-phosphate is probably considered the original slow-release fertilizer. Urea formaldehyde, a common material, is not usually recommended for use on carnations in Colorado. During winter seasons, temperatures are too low to allow sufficient release. As temperatures rise in the spring, excessive nitrogen may be released. Fig. 6-44 shows that

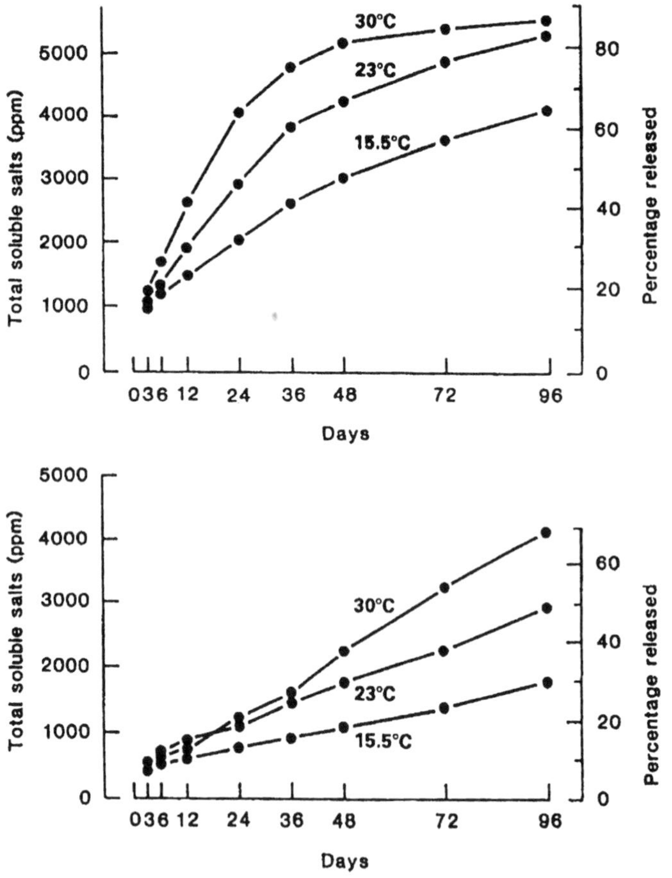

Fig. 6-44. Effect of temperature on release rates of coated fertilizers: **Upper**: Osmocote 14-14-14. **Lower**: Osmocote 18-6-12, 9-12 mo. formulation [Harbaugh and Wilfret, 1982] (With permission of the *Florists' Rev.*).

Table 6-28. Some examples of slow-release fertilizers used in greenhouse culture.

Type	Name	Trade names or acronyms	Factor causing release	Effect of temperature rise on release	Examples Duration
Organic	Urea formaldehyde	Nitro-form, (UF): Several polymers - i.e. methylene-, di-, tri-, and tetramethylene diurea	Microbial activity	High	38% N, few weeks to months
	Iso-butylidene diurea	(IBDU)	Particle size, dissolution	Slight	32% N
	Crotonylidene diurea	Triabon, Crotodur, (CDU)	Microbial activity, particle size	High	28% N, 16-18-12-4[a]
	Diamide of oxalic acid	Oxamide	Particle size, dissolution	Slight	32% N
	Sulfur-coated urea	Gold-In, (SCU): microbicide may be included	Various dissolution rates	Slight	Up to 8 weeks, 6, 5 or 1% per day
Inorganic	Polyolefin-coated	Nutricote	Membrane, pore size and number	High	100-360 dys, 13-13-11, 16-10-10, etc.
	Co-polymers of dicyclopentadine and glycorol ester	Osmocote	Membrane, pore size and number	High	Many formulations, 14-14-14, 15-11-13-2, 18-6-12, 3-9 mo.
	Magnesium-ammonium-phosphate	Magamp (7-40-6-12)[b] , Enmag (6-20-10-8.5)[b]	Particle size	Slight	(Table 6-5)
	Dicalcium phosphate		Particle size	Slight	see Table 6-5
	Frits: K, mainly micronutrients. Compounds fused with sodium silicate	Potassium frit, 29% K FTE 253A (2-B, 2-Cu, 12-Fe, 5-Mn, 0.13-Mo, 4-Zn[c]), #36 (0.5-B, 2-Cu, 9-Fe, 2-Mn, 0.5-Mo, 2-Zn)[c]	Particle size, pH	Slight	Several formulations for micronutrients

[a] Values are percentages of N, P_2O_5, K_2O, and MgO in the formulation.
[b] Values are percentages of N, P_2O_5, K_2O, and Mg in the formulation.
[c] Numbers are percentages of elements in the glass frit.

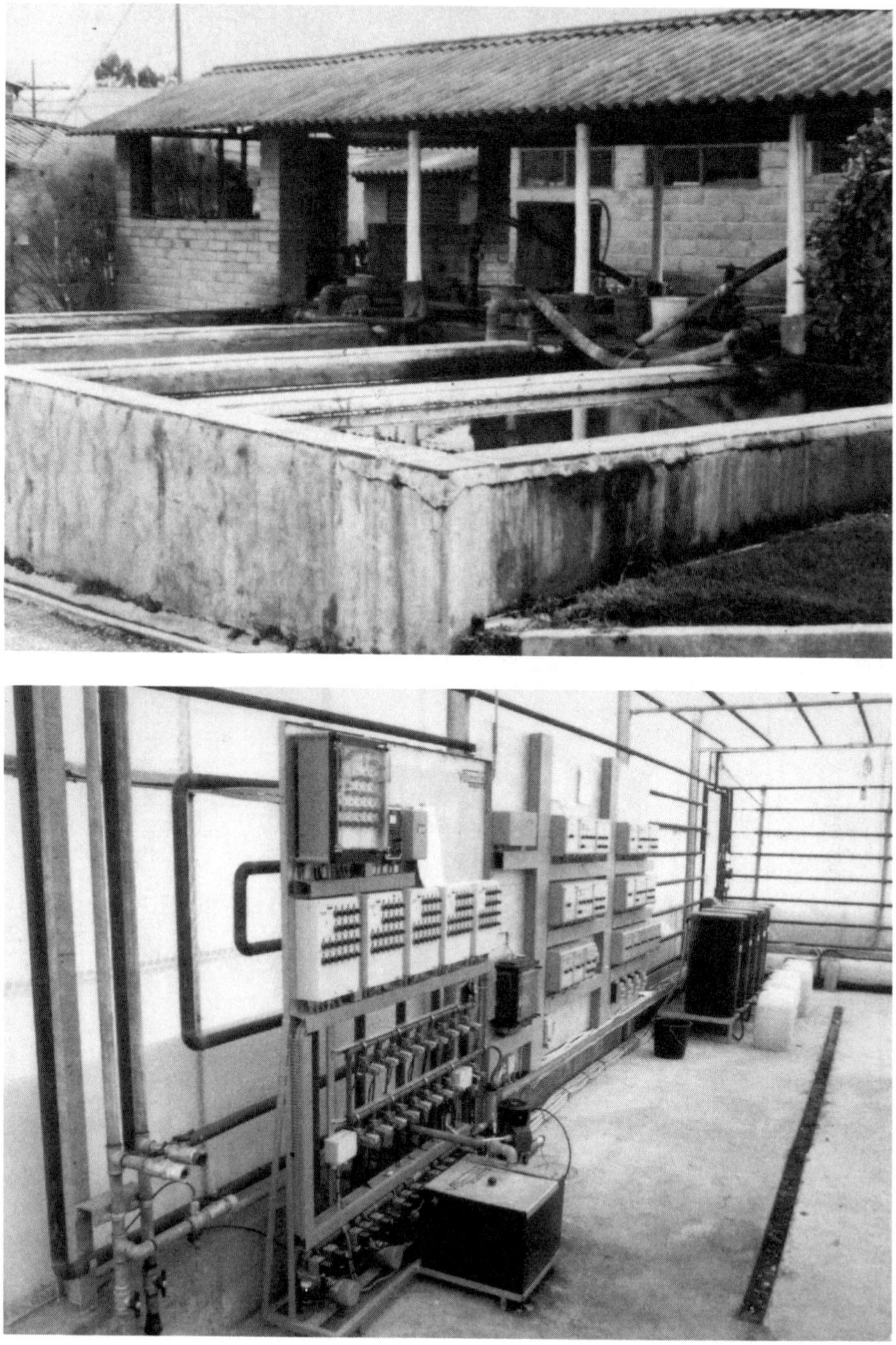

Fig. 6-45. Two examples of fertigation of greenhouse crops. **Upper**: Tanks for mixing fertilizers at the final concentration. **Lower**: System that proportions each individual fertilizer salt as required.

temperature has a marked influence on release from coated fertilizers. Thus, the grower must keep in mind the likelihood of rapid release if substrate temperatures rise excessively. Oertli and Lunt [1962] found release from encapsulated fertilizers to nearly double on a rise from 10 to 20 C. Soil moisture content did not appreciably affect the rate of transfer through the membrane, with an efficiency of recovery from 25 to 45%. Release rates are much slower if the fertilizer is applied as a top dressing. Recommendations are to mix into the substrate prior to planting.

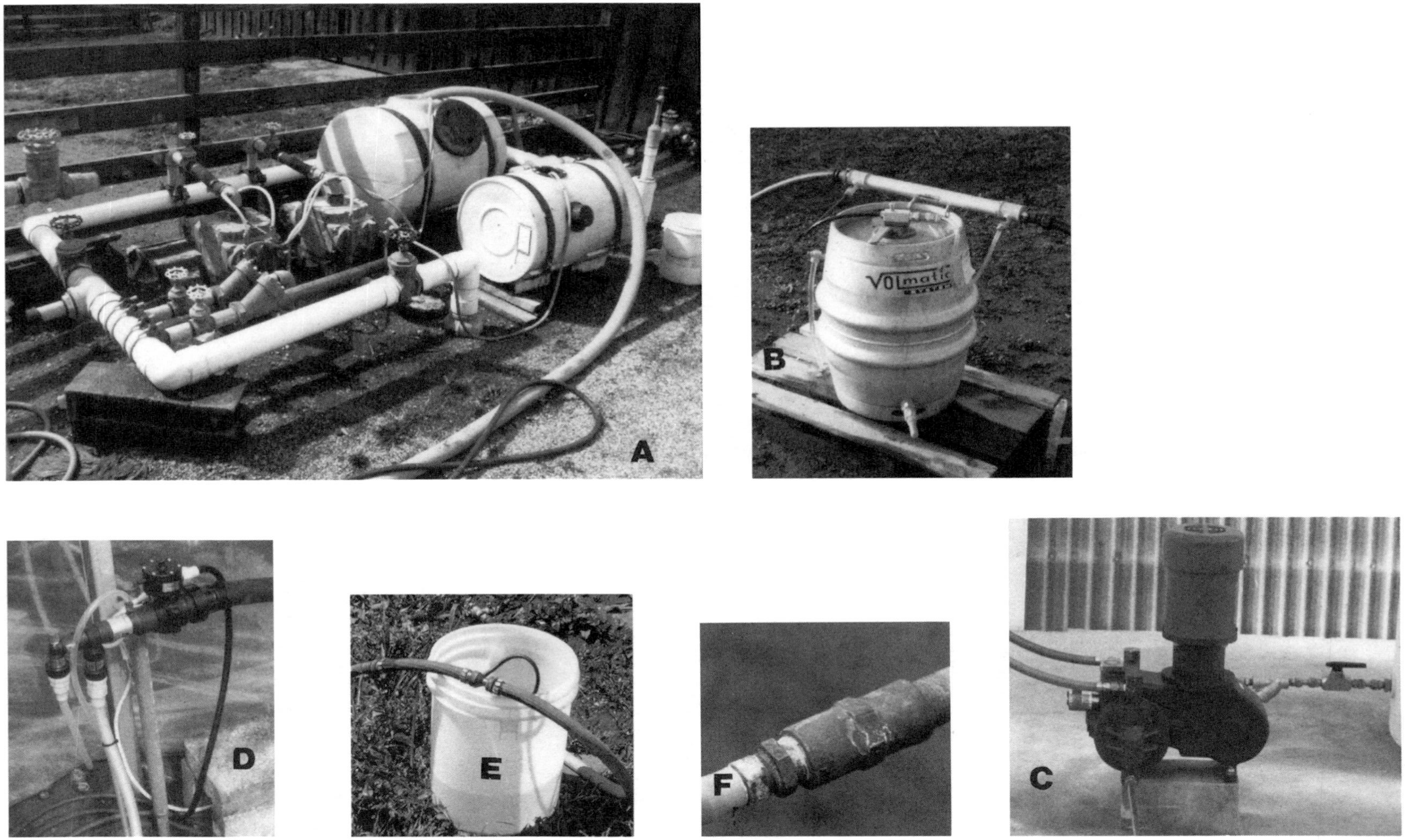

Fig. 6-46. Several examples of proportioning systems for greenhouse production, ranging from water powered to electrical, fixed or variable ratios, and differing costs.

Materials, such as frits, have also been employed in NFT to provide micronutrients. Glass frit 253A (Table 6-28) was recommended at a rate of 25 g per meter of gully –suitable for all trace elements except iron [Hall and Wilson, 1980]. Dolomitic limestone has also been suggested for NFT at 125 g per meter to halt pH drift and stabilize acidity. Use of slow-release fertilizers in container production has been found comparable to liquid feeding [e.g., Harbaugh and Woltz, 1989; Barragry and Morgan, 1978]. According to Roude et al. [1991], however, slow-release fertilizers such as Osmocote had no effect on the longevity of chrysanthemums *per se*. Total nitrogen, regardless of a fertilizer source, had the greatest effect on longevity. In highly porous soils, such as the sandy types in Florida, slow-release fertilizers can maintain sufficient nutrition without rapid loss in recovery efficiency by plants and without significant ground water contamination (see Section 6.VIII).

Nelson [1978] gives rates for a 14-14-14 Osmocote of 7.3 kg m^{-3} in light, course-textured media, and 6.5 kg m^{-3} in medium-textured substrates. The rate is lower for a 19-6-12 formulation, while a medium application for an 18-5-11 is 8.3 kg m^{-3}. The supply for the first two is 3 to 4 months, and 12-14 months for the 18-5-11. About 14 g is suggested by Nelson per 15 cm pot as a top dressing. Mengel and Kirkby [1982] state, however, that nitrogen recovery as measured by the proportion of N taken up by the crop of total N applied, is generally low for slow-release fertilizers as compared to water-soluble compounds. A further disadvantage is their high cost per unit weight of N. Despite these latter comments, such fertilizers are a viable means of nutrient control in greenhouse production, especially as amounts can be added to substrates at rates that would commonly salinize the medium if applied as soluble fertilizers.

4. Fertigation and Solution Formulation

Up to the late 1940s and early 1950s, dry fertilizer applications, either as broadcast or mixed into the substrate, were the principal means of nutrient control. By the late 1950s, it was usual to find some system of placing soluble fertilizers into the irrigation water. There are many systems with varying advantages and disadvantages. The principal advantage of fertigation itself, whatever the method, is that injection into the water supply allows a more consistent fertilizer application that varies with crop growth and season. Thus, it is a semiautomatic program that varies with irrigation frequency and amount. With the ability to use consistent, high-analysis chemicals, the grower may increase or decrease an application rate according to tissue and soil analysis with less likelihood of deficiency or excess. It is applicable to any of the cultural systems employed in greenhouses.

a. Systems

Examples of systems are presented in Figs. 6-45 and 6-46. The upper picture in Fig. 6-45 shows storage tanks in which soluble compounds are dissolved and pumped directly to the crop. It is common to more undeveloped regions where specialized equipment can be difficult to obtain and maintain. Care is required to ensure adequate mixing and complete solution of the fertilizer. In early years, it was usual for growers to modify coal stokers that added fertilizer to the tank under agitation. The lower figure in 6-45 is a highly sophisticated system whereby each nutrient can be added at its own regulated rate, and can be linked to a suitable computer for automatic control. Chapter 8 covers further discussion of automatic fertilizer control systems.

Almost any type of proportional injection system can be purchased that maintains a fixed ratio of chemical to each unit of irrigation water supplied. Fig. 6-46A shows a common system of water-powered injectors that proportion concentrated solutions from the two tanks into the water supply at a fixed 1:200 ratio. Therefore, the concentration in the supply tanks must be 200 times the final dilution. This can lead to problems since the high concentrations can be difficult to fully dissolve in cold water. Greenhouses having a steam supply can inject steam to heat the water. Another problem is that not all fertilizers can be mixed. Phosphorous-containing salts must be mixed separately from those containing calcium. Otherwise, there will be precipitation and the grower winds up with a sludge at the bottom of his tanks. This is the reason for two concentrate tanks in Fig. 6-46A. Pictures 6-46B, D, E, and F are also water-powered injectors. B is common where the fertilizer is injected from the storage tank by displacement and is adequate for small areas. E and F are inexpensive devices. E, usually designated as a "Hoze-on," sucks the fertilizer into the irrigation line by means of a venturi at proportional rates of about 1:24. F is called the "Dole" valve containing a variable diaphragm that changes an orifice diameter with pressure across the membrane. These are usually placed in the suction line of the irrigation pump. Obviously, these (E and F) are much cheaper than the other systems depicted. However, their proportioning rates are much less precise. C is a variable rate, electrically powered injector. The grower may manually set the injection ratio, if the volume of irrigation water through the system is constant. Fig. 6-46D shows a water-powered, inexpensive, pump that wastes part of the water that powers it. D and A can adjust to variations, within limits, in the flow rate of the main water supply. More commercial systems, in recent years, are approaching the versatility of the bottom picture in Fig. 6-45. These newer systems may also be linked with conductivity cells, pH meters, and ion selective sensors for continuous measurement and adjustment of the injected solution on a real-time basis. Several specific ion probes allow continuous measurement of nitrate, ammonium, sulfate, etc. in the irrigation water. Changes in water composition can be automatically compensated [Alberry et al., 1985; Gieling et al., 1988; Gieling and Schurer, 1995]. Measurement of individual ions remains a problem, however. The greenhouse environment is "noisy" to electronic equipment. Some electrical measurements in the greenhouse are difficult to carry out with any accuracy, unless special provisions are made to shield the sensing equipment and its attendant wiring.

b. Solutions

Injection solutions for plant culture are diverse. Where the grower is producing in ground, or in substrates containing soil, a "complete" solution is not often employed. The grower may combine practices by preplant fertilization with lime, superphosphate, or mixing in some slow-release fertilizers and using fertigation to supplement the preplant with such ions as ammonium, nitrate, and potassium. A producer may not always use fertigation with each watering, depending upon soil tests and observation to assess crop performance. Complete solutions of micronutrients are not often employed. Location, substrate type, and plant species may require special additions of one or more trace elements. In Colorado, for example, boron is always added in carnation culture. Zinc is also included in more recent years as piping systems have been replaced with plastics. Many growers purchase complete, proprietary micronutrient mixtures on the idea that if a little is good, more is better. With pure chemicals of the major macronutrients, sufficient trace elements may not be present as contaminants. Hydroponic solutions in good water probably require most of the micronutrients.

Any of the solutions listed in Table 6-22 may be employed to produce acceptable plants. In fact, experience at Colorado State University has shown that the carnation solution can be used to grow any species in any cultural medium. This does not mean that plant response will always be the best. Unfortunately, hydroponic solutions proliferate almost like lemmings. Steiner [1968] reported that more than 300 recipes had been published. If, for example, carnations grow well, it is published as a "carnation" solution. Jones [1983] listed 22 solutions given by various authorities. Cooper [1979], of course, reported his own NFT solution with various modifications. Similar recipes may be found in the publications from research institutions [e.g., Jones, 1983; Ellis et al., 1974; Hanan and Holley, 1974; Maxwell, 1972; Hoagland and Arnon, 1950; etc.]. One finds in the literature, innumerable reports on the effects of individual fertilizers, the use of complete, soluble fertilizers, and the requirements of suitable nutrient balance, depending upon growth stage and particular grower objectives [e.g., Lawton et al., 1989; Harbaugh and Woltz, 1989; Roude et al., 1991; Weston and Zandstra, 1989, Melton and Dufault, 1991; Boertje, 1980; etc.].

One suspects that many of these solutions could be the "best" for the particular conditions under which they were derived. Several investigators in the field have expressed to the author that Hoagland solutions are perfectly adequate for most purposes. Two things are usually neglected in constituting fertigation or hydroponic solutions: 1) individual salts added to water do not always remain the same, regardless of how consumed by plants; and 2) little attention is given about whether a particular solution is physically possible. For 1), reference to Tables 6-25 and 6-26 should be sufficient to be obvious. For 2), Steiner's work [1961; 1966; 1968; 1969] on electrical neutrality and composition, and the development of his "ideal" solution (Table 6-22) was the first to receive adequate attention –although nearly neglected in any practical application. Publication of so-called "ideal" solutions in terms of ppm of elemental nutrients are nearly useless since neither nitrogen nor phosphorous, and some micronutrients, exist as such in nature.

Another factor, often neglected, is the relationship between concentration and rate of supply –or flow rates in hydroponic solutions. It is typical to reduce Hoagland's solutions (Table 6-22) to one-half strength or more in research problems. Typical is Letey et al.'s [1982] determination of minimum nitrate concentrations for lettuce and tomato. The authors neglected, however, to suggest any agitation of their nutrient solutions except in passing that the dense root systems resulting after 60 days probably resulted in insufficient stirring. Their results showed

Table 6-29. Solutions given in Table 6-22 expressed in milliequivalents per ℓ. Micronutrients not considered as necessary in testing electrical neutrality.

Ion	Hoagland 1	Hoagland 2	Modified Hoagland	Johnson's 0.25 strength	Steiner's formula	NFT solution	Carnations	Roses	Tomatoes	Cucumber	Pepper
NH_4^+	---	1.0	0.4	0.5	---	---	2.5	1.0	0.5	0.5	---
K^+	6.0	6.0	2.4	1.5	13.151	7.964	6.0	4.0	7.0	5.5	6.0
Ca^{2+}	10.0	8.0	3.2	2.0	1.644	8.5	3.0	5.0	7.5	7.0	7.5
Mg^{2+}	4.0	4.0	1.6	0.5	1.644	4.154	2.0	2.0	2.0	2.0	2.5
$H_2PO_4^-$	1.0	1.0	0.4	0.5	0.822	2.192	1.1	1.0	1.5	1.25	1.25
NO_3^-	15.0	14.0	5.6	3.5	13.151	14.272	10.4	9.0	10.5	11.75	12.25
SO_4^{2-}	4.0	4.0	1.6	0.6	2.466	4.154	2.0	2.0	5.0	2.0	2.5
Total meq/ℓ	40.0	38	15.2	9.1	32.878	41.236	27.0	24.0	34.0	30.0	32.0
Estimated EC[1]	4.0	3.8	1.5	0.9	3.2	4.1	2.4	2.3	3.9	3.5	3.6

[1] Electrical conductivity in dS m^{-1} estimated from Fig. 6-8.

Table 6-30. Example of setting out a physically possible fertigation or hydroponic solution to ensure initial electrical neutrality, using the carnation solution from Tables 6-22 and 6-29.

Soluble fertilizer salt to be added	Cations						Anions					Milligrams per ℓ required of each fertilizer salt
	H^+	Na^+	K^+	Ca^{2+}	Mg^{2+}	NH_4^+	NO_3^-	$H_2PO_4^-$	SO_4^{2-}	Cl^-	HCO_3^-	
Potassium nitrate (KNO_3)			6.0				6.0					606.0
Calcium nitrate ($Ca(NO_3)_2{\cdot}4H_2O$)				3.0			3.0					354.0
Magnesium sulfate ($MgSO_4{\cdot}7H_2O$)					2.0				2.0			246.0
Ammonium nitrate (NH_4NO_3)						1.4	1.4					112.0
Monoammonium phosphate ($NH_4H_2PO_4$)						1.1		1.1				126.5
Total milliequivalents			6.0	3.0	2.0	2.5	10.4	1.1	2.0			
Total cations = anions						13.5	13.5					

no effect of nitrate concentration on accumulative uptake of the ion until after the first 40 days of growth, even at lowest concentrations of 140 μM ℓ$^{-1}$ NO_3^-. This is expected given the previous discussion in Section 6.III.D.1 (Fig. 6-16), which showed that NO_3^- uptake remained constant down to levels of 3 μM. The relationships between concentration and supply rate are the thrusts of such articles on flow rates by Edwards and Asher [1974] and Ingestad [1974; 1982]. Such studies form the impression that if an ion could be presented directly to the absorbing membrane as required, concentrations of macronutrients in the external solution could well be less than 1 μM ℓ$^{-1}$. But, ensuring minimum diffusion distances even at the outer surfaces of roots is well nigh impossible under commercial conditions. NFT systems are not intended to provide fast flowing solutions to an entire root system of a well-developed crop. In container production, the intense competition between roots in a confined volume requires higher concentrations to ensure adequate nutrient supply.

Although I have shown that a solution for fertigation does not remain the same after chemicals are added –certainly not after being added to a root substrate; there are means to prepare complete solutions that ensure reasonable electrical neutrality and where the solution is physically possible. Assume that a grower is producing carnations in rockwool, so we select the carnation solution in Table 6-22. Table 6-29 shows the same solutions with macronutrient concentrations converted to levels given in equivalencies (meq ℓ$^{-1}$) (Section 6.II.C). Micronutrients are not included since their total concentrations are very small. Also, estimated from Fig. 6-8 is an EC value for the respective solutions based on total meq ℓ$^{-1}$. A second table (6-30) shows the salts necessary to provide the respective ions and the amount necessary to dissolve in a ℓ of solution. In good water, at these concentrations, precipitation should not occur and the pH is estimated to be 7.0. Columns for the ions H^+, Na^+, Cl^-, and HCO_3^- are included since we will have occasion to use a table of this type later in salinity discussion. If a dilution system is being used such as a 1:100 proportioner, then the amounts necessary for 1 ℓ are merely multiplied by 100 times the volume of the concentrate tank. Here, as suggested above, calcium nitrate must be dissolved in a separate tank. Some solutions in Tables 6-22 and 6-29 require the use of calcium sulfate and calcium diphosphate, which are not highly soluble. They cannot be used in a typical proportioning system unless added separately at their final concentration. Note that the amount per ℓ is obtained by multiplying the equivalents required for the fertilizer by its equivalent weight (Table 6-5). The assumption is that each salt will be completely dissociated in solution so that, for example, 6 meq ℓ$^{-1}$ KNO_3 will provide 6 meq ℓ$^{-1}$ K^+ and 6 meq ℓ$^{-1}$ NO_3^- (Eqs. 6.1 and 6.2). As Tables 6-25 and 6-26 show, this is not always the case.

Table 6-30 is highly appropriate for use in a spreadsheet, and the necessary calculations can be easily carried out by the spreadsheet program. A table of this type also presents each macronutrient required by plants, in the form taken up by plants and shows the total concentration in values that can be used to estimate salinity. It eliminates the N-P-K formulae so commonly given in ppm or mg ℓ$^{-1}$, and ensures practicality –i.e., the solution is not likely to precipitate, leading to difficulties with trickle irrigation systems or leaving a part of the fertilizer solution in the dissolving tanks to be thrown out.

5. Examples of Interactions in Container Substrates

Application of a particular fertilizer to supply a known amount of nutrient does not mean that the expected response will occur. Especially in organic container substrates, response may be opposite to that needed. One of the best descriptions of such responses to fertilizers and their interactions is presented in Bunt's Media and Mixtures for Container-Grown Plants [1988]. My purpose here is to show some examples taken from Bunt's text.

Fig. 6-47. Tomato seedlings grown in the winter with equal amounts of ammonium from three sources: **Left-to-right**, ammonium carbonate, urea, and ammonium sulfate [Courtesy of A.C. Bunt].

Fig. 6-48. The effect of superphosphate and nitrogen source on snapdragon growth. From left-to-right, phosphate application rate decreases in both pictures. The top picture are plants with urea as the N source, the bottom picture plants fed with calcium nitrate as the N source [Courtesy of A.C. Bunt].

In his discussion of peat types, Bunt presented several practical examples of the interactions between fertilizer types and plant growth in peat-sand container mixes. When it comes to chemical or organic fertilizers, the results can be startling. Fig. 6-47 shows tomatoes grown in the winter, receiving equal, but from different sources, ammonium. Plants grown with calcium nitrate (not shown) were larger. Free ammonia can occur at pHs close to neutral and increases as pH rises. With an organic fertilizer (urea), the rise in pH can release considerable amounts of toxic ammonia. A second example of interaction between two fertilizer types in peat-sand mixtures is Fig. 6-48. The upper picture shows that greatest response to an organic fertilizer occurs with highest superphosphate rates, but such response to increasing superphosphate does not occur when calcium nitrate is employed. In peat substrates, superphosphate will reduce pH, which slows nitrogen mineralization rates and significantly reduces the free ammonia and nitrite levels.

Fig. 6-49. Tomato plants grown in the winter in four peat-sand mixes. **Upper left**, low N, low $CaCO_3$; **Upper right**, high N, low $CaCO_3$; **Lower left**, low N, high $CaCO_3$; **Lower right**, high N, high $CaCO_3$. N supply 70% from hoof-and-horn meal, 30% from NH_4NO_3. Same treatments in the summer [Courtesy of A.C. Bunt].

The relationship between climate and fertilizers was examined by Bunt as shown for the winter treatment in Fig. 6-49. During the summer, there were essentially no differences between the treatments. The four treatments received calcium carbonate at low or high rates while 70% of the nitrogen was supplied from hoof-and-horn meal and 30% from ammonium nitrate at low or high levels. A part of the reason for the differences between summer and winter results from higher summer temperatures which about double nitrogen mineralization and, with higher radiation, plant growth can be greater with less risk of toxicity. True mineral deficiencies are more quickly observed in summer than in winter. A third example is Fig. 6-50 where urea and calcium nitrate are compared with different sources of calcium. With calcium nitrate as the nitrogen source, growth was similar for all three liming materials. With urea, however, calcium sulfate was superior to calcium carbonate and dolomitic limestone. Bunt stated that the higher pH obtained with $CaCO_3$ caused free ammonia and nitrite production. Dolomitic limestone gave a slightly reduced pH with growth depression correspondingly reduced. Plants grown with ammonium-producing fertilizers are more susceptible to boron deficiency (Fig. 6-51). Boron uptake can also be suppressed by high phosphorous levels. Other examples of trace element interactions were presented earlier (Fig. 6-20).

Interaction between three substrates and various ratios of ammonium-to-nitrate is examined in Fig. 6-52. The pictures are three of several bedding plant species examined by Jeong [1990]. The soil mix (S) consisted

Fig 6-50. The top picture shows tomato growth with the N source urea, the bottom calcium nitrate. From left-to-right, the calcium source is $CaCO_3$, dolomitic limestone, and calcium sulfate [Courtesy of A.C. Bunt].

of 1 part soil (local Ft. Collins clay loam), 3 parts sphagnum peatmoss, and 2 parts by volume perlite. Peat-lite mix (**P**) was a proprietary mixture composed of 75% Canadian sphagnum peatmoss, 24% perlite, and 1% vermiculite with dolomitic limestone and a wetting agent added. All three media, including the rockwool (**R**), had no base fertilizers. The NH_4^+:NO_3^- ratios varied from 0:100, 25:75, 50:50, 75:25, to 100:0 applied in the irrigation water. It can be seen that the response to treatments was much greater in rockwool media as compared to the buffered peat- and soil-containing media. Salvia were much more sensitive to either extreme, particularly 100% NH_4^- and no nitrate. Ageratum response (top), however, suggests some tolerance to 100% ammonium supply. The results point up the inherent criticality of loamless, non-buffered substrates, and the specific responses of each species that were significantly different. Of the 11 species tested, only ageratum showed tolerance to ammonium. Jeong found that pH did not affect growth and tissue composition, reinforcing a previous statement that pH *per se* is not as important as the secondary effects of hydrogen ion concentrations on nutrient relationships.

Fig. 6-51. Boron deficiency on snapdragons grown in peat-sand. **Left**, N source hoof-and-horn meal; **Center**, urea; **Right**, calcium nitrate [Courtesy of A.C. Bunt].

In his Experiment 3, Jeong compared three NH_4^+:NO_3^- ratios of 50:50, 100:0, and 0:100 applied to four bedding plant species grown in peat-lite and rockwool. Each ratio compared solutions applied with no chlorine versus 4 meq ℓ^{-1} Cl^-, all solutions fixed at 50 meq ℓ^{-1} total. Again, responses to these treatments were greater in the unbuffered rockwool. In rockwool, the pictures (Fig. 6-53) suggest that there may have been some beneficial effects from chlorine addition. Still, the absence or presence of Cl^- was not apparent on plants grown in peat-lite. Nukaya et al. [1991] showed that increased Cl^- reduced blossom end rot on tomatoes, but increased gold speck injury. Shelf life decreased with increasing SO_4^{2-} and Cl^-.

These few examples, combined with previous discussion, serve to emphasize that nutrient interactions and corresponding plant growth are not simple concentration manipulations. Much information on specific species is unavailable so a grower is forced to some trial and error.

Fig. 6-52. Effect of five NH_4^+:NO_3^- ratios applied in irrigation water to ageratum (top), celosia and salvia (bottom) grown in a soil mix (**S**), peat-lite (**P**), and rockwool (**R**) [Courtesy of B.R. Jeong].

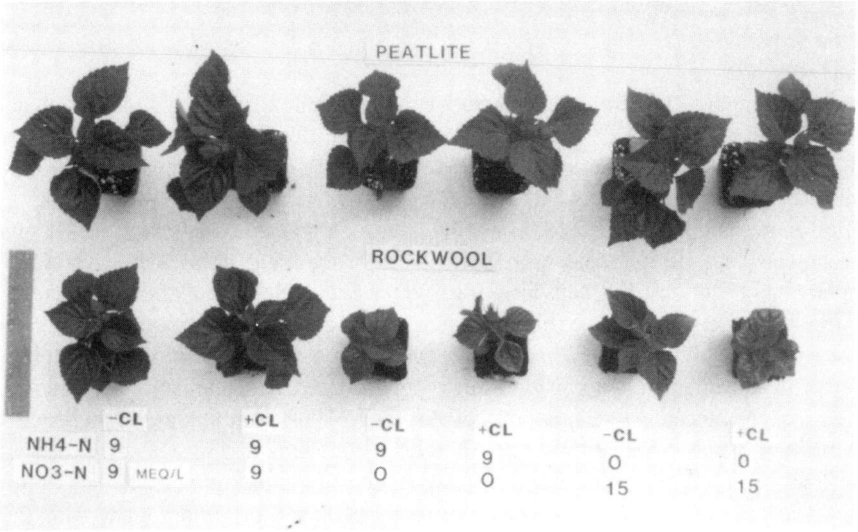

Fig. 6-53. Effect of NH_4^+:NO_3^- ratios with and without 4 meq ℓ^{-1} Cl^- on ageratum (top) and petunia (bottom) grown in two substrates [Courtesy of B.R. Jeong].

F. SOIL ANALYSIS

Soil analysis forms the third leg of nutrient control in greenhouses, the other two being observation and tissue analysis (Section 6.III). The purpose of soil testing is to learn which nutrients are deficient and to estimate how much fertilizer is required to correct a deficiency and increase yield. According to Jones [1985], soil testing is the only means of specifying lime and fertilizer needs. The use of fertilizers without a soil test, and its interpretation, is hazardous to successful crop production. Equally important from the discussion on tissue analysis, soil analyses do not always track tissue concentrations in plants. So there are restrictions on the use of soil analyses to handle nutrient control, just as there are on tissue analyses. There are three aspects to be considered: 1) obtain a representative sample from the crop substrate; 2) obtain a suitable fluid extract for chemical analysis from the sample; and 3) interpret the result.

1. Sampling the Substrate

a. Mineral Soils

Considerable detail for obtaining a representative sample from a field is given by such authors as Jones [1985], Peck and Soltanpour [1990], and James and Wells [1990]. However, most sampling is carried out by the grower, and Peck [1990] states that sampling is the greatest source of error in the testing process. A particular problem is soil heterogeneity horizontally and vertically. Random versus planned sampling has been discussed, and one will find James and Wells' discussion worth reading. In greenhouses, however, the small size (<1 ha) of most establishments, and the fact that a crop is usually growing in a longitudinal bench arrangement, means that most sampling will be to provide a composite sample for each bench. An assumption here is that the soil is uniform, and where a single crop is treated similarly for the entire area, a single composite sample from several benches may be adequate. Although large areas of greenhouse production are in the ground, little attention has been given to possible problems arising from heterogeneity. Since each bench may be analyzed, any variation in soil properties from one location to another may be disguised. Thus, sampling a ground bench randomly, four or more times to obtain a single sample for the bench, is usual for a grower. Long benches, obviously, should be sampled more times, or until the grower has evaluated uniformity in a single bench. The first time a bench is analyzed, individual cores should be analyzed so a grower can determine variation and what sampling method will provide the greatest reliability.

Trickle irrigation systems, particularly if the emitters are 30 cm or more apart, are likely to cause significant variations in nutrient concentrations with distance from the emitter. Such characteristic, concentric patterns of pH, EC, and nutrients, extending in cone shapes through the soil profile, were found by White and Prasad [1980] in tomato culture. Distribution patterns in peat modules were also found. At the end of the growing season, tests in soils revealed the existence of a deep, narrow, strongly leached zone, directly under the emitter, with low pH, Ca, and EC values. Depending upon conditions, evaporation and outward salt movement from the emitter frequently result in salt accumulation at other locations in a bench or container. In peat modules, White and Prasad found no evidence of leaching immediately below the emitter. Under the conditions found in this study, White and Prasad concluded that there was no optimal sampling location. They suggested sampling 15 cm from an emitter, at right angles to the row as the easiest compromise. Such patterns as described –not necessarily the same– can occur in almost any container or ground bed. Problems with an inadequate irrigation system, or with climatic extremes within the greenhouse (i.e., south ends of structures in the winter, etc.), will exacerbate variations in nutrient content of the medium. Growers should allow for these factors when sampling for soil analysis. That is, areas manifestly different in soil moisture or plant response should be sampled separately. Such soil tests may point the way to modifications in the cultural program for greater efficiency and better product quality and yield.

Most analyses for field soils use less than 100 ml soil for extraction. About 0.2 kg dry soil is more than adequate for sending to the laboratory. Most soil testing laboratories will have special containers for the grower to use. This means that one or more cores must be well mixed and a subsample of suitable size taken from the mixture to provide a composite sample. The procedures for mixing and subsampling require care to prevent contamination and to obtain a representative subsample. After mixing several cores, the soil pile is split. One remaining half may be split in succession until the remaining half, the final split, is small enough to place in the laboratory's package.

Depth of a soil core, commonly taken with a stainless steel, hollow tube, 2 to 3 cm diameter, with a sharp edge, is 15 to 20 cm –or the cultivation depth (see Fig. 5-39). Under most greenhouse conditions, such coring tubes can be pressed into the soil, and removed, by one person. Auger-type tools should not be necessary. If the soil is gravelly, a shovel or spade may be required. If the active root system extends below 20 cm (i.e., roses), sampling depth should be extended to include the active root zone. An auger may be required here. In raised soil benches, or pots, a core of the entire depth is taken after first removing the upper half centimeter or so of the soil. Argo and Biernbaum's results [1994] showed the top 2.5 cm in pots growing subirrigated Easter lilies to have nutrient concentrations up to 10 times higher than those measured in the remaining root zone. Use of an evaporation barrier reduced the stratification. In peat modules, two or more samples halfway between center line and the edge, the entire depth, should be adequate. For pots, a smaller tube of 1 to 1.5 cm diameter can be used. Some growers make cheap tubes from rigid electrical conduit. As for sample contamination in micronutrient testing, such sampling tubes are not desirable.

b. Container Mixes and Loamless Media

Most samples for mineral and organic soils are dried, ground, and sieved in preparation for extracting the nutrients for analysis. This is not a good practice where slow-release fertilizers are employed. The general practice is to obtain a moist sample near container capacity, sometimes in amounts up 500 ml. Whereas dry samples are often mixed with extracting solutions on a weight-soil-to-volume-extractant ratio, moist samples, particularly peat-based and loamless media, are extracted based on volume-to-volume. Some laboratories are careful to bring a sample to a known matric suction (i.e., 10 or 60 cm water column) before extraction [van den Ende, 1971; Sonneveld and van den Ende, 1971; Sonneveld et al., 1974]. This is laborious and not often used where large sample numbers are encountered. Markus and Steckel [1980] found that placing moist media samples in sealed plastic bags was adequate although the sample might be in the mail for 2 to 3 days. Individuals dealing with field soil analysis commonly argue that significant changes in chemical composition of moist samples are likely to occur. Extracting moist samples markedly speeds up testing.

Sampling container mixes can use the same tools as outlined above. However, with the marked movement to sample extraction using only water, Wright [1986] discussed the "pour-through" method whereby a liquid sample was obtained from drainage of the container. Containers should be close to, or at, container capacity, and additional water applied sufficient to obtain, for example, a 50 ml aliquot from the drainage. The container is the sample. The grower makes a random selection of containers from the crop for liquid extraction. This procedure has the advantage that root systems are undamaged versus core sampling of the container. For research purposes, Nelson and Faber [1986] discussed a displacement method where a 5 cm diameter plastic column, 60 cm tall, was packed with a ℓ of the sample. About 250 ml of a displacement solution, consisting of 0.5% thiocyanate in 50% ethanol was poured through the column with several samples taken from the drainage until formation of a bright red color from the use of a 0.5% ferric chloride solution showed mixing had occurred. Any drainage after that was discarded. Unfortunately, this method seems laborious for quick testing on a practical basis. Nelson and Faber, I think, did not adequately detail the packing procedure. Given the high percolation rates of some media (Tables 5-9, 5-10), sufficient packing to prevent immediate mixing would be difficult.

A.C. Bunt discussed the problems in analysis of organic and lightweight potting substrates [1986]. He pointed out that drying can cause ingredient separation, reduce amounts of nutrients extracted, and some organic materials (e.g., peatmoss and bark) are difficult to re-wet. Grinding changes medium structure, increasing bulk density. The problem with slow-release fertilizers has been mentioned. With the low bulk density of some media components (0.1 g/cm^3), large errors in analytical interpretation can occur if not corrected. Bunt discouraged the use of displacement methods for any practical testing service.

c. Hydroponic Systems

For hydroponic systems that waste excess water, that is, the nutrients flow once through the inert substrate, sampling for analysis, other than to check the applied nutrient solution, is unnecessary. The same feeding solution is applied at each irrigation. As long as suitable precautions are taken to maintain equipment and ensure proper mixing, there should be no need for numerous analyses. However, as will be noted in Section 6.VII, some means of rapidly measuring pH and total salts are good checks for system performance, and ensuring that the water supply quality has not changed significantly. For recirculating solutions, regular sampling for analysis is an important procedure. Though continuous measurement of pH and EC and some ions (i.e., nitrate) can be done, nutrient uptake varies with each ion and over time. The nutrient solution composition will change markedly from what the grower initially began. Cooper [1979] went into considerable detail on uptake of nutrients in the NFT system. As the crop grows, nutrient uptake increases with many increases or decreases in the main trend that can be attributed to such factors as root death (onset of fruiting or leaf removal in tomatoes), pinching, changes in radiant level, etc. Sampling in either case is merely obtaining a water sample to send to the laboratory. The water sample, here, is analogous to an extracted solution from a substrate sample. The only care is the use of adequate, clean, plastic containers.

d. Handling Samples

James and Wells [1990] emphasized the care required to prevent sample contamination. Common contaminant sources include dirty sampling tools, dirty containers, cigarette or pipe ashes, and preparing samples

Table 6-31. Examples of a number of soil testing procedures in use. The list is not intended to be complete. Procedures in grinding, sieving, preparation of extractant, etc. may be incomplete. Refer to footnotes as necessary.

Name	Conditions and media	Extractant chemicals	Dilution	Shaking time	Nutrients tested	pH of extracted solution	Reference
Morgan universal	All acid, mineral soils, soilless mixes	0.73 M sodium acetate	20 ml dry sample, passing 2 ml sieve in 40 ml solution	5 min and filter	P, K, Ca, Mg, NO_3, NH_4, SO_4, Fe, Cu, Mn, Zn	4.8	Jones, 1990
Morgan-Wolf universal	All acid, mineral and organic soils	0.073 M sodium acetate 0.52 M acetic acid 0.0001 M DTPA	20 ml dry sample passing 2 ml sieve in 40 ml solution	5 min and filter	P, K, Ca, Mg, NO_3, NH_4, SO_4, Fe, B, Cu, Mn, Zn	4.8	Jones, 1990
Modified Morgan	Various potting mixtures, peat-vermiculite, etc.	1.4 N sodium acetate 1.0 N acetic acid	20 ml moist sample, 1:5 ratio	30 min and filter	P, K, Ca, Mg, NO_3, NH_4	4.8	Markus and Steckel, 1980
Mehlich No. 1 universal	Acid, sandy, mineral soils, CEC <10 meq/100 g, O.M. < 5 %	0.05 N HCl 0.025 N H_2SO_4	4 ml dry sample passing 2 mm sieve, 1:5 ratio	5 min and filter	P, K, Ca, Mg. Na, Mn, Zn, NO_3	1.2	Jones, 1990 Markus and Steckel, 1980
Mehlich No. 3 universal	Acid to neutral mineral soils	0.2 N acetic acid 0.25 N NH_4NO_3 0.015 N NH_4F 0.013 N HNO_3 0.001 M EDTA	5 ml dry sample in 50 ml, 1:20 ratio[a]	5 min and filter	P, K, Ca, Mg, B, Cu, Fe, Mn, Zn	1.2	Jones, 1990
AB-DTPA	Alkaline, calcareous, mineral soils, pH >7.5	1M NH_4HCO_3 0.005 M DTPA	8.5 ml dry sample passing 2 mm sieve, in 20 ml	15 min and filter	P, K, Na, Cu, Fe, Mn, Zn	7.6	Jones, 1990 Workman et al.,1988
$NaHCO_3$-DTPA	Potting media, peat-clay mixes	0.5 M $NaHCO_3$ 0.005 M DTPA	?	?	N, K, P, Cu, Ca	?	Alt et al.,1988
Olsen bicarbonate	Alkaline, mineral soils	0.05 $NaHCO_3$	Dry sample, passing 2 mm sieve, 1:20 ratio	30 min and filter	P	8.5	Bates, 1990
Bray-Kurtz P1	Acid, mineral soils	0.025 N HCl 0.03 N NH_4F	Dry sample, passing 2 mm sieve, 1:10 ratio	5 min and filter. 1 min[b]	P, K, Ca, Mg, NO_3	1.2	Bates, 1990 Markus, 1980 Markus and Steckel, 1980
Bray-Kurtz P2	Acid, mineral soils	0.1 N HCl 0.03 NH_4F	Dry sample, passing 2 mm sieve, 1:10 ratio	5 min and filter	P, K, Ca, Mg	1.2	Bates, 1990
DTPA-SME	Potting mixes	0.005 M DTPA, 30 ml in 100 ml soil[c]	Saturate medium after adding DTPA	Mix while saturating, equilibrate 1.5 hr	pH, EC, P, K, Ca, Mg, NO_3, Fe, Mn, B, Cu, Zn		Berghage et al.,1987

Name	Conditions and media	Extractant chemicals	Dilution	Shaking time	Nutrients tested	pH of extracted solution	Reference
Spurway acid extraction	Soils, peat-based potting mixes	0.018 N acetic acid	20 ml moist sample, 1:5 ratio	1 min and filter	P, K, Ca, Mg, NO_3	3.3	Markus, 1986
Dutch 1:5	Soils and soil-based potting mixes	Water, sample brought to 63 cm water matric suction	1:5 ratio by weight (1:25)[d]	15 min	EC, Cl, N, PO_4, K, Mg	---	van den Ende, 1971
Dutch 1:2	Soils, potting mixes, peat- and soil-based	Water	Sample brought to field capacity, 1:2 ratio		EC, PO_4, Cl, N, K, Mg	---	Sonneveld and van den Ende, 1971
Dutch 1:1.5	Potting mixes, peat, bark, clay-sand, etc.	Water, soil brought to 32 cm water matric suction	100 ml soil in 150 ml water, 1:1.5 ratio	15 min, filter	EC, PO_4, NO_3, NH_4, N, K, Cl, Mg	---	Sonneveld et al.,1974
SME (saturated medium extract)	Potting mixes, peat-based, loamless	Water, 500 ml medium brought to saturation	---	Equilibrate 1.5 hr (Overnight)[e]	pH, EC, NO_3, NH_4, P, K, Ca, Mg, Fe, Mn, Cu, Zn	---	Warncke, 1986
Pour-through	Potting mixes, no samples removed, leachate from containers	Water, 50 ml collected	---	---	pH, EC, NO_3, NH_4, P, K, Ca, Mg, Fe, Mn, Cu, Zn	---	Wright, 1986
ADAS (U.K. advisory service)	Potting mixes, loamless media	Saturated $CaSO_4$ solution for EC and NO_3 determination; $NaHCO_3$ for P	Dry sample, passing 2 mm sieve, 20 ml soil, 1:6 ratio	?	P, EC, NO_3, K, Mg	---	Johnson, 1980
Levington (U.K.)	Loamless mixes	Water	1:6 dilution	60 min and filtered	pH, EC, P, K, NO_3, Mg	---	Johnson, 1980
2 mM DTPA	Potting mixes	0.002 M DTPA	Sample matric suction brought to 10 cm water. 25 ml DTPA to 100 ml sample	1.5 hr, intermittant	pH, EC, N, K, P, Ca, Mg, Fe, Cu, B, S, Mn, Zn	---	Handreck, 1991
Trace element	Tropical mineral soils	0.25 N $NaHCO_3$ 0.01 M EDTA 50 ppm Superfloc 127[f] 0.01 N NH_4F	Dry sample, to pass 2 mm sieve, 1:10 ratio	10 min and filter	P, Mn, Zn, Cu	---	Lindsay and Cox, 1985
Hot water	Mineral soils	Hot water	Dry sample to pass 2 mm sieve, 1:2 ratio	5 min and filter	B	---	Lindsay and Cox, 1985
$(NH_4)_2C_2O_4$	Mineral soils	Ammonium oxalate	Dry sample to pass 2 mm sieve, 1:10 ratio	10 hr and filter	Mo	3.3	Lindsay and Cox, 1985

Name	Conditions and media	Extractant chemicals	Dilution	Shaking time	Nutrients tested	pH of extracted solution	Reference
DTPA-TEA	Mineral soils	0.005 M DTPA 0.01 M $CaCl_2$ 0.1 M TEA[g]	Dry sample 1 mm stainless steel sieve, 1:2 ratio (w/v)	2 hr and filter	Zn, Fe, Mn, Cu	7.3	Lindsay and Norvell, 1978
Boron	Mineral soils	0.1 M $CaH_4(PO_4)_2$	Dry sample, 3 mm sieve, 1:5 ratio	10 min and filter	B		Lindsay and Cox, 1985
NaCl	Potting mixes	0.5 N NaCl	20 ml moist sample, 1:5 ratio	30 min and filter	P, K, Ca, Mg, NO_3, NH_4	5.9	Markus and Steckel, 1980

[a] Use only plastic containers with this mixture.
[b] Mixing time used by Markus, 1986.
[c] $LiCl_2$ required at 1000 ppm if emission spectroscopy used for analysis.
[d] P determination on dry sample, shaken 15 min.
[e] See Markus and Steckel, 1980.
[f] Superfloc is an organic flocculating agent manufactured by American Cyanamid.
[g] TEA is triethanolamine $(HOCH_2CH_2)_3N$.

in dusty locations. Obviously, galvanized containers cannot be used if testing for zinc levels is expected. Empty coffee cans can be a major zinc source. Samples to be tested for boron should not be placed on Kraft paper since such paper can be a B contaminant. The best preventive for contamination is use of clean tools, clean plastic buckets, clean plastic bags, and use of containers supplied by the testing laboratories. Tube corers should be stainless steel.

Table 6-32. Standard values for EC and nutrient levels in extracts obtained by the 1:1.5 and SME methods [Adapted from Warncke, 1990] (With permission of the *Amer. Soc. of Agronomy*).

Analytical value	Optimum levels	
	1:1.5	SME
Conductivity[1]	1.3-1.8	2.0-3.5
NO_3[2]	3.7-5.4	7.1-13.2
P	0.48-0.68	0.23-0.42
K	1.5-2.1	4.0-6.0
Ca	---	2.5-5.0
Mg	0.65-0.90	1.5-3.0
Na	---	<3.0
Cl	<3.3	<2.5

[1] Units = dS m^{-1}.
[2] Units = mmol ℓ^{-1}.

2. Analytical Procedures

There is good reason for Table 6-31. Although a grower has no need to know detailed laboratory procedures, he must appreciate the fact that switching willy-nilly from one laboratory to another is likely to cause chaos. Nine of the procedures listed in Table 6-31, such as the Morgan, Mehlich, Bray-Kurtz, and AB-DTPA were devised to test field soils. It is well to remember that most mineral field soil solutions have low concentrations, and the methods allow removal of labile nutrients from the exchange complex that replenish the soil solution. Thus, the tests also suggest the soil's capacity to provide nutrients over the growing season. Most of these procedures deal with acid, mineral soils common to the eastern U.S., whereas the AB-DTPA was devised for alkaline, calcareous mineral soils with pH above 7.0. If the grower producing tomatoes in the southwest U.S. sends his samples to a laboratory using a Mehlich method in New York, the results are not likely to be relevant. Interpretation by someone unfamiliar with local conditions is circumspect.

Table 6-33. Estimated optimum P values based upon cyclamen dry weight for three extracting methods and three media types [Adapted from Prasad et al.,1983].

Test method	Substrate	Optimum level[1]
1:1.5	Peat	0.26-0.28
SME	Peat	0.79-0.86
Spurway	Peat	3.2-3.5
1:1.5	Bark	0.12-0.13
SME	Bark	0.45-0.48
Spurway	Bark	1.4-1.6
1:1.5	Peat + soil	0.045-0.061
SME	Peat + soil	0.074-0.119
Spurway	Peat + soil	0.9-1.2

[1] Units = mmols ℓ^{-1}.

Lindsay and Cox [1985] surveyed procedures used in tropical regions for micronutrients, finding a wide range in practice. Some of their conclusions were that no particular extractant has been superior under all conditions, although several types can be employed successfully. The use of dilute acids is restricted to acid soils. Although chelates have been used mostly on alkaline soils, studies with DTPA suggest it can be used on acid soils as well. Micronutrient testing is particularly difficult as the result of very low concentrations. The preparation procedures, such as sample drying, grinding force and time, sample quantity, shaking type (oscillation versus rotary), and time, all influence the result [Martens and Lindsay, 1990]. So-called "universal" extracting solutions may be misnamed. It can be appreciated that slight variations from one laboratory to another can change the result in so-called duplicate samples. Only in cooperative tests, such as that reported by Warncke [1986] for the SME method, are there likely to be close values from several laboratories.

The higher fertilization rates found in greenhouse practice are likely to result in reports of excessive fertilization when substrate samples are subjected to extraction by procedures suitable for field soils. The laboratory may find it necessary to massively dilute extracts to bring them into the range of their standards. The Spurway system (Table 6-31), using a weak acid solution with minimum shaking, has enjoyed a long period in the U.S. as the principal system for greenhouse substrates. Many laboratories still use it since there is a large background of experience with the system. Because many substrates nowadays are loamless, with low buffering capacity, more laboratories, especially European, are using pure water extraction on moist samples. Even though peat, composted bark, and vermiculite have nutrient-holding abilities, nutrients are held less tightly compared to mineral soils [Warncke, 1990]. With fertigation, the lack of a capacity factor in greenhouse media is less important. The bulk density of container substrates can be variable, and CEC of low-density materials with peatmoss, bark, etc. can appear high when expressed on a weight basis. When expressed on a volume basis, however, CEC values are similar or less than mineral soils. Samples measured on a weight basis require conversion to a volume basis or separate interpretation guidelines.

In the U.S., several laboratories have switched to the saturated media extract (SME) system. This method was first developed by the U.S. Salinity Laboratory [Richards, 1954] for testing total soluble salts. Warncke [1990] states that the method is useful in weakly buffered growth media. The saturated sample is vacuum filtered and all subsequent analyses are performed on the filtrate. Warncke acknowledges that results with SME have been more variable, much of it associated with the difficulty of accurately mixing the medium to the saturation point. With field soils, determining the end-point is relatively easy: the soil begins to flow, the surface glistens, and the soil slides cleanly off a spatula. As the amounts of coarse peat, bark, plastic beads, etc. increase, determining the actual endpoint becomes more difficult. While the SME is more laborious, pH and EC can also be determined from the sample. Other procedures such as the displaced solution method described by Nelson and Faber [1986] and the pour-through [Warncke, 1986] have been described. The detail required for the former procedure precludes its use in routine diagnostic analyses [Warncke, 1990].

In Europe, saturated media extracts and displaced solutions are used in research, but water extracts are employed by service facilities. The Dutch reduce variability by adjusting the moisture tension in the sample to about 32 cm water column before extracting the sample solution at a 1:1.5 ratio. The Levington and ADAS procedures, reported by Johnson [1980], use a 1:6 ratio, which apparently overcomes variability due to failure to account for initial moisture content. Standard test values for the Dutch 1:1.5 and SME are compared in Table 6-32. Table 6-33 compares optimum P levels determined as a function of testing method and substrate composition. As noted, P levels for optimum plant growth vary markedly with root medium composition and test procedure.

For micronutrients, several specific procedures have been included in Table 6-31. Lindsay and Cox's [1985] survey of tropical soil testing showed wide variation. These authors concluded that there is no way to tell from their compiled data the reliability of the various procedures for detecting iron deficiencies in tropical soils. The critical levels of manganese varied from 1 to 28 ppm; zinc, 0.5 to 10.0 ppm; copper, 0.2 to 10.0 ppm; boron, 0.2 to 2.0 ppm; and molybdenum, 0.1 to 0.3 ppm. They also stated that although a particular soil test is used by several laboratories, it is no guarantee that the test is effective.

3. Interpretation of Soil Analyses

Dahnke and Olson [1990] separate understanding soil testing into three steps: 1) correlation, 2) calibration, and 3) interpretation. Correlation is the process to determine if an extracted nutrient and crop response to the added nutrient are so well related that one directly implies the other. There are several means of correlation and calibration, one of which is the graphical Cate-Nelson

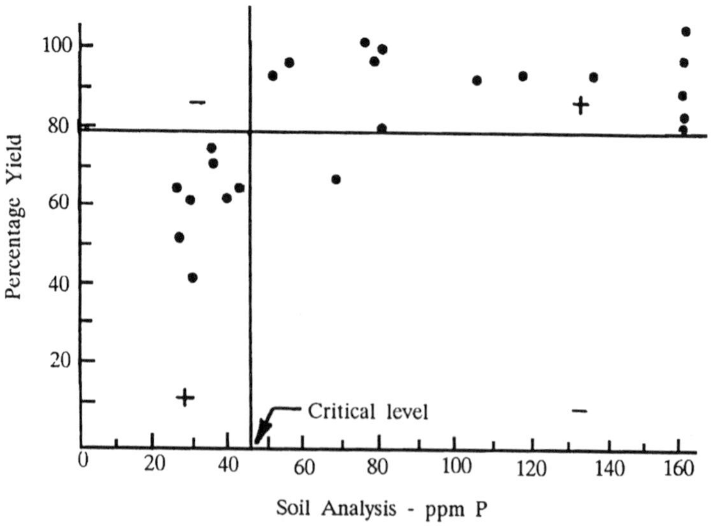

Fig. 6-54. A Cate-Nelson scatter diagram of percentage yield versus soil test P for maize [From Cate and Nelson, 1971] (With permission of the *Amer. Soc. of Agronomy)*.

diagram for P presented in Fig. 6-54. The diagram plots percentage yield against the soil test to give visual indication of soil test reliability. The division between negative and positive quadrants is divided to maximize the number of points in the positive quadrants and minimize those in the negative areas. Many points in the negative quadrants show the soil test is not well suited to the soils of the area, or there is no correlation between soil test values and plant response to the added nutrient. Whether Cate-Nelson diagrams can be applied to greenhouse conditions has not been determined to my knowledge. Calibration finds out the meaning of the soil test in terms of crop response. This allows soils to be placed into response categories such as very low, low, medium, high, and very high concentration ranges [Dahnke and Olson, 1990]. The Cate-Nelson procedure (Fig. 6-54) has at least two advantages over fitting continuous curves to the data. It shows whether there is a good correlation and separates the data into populations likely to respond to added nutrient and those unlikely to respond to added nutrient. The point at which this occurs is the critical level. Laboratories in the U.K., using the ADAS or Levington procedures [Johnson, 1980], assign a numerical value to a range of test values to provide an index. The recommendations are related to the species, growth period, cultural procedures, etc. by assigning an index number that specifies the desirable nutrient test value for each nutrient. Those assigned to the Levington method range from 0 to 9. For tomatoes in modules, Johnson [1980] reported an index of 7 for phosphorous at the start of the crop, dropping to 5 later in the season. These indices corresponded to P levels of 56-75 and 29-40 mg ℓ^{-1}, respectively.

With the background of correlation and calibration, and having selected methods that will best serve the conditions of substrate, species, and cultural procedures, etc., the final step is interpreting and making recommendations. The individual making the recommendations should be thoroughly familiar with the crops, substrates, and cultural conditions of the region. The person making the recommendations must be well-versed in general principles and soil chemistry. The values listed in Table 6-34 are mostly optima, although some are considered as standards or critical levels below which deficiency occurs, or, with NH_4 or Cl, if a potential for toxicity exists. Usually, where constant fertigation is practiced, my experience suggests that soil test values can be lower than those given in Tables 6-32 through 6-34. pH is another parameter that, it appears, may be given more emphasis in the wrong direction than is needful. Black [1992] makes the point that acidity is not the cause of poor growth on acid soils. Furthermore, comments have been made in previous discussion that the hydrogen ion concentration can vary quite widely without direct effects on growth. The secondary effects of H^+ concentration are important: such as the presence of aluminum in acid media, excessive bicarbonate in hydroponics, micronutrient availability, in particular iron availability in well-aerated media with high redox potentials, and the effects H^+ and OH^- may have on microorganisms. Salinity will be discussed in the following section.

In addition to absolute concentrations, nutrient balance can be important. Warncke [1990] cited Geraldson as showing that a good nutrient balance as a percentage of total soluble salts to be: NO_3, 8 to 10; NH_4, <3; K, ll to 13; Ca, 14 to 16; Mg, 4 to 6; Na, < 10; and Cl, <10%. Expressing the nutrient levels as a percentage of total salts helps assess the most limiting nutrient. The values in Table 6-34, with some exceptions, are difficult to use in this fashion. The interpreter can examine: 1) absolute concentrations resulting from the test, 2) crop yields expressed as a percentage of maximum compared with the particular test level, and 3) total soluble salts with test levels expressed as a percentage of that total.

It has been my observation that growers, particularly cut-flower producers, often fail to make full use of soil test analyses. For example, Table 6-35 provides an abbreviated example of results obtained bench-by-bench on roses produced in the ground in a Central American country. Many soil analyses are presented in this fashion, and the grower looks them over, filing them in some remote drawer. This practice is unfortunate. Given the spreadsheets presently available, with graphic capabilities, considerable information can be obtained regardless of whether the analytical procedure is known or understood. Simply looking at such a mass of data cannot provide adequate answers. However, a spreadsheet program allows one to compute the average of all values –for a house, or for a section, where all benches are treated the same– and to calculate the standard deviation of the mean. This at least provides an idea of the average soil test value, and by multiplying the standard deviation by two, a range can be calculated that allows outliers to be identified –e.g., 306 mg ℓ^{-1} P, bench 4; 16 μM ℓ^{-1} Ca, bench 1; etc. Since these samples probably came from one section of an 8 ha range, the outlying data suggest problems in collecting the soil sample.

More important in using soil analysis is the development of a history. If soil analyses are carried out on a

Table 6-34. Recommended soil test values from various sources for different soil test procedures. The majority of tests must be interpreted by specialists familiar with the analytical and cultural procedures as well as the medium and species requirements. Units are ppm, mg ℓ^{-1} or mg kg^{-1} except where noted.

Method	NO_3	NH_4	N	P	K	Ca	Mg	Na	Cl	Fe	Mn	B	Mo	Cu	Zn	Reference
Dutch 1:1.5			3.7-5.4[1]	15-20	1.5-2.1[a]		1.3-1.8[a]		2.3-3.3[a]							Bik and Boertje, 1975
ADAS[b]	51-80	101-150		12-18	101-175		16-25									Bunt, 1988
Levington (Index = 4)	81-130			19-28	176-250		26-35									Johnson, 1980
Spurway[c]	10-20	2		5	20	60-120										Nelson, 1978
	25-100			4-6	30-50	>100										Mastalerz, 1977
	20-80			4-5	25-35	150-200										Hanan et al.,1978
Modified Morgan	76-125			126-250	0.76-1.0[4]	8.1-10[d]	1.3-2.5[d]									Mastalerz, 1977
	100-400			20-45	2.0-4.0[d]	7.5-40[d]	4.0-10.0[d]									Mastalerz, 1977[e]
AB-DTPA[f]				8-11	61-120					>5.0	>0.5			>0.2	>1.50	Soltanpour and Follet, 1985
SME	100-199			6-9	60-149	>200	>70									Warncke, 1980
	8-10	<3			11-13	14-16	4-6	<10	<10							Warncke, 1984
										15	16	0.7		9	14	Berghage et al.,1987
Pour-through			75-100	10-15	30-50	10-15	10-15									Wright, 1986
Mehlich No. 3											3.0[g]			0.3[h]		Martens and Lindsay, 1990
DTPA-TEA										4.5	0.22				0.5-0.8	Martens and Lindsay, 1990
Trace element[i]										3 µg/ml	5 µg/ml			1 µg/ml	2 µg/ml	Lindsay and Cox, 1985
Boron												0.2 µg/ml				Lindsay and Cox, 1985
$(NH_4)_2C_2O_4$													0.2			Lindsay and Cox, 1985
Hot water													>0.3-0.5			Lindsay and Cox, 1985

Method	NO_3	NH_4	N	P	K	Ca	Mg	Na	Cl	Fe	Mn	B	Mo	Cu	Zn	Reference

[a] Units in milliequivalents per ℓ .
[b] Index rating = 3, lower for seedlings, higher for tomatoes in peat modules.
[c] Medium levels of nutrients.
[d] Units in milliequivalents per 100 grams.
[e] Upper values for mixtures containing soils, lower for soilless mixes.
[f] Values for irrigated production in the field.
[g] Units in mg per dm^3.
[h] Value is pH variable, number given is for pH 6.4.
[i] Critical levels for micronutrients.

Table 6-35. An example of a bench-by-bench soil analysis on roses grown in the ground. The average of each value, standard deviation (s^2), and range ($2*s^2$) were calculated.[a]

Bench No.	pH	P (mg/ℓ)	Ca	Mg	K	Fe	Cu	Zn	Mn
1	5.5	110	16	0.9	0.8	69	9	2	11
2	5.5	150	13	0.6	0.6	63	1	1	8
3	5.2	265	6	0.3	0.4	90	6	2	10
4	4.8	306	4	0.3	0.5	68	7	5	15
5	5.3	172	5	0.3	0.3	40	6	5	20
6	5.4	210	4	0.5	0.4	49	4	5	30
7	5.2	270	5	0.4	0.4	84	0.2	3	26
8	5.6	140	6	0.4	0.5	207	1	2	32
9	5.7	153	4	0.8	0.8	74	1	5	13
10	5.4	87	10	0.7	0.6	134	9	1	14
11	5.4	150	7	0.5	0.5	104	2	2	8
2	5.5	135	8	0.5	0.5	81	5	2	10
13	5.5	190	6	0.3	0.5	77	3	4	6
14	5.7	190	8	0.4	0.5	65	1	5	7
15	5.7	123	5	0.3	0.4	78	4	4	9
Average	5.4	177	7	0.5	0.5	68	4	3	15
s^2	0.2	63	4	0.2	0.1	40	3	2	9
Range[b]	5.0-5.9	51-302	0.2-14	0.1-0.9	0.2-0.8	5-166	0-10	0.1-6	0-32

[a] With exception of P and pH, all values in µM ℓ$^{-1}$.
[b] Range calculated by multiplying the standard deviation (s^2) by two and adding to the average.

regular basis (i.e., monthly), than spreadsheet programs (Excel, Lotus 1-2-3) with graphics can be used to develop lines showing actual values. These programs can use a TREND function that predicts the value at some future time from previous results. Obviously, several values are required to establish a trend. The grower can then predict the future concentration level, and make adjustments in his feeding program to lower the rate at which a particular value is increasing or decreasing. Attempts to halt, or reverse, a trend are dangerous unless control has been lost in the feeding program. With fertigation, adjustments can be made incrementally, and this provides greater safety to avoid excesses and high salinity. Once a history has been developed, and is used, the grower can manage his substrate fertility level for maximum profitability.

G. SALINITY AND IRRIGATION WATER QUALITY

To this moment, nutrition has been discussed without reference to irrigation water quality and salinity problems. For one to neglect these two factors in nutrition control can result in serious difficulty. Mistakes made in locating appropriate water sources are difficult to correct –impossible sometimes. Regardless of how good a grower is, or how modern and extensive growth facilities may be, a poor water source makes salinity control difficult, reduces options in nutrient control, and limits yield and quality. A cheap raw material, which growers often ignore, becomes a highly expensive proposition.

1. Salinity

The sum of all cations and anions in the substrate solution is the total soluble salts, which may be expressed in terms of concentration or equivalency. While a complete solution analysis is possible in order to arrive at a figure, measuring the electrical conductivity of a suitable extract from the medium is easier and quicker (Section 6.II.H). This determination is invariably carried out in soil analyses, and it can be made a continuous measurement of flowing solutions such as NFT. The units were given in Table 6-11. The EC of solutions will be employed throughout as a measure of salinity of a substrate or hydroponic solution.

a. Effect of Salinity on Growth

Total salts directly affect osmotic potential as was discussed in Chapter 5 (Section 5.II.A). Fig. 5-14 showed typical examples of excessive soluble salts on commercial greenhouse plants. Besides the osmotic effect, specific ions may be high enough in concentration to cause toxicity and interference in nutrient uptake and balance. Typical of the growth response demonstrated several times for greenhouse crops is Fig. 6-55, showing a continual yield reduction of roses as EC of the soil increases. Such relationships are often curvilinear, and the fact that plants can adjust to increasing salt concentration changes the relationship so expressed. Such responses have been examined by Sonneveld and Mook [1983], Sonneveld and Voogt [1978], Sonneveld and van Beusekom [1974], Maagistad et al. [1943], Hughes [1975], Hughes and Hanan [1977], Schekel [1971], as well as several others. An EC of 4 dS m^{-1} as shown in Fig. 6-55, corresponds to a total salt concentration of about 50 meq ℓ^{-1}, whereas an EC of 8.0 is approximately 90 meq ℓ^{-1} (Fig. 6-7).

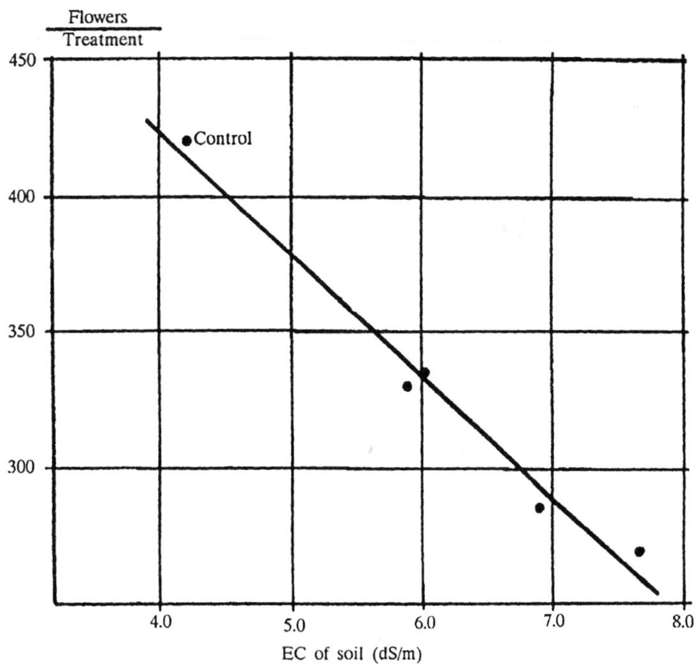

Fig. 6-55. The effect of total soluble salts (EC) on rose growth in the ground . The ordinate is the number of flowers cut per treatment [Reprinted from *Sci. Hort.*, 29:373-385, Fernández-Falcón, M. et al., ©1986. With kind permission of Elsevier Sci.-NL, Sara Burgerhartstraat 25, 1055 KV Amsterdam, The Netherlands].

Effects of high concentrations can be subtle or obvious as shown in Fig. 6-56, depicting a change in color for gerbera and obvious damage on chrysanthemum. The work by Sonneveld and others at Naaldwijk, The Netherlands, showed cucumbers to be highly sensitive, with yields decreasing 4, 7, and 14% for lettuce, tomatoes, and cucumbers, respectively, if the EC of the irrigation water increased 1 dS m^{-1}. An increase of 1 dS m^{-1} in the irrigation water caused an increase of 2 dS m^{-1} in the soil. Cucumbers showed a special sensitivity to excess calcium and magnesium. Most detrimental was the application of sodium bicarbonate. The deleterious effects of bicarbonate have been noted on several greenhouse crops (Fig. 6-11). Spinach, on the other hand, was not affected by the treatments imposed [Sonneveld and van Beusekom, 1974]. Sonneveld and Mook [1983] showed that the incidence of tipburn in lettuce could be attributed to low calcium content versus a high magnesium level. High sodium also promotes tipburn. Thus, some crop responses may be due more to a nutrient imbalance than to high salts.

Common is a salt effect noted in Fig. 6-57, dealing with roses produced in gravel. Treatment 1's applied solution contained the nutrients given in Table 6-22 at an EC of 1.3 dS m^{-1} versus Treatment 2 with 8 meq ℓ^{-1} of HCO_3^- added and an EC of 2.5 dS m^{-1}. There were not, by observation, great differences in visual appearances between treatments. But, the production showed that Treatment 2 reduced peak flowering periods and delayed them to the extent that by early summer peak flower cycles no longer occurred. Thus, excessive and imbalanced salt levels have the practical influence of upsetting timing cycles made by the grower –such as pinching– not

Fig. 6-56. **Upper**: Lighter color in gerbera flowers as the result of high total soluble salts. **Lower**: Fertilizer salt damage to young chrysanthemums.

to mention quality reductions as smaller flowers, or fruit, and shorter stems.

Also common are the effects of internal cell water potentials as influenced by the external solution concentrations. Brouwer's [1963] results with beans are typical regarding transpiration rate and leaf growth as the external concentration increases (Fig. 6-58). As noted in Chapter 5, the water potential of the cell must be lower than the external potential if water is to move into a cell, thereby maintaining sufficient turgor to prevent wilting. An osmotic water potential of 1.0 MPa corresponds approximately to nearly 40 dS m^{-1} EC. A potential of 0.3 MPa is about 8 dS m^{-1} and 0.1 MPa to slightly less than 3 dS m^{-1} [Richards, 1954]. Similar studies have been carried out by many investigators [i.e., Hayward and Spurr, 1944; Nieman and Poulsen, 1967; Bernstein, 1961, 1963; Meiri and Poljakoff-Mayber, 1967; Slatyer, 1961; Wadleigh and Ayers, 1945; etc.]. Bernstein and Hayward [1958] summarized the situation by categorically stating that the effects of excessive salt concentrations

Fig. 6-57. Effect of fertigation treatments on roses in gravel. See Table 6-22 for solution employed in Treatment No. 1. Treatment 2 included 8 meq ℓ^{-1} HCO_3^- and an EC of 2.5 dS m^{-1} versus 1.3 for the former. Note the peak flowering delay of about 2 weeks in No. 2 for the first cycle even through both treatments were pinched for cycling in the late fall at the same time [Hughes and Hanan, 1977].

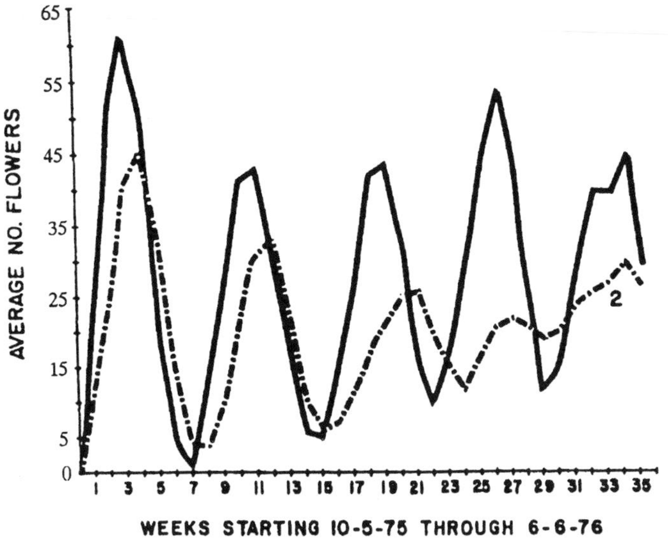

are mediated by osmotic inhibition, by specific effects of the constituent ion, or by a combination of the two.

A particular problem with substrates irrigated at intervals is the increase in solution concentration as the medium dries out. This is illustrated in Fig. 6-59 for selected ions in the soil. In the range of soil matric potentials for most greenhouse substrates, the periodic exposure of root systems to extreme ion concentrations can be frequent. The higher the solution concentration, the more likely acute damage is to occur if the grower allows the medium to dry out. Of course, the ability to maintain high moisture content in ground soils is severely limited by aeration limitations that restrict options available to the grower.

In soils, container substrates, etc., high salt concentrations may also have deleterious effects on soil structure –such as replacement of calcium by sodium on the exchange complex. Most field soil tests also evaluate the salinity hazard with the sodium absorption ratio –which is seldom seen in greenhouse soil tests. For acid soils, the lime requirement is commonly calculated for the grower. In hydroponics, the effects of high concentrations on substrate relationships mentioned above can be ignored. Although high water quality has been emphasized for recirculating culture systems [Anon., 1978b], there have been occasions when the deliberate addition of a salt such as sodium chloride has been found beneficial [Adams, 1987]. Early vigor in tomatoes was controlled by the addition to the NFT solution of 1200 to 1550 mg ℓ^{-1} NaCl (ca. 20 to 27 meq ℓ^{-1}) with an EC of 7.5 to 9 dS m^{-1}. Concentrations of 17 meq ℓ^{-1} (1000 ppm Na) reduced yield, whereas concentrations of about 9 meq ℓ^{-1} (4.6 dS m^{-1}) resulted in the highest yields. Increasing the sodium level improved fruit quality. It is well to remember the very low solar radiation under U.K. winter conditions, and increasing total soluble salts in hydroponic systems is likely to aid in maintenance of acceptable growth and fruit quality. Several years ago, sodium nitrate was often used as a means to increase flower quality during the winter in Colorado. Previous mention has been made of possibilities in crop improvement through the judicious use of such elements as sodium, selenium, etc. There have been few, if any, studies of this kind for greenhouse production. There have been articles dealing with the use of salt water in agricultural production such as the article by Boyko [1967]. Breeding crops for salt tolerance has been a major research thrust [e.g., Saranga et al., 1991; Jones, 1981]. Saranga et al. state that their results suggested the existence of a genetic potential for high salt tolerance in wild tomato germplasm. In either case, given the intensive culture found in greenhouse production, I do not feel that attempting production with salt water, or introducing salt-tolerant species, is economically viable at this time.

b. Ranges of Total Soluble Salts

Recommendations encourage the use of the water-saturated paste as the means to obtain a value for total soluble salts –or EC. However, several laboratories also use 1:2 or 1:5 dilutions to provide a sample for electrical conductivity determination. The latter are convenient for rapid determinations, but reliability depends upon the salts present. For chloride salts, results are only slightly affected by moisture content. However, if sulfate or carbonate salts are present, the apparent amount will depend upon the soil:water ratio. With container mixtures

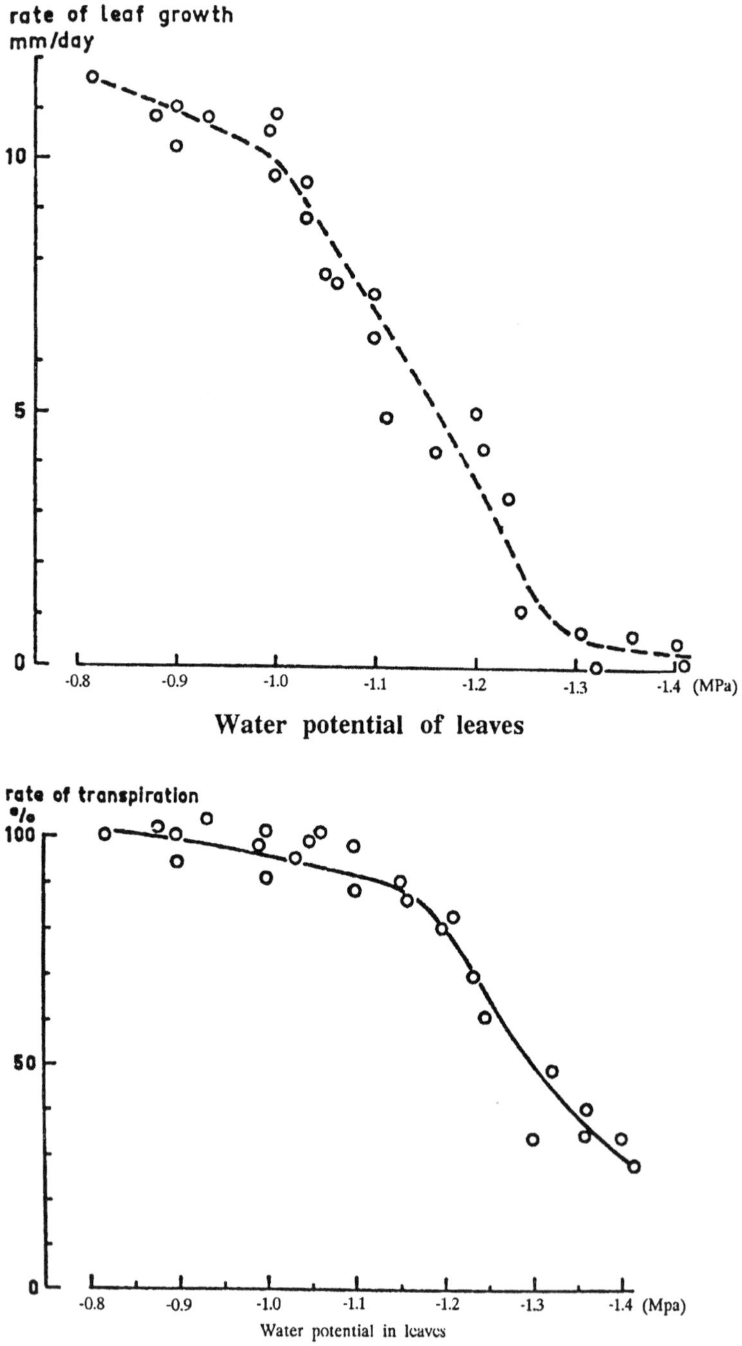

Fig. 6-58. Effect of leaf water potential in beans on leaf growth rate (upper) and transpiration rate (lower) [Brouwer, 1963] (With permission of *Acta Bot. Neerl.,* Blackwell Sci. LTD).

ısing peat and soilless additives, EC readings will be higher at any given moisture content than for mineral clay oams. Sandy loams will give the lowest EC readings [Warncke, 1990]. The U.S. Salinity Laboratory [Richards, 1954] did not recommend such dilutions for arid region, calcareous soils. Waters et al. [1970], in an examination ›f 27 soil media, found that excessive soil moisture in light weight media reduced salinity readings with the lilution method. In the saturated paste method, initial moisture content, sample volume, bulk density, container

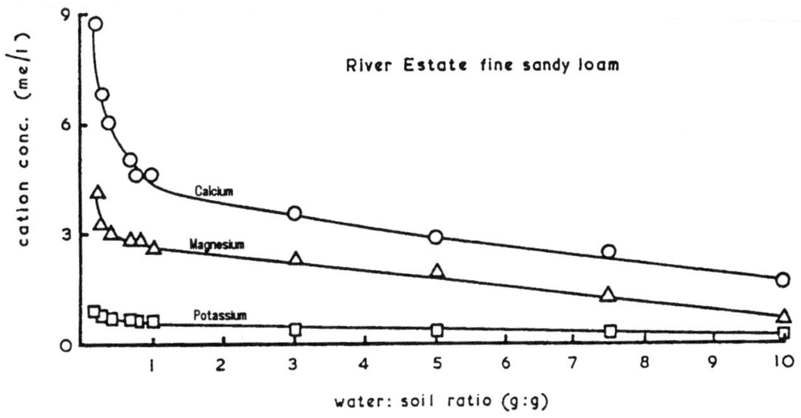

Fig. 6-59. The change in cation concentration of soil solution with a change in moisture content as indicated by the water:soil ratio [From *Plant and Soil*, 28:99-113, Moss, P., ©1963. With kind permission of Kluwer Acad. Publ.]

Table 6-36. Soluble salt guidelines for growth media by three test methods [Warncke, 1990]. Units are dS m^{-1}. (With permission of the *Amer. Soc. of Agronomy*).

Saturated media extract	One part medium to: (v/v)		Interpretation
	Two parts water	Five parts water	
0.0-0.74	0.0-0.24	0.0-0.12	Low nutrient status.
0.75-1.99	0.25-0.49	0.13-0.34	Suitable for seedlings and sensitive plants.
2.00-3.49	0.50-0.99	0.35-0.64	Suitable for most established plants. Upper range may reduce growth of some salt-sensitive plants and seedlings.
3.50-5.00	1.00-1.49	0.65-0.89	Higher than desirable. Loss of vigor in upper range. Okay for high nutrient-requiring plants.
5.00-6.00	1.50-1.99	0.90-1.10	Reduced growth. Wilting and marginal leaf burn.
6.00+	2.00+	1.10+	Severe symptoms, wilting, crop failure.

capacity, or media composition are not critical factors. Distilled or deionized water is added to a soil sample while stirring until the conditions noted earlier are obtained. The sample should be allowed to stand an hour or more and criteria for saturation checked. Free water should not collect on the surface, nor should the paste stiffen or lose its glistening appearance on standing. Richards [1954] suggested a 250 g sample, although the previous section said that 500 ml may be used in the SME analytical procedure. Volumes are better to use for lightweight mixtures. The sample is vacuum filtered. If gypsum is present, the saturated paste should stand for several hours before filtering. Table 6-31 shows overnight standing for some laboratories.

Thus, one finds such tables as 6-36 showing EC values using the different extraction methods and interpretation. Table 6-37 is another example. Optimum EC levels for different analytical extraction methods were given in Table 6-32. Variations can be found in the ranges given and methods by which reported. Table 6-11 shows a plethora of units, with the deciSiemens per meter (dS m^{-1}) being recommended in recent years [e.g., Miller et al., 1981; Anon., 1985]. The approximate relationships between EC and concentrations in milliequivalents per ℓ (meq ℓ^{-1}) were presented in Figs. 6-7 and 6-8. The relationship between osmotic water potential in MegaPascals and EC can be estimated by:

Table 6-37. Interpretation of EC values obtained with a 1:2 soil-water extract [Jones, 1985] (With permission of *Hort. Rev.*).

Specific conductance at 25 C (dS m^{-1})	Interpretation
<0.40	Salinity effects negligible except in beans and carrots.
0.40-0.80	Slightly saline. Yields of salt-sensitive crops such as pepper, lettuce, etc. may be reduced 25-50%.
0.81-1.20	Moderately saline. Yield of salt-sensitive crops restricted. Seedlings injured. Satisfactory for well-drained greenhouse soils.
1.21-1.60	Saline soils. For tolerant crops. Higher than desired for greenhouse soils.
1.61-3.20	Highly saline. Only salt tolerant crops. Leach greenhouse crops with 20 to 40 ℓ s per m^2 water or 0.5 ℓ s per 15 cm pot.
>3.2	Only salt-tolerant species. Very saline.

$$\psi_{\varpi} = 0.36 \times EC \times 0.101 \tag{6.50}$$

where 0.101 is the conversion from atmospheres to MPa and EC is given in dS m^{-1}.

c. Controlling Salinity in Substrates

If, through excessive fertilizer application or failure to water adequately, the grower finds himself faced with high salts, about the only option is to wash excess salts from the soil layer or active root zone. Application of excess water, which itself contains high salts, will not reduce total soluble salts below that contained in the water supply. As will be discussed later, water quality becomes highly important in greenhouse culture.

Table 6-38. Leaching requirement[a] as related to the ECs of irrigation and drainage waters [Richards, 1954].

EC of irrigation waters (dS m^{-1})	Leaching requirement for the indicated maximum values of the EC of the drainage water at the bottom of the root zone			
	4 dS m^{-1}	8 dS m^{-1}	12 dS m^{-1}	16 dS m^{-1}
	Percent	Percent	Percent	Percent
0.1	2.5	1.2	0.8	0.6
0.25	6.2	3.1	2.1	1.6
0.75	18.8	9.4	6.2	4.7
2.25	56.2	28.1	18.8	14.1
5.0		62.5	41.7	31.2

[a] Leaching requirement is the fraction of applied irrigation water that must be leached through the root zone expressed as a percent.

Saline soils are those in which the conductivity of the saturation extract is greater than 4 dS m^{-1} [Richards, 1954]. The previous sections show that EC, under many conditions found in greenhouses, may be much higher. Without salt accumulation from ground water, the EC ranges from 2 to ten times as high as EC of the applied water. This results from moisture extraction by the root system and evaporation. The leaching requirement for field soils as suggested by the U.S. Salinity Laboratory is provided in Table 6-38. The EC of the applied water obviously has a strong influence on water consumption where leaching is required.

Under ideal conditions, when leaching, it is desirable that the applied water not mix with the soil solution, but to "push" the excess salts through the layer with no mixing. This is called "piston" flow since incoming water from the surface acts as an impervious plunger. Even with highly uniform substrates and low infiltration rates, some mixing always occurs. As the infiltration or percolation rate increases, part of the leaching water bypasses small pores and capillaries, leaving some salts behind. Mixing between incoming and outgoing solutions can be very high. This is the reason for the special displacement solution used by Nelson and Faber [1986] in obtaining soil extracts. The situation for container mixes was studied by Kerr and Hanan [1985] (Fig. 6-60), usually

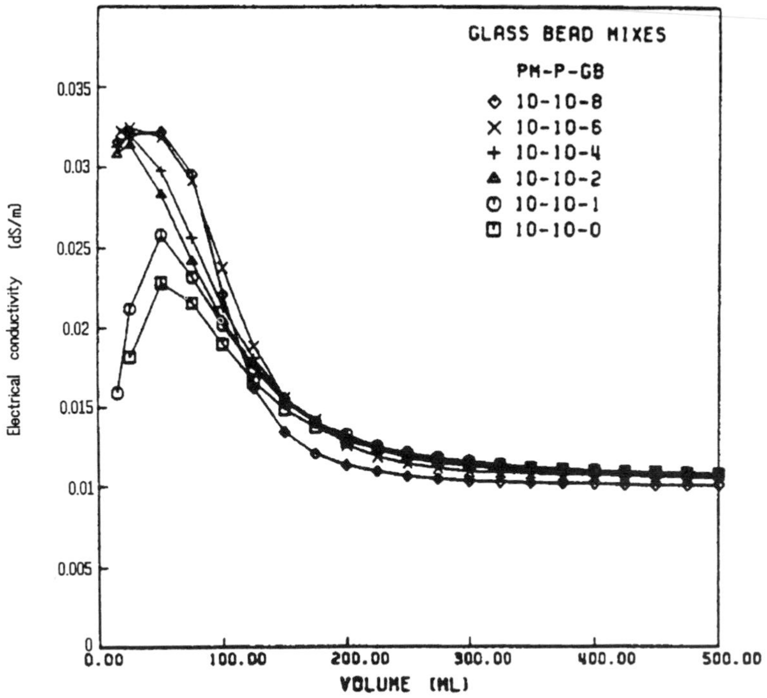

Fig. 6-60. Leaching curves of 15 cm deep columns of mixtures containing various proportions of peatmoss (PM), perlite (P), and glass beads (GB). The surface was flooded with water containing 4 meq ℓ^{-1} $CaCl_2$ and 4 meq ℓ^{-1} NaCl. The columns were salinized prior to leaching with 15 meq ℓ^{-1} each of $CaCl_2$ and NaCl per ℓ medium [Kerr and Hanan, 1985].

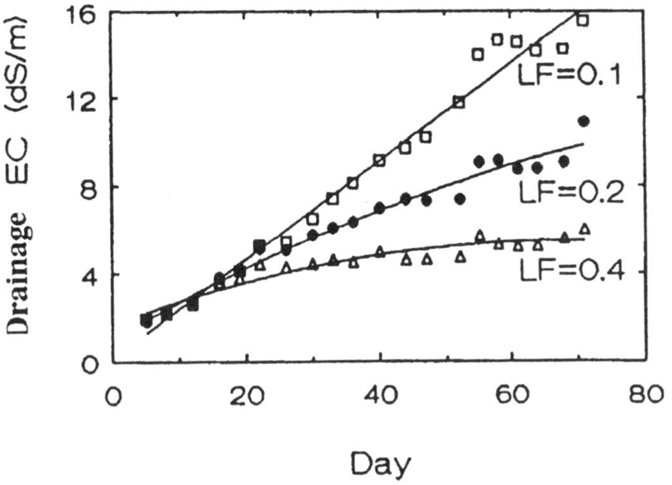

Fig. 6-61. EC of the drainage water from the bottom of containers containing poinsettias during the growing period for different leaching fractions (LF) [Ku and Hershey, 1991] (With permission of the *Amer. Soc. for Hort. Sci.*).

showing an initial increase in EC of the drainage water as water was applied to the sample surface, and then decreasing, approaching a minimum asymptotically as water continued to flow through the sample, regardless of the mixture. The use of glass beads was an attempt by the authors to obtain greater reproducibility. Mixtures with a greater proportion of glass beads resulted in a higher bulk density and greater salt removal rate in initial stages, indicative of lower mixing between soil solution and the leaching water –or closer to piston flow. Regardless of the leaching solution salinity, about the same amount of salt was removed in initial stages, but the higher the EC of the water applied, the higher the final minimum as the removal rates approached zero. Under laboratory conditions, the greatest amount of salt in a mixture was removed when enough water equal to 1.0 to 1.5 container capacities had moved through the sample.

Some authors have examined leaching requirements with plants established [e.g., Ku and Hershey, 1991; 1992; Yelanich and Biernbaum, 1993]. The approach of these authors has been to use the "leaching fraction," which is the volume of solution leached from the container divided by the total solution applied. That is, if 400 ml is the container capacity, an application of 400 ml with 100 ml of drainage, or 200 ml applied with 50 ml leached, would both equal an LF of 0.25. Container capacities would vary since the drainage loss is different. An example of drainage EC over time at different LFs is given in Fig. 6-61. The thrust of this work is to control medium salinity with minimum wastage. Yelanich and Biernbaum state that the recommended LF is 0.1, but most growers use ratios greater than 0.5. Fig. 6-61 indicates an LF of 0.4 is required to maintain EC of the drainage below 8 dS m^{-1}. The latter authors reported LFs greater than 0.35 were required when the commonly recommended 14 mM ℓ^{-1} N was applied to poinsettias grown in 15 cm pots. A total of 1250 ml were required to achieve an LF of 0.5 to 0.6. Adequate fertility levels were maintained using 7 mM ℓ^{-1} N, or less than half the usually recommended amount. If good water is available, drainage produced in poinsettia culture –and consequent fertilizer required and lost– can be reduced significantly.

Unfortunately, water and fertilizer input in greenhouse production has been such a small proportion of total production cost that many growers deliberately apply excess water to be safe. There is also the problem that many container mixes are so porous that trickle irrigation systems may not wet the entire soil mass. This would be particularly true during initial growth stages when root proliferation is least. As root density increases during growth, better wetting from a single trickle emitter would be more likely. Flooding of the medium surface may be necessary at times if efficient leaching is to occur. Another factor to be considered is excessive drying of the substrate between irrigations. Upon watering, most of the water is likely to flow between the container sides and soil mass. Though drainage is occurring, little moves through the soil mass until it has been thoroughly wetted. In raised bench, cut-flower production, "dry" growers find it necessary to water twice over a several hour period to ensure adequate leaching. Of course, such shrinkage may not occur with some substrates, but peat and bark mixtures are very difficult to wet unless a wetting agent has been included in the mixture. Crops watered by sub-irrigation (i.e., capillary mat, ebb-and-flow, constant water table) may experience excessive salts on the medium surface because of evaporation. This can be especially apparent if high fertilizers are included in the water. Overhead irrigation may be necessary.

One cultural procedure that has apparently received insufficient attention is the need to determine container capacity, and water commonly lost between irrigations –exclusive of any drainage from the container. To this time, most growers always supply a surplus, which often exceeds container capacity several fold. As will be emphasized later, some environmental agencies are placing severe restrictions on waste water and fertilizer loss from greenhouse operations. An estimate of container capacity and water consumption for a given crop and season would help answer these new requirements, plus increasing operational efficiency and reducing raw material costs.

H. IRRIGATION WATER QUALITY AND MANIPULATION

The presence of salts in the irrigation water will influence ability to manipulate nutrition. Newer systems for automatic control can take into account variations that may occur in water supply quality. Secondly, water salinity ultimately limits production capability and can markedly reduce profitability –even though the source is cheap compared to labor or fuel.

1. Expectations in Water Quality

Variations in water quality are to be expected, and water analyses are an important adjunct to greenhouse operation. Typical of the analyses performed in Colorado are publications in the *Colorado Greenhouse Growers' Bulletin* [Hanan et al.,1968; Hanan, 1973; 1979] and the discussions on methods for handling saline irrigation supplies [Hanan, 1982a; b; 1988]. Fig. 6-62 graphically presents the concentrations of Mg, SO_4, Ca, and Cl found in 36 shallow wells in the Denver region. Fig. 6-63 illustrates the common salt level found in such wells. Average concentrations for Na and HCO_3 were 4.7 and 5.0 meq ℓ^{-1}, respectively, with 42% of the wells containing sodium between 0 and 3.2 meq ℓ^{-1} and 36% with bicarbonate between 4 and 6 meq ℓ^{-1}. Shallow wells in Colorado –a semiarid region with alkaline soils– are likely to be well above the maximum desirable EC allowable for reasonable nutrient manipulation (Fig. 6-63). Sulfate concentrations can be as high as 25.2 meq

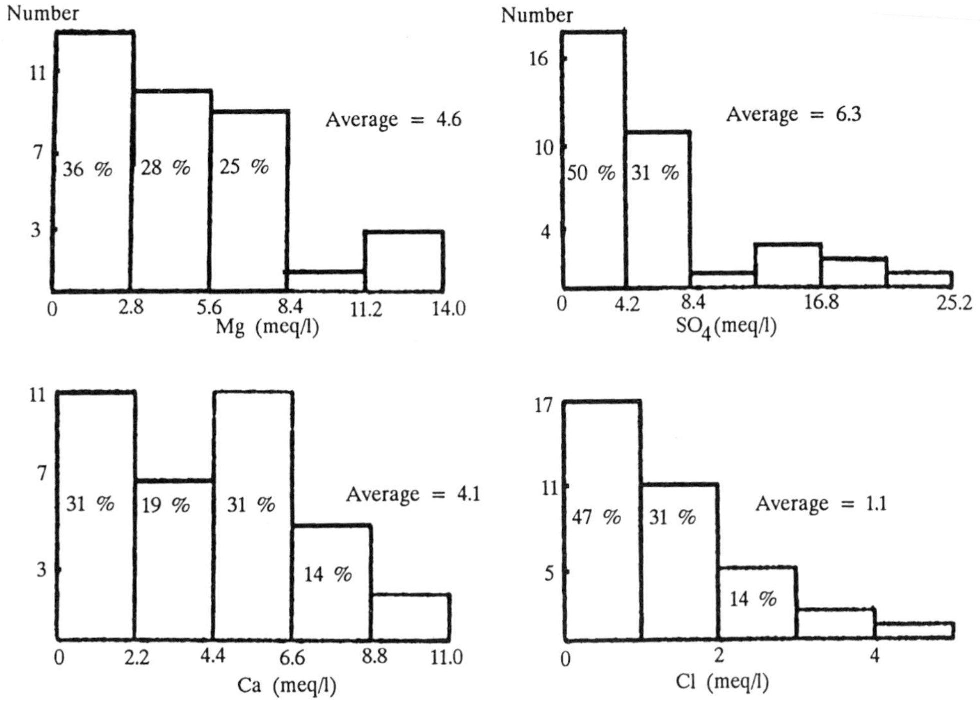

Fig. 6-62. Concentrations of magnesium, calcium, sulfate and chlorine in 36 shallow wells in the Denver, CO, urban area. The vertical axes show the number of samples [Hanan, 1973].

ℓ^{-1}, and I have seen at least one tomato producer attempt culture of tomatoes with such irrigation waters. Calcium and magnesium are likely to be significant in shallow wells (Fig. 6-62). Deep wells in Colorado often contain high sodium and bicarbonate, although some are very high quality (i.e., no salts). Domestic supplies from high mountain regions usually have ECs less than 0.5 dS m^{-1}. Such examples can be found in almost any region of the world. The Netherlands are plagued with high chlorides from the Rhine River. In the Mediterranean region (Spain, Crete, Cyprus, etc.), salt water intrusion is often experienced in shallow wells close to coast lines. The Sea of Galilee contains high chlorides.

The characteristics of irrigation water that are important in determining its quality are: 1) total soluble salt concentration, 2) concentration of individual ions, 3) concentration of boron and other elements that may be toxic, and 4) the bicarbonate concentration as related to concentrations of calcium and magnesium. For substrates containing soil, sodium is particularly important in determining an alkali hazard. Sodium hazard of irrigation water is usually expressed as the sodium absorption ratio (SAR), computed from the formula:

$$SAR = \frac{[Na^{+}]}{\sqrt{\frac{[Ca^{\theta+}] + [Mg^{\theta+}]}{2}}} \qquad (6.51)$$

where concentrations are in meq ℓ^{-1}. Use of water with an SAR greater than 10 must be avoided where soils are part of the substrate. If the mixture contains appreciable amounts of gypsum, an SAR value of 10 can be exceeded. Continued use of high SAR water leads to a breakdown in the soil's physical structure. Adsorbed sodium results in clay dispersion, and the soil becomes hard and compact when dry, with slower water penetration when wet [Follett and Soltanpour, 1985].

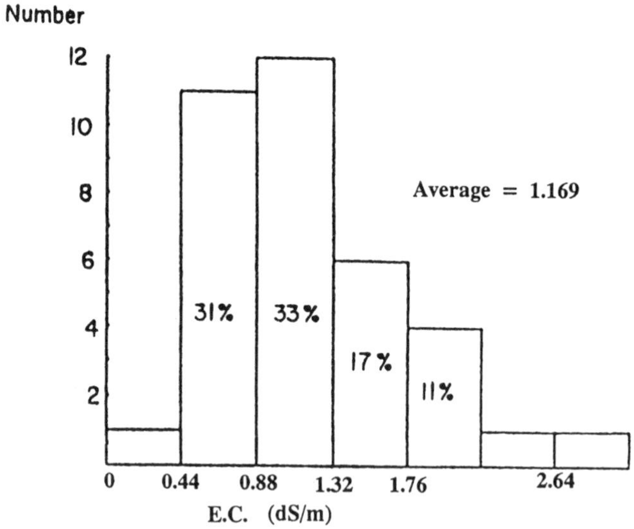

Fig. 6-63. Total soluble salts found in 36 shallow wells in the Denver, CO, region. Vertical axis shows the number of samples. The figures within the bars are percentages of the total samples found in each range [Hanan, 1973].

The U.S. Salinity Laboratory [Richards, 1954] states that water with conductivity values below 0.75 dS m^{-1} is generally safe. However, sensitive species are likely to be affected at ECs above 0.25 dS m^{-1}. The upper limit of 0.75 was chosen by Hanan [1988] as the upper limit for greenhouse production. But, Fig. 6-63 shows that more than 80% of the supplies from shallow wells have ECs ranging from 0.4 to 1.8 dS m^{-1}, and 50% of those are between 0.9 to 1.8 dS m^{-1}. Growers using such water supplies can expect to have difficulty in their nutrition programs to maintain reasonable productivity. Use of an irrigation supply above 2.3 dS m^{-1} is likely to result in crop failure.

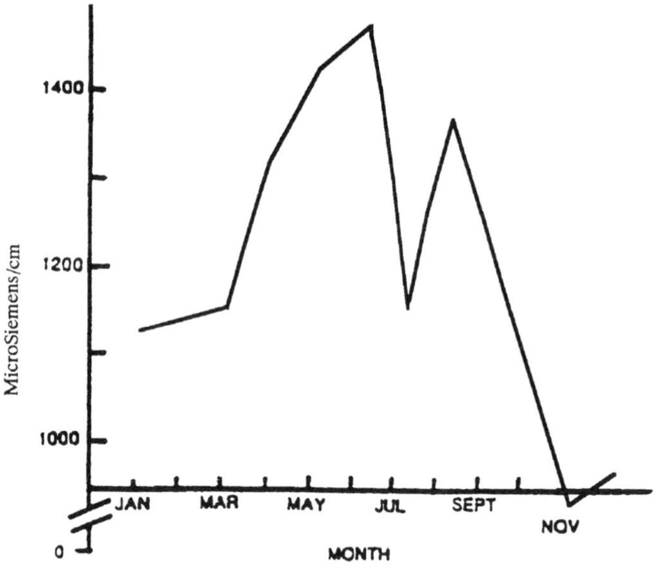

Fig. 6-64. Variation in total soluble salts in water from a shallow well in the South Platte River Valley, CO, over a year. Samples were analyzed once monthly so the curve is likely to conceal small variations probably significant in nutrition practice [Hanan, 1988].

A factor that is seldom emphasized is variation over time in water quality –especially from shallow wells and surface irrigation supplies. Total soluble salts and individual ion concentration can vary markedly with season. A series of tests for a commercial range in Colorado showed that water from a well in the South Platte river basin could vary from about 0.9 to over 1.4 dS m^{-1} (Fig. 6-63), reflecting changes as the result of irrigation, fluctuations in the water table, etc. Similarly, NO_3^- in this supply could fluctuate from less than 1 to more than 2 meq ℓ^{-1}. Calcium and sodium showed even greater differences with season [Hanan, 1988]. Such variations cry out for an automatic system of continuous water analysis so nutrient injection in a fertigation system can be automatically varied as necessary. ECs approaching 1 dS m^{-1} preclude the use of fertigation in crop culture.

There are many laboratories to which one can send a water sample for analysis. Unfortunately, many of these facilities are set up for bacteriological analyses for drinking water suitability, and the results on mineral content are likely to be given in terms of total carbonate, hardness (grains), or softness. The grower is not interested in drinking the water. He needs to know individual ionic content in terms that can be used in manipulating his nutrition program. It has been my observation, in countries requiring mineralogical and bacteriological analyses on bottled water, that often such water would make a very poor irrigation supply. Any water analysis should include, in parts per million or milliequivalents per ℓ, the concentrations of magnesium, calcium, sodium, chloride, bicarbonate, carbonate, sulfate, nitrate, and potassium. Where likely, a test for boron should be included. Presentation of results in meq ℓ^{-1} allows one to check credibility of the analysis by determining if the total cations equal the total anions. The analysis should include electrical conductivity and pH.

Table 6-39. Typical water analyses from four shallow wells with ECs ranging from 0.3 to 1.3 dS m^{-1}. Analyses subjected to MINTEQA2/PRODEFA2 program to calculate equilibrium at CO_2 level of 300 Pa.

Analysis for:	Water sample number 1	2	3	4
Electrical conductivity (dS m^{-1})	0.3	0.5	0.9	1.3
pH	7.7	7.7	7.7	7.6
Ca^{2+} (meq ℓ^{-1})	0.9	4.0	3.8	5.1
Mg^{2+}	0.2	1.6	1.4	2.0
K^+			0.2	0.4
Na^+	1.7	2.0	4.0	5.9
NO_3^-			0.5	0.8
PO_4^{3-}			0.4	0.1
SO_4^{2-}	0.5	2.8	3.2	5.2
Cl^-	0.1	0.4	1.7	3.2
HCO_3^-	2.7	4.4	3.6	4.1
Solids precipitated	None	$CaCO_3$	$CaCO_3$ $Ca_5(PO_4)_3OH$	$CaCO_3$ $Ca_5(PO_4)_3OH$

2. Interpretation and Manipulation of Salty Irrigation Supplies

Previous comments suggest that nutritional control based on a single water analysis can be dangerous. If water quality changes slightly, any program devised with the original analysis in mind becomes improper. As examples of how to deal with irrigation supplies, and their the limits, four water analyses were selected (Table 6-39). The software program MINTEQA2/PRODEFA2 was used to balance the analyses. The program was allowed to calculate the equilibrium pH as a check on the pH actually measured. There were no significant differences between pH, either calculated or measured. These calculations were, of course, highly simplified. A complete analysis of Sample No. 3 showed traces to significant concentrations of aluminum, iron, manganese, copper, zinc, nickel, molybdenum, cadmium, chromium, barium, ammonium, and boron. Inclusion of these in MINTEQA2/PRODEFA2 would have greatly increased complexity. Always, the presence of bicarbonate resulted in alkaline pHs. Secondly, Sample 2 precipitated calcite, and Samples 3 and 4 precipitated both calcite and hydroxyapatite. This suggests that problems might occur with blockage in trickle irrigation systems, whether or not fertilizer is added to the water. Samples 3 and 4 contained phosphates, nitrates, and potassium, showing shallow wells contaminated by nutrients leached from the soils above the water table. This is typical of shallow well supplies. Traces of these nutrients will seldom be found in deep wells (e.g., more than 30 to 60 m deep).

Tables 6-40 and 6-41 show the method for modifying Samples 1 and 2 to reduce salinity and pH. Note the use of phosphoric and nitric acids. Although these acids are highly caustic, they are much better than sulfuric, which can greatly increase sulfate without contributing anything except to acidify the solution. The results of calculation with MINTEQA2/PRODEFA2 are given in Table 6-42. One set of values shows the results of modification according to the water analysis; the other shows the results if the carnation solution as devised in Table 6-30 was simply added to the water supply. Although the estimated EC of the carnation solution in Table 6-30 resulted in a value of 2.4 dS m^{-1}, measurements over several years showed an average of about 1.5 dS m^{-1} for a total of 27 meq ℓ^{-1}. Even with modification of the basic solution, total meq increased to 30.4 meq ℓ^{-1} versus 33.6 meq ℓ^{-1} if the basic solution was simply added without regard to the water analysis. For Sample No. 2, a modified solution resulted in 32.2 meq ℓ^{-1} total versus 42 meq ℓ^{-1} total for the unmodified process. This would probably result in an EC greater than 3.0 dS m^{-1}. In both cases, the use of acids caused pH to be reduced to 6.1

Table 6-40. Manipulation of a water supply with an EC of 0.3 dS m^{-1} and 2.7 meq ℓ^{-1} HCO_3^-. The carnation solution from Tables 6-22 and 6-29 is to be used. Units = meq ℓ^{-1}.

	H^+	K^+	Ca^{2+}	Mg^{2+}	Na^+	NH_4^+	Total cations	NO_3^-	$H_2PO_4^-$	SO_4^{2-}	Cl^-	HCO_3^-	Total anions	mg in 1 ℓ required
Well water analysis			0.9	0.7	1.7		3.3			0.5	0.1	2.7	3.3	
Actual solution desired in good water		6.0	3.0	2.0		2.5	13.5	10.4	1.1	2.0			13.5	
Fertilizers required														
Potassium nitrate		6.0						6.0						606
Monoammonium phosphate						1.1			1.1					127
Nitric acid[1]	0							2.7				0		170
Calcium nitrate			2.1					2.1						248
Magnesium sulfate				1.3						1.3				160
Ammonium sulfate						1.4				1.4				311
Totals	0	6.0	3.0	2.0	1.7	2.5	15.2	10.8	1.1	3.2	0.1	0	15.2	

[1] Nitric acid was used to neutralize bicarbonate so values of H^+ and HCO_3^- were set to zero.

Table 6-41. Manipulation of water supply with 0.5 dS m^{-1} EC, using carnation solution in Tables 6-22 and 6-29. Units = meq ℓ^{-1}.

	H^+	K^+	Ca^{2+}	Mg^{2+}	Na^+	NH_4^+	Total cations	NO_3^-	$H_2PO_4^-$	SO_4^{2-}	Cl^-	HCO_3^-	Total anions	mg fertilizer required per ℓ
Raw water analysis			4.0	1.6	2.0		7.6			2.8	0.4	4.4	7.6	
Desired solution in pure water		6.0	3.0	2.0	0	2.5	13.5	10.4	1.1	2.0	0	0	13.5	
Fertilizers required														
Phosphoric acid									1.1					36
Nitric acid								3.3						208
Potassium nitrate		6.0						6.0						606
Ammonium nitrate						2.5		2.5						200
Totals		6.0	4.0	1.6	2.0	2.5	16.1	11.8	1.1	2.8	0.4		16.1	

Table 6-42. Calculated ionic concentrations ($\mu M\ \ell^{-1}$) in two solutions, one with total soluble salts in the basic water supply of 0.3 dS m^{-1} (Sample 1) and the other with 0.5 dS m^{-1}. With the carnation supply solution, both samples modified to keep total salts as low as possible and neutralize bicarbonates in the solutions. Both water supplies were calculated again with the carnation solution (Tables 6-21, 6-29) added directly to the basic water without modification. Calculations included micronutrients with DTPA chelate, CO_2 concentration at 300 Pa. Computer program was MINTEQA2/PRODEFA2 (Allison et al.,1991). Refer to Tables 6-40 and 6-41 for composition of modified solutions using acids to neutralize bicarbonate. Components having concentrations less than 1 $\mu M\ \ell^{-1}$ ignored. Components with a superscript "o" are neutral, undissociated molecules in solution.

Calculation for:	**Water sample No. 1 (0.3 dS m^{-1})**		**Water sample No. 2 (0.5 dS m^{-1})**	
	Modified	**Unmodified**	**Modified**	**Unmodified**
pH	6.2	7.7	6.1	7.6
NH_4^+	2137	2137	2188	2089
K^+	5248	5248	5248	5248
Ca^{2+}	550	676	813	912
Mg^{2+}	550	724	437	759
NO_3^-	6309	9120	10232	8912
SO_4^{2-}	813	603	692	1096
Na^+	1479	1479	1862	1737
$H_3BO_3^o$	18	18	18	18
Cl^-	87	87	347	347
Mn^{2+}	1		2	
NH_3^o	2	56	1	50
$NH_4SO_4^-$	23	17	19	30
$MgCO_3^o$		3		3
$MgHCO_3^+$		18		17
$MgSO_4^o$	76	74	51	141
$MgH_2PO_4^+$	1		1	
$MgHPO_4^o$	3		2	
$CaCO_3^o$		4		5
$CaSO_4^o$	89	85	112	204
$CaHPO_4^o$	2		2	
$CaH_2PO_4^+$	1		1	
$NaHCO_3^o$		2		2
$NaSO_4^-$	6	5	6	10
KSO_4^-	30	22	25	40
HCO_3^-	66	2089	51	1905
$H_2CO_3^o$	102	102	102	102
HPO_4^{2-}	7		5	
$H_2PO_4^{2-}$	78		71	
$CaDTPA^{3-}$		1		
$FeDTPA^{2-}$	4			
$MnDTPA^{3-}$	2		2	3
Precipitated substances	$MnPO_4 \cdot 1.5H_2O$ α-FeOOH $Ca_5(PO_4)_3OH$	β-MnO_2 α-FeOOH $Ca_5(PO_4)_3OH$	$MnPO_4 \cdot 1.5H_2O$ α-FeOOH $Ca_5(PO_4)_3OH$	β-MnO_2 α-FeOOH $Ca_5(PO_4)_3OH$ $CaCO_3$

versus 7.7 where HCO_3^- was not neutralized.

The total ion concentration for Sample No. 3 (0.9 dS m^{-1}) when modified was 35.4 meq ℓ^{-1}, or about 2.5 dS m^{-1}. The total for Sample No. 4 was estimated above 3.0 dS m^{-1}. The latter value for irrigation supplies is the suggested limit for successful crop production [Richards, 1954]. It is doubtful that basic supplies above 0.75 dS m^{-1}, when fertilizer is injected, will give acceptable growth.

DTPA complexed with Fe^{2+}, Fe^{3+}, Cu^{2+}, Mn^{2+}, Ca^{2+}, and Zn^{2+}. The proportions changed, depending upon pH (see Figs. 6-41 and 6-42); 45 to 52% of the DTPA complexed with iron, and 37 to 44% with manganese in modified solutions, using No. 1 and 2 water samples. In the unmodified, high pH solutions, 24% of the DTPA was complexed with calcium. Formation of MnDTPA utilized 66 to 99% of the manganese. Nearly 100% of the zinc, copper and iron complexed with DTPA. In the unmodified solutions, HPO_4^{2-} and $H_2PO_4^-$ levels dropped below 1 μM ℓ^{-1}, and $MgCO_3^o$, $CaHCO_3^o$ and $NaHCO_3^o$ levels rose above 1 μM. Thus, failure to lower the pH markedly reduced phosphorous supply.

Note (Table 6-42) that solids were calculated to precipitate in all solutions with calcite appearing in the unmodified Sample No. 2. Goethite (α-FeOOH), pyrolusite (β-MnO_2), and hydroxyapatite ($Ca_5(PO_4)_3OH$) were generally the solids to be found.

To the best of my knowledge, these types of calculations have not been published in the technical literature or in the scientific horticultural literature. Fertilizer recommendations in the industry do not commonly acknowledge the influence of basic water composition. Ratios of the different nutrients are considered as though pure water was at hand for all. Cooper [1979] stated that total EC of NFT solutions should never drop below 2.0 dS m^{-1}. My experience suggests the opposite is likely to be more common. Cooper spent considerable effort in dealing with nutrient toxicities and solution control. Unfortunately, computer software, such as applied here, was unavailable when both Cooper and Jones published their manuals on hydroponics [1979; 1983]. These software programs could be incorporated for real-time solution of fertigation problems, given suitable water supplies and instantaneous ion measurement.

3. Improving Water Supplies

When a grower faces a poor water supply, the easiest thing to do is to change to a better supply –even if that costs money. Most domestic supplies that do not depend upon shallow wells will usually be higher quality, i.e., fewer salts. In some countries, however, high quality is not obtainable. If the water supply contains mostly chlorides and sodium, higher concentrations may by lived with, especially if the substrate is loamless or one is using hydroponics. Sodium is not likely to be as dangerous.

The other route to take is directly to improve the water supply by removing salts. Processes for manipulating water supplies are outlined in Table 6-43, except for electrodialysis. Systems such as softening to reduce hardness are of no use in greenhouse culture. Coagulation, sedimentation, and filtration are commonly necessary with any method such as reverse osmosis (RO) or electrodialysis (EDR). Evaporation or distillation systems are not used for greenhouse production, but they are often employed in large-scale plants with a variety of modifications such as vertical tube distillation, multistage flash distillation, multi-effect multistage and vapor compression distillation [e.g., U.S. Dept. Interior, 1962; 1968; 1979; Avissar and Mahrer, 1986]. Reverse osmosis uses a semipermeable membrane through which water is passed by pressurization, leaving the salts behind. Feedwater often requires adjustment to eliminate scaling, $CaSO_4$ precipitation, and some form of biocide to control fouling. Capital costs, given by the Office of Saline Water range from about \$400 per m^{-3} dy^{-1} for an installation of about 40 m^3 daily capacity to \$145 m^{-3} for installations having capacities up to 19000 m^3 daily. Operating costs range from about 20 to 30 ¢ per m^3 for small installations down to 2 to 3 ¢ per m^3 for large operations. There are several variations, depending upon the water supply to be purified. Irreversible membrane fouling is sometimes encountered unless the system is equipped with appropriate safety devices and the operators properly trained.

Table 6-43. Outline of some water and waste treatment processes [U.S. Dept. Interior, 1979]. Electrodialysis not included.

Process	General use or capability	Advantages	Disadvantages
Coagulation, sedimentation, and filtration	Reduction in suspended solids by 90-98%	Low cost, simplicity	Large area required. Does not remove salts or organics
Softening	Reduction in hardness by 95 to 100%	Relatively low in cost, simplicity	Requires frequent regeneration. Does not remove organics (of no use in greenhouses)
Ion exchange	Reduction of salts by 95 to 100%	Can reach very low levels of salinity	Does not remove organics. Requires regeneration
Biological treatment	Reduction of organics by 50 to 90%	Low cost	Subject to upset and variations. Limited to 90% of organic loading
Carbon columns	Reduction of organic loading by 95 to 100%	Good way to remove small amounts of organics	Relatively expensive regeneration. Removes only specific organics
Evaporation	Removes dissolved and suspended solids	Well demonstrated. Handles wide range	High energy use
Reverse osmosis	Reduces nearly all contaminants by 90 to 95%	Simple, low energy utilization	Membrane subject to deterioration. Requires some form of pretreatment. Limited chemical compatibility

Electrodialysis uses DC current to perform separation of charged ions and their removal across suitable membranes [U.S. Dept. Interior, 1979]. EDA is the most advanced of the membrane processes, and it is generally favored for brackish-water conversion. Electrical costs will increase with the feedwater salinity. Capital costs for a polarity reversal plant, for feedwater with 2.3 dS m^{-1} (1500 ppm total dissolved salts), and an output of 0.24 dS m^{-1}, range from $144 to $526 per cubic meter of installed capacity. Operating costs are 10 to 13 ¢ U.S. per m^3 for a system processing about 270 m^3 daily. A polarity reversal system requires no chemical feed or pretreatment.

Packaged reverse osmosis and electrodialysis equipment can be readily purchased in the U.S., in sizes ranging from a few ℓ s per day to several hundred cubic meters per day. Growers producing seedlings are commonly forced to dimineralization when their basic water quality is salty. Cut-flower growers also find it beneficial to use such systems for handling cut flowers after harvest [e.g., Montgomery, 1984; Anon., 1974]. The idea is not to remove all ions, but to reduce concentrations to something near water sample No. 1 in Table 6-39.

Another problem with these systems is disposal of the concentrated brine. Government regulatory agencies are likely to be severe if the grower merely wastes the brine to the most convenient ditch or stream. These methods are not cheap, but acceptable production requires a good water supply.

One way to help surmount the problem is to collect rainwater from the greenhouse roof and store it in a cistern. Some operations have invested in this type of system, especially where high-quality water is required for misting roses or other plants. Deposition of salts on foliage usually causes marked damage, not to mention effects on observable quality. The same water supply can also be used to dilute the poorer water to bring it into a desirable range. Fig. 6-64 is an example of irrigating an ornamental crop with hard water in an African country. Each sprinkler leaves its own pattern in precipitated salts on the screen covering the crop. This obviously will reduce quality.

Far too often, I have seen greenhouse operations started with little attention given to the water supply. The

Fig. 6-65. Example of irrigating a cloth house with water containing high calcium and magnesium carbonates. Each circle is a sprinkler [Courtesy of L. Edstrom].

Table 6-44. Net profits (U.S.$) of reference and closed systems for greenhouse use in The Netherlands [van Os et al., 1991] (©1991, Int. Soc. Hort. Sci., *Acta Hort.*, 294:49-57).

System type	Chrysanthemum	Lettuce	Radish	Cucumber
Reference greenhouse[1]				
Polypropene film under growing layer	+0.56[2]	-2.31	-1.11	-0.62
Gullies				
Nutrient film technique	-1.38	-5.50	-4.46	--
Substrate slabs	--	--	--	-0.45
Slabs and containers	--	--	--	+0.24
Aeroponics	-1.38	--	--	-0.27
Concrete floor[3]	-2.84	-7.40	-7.22	-2.43
Transportable benches[3]	-5.81	-11.68	-11.95	?

[1] Crops in soil except cucumber which is grown in rockwool slabs.
[2] Dutch guilders converted to U.S. dollars.
[3] May use sub-irrigation or ebb-and-flow.

idea is that if one can drink it safely, it is all right to grow plants with it. That is not so. I stated in Chapter 5 that the approach to nutrition in greenhouse production is archaic, and this must change if sufficient return is to be made on the capital investment required for greenhouses.

I. ENVIRONMENTAL CONTAMINATION FROM GREENHOUSES

There are two problems that growers have been forced to deal with in recent years. One is the presence of contaminants, particularly nitrate, in vegetable crops, the second is waste disposal from the greenhouse operation.

Table 6-45. U.S. Health Advisories for chemicals in parts per billion (ppb) [Shimskey, 1988]. HAs are not enforceable by federal government, but may be at the state level (With permission of the *Greenhouse Grower*).

Chemical	ppb	Chemical	ppb	Chemical	ppb
Acifluorfen	9	Aldicarb	10	Ametryn	60
Ammonium sulfamate	1500	Atrazine	3	Bentazon	17.5
Bromacil	80	Butylate	50	Carbaryl	700
Carbofuran	36	Carboxin	700	Chloramben	105
Cyanazine	9	Dalapon	560	DCPA	3500
Diazinon	0.63	Dicamba	9	2,4-D	70
Dimethrin	2100	Dinoseb	7	Diphenamid	200
Disulfoton	0.3	Diuron	14	Endothal	140
Endrin	0.32	Fenamiphos	1.8	Fonofos	14
Glyphosate	700	Hexazinone	210	Maleic hydrazide	3500
MCPA	3.6	Methomyl	175	Methoxychlor	340
Methyl parathion	2	Metolachlor	10	Metribuzin	175
Oxamyl	175	Paraquat	3	Pentachlorophenol	220
Picloram	490	Prometon	100	Pronamide	52
Propachlor	92	Propazine	14	Propham	120
Propoxur	3	Silvex	52	Simazine	35
2,4,5-T	21	Tebuthiuron	35	Terbacil	90
Terbufos	0.18	Trifluralin	2		

In California, the recent drought and effluent polluting runoff in the Half Moon Bay area, have received considerable attention [Hasek et al., 1986; Whitesides, 1989]. In other parts of the country, the varying requirements imposed by states have resulted in grower liability, although their use of chemicals has followed governmental regulations listed on the product [Schmuck and Firth, 1987; Shimskey, 1988]. The Netherlands have imposed severe restrictions on polluting factors [van Os et al.,1991] from greenhouses. The policy plan in The Netherlands, according to van Os et al., is to reduce nitrate and phosphate leaching to surface waters by more than 50%, reduce use of chemical plant protection products by 50%, reduce use of soil disinfection products by 75%, increase energy efficiency by 50%, and reduce CO_2 emission by 5%. The aim is to grow 80% of the greenhouse vegetables and pot plants separately from the soil.

Table 6-44 shows net profits of closed systems simulated by the Dutch, compared with a "reference" operation that produces chrysanthemums, lettuce, and radish in soil and cucumbers in rockwool with free drainage. It can be seen that most of these procedures to reduce pollution also reduce net profit. They are costs that will be transferred to the consumer, and governments will be required to regulate all growers to prevent an advantage to those refusing to carry out the new requirements. This is no different from costs of cleanup for electrical generating, chemical, and manufacturing enterprises. Table 6-45 lists present health advisory limits for several chemicals used in the U.S. by the industry. Regulating will be necessary since there are always a few who attempt to increase profit by polluting the commons.

Keeny [1982] states that nitrates are relatively nontoxic to humans. Acute nitrate poisoning in an adult requires a single oral ingestion of 1 to 2 g NO_3^- –which is far above usual exposure limits. The adverse effect results from reduction of NO_3^- to nitrite (NO_2^-), which can occur in the intestine of some animals and in the human infant during the first few months of life. Maynard et al. [1976] discussed problems of nitrate accumulation in vegetables to some detail. According to these authors, a lethal nitrite dose is about 20 mg NO_2^- per kg body weight. Fatal reactions of this type may be caused by other chemicals. Most cases of acute nitrate toxicity have occurred largely with households having a private well supply. The standard limits in water supply are 10 mg per ℓ NO_3^-. Instances of abortion in cattle have been noted when fed with plant products produced in greenhouses.

As was discussed, nitrate concentration in plants varies widely. Accumulation occurs when the ion is unconverted to other compounds, which may occur with excessive nitrate fertilization, low solar radiation, and high temperatures. Plants subjected to drought tend to accumulate nitrate. There are numerous other sources of nitrate [Maynard et al.,1976]. Comments by Rooda van Eysinga [1984] were to the effect that even if fertilizers are omitted, which would result in 10 to 20% yield reductions, the nitrate contents in lettuce would be reduced by only 10%.

Some investigators have attempted to reduce nitrate in lettuce by transferring the crop to solutions with diluted NO_3^- 2 to 7 days before harvest [Shinohara and Suzuki, 1988]. Hydroponic culture in Japan uses plastic panels floating on the solution, so moving the crop is relatively easy. Eysinga discussed a number of methods to reduce nitrate in lettuce. Nevertheless, he stated that there are no known cases recorded in which methemoglobinemia has been proven in adults from eating vegetables with high nitrates. A causal relationship between vegetable consumption with high nitrate contents and incidence of stomach cancer has never been proved. Some authors have recommended not eating greenhouse-produced vegetables, but only those produced in season and including stinging nettles and sorrel. Eysinga noted concentrations of 6150 and 2900 mg NO_3 per kg, respectively, for these two weeds. This is far above any nitrate levels found in commercial greenhouse vegetables.

One should also keep in mind that not only can nutrients, pesticides, herbicides, and soil chemicals leach into the ground water, but waste materials such as styrofoam beads (used in potting mixtures), plastics, paper, flue gases, etc. can cause pollution. Thus, a beginning can be made in reducing pollution from greenhouses by simply cleaning up the surroundings. Of the individuals immediately concerned with such problems, Biernbaum [1992] discussed methods to reduce runoff from greenhouse property and to lower contamination. Closed systems that recirculate the water would be the most obvious. Experience has shown, however, that such systems are not always the practical solution [Whitesides, 1989]. Even with sterilization of the recirculated solution (Fig. 5-54), a continual increase in salt content usually requires wastage and replenishment with fresh water. Water partially wasted in evaporative pad systems, to prevent salt accumulation on the pads, is a common practice.

Control of salinity in greenhouse substrates has usually involved excess watering, especially where fertigation is continually practiced. Greater care in making sure that fertilizer applications are maintained as low as possible, commensurate with desired results, and measuring the amount of water required by the crop so that application is not excessive, can go far to reducing excess waste to the environment. Pulsed irrigation applications have been suggested as a means to reduce water loss [Biernbaum, 1992], rather than continuous applications to wet the entire substrate volume. Slow-release fertilizers have been found to reduce fertilizer wastage from containers. Biernbaum stated that superabsorbent polyacrylamide gels were ineffective under usual watering practices. Sometimes, growers have placed pans under containers to catch drainage water that can be reused. Manipulation of the irrigation system to ensure uniform watering can go far in eliminating wastage. Whitesides' [1989] account of one California operation cited a 50% reduction in fertilizer usage, a reduction in labor of 4000 man-hours, and 30% less water consumption through nutrient and water control by computer with drip irrigation, variable fertilizer injection, and water consumption correlated with local weather conditions. In all these, storing runoff from greenhouse roofs is encouraged.

In summary, most of the methods to conserve and reduce pollution require greater attention to detail and less sloppiness in cultural procedures. The grower should use all tools available to him since this will often increase efficiency and profitability.

V. REFERENCES

Adams, P. 1987. The test of raised salinity. *The Grower*. 107(2).

Adriano, D.C. et al. 1986. New results in the research of hardly known trace elements and the analytical problems of trace element research. *Proc. Int. Symp*., Budapest, June, 1986. 3-33 pp.

Albery, W.J., B.G.D. Haggett and L.R. Svanberg. 1985. The development of sensors for hydroponics. *Biosensors*. 1:369-397.

Allaway, W.H. 1957. pH, soil acidity and plant growth. *In* Soil, The Yearbook of Agriculture 1957. A. Stefferud, ed. U.S. Government Printing Office, Washington, D.C.

Allison, J.D., D.S. Brown and K.J. Novo-Gradac. 1991. MINTEQA2/PRODEFA2. A geochemical model for environmental systems: Version 3.0 user's manual. EPA/600/3-91/021. U.S. EPA, ORD, Environ. Res. Lab., Athens, GA. 106 pp.

Alt, D., V.T. Gizewski and B. Schroer. 1988. Analysis of substrates by $NAHCO_3$-DTPA method. *Acta Hort.* 221:395-402.

Ames, R.N. and R.G. Linderman. 1977. Vesicular-arbuscular mycorrhizae of Easter lily in the northwestern United States. *Can. J. Microbiol.* 23:1663-1668.

Ames, R.N. and R.G. Linderman. 1978. The growth of Easter lily (*Lilium longiflorum*) as influenced by vesicular-arbuscular mycorrhizal fungi, *Fusarium oxysporum*, and fertility level. *Can. J. Bot.* 56:2773-2783.

Anon. 1974. R.O. helps the flowers grow. *Water Conditioning*. June, 1974, 12-13.

Anon. 1978. Water quality vital for NFT success. *The Grower*. 89(15):880.

Anon. 1985. Soil lab tests help growers adjust soil, define nutrient needs. *Greenhouse Mgr*. Oct., 1985.

Argo, W.R. and J.A. Biernbaum. 1994. Irrigation requirements, root-medium pH, and nutrient concentrations of Easter Lilies grown in five peat-based media with and without an evaporation barrier. *J. Amer. Soc. Hort. Sci.* 119:1151-1156.

Asher, C.J. 1991. Beneficial elements, functional nutrients, and possible new essential elements. *In* Micronutrients in Agriculture. J.J. Mortvedt et al .eds. 2nd edition, No. 4. Soil Sci. Soc. Amer., Madison, WI.

Aung, L.H. 1974. Root-shoot relationships. *In* The Plant Root and Its Environment. E.W. Carson, ed. Univ. Press of VA, Charlottesville. 29-61.

Avissar, R. and Y. Mahrer. 1986. Water desalination in solar earth stills: A numerical study. *Water Resources Res.* 7:1067-1075.

Baldwin, J.P., P.B. Tinker and P.H. Nye. 1972. Uptake of solutes by multiple root systems from soil. II. The theoretical effects of rooting density and pattern on uptake of nutrients from soil. *Plant and Soil.* 36:693-708.

Ball, V., ed. 1985. The Ball Red Book. Reston Publ. Co., Reston, VA. 720 pp.

Barber, D.A. and J.K. Martin. 1976. The release of organic substances by cereal roots into soil. *New Phytol.* 76:69-80.

Barber, S.A. 1962. A diffusion and mass-flow concept of soil nutrient availability. *Soil Sci.* 93:39-49.

Barber, S.A. 1974. Influence of the plant root on ion movement in soil. *In* The Plant Root and Its Environment. E.W. Carson, ed. Univ. Press of VA., Charlottesville. 525-564.

Barker, A.V. and H.A. Mills. 1980. Ammonium and nitrate nutrition of horticultural crops. *Hort. Rev.* 2:395-423.

Barragry, A.R. and J.V. Morgan. 1978. Effect of mineral and slow-release nitrogen combinations on the growth of tomato in a coniferous bark medium. *Acta Hort.* 82:43-53.

Barrows, J.B. and R.W. Roncadori. 1977. Endomychorrizal synthesis by *Gigaspora margarita* in poinsettia. *Mycologia.* 69:1173-1184.

Bates, T.E. 1971. Factors affecting critical nutrient concentrations in plants and their evaluation: A review. *Soil Sci.* 112:116-130.

Bates, T.E. 1990. Prediction of phosphorous availability from 88 Ontario soils using five phosphorous tests. *Commun. in Soil Sci. Plant Anal.* 21:1009-1023.

Bear, F.E. 1957. Toxic elements in soils. *In* Soils. Yearbook of Agriculture 1957. A. Stefferud, ed. U.S. Govern. Publ., Washington, D.C.

Beaufils, E.R. 1971. Physiological diagnosis – A guide for improving maize production based on principles developed for rubber trees. *Fert. Soc. S. Afr.* 1:1-30.

Beaufils, E.R. 1973. Diagnosis and recommendation integrated system (DRIS). A general scheme for experimentation and calibration based on principles developed from research in plant nutrition. *Univ. of Natal Soil Sci. Bull.* 1:1-132.

Berghage, R.D. et al. 1987. Micronutrient testing of plant growth media: Extractant identification and evaluation. *Commun. in Soil Sci. Plant Anal.* 18:1089-1109.

Bernstein, L. 1961. Osmotic adjustment of plants to saline media. I. Steady state. *Amer. J. Bot.* 48:909-918.

Bernstein, L. 1963. Osmotic adjustment of plants to saline media. II. Dynamic phase. *Amer. J. Bot.* 50:360-370.

Bernstein, L. and H.E. Hayward. 1958. Physiology of salt tolerance. *Ann. Rev. Plant Physiol.* 9:25-46.

Biernbaum, J.A. 1992. Root-zone management of greenhouse container-grown crops to control water and fertilizer use. *HortTechnology*. 2:127-132.
Biernbaum, J.A. and W.R. Argo. 1993. Greenhouse peat-based root media: Properties and components. Unpublished MS. Dept. of Horticulture, MI State Univ., East Lansing, MI.
Bik, R.A. and G.A. Boertje. 1975. Fertilizing standards for potting composts based on the 1:1½ volume extraction method of soil testing. *Acta Hort*. 50:153-156.
Black, C.A. 1992. Soil Fertility Evaluation and Control. Lewis Publ., Boca Raton, FL. 746 pp.
Blackman, F.F. 1905. Optima and limiting factors. *Ann. Bot.* 15:185-194.
Blair, G.J, M.H. Miller and W.A. Mitchell. 1970. Nitrate and ammonium as sources of nitrogen for corn and their influence on the uptake of other ions. *Agron. J.* 62:530-532.
Boertje, G.A. 1980. Results of liquid feeding in the production of bedding plants. *Acta Hort*. 99:17-23.
Boodley, J.W. 1962. Fertilization. *In* Snapdragons. R.W. Langhans, ed. Cornell Univ., Ithaca, NY. 185 pp.
Boodley, J.W. 1974. Nutrient content of Paul Mikkelsen poinsettias from juvenile to mature growth. *Florists' Rev*. Dec. 26, 1974.
Boodley, J.W. and J.W. White. 1969. Fertilization. *In* Roses. J.W. Mastalerz and R.W. Langhans eds. PA Flower Growers Assoc., University Park, PA.
Bowe, R. et al. 1969. Gerbera. V.J. Neumann-Neudamm, Berlin. 180 pp.
Boyko, H. 1967. Salt-water agriculture. *Sci. Amer*. 216:89-96.
Boyle, F.W., Jr. and W.L. Lindsay. 1986. Manganese phosphate equilibrium relationships in soils. *Soil Sci. Soc. Amer. J.* 50:588-593.
Brady, N.C. 1974. The Nature and Properties of Soils. MacMillan Publ. Co., New York. 639 pp.
Bray, R.H. 1963. Confirmation of the nutrient mobility concept of soil-plant relationships. *Soil Sci*. 95:124-130.
Brennan, E.G. and J.W. Shive. 1948. Effect of calcium and boron nutrition of the tomato on the relation between these elements in the tissues. *Soil Sci*. 66:65-75.
Brouwer, R. 1963. The influence of the suction tension of the nutrient solutions on growth, transpiration and diffusion pressure deficit of bean leaves (*Phaseolus vulgaris*). *Acta Bot. Neerl*. 12:248-261.
Bunt, A.C. 1976. Modern Potting Composts. Allen and Unwin, London. 277 pp.
Bunt, A.C. 1980. Phosphorous sources for loamless substrates. *Acta Hort*. 99:25-32.
Bunt, A.C. 1986. Problems in the analysis of organic and lightweight potting substrates. *HortScience*. 21:229-231.
Bunt A.C. 1988. Media and Mixes for Container-Grown Plants. Unwin Hyman, London. 309 pp.
Canning, R.E. and P.J. Kramer. 1958. Salt absorption and accumulation in various regions of roots. *Amer. J. Bot.* 45:378-382.
Cate, R.B. and L.A. Nelson. 1971. A simple statistical procedure for partitioning soil test correlation data into two classes. *Soil Sci. Soc. Amer. Proc*. 35:658-660.
Chapman, H.D. 1960. The diagnosis and control of zinc deficiency and excess. *Bull. Res. Council of Israel*. 8D.
Clarkson, D.T. 1985. Factors affecting mineral nutrient acquisition by plants. *Ann. Rev. Plant Physiol*. 36:77-115.
Clement, C.R., M.J. Hopper and L.H.P. Jones. 1978. The uptake of nitrate by *Lolium perenne* from flowing nutrient solution. I. Effect of NO_3^- concentration. *J. Expt. Bot*. 29:453-464.
Colegrave, D. et al. 1973. The Colegrave Manual of Bedding Plants. Grower Books, London. 211 pp.
Cooper, A. 1979. The ABC of NFT. Grower Books, London. 181 pp.
Cregan, P.D., J.R. Hirth and M.K. Conyers. 1989. Amelioration of soil acidity by liming and other amdendments. *In* Soil Acidity and Plant Growth. A.D. Robson, ed. Academic Press, New York. 306 pp.
Crews, C.E., C.R. Johnson and J.N. Joiner. 1978. Benefits of mycorrhizae on growth and development of three woody ornamentals. *HortScience*. 13:429-430.
Crockford, H.D. and S.B. Knight. 1964. Fundamentals of Physical Chemistry. 2nd, ed. J. Wiley & Sons, New York. 415 pp.
Dahnke, W.C. and R.A. Olson. 1990. Soil test correlation, calibration and recommendation. *In* Soil Testing and Plant Analysis. R.L. Westerman, ed. Soil Sci. Soc. Amer. No. 3. Soil Sci. Soc. Amer., Madison, WI.
Dean, L.A. 1957. Plant nutrition and soil fertility. *In* Soil, The Yearbook of Agriculture 1957. A. Stefferud, ed. U.S. Government Printing Office, Washington, D.C.

De Hertogh, A. 1958. Holland Bulb Forcer's Guide. Int. Flower Bulb Centre, Hillegom, The Netherlands. 284 pp.
Dight, R.J.W. 1977. Nutritional requirements of bedding plants. *Expl. Hort.* 29:63-71.
Dole, J.M. and H.F. Wilkins. 1991. Relationship between nodal position and plant age on the nutrient composition of vegetative poinsettia leaves. *J. Amer. Soc. Hort. Sci.* 116:248-252.
Ecke, P., Jr. 1971. The Poinsettia Manual. Paul Ecke Poinsettias, Encinitas, CA. 183 pp.
Editors. 1980. Roses Under Glass. No. 9. Grower Books, London. 78 pp.
Editors. 1980. Cucumbers. No. 15. Grower Books, London. 74 pp.
Editors. 1981. Lettuce Under Glass. No. 21. Grower Books, London. 105 pp.
Edwards, D.G. and C.J. Asher. 1974. The significance of solution flow rate in flowing culture experiments. *Plant and Soil.* 41:161-175.
Elgawhary, S.M., W.L. Lindsay and W.D.Kemper. 1970. Effect of complexing agents and acids on the diffusion of zinc to a simulated root. *Soil Sci. Soc. Amer. Proc.* 34:211-214.
El Kohli, A.F. 1961. An experimental study of the influence of the microelements on the uptake of macro-elements. *Versl. Landbouwk. onderz. NR.* 67.4. Wageningen
Ellis, N.K. et al. 1974. Nutriculture systems. Purdue Univ. Agric. *Expt. Sta. Bull. No.* 44. 20 pp.
Fernández Falcón, M. et al. 1986. The effect of chloride and bicarbonate levels in irrigation water on nutrition content, production and quality of cut roses 'Mercedes'. *Sci. Hort.* 29:373-385.
Follett, R.H. and P.N. Soltanpour. 1985. Irrigation water quality criteria. Service in Action, *CO State Univ. Ext. Service.* No. 506. 2 pp.
Foy, C.D. 1992. Soil chemical factors limiting plant root growth. *Adv. in Soil Sci.* 19:97-149.
Frick, J. And C.A. Mitchell. 1993. Stabilization of pH in solid-matrix hydroponic systems. *HortScience.* 28:981-984.
Geraldson, C.M. 1954. The control of blackheart of celery. *Proc. Amer. Soc. Hort. Sci.* 63:353-358.
Geraldson, C.M. 1955. The use of calcium for control of blossom-end rot of tomatoes. *Proc. FL State Hort. Soc.* 68:197-202.
Geraldson, C.M., G.R. Klacan and O.A. Lorenz. 1973. Plant analys as an aid in fertilizing vegetable crops. *In* Soil Testing and Plant Analysis. L.M. Walsh and J.D. Beaton, eds. Soil Sci. Soc. Amer., Madison, WI.
Gerdemann, J.W. 1974. Mycorrhizae. *In* The Plant Root and Its Environment. E.W. Carson, ed. Univ. Press of VA., Charlottesville. 205-217.
Gieling, T.H., E. van Os and A. de Jager. 1988. The application of chemo-sensors and bio-sensors for soilless culture. *Acta Hort.* 230:357-361.
Gieling, T.H. and K. Schurer. 1995. Sensors and measurement. *In* Greenhouse Climate Control. J.C. Bakker et al., eds. Wageningen Pers.
Gislerød, H.R. and L.M. Mortensen. 1990. Relative humidity and nutrient concentration affect nutrient uptake and growth of *Begonia x hiemalis*. *HortScience.* 25:524-526.
Gislerød, H.R. and L.M. Mortensen. 1991. Air humidity and nutrient concentration affect nutrient uptake and growth of some greenhouse plants. *Acta Hort.* 294:230-235.
Green, J.L. 1967a. Ionic balance and growth of carnations - I. Recent ℓ ature. *CO Flower Growers' Assoc. Res. Bull.* 210:1-4.
Green, J.L. 1967b. Ionic balance and growth of carnation. *CO Flower Growers' Assoc. Res. Bull.* 211:1-5.
Green, J.L., C.V. Cole and W.D. Holley. 1971. Ionic balance and growth of carnations. *Soil Sci.* 111:138-145.
Grinsted, M.J. et al. 1982. Plant-induced changes in the rhizosphere of rape (*Brassica napus* var. Emerald) seedlings. I. pH change and the increase of P concentration in the soil solution. *New Phytol.* 91:19-29.
Hall, D.A. and G. Wilson. 1980. Slow-release mix simple solution for small growers. *The Grower.* 93(5):37-38.
Halvorson, A.D. 1971. Chelation and availability of metal ions in nutrient solutions. Ph.D. Dissertation. Colo. State Univ., Ft. Collins. 134 pp.
Halvorson, A.D. and W.L. Lindsay. 1972. Equilibrium relationships of metal chelates in hydroponic solutions. *Soil Sci. Soc. Amer. Proc.* 36:755-761.
Hanan, J.J. 1964. Oxygen and carbon dioxide concentrations in greenhouse soil-air. *Proc. Amer. Soc. Hort. Sci.* 84:648-652.
Hanan, J.J. 1973. Denver water quality. *CO Flower Growers'Assoc. Res. Bull.* 280:1-3.

Hanan, J.J. 1975. Calcium tissue levels in the carnation. *CO Flower Growers' Assoc. Res. Bull.* 305:1-4.
Hanan, J.J. 1979. Carnation fertilizer injection analysis for the Denver region. *CO Flower Growers' Assoc. Res. Bull.* 343:3-4.
Hanan, J.J. 1982a. Salinity II: A problem in terms. *CO Greenhouse Growers' Assoc. Res. Bull.* 383:1-3.
Hanan, J.J. 1982b. Salinity III: Handling water supplies to minimize salinity problems. *CO Greenhouse Growers' Assoc. Res. Bull.* 384:1-4.
Hanan, J.J. 1984. Thirty years of nutrition studies at Colorado State University. *CO Greenhouse Growers' Assoc. Res. Bull.* 413:1-3.
Hanan, J.J. 1988. Manipulation of nutrition using a salty water. *CO Greenhouse Growers' Assoc. Res. Bull.* 453:1-3.
Hanan, J.J. and W.D. Holley. 1974. Hydroponics. CO State Univ. Expt. Sta. *General Series* 941, 21 pp.
Hanan, J.J., W.D. Holley and K.L. Goldsberry. 1978. Greenhouse Management. Springer-Verlag, Heidelburg. 530 pp.
Hanan, J.J., K. Schekel and W.D. Holley. 1968. Colorado water analysis. *CO Flower Growers' Assoc. Res. Bull.* 222:2-4.
Hansen, M. 1978. Plant specific nutrition and preparation of nutrient solutions. *Acta Hort.* 82:109-112.
Handreck, K.A. 1991. Assessment of available nutrients in potting media using dilute DTPA extractants. *Acta Hort.* 294:250-258.
Harbaugh, B.K. and G.J. Wilfret. 1982. Correct temperature is the key to successful use of Osmocote. *Florists' Review.* 170(4403):21-23.
Harbaugh, B.K. and S.S. Woltz. 1989. Fertilization practice and foliar-bract calcium sprays reduce incidence of marginal bract necrosis of poinsettia. *HortScience.* 24:465-468.
Harter, R.D. 1991. Micronutrient adsorption-desorption reactions in soils. *In* Micronutrients in Agriculture. J.J. Mortvedt et al., eds. 2nd edition, No. 4. Soil Sci. Soc. Amer., Madison, WI.
Hartman, L.D. 1971. Effects of irrigation frequency and nutrient solution concentration on yield of carnation. *CO Flower Growers' Assoc. Res. Bull.* 250:1-5.
Hasek, R.F., R.H. Sciaroni and R.L. Branson. 1986. Water conservation and recycling in ornamentals production. *HortScience.* 21:35-38.
Hayward, H.E. and W.B. Spurr. 1944. Effects of isosmotic concentrations of inorganic and organic substrates on entry of water into corn roots. *Bot. Gaz.* 106:131-139.
Helyar, K.R. and W.M. Porter. 1989. Soil acidification, its measurement and the processes involved. *In* Soil Acidity and Plant Growth. D. Robson, ed. Academic Press, New York. 306 pp.
Hewitt, E.J. 1966. Sand and water culture methods used in the study of plant nutrition. *Tech. Comm. No.* 22 (Revised). Commonwealth Agric. Bur., Maidstone, Kent, England.
Hiatt, A.J. and J.E. Leggett. 1974. Ionic interactions and antagonisms in plants. *In* The Plant Root and Its Environment. E.W. Carson, ed. Univ. Press of VA., Charlottesville. 101-134.
Hoagland, D.R. and D.I. Arnon. 1950. The water-culture method for growing plants without soil. *Calif. Agric. Expt. Sta. Circ.* 347.
Hofstra, J.J. and T. Koch-Bosma. 1970. Organic phosphate in ammonium- and nitrate-fed tomato plants. *Acta Bot. Neerl.* 19:546-552.
Holcomb, E.J. and J.W. White. 1982. Fertilization. *In* Geraniums. J.W. Mastalerz and E.J. Holcomb, eds. *PA Flower Growers*, University Park, PA. 410 pp.
Holevar, C.D. 1966. The effect of vesicular-arbuscular mycorrhiza on the uptake of soil phosphorous by strawberry (*Fragaria* sp. var. Cambridge Favorite). *J. Hort. Sci.* 41:57-64.
Holland, D.A. 1966. The interpretation of leaf analysis. *J. Hort. Sci.* 41:311-329.
Holley, W.D. and R. Baker. 1991. Carnation Production II. Kendall/Hunt Publ. Co., Dubuque, IA. 156 pp..
Hood, T.M., H.A. Mills and P.A. Thomas. 1993. Developmental stage affects nutrient uptake by four snapdragon cultivars. *HortScience.* 28:1008-1010.
Hughes, H. 1975. Progress report. Effect of salinity on roses. *CO Flower Growers' Assoc. Res. Bull.* 304:1-3.
Hughes, H. and J.J. Hanan. 1977. Effects of salinity in water supplies on rose production. Experiment 2. *CO Flower Growers' Assoc. Res. Bull.* 327:1-3.

Hurd, R.G. 1978. The root and its environment in the nutrient film technique of water culture. *Acta Hort.* 82:87-97.

Ingestad, T. 1970. A definition of optimum nutrient requirements in Birch seedlings. *Physiol. Plant.* 23:1127-1138.

Ingestad, T. 1971. A definition of optimum nutrient requirements of birch seedlings. II. *Physiol. Plant.* 24:118-125.

Ingestad, T. 1973b. Mineral nutrient requirements of cucumber seedlings. *Plant Physiol.* 52:332-338.

Ingestad, T. 1974. Towards optimum fertilization. *AMBIO.* 3:49-54.

Ingestad, T. 1982. Relative addition rate and external concentration; driving variables used in plant nutrition research. *Plant, Cell and Environ.* 5:443-453.

James, D.W. and K.L. Wells. 1990. Soil sample collection and handling: Technique based on source and degree of field variability. *In* Soil Testing and Plant Analysis. R.L. Westerman, ed. Soil Sci. Soc. Amer. Series No. 3. Soil Sci. Soc. Amer., Madison, WI.

Jansson, S.L. and J. Persson. 1982. Mineralization and immbolization of soil nitrogen. *In* Nitrogen in Agricultural Soils. F.J. Stevenson, ed. No. 22, Amer. Soc. Agron., Madison, WI. 229-252.

Jeong, B.R. 1990. Ammonium and nitrate nutrition of selected bedding plants. Ph.D. Dissertation, CO State Univ., Ft. Collins. 274 pp.

Johnson, E.W. 1980. Comparison of methods of analysis for loamless composts. *Acta Hort.* 99:197-204.

Johnson, C.R. and C.E. Crews, Jr. 1979. Survival of mychorrizal plants in the landscape. *Amer. Nurseryman.* 150:15, 59.

Joiner, J.N., ed. 1981. Foliage Plant Production. Prentice-Hall, Inglewood Cliffs, NJ. 614 pp.

Joiner, J.N. 1983. Nutrition and fertilization of ornamental greenhouse crops. *Hort. Rev.* 5:317-403.

Joiner, J.N., C.A. Conover and R.T. Poole. 1981. Nutrition and fertilization. *In* Foliage Plant Production. J.N. Joiner, ed.. Prentice-Hall, Englewood, NJ. 614 pp.

Jones, J.B., Jr. 1983. A Guide for the Hydroponic and Soilless Culture Grower. Timber Press, Portland, OR. 124 pp.

Jones, J.B., Jr. 1985. Soil testing and plant analysis: Guides to the fertilization of horticultural crops. *Hort. Rev.* 7:1-68.

Jones, J.B . Jr. 1990. Universal soil extractants: Their composition and use. *Commun. in Soil Sci. Plant Anal.* 21:1091-1101.

Jones, J.B., Jr. 1991. Plant tissue analysis in micronutrients. *In* Micronutrients in Agriculture. J.J. Mortvedt et al., eds. 2nd edition, No. 4. Soil Sci. Soc. Amer., Madison, WI.

Jones, R.G.W. 1981. Plant physiology. *In* McGraw-Hill Yearbook of Science and Technology. McGraw-Hill, New York. 303-306.

Keeney, D.R. 1982. Nitrogen management for maximum efficiency and minimum pollution. *In* Nitrogen in Agricultural Soils. F.J. Stevenson, ed. No. 22, Amer. Soc. Agron., Madison, WI. 605-649,

Kerr, G.P. and J.J. Hanan. 1985. Leaching of container media. *J. Amer. Soc. Hort. Sci.* 110:474-480.

Kingham, H.G., ed. 1973. The U.K. Tomato Manual. Grower Books, London. 223 pp.

Kiplinger, D.C. and R.W. Langhans, eds. 1967. Easter Lilies. N.Y. State College of Agric., Cornell Univ., Ithaca, NY. 158 pp.

Kochian, L.V. 1991. Mechanisms of micronutrient uptake and translocation in plants. *In* Micronutrients in Agriculture. J.J. Mortvedt et al., eds. 2nd edition, No. 4. Soil Sci. Soc. Amer., Madison, WI.

Kofranek, A.M. 1980. Mineral nutrition. *In* Introduction to Floriculture. R.A. Larson, ed.. Academic Press, New York.

Kofranek, A.M. and R.A. Larson eds. 1975. Growing Azaleas Commercially. Publ. 4058., Univ. of CA., Davis, CA. 108 pp.

Kofranek, A.M. and O.R. Lunt. 1975. Mineral nutrition. *In* Growing Azaleas Commercially. A.M. Kofranek and R.A. Larson eds. Publ. No. 4058, Univ. of CA, Davis, CA.

Koide, R.T. and R.P. Schreiner. 1992. Regulation of the vesicular-arbuscular mychorrizal symbiosis. *Ann. Rev. Plant Physiol. and Plant Mol. Biol.* 43:557-581.

Korcak, R.F. 1987. Iron deficiency chlorosis. *Hort. Rev.* 9:133-186.

Ku, C.S.M. and D.R. Hershey. 1991. Leachate electrical conductivity and growth of potted poinsettia with leaching fractions of 0 to 0.4. *J. Amer. Soc. Hort. Sci.* 116:802-806.

Ku, C.S.M. and D.R. Hershey. 1992. Leachate electrical conductivity and growth of potted geranium with leaching fractions of 0 to 0.4. *J. Amer. Soc. Hort. Sci.* 117:893-897.

Lang, H.J., C.L. Rosenfield and D.W. Reed. 1990. Response of *Ficus benjamina* and *Dracaena marginata* to iron stress. *J. Amer. Soc. Hort. Sci.* 115:589-592.

Langhans, R.W., ed. 1962. Snapdragons. NY State Ext. Serv., Cornell Univ., Ithaca, NY. 93 pp.

Langhans, R.W., ed. 1987. Roses. Roses, Inc., Haslett, MI. 372 pp.

Larson, R.A., ed. 1980. Introduction to Floriculture. Academic Press, New York. 607 pp.

Lawton, K.A., G.L. McDaniel and E.T. Graham. 1989. Nitrogen source and calcium supplement affect stem strength of poinsettia. *HortScience*. 24:463-465.

Lee, C.W., J.M. Choi and C.H. Pak. 1996. Micronutrient toxicity in seed geranium (*Pelargonium x hortorum* Bailey). *J. Amer. Soc. Hort. Sci.* 121:77-82.

Legg, J.O. and J.J. Meisinger. 1982. Soil nitrogen budgets. *In* Nitrogen in Agricultural Soils. F.J. Stevensen, ed. No. 22, Amer. Soc. Agron., Madison, WI. 503-566.

Letey, J., W.M. Jarrell and N. Valoras. 1982. Nitrogen and water uptake and growth of plants at various minimum solution nitrate concentrations. *J. Plant Nutr.* 5:73-89.

Lindsay, W.L. 1972a. Zinc in soils and plant nutrition. *Adv. Agron.* 24:147-186.

Lindsay, W.L. 1972b. Influence of the soil matrix on the availability of trace elements to plants. *Ann. NY Acad. Sci.* 199:37-45.

Lindsay, W.L. 1974. Role of chelation in micronutrient availability. *In* The Plant Root and Its Environment. E.W. Carson, ed. Univ. Press of VA., Charlottesville. 507-524.

Lindsay, W.L. 1978. Chemical reactions affecting the availability of micro-nutrients in soils. *In* Mineral Nutrition of Legumes in Tropical and Subtropical Soils. C.S. Andrew and E.J. Kamprath, ed. Proc. Workshop CSIRO, Brisbane, Australia.

Lindsay, W.L. 1979. Chemical Equilibria in Soils. J. Wiley & Sons, New York. 449 pp.

Lindsay, W.L. 1981. Soil chemistry. *In* McGraw-Hill Yearbook of Science and Technology. McGraw-Hill, New York. 358-359.

Lindsay, W.L. 1983. Trace elements and chelation in fertilizers. *10th Ann. Rocky Mtn. Reg. Turfgrass Conf.*, CO State Univ., Ft. Collins. 5 pp.

Lindsay, W.L. 1984. Soil and plant relationships associated with iron deficiency with emphasis on nutrient interactions. *J. Plant Nut.* 7:489-500.

Lindsay, W.L. 1991. Inorganic equlibria affecting micronutrients in soils. *In* Micronutrients in Agriculture. J.J. Mortvedt et al., eds. 2nd edition, No. 4. Soil Sci. Soc. Amer., Madison, WI. 89-112.

Lindsay, W.L. and H.A. Ajwa. 1994. Use of MINTEQA2 for teaching soil chemistry. Unpubl. Ms., CO State Univ., Ft. Collins. 44 pp.

Lindsay, W.L. and F.R. Cox. 1985. Micronutrient soil testing for the tropics. *Fertilizer Res.* 7:169-200.

Lindsay, W.L. and J.D. DeMent. 1961. Effectiveness of some iron phosphates as sources of phosphorous for plants. *Plant and Soil.* 14:118-126.

Lindsay, W.L. and E.C. Moreno. 1960. Phosphate phase equilibria in soils. *Soil Sci. Soc. Amer. Proc.* 24:177-182.

Lindsay, W.L. and W.A. Norvell. 1969. Equilibrium relationships of Zn^{2+}, Fe^{3+}, Ca^{2+}, and H^{+} with EDTA and DTPA in soils. *Soil Sci. Soc. Amer. Proc.* 33:62-68.

Lindsay, W.L. and W.A. Norvell. 1978. Development of a DTPA soil test for zinc, iron, manganese and copper. *Soil Sci. Soc. Amer. Proc.* 42:421-428.

Lindsay, W.L. and A.P. Schwab. 1982. The chemistry of iron in soils and its availability to plants. *J. Plant Nutrition.* 5:821-840.

Lindsay, W.L. and G. Stanford. 1960. What happens to water-soluble phosphate in the soil? *Crops and Soils.* 12:2.

Lindsay, W.L. and A.W. Taylor. 1960. Phosphate reaction products in soil and their availability to plants. *7th Int. Cong. Soil Sci.*, Madison, WI. 3:580-589.

Lindsay, W.L., P.L.G. Vlek and S.H. Chien. 1989. Phosphate minerals. *In* Minerals in Soil Environments, 2nd ed. SSSA Book Series No. 1. Soil Sci. Soc. Amer., Madison, WI. 1089-1130.

Lindsay, W.L., F.S. Watanabe and S.R. Olsen. 1963. Nutritional inter-relationships of phosphorous, zinc and iron. *Proc. 14th Ann. Fert. Conf. Pacific Northwest*, Idaho Falls, ID. 109-115.

Loneragan, J.F. 1968. Nutrient requirement of plants. *Nature*. 220:1307-1308.

Loneragan, J.F. and K. Snowball. 1969a. Calcium requirements of plants. *Austr. J. Agric. Res*. 29:465-478.

Loneragan, J.F. and K. Snowball. 1969b. Rate of calcium absorption by plant roots and its relation to growth. *Austr. J. Agric. Res*. 29:479-490.

Lunt, O.R. and J.J. Oertli. 1962. Controlled release of fertilizer minerals by incapsulating membranes: II. Efficiency of recovery, influence of soil moisture, mode of application, and other considerations related to use. *Soil Sci. Soc. Amer. Proc*. 26:584-587.

Maagistad, O.C. et al. 1943. Effect of salt concentration, kind of salt, and climate on plant growth in sand cultures. *Plant Physiol*. 18:151-166.

Markus, D.K. 1986. Spurway/acid extraction procedures. *HortScience*. 21:217-222.

Markus, D.K. and J.E. Steckel. 1980. Periodical analysis of artificial rooting media and tomato leaf analysis from New Jersey greenhouses. *Acta Hort*. 99:206-217.

Maronek, D.M., J.W. Hendrix and J. Kiernan. 1981. Mycorrhizal fungi and their importance in horticultural crop production. *Hort. Rev*. 3:172-213.

Marschner, H. and V. Römheld. 1983. *In vivo* measurement of root-induced pH changes at the soil-root interface: Effect of plant species and nitrogen source. *Z. Pflanzenphysiol*. 111:241-251.

Martens, D.C. and W.L. Lindsay. 1990. Testing soils for copper, iron, manganese and zinc. *In* Soil Testing and Plant Analysis. Series No. 3. Soil Sci. Soc. Amer., Madison, WI.

Mastalerz, J.W., ed. Bedding Plants. PA Flower Growers, PA State Univ., University Park, PA. 516 pp.

Mastalerz, J.W. 1977. The Greenhouse Environment. Wiley, New York.

Mastalerz, J.W. and E.J. Holcomb, eds. 1982. Geraniums III. PA Flower Growers, PA State Univ., University Park, PA. 410 pp.

Maxwell, M.K. 1972. Soilless culture -hydroponics. Occasional Paper No. 1. Dept. of Plant Sci., Hawkesbury Agric. Col., Richmond, N.S.W., Australia. 10 pp.

Maynard, D.N. et al. 1976. Nitrate accumulation in vegetables. *Adv. in Agron*. 28:71-118.

Meiri, A. and A. Poljakoff-Mayber. 1967. The effect of chlorine salinity on growth of bean leaves in thickness and in area. *Israel J. Bot*. 16:115-123.

Melton, R.R. and R.J. Dufault. 1991. Nitrogen, phosphorous and potassium fertility regimes affect tomato transplant growth. *HortScience*. 26:141-142.

Mengel, K. 1974a. Plant ionic status. *In* The Plant Root and Its Environment. E.W. Carson, ed. Univ. Press of VA, Charlottesville. 63-81.

Mengel, K. 1974b. Ion uptake and translocation. *In* The Plant Root and Its Environment. E.W. Carson, ed. Univ. Press of VA., Charlottesville. 83-100.

Mengel, K. and E.A. Kirkby. 1982. Principles of Plant Nutrition. International Potash Inst., Worblaufen-Bern. 655 pp.

Millar, R.O., D.C. Kiplinger and H.K. Tayama. 1981. Soluble salts: an overview of negative effects. *Florists' Rev*. 168(4364):10, 26-27.

Minotti, P.L., D.C. Williams and W.A. Jackson. 1969. Nitrate uptake by wheat as influenced by ammonium and other cations. *Crop Sci*. 9:9-14.

Montgomery, G. 1984. A low-sodium diet. *Florists' Rev*. 175(4528):52-54.

Moore, D.P. 1974. Physiological effects of pH on roots. *In* The Plant Root and Its Environment. E.W. Carson, ed. Univ. Press of VA., Charlottesville. 135-151.

Moraghan, J.T. and H.J. Mascagni, Jr. 1991. Environmental soil factors affecting micronutrient deficiencies and toxicities. *In* Micronutrients in Agriculture. J.J. Mortvedt et al., eds. No. 4. Soil Sci. Soc. Amer., Madison, WI.

Mortvedt, J.J. 1991. Micronutrient fertilizer technology. *In* Micronutrients in Agriculture. J.J. Mortvedt et al., eds. 2nd edition, No. 4. Soil Sci. Soc. Amer., Madison, WI.

Moss, P. 1963. Some aspects of the cation status of soil moisture. Part I.: The ratio law and soil moisture content. *Plant and Soil*. 28:99-113.

Nelson, P.V. 1971. Copper deficiency in chrysanthemum: Critical level and symptoms. *J. Amer. Soc. Hort. Sci.* 96:661-663.

Nelson, P.V. 1978. Greenhouse Operation and Management. Reston Publ. Co., Reston, VA. 518 pp.

Nelson, P.V. and W.R. Faber. 1986. Bulk solution displacement. *HortScience*. 21:225-227.

Nelson, P.V. and Kuo-Hsien Hsieh. 1971. Ammonium toxicity in chrysanthemum. *Soil Sci. and Plant Anal.* 2:439-448.

Nieman, R.H. and L.L. Poulsen. 1967. Interactive effects of salinity and atmospheric humidity on the growth of bean and cotton plants. *Bot. Gaz.* 128:69-73.

Nobel, P.S. 1991. Physicochemical and Environmental Plant Physiology. Academic Press, New York. 635 pp.

Noggle, J.C. 1966. Ionic balance and growth of sixteen plant species. *Soil Sci. Soc. Amer. Proc.* 30:763-766.

Norvell, W.A. 1970. Solubility of Fe^{3+} in soils. Ph.D. Dissertation. CO State Univ., Ft. Collins. 100 pp.

Norvell, W.A. 1991. Reactions of metal chelates in soils and nutrient solutions. *In* Micronutrients in Agriculture. J.J. Mortvedt et al., eds. 2nd edition, No. 4. Soil Sci. Soc. Amer., Madison, WI.

Norvell, W.A. and W.L. Lindsay. 1969. Reactions of EDTA complexes of Fe, Zn, Mn, and Cu with soils. *Soil Sci. Soc. Amer. Proc.* 33:86-91.

Nukaya, A., W. Voogt and C. Sonneveld. 1991. Effects of NO_3, SO_4 and Cl ratios on tomatoes grown in recirculating system. *Acta Hort.* 294:297-304.

Oertli, J.J. and O.R. Lunt. 1962. Controlled release of fertilizer minerals by incapsulating membranes: I. Factors influencing the rate of release. *Soil Sci. Soc. Amer. Proc.* 26:579-583.

Olsen, C. 1950. The significance of concentration for the rate of ion absorption by higher plants in water culture. *Physiol. Plant.* 3:152-164.

Olsen, C. 1953a. The significance of concentration for the rate of ion absorption by higher plants in water culture. II. Experiments with aquatic plants. *Physiol. Plant.* 6:837-843.

Olsen, C. 1953b. The significance of concentration for the rate of ion absorption by higher plants in water culture. II. The importance of stirring. *Physiol. Plant.* 6:844-847.

Olsen, C. 1953c. The significance of concentration for the rate of ion absorption by higher plants in water culture. IV. The influence of hydrogen ion concentration. *Physiol. Plant.* 6:848-858.

Parent, L.E. and M. Dafir. 1992. A theoretical concept of compositional nutrient diagnosis. *J. Amer. Soc. Hort. Sci.* 117:239-242.

Peaslee, D.E. 1981. Soil phosphorous. *In* McGraw-Hill Yearbook of Science and Technology. McGraw-Hill, New York. 359-361.

Peck, T.R. 1990. Soil testing: Past, present and future. *Commun. in Soil Sci. Plant Anal.* 21:1165-1186.

Peck, T.R. and P.N. Soltanpour. 1990. The principles of soil testing. *In* Soil Testing and Plant Analysis. R.L. Westerman, ed. Soil Sci. Soc. Amer. No. 3. Soil Sci. Soc. Amer., Madison, WI.

Petrucci, R.H. 1977. General Chemistry. 2nd ed. Macmillan Publ. Co., New York. 790 pp.

Poole, R.T., C.A. Connover and J.N. Joiner. 1976. Chemical composition of quality tropical foliage plants. *Proc. FL State Hort. Soc.* 89:307-308.

Prasad, M., R.E. Widmer and R.R. Marshall. 1983. Soil testing of horticultural substrates for cyclamen and poinsettia. *Commun. in Soil Sci.and Plant Anal.* 14:553-573.

Reddy, K.R. and W.H. Patrick, Jr. 1983. Effects of aeration on reactivity and mobility of soil constituents. *In* Chemical Mobility and Reactivity in Soil Systems. D.W. Nelson et al., eds. Soil Sci. Soc. Amer. Special Publ. 11. Madison, WI. 262 pp.

Richards, L.A., ed. 1954. Diagnosis and Improvement of Saline and Alkali Soils. Agric. Handbook No. 60, USDA, Washington, D.C. 160 pp.

Riley, D. and S.A. Barber. 1969. Bicarbonate accumulation and pH changes at the soybean (*Glycine max* (L.) Merr.) root-soil interface. *Soil Sci. Soc. Amer. Proc.* 33:905-908.

Ritchie, G.S.P. 1989. The chemical behavior of aluminum, hydrogen and manganese in acid soils. *In* Soil Acidity and Plant Growth. A.D. Robson, ed. Academic Press, New York. 306 pp.

Robson, A.D. and L.K. Abbott. 1989. The effect of soil acidity on microbial activity in soils. *In* Soil Acidity and Plant Growth. A.D. Robson, ed. Academic Press, New York. 306 pp.

Roude, N., T.A. Nell and J.E. Barrett. 1991. Nitrogen source and concentration, growing medium, and cultivar affect longevity of potted chrysanthemums. *HortScience*. 26:49-52.
Rovira, A.D. and C.B. Davey. 1974. Biology of the rhizosphere. *In* The Plant Root and Its Environment. E.W. Carson, ed. Univ. Press of VA., Charlottesville. 153-204.
Russell, R.S. and V.M. Shorrocks. 1959. The relationship between transpiration and the absorption of inorganic ions by intact plants. *J. Expt. Bot.* 10:201-316.
Sadasiviah, S.P. and W.D. Holley. 1973. Ion balance in nutrition of greenhouse roses. *Roses, Inc. Bull.* (Supplement). Nov., 1973. 27 pp.
Sadiq, M. and W.L. Lindsay. 1989. Electron titration as a technique to study iron and manganese redox transformation in soils. *Soil Sci.* 147:348-356.
Sajwan, K.S. and W.L. Lindsay. 1986. Effect of redox on zinc diffusion in paddy rice. *Soil Sci. Soc. Amer. Jour.*, 50:1264-1269.
Salisbury, F.B. and C.W. Ross. 1985. Plant Physiology. 3rd ed. Wadsworth Publ., Belmont, CA. 540 pp.
Sanchez, C.A., G.H. Snyder and H.W. Burdine. 1991. DRIS evaluation of the nutritional status of crisphead lettuce. *HortScience*. 26:274-276.
Saranga, Y. et al. 1991. Breeding tomatoes for salt tolerance: Field evaluation of *Lycopersicon* germplasm for yield and dry-matter production. *J. Amer. Soc. Hort. Sci.* 116:1067-1071.
Schekel, K.A. 1971. Effects of increased soluble salt concentrations on carnation growth. *CO Flower Growers' Assoc. Res. Bull.* 251:1-5.
Scher, F.M., M. Dupler and R. Baker. Effect of synthetic iron chelates on population densities of *Fusarium oxysporum* and the biological control agent *Pseudomonas putida* in soil. *Can. J. Microbiol.* 30:1271-1275.
Schmuck, D. and K. Firth. 1987. Groundwater contamination raises liability question. *Amer. Vegetable Grower.* 35:14-15.
Schwab, A.P. and W.L. Lindsay. 1983a. Effect of redox on the solubility and availability of iron. *Soil Sci. Soc. Amer. J.* 47:201-205.
Schwab, A.P. and W.L. Lindsay. 1983b. The effect of redox on the solubility and availability of manganese in a calcareous soil. *Soil Sci. Soc. Amer. J.* 47:217-220.
Schwab, A.P. and W.L. Lindsay. 1989. A computer simulation of Fe(III) and Fe(II) complexation in nutrient solution: I. Program development and testing. *Soil Sci. Soc. Amer. Proc.* 53:29-34.
Shimskey, D.S. 1988. Get the facts on groundwater. *Greenhouse Grower*. 6:54-62.
Shinohara, Y. and Y. Suzuki. 1988. Quality improvement of hydroponically grown leaf vegetables. *Acta Hort.* 230:279-286.
Sienko, M.J. and R.A. Plane. 1957. Chemistry. McGraw-Hill, New York. 621 pp.
Simonds, J.N., R. Swidler and H.M. Benedict. 1962. Absorption of chelated iron by soybean roots in nutrient solutions. *Plant Physiol.* 37:460-466.
Slatyer, R.O. 1961. Effects of several osmotic substrates on the water relationships of tomato. *Austr. J. Biol. Sci.* 14:519-541.
Smilde, K.W. and J.P.N.L.R. van Eysinga. 1968. Nutritional Diseases in Glasshouse Tomatoes. Inst. voor Bodemvrucht., Wageningen. 47 pp.
Smith, P.F. 1962. Mineral analysis of plant tissues. *Ann. Rev. Plant Physiol.* 13:81-108.
Soltanpour, P.N. and R.H. Follett. 1985. Soil test explanation. Service in Action No. 502. CO State Univ., Ft. Collins.
Sonneveld, C. and E. Mook. 1983. Lettuce tipburn as related to the cation contents of different plant parts. *Plant and Soil.* 75:29-40.
Sonneveld, C. and J. van Beusekom. 1974. The effect of saline irrigation water on some vegetables under glass. *Acta Hort.* 35:75-86.
Sonneveld, C. and J. van den Ende. Soil analysis by means of a 1:2 volume extract. *Plant and Soil.* 35:505-516.
Sonneveld, C. and S.J. Voogt. 1978. Effects of saline irrigation water on glasshouse cucumbers. *Plant and Soil.* 49:595-606.
Sonnneveld, C. and W. Voogt. 1991. Effects of Ca-stress on blossom-end rot and Mg-deficiency in rockwool grown tomato. *Acta Hort.* 294:81-88.

Sonneveld, C., J. van den Ende and P.A. van Dijk. 1974. Analysis of growing media by means of a 1:1½ volume extract. *Commun. in Soil Sci. Plant Anal.* 5:183-202.

Steiner, A.A. 1961. A universal method for preparing nutrient solutions of a certain desired composition. *Plant and Soil.* 15:134-154.

Steiner, A.A. 1966. The influence of the chemical composition of a nutrient solution on the production of tomato plants. *Plant and Soil.* 24:454-466.

Steiner, A.A. 1968. Soilless culture. *Proc. 6th Colloq. Int. Potash Inst.* Florence, Italy. 324-341.

Steiner, A.A. 1969. Recipe for a universal nutrient solution. *Gostencild Verslag nr.* 35. Vrentrum voor Plantenfys. Oderz., Wageningen. 4 pp.

Stevenson, F.J. 1991. Organic matter-micronutrient reactions in soil. *In* Micronutrients in Agriculture. J.J. Mortvedt et al., eds. 2nd edition, No. 4. Soil Sci. Soc. Amer., Madison, WI.

Sumner, M.E. 1979. Interpretation of foliar analyses for diagnostic purposes. *Agron. J.* 71:343-348.

Sumner, M.E. 1977a. Use of the DRIS system in foliar diagnosis of crops at high yield levels. *Commun. in Soil Sci. and Plant Analysis.* 8:251-268.

Sumner, M.E. 1977b. Effect of corn leaf sampled on N, P, K, Ca and Mg content and calculated DRIS indices. *Commun. in Soil Sci. and Plant Analysis.* 8:269-280.

Sumner, M.E. 1977c. Application of Beaufil's diagnostic indices to maize data published in the literature irrespective of age and conditions. *Plant and Soil.* 46:359-369.

Sumner, M.E. and M.P.W. Farina. 1986. Phosphorous interactions with other nutrients and lime in field cropping systems. *Adv. in Soil Sci.* 5:201-236.

Takahashi, E. and Y. Miyake. 1982. The effects of silicon on the growth of cucumber plant. *In Proc. 9th Intern. Plant Nutrition Colloq.*, M Scaife ed.. Warwick Univ., U.K.

Terman, G.L. and L.A. Nelson. 1976. Comments on the use of multiple regression in plant analysis interpretation. *Agron. J.* 68:148-150.

Tinus, R.W.and S.E. McDonald. 1979. How to grow tree seedlings in containers in greenhouses. *General Tech. Rpt.. RM-60.* Rocky Mtn. Forest and Range Expt. Sta., USDA. Ft. Collins, CO. 256 pp.

Tjia, B., ed. 1984. Commercial Poinsettia Production in Florida. *FL Cooper. Ext. Serv.*, Univ. of FL., Gainesville, FL. 60 pp.

Twigg, M.C. and C.B. Link. 1951. Nutrient deficiency symptoms and leaf analysis of azaleas grown in sand culture. *Proc. Amer. Soc. Hort. Sci.* 57:369-375.

Ulrich, A. 1960. Plant analysis in sugar beet nutrition. *In* Plant Analysis and Fertilizer Problems. W. Reuther, ed. Publ. 8, Amer. Inst. Biol. Sci., Washington, D.C.

Ulrich, A. and F.J. Hill. 1973. Plant analysis as an aid in fertilizing sugar crops: Part 1, Sugar beets. *In* Soil Testing and Plant Analysis. L.M. Walsh and J.D. Beaton, eds. Soil Sci. Soc. Amer., Madison, WI.

Uren, N.C. and H.M. Reisenauer. 1988. The role of root exudates in nutrient acquisition. *Adv. in Plant Nutrition.* B. Tinker and A. Läuchli eds. 3:79-114.

U.S. Dept. of Interior. 1962. Saline water conversion. U.S. Dept. Interior, Office of Saline Water. U.S. Gov. Printing Off., Washington, D.C. 32 pp.

U.S. Dept. of Interior. 1968. The A-B-Seas of desalting. U.S. Dept. Interior, Office of Saline Water. U.S. Gov. Printing Off., Washington, D.C.

U.S. Dept. Interior. 1979. Electrodialysis technology. U.S. Dept. Interior, Office of Water Res. and Tech. U.S. Gov. Printing Off., Washington, D.C. 15 pp.

U.S. Dept. of Interior. 1979. Reverse osmosis. U.S. Dept. Interior, Office of Water Res. and Tech., U.S. Gov. Printing Off., Washington, D.C.

van den Ende, J. 1971. Extraction methods for the determination of major elements in greenhouse soils and potting and culture media. *Acta Hort.* 29:125-140.

van Eysinga, J.P.N.L.R. 1984. Nitrate and glasshouse vegetables. *Fert. Res.* 5:149-156.

Van Os, E.A., M.N.A. Ruijs and P.A. van Well. 1991. Closed business systems for less pollution from greenhouses. *Acta Hort.* 294:49-57.

van Tuil, H.D.W. 1965. Organic salts in plants in relation to nutrition and growth. *Agric. Res. Rpt.* No. 657. Wageningen.

Vlamis, J. and D.E. Williams. 1962. Ion competition in manganese uptake by barley plants. *Plant Physiol.* 37: 650-655.

Wadleigh, C.H. 1957. Growth of plants. *In* Soil, The Yearbook of Agriculture 1957. A. Stefferud, ed. U.S. Government Printing Office, Washington, D.C.

Wadleigh, C.H. and A.D. Ayers. 1945. Growth and biochemical composition of bean plants as conditioned by soil moisture tension and salt concentration. *Plant Physiol.* 20:106-132.

Wadsworth, G.A. 1980. Methods for sampling peat growing bags. *Acta Hort.* 99:57-65.

Wallace, T. 1953. The Diagnosis of Mineral Deficiencies in Plants. Chemical Publ. Co., New York.

Walworth, J.L. and M.E. Sumner. 1987. The diagnosis and recommendation integrated system (DRIS). *Adv. in Soil Sci.* 6:149-188.

Walworth, J.L. and M.E. Sumner. 1988. Foliar diagnosis: A review. *In* Adv. in Plant Nutrition. B. Tinker and A. Läuchi, eds. Volume 3, pp 193-241.

Walworth, J.L., W.S. Letzsch and M.E. Sumner. 1986. Use of boundary lines in establishing diagnostic norms. *Soil Sci. Soc. Amer. J.* 50:123-128.

Warncke, D.D. 1986. Analyzing greenhouse growth media by the saturation extraction method. *HortScience.* 21:223-225.

Warncke, D.D. 1990. Testing artificial growth media and interpreting the results. *In* Soil Testing and Plant Analysis. R.L. Westerman, ed. Soil Sci. Soc. Amer. No. 3. Soil Sci. Soc. Amer., Madison, WI.

Warrick, A.W. 1981. Soil variability. *In* McGraw-Hill Yearbook of Science and Technology. McGraw-Hill, New York. 359-363.

Watanabe, F.S., W.L. Lindsay and S.R. Olsen. 1965. Nutrient balance involving phosphorous, iron, and zinc. *Soil Sci. Soc. Amer. Proc.* 29:562-565.

Waters, W.E. et al. 1973. The interpretation of soluble salt procedures as influenced by salinity testing procedure and soil media. *Proc. Tropical Region, Amer. Soc. Hort. Sci.* 17:397-405.

Webb, R.A. 1972. Use of the boundary line in the analysis of biological data. *J. Hort. Sci.* 47:309-319.

Welch, R.M. et al. 1991, Geographic distribution of trace element problems. *In* Micronutrients in Agriculture. J.J. Mortvedt et al., eds. 2nd edition. No. 4, Soil Sci. Soc. Amer. Madison, WI.

Weston, L.A. and B.H. Zandstra. 1989. Transplant age and N and P nutrition effects on growth and yield of tomatoes. *HortScience.* 24:88-90.

White, J. 1987. Fertilization. *In* Roses. R. Langhans, ed. Roses, Inc., Haslett, MI. 372 pp.

White, J. 1976. Fertilization. *In* Bedding Plants. J.W. Mastalerz, ed. PA Flower Growers, University Park, PA. 516 pp.

White, R.A.J. and M. Prasad. 1980. Nutrient, salt and pH distribution in soil and in peat modules used for tomato growing. *Acta Hort.* 99:167-178.

Whitesides, R. 1989. El Modeno Gardens: Innovative solutions to California's irrigation runoff restrictions. *Grower Talks.* 52:28-36.

Williams, J.W., J.C. Peterson and J.D. Utzinger. 1988. Liming reactions in sphagnum peat-based growing media. *J. Amer. Soc. Hort. Sci.* 113:210-214.

Winsor, G.W. 1973. Nutrition. *In* The U.K. Tomato Manual. H.G. Kingham, ed. Grower Books, London. 223 pp.

Winsor, G.W. amd D.M. Massey. 1978. Some aspects of the nutrition of tomatoes grown in recirculating solution. *Acta Hort.* 82:121-132.

Wittwer, S.H. and S. Honma. 1979. Greenhouse Tomatoes, Lettuce and Cucumbers. MI State Univ. Press., East Lansing, MI. 225 pp.

Workman, S.M., P.N. Soltanpour and R.H. Follett. 1988. Soil testing methods used at Colorado State University for the evaluation of fertility, salinity and trace element toxicity. CO State Univ. *Agric. Expt. Sta. Tech. Bull.* LTB88-2. 29 pp.

Wright, R.D. 1986. The pour-through nutrient extraction procedure. *HortScience.* 21:227-229.

Wright, R.D., K.L. Grueber and C. Leda. 1990. Medium nutrient extraction with the pour-through and saturated medium extract procedures for poinsettia. *HortScience.* 25:658-660.

Yelanich, M.V. and J.A. Biernbaum. 1993. Root-medium nutrient concentration and growth of poinsettia at three fertilizer concentrations and four leaching factors. *J. Amer. Soc. Hort. Sci.* 118:771-776.

CHAPTER 7

CARBON DIOXIDE

I. INTRODUCTION

Plants produced in greenhouses consist mostly of water –on the average around 90% of their fresh weight. Of the remaining 10%, about half that 10% is carbon [Levanon et al., 1986]. This carbon is derived from the CO_2 in the air, which has a very low concentration of about 340 ppm or 34 Pa (0.034%). The principal purpose of photosynthesis is to combine water and carbon in the presence of a suitable energy supply (solar radiation) to form the basic building blocks of life. The converse of photosynthesis, of course, is respiration, occurring in all organisms always. Carbon compounds within plants are broken down, releasing CO_2 so that the organism can grow and replicate.

Many authors, such as Beer [1986], have pointed out that for a plant to grow, a positive carbon balance must be maintained. The photosynthetic rate must exceed the respiratory rate sufficiently so that total respiration during the night and day does not cause a negative balance. Such a situation can occur under northern European conditions of long nights and low solar radiation (Fig. 4-20). As shown by Challa [1976], cucumbers, when subjected to the usual constant night temperatures, can enter a "starvation" phase under Dutch conditions after 12 hours of dark (Chapter 4). A high net photosynthesis, therefore, for a short period in a single leaf is not always indicative of a profitable crop. If, however, net photosynthesis is expressed by net CO_2 uptake per plant or per unit ground area, then a much better idea of potential productivity can be suggested [Bravdo, 1986]. Most of the basic information on the processes involved deal with single leaves for obvious reasons of simplicity in the experimental process. However, there is considerable information on plant performance where either CO_2 uptake by a whole plant, or several plants in a growth chamber or small greenhouse have been examined. Typical examples include work by Hand and Bowman [1969], Enoch et al. [1973], Krizek et al. [1974], Challa [1976], Bierhuizen et al. [1984], plus several others.

One can appreciate that if CO_2 is absent, growth cannot occur. Even if some CO_2 is present in the air around the plant, it may not be at a high enough concentration to result in a positive carbon balance. If the atmospheric concentration of CO_2 can be increased during the day, the photosynthetic rate may be increased –provided no other factors limit the process (i.e., low temperature, low solar radiation, lack of water supply, or inadequate nutrition). CO_2 enrichment in greenhouses is particularly subject to the requirements of the species being grown, and to the restrictions imposed by climatic conditions of the locality.

It is my objective to review what is known of the general relationships between CO_2 supply and plant response, and the practical application of CO_2 enrichment. Usually, the process of photosynthesis can be divided into three major groups: 1) diffusion processes, or the CO_2 supply from bulk air to the chloroplasts within the leaf; 2) biochemical processes, which include incorporation of CO_2 into the plant's structure, and 3) photochemical processes that convert radiant energy to a form usable by the plant. Knowledge of these processes provides a basic understanding of what may be feasible under practical greenhouse conditions.

II. UNITS AND MEASUREMENT FOR CO_2

Although photosynthetic rate can be determined by measuring dry weight accumulation as the plant grows, this is a destructive measurement. More common is measuring the CO_2 concentration in the air under suitable conditions. This has become a standard procedure in greenhouses and the principal means for precision control –although not vitally necessary where CO_2 enrichment is practiced in a "rough-and-ready" manner for many establishments.

A. UNITS OF CONCENTRATION

CO_2 is a colorless, odorless, and noninflammable gas at room temperature, with a boiling point of -78.5 C.

Table 7-1: Conversions and units for CO_2. Sources include Noble [1991], Mortvedt et al. [1991], Enoch and Kimball [1986], and List [1966][a].

To convert Column 1 into Column 2, multiply by	Column 1	Column 2	To convert Column 2 into Column 1, multiply by
	Pressure		
9.90	megaPascal, MPa (10^6 Pa)	atmosphere, atm	0.101
10	megaPascal, MPa (10^6 Pa)	bar, b	0.1
10^6	megaPascal, MPa	Pascal, Pa	10^{-6}
9.871×10^{-6}	Pascal, Pa	atmosphere, atm	1.013×10^5
0.9871	bar, b	atmosphere, atm	1.013
1.033×10^4	atmosphere, atm	kg m^{-2}	9.68×10^{-5}
10^{-3}	Pascal, Pa	kiloPascal, kPa	10^3
10^{-3}	millibar, mb	bar, b	10^3
0.750	millibar, mb	millimeter mercury, mm Hg	1.333
	Concentration		
10^4	percent, %	parts per million, ppm_v	10^{-4}
1.0	microliters per liter, $\mu\ell\ \ell^{-1}$	parts per million, ppm_v microbars per bar, $\mu b\ b^{-1}$ micromoles per mole, $\mu mol\ mol^{-1}$ cubic centimeters per cubic meter, $cm^3\ m^{-3}$ mole fraction x 10^6	1.0
0.1	microliters per liter, $\mu\ell\ \ell^{-1}$	Pascal, Pa	10.0
44.62	microliters per liter, $\mu\ell\ \ell^{-1}$	micromoles per cubic meter, $\mu mol\ m^{-3}$	0.0224
44.0	nanomoles per cubic centimeter, nmol cm^{-3}. millimoles per cubic meter, mmol m^{-3} micromole per liter, $\mu M\ \ell^{-1}$	milligrams per cubic meter, mg m^{-3} nanograms per cubic centimeter, ng cm^{-3}	0.0227
2.24	millimoles per cubic meter, mmol m^{-3}	Pascal, Pa	0.446
22.4	millimoles per cubic meter, mmol m^{-3}	microliters per liter, $\mu\ell\ \ell^{-1}$	0.0446
0.5092	milligrams per cubic meter, mg m^{-3}	microliters per liter, $\mu\ell\ \ell^{-1}$	1.964
0.0554	nanograms per cubic centimeter, ng cm^{-3} milligrams per cubic meter, mg m^{-3}	Pascal, Pa	18.05
0.554	milligrams per cubic meter, mg m^{-3}	microliters per liter, $\mu\ell\ \ell^{-1}$	1.81
0.0446	microbar, μb	millimole per cubic meter, mmol m^{-3}	22.41

To convert Column 1 into Column 2, multiply by	Column 1	Column 2	To convert Column 2 into Column 1, multiply by
		Photosynthetic rates	
0.044		milligrams per square meter-second, mg $m^{-2}s^{-1}$	22.72
1.584		milligrams per square decimeter-hour, mg $dm^{-2}hr^{-1}$	0.6312
1.584		kilograms per hectare-hr, kg $ha^{-1}hr^{-1}$	0.6312
0.1	micromole per square meter-second, μmol $m^{-2}s^{-1}$	nanomoles per square centimeter-second, nmol $cm^{-2}s^{-1}$	10
1.09		kilograms carbohydrate per hectare-hour, kg $ha^{-1}hr^{-1}$	0.92
8.77		cubic millimeters per square centimeter-hour, mm^3 $cm^{-2}hr^{-1}$	0.114
0.158		grams per square meter-hour, g $m^{-2}hr^{-1}$	6.312

[a] Conversions in this table are assumed to be carried out at sea level and STP (0 C, 101.325 kPa).

Molecular weight is 44 g mol^{-1}, with a density at STP of 1.98 kg m^{-3} and 1.83 kg m^{-3} at 20 C. Table 7-1 outlines the many expressions for CO_2 concentration in air, and also units for pressure and photosynthetic rates. Pressure conversions are included since concentration of any gas can be expressed as a partial pressure. That is, as shown for water vapor in Chapter 5, a gas mixed in air behaves the same regardless of the concentration or kinds of other gases that may be present. As a gas at low concentration, treating CO_2 as an ideal gas is common, following the perfect gas law given by Eq.5.13 (Chapter 5) i.e., PV = nRT where **P**ressure times **V**olume equals the number of moles times the perfect gas constant (**R**) times the absolute **T**emperature. If P, V, and T are known, then the molar concentration can be calculated, n = PV/RT. CO_2 levels in air, as noted in Table 7-1, are often expressed in terms of pressure, Pascals (Pa), microbars (μb), as well as microliters (μℓ $ℓ^{-1}$) or percent. As most research is carried out within a few meters of sea level, corrections for variations in total atmospheric pressure (P_t) are seldom made. Furthermore, over the temperature range commonly found in greenhouse practice, corrections for temperature from standard conditions are infrequently calculated. The footnote in Table 7-1 emphasizes that the conversions are for conditions at sea level. The exceptions would be experiments at Colorado State University at an elevation of 1600 m. In the latter case, a 300 μℓ $ℓ^{-1}$ (ppm) concentration at sea level would be 300/22.414 = 13.38 μmol m^{-3}. The value 22.414 μℓ is the volume occupied by 1 μM at standard temperature and pressure (STP). For an average atmospheric pressure of 84.7 kPa at Fort Collins, the correction to the standard volume is 101.315/84.66 = 1.1967, 1.1967 x 22.414 = 26.819 μℓ. The absolute concentration of a 300 μℓ $ℓ^{-1}$ CO_2 level is 300/26.819 = 11.19 μmol m^{-3} at Fort Collins –a difference of about 16% between sea level and 1600 m elevation. Other than this observation, the range of temperatures and pressures to be encountered in greenhouse practice results in trivial changes in absolute CO_2 concentrations.

B. CO_2 MEASUREMENT

Determination of the photosynthetic rate in plants requires a measure of the CO_2 depletion that occurs under suitable radiant energy, or how much CO_2 that must be injected to maintain a specific CO_2 level. Although CO_2

Fig. 7-1. Schematic of an infrared gas analyzer, showing movement of the sample gas and the radiation paths.

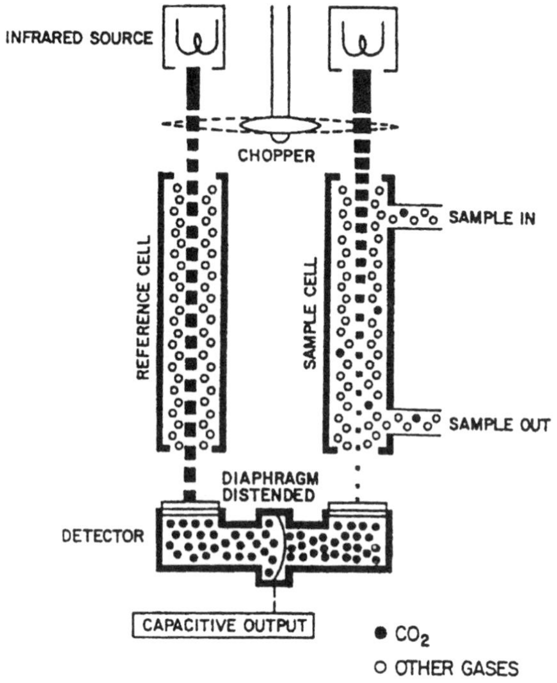

enrichment can be successful without measurement of concentration, newer practices for optimizing greenhouse climate require reliable determination just as one measures temperature, humidity, or radiation [e.g., Schapendonk and van Tilburg, 1984; Schapendonk and Gaastra, 1984a, b; Lake, 1966; Challa and Schapendonk, 1986].

There are numerous methods for measuring carbon dioxide. These include chemical absorption, thermal conductivity, infrared spectroscopy, gas chromatography, and electrical conductivity [Hanan, 1984]. These have been reported in numerous publications, but the principal process now used is infrared spectroscopy or infrared gas analyzers (IRGAs), which are sensitive to the number of CO_2 molecules per unit volume or pressure. The IRGA, according to Pallas [1986], is the most popular and versatile instrument, but more expensive and complicated than instruments based on electrochemistry. They can be used directly with recorders and control devices, and they can be interfaced directly with computers.

Fig. 7-2. Schematic of a nondispersive infrared CO_2 gas analyzer. The system features a solid-state detector and a closed optical pathway.

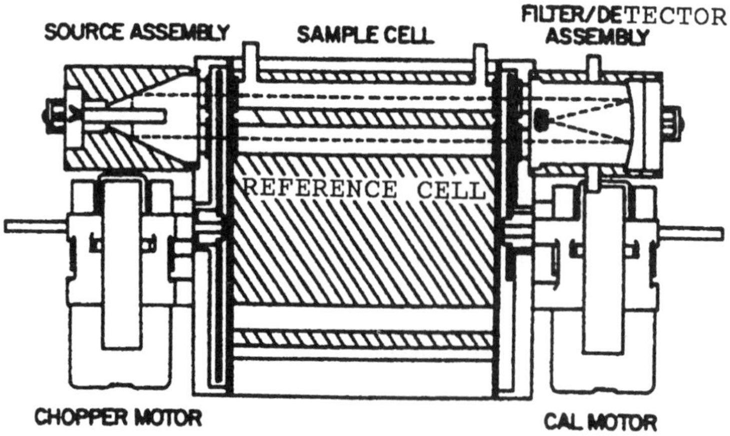

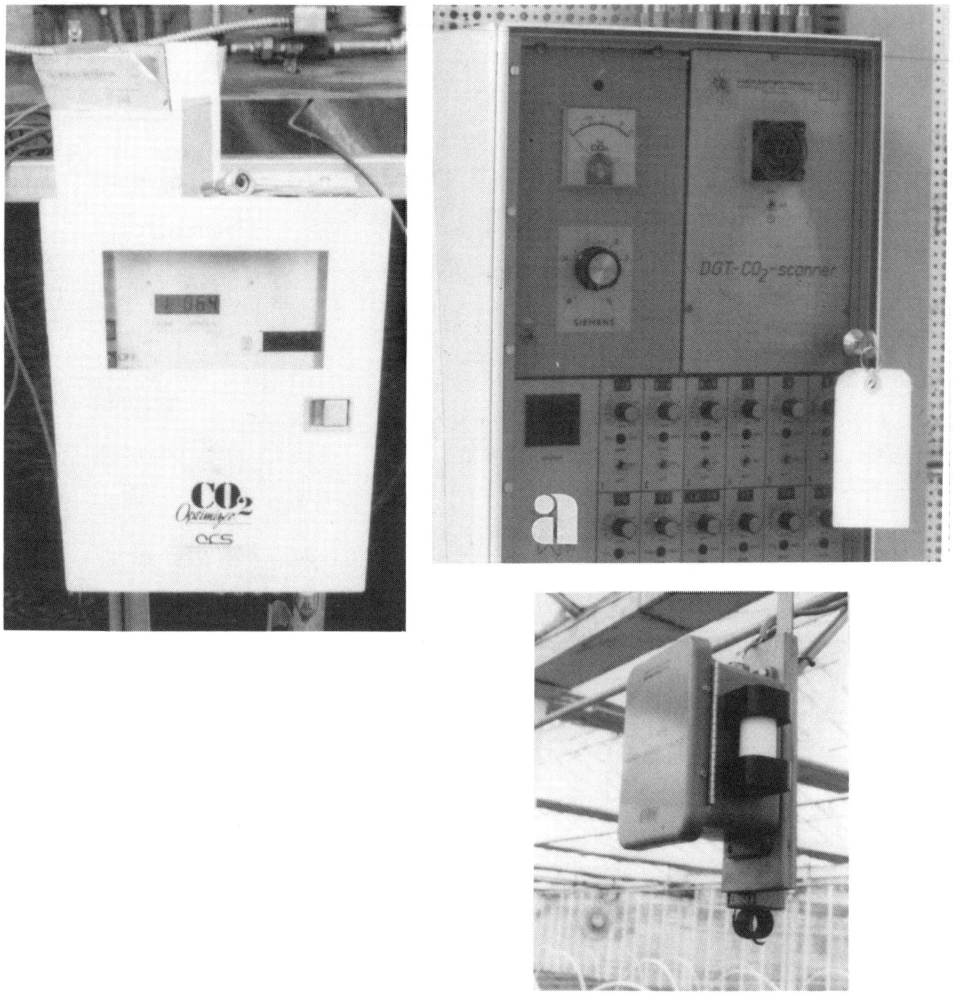

Fig. 7-3. Three examples of CO_2 IRGAs for use in greenhouses. The top left example includes controls for analyses at several locations as well as concentration control. The right picture is another version for measuring several locations. The bottom is a stand-alone analyzer. Location of the latter is not the most desirable from the standpoint of temperature extremes and good sampling.

The principals of operation are shown in Fig. 7-1. There are two balanced tungsten filaments providing infrared radiation, which is modulated by a revolving chopper blade. The radiation passes through two cells, one of which usually contains a fixed reference gas, the other through which the unknown sample to be measured passes. Chambers of the detector cell absorb the remaining radiation that has passed through the cells. The difference in energy due to sample absorption is sensed by a flexible diaphragm, yielding a capacitive output. In some applications, the reference cell may have a different part of the sampling stream passing through it so that the instrument detects the difference between the two. Such a situation could be the fresh air into a plant chamber, the sample cell then passing the air out of the plant chamber. The reduction in concentration would be due to the uptake of CO_2 by the plants in the chamber, and the difference is detected by the IRGA.

The apparatus in Fig. 7-1 employs a mechanical chopper that is subject to wear, and the instrument is prone to calibration shifts due to temperature changes. A more recent variation is an IRGA using a single optical path with two source filaments that are alternately energized (Fig. 7-2). Advantages include electronic versus mechanical modulation with greater stability and selectivity. Most recent instruments have no mechanical parts. Examples of instruments are shown in Fig. 7-3. Although Fig. 7-3 shows an instrument in the greenhouse, most

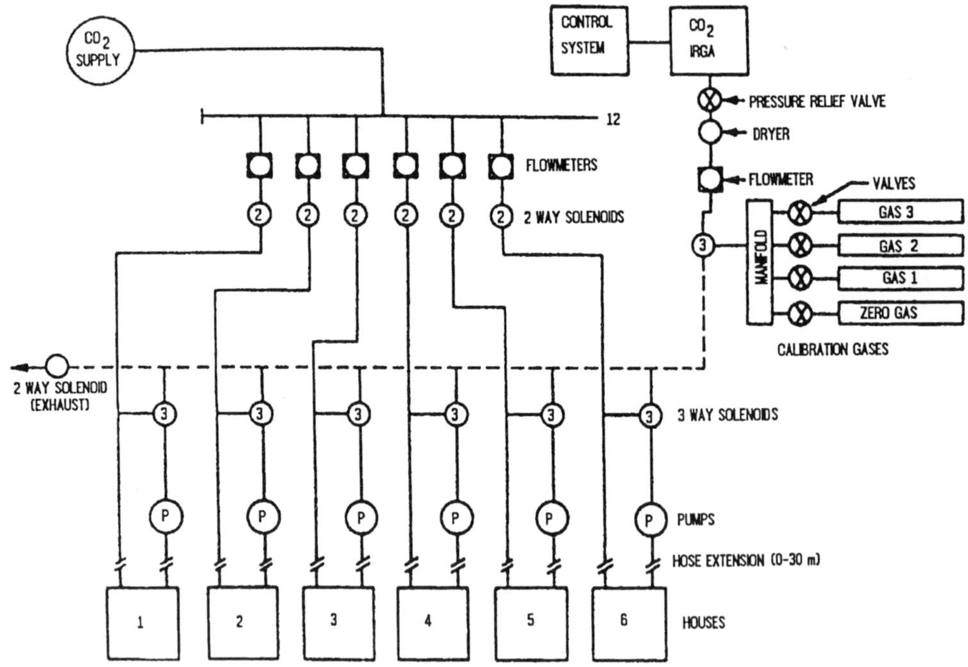

Fig. 7-4. Example of a CO_2 control system [Bailey et al., 1970]. Pumps (P) maintain a continuous air flow from each sampling site, which is wasted unless one of the 3-way valves switches the flow to the dashed line leading to the analyzer. The IRGA is periodically calibrated by a number of known CO_2 concentrations. The 2-way valves inject CO_2 as regulated by the flowmeters upon a suitable signal from control system (With permission of ASAE).

Fig. 7-5. Another example of a CO_2 sampling system. **H1** through **H5** represent sampling sites in separate houses. **E**s are exit ports for exhaust air routed through the 3-way solenoids **SV**s. **SM** is the sampling manifold to the IRGA. **CH** is a switchover valve that switches cells in the IRGA for continuous flowing reference and sampled gases. This reduces calibration drift and the IRGA measures the difference between the **REF** and unknown sample. **M** is a pressure manometer to balance pressures in the two cells of the IRGA. Pumps and filters are not shown [Reprinted from *Agric. and Forest Meteor.*, 40:279-292, Matthews, R.B. et al., ©1987. With kind permission of Elsevier Sci.-NL, Sara Burgerhartstraat 25, 1055 KV Amsterdam, The Netherlands].

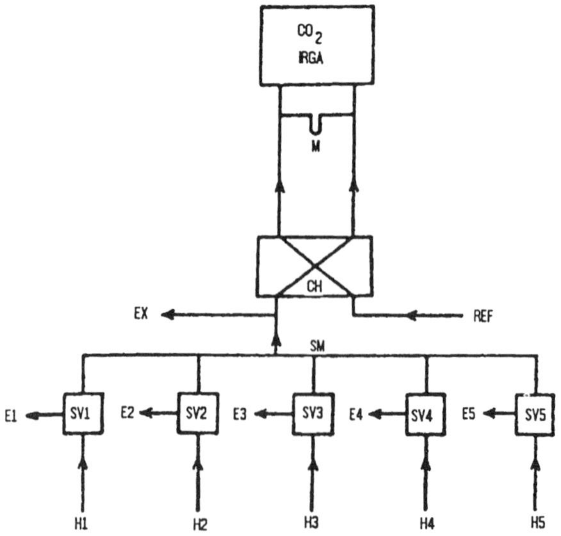

authors do not recommend mounting inside a greenhouse due to temperature extremes and electrical interference. Chemicals employed in pesticide programs may not be suitable for electronic systems. Certainly, sulfur vaporization for mildew control on roses will likely cause damage. Some newer IRGAs are well protected and can often be seen in greenhouses.

Commonly, the analyzer should be in a protected, central location (Fig. 7-3), and the various sites (houses) are sampled by means of pumps and lines to bring the sample to the analyzer, as suggested in Figs. 7-4 and 7-5. Many analyzers contain their own internal pump and filtering system. Nevertheless, where multiple sampling sites are necessary, additional pumps and filters with control valves will be required. Often, recommendations are to dry the air before analysis as water vapor contains infrared absorption bands close to that of CO_2. A desiccant such as silica gel is not recommended as it can both absorb and desorb water. Manufacturers now offer special optical filters [Pallas, 1986; Hand and Bowman, 1969] so the problem of maintaining dryers is eliminated. For precision work, correction for water vapor may be necessary. Pallas recommends against using vacuum pumps to suck the gas stream since this increases leakage danger. Systems can be purchased with simple diaphragm pumps at each sampling location to provide a positive pressure in the sampling lines. As will be noted later, placing the sampling site in the open air above a canopy is not always advisable.

Location and arrangement of sampling lines require special consideration. Most inexpensive plastic tubing is permeable to CO_2, so such material is not recommended, especially where a suction method for moving the gas is employed. Copper tubing of 6 to 8 mm internal diameter is probably the best. If the tubing passes through a temperature region where the gas temperature is reduced below its condensation point, water can condense and block the lines. In sub-zero temperatures, ice can form. Methods to heat and protect the lines become quite elaborate [Matthews et al., 1987] with specialized heating of the tubing to 50 to 60 C. If the rate of air pumping is about 10 ℓ min^{-1}, sufficient to provide a velocity of about 3.0 m s^{-1}, the time for an air sample to travel 100 m will be about 30 sec. Flow rates through an IRGA generally are near 1 ℓ min^{-1}, and 30 sec or so are required for the analyzer to settle before readings are taken. Suitable connection to the average PC allows the timing of valve and analyzer operation to be arranged. The sampling system employed by Nederhoff [1994] consisted of a network of 6 mm nylon tubes with eight inlet points, in the upper half of the plant canopy and connected by a central line to the analyzer. While suitable for research in limited areas, this is complex for a commercial establishment. With the newer controls now available, however, reliable CO_2 measurement and control requires the sampling orifices to be in the main part of the vegetation where interception of solar radiation is the greatest.

Periodic calibration of any type of CO_2 analyzer is required. Small cylinders with known CO_2 concentrations can be purchased. A zero gas (dry air or nitrogen with no CO_2) is necessary, with at least two other calibration gases to provide a curve. IRGA output is not linear so that a single reference, representing the highest concentration, can result in significant error in the midrange. These gases can be connected to the sampling system via pressure-reducing valves that lower pressure to about 25 kPa. The system can recalibrate itself when the operator desires through suitable software and computer interfacing components. Calibration can also be done manually –which may result in neglect. Calibration monthly, if not more often, is mandatory.

III. CO_2 PHYSIOLOGY

The complexity of the photosynthetic process accounts largely for the variable results one may find regarding CO_2 enrichment. Thus, while Kimball [1986b] states that CO_2 enrichment in greenhouses has an overwhelmingly positive effect on yield; his exhaustive evaluation of the many results in Enoch and Kimball's study [Kimball, 1986a] seems confusing. I attribute this to the fact that inclusion of so many reports, coming from so many climatic conditions, under so many variable practices, negates possibility of specific conclusions. The background "noise" is simply too much.

First, there is the problem of CO_2 supply to the leaves in greenhouses where air movement is invariably less than 0.1 m s^{-1} (Section 5.IV.B.3), unless there is deliberate forced-air circulation or the ventilators are wide open and outside wind speed is high. Vegetative canopies are decoupled from the environment during the day with the greenhouse closed (Chapter 5). Second, the biochemical processes are largely influenced by temperature and CO_2 level. Temperature can also influence the photochemical processes, but to a lesser degree. And third, the energy level directly influences the photochemical process, which in turn will affect the CO_2 uptake rate. The multiple reactions must operate in concert. Optimal temperatures for one process may not be optimal for others. There may be water deficits (Chapter 5) that limit CO_2 diffusion through the stomata, or there may be insufficient nutrients available (Chapter 6). One will find contradictory results in application of CO_2 enrichment, and the fact that a particular phenomenon is observed with sunflowers does not mean that the same phenomenon can be observed with cucumbers or roses –even if identical conditions can be imposed. Bravdo [1986] summarized the situation for C_3 plants by stating that the maximum rate of CO_2 assimilation can be obtained in environmental conditions that permit maximum rates of all three major photosynthetic processes.

A. CO_2 FIXATION

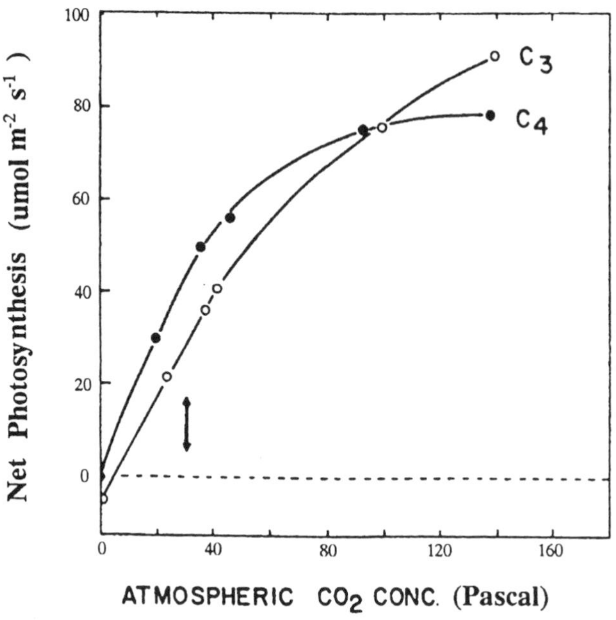

Fig. 7-6. Dependency of net photosynthesis in typical C_3 and C_4 plants on CO_2 concentration. The C_3 curve is dependent also upon O_2 concentration, but not the C_4 curve [adapted from Black, 1986] (Reprinted with permission from Black, C.C. Jr., ©1986, H.Z. Enoch and B.A. Kimball, Eds, Boca Raton, FL).

The photochemical processes and distinction between C_3, C_4 , and CAM plants were briefly discussed in Chapter 3, Section 3.V.B.1. The chapters by Beer, Black and Bravdo [1986] are probably some of the best summaries of carbon fixation to be found in recent years.

All plants use ribulosebiphosphate (RuBP) to fix CO_2 in the chloroplasts to form phosphoglyceric acid (PGA), a three-carbon compound resulting in the name C_3 for plants that use only this pathway. Actually, there are at least six different biochemical pathways known for CO_2 fixation. C_4 plants incorporate CO_2 to form oxaloacetate, malate, or aspartate, which contain four carbons in their skeletons. In separate cells, these compounds are incorporated into the C_3 cycle. A series of reactions are used to move carbon directly to the C_3 chloroplasts. Because of saturating CO_2 concentrations, the enzyme that fixes CO_2 into RuBP does not competitively fix oxygen –which it does in plants with only the C_3 cycle. Therefore, the CO_2 compensation point approaches zero in C_4s; that is, the CO_2 concentration at which the photosynthetic rate equals the respiratory rate (net photosynthesis equals zero). In C_3 plants, the uptake of oxygen results in liberation of CO_2 or "photorespiration," and the CO_2 compensation point must be higher (Table 3-6). Enrichment of the air with CO_2 has a diminishing effect on C_4 plants as suggested in Fig. 7-6. The idea that raised CO_2 levels have insignificant effects on production of C_4 plants important in greenhouse practice, as suggested by some authors [i.e., Ehleringer and Björkman, 1977; Black, 1986; Carlson and Bazzaz, 1982], can be argued and does not appear to be supported by Fig. 7-6. The conclusion that water use efficiency is greater in C_4 plants has no valid application to greenhouse production in view of the discussion in Chapter 5. So-called CAM plants, as indicated in Chapter 3, can fix massive amounts of CO_2 at night as contrasted with others. However, if not subjected to stress, CAM plants can fix CO_2 during the day, especially toward dusk. The large malic acid pool stored in the vacuoles of CAM cells is moved to the cytoplasm the next day and decarboxylated to form CO_2, resulting in a buildup of CO_2 and effectively saturating the CO_2-O_2

enzyme so that photorespiration is low or zero. Black [1986] suggests that CAM plants would benefit from CO_2 enrichment at night. However, I know of no practical studies published on this possibility.

The majority of greenhouse crops are C_3 species. C_4 species grown in greenhouses include *Amaranthus*, *Portulaca*, *Kochia,* and *Gomphrena*. CAM species are more common, including all bromeliads, all cacti, most orchids, and the species *Aloe*, *Kalanchoë,* and *Crassula* [Black, 1986]. With adequate water, some CAM species behave as C_4s. All further information in this chapter deals largely with C_3 species such as tomato, cucumber, lettuce, carnation, roses, etc. Experiments on C_4 and CAM plants are mostly in the European literature, especially the Norwegian work cited in Moe and Mortensen's [1986] publication, the Dutch work by Bierhuizen et al. [1984], and Saxe and Christensen's [1984] on many pot plants.

B. REVIEW OF RADIATION RELATIONSHIPS

Figure 3-49 is a classic illustration from Gaastra's investigations, showing the general effects of environmental factors on photosynthetic rate that hold true for most species in greenhouse production [Gaastra, 1958; 1962]. Variations of this graph can be found in many reviews, texts, and technical publications. Figs. 3-51 through 3-54 and 3-56 are typical for other species, illustrating the effects of temperature, leaf age, and leaf area index (LAI). For the major greenhouse crops, increase of irradiance above the light compensation point (i.e., irradiance where net photosynthesis is zero and CO_2 uptake equals the loss due to "dark" respiration or photorespiration) results in a significant photosynthetic increase at CO_2 levels between 20 to 100 Pascals (Pa).

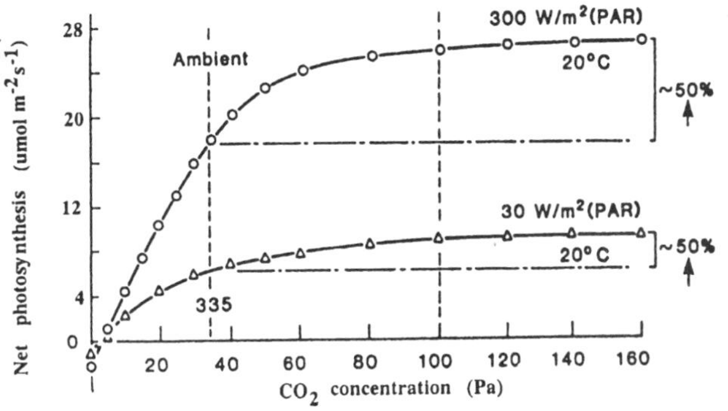

Fig. 7-7. Net photosynthesis versus CO_2 concentration of an attached tomato leaf at two irradiance levels [Hand, 1984] (©1984, Int. Soc. Hort. Sci., *Acta Hort.,* 162:45-63).

At one time, there was some discussion whether CO_2 enrichment would be viable in dark climates. However, most of the species show a variation from Fig. 3-49, as illustrated in Figs. 7-7 and 7-8. Fig. 7-7 shows that for any given energy level, increasing the ambient CO_2 level to about 100 Pa will increase photosynthesis of attached tomato leaves by 50% [Hand, 1984]. Similarly, for roses (Fig. 7-8), as irradiance goes up, the photosynthetic curves for any given CO_2 level begin to deviate almost immediately from each other. Kimball [1986a] emphasized these deviations to be found at low irradiances. While the absolute change at low irradiances may be small, investigation and practical experience have shown CO_2 enrichment to be profitable [e.g., Enoch, 1984; Hand and Cockshull, 1975; Thompson and Hanan, 1975].

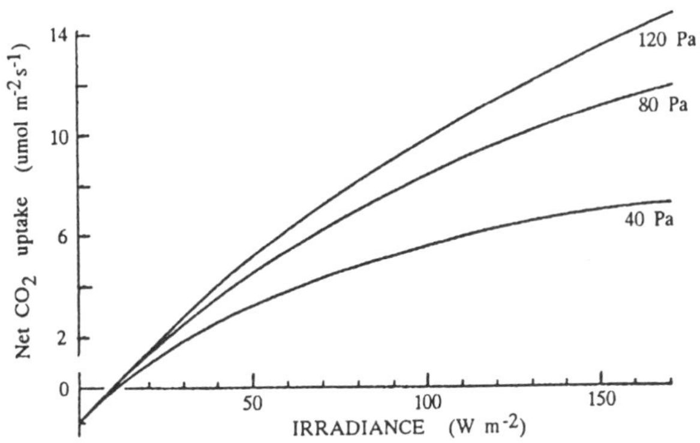

Fig. 7-8. Effect of irradiance and CO_2 level on net photosynthesis in 'Sonia' roses [Calvert and Hand, 1975] (With permission of the British *Grower*).

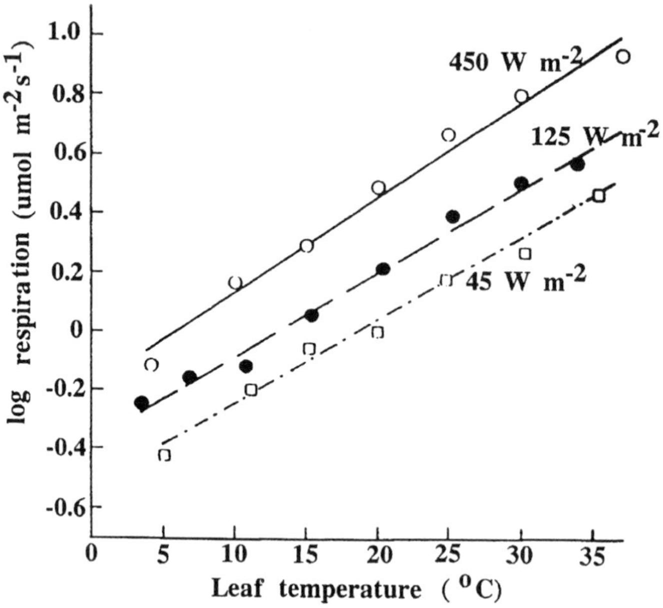

Fig. 7-9. Logarithm of dark respiratory rates in miniature carnation as a function of temperature. The irradiance values were those to which the plants were subjected during the day prior to the measurement the following night [Enoch and Hurd, 1977].

It may be noted that the light compensation point for rose lies below 10 W m^{-2} photosynthetically active radiation (PAR)(46 μmol $m^{-2}s^{-1}$). PAR levels below 30 W m^{-2} are common in northern regions (i.e., northern Europe, eastern U.S., etc.). Hughes and Cockshull [1971] stated that a typical overcast, 8 hr midwinter day in the U.K. provides a total 0.31 MJ m^{-2}. This results in an average 10 W m^{-2} PAR. The compensation point for *Dendranthema*, however, is 0.11 MJ m^{-2}(8 hr dy)$^{-1}$, or an average of 3.8 W m^{-2}. Under these conditions, chrysanthemums are not likely to flower normally. CO_2 enrichment, and supplementary irradiation, are necessary adjuncts to successful culture of such crops as roses, chrysanthemums, carnations, etc. in climates similar to the U.K. The advantages of Mediterranean climates or arid regions as to solar radiation are readily apparent. The saturating irradiance for most species lies between 400 and 600 μmol $m^{-2}s^{-1}$ (80 to 130 W m^{-2}), according to Bravdo [1986]. These values, however, are for single leaves. For a canopy, where the leaf area index (LAI) exceeds 3.0, saturating intensities are not likely below 300 W m^{-2}. As noted in Chapter 3, outside solar irradiances below 300 W m^{-2} in Colorado are likely to be cloudy days. In *Alstroemeria*, Leonardos et al. [1994] found whole-plant saturating intensities to be 1200 μmol $m^{-2}s^{-1}$ (ca. 600 W m^{-2} PAR).

The photosynthetic rate can vary from as low as 4 μmol $m^{-2}s^{-1}$ for *Vriesea* [Bierhuizen et al., 1984] to more than 20 μmol $m^{-2}s^{-1}$ for tomato [Acock et al., 1978], with some plants approaching 60 μmol $m^{-2}s^{-1}$ under suitable conditions. Provided there are no other limiting factors, the relationship between irradiance and CO_2 uptake by most species can be considered as nearly linear and directly proportional to irradiance up to values approaching saturation of the photochemical process.

C. REVIEW OF TEMPERATURE RELATIONSHIPS

The effects of temperature on photosynthesis are more difficult to determine. At low irradiances, the process is limited by energy available, and temperature is unlikely to have much effect (Fig. 3-49). Assuming sufficient energy, particularly at high CO_2 levels, the biochemical process becomes limiting. Temperature becomes an increasingly important factor. However, as pointed out in Chapter 4, there are other processes in crop growth that can directly influence profitability. A temperature optimum for photosynthesis can often be suboptimum for the grower's objectives.

For net photosynthesis, one must take into account respiration, which occurs always. This is often called "dark" respiration to distinguish it from "photorespiration," the latter occurring during the day. The effect of temperature on respiratory rates is direct and fundamental, as noted in Fig. 7-9. There is, in this example with carnation, a pretreatment influence related to the previous day's radiant intensity. That is, the more substrates generated by photosynthesis, the higher the respiratory rate for any given temperature. The classical response, cited in basic physiology texts [e.g., Salisbury and Ross, 1985], is for the respiratory rate to double with a 10 C temperature rise. It can be appreciated that temperatures cannot be increased without compensating action taken

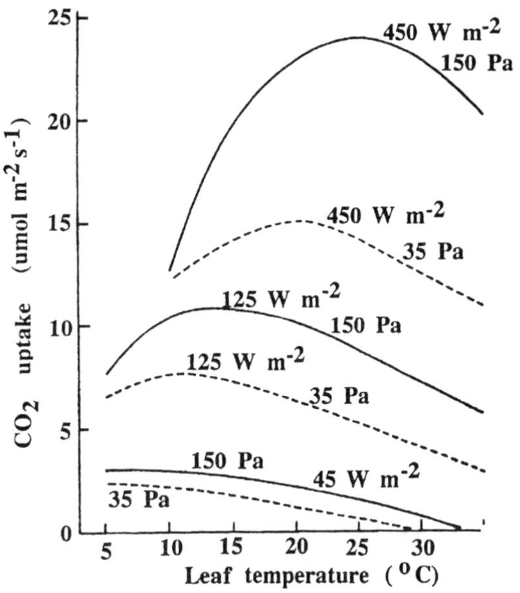

Fig. 7-10. The effect of CO_2 enrichment on net photosynthesis of miniature carnation 'Cerise Royalette' at different PAR irradiances (400-700 nm) [Enoch and Hurd, 1977].

by the grower to improve the total carbohydrates produced –such as CO_2 enrichment, supplementary irradiation, etc.

The carnation is an interesting species, having an unusual growth habit compared to most others, and a photosynthetic rate that is about half other common species [Enoch and Hurd, 1977; Enoch and Sachs, 1978]. Enoch, Hurd, and Sach's efforts also provided one of the best illustrations of the relationships between CO_2 level, temperature, and irradiance; as presented in

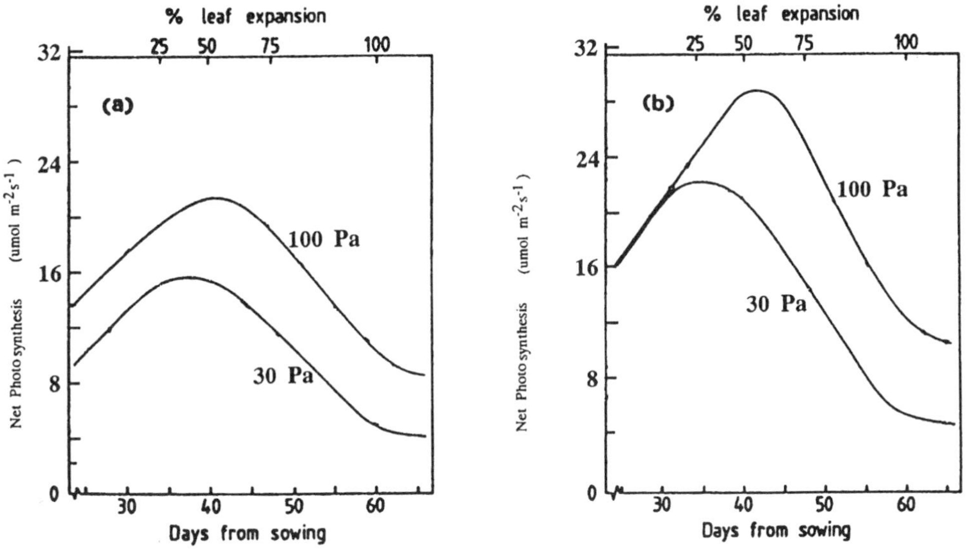

Fig. 7-11. Changes in CO_2 assimilation rate of a tomato leaf at 80 W m^{-2} (**left**) and at 300 W m^{-2} (**right**) at 20 C [Ludwig and Withers, 1984] (Reprinted from *Adv. in Photosynthetic Res.*, Ludwig, L.J. and A.C. Withers, ©1984. 4:217-220. With kind permission of Kluwer Acad. Publ.).

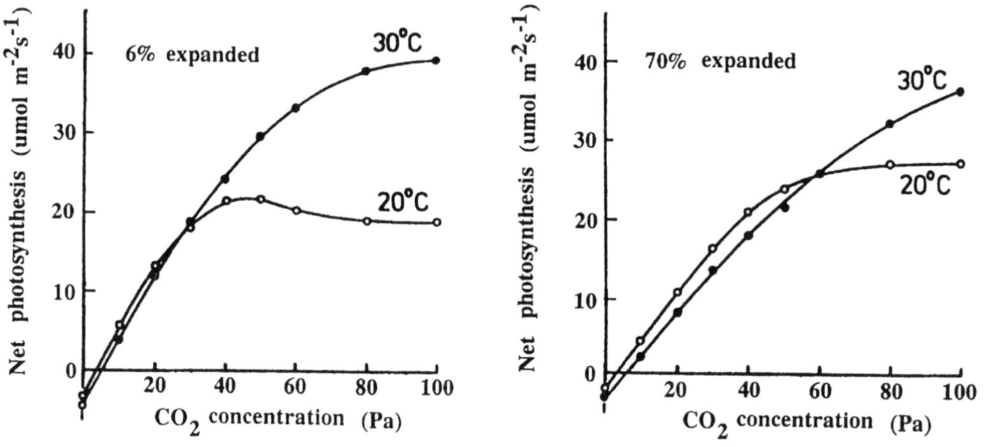

Fig. 7-12. Effect of CO_2 level on assimilation rate of a tomato leaf at two temperatures when the leaf is 6% expanded versus one that has expanded 70% [Ludwig and Withers, 1984] (Reprinted from *Adv. in Photosynthetic Res.*, Ludwig, L.J. and A.C. Withers, ©1984. 4:217-220. With kind permission of Kluwer Acad. Publ.).

Fig. 7-10. For a single leaf, at 45 W m^{-2}, net photosynthesis was highest at the lowest leaf temperature of 5 C, whereas at the highest irradiance, the optimal temperature lay between 20 to 25 C. Temperatures can be increased to enhance growth, especially if CO_2 enrichment can be employed. However, temperatures greater than 20 C, under conditions of high irradiances (summer), with concomitant high CO_2 levels, cannot usually be achieved in commercial greenhouses. It becomes necessary to ventilate and the grower must make do with ambient CO_2 levels of about 34 Pa. Only under suitable climatic conditions of low outside temperatures can increased CO_2 concentrations be achieved under practical conditions. The situation faced by a grower is to achieve sufficient net photosynthesis by his crop so that he may use temperature to obtain suitable growth by using such manipulations as outlined in Chapter 4 (i.e., average daily temperatures (ADT) and difference in temperature (DIF). The ability to do so is closely related to the local climatological conditions.

D. EFFECTS OF LEAF AGE AND NUTRITION

The variations in CO_2 uptake by single leaves as the result of age were discussed earlier (Fig. 3-53). Of particular interest, however, were Ludwig and Withers' results on tomatoes published in 1984. Increasing the CO_2 level from 30 to 100 Pa increased net CO_2 uptake of leaves exposed to 80 W m^{-2} by 4.5 to 6.8 μmol $m^{-2}s^{-1}$ at most stages of leaf development (Fig. 7-11). While CO_2 uptake at 300 W m^{-2} was higher at both concentrations, differences between CO_2 levels were not significant until the 5th leaf had expanded 25%. Afterwards, response was similar to that measured at 80 W m^{-2}. The effects of temperature as a function of CO_2 level are depicted in Fig. 7-12 at two different ages. At 300 W m^{-2}, temperature did not affect net photosynthesis until CO_2 concentrations exceeded about 45 Pa. As the leaves became older, the concentrations at which the two temperature treatments deviated from each other increased. This response was not observed when experiments were conducted at 80 W m^{-2} for either 20 or 30 C treatments, nor in older leaves where photosynthesis increased as CO_2 rose to 100 Pa. As another example, Kelly et al. [1991] found elevated CO_2 enhanced photosynthesis to the greatest extent in middle-aged geranium leaves, whereas it was depressed in the oldest leaves. The effect of nutrition on photosynthesis under elevated CO_2 levels is ambiguous. Thus, Skoye and Toop [1973] concluded that nutrient levels were of minor importance in influencing the effects of temperature or CO_2 level on potted chrysanthemums. Increasing nutrient concentration on tomatoes resulted in lower increases in yield at high CO_2 compared to using a standard nutrient solution [Kimball and Mitchell, 1979]. There appeared to be some difficulty with blossom-end rot and leaf curling during these investigations in the high light Arizona climate.

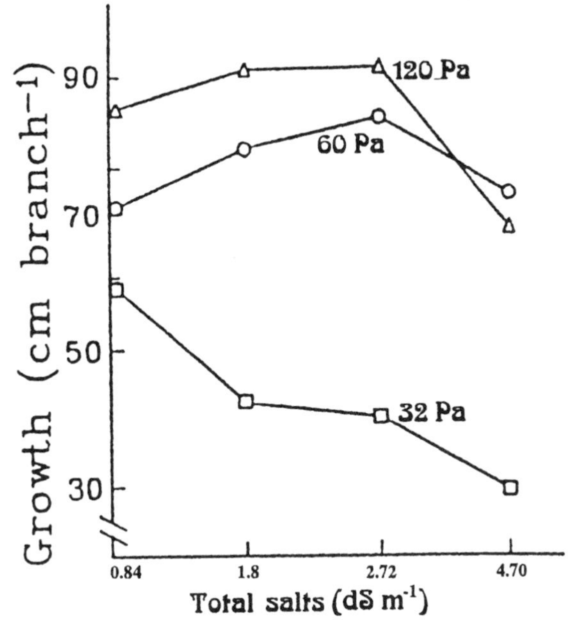

Fig. 7-13. Flowering cane length of 'Sonia' roses at four salinity levels and three CO_2 levels [Zeroni and Gale, 1989] (Reproduced by permission of the *Journal of Horticultural Science*).

Yelle et al. [1987] found increased N and K uptake in tomatoes grown at 80 Pa CO_2 with significant interaction with the root zone temperature. The improvement of shoot growth with CO_2 enrichment increased with increased root zone temperature up to 30 C. Root growth decreased. It also appears that total salinity affects the response of roses to elevated CO_2. In fact, Zeroni and Gale [1989] concluded that 'Sonia' can "tolerate" salinity up to 2.7 dS m^{-1} if CO_2 levels can be increased to at least 120 Pa (Fig. 7-13). Flowering canes on roses grown in solutions of 2.7 dS m^{-1} were 7.1% longer than those grown at high CO_2 alone, and 44.3% longer than those grown in solutions having electrical conductivity levels of 0.84 dS m^{-1} and ambient CO_2 levels.

In my estimation, based upon the discussion on nutrient uptake in Chapter 6, studies involving various macronutrient concentrations and effects of elevated CO_2 levels are likely to be unproductive. Plants adjust nutrient uptake in response to the carbohydrate production rate through active ion uptake and transpiration rate –if the necessary ions are freely available at the root surface. However, the next section deals with some problems of nutrient uptake and actual CO_2 toxicity.

E. EFFECT OF CO_2 CONCENTRATIONS

1. Limiting Values for CO_2

The effect of CO_2 levels on photosynthesis has been shown in many illustrations (e.g., Figs. 7-6 through 7-8, 7-10, and graphs in previous chapters such as Figs. 3-49, 3-64, and 4-10). One notes that the average ambient concentration of CO_2 (34 Pa) lies on that part of the CO_2 uptake curve that is steepest. The portion is linear, or nearly so. Changes in CO_2 level cause a direct, proportional response in photosynthesis that is characteristic for the particular species. Thus, a reduction in CO_2 supply below 34 Pa is likely to have a disproportionate effect on net photosynthesis and consequent growth, especially if available energy approaches saturation. Of course, with C_3 plants, concentrations approaching the CO_2 compensation point of 5 to 6 Pa are an extremely serious problem.

Levels as high as 500 Pa (5000 $\mu\ell\ \ell^{-1}$) have been investigated. Usually, however, such high levels are likely to result in plant injury, and at lower concentrations, to such species as cucumber, tomato, and eggplant [van Berkel and van Uffelen, 1975; van Berkel, 1984b; 1986; Nederhoff, 1994; etc.] and gerbera [van Berkel, 1984a; 1986]. Gerbera may show foliage yellowing at CO_2 levels less than 50 Pa. Van Berkel [1984a] discussed a number of symptoms of excess CO_2 such as leaf roll in tomato, chlorosis and necrosis in chrysanthemum, chlorosis in the oldest leaves of cucumber, and similar problems in the top leaves of tomato. Nederhoff et al. [1992] and Nederhoff [1994] examined the "short leaf syndrome" (SLS) in tomatoes that occurs especially in the summer under Dutch conditions with CO_2 enrichment to ambient concentrations, and leaf-tip chlorosis in eggplant. Nederhoff and Buitelaar [1992] attributed leaf-tip chlorosis (LTC) in eggplant to boron deficiency aggravated by high CO_2 and reduced stomatal conductance. Therefore, the lower transpiration rate reduced boron uptake. Nederhoff [1994] concluded that CO_2 levels should not exceed 70 Pa on spring eggplant crops. Nederhoff and associates suggested that SLS resulted from a calcium deficiency in the apex [1992]. The benefit of CO_2 injection, according to Nederhoff and co-workers, depends upon SLS severity, which can be reduced by

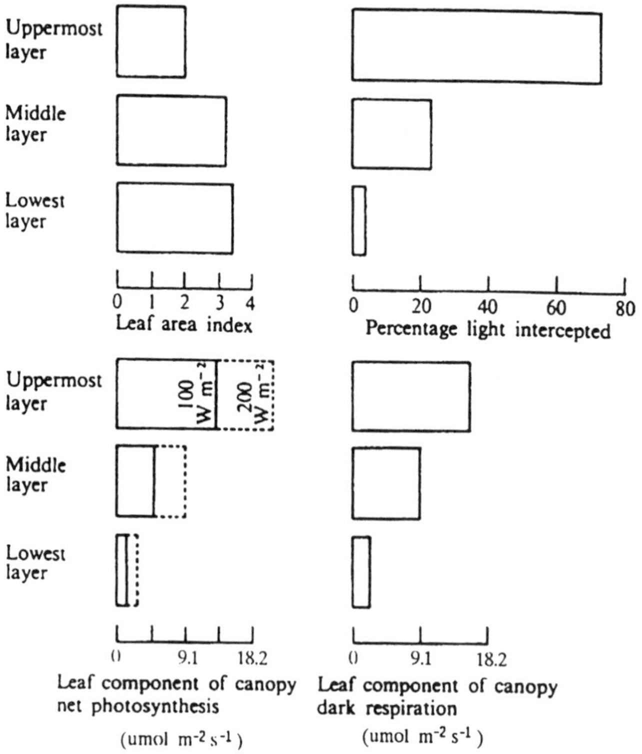

Fig. 7-14. Leaf area index, percentage of irradiance intercepted, net photosynthesis, and dark respiration of three layers of a 1.1 m tall, closed tomato crop with a total LAI of 8.6, divided into three layers [Acock et al., 1978]. The net photosynthetic and respiratory rates are based on a per unit ground area.

increasing plant density during the summer or permitting an extra shoot on the plants in spring and summer. Pfeufer and Krug [1984] also addressed injury at extremely high concentrations of 3000 Pa (3%); 500 Pa (0.5%) CO_2 represents the maximum concentration limit set by the U.S. Occupational Safety and Health Administration [Hanan, 1984]. At 3000 Pa, leaves in spinach were lighter green, corn salad produced small, thick leaves with an upward rolling, radish showed chlorosis beginning at the leaf margins, whereas the leaves in lettuce rolled downward, and one saw in kohlrabi a blackening of thin veins, cellular collapse, chlorosis, and withering. The

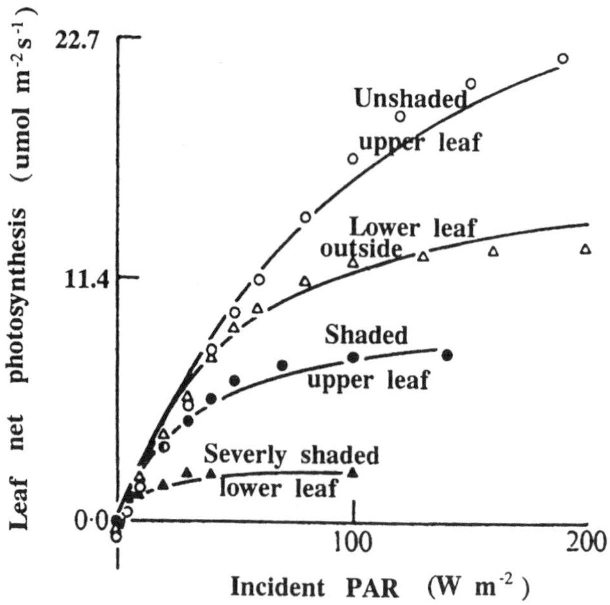

Fig. 7-15. Measured response of net photosynthesis in leaves from various positions to incident irradiance on a closed tomato canopy 1.1 m tall [Acock et al., 1978].

effect was more apparent on warm growing crops compared with the cool crops mentioned. Such high CO_2 levels are utterly uneconomic.

Leaf roll of tomato can be induced by many causes such as high irradiances, viruses and abnormal nutrition plus high CO_2. Van Berkel [1984a] cited the literature to the effect that high CO_2 causing leaf roll was correlated with high starch levels in the chloroplasts. With *Dendranthema*, chlorosis, downward leaf rolling, and necroses have also been associated with excess starch caused by high CO_2. In cucumber, van Berkel stated that damage at 150 Pa CO_2 took place particularly when a spell of overcast weather was followed by bright sunshine, resulting in chlorophyll breakdown and death of the older leaves. Problems of wilting of carnations on bright days following several dark days was mentioned in Chapter 5 –but, this does not usually result in obvious leaf damage. The problem also occurs without CO_2 enrichment. Such difficulties can be found with other species, resulting from severe water deficits as outlined above. Gerbera show particular problems of severe yellowing on the oldest leaves of sensitive cultivars at concentrations above 80 Pa CO_2 concentration. In eggplant, chlorosis begins at the leaf tip, gradually covering the whole leaf with eventual leaf loss [Nederhoff, 1994]. Eggplant has a remarkably sensitive stomatal closure to high CO_2 which might result in less translocation of calcium and boron.

While there have been innumerable studies at CO_2 levels above 100 Pa, and there have been significant responses, it is now considered that 100 Pa (1000 ppm) represents a reasonable, maximum concentration except for such species as gerbera, cucumber and eggplant. This corresponds to the threefold concentration increase recommended by Hand [1984, 1988], and generally accepted as a realistic goal to be achieved under suitable conditions for most species in greenhouse production. But, maintaining these limits with certainty requires suitable analysis and control.

2. CO_2 Diffusion

One is faced with the fact, unfortunately, that CO_2 must diffuse through the boundary air layer surrounding a leaf, through the stomates, and eventually through the water in the cellular tissues to the site of CO_2 fixation in the chloroplasts. The processes regarding water vapor were discussed in Chapter 5. The situation for the surface of a single leaf was suggested in Fig. 5-90. In general, photosynthesis can be described as the result of a gradient in CO_2 concentration between leaf and air, divided by the total resistance to CO_2 diffusion:

$$P = \frac{[CO_2]_{ext} - [CO_2]_{int}}{r_a + r_s + r_m} \tag{7.1}$$

where the brackets ([]) denote the external and internal CO_2 concentrations, and r_a, r_s, and r_m are the boundary, stomatal, and mesophyll resistances (s m^{-1}), respectively. Many variations of this formula may be found in the literature [e.g., Bierhuizen, 1973; Gaastra, 1962; 1966]. Conductances (m s^{-1}) can also be used, and a variation presented by Schapendonk and Gaastra in their seminal 1984 article [1984b] is given in Eq.7.2:

$$P = \frac{\alpha R[CO_2]_{ext} g}{\alpha R + g[CO_2]_{ext}} \tag{7.2}$$

where: P = *gross photosynthesis (mg $m^{-2}s^{-1}$)*
α = *quantum efficiency for CO_2 assimilation (mg J^{-1})*
g = *conductance to CO_2 (m s^{-1})*
R = *irradiance (PAR) (W m^{-2})*
$[CO_2]_{ext}$ = *CO_2 concentration outside the leaf (mg m^{-3})*

As with resistance, conductance includes boundary, stomatal, and mesophyll conductances. Many authors consider mesophyll resistance to be the major impediment in the CO_2 pathway because the diffusion of the gas through water is 10000 times slower than through air. Bierhuizen [1973] stated that r_m could vary from 100 to 1000 s m^{-1}, whereas wide-open stomata have resistances on the order of 100 s m^{-1}. Bravdo [1986] suggested mesophyll values of 400 to 1500 s m^{-1} for common C_3 plants having net photosynthetic rates between 15 to 60 µmol $m^{-2}s^{-1}$. The boundary layer resistance (r_a) is often considered minor compared to r_m [Bravdo, 1986]. Gaastra [1962], for an average leaf, calculated boundary resistances of 330, 170, 90, and 40 s m^{-1} for wind speeds of 0.16, 0.42, 1.0, and 3.0 m s^{-1} respectively. Even with fan-and pad cooling, wind speeds within greenhouses are unlikely to exceed 0.3 m s^{-1} (Section 4.IV.C.2). On the other hand, Shaer and van Bavel [1987] concluded that their boundary layer resistance value was 200 s m^{-1} at my calculated windspeed of 0.04 m s^{-1}. It was decided by Shaer and van Bavel that the boundary layer resistance dominated gas exchange of their tomato plants.

The situation concerning diffusion of vapor was given considerable attention in Section 5.IV.B.3. The low air velocities in greenhouses were emphasized, which, in effect, decouple vegetative canopies from their immediate surroundings. The same general relationship for CO_2 also holds. Most authors have found that increasing CO_2 concentrations tend to close stomata [e.g., Goudriaan and Ajtay, 1979; Shaer and van Bavel, 1987; Bravdo, 1986; Gaastra, 1962; Enoch and Hurd, 1977; etc.]. Fitted equations, calculated by Nederhoff in her dissertation, showed that a 10 Pa CO_2 increase reduced conductivity by 3 to 4% in sweet pepper, cucumber and tomato, and about 11% in eggplant. Reduction in transpiration can be considered as small or negligible at moderate CO_2 levels. At low winter radiation levels, however, any reduction of transpiration as to nutrient uptake could be highly significant. Raschke was cited in Chapter 5 as stating that, if plants were grown under minimal water stress, stomata were likely to be insensitive to CO_2 levels. Also, Figs. 5-94 and 5-95 showed that internal plant water potential must decrease significantly before stomata close to any marked degree. Noble [1991] concluded that although stomatal closure could affect transpiration significantly, the effect on CO_2 uptake would be marginal. In fact, determining resistance to CO_2 directly is difficult. In most cases, boundary layer resistance is determined by exposing a piece of wet paper in the shape of the leaf and calculating the water loss with time. Stomatal resistance to water vapor is determined with suitable instrumentation and the r_m calculated by solving Eq.7.1 for mesophyll resistance. The resistance to CO_2 can be calculated by accepting that the ratio of resistance of water to resistance of CO_2 is 1.56 [Slatyer, 1971]. In other words, resistance to transfer of CO_2 is 1.6 times less compared to vapor.

Nederhoff and de Graaf [1993] explored the effect of CO_2 level on conductance and transpiration from cucumber and tomato. The relative effect of CO_2 increase caused a 3 to 4% decrease in conductance. The effect on transpiration was smaller. In Chapter 5 the general relationship between crop canopy and the greenhouse environment was outlined, showing that a crop is largely decoupled from its environment when dealing with transpiration. In Nederhoff and de Graaf's study, strong decoupling was suggested by the low ratio of relative change in transpiration versus relative change in conductance (0.22). The authors concluded that while the effect of elevated CO_2 on conductance (1/resistance) was significant, the effect on transpiration was small and usually negligible. While there may be some differences to be found in the literature, I think the results affirm my previous conclusions; that in well-watered, full canopies in greenhouses, stomatal resistance is not important and the effect of enhanced CO_2 levels, therefore, does not affect water loss.

As indicated in Chapter 5, the ideal situation for productivity is to maintain stomata open by assuring that internal water potential is as high as possible commensurate with irradiance, temperature, and product quality. Furthermore, increasing air velocity through the plant canopy reduces boundary layer resistance. This is, in effect, an increase in CO_2 concentration. Undoubtedly, the improvement in summer carnation quality in Colorado with the introduction of fan-and-pad cooling resulted, mostly, from improvement of CO_2 supply caused by forced air movement. Such practices as horizontal air flow (HAF) would also enhance CO_2 supply by reducing boundary layer resistance.

F. THE VEGETATIVE CANOPY AND CO_2 UPTAKE

It is one thing to consider CO_2 uptake by a single leaf. It is another thing entirely to encompass the entire canopy of a greenhouse crop. Section 5.III.A briefly discussed radiant interception by canopies at an elemental level. Besides the theoretical considerations to be reviewed in the next chapter, and work with agronomic crops [e.g., Loomis and Williams, 1969; Monteith, 1969; Norman, 1979; etc.], knowledge of radiant interception by the many vegetative canopies to be found in greenhouse production is sadly lacking. The work of Stanghellini, Bot, and others in the Dutch research effort is an exception, as is the outstanding contribution by Acock et al. [1978]. An appreciation of where, in a canopy, the greatest photosynthesis occurs also identifies the largest sink for CO_2 uptake. In view of the recent methods to maintain ambient CO_2 levels even under full ventilation, it would appear that CO_2 should be injected in that same region of the canopy for maximum efficiency.

Acock et al. took a fully mature tomato canopy with an LAI of 8.6 and divided it into three layers, containing 35 plants, 1.1 meters tall, and surrounded by a green, plastic shade to simulate shading by adjacent plants. Fig. 7-14 summarizes the results, showing that the greatest amount of radiation intercepted was confined to the upper third of the canopy, which only had 23% of the total leaf area; 66% of the net CO_2 fixed by the canopy was in the upper third. Removal of the leaves in the lowest third had little effect on dark respiratory or photosynthetic rates. Removal of the middle leaf layer reduced canopy photosynthesis by 33% and dark respiration by 25%. Fig. 7-15 presents photosynthetic curves of individual leaves from different positions in the canopy at 40 Pa CO_2. Results also showed that respiratory rates and conductances of leaves from different levels within the canopy were very different and were associated with the radiant environment rather than differences in leaf age.

As will be noted later, merely injecting CO_2 into an enclosure does not mean that it will be uniformly distributed. Even injection at ground level in benches is no assurance that the gas will affect photosynthetic rates in that part of the canopy where the greatest light interception occurs.

G. CROP RESPONSE TO CO_2 ENRICHMENT

Up to now, discussion has dealt with photosynthetic processes in individual leaves, with some information on whole plant and crop net photosynthesis. It is the carbon balance of the plant that is important, which can be assessed by determining the dry weight increase per unit land area, or kilograms per hectare [Schrader, 1976]. However, few greenhouse species are marketed as dry weight such as cereal crops. The crops are sold fresh where the fresh weight can be the determining yield factor, or, it may be the number of fruits, flowers, size of the pot plant, stem strength of cut flowers, fruit, or flower quality, or the production of a marketable product in as short a time as possible. Furthermore, the crop response to CO_2 enrichment is not immediately apparent, and, unless there is a comparison to an unenriched control, growers may not observe a marked response except as noted in their balance sheets. Given the usual variability in weather, results can change from crop to crop, season to season, and year to year. It is a fact that CO_2 levels approaching 100 Pa can only be economically achieved when the greenhouse is closed. The crop responses to elevated levels are functions of outdoor temperatures, the temperatures required by the crop inside, cultural procedures employed, the method of CO_2 enrichment, and the duration of the period at elevated CO_2 levels.

Much of the practical information on yield responses to CO_2 enrichment comes as percentage differences between enrichment and a non-enriched control, which can range from less than 10% to more than a 100% improvement in whatever parameter compared. Many publications carry summaries of this type of data [e.g., Kimball, 1986b; Heins et al., 1984; Hicklenton, 1988; Shaw and Rogers, 1965; Goldsberry, 1986; etc.]. Table 7-2 lists the economic crops on which reports have been published. Of other crops that may be important, Moe and Mortensen [1986] list no response to raised CO_2 by tulips, and Wilkins [1980] states that CO_2 enrichment for Easter lilies (*Lilium longiflorum*) is not recommended. Rogers [1980] cited the literature to the effect that snapdragon groups I or II resulted in unsalable flowers when treated to elevated CO_2 levels, whereas cultivars in groups III and IV produced excellent cut flowers. Typically, the success of enrichment as to improved yield and growth will depend upon where the work was carried out. Thus, Shaw and Rogers [1965] found that CO_2 enrichment on carnations (a low-temperature crop) in Missouri was not particularly successful as compared to the work conducted in Colorado [Goldsberry, 1963; Holley, 1964; 1970; Holley et al., 1962]. Holley [1970] stated that fall and winter CO_2 additions resulted in yield increases the following March to June. Therefore, CO_2 enrichment in April-May would not be desirable since the resulting crop would be harvested in a low-price period. Yield response to enrichment was not apparent on carnations until about 5 months after the start of enrichment. Sometimes, there have been reports of negative responses [Hesketh and Hellmers, 1973], but not, as far as I know, with commercial greenhouse crops listed in Table 7-2. One has to be careful since problems with CO_2 enrichment may easily have resulted from pollution from the CO_2 source, excessive CO_2 concentrations, or disease as the result of high temperatures combined with high humidity.

Cultivars are listed since there can be extreme variability by different cultivars to identical treatments (Fig. 7-16). Proper selection of varieties grown can make very significant differences in profitability. Those listed in some places of Table 7-2 may no longer be available. For example, 'Better Times' rose is probably not grown to any great extent. Information of this type is not often found in the scientific literature. Commercial breeders may examine response of new cultivars, or the information may be published in regional newsletters and in the local language. It can be appreciated that such information is of immense value to the grower.

Some typical responses are illustrated in Figs. 7-17 and 7-18. If CO_2 enrichment is provided from the

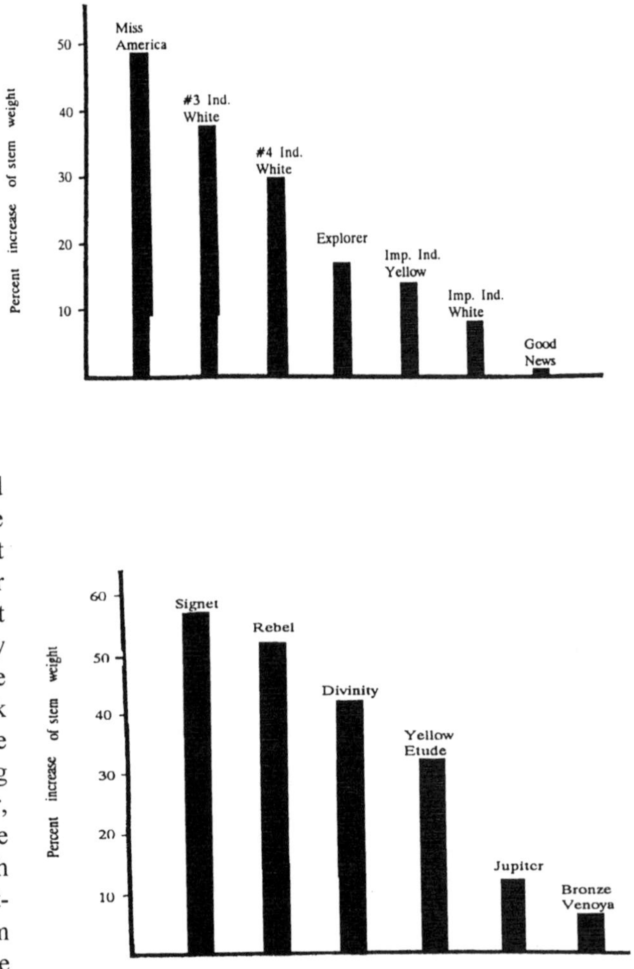

Fig. 7-16. Differences in response of standard (**upper**) and pompon (**lower**) chrysanthemum varieties to CO_2 enrichment expressed as the percentage increase of stem weight compared to an unenriched treatment [Shaw and Rogers, 1965] (With permission of the *Florists' Review*).

seedling stage (pre-planting), through the entire production cycle, the enriched crop will usually begin to produce earlier, and the increased yield will gradually accumulate with time as compared with an unenriched control. Later crops in species such as tomato and cucumbers can show a decline in fruit production although the accumulated yield will remain higher than unenriched crops. In a crop such as lettuce, where the improved photosynthesis is directly related to a harvestable product (leaves), improved yield and earliness due to CO_2 enrichment are remarkable. CO_2 application apparently does not affect lettuce nutrient content, although later harvests showed greater carbohydrate content before an increased growth rate near maturity [Knecht and O'Leary, 1983]. With roses, where production of hybrid teas can be cyclic, the peak yields will be higher and often earlier, so the grower must take into account changes in timing for high-price periods (e.g., Christmas, Easter, etc.). A striking characteristic with roses is the significant reduction in bud abortion with elevated CO_2 [Zieslin and Halevy, 1972]. Cut-flower chrysanthemums will show longer stem length unless the grower makes appropriate changes in pinching and photoperiodic timing. The same is applicable to potted and bedding plants. Growth will be quicker to reach a saleable size. Generally, CO_2 enrichment will increase flower size and improve stem strength. Reports of weak stem strength on carnations in Colorado's winters were usually from those operations that did not employ CO_2 injection. In Europe before the 1960s, the common practice was to allow carnations and roses to go nearly dormant. Hand and Cockshull [1975] addressed the problem, pointing out that newer cultivars such as 'Sonia' greatly improved the ability to economically produce roses in U.K.'s midwinter, followed by greatly improved production with CO_2 enrichment.

Responses to elevated CO_2 will obviously vary with water availability, nutritional status of the crop and the possibility of disease. Some of these have been discussed here and in previous chapters. The increased tightness of new greenhouses, combined with inside combustion to produce CO_2, have resulted in increased problems of disease, especially if the temperature is allowed to rise. Interactions of humidity and CO_2 on growth of chrysanthemums were investigated by Gislerød and Nelson [1989], finding an increase in growth with high CO_2 and high humidity. There have also been reports of adjustment to high CO_2 by plants [Frydrych, 1976; Potter and Breen, 1980; Hicklenton and Jolliffe, 1978, 1980]. A part of this may be due to starch accumulation, which would reduce continued conversion, or an increase in leaf thickness. Continuous CO_2 enrichment usually results in an increase in specific leaf area (SLA), i.e., the amount of leaf area per unit leaf weight. Holley [1964] reported on

Table 7-2. List of economic plants showing positive results to CO_2 enrichment. List is not inclusive. References may not be original sources or the first to be reported for the particular species. Note that equal responses of cultivars to the same treatment is an exception.

Common name	Scientific name	Cultivars	Reference
African violet	*Saintpaulia ionantha*	Artist Touch	Shaw and Rogers, 1965
		Nicole, Lena, Rosa Roccoco	Mortensen, 1986
Alstroemeria	*Alstroemeria aurantiaca*	---	Moe and Mortensen, 1986
		Jacqueline	Leonardos et al., 1994
Aphelandra	*Aphelandra squarrosa*	---	Moe and Mortensen, 1986
Azalea	*Rhododendron indica*	Ingrid	Bierhuizen et al., 1984
Begonia	*Begonia x hiemalis*	---	Bierhuizen et al., 1984
	Begonia semperflorens	---	Moe and Mortensen, 1986
Cabbage	*Brassica capitata*	---	Moe and Mortensen, 1986
Campanula	*Campanula iosphylla*	---	Moe and Mortensen, 1986
Carnation	*Dianthus caryophyllus*	White Pikes Peak	Holley, 1964
		(stock plants)	Goldsberry, 1963
		Safari, White No. 1, Elliott's White, Pink Coquette, CSU Red Sim, Chantilly	Goldsberry, 1986
		Cerise Royalette	Hurd and Enoch, 1976
China aster	*Callistephus chinensis*	Johannistag	Hughes and Cockshull, 1969
Chrysanthemum	*Dendranthema morifolium*	Imp. Princess Anne	Skoye and Toop, 1973
		Miss America, #3 Ind. White, #4 Ind. White, Explorer, Good News,Imp. Ind. Yellow, Imp. Ind. White, Signet, Rebel, Divinity, Yellow Etude, Jupiter, Bronze Lenoya, Star Flite	Shaw and Rogers, 1965
		Bright Golden Anne	Hughes and Cockshull, 1971
		Refour, Dark Flamenco, Cassa	Mortensen, 1986
		Fiesta	Gislerød and Nelson, 1989
		Horim	Mortensen and Moe, 1983a,b
Codiaeum	*Codiaeum variegatum*	---	Moe and Mortensen, 1986
		Phillip Geduldig	Bierhuizen et al., 1984
Cucumber	*Cucumis sativus L.*	Sporu origineel	Challa, 1976
		Burpee Hybrid	Krizek et al., 1974
		Chestnut 5B, Proeftuins, Blackpool, Simone, Kardot, Klack, Kurik, Butcher's Disease Resistant	Wittwer and Robb, 1964
		Tokiwa-Hikori, No. 3-P	Ito, 1978
		Elene	Enoch et al., 1976
		Bet Alpha	Enoch et al., 1970
		Corona	Heij and van Uffelen, 1984a
		Briljant, Toska, Sporu	van Berkel and van Uffelen, 1975
		Jessica, Aramon, Corona, Lucinde, Fairbio, Primio, Tirana	Nederhoff, 1994 Slack, 1983

Common name	Scientific name	Cultivars	Reference
Dieffenbachia	*Dieffenbachia maculata*	Compacta	Bierhuizen et al., 1984 Saxe and Christensen, 1984
		---	Moe and Mortensen, 1986
Dizygotheca	*Dizygotheca 'Pollux'*		Bierhuizen et al., 1984
Eggplant	*Solanum melongena*	Cosmos	Nederhoff, 1994, 1992 Nederhoff and Buitelaar, 1992
Ficus	*Ficus benjamina*	---	Moe and Mortensen, 1986
	Ficus elastica	Robusta	Saxe and Christensen, 1984
Freezia	*Freezia refracta*	---	Moe and Mortensen, 1986
Geranium	*Pelargonium hortorum*	Irene	Shaw and Rogers, 1965
		Ringo Rose	Kelly et al., 1991
Gerbera	*Gerbera Jamesonii*	Markon, Veronica, Appelbloesem, Gosta	van Berkel, 1986
Guzmania	*Guzmania minor*	nr 168	Bierhuizen et al., 1984
Ivy	*Hedera helix*	---	Moe and Mortensen, 1986
		Anne Marie	Saxe and Christensen, 1984 Bierhuizen et al., 1984
	Hedera canariensis	Gloire de Marengo, Montgomery	Saxe and Christensen, 1984
Kalanchoe	*Kalanchoe rauhii*	---	Bierhuizen et al., 1984
	Kalanchoe 'Roodkapje'	---	
	Kalanchoe blossfeldiana	---	Moe and Mortensen, 1986
Lettuce	*Latuca sativa L.*	Grand Rapids	Krizek et al., 1974 Knecht and O'Leary, 1983
		Bibb lettuce	Wittwer and Robb, 1964
		Mildura	Enoch et al., 1970
		Seaqueen, Emerald, Deciso, Amplus, Magiola, Deci-minor, Amanda, Kordaat, Delta, Vitesse, Valentine, Silva, Kerrekt, Tomika, Kweik, Neptune	Johnston, 1972
		Ambassador, Ravel, Renate, Sun	Hand, 1983a
Nephrolepsis	*Nephrolepsis sp.*	---	Moe and Mortensen, 1986
Peperomia	*Peperomia obtusifolia*	USA	Bierhuizen et al., 1984
Pepper	*Capsicum annuum*	---	Vijverberg and van Uffelen, 1977
		Mazurka, Delphine, Rumba	Nederhoff, 1994
		California Wonder	Enoch et al., 1970
Petunia	*Petunia hybrida*	---	Moe and Mortensen, 1986
Poinsettia	*Euphorbia pulcherrima*	---	Moe and Mortensen, 1986

Common name	Scientific name	Cultivars	Reference
Rose	*Rosa hybrida*	Sonia	Zeroni and Gale, 1989 Hand and Cockshull, 1975
		Baccara	Enoch et al., 1973
		Tropicana	Zieslin et al., 1972
		Briarcliff, Supreme, Red Garnette, Rose Elf	Mattson and Widmer, 1971
		Red Delight, Pink Delight, Gorgeous	Holley and Goldsberry, 1961 Dutton et al., 1988
		Better Times	Lindstrom, 1965
		Samantha	Thompson and Hanan, 1976
		Forever Yours, Love Affair	Aikin and Hanan, 1975
Salvia	*Salvia splendens*	---	Mor and Mortensen, 1986
Sinningia	*Sinningia hybride*	v/d Dussen Rood	Bierhuizen et al., 1984
Snapdragon	*Antirrhinum majus*	Debutante, Montezuma, Nevada, Utah, Colorado, Delaware	Shaw and Rogers, 1965
Marigold	*Tagetes erecta*	---	Moe and Mortensen, 1986
Strawberry	*Fragaria chiloensis*	Tioga	Enoch et al., 1976
Tomato	*Lycopersicon esculentum*	Vendor, Carmelo	Yelle et al., 1987
		Michigan-Ohio	Krizek et al., 1974
		WR-7 Globe, Spartan Red, Spartan Pink 10, Ohio Hybrid 1, Ohio Hybrid 0, R-25, Tuckcross 0, Tuckcross M, Heinz 1370, Fireball, Moneymaker	Wittwer and Robb, 1964
		Tropic	Kimball and Mitchell, 1979 Calvert, 1972 Calvert and Slack, 1975
		GCR88 (Miniverella), GCR93, Kingsley Cross, Selsey Cross	Hurd, 1968
		J16, T150	Hicklenton and Jolliffe, 1980
		Vendor	Dutton et al., 1988 Hand and Soffe, 1971
		Campbell 19VF	Nederhoff, 1994
		Craigella, Eurocross B, J172 Calypso, Blizzard Calypso, Counter, Criterion, Marathon	Hand and Slack, 1988
Sweet melon	*Cucumis melo*	---	Moe and Mortensen, 1986
Vriesa	*Vriesa splendens*	---	Bierhuizen et al., 1984

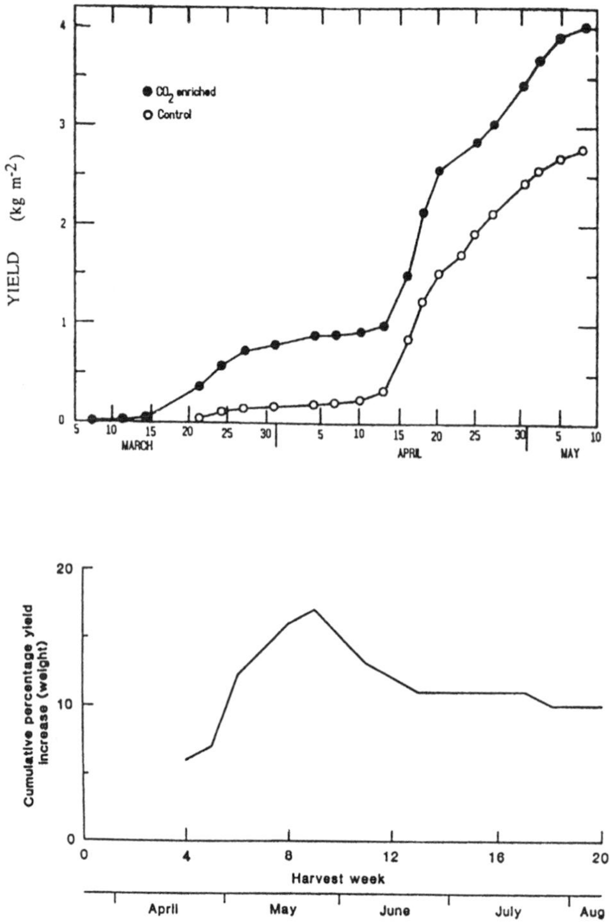

Fig. 7-17. Two examples of yield in cucumbers subjected to CO_2 enrichment. **Upper**: Cumulative yield for a spring crop in Israel [Enoch et al., 1970] (With permission of the author). **Lower**: Cumulative percentage yield for a spring crop in the U.K. subjected to partial enrichment, i.e., ventilators open with injection to maintain ambient concentration [Slack, 1983] (With permission of the British *Grower*).

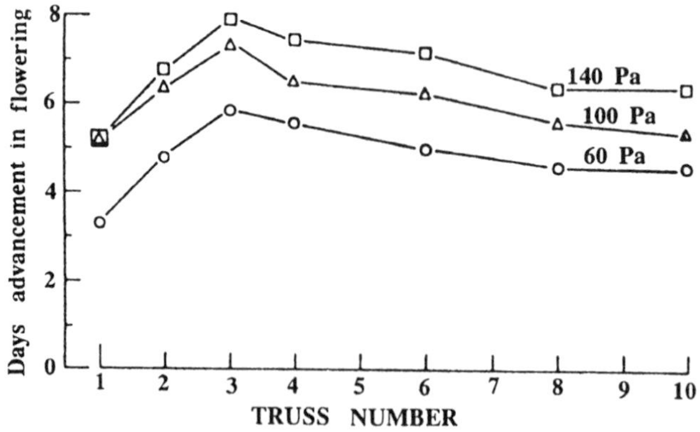

Fig. 7-18. Effect of CO_2 enrichment on flowering time of tomatoes as a function of truss number [Calvert and Slack, 1975] (With permission of the *Journal of Horticultural Science*).

Fig. 7-19. Smoothed weekly production of 'Forever Yours' roses, beginning in October at two ventilation temperatures and two CO_2 levels [Hanan, 1973].

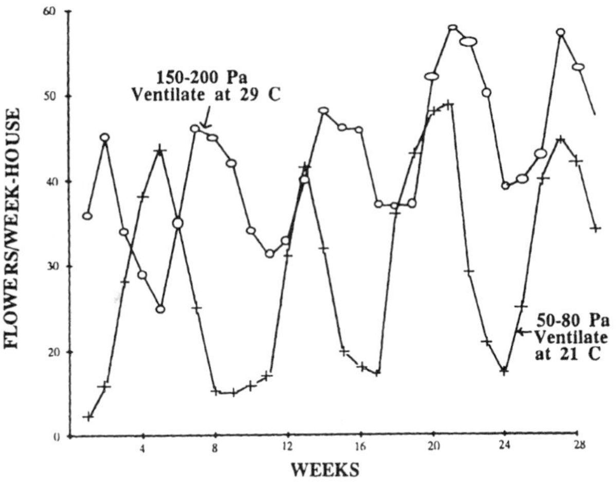

improved growth under fiberglass compared to glass. I think improved growth under fiber-reinforced plastic (FRP) in clear-day regimes results from improved radiation distribution (Fig. 3-19).

IV. CO_2 SOURCES AND PROBLEMS WITH POLLUTION

There are two main CO_2 sources, pure gas and that produced by fuel combustion. Organic matter decomposition can be considered as a form of combustion.

A. PURE CO_2

Pure CO_2 gas is the most expensive source, but also the safest and most easily controlled. Three examples of storage facilities are provided in Fig. 7-20. The top picture shows a tank for dry ice, the middle and lower pictures are for compressed gas. The middle tank is interesting since it can be weighed. This opens the possibility of continuous gas consumption recording, that with suitable systems for determining ventilation rates of the structure, should permit a measure of photosynthetic rates for the crop. This has been done on a small experimental scale and will be discussed later. According to Hand [1984], bulk storage of the types shown in Fig. 7-20 is difficult to justify unless the area to be enriched is at least 4000 m^2. If partial enrichment during ventilation periods is employed, the reason for use of pure CO_2 is strengthened. For small areas (<2000 m^2), a system based upon one or more banks of wall-mounted cylinders can be used [Hand, 1984; 1986; Calvert and Hand, 1975]. Handling such cylinders can be laborious and they are seldom seen in commercial ranges in the U.S. The capacity of standard cylinders is 22.6 kg CO_2 at a working pressure of about 5500 kPa (U.K.). For small areas, the tanks can be nonsyphon with the gas drawn directly from the top of the cylinder. Larger areas require liquid CO_2 withdrawn from cylinders equipped with internal dip tubes, and then converted to a gas by a suitable heating system. Large storage tanks of the type shown in Fig. 7-20 (bottom) are kept at pressures below 2100 kPa by small refrigerators. Alarms and relief valves are necessary equipment. Bulk storage capacities range from 6000 to 30000 kg for horizontal vessels, upwards to 52000 for vertical storages. In the U.K., costs range from U.S. \$28000 to \$80000 (£ = \$1.49) for storages between 6000 to 52000 kg capacity, including vaporizer and pressure-reducing valve [Hand, 1986]. Delivery costs in bulk range from \$0.12 to \$0.19 kg^{-1}. Pressure is usually reduced to 70 kPa for distribution, and to about 30 kPa in the greenhouse.

In Scandinavia, particularly Norway, pure CO_2 is the principal source [Moe and Mortensen, 1986]. CO_2 may be delivered in high-pressure cylinders, or in larger units with built-in cooling units and insulation. A single steel cylinder contains 30 kg CO_2. At present, the local supplier in Norway provides a unit of 12 cylinders, each with 20 kg CO_2 and a total weight of 900 kg. The grower pays a daily rent of \$0.10 per cylinder. Larger containers are rented at \$3000 per year where the CO_2 use is 15000 to 20000 kg per year. Moe and Mortensen state that the limiting distance for bulk delivery is 500 km, with an increase in cost of \$0.01 kg^{-1} $(100\ km)^{-1}$ from the factory. Base prices are about \$0.30 kg^{-1} CO_2. Additional costs include about \$150 per operation for automatic valves, flow meter, and ventilator controller.

Gent [1982] gave a cost of \$0.16 kg^{-1} in the Eastern U.S., which was competitive in price with propane at

Fig. 7-20. Three types of bulk storage for pure CO_2. The tanks in the top picture use dry ice. In the center picture is a tank that can be continuously weighed, thereby determining CO_2 utilization over short periods. The lower picture is a typical 4.5 ton bulk storage used in the U.S. This type includes a refrigeration compressor to maintain temperature at about −18 C. A vaporizer is included. All types include alarm and pressure safety valves.

$0.34 ℓ$^{-1}$. Prices for large tanks in the U.S. range from $20000 to $50000, although rental or lease arrangements are available. These latter arrangements can vary from $400 per month for 2.5 ton vessels to more than $1000 per month. The average North American price in 1988 was about $0.20 kg^{-1} [Hicklenton, 1988]. How much CO_2 is required depends upon the uptake rate of the crop, loss due to infiltration from the structure, and the CO_2 level needed. Under U.K. conditions, at the winter solstice, Hand [1984] suggested maximum rates of photosynthesis for tomatoes of 22 kg ha^{-1}hr^{-1}, and 47 kg ha^{-1}hr^{-1} for infiltration losses when maintaining a threefold enrichment. For an infiltration exchange of 1 air change per hour at 100 Pa, with no plants, the requirement would be 42 kg ha^{-1}hr^{-1}. An air infiltration rate of 1 hr^{-1} is high, and the CO_2 injection, or amount of fuel burned, increases asymptotically with increasing air exchange. At exchange rates less than 0.1 hr^{-1}, injection requirements for common CO_2 concentrations become nearly constant [Hanan, 1972]. As noted in Chapter 4, infiltration rates will vary with structural type, cover and heating method. Forced air circulation will increase infiltration. At full ventilation, maintaining CO_2 levels above the ambient is not possible. Hicklenton [1988] states that, where control is unavailable, it is customary to inject at 5.6 g m^{-2}hr^{-1} of growing area. However, under Israeli conditions, 12 g m^{-2}hr^{-1} CO_2 injected at 390 µmol m^{-2}s^{-1} PAR (90 W m^{-2}) on roses was insufficient to prevent concentration from declining to 20 Pa (120 kg CO_2 ha^{-1}hr^{-1}) [Enoch, 1984]. I would expect that CO_2 injection rates for north European midwinter conditions underestimate such requirements for crops grown in more southerly areas, particularly those with clear-day regimes (i.e., Israel, southwest U.S., etc.), by more than a factor of three. The types of investigations carried out in the U.K., The Netherlands, Scandinavia, etc. on a representative, fully developed crop have seldom been executed elsewhere.

Table 7-3. Approximate conversions for natural gas and LPG to µmol m^{-2}s^{-1} [Enoch and Kimball, 1986].

Original units	Conversion factor for natural gas[a]	Conversion factor for propane[b]
mg m^{-2}s^{-1}	63	
g m^{-2}hr^{-1}	17.4	19.1
mℓ m^{-2}s^{-1}	50	
µℓ m^{-2}s^{-1}		40.5
ℓ m^{-2}hr^{-1}	13.8	11200
W m^{-2}	1.27	1.43
kcal m^{-2}hr^{-1}	1.47	1.66
kJ m^{-2}hr^{-1}	0.35	0.395

[a] Conversions vary with fuel source, based on molecular weight of 17.7 g mol^{-1}, production of 1.11 mol CO_2 per mole of natural gas and an energy equivalence of 49.5 MJ kg^{-1}.

[b] Conversions vary with fuel source, based on equal concentrations of propane and butane with molecular weight of 51 g mol^{-1}, production of 3.5 mol CO_2 per mole of LP, a density of 590 kg m^{-3}, and an energy equivalence of 48 MJ kg^{-1}.

Table 7-4. Composition of natural gas from three sources (percent by volume).

Constituent[a]	North Sea[b] (British)	Groningen field[b] (Dutch)	Colorado[c3]
Methane (CH_4)	90.43	81.40	85.6
Ethane (C_2H_6)	3.76	2.86	5.51
Propane (C_3H_8)	0.94	0.39	1.40
Butane (C_4H_{10})	0.30	0.13	0.09
Pentane (C_5H_{12})	0.14	0.03	trace
Nitrogen (N_2)	3.86	14.15	6.17
CO_2	0.26	0.95	1.00

[a] Natural gas supplies may also contain hexanes and higher paraffins, as well as helium and benzene.
[b] From Hand, 1986.
[c] From Hanan, 1972.

Commonly, in the U.S., pure CO_2 is distributed in the center of a house or section, it being the assumption that it will readily diffuse through the house as the result of convective air movements. The fact that CO_2 is heavier than air, and theoretically could accumulate at ground level, is not a particular problem. As will be seen later, for a variety of reasons, the CO_2 should be distributed directly into the principal part of the vegetative canopy by means of small tubes such as Tygon tubing pierced by small holes (0.1 mm) about every 30 cm. Five cm, layflat polyethylene can also be installed, one to a bench, pierced at 30 cm

Fig. 7-21. Four types of burners often utilized in CO_2 production. **A** is the most common for small operations. **B** is an atomizing kerosene burner with fan. **C** is a unit heater that many growers in the U.S. adapted for CO_2 enrichment by merely removing the exhaust stack. Such types are frowned upon by public service companies and can lead to serious crop damage if not properly fitted. **D** is a smaller, open burner.

intervals with 1 mm diameter holes. This is for the Dutch system with a high volume delivery rate. Control of CO_2 will be considered in Section 7.VI.

B. COMBUSTION

Complete combustion of any fuel results in generation of CO_2 and water. The principal fuels are natural gas, propane or butane (liquified petroleum gas, LPG), and kerosene. Koths [1964] reported on the use of ethanol and methanol. A liter of 95% ethanol will produce about 1.4 kg CO_2 or 0.7 m^3. Similarly, a liter of 99.5% methanol burns to produce about 1 kg or 0.6 m^3 CO_2. These two alcohols are not used extensively although they are considered "clean" fuels. Isopropyl alcohol has also been discussed. All fuels require that sulfur content not exceed 200 $\mu g\ g^{-1}$ [Hand, 1986], otherwise plant damage from SO_2 is likely to occur. Kerosene with sulfur contents up to 600 $\mu g\ g^{-1}$ (0.06%) has been employed successfully if the CO_2 concentration is limited to 100 Pa and is uniformly distributed. Table 7-3 provides many conversions for natural gas and propane/butane. Propane and natural gas composition can vary markedly, as illustrated for natural gas from three sources in Table 7-4.

Commercial grades of propane can vary from 30 to 100% propane and 70 to 0% propylene. In the U.K., sulfur content of LPG is usually below 60 μg g^{-1}. Natural gas sources, though they may contain potentially harmful contaminants, are seldom in high enough concentrations to cause damage. McKeag [1965] reported gas supplies in Colorado to contain a maximum 9.16 μg ℓ^{-1} sulfur. Van Berkel [1984b] reported a value of 8 μg ℓ^{-1} sulfur for Dutch sources. Where available, natural gas is the best fuel for both heat and CO_2. Liquified petroleum gases (propane and butane) are common where natural gas is unavailable. While gaseous at atmospheric pressure, they can be readily stored in the liquid state in portable containers or fixed tanks. For CO_2 production, the fuel gases are piped from the external supply to open-flame burners suspended over the crop. Kerosene is probably the cheapest fuel and most commonly employed in lettuce production.

Fuels burned for CO_2 supply have two outstanding advantages: 1) they are cheap compared with pure CO_2 with minimum capital investment, and 2) they can also supply heat simultaneously. However, they can be a serious disadvantage under conditions where no heat is required (sunny days and/or warm temperatures). There is another important disadvantage, especially if the fuel is burned inside the greenhouse –that is, the possibility of plant damage resulting from fuel contaminants, improper combustion (maladjustment of burners), and poor distribution of the combustion products in the structure. Pure CO_2, in contrast, has two outstanding advantages: 1) there are no contaminants and 2), it can be controlled precisely with present-day computer systems.

Examples of the burner types available for in-house CO_2 generation are depicted in Fig. 7-21. Picture **A** shows a type of burner most commonly utilized in small operations. These can burn natural gas or propane. They consume between 1 and 1.6 kg propane per hour, producing between 3 and 4.8 kg CO_2; 1 kg of propane yields 3 kg CO_2. Using Hinklenton's previous guideline for injection, a minimum of 12 burners is required per hectare. Modern burners, which bring fresh air directly from the outside, have fuel consumption rates up to 7 kg hr^{-1}. Burners using natural gas will consume about 2 m^3 hr^{-1}, requiring 15 burners per hectare to attain 100 Pa CO_2 concentration in a tight greenhouse. On combustion, 1 m^3 natural gas produces an equivalent amount of CO_2. To maintain 100 Pa, gas must be burned at 30.4 m^3 $ha^{-1}hr^{-1}$ to produce 56 kg CO_2 $ha^{-1}hr^{-1}$. Natural gas is often supplied to the greenhouse from a high-pressure main at 175 to 350 kPa. In cities, the pressure may be at 3.5 kPa, and is usually piped around the nursery at pressures below 3.5 kPa. Fig. 7-21B is a kerosene burner common in Europe.

Care is required since kerosene sources may have excessive sulfur. On combustion, the sulfur is oxidized to SO_2, which can be a serious pollutant in closed greenhouses. The atomizing type is preferable to vaporizing types. Hand [1986] provided prices of $0.33 to $0.42 per liter or $0.13 to $0.16 per kg CO_2. Van Berkel [1986] gave costs for kerosene of about $0.33 m^{-2} greenhouse area with a rate of combustion of 5 g $m^{-2}hr^{-1}$.

Of particular interest is the Dutch system of using flue gases from centralized boiler systems [van Berkel, 1975; van Berkel and Verveer, 1984]. Fig. 7-22 diagrammatically outlines the system. The two photographs below the diagram illustrate the control system in the boiler room and the main distribution tube from the secondary fan to the greenhouses. Natural gas is supplied in Holland at very competitive prices –$0.15 m^{-3}, which is related to the price of heavy fuel oil as compared to the more expensive light fuel oil in the U.K. [Hand, 1986]. Part of the boiler combustion gas is extracted from the flue with a primary fan. A secondary fan adds air and feeds the mixtures, containing 2 to 3% CO_2, into the main supply duct. The problem with this method is the heat generated under high radiation that might cause the ventilators to open. The situation is particularly a problem in arid, clear-day regions as contrasted to the northern European climate. To overcome this problem of no CO_2 under bright-light conditions, the Dutch have developed burners that are capable of modulation down to 25 m^3 hr^{-1} natural gas combustion. Or, a second smaller burner may be used; 25 $m^3ha^{-1}hr^{-1}$ is sufficient to maintain 100 Pa CO_2 with ventilators closed, and to maintain ambient levels during ventilation. The heat is stored, partially within the hot water boiler itself and into an additional tank with a capacity of about 30000 liters or more. Various combinations with pure CO_2, single or double burner, storage, etc. were outlined by van Berkel and Verweer [1984]. Capital costs could range upward to $18000, with annual operating costs ranging from about $3000 to $8600.

The main valve in the system (Fig 7-22) closes when the boiler burner is off, or is starting, when the secondary fan is not operating, or the temperature exceeds a maximum setpoint. The latter is to prevent softening of the main PVC duct at temperatures above 70 C. Because of pressure drops and volume reduction as the gas cools and water condenses, duct diameter has to be adjusted with distance from the boiler. Laterals from the main

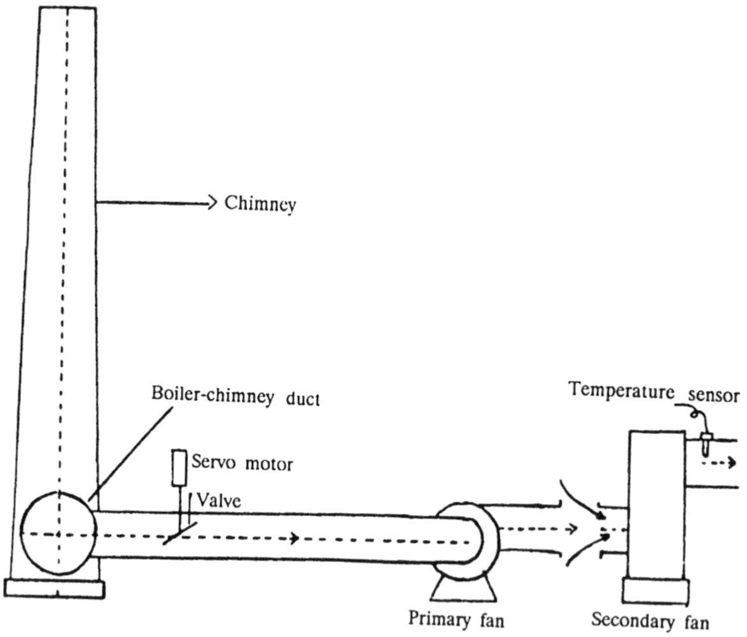

Fig. 7-22. Dutch system for providing CO_2 from boilers burning natural gas. The left, bottom picture indicates the control system, while the right photograph shows the main distribution tube. Flue gases are diluted by the secondary fan at a ratio of 1:2 parts air [Diagram from van Berkel, 1984b].

duct are 5 cm layflat PE tubing with l mm diameter holes every 30 cm. These are spaced 3 m apart with maximum lengths up to 35 m, and laid on the ground, below the crop. Van Berkel [1986] showed that this distribution provided a fairly homogenous CO_2 level in greenhouses.

The rates of fuel combustion and CO_2 injection reviewed above deal mostly with north European conditions, which are comparable with the northern U.S. In Mediterranean climates, or in the southwest U.S., these rates are low when the greenhouse can be tightly closed. When the temperature outside is about 20 C lower than the inside, a single-layer greenhouse can be tightly closed even on clear days. This situation occurs from approximately November 1 through March in northern Colorado [Hanan, 1973]. For warm crops such as roses, tomatoes, etc., this provides an outstanding climatic advantage.

C. ORGANIC MATTER DECOMPOSITION

Decomposition of any organic matter produces CO_2. In fact, before the 1960s, considerable CO_2 was generated unknowingly through the common use of manures and straw composts in greenhouses. Any organic matter added to the soil, or spread on its surface, is subject to degradation. CO_2 from the soil ranges from 0.2 to 25 g $m^{-2}dy^{-1}$, with an average of about 2 kg $ha^{-1}dy^{-1}$. With the movement toward soilless culture, however, problems in the lack of CO_2, especially where the ground was covered, became more apparent. With the requirements for dependable and controllable CO_2 sources, decomposition is seldom practiced in modern greenhouses. The review by Levanon et al. [1986] and the discussion by Hinklenton [1988] represent the most detailed expositions of organic matter use in recent years.

Table 7-5. Scheme for cucumber production on straw bales, generally producing sufficient CO_2 for an autumn crop[a] [Hicklenton, 1988].

Days after placing bales[b]	Fertilizer
4	7.5 kg Ammonium nitrate
7	3.8 kg Ammonium nitrate
10	3.7 kg Ammonium nitrate
	11.5 kg Triple superphosphate
	20.5 kg Potassium nitrate
Water bales daily until day 11, apply peat and plant	

[a] Weight of fertilizer per ton of organic material.
[b] Bales (ca 100 ton ha^{-1}) are placed 2 to 3 weeks before planting, thoroughly wetted and fertilized to initiate decomposition. Air temperature kept at 15 C.

Decomposition is carried out by microorganisms that use organic materials as energy and carbon sources. The process most interesting for greenhouse use is aerobic decomposition, as contrasted with anaerobic decomposition, which produces methane plus CO_2 and other by-products (biogas). Composting for CO_2 production can be "batch"or "windrow" processing. The latter is a long row about 1.5 to 1.8 m high and 2.4 m wide. This is turned every 2 to 3 days. Sufficient aeration and self-heating are required to ensure elimination of harmful insects and microorganisms. This means temperatures near 60 to 70 C. If the temperatures rise too high, the pile could ignite. Temperatures of 70 to 80 C result in caramelization of sugars, causing a decrease in dry weight without CO_2 production. The windrow can also be force ventilated with special blowers. A decline in the oxygen level much below 0.5% O_2 will lead to anaerobic conditions in the center of the pile, producing organic vapors with bad odors. Compost particles should be about 0.5 to 5 cm diameter. The least water content must be no less than 40%. Good composting is achieved with water contents of 75 to 85% for straw and plant wastes. With manures, on the other hand, water content should not exceed 55 to 65%.

Microorganisms are commonly present. There is no need to make special inoculations. The quantitative ratio between macroelements in microbial cells is 100:10:1:1 for C:N:P:S. The ratio between N, P, and S in the raw materials must be higher than the ratio in the microorganisms. Optimal C/N ratios for composts is 26 to 35. A higher ratio slows the process, and a lower ratio will result in more gases such as ammonia. The best estimate, according to Levanon et al., is that a dry weight loss of 1.0 kg in the compost results in about 1.5 kg CO_2. For an average enrichment rate of 5 to 10 g CO_2 $m^{-2}hr^{-1}$, the authors estimate that for each m^2 greenhouse area, 7 to 14 kg of wet compost will supply sufficient CO_2 for 20 days.

Production of cucumbers on straw bales will usually provide sufficient CO_2 for the first crop, and is an

Fig. 7-23. Examples of sulfur dioxide damage to cucumber (**top**) and tomato (**bottom**) [bottom picture courtesy of McKeag, 1965].

example of a "batch" system. Hinklenton [1988] provided a recipe for straw bale production, which is reproduced in Table 7-5. Another example of a batch process was described for a commercial greenhouse in the U.K. by Hayman [1987]. In this case, polyethylene tunnels were erected, each tunnel containing 1500 straw bales. After an initial wetting, 150 mM ℓ^{-1} of ammonium nitrate in approximately 1000 liters were applied to each meter length with a sprayline. About 500 kg CO_2 were produced from each ton of straw at a cost of about \$0.05 to \$0.06 kg^{-1}. The gas was blown into the greenhouse from the tunnels.

Whereas decomposition is an inexpensive method, and there are any number of organic materials that can be used, there are several difficulties. One matter, of course, is that the process cannot be shut off, with CO_2 production gradually declining with time. A second problem is the possibility of introducing harmful organisms that are not eliminated in the decomposition process. A third aspect is that greenhouses may not be within reasonable distance of the compost supply –i.e., sewage plants, garbage processing centers, etc. Most vegetative wastes are returned to the land. Except for Levanon et al.'s report on work in Israel, I am not aware of extensive commercial trials with CO_2 supply from organic matter. Anaerobic decomposition for the production of biogas is much more expensive. In addition, pollutants such as ethylene, hydrogen sulfide, etc. may be present.

Any fuel is a potential CO_2 source. It is the contaminants, either in the fuel or as the result of improper combustion, that can limit their use. For example, Rustad et al. [1984] examined a charcoal-based system for CO_2 enrichment in Norway. Nitrous oxides were a problem. Extending operation into the summer, as with the Dutch method, requires an alternative heat storage facility. I do not believe that this system type has been carried out on a large scale.

D. PROBLEMS OF POLLUTANTS FROM COMBUSTION SOURCES

Any combustion in the confines of an enclosed building has inherent dangers. Among the possible pollutants are sulfur dioxide (SO_2), nitrous oxides (NO, NO_2), ethylene (C_2H_4), and leakage from pipe joints of propylene and natural gas. Hand [1986; 1983] reported on propylene, which is a major constituent of propane. Equipment in the U.K. has often been poorly installed with leaky pipe joints. Propylene mimics ethylene and has been found in greenhouses in 5 to 100 ppm concentrations, resulting from leaks from propane-fired CO_2 producers. A level of 10 to 50 ppm will cause epinasty (downward curvature of the petioles) in tomato, similar to 100 to 500 ppb ethylene. The gas can also interfere with photoperiodic responses in CO_2-enriched chrysanthemums.

Inefficient combustion of petroleum fuels can lead to formation of carbon monoxide (CO), formaldehyde (HCHO), and acrolein (CH_2CHCHO). Nevertheless, if the CO level is below 50 ppm (upper limit for human safety), CO is unlikely to cause serious problems. The other two, HCHO and CH_2CHCHO, are potentially harmful to crops at 1 ppm or less. Improper combustion also increases the release of ethylene, and it is a particular problem in severe climates such as Colorado and northern, continental U.S. Outside temperatures much below -5 to -10 C will cause cracks to freeze, greatly reducing air infiltration. Inevitably, two or three episodes of pollution will occur in Colorado in the winter where the grower is using petroleum fuels for CO_2 production or for heating the greenhouse. Recommendations are for heaters to have separate air supplies from outside the structure. It has been my observation that, as an industry develops, growers eventually move heaters to outside the structure to ensure proper air supply and combustion. Work by Hanan [1972] resulted in a recommendation of a 10 cm^2 free-air opening for each kW output of the burner (1 kJ s^{-1}). Unfortunately, the fact that there is 1 m^2 of opening in one end of the structure does not mean that the burners within the house will obtain sufficient air. Units producing heat or CO_2, even if equipped with fans, tend to set up independent circulation patterns that may, or may not, draw fresh air from the aperture.

Sulfur dioxide, as a contaminant in the fuel supply, has been mentioned, with the maximum concentrations permitted in fuels. SO_2 was also discussed in Chapter 6 as a sulfur source in nutrition. Pollution control has gradually resulted in lowering the SO_2 concentrations commonly prevalent in the air a decade ago. Fig. 7-23 shows examples of SO_2 injury to cucumber and tomato. The ivory-colored necrotic areas are typical on many species. This injury can occur from fuel oil heaters in which the combustion chamber has cracked or rusted, or the exhaust stack has been damaged. Periodic inspection of heater chambers, air supply, and exhaust is good preventive maintenance.

Nitrous oxides are the inevitable product of combustion regardless of the burner type. Nitrous oxide (NO) is not derived from the nitrogen content of the fuel, but the combination of atmospheric N_2 and O_2 in the flame to form nitrogen monoxide, which is subsequently oxidized to nitrogen dioxide (NO_2). The heat of the burning fuel promotes formation of NO. Modern CO_2 burners have flame temperatures well above the critical temperature for NO formation (ca. 1100 C). The higher the temperature, the more NO formed [Hand, 1986]. There is then a spontaneous reaction between NO and oxygen to form NO_2, the rate of which depends on the square of the NO level and is accelerated by photochemical reactions involving the absorption of sunlight and reactions with hydrocarbons and ozone. Due to the greenhouse cover, which filters most of the ultraviolet from sunshine, photochemical reactions continue more slowly inside a greenhouse. Much of the work on NO/NO_2 pollution has been conducted by Capron and Mansfield [1975; 1976; 1977] and Mansfield and Murray [1984]. When burners are turned on for CO_2 production, the levels of nitrous oxides (NO_x) will increase steadily, as shown in Fig. 7-24 for a kerosene burner. Phytotoxicity, especially to tomatoes, can result from concentrations in the parts per billion range (ppb). The main problem is photosynthetic inhibition that is not usually observable by the grower unless there is some means of comparison with an uninhibited control. Both NO and NO_2 act to reduce photosynthesis significantly at 250 ppb [Capron and Mansfield, 1976]. High concentrations did produce necrotic areas on tomatoes, with severe growth reduction at 500 ppb NO_x [Capron and Mansfield, 1977]. Depending upon conditions (temperatures, burner types, etc.), Hand [1986] concluded that NO_x levels in excess of 1000 ppb (1 ppm = 1 $\mu\ell\ \ell^{-1}$), and even 5 ppm (5000 ppb) under severe wintry conditions, can occur. Injury severity is related to actual uptake by the plant, rather than the time x concentration dose relationship found with ethylene. Lettuce is another crop, that shows high sensitivity to NO_x, with long-term exposure at 500 ppb showing significant growth inhibition. Saxe and Christensen [1984] examined several decorative foliage plant species. High CO_2 stimulated growth in the eight species examined. The effect of NO fumigation was to decrease the advantage of CO_2 enrichment. The most visible effect of NO was leaf scorching of *Dieffenbachia*. The authors concluded by stating that it is difficult to convince growers of the danger of combustion for CO_2 production without any obvious NO_x damage on most plant species.

Complex reactions occur during combustion that can produce unwanted hydrocarbons, such as ethylene. C_2H_4 is an endogenous growth regulator and concentrations as low as 20 ppb can have disastrous effects on orchids such as *Cattleya* species ("dry sepal" damage).

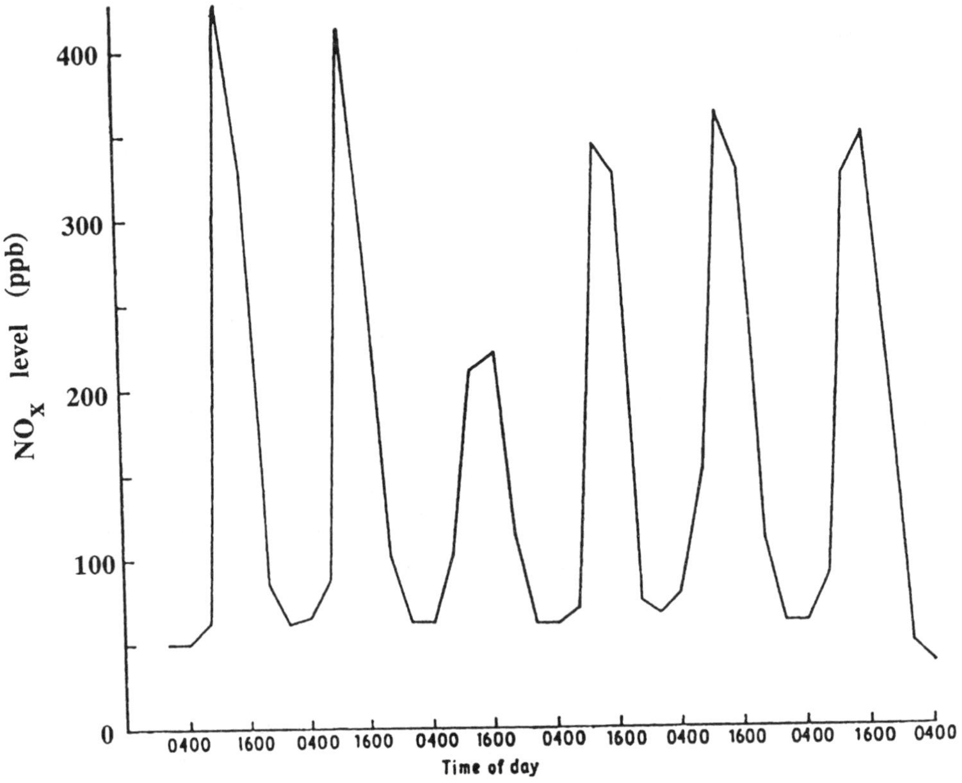

Fig. 7-24. NO_x produced from kerosene burners and time of day. Burners switched on at 0800 and turned off at 1600 hr each day from Feb. 24 to Mar. 2 [Capron and Mansfield, 1975].

Symptoms include reduced growth, decreased apical dominance, internode shortening, leaf epinasty, premature senescence and abscission of leaves and flowers, delayed flowering, and malformed flowers [Hand, 1983b; 1986]. Tomato flowers are particularly susceptible to damage from bud opening to anthesis. Levels of 500 ppb are sufficient to induce complete flower abortion. Examples of ethylene damage are given in Fig. 7-25. Because low concentrations, given for a long period, are just as effective as short exposures at high levels, concentrations below 50 ppb can be highly dangerous to sensitive species.

One can detect ethylene in any combustion process with safety to be obtained by dilution, use of well-designed burners, and complete combustion. The burner shown in Fig. 7-21A is safe if there is sufficient air infiltration. However, the heaters in Fig. 7-21C were noted for the damage to carnations shown in Fig. 7-25. Periodic internode stacking always followed periods of severe cold, especially when the heaters in Fig. 7-21C were mounted to fan across the house or, if the heaters blew directly into the air versus those mounting polyethylene tubes [Hanan, 1972]. From sampling several commercial ranges, Hanan found that heaters used for CO_2 production must have distribution tubes the length of the house in order to force complete circulation and bring fresh air to the burner. This was confirmed when tube removal caused an immediate ethylene concentration increase in the exhaust gases from the tubeless heater. One may also expect an ethylene burst from cold heaters when first started.

Measurement of ethylene requires sophisticated analytical equipment. In a personal communication to the author, Rosenberg [1971] stated that presence of CO is no guarantee that C_2H_4 is present. That is, reactions leading to CO formation are different from those producing ethylene. So it was Hanan's conclusion that analysis

Fig. 7-25. Three examples of ethylene damage. **Left**: Internode stacking on carnation. Very high C_2H_4 levels can make the distance between nodes almost indistinguishable. **Top right**: Flower abscission on azalea. This is fairly typical on those species having multiple blooms or florets. **Bottom right**: Shortening on the right poinsettia. Epinasty (downward petiole curvature) is a characteristic symptom of ethylene damage on numerous species.

of CO for determining C_2H_4 was not reliable. This is contrary to the Dutch practice [van Berkel, 1984] where measurements of CO and C_2H_4 on many burners, under a variety of conditions, showed linear relationships between these two gases. The Dutch concluded that the C_2H_4 level would be less than 0.1 of the critical 50 ppb, if the CO content of the undiluted flue gases did not exceed 50 ppm and the CO_2 dispersed into the greenhouses was diluted to 100 Pa (1000 ppm). Since CO monitors are readily available and much cheaper, CO analysis is used to detect the status of possible pollution in Dutch greenhouses.

It is obvious from the previous discussion that combustion inside greenhouses, whether for CO_2 production or heating, is fraught with many dangers to healthy crop production. Where heaters are the principal means of heating, they should be equipped with fresh air supplies from outside, suitable exhaust stacks to outside, and electronic ignition. Obviously, these burners are much more expensive. Although CO_2 distribution from central boilers has a principal advantage of low cost, I do not like the procedure. As will be noted in the Section **7.VI**, the heat produced during warm periods requires specialized adjustment and increased capital expenditure (see above).

One must keep in mind that combustion also produces water vapor, which will raise humidity levels. This may be undesirable, especially in tight structures. Disease problems may be exacerbated, especially if lower temperatures are employed at night. With little air movement, condensation is likely to occur on foliage, which can lead to problems as outlined in Chapter 5. Dutch publications do not emphasize this aspect, which may be the result of their climatological conditions.

V. CO_2 DISTRIBUTION AND CONCENTRATION IN GREENHOUSES

Measurements within greenhouses without CO_2 enrichment have shown that concentrations during the day are invariably below ambient outside levels of 34 Pa. An example of typical concentrations in Dutch greenhouses is presented in Fig. 7-26. Similar results have been obtained by Heij and Schapendonk [1984], Heij and de Lint [1984], and Goldsberry and Holley [1962]. Goldsberry and Holley, under Colorado conditions, found that CO_2 levels in a carnation house, with no CO_2 injection, invariably were below ambient when ventilators were closed. Heij and de Lint [1984], in Fig. 7-26, showed that concentrations could be as low as 10 Pa –which is approaching the CO_2 compensation point. At night, as the result of respiration and soil loss, CO_2 levels may exceed 100 Pa. This always drops when the plants begin photosynthesizing in the mornings. Unfortunately, even if ventilators are open wide, or fan-and-pad cooling systems are operating maximally, concentrations within the vegetative canopy can be 5+ Pa below ambient (Fig. 7-27). The situation can be aggravated by calm, clear conditions where air movement within the greenhouse will depend largely on convective flow. Thus, in recent years, considerable attention has been given to summer CO_2 enrichment with injection into the canopy

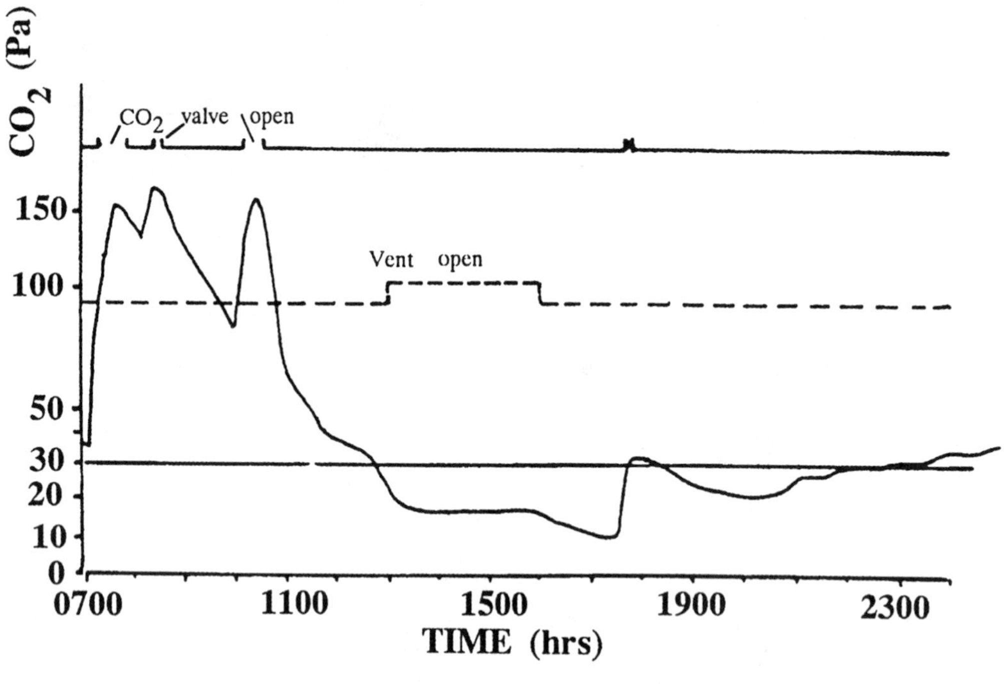

Fig. 7-26. CO_2 concentrations on April 24 in a Dutch widespsan house growing chrysanthemums. Even with ventilators open, CO_2 remained below ambient [Heij and de Lint, 1984] (©1984, Int. Soc. Hort. Sci., *Acta Hort.*, 162:93-100).

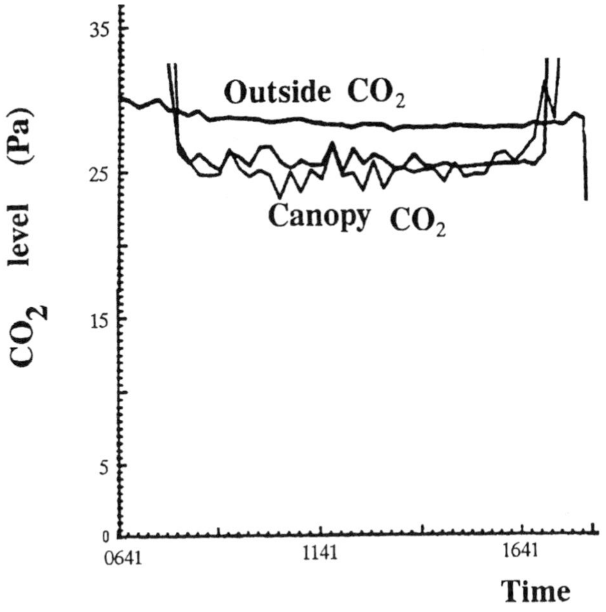

Fig. 7-27. CO_2 concentrations within a mature rose canopy (2 m tall) under maximum pad-and-fan cooling. Wind speed about 0.3 m s^{-1} with slight leaf fluttering on the outside canopy edges. Benches arranged perpendicular to wind flow direction [Hanan, 1988].

to maintain levels equal to that outside[1]. The methods will be reviewed in Section 7.VI. An example of the year-round CO_2 levels to be expected is presented in Fig. 7-28. The CO_2 curve is opposite to the radiation curve. Increasing the CO_2 level under high radiation would be more desirable if that were possible. The criticality of maintaining at least ambient levels in the canopy is suggested in Fig. 7-29, where van Berkel showed that growth rates are likely to slow remarkably with slight CO_2 decreases below 33 to 34 Pa. Van Berkel [1986] calculated from Heij and Schapendonk's data [1984], that CO_2 depletion decreased net photosynthetic rate nearly 50 kg $ha^{-1}dy^{-1}$, regardless of the irradiance (184 and 920 µmol $m^{-2}s^{-1}$) (94 and 419 W m^{-2} global). From measurements made in many commercial ranges, CO_2 levels fell below 34 Pa an average 0.5 hours in February, increasing to 10 hours daily in June. Maintaining the CO_2 level at ambient, with ventilators open, would require 2 to 2.5 m^3 natural gas per 1000 m^2 hr^{-1}, or 3.5 to 4.5 kg pure CO_2 [van Berkel, 1986].

When CO_2 first began to be introduced in greenhouses, it was common in the U.S. to run a 5 to 6 mm diameter tube down the length of the bench to disburse the gas. With continued practice, particularly where fuel combustion was employed, it was felt that the gas would be completely distributed throughout the structure. This is not at all the case. In structures covering large areas, outside wind is likely to cause unequal CO_2 levels, depending upon the wind direction –just as with temperature control as outlined in Chapter 4. Gormley and Walshe [1979] found a threefold concentration difference in a house covering 1.21 ha. Overall, these authors concluded that distribution in structures needed to be monitored.

One of the best expositions of CO_2 distribution in greenhouses was van Berkel's [1975] study in commercial Dutch ranges. The houses were multispan with bay width of 3.2 m, and a gutter height of about 2.5 m. CO_2 was transported via a main duct suspended above the main path, or via two ducts along opposite gable ends. Final

---

[1]The ventilation rate required to maintain a CO_2 level close to ambient concentrations (34 Pa) is astounding. For example, assume a crop with an LAI of 3, which is taking up CO_2 at a rate of 25 µmol $s^{-1}m^{-2}$ leaf surface. Or, total uptake for 1 m^2 of ground area is 75 µmol s^{-1}. In a 30x30 m greenhouse, with 810 m^2 in production, the uptake would remove 0.061 mol s^{-1} from the greenhouse air. If the greenhouse volume is 2700 m^3, then at 34 Pa, the greenhouse contains 40.9 mol CO_2. If one limits reduction to 30 Pa, the CO_2 that can be removed between 30 and 34 Pa amounts to 4.8 mols. So, 4.8/0.061 = 78.7 sec, the time required to reduce the greenhouse air volume to 30 Pa CO_2, assuming no change in CO_2 uptake rate . That is, the greenhouse air volume must be completely exchanged every 1.3 minutes if CO_2 is not to drop below 30 Pa, or nearly 50 exchanges hr^{-1}. Providing adequate CO_2 to a full vegetative canopy even at full ventilation is not simple.

Fig. 7-28. Yearly course of CO_2 concentration and global radiation in a widespan chrysanthemum house in Holland [Hiej and de Lint, 1984] (©1984, Int. Soc. Hort. Sci., *Acta Hort.*, 162:93-100).

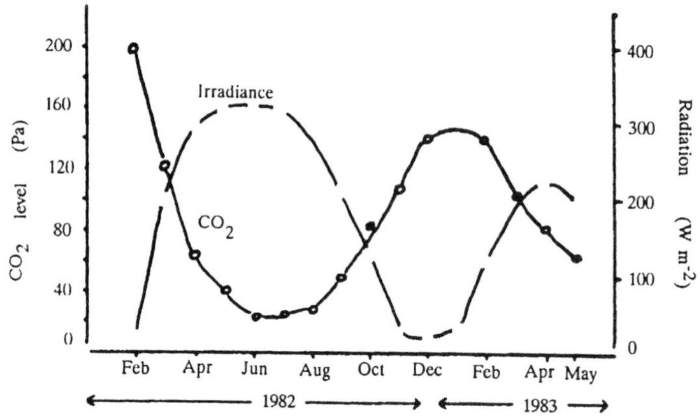

Fig. 7-29. Growth rate as a percentage of 100 at ambient CO_2 level. In the small interval between ambient and the CO_2 compensation level (5 Pa), growth rate drops to zero [van Berkel, 1984b] (©1894, Int. Soc. Hort. Sci., *Acta Hort.*, 162:197-205).

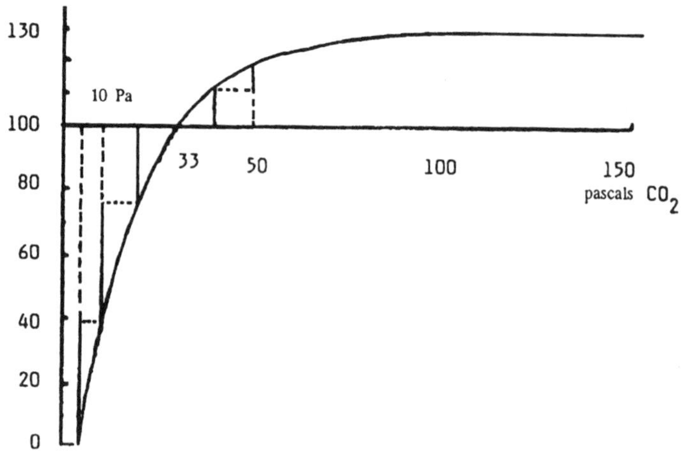

distribution used lay flat tubing, 35 m long, at a spacing of one tube per bay. The results are shown in Figs. 7-30 through 7-32 . The variations shown in Fig. 7-30 were measured 1.8 m above the ground for a tomato crop with open ventilators. A central main duct (right) resulted in less variation horizontally as compared to two gable mounted main ducts (left), with highest concentrations at the gable ends for gable-mounted main ducting and reversed for a single main duct. Van Berkel cites others as showing that CO_2 distributed through layflat tubes, at an initial pressure of 3 cm water column, decreases with length. This was the situation in Fig. 7-31.

If the CO_2 was released from the main ducts at 10 and 20 cm above the crop, or into the crop at half the height of the crop, CO_2 levels could vary markedly (Fig. 7-31). The variation was less in the chrysanthemum crop, either with layflat tubing or injecting the CO_2 1.2 m above the crop. No clear difference with height above the ground was found in tomatoes (Fig. 7-32). For chrysanthemums (right), the CO_2 level in the crop at 10 cm height was clearly higher than at other points above the crop. The author attributed these differences as probably caused by plant density and heating pipe position. The heating pipes in tomatoes were located at 30 to 60 cm, as contrasted with a height of 1.8 m for chrysanthemums. Air circulation was improved by placing heating pipes low in the canopy. Tomatoes were planted at 2.7 m^{-2} as contrasted to chrysanthemums at 58 to 62 plants per m^2. Thus, air movement within a cut-flower chrysanthemum canopy would likely be very slow, if not stagnant.

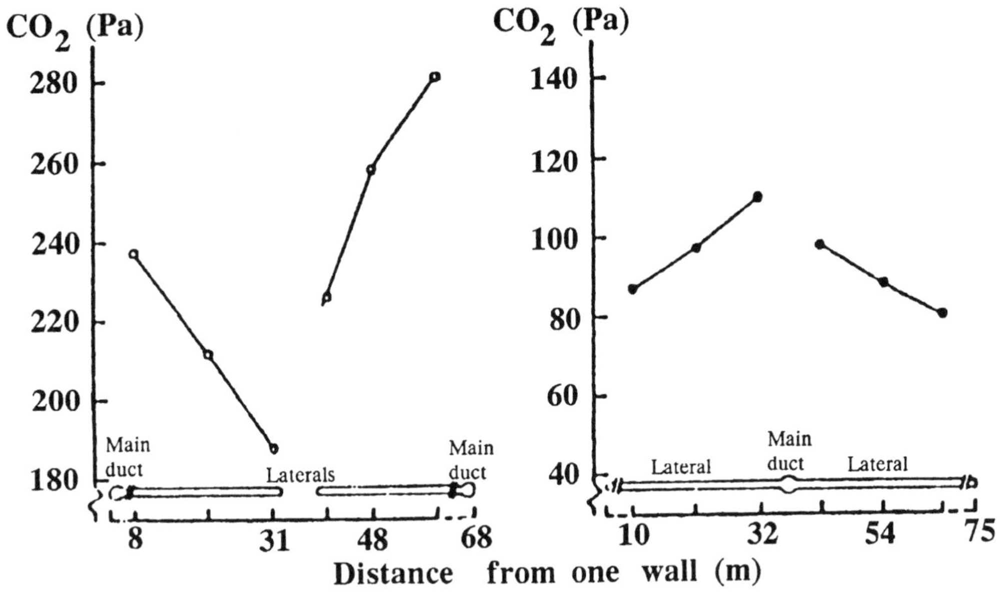

Fig. 7-30. Gradient of CO_2 levels with distance from one wall in Dutch greenhouses, comparing two arrangements of the main supply ducts [van Berkel, 1975] (With permission of the *Netherlands J. of Agric. Sci.*).

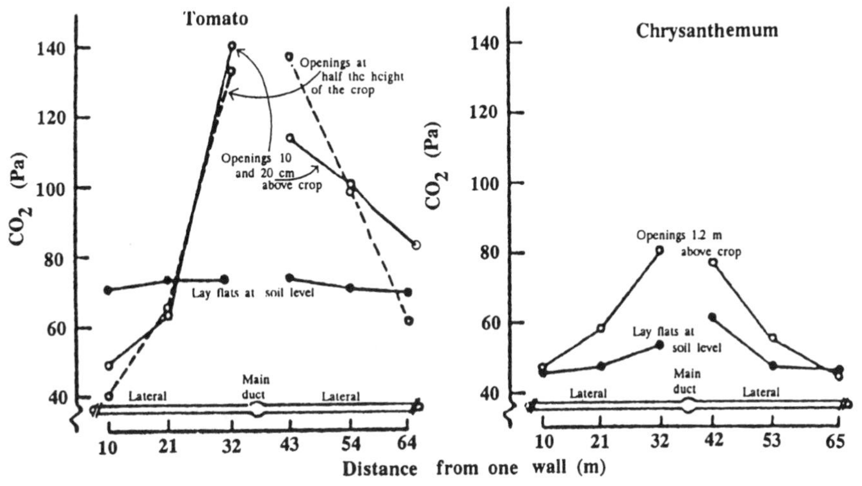

Fig. 7-31. Horizontal CO_2 distribution for a tomato crop (**left**) versus a chrysanthemum crop (**right**) in commercial Dutch greenhouses. The tomato crop compares CO_2 distributed through layflat tubing on the ground versus air blowing from the main duct into the houses at 10 and 20 cm above the crop or at one-half the height of the crop. For chrysanthemums, the openings from the main duct were 1.2 m above the crop [van Berkel, 1975] (With permission of the *Netherlands J. of Agric. Sci.*).

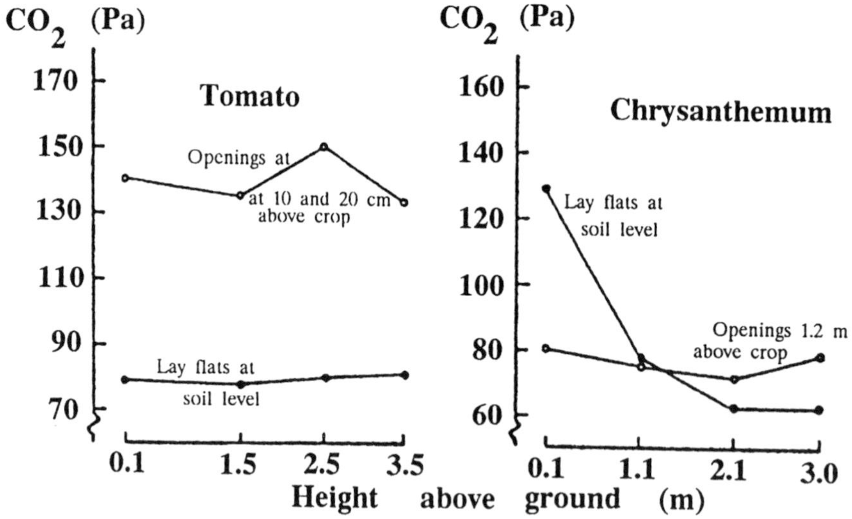

Fig. 7-32. CO_2 concentrations with height in a tomato crop (**left**) and in a chrysanthemum crop (**right**). A lay flat tubing distribution system is compared with direct supply from a main duct at 10 and 20 cm above the crop (**left**), and 1.2 m above the crop (**right**) [van Berkel, 1975] (With permission of the *Netherlands J. of Agric. Sci.*).

Van Berkel's study is the only comprehensive exposition of CO_2 distribution in greenhouses of which I am aware. One should remember that the results pertain to the Dutch system where a relatively large air volume is distributed through canopies at ground level. What the situation would be with pure CO_2 distribution is unknown. However, the previous discussion (Section 7.III.F) suggests that pure CO_2 should use a distribution system in the upper part of the crop canopy (e.g., 6 mm diameter tubing with 0.1 mm diameter holes every 30 cm). I do not know the availability of this type of tubing.

In summary, in the modern, tight greenhouse, CO_2 levels will be below ambient concentrations during day if the house is unenriched. The situation will be worse on clear days, and the crop can be severely delayed if not damaged. Deficient levels will occur in a mature crop even with full ventilation. Secondly, CO_2 injection should be made into the upper part of the vegetative canopy. Simply admitting pure CO_2 or flue gases into the house is likely to result in uneven concentrations with location. I have not seen any studies on evenness of CO_2 levels from open-flame CO_2 burners, which have the disadvantage of releasing flue gases above the crop as contrasted to mixing it into the canopy.

VI. CO_2 CONTROL

It appears from the previous discussion that there are several options a grower can use to supplement CO_2 in the greenhouse. One can simply inject at a constant rate when the greenhouse is closed, using the various recommendations as to weight (or volume) of pure CO_2 per unit area-time, or burn the appropriate fuel quantities recommended per unit area-time. This is the simplest procedure and requires no fancy equipment beyond a time clock and interlock with the ventilation system. However, there are obvious disadvantages for the sake of a cheap system. These can be, for example, a limitation to the period when the greenhouse is closed, oversupply of CO_2 in dark weather that is inefficient and likely to cause damage to some species, and does not prevent deficiencies under full ventilation conditions. Or, the grower can invest in suitable analytical equipment and interface this with a climate-control computer. This approach, while more expensive, can save CO_2

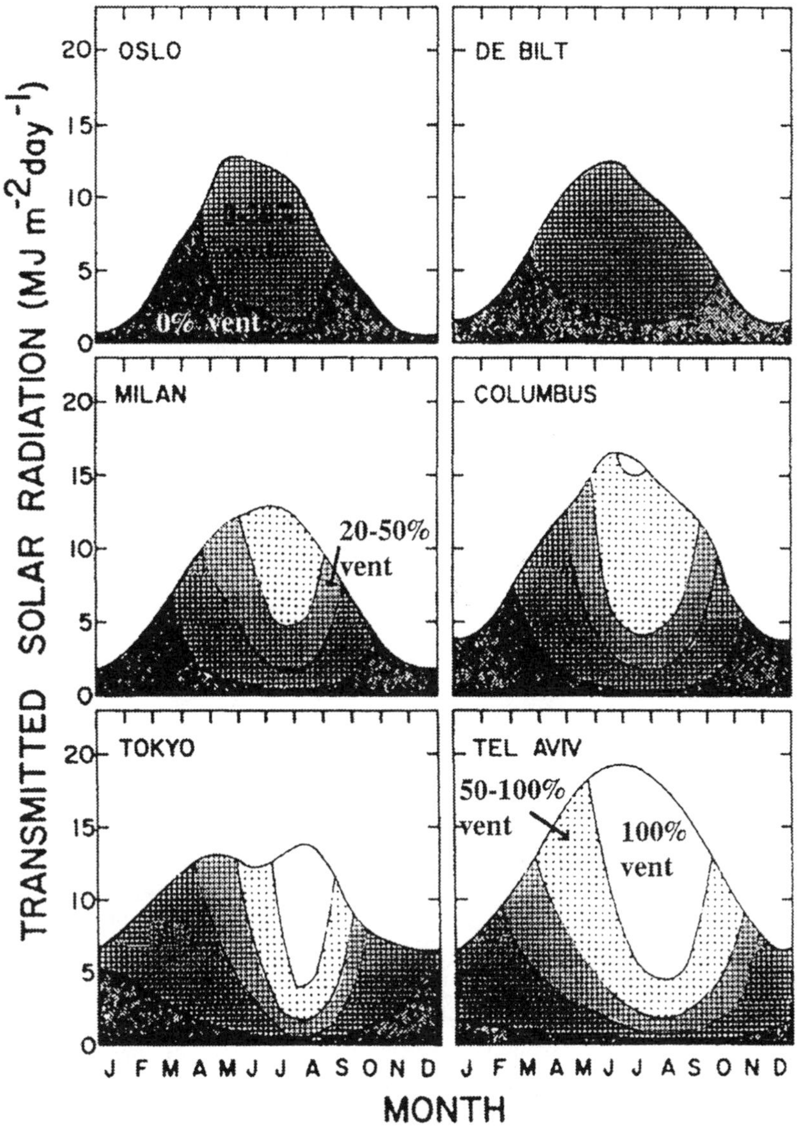

Fig. 7-33. Daily solar radiation inside greenhouses during various ventilation time fractions for the year at six locations. Vegetation temperature at 30 C, combustion type CO_2 generator and assumed transmittance set at 0.7. [Reprinted from Bellamy, L.A. and B.A. Kimball, ©1986. In *Carbon Dioxide Enrichment of Greenhouse Crops*, CRC Press, Boca Raton, FL].

costs, prevent crop damage as the result of CO_2 excess, and improve net return to the operation. Software programs can vary as to complexity and sophistication, and unless the grower has suitable, in-house technical support, he is dependent upon the computer and software suppliers. There are, to my knowledge, no standards upon which operators can depend, other than the reliability of the equipment supplier. The situation in The Netherlands, with their close communication and high density, is completely different from, say, the U.S. The

Table 7-6. Annual relative yields and CO_2 enrichment response for six locations as calculated by Bellamy and Kimball. See Fig. 7-23. [Reprinted from Bellamy, L.A. and B.A. Kimball, ©1986. In *Carbon Dioxide Enrichment of Greenhouse Crops*. CRC Press, Boca Raton, FL].

Ventilation class	Location					
	Oslo	De Bilt	Milan	Columbus	Tokyo	Tel Aviv
Yield increment due to CO_2 enrich ment during various ventilation classes						
0%	0.26	0.21	0.17	0.21	0.16	0.05
0-20%	0.22	0.30	0.16	0.21	0.30	0.24
20-50%	0	0	0.09	0.10	0.08	0.12
50-100%	0	0	0.03	0.06	0.03	0.07
0% + (1-20%)	0.48	0.51	0.33	0.42	0.46	0.29
Total	0.48	0.51	0.45	0.58	0.57	0.48
Percent yield increase due to CO_2 enrichment						
0%	26	19	14	13	9	2
0% + (0-20%)	48	47	28	26	27	13
Total	48	47	38	35	34	22

objective in this section is to look at various aspects influencing supply that growers should keep in mind, leaving to the next chapter some problems with actual climate control and CO_2 management.

A. THE CLIMATIC INFLUENCE

In the early years of CO_2 enrichment, the common procedure was to estimate the required quantity with time, and to inject that amount during the daylight hours when there was no ventilation. Often, the quantity was sufficient, for example, to maintain 100 Pa on clear days. This, of course, resulted in concentrations often exceeding 500 Pa on cloudy days, which was wasteful and likely to be damaging to the crop –especially with combustion sources within the structure.

Apparently, certain locations provided greater advantages for CO_2 injection. Examples were northern Europe, particularly Scandinavia, and those regions having clear skies with low outside temperatures (e.g., Colorado). Locations such as the Mediterranean and southern California found that greenhouses could not be kept closed long enough for significant crop enhancement if the crop required temperatures ranging from 5 to 10 C. (e.g., lettuce, radish, carnations, etc.). Opportunities were available for warm crops such as roses, tomatoes, cucumbers, and others during midwinter. An exhaustive analysis of climatic influence was made by Bellamy and Kimball [1986] in which they selected certain locations and analyzed ventilation requirements (Fig. 7-33). Based upon the assumption that elevated CO_2 levels could be maintained economically even with 20% ventilation [Challa and Schapendonk, 1986], the authors calculated, by means of a model, the percent increase in yield due to CO_2 enrichment during different ventilation classes (Table 7-6). It can be noted from Fig. 7-33 that locations such as Oslo and de Bilt did not usually exceed the ventilation class 0 to 20%, so that the crop response from enrichment was about 47 to 48% annually. This contrasts with Tel Aviv, where the annual improvement was 22% –even though, as the result of high radiation, the yield potential at Tel Aviv was more than twice that found for Oslo. The advantages of clear-day climatic regimes were emphasized by Hanan [1968]. If coupled with low outside temperatures, the period for enrichment is extended.

Fig. 7-34. Net monetary gain per square meter during summer months with increasing CO_2 concentration and ventilation at 21 C for tomatoes grown in the U.K. [Hand and Slack, 1988]. Exchange: £ = $1.49 (With permission of the British *Grower*).

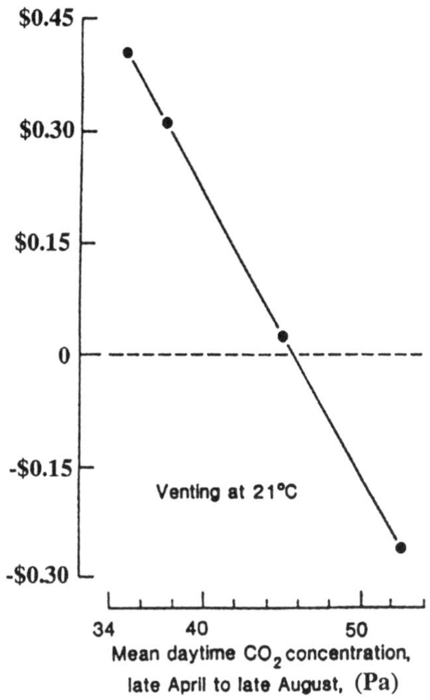

B. RAISING TEMPERATURES

There are ways of increasing the non-ventilation time, allowing for more CO_2 enrichment. One of the most obvious is to raise the ventilation temperature, based upon the idea that optimum temperature with raised CO_2 levels will be higher, and that with greater solar radiation, CO_2 enrichment will be more beneficial. Again, the ability to do this successfully depends upon the species and the climatic location. Lindstrom [1965] attempted this on roses in Michigan. However, this was not as successful as in Colorado [Hanan, 1973]. Fig. 7-18 (lower) showed an example of raising ventilation temperatures on roses, showing a faster growth rate with higher production. The internal ventilation temperature, above the basic day setpoint of 21 C, was 29 C, and actual temperature varied between the minimum and maximum, depending upon the amount of solar radiation. This might be considered as a semiautomatic temperature adjustment. But, the CO_2 levels must be correspondingly higher. Such practice was not successful in the summer [Mathis, 1972] when high ventilation rates reduced CO_2 levels to ambient concentrations, resulting in high production but very poor quality. While night and day temperatures can be raised slightly on carnations (ca. 1 C) subjected to enrichment, attempting to raise ventilation temperatures is likely to be counterproductive, even in the winter for carnations.

Other authors have attempted higher ventilation temperatures on warm vegetable crops [e.g., Schapendonk and Gaastra, 1984a; Hand and Slack, 1988], extending the enrichment season even into summer conditions where the ventilators are normally open. Hand and Slack examined tomatoes in the summer under conditions of ventilation at 21 and 26 C. While venting at 26 C reduced the CO_2 necessary to maintain levels of 37.5, 45.0, and 52.5 Pa, respectively, the percentages of marketable fruit and fruit yield were lowered markedly. There was a significant increase in net monetary gain at 21 C ventilation temperature, as seen in Fig. 7-34. According to Hand and Slack, enriching a greenhouse just to maintain CO_2 concentrations of 34 Pa can increase output and value of long-season tomato crops by as much as 10%, with reports of yield increases amounting to 4 kg m^{-2}. It can be shown that for every additional kilogram of CO_2 fixed by the crop, there is a potential for an extra 5 to 6 kg of saleable fruit at the correct temperature. Raising temperatures will affect DIF and ADT, as discussed in Chapter 4. The grower will find it necessary to reformulate his cultural practices. I am not aware of any information in the U.S. literature about this aspect of CO_2 on crops where ADT and DIF are important.

One must use care in raising ventilation temperatures. If the outside temperature is high enough, little advantage will be gained and the grower's objectives will not be met. Studies have been made on automatically controlling temperatures according to solar radiation [Hand and Soffe, 1971; Rudd-Jones et al., 1978]; that is, raising the minimum "heat-to" air temperature as solar radiation increases. The conclusions are contradictory. The results of Hand and Soffe caused tomatoes to reach the half-harvest date 5 days ahead of those grown in a steady-temperature regime. The success of the treatment was attributed to warmer daily mean air temperature and markedly extended periods of CO_2 enrichment. There was a greater incidence of *Botrytis* in the light-modulated treatments. For Rudd-Jones et al.'s experiments, the results suggested no financial advantage gained from any adjustment of the night or day temperature according to radiation integral. These authors pointed out that the greenhouse itself acts as a radiation-dependent controller on conventional thermostatic control under conditions

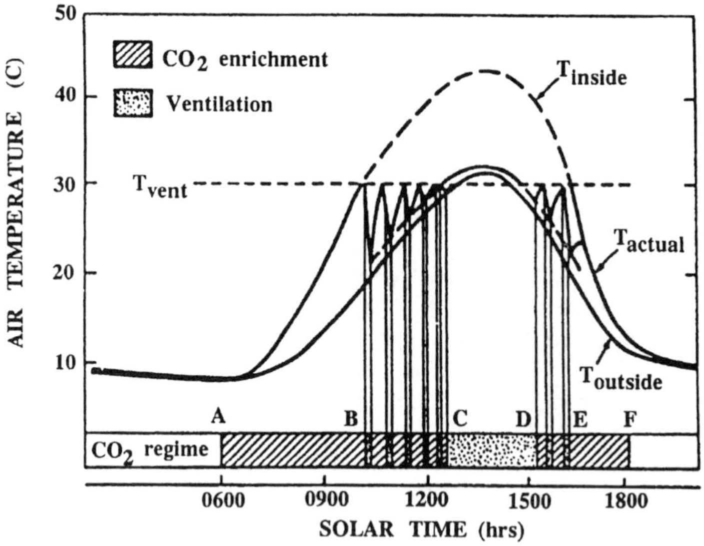

Fig. 7-35. Illustration of actual greenhouse air temperature and CO_2 enrichment. Periods A-B and E-F on the bottom bar represent periods of continuous injection. In periods B-C and D-E, the greenhouse is allowed to cool several degrees and then CO_2 is injected. From C to D, conditions are such as to require continuous ventilation [Enoch, 1984] (©1984, Int. Soc. Hort. Sci., *Acta Hort.*, 162:137-147).

of high radiation. The latter situation is similar to that found by Hanan [1973]. A more relevant point was emphasized by Hand and Soffe: allowing the air temperature to rise with higher solar radiation results in an increasingly uncomfortable environment for workers. This situation is particularly apparent if misting is employed, as in rose growing. Human comfort is a subjective quality, but highly important so that staff can carry out operations efficiently. Thus, arbitrarily raising ventilation temperatures can be nonprofitable.

C. INTERMITTENT CO_2 INJECTION

Another approach is to pulse or inject CO_2 at intervals so that some benefit is gained but the greenhouse can be ventilated to avoid excessive temperatures. This was described by Enoch [1984] and illustrated in Fig. 7-35. Since the inside temperature is below the ventilation setpoint, of course, continuous enrichment can go on (sections AB and EF). If the ventilators were closed all day, the temperature would rise, as indicated by the broken line. During the periods BC and DE, the house is allowed to ventilate, reducing the temperature, and then CO_2 was injected until inside temperature again reaches the ventilation setpoint. During the period CD, outside air temperatures and radiation are such as to require continuous ventilation. This procedure was tested by Mortensen [1986] under Norwegian conditions on African violet and chrysanthemum. Mortensen found continuous enrichment to be superior to intermittent treatments. One must also keep in mind the difference between conditions in Norway versus those in Israel.

One may totally enclose the greenhouse with an internal cooling system, as was examined by Kimball and Mitchell [1979] under Arizona conditions. While there was a highly significant yield increase on tomatoes in an unventilated house with an internal cooling system, such a system would be much too expensive on a commercial scale. However, Kimball and Mitchell calculated that even in a conventional structure, growers could expect a 10% yield increase from enrichment at Phoenix, AZ, in weather cool enough to permit injection. They concluded that CO_2 from combustion would be profitable, but not the use of pure CO_2.

D. DUTCH GUIDELINES

Using a model, Schapendonk and Gaastra [1984a] calculated relationships between radiation levels and CO_2 enrichment under Dutch conditions. Up to irradiance integrals of 1.25 MJ $m^{-2}dy^{-1}$, the CO_2 supplied by the heating system was sufficient to prevent CO_2 depletion. From 1.25 to 2.30 MJ $m^{-2}dy^{-1}$, requirements for heat decreased; consequently, CO_2 levels also decreased though there was no ventilation. As irradiance increased further, photosynthetic assimilation losses increased dramatically –calculated at 69 kg $ha^{-1}dy^{-1}$ at 3.30 MJ $m^{-2}dy^{-1}$

and a ventilation setpoint of 30 C. The authors divided production losses due to CO_2 depletion into four major classes:

1. Irradiance below 1.25 MJ $m^{-2}dy^{-1}$, excess CO_2 from heating system
2. From 1.25 to 2.30 MJ $m^{-2}dy^{-1}$, shortage of CO_2, ventilators still closed, less heat required
3. From 2.30 to 3.30 MJ $m^{-2}dy^{-1}$, severe CO_2 depletion as air exchange through ventilators insufficient
4. Above 3.30 MJ $m^{-2}dy^{-1}$, CO_2 depletion limited by high air exchange rates; assimilation still inhibited about 20%

Schapendonk and Gaastra assessed the monetary loss for 69 kg $ha^{-2}dy^{-1}$ as equal to \$630.00 $ha^{-1}dy^{-1}$ (Dfl = \$0.33), which an extra 39 m^3 of natural gas at \$5.60 would be sufficient to correct. At an irradiance of 4.82 MJ $m^{-2}dy^{-1}$, and a midday ventilation rate of 50 m^3 m^{-2} hr^{-1}, the simulated production at 33 Pa was 186 kg ha^{-1}. This was raised to 222 kg ha^{-1} when CO_2 was injected to 100 Pa every 30 minutes, providing another \$329.00 ha^{-1} at a cost of \$45.00 for 306 m^3 natural gas. There were a number of aspects of the mathematical model that were unrealistic –e.g., a ventilation setpoint of 30 C. If we assume an 8-hour day, a daily irradiance of 1.25 MJ m^{-2} dy^{-1} would be an average 43 W m^{-2}, 3.30 equals 115 and 4.82 MJ yields an average 167 W m^{-2}. In Chapter 3, I stated that, under Colorado conditions, radiation below 300 W m^{-2} would suggest a cloudy situation. Climatic conditions in Holland are much more overcast with higher outside temperatures, limiting the period of CO_2 enrichment above ambient levels. This discussion emphasizes climatic problems and the fact that any CO_2 depletion below 34 Pa represents a marked monetary loss to the grower[2].

E. HEAT FROM COMBUSTION

Fig. 7-36. Heat requirement and heat produced by CO_2 enrichment in The Netherlands for a standard Venlo structure [Nawrocki, 1984] (©1984, Int. Soc. Hort. Sci., *Acta Hort.*, 162:233-236).

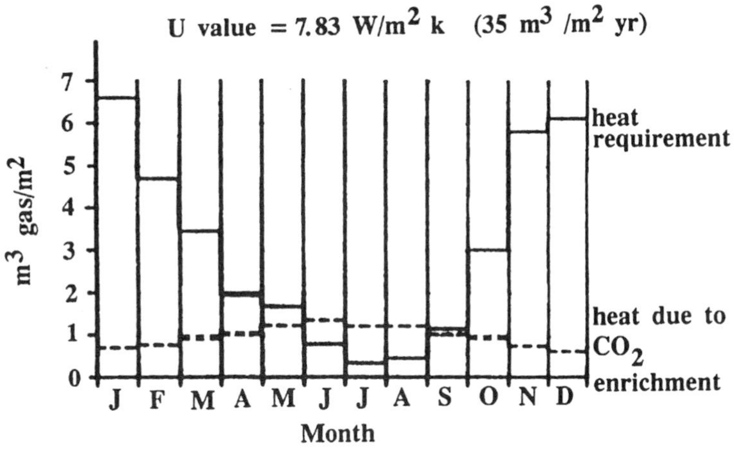

The grower has a problem where CO_2 is provided from combustion. While fuel cost is much lower than pure CO_2, the heat produced from combustion during the winter can be considered as an advantage. In fact, growers in Holland and the U.K. noted the benefits from the use of kerosene burners to prevent frost on lettuce. The

[2]Some reviewers of this text have the desire for radiation values to be expressed in mol $m^{-2}dy^{-1}$. Therefore, 1.25 MJ $m^{-2}dy^{-1}$, when converted according to Tables 3-3 and 3-4, would be equivalent to 2.45 mol and 3.30 MJ equivalent to 6.47 mol $m^{-2}dy^{-1}$ PAR. In view of the fact, however, that the energy utilized in photosynthesis can be considered as hidden in the error common to greenhouse energy balances, and that most of the energy evaporates water; total energy terms in units of MJ or kW-hrs are more meaningful. Otherwise, one is neglecting more than 70% of the energy important in heating, cooling and water loss.

common situation in Holland, given their low gas prices, is to inject CO_2 from the central boiler system at a minimum combustion rate of 25 m^3 ha^{-2} hr^{-1}, storing the additional heat in the boiler or in well-insulated, subsidiary tanks. Except in very severe weather, minimum temperatures in northern Europe are not what can be experienced in the mid-continent or the continental U.S. The relationship between heat required and heat produced due to enrichment was calculated by Nawrocki [1984] as shown in Fig. 7-36 for a standard Dutch Venlo structure. It will be noted that Nawrocki used a single-glazed roof with thermal screens and double-glazed walls. The total fuel consumption here was 35 m^3 gas $m^{-2}yr^{-1}$. In a total 4747 hours per year, 1548 hours of CO_2 heat can be used for heating the structure. The remaining 3199 hours will be lost unless heat-storage facilities are in place.

F. DAILY INJECTION PERIODS

The general procedure is to inject CO_2 from dawn to dusk. Attempts have been made to reduce CO_2 costs by restricting injection to the middle hours of the day. Hand [1983a] compared benefits of 100% (sunrise to sunset) versus regimes using 75, 50, and 25% of the time on lettuce. For an 8hour day at the winter solstice, this involved delaying the start of enrichment 1, 2, or 3 hours after sunrise and ending the same number of hours before sunset. Where the CO_2 supply is liquid, such regimes may be economical. Hand concluded that enriching continuously on lettuce in the U.K. was unnecessary. He figured that enrichment up to 50% during midday of the period between sunrise and sunset would be most efficient. Nevertheless, the usual practice is to enrich throughout the daylight hours despite irradiance. The possibility of failure to achieve maximal return from CO_2 injection is too high.

G. ENRICHMENT AT FULL VENTILATION

Throughout this section, references have been made to enrichment even during periods of full ventilation. Previous discussion has shown that CO_2 depletion can occur under high radiation even with the forced circulation of fan-and-pad cooling (Fig. 7-27). There are two ways of looking at the situation. One is to calculate the ventilation rate of the structure and inject only that amount required to maintain outside CO_2 concentrations of about 34 Pa. The second is to assume that if the inside CO_2 level is maintained at the outside level, only that amount necessary will be injected into the canopy and there will be no loss of CO_2 to the outside. Slack [1983] cited results to show that the photosynthetic rate of tomato leaves will be halved when the concentration falls to 17 to 18 Pa. Cucumber leaves can be expected to behave similarly. The answer, according to Slack, is to provide partial enrichment to avoid CO_2 levels below ambient outside concentrations. In a February 3 cucumber planting, there was a significant enhancement of yield at the fourth growth stage with 35 to 40 Pa CO_2. The biggest increases were in May (Fig. 7-17(upper)). Partial CO_2 enrichment from the 3rd to 20th week increased fruit numbers by 8%, total fruit weight by 11%, and gross monetary value by 12%. CO_2 costs for the 17-week period were \$0.25 m^{-2}. After deduction of recurrent costs, the increased dollar value was \$1.80 m^{-2}. Mean CO_2 usage for partial enrichment was about 23.9 kg $ha^{-1}hr^{-1}$, or 2.5 tons $ha^{-1}wk^{-1}$.

Nederhoff [1994] cites a publication from the experiment station at Naaldwijk (in Dutch) for actual CO_2 setpoints through the year in Holland. The diagram is reproduced in Fig. 7-37 . Under conditions of heating and no ventilation, the CO_2 setpoint is high (80 Pa). When there is almost no ventilation, and no heating, the setpoint is reduced to 50 Pa. Of course, if there is no heat input from the CO_2 source, this level could be adjusted, depending upon radiation and outside temperature. Beyond, say, 20% ventilation, the setpoint can be reduced to 35 Pa. In a more recent publication, van Meurs and Nederhoff [1995] have modified Fig. 7-37, using an algorithm that calculates the instantaneous CO_2 setpoint as the result of heat demand, radiation, wind speed, and ventilator position. This system requires reliable CO_2 measurement and a software program to adjust the various parameters, depending upon numerical values input to the program. An IRGA can be equipped with an on/off control if there is no climate-control computer. There are many publications detailing methods for CO_2 control and adjustment [e.g., Matthew et al., 1987; Harper et al., 1979; Hand and Bowman, 1969; Pettibone et al., 1970; Bowman, 1968; Dutton et al., 1988]. The procedure can be on/off for a given setpoint with suitable adjustment in flow rate and time to meet the setpoint.

Based upon a proposal by Lake [1966], Hand and Bowman [1969] computed air infiltration rates, or ventilation, of small greenhouses by measuring the diffusion of nitrous oxide (N_2O), which has a density and

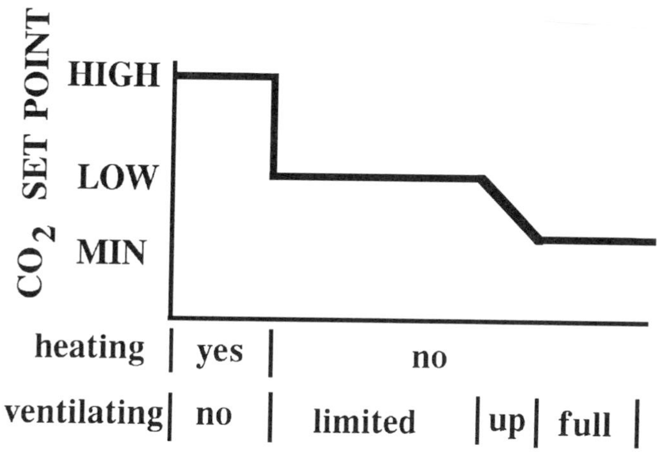

Fig. 7-37. Example of a method to control CO_2 based upon conditions in Dutch greenhouses [cited by Nederhoff, 1994, from PTG, 1992].

molecular weight similar to CO_2. By actually measuring weight changes in small cylinders of N_2O and CO_2, the exchanges due to ventilation and CO_2 uptake by the crop could be recorded. This represented a complex system for commercial establishments, although the middle photograph in Fig. 7-19 suggests the possibility of continuous measurement of CO_2 use. There is also the possibility of employing an IRGA in a differential mode. That is, an outside air sample goes through one cell of the IRGA, and another sample from inside the greenhouse is pumped through the other cell (see Fig. 7-1). The system can be programmed to reduce any difference detected by the analyzer by increasing or decreasing the CO_2 injection rate. Matthews et al. [1987] employed an IRGA in the differential mode for this purpose. The method has not, to my knowledge, been used on a commercial scale. It would reduce the possibility of excess CO_2 enrichment under conditions of maximum cooling. Where, as in the Dutch practice, excess heat from the central burner is stored, there is also the possibility of cryoscopic pumping at night to store liquid CO_2 that is released the following day. Thus, burner operation might not be required during the day. Cryoscopic pumping has been used on a small scale to produce high-pressure cylinders of known gas concentrations for experimentation [Hanan, 1984]. Essentially, one immerses a high-pressure cylinder in liquid nitrogen. Since the boiling point of CO_2 is -78.5 C versus -195.8 C for liquid N_2, CO_2 in an air stream passing through the cylinder will be condensed, particularly if there is a suitable absorbent such as crushed firebrick in the cylinder. Water vapor, of course, must be removed. This procedure has not been examined on any scale. Given solution of the problems (i.e., handling liquid N_2), this might be a method for an inexhaustible CO_2 supply in greenhouses.

It is surprising that studies of continuous CO_2 enrichment have not been published in the U.S. horticultural literature –though English literature from Europe suggests common use since the early 1980s. The two-volume work on CO_2, edited by Kimball and Enoch, makes little mention of enrichment to ambient CO_2 levels in the summer. Knowledge of the CO_2 uptake rate in a crop is important information required for optimizing the greenhouse climate. In particular, there have been many publications on optimizing CO_2 levels from the standpoint of maximizing net income [e.g., Challa and Schapendonk, 1986; Longuenesse et al., 1993; Zorn and Knoppik, 1993; Seginer et al., 1986; etc.]. These models and their validation are complex, requiring knowledge of higher mathematics. The purpose is to simulate crop response to CO_2 levels dynamically, enabling the grower to predict crop response; thereby, he can adjust in real time the climate controls to maximize net income. Rather than attempting to outline some of these procedures in this chapter, their general descriptions will be given in Chapter 8, control of the greenhouse environment.

VII. REFERENCES

Acock, B. et al. 1978. The contribution of leaves from different levels within a tomato crop to canopy net photosynthesis: An experimental examination of two canopy models. *J. Expt. Bot.* 29:815-827.

Aikin, W.J. and J.J. Hanan. 1975. Photosynthesis in the rose: Effect of light intensity, water potential and leaf age. *J. Amer. Soc. Hort. Sci.* 100:551-553.

Bailey, W.A. et al. 1970. CO_2 systems for growing plants. *Trans. ASAE.* 13:263-268.

Beer, S. 1986. The fixation of inorganic carbon in plant cells. *In* Carbon Dioxide Enrichment of Greenhouse Crops. Vol. II. H.Z. Enoch and B.A. Kimball, eds. CRC Press, Boca Raton, FL.

Bellamy, L.A. and B.A. Kimball. 1986. CO_2 enrichment duration and heating credit as determined by climate. *In* Carbon Dioxide Enrichment of Greenhouse Crops. H.Z. Enoch and B.A. Kimball, eds. Vol. II. CRC Press, Boca Raton, FL.

Bierhuizen, J.F. 1973. Carbon dioxide supply and net photosynthesis. *Acta Hort*. 32:119-126.

Bierhuizen, J.F., J.M. Bierhuizen and G.F.P. Martakis. 1984. The effect of light and CO_2 on photosynthesis of various pot plants. *Gartenbauwissenschaft*. 49:251-257.

Black, C.C., Jr. 1986. Effects of CO_2 concentration on photosynthesis and respiration of C_4 and CAM plants. *In* Carbon Dioxide Enrichment of Greenhouse Crops. H.Z. Enoch and B.A. Kimball, eds. CRC Press, Boca Raton, FL.

Bowman, G.E. 1968. The control of carbon dioxide concentration in plant enclosures. *In* Functioning of terrestrial ecosystems at the primary production level. *Proc. Copenhagen Symp*., 335-343. UNESCO, Paris.

Bravdo, B. 1986. Effect of CO_2 enrichment on photosynthesis of C_3 plants. *In* Carbon Dioxide Enrichment of Greenhouse Crops. Vol. II. H.Z. Enoch and B.A. Kimball, eds. CRC Press, Boca Raton, FL.

Calvert, A. 1972. Effects of day and night temperatures and carbon dioxide enrichment on yield of glasshouse tomatoes. *J. Hort. Sci*. 47:231-247.

Calvert, A. and D.W. Hand. 1975. CO_2 enrichment is still important. *The Grower*. 84(4):617-620.

Calvert, A. and G. Slack. 1975. Effects of carbon dioxide enrichment on growth, development and yield of glasshouse tomatoes. I. Responses to controlled concentrations. *J. Hort. Sci*. 50:61-71.

Calvert, A. and G. Slack. 1976. Effect of carbon dioxide on growth, development and yield of glasshouse tomatoes. II. The duration of daily periods of enrichment. *J. Hort. Sci*. 51:401-409.

Capron, T.M. and T.A. Mansfield. 1975. Generation of nitrogen oxide pollutants during CO_2 enrichment of glasshouse atmospheres. *J. Hort. Sci*. 50:233-238.

Capron, T.M. and T.A. Mansfield. 1976. Inhibition of net photosynthesis in tomato in air polluted with NO and NO_2. *J. Exptl. Bot*. 27:1181-1186.

Capron, T.M. and T.A. Mansfield. 1977. Inhibition of growth in tomato by air polluted with nitrogen oxides. *J. Exptl. Bot.* 27:112-116.

Carlson, R.W. and F.A. Bazzaz. 1982. Photosynthetic and growth response to fumigation with SO_2 at elevated CO_2 for C_3 and C_4 plants. *Oecologia*. 54:50-54.

Challa, H. 1976. An analysis of the diurnal course of growth, carbon dioxide exchange and carbohydrate reserve content of cucumber. Ph.D. Dissertation., Wageningen. 88 pp.

Challa, H. and A.H.C.M. Schapendonk. 1986. Dynamic optimization of CO_2 concentration in relation to climate control in greenhouses. *In* Carbon Dioxide Enrichment of Greenhouse Crops. Vol. I. H.Z. Enoch and B.A. Kimball, eds. CRC Press, Boca Raton, FL.

Dutton, R.G. et al. 1988. Whole plant CO_2 exchange measurements for nondestructive estimation of growth. *Plant Physiol*. 86:355-358.

Ehleringer, J. and O. Björkman. 1977. Quantum yields for CO_2 uptake in C_3 and C_4 plants. *Plant Physiol*. 59:86-90.

Enoch, H.Z. 1976. CO_2 enrichment of strawberry and cucumber plants grown in unheated greenhouses in Israel. *Sci. Hort*. 5:33-41.

Enoch, H.Z. 1984. Carbon dioxide uptake efficiency in relation to crop-intercepted solar radiation. *Acta Hort*. 162:137-147.

Enoch, H.Z. and R.G. Hurd. 1977. Effect of light intensity, carbon dioxide concentration and leaf temperature on gas exchange of spray carnation plants. *J. Exptl. Bot*. 28:84-95.

Enoch, H.Z. and B.A. Kimball, eds. 1986. Carbon Dioxide Enrichment of Greenhouse Crops. Vol. I. CRC Press, Boca Raton, FL.

Enoch, H.Z. and J.M. Sachs. 1978. An empirical model of CO_2 exchange of a C_3 plant in relation to light, CO_2 concentration and temperature. *Photosynthetica*. 12:150.

Enoch, H.Z., I. Rylski and Y. Samish. 1970. CO_2 enrichment to cucumber, lettuce and sweet pepper plants grown in low plastic tunnels in a subtropical climate. *Israel J. Agric. Res*. 20:63-69.

Enoch, H.Z. et al. 1973. Principles of CO_2 nutrition research. *Acta Hort.* 32:97-118.
Frydrych, J. 1976. Photosynthetic characteristics of cucumber seedlings grown under two levels of carbon dioxide. *Photosynthetica.* 10:335-338.
Gaastra, P. 1958. Light energy conversion in field crops in comparison with the photosynthetic efficiency under laboratory conditions. *Med. Landbouw*. Wageningen. 58:1-12.
Gaastra, P. 1962. Photosynthesis of leaves and field crops. *Neth. J. Agric. Sci.* 10:311-324.
Gaastra, P. 1966. Some physiological aspects of CO_2-application in glasshouse culture. *Acta Hort.* 4:111-116.
Gent, M.P.N. 1982. Controlled injection of carbon dioxide for rose production. *CT Greenhouse Newsletter.* 111:2-6.
Gislerød, H.R. and P.V. Nelson. 1989. The interaction of relative air humidity and carbon dioxide enrichment in the growth of *Chrysanthemum X morifolium* Ramat. *Sci. Hort.* 38:305-313.
Goldsberry, K.L. 1963. Growth of carnation increase blocks with supplementary CO_2. *CO Flower Growers' Assoc. Res. Bull.* 164:1-2.
Goldsberry, K.L. 1986. CO_2 fertilization of carnations and some other flower crops. *In* Carbon Dioxide Enrichment of Greenhouse Crops. H.Z. Enoch and B.A. Kimball, eds. CRC Press, Boca Raton, FL.
Goldsberry, K.L. and W.D. Holley. 1962. Carbon dioxide research on roses at Colorado State University. *CO Flower Growers' Assoc. Res. Bull.* 151:1-6.
Gormley, T.R. and P.E. Walshe. 1979. Carbon dioxide distribution in glasshouses. *Irish J. Agric. Res.* 18:45-53.
Goudriaan, J. and G.L. Atjay. 1979. The possible effects of increased CO_2 on photosynthesis. *In* The Global Carbon Cycle (SCOPE 13). B. Bolin et al., eds. Wiley, New York.
Hanan, J.J. 1968. The advantages of clear-day climatic regions for greenhouse production. *CO Flower Growers' Assoc. Res. Bull.* 217:1-4.
Hanan, J.J. 1972. Use of natural gas for CO_2 production in greenhouses. *CO Flower Growers' Assoc. Res. Bull.* 262:1-5.
Hanan, J.J. 1973. Ventilation temperatures, CO_2 levels and rose production. *CO Flower Growers' Assoc. Res. Bull.* 279:1-5.
Hanan, J.J. 1984. Plant Environmental Measurement. Bookmakers Guild, Longmont, CO.
Hanan, J.J. 1988. Summer greenhouse climate in Colorado. *CO Greenhouse Growers' Assoc. Res. Bull.* 461:1-6.
Hand, D.W. 1983a. Pruning CO_2 costs to maximize profits. *The Grower* (Supplement), 89-93. Feb. 17, 1983.
Hand, D.W. 1983b. On guard for air pollution. *The Grower.* 100:23-28.
Hand, D.W. 1984. Crop responses to winter and summer CO_2 enrichment. *Acta Hort.* 162:45-63.
Hand, D.W. 1986. CO_2 sources and problems in burning hydrocarbon fuels for CO_2 enrichment. *In* Carbon Dioxide Enrichment of Greenhouse Crops. Vol. I. H.Z. Enoch and B.A. Kimball, eds. CRC Press, Boca Raton, FL.
Hand, D.W. and G. Slack. 1988. What price summer CO_2 enrichment? *The Grower.* 109:27-31.
Hand, D.W. and G.E. Bowman. 1969. Carbon dioxide assimilation in a controlled environment greenhouse. *J. Agric. Eng. Res.* 14:92-99.
Hand, D.W. and K.E. Cockshull. 1975. Roses I: The effects of CO_2 enrichment on winter bloom production. *Hort. Sci.* 50:183-192.
Hand, D.W. and R.W. Soffe. 1971. Light-modulated temperature control and the response of greenhouse tomatoes to different CO_2 regimes. *J. Hort. Sci.* 46:381-396.
Harper, L.A., B.W. Mitchell and J.E. Pallas, Jr. 1979. A CO_2 controller for greenhouses. *Trans. ASAE.* 22:649-652.
Hayman, G. 1987. A success with straw. *The Grower.* 108:19-22.
Heij, G. and P.J.A.L. de Lint. 1984. Prevailing CO_2 concentrations in glasshouses. *Acta Hort.* 162:93-100.
Heij, G. and A.H.C.M. Schapendonk. 1984. CO_2 depletion in greenhouses. *Acta Hort.* 148:351-358.
Hiej, G. and J.A.M. van Uffelen. 1984a. Effects of CO_2 concentration on growth and production of glasshouse vegetable crops. *Acta Hort.* 148:591-595.
Heij, G. and J.A.M. van Uffelen. 1984b. Effects of CO_2 concentration on growth of glasshouse cucumber. *Acta Hort.* 162:29-36.
Heins, R.D. et al. 1984. Interaction of CO_2 and environmental factors on crop responses. *Acta Hort.* 162:21-28.

Hesketh, J.D. and H. Hellmers. 1973. Floral initiation in four plant species growing in CO_2 enriched air. *Environ. Control in Biol.* 11:51-53.

Hicklenton, P.R. 1988. CO_2 Enrichment in the Greenhouse. Growers Handbook Series, Vol. 2. A.M. Armitage, ed. Timber Press, Portland, OR. 58 pp.

Hicklenton, P.R. and P.A. Jolliffe. 1978. Effects of greenhouse CO_2 enrichment on the yield and photosynthetic physiology of tomato plants. *Can. J. Plant Sci.* 58:801-817.

Hicklenton, P.R. and P.A. Jolliffe. 1980. Alterations in the physiology of CO_2 exchange in tomato plants grown in CO_2-enriched atmospheres. *Can. J. Bot.* 58:2181-2189.

Holley, W.D. 1964. Type of greenhouse covering may affect CO_2 utilization by carnations. *CO Flower Growers' Assoc. Res. Bull.* 172:1-3.

Holley, W.D. 1970. CO_2 enrichment for flower production. *Trans. ASAE.* 13:257-258.

Holley, W.D. and K.L. Goldsberry. 1961. Carbon dioxide increases growth of greenhouse roses. *CO Flower Growers' Assoc. Res. Bull.* 139:1-3.

Holley, W.D. and K.L. Goldsberry. 1963. CO_2 and temperature requirements. CO_2 required per 1,000 square feet. *CO Flower Growers' Assoc. Res. Bull.* 164:3.

Holley, W.D., C.H. Korns and K.L. Goldsberry. 1962. The use of carbon dioxide on carnations. *CO Flower Growers' Assoc. Res. Bull.* 149:1-4.

Hughes, A.P. and K.E. Cockshull. 1969. Effects of carbon dioxide concentration on the growth of *Callistephus chinensis* cultivar Johannistag. *Ann. Bot.* 33:351-365.

Hughes, A.P. and K.E. Cockshull. 1971. The effects of light intensity and carbon dioxide concentration on the growth of *Chrysanthemum morifolium* cv. Bright Golden Anne. *Ann. Bot.* 35:899-914.

Hurd, R.G. 1968. Effects of CO_2-enrichment on the growth of young tomato plants in low light. *Ann. Bot. (N.S.).* 32:531-542.

Hurd, R.G. and H.Z. Enoch. 1976. Effect of night temperature on photosynthesis, transpiration and growth of spray carnations. *J. Exptl. Bot.* 27:695-703.

Ito, T. 1978. Physiological aspects of carbon dioxide enrichment to cucumber plants grown in greenhouses. *Acta Hort.* 87:139-146.

Jones, P. et al. 1985. Photosynthesis and transpiration responses of soybean canopies to short-and long-term CO_2 treatments. *Agron. J.* 77:119-126.

Johnston, R.E. 1972. A trial of glasshouse winter lettuce in Scotland. *Hort. Res.* 12:149-152.

Kelly, D.W., P.R. Hicklenton and E.G. Reekie. 1991. Photosynthetic response of geranium to elevated CO_2 as affected by leaf age and time of CO_2 exposure. *Can. J. Bot.* 69:2482-2488.

Kimball, B.A. 1986a. CO_2 stimulation of growth and yield under environmental restraints. *In* Carbon Dioxide Enrichment of Greenhouse Crops. Vol. II. H.Z. Enoch and B.A. Kimball, eds. CRC Press, Boca Raton, FL.

Kimball, B.A. 1986b. Influence of elevated CO_2 on crop yield. *In* Carbon Dioxide Enrichment of Greenhouse Crops. Vol. II. H.Z. Enoch and B.A. Kimball, eds. CRC Press, Boca Raton, FL.

Kimball, B.A. and S.T. Mitchell. 1979. Tomato yields from CO_2-enrichment in unventilated and conventionally ventilated greenhouses. *J. Amer. Soc. Hort. Sci.* 104:515-520.

Knecht, G.N. and J.W. O'Leary. 1983. The influence of carbon dioxide on the growth, pigment, protein, carbohydrate and mineral status of lettuce. *J. Plant Nutrition.* 6:301-312.

Koths, J.S. 1964. Alcohols as sources of CO_2 for greenhouse plants. Extension Reports No. 73. Univ. of CT, Storrs, CT. 2 pp.

Krizek, D.T. et al. 1974. Maximizing growth of vegetable seedlings in controlled environments at elevated temperature, light and CO_2. *Acta Hort.* 39:89-102.

Lake, J.V. 1966. Measurement and control of the rate of carbon dioxide assimilation by glasshouse crops. *Nature.* 209:97-98.

Leonardos, E.D., M.J. Tsujita and B. Grodzinski. 1994. Net carbon dioxide exchange rates and predicted growth patterns in *Alstroemeria* 'Jacqueline' at varying irradiances, carbon dioxide concentrations and air temperatures. *J. Amer. Soc. Hort. Sci.* 119:1265-1275.

Levanon, D., B. Motro and U. Marchaim. 1986. Organic materials degradation for CO_2 enrichment of greenhouse crops. *In* Carbon Dioxide Enrichment of Greenhouse Crops. Vol. I. H.Z. Enoch and B.A. Kimball, eds. CRC Press, Boca Raton, FL.

Lindstrom, R.S. 1965. Carbon dioxide and its effect on the growth of roses. *Proc. Amer. Soc. Hort. Sci.* 87:521-524.

List, R.J. 1966. Smithsonian Meteorological Tables. 6th ed. Smith. Misc. Coll. 114, Smithsonian Inst., Washington, D.C.

Longuenesse, J.J., C. Gary and M. Tchamitchian. 1993. Modelling CO_2 exchanges of greenhouse crops: A matter of scales and boundaries. *Acta Hort.* 328:33-47.

Loomis, R.S. and W.A. Williams. 1969. Productivity and the morphology of crop stands: Patterns with leaves. *In* Physiological Aspects of Crop Yield. J.D. Eastin et al., eds. Amer. Soc. Agron., Madison, WI.

Ludwig, L.J. and A.C. Withers. 1984. Photosynthetic responses to CO_2 in relation to leaf development in tomato. *Adv. in Photosynthetic Res.* 4:217-220.

Mansfield, T.A. and A.J.S. Murray. 1984. Pollutants generated in greenhouses during CO_2 enrichment. *Acta Hort.* 162:171-178.

Mathis, V. 1972. The effect of high day temperatures on new plantings of Forever Yours, Love Affair and Cara Mia roses. *CO Flower Growers' Assoc. Res. Bull.* 270:1-3.

Matthews, R.B. et al. 1987. Computer control of carbon dioxide concentration in experimental glasshouses and its use to estimate net canopy photosynthesis. *Agric. and Forest Meteor.* 40:279-292.

Mattson, R.H. and R.E. Widmer. 1971. Year around effects of carbon dioxide supplemented atmospheres on greenhouse rose (*Rosa hybrida*) production. *J. Amer. Soc. Hort. Sci.* 96:487-488.

McKeag, R. 1965. Natural gas as a source of CO_2 for greenhouse plants. *CO Flower Growers' Assoc. Res. Bull.* 185:1-4.

Moe, R. and L.M. Mortensen. 1986. CO_2 enrichment in Norway. *In* Carbon Dioxide Enrichment of Greenhouse Crops. Vol. I. H.Z. Enoch and B.A. Kimball, eds. CRC Press, Boca Ration, FL.

Monteith, J.L. 1969. Light interception and radiative exchange in crop stands. *In* Physiological Aspects of Crop Yields. J.D. Eastin et al., eds. Amer. Soc. Agron., Madison, WI.

Mortensen, L.M. 1986. Effect of intermittent as compared to contiuous CO_2 enrichment on growth and flowering of *Chrysanthemum X morifolium* Ramat. and *Saintpaulia ionantha* H. Wendl. *Sci. Hort.* 29:283-289.

Mortensen, L.M. and R. Moe. 1983a. Growth responses of some greenhouse plants to environment. V. Effect of CO_2, O_2 and light on net photosynthetic rate in *Chrysanthemum morifolium* Ramat. *Sci. Hort.* 19:133-140.

Mortensen, L.M. and R. Moe. 1983b. Growth responses of some greenhouse plants to environment. VI. Effect of CO_2 and artificial light on growth of *Chrysanthemum morifolium* Ramat. *Sci. Hort.* 19:141-147.

Mortvedt, J.J. et al. eds. 1991. Micronutrients in Agriculture. 2nd ed. Soil Sci. Soc. Amer., Madison, WI.

Nawrocki, K.R. 1984. Greenhouse heat balance as influenced by CO_2-enrichment. *Acta Hort.* 162:233-236.

Nederhoff, E.M. 1992. Effects of CO_2 on greenhouse grown eggplant (*Solanum melongena* L.) I. Leaf conductance. *J. Hort. Sci.* 67:795-803.

Nederhoff, E.M. 1994. Effects of CO_2 concentrations on photosynthesis, transpiration and production of greenhouse fruit vegetable crops. Ph.D. Dissertation. Wageningen. 213 pp.

Nederhoff, E.M. and K. Buitelaar. 1992. Effects of CO_2 on greenhouse grown eggplant (*Solanum melongena* L.) II. Leaf tip chlorosis and fruit production. *J. Hort. Sci.* 67:805-812.

Nederhoff, E.M. and R. de Graaf. 1993. Effects of CO_2 on leaf conductance and canopy transpiration of greenhouse grown cucumber and tomato. *J. Hort Sci.* 68:925-937.

Nederhoff, E.M., A.N.M. de Koning and A.A. Rijsdijk. 1992. Leaf deformation and fruit production of glasshouse grown tomato (*Lycopersicon esculentum* Mill.) as affected by CO_2, plant density and pruning. *J. Hort. Sci.* 67:411-420.

Nobel, P.S. 1991. Physicochemical and Environmental Plant Physiology. Academic Press, New York. 635 pp.

Norman, J.M. 1979. Modeling the complete crop canopy. *In* Modification of the Aerial Environment of Crops. B.J. Barfield and J.F. Gerber, eds. Amer. Soc. Agric. Engin., St. Joseph, MI.

Pallas, J.E., Jr. 1986. CO_2 measurement and control. *In* Carbon Dioxide Enrichment of Greenhouse Crops. Vol. I. H.Z. Enoch and B.A. Kimball, eds. CRC Press, Boca Raton, FL.

Pettibone, C.A. et al. 1970. The control and effects of supplemental carbon dioxide in air-supported plastic greenhouses. *Trans. ASAE.* 13:259-262, 268.

Pfeufer, B. and H. Krug. 1984. Effects of high CO_2 concentrations on vegetables. *Acta Hort.* 162:37-44.

Potter, J.R. and P.J. Breen. 1980. Maintenance of high photosynthetic rates during the accumulation of high leaf starch levels in sunflower and soybean. *Plant Physiol.* 66:528-531.

PTG. 1992. Terminologie geautomatiseerde kasklimaatregeling. Informatiereeks no. 102. Proefstation voor Tuinbouw onder Glas, Naaldwijk. 32 pp. (Cited in Nederhoff, 1994).

Rogers, M.N. 1980. Snapdragons. *In* Introduction to Floriculture. R.A. Larson, ed. Academic Press, New York.

Rudd-Jones, D., A. Calvert and G. Slack. 1978. CO_2 enrichment and light-dependent temperature control in glasshouse tomato production. *Acta Hort.* 87:147-155.

Rustad, S., T. Olsen and P. Thoresen. 1984. A technical and economical evaluation of a charcoal-based combustion system for CO_2-enrichment. *Acta Hort.* 162:189-196.

Salisbury, F.B. and C.W. Ross. 1985. Plant Physiology. Wadsworth Publ. Co., Belmont, CA. 540 pp.

Saxe, H. and O.V. Christensen. 1984. Effects of carbon dioxide with and without nitric oxide pollution on growth, morphogenesis and production time of potted plants. *Acta Hort.* 162:179-186.

Schapendonk, A.H.C.M. and P. Gaastra. 1984a. Physiological aspects of optimal CO_2-control in protected cultivation. *Acta Hort.* 148:477-484.

Schapendonk, A.H.C.M. and P. Gaastra. 1984b. A simulation study of CO_2 concentration in protected cultivation. *Sci. Hort.* 23:217-229.

Schapendonk, A.H.C.M. and W. van Tilburg. 1984. The CO_2 factor in modelling photosynthesis and growth of greenhouse crops. *Acta Hort.* 162:83-92.

Schrader, L.E. 1976. CO_2 metabolism and productivity in C_3 plants. *In* CO_2 Metabolism and Plant Productivity. R.H. Burris and C.C. Black, eds. University Park Press, Baltimore, MD.

Seginer, I. et al. 1986. Optimal CO_2 enrichment strategy for greenhouses: a simulation study. *J. Agric. Eng. Res.* 34:285-304.

Shaer, Y.A. and C.H.M. van Bavel. 1987. Relative role of stomatal and aerodynamic resistances in transpiration of a tomato crop in a CO_2-enriched greenhouse. *Agric. and Forest Meteor.* 41:77-85.

Shaw, R.J. and M.N. Rogers. 1965. Interactions between elevated carbon dioxide levels and greenhouse temperatures on the growth of roses, chrysanthemums, carnations, geraniums, snapdragons and african violets. *Florists' Rev.* 135(3486):23-24, 88-89; 135(3487):21-22, 82; 135(3488):73-74, 95-96; 135(3489):21, 59-60; 135(3491):19, 37-39.

Skoye, D.A. and E.W. Toop. 1973. Relationship of temperature and mineral nutrition to carbon dioxide enrichment in the forcing of pot chrysanthemums. *Can. J. Plant Sci.* 53:609-614.

Slack, G. 1983. CO_2: A new technique for an old commodity. *The Grower.* Mar. 17, 1983.

Slatyer, R.O. 1971. Effect of errors in measuring leaf temperature and ambient gas concentration on calculated resistance to CO_2 and water vapor exchanges in plant leaves. *Plant Physiol.* 47:269-274.

Thompson, C.J. and J.J. Hanan. 1975. Effect of CO_2 concentrations on roses: I. CO_2 uptake by individual leaves. *CO Flower Growers' Assoc. Res. Bull.* 306:1-2.

Thompson, C.J. and J.J. Hanan. 1976. Effect of CO_2 concentrations on roses: II. Yield. *CO Flower Growers' Assoc. Res. Bull.* 307:1-2.

van Berkel, N. 1975. CO_2 from gas-fired heating boilers - its distribution and exchange rate. *Neth. J. Agric. Sci.* 23:202-210.

van Berkel, N. 1984a. Injurious effects of high CO_2 concentrations on cucumber, tomato, chrysanthemum and gerbera. *Acta Hort.* 162:101-112.

van Berkel, N. 1984b. CO_2 enrichment in the Netherlands. *Acta Hort.* 162:197-205.

van Berkel, N. 1986. CO_2 enrichment in the Netherlands. *In* Carbon Dioxide Enrichment of Greenhouse Crops. Vol. I. H.Z. Enoch and B.A. Kimball, eds. CRC Press, Boca Raton, FL.

van Berkel, N. and J.A.M. van Uffelen. 1975. CO_2 nutrition of spring cucumbers in The Netherlands. *Acta Hort.* 51:213-219.

van Berkel, N. and J.B. Verveer. 1984. Methods of CO_2 enrichment in the Netherlands. *Acta Hort.* 162:227-231.

van Meurs, W.T.M. and E.M. Nederhoff. 1995. CO_2 control. *In* Greenhouse Climate Control. J.C. Bakker et al., eds. Wageningen Pers.

Vijverberg, A.J. and J.A.M. van Uffelen. 1977. The application of CO_2 in the culture of sweet peppers. *Acta Hort*. 58:293-296.

Wilkins, H.F. 1980. Easter lilies. *In* Introduction to Floriculture. R.A. Larson, ed. Academic Press, New York.

Wittwer, S.H. and W. Robb. 1964. Carbon dioxide enrichment of greenhouse atmospheres for food crop production. *Econ. Bot*. 18:34-56.

Yelle, S., A. Gosselin and M.J. Trudel. 1987. Effect of atmospheric CO_2 concentration and root-zone temperature on growth, mineral nutrition and nitrate reductase activity of greenhouse tomato. *J. Amer. Soc. Hort. Sci.* 112:1036-1040.

Zeroni, M. and J. Gale. 1989. Response of 'Sonia' roses to salinity at three levels of ambient CO_2. *J. Hort. Sci.* 64:503-511.

Zieslin, N. and A.H. Halevy. 1972. The role of CO_2 in increasing the yield of 'Baccara' roses. *Hort. Res*. 12:97-100.

Zorn, W. and D. Knoppik. 1993. Optimizing crop growth model input data acquisition - Computer simulation of dynamic gas exchange measurement. *Acta Hort*. 328:163-170.

CHAPTER 8

CLIMATE CONTROL

I. INTRODUCTION

Previous chapters each dealt with units, definitions and measurement. From time to time, discussion on control of the main factors (temperature, radiation, water, nutrition, and CO_2) has been included. The proper adjustment for purposes of maximizing profitability is not simple. In this chapter, one finds the greatest revolution in methodology that has ever occurred in greenhouse history. In the last decade, we have seen the digital computer reach a stage where enormous computational power is cheaply and readily available. Coincident with this, now, is the application of control theory and powerful mathematical procedures that will eventually help a greenhouse operator to maximize net income continuously.

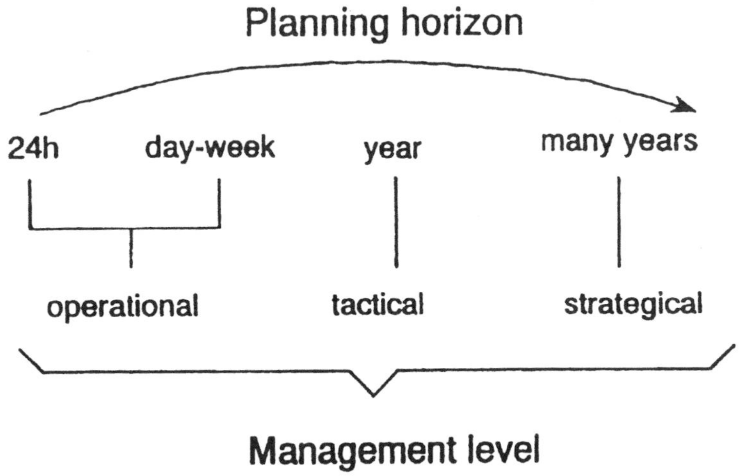

Fig. 8-1. Planning horizons for the greenhouse with suggested time periods as outlined by Challa and van Straten [1993]. The operational level is the management level dealing most directly with climate control (With permission of Academic Press, Inc.).

In the management of any business, the operator not only works with short-term problems, but he must continually plan over several periods of varying length. One of the better outlines of the process is Challa and van Straten's diagram in Fig. 8-1. Over several years, decisions on the strategic level determine the technical possibilities for climate control and the long-term nursery's policies on such factors as marketing, product quality, etc. At the tactical level, one formulates the crops, cultivars, timing, and so-forth in the expectation of **average weather**. Deviations in climate, market, and crop behavior can occur that require changes at the operational level with a duration ranging from a few minutes to as long as a week. A similar and more detailed approach for simplification has been the hierarchical levels outlined by Udink ten Cate [1983], Tantau [1990], and Bailey and Chalabi [1990]. The top, third level in Fig. 8-2 is concerned with the basic decision of crop planning, corresponding to the tactical level in Fig. 8-1. The middle and bottom parts of Fig. 8-2 are 8-1's operational level. The bottom level has a time span of minutes, or nearly instantaneous. The complexity of level two suggests the need for a digital computer of sufficient capacity and speed. While some attention will be given to level two, level one is the region with which we will be mostly concerned in this chapter. The main factors that the grower directly deals with on the first level are set out in Table 8-1.

Hammer and Langhans [1978] suggested more than 24 environmental parameters can be controlled. If one adds the possible subdivisions in Table 8-1, there are at least 33 separate and independent measurements (including individual nutrients) that can be used to characterize conditions inside and outside the greenhouse, which may be necessary to control the internal environment adequately. However, the table does not mention other information that can be important. These are position of, and opening and closure of, thermal and shade

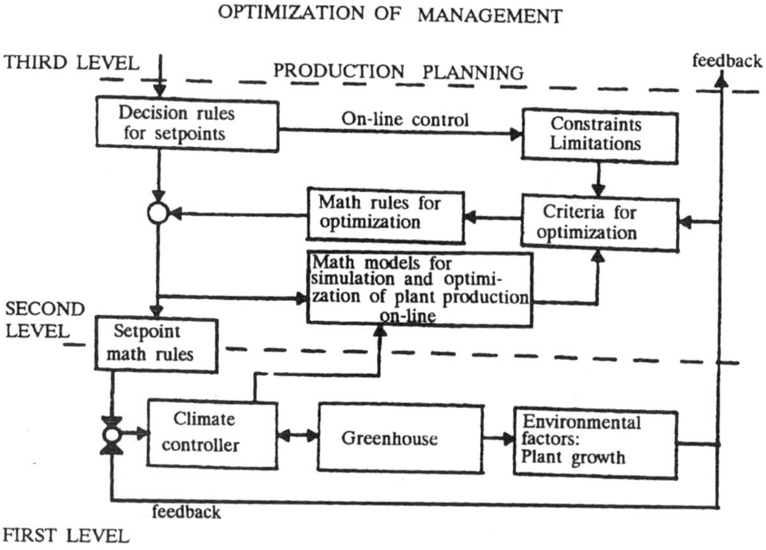

Fig. 8-2. Possible block structure of an on-line optimal climate control system as envisioned by Bailey and Chalabi [1990] (With permission of the ASAE) and Tantau (1990) (©1990, Int. Soc. Hort. Sci., *Acta Hort.*, 106:49-54). Control is exercised at three levels with the actual climate controller at the bottom level, supervised by a process computer at the middle level. Some of the terminology will be discussed later. Feedback is required from each level as noted on the right.

screens for energy conservation, photoperiodic control, and reduction of sunlight. Power consumption of photoperiodic lighting and supplementary irradiation could be recorded and used as a means to decide proper functioning. Such measurements would be valuable for appropriate operation of critical pumps, valves, and injection systems. Records of fuel and water use would be important to the grower in assessing costs. Table 8-1 does not show that the ultimate harvestable yield of the crop –which the grower may keep separately in his business account– could be used in maximizing profitability in climate control. These inputs could be weight, quality, and numbers of vegetable fruits, number and quality of cut flowers, growth of pot plants and final number and quality, actual product prices, and many other inputs as to crop status. With present programmable computers, there is sufficient power eventually to bring all these together into one system under grower direction.

The ability to regulate any of the main factors listed in Table 8-1 is inexorably related to:

1). Local climate
2). Structure in which control is to be accomplished
3). Technical knowledge available
4). Equipment availability

In particular, Chapters 2 through 4 explored the effects of climate in some detail. One can appreciate that fancy control equipment is of small use where there is little need for heating, or, the structure does not warrant an expensive outlay. Since increasing the number of parts to a system follows "Murphy's Law" –anything that can go wrong will go wrong– the available technical knowledge and parts supply will be important. However, this does not mean that present computers and related information-gathering equipment are of no use. Possibilities in this area will be discussed later.

Temperature control is a dominant factor, followed by humidity and CO_2. Control of the former will be found in almost any structure where outside freezing conditions are likely to occur. The control can be simply an on-off thermostat to turn on and turn off heaters. Temperate climatic regions often include some modicum of humidity and CO_2 management. Often, radiation, outside temperature and humidity, wind, and precipitation can be considered as disturbances for which suitable correction must be made internally. Water potential and photosynthesis, which are closely connected to most of the other main factors, cannot be measured in a meaningful fashion for

Table 8-1. Environmental factors measured for greenhouse control, and subject to control, depending upon external climate. Feasibility, accuracy, and precision vary widely. Some main factors can only be determined indirectly at present.

Main factor	Units of measurement	Subdivisions	Instruments	Actuators	Remarks
Radiation	Watts m^{-2} μmols m^{-2} s^{-1}	Outside global radiation Net radiation PAR, sky temperature	Pryanometer, net radiometer, quantum sensor, prygeometer	Shade screens, day-night settings, supplemental irradiation	Assuming radiation transfer characteristics of greenhouse known, outside measurement simplest. Instantaneous and accumulative values required. Interaction with temperature and CO_2 level. PAR measurement not vital.
Temperature	° Celsius	Outside, inside, crop, and root zone. Water and pipe temperatures if hot water system in use. Split zones (soil-air) require independent measurement of pipe temperature	Resistor, thermistor, thermocouple, liquid-in-glass, infrared thermometry	Boilers, unit heaters, ventilators, evaporative pads, air mixing systems, thermal screens, exhaust fans, actuating and mixing valves	Maximum and minimum temperatures must be included. Instantaneous, average and accumulative values required. Interaction with radiation and humidity. Separation between day and night temperature regimes. Sky temperatures necessary for energy balance.
Humidity	kilopascals (kPa), % relative humidity, dew point (C)	Inside and outside	Hair element, capacitance probe, psychrometer	Misting systems, ventilators, heating systems	Vapor pressure deficits between 0.2 and 1.0 kPa generally have slight effect on growth. Below 0.2 kPa VPD, % RH approaches 100, and considered undesirable. Above 1.0 kPa VPD, RH often below 60%, depending upon temperature. Considerable interaction with heating and ventilation systems. Instantaneous and average values required.
Carbon dioxide	Pascals, μℓ $ℓ^{-1}$, ppm, mmol m^{-3}	Inside and outside. Measurements should be within vegetative canopy inside	Infrared gas analyzer (IRGA)	CO_2 injection, which may include liquid or combustion systems	Strong interaction with radiation and temperature. Levels, when possible, not less than atmospheric to about 100 Pa (1000 ppm). Instantaneous and average values required.
Wind	m s^{-1}, km hr^{-1}	Outside, velocity always measured, direction usually	Cup or windmill anemometer. May be combined with wind vane	Ventilator system	Interaction with temperature and with ventilator position. Average values and extremes desired.

Main factor	Units of measurement	Subdivisions	Instruments	Actuators	Remarks
Precipitation	Presence or absence, may include rain gauge	Outside. (May be utilized for mist system control in propagation)	Resistance or optical device to sense rainfall	Ventilator system, also heating system in cases of snow	Utilized to position ventilators to exclude rain, or with heating pipes under gutters to melt snow. Use rain gages to provide totals.
Water	$m^3 m^{-2}$	Plant consumption	Numerous devices, depending upon cultivation system. Tensiometer, psychrometer	Irrigation systems, control in hydroponic culture	Not very common in practice due to numerous difficulties in measurement and assessing water loss over short time periods. Often ignored where water freely available to roots.
Nutrition	Individual nutrients mmol ℓ^{-1}, ppm, pH	In the root substrate, water supply	Selective electrodes, pH meters, conductivity bridges	Nutrient injection systems	Depends strongly on cultivation system. Strong interaction with water supply quality. Capability of measuring all essential nutrients in real time not available.
Water potential	MPa	Plants	Several experimental procedures on one or more plants	Humidity, irrigation systems	Mostly experimental, but a highly desirable measurement for evaluating internal plant stress, water demand and supply, etc.
Photosynthesis	$\mu mol\ cm^{-2}s^{-1}$, CO_2 uptake	Plants	IRGA. Requires knowledge of CO_2 uptake by canopy	CO_2 injection systems, for purposes of optimal solutions in control	Computers required with suitable software and attendant, related sensing units.

commercial application at present. Nevertheless, water potential (stress) and photosynthetic rate represent what is going on **in** the crop, and their indirect measurement is being rigorously perused at many research stations. Water supply, in hydroponics, can be bypassed. Nutrition control, again in hydroponics, is being continuously improved with new sensing elements that permit individual ion analysis and appropriate modifications of the solution.

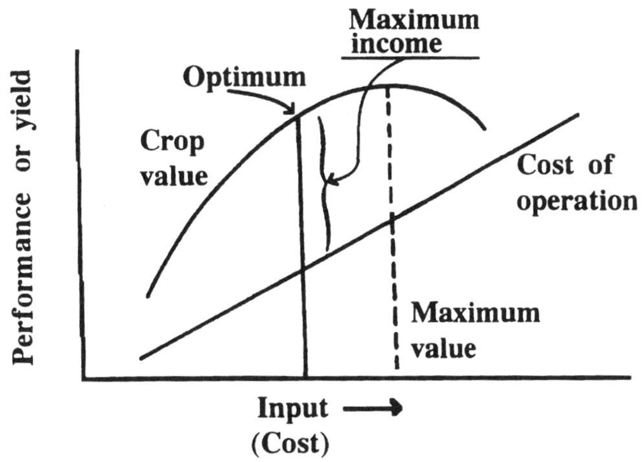

Fig. 8-3. Hypothetical relationship between cost of one or more operations (heating, cooling, CO_2 injection, etc.) and performance of the crop in cash value. Performance might be timing, weight gain, yield, etc. The particular diagram was modified after Bailey and Chalibi [1990] (With permission of the ASAE), but may be found in slightly different form in articles such as Challa et al. [1988] and Nederhoff [1988].

The first objective of a greenhouse operator is profitability. A well-defined objective is, according to Arnold [1979], necessary for automation; and, there must be included a plan by which that objective can be obtained. The situation one desires is diagrammed in Fig. 8-3, which depicts the difference between yield in economic return and costs of growing the crop. Maximum crop value is not necessarily the point that will provide the largest profit margin. Usually, these types of optimization as depicted in Fig. 8-3 have been applied to temperature control versus heating costs, largely for vegetable production. The literature is extensive in outlining the various approaches that include higher mathematics, statistics, linear programming, etc. plus the digital equipment required. Control theory is the subject of many texts [e.g., Bollinger and Duffie, 1988; Kuo, 1987; Ogata, 1970; etc.], requiring, especially, knowledge of calculus and matrix algebra. Classical and modern control theories make extensive use of the Laplace, and Ƶ transforms. With higher mathematics, such as Lagrange multipliers [Seginer et al., 1986; Gal et al., 1984] and Hamiltonian functions [Challa, 1993; Seginer, 1993; Pontryagin et al., 1962], optimization of the entire greenhouse operation, at least through the tactical management level (Fig. 8-1), is being addressed. Application of these methods, with the aid of high processing capability, means that this chapter will be limited in its discussion –as the author is no mathematician.

The ability to apply these approaches via computerization requires knowledge of basic greenhouse physics and plant physiology. Effort expended as shown by the number of publications has made climatic greenhouse modeling one of the "buzz" areas for the past decade. In particular, the Dutch have been most forward with many dissertations from Wageningen [e.g., Bot, 1983; Udink ten Cate, 1983; Stanghellini, 1987; Nederhoff, 1994; etc.], culminating with the recent publication of Greenhouse Climate Control [1995] with Bakker, Bot, Challa, and van de Braak as editors. Many contributions in this area have also been made by the Israelis [e.g., Seginer, 1980; 1981; 1989; Seginer and Sher, 1993; etc.], Japanese [e.g., Takakura, 1991, 1994, etc.], Germans [e.g., Tantau, 1980; 1985; 1990; 1993; etc.], as well as numerous other investigators from European countries. There is an even greater body of scientific and practical literature in other languages. The U.S. contribution in this area –with few exceptions– is minuscule in comparison.

Interestingly, many of these mathematical procedures were devised in the 18th and 19th centuries for dealing with the physical sciences. Pontryagin et al.'s [1962] development begins with a discussion of determining the optimal path for intercepting an object traveling in space –a rather warlike application. To see this terminology introduced into horticultural literature as though their definition and benefit are obvious to anyone of median intelligence is disconcerting to horticulturists, leading to a high frustration level. Publication quality in this area is unsatisfactory, with so much jargon, sometimes, as to make the author(s) unintelligible. As can be deduced from the terminology introduced above, mathematicians are better than horticulturists in disguising their trade.

II. BACKGROUND AND DEVELOPMENT

Throughout the 1940s and into the 1950s, control systems were primitive, consisting mostly of on-off control of temperature with heating and ventilation determined by separate thermostats. Examples of such devices and their mountings are typified in Figs. 4-4 and 8-4. Such control inevitably results in a temperature difference between the on and off points. Depending upon the particular situation, temperature variations can be much more than a one to 2 degree thermostat range. Improper heating system balance, unsuitable thermostat location and protection, or rapid changes in radiation or external temperature can result in unstable control with on-off temperature variation sufficient to reduce crop quality significantly. Suitably designed on-off systems, however, can be quite good and economically feasible.

As the result of development from the Second World War, the use of electrical circuitry to set inputs from measurements of the environment, and automatically to calculate the necessary output to actuating machinery, was gradually developed. This system is called "analog control" because a continuously varying environmental signal is subjected to modification to provide actuating signals to the heating and cooling systems [Carr, 1984]. Possibly the first discussion of these controls in the English horticultural literature was Winspear and Morris' 1965 publication, followed by Winspear in 1968. Two examples of such systems are pictured in Fig. 8-5. Control values were determined by experiment, and devoted to providing a set of "blueprint" values. The application of movable screens, supplemental irradiation, CO_2 injection, with multiple input variables, etc., however, led to increasingly complicated systems with numerous setpoints –the setting of which were not particularly user friendly. Cost began to exceed what could be obtained with digital systems. Nevertheless, many of these devices are still being manufactured and installed, although designed 20 years ago, since their use remains economically desirable, depending upon the particular situation.

Fig. 8-4. Typical thermostat mountings for temperature control in greenhouses prior to the 1960s. See also Fig. 4-4. These types may still be found in numerous installations. Application of sophisticated control systems, using these types of mountings, is ridiculous.

In the 1970s, computers were available for climate control in Holland [Gieling, 1980]. Within 2 years of 1974, five firms in the Netherlands were selling computer systems to greenhouses, increasing to nine by 1979, with more than 400 systems sold in that year. By 1986, the number of computer systems had increased to 4000 [Kooistra, 1986]. This development, much to the surprise of many individuals, resulted from the increased versatility of digital systems, labor saving, and the ability to upgrade the systems from time to time with the latest software. With continued development, systems became available that could receive inputs from and control more than 50 separate environments, all under the supervision of one centrally located programmable computer (PC). In 1989, the average state-of-the-art PC had a 386-Intel microprocessor, 2 megabytes (MB) of random access memory (RAM), a 50 MB hard drive, 13.5 or 8.9 cm floppy drives, and a 2400 bits per second (bps) modem. By 1995, the Pentium microprocessor was available with RAMs more than 30 MB, hard drives greater than 5 gigabytes (GB), with modems capable of transmitting up to 28800 bps. Historically, a complete revolution has occurred at least every 6 months in the computer industry in the past 2 decades, resulting in, Norton et al. [1995] states, the most powerful tool man has ever created. An outline of computer development is provided in Table 8-2.

Table 8-2. An abridged outline of the evolution of computing devices from Swetz [1994].

Date	Accomplishment or event
ca. 600 BC	Abacus used in classical Greece
AD 1614	Logarithms developed by John Napier
1620	Logarithmic scale basis for slide rule by Edmund Gunter
1623	Wilhelm Schickard invents machine that can perform four operations
1642	Pascal builds gear-driven computer to add and subtract with six-digit numbers
1673	Samuel Morland invents multiplying machine
1805	Jacquard develops card punch for textile looms
1830	Babbage conceives of great computing engines whose designs incorporate specifications for modern digital computers
1875	Frank Baldwin obtains patent for calculating machine
1945	Electronic Numerical Integrator and Computer (ENIAC) begins operation at the University of Pennsylvania
1947	Transistor developed at Bell Laboratories
1951	U.S. Census Bureau accepts delivery of Remington Rand UNIVAC 1 computer
1953	Magnetic core memory introduced
1957	FORTRAN programming language introduced
1959	Concept of integrated circuits conceived by Noyce
1960	COBOL language introduced
1964	IBM 360 marketed with binary addressing, cheap, feasible time-sharing and virtual memory BASIC language introduced
1969	UNIX operating system introduced Intel develops microprocessor
1970	Floppy disc introduced
1971	PASCAL language introduced First pocket calculators appear
1975	Microcomputers marketed
1976	Cray 1 supercomputer is operational
1980	ADA language introduced
1985	The Connection Machine, a highly parallel supercomputer with 65536 processors
1988	Computer networking well established
1995	Internet well established; PCs readily available to consumers

The thrust toward computerization in Europe, especially Holland, has been fostered by increasing governmental regulation to reduce pollution. Some of these developments have been discussed by van den Berg and Bogers [1995] and Bot [1991]. These regulations have moved toward completely enclosed systems by the year 2000 with pesticide use cut in half, reduced chemical use, reduced CO_2 and NO_x emissions, and all drainage water being reused. Bot pointed out that a closed system means that natural ventilation is not permitted and sensors will be required for monitoring air composition in addition to CO_2. The problems with nutrition control are formidable. The dispersed nature of greenhouses in the U.S., combined with lower economic importance, has, so far, resulted in few broad policies by government to enforce the changes seen in Europe. In developing countries, such as South and Central America, parts of Africa, and the Far East, enforcement of safety and pollution control measures remain nearly nonexistent. An enforced, greater economic burden in developed regions means that the local industry must continually strive to increase yields and quality while keeping unit cost

Fig. 8-5: Two examples of analog controllers still being installed in greenhouses. Note that thermostats may be used in combination. Various modulating, proportional controllers may be in use.

as low as possible. Appropriate computer application can be highly significant in meeting these objectives.

III. GREENHOUSE CLIMATE AND CONTROL

From a horticulturist's point of view, the introduction of so many new ideas in greenhouse climate control leaves the plantsman feeling dismayed. The recent publication by Bakker et al. [1995] starts at an advanced level –much beyond what the author taught in 1989 in greenhouse management. Bakker et al. assume an advanced mathematical background of the reader, as well as knowledge in plant physiology and engineering. Because the authors cover physiology and also control, many sections are brief. Nevertheless, this publication represents the most recent thinking of the Dutch research establishment, dealing with the standard Venlo greenhouse structure under north European climatic conditions and practice. The very good thing about the book is the fact it is in English, well representing the present status of climate control and its future.

Development in this chapter deals with some basics and a brief description of equipment available, following the author's learning process in reviewing the literature. The sections deal with:

1). Requirements of a control system
2). A brief summary of the greenhouse climate discussed in previous chapters
3). The difference between analog and digital recording and operation, and the use of feedback and feed-forward systems
4). The advantages of digital computer systems, with a brief outline of computer hardware and languages
5). A review of the instruments necessary for climate measurement and some of the problems involved

6). Modeling of the control process, plus discussion of the various control options at the operational level (Fig. 8-2)
7). Other models to be found in the literature that deal with growth, photosynthesis, and water

A. CONTROL REQUIREMENTS

There are at least four important attributes required of a control system: 1) stability, 2) accuracy, 3) sensitivity, and 4) reliability.

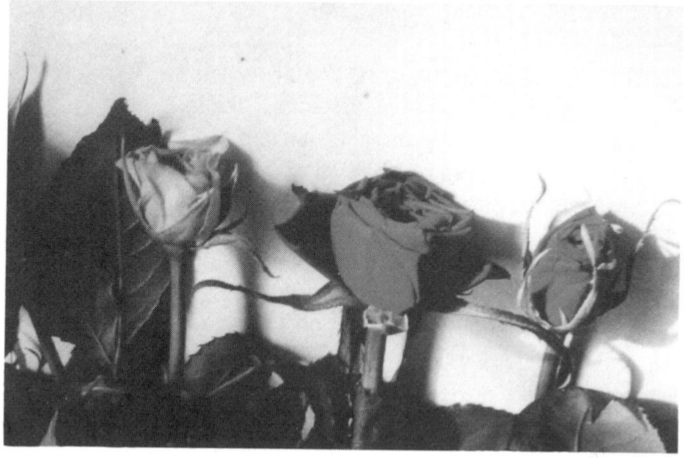

Fig. 8-6. An example of the damage an unstable heating system may cause to roses. Rapid, extreme temperature variation as the system cycled from full cooling to full heating resulted in abnormal flower development in a relatively short period.

1. Stability

Stability is the notion that the system will be able to follow an input command, and it is unstable if output is out of control or increases without bound. A greenhouse example is temperature "hunting." The control system reverses from extreme cooling to extreme heating. Fig. 8-6 is an example of the damage that can occur with a sensitive crop where hunting occurs –usually as the result of a narrow modulating range, coupled with slow equipment reaction such as ventilator movement. Many mathematical procedures can be used to decide whether a system

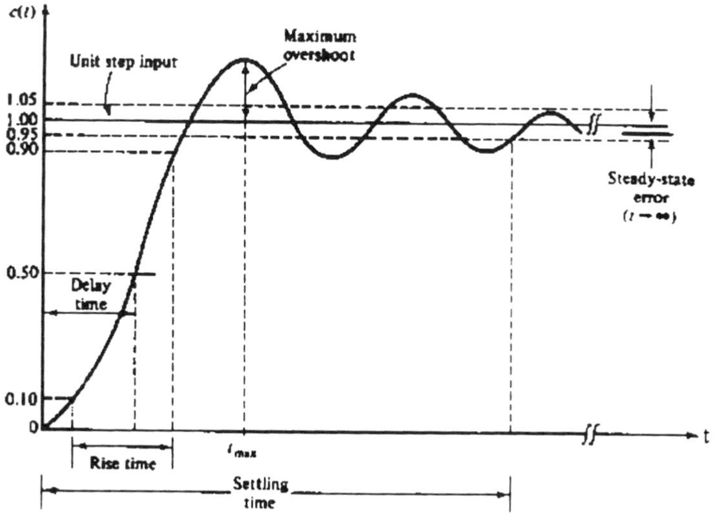

Fig. 8-7. Typical response of a control system to a change in input or a unit step input [Kuo, 1987]. The system is "linear" and the performance is stated to be in the "time domain". The "delay" time is that required for the step response to reach 50% of its final value, while "rise" time is that required for the response to go from 10 to 90% of its final value. The "settling" time is time required for the step response to stay within a specified percentage of its final value (With permission of Prentice-Hall).

is stable or how much and when it may be unstable. In any case, an unstable system is a useless system.

Fig. 8-8. Abnormal carnation flower development during dark, cold, winter weather when the system failed to raise the day temperature sufficiently. The converse is failure to lower the night temperature, resulting in weak flower stems in less than a week of such treatment. Other crops often behave similarly, showing the need of a regular program to oversee control system operation.

2. Accuracy

Accuracy refers to the ability of the system's output to achieve the desired input –or to "track," within a reasonable error margin, the input. An example of a typical response to a "step" input (reaction to a sudden change) is illustrated in Fig. 8-7. When the change is made such as when the thermostat setting is varied, there is a period required for the temperature to respond in a greenhouse. The temperature may initially overshoot the setting and then oscillate for a period (settling time) until reaction ceases and the temperature, then, remains steady at some value different from the desired setting. This is the steady-state error. Of course, the most desirable situation is immediate response with no oscillation and no error. But, this is impractical and may not be especially desirable if one is to meet other requirements such as stability. In a complete system, there may be several devices in the signal path that add to error. There are errors in the measuring devices, errors in the necessary analog-digital, digital-analog conversions, and there may be restrictions imposed by the number of bits chosen to represent a number in the digital computer. As a rule, the final error is the sum of all the errors in the system. In control theory there are computational means to assess error and its significance [i.e., Kuo, 1987; Ogata, 1970; Bollinger and Duffie, 1988].

3. Sensitivity

For measuring instruments, sensitivity shows the smallest change in the parameter being measured necessary to achieve a detectable output. In control systems, however, other parts of the system, besides sensors, may change over time. As temperature rises during operation of an electrical motor, for example, winding resistance changes. Or, the output of a sensor in the greenhouse may be sensitive to related environmental parameters –as well as the one being measured. In a good control system, the output should be insensitive to these latter changes while still able to follow the command responsively. In this respect, as we shall see later, the primary purpose of "feedback" is to reduce the sensitivity of the system to parameter variations and to unwanted disturbances.

4. Reliability

Reliability deals with the problem that output of the control process is what it should be over time. The system operates as it should without breakdown, or it is "robust." Overall, present hardware of computer systems is extremely reliable since it is solid-state. Hard discs eventually wear out, and the main actuators (ventilators, relays, motors, etc.) are likely to wear or break before any problems occur with the computer hardware. A discussion later deals with software, guiding the computer, which is likely to cause problems as it becomes more complex. The reliability of software is a serious problem in computerization. Another serious problem is interference by unwanted programs if the system happens to be in communication with others (i.e., Internet, viruses, hackers, etc.). A good example of unreliability is Fig. 8-8, showing what happens to a carnation flower if the system fails to act as it should –either because of system breakdown or the grower attempts a misguided practice to conserve energy.

5. Other Attributes

Of a system's performance, attributes include bandwidth (i.e., the ability of a system to respond to a wide range of frequencies); overall gain, which may be less or more than the input; and impedance, or resistance to passage of different frequencies. The importance of these latter attributes is less in greenhouse control as compared with some industrial processes. Although Winspear [1968] considered the response of greenhouses to outside disturbances (wind, radiation, etc.) as rapid, it depends upon the comparison. Regarding many aspects of modern processes (i.e., rocket control), a greenhouse, despite its small mass and high heat transmission, may require several minutes to hours to change a parameter such as temperature. So the considerations of impedance and bandwidth are not particularly germane because the various parameters change slowly and frequency will generally be less than 0.01 Hz (see Table 4-10) or essentially DC from the sensors.

However, it is the nature of present computers that the machine can only work with one variable at a time. Thus, Bollinger and Duffie [1988] show the need to consider control systems in terms of "discrete" modeling for which "difference" equations provide the mathematical basis versus "differential" equations used in analog control.

6. Electrical Interference

A particular problem that Gieling and Schurer [1994; 1995] discuss is electrical shielding from radio frequency interference (RFI) and protection of the system from lightning or lightning electromagnetic pulses (LEMP). Greenhouses are noisy in regard to lines carrying low signal voltages (mV or less). Relays may be opening or closing, motors starting and running, etc., each action probably generating radio frequencies that can be imposed upon unshielded lines. LEMP can be disaster to computer systems. Gieling and Schurer [1995] give an estimation of a 100% loss of tomatoes in soilless culture in 8 minutes, and 360 minutes for the same crop in soil, upon system failure. Failure of the first type was presented pictorially in Fig. 5-8. My experience with such failures has shown a longer period before visible damage is manifest. There are many precautions to be taken, whether or not a computer system is employed. There should be filters on incoming cables to short-circuit all transient voltages to a central ground. Small, inexpensive devices can be purchased that limit voltages on lines to a reasonable value –e.g., 12 Volts on a humidity sensor power supply. All electrically conducting metallic parts should be connected to each other and to a central ground electrode. This includes all piping and heating installations. Greenhouses themselves should have all construction parts provided with a galvanic interconnection, in turn connected to the central ground. Gieling and Schurer recommend double shielding on cables to sensors and actuators. The inner shield is connected to the main ground terminal of the computer electronics as protection from RFI, with the outer shield grounded at both ends to protect against LEMP. Lightning rods on meteorological masts, and on buildings in which the electronic systems are located, are necessary insurance. Sensors that have high impedance are more susceptible to RFI [Gieling and Schurer, 1995; Alberry et al., 1985]. Appropriate shielding becomes extremely important if uncorrupted signals are to be obtained.

7. System Failure (Reliability)

The possibility of system failure –either partially or completely– is always present no matter how simple the system. A thermostat can break, there can be a sudden power loss, or a boiler return pump can cease functioning. This means that any control system must have a suitable alarm and report generation. In 30 years of experience, it is remarkable how many growers have suffered disaster because they had no warning of what was going on in the greenhouse. Warnings and reports of parameter extremes (temperature, CO_2, humidity, etc.) are easy to program on computer systems. Lock-up of the computer due to mistakes in the software, failure of a component, or power failure is more difficult. Most growers of any size maintain standby generators capable of running essential equipment (e.g., boilers, circulating pumps, minimum exhaust fans, etc.), which start automatically if there is power failure. The computer system must be capable of re-establishing control when power returns [e.g., Gieling and van Meurs, 1979]. At least, appropriate authority must receive timely warning that something is amiss in an expensive establishment. Among the types of failure one may deal with in central computer-based control systems, Mitchell [1991] listed problems with input/output (I/O) devices (disc drives, monitor, printer, serial or parallel I/O, etc.), computer failure, remote I/O failure, individual input or output failure, and failure diagnosis with expert systems, as well as power failure. One may install manual and automatic switchover to conventional controls. During the early 1980s, Willits et al. [1980] dealt with the problem of lockup of the CPU on low voltages. They solved this by operating on a battery. Such solutions are no longer necessary, nor is the timing delay for power restoration to the CPU as outlined by Mitchell. Delays to high-output lamps and large motors can be necessary. Although, with present computer systems, warning systems can be programmed easily, my personal predilection is to include at least a basic temperature and power failure alarm entirely separate from the main control system. The simplicity and redundancy afforded assures safety.

B. GREENHOUSE CLIMATE

In the first chapter (1.II.F), the effect of the cover on air movement in a greenhouse was suggested as the most significant factor in determining internal climate. In particular, Section 4.IV.B.4 outlined the effects of energy conservation on air infiltration, CO_2 level, and humidity. Conservation measures will reduce air movement, reduce CO_2 during the day, and increase humidity. These factors all have significant impact on climate control. Thus, a summary of internal climate compared to the outside includes:

1. Wind speeds that are usually less than 0.1 m s^{-1}, frequently less than 0.04 m s^{-1}, whereas outside winds are ordinarily more than 1 m s^{-1}. Wind speeds in excess of 0.3 m s^{-1} are seldom encountered within greenhouses unless forced air cooling or hot air heaters are employed. As a result, energy and mass transfer (water vapor) will be reduced or significantly modified.

2. With full vegetative canopies (LAI > 3.0), humidity will invariably be higher than outside. Relative humidity can frequently exceed 90%, reaching 100% (saturation), especially when temperatures decline at sunset. The vapor pressure deficit (VPD) will invariably be less than 0.5 kPa. Values less than 0.3 kPa can have adverse effects on crop production. Special provisions are often required to reduce or prevent excess humidity (Section 5.IV) and to eliminate condensation on sensitive plant material.

3. Carbon dioxide concentration will decrease during the day, more so in tight greenhouses, to levels approaching 100 ppm (10 Pa) –unless CO_2 injection is practiced. Even with evaporative cooling, CO_2 levels within a dense canopy, on bright days, are commonly below the ambient CO_2 concentration (Chapter 7, Fig. 7-26) by several Pascals. With CO_2 injection, levels near, or much above, 100 Pa (1000 ppm) are encountered, depending upon the control system. Other gaseous components are more likely to be at high concentrations compared to outside. These could include boiler gases (CO, NO_x) and ethylene, plus vaporized products from plastic materials.

4. Radiation is invariably lower than outside (usually between 60 to 70% of the outside), so the greenhouse structure and cover, including internal components, should be chosen to maximize incoming radiant energy. Supplemental irradiation may provide additional energy on dark days.

5. Temperatures generally range between 5 and 30 C. Only in unheated houses in Mediterranean-type climates are temperatures below 5 C experienced during frost episodes. Much of the research related to temperature effects on plants is only valid in the range between 15 and 25 C.

1. Climate Variability in Structures

Sections 3.IV.B, 3.V, 4.IV.B.1, and 4.IV.D dealt with radiation and temperature characteristics inside greenhouses. For example, Figs. 3-25 through 28, 3-41, and 3-42 show the radiation differences that can exist internally depending upon structural type, season, and solar position. Where radiation interception on the supporting structure or plant canopy is the highest, there also will temperature be the highest unless evaporation is occurring as from a well-watered plant. Sensible heat is converted to latent heat, and those locations where radiant interception is the greatest will dry out quicker. Similarly, Figs. 4-86,4- 87, 4-89 through 4-92 and 4-94 illustrate examples of temperature variations as the result of the heating and cooling systems employed and the kinds of air circulation patterns one may expect. Vertical and horizontal temperature gradients are to be expected that vary with heating and cooling load and radiation intensity. Although many investigators make a basic assumption in their mathematical modeling of the greenhouse climate that the interior is a perfectly stirred environment (Chapter 1), **this is the exception**. Typically, the grower acknowledges this problem by suitable positioning of varieties and species to account for locations that may be above or below the "normal" or "average" temperature level.

In the European situation, the attempt is made to reduce vertical and horizontal temperature gradients by appropriate positioning of the heating pipe across the house near ground level, with continuous circulation of hot water. Lines are also positioned around the periphery and overhead –although the latter reduces incoming radiation. The ideal is to input only as much energy as required to maintain the set temperature. This can be difficult to do with steam or hot air convectors –the latter may be on or off with no modulation. Climate may be such as to require a high heat input that will tend to increase gradients in the structure. Although Holland is on a latitude higher than the U.S. northern boundary, the maritime influence tends to ameliorate extremes; whereas, in the continental Colorado climate, extremes are to be expected with minimum temperatures as low as -30 C and maximums above 40 C. These extremes will invariably increase temperature, CO_2, and humidity gradients within the structure.

The reason for the necessity to account for climate variations from one place to another within structures arises from the fact that in any one "environment" only one sensor may be present. Shielding helps reduce errors caused by radiation and increase sensitivity of the measuring instrument (Section 4.II, Fig. 4-1). Fig. 8-9 illustrates differences between shielded and unshielded temperature sensors. Note the higher temperatures and rapid fluctuations during the day in the unshielded detector, and the reduced sensitivity to on-off heater fluctuations at night. The fact remains that instrument location in a greenhouse is critical, varying with the crop species and its growth, and with the kinds of climate controllers utilized. I think multiple temperature and humidity sensors would afford better estimation of the "average" climate in the particular "environment" being controlled, and reduce the significance of instrument placement. Surprisingly, no one has investigated the utility of this idea where very expensive installations are likely. The expense of recommended sensors, their calibration, and maintenance mitigates against having more than one. Furthermore, few growers have any idea of how the internal climate varies unless they have observed obvious crop differences over a long time. The Dutch, with well-standardized structural designs, have compensated to a fair degree when using single measuring stations.

C. SOME BASICS OF CONTROL

Referring to Fig. 8-2, this Section deals with Level l with the setpoints already entered. The controller manipulates the signal outputs to the actuators to achieve the desired setpoint. We assume the grower enters the various setpoints manually. There are many other factors that are discussed here.

1. Analog versus Digital Signals

In the greenhouse, nearly all signals from sensors will be continuously varying in amplitude and over time, whether current or voltage. Many of the present simple control systems operate on Level 1 of Fig. 8-2. Analog signals can be amplified, added, compared with other signals, subtracted, multiplied, etc. to provide an appropriate output to correct or change the measured environment [e.g., Carr, 1984; Arnold, 1979]. The signal can be digitized at the sensor in most cases –which has advantages in transfer of uncorrupted information. An example of an analog signal is "A" in Fig. 8-10. While this example shows an equal and regular variation above and below the zero reference line, the real situation may be highly irregular. Only where signals deal with, say, ventilator or screen operations, is there likely to be a simple presence or absence of a voltage, indicating position, or a step change in voltage or current when, for example, a ventilator is closed or fully opened. Such on-off signals are often necessary to show onset or completion of an operation in climate contro!. Unfortunately, as the number of inputs and outputs grow, analog devices become cumbersome and complicated, with many settings that the grower may be unable to service correctly. Analog systems do have the advantage that "saturation" (extremes of output) is limited. Especially when numerous calculations are required, and there may be periodic changes, the digital computer is without parallel.

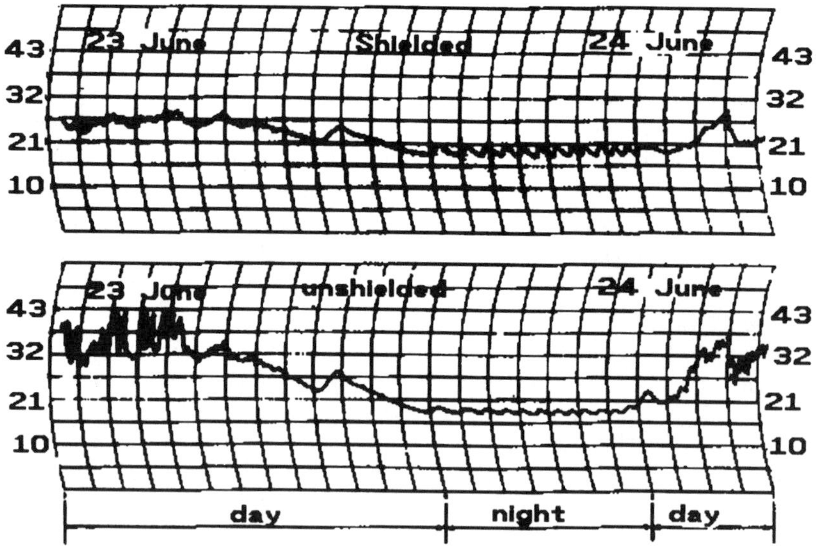

Fig. 8-9. Comparison of temperature recordings from shielded and unshielded temperature sensors [Adapted from Winspear, 1968]. Note that as the sun heats the sensor, temperature is much higher. Under these conditions, the control system would attempt to markedly reduce air temperature either by fully opening ventilators or turning on all exhaust fans and evaporative pad. At night, the unaspirated sensor indicates an average lower temperature without the sharp variations exhibited by the more sensitive aspirated and shielded detector (©1968, Int. Soc. Hort. Sci., *Acta Hort.*, 6:62-78).

However, when one employs digital machines to do the calculations, then problems arise. Digital computers operate as an on-off system, and their whole structure is to detect whether a "bit" of information is zero or one, to combine bits in appropriate words or bytes, and to operate upon them in a serial fashion. This means that the actual signal from the sensor can only be determined at some particular moment in time. Assuming a periodic signal as in "D" of Fig. 8-10, one samples the analog signal as suggested in "B," and an analog-to-digital (A/D) converter changes the information into the form upon which the PC can work. The value of the signal is known precisely only at certain instants in time. In Fig. 8-10, "C," the level will be held constant until the next signal to sample the sensor output. Obviously, the more rapid the sampling clock signal, the closer the digital representation comes to the actual analog input. Fortunately, with the slow response of environmental parameters in greenhouses, a sample every 30 seconds, or even every 3 to 5 minutes, can be sufficient to adequately represent temperature or CO_2 levels. Van Henten [1996], in his simulations of control, used 2 minute measurements with lettuce as the crop. Van Meurs [1995], however, said that all sensors were read every 15 sec, with setpoints recalculated each minute. Bot [1980] stated the sampling of radiation sensors as every second to get an accurate mean value in an 8 minute period. Bot said the time constant of the greenhouse system under examination was about 25 minutes when radiation changed. Generally, the rate of sampling should be at least twice the signal's highest frequency. This is a minimal rate, not necessarily optimal. Much higher sampling rates are often found, for example, in electrocardiographs, etc.

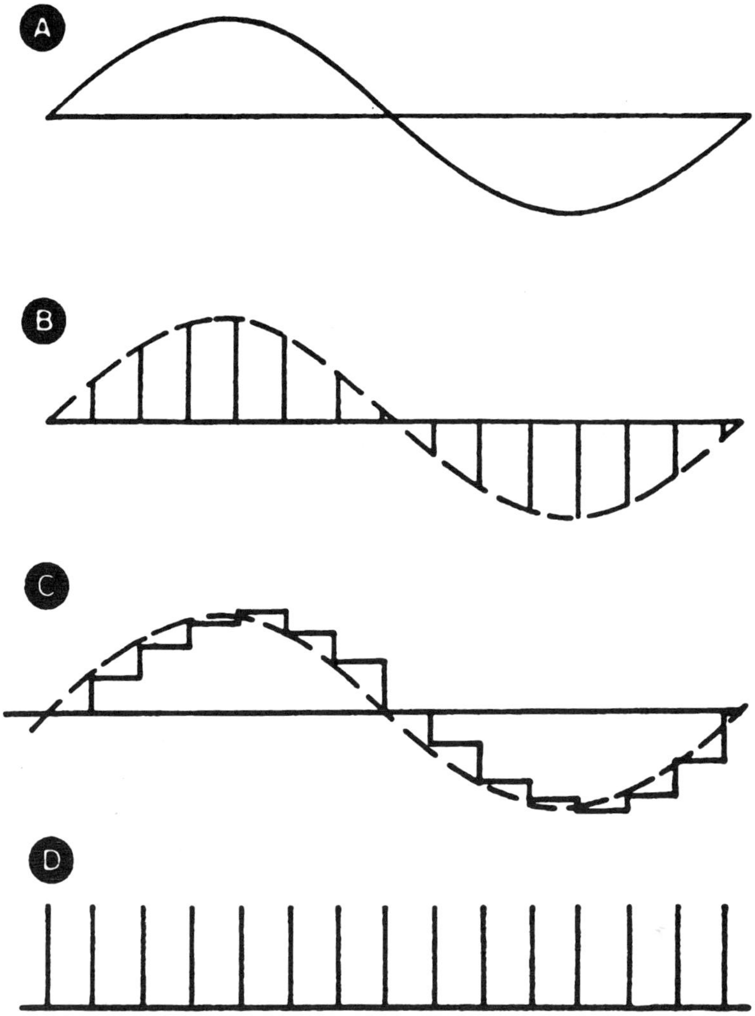

Fig. 8-10. Examples of signals: **A** analog, **B** a sampled analog, **C** a sampled and held signal, and **D** the sampling clock signal.

When the sample signal has been stored, printed, and manipulated appropriately, then it must be output in a fashion to operate a motorized hot water or steam valve or ridge ventilators. This will require a digital-to-analog (D/A) converter to output from a digital word the necessary voltage or current level to tell the valve to open or close correspondingly. With the speeds and capabilities of present PCs, the action is, to the human observer, instantaneous. One can diagram this system as in Fig. 8-11, showing sensors, A/D converter, computer with operator input, the D/A converter, and the actuators. Level 1 of Fig. 8-2 has been expanded to show a typical "feedback" system with the greenhouse as the process or "plant." In this system, the computer compares the sensor input to setpoints in memory to correct and change an output signal to actuators in a way that brings the environment back to the desired values.

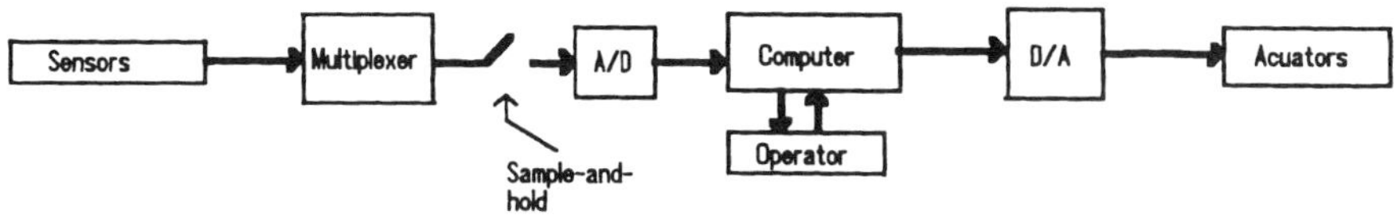

Fig. 8-11. A basic feedback control system. The sensors send an analog signal to the A/D converter under control of the computer to output to the computer an appropriate byte or word. For several inputs, there may be a "multiplexer" between the sensors and converter whereby the computer can switch from one signal line to another as appropriate. The interactive "operator" would include a monitor, keyboard, printer, and storage facilities. The revised signal is output to the D/A converter to control the actuators in the greenhouse. The operator enters the setpoints and the computer compares these setpoints with sensor input. The break in the flow line indicates a "sample-and-hold" device. That is, the signal level is held constant until initiation of the next measurement and hold cycle.

Fig. 8-12. Example of a closed-loop feedback control system. The circle indicates a comparison point, in this case an error detector in which the feedback is subtracted from the setpoint. Disturbances might include radiation level, outside temperature, humidity, etc.

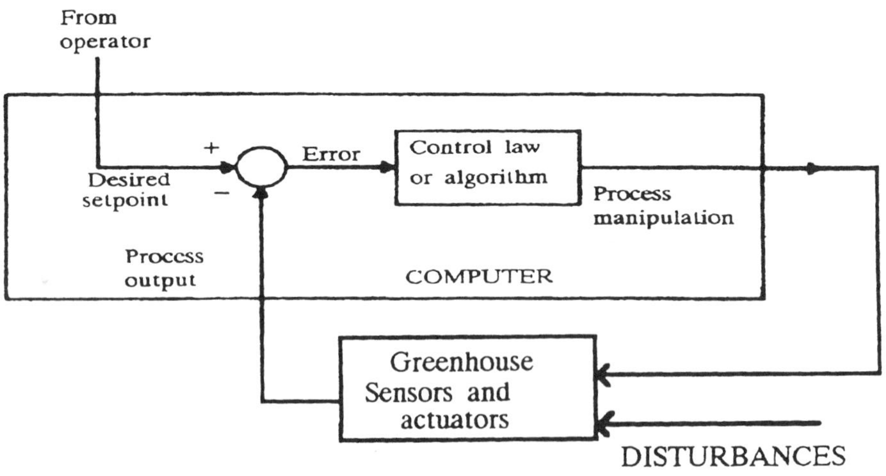

2. Feedback and Feedforward

Another way of representing Fig. 8-11 is given in Fig. 8-12, which is more nearly the manner in which control systems are presented in control textbooks. The greenhouse, in this case, is called the "process" or "plant." The signal from the greenhouse (e.g., temperature, humidity, etc.) is fed backward to an error detector where the error signal to the controller is simply the setpoint minus the actual output. In this fashion, the system continuously monitors and attempts to correct the difference in the signals to actuators in the greenhouse. It reduces the error between the reference input and the system output. This is the essence of "closed-loop" or feedback control.

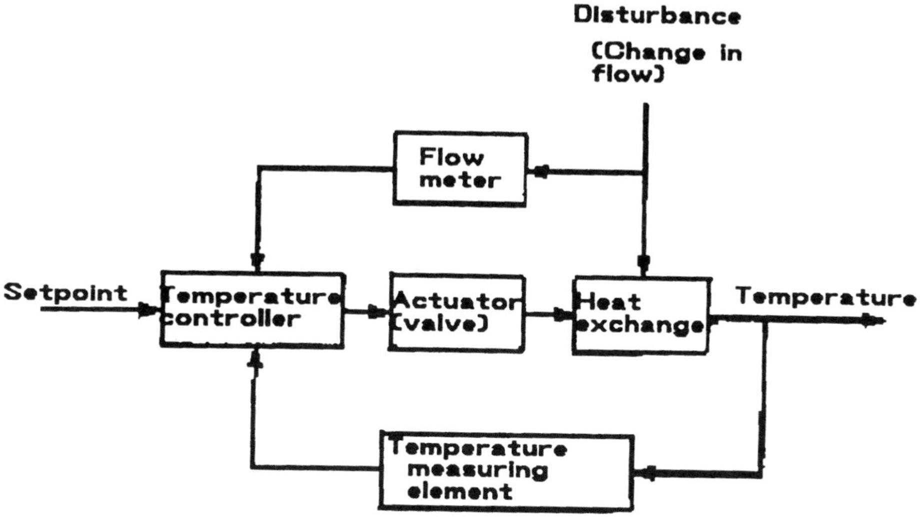

Fig. 8-13. A feedforward system combined with feedback. This could be a hot water heater. The change in water flow rate is measured and fed to the controller and added to the setpoint, correcting for an increase in flow rate [Adapted from Ogata, 1970] (With permission of Prentice-Hall).

The effects of feedback are profound. Whenever there is a closed sequence of cause-and-effect relationships among the variables of a system, feedback exists. Although feedback can stabilize an unstable system, it is a two-edged sword. Besides the manipulation by the control law (controller) in the signals to the greenhouse, the feedback can also be modified to cause a larger or smaller error signal for any given setpoint. Typically, feedback may be used to reduce the effect of noise on system performance. Feedback also has effects on system characteristics of bandwidth, impedance, transient response, and frequency response. There may be more than one feedback (Fig. 8-2), or they may be nested, depending upon the system.

In Fig. 8-12, the greenhouse is "disturbed" by such uncontrollable factors as radiation, outside temperature, wind speed, etc. If such disturbances are measurable, feedforward control is a means of canceling their effects upon system output. For example, an outside air temperature measurement can be used to correct the control signal, compensating an outside temperature reduction by increasing the output from the controller. The usual feedback starts corrective action only after the output has been affected. In essence, feedforward minimizes transient error caused by measurable disturbances, but it will not cancel effects of unmeasurable disturbances. A feedforward control must include its own feedback loop. An example is Fig. 8-13, which could be a means of controlling water temperature in a heating system in which flow rate changes as the result of mixing valve operation. The change in flow rate is the disturbance, and its measurement is a means to modify controller output so temperature is maintained at its set level. Another interesting system is the "cascade" discussed by Bontsema [1995], in which the effect of radiation on the heating system is compensated. Hot water systems have the unfortunate characteristic that they heat up fast but cool off slowly. Measurement of the incoming radiation allows the system to compensate quicker for the radiant heat load. Contrary to this situation in Holland is the use of radiant heat in Colorado, during low outside temperatures, to raise inside temperature, taking advantage of elevated CO_2 and permitting a faster photosynthetic rate. In this case, both outside temperature and radiation would be combined to raise the ventilation setpoint. Elevation of temperature setpoints with increased sunlight has been examined by Rudd-Jones et al. [1978], Hand and Soffe [1971], and Bowman and Weaving [1970]. The systems

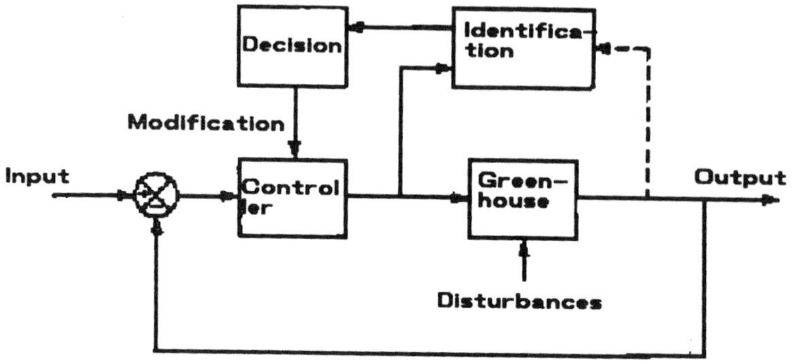

Fig. 8-14. An adaptive control system wherein the dynamic characteristics of the greenhouse are measured and compared with the desired characteristics, using the difference to adjust controller characteristics, or to generate a signal so that optimal performance can be maintained regardless of environmental changes [Ogata, 1970]. Although the literature discusses adaptive systems, use in commercial greenhouses has not been achieved (With permission of Prentice-Hall).

designed used analog equipment. The benefits depend upon the species and can be limited by the fact that there are other plant processes affected by temperature that may be more important than photosynthetic rate.

In some instances, an "adaptive" controller may be employed (Fig. 8-14) [e.g., Hooper and Davis, 1985]. Here, there is an identification of the dynamic characteristics of the greenhouse, a decision-making process based on the identification, and modification based on the decision made. If the greenhouse is imperfectly "known," then it becomes necessary to carry out the previous three functions continuously or at intervals. In other words, there is a constant redesign of the system to compensate for unpredictable changes in the greenhouse. In Fig. 8-14, the greenhouse is identified and the performance index measured periodically. The performance index is compared with the optimal and a decision made based on the findings about how to modify the actuating signal. Test signals may be necessary to make the identification. Udink ten Cate [1980], however, pointed out that all adaptive approaches require greenhouse identification to be done in models that are linearized around a working point that is not known and varies with time. In industry, adaption is based upon known disturbances or external influences –or feedforward control. The problem in greenhouses is that it is difficult to determine beforehand the effects of all the influences, nor are their relationships all that well known. Their relationships are usually statistical as contrasted to dynamic. When, as in greenhouses, there are many variables, strongly interrelated, classical control theory does not usually lead to satisfactory designs.

3. Closed- and Open-loop Systems

The first paragraph in the preceding Section mentioned that systems with feedback are considered "closed loops." A closed-loop system has a direct effect upon control action. The term implies the use of feedback to reduce system error. With "open-loop" systems, the output has no effect upon the control action. Nothing is fed back for comparison. Here, an example could be CO_2 injection that occurs during the day and when there is no ventilation. The injection rate is fixed. In older American literature, it was common to fix the CO_2 injection rate at a level to maintain the desired CO_2 concentration on a bright day. This leads to undesirable excessive CO_2 levels on cloudy days as CO_2 uptake by the crop would be reduced correspondingly. Open-loop control can be used only if the relationship between input and output is known and there are neither internal nor external disturbances. Any system that operates on a time basis is open loop –i.e., automatic washing machines and simple traffic signals.

4. Controller Functions

One should say here that powerful mathematical calculations are not necessary for successful climate control. The grower may use simple, direct-acting devices to control the environment quite successfully. A published example is Hanan et al.'s [1987] description for control of four 6 x 15 m research greenhouses. Transfer functions as found in control theory were not employed. Temperature, averaged from three stations in each house, was less than ±1 C at night and about ±1 C during the day. Problems of stability were small. What one finds is that present-day computers are wonderfully versatile machines, allowing an individual to program and control many environments in any fashion he sees fit; and to include as much, or as little, of the control as may be feasible. A second advantage is that new programs and changes in existing software may be made quickly whenever necessary. Once the initial capital investment has been made, upgrades can be introduced at nominal cost.

a. On-Off Control

This is the most common and least expensive system to install. It is used, obviously, where several unit heaters or convectors are employed, and with fan-and-evaporative cooling. In this case, the individual units are staged over a temperature range that is sufficiently wide to prevent continuous on-off operation –otherwise, the wear and tear on equipment is severe. The published articles on control, unfortunately, deal mostly with small research greenhouses in which environmental changes are likely to be rapid. For example, Saffell and Marshall [1983] described a system for five 4.7 x 10.1 m houses, combining an analog controller with a digital system, in which the mean daily temperature never deviated more than 0.1 C from the intended value. The authors' reason for using a hot-air system was its greater responsiveness.

A publication by Hesketh et al. [1986] described a dynamic model wherein the digital control provided a more closely controlled temperature than a simple proportional controller. Their article represents one of the few published greenhouse systems to be mathematically modeled, following general control theory with difference equations. The greenhouse in which this system was tested covered 120 m^2. Another interesting facet was the normalization of all signal values, i.e., the range was -1 to 1, with the values based upon units of ground area, with the means close to zero. There are a number of advantages to normalization, such as software simplification and easier mathematical computation. Normalization is a common procedure in control. An example of Hesketh et al's simulation is provided in Fig. 8-15. Over an approximate 2 hour period, their data on heater operation (**c**) suggest a cycling rate of about once every 4 minutes. While this might be sufficient to maintain temperature within about ±0.3 C, it would seem excessive cycling to me. A large greenhouse range would take much longer to respond, and attempts to maintain such close control would be self-defeating. Graph (**d**) shows the typical temperature cycling for this operation. Such cycling is to be expected even with staged systems as illustrated in Figs. 4-42, 4-43, 4-46, 4-47 and 4-77.

It appears that with on-off requirements so common, an adaptive system to control on-off cycles according to demand, either by adjusting the setpoint or the range between on and off, would be a good approach for commercial ranges. There are many situations requiring on-off besides temperature control. Ventilator positioning, screen positioning, CO_2 injection, supplementary and photoperiodic radiation, etc. are among the candidates.

Note that the setpoints between heating and cooling are almost invariably separated partially. This is to avoid simultaneous heating and ventilation except in the interest of reducing humidity. The temperature lift that may be allowed between heating and cooling varies with the system and plant species. In cool-requiring crops, temperature cannot be allowed to rise far (1-2 C), whereas with warm-requiring crops, the temperature might be allowed to rise more than 5 C before cooling begins. Although allowing temperature rise to occur before cooling affords advantages in CO_2 control, there is a limit in terms of worker comfort. Warm environments with high humidity may give difficulty in labor efficiency –unless all manual operations on the crop can be carried out in another enclosure.

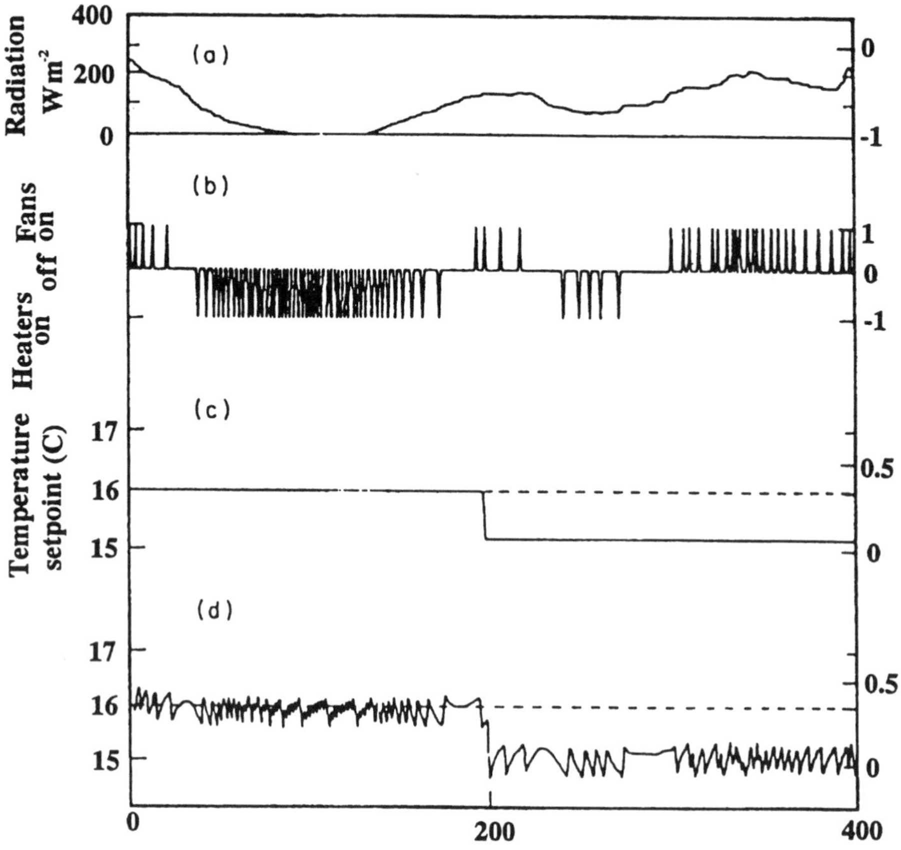

Fig. 8-15. Simulation of on-off control of a small greenhouse showing smoothed solar radiation (**a**), heater and cooling fan actuation (**b**), actual setpoint (**c**), and actual greenhouse temperature (**d**) [Hesketh et al., 1986].

b. Proportional, Integral and Derivative

P, I, and D control systems were originally applied with analog devices, and later used directly with digital computers. While there are reasons to apply different procedures to digital systems [Bollinger and Duffie, 1988; Young et al., 1993], these will be briefly discussed later. P, I, and D controllers in combination or singly are quite common where hot water systems are employed. Among those first publishing on such control systems for greenhouses were Udink ten Cate [1983; 1987], Udink ten Cate and van de Vooren [1984a; b] and Tantau [1980; 1985]. P, I, and D controllers may also be used where staging of several exhaust fans, unit heaters, or ventilators is required [Jones, 1994]. I am not sure that the latter have been carried out to any great degree. Software sold by companies supplying such control systems is undoubtedly considered proprietary. Publications on such subjects are few.

Eqs. 8.1 through 8.3 show the formulae for each of the types of control:

$$u(t) = K_p e(t) \tag{8.1}$$

$$u(t) = K_i \int_0^t e(\tau)d\tau \tag{8.2}$$

$$u(t) = K_d \frac{de(t)}{dt} \tag{8.3}$$

where: $u(t)$ *= output of the controller or input to the process*
K_p, K_i, K_d *= controller proportionality constants or proportional gain, integral controller gain and derivative controller gain respectively*
t *= time*
τ *= a dummy variable in control theory, but it is listed as time constant of the greenhouse in Chapter 5 of Bakker et al's text [1995]*
e *= error input signal resulting from comparison of the setpoint and feedback*

If the Laplace transform is used, as in an analog control, the following represents an approximation of the majority of processes [Bontsema, 1995]:

$$H(s) = \frac{K_p e^{-t_d}}{\tau s + 1} \tag{8.4}$$

where: t_d *= dead time, or the time for a response to occur*
e *= natural logarithm*
s *= the Laplace operator*

Fig. 8-16 illustrates the usual output from a proportional controller, combined with integral and then with derivative control, or the PID controller. For analog systems, the general formula is:

$$u(t) = K_c\{e(t) + \tau_I \int_0^t e(t)d(t) + \tau_D \frac{de(t)}{dt}\} \tag{8.5}$$

where: τ_I *= integral time (reset time)*
τ_D *= the derivative time (rate time)*
K_c *= proportional gain*

In P (proportional) processes, the actuator (or regulator) is varied directly with the deviation from the setpoint –also called modulation. The error between setpoint and actual output is multiplied by the gain, which can lead to an offset (Fig. 8-16). With I (integral) control, the value of the controller output is changed at a rate proportional to the integral of the error signal. If e(t) is doubled, u(t) varies twice as fast. Integral is sometimes called "reset" control. As noted in Fig. 8-16, the combination of P + I results in the error signal eventually being reduced to zero. Integral, however, tends to increase oscillation of the controller output. In derivative (D) control, the actuator is varied directly with the rate of change of the error signal. The derivative time constant is the interval by which the rate action advances the effect of proportional control action. D control has an anticipatory character. Unfortunately, it amplifies noise and can cause a saturation effect in the actuator –i.e., wind-up or the output goes to infinity. Due to the large time constants of greenhouses, Tantau [1980] stated that analog systems are difficult to adjust (tune). The time constants of the control system must be very large. It is much

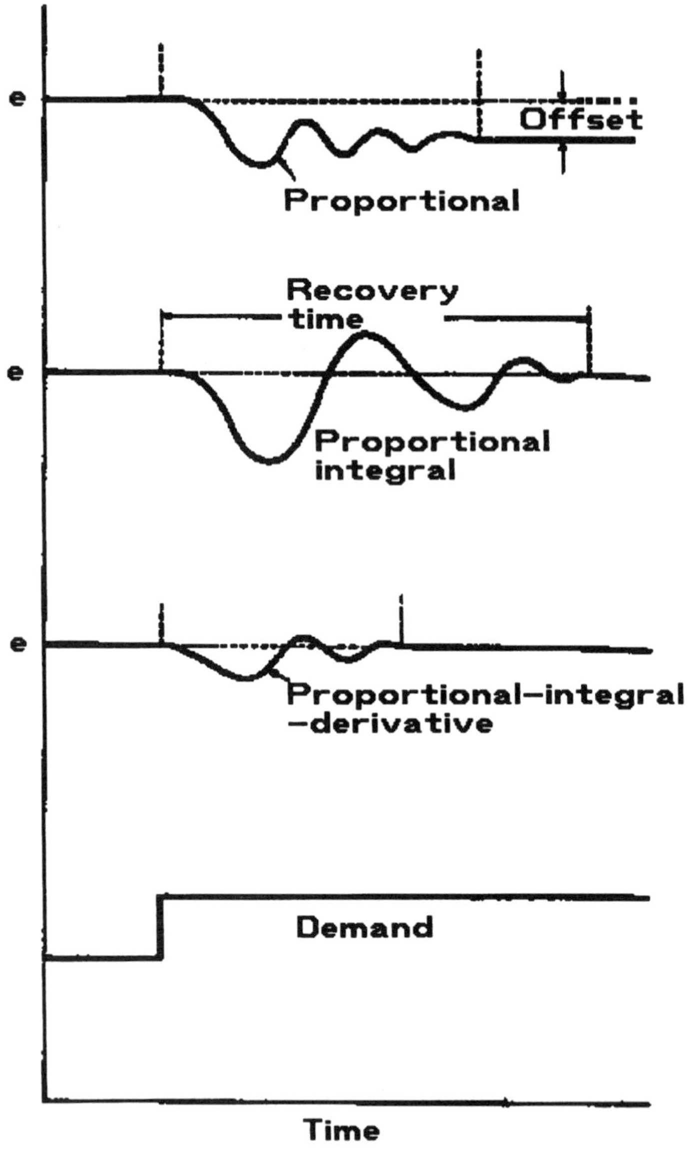

Fig. 8-16. Response of proportional, PI, and PID control to a step change in temperature [Winspear and Morris, 1965] (©1965, Int. Soc. Hort. Sci., *Acta Hort.*, 2:61-70).

easier with digital processors. Proportional control is considered rather slow, although including integral and derivative control can speed up the process.

Although PID controllers are quite common, other combinations such as P, PI and PD may be chosen for the particular problem. There is the necessity of tuning for the particular greenhouse by which the values of K_c, τ_I and τ_D are chosen. Performance criteria are necessary such as small maximum error, short settling time, minimal integrated error, minimum or no overshoot, small rise time, decay ratio, etc. [Bontsema, 1995]. Bontsema gives a brief outline of the procedures for P, PI, and PID tuning which can be carried out with the computer system and its recording facilities.

P, I, and D controllers in greenhouses are used mainly for hot water systems. The actuating valves are modulating, illustrated by Fig. 4-36. Beginning in the 80s with Udink ten Cate's dissertation [1983], numerous articles have been published [e.g., Udink ten Cate, 1985; 1987; Udink ten Cate and van Zeeland, 1981; Udink ten Cate and van de Vooren, 1984; Udink ten Cate et al., 1978; Bot et al., 1977; Valentin and van Zeeland, 1980; van Meurs and Stanghellini, 1989; Bakker, 1985; Challa et al., 1988, etc.]. In a recent publication [Bakker et al., 1995], van Meurs summarized present thinking. Figs. 8-17 and 8-18, for example, show temperature and ventilator control with a split range for the former for low and high pipe locations. Van Meurs states that well tuned controllers have an accuracy of better than ±0.2 C with over- and undershoots less than 25% of the temperature step.

In the same article, van Meurs discussed a combination of feedforward and feedback, with the former reducing expected inside changes as the result of changes outside. In this system the lower pipes are brought into use first, the upper pipes only if required to meet the heating load. The controller for the split-range calculates the individual setpoints of the hot water system, taking into account minimum and maximum temperatures of each system. Valentin and van Zeeland proposed this system in 1980. For digital systems, the discrete equation for a PI algorithm is:

$$u(k) = K_c e(k) + K_i t_s \sum_{j=0}^{k-1} e(k-j) \tag{8.6}$$

where: K_c = *proportional gain, the first group on the right-hand side being the proportional action*
K_i = $K_c x\ \tau_I$ = *integral gain, the right-most group*
$e(k)$ = $T_{sp}(k)$ - $T_I(k)$, *setpoint minus measured, integral, temperature (C)*
t_s = *the sampling time interval*
k = *an integer index*

Regardless of the system, changing the setpoint should be done smoothly, especially in the mornings, otherwise the boiler cannot meet demands. Saturation occurs in the signal and the integral part goes to infinity –or winds up. A sloping change diminishes the wind-up. This temperature change may start before sunrise, and reach the day setpoint some time after sunrise (a period of 1 to 3 hours). Of course, sunlight may hasten the warming. Maximum and minimums are necessary to keep saturation within limits. Udink ten Cate and van Zeeland [1981] proposed an anti-wind-up approach that keeps u(k) within limits of the pipe temperature. Eq. 8.6 becomes:

$$\begin{aligned} u(k) &= u(k-1) + K_c[e(k) - e(k-1) + K_i^* e(k)] \\ &\text{with } K_i^* = K_c K_i \text{ and subject to} \\ T_h(k-1) - c &\le u(k-1) \le T_h(k-1) + c \end{aligned} \tag{8.7}$$

where: T_h = *heating pipe temperature*
c = *a constant* ($\approx$ *5 C*)

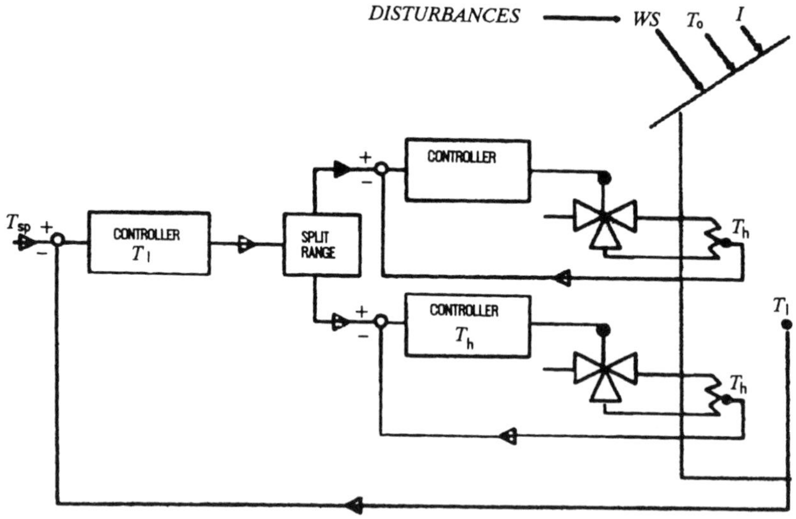

Fig. 8-17. Split range control for two hot water piping systems in Venlo structures [van Meurs, 1995]. T_I is the greenhouse air temperature integral, T_h the pipe temperatures, both fed back to comparators (Courtesy of Wageningen Pers).

The temperature at sunset should be allowed to return to the night setpoint gradually, similar to the day rise.

Another method to increase accuracy and stability is to calculate the heat load of the greenhouse from the internal and external temperature difference, wind speed, and radiation. This can be used to calculate a provisional setpoint for water temperature:

$$T_h^{/} = T_{hmin} + \frac{T_{omax} - T_o}{T_{omax} - T_{omin}}(T_{hmax} - T_{hmin}) + aWS - bR \tag{8.8}$$

where: T_o = *maximum and minimum outside temperatures at which pipe temperatures are maximum and minimum respectively,* T_o *being the outside temperature*

WS = wind speed

R = global radiation inside

a, b = adjustable coefficients

Van Meurs [1995] commented upon heated floors that have been examined by Verwaaijen et al. [1985] (see also Section 4.IV.B.4). As the response time of a concrete floor is between 5 and 8 hours, the floor can only be used as a base heating system, limited to 26 C (≈45 C water temperature). Problems with temperature control, using floor heating are obvious. If there is cloudy weather, the system will always be much behind air temperature response as radiant heat load changes. This might place a severe load on the pipe system to meet sudden radiation drops, or to overheating with a sudden increase in radiant load. Under clear-day climates, the use of floor heating requires a highly sophisticated control system that could be impractical from the standpoint of accuracy and

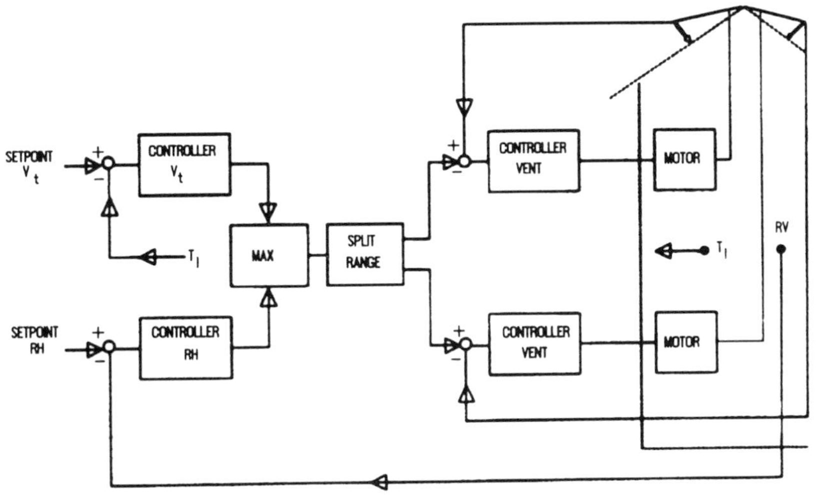

Fig. 8-18. Greenhouse ventilation and humidity control as given by van Meurs [1995]. T_I is the internal temperature integral and RV the humidity measurement. Note the use of relative humidity (RH) as contrasted with the use of an absolute measurement of vapor pressure deficit (VPD). The setpoints are V_t and RH (Courtesy of Wageningen Pers).

stability. The efficiency of a heated floor is stated as "very low" [van Meurs, 1995].

As with heating, ventilation in Venlo houses is controlled by a proportional algorithm (Fig. 8-18). Ventilator position, according to van Meurs, is calculated every time the system executes, rather than using potentiometers to measure position. Ventilator position depends upon wind speed and temperature difference between outside and inside and the proportional band. Control of relative humidity is included since only by opening the vents can humidity be efficiently lowered. The proportional band for humidity is wider than that for temperature. A combination of heating and ventilation is used when the outside-to-inside temperature difference is small and it is cloudy. A minimum pipe temperature is set, and the greenhouse temperature may rise above the ventilator setpoint so ventilators are forced open. The Dutch state that this also stimulates transpiration, which reduces the problems caused by low water demand discussed in Section 5.II.B. As RH increases, the vent setpoint will decrease and vice versa. The program also decides which ventilator to open first, based upon wind direction. Rain or snow may also cause a limit on vent position, sufficient to allow drainage.

The above contrasts with situations where fan-and-pad cooling is employed. Hanan et al. [1987] measured VPD, which was used as a setpoint to force first-stage cooling on at a designated VPD and opening the fan-jet louvers. The temperature setpoint was not changed. Under the low humidities found in Colorado, it was seldom necessary to force cooling and heating beyond a few early morning hours in the spring and autumn. Temperature reduction at sunset would cause RH to rise, but unless, in this case, the minimum VPD was less than the setpoint VPD, first-stage fan operation would not occur. When the system changed to the night setpoint, cooling fans were generally locked out and the inside temperature allowed to drift downward to the night setpoint, unless the outside temperature exceeded the day setpoint.

c. Other Control Functions

Other functions include thermal and shade screens, and photoperiodic and supplemental irradiation. These require, in addition to the times and conditions of opening and closing, suitable delay preventing continuous operation or sudden reversals. High-intensity lamps require a period of about 20 minutes to cool off before being switched on again. A minimum on time of about 15 min is necessary. Recording of power consumption by the irradiation system is not only needed for such optimizing algorithms discussed in Chapter 3, but also to give indication of burn-out of one or more lamps.

In Chapter 7 (Fig. 7-36), possible setpoints for CO_2 as outlined by Nederhoff [1994] are illustrated under a variety of conditions. In 1995, van Meurs and Nederhoff suggested the use of an algorithm to set instantaneous setpoint for concentration based upon heat demand, radiation, wind speed, and ventilator position. The use of a central, natural gas heating system provides a cheap source of CO_2 that can be blown into the canopy through perforated tubes. According to these authors, the maximum CO_2 setpoint (ca. 100 Pa) is used as long as heating is required independent of ventilation and radiation. A lower, medium setpoint is used when no heat is required and the radiation exceeds a certain level. A low CO_2 setpoint is used with no heat and radiation below a given level. Both the medium- and low-level setpoints are gradually reduced to the minimum level (ca. 34 Pa) as ventilation increases. As wind speed increases, both of these setpoints are shifted to the left on a graph of ventilator aperture versus CO_2 concentration. Heat generation as a consequence of CO_2 generation can be stored in suitable hot water tanks. Van Meurs and Nederhoff say that there is at least one commercial system that optimizes CO_2 supply and takes into account expected heat demand during the night on the basis of local weather forecasts.

A problem remains in that even with full, forced ventilation, the CO_2 level is likely to be below ambient in a dense crop in full sunlight. In 1966, Lake proposed control of CO_2 by injecting nitrogen dioxide (N_2O) into the house and measuring its decline with time. Ventilation rate could be determined and CO_2 levels better controlled. A procedure, based upon Lake's thoughts, was published by Hand and Bowman [1969], followed later by Nederhoff et al. [1984; 1985]. The latter authors determined ventilation rates by measuring inside and outside CO_2 levels. If ventilation rates are known, canopy CO_2 may be kept at least at ambient outside levels. A number of possible algorithms have been published by the Dutch [e.g., Nederhoff, 1988; Schapendonk and Gaastra, 1984a; Nederhoff et al., 1989; Bakker, 1985] as well as others [e.g., Matthews et al., 1987]. The inexpensive CO_2 supply available to the Dutch means that injection can continue even at high ventilation rates if the gas is distributed properly in the vegetative canopy. A sophisticated optimization program was proposed by Challa and Schapendonk in 1986. It appears that inside and outside CO_2 determinations should be made regularly. Not only would this influence CO_2 injection, but the process would be important in determination of sensible heat transfer to and from the structure, and lead to a real-time determination of photosynthetic rate.

So far, very little has been said about nutrition control. Generally, nutrition and water control have been considered separately from procedures outlined above. Systems as described by van den Vlekkert et al. [1992], van den Vlekkert and Kouwen-hoven [1992], and Fynn et al. [1991; 1994] are approaches for enclosed, continuous recirculating setups in which specific nutrients are added as required to the solution. Gieling and Schurer indicated that up to 14 stock solutions may be used to dispense individual concentrates into the circulating solution. Fig. 8-19 is an example of such a system that obviously uses feedback to control nutrient level. If only pH and electrical conductivity are used as feedback signals, then excess water to plants is wasted. As will be noted, sensors for ion concentration measurement are still in the developmental process. Measurement of soil moisture was briefly covered in Section 5.III.E. Gieling and Schurer mention other sensors under development, but given the thrust toward unlimited water supply and recirculating systems, I am not optimistic for computer control in handling soil moisture in greenhouses where plants are grown in the ground.

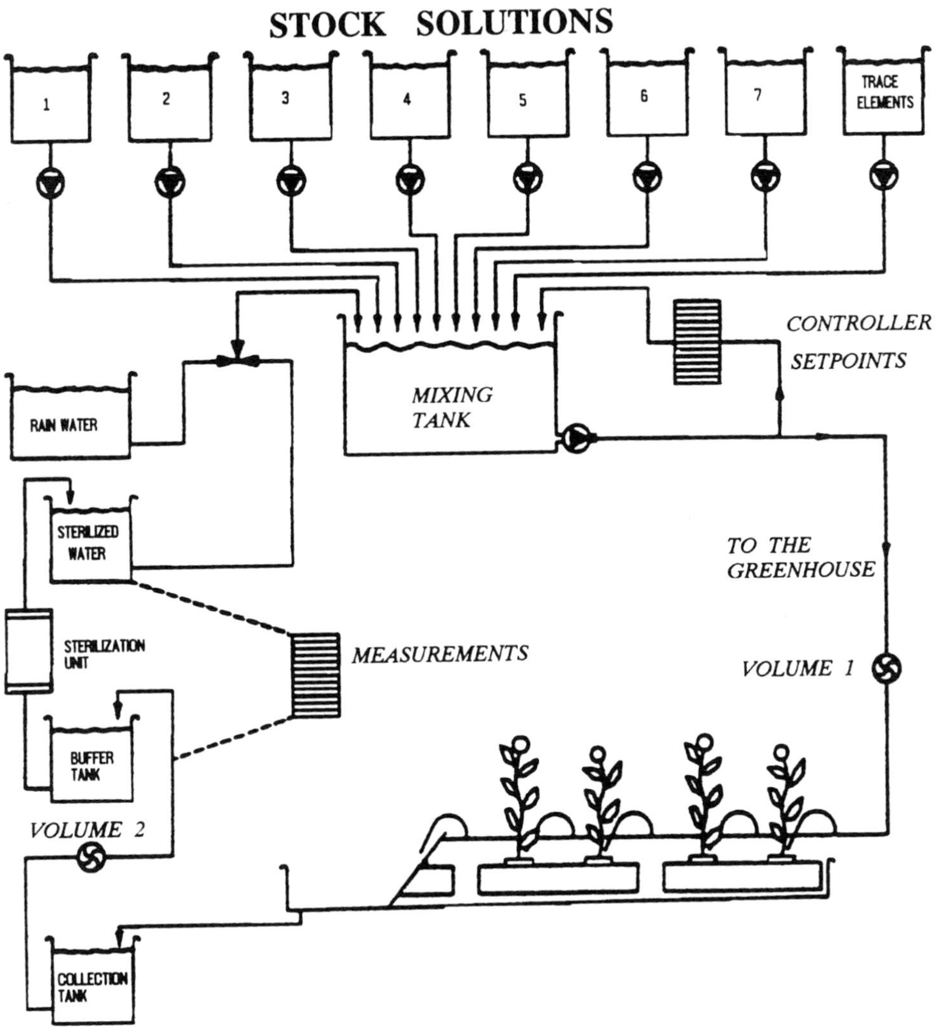

Fig. 8-19. Schematic of a proposed closed-loop nutrient control system. Van den Vlekkert et al. [1992] proposed measurement and control of pH, EC, K^+, Mg^{2+}, Ca^{2+}, Na^+, NO_3^-, SO_4^{2-}, PO_4^{3-}, and Cl (©1992, Int. Soc. Hort. Sci., *Acta Hort.*, 304:309-320).

d. Summary

Bot [1995] summarized the present situation as: "a modern complex version of a digitalized collection of analog controllers." In spite of high-powered computers, the systems outlined above are complicated to operate (Level 1, Fig. 8-2). As many as 300 to 400 settings for a modern installation are common. These settings may not always be clear to the grower, given the problems of conflicting requirements. The problems of interacting control loops have not been solved, along with a concentration on the physical environment as contrasted with

Table 8-3. Setpoints and relationships for greenhouse climate control (see Level 1, Fig. 8-2).

Main setpoints	Disturbances	Equipment operated	Interactions
T_I = Air temperature integral inside T_h = Pipe temperature(s) T_v = Ventilation temperature T_d = Day temperature T_n = Night temperature	T_o = Outside air temperature R_o = Outside global radiation WS = Outside wind speed WD = Outside wind direction	Boiler or heater operation Vent or fan operation Valve operation, fan staging, heat storage, thermal screen	VPD,[1] CO_2 R_s,[2] Cogen,[3] DIF,[4] Integral,[5] T^{*}[6], T_c,[7] ΔT,[8] VR,[9] T_r,[10] HAF,[11] T_s[12]
R_t = Minimum radiation for thermal screen closure R_s = Minimum radiation for supplementary irradiation R_h = Maximum radiation for shade screen	R_o = Outside global radiation T_o = Outside air temperature WS = Outside wind speed	Thermal screen, shade screen, photoperiodic control (shade or light), supplementary irradiation	T_I, CO_2, ET,[13] P[14] Minimum periods for lights and screen movement
VPD_h = Minimum vapor pressure deficit VPD_l = Maximum vapor pressure deficit	VPD_o = Outside vapor pressure deficit T_o = Outside air temperature T_i = Inside air temperature	Vent or fan operation, misting	T_h, ET, HAF
CO_{2h} = Maximum CO_2 concentration CO_{2m} = Rate of CO_2 change with ventilation CO_{2l} = Minimum CO_2 concentration	CO_{2o} = Outside CO_2 concentration R_o = Outside global radiation WS = Outside wind speed	CO_2 injection system, heat storage if CO_2 is from boiler	P, T_i, VR, R_o
WC = Water circulation SWP = Soil water potential (suction)	VPG_i = Vapor pressure gradient inside R_o = Outside global radiation	Water circulation pump, water valve, irrigation system	ET, EC

Main setpoints	Disturbances	Equipment operated	Interactions
EC = Electrical conductivity limit pH = Nutrient solution pH	Makeup water EC	Waste water release, water valve, acid-base injection system	Nutrient ions
NO_3^-, NH_4^+, Ca^{2+}, Mg^{2+}, K^+, Na^+, Cl^-, PO_4^{3+}, SO_4^{2-} = Ion setpoints	Makeup water quality, EC, pH	Nutrient injection system	pH, EC, ion interaction
CH = Day-night changeover (Rate of change) R_t = Radiation toggle setpoint	DL = Daylength R_o = Outside global radiation	Heating or cooling operation	Screens, photoperiod, supplementary irradiation

[1]VPD = Vapor pressure deficit (Pressure difference between pressure at saturation and actual at existing temperature).
[2]R_s = Radiation for temperature adjustment T_I.
[3]Cogen = Cogeneration when present, includes heat distribution from generators, electrical consumption.
[4]DIF = Temperature difference between day and night.
[5]Integral = Temperature integral (T_I), one or more 24 hour periods.
[6]T^* = Outside temperature for opening or closing of thermal screens.
[7]T_c = Temperature adjustment for CO_2 concentration.
[8]ΔT =Rate of temperature change.
[9]VR = Ventilation rate.
[10]T_r = Root temperature.
[11]HAF = Horizontal air flow.
[12]T_s = Heat storage temperature.
[13]ET = Evapotranspiration, water utilization.
[14]P = Photosynthesis.

Table 8-4. Constraints on environmental parameters for digital control. Several adapted from Bakker [1995]. See also van Henten [1994] and van Henten and Botsema [1991].

Parameter	Limitations	Remarks
Temperature	T_l = Maximum and minimum T_h = Water temperature limits ΔT = Temperature rate of change	ca. 5 to 30 C maximum range, often less DIF and Integrals may be limited
Radiation	R_l = Maximum and minimum (daily integral and total) DL = Daylength	Supplemental irradiation at R_s Shade screens at R_h Photoperiod (shade and lights)
CO_2	CO_{2h} = Maximum CO_2 (100 Pa) CO_{2l} = Minimum CO_2 (34 Pa)	ΔP = Photosynthetic rate Transpiration and water status
Humidity	VPD_l = Minimum deficit (0.2 kPa) VPD_h = Maximum vapor pressure deficit (1.0 kPa)	Force ventilation and heating below VPD_h, misting above VPD_l, water status, mineral uptake, and disease
Nutrition	EC_{max} = Maximum salinity Individual ion concentrations (minimums, maximums) pH = Limitations (ca. 6.0 to 7.0 for most)	Complex with problems in water status, morphology, and photosynthesis
Human constraints	15-25 C and 65-90% relative humidity Maximum pipe temperatures, safe operation of fans heaters, etc.	May be bypassed if manual operations on crop carried out in separate enclosure

an appreciation of plant response. Theoretical and practical approaches to a true digital system may be found in such texts as Bollinger and Duffie [1988], and more recently, in articles such as Young et al.'s [1993]. Young et al. make a number of comments, in addition to giving a highly technical outline of their approach. They name their system as PIP for Proportional-Integral-Plus, touting it as True Digital Control, in contrast to some form of Direct Digital Control or digitized versions of PI or PID controllers. Bot pointed out that present systems do not show consequences of proposed settings regarding product yield, risk, energy costs, etc. The systems outlined above have not been implemented to any great degree in the U.S.

5. Setpoints and Constraints

As a summation of the previous discussion, Table 8-3 suggests the possible setpoints, equipment operated, and interactions that have to be considered. Undoubtedly, not everything that one may run across is included, but the table serves to emphasize Bot's statements as to complexity of existing climate control systems. Usually, these are factors that are to be considered as immediate in climate control. As we will see later, a computer system can be utilized to modify these setpoints and to relieve the grower of the need to continuously monitor them. A computer system should also protect managers from inadvertent mistakes that conflict with the numerous interactions and disturbances that could crash operation or prevent the grower from maximizing profitability. In the possibility of upper-level control failure (a central computer failure), the system should be capable of a minimal control with, perhaps, a reduced set of algorithms and setpoints that the grower could operate manually.

Constraints on systems can be viewed from two points. One view is the fact that contrary to analog systems, digital systems require limits that prevent unbounded output to actuating equipment. Analog systems have one advantage in that their output is limited by their voltage supply. A digital output, on the other hand, can grossly exceed equipment capability, and limits are required in the software program. The actual constraints may be stated:

$$x_{i,min}(t) \leq x_i(t) \geq x_{i,max}(t) \tag{8.9}$$

where: x_i = *climate state variable, i, such as temperature, and must lie within the range expressed by* $x_{i,min}$ *and* $x_{i,max}$

Limits may also be required on control inputs:

$$u_{i,min}(t) \leq u_i(t) \leq u_{i,max}(t) \qquad (8.10)$$

where: u_i = *a control input, i, such as valve position*

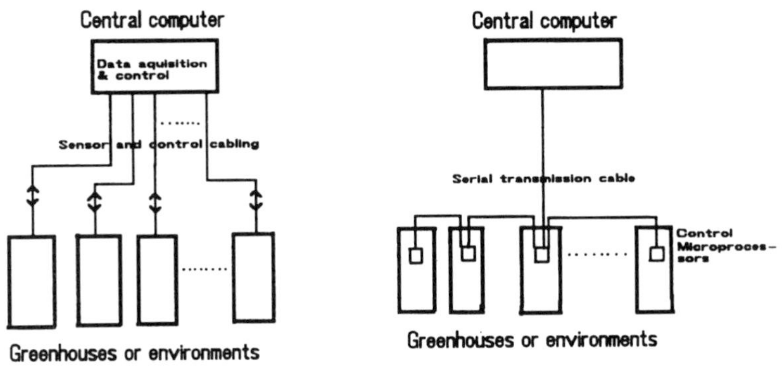

Fig. 8-20. Diagrammatic representation of a centralized climate control system (**left**) and a distributed system (**right**). The distributed system is connected to local processors by means of a serial transmission cable that allows each local processor to be polled and information and commands passed digitally. The centralized system requires shielded cable and control lines to each environment.

There are also requirements that allow determination of sensor failure, and these may be in terms of limits on the output. The second viewpoint to consider is constraints on the crop in the manner discussed by Bakker [1995], which are modified and partially included in Table 8-4. In Bakker's discussion, the crop response can be described by the minimum, optimum, and maximum characteristic levels. There are also "threshold" doses (level times duration) that one may be required to consider. Here, inclusion of thresholds as constraints requires knowledge of crop history. In terms of priority, plant temperatures regarding energy balance (injury from cold or heat) and concerning the dewpoint (condensation on plant tissue) are considered first. Water potential in the crop, optimization of CO_2 efficiency, long-term crop water balance, daylength or phase transitions (flowering and fruiting), dry matter distribution in the crop, and constraints on plant height or internode length follow. These latter constraints are limited in their application at this time, but subject to an intensive research effort.

6. Centralized versus Distributed Systems

There are two basic arrangements for control systems as illustrated in Fig. 8-20. The figure on the left denotes a system in which the computer, its I/O accessories, and the data acquisition and control device for interfacing between sensors, actuators, and computer are all under one roof. Sensor cabling and signal lines are run separately to each "environment." There are advantages since sensitive electronic equipment can be removed from greenhouse environments and placed in protected conditions. Greenhouses are not the best place for instrumentation unless well-protected from heat, humidity, and aerial pollutants that may be present from spraying and fumigation operations. Published examples of such systems are Hanan et al. [1987] and Willits et al. [1980a, b]. The disadvantage of centralized systems is the necessity to run well shielded, multiple strand cabling from the central location to each controlled location. There is not only the basic expense in shielded cables, but the increased chance of electrical interference, damage to the lines, and lightning strikes. Data is more likely to be corrupted and grounding problems increase.

The figure on the right (Fig. 8-20) is a distributed system that has gained favor in recent years. In this case, each environment is equipped with its own microprocessor with a limited memory capacity, keypad, and display. Or, a portable keyboard and display may be plugged into the local processor as required. Setpoints are entered with the local processor doing most of the immediate control, and reading the relevant sensors as necessary. The central computer polls each local processor in succession, obtains the latest information on climate, and recalculates the setpoints if required. The operator can enter commands locally by placing the processor into a local mode. In case of central computer failure, the local processor can operate the system to a limited extent. Examples of these systems are the detailed publications by Gieling and van Meurs [1979], van Meurs [1980], and the description of the Naaldwijk research facility [van de Vooren and Koppe, 1975]. According to these authors, noise problems are less, and cabling cost is minimum with the use of optical isolation. The latter devices physically separate circuitry, reducing interference and unwanted signals. Serial communication for the Dutch system is over a 20 mA full duplex line with two, shielded, twisted pairs that allow up to 1 km between the local microprocessor and the multiplexing processor at the central computer. Miles [1994] stated that, for up to 50 m distance, the standard RS-232 communication protocol can be used. Beyond 50 m, modems (modulator-demodulators) are necessary. They are used commonly in telephone line connections. Modems can also be utilized for remote radio communication. Fiber optics, or coaxial cable, permit data to be transmitted at rates up to 10 MBaud or Mega bits per second. Miles suggested that pictures of plants, tacky pest control strips, etc. might eventually be transmitted, as well as permitting a salesman to call up live plant images to show prospective customers in other regions. Such procedures can now be found on the Internet, and one can order flowers if he so wishes. Udink ten Cate [1985] considered the main trend to be distributed control.

D. SENSORS IN THE GREENHOUSE

Discussion on sensors was presented in previous chapters:

Radiation: Section 3.II.E, Fig. 3-5
Temperature: Section 4.II, Figs. 4-1, 4-6
Water: Section 5.II.C and 5.III.D, Figs. 5-16, 5-17, 5-40 through 5-42
Humidity: Section 5.IV.B.2, Figs. 5-82 through 5-84
Electrical conductivity: Section 6.II.H
pH: Section 6.IV.C.3
CO_2: Section 7.II.B, Figs. 7-1 though 7-3

It is well to review salient features of sensor use since these devices are the basis of climate control, providing information necessary to optimization. One of the better expositions is Gieling and Schurer's presentation at Rutgers in 1994, some based upon information from the Honeywell corporation and their more recent publication [1995]. They classified functional tasks for sensors as:

Presence or absence (operation or closure, etc.)
Positioning (valve, ventilator, etc. position)
Inspection (proper operation of equipment)
Condition measurement (state of equipment)
Identification (what is happening)
The sensor interface (connection to other equipment)
The operator machine interface (keyboard, monitor, etc.)

The application is divided into three areas: monitoring, qualification, and control. The first is defined as generating data about a procedure or piece of equipment that is being monitored, while qualification is a procedural function where results of the procedure are checked to see if they meet established requirements. Control is making something happen as the result of data gathered by sensors. One can see that sensors may do more than merely measure the parameters listed in Table 8-1.

Some general properties of sensors are interchangeability, repeatability, producibility, output signal compatibility, and "usability by growers." From Gieling and Schurer [1994], levels of sensing sophistication are: 1) conversion, or the ability to convert the physical condition to a measurable and interpretable signal; 2) environmental compensation, or the ability to correct for changes in, or protection against, the operating environment; 3) communication, or the ability to interface with the host system understandably and acceptably; 4) diagnostics, or ability to inform the host system of operating problems and sensor failure; and 5) logic/actuation, which is the ability to make decisions and perform control actions. The majority of sensors in use employ basic technologies of electromechanical, semiconductor, and optoelectronics.

1. Temperature

Dutch authors advocate use of resistive temperature measuring devices such as standard ceramic thermistors (NTC) or platinum resistance thermometers (Pt_{100}, Pt_{500}, and Pt_{1000}). The subscript shows resistance value at zero C. Contrary to thermistors, platinum devices are linear, highly stable and with an uncertainty of ±0.3 C. While two-line resistive devices are common in the U.S., most published recommendations indicate the use of four-wire shielded cable for connections. Only two lines means that resistance of the lines, which varies with temperature, are included in the measurement. For four lines, two are necessary for the power required in the measurement, limited to 1 mA for platinum and 0.2 mA for high-resistance NTC sensors. High currents can cause self-heating in resistive devices. It is common to switch current on for a few milliseconds during measurement. The other pairs of lines are used to connect the measured voltage across a high input impedance, differential amplifier. Temperature sensors should be covered with thin-walled glass or ceramic tubing because these materials conduct less heat compared to the metal casing of heavy-duty industrial sensors. Calibration can be at two points (0 C and higher) annually. Thermocouples are the least expensive devices to be found, but they require a suitable temperature-controlled reference with thermocouple cabling, which increases cost.

2. Humidity

Psychrometric measurements of humidity are usually recommended in the literature, but this calls for rigorous maintenance, with uncertainty in temperature sensors less than ±0.1 C. In my own experience, psychrometric devices were unsatisfactory over long periods. Work at Colorado State University showed some types of capacitance probes to be highly stable outdoors, maintaining calibration over a year. Inside greenhouses, sensors were subjected to monthly recalibration, rotating them with a replacement unit as necessary. Certain designs allow for simple recalibration with a drying agent for zero RH and above a saturated KCl solution for the high point. A protective cap or filter is required on the sensor, and they should be protected from misting or irrigation systems. The high ventilation rate (3-4 m s^{-1}) required for psychrometric sensors is not necessary for capacitance probes –0.5 to 1 m s^{-1} through the shelter is adequate. Dutch investigators have spent considerable effort on improving psychrometric sensors. Their experience with capacitive devices has not been acceptable due to drift at high humidities. That situation is changing as continual improvements are made in newer capacitance probes.

Fig. 8-21. A vastly enlarged cross-section of an ion-sensitive field effect transistor (ISFET) compared to the common metal oxide semiconductor field effect transistor (MOSFET), the latter found in all computing equipment. A polymeric membrane, with ion-sensitive ionophores, is chemically bonded to the ISFET surface [From van den Vlekkert et al., 1992] (©1992, Int. Soc. Hort. Sci., *Acta Hort.*, 304:309-320).

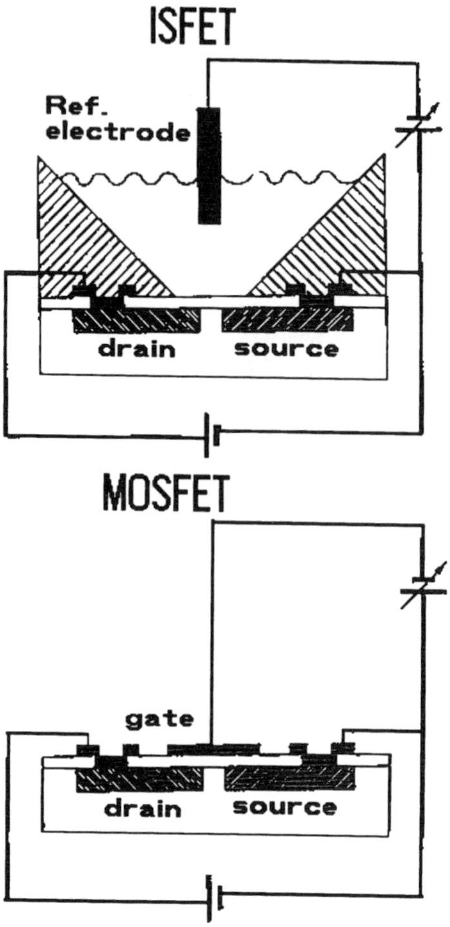

3. Carbon Dioxide

CO_2 devices were well covered in Chapter 7. In view of analyzer expense, a multiplexed sampling system with one IRGA is common (Figs. 7-4 and 7-5). Sampling lines of nylon or high-density polyethylene are recommended, with sufficient diameter to prevent a marked pressure drop over the required length. Inlets, in my experience, should be located within a dense canopy, or at least in close juxtaposition to the main leaf area of a crop. Sampling lines should not be subjected to low temperatures since this may cause condensation internally, leading to blockage. Some systems place individual pumps at the sampling inlet, or a single pump at the instrument location. It is usual to find single instruments placed within the greenhouse. The latter can be very expensive where many environments are controlled. Also, the sampling point is not always in the proper location. The output from infrared analyzers is not linear, and at least three points are required for reasonable calibration accuracy. Gieling and Schurer recommend calibration every 3 to 6 months, but I would suggest monthly calibrations. Some manufacturers supply small, pressurized canisters of calibrating gas, or larger tanks of certified concentrations can be obtained. Simple software programs can be utilized to perform calibration.

4. Radiation

The common instrument is a broadband thermal receiver covered with 1 or 2 glass envelopes, normally located outside the greenhouse (pyranometer). This avoids problems of finding a suitable and representative location inside the structure. For standard houses such as the Venlo, software programs are available that take into account solar position, time of day, etc. for calculating transmittance and solar intensity at crop level. It is common to find in the literature a single transmittance coefficient, such as 60 to 70%, used despite structure and the sun's position. This is not at all valid for precision analysis of the internal climate. Due to the climatic conditions in Europe, diffuse radiation is more common, which is sometimes measured with a shadow band mounted above the receiver. This is seldom carried out commercially. Since the total radiation is important for calculating energy flow, PAR measurements are seldom included. Bot and van de Braak [1995] state that in the energy balance equation, the energy converted to photosynthesis can be neglected. A special problem is the radiative exchange between the cover and the hemisphere of the sky. This is an important factor in the energy balance equation. An instrument for measuring sky temperature directly is the pyrgeometer that measures net solar, terrestrial, and atmospheric radiation on an upward-facing surface. I have not seen any instruments of this type in commercial use. The uncertainty of measuring shortwave, global radiation is ±0.5%, which does not accord with remarks on measurement made in Chapter 3. The outer dome of the pyranometer is cleaned monthly, with recalibration every 2 years [Gieling and Schurer, 1995].

Silicon cell pyranometers are available, and are usually cheaper than standard thermal devices. Gieling and Schurer [1994] consider these devices as inferior substitutes for the thermal instrument. Light meters, illumination meters, photometers, etc. are very common in the industry, mainly employed to assess intensity from supplementary irradiation lamps and photoperiod lights. These are not suitable for advanced climate control. It is common to state the level of additional lighting in greenhouses in lux rather than PAR terms of $\mu mol\ m^{-2}s^{-1}$. The photosynthetic effect of radiation from high-pressure sodium lamps in comparison to daylight is overestimated nearly 50% when using a photometer. Modern lamps are designed to be efficient sources of light rather than PAR.

5. Nutrition

Chemosensors provide an electrical signal proportional to the number particles in a fluid or gas that can be atoms, molecules, or ions. Their development is becoming particularly important in the analysis of nutrient solutions. There are many different types. Ion selective electrodes (ISE) use membranes permeable to certain ions in solution, such as H^+ in pH measurements. Alberry et al. [1985] published their work on a system for measuring six ions (H^+, K^+, Na^+, Ca^{2+}, Cl^-, NO_3^-), as well as Hashimoto et al. [1989] who measured Ca^{2+}, K^+, and NO_3^-, including EC and pH. Heinen and Harmanny [1992] evaluated ISE performance, concluding that continuous use of membrane and glass electrodes is possible when temperatures are maintained constant with daily cleaning. Gieling and Schurer [1995] concluded that since ISE electrodes measure the logarithm of activity, the overall uncertainty of ISE measurements is not very good. Such electrodes have high output impedance (ca. 10^9 ohms) so that considerable care is required in shielding against electronic interference with an impedance converter on each electrode; 1 nA of current, such as generated by waving a hand in the vicinity, will generate 1 Volt of noise. Alberry et al. also found that locating computers in the greenhouse was not a suitable practice.

Of more interest are ISFETs or ion sensitive field effect transistors. Transistors are, of course, the basis of present-day computers. The type of field effect transistor modified for ionic analysis is the MOSFET or metal oxide semiconductor field effect transistor. There are a number of variations in the MOSFET catalog, but for ISFETs, the metal ohmic contact with the gate is replaced by a reference electrode located in the fluid to be analyzed. An ion selective material is deposited over the gate insulator, containing a compound that facilitates transmission of an appropriate ion across a lipid barrier by combining with the ion or increasing the permeability of the barrier to the ion. Such compounds are called ionophores. An example of the device is given in Fig. 8-21, with the transistor greatly enlarged for clarity. Van den Vlekkert [1992] and van den Vlekkert et al. [1992] have published on the special application of ISFETs in closed, recirculating nutrient solutions. The latter authors state that, with an appropriate calibration, high accuracy can be achieved over long periods (±10%, concentrations in mMol ℓ^{-1}). To overcome most problems, the membrane is chemically attached to the ISFET surface and the ionophore to the membrane. To increase stability, a hydrogel is introduced between ISFET and the membrane. The distinct advantage of ISFETs over ISEs is the formers' low-cost, mass producible semiconductor component. It can be produced as a small, reinforced device such as a dipstick or flow-through cell. ISFETs can be integrated with electronics for amplification and data handling on the same chip –smart sensors– and selective sensors for several ions can be incorporated on one unit (NO_3^-, K^+, Ca^{2+}, SO_4^{2-}, Na^+, Cl^-, NH_4^+). Gieling and Schurer [1995] state that commercial breakthrough for ISFETs is still limited by accuracy and life expectancy.

If part of the sensor consists of a biological material such as an enzyme, bacteria, or whole cells, the device is called a "biosensor." Development here is the forefront of sensing technology. Many commercial applications are in use, such as glucose and lactate analysis in the medical field. Potential for analysis in odors, taste, mutagenicity, allergenicity, biological oxygen demand, etc. are probable, with further progress by applying tailor-made enzymes, antibodies, and neuronal networks [Schubert and Scheller, 1992]. Basic types of microsensors have been ISFETs, metal oxide semiconductor (MOS) capacitors, and thin-film electrodes. Thermistors, integrated optical sensors, and surface acoustic devices have also been introduced.

pH sensing still uses the standard combination, glass units with a life of about 1 year. Two sensors are used as a check, and further checked with biweekly laboratory analyses. Electrical conductivity (EC), or salinity, is measured in Holland with three, ring-shaped electrodes inside the water pipe: 1 Volt AC, ranging from 400 Hz to 50 kHz is applied between the central electrode and the two interconnected and grounded end electrodes. The two currents resulting are summed to eliminate parasitic effects (e.g., water flow). Grounding the two end electrodes allows sequential or parallel use of more EC electrodes in the water supply system. The sensor output must be compensated by temperature measurement to modify the AC voltage applied. In practice, two distinct EC electrodes are used in parallel, enabling a check on functioning. Biweekly laboratory analysis should be carried out on EC as well [Gieling and Schurer, 1995].

6. Miscellaneous Measurements and Meteorological Masts

Instruments for measuring soil moisture were considered in Chapter 5, along with the problems common to such measurements such as suitable location of the instrument. Even newer instruments such as TDRs (time

Table 8-5. General outline of possible inputs to processor from greenhouse sensors.

Analog-digital	Input	Transducer	Number used	Range	Resolution	Unit
Analog	Air temperature	Pt_{100}	1 or more	0-50	0.1	C
"	Pipe temperature	Pt_{100}	2 or more	0-100	0.1	C
"	Soil temperature	Pt_{100}	1 or more	0-50	0.1	C
"	Crop temperature	IR thermometer	1	0-50	0.5	C
"	Cover temperature	Pt_{100}	1 or more	-30-70	0.1	C
"	Humidity	Capacitance probe	1 or more	0-100	1%	RH
"	Heating valve positions	Potentiometric	2 or more	0-100	0.1	% of max.
"	Soil moisture	Tensiometric	1 or more	0-800	1	kPa
"	Electrical conductivity	EC electrode	1 or more	0-20	1	dS m^{-1}
"	pH	pH electrode	1 or more	0-14	1	pH units
"	CO_2	IRGA	1	0-200	1	Pascal
"	Ventilator position	Potentiometric	2 or more	0-100	0.1	% of max.
"	Curtain position	Potentiometric	1 or more	0-100	0.1	% of max.
Digital	Exhaust fan operation		1 or more			
"	Pump operation		1 or more			
"	Curtains open		1 or more			
"	Curtains closed		1 or more			
"	Ventilator full open		1 or more			
"	Ventilator full closed		1 or more			
"	Mixing valve upper		1			
"	Mixing valve lower		1			
"	Heater unit on		1 or more			
"	HAF on		1			

domain reflectometry) suffer from the problem of representative location. In my opinion, the most reliable instrument, and least expensive, remains the tensiometer, limited mostly to substrates with significant soil content. In soilless and solution cultures, the status of water potential can be assessed by EC measurements or the use of porous, ceramic-enclosed thermocouple psychrometers.

Outdoor wind speed is usually measured with a 3-cup anemometer that has a tendency to errors at low wind velocities, and may have a minimum threshold wind speed value, resulting from friction in the moving parts. Instruments capable of measuring the low wind speeds common inside greenhouses are generally too delicate. Wind direction uses the standard wind vane. If propeller anemometers are employed, wind direction is obtained with the same instrument. Rain detectors are standard devices to generate a yes/no signal as to presence of rain. For quantitative measurement, tipping-bucket instruments are employed. These have not been seen in commercial use.

One of the most important factors in plant growth is water potential inside the crop. As pointed out in Chapter 5, unfortunately, sensors available are mostly for research purposes. There remains the problem of obtaining representative samples with those instruments available (Figs. 5-16, 5-17). Measurements of the crop have been investigated by Hashimoto [1980; 1989; 1993] as the "speaking plant" concept. Despite considerable effort, there has not been commercial use since the technology and sampling procedures are far from commercial acceptability. Photosynthesis rate is capable of solution through indirect means. That is, CO_2 consumption, where ventilation rate can be evaluated, is a possibility for measurement in real time of crop photosynthesis. Both water potential and photosynthetic measurement are undergoing rigorous investigation.

There are also problems in determining growth rate, fruit development, etc. Some solution of these plant factors may be susceptible to video procedures as investigated by van Henten and Bontsema [1995]. Their application to determine canopy area is discussed in the last section.

In view of the number of disturbances in climate control, an external set of meteorological instruments is required. These include air temperature, global radiation (Chapter 3), precipitation, humidity, CO_2, wind speed, and wind direction. Psychrometers cannot be used outside at low temperatures so a capacitance probe is the likely candidate. The CO_2 sample can be obtained easily if a multiplex sampling system is utilized. Eventually, pyrgeometers may also be included. These instruments can be mounted on a support on the top of the greenhouse, or on a meteorological mast near the greenhouse. The instrument group must be readily accessible for maintenance. For climate control purposes, instruments should be mounted at the height of the greenhouse roof. Standard heights for usual meteorological instruments are not valid. Disturbing influences should be minimized –such as locating near chimneys, ventilators, or nearby buildings or trees. Shading from such is not allowable. Lightning protection is a necessity.

7. Summary

Table 8-5 suggests the possible inputs from sensors to a local processor or to the central computer, depending upon whether the system is distributed or centralized. Not all these may be required, or wanted, and they will vary with the systems employed (e.g., fan-and-pad cooling versus natural ventilation, unit heaters versus steam or hot water). One can appreciate that if a full complement of sensors is employed, even in one environment, a computer is a necessity.

The grower should be careful of a false sense of security. Usually, he knows little about the equipment, especially if connected to a sophisticated computer. Special means should be incorporated to detect sensor failure. A regular, timely procedure of calibration, cleaning, and inspection is required, with suitable records maintained. Whereas, even 10 years ago, a grower might walk through his establishment to assess growth and proper operation, such a procedure with a computer climate control system is fraught with danger. Since the whole purpose of computerization is to increase net profit, the grower should establish those methods that ensure appropriate operation.

IV. THE COMPUTER

The present programmable or personal computer is an astounding device. The basic design is shown in Fig. 8-22. Its information is encoded in "bits," the smallest amount of meaningful, distinguishable information possible. That is, on or off, true or false, presence or absence, zero or one. The second remarkable aspect of this use of bits is the fact that changes, calculations, comparisons, etc. are done mostly in serial fashion –one after another. The fact that something comes out in reasonable time results from miniaturization and the blinding speed of instruction execution. The original Intel 8088 microprocessor ran at 4.77 MHz with 29000 transistors, whereas the present Intel Pentium has more than 3.1 million transistors and can operate at clock speeds greater than 100 MHz, or more than 150 million instructions (machine language) per second (MIPS). The miniaturization process has gone from dimensions of 3 μm for a transistor to 0.35 μm, and these dimensions are rapidly approaching fundamental limits regarding electron movement. Speed is further enhanced by causing the CPU to execute two or more instructions for each clock cycle, with the onboard memory being utilized to actually anticipate instructions. Newer processors are also capable of "pipelining." That is, instructions are pre-fetched from memory in a linear manner and stored linearly in a memory cache, later loading into the "pipeline" for superscalar execution or several executions per clock cycle. Several pipelines may be included that in effect, turns the processor into a parallel system –as contrasted with a series execution design. These and other enhancements to the computing system provide an individual user with more power than was available on mainframe computers a few years ago.

One of the greatest attributes of the PC is its interrupt ability. It enables the computer to respond to the unpredictable variety of work that may come its way. Interrupts are required for response to pressing and releasing a keyboard button, the need to access the hard disc to load or record a file, to draw a rectangle on the monitor, and to respond to external interrupts such as the printer running out of paper or the need to access and read sensors or download data from slave microprocessors in an environment. There are also software interrupts that call for special handling in the software program. In internal interrupts, at least, the CPU is capable of servicing the interrupt and returning to where it left off in its normal computations. The CPU saves the instructions for resuming its task. Interrupts are generally ranked according to importance, and the operation speed is sufficient to handle several interrupts and still get on with the main program. A second attribute of PCs is a variety of instructions called computer logic. In addition to the usual mathematical computations of addition, subtraction, multiplication, and division, the CPU is capable of "conditional" branching. That is, from two or more possible courses of action, the processor can make a selection. These decisions may involve controlling a loop, repeating an action based upon another condition, exiting to a subroutine, or calling another program. One might say that an intelligent process is in use. However, despite the heralded approaches to artificial intelligence, the PC is stupid. It does what the operator tells it to do, and if garbage is put in, whether in software or data, garbage is what one gets back out.

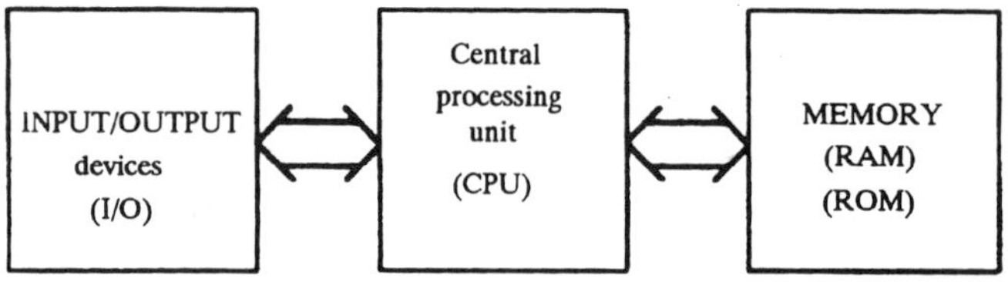

Fig. 8-22. The basic computer components. The I/O devices include keyboard, monitor, acquisition system for data, and control output, printers, recording devices, and modems for external communications. Memory, which commonly now exceeds 20 MB, includes random access and read only memories.

It is not my purpose here to use much detail in computer operation. That would require a large tome such as Norton et al.'s *Inside the PC* [1995]. Even Norton does not deal with complete detail of present PCs. The factors to be discussed in this Section are the advantages of using computers, and problems with software and networking.

A. ADVANTAGES OF COMPUTERS AND POSSIBILITIES

The previous sections, with their tables, show that a sophisticated computational system is required if the grower is to avoid confusion. A fast machine is needed merely to read and record the data input from the many sensors. The attempt to include the numerous features in analog systems is so expensive that computers are economically justified [Bot, 1995]. Initially, the primary objective of computers was to save labor, particularly where equipment operation requires continual supervision (i.e., heating, cooling, ventilator operation, etc.) [Bakker and Challa, 1995]. Growers have increasingly come to rely on automatic control systems to provide consistent, favorable environmental conditions.

Table 8-6. Comparisons between binary, decimal, hexadecimal, and octal numbering systems.

Binary (base 2)	Decimal (base 10)	Hexadecimal (base 16)	Octal (base 8)
0000	0	0	0
0001	1	1	1
0010	2	2	2
0011	3	3	3
0100	4	4	4
0101	5	5	5
0110	6	6	6
0111	7	7	7
1000	8	8	10
1001	9	9	11
1010	10	A	12
1011	11	B	13
1100	12	C	14
1101	13	D	15
1110	14	E	16
1111	15	F	17

Pierce Jones [1994] has given several examples of actual benefits from proper use of computer control. In one instance, a rose grower could delay flower development by lowering temperature to ship more flowers on a Monday, as opposed to running the same temperatures over the preceding weekend. This example was used by Jones to illustrate the considerable versatility available in such systems. By suitable use of PID algorithms, a Florida grower could control ventilation (exhaust fans) in a manner to provide a 15% reduction in fan operation (and his electrical bill). There are numerous instances of alarms that warn growers of equipment failure, or the fact that someone has left the wrong valve opened or closed, as the case may be. It can be appreciated that internal sensors and control programs are nearly useless when plants are grown in climatic regions where the greenhouse may be similar to those shown in Figs. 2-9, 2-19, 2-30 and 2-31. Nevertheless, Jones has provided examples of growers looking at climate data to determine improper operation or the reasons for a particular disease outbreak such as botrytis. Thus, a simple meteorological recording system could be used by a grower to assess crop performance and potential as the result of local variations in weather in tropical or Mediterranean-like climates. This is in contrast to Knight's conclusion [1985] that if one does not use a thermostat or powered ventilators, or has less than 300 m^2 area, he does not need a computer. The latter might be true since a computer would be a very large capital input per square meter, representing a cost that could not be retrieved effectively on a small total area. However, Tantau [1985] pointed out that the average greenhouse holding in Germany is less than 1000 m^2. Nevertheless, a computer system has been developed for commercial application.

Marsh, in 1987, pointed out means by which computer control in the U.K. could cut fuel costs by looking at the trade off between cost of heat and worth of the corresponding crop growth. As weather conditions vary from one location to another, the economically optimum environment is crop and site specific. An anonymous report in the British *Grower* [1988] showed that the use of well-designed computer programs by the Dutch had aided in the reduction of gas consumption from 47 to 29 m^3m^{-2} over a period of 4 years. In the same period (1979 to 1983), crop yields had increased significantly. Knight [1985] pointed out some things a properly programmed computer can do: such as "knowing" what all systems are doing, coordinating all the operating systems, giving alarms, conserving energy, reporting on current and past conditions, etc.

Often, a distinction is made between the usual PC that may be used for financial, publication, or inventory purposes, and so-called "environmental control" (EC) computers. I think that such differences are no longer germane. The present PC machine can handle most environmental control jobs –it is a matter of the program instructions employed, and connection (interfacing) between sensors, PC, and control signals to the actuating equipment. With a distributed system, a modem can be employed to interact with the various microprocessors in each environment. Or, in a centralized system, a digital acquisition and control module is present, which may be connected to the PC with an RS-232 interface or parallel port. All of the necessary hardware is readily available and well standardized.

In the future we may see holographic memory storage [Psaltis and Mok, 1995] which will turn the present optically recorded CD-ROMs into super storage devices. A compact disc can hold about 640 x 10^6 bytes, or more than 300000 pages of double-spaced, typewritten text. Existing schemes are expected to push CD capacity into the tens of billions of bytes within 5 years. Holographic memories could conceivably store hundreds of billions, transfer data at a rate of a billion or more bits per second, and select a randomly chosen data element in 100 μsec or less. Others have described quantum-mechanical computers in the process of development. Advanced lithographic techniques can yield parts 100 times smaller than that currently available. Nevertheless, at this scale where bulk matter reveals itself as a crowd of individual atoms, integrated circuits barely function. In the next decade, the present PC will be unrecognizable. However, the present, off-the-shelf PC can control most environmental systems. The control functions can be integrated with the economic functions of the enterprise in the same machine or connected to a communication system by which the "EC" PC can interact with the grower's financial, inventory, and marketing system.

B. SOFTWARE: THE GUIDING REINS

Usually, computer hardware available is reliable and standardized. The ROM-BIOS (read only memory-basic input-output system), as the underlying pinnings to PC operation, is reliable. It is the software that determines the course of execution, and what the PC does with respect to climate control. Surprisingly, this aspect of computer control is seldom addressed in existing horticultural literature. Schmidt [1986] stated the general conclusion that error-free software does not exist, whereas the common hardware circuitry executes thousands to millions of instructions with extreme reliability. Software, as pointed out by Schmidt, does not age or rust, does not deform under influence of heat or cold, and it is not exposed to physical strain. Software usually escapes objective scrutiny, but the DIN (Deutsches Institut für Normung e.V.) in Berlin has published guidelines for software quality. Similar efforts in the U.S. are not well publicized in available literature. Most software is considered proprietary, and the grower is dependent upon the company's expertise in programming and its reliability. The desirable properties of software are completeness, correctness and accuracy, sturdiness, consistency, and informing the user in case of failure. In the final Section of this chapter, the modeling of the greenhouse operation will be briefly discussed. This represents one of the greatest opportunities in climate control in the next few years. Still, any standardization in the translation of the requirements into reliable software has yet to be discussed in a manner suitable for the greenhouse operator. It, translation, is generally dismissed as though the user will be presented with a delicious cookie, free of charge. The grower generally has no knowledge of the program concerning what it does, how it does it, or what, if any, "bugs" are in it.

It is reality that computers execute only those instructions given to them in "machine" language, or bits. The number of bits required for the word "hexadecimal" is 88 –a cumbersome business to write out with no mistakes [Norton et al., 1995]. A programmer who writes in machine language is writing out, one by one, the detailed instructions for the computer to follow, or the binary numbering system based upon the numeral 2 (2^0, 2^1, 2^2, . . . 2^6, etc.). Whether the "byte" represents numbers or characters depends on what the programmer wants done.

As a compromise, permitting the programmer to deal with 8 bit, 16 bit, 32 bit, etc. representations of the machine language, the hexadecimal numbering system permits a set of 4 bits (0000) to be represented by a single digit. Since there are 16 possible combinations in a 4 bit byte, there are 16 numbers in the "hex" system or 0 through 9 and the letters A through F. Thus, an 8 bit byte, or word, can be presented by two digits. The use of this numerical system is the basis of "assembly" language as contrasted with "machine" language. A comparison of binary, hex, decimal, and octal numbering systems is provided in Table 8-6. Assembly language forms the instructions translated into the individual instruction bits used by the computer. Conversions from one numbering system to another can be found in any number of textbooks. Rather than doing it longhand, software programs will do conversions.

However, even assembly language can be difficult. Although easier than machine language, it is still a tedious and long process. This problem has resulted in the development of "high-level" languages, of which there are probably more than a thousand floating around. Table 8-7 provides a list of many high-level languages that can be found in the horticultural and agricultural literature. Commonly, the use of a particular language is what the programmer is familiar with and not necessarily because the language is the best for the job. According to Bloor [1994], the most recent languages such as Smalltalk, VisualBasic, Forte, SNAP, etc. are considered fifth-generation languages. The first generation was the actual machine instructions, the second the assembler languages, and the third level consists of such languages as Cobol and Fortran. The distinguishing characteristic between 3rd and 4th generations was the use of a data dictionary to define database formats and field presentation defaults and various screen-formatting and report-generation features that were highly productive. The attempt by European and Japanese governments to throw money at software development has not been particularly productive in Bloor's opinion. Of the languages listed in Table 8-7, Fortran, Pascal, C++, VisualBasic and Smalltalk probably represent the most commonly utilized. The latter two are geared more to PCs. An example of a short Section of code in HP's Basic is a simple routine to control evaporative pads:

```
1692 !
1694 IF NOT Flag3 THEN
1696   IF T_out>Const_set(7) AND Rad_out>Const_set(8) THEN
1698    OUTPUT 709; "DC12,0,1,2,3"
1700   ELSE
1702    OUTPUT 709; "DO12,0,1,2,3"
1704   END IF
1706 ELSE
1708   OUTPUT 709; "DO12,0,1,2,3"
1710 END IF
1712 !
1714 RETURN
```

Flag3 in line 1694 signals nighttime (1) or daytime (0), which is set or cleared in another part of the program. If the flag is 1, then the program goes to line 1706, which sends a signal to the digital acquisition/control system (address 709) to open relays for the pad pumps in each of four separate environments, and stop the water pumps. If Flag3 = 0 (daytime), the execution goes to line 1696 where the outside temperature is compared to a constant and the outside radiation is compared to a constant. If both temperature and radiation are larger than the constants, then a signal is sent to the control system to close the pad pump relays and put water into the pads. The program switches to 1714 and returns to the main routine. If one or both of the comparisons are false, line 1702 is executed, shutting off the pumps, and then going to line 1714. Note that two IF statements are nested. There are many other variations of such conditional branching, depending upon the programmer's objectives and the particular language being used.

Before any program, regardless of language, can be executed, the language must be translated into the only thing the computer can actually execute –machine language instructions. Translation may be one of three ways: interpreting, assembling, or compiling. Interpreting is a kind of translation by which the program is translated into machine language as the program is executed. The computer is actually running the interpreter that executes the program step by step. The result is a slow and inefficient process, but flexible. A particular instruction may have to be interpreted numerous times, which obviously will slow program execution. Basic, which comes with DOS (disk operating system) in PCs is interpreted, as is VisualBasic. Assembly and compiling, however, translate the programs before executing them. Translation is part of the program development process. Once accomplished, one does not have to waste time translating the program again, nor does the translating software have to be available. For these reasons, programmers use a specific implementation of the general programming language such as TurboPascal or Lightspeed C. With some compilers, a program can be executed immediately after translation. The compiler has a demanding task of deciding what kind of machine language instructions will be used to accomplish the particular higher-level instruction. On the other hand, an assembler performs a mechanical conversion of the programmer's instructions into equivalent machine instructions. Although a faultless program is provided in Fortran, for example, considerable effort is still required in translating and assembling the code into a finished product. Unfortunately, many, small modifications to the code may require excessive recompilations, whereas interpreter methods can be instantly modified during execution.

Even should one or more of the above languages be highly suitable for environmental control, nothing is stated as to reliability and efficiency of the code that may be written. These problems are likely to cause serious difficulty as the number of instruction lines and complexity increases. Two notable examples of software

Table 8-7. Representative selection of high level programming languages that may be found in use in greenhouse climate control, modeling, and simulation.

Language	Remarks
Ada	Language mandated by U.S. Department of Defense to provide uniform codes in federal contracts. Some problems with compilation.
Alloy	Parallel programming language for massively parallel computing systems [Morin, 1994a; b]
Basic	First used in IBM machines, undergone numerous revisions, with latest known as "VisualBasic," considered a 4th or 5th generation language (object oriented). Closest one to a universal language for the PC [Norton et al., 1995]. Numerous variations over the years. Slow compared to others since it is usually interpreted as contrasted to compiled. WordBasic considered one of the most powerful macro languages [Prosise, 1993]. VisualBasic resembles C or Pascal [Spencer, 1994].
C	C++, an enhanced C, ideally suited, along with Pascal, for professional programming [Norton et al., 1995]. Highly efficient. Some problems with secrecy and privacy [Waldo, 1994]. One of the most popular languages over the past decade in the U.S.. Problems with suitable compilation [Coffee, 1994a].
C++SIM	Set of C++ class definitions that mimic process-based simulation facilities of SIMULA and SIMSET [Morin, 1994a; b].
Cobol	Used mainly in financial and business pursuits, introduced in 1960. Considered by some as outmoded [Cunningham, 1994; Machrone, 1994; Coffee, 1993a].
CSMP	Continuous Simulation Modeling Program, originally supported by IBM, sometimes used in conjunction with Fortran [Goudriaan, 1982; van Bavel et al., 1981]. Also THSim [Bot, 1980], TRANSYS [Cooper and Fuller, 1983] and numerous others.
Forth	Industrial language utilized by some American climate control companies.
Fortran	Introduced in 1957, first widely disseminated language in use at universities. Continuously updated, numerous other codes based on Fortran [Coffee, 1993b].
GPSS	Discrete simulation language as contrasted to CSMP as a "continuous" simulation language [Takakura, 1993].
Lisp	Second oldest artificial intelligence language (AI) [Newquist, 1994]. Basis of a number of newer languages. Utilized in expert systems.
Maisie	Process oriented parallel programming language [Morin, 1994a; b].
Modula-2	Also Modula-3 and Oberon, offshoots of Pascal and somewhat similar [Morin, 1993].
Pascal	Designed originally as an education programming language. Utilized heavily in computer sciences. A number of disadvantages (Morin, 1993). No compiler for PCs, but interpreted (See below) (Norr, 1994).
Prolog	One of best languages for programs that involve satisfying constraints (Lane, 1994).
Smalltalk	Somewhat similar to VisualBasic, both a language and a programming environment, requiring a 486 Intel CPU with 12 MB of RAM [Reid, 1994] (object oriented). Highly touted for greenhouses by Gauthier [1991].

difficulties are the English tunnel (Chunnel) and the Denver International Airport (DIA) [Coffee, 1994b; Collins, 1994]. The tunnel was delayed 1 year by problems with more than 3 million original lines of code. A total of 8 million lines supports Chunnel operation. Even an error rate of less than 1 defect per 1000 code lines could represent a significant failure in critical safety equipment. The DIA imbroglio with the baggage system software resulted in mangled baggage. Without proper software, an airport is a large parking lot and a tunnel simply a hole in the ground.

Problems in other areas relating to improper software have been addressed by Petroski [1994] and Littlewood and Strigini [1992]. Programming bugs have disrupted telephone service and delayed shuttle launches. Petroski went into considerable detail on the loss of a massive Norwegian oil platform in the North Sea. Here, there was improper modeling of the stresses to which the platform would be subjected. Radiation overdoses in medical devices have been attributed to the situation where the typist was too fast. In Reagan's Star Wars initiative several years ago, the point was made that there was no way to test the final system. Hayes [1995] pointed out that debugging tools need to be improved considerably. It might be well that providers of software should include debugging programs that will allow the user to fix and tune his system for his location. At least with climate control computers for greenhouses, an alert grower can usually prevent a catastrophe. Nevertheless, as software complexity increases problems of code defects will also increase. Seldom have I seen specific measures to address these concerns about obtaining reliable software. The grower can protect himself to some extent by requiring the seller to provide adequate documentation, user-friendly debugging procedures, and proper backup for problems that might arise in execution.

C. INTERCOMMUNICATION

In recent years, the development of computer communications has exploded. The common publications such as *Time* and *U.S. News & World Report* nearly always carry some article on "cyberspace" each week, along with special issues [*Time*, Nov. 13, 1995]. The *National Geographic* magazine devoted considerable space to the information revolution [Swerdlow, 1995]. Local newspapers are often seen to tout the use of Internet. The potential for obtaining information and communication on the PC holds marked promise, particularly in passage of the new telecommunications law in the U.S..

In the development of optimal climate control (second and third levels of Fig. 8-2), information from other sources such as weather predictions and financial data are required. This means a communications link with other data sources. Ammerlaan [1984] went into considerable detail of the kinds of information contemplated in The Netherlands. He stated that a necessary condition for any properly working management information system is adequate recording of actual greenhouse data –meaning financial data, plus production and marketing. Action of the Dutch auction could be made available to the grower's machine, and this information used in price predictions and product requirements. Growers' data would be used by research organizations and the advisory service to improve actual firm management, with the possibility that an operator could access a huge data base with programs for optimizing production requiring computer services that have greater capacity than the firm's local PC. Steinbuch [1986] also discussed these possibilities using collected data from climate measurements for use in management programs, and the possibility of calculating new setpoints based on actual fuel prices, market situation, and weather forecasts. Whether the individualistic American grower will be willing to share proprietary information remains to be seen.

A degree of centralization in a common information system is required –which is forced upon Dutch growers by the fact that most marketing in the Netherlands is conducted through auction, and research and advisory organizations are closely related, as contrasted with the U.S. situation. However, opening communication lines between individual operations and centralized data gathering points leads to problems of adequate security. The existing situation with the Internet has been addressed by several authors [e.g., Shiller, 1994; Mann, 1995; Wallich, 1994]. It is a fact in the intercollegiality of the Internet that few items are secure in privacy terms. Most people know that e-mail messages can be read by many people other than the intended recipients, and such messages can be forged with no one sure that the message came from the ostensible sender. Electronic impersonators can commit slander or solicit criminal acts in someone else's name. They can even masquerade as a trusted colleague to convince someone to reveal sensitive personal or business information. Anyone who has a PC, a modem, and $20.00 per month in connection fees can have a direct link to Internet and be subjected to break-ins –or launch attacks on others. In fact, Wallich [1994] states that, in European countries, computer intrusion is not necessarily a crime. There are instances of computer science professors who assign their students sites on the Internet to break into, and files to bring back as proof they understand the protocols involved. In a recent newspaper article [Johnson, 1996], an on-line service's computer files were obtained by the police in a murder investigation. In still another example, Elmer-Dewitt [1995] reported on the problems Netscape was having in eliminating security flaws. The new Internet software such as "Java" had loopholes that permitted software viruses to enter –which was caught before any harm occurred.

There are examples of secure distributed computing [e.g., Schiller, 1994], such as the system used at the Massachusetts Institute of Technology. Encryption programs have been written that are completely secure, but their use in the U.S. has been restricted by the government. If the server machine provides it, encryption at 128 digits is permitted on systems such as Netscape. Beyond this number (e.g., 256, 512, etc.), possibilities of breaking into the encryption are so difficult that the U.S. government prohibits their use. Thus, in any system of intercommunication, security standards should be thoroughly thought out and set up before the industry can be expected to open their computing facilities for sharing data, utilizing special programs, or downloading market and financial information. The possibilities that exist in this area for significant improvement in industry profitability are enormous.

V. OPTIMIZATION AND MODELING

A. REQUIREMENTS

In achieving the ultimate –optimization of climate control by determining the most profitable setpoints that can be achieved for the climate and structure– a number of points should be mentioned:

1. The necessary setpoints obviously depend on the species being grown. Preliminary information has to be entered by the grower as to crop requirements in terms of optimal setpoints given an **average** weather outlook and market, boundary conditions (i.e., maximum and minimums), stage, or stages, of crop growth that will be encountered, and the duration of each stage given an **average** crop history.

2. Short-term actions generally occur within minutes, depending upon the response time of the greenhouse and the crop within it. Photosynthesis is a typical crop process that changes rapidly according to environmental changes. Transpiration and internal water potential is another aspect, along with respiration, translocation, etc. Although the results may not be readily apparent, their cumulative effect is highly significant. The short-term setpoints need to be optimized so that the greenhouse climate, at least, does not reduce growth or miss timing. The cost of producing the necessary climate is reduced.

3. The growth of the crop occurs over a long period. Production may be continuous once the crop reaches production, or it may be sold within a short period. While one can estimate, from previous years' market performance, the probable returns, these change with time. Now, the grower must be able to recalculate, based on most recent estimates, the possible return, and to calculate optimal setpoints as the result of new information. Thus, there is a long-term optimization problem. It may be necessary for the grower to "force" certain trends through manual manipulation of climate variables, and the possible results should be given to the grower for his ultimate decision.

4. In solution to climate control, it is necessary to account for the various parts of the greenhouse and its contents. This includes the energy balances and response of the crop, the greenhouse air, the cover or "hull" of the structure, and the greenhouse soil [Bot and van de Braak, 1995]. The hemisphere above the greenhouse, of course, influences energy and mass exchanges. This is a simple division of the problem of calculating energy and mass (vapor) exchanges between the various components. Where a thermal curtain is employed, the space above the crop may have to be further subdivided –below the curtain

and between the curtain and greenhouse hull. The soil is a storage system, requiring a measurement of the surface temperature and the flux of energy into and out of the soil volume. Sometimes, investigators will divide the soil volume into as many as seven layers. The crop may be divided into layers. Heat storage capacity of the crop, greenhouse air, and cover are generally neglected. Most divisions of the greenhouse found in the literature use similar or slight variations of the ones introduced above. I have yet to see any three-dimensional considerations of the problem. Energy and mass exchange are considered one-dimensional in vertical directions.

These problems of the crop, long-term and short-term requirements, have been the subject of many authors in the past 2 decades [e.g., Challa and van de Vooren, 1980; Udink ten Cate and Challa, 1984; Challa, 1993; Bailey, 1995; Bailey and Chalabi, 1991; Tantau, 1993; Seginer, 1993; Seginer et al., 1991; etc.]. Some of these will be discussed at a later point.

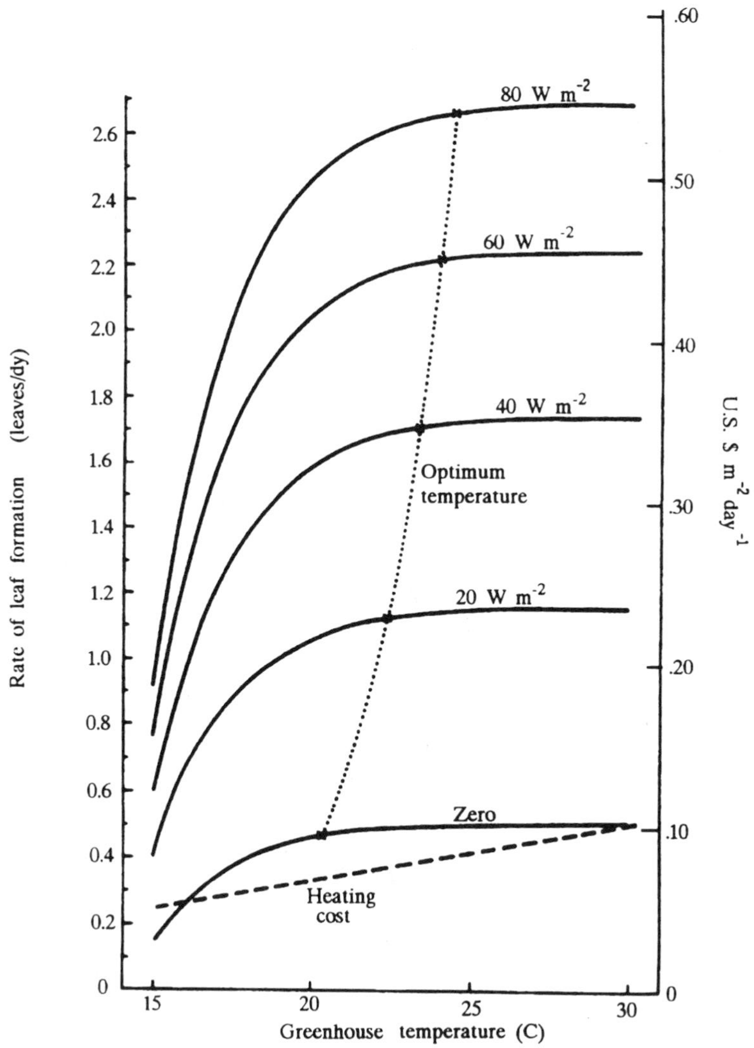

Fig. 8-23. An example of the relationship between rate of leaf formation, radiation, and temperature during the first stage of cucumber growth. The leaf formation rate determines earliness in fruit production. Optimization is calculated by comparing the costs of heating with the expected economic returns (right ordinate) [Challa and van de Vorren, 1980] (©1980, Int. Soc. Hort. Sci., *Acta Hort.*, 106:159-163).

B. EXAMPLES OF LIMITED APPROACHES

In Chapter 3, work by Heuvelink and Challa [1989] to optimize the use of supplementary irradiation by comparing the dry weight increase in photosynthesis versus the marginal costs of supplemental irradiation was briefly outlined (Fig. 3-66). Similar restricted programs for energy conservation by Bailey [1990], Hurd and Graves [1984], and Butters [1977] were cited in Chapter 4. Aikman and Picken [1989] examined a means to relate the mean temperature integral with wind for the purposes of energy conservation. Hand and Soffee [1971] and Rudd-Jones et al. [1978] considered temperature adjustment according to radiant intensity early in control development. These maximizations of photosynthesis are not always the most desirable procedures because there are other temperature-sensitive plant reactions that can be more important (Chapter 4). Thermal screen optimization was also mentioned [Seginer and Albright, 1980] (Table 4-16), in which the tradeoffs between photosynthesis, respiration, and fuel requirement were investigated. In Chapter 7, work by Schapendonk and Gaastra [1984] on costs of CO_2 enrichment emphasized the importance of CO_2 in production. The literature in the area of CO_2 optimization and costs-versus-benefits is extensive [e.g., Challa and Schapendonk, 1986; Bakker, 1985; Nederhoff, 1988; Nederhoff et al., 1989; Rijsdijk and Houter, 1993; etc.].

Usually, the situations cited above are limited to 1 or 2 variables. In most of the examples, the control is on-off, or "bang-bang" by the jargonists. There is a whole class of such operations to be found in greenhouse climate control (e.g., thermal and shade screens, supplementary and photoperiod lighting, evaporative pad pumps, etc.). Several of these manipulations of setpoints and control actions could be carried out in computer systems with subroutines in a program to provide the result as diagrammed in Fig. 8-2 or 8-23, where the cost of obtaining some increase in crop response is set against the monetary return. Unfortunately, as stated, these are limited in scope and beg the question of a truly integrated climate control system.

C. MODELS

In order for the computer to have something upon which to work, the relationships between the environment, the grower's wishes, and the required setpoints to achieve the objectives of maximizing profit must be presented in some way that relates them. This can be done by patterns, called "models" which are, in our case, mathematical representations of the real world. The greenhouse, its crop, and the management thereof, is a system. A model is a simplified representation of this system. In this area, the proliferation of kinds of models, and the mathematical methods and complexity become an art that only someone thoroughly familiarized can succeed.

Seginer [1993] outlined some approaches to the problem of determining control:

1) The traditional approach where experiments are designed to test a range of suggested setpoint rules. While this method has been most common, interaction and complexity of climate control has increased to the point that the method is very expensive, requiring an extensive testing facility. Setpoints derived from such studies are, according to Seginer, incapable of weighing short-term, conflicting interests (e.g., simultaneous ventilation and CO_2 injection). Examples would be investigations of crop response to different temperatures imposed in a number of chambers or greenhouse sections.
2) The expert system approach where rules are extracted by communication with an expert grower. There are many examples of this type, such as Fynn et al.'s system for nutrient selection [1991].
3) A learning system (pattern recognition approach), where the state of the environment and the actions of the expert grower are monitored. The data obtained are used for automatic rule formation. This could be considered as the pattern employed by Sigrimis [1989] in which the variables were "normalized" (range zero to one) and the values required were obtained by recording actual system data over a 24-hour period. These were then used as the required parameters that could be "tuned."
4) A system model approach, where mathematical algorithms are used in an optimization scheme to produce individual control decisions. This approach is being actively persued by many authors [e.g., Seginer, 1991; Gal et al., 1984; Challa and van Straten, 1991; van Straten and Challa, 1995; etc.].

When it comes to actual models, there is a plethora, including:

Static models
Steady-state models
Black box models
Dynamic models
Mechanistic models
Stochastic models
Heuristic models
Simulation models
Descriptive models
Explanatory models
State-variable models, etc.

Static models are a set of equations relating the various aspects of heat loss, heat input, ventilation, etc. that can be solved for an instant in time when, essentially, the system is in equilibrium. The same thing may also be accomplished for crop growth. As such, one may consider static models as **steady-state** models. Equations are based upon physical laws, and are considered **mechanistic** models. For a digital computer system, which samples at instants in time (t = 0, t = t + 1, t = t + 2, etc.), and which is fast, it is possible to solve the necessary equations and to signal actuation equipment before sampling again and recalculating. Most of the equations in this book are static equations. The simple Eq. 4.13 for calculating the sensible heat exchange by convection and conduction (H_{sg}) could be solved at each execution of the program, and then utilized in other equations to determine the structure's heat requirement. More detailed models, some dynamic, for thermal analyses include those by Bot [1983], Schokert and von Zabeltitz [1980], Chandra et al. [1981], Albright et al. [1985], Albright [1984], Maher and O'Flaherty [1973], etc. Other examples include those to calculate radiant energy distribution in a canopy or the relationship between respiration and growth. Static models do not involve time, as contrasted with dynamic relationships that do consider changes with time. If the results are placed in terms of cost to maintain a particular setpoint, then adjustments could be made based upon the predicated plant response. Examples of static models for predicting greenhouse energy relations are Jolliet et al's [1991] "HORTICERN," Tantau's [1985] simplified model for German conditions, Cooper and Fuller's [1983] transient model, and Liebig and Alscher [1993] and Garzoli [1985] –to mention a few.

Since several inputs and outputs are generally required, a set of simultaneous equations result that are often solved by using matrix algebra. Sigrimis' [1989] equation is an example of this. It is also a **stochastic** model:

$$\begin{bmatrix} T_i \\ H_i \\ T_s \end{bmatrix}(k+1) = \begin{bmatrix} a_{11} & 0 & a_{13} \\ a_{21} & a_{22} & 0 \\ a_{31} & 0 & a_{33} \end{bmatrix}\begin{bmatrix} T_i \\ H_i \\ T_s \end{bmatrix}(k) + \begin{bmatrix} b_{11} & b_{12} & b_{13} & b_{14} & b_{15} \\ 0 & b_{22} & 0 & b_{24} & b_{25} \\ 0 & b_{32} & b_{33} & b_{34} & 0 \end{bmatrix}\begin{bmatrix} U_1 \\ U_2 \\ U_3 \\ U_4 \\ U_5 \end{bmatrix}(k) + \begin{bmatrix} c_{11} & c_{12} & 0 & c_{14} \\ 0 & 0 & c_{23} & c_{24} \\ c_{31} & c_{32} & 0 & c_{34} \end{bmatrix}\begin{bmatrix} S_o \\ T_o \\ H_o \\ V_t \end{bmatrix}(k) + \begin{bmatrix} V_1 \\ V_2 \\ V_3 \end{bmatrix} \tag{8.11}$$

where: k = a sample counter t = kT T = intersampling period
T_i = inside air temperature,

H_i = inside relative humidity
T_s = soil temperature
U_1 through U_5 = control inputs for space heater, ventilator openings, fan ventilation, ground heater and mist
S_o = solar radiation
T_o = outside temperature
H_o = outside relative humidity
V_t = inside evapotranspiration rate
V_1 through V_3 = measured, random system noise
a, b and c = parametric constants indicated by subscripts, usually determined by statistical methods

Use of matrix equations is common where the solution to multiple inputs and outputs is to be solved. Complex models, of course, increase the demand on a computer system. The ability to recalculate constants between each execution may not be possible. For example, Ritchie [1986] suggested that versions of the CERES crop model for predicting yield of agronomic plants required 30 seconds on 1985 versions of IBM-PC-AT computers. In comparison with present computers, 30 seconds is extremely slow. The solution could probably be achieved in a few seconds or less today.

One thinks of a **black box** as something mysterious. A general idea is some removable, electronic device. The term is an informal one meaning that the output of such a device is the result of unknown factors. Yet the term is used in modeling to denote a system that sometimes depends upon statistical relationships to provide an output from some input. This might be a straightforward linearization of a variable such as inside temperature with outside radiation by least squares regression. Udink ten Cate [1983] defined a black box model as a working point (steady-state situation) that is the result of one or more variables acting upon the greenhouse climate. The output is not dependent upon equations that follow a particular physical law. Sigrimis' Eq. 8.11 is an example. The selection of a "difference" equation, or process model, is called "identification."

The terms **heuristic** and **stochastic**, which are well-loved by many authors, refer to the means by which models are solved. Heuristic models are those in which problem-solving is by exploration or trial and error. One tries one method, based upon experience. If that does not work, one tries another way that may be as much intuition as clear reasoning. Stochastic models, on the other hand, deal with random variables. The problem-solving process depends upon statistical methods as opposed to fixed rules or relations. Statistical methods are used to reduce and predict random variables by optimizing the system design. A stochastic model could be considered a black box model as the exact nature of the process may be unknown [Bollinger and Duffie, 1988].

As with static equations, "dynamic" equations follow, wherever possible, well-known physical laws of energy, mass transfer, and plant physiological processes. With the digital computer, it is possible to turn static equations into a dynamic process since a new set of conditions can be calculated each time the computer cycles through its routine. Usually, however, most think in terms of differential equations such as Nederhoff's [1994] for solving the CO_2 balance of greenhouse air:

$$\frac{dC(t)}{dt} = \frac{G_c + S_c - P - L_c}{h\rho_c} \tag{8.12}$$

where: $dC(t)/dt$ = an infinitesimal change in CO_2 concentration with an infinitesimal change in time ($C(t)$ = $\mu mol\ mol^{-1}$)
G_c = rate of CO_2 emission from the ground
P = rate of CO_2 uptake (photosynthesis) ($g\ m^{-2}hr^{-1}$)
S_c = rate of CO_2 supply ($g\ m^{-2}hr^{-1}$)
L_c = rate of loss of CO_2 due to ventilation ($g\ m^{-2}hr^{-1}$)

h	= *average height of greenhouse (m)*
ρ_c	= CO_2 *density (1.83 kg m*$^{-3}$ *at NPT)*

There are other equations necessary for providing the values of P, S_c, G_c, and L_c. The equation is considered "first-order" linear, as contrasted with a higher-order polynomial or nonlinear equation. Most processes in the greenhouse are nonlinear (i.e., Fig. 8-24), but these are extremely difficult to solve. Typically, both static and dynamic relationships may be found in greenhouse models. **Explanatory** models may be static or dynamic. A descriptive model might be merely a file of data on the greenhouse climate.

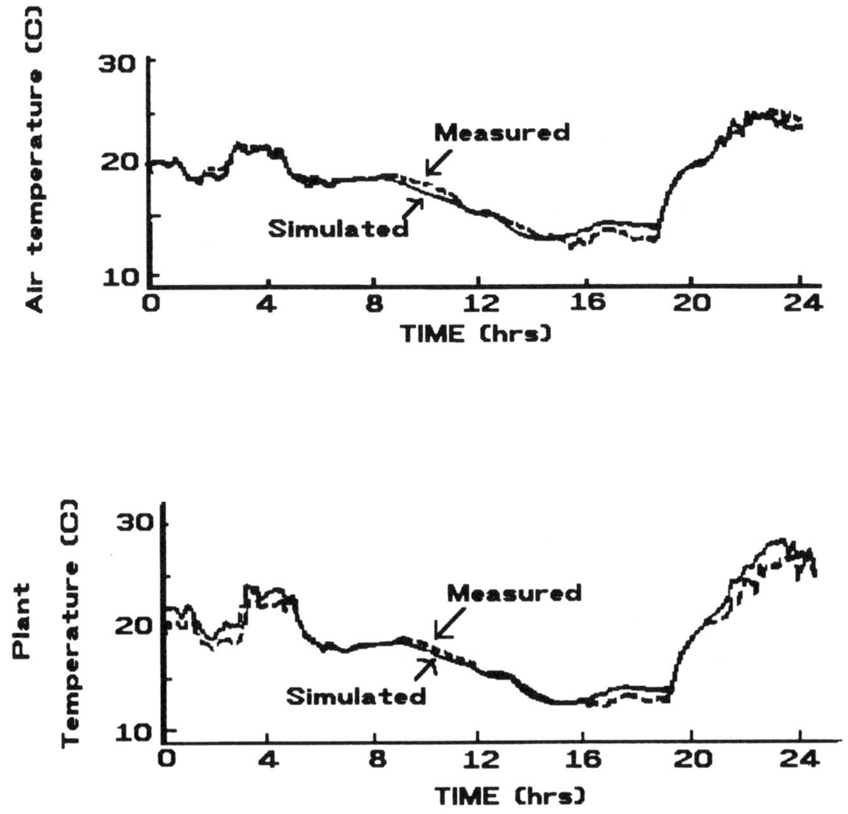

Fig. 8-24. An example of a simulation run comparing air and plant temperatures as predicted by the model with actual measured temperatures for the same period [Bot, 1989b]. For Venlo-type houses, Wageningen. The day had heavy cloudiness, followed by a night with decreasing cloudiness and a clear day (©1989, Int. Soc. Hort. Sci., *Acta Hort.*, 245:389-396).

Simulation models might consist of any of the above. The method is to reproduce, in this case, the greenhouse climate and crop response to one or more variables such as temperature, radiation, CO_2, etc. Simulation is an inexpensive process for studying climate or crop response without the cost of building a greenhouse and testing crop growth within it. The many variables that influence growth make actual testing very expensive (refer to Seginer's type 1 approach). The results of simulation can be run in a short time, and model validity can be compared with what happens under real conditions if the investigator feels it necessary to do so. Unfortunately, one finds in the literature many statements of agreement between simulation and actual measured results modified by the words "good," "close," "reasonably well," etc. without any definition of what the author means by "close." Examples of simulation studies include publications by Jones et al. [1995; 1990], Seginer et al. [1986], van Bavel et al. [1981], Mahrer and Avissar [1984], Giniger et al. [1988], etc., including the numerous simulation monographs from Wageningen, of which the ones edited by Penning de Vries and van Laar [1982] and Rabbinge et al. [1989; 1990] are representative. Bot [1993] has done a considerable amount of simulation for which typical results of some of his earlier work are illustrated in Fig. 8-24.

De Wit [1982], in his lucid description of modeling, pointed out that explanatory models in biology are rudimentary –although improvements have been made since his publication. More to the point is de Wit's explanation of the "state-variable" approach. State-variable models are based on the assumption that the state of each system at any moment can be quantified. Changes in the state can be described by mathematical equations. State variables in a greenhouse model can be temperature, CO_2 level, and humidity, or for the plant, leaf area index, assimilate distribution, or dry weight [Challa, 1993]. These are variables that can be measured as though time stood still. Driving variables, or forcing functions, characterize the effect of the environment, and their value must be measured continuously –i.e., external wind, temperature, irradiation, etc. Each state variable is associated with rate variables that characterize their rate of change at a certain instant. These variables represent energy or material flows between state variables. The values are not dependent upon statistical analysis, but represent rules based upon knowledge of the physical, chemical, and biological processes taking place. When the rate variables have been determined, these are used to calculate the state variable at time $t + \Delta t$, which equals the state variable at time t plus the rate at time t multiplied by Δt. This is integration that gives new values of the state variables. The interval, Δt, is small enough that the rates do not change appreciably. The principal factor to consider is that rates are not dependent upon each other. Each rate depends at each moment on state and forcing variables only, and is computed independently of any other rate. It is never necessary to solve most simultaneous equations with several unknowns.

Another explanation of state variables is Voland's [1986], in which he refers to the term "state space." State space is a multidimensional space in which state variables act as coordinates, "n" being equal to the number of simultaneous values for the "n" state variables. The number "n" of state variables is the minimum number of linearly independent coordinates in state space that must be known to specify the state of the system. If the values of these state variables are known at any given time, t_o, the behavior of the system can be predicted for any time greater than t_o. The n state variables ($x_1, x_2, \ldots, x_n$) form the coordinates of the state vector **x**. One may also define a multidimensional control vector **u**, with k dimensions, which are the system inputs. Finally, a function vector, **f**, can be formulated as the mathematical model of the system. Vorland points out that a set of state variables that form the mathematical model is not unique. Other sets can be formulated. Any feasible set of state variables, $x_1, x_2, \ldots, x_n$, must contain only linearly independent x_i. That is, the relationship:

$$a_1x_1 + a_2x_2 + \ldots + a_nx_n = 0 \qquad (8.13)$$

must not be satisfied by any constants, a_i, other than zero.

Note that the distinction between these model types can be "murky." Usually, however, one finds the state-variable approach, using a combination of static and dynamic relationships, to be the major thrust in combining levels 2 and 3 of Fig. 8-2.

D. COMPLETE MODELS

There are two main objectives to be attained in modeling the greenhouse operation: 1) the energy budget of the structure is to be solved, and 2) the crop production output is to be determined. In general, the physical laws of energy exchange for the structure are well known, and physiological models are more often estimates.

1. Energy Balance of the Structure

Overall, the rate of energy exchange for any volume has been stated simply by Bot and van de Braak [1995]:

$$\frac{dQ_h}{dt} = q_{in,h} - q_{out/h} + P_h \tag{8.14}$$

where: dQ_h/dt = *the rate of energy exchange* $(J\ s^{-1})$
$q_{in,h}$ = *the amount of energy entering the structure,* $(J\ s^{-1})$
$q_{out,h}$ = *the amount of energy leaving the structure,* $(J\ s^{-1})$
P_h = *energy produced within the structure* $(J\ s^{-1})$

As energy is related to temperature (T), the right side of Eq. 8.14 can be equated to Cap_hT, where Cap_h is the thermal capacity of the air ($J\ K^{-1}$). Joules per second, of course, equal Watts. The same relationship can be written for mass transfer, or the exchange of water vapor:

$$\frac{dQ_m}{dt} = q_{in,m} - q_{out,m} + p_m \tag{8.15}$$

where: dQ_m/dt = *the rate of mass (water vapor) exchange* $(kg\ s^{-1})$
$q_{in,m}$ = *rate of vapor entering the structure* $(kg\ s^{-1})$
$q_{out,m}$ = *rate of vapor exiting the structure* $(kg\ s^{-1})$
p_m = *mass of vapor produced in the structure* $(kg\ s^{-1})$

$Q_m = Vc_m$ where c_m is the vapor concentration ($kg\ m^{-3}$) and V = volume of the structure (m^3). As with Eq. 8.14, Vdc_m/dt equals the right-hand side of Eq. 8.15.

There is much more to be done to achieve the solution of these differential equations. Without going into the caveats for the relationships shown, or Bot and van de Braak's discussion of dimensionless numbers that help solve energy and mass transfer, the net energy and mass flow can be equated to the differences in temperature and concentration between inside and outside:

$$q_h = q_v \rho C_p(T_i - T_o) \tag{8.16}$$
$$q_m = q_v(c_i - c_o) \tag{8.17}$$

where: q_h *and* q_m = *energy exchange* $(J\ s^{-1})$ *and vapor exchange* $(kg\ s^{-1})$ *by ventilation*
q_v = *volumetric exchange by ventilation or infiltration* $(m^3 s^{-1})$
ρC_p = *volumetric specific heat* $(J\ m^{-3} K^{-1})$
T_i *and* T_o = *inside and outside temperature (K)*
c_i *and* c_o = *inside and outside vapor concentration* $(kg\ m^{-3})$

Eqs. 8.16 and 8.17 account for exchange due to ventilation and infiltration, with Eq. 8.17 also applicable in some respects to CO_2 exchange. In the Dutch experience [Bot and van de Braak, 1995; Nederhoff et al., 1984; Fernandez and Bailey, 1992; Bot, 1983; Goedhart et al., 1984], the ventilation number $G_v = q_v(uA_{op})^{-1}$. The symbol "u" = wind speed and A_{op} = area of ventilator opening. Of course, with fan ventilation, if the fan capacity is known, the ventilation exchange is easy to obtain.

Heat and vapor exchange by conduction and convection are more difficult in several respects. One can use the relationships discussed in Chapter 4, Eq. 4.13. The precision of that equation can vary remarkably with the transfer coefficient, which also varies with such factors as sky conditions and condensation as well as wind speed (Table 4-5, Fig. 4-31). The addition of heat to the inside can be more difficult, depending upon the heating system and its arrangement [Stanghellini, 1983; Bot, 1983]. The crop is also a major absorber of radiation that is largely converted to water vapor. The energy storage of the crop is usually neglected, but the energy storage of the soil determines dynamics of the greenhouse system on a daily basis [Bot, 1989a]. Thus, the absorption and release of energy from the soil needs to be determined.

Radiation in greenhouses was discussed in Chapter 3. The transmission is the relation between the useful irradiation falling on the total ground area of the greenhouse inside and that falling on the same area without the structure. This requires determination of greenhouse transmission (τ). The outside pyranometer provides total global radiation, diffuse and direct, falling on a horizontal surface per unit area. For large, contiguous structures, the effect of sidewalls can be ignored when one considers the usual error in radiation measurement [van den Kiebom and Stoffers, 1985; Tchamitchian, 1993]. For Venlo-type houses with limited distance from gutters to ridge, one could consider the area of the range as a flat plate. It is necessary to calculate the direct solar radiation as a function of the sun's position, and where the facing roof is large, orientation and roof angle should probably be taken into account. Under European conditions, the largest part of radiation is diffuse, particularly in the winter. The paper by Spitters et al. [1986] dealt with determining the relative proportions of diffuse versus direct radiation at the earth's surface. Differences between diffuse and direct require different treatment in radiation interception by the plant canopy. Bot [1983] went into considerable detail for determining transmission for Venlo structures. Iqbal [1983] has a good discussion on some terminology in dealing with solar position. It can be appreciated from Fig. 2-7, that the equations necessary to calculate the angle of incidence on the roof for direct solar radiation will vary with the greenhouse type and orientation. Bot's development, while informative, will not be germane for all structures. Figs. 3-14 through 3-16 and 3-21 show that transmissivity varies not only with the angle of incidence, but also with the cover type and materials and construction of the range. Once the transmission coefficient has been determined, then one may arrive at the radiation available for photosynthesis, and sensible and latent heat exchanges inside. Also, note the differences in the heat transfer coefficient as the result of cloud conditions in Table 4-5.

There are also radiative exchanges between plants, structure and the heating system, and between the cover and sky, especially in thermal radiation. From the Stefan-Boltzmann relation, and ignoring possible multiple reflections, the net heat flow (flux) ($q_{rad,ij}$) from surface "i" to surface "j" can be written as [Bot and van de Braak, 1995]:

$$q_{rad,ij} = A_i \epsilon_i \epsilon_j F_{ij} \sigma (T_i^4 - T_j^4) \tag{8.18}$$

where: $q_{rad,ij}$ = *J s-1,*
A_i = *Area of surface "i" (e.g., pipe surface, cover of greenhouse, etc)*
ϵ_i, ϵ_j = *emissivities of surfaces "i" and "j"*
F_{ij} = *view factor between surfaces "i" and "j", determining which of radiation from "I" actually falls on "j"*
σ = *Stefan-Boltzmann constant*
T_i, T_j = *absolute temperatures of the two surfaces*

For real-time information, the sky temperature is important in determining energy exchange between the greenhouse roof and sky. Outdoor camping experts are well aware of the effects of clear skies, particularly at high altitudes. Sky temperature can be measured with a pyrgeometer, but I know of no commercial operation in the U.S. that is making such measurements.

Measurement of vapor generation within a greenhouse, resulting largely from transpiration, is subject to some difficulty since it involves physiological aspects of the crop. Section 5.IV discussed many items that determine crop transpiration. Eq. 5.27 showed Stanghellini's [1987] formula for computing "E." Efforts to refine the transpiration model by the Dutch have continued [e.g., de Graaf, 1988; van Meurs and Stanghellini, 1989]. The most recent discussion has been provided by Stanghellini [1995]. One problem is the resistance to vapor transfer between canopy and air (see Table 5-32). In darkness, resistance through the stomata increases to very high levels (conductance close to zero); whereas, in daylight hours, stomata can be considered fully opened in non-stressed plants, and the major resistance occurs between the canopy and surrounding bulk air. Stanghellini holds that leaf (stomatal) resistance has to be known, along with net radiation and boundary layer resistance. Parenthetically, although stomata may be closed at night, the 16 hour dark periods during European winters can result in greater total water loss than during the short hours of an overcast day (see Chapters 3 and 5). Stanghellini concludes that the boundary layer resistance in Venlo houses is about a few hundred s m^{-1}, and the value can be considered a constant. If radiation, humidity and temperature are known, Eq. 5.27 can be solved without difficulty. Data for other structures (i.e., American) are not available.

Condensation of 1 g vapor on the inside roof of a greenhouse will warm a cubic meter of air nearly 2 C. The amount of water can be calculated [Stanghellini, 1995]:

$$C = g_{cnd}(\chi_a - \chi_r^*) \tag{8.19}$$

where: C = *condensation (kg $m^{-2}s^{-1}$)*
χ_a = *ambient absolute humidity (kg m^{-3})*
χ_r^* = *absolute humidity at the cover (kg m^{-3})*
g_{cnd} = *conductance (m s^{-1})*

g_{cnd} can be calculated:

$$g_{cnd} = \frac{A_r}{A_g} 1.64x10^{-3}(\check{T}_a - \check{T}_r)^{1/3} \tag{8.20}$$

where: A_r = *cover area (m^2)*
A_g = *ground area (m^2)*
$\check{T}_a$ = *virtual temperature of the air*
$\check{T}_r$ = *virtual temperature of the cover surface at T = inside cover temperature (see Section 5.IV.B.2 for discussion and definition of virtual temperature)*

For a Venlo house, A_r/A_g = 1.1, and when there is condensation, the far right factor will be between 1 and 2.5. g_{cnd} is about 3 x 10^{-3} m s^{-1}. Of course, if χ_a is less than χ_r, there will be no condensation. There will be no vapor exchange from, or to, the roof.

The vapor balance of the greenhouse air can be written:

$$h\frac{d\chi_a}{dt} = g_{tr}(\chi_{eff} - \chi_a) - g_{cnd}(\chi_a - \chi_r^*) - g_{vnt}(\chi_a - \chi_o) \tag{8.21}$$

where: h = *ratio of greenhouse volume to ground area, or the mean height of the house (m),*
$d\chi_a/dt$ = *the change in absolute humidity,*
g_{tr} = *conductance of the canopy,*
χ_e = *effective absolute humidity in the crop*

g_{vnt} = *conductance of the ventilation*
χ_o = *outside absolute humidity*

The summation of all conductances gives a total conductance of the house (g_{tot}) and:

$$g_{tot} = \frac{2LAI}{\left(1 + \frac{\Delta}{\gamma}\right) r_b + r_l} + 1.64x10^{-3}\frac{A_r}{A_g}\left(T_a - T_r\right)^{1/3} + h\frac{n}{3600} \tag{8.22}$$

The extreme right-hand term above represents g_{vnt}, and n is the volume changes per hour.

The time constant of this system, according to Stanghellini [1995], is $\tau = h/g_{tot}$. During the day, under full sunshine and a mean greenhouse height of 3.6 m, τ is between 2 to 10 minutes.

From Nederhoff [1995], the supply rate of CO_2 must compensate that taken up by photosynthesis plus that gained or lost by air exchange (ventilation and infiltration):

$$\phi_{c,in} = \phi_{c,p} + \phi_{c,vnt} \tag{8.23}$$

where the subscript c refers to CO_2, and the subscripts "in," "p," and "vnt" refer to CO_2 entering the house, i.e., that taken up by photosynthesis and CO_2 exchanged by ventilation, respectively. ϕ is in units of kg $m^{-2}s^{-1}$. Nederhoff cites the literature to the effect that the recommended, minimum CO_2 supply rate is 4.5 g $m^{-2}hr^{-1}$, or the combustion of 25 m^3 ha^{-1} hr^{-1} natural gas. This is sufficient to maintain 100 Pa in a closed Venlo under European conditions, and prevent severe CO_2 depletion where ventilation occurs. I have not seen figures for other structures, especially in clear-day climates.

The energy and mass balances for the environment, briefly outlined above, result in a set of first-order differential equations, describing the state variables temperature and mass concentration as functions of time with given initial and boundary conditions [Bot and van de Braak, 1995]. There are a number of climate models, of which three were compared by van Bavel et al. in 1985. These were simulation models that were tested using weather data from three locations: Lubbock, TX; Tokyo; and Wageningen. The general conclusions were that the three models produced essentially the same results although significant differences were found regarding evaporative cooling and estimated daily transpiration. Since this 1985 comparison, there has been considerable improvement in understanding physical processes in greenhouses, along with the significant advance in hardware available. Nevertheless, the result shows that models may vary in their parts and still give acceptable performance. Among those who have published on modeling of greenhouses –besides Bot [1983; 1993]– are Avissar and Mahrer [1982], Kindelin [1980], Duncan et al. [1981] and Seginer and Kantz [1985], as well as many others. A detailed, 356 page, energy balance program was published by Kimball in 1986. A computer program consisting of parts of a simulation program TRNSYS, written in Fortran, was provided.

If the energy and power requirements for maintaining a given setpoint (temperature, humidity, CO_2) are expressed in monetary terms, then calculation of operating costs can be made. While these costs are true at the moment of calculation, these same calculations can be made to predict costs based upon the average predicted weather for the locality through the crop production period, and updated every execution based upon the measured weather at that time. Given the ability of most crops to integrate temperature over time, there is no reason the mean temperatures, or the DIF (T_{dy}-T_n), cannot be manipulated over periods of 1 day or more to provide setpoint changes commensurate with minimum cost. Thus, we begin to see the management power that computer control can provide the grower.

2. Models of Crop Growth

In looking at crop growth, Challa and Bakker [1995] point out three items one wishes to know: the total amount produced, onset and time course of production, and the quality of the produce. Unfortunately, not only is the crop influenced by the weather conditions (radiation especially), but the crop itself influences the greenhouse climate, as discussed in Section 5.IV.B.3. The process of crop growth may be divided into short and long term, according to the growth stage of the crop. The short terms are carbohydrate production and the water status. Long term is characterized by dry matter accumulation and development. According to Penning de Vries [1990], one wishes to simulate the behavior of an agricultural crop, but economic and sociological factors are no real part of crop models; however, crop models can provide information that improve decision-making where economic constraints are important.

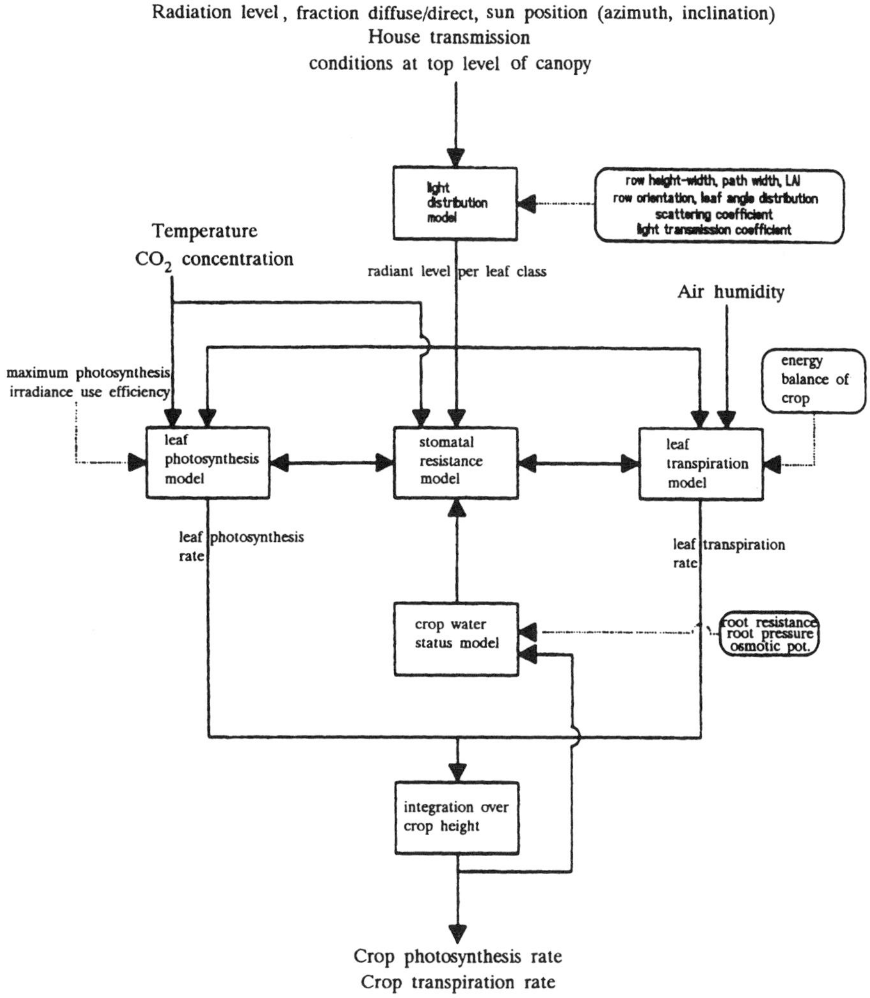

Fig. 8-25. General outline of a short-term crop growth model (< 1 dy) as envisioned by Challa [1989] (©1989, Int. Soc. Hort. Sci., *Acta Hort.*, 248:209-215).

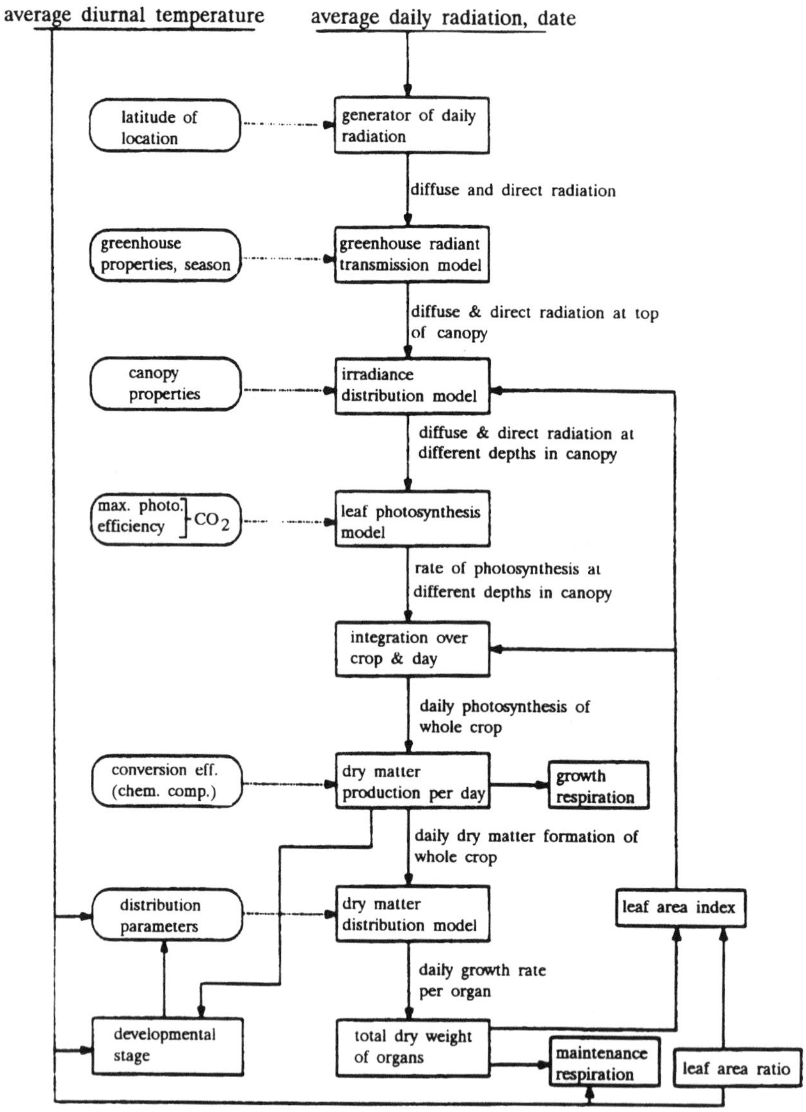

Fig. 8-26. General outline of a long-term crop growth model as suggested by Challa [1989] (©1989, Int. Soc. Hort. Sci., *Acta Hort.*, 248:209-215).

The complexity of the problem one is faced with is exemplified by Figs. 8-25 and 8-26. Fig. 8-25 is an outline of the short-term requirement (1 day or less) for determining crop photosynthesis and transpiration as based upon the SUCROS87 model [Spitters et al., 1989] presented by Challa [1989]. The figure shows, actually, the use of several models, parameter requirements, and the interactions that must be dealt with. The transpiration model was discussed in the previous Section (Eq 5.27). The photosynthetic model may be one of several with Faquhar et al.'s [1980] presentation the one favored by Gijzen [1995a]. The Acock et al. [1978] model, which enjoyed a period of popularity, was stated by Jones et al. [1989) as too simple for their conditions of complex row planting and spacing. There were no corrections for plant age effects. Farquhar et al.'s model seeks to relate the biochemical processes in C_3 species that can limit photosynthesis. The overall relation is given by a seemingly simple relationship [Gijzen, 1995a]:

$$\frac{dW}{dt} = C_f(30/44\,P_{gc,d} - R_{m,d}) \tag{8.24}$$

where: dW/dt = *the rate of dry matter production* ($g\ m^{-2}dy^{-1}$)
C_f = *conversion efficiency of dry weight formed per gram of carbohydrates (assimilates)*
$P_{gc,d}$ = *daily rate of gross photosynthesis per unit greenhouse area* ($CO_2\ g\ m^{-2}dy^{-1}$)
$R_{m,d}$ = *daily rate of maintenance respiration* ($CH_2O\ g\ m^{-2}dy^{-1}$)

Gijzen [1995a] and van de Sanden [1995] provide a good general discussion of photosynthetic and water status relationships as influenced by the interior climate and radiation. The simulation model, described by Gijzen was based upon his earlier work, Spitters et al. and a detailed biochemical model published by Farquhar et al. [1980]. Although nearly all greenhouse products are sold fresh, the fresh weight values can be obtained by multiplying the dry weight by suitable coefficients. The transpiration model is vital since this influences the internal crop water status (i.e., size, stem length, etc.). This short-term model allows balancing the immediate crop requirements with the engineering models discussed in the previous section. Gijzen [1995b] spends considerable space on the interactions between CO_2 uptake and water loss as influenced by stomatal and boundary resistance. However, both this and the long-term model could be simplified –such as assuming that the stomatal and boundary resistances are constants during the day.

The interception of radiant energy by the crop, of course, must be determined. This can be simple as Stanghellini's “big” leaf (2LAI), or as complex as Spitter's [1986] calculation of a closed canopy (LAI = 5) photosynthesis. Not only did Spitters differentiate between diffuse and direct energy interception, but the canopy, in one version, was divided into 30 layers, taking into account the leaf angle and distribution. More often, the investigator considers three layers as representative [Spitters, 1986; Acock et al., 1978]. One can appreciate that models can become very complex. Often, the operator must choose between the complex details and some more simplified model that yields reasonable results.

In the long-term process, one wishes to know the quantity produced and the moment of harvest. The short-term response is superimposed on long-term reactions that become apparent after a period of more than one to several days [Challa et al., 1995]. Growth has a delayed response to the environment. At this point, one begins to run into terms such as LAI, NAR, RGR, etc. These are defined in Appendix A with some discussion in Chapter 3. A short definition and units are given below:

DVR = rate of development (dy^{-1})
LAI = leaf area index (m^2m^{-2})
LAR = leaf area ratio –leaf area per unit plant dry weight (m^2g^{-1})
LWR = leaf weight ratio –leaf dry weight per plant dry weight
NAR = net assimilation rate ($g\ m^{-2}dy^{-1}$)
RGR = relative growth rate ($g\ g^{-1}dy^{-1}$)
SLA = specific leaf area –leaf area per unit leaf dry weight

With young plants, growth is an amplifying process, which is not constant. Absolute growth is not a suitable criterion, and one uses RGR [Challa et al., 1995]:

$$RGR = \frac{[\ln(W_{t2}) - \ln(W_{t1})]}{(t_2 - t_1)} \tag{8.25}$$

where: RGR = *relative growth rate* ($g\ g^{-1} dy^{-1}$)
t_1 *and* t_2 = *times*
W_t = *crop dry weight* ($g\ m^{-2}$ *at time t*)

At a constant RGR, growth is exponential in young plants. The leaf area ratio (LAR) shows the energy required in leaf area formation: $A_t/W_{p,t}$, where A_t is the leaf area per plant at time t, and $W_{p,t}$ is the plant dry weight at time t. When comparing young plants of a given weight, the crop with a higher LAR will have a higher LAI (leaf area index), and a higher photosynthetic rate. The LAR_t is:

$$LAR_t = SLA_t\, x\, LWR_t \tag{8.26}$$

where: LWR_t = *leaf weight ratio, which expresses the dry matter distribution over leaves and other organs* ($g\ g^{-1}$)
SLA_t = *specific leaf area, or the leaf area per unit dry weight in the leaves* ($m^2 g^{-1}$)

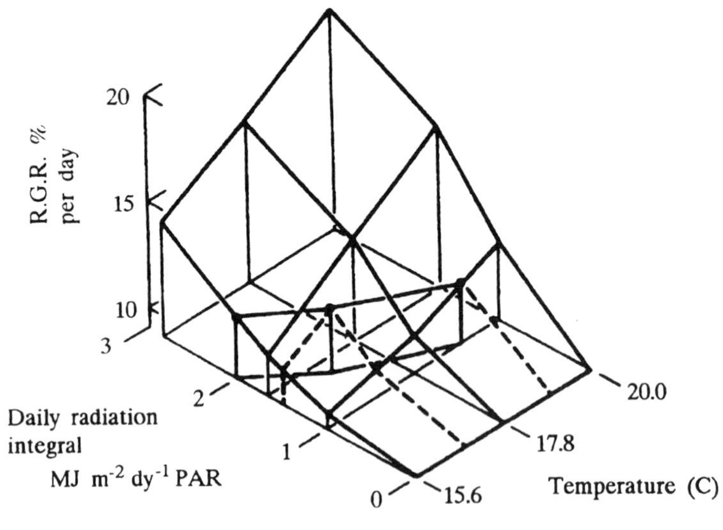

Fig. 8-27. Cockshull's (1988) rendering of A. Calvert's data on tomato, showing the effect of total daily radiation and 24 hour temperature average on relative growth rate (©1988, Int. Soc. Hort. Sci., *Acta Hort.*, 229:113-123).

Once the canopy becomes closed (ca. LAI = 3), the crop shifts to a new phase, and leaf area expansion becomes minor in importance. A further increase in LAI has only marginal effect. Challa et al. [1995] state that it is more meaningful to consider absolute growth at this point: GR = $\Delta W/\Delta t$ (g $m^{-2}dy^{-1}$), and provide an approximation:

$$P_{gc,d} = \alpha_c I_\Sigma (1 - \rho)(1 - e^{-K x LAI}) \tag{8.27}$$

where: $P_{gc,d}$ *= crop photosynthesis per unit house area (g CO_2 $m^{-2}dy^{-1}$)*
α_c *= average crop PAR use efficiency for CO_2 uptake, (g MJ^{-1})*
I_Σ *= daily integral of PAR at the top of the canopy (MJ $m^{-2}dy^{-1}$)*
ρ *= reflection coefficient of the canopy*
K *= extinction coefficient of PAR in the canopy*
LAI = leaf area index

The point at which shifts occur in the growth process is critical and sometimes difficult to learn. Some of these growth phases, and the effect of preceding stages, were discussed in Chapter 4. Cockshull [1988] pointed out, for example, that when chrysanthemum is grown under controlled conditions, yield (in terms of flower weight) is directly proportional to the daily light integral (total daily radiation). In the initial period, dry matter is distributed between the leaf, stem, and root. Once flower initiation and development has started, flower buds gradually develop to the point that they are significant sinks for assimilates (carbohydrates). The pattern of assimilate partitioning changes with time, and the flower size will be influenced by when a change in the radiant quantity occurs. Low radiation in the first weeks will not affect flower size (may affect flower number), but will reduce size (yield) if given in the last few weeks.

On the other hand, with cucumber, Cockshull pointed out that a relatively constant rate of partitioning is maintained once fruiting commences. The weight of fruit produced is proportional to energy received. When cumulative radiation is compared with cumulative yield, a 1% reduction in solar energy causes a 1% yield loss **if** the line, showing the relationship between cumulative yield and cumulative energy, passes through the origin. The percentage loss will vary with where the cumulative yield intercepts the axis above or below the line intersecting the origin. If the line intercepts the cumulative radiation axis, the proportional reduction in yield with reduced radiation will be greater, and smaller if the line intersects the cumulative yield above the origin. The interactions between radiation and temperature on another continuous production crop such as tomato is shown in Fig. 8-27. At 15.6 C, the three-dimensional graph shows that RGR will fall if the daily integral drops from 2 to 1.5 MJ $m^{-2}dy^{-1}$, but can be restored if the average 24-hr temperature is raised to 17.2 C.

Challa et al. [1995b] spend some time on the above subjects and phase transitions. Thus, one has germination and sprouting of bulbs and corms, flowering (photoperiodic signals in combination with temperature and radiation), formation of storage organs, and control of size and shape. A simplified presentation of some growth relationships during the "young" phase is insufficient. An important distinction is between energy fixation for photosynthesis in the production of assimilates, and the use of the assimilates for growth and production of harvestable product [Challa and Bakker, 1995b]. Thus, the outline in Fig. 8-26 is a simplified presentation of the relationships that one must contend with in modeling crop growth. Controlling climate according to predefined setpoints does not adequately meet grower requirements even though Heins [personal communication, 1996] says that many growers do this successfully.

The preceding offers an insight into "dynamic" modeling of crop growth, using known physical and biochemical laws. As may be appreciated, much is yet to be learned before such systems become common commercially. Most effort is spent on the reactions of single leaves or plants, and the scale-up to whole crops is still in its infancy. Among those who have published on the subject are Marsh and Albright [1991a; b] and Seginer et al. [1991], dealing with optimal temperature setpoints for lettuce; and the work by Jones et al. [1989a; b; 1990; 1991], using the TOMGRO model for tomatoes under Florida and Israeli conditions.

Many investigators turn to simpler models for limited purposes such as those given earlier in Chapter 4 (Eqs.

4.6 through 4.12). These represent special mathematical means of relating some variable such as temperature to a particular plant response such as days to flower, internode length, number of flowers, etc. Such models do not attempt to "explain," but are descriptive in that they show what happened. Use of such models is obviously restricted to a narrow range of environmental conditions. Application of these models by growers is predicated on the assumption that the grower's conditions are similar to those under which the model was derived –and the derivation was obtained in a "normal" season. A more thorough discussion of plant modeling of these types was provided by several authors [Chanter, 1981; Ross, 1981; Hunt, 1981] in Rose and Charles-Edwards' *Mathematics and Plant Physiology*. Charles-Edwards concluded his introduction to *Techniques of Plant Modeling* by stating that:

State = Initial State + Integral (Rate)

That is, the state of the system (plant) is obtained by integrating rates over time from a set of initial conditions. The essential problem in the mathematical description of control processes is to set up the physiological problem in the form of the above equation. In this area, there are models subjected to statistical methods that may be:

Linear, using linear regression methods
Nonlinear, using multiple, stepwise regression methods
Richards functions
Gompertz relations
Splines
Weibull functions
Feldmann equations
The Mitscherlich equation, etc.

All these are mathematical procedures that allow one to provide a curve that best explains the relationship between some plant response and one or more independent variables. They are limited to the range of data from which the relationship was originally derived, and do not attempt to explain the fundamental physical relationships that might underlie the response. They are essentially a means to an end. They may be variously modified from that given by the original author [e.g., Liebig and Lederle, 1985; Leibig, 1989; Reinisch et al., 1989]. Various combinations of these mathematical relations that have no relation to physical laws, and static and dynamic models, may be employed where there is insufficient knowledge and the author needs a certain end. Challa [1988], however, pointed out that while the time required to developed statistical models is far less than simulation models; extrapolation outside the range of conditions of the original experiment is not permitted.

3. Optimization

Throughout this chapter, we have referred to "optimization" of climate control. That is, to control the setpoints, constraints, and limits such as to maximize the net profit or reduce the operational cost (objective functions). Examples of such systems have been given earlier in the book, and Figs. 8-2 and 8-23 were generic and specific examples, respectively. If a set of equations to explain what is happening has been outlined, then how may these be used to maximize profit? As we have seen, there is a complex set of many processes, taking place simultaneously, with different response times and different response patterns and characterized by many interactions [Challa and van Straten, 1991; see van Henten, 1995, in particular]. The grower may have several objectives that conflict. Beginning in the 1990s, Challa [Challa and van Straten, 1991; Challa and van Straten, 1993; van Straten and Challa, 1995] and Seginer [Bailey and Seginer, 1989; Seginer, 1989; Seginer, 1993; Seginer et al., 1991] turned to the use of the "Pontryagin Maximum Principle" [Pontryagin et al., 1962], which allows the use of a single value that can be maximized to show optimal climate control.

Assuming ordinary differential equations represented by a "vector" **x**, the dynamics of the greenhouse and crop can be described in general terms:

$$\frac{dx_c}{dt} = f(x_c, x_p, u_e, u_c) \text{ and} \tag{8.28}$$

$$\frac{dx_p}{dt} = g(x_c, x_p, u_e) \tag{8.29}$$

where: $\mathbf{x}_c$ = *the indoor climate state variables (e.g., temperature, CO_2 level, humidity, etc.)*
$\mathbf{x}_p$ = *the plant state variables (e.g., LAI, weight, assimilate distribution, etc.)*
$\mathbf{u}_e$ = *the external inputs (e.g., solar radiation, outdoor temperature, outdoor humidity, etc.)*
$\mathbf{u}_c$ = *the control inputs (e.g., heat supply, ventilation, CO_2 injection, etc.)*

In effect, these generic equations state that the change in $\mathbf{x}_c$ and $\mathbf{x}_p$ are functions of the right-hand terms that are vectors, which represent the various mathematical formulae, with dimensions of n_{xc} and n_{xp} (see Eq. 8.11 for examples of a single-column "vector"). Photosynthesis, respiration, transpiration, etc. are not shown here, nor are the constants (parameters). The functions "f" and "g" are the mathematician's shorthand for a value determined by the following variables in the parentheses.

The basic idea is to generate a sequence of controls ($\mathbf{u}_c$) such that the "path" ($\mathbf{x}$(t)), given by Eqs. 8.28 and 8.29, is optimal. Separation into Eqs. 8.28 and 8.29 is useful since the function "f" has fast dynamics while "g" has slow dynamics [Challa and van Straten, 1993]. The economically optimal control of the system expressed by Eqs. 8.28 and 8.29 requires the formulation of a goal function. A general relationship is to find the control time pattern $\mathbf{u}_c$ such that the goal "J" is maximized, subject to the constraints and limitations that may be imposed:

$$J = \phi(x_p(t_f)) - \int_0^{t_f} L(x_c, x_p, u_e, u_c) \tag{8.30}$$

where: J = *the net profit to be maximized (\$)*
$\phi(x_p(t_f))$ = *the termination value of the system that may be the price received for the final harvest (e.g., bedding plants)*
L = *the control costs at any instant in time*
t_f = *the termination time, or harvest, over which the system is integrated*

The control costs include prices paid for energy and CO_2 supply, and may also include benefits such as earliness (e.g., lettuce).

The task is to construct an optimal "path" of both climatic factors ($\mathbf{x}_c$) and steering controls ($\mathbf{u}_c$) over the period t_o to t_f that maximizes J. The global, or over the growing period, of J can be transformed to a problem of maximizing locally (for every time t), following Pontryagin et al.'s procedure. This is the formulation of the Hamiltonian function:

$$H(x, u_c) = L(x, u_e, u_c) + p_x f(x, u_e, u_c) \tag{8.31}$$

where: H = *Hamiltonian function to be maximized (\$)*
p_x = *a co-state variable or function (vector), representing the value attached to the marginally produced unit of* $\mathbf{x}$ *(i.e., \$ kg^{-1}), similar to Lagrange multipliers for static optimization*

Each of the state variables in $\mathbf{x}$ has a corresponding co-state variable. The first term on the right-hand side represents the current rate of net gain, while the second term $\mathbf{p}_x\mathbf{f}$ is the rate of investment in the future, or the rate at which the potential for future gains is created.

Once the co-state variables are known, which may be difficult to learn, H becomes a control generating function (control rule), because at any given time, and any given state (**x**) and boundary conditions ($\mathbf{u}_e$), it is possible to find a control ($\mathbf{u}_c$) that maximizes the Hamiltonian. One must take other considerations into account. Van Henten [1995] showed for lettuce that the co-state patterns related to structural and nonstructural dry weight are very similar for various weather patterns. As the co-states represent a shadow price, the information can be used to solve the problem of minute-by-minute optimal control where instantaneous weather variations can be taken into account. What van Straten and Challa propose is "receding horizon predictive control" [RHPC), where the optimization is calculated over a short period (24 hr maximum), and then repeated another period with only the first optimal control sequence applied. System behavior is then observed with another optimization, starting with the observed state. The limit (t_f) continually recedes at the same rate as the control system cycles. Comments on the application of this system in Holland were interesting with about a 20% larger harvest weight, 8% lower CO_2 costs, and about 30% lower energy costs for lettuce. Results showed that if the "horizon," or period for optimization, was too short, solutions were not optimal. The authors concluded that a prediction horizon of about 1 hr gave solutions very close to the optimum. Furthermore, using as a forecast that the next hour weather will be the same as the present weather, and updating this forecast every minute, gave almost optimal results.

The Hamiltonian procedure has been examined by Challa and others as cited above, Marsh and Albright [1991a; b], Seginer et al. [1991] and particularly by van Henten [1995]. In most of the investigations, the crop has been lettuce. This is an obvious choice since the growth of the crop follows a well-defined pattern to reach a required small weight, and the crop is harvested over a short period. It should be possible to apply the procedure for lettuce to other single-harvest crops such as poinsettias, Easter lilies, etc., although these are not sold based on dry weight. In van Henten and Bontsema's [1996] approach, the control problem was divided into a fast and a slow model:

$$\frac{dx}{dt} = f(x,z,u,v,t) \tag{8.32}$$

$$\epsilon\frac{dz}{dt} = g(x,z,u,v,t) \tag{8.33}$$

where: x = *slow state variables such as crop dry weight*
z = *fast state variables such as air temperature, CO_2 level and humidity*
u = *energy inputs, CO_2 inputs, and ventilation rate*
v = *outside disturbances (temperature, radiation, wind speed, CO_2)*
t = *time with t_o = planting date; and x_o, z_o = initial states of the process*
ϵ = *time-scaling constant, or the ratio between fast and slow system responses*

The Pontryagin procedure was followed for Eq. 8.32 and for Eq. 8.33 to find **u** that maximized **J** (fast and slow values). A Hamiltonian was not considered in this paper [1996]. Comparison of dry weight of lettuce, harvested at regular intervals, with simulation of the dry weight over the same period usually resulted in the simulated results being within the 95% confidence interval [van Henten, 1994; ven Henten and Bontsema, 1995]. In actuality, the confidence intervals for dry weight were wide due to crop variability. Fig. 8-28 is a scheme proposed by van Henten and Bontsema [1996], containing an outer loop for the control of the slow crop growth and an inner loop controlling the fast greenhouse dynamics.

The plan outlined in Fig. 8-28 is different from the three-level system given in Fig. 8-2. Apparently, the authors feel this to represent a considerable simplification of the control system compared with what has been discussed. Its achievement may be difficult, and van Henten's dissertation [1995] does not go into detail on this aspect.

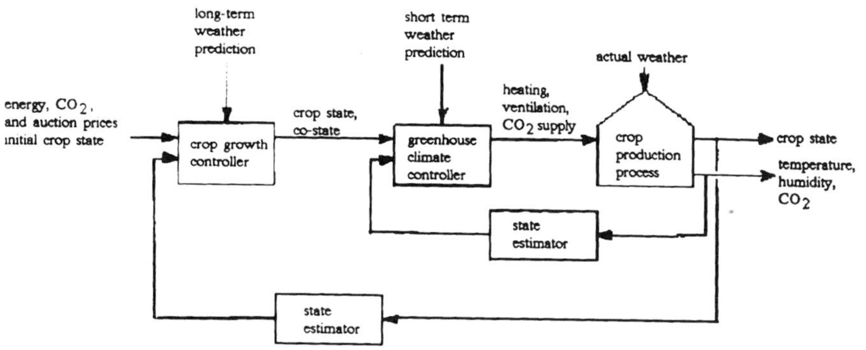

Fig. 8-28. Van Henten and Bontsema's [1996] proposed outline of a climate manager for greenhouses. The state estimator in the outer feedback loop caculates the state and co-state change while the inner loop state estimator calculates the greenhouse climate dynamics. This system uses a receding horizon optimal control similar to the RHPC discussed above.

There are some 26 parameters or constants used in the formulae for calculating optimal setpoints in these state-space models. Van Henten's dissertation lists 62 in his appendix. Of these, van Henten and Bontsema [1994] suggest, from their sensitivity analyses, that about five appear critical. Their influence, as one might expect, varies with growth stage. They were: a "yield factor" that indicates the respiratory and synthesis losses of carbohydrates as growth proceeds, the maximum growth rate at 20 C, the light use efficiency at high CO_2, and the leaf area ratio (LAR) as defined above. Given the number of possible constants, inputs, and outputs (Eq. 8.32, 8.33), the equations can be complex.

The efforts of the Europeans in climate control are exciting, even more so with Seginer's [1993] suggestion of variable bias dials on the control system by which the grower could directly affect the co-state variables. That is, there could be a node-initiating dial, a photosynthesis dial, etc. The grower would not be faced with the bewildering task of changing setpoints, but could vary the optimization program to achieve a desired end-result by rotating the appropriate dial.

Other variations and modifications of models have been published. For example, Gauthier and Guay [1990] and Gauthier [1991: 1995] have published on the use of Smalltalk in setting up a control system, using several different procedures in the artificial intelligence area. The model TOMGRO was mentioned previously, and that plant growth model and its employment have been examined by Jones et al. [1989a; b; 1991] and Jones et al. [1990]. Their work represents one of the few full-scale, detailed examinations of greenhouse modeling and climate control in the English literature –especially the BARD report published in 1989. As a comment, Seginer's experience with Israeli conditions in mild climates shows in his publications where the emphasis is on on-off control compared with the European situation of modulated hot water heating. One will note that most of the publications cited deal with vegetable crops, and except a few papers such as by Lieth and Pasian [1988; 1993] and by Tsujita [1988] on rose production, there are practically no complete models available for ornamental crops –especially in the American literature.

4. Artificial Intelligence

Artificial intelligence in computer sciences is the development and use of computer systems that have some resemblance to human intelligence, including such operations as natural language recognition and use, problem-solving, selection from alternatives, pattern recognition, and generalization based on experience [Morris, 1992].

Webster states that intelligence is the ability to learn or understand from experience; ability to find and retain knowledge; mental ability; the ability to respond quickly and successfully to a new situation; or the use of the faculty of reason in solving problems, directing conduct, etc. It can be appreciated from this complex definition, that attempting to make computers intelligent is not easy —especially as the definition is not very explicit. Rather than addressing the issue of what intelligence is, AI researchers are more concerned with recognizing intelligence and developing systems that behave intelligently [Zazueta, 1994]. AI rests on the notion that a machine can be built that emulates human behavior. Some areas that are useful in greenhouse climate control are robotics, vision, natural language, understanding, artificial neural systems, speech, and expert systems.

Expert, or knowledge, systems use a collection of independent techniques to develop programs that use specialized knowledge. A human expert is an individual who can solve problems in a restricted area of expertise that most people cannot solve at all. A computer expert system includes a knowledge base (or set of rules), a working memory (facts), an inference engine (agenda), an explanation facility, a knowledge acquisition facility, and the user interface. Expert systems may be the best alternative if problems are ill-structured with no mathematical solution evident. The domain of the problem must be clearly recognized and well bounded. The solutions to the problem must be encodable into production rules. Assuming the limitations have been met, ES can be used in configuration, diagnosis, instruction, interpretation, monitoring, planning, prognosis, remedy, and control. Examples of an expert system for greenhouse application have been Fynn et al.'s [1991; 1994] development for nutrient selection, finding that the extraction of expertise from the experts was a major difficulty. With Jones et al.'s [1986; 1988] development, the idea was that an expert system could mimic a grower's perceptions. Jones et al. implemented a misting system in propagation, based upon the perceived optimal strategy of an experienced grower. Expert systems may also be considered as "knowledge based," as suggested by Kozai [1985]. The SERRISTE system discussed by Martin-Clouaire et al. [1993] represents one of the more thorough expert systems about which I have read. The situation from these few citations shows that expert systems are restricted if they are to be successful.

Another aspect of AI in greenhouses is the use of image processing to recognize diseases or measure growth and well-being of plant material [Hack, 1989; Hashimoto et al., 1981; Hatou et al., 1995]. Video imaging requires huge amounts of memory and very fast data transfer. Thus, practical implementation of vision has been limited to advances in hardware. According to Hatou et al., the video system, when combined with an expert system, could be used to avoid growth inhibition in stems and blossom-end rot on tomatoes in hydroponics. Hashimoto et al. used IR thermometers to measure leaf temperatures and to digitize leaf temperature patterns to suggest plant-water relationships. Hack used image processing to allow nondestructive measurement of plant growth. As long as the image really represented leaf area, the method gave good correlation between measured values and fresh weight in lettuce. Another similar approach has been van Henten and Bontsema's [1995] use of an image camera mounted above the lettuce crop in order to evaluate canopy coverage. The total leaf area is a critical factor in their growth models. Such a procedure avoids destructive testing during crop growth. According to van Henten and Bontsema, the model, used with the image camera, can estimate plant dry weight from relative soil coverage to within 5%.

Commonly, the information available to the control system is inaccurate, and the handling of this inaccuracy leads to the idea of "fuzzy" logic, which is a branch of fuzzy set theory. The control system may be difficult to describe precisely, and the objectives are not clearly defined. On the other hand, human expertise on how to smoothly run these systems is available. The knowledge can be structured as "if-then" rules containing terms such as *low* pressure, *medium* temperature, etc. Pedrycez [1991] stated that the role of fuzzy sets is to capture ill-defined objectives specified in decision-making procedures, and allow reasoning in the presence of uncertainty. Fuzzy controllers are somewhat similar to expert systems. Kurata and Eguchi [1990] pointed out that a grower's behavior cannot be considered as accurate as that of machines, and this involves "fuzziness." Even though a computer has "learned" the grower rules for managing the system, the inaccuracies require a means to account for the fuzziness. Examples where fuzzy logic, in a broad sense, has been applied include medical diagnosis systems, autofocusing systems in cameras, automobile transmissions, rice cookers, and vacuum cleaners.

Neural networks, which have seldom been mentioned in the horticultural literature, attempt to mimic the structure of nerve cells (neurons) in the human brain. A basic network has three layers: an input, an output, and a "hidden" layer between. Each processor is connected to every other in the network by a system of "synapses". When the amount of the input exceeds a critical level, an output signal can be obtained, or the synapse "fires." The chief characteristic of neural networks is their ability to sum large amounts of imprecise data and decide whether they match a pattern or not. Networks of this type are used in developing robot vision, matching fingerprints, and analyzing flucuations in the stock market. A properly constructed neural network can "learn," and some requirements discussed above might be used in such a network for the climate control system to learn the optimum environment –which may call for both an expert system and fuzzy logic.

Successes of these few examples of AI for climate control have not been remarkable. Still, the possibility exists to supplement the systems outlined above with well-defined types of AI that would enhance the control and relieve the grower of the time necessary to revise and introduce setpoints and mechanisms for control operation. In particular, expert systems have been applied to disease recognition and diagnoses, and these could be used in climate control to enhance disease prevention and control. There is no limit to the possibilities that exist in precision greenhouse climate manipulation.

VI. SUMMARY

This chapter is peculiar since, particularly in the last section, the ideas have yet to be used to any extent in the greenhouse industry. Nevertheless, the possibilities of advanced growth facilities in protected horticulture, combined with sophisticated climate control, will be waiting for those able to use them for their enhanced profitability. There is no foreseeing which group or region will be the first to take advantage of these advanced concepts. Nor can one predict the final shape of what has been presented here. It remains to conclude, as pointed out in Chapter 1, that the idea that everything has been discovered, explained, and finished is false. Such an idea is likely to leave their prognosticators in their own dust.

VII. REFERENCES

Acock, B. et al. 1978. The contribution of leaves from different levels within a tomato crop to canopy net photosynthesis: An experimental examination of two canopy models. *J. Expt. Bot.* 29:815-827.

Aikman, D.F. and A.J.F. Picken. 1989. Wind-related temperature setting in glasshouses. *J. Hort. Sci.* 64:649-654.

Alberry, W.J. B.G.D. Haggett and L.R. Svanberg. 1985. The development of sensors for hydroponics. *Biosensors*. 1:369-397.

Albright, L.D. 1984. Modeling thermal mass effects in commercial greenhouses. *Acta Hort.* 148:359-368.

Albright, L.D. et al. 1985. *In situ* thermal calibration of unventilated greenhouses. *J. Agric. Eng. Res.* 31:265-281.

Ammerlaan, J.C.J. 1984. Development of management information systems in the glasshouse horticulture in The Netherlands. *Acta Hort.* 155:235-241.

Anon. 1988. Committed to energy productivity. *The Grower*. 109(4):(Supplement). Jan. 18, 1988.

Arnold, J.T. 1979. Simplified Digital Automation with Microprocessors. Academic Press, New York. 267 pp.

Avissar, R. and Y. Mahrer. 1982. Verification study of a numerical greenhouse microclimate model. *Trans. ASAE*. 25:1711-1721.

Bailey, B.J. 1995. Greenhouse climate control - New challenges. *Acta Hort.* 399:13-23.

Bailey, B.J. and Z.S. Chalabi. 1991. Optimization of controlled plant environments. *Proc. 1991 Symp. Automated Agriculture for the 21st Century*. ASAE, St. Joseph, MI.

Bailey, B.J. and I. Seginer. 1989. Optimum control of greenhouse heating. *Acta Hort.* 245:512-518.

Bakker, J.C. 1985. A CO_2 control algorithm based on simulated photosynthesis and ventilation rate. *Acta Hort.* 174:387-392.

Bakker, J.C. 1995. Greenhouse climate control: Constraints and limitations. *Acta Hort.* 399:25-33.

Bakker, J.C. and H. Challa. 1995. Aim and approach of this book. *In* Greenhouse Climate Control. J.C. Bakker et al., eds. Wageningen Pers.

Bakker, J.C. et al., eds. 1995. Greenhouse Climate Control. Wageningen Pers. 279 pp.

Bloor, R. 1994. New kids on the block: Still wondering what the future 5GLs will look like? They're already here. *DBMS*. 7:14-16.

Bolinger, J.G. and N.A. Duffie. 1988. Computer Control of Machines and Processes. Addison-Wesley, New York. 613 pp.

Bontsema, J. 1995. Greenhouse climate control. Control principles. *In* Greenhouse Climate Control. J.C. Bakker et al., eds. Wageningen Pers.

Bot, G.P.A. 1980. Validation of a dynamical model of greenhouse climate. *Acta Hort*. 106:149-158.

Bot, G.P.A. 1983. Greenhouse climate: From physical processes to a dynamic model. Ph.D. Dissertation. Landb. Wageningen. 240 pp.

Bot, G.P.A. 1989a. Greenhouse simulation models. *Acta Hort*. 245:315-325.

Bot, G.P.A. 1989b. A validated physical model of greenhouse climate. *Acta Hort*. 245:389-396.

Bot, G.P.A. 1991. Greenhouse growing: *World Outlook 2000*. Sensor workshop, Nordurgherherid. 6 pp.

Bot, G.P.A. 1993. Physical modelling of greenhouse climate. *In* The Computerized Greenhouse. Y. Hasimoto et al., eds.. Academic Press, San Diego.

Bot, G.P.A. 1995. Conclusions. *In* Greenhouse Climate Control. J.C. Bakker et al. eds. Wageningen Pers.

Bot, G.P.A. and N.J. van de Braak. 1995. Physics of greenhouse climate. Basic principles. *In* Greenhouse Climate Control. J.C. Bakker et al. eds. Wageningen Pers.

Bot, G.P.A., J.J. van Dixhoorn and A.J. Udink ten Cate. 1977. Dynamic modelling of greenhouse climate and the application to greenhouse climate control. *ICHMT Seminar*, Dubrovnik, Yugoslavia. 6 pp.

Bowman, G.E. and G.S. Weaving. 1970. A light-modulated greenhouse control system. *J. Agric. Engin. Res*. 15:255-264.

Carr, J.J. 1984. Interfacing Your Microcomputer to Virtually Anything. TAB Books, Inc., Blue Ridge Summit, PA. 325 pp.

Challa, H. 1988. Prediction of production: Requisite of an integrated approach. *Acta Hort* 229:133-141.

Challa, H. 1989. Modelling for crop growth control. *Acta Hort*.248:209-215.

Challa, H. 1993. Optimal diurnal climate control in greenhouses as related to greenhouse management and crop requirements. *In* The Computerized Greenhouse. Y. Hashimoto et al., eds. Academic Press, San Diego.

Challa, H. and J.C. Bakker. 1995a. Aim and approach of this book. *In* Greenhouse Climate Control. J.C. Bakker et al., eds. Wageningen Pers.

Challa, H. and J..C Bakker. 1995b. Introduction. *In* Greenhouse Climate Control. J.C.Bakker et al., eds. Wageningen Pers.

Challa, H. and A.H.C.M. Schapendonk. 1986. Dynamic optimalization of CO_2 concentration in relation to climate control in greenhouses. *In* Carbon Dioxide Enrichment of Greenhouse Crops. H.Z. Enoch and B.A. Kimball, eds. CRC Press, Boca Raton, Fl.

Challa, H. and J. van de Vooren. 1980. A strategy for climate control in greenhouses in early winter production. *Acta Hort*. 106:159-163.

Challa, H. and G. van Straten. 1991. Reflections about optimal climate control in greenhouse cultivation. *IFAC Mathematical and Control Applications in Agric. and Hort.*, Matsuyama, Japan. 13-18.

Challa, H. and G. van Straten. 1993. Optimal diurnal climate control in greenhouses as related to greenhouse management and crop requirements. *In* The Computerized Greenhouse. Y. Hashimoto et al., eds. Academic Press, New York. 340 pp.

Challa, H., E. Heuvelink and U. van Meeteren. 1995. Crop growth and development. *In* Greenhouse Climate Control. J.C. Bakker et al., eds. Wageningen Pers.

Challa, H. et al. 1988. Greenhouse climate control in the nineties. *Acta Hort*. 230:459-470.

Chandra, P., L.D. Albright and N.R. Scott. 1981. A time dependent analysis of greenhouse thermal environment. *Trans. ASAE*. 24:442-449.

Chanter, D.O. 1981. The use and misuse of linear regression methods in crop modelling. *In* Mathematics and Plant Physiology. D.A. Rose and D.A. Charles-Edwards, eds. Academic Press, New York.

Charles-Edwards, D.A. 1981. The mathematical description of control processes. *In* Mathematics and Plant Physiology. D.A. Rose and D.A. Charles-Edwards, eds. Academic Press, New York.

Cockshull, K.E. 1988. The integration of plant physiology with physical changes in the greenhouse climate. *Acta Hort.* 229:113-123.

Coffee, P. 1993a. 'Dead' languages still sources of critical skills. *PC Week.* 10:68.

Coffee, P. 1993b. FORTRAN retains its industrial strength. *PC Week.* 10:46-47.

Coffee, P. 1994a. C++ programming language is still a work in progress. *PC Week.* 11:73-74.

Coffee, P. 1994b. Bugs flourish as bane of grand public works. *PC Week.* 11:63-64.

Collins, T. 1994. Channel tunnel code faces timetable crisis. *Computer Weekly.* Feb. 24, 1994.

Cooper, P.I. and R.J. Fuller. 1983. A transient model of the interaction between crop, environment and greenhouse structure for predicting crop yield and energy consumption. *J. Agric. Eng. Res.* 28:401-417.

Cunningham, J. 1994. Language brouhaha: Is Cobol dead? *Computer World.* 28:113-115.

de Graff, R. 1988. Automation of the water supply of glasshouse crops by means of calculating the transpiration and measuring the amount of drainage water. *Acta Hort.* 229:219-231.

de Wit, C.T. 1982. Simulation of living systems. *In* Simulation of Plant Growth and Crop Production. F.W.T. Penning de Vries and H.H. van Laar, eds. Simulation Monograph. Pudoc, Wageningen.

Duncan, G.A., O.J. Loewer, Jr. and D.G. Colliver. 1981. Simulation of energy flows in a greenhouse: Magnitudes and conservation potential. *Trans. ASAE.* 24:1014-1021.

Elmer-Dewitt, P. 1955. Bugs bounty. *Time* 146(17):86.

Farquhar, G.D., S. von Caemmerer and J.A. Berry. 1980. A biochemical model of photosynthetic CO_2 assimilation in leaves of C_3 species. *Planta.* 149:78-90.

Fernandez, J.E. and B.J. Bailey. 1993. Predicting greenhouse ventilation rates. *Acta Hort.* 328:107-114.

Fynn, R.P., W.L. Bauerle and W.L. Roller. 1991. Expert system model for nutrient selection in a drip irrigation system. *Proc. 1991 Symp. Automated Agriculture for the 21st Century.* ASAE, St. Joseph, MI.

Fynn, R.P., W.L. Bauerle and W.L. Roller. 1994. Implementing a decision and expert system model for individual nutrient selection. *Agric. Systems.* 44:125-142.

Gal, S., A. Angel and I. Seginer. 1984. Optimal control of greenhouse climate: Methodology. *European J. Operational Res.* 17:45-56.

Garzoli, K. 1985. A simple greenhouse climate model. *Acta Hort.* 174:393-400.

Gauthier, L. 1991. Greenhouse environment control using a knowledge-based approach. *Proc. 1991 Symp. Automated Agriculture for the 21st Century.* ASAE, St. Joseph, MI.

Gauthier, L. 1995. A multi-agent approach for the dynamic optimization of greenhouse environments. *Acta Hort.* 399:61-71.

Gauthier, L. and R. Guay. 1990. An object-oriented design for a greenhouse climate control system. *Trans. ASAE.* 33:999-1004.

Gieling, T.H. 1980. Commercial greenhouse computersystems: Investigation of implemented research, carried out by commercial firms. *Acta Hort.* 106:59-66.

Gieling, T.H. and A. de Jager. 1988. The application of chemisensors and bio-sensors for soilless cultures. *Acta Hort.* 230:357-361.

Gieling, T.H. and K. Schurer. 1994. Sensors for information acquisition and control. *In* Greenhouse Systems. Automation, Culture and Control. NRAES-72. Rutgers Univ., New Brunswick, NJ.

Gieling, T.H. and K. Schurer. 1995. Greenhouse climate control. Sensors and measurement. *In* Greenhouse Climate Control. J.C. Bakker et al., eds. Wageningen Pers.

Gieling, T.H. and W.T.M. van Meurs. 1979. Specifications for environmental control. The PDP-11 to MuP/KuP interface. Res. Rpt.79-5. IMAG. Wageningen. 42 pp.

Gijzen, H. 1995a. CO_2 uptake by the crop. *In* Greenhouse Climate Control. J.C. Bakker et al., eds. Wageningen Pers.

Gijzen, H. 1995b. Interaction between CO_2 uptake and water loss. *In* Greenhouse Climate Control. J.C. Bakker et al., eds. Wageningen Pers.
Giniger, M.S. et al. 1988. Computer simulation of a single truss tomato cropping system. *Trans. ASAE.* 31:1176-1179.
Goedhart, M. et al. 1984. Methods and instruments for ventilation rate measurements. *Acta Hort.* 148:393-400.
Goudriaan, J. 1982. Some techniques in dynamic simulation. *In* Simulation of Plant Growth and Crop Production. F.W.T. Penning de Vries and H.H. van Laar, eds. Simulation Monograph. Pudoc, Wageningen.
Hack, G.R. 1989. The use of image processing under greenhouse conditions for growth and climate control. *Acta Hort.* 245:455-461.
Hammer, P.A. and R.W. Langhans. 1978. Modeling of plant growth in horticulture. *HortScience.* 13:456-458.
Hanan, J.J., F.A. Coker and K.L. Goldsberry. 1987. A climate control system for greenhouse research. *HortScience.* 22:704-708.
Hand, D.W. and G.E. Bowman. 1969. Carbon dioxide assimilation in a controlled environment greenhouse. *J. Agric. Eng. Res.* 14:92-99.
Hand, D.W. and R.W. Soffe. 1971. Light-modulated temperature control and the response of greenhouse tomatoes to different CO_2 regimes. *J. Hort. Sci.* 46:381-396.
Hashimoto, Y. 1980. Computer control of short term plant growth by monitoring leaf temperature. *Acta Hort.* 106:139-146.
Hashimoto, Y. 1989. Recent strategies of optimal growth regulation by the speaking plant concept. *Acta Hort.* 260:115-121.
Hashimoto, Y. 1993. Computer integrated system for the cultivating process in agriculture and horticulture. *In* The Computerized Greenhouse. Y. Hashimoto et al., eds. Academic Press, New York. 340 pp.
Hashimoto, Y., T. Morimoto and S. Funada. 1981. Computer processing of speaking plant for climate control and computer aided plantation (computer aided cultivation). *Acta Hort.* 115:317-325.
Hatou, K. et al. 1995. Physiological diagnosis of tomato plants grown in hydroponic culture by using image analysis. *Acta Hort.* 399:225-232.
Hayes, B. 1995. Debugging myself. *Amer. Sci.* 83:404-408.
Heinen, M. and K. Harmanny. 1992, Evaluation of the performance of ion-selective electrodes in an automated NFTS system. *Acta Hort.* 304:273-280.
Hesketh, T., R.A. Skilton and C.J. Studman. 1986. Advanced digital control for New Zealand glasshouses. *J. Agric. Eng. Res.* 34:207-218.
Heuvelink, E. and H. Challa. 1989. Dynamic optimization of artificial lighting in greenhouses. *Acta Hort.* 260:401-412.
Hooper, A.W. and P.F. Davis. 1985. Control of greenhouse air temperature with an adaptive control alogrithm. *Acta Hort.* 174:407-412.
Hunt, R. 1981. The fitted curve in plant growth studies. *In* Mathematics and Plant Physiology. D.A. Rose and D.A. Charles-Edwards, eds. Academic Press, New York.
Hurd, R.G. and C.J. Graves. 1984. The influence of different temperature patterns having the same integral on the earliness and yield of tomatoes. *Acta Hort.* 148:547-554.
Iqbal, M. 1983. An Introduction to Solar Radiation. Academic Press, New York. 390 pp.
Johnson, L.A. 1996. Internet chat room linked to New Jersey homicide. *Coloradoan.* Feb. 5, 1996.
Jolliet, O. et al. 1991. HORTICERN: An improved static model for predicting the energy consumption of a greenhouse. *Agric. and Forest Meteor.* 55:265-294.
Jones, J.W., Y.K. Hwang and I. Seginer. 1995. Simulation of greenhouse crops, environments and control systems. *Acta Hort.* 399:73-84.
Jones, J.W. et al. 1989a. On-line computer control system for greenhouses under high radiation and temperature zones. *Final Res. Rpt., BARD Project US-871-84.* Agric. Eng. Dept., Univ. of FL, Gainesville, FL.
Jones, J.W. et al. 1989b. Modeling tomato growth for optimizing greenhouse temperatures and carbon dioxide concentrations. *Acta Hort.* 248:285-294.
Jones, J.W. et al. 1991. A dynamic tomato growth and yield model (TOMGRO). *Trans. ASAE.* 34:663-672.

Jones, P. 1994. Environmental controls: Thermostats to computers. *In* Greenhouse Systems. Automation, Culture, and Environment. NRAES-72. Rutgers Univ., New Brunswick, NJ.

Jones, P., B.K. Jacobson and J.W. Jones. 1988. Applying expert systems concepts to real-time greenhouse controls. *Acta Hort.* 230:201-208.

Jones, P, J.W. Jones and Y. Hwang. 1990. Simulation for determining greenhouse temperature setpoints. *Trans. ASAE.* 33:1722-1728.

Jones, P. et al. 1986. Real time greenhouse monitoring and control with an expert system. *ASAE Paper No. 86-4515.* ASAE Ann. Mtg., Chicago, IL. 16 pp.

Kimball, 1986. A modular energy balance program including subroutines for greenhouses and other latent heat devices. *USDA, ARS, ARS-33.* Superintendent of Documents, Washington, D.C. 356 pp.

Kindelan, M. 1980. Dynamic modeling of greenhouse environment. *Trans ASAE.* 23:1232-1239.

Knight, J. 1985. Computers for environmental control. *Canadian Greenhouse Conf.*, Univ. Guelph, Guelph, Ontario.

Kooistra, E. 1986. Developments in the control of growing conditions in Dutch glasshouse horticulture. *Netherlands J. Agric. Sci.* 3:381-385.

Kozai, T. 1985. Ideas of greenhouse climate control based on knowledge engineering techniques. *Acta Hort.* 174:365-373.

Kuo, B.C. 1987. Automatic Control Systems. 5th ed. Prentice-Hall, Inglewood Cliffs, NJ. 720 pp.

Kurata, K. and N. Eguchi. 1990. Learning of fuzzy rules for crop management in protected cultivation. *Trans. ASAE.* 33:1360-1368.

Lane, A. 1994. More basic than BASIC. *Computer Shopper.* 14:622-624.

Lane, A. 1994. A paean to Prolog. *AI Expert.* 9:13-26.

Lake, J.V. 1966. Measurement and control of the rate of carbon dioxide assimilation by glasshouse crops. *Nature.* 209:97-98.

Leung, C.M. et al. 1985. A knowledge based system for the control of greenhouse environment. *Acta Hort.* 174:425-431.

Liebig, H.P. 1989. Growth and yield models as an aid for decision making in protected crop production control. *Acta Hort.* 260:99-113.

Liebig, H.P. and G. Alscher. 1993. Combination of growth models for optimized CO_2- and temperature-control of lettuce. *Acta Hort.* 328:155-162.

Liebig, H.P. and E. Lederle. 1985. Strategy for modelling plant growth. *Acta Hort.* 174:177-192.

Lieth, J.H. and C.C. Pasian. 1988. Automated optimization of rose production: A mathematical model for partitioning of carbohydrates in roses. *Roses, Inc. Bull.* Nov., 1988. 50-52.

Lieth, J.H. and C.C. Pasian. 1993. Development of a crop simulation model for cut-flower roses. *Acta Hort.* 328:179-184.

Littlewood, B. and L. Strigini. 1992. The risks of software. *Sci. Amer.* 267:62-75.

Machrone, B. 1994. A life after COBOL? Yes, with training. *PC Week.* 11:97-98.

Maher, M.J. and T. O'Flaherty. 1973. An analysis of greenhouse climate. *J. Agric. Eng. Res.* 18:197-203.

Mahrer, Y. and R. Avissar. 1984. A numerical simulation of the greenhouse microclimate. *Mathematics and Computers in Simulation.* 24:218-228.

Mann, C.C. 1995. Is the Internet doomed? *Inc. Technology.* 17:47-54.

Marsh, L. 1987. Computer control cuts fuel costs. *Amer. Veg. Grower.* 35(1):44.

Marsh, L.S. and L.D. Albright. 1991. Economically optimum day temperatures for greenhouse hydroponic lettuce production., Part I: A computer model. *Trans. ASAE.* 34:550-556.

Marsh, L.S. and L.D. Albright. 1991. Economically optimum day temperatures for greenhouse hydroponic lettuce production. Part II: Results and simulations. *Trans. ASAE.* 34:557-562.

Martin-Clouaire, R. et al. 1993. Using empirical knowledge for the determination of climatic setpoints: An artificial intelligence approach. *In* The Computerized Greenhouse. Y. Hashimoto et al., eds. Academic Press, San Diego.

Matthews, R.B. and R.A. Saffell. 1986. Computer control of humidity in experimental greenhouses. *J. Agric. Eng. Res.* 33:213-221.
Matthews, R.B. et al. 1987. Computer control of carbon dioxide concentration in experimental greenhouses and its use to estimate net canopy photosynthesis. *Agric. and Forest Meteor.* 40:279-292.
Miles, G.E. 1994. Automation basics: Perception, reasoning, communication, planning, and implementation. *In* Greenhouse Systems. Automation, Culture, and Environment. NRAES-72. Rutgers Univ., New Brunswick, NJ.
Mitchell, B.W. 1991. Failure detection and redundancy for central computer - based control systems. *Proc. 1991 Symp., Automated Agriculture for the 21st Century.* ASAE, St. Joseph, MI
Morin, R. 1993. Pascal's progeny. *UNIX Rev.* 11:167-169.
Morin, R. 1994a. Parallel programming languages. Part I. *UNIX Rev.* 12:85-87.
Morin, R. 1994b. Parallel programming languages, Part II. *UNIX Rev.* 12:109-111.
Nederhoff, E.M. 1988. Dynamic optimilization of the CO_2 concentration in greenhouses: An experiment with cucumber (*Cucumis sativus* L). *Acta Hort.* 229:341-348.
Nederhoff, E.M. 1994. Effects of CO_2 concentration on photosynthesis, transpiration and production of greenhouse fruit vegetable crops. Ph.D. Dissertation. Landbw. Wageningen. 214 pp.
Nederhoff, E.M. 1995. Physics of greenhouse climate. Carbon dioxide balance. *In* Greenhouse Climate Control. J.C. Bakker et al., eds. Wageningen Pers.
Nederhoff, E.M., J. van de Vorren and A.J. Udink ten Cate. 1984. A method to determine ventilation in greenhouses. *Acta Hort.* 148:345-350.
Nederhoff, E.M., J. van de Vooren and A.J. Udink ten Cate. 1985. A practical tracer gas method to determine ventilation in greenhouses. *J. Agric. Eng. Res.* 31:309-319.
Nederhoff, E.M. et al. 1989. Dynamic model for greenhouse crop photosynthesis: Validation by measurements and application for CO_2 optimization. *Acta Hort.* 260:137-147.
Newquist, H.P. III. 1994. Lisp for lunch. *AI Expert.* 9:11-13.
Norr, H. 1994. Pascal problem shows pitfalls in tool effort. *MacWEEK.* 8:38-39.
Norton, P., L.C. Eggebrecht and S.H.A. Clark. 1995. Inside the PC. 6th edition. Sams Publ., Indianapolis. 627 pp.
Ogata, K. 1970. Modern Control Engineering. Prentice-Hall, Englewood Cliffs, NJ. 836 pp.
Pedrycz, W. 1991. Application of fuzzy logic in control systems. *Proc. 1991 Symp. Automated Agriculture for the 21st Century.* ASAE, St. Joseph, MI.
Penning de Vries, F.W.T. 1990. Can crop models contain economic factors? *In* Theoretical Production Ecology: Reflections and Prospects. R. Rabbinge et al., eds. Simulation Monograph 34. Pudoc, Wageningen.
Penning de Vries, F.W.T. and H.H. van Laar. 1982. Simulation of growth processes and the model BACROS. *In* Simulation of Plant Growth and Crop Production. F.W.T. Penning de Vries and H.H. van Laar, eds. Simulation Monograph. Pudoc, Wageningen.
Petroski, H. 1994. Failed promises. *Amer. Sci.* 82:6-9.
Potryagin, L.S. et al. 1962. The Mathematical Theory of Optimal Processes. John Wiley Interscience, New York. 360 pp.
Prosise, J. 1994. The WordBasic programming language. *PC Magazine.* 12:383-387.
Psaltis, D and F. Mok. 1995. Holographic memories. *Sci. Amer.* 273:70-76.
Rabbinge, R., S.A. Ward and H.H. van Laar, eds. 1989. Simulation and Systems Management in Crop Protection. *Simulation Monograph 32.* Wageningen. 420 pp.
Rabbinge, R. et al., eds. 1990. Theoretical Production Ecology: Reflections and Prospects. *Simulation Monograph 34.* Wageningen. 301 pp.
Reid, G. 1994. Weighing the merits of Smalltalk and C++. *Computing Canada.* 28:28-29.
Reinisch, K. et al. 1989. Development of strategies for temperature and CO_2 control in the greenhouse production of cucumbers and tomatoes based on modelbuilding and optimization. *Acta Hort.* 260:67-75.
Rijsdijk, A.A. and G. Houter. 1993. Validation of a model for energy consumption, CO_2 consumption and crop production (ECP-model). *Acta Hort.* 328:125-131.

Ritchie, J.T. 1986. Process control and monitoring in crop production. *In* Microelectronics in Agriculture - Facts and Trends. *Deutsche Landwirkschaffs Gesellschaft*, Frankfurt. 27-41.

Ross, G.J.S. 1981. The use of non-linear regression methods in crop modelling. *In* Mathematics and Plant Physiology. D.A. Rose and D.A. Charles-Edwards, eds. Academic Press, New York.

Rudd-Jones, D., A. Calvert and G. Slack. 1978. CO_2 enrichment and light-dependent temperature control in glasshouse tomato production. *Acta Hort.* 87:147-155.

Saffell, R.A. and B. Marshall. 1983. Computer control of air temperature in a glasshouse. *J. Agric. Eng. Res.* 28:469-477.

Schapendonk, A.H.C.M. and P. Gaastra. 1984. Physiological aspects of optimal CO_2-control in protected cultivation. *Acta Hort.* 148:477-484.

Schiller, J.I. 1994. Secure distributed computing. *Sci. Amer.* 271:72-76.

Schmidt, W. 1986. Test and maintenance of quality standards for agricultural software. *In* Microelectronics in Agriculture - Facts and Trends. *Deutsche Landwirtschafts Gesellschaft*, Frankfurt. 197-208.

Schockert, K. and C. von Zabeltitz. 1980. Energy consumption of greenhouses. *Acta Hort.* 106:21-26.

Schubert, F. and F. Scheller. 1992. Biosensors. *Acta Hort.* 304:71-77.

Seginer, I. 1980. Optimizing greenhouse operation for best aerial environment. *Acta Hort.* 106:169-178.

Seginer, I. 1981. Economic greenhouse temperatures. *Acta Hort.* 115:439-452.

Seginer, I. 1989. Optimal greenhouse production under economic constraints. *Agric. Systems.* 29:17-80.

Seginer, I. 1993. Crop models in greenhouse climate control. *Acta Hort.* 328:79-97.

Seginer, I. and L.D. Albright, 1980. Rational operation of greenhouse thermal curtains. *Trans. ASAE.* 23:1240-1245.

Seginer, I. and D. Kantz. 1986. *In situ* determination of transfer coefficients for heat and water vapour in a small greenhouse. *J. Agric. Eng. Res.* 35:39-54.

Seginer, I. and A. Sher. 1993. Optimal greenhouse temperature trajectories for a multi-state-variable tomato model. *In* The Computerized Greenhouse. Y. Hashimoto et al., eds. Academic Press, New York. 340 pp.

Seginer, I. et al. 1986. Optimal CO_2 enrichment strategy for greenhouses: A simulation study. *J. Agric. Eng. Res.* 34:285-304.

Seginer, I. et al. 1991. Optimal temperature setpoints for greenhouse lettuce. *J. Agric. Eng. Res.* 49:209-226.

Sigrimis, N. 1989. Greenhouse control optimization. *ASAE Paper No. 89-4017.* ASAE, Quebec, Canada. 25 pp.

Spencer, K.L. 1994. Visual Basic: a new development in productivity. *DEC Professional.* 13:48-54.

Spitters, C.J.T. 1986. Separating the diffuse and direct component of global radiation and its implications for modeling canopy photosynthesis. Part II. Calculation of canopy photosynthesis. *Agric. and Forest Meteor.* 38:231-242.

Spitters, C.J.T., H.A.J.M. Toussaint and J. Goudriaan. 1986. Separating the diffuse and direct component of global radiation and its implications for modeling canopy photosynthesis. Part I. Components of incoming radiation. *Agric. and Forest Meteor.* 38:217-229.

Spitters, C.J.T., H. van Keulen and D.W.G. van Kraalingen. 1989. A simple and universal crop growth simulator: SUCROS87. *In* Simulation and Systems Management in Crop Protection. R. Rabbinge, S.A. Ward and H.H. van Laar, *eds. Simulation Monograph 32.* Pudoc, Wageningen.

Stanghellini, C. 1983. Radiation absorbed by a tomato crop in a greenhouse. *Res. Rpt. 83-5.* IMAG, Wageningen. 23 pp.

Stanghellini, C. 1987. Transpiration of greenhouse crops. An aid to climate management. *Ph.D. Dissertation.* IMAG, Wageningen. 150 pp.

Stanghellini, C. 1995. Physics of greenhouse climate. Vapour balance. *In* Greenhouse Climate Control. J.C. Bakker et al., eds. Wageningen Pers.

Steinbuch, F. 1986. Microelectronics for climate control in greenhouses. Microelectronics in Agriculture - Facts and Trends. *Deutsche Landwirtschafts Gesellschaft.* Frankfurt. 64-67.

Swaine, M. 1994. Forth and standards and chaos and life. *Dr. Dobb's Journal.* 18:107-109.

Swerdlow, J.L. 1995. Information revolution. *Nat. Geogr.* 188:5-36.

Swetz, F.J. ed. 1994. From Five Fingers to Infinity. Open Court, Chicago. 770 pp.

Takakura, T. 1991. Environmental control systems for greenhouses. *Proc. 1991 Symp. Automated Agriculture for the 21st Century.* ASAE, St. Joseph, MI.

Takakura, T. 1993. Climate Under Cover. Kluwer Academic Publ., London. 155 pp.

Takakura, T. 1994. Temperature, humidity and ventilation. - Interrelationships. *In* Greenhouse Systems. Automation, Culture, and Environment. NRAES-72. Rutgers Univ., New Brunswick, NJ.

Tantau, H.J. 1980. Climate control algorithms. *Acta Hort.* 106:49-54.

Tantau, H.J. 1985. The ITG digital greenhouse climate control system for energy saving. Unpubl. Manuscript., Inst. for Hort. Eng., Hannover. 7 pp.

Tantau, H.J. 1990. Automatic control application in greenhouse. *Int. Fed. of Automatic Control.* World Congress. 6:277-280.

Tantau, H.J. 1993. Optimal control for plant production in greenhouses. *In* The Computerized Greenhouse. Y Hashimoto et al., eds. Academic Press, New York. 340 pp.

Tchamitchian, M. 1993. Including the greenhouse side walls in a light interception model. *Acta Hort.* 378:133-140.

Tsujita, M.J. 1988. Phase I. Environmental parameters for computer-controlled greenhouses. Phase II. Optimum temperature for roses as influenced by light levels. *Roses, Inc. Bul.* Nov., 1988. 53-61.

Udink ten Cate, A.J. 1980. Remarks on greenhouse climate control models. *Acta Hort.* 106:43-47.

Udink ten Cate, A.J. 1983. Modeling and (adaptive) control of greenhouse climates. Ph.D. Dissertation. Landbw. Wageningen. 159 pp.

Udink ten Cate, A.J. 1985. Advances in greenhouse climate control. *Acta Hort.* 174:361-363.

Udink ten Cate, A.J. 1987. Analysis and synthesis of greenhouse climate controllers. *In* Computer Applications in Agricultural Environments. J.A. Clark et al., eds. Butterworths, London. Chap. 1, pp 1-20.

Udink ten Cate, A.J. and J. van de Vooren. 1984a. New models for greenhouse climate control. *Acta Hort.* 148:277-285.

Udink ten Cate, A.J. and J. van de Vooren. 1984b. Modelling of greenhouse temperatures using time-series analysis techniques. *9th IFAC World Congress*, Budapest. 1:272-276.

Udink ten Cate, A.J. and J. van Zeeland. 1981. A modified PI-algorithm for a glasshouse heating system. *Acta Hort.* 115:351-358.

Udink ten Cate, A.J., G.P.A. Bot and J.J. van Dixhoorn. 1978. Computer control of greenhouse climates. *Acta Hort.* 87:265-272.

Valentin, J. and J. van Zeeland. 1980. Adaptive split-range control of a glasshouse heating system. *Acta Hort.* 106:109-115.

van Bavel, C.H.M., J. Damagnez and E.J. Sadler. 1981. The fluid-roof solar greenhouse: Energy budget analysis by simulation. *Agric. Meteor.* 23:61-76.

van Bavel, C.H.M., T. Takakura and G.P.A. Bot. 1985. Global comparison of three greenhouse climate models. *Acta Hort.* 174:21-33.

van de Sanden, P.A.C.M. 1995. Water balance. *In* Greenhouse Climate Control. J.C. Bakker et al., eds. Wageningen Pers.

van de Vooren, J. and R. Koppe. 1975. The climate greenhouse at Naaldwijk. *Netherlands J. Agric. Sci.* 23:238-247.

van den Berg, G.A. and R.J. Bogers. 1995. Production planning in closed floriculture systems. *Acta Hort.* 399:95-100.

van den Kieboom, A.M.G. and J.A. Stoffers. 1985. Light transmittance under diffuse radiation circumstances. *Acta Hort.* 174:67-74.

van den Vlekkert, H.H. 1992. Ion-sensitive field effect transistors. *Acta Hort.* 304:113-125.

van den Vlekkert, H.H., J.P.M. Kouwenhoven and A.A.M. van Wingerden. 1992. Application of ISFETs in closed-loop systems for horticulture. *Acta Hort.* 304:309-320.

van Henten, E.J. 1994. Greenhouse climate management: An optimal control approach. Ph.D. Dissertation. Wageningen. 329 pp.

van Henten, E.J. and J. Bontsema. 1994. Sensitivity analysis of a dynamic growth model of lettuce. *J. Agric. Eng. Res.* 59:19-31.
van Henten, E.J. and J. Bontsema. 1995. Non-destructive crop measurements by image processing for crop growth control. *J. Agric. Eng. Res.* 61:97-105.
van Henten, E.J. and J. Bontsema, 1996. Greenhouse climate control: A two time-scale approach. *Acta Hort.* In press.
van Meurs, W.T.M. 1980. The climate control computersystem at the IMAG, Wageningen. *Acta Hort.* 106:77-83.
van Meurs, W.T.M. 1995. Current implementation of hardware and software. *In* Greenhouse Climate Control. J.C. Bakker et al., eds. Wageningen Pers.
van Meurs, W.T.M. and E.M. Nederhoff. 1995. Greenhouse climate control. CO_2 control. *In* Greenhouse Climate Control. J.C. Bakker et al., eds. Wageningen .
van Meurs, W.T.M. and C. Stanghellini. 1989. A transpiration-based climate control algorithm. *Acta Hort.* 245:476-481.
van Straten, G. and H. Challa. 1995. Towards integration. Requirements for intelligent control systems of the future. *In* Greenhouse Climate Control. J.C. Bakker et al., eds. Wageningen Pers.
Verwaaijen, P.W.T., T.H. Gieling and W.T.M. van Meurs. 1985. Measurement and control of the climate in a greenhouse, where a heated concrete-floor is used as low temperature energy source. *Acta Hort.* 174:469-475.
Voland, G. 1986. Control Systems Modeling and Analysis. Prentice-Hall, Engelwood Cliffs, NJ. 266 pp.
Waldo, J. 1994. Secrets and privacy in C++. *UNIX Rev.* 12:63-66.
Wallich, P. 1994. Wire pirates. *Sci. Amer.* 270:90-101.
Willits, D.H., T.K. Karnoski and W.F. McClure. 1980. A microprocessor-based control system for greenhouse research: Part I. Hardware. *Trans. ASAE.* 5:688-692.
Willits, D.H., T.K. Karnoski and E.H. Wiser. 1980. A microprocessor-based control system for greenhouse research. Part II. Software. *Trans. ASAE.* 5:693-698.
Winspear, K.W. 1968. Control of heating and ventilation in glasshouses. *Acta Hort.* 6:62-78.
Winspear, K.W. and L.G. Morris. 1965. Automation and control in glasshouses. *Acta Hort.* 2:61-70.
Young, P., A. Chotai and W. Tych. 1993. Identification, estimation and true digital control of glasshouse systems. *In* The Computerized Greenhouse. Y. Hashimoto et al., eds. Academic Press, San Diego.
Zazueta, F.S. 1994. Rapid prototyping of expert systems. *1994 Int. Cong. on Computers in Agric.*, Orlando, FL.

CONCLUSIONS

In retrospect, the finish of Chapter 8 leaves something to be desired. That is, what are the major areas in greenhouse technology where rapid advances and improvements might be possible? One might think this as equivalent to fortune telling. However, the massive collation of the available English information in the field, combined with the author's personal experience, points to areas that are deficient with some possibilities of advances that would require minimal new equipment. Among those areas are:

1. The accumulation of additional information on tissue analyses as outlined in Chapter 6, whereby boundaries can be established, indicating general nutrient ranges and optima (Figs. 6-13, 6-14 and 6-25). Much information could be collated from existing data, reducing the amount of actual experimentation required. This would be a first step toward analysis of data using the DRIS system or similar methods. Such work could be extended to soil and nutrient solution analysis. The work requires no additional technical equipment beyond what most experimental stations already have in place. Thus, expense would be minimal. Assuming sufficient data over a wide range of conditions, there seems to be a good possibility of determining universal tissue nutrient levels, applicable to a variety of cultural procedures and climate locations.

2. Also, in Tables 6-25 and 6-26, it was shown that nutrient solutions are anything but simple. Table 6-42 showed the effects of water quality on fertigation supplies and some of the limitations. Software programs, such as MINTEQA2/PRODEFA2, greatly expand potentials for precision manipulation of fertilizers, opening new advances in nutrition that can be integrated with exisiting computer control programs. Continued advances in ion selective probes will occur. These can be used in experimental programs to manipulate nutrient supply and for commercial application. Investment in capital equipment to undertake program generation and experimentation does not appear great. If the industry and investigators can eventually utilize proper terminology in speaking of nutrition relationships as outlined in the first part of Chapter 6, advances in this area would be expedited.

3. Chapter 8 presented the possibility of on-line optimization of greenhouse production. Computers also are useful for simulation (e.g., Fig. 8-24), thereby eliminating the need for extensive experimental facilities. Given sufficiently low error, the simulation results can be integrated with environmental control programs. However, experience has shown that as program complexity increases, possibilities of program corruption also increase. There are no good standards, at least in the U.S., to assure robustness in existing programming languages, as well as protection against outside, undesirable code being introduced. The control industry needs to ensure that a grower, if necessary, can dump existing programs and re-install uncorrupted programs easily and when necessary. Guidelines need to be established for procedures to protect existing software, reduce errors, install upgrades, and re-establish control in the event of failure of all or a portion of the program. The production of good software for control and optimization seems to have high potential without great outlay for equipment. Programs can be tested in conjunction with those greenhouse operators having good equipment.

4. Considerable progress in temperature and radiation manipulation was outlined in Chapters 3 and 4. However, the results, while useful, are limited in range and climatic region. It seems to me that the great influence of internal water potential has not been given sufficient attention (Chapter 5). As pointed out, the ability to measure water potential in a meaningful fashion would allow integration of this factor with the data available for DIF, temperature integral, photosynhesis, and morphogenesis. The lack of information on stress and its measurement is the result of the difficulty in existing procedures. A concerted effort to provide reliable measurement of crop stress would result in an outstanding advance in greenhouse production.

APPENDIX A

SYMBOLISM, UNITS, AND PHYSICAL CONSTANTS

Symbol	Definition	Units	Value
α	Absorption ratio	Decimal ratio	
α_c	Average crop radiation use efficiency for CO_2 uptake	g MJ^{-1}	
β	Radiance	W $sr^{-1}m^{-2}$	
	Bowen ratio	H/LE	
β	Thermal expansion coefficient of air	K^{-1}	3.6×10^{-2}
	Ratio floor area to cover surface area	Decimal ratio	
β-TCP	β-tricalcium phosphate		
Δ	Difference	Misc.	
Δq	Specific humidity difference	kg kg^{-1}	
$\Delta\mu$	Energy difference	ergs mol^{-1}	
ϵ	Emissivity	Decimal ratio	
ϵ	Time scaling constant or ratio between fast and slow	Decimal ratio	
θ_m	Soil water content	g g^{-1} dry soil	
θ_v	Soil water ratio	cm^3 cm^{-3}	
λ	Radiation wavelength	nanometers	
	Thermal conductivity of air	W $m^{-1}K^{-1}$	2.2×10^{-2}
μ	Ratio air volume to floor area	Decimal	
μ_w	Chemical potential of water in a system	ergs mol^{-1}	
μ_w^o	Chemical potential of pure, free water at T	ergs mol^{-1}	
μmho cm^{-1}	Electrical conductivity	micromho cm^{-1}	
σ	Stefan-Boltzmann constant	W $m^{-2}K^{-4}$	5.673×10^{-6}
π	Pi		3.141
ρ	Reflection coefficient of canopy	Decimal ratio	
ρ	Air density	kg m^{-3}	1.29 at NPT
ρ_c	CO_2 density	kg m^{-3}	1.83 at NPT
ρ_s	Shortwave reflectivity	Decimal ratio	
ϕ	CO_2 uptake	kg $m^{-2}s^{-1}$	
ϕ_w	Chemical activity of water	Decimal ratio	
$\phi(x_p(t_f))$	Termination value of the system	$	
χ	Absolute humidity	g m^{-3}, kg m^{-3}	
χ_a	Ambient absolute humidity	kg m^{-3}	
χ_o	Outside absolute humidity	kg m^{-3}	
χ_r	Absolute humidity at cover surface	kg m^{-3}	
χ_{eff}	Effective absolute humidity in the crop	kg m^{-3}	
γ	Psychrometric constant	Pa K^{-1}	0.66
τ	Transmissivity	Decimal ratio or percent	
	Response time	seconds or minutes	Venlo, 2-10 min
	Dummy variable in control theory		
ψ	Chemical potential of water	J m^{-1}	
ψ_t	Total plant water potential	MJ	
ψ_m	Matric water potential	MJ	
ψ_o	Osmotic water potential	MJ	
$\psi_{(r,s,l,a)}$	Water potential, root, soil, leaf, or air	MJ	
ε	Mol weight water/Mol weight air		0.622
	Time scaling response or ratio between fast and slow		

Symbol	Definition	Units	Value
A	Amperes, current		
	Solar altitude	Degrees	
	Area	km^2, m^2, etc.	
	Avogadro's Number	$mole^{-1}$	6.02×10^{23}
A_g	Ground area	m^2	
A_r	Greenhouse cover area	m^2	
AI	Artificial intelligence		
Al	Aluminum		
ADT	Average daily temperature, integral	Degrees C	
A/D	Analog to digital conversion		
ASAE	American Society of Agricultural Engineers		
ASCE	American Society of Civil Engineers		
ASHRAE	American Society of Heating, Refrigeration and Air Conditioning Engineers		
a	Parametric constant		
a	Solar altitude	Degrees	
a_w	Chemical activity of water	Decimal ratio	
B	Boron		
BD	Bulk density	$g\ cm^{-3}$	
b	Parametric constant		
C	Roof shape factor	Decimal	
	Carbon		
	Degrees Celsius		
	Condensation	$kg\ m^{-2}s^{-1}$	
C_f	Conversion efficiency of dry weight formed per gram of carbohydrates (assimilates)	Decimal ratio	
Ca	Calcium		
Cl	Chlorine		
Co	Cobalt		
Cu	Copper		
C_p	Specific heat of air	$J\ kg^{-1}K^{-1}$	1.01×10^3
CV	Coefficient of variability		
Cap_h	Thermal capacity of air	$J\ K^{-1}$	1.01×10^1
CEC	Cation excange capacity	$meq\ 100g^{-1}$	
CPU	Computer processer unit		
CTV	Canopy temperature variability		
CVA	Critical value approach		
CO_2	Carbon dioxide		
CO_3^{2-}	Carbonate		
CP3	U.K. Greenhouse Design Standards		
COP	Coefficient of performance	Decimal	
c	Radiation velocity	$cm\ s^{-1}$	2.998×10^{10}
	Parametric constant		

Symbol	Definition	Units	Value
c_m	Vapor concentration	kg m^{-3}	
c_i	Vapor concentration inside	kg m^{-3}	
c_o	Vapor concentration outside	kg m^{-3}	
cd	Candela, unit of luminous intensity		
DM	Dry matter		
DT	Day temperature	C	
DIF	Difference (DT-NT)	C	
DIN	Deutsches Institut für Normung e.V.		
DOS	Disc operating system		
DVR	Plant rate of development	dy^{-1}	
DCPA	Dicalcium phosphate anhydrous	$CaHPO_4$	
DCPD	Dicalcium phosphate dihydrate	$CaHPO_4.2H_2O$	
DRIS	Diagnosis and recommendation integrated system		
DTPA	Fe 330 chelate		
D/A	Digital to analog conversion		
d_c	Discount coefficient		
dS m^{-1}	Electrical conductivity	deciSiemans per meter	
E	Energy	Joules	
ET	Evapotranspiration	mm water column	
E_p	Annual electrical production, cogeneration	kW-hr	
E_t	Annual thermal energy used in house	kW-hr	
E_f	Annual fuel energy utilized, cogeneration	kW-hr	
EC	Electrical conductivity	dS m^{-1}	
ECV	Equilibrium capacity variable		
EDR	Electrodialysis		
EDTA	Iron sequestrene chelate		
EVA	Ethyl vinyl acetate		
EDDHA	Fe 138 chelate		
e	Natural logarithm Error value		2.71828
e_o	Vapor pressure at evaporating surface	Pascal	
e_a	Vapor pressure of air	Pascal	
e_s	Saturation vapor pressure at T	Pascal	
F	Flux Air flow	Watts or lumens m^3s^{-1}	
F_{ij}	View factor from surface 'i' to 'j'	Decimal ratio	
Fe	Iron		
FA	Fluorapatite		
FR	Far red radiation	725 to 735 nm	
FRP	Fiber reinforced plastic		
G_c	Rate of CO_2 emission from ground	kg $m^{-2}s^{-1}$	
Gr	Grashof number = $g\beta\Delta T\ell^3 v^2$	Dimensionless	
GR	Absolute growth	g $m^{-2}dy^{-1}$	
g	Acceleration of gravity	m s^{-2}	0.980

Symbol	Definition	Units	Value
g_{cnd}	Conductance	m s^{-1}	
g_{tot}	Total conductance	m s^{-1}	
g_{tr}	Canopy conductance	m s^{-1}	
g_{vnt}	Ventilation conductance	m s^{-1}	
H	Hydrogen		
	Hamiltonian function		
H_s	Height	meters, centimeters, etc.	
H_w	Sensible heat exchange	W m^{-2}	
H_{sp}	Latent heat exchange, condensation	W m^{-2}	
H_{sg}	Sensible heat from heating system	W m^{-2}	
	Sensible heat exchange by conduction and convection	W m^{-2}	
H_v	Sensible and latent heat exchange through ventilators	W m^{-2}	
HA	Hydroxyapatite		
HG	Clear mercury lamp		
HCO_3^-	Bicarbonate		
HID	High-intensity discharge lamp		
HPS	High-pressure sodium lamp		
Hz	Hertz (frequency)	Cycles per second	
HEDTA	Versonal chelate		
$H_2PO_4^-$	Phosphate ion		
h	Plank's constant	J-s	6.62×10^{-34}
	Height	m, cm, etc.	
	Heat transmission coefficient	W $m^{-2}K^{-1}$, J $s^{-1}m^{-2}K^{-1}$	
	Ratio of volume to ground area or mean height		
ha	Hectare (area)		
h_{inf}	Infiltration transfer coefficient	kJ $m^{-2}hr^{-1}K^{-1}$	
I	Intensity	Watts or lumens	
I_d	Insulation index	Decimal	
I_o	Irradiance at the top of the canopy	W m^{-2}	
I_Σ	Daily integral of PAR at canopy top	MJ $m^{-2}dy^{-1}$	
IR	Infrared radiation	700 to 1100 nm	
IRR	Internal rate of return		
ISE	Ion selective electrode		
ISFET	Ion sensitive field effect transistor		
i	Discount rate	Percent as a decimal	
J	Joules, unit of energy		
	Goal to be maximized = income		
K	Potassium		
	Kelvin, absolute temperature scale	Degrees 0 C = ca. 273 K	
	Extinction coefficient	Decimal ratio	
	Extinction coefficient of PAR in the canopy	Decimal ratio	
K°	Equilbrium constant for a chemical reaction		
KW	kiloWatts, power	1000 Joules per second	
L	Distance, length	meters, centimeters, etc.	
	Latent heat of vaporization	MJ kg^{-1}	2.45 at 20 C
	Constant costs at any instant in time		
L_e	Illuminance at angle **a** above horizon	Lumens	

Symbol	Definition	Units	Value
L_Z	Illuminance at the zenith	Lumens	
Le	Lewis number = $PrSc^{-1}$	Dimensionless	
LE	Latent heat times evapotranspiration	$W\ m^{-2}$	
LF	Leaching fraction	Water drained/water applied	
LAI	Leaf area index	m^2m^{-2} ground surface	
LAR	Leaf area ratio	Leaf area / plant dry weight (m^2kg^{-1})	
LDP	Long-Day Plant		
LPS	Low-pressure sodium lamp		
LTC	Leaf tip chlorosis		
LWR	Leaf weight ratio	Leaf dry weight per plant dry weight	
LEMP	Lightning electromagnetic pulse		
ℓ	Characteristic length Liter	meter	
lm	Unit of illuminance	Lumen	
ln	Natural logarithm		2.71828
lx	Luminous intensity	lux ($lm\ m^{-2}$)	
M	Moles per liter (concentration)(molality)	$M\ \ell^{-1}$	
Mg	Magnesium		
Mn	Manganese		
Mo	Molybdenum		
M_v	Molecular weight of water	$kg\ mol^{-1}$	0.018
MH	Metal halide lamp		
MOS	Metal oxide semiconductor		
MIPS	Million instructions per second		
MOSFET	Metal oxide semiconductor field effect transistor		
m	Moles per kilogram (concentration)(molarity) Meter (length) Leaf transmission coefficient	$m\ kg^{-1}$ Decimal ratio	
$meq\ \ell^{-1}$	Milliequivalents per liter (concentration)		
n	Number (units of years, sets, etc.)		
N	Nitrogen Newton, unit of force whose mass is 1 kg which would experience an acceleration of 1 m^{-2}. Air exchange per hour One equivalent weight per liter (normality)		
Na	Sodium		
Nu	Nusselt number = $\alpha_h \ell \lambda^{-1}$	Dimensionless	
NT	Night temperature	C	
NAR	Net assimilation rate	$gm\ m^{-2}dy^{-1}$	
NFT	Nutrient film technique		
NTC	Standard, ceramic thermistor		

Symbol	Definition	Units	Value
NCER	Net carbon exchange rate	µmol. $m^{-2}s^{-1}$	
NEN3859	Dutch Design Standards, "Venlo"		
NMGA	National Manufacturers Greenhouse Association		
NPT	Normal pressure and temperature (also STP)	101.3 kPa and 20 C	
NH_4^+	Ammonium		
NO_3^-	Nitrate		
NR	Near red radiation	655 to 665 nm	
N_w	Mole fraction of water	Decimal ratio	
O	Oxygen		
OCP	Octocalcium phosphate		
P	Phosphorus Atmospheric pressure Photosynthesis energy Photosynthetic rate	 kPa W m^{-2} mmol $m^{-2}s^{-1}$	 101.3 NPT
Pr	Prandtl number = vD^{-1}	Dimensionless	
$P_{gc,d}$	Crop photosynthesis per unit area	g CO_2 $m^{-2}dy^{-1}$	
PC	Polycarbonate Programmable computer		
PE	Polyethylene		
PAR	Photosynthetic active radiation	µmol. $m^{-2}s^{-1}$, 400 to 700 nm	
PMMA	Polymethylmetacetate		
PO_4^{3-}	Phosphate		
PPF	Photosynthetic Photon Flux	µmol. $m^{-2}s^{-1}$	
PVC	Poly vinyl chloride		
PVF	Poly vinyl fluoride		
pe	Oxidation-reduction potential	-log electron activity (mols)	
pH	Hydrogen ion activity	-log ion activity (mols)	
$\mathbf{p_x}$	A costate variable or vector value attached to a marginally produced unit of 'x'	\$ kg^{-1}	
p_h p_m	Energy produced in the greenhouse Vapor produced in the greenhouse	J s^{-1} kg s^{-1}	
p_{xf}	Rate of investment in the future. Rate of future gains	Percentage	
Q	Radiant flux at a surface Water flux in a plant	W m^{-2} cm^3s^{-1}	
dQ_h/dt dQ_m/dt	Rate of energy exchange of greenhouse Rate of mass (vapor) exchange of greenhouse	J s^{-1} kg s^{-1}	
q q_i q_g	Specific humidity Specific humidity inside Specific humidity outside	g kg^{-1} moist air g kg^{-1} g kg^{-1}	
$q_{h,m,v}$	Energy, vapor or volumetric exchange by ventilation and infiltration	J, kg, m^3 s^{-1}	
$q_{(in,h,\ in,m)}$ $q_{(out,h,\ out,m)}$ $q_{rad,ij}$	Energy or vapor exchange into greenhouse Energy or vapor exchange leaving greenhouse Radiation exchange between surfaces 'i' and 'j'	J, kg s^{-1} J, kg s^{-1} J s^{-1}	

Symbol	Definition	Units	Value
R	Perfect gas constant Radiant emittance Richard's function	ergs $mol^{-1}K^{-1}$ W m^{-2}	8.314 x 10^7
Re	Reynold's number = $u\ell v^{-1}$	Dimensionless	
R_{dir}	Direct radiation	W m^{-2}	
R_{dif}	Diffuse radiation	W m^{-2}	
R_{lw}	Longwave radiation	W m^{-2}	
R_n	Net radiation	W m^{-2}	
R_{ref}	Reflected radiation	W m^{-2}	
R_{tot}	Total radiation	W m^{-2}	
R_{lwg}	Long wave radiation exchange between cover and sky	W m^{-2}	
R_{lwp}	Long wave radiation from heating system	W m^{-2}	
$R_{m,d}$	Daily rate of maintenance respiration	CH_2O g $m^{-2}dy^{-1}$	
RH	Relative humidity	Decimal ratio (e_a/e_s)	
RO	Reverse osmosis		
RFI	Radio frequency interference		
RGR	Relative growth rate	g $g^{-1}dy^{-1}$	
RHPC	Receding horizon predictive control		
ROM-BIOS	Read only memory - basic input-output system		
r	Mixing ratio	g kg^{-1} dry air	
$r_{(s,r,l,a)}$	Resistance of soil, root, leaf or air	s m^{-1}	
r_H	Resistance to heat transfer	s m^{-1}	
S	Sulfur Energy flux from soil Energy storage in soil, canopy, etc.	 W m^{-2} J kg^{-1}, m^{-2}	
Sc	Schmidt number = aD^{-1}	Dimensionless	
Se	Selenium		
Sh	Sherwood number = $k_m\ell D^{-1}$	Dimensionless	
Si	Silicon		
SR	Sufficiency range		
SDD	Stress degree days		
SDP	Short Day Plant		
SLA	Specific leaf area	Leaf area per unit leaf dry weight	
SLS	Short leaf syndrome		
SLW	Specific leaf weight	Leaf weight per plant dry weight	
SME	Saturated medium extract (Soil test)		
SO_4^{2-}	Sulfate		
s	Second Standard deviation of a sample Laplace transform		
s^2	Variance of a mean		
sr	Steradian	Unit of solid angle	

Symbol	Definition	Units	Value
T	Temperature	C or K	
	Transmissivity	Decimal ratio	
T_p	Plant temperature (or radiating surface)	C or K	
T_a	Air temperature (or absorbing surface)	C or K	
T_{wb}	Wet bulb temperature	C or K	
T_{db}	Dry bulb temperature (air temperature)	C or K	
T_{dp}	Dew point temperature	C or K	
$T_{i,o}$	Inside or outside temperature	C or K	
T_v	Virtual temperature	C or K	
$\bar{T}_a$	Virtual temperature of the air	C or K	
$\bar{T}_r$	Virtual temperature at a cover surface at T	C or K	
TDR	Time domain reflectometry		
TSD	Temperature stress day		
t	Time	s, hr, dy, etc.	
UV	Ultraviolet radiation	100 to 380 nm	
UV-A	Ultraviolet radiation, A band	320 to 380 nm	
UV-B	Ultraviolet radiation, B band	280 to 320 nm	
UV-C	Ultraviolet radiation, C band	100 to 280 nm	
$u_{c,e}$	External or control inputs to control system		
V	Luminous efficiency (output/input)	decimal	
	Volume	m^3	
VA	Vesicular-arbuscular mychorrizae		
V_g	Air change rate	$m^3\ m^{-2}\ s^{-1}$	
$\underline{V}_w$	Molal volume of water	$cm^3\ mol^{-1}$	≈18.0
VPD	Vapor pressure deficit	kPa	
v	Radiation waves per second	s^{-1}	
	Wind speed	$m\ s^{-1}$	
	Kinematic viscosity of air	m^2s^{-1}	$1.4x10^{-5}$
	Outside disturbances		
	Random noise		
v'	Radiation wave number	Waves per centimeter	
W	Unit of power	Watt ($J\ s^{-1}$)	
W_t	Crop dry weight at time 't'	$g\ m^{-2}$	
$W\ sr^{-1}$	Watt per steradian		
YPF	Yield Photon Flux	$\mu mol.\ m^{-2}s^{-1}$	
x, $x_{c,p}$	Slow, indoor climatic, or plant state variable		
Zn	Zinc		
z	Fast variables		
ω	Solid angle	Degrees	
Z	'Z' transform		

APPENDIX B

CONVERSIONS METRIC TO ENGLISH
(UNITS GENERALLY ROUNDED TO THREE SIGNIFICANT DIGITS)

From	To	Multiply by
candela	candle	0.981
candela per square meter	foot candle	0.292
candela per square meter	lambert	3.145×10^{-4}
Celsius	Fahrenheit	F = (9/5)(C+32)
centimeter	foot	3.28×10^{-2}
centimeter	inch	0.394
centimeter	yard	0.011
centimeter per second	foot per second	3.279×10^{-2}
cents per liter	cents per gallon	3.788
cubic centimeter	cubic inch	6.10×10^{-2}
cubic meter	ounce (U.S. fluid)	3.378×10^{4}
cubic meter	bushel (U.S.)	28.409
cubic meter	barrel (42 U.S. gallons)	6.289
cubic meter	liter	1000
cubic meter	quart (U.S. liquid)	1.057×10^{3}
cubic meter	pint (U.S. liquid)	2.114×10^{3}
cubic meter	cubic yard	1.307
cubic meter	acre-foot	8.13×10^{-4}
cubic meter	gallon (U.S. liquid)	2.639×10^{2}
cubic meter per second	cubic foot per second	35.336
cubic meter per second	cubic foot per minute	2.119×10^{3}
cubic meter per second	liter per minute	5.988×10^{4}
cubic meter per second	liter per second	1×10^{3}
cubic meter per second	gallon per minute	1.585×10^{3}
dollar per gigajoule	cents per gallon (no. 2 fuel oil)	14.77
dollar per gigajoule	cents per kilowatt-hour	0.360
dollar per gigajoule	cents per gallon (propane)	8.85
dollar per kilogram	dollar per pound	0.455
dollar per square meter	dollar per square foot	0.093
erg per second-square centimeter	British thermal unit per hour-square foot	6×10^{-9}
erg per second-square centimeter	gram-calorie per minute-square centimeter	1.43×10^{-6}
erg per second-square centimeter	langley per minute	1.43×10^{-6}
gram	pound	2.205×10^{-3}
gram	ounce (mass-avoirdupois)	0.0353

From	To	Multiply by
gram	grain	15.385
gram per cubic centimeter	pound per cubic foot	62.5
hectare	acre	2.469
joule	electron Volt	6.25×10^{18}
joule	British thermal unit	9.43×10^{-4}
joule	Watt-hour	2.28×10^{-4}
joule	calorie (mean)	0.239
joule	kilocalorie (mean)	2.387×10^{-4}
joule	Watt-second	1.0
joule per square meter	calorie per square centimeter	2.387×10^{-5}
joule per square meter	langley	2.392×10^{-5}
Kelvin	Fahrenheit	K=(9/5)(F+273.18)
kilogram	pound	2.222
kilogram	ounce (mass-avoirdupois)	35.274
kilogram	ton (short)	1.10×10^{-3}
kilogram	ton (long)	9.80×10^{-4}
kilogram	ounce (mass-troy)	32.151
kilogram per cubic meter	pound per cubic foot	0.063
kilogram per cubic meter	pound per cubic yard	1.686
kilojoule	British thermal unit	0.948
kilojoule per cubic meter, joule per liter	Briitish thermal unit per cubic foot	2.68×10^{-2}
kilometer	mile	0.621
kiloPascal	pound per square inch	0.145
kiloPascal	bar	0.01
liter	cubic foot	3.534×10^{-2}
liter	barrel (42 U.S. gallons)	6.29×10^{-3}
liter per minute	cubic foot per minute	3.534×10^{-2}
liter per second	gallon per minute	15.848
liter per second	cubic foot per seond	3.53×10^{-2}
lumen	candela per steradian	1.00
lumen per square meter	foot candle	9.259×10^{-2}
lux	lumen per square centimeter	10^{-4}
lux	foot candle	9.259×10^{-2}
megajoule	therm (U.S.)	9.479×10^{-3}
megajoule	kilowatt-hour	0.278
megaPascal	bar	1
megaPascal	atmosphere	9.90

From	To	Multiply by
meter	micron	10^{6}
meter	foot	3.279
meter	nautical mile (U.S.)	5.405×10^{-4}
meter	light year	1.057×10^{-16}
meter	inch	39.370
meter	fathom	0.546
meter	mile (U.S. statute)	6.25×10^{-4}
meter per second	mile per hour	2.237
meter per second	knot	1.946
meter per second	foot per minute	196.85
meter per second-second	foot per second-second	3.279
metric ton	ton (short)	1.103
millibar	inch of mercury column	2.95×10^{-2}
millibar	atmosphere	9.90×10^{-4}
millimeter	foot	3.28×10^{-3}
milllimeter	inch	3.937×10^{-2}
newton per square meter	pound per square inch	1.451×10^{-4}
newton per square meter	Pascal	1.0
newton per square meter	atmosphere	9.90×10^{-6}
newton per square meter	millimeter mercury column	7.519×10^{-3}
newton per square meter	millibar	10^{-2}
newton per square meter	centimeter of water column	1.020×10^{-2}
newton per square meter	inch of water column	4.016×10^{-3}
newton per square meter	bar	1×10^{-5}
newton per square meter	inch of mercury column	2.95×10^{-4}
square centimeter	square inch	0.155
square kilometer	square mile	0.385
square meter	section	4×10^{-7}
square meter	square inch	1.550×10^{3}
square meter	acre	2.46×10^{-4}
square meter	square mile (U.S. statute)	3.86×10^{-7}
square meter	square foot	10.764
square meter	square yard	1.196
square meter-Kelvin per Watt (thermal resistance, R)	square foot-hour-fahrenheit per British thermal unit	5.682
square millimeter	square inch	1.55×10^{-3}
ton (metric)	ton (short)	1.103

From	To	Multiply by
Watt	calorie per minute	14.347
Watt	horsepower (550 ft-lbf sec^{-1})	1.340×10^{-3}
Watt	horsepower (electric)	1.340×10^{-3}
Watt	horsepower (boiler)	1.019×10^{-4}
Watt	British thermal unit per minute	5.68×10^{-2}
Watt per square centimeter	langley per minute	14.306
Watt per square meter	British thermal unit per hour-square foot	0.317
Watt per square meter	watt per square foot	0.093
Watt per square meter	British thermal unit per hour-square foot	3.175
Watt per square meter	calorie per square centimeter-minute	1.433×10^{-3}
Watt per square meter-Kelvin	British thermal unit per hour-square foot-fahrenheit	0.176
Watt-hour per square meter	gram-calorie per square centimeter	6.606×10^{-2}
Watt-hour per square meter	British thermal unit per square foot	0.317

INDEX

A

E

H

I

J

K

L

O

P

Q

R

S

T

U

V

W

X

Y

Z

DATE			
MAY 18 2000			
OCT 12 2000			